Chapter 5, continued

$C^\infty[a, b]$, space of real-valued functions with continuous derivatives of all orders on the closed interval $[a, b]$

$C^\infty(a, b)$, space of real-valued functions with continuous derivatives of all orders on the open interval (a, b)

span(S), space of all linear combinations of the vectors in the set S

P_n, space of polynomials of degree $\leq n$

$W(x)$, Wronskian

dim(V), dimension of a finite-dimensional vector space V

rank(A), common dimension of the row and column spaces of the matrix A

nullity(A), dimension of the nullspace of an $m \times n$ matrix A

$(\mathbf{v})_S = (c_1, c_2, \ldots, c_n)$, coordinate vector of $\mathbf{v}$ relative to a basis S

CHAPTER 6

$\langle \mathbf{u}, \mathbf{v} \rangle$, inner product of vectors $\mathbf{u}$ and $\mathbf{v}$

$\|\mathbf{u}\|$, norm of a vector $\mathbf{u}$

$d(\mathbf{u}, \mathbf{v})$, distance between vectors $\mathbf{u}$ and $\mathbf{v}$

$W^\perp$, orthogonal complement of a subspace W

$\dfrac{1}{\|\mathbf{v}\|}\mathbf{v}$, normalized vector

$\text{proj}_W \mathbf{u}$, orthogonal projection of the vector $\mathbf{u}$ on the subspace W

$A^{-1} = A^T$, orthogonal matrix

$[\mathbf{v}]_S$, coordinate matrix of $\mathbf{v}$ relative to the basis S

CHAPTER 7

$\det(\lambda I - A) = 0$, characteristic equation of a square matrix A

D, symbol usually used to denote a diagonal matrix

CHAPTER 8

$T: V \to W$, linear transformation from a vector space V into a vector space W

ker(T), kernel of the linear transformation T

$R(T)$, range of the linear transformation T

rank(T), rank of the linear transformation T

nullity(T), nullity of a linear transformation

$[T]_{B', B}$, matrix for T with respect to bases B and B'

CHAPTER 9

$Y' = AY$, linear system of first-order differential equations

$\displaystyle\int_a^b [f(x) - g(x)]^2\, dx$, mean square error

CHAPTER 10

i, symbol used to denote $\sqrt{-1}$

$z = a + bi$, a general complex number

(a, b), alternative symbol for a complex number

$\bar{z} = a - bi$, conjugate of a complex number z

Re(z), real part of a complex number z

Im(z), imaginary part of a complex number z

arg z, argument of a complex number z

Arg z, principal argument of a complex number z

$|z| = \sqrt{a^2 + b^2}$, modulus of a complex number z

$z = r(\cos\theta + i\sin\theta)$, polar form of a complex number z

$z = re^{i\theta}$, alternative polar form of a complex number z

C^n, space of n-tuples of complex numbers (complex Euclidean n-space)

complex $C(-\infty, \infty)$, vector space of complex-valued functions of a real variable that are continuous on the interval $(-\infty, \infty)$

complex $C[a, b]$, vector space of complex-valued functions of a real variable that are continuous on the closed interval $[a, b]$

$\mathbf{u} \cdot \mathbf{v} = u_1\bar{v}_1 + u_2\bar{v}_2 + \cdots + u_n\bar{v}_n$, complex Euclidean inner product

$A^* = A$, Hermitian matrix

$A^* = -A$, skew-Hermitian matrix

$A^* = A^{-1}$, unitary matrix

$AA^* = A^*A$, normal matrix

Elementary Linear Algebra

Applications Version

Elementary
Linear
Algebra

Applications Version   **Eighth Edition**

HOWARD ANTON
CHRIS RORRES

Drexel University

John Wiley & Sons, Inc.
New York Chichester Weinheim Brisbane Toronto Singapore

MATHEMATICS EDITOR   Barbara Holland
PRODUCTION EDITOR   Ken Santor
DESIGNER   Dawn L. Stanley
PHOTO EDITOR   Hilary Newman
ILLUSTRATION COORDINATOR   Techsetters/Sigmund Malinowski
COVER PHOTO   ©Joao Paulo/The Image Bank

This book was set in Times New Roman PS by Techsetters, Inc. and printed and bound by Von Hoffman Press, Inc. The cover was printed by Lehigh Press, Inc.

This book is printed on acid-free paper.∞

The paper in this book was manufactured by a mill whose forest management programs include sustained yield harvesting of its timberlands. Sustained yield harvesting principles ensure that the numbers of trees cut each year does not exceed the amount of new growth. See section 11.10, "Forest Management," to learn how linear algebra is applied to the problem of sustained yield harvesting.

Library of Congress Cataloging-in-Publication Data:
Anton, Howard.
        Elementary linear algebra : applications version / Howard Anton,
    Chris Rorres.—8th ed.
            p.   cm.
        Includes index.
        ISBN 0-471-17052-6 (cl.: alk. paper)
        1. Algebras, Linear.   I. Rorres, Chris.   II. Title.

QA184.A57   2000
512′.5—dc21   99-44870
                                                                          CIP

Printed in the United States of America

10 9 8 7 6 5 4 3

To my wife Pat
and my children
Brian, David, and Lauren
HA

To Billie
CR

Preface

This textbook is an expanded version of *Elementary Linear Algebra*, Eighth Edition, by Howard Anton. The first ten chapters of this book are identical to the first ten chapters of that text; the eleventh chapter consists of 21 applications of linear algebra drawn from business, economics, engineering, physics, computer science, approximation theory, ecology, sociology, demography, and genetics. The applications are, with one exception, independent of one another and each comes with a list of mathematical prerequisites. Thus, each instructor has the flexibility to choose those applications that are suitable for his or her students and to incorporate each application anywhere in the course after the mathematical prerequisites have been satisfied.

This edition, in the spirit of its predecessors, gives an elementary treatment of linear algebra and its applications that is suitable for students in their freshman or sophomore year. The aim is to present the fundamentals of linear algebra and its applications in the clearest possible way—pedagogy is the main consideration. Calculus is not a prerequisite, but there are clearly labeled exercises and examples for students with calculus backgrounds. Those exercises can be omitted without loss of continuity. Technology is also not required, but for those who would like to use MATLAB, Maple, *Mathematica*, or calculators with linear algebra capabilities, exercises have been included at the ends of the chapters for that purpose.

SUMMARY OF CHANGES IN THIS EDITION

This edition is a refinement of the previous edition. We have tried to maintain the clarity and style of the earlier editions, yet accommodate the needs of a new generation of students. Here is a summary of the changes:

- **Addition of Technology Exercises:** A set of technology exercises has been placed at the ends of chapters, but they are categorized according to section, so they can be assigned as part of the regular section exercises. The technology exercises are designed to acquaint the student with most of the basic commands that are required to solve problems in linear algebra using technology. The exercises are written in a generic, syntax-free form with the understanding that the student's own documentation will be the resource for specific commands and procedures. To relieve students of the burden of data entry, data for the technology exercises are provided in MATLAB, Maple, and *Mathematica* formats. These data can be downloaded from **www.wiley.com/college/anton**.

- **Addition of Discussion and Discovery Exercises:** A new category of exercises, identified as *Discussion and Discovery*, has been added to many exercise sets. In keeping with modern pedagogical trends, these exercises are more open-ended than others in the set. They typically include true/false, conjecture, discovery, and informal explanations of how conclusions are reached.

- **Refinement of Exposition:** Parts of the exposition have been refined and improved, but no substantial changes in style, organization, or content have been made, except as previously noted.

• **New Application on Warps and Morphs:** This new application at the end of Chapter 11 provides a look at some recent image-manipulation techniques available for computer graphics.

Hallmark Features

• **Relationships Between Concepts:** One of the important goals of a course in linear algebra is to establish the intricate thread of relationships between systems of linear equations, matrices, determinants, vectors, linear transformations, and eigenvalues. That thread of relationships is developed through the following crescendo of theorems that link each new idea with ideas that preceded it: 1.5.3, 1.6.4, 2.3.6, 4.3.4, 5.6.9, 6.2.7, 6.4.5, 7.1.5. These theorems bring a coherence to the linear algebra landscape and also serve as a constant source of review.

• **Smooth Transition to Abstraction:** The transition from R^n to general vector spaces is traumatic for most students, so that transition has been smoothed out by emphasizing the underlying geometry and developing key ideas in R^n before proceeding to general vector spaces.

• **Early Exposure to Linear Transformations and Eigenvalues:** To ensure that the material on linear transformations and eigenvalues does not get lost at the end of the course, some of the basic concepts relating to those topics are developed early in the text and then reviewed when the topic is treated in more depth later in the text. For example, characteristic equations are discussed briefly in the section on determinants, and linear transformations from R^n to R^m are discussed immediately after R^n is introduced, then reviewed later in the context of general linear transformations.

About the Exercises

Each section exercise set begins with routine drill problems, progresses to problems with more substance, and concludes with theoretical problems. In most sections, the main part of the exercise set is followed by the *Discussion and Discovery* problems described above. Most chapters end with a set of supplementary exercises that tend to be more challenging and force the student to draw on ideas from the entire chapter rather than a specific section. The technology exercises follow the supplementary exercises and are classified according to the section in which we suggest that they be assigned. Data for these exercises can be downloaded from **www.wiley.com/college/anton**.

About Chapter 11

This chapter consists of 21 applications of linear algebra. With one clearly marked exception, each application is in its own independent section, so that sections can be deleted or permuted freely to fit individual needs and interests. Each topic begins with a list of linear algebra prerequisites so that a reader can tell in advance if he or she has sufficient background to read the section.

Because the topics vary considerably in difficulty, we have included a subjective rating of each topic—easy, moderate, more difficult. (See "A Guide for the Instructor" following this preface.) Our evaluation is based more on the intrinsic difficulty of the material rather than the number of prerequisites; thus, a topic requiring fewer mathematical prerequisites may be rated harder than one requiring more prerequisites.

Because our primary objective is to present applications of linear algebra, proofs are often omitted. We assume that the reader has met the linear algebra prerequisites and whenever results from other fields are needed, they are stated precisely (with motivation where possible), but usually without proof.

Since there is more material in this book than can be covered in a one-semester or one-quarter course, the instructor will have to make a selection of topics. Help in making this selection is provided in the Guide for the Instructor below.

Supplementary Materials for Students

Student Solutions Manual to Accompany Elementary Linear Algebra Applications Version, Eighth Edition, by Elizabeth M. Grobe, Charles A. Grobe, Jr. (Bowdoin College), and Chris Rorres (Drexel University). This supplement provides detailed solutions to most theoretical exercises and to at least one nonroutine exercise of every type. (ISBN 0-471-38248-5)

Linear Algebra Applications Software by Howard Anton (Drexel University), Chris Rorres (Drexel University), and Intellipro, Inc. This is a set of ten modules, each of which focuses on an application of linear algebra. Each module has four parts:

1. A statement of the problem, supported by graphics and animations.
2. A discussion of the ideas relevant to the solution.
3. Interactive practice sessions that allow students to perform "what if" analyses by varying parameters and observing resulting simulations.
4. Exercises.

This software, which is provided for Windows 95 and NT, is available on CD or can be downloaded from **www.wiley.com/college/anton**

Data for Technology Exercises at the ends of Chapters is provided in MATLAB, Maple, and *Mathematica* formats. These data can be downloaded from **www.wiley.com/college/anton**

Supplementary Materials for Instructors

Linear Algebra Test Bank: *Randy Schwartz (Schoolcraft College).* This includes approximately 50 free-form questions, five essay questions for each chapter, and a sample cumulative final examination. Worked out solutions are given for each question. This manual is available in hard copy form or can be downloaded from **www.wiley.com/college/anton**

Wiley Web Tests for Linear Algebra: This is a system for assigning homework or giving examinations over the World Wide Web. Questions are multiple choice, free response, and true/false. The system can be used to administer tests in either practice or proctored modes. Students receive immediate feedback on grades. This powerful testing and homework management system is available to instructors upon adoption of this text.

Web Resources: More information about this text and its resources can be obtained from your Wiley representative or from **www.wiley.com/college/anton**

A GUIDE FOR THE INSTRUCTOR

Linear algebra courses vary widely between institutions in content and philosophy, but most courses fall into two categories: those with about 35–40 lectures (excluding tests and reviews) and those with about 25–30 lectures (excluding tests and reviews). Accordingly, long and short templates have been created as possible starting points for constructing a course outline. In the long template it is assumed that all sections in the indicated chapters are covered, and in the short template it is assumed that instructors will make selections from the chapters to fit the available time. Of course, these are just guides and you may want to customize them to fit your local interests and requirements.

The organization of the text has been carefully designed to make life easier for instructors working under time constraints: A brief introduction to eigenvalues and eigenvectors occurs in Sections 2.3 and 4.3, and linear transformations from R^n to R^m are discussed in Chapter 4. This makes it possible for all instructors to cover these topics at a basic level when the time available for their more extensive coverage in Chapters 7 and 8 is limited. Also, note that Chapter 3 can be omitted without loss of continuity for students who are already familiar with the material.

	Long Template	**Short Template**
Chapter 1	7 lectures	6 lectures
Chapter 2	4 lectures	3 lectures
Chapter 4	3 lectures	3 lectures
Chapter 5	8 lectures	7 lectures
Chapter 6	6 lectures	3 lectures
Chapter 7	4 lectures	3 lectures
Chapter 8	6 lectures	2 lectures
Total	38 lectures	27 lectures

Variations in the Standard Course

Many variations in the long template are possible. For example, one might create an alternative long template by following the time allocations in the short template and devoting the remaining 11 lectures to some of the topics in Chapters 9, 10, and 11.

An Applications-Oriented Course

Once the necessary core material is covered, the instructor can choose applications from Chapter 9 or Chapter 11. The following table classifies each of the 21 sections in Chapter 11 according to difficulty:

Easy—The average student who has met the stated prerequisites should be able to read the material with no help from the instructor.

Moderate—The average student who has met the stated prerequisites may require a little help from the instructor.

More Difficult—The average student who has met the stated prerequisites will probably need help from the instructor.

	1	2	3	4	5	6	7	8	9	10	11	12	13	14	15	16	17	18	19	20	21
EASY	●	●																			
MODERATE			●	●		●	●	●	●	●	●	●				●				●	●
MORE DIFFICULT					●								●	●	●		●	●	●		

Acknowledgments

We express our appreciation for the helpful guidance provided by the following people:

REVIEWERS AND CONTRIBUTORS TO EARLIER EDITIONS

Steven C. Althoen, *University of Michigan–Flint*
C. S. Ballantine, *Oregon State University*
Erol Barbut, *University of Idaho*
George Bergman, *University of California–Berkeley*
William A. Brown, *University of Maine*
Joseph Buckley, *Western Michigan University*
Thomas Cairns, *University of Tulsa*
Douglas E. Cameron, *University of Akron*
Bomshik Chang, *University of British Columbia*
Peter Colwell, *Iowa State University*
Carolyn A. Dean, *University of Michigan*
Ken Dunn, *Dalhousie University*
Bruce Edwards, *University of Florida*
Murray Eisenberg, *University of Massachusetts*
Harold S. Engelsohn, *Kingsborough Comm. College*
Garret Etgen, *University of Houston*
Marjorie E. Fitting, *San Jose State University*
Dan Flath, *University of South Alabama*
David E. Flesner, *Gettysburg College*
Mathew Gould, *Vanderbilt University*
Ralph P. Grimaldi, *Rose–Hulman Institute*
William W. Hager, *University of Florida*
Collin J. Hightower, *University of Colorado*
Joseph F. Johnson, *Rutgers University*
Robert L. Kelley, *University of Miami*
Arlene Kleinstein
Myren Krom, *California State University*
Lawrence D. Kugler, *University of Michigan*
Charles Livingston, *Indiana University*
Nicholas Macri, *Temple University*

Roger H. Marty, *Cleveland State University*
Patricia T. McAuley, *SUNY–Binghamton*
Robert M. McConnel, *University of Tennessee*
Douglas McLeod, *Drexel University*
Michael R. Meck, *Southern Connecticut State Univ.*
Craig Miller, *University of Pennsylvania*
Donald P. Minassian, *Butler University*
Hal G. Moore, *Brigham Young University*
Thomas E. Moore, *Bridgewater State College*
Robert W. Negus, *Rio Hondo Junior College*
Bart S. Ng, *Purdue University*
James Osterburg, *University of Cincinnati*
Michael A. Penna, *Indiana–Purdue University*
Gerald J. Porter, *University of Pennsylvania*
F. P. J. Rimrott, *University of Toronto*
C. Ray Rosentrater, *Westmont College*
Kenneth Schilling, *University of Michigan–Flint*
William Scott, *University of Utah*
Donald R. Sherbert, *University of Illinois*
Bruce Solomon, *Indiana University*
Mary T. Treanor, *Valparaiso University*
William F. Trench, *Trinity University*
Joseph L. Ullman, *University of Michigan*
W. Vance Underhill, *East Texas State University*
James R. Wall, *Auburn University*
Arthur G. Wasserman, *University of Michigan*
Evelyn J. Weinstock, *Glassboro State College*
Rugang Ye, *Stanford University*
Frank Zorzitto, *University of Waterloo*
Daniel Zwick, *University of Vermont*

Reviewers and Contributors to the Seventh Edition

Mark B. Beintema, *Southern Illinois University*
Paul Wayne Britt, *Louisiana State University*
David C. Buchthal, *University of Akron*
Keith Chavey, *University of Wisconsin–River Falls*

Stephen L. Davis, *Davidson College*
Blaise DeSesa, *Drexel University*
Dan Flath, *University of South Alabama*
Peter Fowler, *California State University*

Marc Frantz, *Indiana–Purdue University*
Sue Friedman, *Bernard M. Baruch College, CUNY*
William Golightly, *College of Charleston*
Hugh Haynsworth, *College of Charleston*
Tom Hern, *Bowling Green State University*
J. Hershenov, *Queens College, CUNY*
Steve Humphries, *Brigham Young University*
Steven Kahan, *Queens College, CUNY*
Andrew S. Kim, *Westfield State College*
John C. Lawlor, *University of Vermont*
M. Malek, *California State University at Hayward*
J. J. Malone, *Worcester Polytechnic Institute*

William McWorter, *Ohio State University*
Valerie A. Miller, *Georgia State University*
Hal G. Moore, *Brigham Young University*
S. Obaid, *San Jose State University*
Ira J. Papick, *University of Missouri–Columbia*
Donald Passman, *University of Wisconsin*
Robby Robson, *Oregon State University*
David Ryeburn, *Simon Fraser University*
Ramesh Sharma, *University of New Haven*
David A. Sibley, *Pennsylvania State University*
Donald Story, *University of Akron*
Michael Tarabek, *Southern Illinois University*

Problem Solutions, Proofreading, and Index

Michael Dagg, *Numerical Solutions, Inc.*
Susan L. Friedman, *Bernard M. Baruch*
 College, CUNY

Maureen Kelley, *Northern Essex Community College*
Randy Schwartz, *Schoolcraft College*
Daniel Traster (*Student*), *Yale University*

Supplements

Benny Evans, *Oklahoma State University*
Charles A. Grobe, Jr., *Bowdoin College*
Elizabeth M. Grobe

IntelliPro, Inc.
Jerry Johnson, *Oklahoma State University*
Randy Schwartz, *Schoolcraft College*

Other Contributions   Special thanks to the following professors who read the text material in depth and made significant contributions to the quality of the mathematics and the exposition:

George Bergman, *University of California–Berkeley*
Stephen Davis, *Davidson College*
Blaise DeSesa, *Drexel University*
Dan Flath, *University of South Alabama*
Marc Frantz, *Indiana–Purdue University*

William McWorter, *Ohio State University*
Donald Passman, *University of Wisconsin*
David Ryeburn, *Simon Fraser University*
Lois Craig Stagg, *University of Wisconsin–Milwaukee*

REVIEWERS AND CONTRIBUTORS TO THE EIGHTH EDITION

Richard Alfaro, *University of Michigan–Flint*
Stuart Boersma, *Alfred University*
Scott Chapman, *Trinity University*
KarabgDatta, *Northern Illinois University*
Mark Davis, *City College of San Francisco*
Alberto L. Delgado, *Kansas State University*
Willy Hereman, *Colorado School of Mines*
Chandanie Hetti-Arachchige, *Northern Illinois University*
Farhad Jafari, *University of Wyoming*
Eugene W. Johnson, *University of Iowa*
John Johnson, *George Fox College*

Steven Kahan, *Queens College*
John W. Krussel, *Lewis & Clark College*
Steffen Lempp, *University of Wisconsin*
Thomas A. Metzger, *University of Pittsburgh*
Gary L. Mullen, *Pennsylvania State University–University Park*
Sheldon Rothman, *Long Island University–C.W. Post*
Mark Sepanski, *Baylor University*
Sally Shao, *Cleveland State*
Evelyn Weinstock, *Rowan University*
T. J. Ypma, *Colorado School of Mines*

Answers and Solutions

Special thanks to the team of people who assisted with answers and solutions. Their work was outstanding, and we are grateful for their professionalism and attention to detail.

Charles A. Grobe, Jr., *Bowdoin College*
Elizabeth M. Grobe
Michael A. Carchidi, *Drexel University*
Scott Chapman, *Trinity University*
Mark Davis, *City College of San Francisco*

Herbert Kreyszig, M.B.A., *Columbia University*
Dr. Erwin Kreyszig, *Carleton University*
Donald Passman, *University of Wisconsin*
Jean Springer, *Mt. Royal College*

Mathematical Advisors

Special thanks are also due to two very talented mathematicians who read the manuscript in detail for technical accuracy and provided us with excellent advice on numerous pedagogical and mathematical matters.

Dean Hickerson
David Ryeburn, *Simon Fraser University*

Special Contributions

A special debt of gratitude to:

Barbara Holland—for working so hard and solving the many problems that were required to make this new edition a reality.

Ken Santor—for his attention to detail and his superb job in managing this project.

The Team at Techsetters—for their beautiful typesetting of a complex design and for making a very tight schedule go so smoothly.

Lilian Brady—for her usual unerring eye for typography and aesthetics.

Hilary Newman—for her remarkable ability to unearth the most obscure photographic and historical materials.

Roseann Zappia—for her constant good cheer and outstanding job in coordinating the work on the answers and Solutions Manual.

Dawn Stanley—for a beautiful design and cover.

The Wiley Production Staff—with special thanks to Lucille Buonocore, Maddy Lesure, Sigmund Malinowski, and Ann Berlin for their efforts behind the scenes and for their support on many books over the years.

HOWARD ANTON
CHRIS RORRES

Contents

CHAPTER 10   COMPLEX VECTOR SPACES   487

CHAPTER 11   APPLICATIONS OF LINEAR ALGEBRA   531

11.1 CONSTRUCTING CURVES AND SURFACES THROUGH SPECIFIED POINTS (p.532)

One of the problems that Isaac Newton considered in his monumental work *Mathematical Principles of Natural Philosophy (Principia Mathematica)* was the construction of an ellipse through five given points, illustrating how to find the orbit of a comet or planet with five observations. Determinants can be used to solve this problem analytically, in contrast to Newton's geometric procedure.

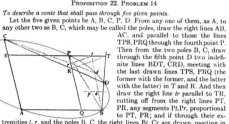

PROPOSITION 22. PROBLEM 14

To describe a conic that shall pass through five given points.

Let the five given points be A, B, C, P, D. From any one of them, as A, to any other two as B, C, which may be called the poles, draw the right lines AB, AC, and parallel to those the lines TPS, PRQ through the fourth point P. Then from the two poles B, C, draw through the fifth point D two indefinite lines BDT, CRD, meeting with the last drawn lines TPS, PRQ (the former with the former, and the latter with the latter) in T and R. And then draw the right line *tr* parallel to TR, cutting off from the right lines PT, PR, any segments P*t*,P*r*, proportional to PT, PR; and if through their extremities *t*, *r*, and the poles B, C, the right lines B*l*, C*r* are drawn, meeting in *d*, that point *d* will be placed in the conic required. For (by Lem. 20) that point *d* is placed in a conic section passing through the four points A, B, C, P; and the lines R*r*, T*t* vanishing, the point *d* comes to coincide with the point D. Wherefore the conic section passes through the five points A, B, C, P, D.

Q.E.D.

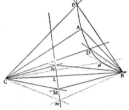

The same otherwise.

Of the given points join any three, as A, B, C; and about two of them B, C, as poles, making the angles ABC, ACB of a given magnitude to revolve, apply the legs BA, CA, first to the point D, then to the point P, and mark the points M, N, in which the other legs BL, CL intersect each other in both cases. Draw the indefinite right line MN, and let those movable angles revolve about their poles B, C, in such manner that the intersection, which is now supposed to be *m*, of the legs BL, CL, or BM, CM, may always fall in that indefinite right line MN; and the intersection, which is now supposed to be *d*, of the legs BA, CA, or BD, CD, will describe the conic required, PAD*d*B. For (by Lem. 21) the point *d* will be placed in a conic section passing through the points B, C; and when the point *m* comes to coincide with the points L, M, N, the point *d* will (by construction) come to coincide with the points A, D, P. Wherefore a conic section will be described that shall pass through the five points A, B, C, P, D.

Q.E.F.

COR. I. Hence a right line may be readily drawn which shall be a tangent to

11.2 ELECTRICAL NETWORKS (p. 538)

In the circuit board shown, the components labeled with the letter "R" are resistors, which limit the flow of electrical current. Circuits that contain only resistors and voltage supplies can be analyzed using systems of linear equations that result from the basic laws of circuit theory.

11.3 GEOMETRIC LINEAR PROGRAMMING (p. 542)

A common problem treated in the field of Linear Programming is determining the proportions of ingredients in a mixture in order to minimize its cost when the proportions may vary within certain limits. In business and industry an immense amount of computer time is devoted to linear programming problems.

11.4 THE ASSIGNMENT PROBLEM [p. 554]

Relocation of personnel and resources in a cost-efficient way is an important problem in industry. For example, a construction company may want to choose routes for moving bulldozers from its warehouses to its construction sites in a way that minimizes the total distance traveled.

11.5 CUBIC SPLINE INTERPOLATION [p. 565]

PostScript™ and TrueType™ fonts used for computer displays and printers are defined by piecewise polynomial curves called "splines". The parameters determining these splines are stored in the computer's memory, one set of parameters for each of the letters constituting a particular typeface.

11.6 MARKOV CHAINS [p. 576]

Weather records in a specific location can be used to estimate the probability that it will snow one day based on whether it snowed or not on the preceding day. The theory of Markov chains can utilize such data to predict the long term probabilities of a snow day in the location.

11.7 GRAPH THEORY [p. 587]

The social rankings among a group of animals is a relationship that can be described and analyzed through graph theory. Graph theory also has applications to such diverse problems as airline routing and analysis of voting patterns.

11.8 GAMES OF STRATEGY (p. 598)

In the game of roulette a player makes a move by placing a bet, and the casino makes a counter move by spinning a wheel; a payoff to the player or casino is determined from the two moves. These are the basic ingredients in a variety of games that contain elements of both strategy and chance. Matrix methods can be used to develop optimal strategies for the players.

11.9 LEONTIEF ECONOMIC MODELS (p. 608)

In a simple economic system, a coal mine, a railroad, and a power plant need portions of each other's output to run themselves and to supply other consumers of their product. Leontief production models can be used to determine the level of output of the three industries required to support the economic system.

11.10 FOREST MANAGEMENT (p. 618)

A manager of a Christmas tree farm wants to plant and harvest trees in a way that will perpetuate the forest configuration from year to year. The manager also seeks to maximize revenues, which depend on the numbers and sizes of the trees harvested. Matrix techniques can quantify this problem and help the manager select an appropriate sustainable harvesting policy.

11.11 COMPUTER GRAPHICS [p. 626]

Flight simulation is one of the most useful applications of computer graphics. Matrices provide a convenient way of handling the enormous bookkeeping required to construct and animate the three-dimensional objects used by flight simulators to depict a moving scene.

11.12 EQUILIBRIUM TEMPERATURE DISTRIBUTIONS [p. 636]

Determining the temperature distributions of objects, such as the steel coming out of a furnace, is a basic scientific and engineering task that can be reduced to solving a linear system of equations through iterative matrix techniques.

11.13 COMPUTED TOMOGRAPHY [p. 647]

The development of noninvasive methods such as CAT scans (Computer Aided Tomography) and MRI (Magnetic Resonance Imaging) to obtain images of cross sections of the human body has been a major advancement in medical diagnosis. Methods of linear algebra can be used to accomplish image reconstruction from CAT scan X-rays.

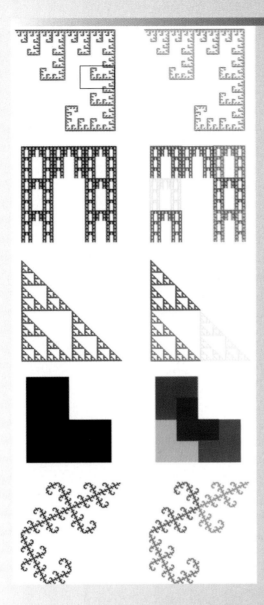

11.14 FRACTALS [p. 659]

Each of the black sets left can be broken into congruent scaled-down versions of the original set. This property is typical of sets called "fractals", which have recently been applied to the compression of computer data. Methods of linear algebra can be used to construct and classify fractals.

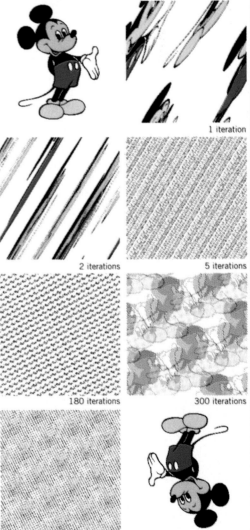

1 iteration

2 iterations

5 iterations

180 iterations

300 iterations

350 iterations

375 iterations

11.15 CHAOS [p. 678]

The pixels constituting the pixel map of Mickey Mouse undergo identical repeated shufflings in an attempt to randomize them. However, undesirable patterns keep coming up in the shuffling process. The matrix mapping that describes the shuffling process illustrates both the order and disorder that characterizes such "chaotic" processes.

11.16 CRYPTOGRAPHY [p. 692]

During World War II, American and British codebreakers succeeded in breaking enemy military codes using sophisticated mathematical techniques and machines. Today, computer and tele-communication confidentiality is the main impetus for developing secure codes.

11.17 GENETICS [p. 705]

The rulers of ancient Egypt practiced brother-sister marriages to keep the royal line pure. This propagated and accentuated certain genetic traits through many generations. Matrix theory provides a mathematical framework to examine the general propagation of genetic traits.

11.18 AGE-SPECIFIC POPULATION GROWTH [p. 716]

Given the age-specific birth and death rates of a population, matrix algebra can be applied to project the population configuration forward in time. The long-term evolution of the population depends on mathematical characteristics of a projection matrix that contains the population's demographic parameters.

11.19 HARVESTING OF ANIMAL POPULATIONS [p. 727]

Sustainable harvesting of animal populations requires knowledge of the demographics of the population. To maximize the yield of a periodic harvest, different sustainable harvesting strategies can be compared through matrix techniques that describe the population's growth dynamics.

11.20 A LEAST-SQUARES MODEL FOR HUMAN HEARING [p. 735]

The inner ear contains a structure with thousands of hairlike sensory receptors. These receptors, driven by the vibrations of the eardrum, respond to different frequencies according to their locations and produce electrical impulses that travel to the brain through the auditory nerve. In this way the inner ear acts as a signal processor that decomposes a complicated sound wave into a spectrum of different frequencies.

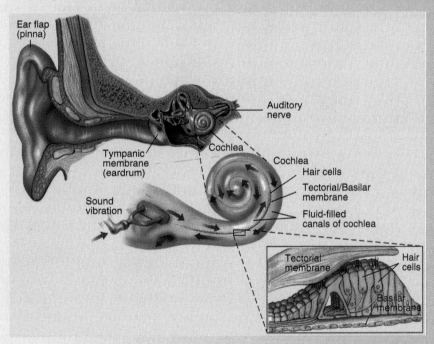

WARPS AND MORPHS [p. 742]

Of the sixeen images displayed showing a woman over a fifty-year period, only the four diagonal ones from top-left to bottom–right are actual photographs. The others are computer–generated images called "morphs", which are blends of the actual photographs. Such image-manipulation techniques have found countless applications in the medical, scientific, and entertainment industries.

Systems of Linear Equations and Matrices

Chapter Contents

INTRODUCTION: Information in science and mathematics is often organized into rows and columns to form rectangular arrays, called "matrices" (plural of "matrix"). Matrices are often tables of numerical data that arise from physical observations, but they also occur in various mathematical contexts. For example, we shall see in this chapter that to solve a system of equations such as

$$5x + y = 3$$
$$2x - y = 4$$

all of the information required for the solution is embodied in the matrix

$$\begin{bmatrix} 5 & 1 & 3 \\ 2 & -1 & 4 \end{bmatrix}$$

and that the solution can be obtained by performing appropriate operations on this matrix. This is particularly important in developing computer programs to solve systems of linear equations because computers are well suited for manipulating arrays of numerical information. However, matrices are not simply a notational tool for solving systems of equations; they can be viewed as mathematical objects in their own right, and there is a rich and important theory associated with them that has a wide variety of applications. In this chapter we will begin the study of matrices.

1.1 INTRODUCTION TO SYSTEMS OF LINEAR EQUATIONS

Systems of linear algebraic equations and their solutions constitute one of the major topics studied in the course known as "linear algebra." In this first section we shall introduce some basic terminology and discuss a method for solving such systems.

Linear Equations   Any straight line in the xy-plane can be represented algebraically by an equation of the form

$$a_1 x + a_2 y = b$$

where a_1, a_2, and b are real constants and a_1 and a_2 are not both zero. An equation of this form is called a linear equation in the variables x and y. More generally, we define a ***linear equation*** in the n variables $x_1, x_2, \ldots, x_n$ to be one that can be expressed in the form

$$a_1 x_1 + a_2 x_2 + \cdots + a_n x_n = b$$

where $a_1, a_2, \ldots, a_n$, and b are real constants. The variables in a linear equation are sometimes called ***unknowns***.

EXAMPLE 1   Linear Equations

The equations

$$x + 3y = 7, \quad y = \tfrac{1}{2}x + 3z + 1, \quad \text{and} \quad x_1 - 2x_2 - 3x_3 + x_4 = 7$$

are linear. Observe that a linear equation does not involve any products or roots of variables. All variables occur only to the first power and do not appear as arguments for trigonometric, logarithmic, or exponential functions. The equations

$$x + 3\sqrt{y} = 5, \quad 3x + 2y - z + xz = 4, \quad \text{and} \quad y = \sin x$$

are *not* linear. ◆

A ***solution*** of a linear equation $a_1 x_1 + a_2 x_2 + \cdots + a_n x_n = b$ is a sequence of n numbers $s_1, s_2, \ldots, s_n$ such that the equation is satisfied when we substitute $x_1 = s_1$, $x_2 = s_2, \ldots, x_n = s_n$. The set of all solutions of the equation is called its ***solution set*** or sometimes the ***general solution*** of the equation.

EXAMPLE 2   Finding a Solution Set

Find the solution set of (a) $4x - 2y = 1$, and (b) $x_1 - 4x_2 + 7x_3 = 5$.

Solution (a).   To find solutions of (a), we can assign an arbitrary value to x and solve for y, or choose an arbitrary value for y and solve for x. If we follow the first approach and assign x an arbitrary value t, we obtain

$$x = t, \quad y = 2t - \tfrac{1}{2}$$

These formulas describe the solution set in terms of an arbitrary number t, called a **parameter**. Particular numerical solutions can be obtained by substituting specific values for t. For example, $t = 3$ yields the solution $x = 3$, $y = \frac{11}{2}$; and $t = -\frac{1}{2}$ yields the solution $x = -\frac{1}{2}$, $y = -\frac{3}{2}$.

If we follow the second approach and assign y the arbitrary value t, we obtain

$$x = \tfrac{1}{2}t + \tfrac{1}{4}, \qquad y = t$$

Although these formulas are different from those obtained above, they yield the same solution set as t varies over all possible real numbers. For example, the previous formulas gave the solution $x = 3$, $y = \frac{11}{2}$ when $t = 3$, while the formulas immediately above yield that solution when $t = \frac{11}{2}$.

Solution (b). To find the solution set of (b) we can assign arbitrary values to any two variables and solve for the third variable. In particular, if we assign arbitrary values s and t to x_2 and x_3, respectively, and solve for x_1, we obtain

$$x_1 = 5 + 4s - 7t, \qquad x_2 = s, \qquad x_3 = t \qquad\qquad ◆$$

Linear Systems   A finite set of linear equations in the variables $x_1, x_2, \ldots, x_n$ is called a **system of linear equations** or a **linear system**. A sequence of numbers $s_1, s_2, \ldots, s_n$ is called a **solution** of the system if $x_1 = s_1, x_2 = s_2, \ldots, x_n = s_n$ is a solution of every equation in the system. For example, the system

$$4x_1 - x_2 + 3x_3 = -1$$
$$3x_1 + x_2 + 9x_3 = -4$$

has the solution $x_1 = 1$, $x_2 = 2$, $x_3 = -1$ since these values satisfy both equations. However, $x_1 = 1$, $x_2 = 8$, $x_3 = 1$ is not a solution since these values satisfy only the first of the two equations in the system.

Not all systems of linear equations have solutions. For example, if we multiply the second equation of the system

$$x + y = 4$$
$$2x + 2y = 6$$

by $\frac{1}{2}$, it becomes evident that there are no solutions since the resulting equivalent system

$$x + y = 4$$
$$x + y = 3$$

has contradictory equations.

A system of equations that has no solutions is said to be **inconsistent**; if there is at least one solution of the system, it is called **consistent**. To illustrate the possibilities that can occur in solving systems of linear equations, consider a general system of two linear equations in the unknowns x and y:

$$a_1 x + b_1 y = c_1 \quad (a_1, b_1 \text{ not both zero})$$
$$a_2 x + b_2 y = c_2 \quad (a_2, b_2 \text{ not both zero})$$

The graphs of these equations are lines; call them l_1 and l_2. Since a point (x, y) lies on a line if and only if the numbers x and y satisfy the equation of the line, the solutions of the system of equations correspond to points of intersection of l_1 and l_2. There are three possibilities illustrated in Figure 1.1.1:

- The lines l_1 and l_2 may be parallel, in which case there is no intersection and consequently no solution to the system.

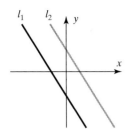

(a) No solution

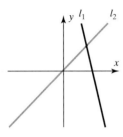

(b) One solution

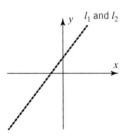

(c) Infinitely many solutions

Figure 1.1.1

• The lines l_1 and l_2 may intersect at only one point, in which case the system has exactly one solution.

• The lines l_1 and l_2 may coincide, in which case there are infinitely many points of intersection and consequently infinitely many solutions to the system.

Although we have considered only two equations with two unknowns here, we will show later that the same three possibilities hold for arbitrary linear systems:

Every system of linear equations has either no solutions, exactly one solution, or infinitely many solutions.

An arbitrary system of m linear equations in n unknowns can be written as

$$a_{11}x_1 + a_{12}x_2 + \cdots + a_{1n}x_n = b_1$$
$$a_{21}x_1 + a_{22}x_2 + \cdots + a_{2n}x_n = b_2$$
$$\vdots \qquad \vdots \qquad \qquad \vdots \qquad \vdots$$
$$a_{m1}x_1 + a_{m2}x_2 + \cdots + a_{mn}x_n = b_m$$

where $x_1, x_2, \ldots, x_n$ are the unknowns and the subscripted a's and b's denote constants. For example, a general system of three linear equations in four unknowns can be written as

$$a_{11}x_1 + a_{12}x_2 + a_{13}x_3 + a_{14}x_4 = b_1$$
$$a_{21}x_1 + a_{22}x_2 + a_{23}x_3 + a_{24}x_4 = b_2$$
$$a_{31}x_1 + a_{32}x_2 + a_{33}x_3 + a_{34}x_4 = b_3$$

The double subscripting on the coefficients of the unknowns is a useful device that is used to specify the location of the coefficient in the system. The first subscript on the coefficient a_{ij} indicates the equation in which the coefficient occurs, and the second subscript indicates which unknown it multiplies. Thus, a_{12} is in the first equation and multiplies unknown x_2.

Augmented Matrices   If we mentally keep track of the location of the $+$'s, the x's, and the $=$'s, a system of m linear equations in n unknowns can be abbreviated by writing only the rectangular array of numbers:

$$\begin{bmatrix} a_{11} & a_{12} & \cdots & a_{1n} & b_1 \\ a_{21} & a_{22} & \cdots & a_{2n} & b_2 \\ \vdots & \vdots & & \vdots & \vdots \\ a_{m1} & a_{m2} & \cdots & a_{mn} & b_m \end{bmatrix}$$

This is called the **augmented matrix** for the system. (The term *matrix* is used in mathematics to denote a rectangular array of numbers. Matrices arise in many contexts, which we will consider in more detail in later sections.) For example, the augmented matrix for the system of equations

$$x_1 + x_2 + 2x_3 = 9$$
$$2x_1 + 4x_2 - 3x_3 = 1$$
$$3x_1 + 6x_2 - 5x_3 = 0$$

is

$$\begin{bmatrix} 1 & 1 & 2 & 9 \\ 2 & 4 & -3 & 1 \\ 3 & 6 & -5 & 0 \end{bmatrix}$$

REMARK. When constructing an augmented matrix, the unknowns must be written in the same order in each equation and the constants must be on the right.

The basic method for solving a system of linear equations is to replace the given system by a new system that has the same solution set but which is easier to solve. This new system is generally obtained in a series of steps by applying the following three types of operations to eliminate unknowns systematically.

1. Multiply an equation through by a nonzero constant.
2. Interchange two equations.
3. Add a multiple of one equation to another.

Since the rows (horizontal lines) of an augmented matrix correspond to the equations in the associated system, these three operations correspond to the following operations on the rows of the augmented matrix.

1. Multiply a row through by a nonzero constant.
2. Interchange two rows.
3. Add a multiple of one row to another row.

Elementary Row Operations These are called *elementary row operations*. The following example illustrates how these operations can be used to solve systems of linear equations. Since a systematic procedure for finding solutions will be derived in the next section, it is not necessary to worry about how the steps in this example were selected. The main effort at this time should be devoted to understanding the computations and the discussion.

EXAMPLE 3 Using Elementary Row Operations

In the left column below we solve a system of linear equations by operating on the equations in the system, and in the right column we solve the same system by operating on the rows of the augmented matrix.

$$\begin{array}{rcl} x + y + 2z &=& 9 \\ 2x + 4y - 3z &=& 1 \\ 3x + 6y - 5z &=& 0 \end{array} \qquad \begin{bmatrix} 1 & 1 & 2 & 9 \\ 2 & 4 & -3 & 1 \\ 3 & 6 & -5 & 0 \end{bmatrix}$$

Add -2 times the first equation to the second to obtain

Add -2 times the first row to the second to obtain

$$\begin{array}{rcr} x + y + 2z &=& 9 \\ 2y - 7z &=& -17 \\ 3x + 6y - 5z &=& 0 \end{array} \qquad \begin{bmatrix} 1 & 1 & 2 & 9 \\ 0 & 2 & -7 & -17 \\ 3 & 6 & -5 & 0 \end{bmatrix}$$

Add -3 times the first equation to the third to obtain

Add -3 times the first row to the third to obtain

$$\begin{array}{rcr} x + y + 2z &=& 9 \\ 2y - 7z &=& -17 \\ 3y - 11z &=& -27 \end{array} \qquad \begin{bmatrix} 1 & 1 & 2 & 9 \\ 0 & 2 & -7 & -17 \\ 0 & 3 & -11 & -27 \end{bmatrix}$$

Multiply the second equation by $\frac{1}{2}$ to obtain

$$\begin{aligned} x + y + 2z &= 9 \\ y - \tfrac{7}{2}z &= -\tfrac{17}{2} \\ 3y - 11z &= -27 \end{aligned}$$

Multiply the second row by $\frac{1}{2}$ to obtain

$$\begin{bmatrix} 1 & 1 & 2 & 9 \\ 0 & 1 & -\tfrac{7}{2} & -\tfrac{17}{2} \\ 0 & 3 & -11 & -27 \end{bmatrix}$$

Add -3 times the second equation to the third to obtain

$$\begin{aligned} x + y + 2z &= 9 \\ y - \tfrac{7}{2}z &= -\tfrac{17}{2} \\ -\tfrac{1}{2}z &= -\tfrac{3}{2} \end{aligned}$$

Add -3 times the second row to the third to obtain

$$\begin{bmatrix} 1 & 1 & 2 & 9 \\ 0 & 1 & -\tfrac{7}{2} & -\tfrac{17}{2} \\ 0 & 0 & -\tfrac{1}{2} & -\tfrac{3}{2} \end{bmatrix}$$

Multiply the third equation by -2 to obtain

$$\begin{aligned} x + y + 2z &= 9 \\ y - \tfrac{7}{2}z &= -\tfrac{17}{2} \\ z &= 3 \end{aligned}$$

Multiply the third row by -2 to obtain

$$\begin{bmatrix} 1 & 1 & 2 & 9 \\ 0 & 1 & -\tfrac{7}{2} & -\tfrac{17}{2} \\ 0 & 0 & 1 & 3 \end{bmatrix}$$

Add -1 times the second equation to the first to obtain

$$\begin{aligned} x + \tfrac{11}{2}z &= \tfrac{35}{2} \\ y - \tfrac{7}{2}z &= -\tfrac{17}{2} \\ z &= 3 \end{aligned}$$

Add -1 times the second row to the first to obtain

$$\begin{bmatrix} 1 & 0 & \tfrac{11}{2} & \tfrac{35}{2} \\ 0 & 1 & -\tfrac{7}{2} & -\tfrac{17}{2} \\ 0 & 0 & 1 & 3 \end{bmatrix}$$

Add $-\frac{11}{2}$ times the third equation to the first and $\frac{7}{2}$ times the third equation to the second to obtain

$$\begin{aligned} x &= 1 \\ y &= 2 \\ z &= 3 \end{aligned}$$

Add $-\frac{11}{2}$ times the third row to the first and $\frac{7}{2}$ times the third row to the second to obtain

$$\begin{bmatrix} 1 & 0 & 0 & 1 \\ 0 & 1 & 0 & 2 \\ 0 & 0 & 1 & 3 \end{bmatrix}$$

The solution $x = 1$, $y = 2$, $z = 3$ is now evident. ◆

Exercise Set 1.1

1. Which of the following are linear equations in x_1, x_2, and x_3?

(a) $x_1 + 5x_2 - \sqrt{2}x_3 = 1$ (b) $x_1 + 3x_2 + x_1x_3 = 2$ (c) $x_1 = -7x_2 + 3x_3$

(d) $x_1^{-2} + x_2 + 8x_3 = 5$ (e) $x_1^{3/5} - 2x_2 + x_3 = 4$ (f) $\pi x_1 - \sqrt{2}x_2 + \tfrac{1}{3}x_3 = 7^{1/3}$

2. Given that k is a constant, which of the following are linear equations?

(a) $x_1 - x_2 + x_3 = \sin k$ (b) $kx_1 - \dfrac{1}{k}x_2 = 9$ (c) $2^k x_1 + 7x_2 - x_3 = 0$

3. Find the solution set of each of the following linear equations.

(a) $7x - 5y = 3$ (b) $3x_1 - 5x_2 + 4x_3 = 7$

(c) $-8x_1 + 2x_2 - 5x_3 + 6x_4 = 1$ (d) $3v - 8w + 2x - y + 4z = 0$

4. Find the augmented matrix for each of the following systems of linear equations.

(a) $\begin{aligned} 3x_1 - 2x_2 &= -1 \\ 4x_1 + 5x_2 &= 3 \\ 7x_1 + 3x_2 &= 2 \end{aligned}$ (b) $\begin{aligned} 2x_1 \qquad + 2x_3 &= 1 \\ 3x_1 - x_2 + 4x_3 &= 7 \\ 6x_1 + x_2 - x_3 &= 0 \end{aligned}$ (c) $\begin{aligned} x_1 + 2x_2 \qquad - x_4 + x_5 &= 1 \\ 3x_2 + x_3 \qquad - x_5 &= 2 \\ x_3 + 7x_4 \qquad &= 1 \end{aligned}$ (d) $\begin{aligned} x_1 \qquad &= 1 \\ x_2 \quad &= 2 \\ x_3 &= 3 \end{aligned}$

5. Find a system of linear equations corresponding to the augmented matrix.

(a) $\begin{bmatrix} 2 & 0 & 0 \\ 3 & -4 & 0 \\ 0 & 1 & 1 \end{bmatrix}$
(b) $\begin{bmatrix} 3 & 0 & -2 & 5 \\ 7 & 1 & 4 & -3 \\ 0 & -2 & 1 & 7 \end{bmatrix}$

(c) $\begin{bmatrix} 7 & 2 & 1 & -3 & 5 \\ 1 & 2 & 4 & 0 & 1 \end{bmatrix}$
(d) $\begin{bmatrix} 1 & 0 & 0 & 0 & 7 \\ 0 & 1 & 0 & 0 & -2 \\ 0 & 0 & 1 & 0 & 3 \\ 0 & 0 & 0 & 1 & 4 \end{bmatrix}$

6. (a) Find a linear equation in the variables x and y that has the general solution $x = 5 + 2t$, $y = t$.

(b) Show that $x = t$, $y = \frac{1}{2}t - \frac{5}{2}$ is also the general solution of the equation in part (a).

7. The curve $y = ax^2 + bx + c$ shown in the accompanying figure passes through the points (x_1, y_1), (x_2, y_2), and (x_3, y_3). Show that the coefficients a, b, and c are a solution of the system of linear equations whose augmented matrix is

$\begin{bmatrix} x_1^2 & x_1 & 1 & y_1 \\ x_2^2 & x_2 & 1 & y_2 \\ x_3^2 & x_3 & 1 & y_3 \end{bmatrix}$

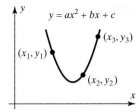

Figure Ex-7

8. Consider the system of equations

$$x + y + 2z = a$$
$$x \quad\;\; + z = b$$
$$2x + y + 3z = c$$

Show that for this system to be consistent, the constants a, b, and c must satisfy $c = a + b$.

9. Show that if the linear equations $x_1 + kx_2 = c$ and $x_1 + lx_2 = d$ have the same solution set, then the equations are identical.

Discussion and Discovery

10. For which value(s) of the constant k does the system

$$x - y = 3$$
$$2x - 2y = k$$

have no solutions? Exactly one solution? Infinitely many solutions? Explain your reasoning.

11. Consider the system of equations

$$ax + by = k$$
$$cx + dy = l$$
$$ex + fy = m$$

What can you say about the relative positions of the lines $ax + by = k$, $cx + dy = l$, and $ex + fy = m$ when

(a) the system has no solutions;

(b) the system has exactly one solution;

(c) the system has infinitely many solutions?

12. If the system of equations in Exercise 11 is consistent, explain why at least one equation can be discarded from the system without altering the solution set.

13. If $k = l = m = 0$ in Exercise 11, explain why the system must be consistent. What can be said about the point of intersection of the three lines if the system has exactly one solution?

1.2 GAUSSIAN ELIMINATION

In this section we shall develop a systematic procedure for solving systems of linear equations. The procedure is based on the idea of reducing the augmented matrix of a system to another augmented matrix that is simple enough that the solution of the system can be found by inspection.

Echelon Forms   In Example 3 of the last section, we solved a linear system in the unknowns x, y, and z by reducing the augmented matrix to the form

$$\begin{bmatrix} 1 & 0 & 0 & 1 \\ 0 & 1 & 0 & 2 \\ 0 & 0 & 1 & 3 \end{bmatrix}$$

from which the solution $x = 1$, $y = 2$, $z = 3$ became evident. This is an example of a matrix that is in **reduced row-echelon form**. To be of this form a matrix must have the following properties:

  1. If a row does not consist entirely of zeros, then the first nonzero number in the row is a 1. We call this a **leading 1**.

  2. If there are any rows that consist entirely of zeros, then they are grouped together at the bottom of the matrix.

  3. In any two successive rows that do not consist entirely of zeros, the leading 1 in the lower row occurs farther to the right than the leading 1 in the higher row.

  4. Each column that contains a leading 1 has zeros everywhere else.

A matrix that has the first three properties is said to be in **row-echelon form**. (Thus, a matrix in reduced row-echelon form is of necessity in row-echelon form, but not conversely.)

EXAMPLE 1 Row-Echelon and Reduced Row-Echelon Form

The following matrices are in reduced row-echelon form.

$$\begin{bmatrix} 1 & 0 & 0 & 4 \\ 0 & 1 & 0 & 7 \\ 0 & 0 & 1 & -1 \end{bmatrix}, \quad \begin{bmatrix} 1 & 0 & 0 \\ 0 & 1 & 0 \\ 0 & 0 & 1 \end{bmatrix}, \quad \begin{bmatrix} 0 & 1 & -2 & 0 & 1 \\ 0 & 0 & 0 & 1 & 3 \\ 0 & 0 & 0 & 0 & 0 \\ 0 & 0 & 0 & 0 & 0 \end{bmatrix}, \quad \begin{bmatrix} 0 & 0 \\ 0 & 0 \end{bmatrix}$$

The following matrices are in row-echelon form.

$$\begin{bmatrix} 1 & 4 & -3 & 7 \\ 0 & 1 & 6 & 2 \\ 0 & 0 & 1 & 5 \end{bmatrix}, \quad \begin{bmatrix} 1 & 1 & 0 \\ 0 & 1 & 0 \\ 0 & 0 & 0 \end{bmatrix}, \quad \begin{bmatrix} 0 & 1 & 2 & 6 & 0 \\ 0 & 0 & 1 & -1 & 0 \\ 0 & 0 & 0 & 0 & 1 \end{bmatrix}$$

We leave it for you to confirm that each of the matrices in this example satisfies all of the requirements for its stated form.   ◆

EXAMPLE 2 More on Row-Echelon and Reduced Row-Echelon Form

As the last example illustrates, a matrix in row-echelon form has zeros below each leading 1, whereas a matrix in reduced row-echelon form has zeros below *and above* each leading 1. Thus, with any real numbers substituted for the $*$'s, all matrices of the following types are in row-echelon form:

$$\begin{bmatrix} 1 & * & * & * \\ 0 & 1 & * & * \\ 0 & 0 & 1 & * \\ 0 & 0 & 0 & 1 \end{bmatrix}, \quad \begin{bmatrix} 1 & * & * & * \\ 0 & 1 & * & * \\ 0 & 0 & 1 & * \\ 0 & 0 & 0 & 0 \end{bmatrix},$$

$$\begin{bmatrix} 1 & * & * & * \\ 0 & 1 & * & * \\ 0 & 0 & 0 & 0 \\ 0 & 0 & 0 & 0 \end{bmatrix}, \quad \begin{bmatrix} 0 & 1 & * & * & * & * & * & * & * \\ 0 & 0 & 0 & 1 & * & * & * & * & * \\ 0 & 0 & 0 & 0 & 1 & * & * & * & * \\ 0 & 0 & 0 & 0 & 0 & 1 & * & * & * \\ 0 & 0 & 0 & 0 & 0 & 0 & 0 & 1 & * \end{bmatrix}$$

Moreover, all matrices of the following types are in reduced row-echelon form:

$$\begin{bmatrix} 1 & 0 & 0 & 0 \\ 0 & 1 & 0 & 0 \\ 0 & 0 & 1 & 0 \\ 0 & 0 & 0 & 1 \end{bmatrix}, \quad \begin{bmatrix} 1 & 0 & 0 & * \\ 0 & 1 & 0 & * \\ 0 & 0 & 1 & * \\ 0 & 0 & 0 & 0 \end{bmatrix},$$

$$\begin{bmatrix} 1 & 0 & * & * \\ 0 & 1 & * & * \\ 0 & 0 & 0 & 0 \\ 0 & 0 & 0 & 0 \end{bmatrix}, \quad \begin{bmatrix} 0 & 1 & * & 0 & 0 & 0 & * & * & 0 & * \\ 0 & 0 & 0 & 1 & 0 & 0 & * & * & 0 & * \\ 0 & 0 & 0 & 0 & 1 & 0 & * & * & 0 & * \\ 0 & 0 & 0 & 0 & 0 & 1 & * & * & 0 & * \\ 0 & 0 & 0 & 0 & 0 & 0 & 0 & 0 & 1 & * \end{bmatrix} \quad \blacklozenge$$

If, by a sequence of elementary row operations, the augmented matrix for a system of linear equations is put in reduced row-echelon form, then the solution set of the system will be evident by inspection or after a few simple steps. The next example illustrates this situation.

EXAMPLE 3 **Solutions of Four Linear Systems**

Suppose that the augmented matrix for a system of linear equations has been reduced by row operations to the given reduced row-echelon form. Solve the system.

(a) $\begin{bmatrix} 1 & 0 & 0 & 5 \\ 0 & 1 & 0 & -2 \\ 0 & 0 & 1 & 4 \end{bmatrix}$
(b) $\begin{bmatrix} 1 & 0 & 0 & 4 & -1 \\ 0 & 1 & 0 & 2 & 6 \\ 0 & 0 & 1 & 3 & 2 \end{bmatrix}$

(c) $\begin{bmatrix} 1 & 6 & 0 & 0 & 4 & -2 \\ 0 & 0 & 1 & 0 & 3 & 1 \\ 0 & 0 & 0 & 1 & 5 & 2 \\ 0 & 0 & 0 & 0 & 0 & 0 \end{bmatrix}$
(d) $\begin{bmatrix} 1 & 0 & 0 & 0 \\ 0 & 1 & 2 & 0 \\ 0 & 0 & 0 & 1 \end{bmatrix}$

Solution (*a*). The corresponding system of equations is

$$\begin{aligned} x_1 & & & = 5 \\ & x_2 & & = -2 \\ & & x_3 & = 4 \end{aligned}$$

By inspection, $x_1 = 5$, $x_2 = -2$, $x_3 = 4$.

Solution (b). The corresponding system of equations is

$$
\begin{aligned}
x_1 \quad\quad\quad\quad + 4x_4 &= -1 \\
x_2 \quad\quad + 2x_4 &= 6 \\
x_3 + 3x_4 &= 2
\end{aligned}
$$

Since x_1, x_2, and x_3 correspond to leading 1's in the augmented matrix, we call them *leading variables*. The nonleading variables (in this case x_4) are called *free variables*. Solving for the leading variables in terms of the free variable gives

$$
\begin{aligned}
x_1 &= -1 - 4x_4 \\
x_2 &= 6 - 2x_4 \\
x_3 &= 2 - 3x_4
\end{aligned}
$$

From this form of the equations we see that the free variable x_4 can be assigned an arbitrary value, say t, which then determines the values of the leading variables x_1, x_2, and x_3. Thus there are infinitely many solutions, and the general solution is given by the formulas

$$
x_1 = -1 - 4t, \quad x_2 = 6 - 2t, \quad x_3 = 2 - 3t, \quad x_4 = t
$$

Solution (c). The row of zeros leads to the equation $0x_1 + 0x_2 + 0x_3 + 0x_4 + 0x_5 = 0$, which places no restrictions on the solutions (why?). Thus, we can omit this equation and write the corresponding system as

$$
\begin{aligned}
x_1 + 6x_2 \quad\quad\quad + 4x_5 &= -2 \\
x_3 \quad\quad + 3x_5 &= 1 \\
x_4 + 5x_5 &= 2
\end{aligned}
$$

Here the leading variables are x_1, x_3, and x_4, and the free variables are x_2 and x_5. Solving for the leading variables in terms of the free variables gives

$$
\begin{aligned}
x_1 &= -2 - 6x_2 - 4x_5 \\
x_3 &= 1 - 3x_5 \\
x_4 &= 2 - 5x_5
\end{aligned}
$$

Since x_5 can be assigned an arbitrary value, t, and x_2 can be assigned an arbitrary value, s, there are infinitely many solutions. The general solution is given by the formulas

$$
x_1 = -2 - 6s - 4t, \quad x_2 = s, \quad x_3 = 1 - 3t, \quad x_4 = 2 - 5t, \quad x_5 = t
$$

Solution (d). The last equation in the corresponding system of equations is

$$
0x_1 + 0x_2 + 0x_3 = 1
$$

Since this equation cannot be satisfied, there is no solution to the system. ◆

Elimination Methods

We have just seen how easy it is to solve a system of linear equations once its augmented matrix is in reduced row-echelon form. Now we shall give a step-by-step *elimination* procedure that can be used to reduce any matrix to reduced row-echelon form. As we state each step in the procedure, we shall illustrate the idea by reducing the following matrix to reduced row-echelon form.

$$
\begin{bmatrix}
0 & 0 & -2 & 0 & 7 & 12 \\
2 & 4 & -10 & 6 & 12 & 28 \\
2 & 4 & -5 & 6 & -5 & -1
\end{bmatrix}
$$

Step 1. Locate the leftmost column that does not consist entirely of zeros.

$$\begin{bmatrix} 0 & 0 & -2 & 0 & 7 & 12 \\ 2 & 4 & -10 & 6 & 12 & 28 \\ 2 & 4 & -5 & 6 & -5 & -1 \end{bmatrix}$$

    ↑
└── **Leftmost nonzero column**

Step 2. Interchange the top row with another row, if necessary, to bring a nonzero entry to the top of the column found in Step 1.

$$\begin{bmatrix} 2 & 4 & -10 & 6 & 12 & 28 \\ 0 & 0 & -2 & 0 & 7 & 12 \\ 2 & 4 & -5 & 6 & -5 & -1 \end{bmatrix}$$ ←—— The first and second rows in the preceding matrix were interchanged.

Step 3. If the entry that is now at the top of the column found in Step 1 is a, multiply the first row by $1/a$ in order to introduce a leading 1.

$$\begin{bmatrix} 1 & 2 & -5 & 3 & 6 & 14 \\ 0 & 0 & -2 & 0 & 7 & 12 \\ 2 & 4 & -5 & 6 & -5 & -1 \end{bmatrix}$$ ←—— The first row of the preceding matrix was multiplied by $\frac{1}{2}$.

Step 4. Add suitable multiples of the top row to the rows below so that all entries below the leading 1 become zeros.

$$\begin{bmatrix} 1 & 2 & -5 & 3 & 6 & 14 \\ 0 & 0 & -2 & 0 & 7 & 12 \\ 0 & 0 & 5 & 0 & -17 & -29 \end{bmatrix}$$ ←—— -2 times the first row of the preceding matrix was added to the third row.

Step 5. Now cover the top row in the matrix and begin again with Step 1 applied to the submatrix that remains. Continue in this way until the *entire* matrix is in row-echelon form.

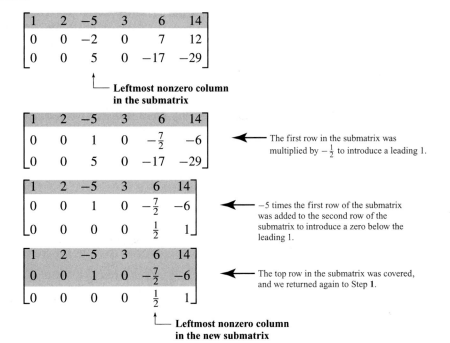

$$\begin{bmatrix} 1 & 2 & -5 & 3 & 6 & 14 \\ 0 & 0 & -2 & 0 & 7 & 12 \\ 0 & 0 & 5 & 0 & -17 & -29 \end{bmatrix}$$

    ↑
└── **Leftmost nonzero column**
      in the submatrix

$$\begin{bmatrix} 1 & 2 & -5 & 3 & 6 & 14 \\ 0 & 0 & 1 & 0 & -\frac{7}{2} & -6 \\ 0 & 0 & 5 & 0 & -17 & -29 \end{bmatrix}$$ ←—— The first row in the submatrix was multiplied by $-\frac{1}{2}$ to introduce a leading 1.

$$\begin{bmatrix} 1 & 2 & -5 & 3 & 6 & 14 \\ 0 & 0 & 1 & 0 & -\frac{7}{2} & -6 \\ 0 & 0 & 0 & 0 & \frac{1}{2} & 1 \end{bmatrix}$$ ←—— -5 times the first row of the submatrix was added to the second row of the submatrix to introduce a zero below the leading 1.

$$\begin{bmatrix} 1 & 2 & -5 & 3 & 6 & 14 \\ 0 & 0 & 1 & 0 & -\frac{7}{2} & -6 \\ 0 & 0 & 0 & 0 & \frac{1}{2} & 1 \end{bmatrix}$$ ←—— The top row in the submatrix was covered, and we returned again to Step **1**.

    ↑
└── **Leftmost nonzero column**
      in the new submatrix

$$\begin{bmatrix} 1 & 2 & -5 & 3 & 6 & 14 \\ 0 & 0 & 1 & 0 & -\frac{7}{2} & -6 \\ 0 & 0 & 0 & 0 & 1 & 2 \end{bmatrix}$$

⟵ The first (and only) row in the new submatrix was multiplied by 2 to introduce a leading 1.

The *entire* matrix is now in row-echelon form. To find the reduced row-echelon form we need the following additional step.

Step 6. Beginning with the last nonzero row and working upward, add suitable multiples of each row to the rows above to introduce zeros above the leading 1's.

$$\begin{bmatrix} 1 & 2 & -5 & 3 & 6 & 14 \\ 0 & 0 & 1 & 0 & 0 & 1 \\ 0 & 0 & 0 & 0 & 1 & 2 \end{bmatrix}$$

⟵ $\frac{7}{2}$ times the third row of the preceding matrix was added to the second row.

$$\begin{bmatrix} 1 & 2 & -5 & 3 & 0 & 2 \\ 0 & 0 & 1 & 0 & 0 & 1 \\ 0 & 0 & 0 & 0 & 1 & 2 \end{bmatrix}$$

⟵ -6 times the third row was added to the first row.

$$\begin{bmatrix} 1 & 2 & 0 & 3 & 0 & 7 \\ 0 & 0 & 1 & 0 & 0 & 1 \\ 0 & 0 & 0 & 0 & 1 & 2 \end{bmatrix}$$

⟵ 5 times the second row was added to the first row.

The last matrix is in reduced row-echelon form.

If we use only the first five steps, the above procedure produces a row-echelon form and is called **Gaussian elimination**. Carrying the procedure through to the sixth step and producing a matrix in reduced row-echelon form is called **Gauss–Jordan elimination**.

REMARK. It can be shown that *every matrix has a unique reduced row-echelon form*; that is, one will arrive at the same reduced row-echelon form for a given matrix no matter how the row operations are varied. (A proof of this result can be found in the article "The Reduced Row Echelon Form of a Matrix Is Unique: A Simple Proof," by Thomas Yuster, *Mathematics Magazine*, Vol. 57, No. 2, 1984, pp. 93–94.) In contrast, *a row-echelon form of a given matrix is not unique*: different sequences of row operations can produce different row-echelon forms.

EXAMPLE 4 Gauss–Jordan Elimination

Solve by Gauss–Jordan elimination.

$$\begin{aligned} x_1 + 3x_2 - 2x_3 \quad\quad\quad + 2x_5 \quad\quad\quad &= 0 \\ 2x_1 + 6x_2 - 5x_3 - 2x_4 + 4x_5 - 3x_6 &= -1 \\ 5x_3 + 10x_4 \quad\quad + 15x_6 &= 5 \\ 2x_1 + 6x_2 \quad\quad + 8x_4 + 4x_5 + 18x_6 &= 6 \end{aligned}$$

Solution.

The augmented matrix for the system is

$$\begin{bmatrix} 1 & 3 & -2 & 0 & 2 & 0 & 0 \\ 2 & 6 & -5 & -2 & 4 & -3 & -1 \\ 0 & 0 & 5 & 10 & 0 & 15 & 5 \\ 2 & 6 & 0 & 8 & 4 & 18 & 6 \end{bmatrix}$$

Karl Friedrich Gauss

Wilhelm Jordan

Karl Friedrich Gauss (1777–1855) was a German mathematician and scientist. Sometimes called the "prince of mathematicians," Gauss ranks with Isaac Newton and Archimedes as one of the three greatest mathematicians who ever lived. In the entire history of mathematics there may never have been a child so precocious as Gauss—by his own account he worked out the rudiments of arithmetic before he could talk. One day, before he was even three years old, his genius became apparent to his parents in a very dramatic way. His father was preparing the weekly payroll for the laborers under his charge while the boy watched quietly from a corner. At the end of the long and tedious calculation, Gauss informed his father that there was an error in the result and stated the answer, which he had worked out in his head. To the astonishment of his parents, a check of the computations showed Gauss to be correct!

In his doctoral dissertation Gauss gave the first complete proof of the fundamental theorem of algebra, which states that every polynomial equation has as many solutions as its degree. At age 19 he solved a problem that baffled Euclid, inscribing a regular polygon of seventeen sides in a circle using straightedge and compass; and in 1801, at age 24, he published his first masterpiece, *Disquisitiones Arithmeticae*, considered by many to be one of the most brilliant achievements in mathematics. In that paper Gauss systematized the study of number theory (properties of the integers) and formulated the basic concepts that form the foundation of the subject.

Among his myriad achievements, Gauss discovered the Gaussian or "bell-shaped" curve that is fundamental in probability, gave the first geometric interpretation of complex numbers and established their fundamental role in mathematics, developed methods of characterizing surfaces intrinsically by means of the curves that they contain, developed the theory of conformal (angle-preserving) maps, and discovered non-Euclidean geometry 30 years before the ideas were published by others. In physics he made major contributions to the theory of lenses and capillary action, and with Wilhelm Weber he did fundamental work in electromagnetism. Gauss invented the heliotrope, bifilar magnetometer, and an electrotelegraph.

Gauss was deeply religious and aristocratic in demeanor. He mastered foreign languages with ease, read extensively, and enjoyed mineralogy and botany as hobbies. He disliked teaching and was usually cool and discouraging to other mathematicians, possibly because he had already anticipated their work. It has been said that if Gauss had published all of his discoveries, the current state of mathematics would be advanced by 50 years. He was without a doubt the greatest mathematician of the modern era.

Wilhelm Jordan (1842–1899) was a German engineer who specialized in geodesy. His contribution to solving linear systems appeared in his popular book, *Handbuch der Vermessungskunde* (*Handbook of Geodesy*), in 1888.

Adding -2 times the first row to the second and fourth rows gives

$$\begin{bmatrix} 1 & 3 & -2 & 0 & 2 & 0 & 0 \\ 0 & 0 & -1 & -2 & 0 & -3 & -1 \\ 0 & 0 & 5 & 10 & 0 & 15 & 5 \\ 0 & 0 & 4 & 8 & 0 & 18 & 6 \end{bmatrix}$$

Multiplying the second row by -1 and then adding -5 times the new second row to the third row and -4 times the new second row to the fourth row gives

$$\begin{bmatrix} 1 & 3 & -2 & 0 & 2 & 0 & 0 \\ 0 & 0 & 1 & 2 & 0 & 3 & 1 \\ 0 & 0 & 0 & 0 & 0 & 0 & 0 \\ 0 & 0 & 0 & 0 & 0 & 6 & 2 \end{bmatrix}$$

Interchanging the third and fourth rows and then multiplying the third row of the resulting matrix by $\frac{1}{6}$ gives the row-echelon form

$$\begin{bmatrix} 1 & 3 & -2 & 0 & 2 & 0 & 0 \\ 0 & 0 & 1 & 2 & 0 & 3 & 1 \\ 0 & 0 & 0 & 0 & 0 & 1 & \frac{1}{3} \\ 0 & 0 & 0 & 0 & 0 & 0 & 0 \end{bmatrix}$$

Adding -3 times the third row to the second row and then adding 2 times the second row of the resulting matrix to the first row yields the reduced row-echelon form

$$\begin{bmatrix} 1 & 3 & 0 & 4 & 2 & 0 & 0 \\ 0 & 0 & 1 & 2 & 0 & 0 & 0 \\ 0 & 0 & 0 & 0 & 0 & 1 & \frac{1}{3} \\ 0 & 0 & 0 & 0 & 0 & 0 & 0 \end{bmatrix}$$

The corresponding system of equations is

$$\begin{aligned} x_1 + 3x_2 \quad\quad + 4x_4 + 2x_5 \quad\quad &= 0 \\ x_3 + 2x_4 \quad\quad &= 0 \\ x_6 &= \tfrac{1}{3} \end{aligned}$$

(We have discarded the last equation, $0x_1 + 0x_2 + 0x_3 + 0x_4 + 0x_5 + 0x_6 = 0$, since it will be satisfied automatically by the solutions of the remaining equations.) Solving for the leading variables, we obtain

$$\begin{aligned} x_1 &= -3x_2 - 4x_4 - 2x_5 \\ x_3 &= -2x_4 \\ x_6 &= \tfrac{1}{3} \end{aligned}$$

If we assign the free variables x_2, x_4, and x_5 arbitrary values r, s, and t, respectively, the general solution is given by the formulas

$$x_1 = -3r - 4s - 2t, \quad x_2 = r, \quad x_3 = -2s, \quad x_4 = s, \quad x_5 = t, \quad x_6 = \tfrac{1}{3} \quad \blacklozenge$$

Back-Substitution   It is sometimes preferable to solve a system of linear equations by using Gaussian elimination to bring the augmented matrix into row-echelon form without continuing all the way to the reduced row-echelon form. When this is done, the corresponding system of equations can be solved by a technique called ***back-substitution***. The next example illustates the idea.

EXAMPLE 5 Example 4 Solved by Back-Substitution

From the computations in Example 4, a row-echelon form of the augmented matrix is

$$\begin{bmatrix} 1 & 3 & -2 & 0 & 2 & 0 & 0 \\ 0 & 0 & 1 & 2 & 0 & 3 & 1 \\ 0 & 0 & 0 & 0 & 0 & 1 & \frac{1}{3} \\ 0 & 0 & 0 & 0 & 0 & 0 & 0 \end{bmatrix}$$

To solve the corresponding system of equations

$$
\begin{aligned}
x_1 + 3x_2 - 2x_3 \quad\;\; + 2x_5 \quad\quad\;\; &= 0 \\
x_3 + 2x_4 \quad\quad + 3x_6 &= 1 \\
x_6 &= \tfrac{1}{3}
\end{aligned}
$$

we proceed as follows:

Step 1. Solve the equations for the leading variables.

$$
\begin{aligned}
x_1 &= -3x_2 + 2x_3 - 2x_5 \\
x_3 &= 1 - 2x_4 - 3x_6 \\
x_6 &= \tfrac{1}{3}
\end{aligned}
$$

Step 2. Beginning with the bottom equation and working upward, successively substitute each equation into all the equations above it.

Substituting $x_6 = \tfrac{1}{3}$ into the second equation yields

$$
\begin{aligned}
x_1 &= -3x_2 + 2x_3 - 2x_5 \\
x_3 &= -2x_4 \\
x_6 &= \tfrac{1}{3}
\end{aligned}
$$

Substituting $x_3 = -2x_4$ into the first equation yields

$$
\begin{aligned}
x_1 &= -3x_2 - 4x_4 - 2x_5 \\
x_3 &= -2x_4 \\
x_6 &= \tfrac{1}{3}
\end{aligned}
$$

Step 3. Assign arbitrary values to the free variables, if any.

If we assign x_2, x_4, and x_5 the arbitrary values r, s, and t, respectively, the general solution is given by the formulas

$$
x_1 = -3r - 4s - 2t, \quad x_2 = r, \quad x_3 = -2s, \quad x_4 = s, \quad x_5 = t, \quad x_6 = \tfrac{1}{3}
$$

This agrees with the solution obtained in Example 4. ◆

REMARK. The arbitrary values that are assigned to the free variables are often called **parameters**. Although we shall generally use the letters $r, s, t, \ldots$ for the parameters, any letters that do not conflict with the variable names may be used.

EXAMPLE 6 Gaussian Elimination

Solve

$$
\begin{aligned}
x + \;\; y + 2z &= 9 \\
2x + 4y - 3z &= 1 \\
3x + 6y - 5z &= 0
\end{aligned}
$$

by Gaussian elimination and back-substitution.

Solution.

This is the system in Example 3 of Section 1.1. In that example we converted the augmented matrix

$$\begin{bmatrix} 1 & 1 & 2 & 9 \\ 2 & 4 & -3 & 1 \\ 3 & 6 & -5 & 0 \end{bmatrix}$$

to the row-echelon form

$$\begin{bmatrix} 1 & 1 & 2 & 9 \\ 0 & 1 & -\frac{7}{2} & -\frac{17}{2} \\ 0 & 0 & 1 & 3 \end{bmatrix}$$

The system corresponding to this matrix is

$$\begin{aligned} x + y + 2z &= 9 \\ y - \tfrac{7}{2}z &= -\tfrac{17}{2} \\ z &= 3 \end{aligned}$$

Solving for the leading variables yields

$$\begin{aligned} x &= 9 - y - 2z \\ y &= -\tfrac{17}{2} + \tfrac{7}{2}z \\ z &= 3 \end{aligned}$$

Substituting the bottom equation into those above yields

$$\begin{aligned} x &= 3 - y \\ y &= 2 \\ z &= 3 \end{aligned}$$

and substituting the second equation into the top yields $x = 1, y = 2, z = 3$. This agrees with the result found by Gauss–Jordan elimination in Example 3 of Section 1.1. ◆

Homogeneous Linear Systems

A system of linear equations is said to be **homogeneous** if the constant terms are all zero; that is, the system has the form

$$\begin{aligned} a_{11}x_1 + a_{12}x_2 + \cdots + a_{1n}x_n &= 0 \\ a_{21}x_1 + a_{22}x_2 + \cdots + a_{2n}x_n &= 0 \\ \vdots \qquad \vdots \qquad\qquad \vdots \quad \vdots \\ a_{m1}x_1 + a_{m2}x_2 + \cdots + a_{mn}x_n &= 0 \end{aligned}$$

Every homogeneous system of linear equations is consistent, since all such systems have $x_1 = 0, x_2 = 0, \ldots, x_n = 0$ as a solution. This solution is called the **trivial solution**; if there are other solutions, they are called **nontrivial solutions**.

Because a homogeneous linear system always has the trivial solution, there are only two possibilities for its solutions:

- The system has only the trivial solution.
- The system has infinitely many solutions in addition to the trivial solution.

In the special case of a homogeneous linear system of two equations in two unknowns, say

$$\begin{aligned} a_1 x + b_1 y &= 0 \quad (a_1, b_1 \text{ not both zero}) \\ a_2 x + b_2 y &= 0 \quad (a_2, b_2 \text{ not both zero}) \end{aligned}$$

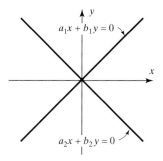

(a) Only the trivial solution

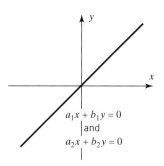

(b) Infinitely many solutions

Figure 1.2.1

the graphs of the equations are lines through the origin, and the trivial solution corresponds to the point of intersection at the origin (Figure 1.2.1).

There is one case in which a homogeneous system is assured of having nontrivial solutions, namely, whenever the system involves more unknowns than equations. To see why, consider the following example of four equations in five unknowns.

EXAMPLE 7 Gauss–Jordan Elimination

Solve the following homogeneous system of linear equations by using Gauss–Jordan elimination.

$$\begin{aligned}
2x_1 + 2x_2 - x_3 \quad\quad + x_5 &= 0 \\
-x_1 - x_2 + 2x_3 - 3x_4 + x_5 &= 0 \\
x_1 + x_2 - 2x_3 \quad\quad - x_5 &= 0 \\
x_3 + x_4 + x_5 &= 0
\end{aligned} \tag{1}$$

Solution.

The augmented matrix for the system is

$$\begin{bmatrix}
2 & 2 & -1 & 0 & 1 & 0 \\
-1 & -1 & 2 & -3 & 1 & 0 \\
1 & 1 & -2 & 0 & -1 & 0 \\
0 & 0 & 1 & 1 & 1 & 0
\end{bmatrix}$$

Reducing this matrix to reduced row-echelon form, we obtain

$$\begin{bmatrix}
1 & 1 & 0 & 0 & 1 & 0 \\
0 & 0 & 1 & 0 & 1 & 0 \\
0 & 0 & 0 & 1 & 0 & 0 \\
0 & 0 & 0 & 0 & 0 & 0
\end{bmatrix}$$

The corresponding system of equations is

$$\begin{aligned}
x_1 + x_2 \quad\quad + x_5 &= 0 \\
x_3 \quad + x_5 &= 0 \\
x_4 \quad &= 0
\end{aligned} \tag{2}$$

Solving for the leading variables yields

$$\begin{aligned}
x_1 &= -x_2 - x_5 \\
x_3 &= -x_5 \\
x_4 &= 0
\end{aligned}$$

Thus, the general solution is

$$x_1 = -s - t, \quad x_2 = s, \quad x_3 = -t, \quad x_4 = 0, \quad x_5 = t$$

Note that the trivial solution is obtained when $s = t = 0$. ◆

Example 7 illustrates two important points about solving homogeneous systems of linear equations. First, none of the three elementary row operations alters the final column of zeros in the augmented matrix, so that the system of equations corresponding to the reduced row-echelon form of the augmented matrix must also be a homogeneous system [see system (2)]. Second, depending on whether the reduced row-echelon form of the augmented matrix has any zero rows, the number of equations in the reduced system is the same as or less than the number of equations in the original system [compare systems (1) and (2)]. Thus, if the given homogeneous system has m equations in n unknowns with $m < n$, and if there are r nonzero rows in the reduced row-echelon form of the augmented matrix, we will have $r < n$. It follows that the system of equations corresponding to the reduced row-echelon form of the augmented matrix will have the form

$$
\begin{aligned}
\cdots x_{k_1} && + \Sigma(\ ) &= 0 \\
\cdots x_{k_2} && + \Sigma(\ ) &= 0 \\
\cdots \ddots && \vdots & \\
x_{k_r} && + \Sigma(\ ) &= 0
\end{aligned}
\tag{3}
$$

where $x_{k_1}, x_{k_2}, \ldots, x_{k_r}$ are the leading variables and $\Sigma(\ )$ denotes sums (possibly all different) that involve the $n - r$ free variables [compare system (3) with system (2) above]. Solving for the leading variables gives

$$
\begin{aligned}
x_{k_1} &= -\Sigma(\ ) \\
x_{k_2} &= -\Sigma(\ ) \\
&\vdots \\
x_{k_r} &= -\Sigma(\ )
\end{aligned}
$$

As in Example 7, we can assign arbitrary values to the free variables on the right-hand side and thus obtain infinitely many solutions to the system.

In summary, we have the following important theorem.

Theorem 1.2.1

A homogeneous system of linear equations with more unknowns than equations has infinitely many solutions.

REMARK. Note that Theorem 1.2.1 applies only to homogeneous systems. A nonhomogeneous system with more unknowns than equations need not be consistent (Exercise 28); however, if the system is consistent, it will have infinitely many solutions. This will be proved later.

Computer Solution of Linear Systems   In applications it is not uncommon to encounter large linear systems that must be solved by computer. Most computer algorithms for solving such systems are based on Gaussian elimination or Gauss–Jordan elimination, but the basic procedures are often modified to deal with such issues as

- Reducing roundoff errors
- Minimizing the use of computer memory space
- Solving the system with maximum speed

Some of these matters will be considered in Chapter 9. For hand computations fractions are an annoyance that often cannot be avoided. However, in some cases it is possible to avoid them by varying the elementary row operations in the right way. Thus, once the methods of Gaussian elimination and Gauss–Jordan elimination have been mastered, the reader may wish to vary the steps in specific problems to avoid fractions (see Exercise 18).

REMARK. Since Gauss–Jordan elimination avoids the use of back-substitution, it would seem that this method would be the more efficient of the two methods we have considered.

It can be argued that this statement is true when solving small systems by hand since Gauss–Jordan elimination actually involves less writing. However, for large systems of equations, it has been shown that the Gauss–Jordan elimination method requires about 50% more operations than Gaussian elimination. This is an important consideration when working on computers.

Exercise Set 1.2

1. Which of the following 3×3 matrices are in reduced row-echelon form?

(a) $\begin{bmatrix} 1 & 0 & 0 \\ 0 & 1 & 0 \\ 0 & 0 & 1 \end{bmatrix}$ (b) $\begin{bmatrix} 1 & 0 & 0 \\ 0 & 1 & 0 \\ 0 & 0 & 0 \end{bmatrix}$ (c) $\begin{bmatrix} 0 & 1 & 0 \\ 0 & 0 & 1 \\ 0 & 0 & 0 \end{bmatrix}$ (d) $\begin{bmatrix} 1 & 0 & 0 \\ 0 & 0 & 1 \\ 0 & 0 & 0 \end{bmatrix}$ (e) $\begin{bmatrix} 1 & 0 & 0 \\ 0 & 0 & 0 \\ 0 & 0 & 1 \end{bmatrix}$

(f) $\begin{bmatrix} 0 & 1 & 0 \\ 1 & 0 & 0 \\ 0 & 0 & 0 \end{bmatrix}$ (g) $\begin{bmatrix} 1 & 1 & 0 \\ 0 & 1 & 0 \\ 0 & 0 & 0 \end{bmatrix}$ (h) $\begin{bmatrix} 1 & 0 & 2 \\ 0 & 1 & 3 \\ 0 & 0 & 0 \end{bmatrix}$ (i) $\begin{bmatrix} 0 & 0 & 1 \\ 0 & 0 & 0 \\ 0 & 0 & 0 \end{bmatrix}$ (j) $\begin{bmatrix} 0 & 0 & 0 \\ 0 & 0 & 0 \\ 0 & 0 & 0 \end{bmatrix}$

2. Which of the following 3×3 matrices are in row-echelon form?

(a) $\begin{bmatrix} 1 & 0 & 0 \\ 0 & 1 & 0 \\ 0 & 0 & 1 \end{bmatrix}$ (b) $\begin{bmatrix} 1 & 2 & 0 \\ 0 & 1 & 0 \\ 0 & 0 & 0 \end{bmatrix}$ (c) $\begin{bmatrix} 1 & 0 & 0 \\ 0 & 1 & 0 \\ 0 & 2 & 0 \end{bmatrix}$ (d) $\begin{bmatrix} 1 & 3 & 4 \\ 0 & 0 & 1 \\ 0 & 0 & 0 \end{bmatrix}$

(e) $\begin{bmatrix} 1 & 5 & -3 \\ 0 & 1 & 1 \\ 0 & 0 & 0 \end{bmatrix}$ (f) $\begin{bmatrix} 1 & 2 & 3 \\ 0 & 0 & 0 \\ 0 & 0 & 1 \end{bmatrix}$

3. In each part determine whether the matrix is in row-echelon form, reduced row-echelon form, both, or neither.

(a) $\begin{bmatrix} 1 & 2 & 0 & 3 & 0 \\ 0 & 0 & 1 & 1 & 0 \\ 0 & 0 & 0 & 0 & 1 \\ 0 & 0 & 0 & 0 & 0 \end{bmatrix}$ (b) $\begin{bmatrix} 1 & 0 & 0 & 5 \\ 0 & 0 & 1 & 3 \\ 0 & 1 & 0 & 4 \end{bmatrix}$ (c) $\begin{bmatrix} 1 & 0 & 3 & 1 \\ 0 & 1 & 2 & 4 \end{bmatrix}$

(d) $\begin{bmatrix} 1 & -7 & 5 & 5 \\ 0 & 1 & 3 & 2 \end{bmatrix}$ (e) $\begin{bmatrix} 1 & 3 & 0 & 2 & 0 \\ 1 & 0 & 2 & 2 & 0 \\ 0 & 0 & 0 & 0 & 1 \\ 0 & 0 & 0 & 0 & 0 \end{bmatrix}$ (f) $\begin{bmatrix} 0 & 0 \\ 0 & 0 \\ 0 & 0 \end{bmatrix}$

4. In each part suppose that the augmented matrix for a system of linear equations has been reduced by row operations to the given reduced row-echelon form. Solve the system.

(a) $\begin{bmatrix} 1 & 0 & 0 & -3 \\ 0 & 1 & 0 & 0 \\ 0 & 0 & 1 & 7 \end{bmatrix}$ (b) $\begin{bmatrix} 1 & 0 & 0 & -7 & 8 \\ 0 & 1 & 0 & 3 & 2 \\ 0 & 0 & 1 & 1 & -5 \end{bmatrix}$

(c) $\begin{bmatrix} 1 & -6 & 0 & 0 & 3 & -2 \\ 0 & 0 & 1 & 0 & 4 & 7 \\ 0 & 0 & 0 & 1 & 5 & 8 \\ 0 & 0 & 0 & 0 & 0 & 0 \end{bmatrix}$ (d) $\begin{bmatrix} 1 & -3 & 0 & 0 \\ 0 & 0 & 1 & 0 \\ 0 & 0 & 0 & 1 \end{bmatrix}$

5. In each part suppose that the augmented matrix for a system of linear equations has been reduced by row operations to the given row-echelon form. Solve the system.

(a) $\begin{bmatrix} 1 & -3 & 4 & 7 \\ 0 & 1 & 2 & 2 \\ 0 & 0 & 1 & 5 \end{bmatrix}$
(b) $\begin{bmatrix} 1 & 0 & 8 & -5 & 6 \\ 0 & 1 & 4 & -9 & 3 \\ 0 & 0 & 1 & 1 & 2 \end{bmatrix}$

(c) $\begin{bmatrix} 1 & 7 & -2 & 0 & -8 & -3 \\ 0 & 0 & 1 & 1 & 6 & 5 \\ 0 & 0 & 0 & 1 & 3 & 9 \\ 0 & 0 & 0 & 0 & 0 & 0 \end{bmatrix}$
(d) $\begin{bmatrix} 1 & -3 & 7 & 1 \\ 0 & 1 & 4 & 0 \\ 0 & 0 & 0 & 1 \end{bmatrix}$

6. Solve each of the following systems by Gauss–Jordan elimination.

(a)
$$\begin{aligned} x_1 + x_2 + 2x_3 &= 8 \\ -x_1 - 2x_2 + 3x_3 &= 1 \\ 3x_1 - 7x_2 + 4x_3 &= 10 \end{aligned}$$
(b)
$$\begin{aligned} 2x_1 + 2x_2 + 2x_3 &= 0 \\ -2x_1 + 5x_2 + 2x_3 &= 1 \\ 8x_1 + x_2 + 4x_3 &= -1 \end{aligned}$$

(c)
$$\begin{aligned} x - y + 2z - w &= -1 \\ 2x + y - 2z - 2w &= -2 \\ -x + 2y - 4z + w &= 1 \\ 3x \qquad\quad - 3w &= -3 \end{aligned}$$
(d)
$$\begin{aligned} - 2b + 3c &= 1 \\ 3a + 6b - 3c &= -2 \\ 6a + 6b + 3c &= 5 \end{aligned}$$

7. Solve each of the systems in Exercise 6 by Gaussian elimination.

8. Solve each of the following systems by Gauss–Jordan elimination.

(a)
$$\begin{aligned} 2x_1 - 3x_2 &= -2 \\ 2x_1 + x_2 &= 1 \\ 3x_1 + 2x_2 &= 1 \end{aligned}$$
(b)
$$\begin{aligned} 3x_1 + 2x_2 - x_3 &= -15 \\ 5x_1 + 3x_2 + 2x_3 &= 0 \\ 3x_1 + x_2 + 3x_3 &= 11 \\ -6x_1 - 4x_2 + 2x_3 &= 30 \end{aligned}$$

(c)
$$\begin{aligned} 4x_1 - 8x_2 &= 12 \\ 3x_1 - 6x_2 &= 9 \\ -2x_1 + 4x_2 &= -6 \end{aligned}$$
(d)
$$\begin{aligned} 10y - 4z + w &= 1 \\ x + 4y - z + w &= 2 \\ 3x + 2y + z + 2w &= 5 \\ -2x - 8y + 2z - 2w &= -4 \\ x - 6y + 3z \qquad &= 1 \end{aligned}$$

9. Solve each of the systems in Exercise 8 by Gaussian elimination.

10. Solve each of the following systems by Gauss–Jordan elimination.

(a)
$$\begin{aligned} 5x_1 - 2x_2 + 6x_3 &= 0 \\ -2x_1 + x_2 + 3x_3 &= 1 \end{aligned}$$
(b)
$$\begin{aligned} x_1 - 2x_2 + x_3 - 4x_4 &= 1 \\ x_1 + 3x_2 + 7x_3 + 2x_4 &= 2 \\ x_1 - 12x_2 - 11x_3 - 16x_4 &= 5 \end{aligned}$$
(c)
$$\begin{aligned} w + 2x - y &= 4 \\ x - y &= 3 \\ w + 3x - 2y &= 7 \\ 2u + 4v + w + 7x &= 7 \end{aligned}$$

11. Solve each of the systems in Exercise 10 by Gaussian elimination.

12. Without using pencil and paper, determine which of the following homogeneous systems have nontrivial solutions.

(a)
$$\begin{aligned} 2x_1 - 3x_2 + 4x_3 - x_4 &= 0 \\ 7x_1 + x_2 - 8x_3 + 9x_4 &= 0 \\ 2x_1 + 8x_2 + x_3 - x_4 &= 0 \end{aligned}$$
(b)
$$\begin{aligned} x_1 + 3x_2 - x_3 &= 0 \\ x_2 - 8x_3 &= 0 \\ 4x_3 &= 0 \end{aligned}$$

(c)
$$\begin{aligned} a_{11}x_1 + a_{12}x_2 + a_{13}x_3 &= 0 \\ a_{21}x_1 + a_{22}x_2 + a_{23}x_3 &= 0 \end{aligned}$$
(d)
$$\begin{aligned} 3x_1 - 2x_2 &= 0 \\ 6x_1 - 4x_2 &= 0 \end{aligned}$$

13. Solve the following homogeneous systems of linear equations by any method.

(a)
$$\begin{aligned} 2x_1 + x_2 + 3x_3 &= 0 \\ x_1 + 2x_2 &= 0 \\ x_2 + x_3 &= 0 \end{aligned}$$
(b)
$$\begin{aligned} 3x_1 + x_2 + x_3 + x_4 &= 0 \\ 5x_1 - x_2 + x_3 - x_4 &= 0 \end{aligned}$$
(c)
$$\begin{aligned} 2x + 2y + 4z &= 0 \\ w - y - 3z &= 0 \\ 2w + 3x + y + z &= 0 \\ -2w + x + 3y - 2z &= 0 \end{aligned}$$

14. Solve the following homogeneous systems of linear equations by any method.

(a) $2x - y - 3z = 0$
$-x + 2y - 3z = 0$
$x + y + 4z = 0$

(b) $v + 3w - 2x = 0$
$2u + v - 4w + 3x = 0$
$2u + 3v + 2w - x = 0$
$-4u - 3v + 5w - 4x = 0$

(c) $x_1 + 3x_2 + x_4 = 0$
$x_1 + 4x_2 + 2x_3 = 0$
$-2x_2 - 2x_3 - x_4 = 0$
$2x_1 - 4x_2 + x_3 + x_4 = 0$
$x_1 - 2x_2 - x_3 + x_4 = 0$

15. Solve the following systems by any method.

(a) $2I_1 - I_2 + 3I_3 + 4I_4 = 9$
$I_1 - 2I_3 + 7I_4 = 11$
$3I_1 - 3I_2 + I_3 + 5I_4 = 8$
$2I_1 + I_2 + 4I_3 + 4I_4 = 10$

(b) $Z_3 + Z_4 + Z_5 = 0$
$-Z_1 - Z_2 + 2Z_3 - 3Z_4 + Z_5 = 0$
$Z_1 + Z_2 - 2Z_3 - Z_5 = 0$
$2Z_1 + 2Z_2 - Z_3 + Z_5 = 0$

16. Solve the following systems, where a, b, and c are constants.

(a) $2x + y = a$
$3x + 6y = b$

(b) $x_1 + x_2 + x_3 = a$
$2x_1 + 2x_3 = b$
$3x_2 + 3x_3 = c$

17. For which values of a will the following system have no solutions? Exactly one solution? Infinitely many solutions?

$$x + 2y - 3z = 4$$
$$3x - y + 5z = 2$$
$$4x + y + (a^2 - 14)z = a + 2$$

18. Reduce

$$\begin{bmatrix} 2 & 1 & 3 \\ 0 & -2 & -29 \\ 3 & 4 & 5 \end{bmatrix}$$

to reduced row-echelon form without introducing any fractions.

19. Find two different row-echelon forms of

$$\begin{bmatrix} 1 & 3 \\ 2 & 7 \end{bmatrix}$$

20. Solve the following system of nonlinear equations for the unknown angles α, β, and γ, where $0 \le \alpha \le 2\pi$, $0 \le \beta \le 2\pi$, and $0 \le \gamma < \pi$.

$$2 \sin \alpha - \cos \beta + 3 \tan \gamma = 3$$
$$4 \sin \alpha + 2 \cos \beta - 2 \tan \gamma = 2$$
$$6 \sin \alpha - 3 \cos \beta + \tan \gamma = 9$$

21. Show that the following nonlinear system has 18 solutions if $0 \le \alpha \le 2\pi$, $0 \le \beta \le 2\pi$, and $0 \le \gamma < 2\pi$.

$$\sin \alpha + 2 \cos \beta + 3 \tan \gamma = 0$$
$$2 \sin \alpha + 5 \cos \beta + 3 \tan \gamma = 0$$
$$- \sin \alpha - 5 \cos \beta + 5 \tan \gamma = 0$$

22. For which value(s) of λ does the system of equations

$$(\lambda - 3)x + y = 0$$
$$x + (\lambda - 3)y = 0$$

have nontrivial solutions?

23. Solve the system

$$2x_1 - x_2 \qquad = \lambda x_1$$
$$2x_1 - x_2 + x_3 = \lambda x_2$$
$$-2x_1 + 2x_2 + x_3 = \lambda x_3$$

for x_1, x_2, and x_3 in the two cases $\lambda = 1$, $\lambda = 2$.

24. Solve the following system for x, y, and z.

$$\frac{1}{x} + \frac{2}{y} - \frac{4}{z} = 1$$

$$\frac{2}{x} + \frac{3}{y} + \frac{8}{z} = 0$$

$$-\frac{1}{x} + \frac{9}{y} + \frac{10}{z} = 5$$

25. Find the coefficients a, b, c, and d so that the curve shown in the accompanying figure is the graph of the equation $y = ax^3 + bx^2 + cx + d$.

26. Find coefficients a, b, c, and d so that the curve shown in the accompanying figure is given by the equation $ax^2 + ay^2 + bx + cy + d = 0$.

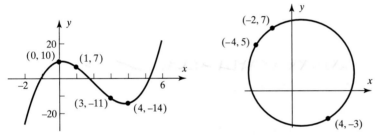

Figure Ex-25   **Figure Ex-26**

27. (a) Show that if $ad - bc \neq 0$, then the reduced row-echelon form of

$$\begin{bmatrix} a & b \\ c & d \end{bmatrix} \quad \text{is} \quad \begin{bmatrix} 1 & 0 \\ 0 & 1 \end{bmatrix}$$

(b) Use part (a) to show that the system

$$ax + by = k$$
$$cx + dy = l$$

has exactly one solution when $ad - bc \neq 0$.

28. Find an inconsistent linear system that has more unknowns than equations.

Discussion and Discovery

29. Discuss the possible reduced row-echelon forms of

$$\begin{bmatrix} a & b & c \\ d & e & f \\ g & h & i \end{bmatrix}$$

30. Consider the system of equations

$$ax + by = 0$$
$$cx + dy = 0$$
$$ex + fy = 0$$

Discuss the relative positions of the lines $ax + by = 0$, $cx + dy = 0$, and $ex + fy = 0$ when (a) the system has only the trivial solution, and (b) the system has nontrivial solutions.

31. Indicate whether the statement is always true or sometimes false. Justify your answer by giving a logical argument or a counterexample.

(a) If a matrix is reduced to reduced row-echelon form by two different sequences of elementary row operations, the resulting matrices will be different.

(b) If a matrix is reduced to row-echelon form by two different sequences of elementary row operations, the resulting matrices might be different.

(c) If the reduced row-echelon form of the augmented matrix for a linear system has a row of zeros, then the system must have infinitely many solutions.

(d) If three lines in the xy-plane are sides of a triangle, then the system of equations formed from their equations has three solutions, one corresponding to each vertex.

32. Indicate whether the statement is always true or sometimes false. Justify your answer by giving a logical argument or a counterexample.

(a) A linear system of three equations in five unknowns must be consistent.

(b) A linear system of five equations in three unknowns cannot be consistent.

(c) If a linear system of n equations in n unknowns has n leading 1's in the reduced row-echelon form of its augmented matrix, then the system has exactly one solution.

(d) If a linear system of n equations in n unknowns has two equations that are multiples of one another, then the system is inconsistent.

1.3   MATRICES AND MATRIX OPERATIONS

Rectangular arrays of real numbers arise in many contexts other than as augmented matrices for systems of linear equations. In this section we begin our study of matrix theory by giving some of the fundamental definitions of the subject. We shall see how matrices can be combined through the arithmetic operations of addition, subtraction, and multiplication.

Matrix Notation and Terminology   In Section 1.2 we used rectangular arrays of numbers, called *augmented matrices*, to abbreviate systems of linear equations. However, rectangular arrays of numbers occur in other contexts as well. For example, the following rectangular array with three rows and seven columns might describe the number of hours that a student spent studying three subjects during a certain week:

	Mon.	Tues.	Wed.	Thurs.	Fri.	Sat.	Sun.
Math	2	3	2	4	1	4	2
History	0	3	1	4	3	2	2
Language	4	1	3	1	0	0	2

If we suppress the headings, then we are left with the following rectangular array of numbers with three rows and seven columns called a "matrix":

$$\begin{bmatrix} 2 & 3 & 2 & 4 & 1 & 4 & 2 \\ 0 & 3 & 1 & 4 & 3 & 2 & 2 \\ 4 & 1 & 3 & 1 & 0 & 0 & 2 \end{bmatrix}$$

More generally, we make the following definition.

Definition

A **matrix** is a rectangular array of numbers. The numbers in the array are called the **entries** in the matrix.

EXAMPLE 1 Examples of Matrices

Some examples of matrices are

$$\begin{bmatrix} 1 & 2 \\ 3 & 0 \\ -1 & 4 \end{bmatrix}, \quad [2 \quad 1 \quad 0 \quad -3], \quad \begin{bmatrix} e & \pi & -\sqrt{2} \\ 0 & \frac{1}{2} & 1 \\ 0 & 0 & 0 \end{bmatrix}, \quad \begin{bmatrix} 1 \\ 3 \end{bmatrix}, \quad [4] \quad \blacklozenge$$

The **size** of a matrix is described in terms of the number of rows (horizontal lines) and columns (vertical lines) it contains. For example, the first matrix in Example 1 has three rows and two columns, so its size is 3 by 2 (written 3×2). In a size description, the first number always denotes the number of rows and the second denotes the number of columns. The remaining matrices in Example 1 have sizes 1×4, 3×3, 2×1, and 1×1, respectively. A matrix with only one column is called a **column matrix** (or a **column vector**), and a matrix with only one row is called a **row matrix** (or a **row vector**). Thus, in Example 1 the 2×1 matrix is a column matrix, the 1×4 matrix is a row matrix, and the 1×1 matrix is both a row matrix and a column matrix. (The term *vector* has another meaning that we will discuss in subsequent chapters.)

REMARK. It is common practice to omit the brackets on a 1×1 matrix. Thus, we might write 4 rather than [4]. Although this makes it impossible to tell whether 4 denotes the number "four" or the 1×1 matrix whose entry is "four," this rarely causes problems, since it is usually possible to tell which is meant from the context in which the symbol appears.

We shall use capital letters to denote matrices and lowercase letters to denote numerical quantities; thus, we might write

$$A = \begin{bmatrix} 2 & 1 & 7 \\ 3 & 4 & 2 \end{bmatrix} \quad \text{or} \quad C = \begin{bmatrix} a & b & c \\ d & e & f \end{bmatrix}$$

When discussing matrices, it is common to refer to numerical quantities as **scalars**. Unless stated otherwise, *scalars will be real numbers*; complex scalars will be considered in Chapter 10.

The entry that occurs in row i and column j of a matrix A will be denoted by a_{ij}. Thus, a general 3×4 matrix might be written as

$$A = \begin{bmatrix} a_{11} & a_{12} & a_{13} & a_{14} \\ a_{21} & a_{22} & a_{23} & a_{24} \\ a_{31} & a_{32} & a_{33} & a_{34} \end{bmatrix}$$

and a general $m \times n$ matrix as

$$A = \begin{bmatrix} a_{11} & a_{12} & \cdots & a_{1n} \\ a_{21} & a_{22} & \cdots & a_{2n} \\ \vdots & \vdots & & \vdots \\ a_{m1} & a_{m2} & \cdots & a_{mn} \end{bmatrix} \tag{1}$$

When compactness of notation is desired, the preceding matrix can be written as

$$[a_{ij}]_{m \times n} \quad \text{or} \quad [a_{ij}]$$

the first notation being used when it is important in the discussion to know the size and the second when the size need not be emphasized. Usually, we shall match the letter denoting a matrix with the letter denoting its entries; thus, for a matrix B we would generally use b_{ij} for the entry in row i and column j and for a matrix C we would use the notation c_{ij}.

The entry in row i and column j of a matrix A is also commonly denoted by the symbol $(A)_{ij}$. Thus, for matrix (1) above, we have

$$(A)_{ij} = a_{ij}$$

and for the matrix

$$A = \begin{bmatrix} 2 & -3 \\ 7 & 0 \end{bmatrix}$$

we have $(A)_{11} = 2$, $(A)_{12} = -3$, $(A)_{21} = 7$, and $(A)_{22} = 0$.

Row and column matrices are of special importance, and it is common practice to denote them by boldface lowercase letters rather than capital letters. For such matrices double subscripting of the entries is unnecessary. Thus, a general $1 \times n$ row matrix **a** and a general $m \times 1$ column matrix **b** would be written as

$$\mathbf{a} = [a_1 \quad a_2 \quad \cdots \quad a_n] \quad \text{and} \quad \mathbf{b} = \begin{bmatrix} b_1 \\ b_2 \\ \vdots \\ b_m \end{bmatrix}$$

A matrix A with n rows and n columns is called a ***square matrix of order n***, and the shaded entries $a_{11}, a_{22}, \ldots, a_{nn}$ in (2) are said to be on the ***main diagonal*** of A.

$$\begin{bmatrix} a_{11} & a_{12} & \cdots & a_{1n} \\ a_{21} & a_{22} & \cdots & a_{2n} \\ \vdots & \vdots & & \vdots \\ a_{n1} & a_{n2} & \cdots & a_{nn} \end{bmatrix} \tag{2}$$

Operations on Matrices   So far, we have used matrices to abbreviate the work in solving systems of linear equations. For other applications, however, it is desirable to develop an "arithmetic of matrices" in which matrices can be added, subtracted, and multiplied in a useful way. The remainder of this section will be devoted to developing this arithmetic.

Definition

Two matrices are defined to be ***equal*** if they have the same size and their corresponding entries are equal.

In matrix notation, if $A = [a_{ij}]$ and $B = [b_{ij}]$ have the same size, then $A = B$ if and only if $(A)_{ij} = (B)_{ij}$, or equivalently, $a_{ij} = b_{ij}$ for all i and j.

EXAMPLE 2 Equality of Matrices

Consider the matrices

$$A = \begin{bmatrix} 2 & 1 \\ 3 & x \end{bmatrix}, \quad B = \begin{bmatrix} 2 & 1 \\ 3 & 5 \end{bmatrix}, \quad C = \begin{bmatrix} 2 & 1 & 0 \\ 3 & 4 & 0 \end{bmatrix}$$

If $x = 5$, then $A = B$, but for all other values of x the matrices A and B are not equal, since not all of their corresponding entries are equal. There is no value of x for which $A = C$ since A and C have different sizes. ◆

Definition

If A and B are matrices of the same size, then the **sum** $A + B$ is the matrix obtained by adding the entries of B to the corresponding entries of A, and the **difference** $A - B$ is the matrix obtained by subtracting the entries of B from the corresponding entries of A. Matrices of different sizes cannot be added or subtracted.

In matrix notation, if $A = [a_{ij}]$ and $B = [b_{ij}]$ have the same size, then

$$(A + B)_{ij} = (A)_{ij} + (B)_{ij} = a_{ij} + b_{ij} \quad \text{and} \quad (A - B)_{ij} = (A)_{ij} - (B)_{ij} = a_{ij} - b_{ij}$$

EXAMPLE 3 Addition and Subtraction

Consider the matrices

$$A = \begin{bmatrix} 2 & 1 & 0 & 3 \\ -1 & 0 & 2 & 4 \\ 4 & -2 & 7 & 0 \end{bmatrix}, \quad B = \begin{bmatrix} -4 & 3 & 5 & 1 \\ 2 & 2 & 0 & -1 \\ 3 & 2 & -4 & 5 \end{bmatrix}, \quad C = \begin{bmatrix} 1 & 1 \\ 2 & 2 \end{bmatrix}$$

Then

$$A + B = \begin{bmatrix} -2 & 4 & 5 & 4 \\ 1 & 2 & 2 & 3 \\ 7 & 0 & 3 & 5 \end{bmatrix} \quad \text{and} \quad A - B = \begin{bmatrix} 6 & -2 & -5 & 2 \\ -3 & -2 & 2 & 5 \\ 1 & -4 & 11 & -5 \end{bmatrix}$$

The expressions $A + C$, $B + C$, $A - C$, and $B - C$ are undefined. ◆

Definition

If A is any matrix and c is any scalar, then the **product** cA is the matrix obtained by multiplying each entry of the matrix A by c. The matrix cA is said to be a **scalar multiple** of A.

In matrix notation, if $A = [a_{ij}]$, then

$$(cA)_{ij} = c(A)_{ij} = ca_{ij}$$

EXAMPLE 4 Scalar Multiples

For the matrices

$$A = \begin{bmatrix} 2 & 3 & 4 \\ 1 & 3 & 1 \end{bmatrix}, \qquad B = \begin{bmatrix} 0 & 2 & 7 \\ -1 & 3 & -5 \end{bmatrix}, \qquad C = \begin{bmatrix} 9 & -6 & 3 \\ 3 & 0 & 12 \end{bmatrix}$$

we have

$$2A = \begin{bmatrix} 4 & 6 & 8 \\ 2 & 6 & 2 \end{bmatrix}, \qquad (-1)B = \begin{bmatrix} 0 & -2 & -7 \\ 1 & -3 & 5 \end{bmatrix}, \qquad \tfrac{1}{3}C = \begin{bmatrix} 3 & -2 & 1 \\ 1 & 0 & 4 \end{bmatrix}$$

It is common practice to denote $(-1)B$ by $-B$. ◆

If $A_1, A_2, \ldots, A_n$ are matrices of the same size and $c_1, c_2, \ldots, c_n$ are scalars, then an expression of the form

$$c_1 A_1 + c_2 A_2 + \cdots + c_n A_n$$

is called a ***linear combination*** of $A_1, A_2, \ldots, A_n$ with ***coefficients*** $c_1, c_2, \ldots, c_n$. For example, if A, B, and C are the matrices in Example 4, then

$$2A - B + \tfrac{1}{3}C = 2A + (-1)B + \tfrac{1}{3}C$$

$$= \begin{bmatrix} 4 & 6 & 8 \\ 2 & 6 & 2 \end{bmatrix} + \begin{bmatrix} 0 & -2 & -7 \\ 1 & -3 & 5 \end{bmatrix} + \begin{bmatrix} 3 & -2 & 1 \\ 1 & 0 & 4 \end{bmatrix} = \begin{bmatrix} 7 & 2 & 2 \\ 4 & 3 & 11 \end{bmatrix}$$

is the linear combination of A, B, and C with scalar coefficients 2, -1, and $\tfrac{1}{3}$.

Thus far we have defined multiplication of a matrix by a scalar but not the multiplication of two matrices. Since matrices are added by adding corresponding entries and subtracted by subtracting corresponding entries, it would seem natural to define multiplication of matrices by multiplying corresponding entries. However, it turns out that such a definition would not be very useful for most problems. Experience has led mathematicians to the following more useful definition of matrix multiplication.

Definition

If A is an $m \times r$ matrix and B is an $r \times n$ matrix, then the ***product*** AB is the $m \times n$ matrix whose entries are determined as follows. To find the entry in row i and column j of AB, single out row i from the matrix A and column j from the matrix B. Multiply the corresponding entries from the row and column together and then add up the resulting products.

EXAMPLE 5 Multiplying Matrices

Consider the matrices

$$A = \begin{bmatrix} 1 & 2 & 4 \\ 2 & 6 & 0 \end{bmatrix}, \qquad B = \begin{bmatrix} 4 & 1 & 4 & 3 \\ 0 & -1 & 3 & 1 \\ 2 & 7 & 5 & 2 \end{bmatrix}$$

Since A is a 2×3 matrix and B is a 3×4 matrix, the product AB is a 2×4 matrix. To determine, for example, the entry in row 2 and column 3 of AB, we single out row 2 from A and column 3 from B. Then, as illustrated below, we multiply corresponding entries together and add up these products.

$$\begin{bmatrix} 1 & 2 & 4 \\ 2 & 6 & 0 \end{bmatrix} \begin{bmatrix} 4 & 1 & 4 & 3 \\ 0 & -1 & 3 & 1 \\ 2 & 7 & 5 & 2 \end{bmatrix} = \begin{bmatrix} \square & \square & \square & \square \\ \square & \square & 26 & \square \end{bmatrix}$$

$$(2 \cdot 4) + (6 \cdot 3) + (0 \cdot 5) = 26$$

The entry in row 1 and column 4 of AB is computed as follows.

$$\begin{bmatrix} 1 & 2 & 4 \\ 2 & 6 & 0 \end{bmatrix} \begin{bmatrix} 4 & 1 & 4 & 3 \\ 0 & -1 & 3 & 1 \\ 2 & 7 & 5 & 2 \end{bmatrix} = \begin{bmatrix} \square & \square & \square & 13 \\ \square & \square & \square & \square \end{bmatrix}$$

$$(1 \cdot 3) + (2 \cdot 1) + (4 \cdot 2) = 13$$

The computations for the remaining products are

$$(1 \cdot 4) + (2 \cdot 0) + (4 \cdot 2) = 12$$
$$(1 \cdot 1) - (2 \cdot 1) + (4 \cdot 7) = 27$$
$$(1 \cdot 4) + (2 \cdot 3) + (4 \cdot 5) = 30$$
$$(2 \cdot 4) + (6 \cdot 0) + (0 \cdot 2) = 8$$
$$(2 \cdot 1) - (6 \cdot 1) + (0 \cdot 7) = -4$$
$$(2 \cdot 3) + (6 \cdot 1) + (0 \cdot 2) = 12$$

$$AB = \begin{bmatrix} 12 & 27 & 30 & 13 \\ 8 & -4 & 26 & 12 \end{bmatrix} \qquad ♦$$

The definition of matrix multiplication requires that the number of columns of the first factor A be the same as the number of rows of the second factor B in order to form the product AB. If this condition is not satisfied, the product is undefined. A convenient way to determine whether a product of two matrices is defined is to write down the size of the first factor and, to the right of it, write down the size of the second factor. If, as in (3), the inside numbers are the same, then the product is defined. The outside numbers then give the size of the product.

$$\begin{array}{ccccccc} A & & B & & AB \\ m \times r & & r \times n & = & m \times n \\ & \text{Inside} & & & \\ & \text{Outside} & & & \end{array} \qquad (3)$$

EXAMPLE 6 Determining Whether a Product Is Defined

Suppose that A, B, and C are matrices with the following sizes:

$$\begin{array}{ccc} A & B & C \\ 3 \times 4 & 4 \times 7 & 7 \times 3 \end{array}$$

Then by (3), AB is defined and is a 3×7 matrix; BC is defined and is a 4×3 matrix; and CA is defined and is a 7×4 matrix. The products AC, CB, and BA are all undefined. ♦

In general, if $A = [a_{ij}]$ is an $m \times r$ matrix and $B = [b_{ij}]$ is an $r \times n$ matrix, then as illustrated by the shading in (4),

$$AB = \begin{bmatrix} a_{11} & a_{12} & \cdots & a_{1r} \\ a_{21} & a_{22} & \cdots & a_{2r} \\ \vdots & \vdots & & \vdots \\ a_{i1} & a_{i2} & \cdots & a_{ir} \\ \vdots & \vdots & & \vdots \\ a_{m1} & a_{m2} & \cdots & a_{mr} \end{bmatrix} \begin{bmatrix} b_{11} & b_{12} & \cdots & b_{1j} & \cdots & b_{1n} \\ b_{21} & b_{22} & \cdots & b_{2j} & \cdots & b_{2n} \\ \vdots & \vdots & & \vdots & & \vdots \\ b_{r1} & b_{r2} & \cdots & b_{rj} & \cdots & b_{rn} \end{bmatrix} \tag{4}$$

the entry $(AB)_{ij}$ in row i and column j of AB is given by

$$(AB)_{ij} = a_{i1}b_{1j} + a_{i2}b_{2j} + a_{i3}b_{3j} + \cdots + a_{ir}b_{rj} \tag{5}$$

Partitioned Matrices   A matrix can be subdivided or *partitioned* into smaller matrices by inserting horizontal and vertical rules between selected rows and columns. For example, below are three possible partitions of a general 3×4 matrix A— the first is a partition of A into four *submatrices* A_{11}, A_{12}, A_{21}, and A_{22}; the second is a partition of A into its row matrices $\mathbf{r}_1$, $\mathbf{r}_2$, and $\mathbf{r}_3$; and the third is a partition of A into its column matrices $\mathbf{c}_1$, $\mathbf{c}_2$, $\mathbf{c}_3$, and $\mathbf{c}_4$:

$$A = \left[\begin{array}{ccc|c} a_{11} & a_{12} & a_{13} & a_{14} \\ a_{21} & a_{22} & a_{23} & a_{24} \\ \hline a_{31} & a_{32} & a_{33} & a_{34} \end{array} \right] = \begin{bmatrix} A_{11} & A_{12} \\ A_{21} & A_{22} \end{bmatrix}$$

$$A = \left[\begin{array}{cccc} a_{11} & a_{12} & a_{13} & a_{14} \\ \hline a_{21} & a_{22} & a_{23} & a_{24} \\ \hline a_{31} & a_{32} & a_{33} & a_{34} \end{array} \right] = \begin{bmatrix} \mathbf{r}_1 \\ \mathbf{r}_2 \\ \mathbf{r}_3 \end{bmatrix}$$

$$A = \left[\begin{array}{c|c|c|c} a_{11} & a_{12} & a_{13} & a_{14} \\ a_{21} & a_{22} & a_{23} & a_{24} \\ a_{31} & a_{32} & a_{33} & a_{34} \end{array} \right] = \begin{bmatrix} \mathbf{c}_1 & \mathbf{c}_2 & \mathbf{c}_3 & \mathbf{c}_4 \end{bmatrix}$$

Matrix Multiplication by Columns and by Rows   Sometimes it may be desirable to find a particular row or column of a matrix product AB without computing the entire product. The following results, whose proofs are left as exercises, are useful for that purpose:

$$j\text{th column matrix of } AB = A[\,j\text{th column matrix of } B] \tag{6}$$

$$i\text{th row matrix of } AB = [i\text{th row matrix of } A]B \tag{7}$$

EXAMPLE 7 Example 5 Revisited

If A and B are the matrices in Example 5, then from (6) the second column matrix of AB can be obtained by the computation

$$\begin{bmatrix} 1 & 2 & 4 \\ 2 & 6 & 0 \end{bmatrix} \begin{bmatrix} 1 \\ -1 \\ 7 \end{bmatrix} = \begin{bmatrix} 27 \\ -4 \end{bmatrix}$$

$$\uparrow \qquad\qquad \uparrow$$

Second column   Second column
of B of AB

and from (7) the first row matrix of AB can be obtained by the computation

$$\underset{\text{First row of } A}{\underbrace{\left[1 \quad 2 \quad 4\right]}} \begin{bmatrix} 4 & 1 & 4 & 3 \\ 0 & -1 & 3 & 1 \\ 2 & 7 & 5 & 2 \end{bmatrix} = \underset{\text{First row of } AB}{\underbrace{\left[12 \quad 27 \quad 30 \quad 13\right]}} \qquad ◆$$

If $\mathbf{a}_1, \mathbf{a}_2, \ldots, \mathbf{a}_m$ denote the row matrices of A and $\mathbf{b}_1, \mathbf{b}_2, \ldots, \mathbf{b}_n$ denote the column matrices of B, then it follows from Formulas (6) and (7) that

$$AB = A[\mathbf{b}_1 \quad \mathbf{b}_2 \quad \cdots \quad \mathbf{b}_n] = [A\mathbf{b}_1 \quad A\mathbf{b}_2 \quad \cdots \quad A\mathbf{b}_n] \tag{8}$$

(*AB* computed column by column)

$$AB = \begin{bmatrix} \mathbf{a}_1 \\ \mathbf{a}_2 \\ \vdots \\ \mathbf{a}_m \end{bmatrix} B = \begin{bmatrix} \mathbf{a}_1 B \\ \mathbf{a}_2 B \\ \vdots \\ \mathbf{a}_m B \end{bmatrix} \tag{9}$$

(*AB* computed row by row)

REMARK. Formulas (8) and (9) are special cases of a more general procedure for multiplying partitioned matrices (see Exercises 15–17).

Matrix Products as Linear Combinations

Row and column matrices provide an alternative way of thinking about matrix multiplication. For example, suppose that

$$A = \begin{bmatrix} a_{11} & a_{12} & \cdots & a_{1n} \\ a_{21} & a_{22} & \cdots & a_{2n} \\ \vdots & \vdots & & \vdots \\ a_{m1} & a_{m2} & \cdots & a_{mn} \end{bmatrix} \quad \text{and} \quad \mathbf{x} = \begin{bmatrix} x_1 \\ x_2 \\ \vdots \\ x_n \end{bmatrix}$$

Then

$$A\mathbf{x} = \begin{bmatrix} a_{11}x_1 + a_{12}x_2 + \cdots + a_{1n}x_n \\ a_{21}x_1 + a_{22}x_2 + \cdots + a_{2n}x_n \\ \vdots \qquad \vdots \qquad \qquad \vdots \\ a_{m1}x_1 + a_{m2}x_2 + \cdots + a_{mn}x_n \end{bmatrix} = x_1 \begin{bmatrix} a_{11} \\ a_{21} \\ \vdots \\ a_{m1} \end{bmatrix} + x_2 \begin{bmatrix} a_{12} \\ a_{22} \\ \vdots \\ a_{m2} \end{bmatrix} + \cdots + x_n \begin{bmatrix} a_{1n} \\ a_{2n} \\ \vdots \\ a_{mn} \end{bmatrix} \tag{10}$$

In words, (10) tells us that *the product $A\mathbf{x}$ of a matrix A with a column matrix $\mathbf{x}$ is a linear combination of the column matrices of A with the coefficients coming from the matrix $\mathbf{x}$.* In the exercises we ask the reader to show that *the product $\mathbf{y}A$ of a $1 \times m$ matrix $\mathbf{y}$ with an $m \times n$ matrix A is a linear combination of the row matrices of A with scalar coefficients coming from $\mathbf{y}$.*

EXAMPLE 8 Linear Combinations

The matrix product

$$\begin{bmatrix} -1 & 3 & 2 \\ 1 & 2 & -3 \\ 2 & 1 & -2 \end{bmatrix} \begin{bmatrix} 2 \\ -1 \\ 3 \end{bmatrix} = \begin{bmatrix} 1 \\ -9 \\ -3 \end{bmatrix}$$

can be written as the linear combination of column matrices

$$2 \begin{bmatrix} -1 \\ 1 \\ 2 \end{bmatrix} - 1 \begin{bmatrix} 3 \\ 2 \\ 1 \end{bmatrix} + 3 \begin{bmatrix} 2 \\ -3 \\ -2 \end{bmatrix} = \begin{bmatrix} 1 \\ -9 \\ -3 \end{bmatrix}$$

The matrix product

$$\begin{bmatrix} 1 & -9 & -3 \end{bmatrix} \begin{bmatrix} -1 & 3 & 2 \\ 1 & 2 & -3 \\ 2 & 1 & -2 \end{bmatrix} = \begin{bmatrix} -16 & -18 & 35 \end{bmatrix}$$

can be written as the linear combination of row matrices

$$1 \begin{bmatrix} -1 & 3 & 2 \end{bmatrix} - 9 \begin{bmatrix} 1 & 2 & -3 \end{bmatrix} - 3 \begin{bmatrix} 2 & 1 & -2 \end{bmatrix} = \begin{bmatrix} -16 & -18 & 35 \end{bmatrix} \quad \blacklozenge$$

It follows from (8) and (10) that *the jth column matrix of a product AB is a linear combination of the column matrices of A with the coefficients coming from the jth column of B.*

EXAMPLE 9 Columns of a Product AB as Linear Combinations

We showed in Example 5 that

$$AB = \begin{bmatrix} 1 & 2 & 4 \\ 2 & 6 & 0 \end{bmatrix} \begin{bmatrix} 4 & 1 & 4 & 3 \\ 0 & -1 & 3 & 1 \\ 2 & 7 & 5 & 2 \end{bmatrix} = \begin{bmatrix} 12 & 27 & 30 & 13 \\ 8 & -4 & 26 & 12 \end{bmatrix}$$

The column matrices of AB can be expressed as linear combinations of the column matrices of A as follows:

$$\begin{bmatrix} 12 \\ 8 \end{bmatrix} = 4 \begin{bmatrix} 1 \\ 2 \end{bmatrix} + 0 \begin{bmatrix} 2 \\ 6 \end{bmatrix} + 2 \begin{bmatrix} 4 \\ 0 \end{bmatrix}$$

$$\begin{bmatrix} 27 \\ -4 \end{bmatrix} = \begin{bmatrix} 1 \\ 2 \end{bmatrix} - \begin{bmatrix} 2 \\ 6 \end{bmatrix} + 7 \begin{bmatrix} 4 \\ 0 \end{bmatrix}$$

$$\begin{bmatrix} 30 \\ 26 \end{bmatrix} = 4 \begin{bmatrix} 1 \\ 2 \end{bmatrix} + 3 \begin{bmatrix} 2 \\ 6 \end{bmatrix} + 5 \begin{bmatrix} 4 \\ 0 \end{bmatrix}$$

$$\begin{bmatrix} 13 \\ 12 \end{bmatrix} = 3 \begin{bmatrix} 1 \\ 2 \end{bmatrix} + \begin{bmatrix} 2 \\ 6 \end{bmatrix} + 2 \begin{bmatrix} 4 \\ 0 \end{bmatrix} \quad \blacklozenge$$

Matrix Form of a Linear System

Matrix multiplication has an important application to systems of linear equations. Consider any system of m linear equations in n unknowns.

$$\begin{aligned} a_{11}x_1 + a_{12}x_2 + \cdots + a_{1n}x_n &= b_1 \\ a_{21}x_1 + a_{22}x_2 + \cdots + a_{2n}x_n &= b_2 \\ \vdots \qquad \vdots \qquad\quad \vdots \qquad \vdots \\ a_{m1}x_1 + a_{m2}x_2 + \cdots + a_{mn}x_n &= b_m \end{aligned}$$

Since two matrices are equal if and only if their corresponding entries are equal, we can replace the m equations in this system by the single matrix equation

$$\begin{bmatrix} a_{11}x_1 + a_{12}x_2 + \cdots + a_{1n}x_n \\ a_{21}x_1 + a_{22}x_2 + \cdots + a_{2n}x_n \\ \vdots \qquad \vdots \qquad\qquad \vdots \\ a_{m1}x_1 + a_{m2}x_2 + \cdots + a_{mn}x_n \end{bmatrix} = \begin{bmatrix} b_1 \\ b_2 \\ \vdots \\ b_m \end{bmatrix}$$

The $m \times 1$ matrix on the left side of this equation can be written as a product to give

$$\begin{bmatrix} a_{11} & a_{12} & \cdots & a_{1n} \\ a_{21} & a_{22} & \cdots & a_{2n} \\ \vdots & \vdots & & \vdots \\ a_{m1} & a_{m2} & \cdots & a_{mn} \end{bmatrix} \begin{bmatrix} x_1 \\ x_2 \\ \vdots \\ x_n \end{bmatrix} = \begin{bmatrix} b_1 \\ b_2 \\ \vdots \\ b_m \end{bmatrix}$$

If we designate these matrices by A, $\mathbf{x}$, and $\mathbf{b}$, respectively, the original system of m equations in n unknowns has been replaced by the single matrix equation

$$A\mathbf{x} = \mathbf{b}$$

The matrix A in this equation is called the ***coefficient matrix*** of the system. The augmented matrix for the system is obtained by adjoining $\mathbf{b}$ to A as the last column; thus the augmented matrix is

$$[A \mid \mathbf{b}] = \begin{bmatrix} a_{11} & a_{12} & \cdots & a_{1n} & b_1 \\ a_{21} & a_{22} & \cdots & a_{2n} & b_2 \\ \vdots & \vdots & & \vdots & \vdots \\ a_{m1} & a_{m2} & \cdots & a_{mn} & b_m \end{bmatrix}$$

Transpose of a Matrix   We conclude this section by defining two matrix operations that have no analogs in the real numbers.

Definition

If A is any $m \times n$ matrix, then the ***transpose of A***, denoted by A^T, is defined to be the $n \times m$ matrix that results from interchanging the rows and columns of A; that is, the first column of A^T is the first row of A, the second column of A^T is the second row of A, and so forth.

EXAMPLE 10 Some Transposes

The following are some examples of matrices and their transposes.

$$A = \begin{bmatrix} a_{11} & a_{12} & a_{13} & a_{14} \\ a_{21} & a_{22} & a_{23} & a_{24} \\ a_{31} & a_{32} & a_{33} & a_{34} \end{bmatrix}, \quad B = \begin{bmatrix} 2 & 3 \\ 1 & 4 \\ 5 & 6 \end{bmatrix}, \quad C = [1 \;\; 3 \;\; 5], \quad D = [4]$$

$$A^T = \begin{bmatrix} a_{11} & a_{21} & a_{31} \\ a_{12} & a_{22} & a_{32} \\ a_{13} & a_{23} & a_{33} \\ a_{14} & a_{24} & a_{34} \end{bmatrix}, \quad B^T = \begin{bmatrix} 2 & 1 & 5 \\ 3 & 4 & 6 \end{bmatrix}, \quad C^T = \begin{bmatrix} 1 \\ 3 \\ 5 \end{bmatrix}, \quad D^T = [4] \quad \blacklozenge$$

Observe that not only are the columns of A^T the rows of A, but the rows of A^T are columns of A. Thus, the entry in row i and column j of A^T is the entry in row j and column i of A; that is,

$$(A^T)_{ij} = (A)_{ji} \tag{11}$$

Note the reversal of the subscripts.

In the special case where A is a square matrix, the transpose of A can be obtained by interchanging entries that are symmetrically positioned about the main diagonal. In (12) it is shown that A^T can also be obtained by "reflecting" A about its main diagonal.

$$A = \begin{bmatrix} 1 & -2 & 4 \\ 3 & 7 & 0 \\ -5 & 8 & 6 \end{bmatrix} \rightarrow \begin{bmatrix} 1 & -2 & 4 \\ 3 & 7 & 0 \\ -5 & 8 & 6 \end{bmatrix} \rightarrow A^T = \begin{bmatrix} 1 & 3 & -5 \\ -2 & 7 & 8 \\ 4 & 0 & 6 \end{bmatrix} \tag{12}$$

Interchange entries that are symmetrically positioned about the main diagonal.

Definition

If A is a square matrix, then the ***trace of*** A, denoted by $\text{tr}(A)$, is defined to be the sum of the entries on the main diagonal of A. The trace of A is undefined if A is not a square matrix.

EXAMPLE 11 Trace of a Matrix

The following are examples of matrices and their traces.

$$A = \begin{bmatrix} a_{11} & a_{12} & a_{13} \\ a_{21} & a_{22} & a_{23} \\ a_{31} & a_{32} & a_{33} \end{bmatrix}, \qquad B = \begin{bmatrix} -1 & 2 & 7 & 0 \\ 3 & 5 & -8 & 4 \\ 1 & 2 & 7 & -3 \\ 4 & -2 & 1 & 0 \end{bmatrix}$$

$$\text{tr}(A) = a_{11} + a_{22} + a_{33} \qquad \text{tr}(B) = -1 + 5 + 7 + 0 = 11 \qquad \blacklozenge$$

Exercise Set 1.3

1. Suppose that A, B, C, D, and E are matrices with the following sizes:

$$\begin{array}{ccccc} A & B & C & D & E \\ (4 \times 5) & (4 \times 5) & (5 \times 2) & (4 \times 2) & (5 \times 4) \end{array}$$

Determine which of the following matrix expressions are defined. For those which are defined, give the size of the resulting matrix.

(a) BA
(b) $AC + D$
(c) $AE + B$
(d) $AB + B$
(e) $E(A + B)$
(f) $E(AC)$
(g) $E^T A$
(h) $(A^T + E)D$

2. Solve the following matrix equation for a, b, c, and d.

$$\begin{bmatrix} a-b & b+c \\ 3d+c & 2a-4d \end{bmatrix} = \begin{bmatrix} 8 & 1 \\ 7 & 6 \end{bmatrix}$$

3. Consider the matrices

$$A = \begin{bmatrix} 3 & 0 \\ -1 & 2 \\ 1 & 1 \end{bmatrix}, \quad B = \begin{bmatrix} 4 & -1 \\ 0 & 2 \end{bmatrix}, \quad C = \begin{bmatrix} 1 & 4 & 2 \\ 3 & 1 & 5 \end{bmatrix}, \quad D = \begin{bmatrix} 1 & 5 & 2 \\ -1 & 0 & 1 \\ 3 & 2 & 4 \end{bmatrix}, \quad E = \begin{bmatrix} 6 & 1 & 3 \\ -1 & 1 & 2 \\ 4 & 1 & 3 \end{bmatrix}$$

Compute the following (where possible).

(a) $D + E$ (b) $D - E$ (c) $5A$ (d) $-7C$
(e) $2B - C$ (f) $4E - 2D$ (g) $-3(D + 2E)$ (h) $A - A$
(i) $\operatorname{tr}(D)$ (j) $\operatorname{tr}(D - 3E)$ (k) $4\operatorname{tr}(7B)$ (l) $\operatorname{tr}(A)$

4. Using the matrices in Exercise 3, compute the following (where possible).

(a) $2A^T + C$ (b) $D^T - E^T$ (c) $(D - E)^T$ (d) $B^T + 5C^T$
(e) $\frac{1}{2}C^T - \frac{1}{4}A$ (f) $B - B^T$ (g) $2E^T - 3D^T$ (h) $(2E^T - 3D^T)^T$

5. Using the matrices in Exercise 3, compute the following (where possible).

(a) AB (b) BA (c) $(3E)D$ (d) $(AB)C$
(e) $A(BC)$ (f) CC^T (g) $(DA)^T$ (h) $(C^TB)A^T$
(i) $\operatorname{tr}(DD^T)$ (j) $\operatorname{tr}(4E^T - D)$ (k) $\operatorname{tr}(C^TA^T + 2E^T)$

6. Using the matrices in Exercise 3, compute the following (where possible).

(a) $(2D^T - E)A$ (b) $(4B)C + 2B$ (c) $(-AC)^T + 5D^T$
(d) $(BA^T - 2C)^T$ (e) $B^T(CC^T - A^TA)$ (f) $D^TE^T - (ED)^T$

7. Let

$$A = \begin{bmatrix} 3 & -2 & 7 \\ 6 & 5 & 4 \\ 0 & 4 & 9 \end{bmatrix} \quad \text{and} \quad B = \begin{bmatrix} 6 & -2 & 4 \\ 0 & 1 & 3 \\ 7 & 7 & 5 \end{bmatrix}$$

Use the method of Example 7 to find

(a) the first row of AB (b) the third row of AB (c) the second column of AB
(d) the first column of BA (e) the third row of AA (f) the third column of AA

8. Let A and B be the matrices in Exercise 7.

(a) Express each column matrix of AB as a linear combination of the column matrices of A.
(b) Express each column matrix of BA as a linear combination of the column matrices of B.

9. Let

$$\mathbf{y} = [y_1 \quad y_2 \quad \cdots \quad y_m] \quad \text{and} \quad A = \begin{bmatrix} a_{11} & a_{12} & \cdots & a_{1n} \\ a_{21} & a_{22} & \cdots & a_{2n} \\ \vdots & \vdots & & \vdots \\ a_{m1} & a_{m2} & \cdots & a_{mn} \end{bmatrix}$$

Show that the product $\mathbf{y}A$ can be expressed as a linear combination of the row matrices of A with the scalar coefficients coming from $\mathbf{y}$.

10. Let A and B be the matrices in Exercise 7.

(a) Use the result in Exercise 9 to express each row matrix of AB as a linear combination of the row matrices of B.
(b) Use the result in Exercise 9 to express each row matrix of BA as a linear combination of the row matrices of A.

11. Let C, D, and E be the matrices in Exercise 3. Using as few computations as possible, determine the entry in row 2 and column 3 of $C(DE)$.

12. (a) Show that if AB and BA are both defined, then AB and BA are square matrices.

(b) Show that if A is an $m \times n$ matrix and $A(BA)$ is defined, then B is an $n \times m$ matrix.

13. In each part find matrices A, $\mathbf{x}$, and $\mathbf{b}$ that express the given system of linear equations as a single matrix equation $A\mathbf{x} = \mathbf{b}$.

(a) $2x_1 - 3x_2 + 5x_3 = 7$
    $9x_1 - x_2 + x_3 = -1$
    $x_1 + 5x_2 + 4x_3 = 0$

(b) $4x_1 \quad - 3x_3 + x_4 = 1$
    $5x_1 + x_2 \quad - 8x_4 = 3$
    $2x_1 - 5x_2 + 9x_3 - x_4 = 0$
    $3x_2 - x_3 + 7x_4 = 2$

14. In each part, express the matrix equation as a system of linear equations.

(a) $\begin{bmatrix} 3 & -1 & 2 \\ 4 & 3 & 7 \\ -2 & 1 & 5 \end{bmatrix} \begin{bmatrix} x_1 \\ x_2 \\ x_3 \end{bmatrix} = \begin{bmatrix} 2 \\ -1 \\ 4 \end{bmatrix}$

(b) $\begin{bmatrix} 3 & -2 & 0 & 1 \\ 5 & 0 & 2 & -2 \\ 3 & 1 & 4 & 7 \\ -2 & 5 & 1 & 6 \end{bmatrix} \begin{bmatrix} w \\ x \\ y \\ z \end{bmatrix} = \begin{bmatrix} 0 \\ 0 \\ 0 \\ 0 \end{bmatrix}$

15. If A and B are partitioned into submatrices, for example,

$$A = \left[\begin{array}{c|c} A_{11} & A_{12} \\ \hline A_{21} & A_{22} \end{array}\right] \quad \text{and} \quad B = \left[\begin{array}{c|c} B_{11} & B_{12} \\ \hline B_{21} & B_{22} \end{array}\right]$$

then AB can be expressed as

$$AB = \left[\begin{array}{c|c} A_{11}B_{11} + A_{12}B_{21} & A_{11}B_{12} + A_{12}B_{22} \\ \hline A_{21}B_{11} + A_{22}B_{21} & A_{21}B_{12} + A_{22}B_{22} \end{array}\right]$$

provided the sizes of the submatrices of A and B are such that the indicated operations can be performed. This method of multiplying partitioned matrices is called ***block multiplication***. In each part compute the product by block multiplication. Check your results by multiplying directly.

(a) $A = \left[\begin{array}{cc|cc} -1 & 2 & 1 & 5 \\ 0 & -3 & 4 & 2 \\ \hline 1 & 5 & 6 & 1 \end{array}\right]$, $B = \left[\begin{array}{cc|c} 2 & 1 & 4 \\ -3 & 5 & 2 \\ \hline 7 & -1 & 5 \\ 0 & 3 & -3 \end{array}\right]$

(b) $A = \left[\begin{array}{ccc|c} -1 & 2 & 1 & 5 \\ 0 & -3 & 4 & 2 \\ 1 & 5 & 6 & 1 \end{array}\right]$, $B = \left[\begin{array}{cc|c} 2 & 1 & 4 \\ -3 & 5 & 2 \\ 7 & -1 & 5 \\ \hline 0 & 3 & -3 \end{array}\right]$

16. Adapt the method of Exercise 15 to compute the following products by block multiplication.

(a) $\left[\begin{array}{cc|c} 3 & -1 & 0 & -3 \\ 2 & 1 & 4 & 5 \end{array}\right] \left[\begin{array}{ccc} 2 & -4 & 1 \\ 3 & 0 & 2 \\ \hline 1 & -3 & 5 \\ 2 & 1 & 4 \end{array}\right]$

(b) $\left[\begin{array}{c|c} 2 & -5 \\ 1 & 3 \\ 0 & 5 \\ \hline 1 & 4 \end{array}\right] \left[\begin{array}{cc|cc} 2 & -1 & 3 & -4 \\ 0 & 1 & 5 & 7 \end{array}\right]$

(c) $\left[\begin{array}{ccc|cc} 1 & 0 & 0 & 0 & 0 \\ 0 & 1 & 0 & 0 & 0 \\ 0 & 0 & 1 & 0 & 0 \\ \hline 0 & 0 & 0 & 2 & 0 \\ 0 & 0 & 0 & -1 & 2 \end{array}\right] \left[\begin{array}{cc} 3 & 3 \\ -1 & 4 \\ 1 & 5 \\ \hline 2 & -2 \\ 1 & 6 \end{array}\right]$

17. In each part determine whether block multiplication can be used to compute AB from the given partitions. If so, compute the product by block multiplication.

(a) $A = \left[\begin{array}{ccc|c} -1 & 2 & 1 & 5 \\ 0 & -3 & 4 & 2 \\ \hline 1 & 5 & 6 & 1 \end{array}\right]$, $B = \left[\begin{array}{cc|c} 2 & 1 & 4 \\ -3 & 5 & 2 \\ 7 & -1 & 5 \\ \hline 0 & 3 & -3 \end{array}\right]$

(b) $A = \left[\begin{array}{cccc} -1 & 2 & 1 & 5 \\ \hline 0 & -3 & 4 & 2 \\ \hline 1 & 5 & 6 & 1 \end{array}\right]$, $B = \left[\begin{array}{c|c|c} 2 & 1 & 4 \\ -3 & 5 & 2 \\ 7 & -1 & 5 \\ 0 & 3 & -3 \end{array}\right]$

18. (a) Show that if A has a row of zeros and B is any matrix for which AB is defined, then AB also has a row of zeros.

(b) Find a similar result involving a column of zeros.

19. Let A be any $m \times n$ matrix and let 0 be the $m \times n$ matrix each of whose entries is zero. Show that if $kA = 0$, then $k = 0$ or $A = 0$.

20. Let I be the $n \times n$ matrix whose entry in row i and column j is

$$\begin{cases} 1 & \text{if } i = j \\ 0 & \text{if } i \neq j \end{cases}$$

Show that $AI = IA = A$ for every $n \times n$ matrix A.

21. In each part find a 6×6 matrix $[a_{ij}]$ that satisfies the stated condition. Make your answers as general as possible by using letters rather than specific numbers for the nonzero entries.

(a) $a_{ij} = 0$ if $i \neq j$ (b) $a_{ij} = 0$ if $i > j$ (c) $a_{ij} = 0$ if $i < j$ (d) $a_{ij} = 0$ if $|i - j| > 1$

22. Find the 4×4 matrix $A = [a_{ij}]$ whose entries satisfy the stated condition.

(a) $a_{ij} = i + j$ (b) $a_{ij} = i^{j-1}$ (c) $a_{ij} = \begin{cases} 1 & \text{if } |i - j| > 1 \\ -1 & \text{if } |i - j| \leq 1 \end{cases}$

23. Prove: If A and B are $n \times n$ matrices, then $\text{tr}(A + B) = \text{tr}(A) + \text{tr}(B)$.

Discussion and Discovery

24. Describe three different methods for computing a matrix product, and illustrate the methods by computing some product AB three different ways.

25. How many 3×3 matrices A can you find such that

$$A \begin{bmatrix} x \\ y \\ z \end{bmatrix} = \begin{bmatrix} x + y \\ x - y \\ 0 \end{bmatrix}$$

for all choices of x, y, and z?

26. How many 3×3 matrices A can you find such that

$$A \begin{bmatrix} x \\ y \\ z \end{bmatrix} = \begin{bmatrix} xy \\ 0 \\ 0 \end{bmatrix}$$

for all choices of x, y, and z?

27. A matrix B is said to be a ***square root*** of a matrix A if $BB = A$.

(a) Find two square roots of $A = \begin{bmatrix} 2 & 2 \\ 2 & 2 \end{bmatrix}$.

(b)   How many different square roots can you find of $A = \begin{bmatrix} 5 & 0 \\ 0 & 9 \end{bmatrix}$?

(c)   Do you think that every 2×2 matrix has at least one square root? Explain your reasoning.

28.   Let 0 denote a 2×2 matrix, each of whose entries is zero.

(a)   Is there a 2×2 matrix A such that $A \neq 0$ and $AA = 0$? Justify your answer.

(b)   Is there a 2×2 matrix A such that $A \neq 0$ and $AA = A$? Justify your answer.

29.   Indicate whether the statement is always true or sometimes false. Justify your answer with a logical argument or a counterexample.

(a)   The expressions $\text{tr}(AA^T)$ and $\text{tr}(A^TA)$ are always defined, regardless of the size of A.

(b) $\text{tr}(AA^T) = \text{tr}(A^TA)$ for every matrix A.

(c)   If the first column of A has all zeros, then so does the first column of every product AB.

(d)   If the first row of A has all zeros, then so does the first row of every product AB.

30.   Indicate whether the statement is always true or sometimes false. Justify your answer with a logical argument or a counterexample.

(a)   If A is a square matrix with two identical rows, then AA has two identical rows.

(b)   If A is a square matrix and AA has a column of zeros, then A must have a column of zeros.

(c)   If B is an $n \times n$ matrix whose entries are positive even integers, and if A is an $n \times n$ matrix whose entries are positive integers, then the entries of AB and BA are positive even integers.

(d)   If the matrix sum $AB + BA$ is defined, then A and B must be square.

1.4   INVERSES; RULES OF MATRIX ARITHMETIC

In this section we shall discuss some properties of the arithmetic operations on matrices. We shall see that many of the basic rules of arithmetic for real numbers also hold for matrices but a few do not.

Properties of Matrix Operations   For real numbers a and b, we always have $ab = ba$, which is called the *commutative law for multiplication*. For matrices, however, AB and BA need not be equal. Equality can fail to hold for three reasons: It can happen that the product AB is defined but BA is undefined. For example, this is the case if A is a 2×3 matrix and B is a 3×4 matrix. Also, it can happen that AB and BA are both defined but have different sizes. This is the situation if A is a 2×3 matrix and B is a 3×2 matrix. Finally, as Example 1 shows, it is possible to have $AB \neq BA$ even if both AB and BA are defined and have the same size.

EXAMPLE 1   *AB and BA Need Not Be Equal*

Consider the matrices

$$A = \begin{bmatrix} -1 & 0 \\ 2 & 3 \end{bmatrix}, \qquad B = \begin{bmatrix} 1 & 2 \\ 3 & 0 \end{bmatrix}$$

Multiplying gives

$$AB = \begin{bmatrix} -1 & -2 \\ 11 & 4 \end{bmatrix}, \qquad BA = \begin{bmatrix} 3 & 6 \\ -3 & 0 \end{bmatrix}$$

Thus, $AB \neq BA$.   ◆

Although the commutative law for multiplication is not valid in matrix arithmetic, many familiar laws of arithmetic are valid for matrices. Some of the most important ones and their names are summarized in the following theorem.

Theorem 1.4.1   **Properties of Matrix Arithmetic**

Assuming that the sizes of the matrices are such that the indicated operations can be performed, the following rules of matrix arithmetic are valid.

(*a*) $A + B = B + A$ (**Commutative law for addition**)

(*b*) $A + (B + C) = (A + B) + C$ (**Associative law for addition**)

(*c*) $A(BC) = (AB)C$ (**Associative law for multiplication**)

(*d*) $A(B + C) = AB + AC$ (**Left distributive law**)

(*e*) $(B + C)A = BA + CA$ (**Right distributive law**)

(*f*) $A(B - C) = AB - AC$ (*j*) $(a + b)C = aC + bC$

(*g*) $(B - C)A = BA - CA$ (*k*) $(a - b)C = aC - bC$

(*h*) $a(B + C) = aB + aC$ (*l*) $a(bC) = (ab)C$

(*i*) $a(B - C) = aB - aC$ (*m*) $a(BC) = (aB)C = B(aC)$

To prove the equalities in this theorem we must show that the matrix on the left side has the same size as the matrix on the right side and that corresponding entries on the two sides are equal. With the exception of the associative law in part (*c*), the proofs all follow the same general pattern. We shall prove part (*d*) as an illustration. The proof of the associative law, which is more complicated, is outlined in the exercises.

Proof (*d*). We must show that $A(B + C)$ and $AB + AC$ have the same size and that corresponding entries are equal. To form $A(B + C)$, the matrices B and C must have the same size, say $m \times n$, and the matrix A must then have m columns, so its size must be of the form $r \times m$. This makes $A(B + C)$ an $r \times n$ matrix. It follows that $AB + AC$ is also an $r \times n$ matrix and, consequently, $A(B + C)$ and $AB + AC$ have the same size.

Suppose that $A = [a_{ij}]$, $B = [b_{ij}]$, and $C = [c_{ij}]$. We want to show that corresponding entries of $A(B + C)$ and $AB + AC$ are equal; that is,

$$[A(B + C)]_{ij} = [AB + AC]_{ij}$$

for all values of i and j. But from the definitions of matrix addition and matrix multiplication we have

$$[A(B + C)]_{ij} = a_{i1}(b_{1j} + c_{1j}) + a_{i2}(b_{2j} + c_{2j}) + \cdots + a_{im}(b_{mj} + c_{mj})$$
$$= (a_{i1}b_{1j} + a_{i2}b_{2j} + \cdots + a_{im}b_{mj}) + (a_{i1}c_{1j} + a_{i2}c_{2j} + \cdots + a_{im}c_{mj})$$
$$= [AB]_{ij} + [AC]_{ij} = [AB + AC]_{ij} \qquad\blacksquare$$

REMARK. Although the operations of matrix addition and matrix multiplication were defined for pairs of matrices, associative laws (*b*) and (*c*) enable us to denote sums and products of three matrices as $A + B + C$ and ABC without inserting any parentheses. This is justified by the fact that no matter how parentheses are inserted, the associative laws guarantee that the same end result will be obtained. In general, *given any sum or any product of matrices, pairs of parentheses can be inserted or deleted anywhere within the expression without affecting the end result.*

EXAMPLE 2 Associativity of Matrix Multiplication

As an illustration of the associative law for matrix multiplication, consider

$$A = \begin{bmatrix} 1 & 2 \\ 3 & 4 \\ 0 & 1 \end{bmatrix}, \quad B = \begin{bmatrix} 4 & 3 \\ 2 & 1 \end{bmatrix}, \quad C = \begin{bmatrix} 1 & 0 \\ 2 & 3 \end{bmatrix}$$

Then

$$AB = \begin{bmatrix} 1 & 2 \\ 3 & 4 \\ 0 & 1 \end{bmatrix} \begin{bmatrix} 4 & 3 \\ 2 & 1 \end{bmatrix} = \begin{bmatrix} 8 & 5 \\ 20 & 13 \\ 2 & 1 \end{bmatrix} \quad \text{and} \quad BC = \begin{bmatrix} 4 & 3 \\ 2 & 1 \end{bmatrix} \begin{bmatrix} 1 & 0 \\ 2 & 3 \end{bmatrix} = \begin{bmatrix} 10 & 9 \\ 4 & 3 \end{bmatrix}$$

Thus,

$$(AB)C = \begin{bmatrix} 8 & 5 \\ 20 & 13 \\ 2 & 1 \end{bmatrix} \begin{bmatrix} 1 & 0 \\ 2 & 3 \end{bmatrix} = \begin{bmatrix} 18 & 15 \\ 46 & 39 \\ 4 & 3 \end{bmatrix}$$

and

$$A(BC) = \begin{bmatrix} 1 & 2 \\ 3 & 4 \\ 0 & 1 \end{bmatrix} \begin{bmatrix} 10 & 9 \\ 4 & 3 \end{bmatrix} = \begin{bmatrix} 18 & 15 \\ 46 & 39 \\ 4 & 3 \end{bmatrix}$$

so $(AB)C = A(BC)$, as guaranteed by Theorem 1.4.1c.   ◆

Zero Matrices   A matrix, all of whose entries are zero, such as

$$\begin{bmatrix} 0 & 0 \\ 0 & 0 \end{bmatrix}, \quad \begin{bmatrix} 0 & 0 & 0 \\ 0 & 0 & 0 \\ 0 & 0 & 0 \end{bmatrix}, \quad \begin{bmatrix} 0 & 0 & 0 & 0 \\ 0 & 0 & 0 & 0 \end{bmatrix}, \quad \begin{bmatrix} 0 \\ 0 \\ 0 \\ 0 \end{bmatrix}, \quad [0]$$

is called a ***zero matrix***. A zero matrix will be denoted by 0; if it is important to emphasize the size, we shall write $0_{m \times n}$ for the $m \times n$ zero matrix. Moreover, in keeping with our convention of using boldface symbols for matrices with one column, we will denote a zero matrix with one column by **0**.

If A is any matrix and 0 is the zero matrix with the same size, it is obvious that $A + 0 = 0 + A = A$. The matrix 0 plays much the same role in these matrix equations as the number 0 plays in the numerical equations $a + 0 = 0 + a = a$.

Since we already know that some of the rules of arithmetic for real numbers do not carry over to matrix arithmetic, it would be foolhardy to assume that all the properties of the real number zero carry over to zero matrices. For example, consider the following two standard results in the arithmetic of real numbers.

- If $ab = ac$ and $a \neq 0$, then $b = c$. (This is called the *cancellation law*.)
- If $ad = 0$, then at least one of the factors on the left is 0.

As the next example shows, the corresponding results are not generally true in matrix arithmetic.

EXAMPLE 3 The Cancellation Law Does Not Hold

Consider the matrices

$$A = \begin{bmatrix} 0 & 1 \\ 0 & 2 \end{bmatrix}, \quad B = \begin{bmatrix} 1 & 1 \\ 3 & 4 \end{bmatrix}, \quad C = \begin{bmatrix} 2 & 5 \\ 3 & 4 \end{bmatrix}, \quad D = \begin{bmatrix} 3 & 7 \\ 0 & 0 \end{bmatrix}$$

You should verify that

$$AB = AC = \begin{bmatrix} 3 & 4 \\ 6 & 8 \end{bmatrix} \quad \text{and} \quad AD = \begin{bmatrix} 0 & 0 \\ 0 & 0 \end{bmatrix}$$

Thus, although $A \neq 0$, it is *incorrect* to cancel the A from both sides of the equation $AB = AC$ and write $B = C$. Also, $AD = 0$, yet $A \neq 0$ and $D \neq 0$. Thus, the cancellation law is not valid for matrix multiplication, and it is possible for a product of matrices to be zero without either factor being zero. ♦

In spite of the above example, there are a number of familiar properties of the real number 0 that *do* carry over to zero matrices. Some of the more important ones are summarized in the next theorem. The proofs are left as exercises.

Theorem 1.4.2 Properties of Zero Matrices

Assuming that the sizes of the matrices are such that the indicated operations can be performed, the following rules of matrix arithmetic are valid.

(*a*) $A + 0 = 0 + A = A$
(*b*) $A - A = 0$
(*c*) $0 - A = -A$
(*d*) $A0 = 0; \quad 0A = 0$

Identity Matrices

Of special interest are square matrices with 1's on the main diagonal and 0's off the main diagonal, such as

$$\begin{bmatrix} 1 & 0 \\ 0 & 1 \end{bmatrix}, \quad \begin{bmatrix} 1 & 0 & 0 \\ 0 & 1 & 0 \\ 0 & 0 & 1 \end{bmatrix}, \quad \begin{bmatrix} 1 & 0 & 0 & 0 \\ 0 & 1 & 0 & 0 \\ 0 & 0 & 1 & 0 \\ 0 & 0 & 0 & 1 \end{bmatrix}, \quad \text{and so on.}$$

A matrix of this form is called an ***identity matrix*** and is denoted by I. If it is important to emphasize the size, we shall write I_n for the $n \times n$ identity matrix.

If A is an $m \times n$ matrix, then, as illustrated in the next example,

$$AI_n = A \quad \text{and} \quad I_m A = A$$

Thus, an identity matrix plays much the same role in matrix arithmetic as the number 1 plays in the numerical relationships $a \cdot 1 = 1 \cdot a = a$.

EXAMPLE 4 Multiplication by an Identity Matrix

Consider the matrix

$$A = \begin{bmatrix} a_{11} & a_{12} & a_{13} \\ a_{21} & a_{22} & a_{23} \end{bmatrix}$$

Then

$$I_2A = \begin{bmatrix} 1 & 0 \\ 0 & 1 \end{bmatrix} \begin{bmatrix} a_{11} & a_{12} & a_{13} \\ a_{21} & a_{22} & a_{23} \end{bmatrix} = \begin{bmatrix} a_{11} & a_{12} & a_{13} \\ a_{21} & a_{22} & a_{23} \end{bmatrix} = A$$

and

$$AI_3 = \begin{bmatrix} a_{11} & a_{12} & a_{13} \\ a_{21} & a_{22} & a_{23} \end{bmatrix} \begin{bmatrix} 1 & 0 & 0 \\ 0 & 1 & 0 \\ 0 & 0 & 1 \end{bmatrix} = \begin{bmatrix} a_{11} & a_{12} & a_{13} \\ a_{21} & a_{22} & a_{23} \end{bmatrix} = A \qquad \blacklozenge$$

As the next theorem shows, identity matrices arise naturally in studying reduced row-echelon forms of *square* matrices.

Theorem 1.4.3

If R is the reduced row-echelon form of an n × n matrix A, then either R has a row of zeros or R is the identity matrix I_n.

Proof. Suppose that the reduced row-echelon form of A is

$$R = \begin{bmatrix} r_{11} & r_{12} & \cdots & r_{1n} \\ r_{21} & r_{22} & \cdots & r_{2n} \\ \vdots & \vdots & & \vdots \\ r_{n1} & r_{n2} & \cdots & r_{nn} \end{bmatrix}$$

Either the last row in this matrix consists entirely of zeros or it does not. If not, the matrix contains no zero rows, and consequently each of the n rows has a leading entry of 1. Since these leading 1's occur progressively further to the right as we move down the matrix, each of these 1's must occur on the main diagonal. Since the other entries in the same column as one of these 1's are zero, R must be I_n. Thus, either R has a row of zeros or $R = I_n$. ∎

Definition

If A is a square matrix, and if a matrix B of the same size can be found such that $AB = BA = I$, then A is said to be ***invertible*** and B is called an ***inverse*** of A. If no such matrix B can be found, then A is said to be ***singular***.

EXAMPLE 5 Verifying the Inverse Requirements

The matrix

$$B = \begin{bmatrix} 3 & 5 \\ 1 & 2 \end{bmatrix} \quad \text{is an inverse of} \quad A = \begin{bmatrix} 2 & -5 \\ -1 & 3 \end{bmatrix}$$

since

$$AB = \begin{bmatrix} 2 & -5 \\ -1 & 3 \end{bmatrix} \begin{bmatrix} 3 & 5 \\ 1 & 2 \end{bmatrix} = \begin{bmatrix} 1 & 0 \\ 0 & 1 \end{bmatrix} = I$$

and

$$BA = \begin{bmatrix} 3 & 5 \\ 1 & 2 \end{bmatrix} \begin{bmatrix} 2 & -5 \\ -1 & 3 \end{bmatrix} = \begin{bmatrix} 1 & 0 \\ 0 & 1 \end{bmatrix} = I \qquad \blacklozenge$$

EXAMPLE 6 **A Matrix with No Inverse**

The matrix

$$A = \begin{bmatrix} 1 & 4 & 0 \\ 2 & 5 & 0 \\ 3 & 6 & 0 \end{bmatrix}$$

is singular. To see why, let

$$B = \begin{bmatrix} b_{11} & b_{12} & b_{13} \\ b_{21} & b_{22} & b_{23} \\ b_{31} & b_{32} & b_{33} \end{bmatrix}$$

be any 3×3 matrix. The third column of BA is

$$\begin{bmatrix} b_{11} & b_{12} & b_{13} \\ b_{21} & b_{22} & b_{23} \\ b_{31} & b_{32} & b_{33} \end{bmatrix} \begin{bmatrix} 0 \\ 0 \\ 0 \end{bmatrix} = \begin{bmatrix} 0 \\ 0 \\ 0 \end{bmatrix}$$

Thus,

$$BA \neq I = \begin{bmatrix} 1 & 0 & 0 \\ 0 & 1 & 0 \\ 0 & 0 & 1 \end{bmatrix} \qquad \blacklozenge$$

Properties of Inverses It is reasonable to ask whether an invertible matrix can have more than one inverse. The next theorem shows that the answer is no—*an invertible matrix has exactly one inverse.*

Theorem 1.4.4

If B and C are both inverses of the matrix A, then $B = C$.

Proof. Since B is an inverse of A, we have $BA = I$. Multiplying both sides on the right by C gives $(BA)C = IC = C$. But $(BA)C = B(AC) = BI = B$, so that $C = B$. ∎

As a consequence of this important result, we can now speak of "the" inverse of an invertible matrix. If A is invertible, then its inverse will be denoted by the symbol A^{-1}. Thus,

$$AA^{-1} = I \quad \text{and} \quad A^{-1}A = I$$

The inverse of A plays much the same role in matrix arithmetic that the reciprocal a^{-1} plays in the numerical relationships $aa^{-1} = 1$ and $a^{-1}a = 1$.

In the next section we shall develop a method for finding inverses of invertible matrices of any size; however, the following theorem gives conditions under which a 2×2 matrix is invertible and provides a simple formula for the inverse.

Theorem 1.4.5

The matrix

$$A = \begin{bmatrix} a & b \\ c & d \end{bmatrix}$$

is invertible if $ad - bc \neq 0$, in which case the inverse is given by the formula

$$A^{-1} = \frac{1}{ad - bc} \begin{bmatrix} d & -b \\ -c & a \end{bmatrix} = \begin{bmatrix} \dfrac{d}{ad - bc} & -\dfrac{b}{ad - bc} \\ -\dfrac{c}{ad - bc} & \dfrac{a}{ad - bc} \end{bmatrix}$$

Proof. We leave it for the reader to verify that $AA^{-1} = I_2$ and $A^{-1}A = I_2$. ∎

Theorem 1.4.6

If A and B are invertible matrices of the same size, then AB is invertible and

$$(AB)^{-1} = B^{-1}A^{-1}$$

Proof. If we can show that $(AB)(B^{-1}A^{-1}) = (B^{-1}A^{-1})(AB) = I$, then we will have simultaneously shown that the matrix AB is invertible and that $(AB)^{-1} = B^{-1}A^{-1}$. But $(AB)(B^{-1}A^{-1}) = A(BB^{-1})A^{-1} = AIA^{-1} = AA^{-1} = I$. A similar argument shows that $(B^{-1}A^{-1})(AB) = I$. ∎

Although we will not prove it, this result can be extended to include three or more factors; that is,

A product of any number of invertible matrices is invertible, and the inverse of the product is the product of the inverses in the reverse order.

EXAMPLE 7 Inverse of a Product

Consider the matrices

$$A = \begin{bmatrix} 1 & 2 \\ 1 & 3 \end{bmatrix}, \qquad B = \begin{bmatrix} 3 & 2 \\ 2 & 2 \end{bmatrix}, \qquad AB = \begin{bmatrix} 7 & 6 \\ 9 & 8 \end{bmatrix}$$

Applying the formula in Theorem 1.4.5, we obtain

$$A^{-1} = \begin{bmatrix} 3 & -2 \\ -1 & 1 \end{bmatrix}, \qquad B^{-1} = \begin{bmatrix} 1 & -1 \\ -1 & \frac{3}{2} \end{bmatrix}, \qquad (AB)^{-1} = \begin{bmatrix} 4 & -3 \\ -\frac{9}{2} & \frac{7}{2} \end{bmatrix}$$

Also,

$$B^{-1}A^{-1} = \begin{bmatrix} 1 & -1 \\ -1 & \frac{3}{2} \end{bmatrix} \begin{bmatrix} 3 & -2 \\ -1 & 1 \end{bmatrix} = \begin{bmatrix} 4 & -3 \\ -\frac{9}{2} & \frac{7}{2} \end{bmatrix}$$

Therefore, $(AB)^{-1} = B^{-1}A^{-1}$ as guaranteed by Theorem 1.4.6. ◆

Powers of a Matrix

Next, we shall define powers of a square matrix and discuss their properties.

Definition

If A is a square matrix, then we define the nonnegative integer powers of A to be

$$A^0 = I \qquad A^n = \underbrace{AA \cdots A}_{n \text{ factors}} \qquad (n > 0)$$

Moreover, if A is invertible, then we define the negative integer powers to be

$$A^{-n} = (A^{-1})^n = \underbrace{A^{-1}A^{-1} \cdots A^{-1}}_{n \text{ factors}}$$

Because this definition parallels that for real numbers, the usual laws of exponents hold. (We omit the details.)

Theorem 1.4.7 **Laws of Exponents**

If A is a square matrix and r and s are integers, then

$$A^r A^s = A^{r+s}, \qquad (A^r)^s = A^{rs}$$

The next theorem provides some useful properties of negative exponents.

Theorem 1.4.8 **Laws of Exponents**

If A is an invertible matrix, then:

(a) A^{-1} *is invertible and* $(A^{-1})^{-1} = A$.
(b) A^n *is invertible and* $(A^n)^{-1} = (A^{-1})^n$ *for* $n = 0, 1, 2, \ldots$.
(c) *For any nonzero scalar k, the matrix kA is invertible and* $(kA)^{-1} = \dfrac{1}{k}A^{-1}$.

Proof.

(a) Since $AA^{-1} = A^{-1}A = I$, the matrix A^{-1} is invertible and $(A^{-1})^{-1} = A$.
(b) This part is left as an exercise.
(c) If k is any nonzero scalar, results (l) and (m) of Theorem 1.4.1 enable us to write

$$(kA)\left(\frac{1}{k}A^{-1}\right) = \frac{1}{k}(kA)A^{-1} = \left(\frac{1}{k}k\right)AA^{-1} = (1)I = I$$

Similarly, $\left(\dfrac{1}{k}A^{-1}\right)(kA) = I$ so that kA is invertible and $(kA)^{-1} = \dfrac{1}{k}A^{-1}$. ∎

EXAMPLE 8 Powers of a Matrix

Let A and A^{-1} be as in Example 7, that is,

$$A = \begin{bmatrix} 1 & 2 \\ 1 & 3 \end{bmatrix} \quad \text{and} \quad A^{-1} = \begin{bmatrix} 3 & -2 \\ -1 & 1 \end{bmatrix}$$

Then

$$A^3 = \begin{bmatrix} 1 & 2 \\ 1 & 3 \end{bmatrix} \begin{bmatrix} 1 & 2 \\ 1 & 3 \end{bmatrix} \begin{bmatrix} 1 & 2 \\ 1 & 3 \end{bmatrix} = \begin{bmatrix} 11 & 30 \\ 15 & 41 \end{bmatrix}$$

$$A^{-3} = (A^{-1})^3 = \begin{bmatrix} 3 & -2 \\ -1 & 1 \end{bmatrix} \begin{bmatrix} 3 & -2 \\ -1 & 1 \end{bmatrix} \begin{bmatrix} 3 & -2 \\ -1 & 1 \end{bmatrix} = \begin{bmatrix} 41 & -30 \\ -15 & 11 \end{bmatrix}$$ ◆

Polynomial Expressions Involving Matrices

If A is a square matrix, say $m \times m$, and if

$$p(x) = a_0 + a_1 x + \cdots + a_n x^n \tag{1}$$

is any polynomial, then we define

$$p(A) = a_0 I + a_1 A + \cdots + a_n A^n$$

where I is the $m \times m$ identity matrix. In words, $p(A)$ is the $m \times m$ matrix that results when A is substituted for x in (1) and a_0 is replaced by $a_0 I$.

EXAMPLE 9 Matrix Polynomial

If

$$p(x) = 2x^2 - 3x + 4 \quad \text{and} \quad A = \begin{bmatrix} -1 & 2 \\ 0 & 3 \end{bmatrix}$$

then

$$p(A) = 2A^2 - 3A + 4I = 2 \begin{bmatrix} -1 & 2 \\ 0 & 3 \end{bmatrix}^2 - 3 \begin{bmatrix} -1 & 2 \\ 0 & 3 \end{bmatrix} + 4 \begin{bmatrix} 1 & 0 \\ 0 & 1 \end{bmatrix}$$

$$= \begin{bmatrix} 2 & 8 \\ 0 & 18 \end{bmatrix} - \begin{bmatrix} -3 & 6 \\ 0 & 9 \end{bmatrix} + \begin{bmatrix} 4 & 0 \\ 0 & 4 \end{bmatrix} = \begin{bmatrix} 9 & 2 \\ 0 & 13 \end{bmatrix}$$ ◆

Properties of the Transpose

The next theorem lists the main properties of the transpose operation.

Theorem 1.4.9   **Properties of the Transpose**

If the sizes of the matrices are such that the stated operations can be performed, then

(a) $((A)^T)^T = A$
(b) $(A + B)^T = A^T + B^T$ and $(A - B)^T = A^T - B^T$
(c) $(kA)^T = kA^T$, *where k is any scalar*
(d) $(AB)^T = B^T A^T$

Keeping in mind that transposing a matrix interchanges its rows and columns, parts (a), (b), and (c) should be self-evident. For example, part (a) states that interchanging rows and columns twice leaves a matrix unchanged; part (b) asserts that adding and then interchanging rows and columns yields the same result as first interchanging rows and columns, then adding; and part (c) asserts that multiplying by a scalar and then interchanging rows and columns yields the same result as first interchanging rows and columns, then multiplying by the scalar. Part (d) is not so obvious, so we give its proof.

Proof (***d***). Let $A = [a_{ij}]_{m \times r}$ and $B = [b_{ij}]_{r \times n}$ so that the products AB and $B^T A^T$ can both be formed. We leave it for the reader to check that $(AB)^T$ and $B^T A^T$ have the same size, namely $n \times m$. Thus, it only remains to show that corresponding entries of $(AB)^T$ and $B^T A^T$ are the same; that is,

$$\left((AB)^T\right)_{ij} = (B^T A^T)_{ij} \tag{2}$$

Applying Formula (11) of Section 1.3 to the left side of this equation and using the definition of matrix multiplication, we obtain

$$\left((AB)^T\right)_{ij} = (AB)_{ji} = a_{j1}b_{1i} + a_{j2}b_{2i} + \cdots + a_{jr}b_{ri} \tag{3}$$

To evaluate the right side of (2) it will be convenient to let a'_{ij} and b'_{ij} denote the ijth entries of A^T and B^T, respectively, so

$$a'_{ij} = a_{ji} \quad \text{and} \quad b'_{ij} = b_{ji}$$

From these relationships and the definition of matrix multiplication we obtain

$$\begin{aligned}
(B^T A^T)_{ij} &= b'_{i1}a'_{1j} + b'_{i2}a'_{2j} + \cdots + b'_{ir}a'_{rj} \\
&= b_{1i}a_{j1} + b_{2i}a_{j2} + \cdots + b_{ri}a_{jr} \\
&= a_{j1}b_{1i} + a_{j2}b_{2i} + \cdots + a_{jr}b_{ri}
\end{aligned}$$

This, together with (3), proves (2). ∎

Although we shall not prove it, part (*d*) of this theorem can be extended to include three or more factors; that is,

> *The transpose of a product of any number of matrices is equal to the product of their transposes in the reverse order.*

REMARK. Note the similarity between this result and the result following Theorem 1.4.6 about the inverse of a product of matrices.

Invertibility of a Transpose

The following theorem establishes a relationship between the inverse of an invertible matrix and the inverse of its transpose.

Theorem 1.4.10

If A is an invertible matrix, then A^T is also invertible and

$$(A^T)^{-1} = (A^{-1})^T \tag{4}$$

Proof. We can prove the invertibility of A^T and obtain (4) by showing that

$$A^T(A^{-1})^T = (A^{-1})^T A^T = I$$

But from part (*d*) of Theorem 1.4.9 and the fact that $I^T = I$ we have

$$\begin{aligned}
A^T(A^{-1})^T &= (A^{-1}A)^T = I^T = I \\
(A^{-1})^T A^T &= (AA^{-1})^T = I^T = I
\end{aligned}$$

which completes the proof. ∎

EXAMPLE 10 Verifying Theorem 1.4.10

Consider the matrices

$$A = \begin{bmatrix} -5 & -3 \\ 2 & 1 \end{bmatrix}, \quad A^T = \begin{bmatrix} -5 & 2 \\ -3 & 1 \end{bmatrix}$$

Applying Theorem 1.4.5 yields

$$A^{-1} = \begin{bmatrix} 1 & 3 \\ -2 & -5 \end{bmatrix}, \quad (A^{-1})^T = \begin{bmatrix} 1 & -2 \\ 3 & -5 \end{bmatrix}, \quad (A^T)^{-1} = \begin{bmatrix} 1 & -2 \\ 3 & -5 \end{bmatrix}$$

As guaranteed by Theorem 1.4.10, these matrices satisfy (4).   ◆

Exercise Set 1.4

1. Let

$$A = \begin{bmatrix} 2 & -1 & 3 \\ 0 & 4 & 5 \\ -2 & 1 & 4 \end{bmatrix}, \quad B = \begin{bmatrix} 8 & -3 & -5 \\ 0 & 1 & 2 \\ 4 & -7 & 6 \end{bmatrix}, \quad C = \begin{bmatrix} 0 & -2 & 3 \\ 1 & 7 & 4 \\ 3 & 5 & 9 \end{bmatrix}, \quad a = 4, \quad b = -7$$

Show that

(a) $A + (B + C) = (A + B) + C$ (b) $(AB)C = A(BC)$ (c) $(a+b)C = aC + bC$

(d) $a(B - C) = aB - aC$

2. Using the matrices and scalars in Exercise 1, verify that

(a) $a(BC) = (aB)C = B(aC)$ (b) $A(B - C) = AB - AC$ (c) $(B + C)A = BA + CA$

(d) $a(bC) = (ab)C$

3. Using the matrices and scalars in Exercise 1, verify that

(a) $(A^T)^T = A$ (b) $(A + B)^T = A^T + B^T$ (c) $(aC)^T = aC^T$ (d) $(AB)^T = B^T A^T$

4. Use Theorem 1.4.5 to compute the inverses of the following matrices.

(a) $A = \begin{bmatrix} 3 & 1 \\ 5 & 2 \end{bmatrix}$ (b) $B = \begin{bmatrix} 2 & -3 \\ 4 & 4 \end{bmatrix}$ (c) $C = \begin{bmatrix} 6 & 4 \\ -2 & -1 \end{bmatrix}$ (d) $D = \begin{bmatrix} 2 & 0 \\ 0 & 3 \end{bmatrix}$

5. Use the matrices A and B in Exercise 4 to verify that

(a) $(A^{-1})^{-1} = A$ (b) $(B^T)^{-1} = (B^{-1})^T$

6. Use the matrices A, B, and C in Exercise 4 to verify that

(a) $(AB)^{-1} = B^{-1}A^{-1}$ (b) $(ABC)^{-1} = C^{-1}B^{-1}A^{-1}$

7. In each part use the given information to find A.

(a) $A^{-1} = \begin{bmatrix} 2 & -1 \\ 3 & 5 \end{bmatrix}$ (b) $(7A)^{-1} = \begin{bmatrix} -3 & 7 \\ 1 & -2 \end{bmatrix}$

(c) $(5A^T)^{-1} = \begin{bmatrix} -3 & -1 \\ 5 & 2 \end{bmatrix}$ (d) $(I + 2A)^{-1} = \begin{bmatrix} -1 & 2 \\ 4 & 5 \end{bmatrix}$

8. Let A be the matrix

$$\begin{bmatrix} 2 & 0 \\ 4 & 1 \end{bmatrix}$$

Compute A^3, A^{-3}, and $A^2 - 2A + I$.

9. Let A be the matrix

$$\begin{bmatrix} 3 & 1 \\ 2 & 1 \end{bmatrix}$$

In each part find $p(A)$.

(a) $p(x) = x - 2$ (b) $p(x) = 2x^2 - x + 1$ (c) $p(x) = x^3 - 2x + 4$

10. Let $p_1(x) = x^2 - 9$, $p_2(x) = x + 3$, and $p_3(x) = x - 3$.

(a) Show that $p_1(A) = p_2(A)p_3(A)$ for the matrix A in Exercise 9.

(b) Show that $p_1(A) = p_2(A)p_3(A)$ for any square matrix A.

11. Find the inverse of

$$\begin{bmatrix} \cos\theta & \sin\theta \\ -\sin\theta & \cos\theta \end{bmatrix}$$

12. Find the inverse of

$$\begin{bmatrix} \frac{1}{2}(e^x + e^{-x}) & \frac{1}{2}(e^x - e^{-x}) \\ \frac{1}{2}(e^x - e^{-x}) & \frac{1}{2}(e^x + e^{-x}) \end{bmatrix}$$

13. Consider the matrix

$$A = \begin{bmatrix} a_{11} & 0 & \cdots & 0 \\ 0 & a_{22} & \cdots & 0 \\ \vdots & \vdots & & \vdots \\ 0 & 0 & \cdots & a_{nn} \end{bmatrix}$$

where $a_{11}a_{22}\cdots a_{nn} \neq 0$. Show that A is invertible and find its inverse.

14. Show that if a square matrix A satisfies $A^2 - 3A + I = 0$, then $A^{-1} = 3I - A$.

15. (a) Show that a matrix with a row of zeros cannot have an inverse.

(b) Show that a matrix with a column of zeros cannot have an inverse.

16. Is the sum of two invertible matrices necessarily invertible?

17. Let A and B be square matrices such that $AB = 0$. Show that if A is invertible, then $B = 0$.

18. Let A, B, and 0 be 2×2 matrices. Assuming that A is invertible, find a matrix C so that

$$\left[\begin{array}{c|c} A^{-1} & 0 \\ \hline C & A^{-1} \end{array} \right]$$

is the inverse of the partitioned matrix

$$\left[\begin{array}{c|c} A & 0 \\ \hline B & A \end{array} \right]$$

(See Exercise 15 of the preceding section.)

19. Use the result in Exercise 18 to find the inverses of the following matrices.

(a) $\begin{bmatrix} 1 & 1 & 0 & 0 \\ -1 & 1 & 0 & 0 \\ 1 & 1 & 1 & 1 \\ 1 & 1 & -1 & 1 \end{bmatrix}$ (b) $\begin{bmatrix} 1 & 1 & 0 & 0 \\ 0 & 1 & 0 & 0 \\ 0 & 0 & 1 & 1 \\ 0 & 0 & 0 & 1 \end{bmatrix}$

20. (a) Find a nonzero 3×3 matrix A such that $A^T = A$.

(b) Find a nonzero 3×3 matrix A such that $A^T = -A$.

21. A square matrix A is called *symmetric* if $A^T = A$ and *skew-symmetric* if $A^T = -A$. Show that if B is a square matrix, then

(a) BB^T and $B + B^T$ are symmetric   (b) $B - B^T$ is skew-symmetric

22. If A is a square matrix and n is a positive integer, is it true that $(A^n)^T = (A^T)^n$? Justify your answer.

23. Let A be the matrix

$$\begin{bmatrix} 1 & 0 & 1 \\ 1 & 1 & 0 \\ 0 & 1 & 1 \end{bmatrix}$$

Determine whether A is invertible, and if so, find its inverse. [**Hint.** Solve $AX = I$ by equating corresponding entries on the two sides.]

24. Prove:

(a) part (b) of Theorem 1.4.1   (b) part (i) of Theorem 1.4.1   (c) part (m) of Theorem 1.4.1

25. Apply parts (d) and (m) of Theorem 1.4.1 to the matrices A, B, and $(-1)C$ to derive the result in part (f).

26. Prove Theorem 1.4.2.

27. Consider the laws of exponents $A^r A^s = A^{r+s}$ and $(A^r)^s = A^{rs}$.

(a) Show that if A is any square matrix, these laws are valid for all nonnegative integer values of r and s.

(b) Show that if A is invertible, these laws hold for all negative integer values of r and s.

28. Show that if A is invertible and k is any nonzero scalar, then $(kA)^n = k^n A^n$ for all integer values of n.

29. (a) Show that if A is invertible and $AB = AC$, then $B = C$.

(b) Explain why part (a) and Example 3 do not contradict one another.

30. Prove part (c) of Theorem 1.4.1. [**Hint.** Assume that A is $m \times n$, B is $n \times p$, and C is $p \times q$. The ijth entry on the left side is $l_{ij} = a_{i1}[BC]_{1j} + a_{i2}[BC]_{2j} + \cdots + a_{in}[BC]_{nj}$ and the ijth entry on the right side is $r_{ij} = [AB]_{i1}c_{1j} + [AB]_{i2}c_{2j} + \cdots + [AB]_{ip}c_{pj}$. Verify that $l_{ij} = r_{ij}$.]

Discussion and Discovery

31. Let A and B be square matrices with the same size.

(a) Give an example in which $(A + B)^2 \neq A^2 + 2AB + B^2$.

(b) Fill in the blank to create a matrix identity that is valid for all choices of A and B.
$(A + B)^2 = A^2 + B^2 + $ _____.

32. Let A and B be square matrices with the same size.

(a) Give an example in which $(A + B)(A - B) \neq A^2 - B^2$.

(b) Let A and B be square matrices with the same size. Fill in the blank to create a matrix identity that is valid for all choices of A and B. $(A + B)(A - B) = $ _____.

33. In the real number system the equation $a^2 = 1$ has exactly two solutions. Find at least eight different 3×3 matrices that satisfy the equation $A^2 = I_3$. [**Hint.** Look for solutions in which all entries off the main diagonal are zero.]

34. A statement of the form "If p, then q" is logically equivalent to the statement "If not q, then not p." (The second statement is called the **logical contrapositive** of the first.) For example, the logical contrapositive of the statement "If it is raining, then the ground is wet" is "If the ground is not wet, then it is not raining."

(a) Find the logical contrapositive of the following statement: If A^T is singular, then A is singular.

(b) Is the statement true or false? Explain.

35. Let A and B be $n \times n$ matrices. Indicate whether the statement is always true or sometimes false. Justify each answer.

(a) $(AB)^2 = A^2 B^2$ (b) $(A - B)^2 = (B - A)^2$

(c) $(AB^{-1})(BA^{-1}) = I_n$ (d) $AB \neq BA$

1.5 ELEMENTARY MATRICES AND A METHOD FOR FINDING A^{-1}

In this section we shall develop an algorithm for finding the inverse of an invertible matrix. We shall also discuss some of the basic properties of invertible matrices.

We begin with the definition of a special type of matrix that can be used to carry out an elementary row operation by matrix multiplication.

> **Definition**
>
> An $n \times n$ matrix is called an ***elementary matrix*** if it can be obtained from the $n \times n$ identity matrix I_n by performing a single elementary row operation.

EXAMPLE 1 Elementary Matrices and Row Operations

Listed below are four elementary matrices and the operations that produce them.

$$
\begin{bmatrix} 1 & 0 \\ 0 & -3 \end{bmatrix} \qquad
\begin{bmatrix} 1 & 0 & 0 & 0 \\ 0 & 0 & 0 & 1 \\ 0 & 0 & 1 & 0 \\ 0 & 1 & 0 & 0 \end{bmatrix} \qquad
\begin{bmatrix} 1 & 0 & 3 \\ 0 & 1 & 0 \\ 0 & 0 & 1 \end{bmatrix} \qquad
\begin{bmatrix} 1 & 0 & 0 \\ 0 & 1 & 0 \\ 0 & 0 & 1 \end{bmatrix}
$$

Multiply the second row of I_2 by -3.　Interchange the second and fourth rows of I_4.　Add 3 times the third row of I_3 to the first row.　Multiply the first row of I_3 by 1. ◆

When a matrix A is multiplied on the *left* by an elementary matrix E, the effect is to perform an elementary row operation on A. This is the content of the following theorem, the proof of which is left for the exercises.

> **Theorem 1.5.1** **Row Operations by Matrix Multiplication**
>
> *If the elementary matrix E results from performing a certain row operation on I_m and if A is an $m \times n$ matrix, then the product EA is the matrix that results when this same row operation is performed on A.*

EXAMPLE 2 Using Elementary Matrices

Consider the matrix

$$
A = \begin{bmatrix} 1 & 0 & 2 & 3 \\ 2 & -1 & 3 & 6 \\ 1 & 4 & 4 & 0 \end{bmatrix}
$$

and consider the elementary matrix

$$E = \begin{bmatrix} 1 & 0 & 0 \\ 0 & 1 & 0 \\ 3 & 0 & 1 \end{bmatrix}$$

which results from adding 3 times the first row of I_3 to the third row. The product EA is

$$EA = \begin{bmatrix} 1 & 0 & 2 & 3 \\ 2 & -1 & 3 & 6 \\ 4 & 4 & 10 & 9 \end{bmatrix}$$

which is precisely the same matrix that results when we add 3 times the first row of A to the third row. ◆

REMARK. Theorem 1.5.1 is primarily of theoretical interest and will be used for developing some results about matrices and systems of linear equations. Computationally, it is preferable to perform row operations directly rather than multiplying on the left by an elementary matrix.

If an elementary row operation is applied to an identity matrix I to produce an elementary matrix E, then there is a second row operation that, when applied to E, produces I back again. For example, if E is obtained by multiplying the ith row of I by a nonzero constant c, then I can be recovered if the ith row of E is multiplied by $1/c$. The various possibilities are listed in Table 1. The operations on the right side of this table are called the ***inverse operations*** of the corresponding operations on the left.

TABLE 1

Row Operation on *I* That Produces *E*	Row Operation on *E* That Reproduces *I*
Multiply row i by $c \neq 0$	Multiply row i by $1/c$
Interchange rows i and j	Interchange rows i and j
Add c times row i to row j	Add $-c$ times row i to row j

EXAMPLE 3 Row Operations and Inverse Row Operations

In each of the following, an elementary row operation is applied to the 2×2 identity matrix to obtain an elementary matrix E, then E is restored to the identity matrix by applying the inverse row operation.

$$\begin{bmatrix} 1 & 0 \\ 0 & 1 \end{bmatrix} \longrightarrow \begin{bmatrix} 1 & 0 \\ 0 & 7 \end{bmatrix} \longrightarrow \begin{bmatrix} 1 & 0 \\ 0 & 1 \end{bmatrix}$$

Multiply the second row by 7.   Multiply the second row by $\frac{1}{7}$.

$$\begin{bmatrix} 1 & 0 \\ 0 & 1 \end{bmatrix} \longrightarrow \begin{bmatrix} 0 & 1 \\ 1 & 0 \end{bmatrix} \longrightarrow \begin{bmatrix} 1 & 0 \\ 0 & 1 \end{bmatrix}$$

↑   ↑

Interchange the first   Interchange the first
and second rows.   and second rows.

$$\begin{bmatrix} 1 & 0 \\ 0 & 1 \end{bmatrix} \longrightarrow \begin{bmatrix} 1 & 5 \\ 0 & 1 \end{bmatrix} \longrightarrow \begin{bmatrix} 1 & 0 \\ 0 & 1 \end{bmatrix}$$

↑   ↑

Add 5 times the   Add -5 times the
second row to the   second row to the
first.   first.   ◆

The next theorem gives an important property of elementary matrices.

Theorem 1.5.2

Every elementary matrix is invertible, and the inverse is also an elementary matrix.

Proof. If E is an elementary matrix, then E results from performing some row operation on I. Let E_0 be the matrix that results when the inverse of this operation is performed on I. Applying Theorem 1.5.1 and using the fact that inverse row operations cancel the effect of each other, it follows that

$$E_0 E = I \quad \text{and} \quad E E_0 = I$$

Thus, the elementary matrix E_0 is the inverse of E. ■

The next theorem establishes some fundamental relationships between invertibility, homogeneous linear systems, reduced row-echelon forms, and elementary matrices. These results are extremely important and will be used many times in later sections.

Theorem 1.5.3   Equivalent Statements

If A is an $n \times n$ matrix, then the following statements are equivalent, that is, all true or all false.

(a) A is invertible.
(b) $A\mathbf{x} = \mathbf{0}$ has only the trivial solution.
(c) The reduced row-echelon form of A is I_n.
(d) A is expressible as a product of elementary matrices.

Proof. We shall prove the equivalence by establishing the chain of implications: $(a) \Rightarrow (b) \Rightarrow (c) \Rightarrow (d) \Rightarrow (a)$.

(a) $\Rightarrow$ (b). Assume A is invertible and let $\mathbf{x}_0$ be any solution of $A\mathbf{x} = \mathbf{0}$; thus, $A\mathbf{x}_0 = \mathbf{0}$. Multiplying both sides of this equation by the matrix A^{-1} gives $A^{-1}(A\mathbf{x}_0) = A^{-1}\mathbf{0}$, or $(A^{-1}A)\mathbf{x}_0 = \mathbf{0}$, or $I\mathbf{x}_0 = \mathbf{0}$, or $\mathbf{x}_0 = \mathbf{0}$. Thus, $A\mathbf{x} = \mathbf{0}$ has only the trivial solution.

$(b) \Rightarrow (c)$. Let $A\mathbf{x} = \mathbf{0}$ be the matrix form of the system

$$
\begin{aligned}
a_{11}x_1 + a_{12}x_2 + \cdots + a_{1n}x_n &= 0 \\
a_{21}x_1 + a_{22}x_2 + \cdots + a_{2n}x_n &= 0 \\
\vdots \qquad \vdots \qquad\qquad \vdots \qquad \vdots \\
a_{n1}x_1 + a_{n2}x_2 + \cdots + a_{nn}x_n &= 0
\end{aligned}
\tag{1}
$$

and assume that the system has only the trivial solution. If we solve by Gauss–Jordan elimination, then the system of equations corresponding to the reduced row-echelon form of the augmented matrix will be

$$
\begin{aligned}
x_1 \qquad\qquad\qquad &= 0 \\
x_2 \qquad\qquad &= 0 \\
\ddots \\
x_n &= 0
\end{aligned}
\tag{2}
$$

Thus, the augmented matrix

$$
\begin{bmatrix}
a_{11} & a_{12} & \cdots & a_{1n} & 0 \\
a_{21} & a_{22} & \cdots & a_{2n} & 0 \\
\vdots & \vdots & & \vdots & \vdots \\
a_{n1} & a_{n2} & \cdots & a_{nn} & 0
\end{bmatrix}
$$

for (1) can be reduced to the augmented matrix

$$
\begin{bmatrix}
1 & 0 & 0 & \cdots & 0 & 0 \\
0 & 1 & 0 & \cdots & 0 & 0 \\
0 & 0 & 1 & \cdots & 0 & 0 \\
\vdots & \vdots & \vdots & & \vdots & \vdots \\
0 & 0 & 0 & \cdots & 1 & 0
\end{bmatrix}
$$

for (2) by a sequence of elementary row operations. If we disregard the last column (of zeros) in each of these matrices, we can conclude that the reduced row-echelon form of A is I_n.

$(c) \Rightarrow (d)$. Assume that the reduced row-echelon form of A is I_n, so that A can be reduced to I_n by a finite sequence of elementary row operations. By Theorem 1.5.1 each of these operations can be accomplished by multiplying on the left by an appropriate elementary matrix. Thus, we can find elementary matrices $E_1, E_2, \ldots, E_k$ such that

$$
E_k \cdots E_2 E_1 A = I_n
\tag{3}
$$

By Theorem 1.5.2, $E_1, E_2, \ldots, E_k$ are invertible. Multiplying both sides of Equation (3) on the left successively by $E_k^{-1}, \ldots, E_2^{-1}, E_1^{-1}$ we obtain

$$
A = E_1^{-1} E_2^{-1} \cdots E_k^{-1} I_n = E_1^{-1} E_2^{-1} \cdots E_k^{-1}
\tag{4}
$$

By Theorem 1.5.2, this equation expresses A as a product of elementary matrices.

$(d) \Rightarrow (a)$. If A is a product of elementary matrices, then from Theorems 1.4.6 and 1.5.2 the matrix A is a product of invertible matrices, and hence is invertible. ∎

Row Equivalence
If a matrix B can be obtained from a matrix A by performing a finite sequence of elementary row operations, then obviously we can get from B back to A by performing the inverses of these elementary row operations in reverse order. Matrices that can be obtained from one another by a finite sequence of elementary row operations are said to be ***row equivalent***. With this terminology it follows from

parts (*a*) and (*c*) of Theorem 1.5.3 that an $n \times n$ matrix A is invertible if and only if it is row equivalent to the $n \times n$ identity matrix.

A Method for Inverting Matrices

As our first application of Theorem 1.5.3, we shall establish a method for determining the inverse of an invertible matrix. Multiplying (3) on the right by A^{-1} yields

$$A^{-1} = E_k \cdots E_2 E_1 I_n \tag{5}$$

which tells us that A^{-1} can be obtained by multiplying I_n successively on the left by the elementary matrices $E_1, E_2, \ldots, E_k$. Since each multiplication on the left by one of these elementary matrices performs a row operation, it follows, by comparing Equations (3) and (5), that *the sequence of row operations that reduces A to I_n will reduce I_n to A^{-1}.* Thus, we have the following result:

> *To find the inverse of an invertible matrix A, we must find a sequence of elementary row operations that reduces A to the identity and then perform this same sequence of operations on I_n to obtain A^{-1}.*

A simple method for carrying out this procedure is given in the following example.

EXAMPLE 4 Using Row Operations to Find A^{-1}

Find the inverse of

$$A = \begin{bmatrix} 1 & 2 & 3 \\ 2 & 5 & 3 \\ 1 & 0 & 8 \end{bmatrix}$$

Solution.

We want to reduce A to the identity matrix by row operations and simultaneously apply these operations to I to produce A^{-1}. To accomplish this we shall adjoin the identity matrix to the right side of A, thereby producing a matrix of the form

$$[A \mid I]$$

Then we shall apply row operations to this matrix until the left side is reduced to I; these operations will convert the right side to A^{-1}, so that the final matrix will have the form

$$[I \mid A^{-1}]$$

The computations are as follows:

$$\begin{bmatrix} 1 & 2 & 3 & 1 & 0 & 0 \\ 2 & 5 & 3 & 0 & 1 & 0 \\ 1 & 0 & 8 & 0 & 0 & 1 \end{bmatrix}$$

$$\begin{bmatrix} 1 & 2 & 3 & 1 & 0 & 0 \\ 0 & 1 & -3 & -2 & 1 & 0 \\ 0 & -2 & 5 & -1 & 0 & 1 \end{bmatrix}$$ ⟵ We added -2 times the first row to the second and -1 times the first row to the third.

$$\begin{bmatrix} 1 & 2 & 3 & 1 & 0 & 0 \\ 0 & 1 & -3 & -2 & 1 & 0 \\ 0 & 0 & -1 & -5 & 2 & 1 \end{bmatrix}$$ ⟵ We added 2 times the second row to the third.

$$\left[\begin{array}{ccc|ccc} 1 & 2 & 3 & 1 & 0 & 0 \\ 0 & 1 & -3 & -2 & 1 & 0 \\ 0 & 0 & 1 & 5 & -2 & -1 \end{array}\right]$$ ⟵ We multiplied the third row by -1.

$$\left[\begin{array}{ccc|ccc} 1 & 2 & 0 & -14 & 6 & 3 \\ 0 & 1 & 0 & 13 & -5 & -3 \\ 0 & 0 & 1 & 5 & -2 & -1 \end{array}\right]$$ ⟵ We added 3 times the third row to the second and -3 times the third row to the first.

$$\left[\begin{array}{ccc|ccc} 1 & 0 & 0 & -40 & 16 & 9 \\ 0 & 1 & 0 & 13 & -5 & -3 \\ 0 & 0 & 1 & 5 & -2 & -1 \end{array}\right]$$ ⟵ We added -2 times the second row to the first.

Thus,

$$A^{-1} = \begin{bmatrix} -40 & 16 & 9 \\ 13 & -5 & -3 \\ 5 & -2 & -1 \end{bmatrix}$$ ◆

Often it will not be known in advance whether a given matrix is invertible. If an $n \times n$ matrix A is not invertible, then it cannot be reduced to I_n by elementary row operations [part (c) of Theorem 1.5.3]. Stated another way, the reduced row-echelon form of A has at least one row of zeros. Thus, if the procedure in the last example is attempted on a matrix that is not invertible, then at some point in the computations a row of zeros will occur on the *left side*. It can then be concluded that the given matrix is not invertible, and the computations can be stopped.

EXAMPLE 5 Showing That a Matrix Is Not Invertible

Consider the matrix

$$A = \begin{bmatrix} 1 & 6 & 4 \\ 2 & 4 & -1 \\ -1 & 2 & 5 \end{bmatrix}$$

Applying the procedure of Example 4 yields

$$\left[\begin{array}{ccc|ccc} 1 & 6 & 4 & 1 & 0 & 0 \\ 2 & 4 & -1 & 0 & 1 & 0 \\ -1 & 2 & 5 & 0 & 0 & 1 \end{array}\right]$$

$$\left[\begin{array}{ccc|ccc} 1 & 6 & 4 & 1 & 0 & 0 \\ 0 & -8 & -9 & -2 & 1 & 0 \\ 0 & 8 & 9 & 1 & 0 & 1 \end{array}\right]$$ ⟵ We added -2 times the first row to the second and added the first row to the third.

$$\left[\begin{array}{ccc|ccc} 1 & 6 & 4 & 1 & 0 & 0 \\ 0 & -8 & -9 & -2 & 1 & 0 \\ 0 & 0 & 0 & -1 & 1 & 1 \end{array}\right]$$ ⟵ We added the second row to the third.

Since we have obtained a row of zeros on the left side, A is not invertible. ◆

EXAMPLE 6 A Consequence of Invertibility

In Example 4 we showed that

$$A = \begin{bmatrix} 1 & 2 & 3 \\ 2 & 5 & 3 \\ 1 & 0 & 8 \end{bmatrix}$$

is an invertible matrix. From Theorem 1.5.3 it follows that the homogeneous system

$$\begin{aligned} x_1 + 2x_2 + 3x_3 &= 0 \\ 2x_1 + 5x_2 + 3x_3 &= 0 \\ x_1 \quad\quad + 8x_3 &= 0 \end{aligned}$$

has only the trivial solution. ◆

Exercise Set 1.5

1. Which of the following are elementary matrices?

(a) $\begin{bmatrix} 1 & 0 \\ -5 & 1 \end{bmatrix}$ (b) $\begin{bmatrix} -5 & 1 \\ 1 & 0 \end{bmatrix}$ (c) $\begin{bmatrix} 1 & 0 \\ 0 & \sqrt{3} \end{bmatrix}$ (d) $\begin{bmatrix} 0 & 0 & 1 \\ 0 & 1 & 0 \\ 1 & 0 & 0 \end{bmatrix}$

(e) $\begin{bmatrix} 1 & 1 & 0 \\ 0 & 0 & 1 \\ 0 & 0 & 0 \end{bmatrix}$ (f) $\begin{bmatrix} 1 & 0 & 0 \\ 0 & 1 & 9 \\ 0 & 0 & 1 \end{bmatrix}$ (g) $\begin{bmatrix} 2 & 0 & 0 & 2 \\ 0 & 1 & 0 & 0 \\ 0 & 0 & 1 & 0 \\ 0 & 0 & 0 & 1 \end{bmatrix}$

2. Find a row operation that will restore the given elementary matrix to an identity matrix.

(a) $\begin{bmatrix} 1 & 0 \\ -3 & 1 \end{bmatrix}$ (b) $\begin{bmatrix} 1 & 0 & 0 \\ 0 & 1 & 0 \\ 0 & 0 & 3 \end{bmatrix}$ (c) $\begin{bmatrix} 0 & 0 & 0 & 1 \\ 0 & 1 & 0 & 0 \\ 0 & 0 & 1 & 0 \\ 1 & 0 & 0 & 0 \end{bmatrix}$ (d) $\begin{bmatrix} 1 & 0 & -\frac{1}{7} & 0 \\ 0 & 1 & 0 & 0 \\ 0 & 0 & 1 & 0 \\ 0 & 0 & 0 & 1 \end{bmatrix}$

3. Consider the matrices

$$A = \begin{bmatrix} 3 & 4 & 1 \\ 2 & -7 & -1 \\ 8 & 1 & 5 \end{bmatrix}, \quad B = \begin{bmatrix} 8 & 1 & 5 \\ 2 & -7 & -1 \\ 3 & 4 & 1 \end{bmatrix}, \quad C = \begin{bmatrix} 3 & 4 & 1 \\ 2 & -7 & -1 \\ 2 & -7 & 3 \end{bmatrix}$$

Find elementary matrices, E_1, E_2, E_3, and E_4 such that

(a) $E_1 A = B$ (b) $E_2 B = A$ (c) $E_3 A = C$ (d) $E_4 C = A$

4. In Exercise 3 is it possible to find an elementary matrix E such that $EB = C$? Justify your answer.

In Exercises 5–7 use the method shown in Examples 4 and 5 to find the inverse of the given matrix if the matrix is invertible and check your answer by multiplication.

5. (a) $\begin{bmatrix} 1 & 4 \\ 2 & 7 \end{bmatrix}$ (b) $\begin{bmatrix} -3 & 6 \\ 4 & 5 \end{bmatrix}$ (c) $\begin{bmatrix} 6 & -4 \\ -3 & 2 \end{bmatrix}$

6. (a) $\begin{bmatrix} 3 & 4 & -1 \\ 1 & 0 & 3 \\ 2 & 5 & -4 \end{bmatrix}$ **(b)** $\begin{bmatrix} -1 & 3 & -4 \\ 2 & 4 & 1 \\ -4 & 2 & -9 \end{bmatrix}$ **(c)** $\begin{bmatrix} 1 & 0 & 1 \\ 0 & 1 & 1 \\ 1 & 1 & 0 \end{bmatrix}$ **(d)** $\begin{bmatrix} 2 & 6 & 6 \\ 2 & 7 & 6 \\ 2 & 7 & 7 \end{bmatrix}$ **(e)** $\begin{bmatrix} 1 & 0 & 1 \\ -1 & 1 & 1 \\ 0 & 1 & 0 \end{bmatrix}$

7. (a) $\begin{bmatrix} \frac{1}{5} & \frac{1}{5} & -\frac{2}{5} \\ \frac{1}{5} & \frac{1}{5} & \frac{1}{10} \\ \frac{1}{5} & -\frac{4}{5} & \frac{1}{10} \end{bmatrix}$ **(b)** $\begin{bmatrix} \sqrt{2} & 3\sqrt{2} & 0 \\ -4\sqrt{2} & \sqrt{2} & 0 \\ 0 & 0 & 1 \end{bmatrix}$ **(c)** $\begin{bmatrix} 1 & 0 & 0 & 0 \\ 1 & 3 & 0 & 0 \\ 1 & 3 & 5 & 0 \\ 1 & 3 & 5 & 7 \end{bmatrix}$

(d) $\begin{bmatrix} -8 & 17 & 2 & \frac{1}{3} \\ 4 & 0 & \frac{2}{5} & -9 \\ 0 & 0 & 0 & 0 \\ -1 & 13 & 4 & 2 \end{bmatrix}$ **(e)** $\begin{bmatrix} 0 & 0 & 2 & 0 \\ 1 & 0 & 0 & 1 \\ 0 & -1 & 3 & 0 \\ 2 & 1 & 5 & -3 \end{bmatrix}$

8. Find the inverse of each of the following 4×4 matrices, where $k_1, k_2, k_3, k_4,$ and k are all nonzero.

(a) $\begin{bmatrix} k_1 & 0 & 0 & 0 \\ 0 & k_2 & 0 & 0 \\ 0 & 0 & k_3 & 0 \\ 0 & 0 & 0 & k_4 \end{bmatrix}$ **(b)** $\begin{bmatrix} 0 & 0 & 0 & k_1 \\ 0 & 0 & k_2 & 0 \\ 0 & k_3 & 0 & 0 \\ k_4 & 0 & 0 & 0 \end{bmatrix}$ **(c)** $\begin{bmatrix} k & 0 & 0 & 0 \\ 1 & k & 0 & 0 \\ 0 & 1 & k & 0 \\ 0 & 0 & 1 & k \end{bmatrix}$

9. Consider the matrix

$$A = \begin{bmatrix} 1 & 0 \\ -5 & 2 \end{bmatrix}$$

(a) Find elementary matrices E_1 and E_2 such that $E_2 E_1 A = I$.

(b) Write A^{-1} as a product of two elementary matrices.

(c) Write A as a product of two elementary matrices.

10. In each part perform the stated row operation on

$$\begin{bmatrix} 2 & -1 & 0 \\ 4 & 5 & -3 \\ 1 & -4 & 7 \end{bmatrix}$$

by multiplying A on the left by a suitable elementary matrix. Check your answer in each case by performing the row operation directly on A.

(a) Interchange the first and third rows.

(b) Multiply the second row by $\frac{1}{3}$.

(c) Add twice the second row to the first row.

11. Express the matrix

$$A = \begin{bmatrix} 0 & 1 & 7 & 8 \\ 1 & 3 & 3 & 8 \\ -2 & -5 & 1 & -8 \end{bmatrix}$$

in the form $A = EFGR$, where $E, F,$ and G are elementary matrices, and R is in row-echelon form.

12. Show that if

$$A = \begin{bmatrix} 1 & 0 & 0 \\ 0 & 1 & 0 \\ a & b & c \end{bmatrix}$$

is an elementary matrix, then at least one entry in the third row must be a zero.

13. Show that

$$A = \begin{bmatrix} 0 & a & 0 & 0 & 0 \\ b & 0 & c & 0 & 0 \\ 0 & d & 0 & e & 0 \\ 0 & 0 & f & 0 & g \\ 0 & 0 & 0 & h & 0 \end{bmatrix}$$

is not invertible for any values of the entries.

14. Prove that if A is an $m \times n$ matrix, there is an invertible matrix C such that CA is in reduced row-echelon form.

15. Prove that if A is an invertible matrix and B is row equivalent to A, then B is also invertible.

16. (a) Prove: If A and B are $m \times n$ matrices, then A and B are row equivalent if and only if A and B have the same reduced row-echelon form.

(b) Show that A and B are row equivalent, and find a sequence of elementary row operations that produces B from A.

$$A = \begin{bmatrix} 1 & 2 & 3 \\ 1 & 4 & 1 \\ 2 & 1 & 9 \end{bmatrix}, \quad B = \begin{bmatrix} 1 & 0 & 5 \\ 0 & 2 & -2 \\ 1 & 1 & 4 \end{bmatrix}$$

17. Prove Theorem 1.5.1.

Discussion and Discovery

18. Suppose that A is some unknown invertible matrix, but you know of a sequence of elementary row operations that produces the identity matrix when applied in succession to A. Explain how you can use the known information to find A.

19. Indicate whether the statement is always true or sometimes false. Justify your answer with a logical argument or a counterexample.

(a) Every square matrix can be expressed as a product of elementary matrices.

(b) The product of two elementary matrices is an elementary matrix.

(c) If A is invertible and a multiple of the first row of A is added to the second row, then the resulting matrix is invertible.

(d) If A is invertible and $AB = 0$, then it must be true that $B = 0$.

20. Indicate whether the statement is always true or sometimes false. Justify your answer with a logical argument or a counterexample.

(a) If A is a singular $n \times n$ matrix, then $A\mathbf{x} = \mathbf{0}$ has infinitely many solutions.

(b) If A is a singular $n \times n$ matrix, then the reduced row-echelon form of A has at least one row of zeros.

(c) If A^{-1} is expressible as a product of elementary matrices, then the homogeneous linear system $A\mathbf{x} = \mathbf{0}$ has only the trivial solution.

(d) If A is a singular $n \times n$ matrix, and B results by interchanging two rows of A, then B may or may not be singular.

21. Do you think that there is a 2×2 matrix A such that

$$A \begin{bmatrix} a & b \\ c & d \end{bmatrix} = \begin{bmatrix} b & d \\ a & c \end{bmatrix}$$

for all values of a, b, c, and d? Explain your reasoning.

1.6 FURTHER RESULTS ON SYSTEMS OF EQUATIONS AND INVERTIBILITY

In this section we shall establish more results about systems of linear equations and invertibility of matrices. Our work will lead to a new method for solving n equations in n unknowns.

A Basic Theorem   In Section 1.1 we made the statement (based on Figure 1.1.1) that every linear system has either no solutions, one solution, or infinitely many solutions. We are now in a position to prove this fundamental result.

> **Theorem 1.6.1**
>
> *Every system of linear equations has either no solutions, exactly one solution, or infinitely many solutions.*

Proof. If $A\mathbf{x} = \mathbf{b}$ is a system of linear equations, exactly one of the following is true: (a) the system has no solutions, (b) the system has exactly one solution, or (c) the system has more than one solution. The proof will be complete if we can show that the system has infinitely many solutions in case (c).

Assume that $A\mathbf{x} = \mathbf{b}$ has more than one solution, and let $\mathbf{x}_0 = \mathbf{x}_1 - \mathbf{x}_2$, where $\mathbf{x}_1$ and $\mathbf{x}_2$ are any two distinct solutions. Because $\mathbf{x}_1$ and $\mathbf{x}_2$ are distinct, the matrix $\mathbf{x}_0$ is nonzero; moreover,

$$A\mathbf{x}_0 = A(\mathbf{x}_1 - \mathbf{x}_2) = A\mathbf{x}_1 - A\mathbf{x}_2 = \mathbf{b} - \mathbf{b} = \mathbf{0}$$

If we now let k be any scalar, then

$$A(\mathbf{x}_1 + k\mathbf{x}_0) = A\mathbf{x}_1 + A(k\mathbf{x}_0) = A\mathbf{x}_1 + k(A\mathbf{x}_0)$$
$$= \mathbf{b} + k\mathbf{0} = \mathbf{b} + \mathbf{0} = \mathbf{b}$$

But this says that $\mathbf{x}_1 + k\mathbf{x}_0$ is a solution of $A\mathbf{x} = \mathbf{b}$. Since $\mathbf{x}_0$ is nonzero and there are infinitely many choices for k, the system $A\mathbf{x} = \mathbf{b}$ has infinitely many solutions.   ∎

Solving Linear Systems by Matrix Inversion   Thus far, we have studied two methods for solving linear systems: Gaussian elimination and Gauss–Jordan elimination. The following theorem provides a new method for solving certain linear systems.

> **Theorem 1.6.2**
>
> *If A is an invertible $n \times n$ matrix, then for each $n \times 1$ matrix $\mathbf{b}$, the system of equations $A\mathbf{x} = \mathbf{b}$ has exactly one solution, namely, $\mathbf{x} = A^{-1}\mathbf{b}$.*

Proof. Since $A(A^{-1}\mathbf{b}) = \mathbf{b}$, it follows that $\mathbf{x} = A^{-1}\mathbf{b}$ is a solution of $A\mathbf{x} = \mathbf{b}$. To show that this is the only solution, we will assume that $\mathbf{x}_0$ is an arbitrary solution and then show that $\mathbf{x}_0$ must be the solution $A^{-1}\mathbf{b}$.

If $\mathbf{x}_0$ is any solution, then $A\mathbf{x}_0 = \mathbf{b}$. Multiplying both sides by A^{-1}, we obtain $\mathbf{x}_0 = A^{-1}\mathbf{b}$.   ∎

EXAMPLE 1 Solution of a Linear System Using A^{-1}

Consider the system of linear equations

$$
\begin{aligned}
x_1 + 2x_2 + 3x_3 &= 5 \\
2x_1 + 5x_2 + 3x_3 &= 3 \\
x_1 \qquad\quad + 8x_3 &= 17
\end{aligned}
$$

In matrix form this system can be written as $A\mathbf{x} = \mathbf{b}$, where

$$
A = \begin{bmatrix} 1 & 2 & 3 \\ 2 & 5 & 3 \\ 1 & 0 & 8 \end{bmatrix}, \qquad \mathbf{x} = \begin{bmatrix} x_1 \\ x_2 \\ x_3 \end{bmatrix}, \qquad \mathbf{b} = \begin{bmatrix} 5 \\ 3 \\ 17 \end{bmatrix}
$$

In Example 4 of the preceding section we showed that A is invertible and

$$
A^{-1} = \begin{bmatrix} -40 & 16 & 9 \\ 13 & -5 & -3 \\ 5 & -2 & -1 \end{bmatrix}
$$

By Theorem 1.6.2 the solution of the system is

$$
\mathbf{x} = A^{-1}\mathbf{b} = \begin{bmatrix} -40 & 16 & 9 \\ 13 & -5 & -3 \\ 5 & -2 & -1 \end{bmatrix} \begin{bmatrix} 5 \\ 3 \\ 17 \end{bmatrix} = \begin{bmatrix} 1 \\ -1 \\ 2 \end{bmatrix}
$$

or $x_1 = 1$, $x_2 = -1$, $x_3 = 2$. ♦

REMARK. Note that the method of Example 1 applies only when the system has as many equations as unknowns and the coefficient matrix is invertible.

Linear Systems with a Common Coefficient Matrix

Frequently, one is concerned with solving a sequence of systems

$$
A\mathbf{x} = \mathbf{b}_1, \quad A\mathbf{x} = \mathbf{b}_2, \quad A\mathbf{x} = \mathbf{b}_3, \dots, \quad A\mathbf{x} = \mathbf{b}_k
$$

each of which has the same square coefficient matrix A. If A is invertible, then the solutions

$$
\mathbf{x}_1 = A^{-1}\mathbf{b}_1, \quad \mathbf{x}_2 = A^{-1}\mathbf{b}_2, \quad \mathbf{x}_3 = A^{-1}\mathbf{b}_3, \dots, \quad \mathbf{x}_k = A^{-1}\mathbf{b}_k
$$

can be obtained with one matrix inversion and k matrix multiplications. However, a more efficient method is to form the matrix

$$
[A \mid \mathbf{b}_1 \mid \mathbf{b}_2 \mid \cdots \mid \mathbf{b}_k] \tag{1}
$$

in which the coefficient matrix A is "augmented" by all k of the matrices $\mathbf{b}_1, \mathbf{b}_2, \dots, \mathbf{b}_k$. By reducing (1) to reduced row-echelon form we can solve all k systems at once by Gauss–Jordan elimination. This method has the added advantage that it applies even when A is not invertible.

EXAMPLE 2 Solving Two Linear Systems at Once

Solve the systems

$$
\begin{array}{ll}
\text{(a)} \quad \begin{aligned} x_1 + 2x_2 + 3x_3 &= 4 \\ 2x_1 + 5x_2 + 3x_3 &= 5 \\ x_1 \qquad\quad + 8x_3 &= 9 \end{aligned} &
\text{(b)} \quad \begin{aligned} x_1 + 2x_2 + 3x_3 &= 1 \\ 2x_1 + 5x_2 + 3x_3 &= 6 \\ x_1 \qquad\quad + 8x_3 &= -6 \end{aligned}
\end{array}
$$

Solution.

The two systems have the same coefficient matrix. If we augment this coefficient matrix with the columns of constants on the right sides of these systems, we obtain

$$\begin{bmatrix} 1 & 2 & 3 & 4 & 1 \\ 2 & 5 & 3 & 5 & 6 \\ 1 & 0 & 8 & 9 & -6 \end{bmatrix}$$

Reducing this matrix to reduced row-echelon form yields (verify)

$$\begin{bmatrix} 1 & 0 & 0 & 1 & 2 \\ 0 & 1 & 0 & 0 & 1 \\ 0 & 0 & 1 & 1 & -1 \end{bmatrix}$$

It follows from the last two columns that the solution of system (a) is $x_1 = 1$, $x_2 = 0$, $x_3 = 1$ and of system (b) is $x_1 = 2$, $x_2 = 1$, $x_3 = -1$.   ◆

Properties of Invertible Matrices

Up to now, to show that an $n \times n$ matrix A is invertible, it has been necessary to find an $n \times n$ matrix B such that

$$AB = I \quad \text{and} \quad BA = I$$

The next theorem shows that if we produce an $n \times n$ matrix B satisfying *either* condition, then the other condition holds automatically.

Theorem 1.6.3

Let A be a square matrix.

(a) If B is a square matrix satisfying $BA = I$, then $B = A^{-1}$.
(b) If B is a square matrix satisfying $AB = I$, then $B = A^{-1}$.

We shall prove part (*a*) and leave part (*b*) as an exercise.

Proof (a). Assume that $BA = I$. If we can show that A is invertible, the proof can be completed by multiplying $BA = I$ on both sides by A^{-1} to obtain

$$BAA^{-1} = IA^{-1} \quad \text{or} \quad BI = IA^{-1} \quad \text{or} \quad B = A^{-1}$$

To show that A is invertible, it suffices to show that the system $A\mathbf{x} = \mathbf{0}$ has only the trivial solution (see Theorem 1.5.3). Let $\mathbf{x}_0$ be any solution of this system. If we multiply both sides of $A\mathbf{x}_0 = \mathbf{0}$ on the left by B, we obtain $BA\mathbf{x}_0 = B\mathbf{0}$ or $I\mathbf{x}_0 = \mathbf{0}$ or $\mathbf{x}_0 = \mathbf{0}$. Thus, the system of equations $A\mathbf{x} = \mathbf{0}$ has only the trivial solution.   ■

We are now in a position to add two more statements that are equivalent to the four given in Theorem 1.5.3.

Theorem 1.6.4   Equivalent Statements

If A is an $n \times n$ matrix, then the following are equivalent.

(a)   A is invertible.
(b) $A\mathbf{x} = \mathbf{0}$ has only the trivial solution.
(c)   The reduced row-echelon form of A is I_n.
(d)   A is expressible as a product of elementary matrices.
(e) $A\mathbf{x} = \mathbf{b}$ is consistent for every $n \times 1$ matrix $\mathbf{b}$.
(f) $A\mathbf{x} = \mathbf{b}$ has exactly one solution for every $n \times 1$ matrix $\mathbf{b}$.

Proof. Since we proved in Theorem 1.5.3 that (a), (b), (c), and (d) are equivalent, it will be sufficient to prove that $(a) \Rightarrow (f) \Rightarrow (e) \Rightarrow (a)$.

$(a) \Rightarrow (f)$. This was already proved in Theorem 1.6.2.

$(f) \Rightarrow (e)$. This is self-evident: If $A\mathbf{x} = \mathbf{b}$ has exactly one solution for every $n \times 1$ matrix $\mathbf{b}$, then $A\mathbf{x} = \mathbf{b}$ is consistent for every $n \times 1$ matrix $\mathbf{b}$.

$(e) \Rightarrow (a)$. If the system $A\mathbf{x} = \mathbf{b}$ is consistent for every $n \times 1$ matrix $\mathbf{b}$, then in particular, the systems

$$A\mathbf{x} = \begin{bmatrix} 1 \\ 0 \\ 0 \\ \vdots \\ 0 \end{bmatrix}, \qquad A\mathbf{x} = \begin{bmatrix} 0 \\ 1 \\ 0 \\ \vdots \\ 0 \end{bmatrix}, \dots, \qquad A\mathbf{x} = \begin{bmatrix} 0 \\ 0 \\ 0 \\ \vdots \\ 1 \end{bmatrix}$$

are consistent. Let $\mathbf{x}_1, \mathbf{x}_2, \dots, \mathbf{x}_n$ be solutions of the respective systems, and let us form an $n \times n$ matrix C having these solutions as columns. Thus, C has the form

$$C = [\mathbf{x}_1 \mid \mathbf{x}_2 \mid \cdots \mid \mathbf{x}_n]$$

As discussed in Section 1.3, the successive columns of the product AC will be

$$A\mathbf{x}_1, A\mathbf{x}_2, \dots, A\mathbf{x}_n$$

Thus,

$$AC = [A\mathbf{x}_1 \mid A\mathbf{x}_2 \mid \cdots \mid A\mathbf{x}_n] = \begin{bmatrix} 1 & 0 & \cdots & 0 \\ 0 & 1 & \cdots & 0 \\ 0 & 0 & \cdots & 0 \\ \vdots & \vdots & & \vdots \\ 0 & 0 & \cdots & 1 \end{bmatrix} = I$$

By part (b) of Theorem 1.6.3 it follows that $C = A^{-1}$. Thus, A is invertible. ∎

We know from earlier work that invertible matrix factors produce an invertible product. The following theorem, which will be proved later, looks at the converse: It shows that if the product of square matrices is invertible, then the factors themselves must be invertible.

Theorem 1.6.5

Let A and B be square matrices of the same size. If AB is invertible, then A and B must also be invertible.

In our later work the following fundamental problem will occur frequently in various contexts.

A Fundamental Problem. Let A be a fixed $m \times n$ matrix. Find all $m \times 1$ matrices $\mathbf{b}$ such that the system of equations $A\mathbf{x} = \mathbf{b}$ is consistent.

If A is an invertible matrix, Theorem 1.6.2 completely solves this problem by asserting that for *every* $m \times 1$ matrix $\mathbf{b}$, the linear system $A\mathbf{x} = \mathbf{b}$ has the unique solution $\mathbf{x} = A^{-1}\mathbf{b}$. If A is not square, or if A is square but not invertible, then Theorem 1.6.2 does not apply. In these cases the matrix $\mathbf{b}$ must usually satisfy certain conditions in order for $A\mathbf{x} = \mathbf{b}$ to be consistent. The following example illustrates how the elimination methods of Section 1.2 can be used to determine such conditions.

EXAMPLE 3 Determining Consistency by Elimination

What conditions must b_1, b_2, and b_3 satisfy in order for the system of equations

$$x_1 + x_2 + 2x_3 = b_1$$
$$x_1 \qquad + \ x_3 = b_2$$
$$2x_1 + x_2 + 3x_3 = b_3$$

to be consistent?

Solution.

The augmented matrix is

$$\begin{bmatrix} 1 & 1 & 2 & b_1 \\ 1 & 0 & 1 & b_2 \\ 2 & 1 & 3 & b_3 \end{bmatrix}$$

which can be reduced to row-echelon form as follows.

$$\begin{bmatrix} 1 & 1 & 2 & b_1 \\ 0 & -1 & -1 & b_2 - b_1 \\ 0 & -1 & -1 & b_3 - 2b_1 \end{bmatrix}$$ ← -1 times the first row was added to the second and -2 times the first row was added to the third.

$$\begin{bmatrix} 1 & 1 & 2 & b_1 \\ 0 & 1 & 1 & b_1 - b_2 \\ 0 & -1 & -1 & b_3 - 2b_1 \end{bmatrix}$$ ← The second row was multiplied by -1.

$$\begin{bmatrix} 1 & 1 & 2 & b_1 \\ 0 & 1 & 1 & b_1 - b_2 \\ 0 & 0 & 0 & b_3 - b_2 - b_1 \end{bmatrix}$$ ← The second row was added to the third.

It is now evident from the third row in the matrix that the system has a solution if and only if b_1, b_2, and b_3 satisfy the condition

$$b_3 - b_2 - b_1 = 0 \ \text{ or } \ b_3 = b_1 + b_2$$

To express this condition another way, $A\mathbf{x} = \mathbf{b}$ is consistent if and only if $\mathbf{b}$ is a matrix of the form

$$\mathbf{b} = \begin{bmatrix} b_1 \\ b_2 \\ b_1 + b_2 \end{bmatrix}$$

where b_1 and b_2 are arbitrary. ♦

EXAMPLE 4 Determining Consistency by Elimination

What conditions must b_1, b_2, and b_3 satisfy in order for the system of equations

$$x_1 + 2x_2 + 3x_3 = b_1$$
$$2x_1 + 5x_2 + 3x_3 = b_2$$
$$x_1 \qquad + 8x_3 = b_3$$

to be consistent?

Solution.

The augmented matrix is

$$\begin{bmatrix} 1 & 2 & 3 & b_1 \\ 2 & 5 & 3 & b_2 \\ 1 & 0 & 8 & b_3 \end{bmatrix}$$

Reducing this to reduced row-echelon form yields (verify)

$$\begin{bmatrix} 1 & 0 & 0 & -40b_1 + 16b_2 + 9b_3 \\ 0 & 1 & 0 & 13b_1 - 5b_2 - 3b_3 \\ 0 & 0 & 1 & 5b_1 - 2b_2 - b_3 \end{bmatrix} \qquad (2)$$

In this case there are no restrictions on b_1, b_2, and b_3; that is, the given system $A\mathbf{x} = \mathbf{b}$ has the unique solution

$$x_1 = -40b_1 + 16b_2 + 9b_3, \quad x_2 = 13b_1 - 5b_2 - 3b_3, \quad x_3 = 5b_1 - 2b_2 - b_3 \qquad (3)$$

for all **b**. ◆

REMARK. Because the system $A\mathbf{x} = \mathbf{b}$ in the preceding example is consistent for all **b**, it follows from Theorem 1.6.4 that A is invertible. We leave it for the reader to verify that the formulas in (3) can also be obtained by calculating $\mathbf{x} = A^{-1}\mathbf{b}$.

Exercise Set 1.6

In Exercises 1–8 solve the system by inverting the coefficient matrix and using Theorem 1.6.2.

1. $\begin{aligned} x_1 + x_2 &= 2 \\ 5x_1 + 6x_2 &= 9 \end{aligned}$

2. $\begin{aligned} 4x_1 - 3x_2 &= -3 \\ 2x_1 - 5x_2 &= 9 \end{aligned}$

3. $\begin{aligned} x_1 + 3x_2 + x_3 &= 4 \\ 2x_1 + 2x_2 + x_3 &= -1 \\ 2x_1 + 3x_2 + x_3 &= 3 \end{aligned}$

4. $\begin{aligned} 5x_1 + 3x_2 + 2x_3 &= 4 \\ 3x_1 + 3x_2 + 2x_3 &= 2 \\ x_2 + x_3 &= 5 \end{aligned}$

5. $\begin{aligned} x + y + z &= 5 \\ x + y - 4z &= 10 \\ -4x + y + z &= 0 \end{aligned}$

6. $\begin{aligned} -x - 2y - 3z &= 0 \\ w + x + 4y + 4z &= 7 \\ w + 3x + 7y + 9z &= 4 \\ -w - 2x - 4y - 6z &= 6 \end{aligned}$

7. $\begin{aligned} 3x_1 + 5x_2 &= b_1 \\ x_1 + 2x_2 &= b_2 \end{aligned}$

8. $\begin{aligned} x_1 + 2x_2 + 3x_3 &= b_1 \\ 2x_1 + 5x_2 + 5x_3 &= b_2 \\ 3x_1 + 5x_2 + 8x_3 &= b_3 \end{aligned}$

9. Solve the following general system by inverting the coefficient matrix and using Theorem 1.6.2.

$$\begin{aligned} x_1 + 2x_2 + x_3 &= b_1 \\ x_1 - x_2 + x_3 &= b_2 \\ x_1 + x_2 &= b_3 \end{aligned}$$

Use the resulting formulas to find the solution if

(a) $b_1 = -1$, $b_2 = 3$, $b_3 = 4$ (b) $b_1 = 5$, $b_2 = 0$, $b_3 = 0$ (c) $b_1 = -1$, $b_2 = -1$, $b_3 = 3$

10. Solve the three systems in Exercise 9 using the method of Example 2.

In Exercises 11–14 use the method of Example 2 to solve the systems in all parts simultaneously.

11. $x_1 - 5x_2 = b_1$
$3x_1 + 2x_2 = b_2$

(a) $b_1 = 1,\ \ b_2 = 4$
(b) $b_1 = -2,\ \ b_2 = 5$

12. $-x_1 + 4x_2 +\ \ x_3 = b_1$
$x_1 + 9x_2 - 2x_3 = b_2$
$6x_1 + 4x_2 - 8x_3 = b_3$

(a) $b_1 = 0,\ \ b_2 = 1,\ \ b_3 = 0$
(b) $b_1 = -3,\ \ b_2 = 4,\ \ b_3 = -5$

13. $4x_1 - 7x_2 = b_1$
$x_1 + 2x_2 = b_2$

(a) $b_1 = 0,\ \ b_2 = 1$
(b) $b_1 = -4,\ \ b_2 = 6$
(c) $b_1 = -1,\ \ b_2 = 3$
(d) $b_1 = -5,\ \ b_2 = 1$

14. $x_1 + 3x_2 + 5x_3 = b_1$
$-x_1 - 2x_2 \ = b_2$
$2x_1 + 5x_2 + 4x_3 = b_3$

(a) $b_1 = 1,\ \ b_2 = 0,\ \ b_3 = -1$
(b) $b_1 = 0,\ \ b_2 = 1,\ \ b_3 = 1$
(c) $b_1 = -1,\ \ b_2 = -1,\ \ b_3 = 0$

15. The method of Example 2 can be used for linear systems with infinitely many solutions. Use that method to solve the systems in both parts at the same time.

(a) $x_1 - 2x_2 +\ \ x_3 = -2$
$2x_1 - 5x_2 +\ \ x_3 =\ \ \ 1$
$3x_1 - 7x_2 + 2x_3 = -1$

(b) $x_1 - 2x_2 +\ \ x_3 =\ \ \ 1$
$2x_1 - 5x_2 +\ \ x_3 = -1$
$3x_1 - 7x_2 + 2x_3 =\ \ \ 0$

In Exercises 16–19 find conditions that b's must satisfy for the system to be consistent.

16. $6x_1 - 4x_2 = b_1$
$3x_1 - 2x_2 = b_2$

17. $x_1 - 2x_2 + 5x_3 = b_1$
$4x_1 - 5x_2 + 8x_3 = b_2$
$-3x_1 + 3x_2 - 3x_3 = b_3$

18. $x_1 - 2x_2 -\ \ x_3 = b_1$
$-4x_1 + 5x_2 + 2x_3 = b_2$
$-4x_1 + 7x_2 + 4x_3 = b_3$

19. $x_1 -\ \ x_2 + 3x_3 + 2x_1 = b_1$
$-2x_1 +\ \ x_2 + 5x_3 +\ \ x_1 = b_2$
$-3x_1 + 2x_2 + 2x_3 -\ \ x_1 = b_3$
$4x_1 - 3x_2 +\ \ x_3 + 3x_1 = b_4$

20. Consider the matrices

$$A = \begin{bmatrix} 2 & 1 & 2 \\ 2 & 2 & -2 \\ 3 & 1 & 1 \end{bmatrix} \quad \text{and} \quad \mathbf{x} = \begin{bmatrix} x_1 \\ x_2 \\ x_3 \end{bmatrix}$$

(a) Show that the equation $A\mathbf{x} = \mathbf{x}$ can be rewritten as $(A - I)\mathbf{x} = \mathbf{0}$ and use this result to solve $A\mathbf{x} = \mathbf{x}$ for $\mathbf{x}$.
(b) Solve $A\mathbf{x} = 4\mathbf{x}$.

21. Solve the following matrix equation for X.

$$\begin{bmatrix} 1 & -1 & 1 \\ 2 & 3 & 0 \\ 0 & 2 & -1 \end{bmatrix} X = \begin{bmatrix} 2 & -1 & 5 & 7 & 8 \\ 4 & 0 & -3 & 0 & 1 \\ 3 & 5 & -7 & 2 & 1 \end{bmatrix}$$

22. In each part determine whether the homogeneous system has a nontrivial solution (without using pencil and paper); then state whether the given matrix is invertible.

(a) $2x_1 +\ \ x_2 - 3x_3 +\ \ x_4 = 0$
$5x_2 + 4x_3 + 3x_4 = 0$
$x_3 + 2x_4 = 0$
$3x_4 = 0$

$\begin{bmatrix} 2 & 1 & -3 & 1 \\ 0 & 5 & 4 & 3 \\ 0 & 0 & 1 & 2 \\ 0 & 0 & 0 & 3 \end{bmatrix}$

(b) $5x_1 + x_2 + 4x_3 +\ \ x_4 = 0$
$2x_3 -\ \ x_4 = 0$
$x_3 +\ \ x_4 = 0$
$7x_4 = 0$

$\begin{bmatrix} 5 & 1 & 4 & 1 \\ 0 & 0 & 2 & -1 \\ 0 & 0 & 1 & 1 \\ 0 & 0 & 0 & 7 \end{bmatrix}$

23. Let $A\mathbf{x} = \mathbf{0}$ be a homogeneous system of n linear equations in n unknowns that has only the trivial solution. Show that if k is any positive integer, then the system $A^k\mathbf{x} = \mathbf{0}$ also has only the trivial solution.

24. Let $A\mathbf{x} = \mathbf{0}$ be a homogeneous system of n linear equations in n unknowns, and let Q be an invertible $n \times n$ matrix. Show that $A\mathbf{x} = \mathbf{0}$ has just the trivial solution if and only if $(QA)\mathbf{x} = \mathbf{0}$ has just the trivial solution.

25. Let $A\mathbf{x} = \mathbf{b}$ be any consistent system of linear equations, and let $\mathbf{x}_1$ be a fixed solution. Show that every solution to the system can be written in the form $\mathbf{x} = \mathbf{x}_1 + \mathbf{x}_0$, where $\mathbf{x}_0$ is a solution to $A\mathbf{x} = \mathbf{0}$. Show also that every matrix of this form is a solution.

26. Use part (*a*) of Theorem 1.6.3 to prove part (*b*).

Discussion and Discovery

27. (a) If A is an $n \times n$ matrix and if $\mathbf{b}$ is an $n \times 1$ matrix, what conditions would you impose to ensure that the equation $\mathbf{x} = A\mathbf{x} + \mathbf{b}$ has a unique solution for $\mathbf{x}$?

   (b) Assuming that your conditions are satisfied, find a formula for the solution in terms of an appropriate inverse.

28. Suppose that A is an invertible $n \times n$ matrix. Must the system of equations $A\mathbf{x} = \mathbf{x}$ have a unique solution? Explain your reasoning.

29. Is it possible to have $AB = I$ without B being the inverse of A? Explain your reasoning.

30. Create a theorem by rewriting Theorem 1.6.5 in contrapositive form (see Exercise 34 of Section 1.4).

1.7 DIAGONAL, TRIANGULAR, AND SYMMETRIC MATRICES

In this section we shall consider certain classes of matrices that have special forms. The matrices that we study in this section are among the most important kinds of matrices encountered in linear algebra and will arise in many different settings throughout the text.

Diagonal Matrices   A square matrix in which all the entries off the main diagonal are zero is called a **diagonal matrix**. Here are some examples.

$$\begin{bmatrix} 2 & 0 \\ 0 & -5 \end{bmatrix}, \quad \begin{bmatrix} 1 & 0 & 0 \\ 0 & 1 & 0 \\ 0 & 0 & 1 \end{bmatrix}, \quad \begin{bmatrix} 6 & 0 & 0 & 0 \\ 0 & -4 & 0 & 0 \\ 0 & 0 & 0 & 0 \\ 0 & 0 & 0 & 8 \end{bmatrix}$$

A general $n \times n$ diagonal matrix D can be written as

$$D = \begin{bmatrix} d_1 & 0 & \cdots & 0 \\ 0 & d_2 & \cdots & 0 \\ \vdots & \vdots & & \vdots \\ 0 & 0 & \cdots & d_n \end{bmatrix} \tag{1}$$

A diagonal matrix is invertible if and only if all of its diagonal entries are nonzero; in this case the inverse of (1) is

$$D^{-1} = \begin{bmatrix} 1/d_1 & 0 & \cdots & 0 \\ 0 & 1/d_2 & \cdots & 0 \\ \vdots & \vdots & & \vdots \\ 0 & 0 & \cdots & 1/d_n \end{bmatrix}$$

The reader should verify that $DD^{-1} = D^{-1}D = I$.

Powers of diagonal matrices are easy to compute; we leave it for the reader to verify that if D is the diagonal matrix (1) and k is a positive integer, then

$$D^k = \begin{bmatrix} d_1{}^k & 0 & \cdots & 0 \\ 0 & d_2{}^k & \cdots & 0 \\ \vdots & \vdots & & \vdots \\ 0 & 0 & \cdots & d_n{}^k \end{bmatrix}$$

EXAMPLE 1 Inverses and Powers of Diagonal Matrices

If

$$A = \begin{bmatrix} 1 & 0 & 0 \\ 0 & -3 & 0 \\ 0 & 0 & 2 \end{bmatrix}$$

then

$$A^{-1} = \begin{bmatrix} 1 & 0 & 0 \\ 0 & -\frac{1}{3} & 0 \\ 0 & 0 & \frac{1}{2} \end{bmatrix}, \quad A^5 = \begin{bmatrix} 1 & 0 & 0 \\ 0 & -243 & 0 \\ 0 & 0 & 32 \end{bmatrix}, \quad A^{-5} = \begin{bmatrix} 1 & 0 & 0 \\ 0 & -\frac{1}{243} & 0 \\ 0 & 0 & \frac{1}{32} \end{bmatrix}$$

$\blacklozenge$

Matrix products that involve diagonal factors are especially easy to compute. For example,

$$\begin{bmatrix} d_1 & 0 & 0 \\ 0 & d_2 & 0 \\ 0 & 0 & d_3 \end{bmatrix} \begin{bmatrix} a_{11} & a_{12} & a_{13} & a_{14} \\ a_{21} & a_{22} & a_{23} & a_{24} \\ a_{31} & a_{32} & a_{33} & a_{34} \end{bmatrix} = \begin{bmatrix} d_1 a_{11} & d_1 a_{12} & d_1 a_{13} & d_1 a_{14} \\ d_2 a_{21} & d_2 a_{22} & d_2 a_{23} & d_2 a_{24} \\ d_3 a_{31} & d_3 a_{32} & d_3 a_{33} & d_3 a_{34} \end{bmatrix}$$

$$\begin{bmatrix} a_{11} & a_{12} & a_{13} \\ a_{21} & a_{22} & a_{23} \\ a_{31} & a_{32} & a_{33} \\ a_{41} & a_{42} & a_{43} \end{bmatrix} \begin{bmatrix} d_1 & 0 & 0 \\ 0 & d_2 & 0 \\ 0 & 0 & d_3 \end{bmatrix} = \begin{bmatrix} d_1 a_{11} & d_2 a_{12} & d_3 a_{13} \\ d_1 a_{21} & d_2 a_{22} & d_3 a_{23} \\ d_1 a_{31} & d_2 a_{32} & d_3 a_{33} \\ d_1 a_{41} & d_2 a_{42} & d_3 a_{43} \end{bmatrix}$$

In words, *to multiply a matrix A on the left by a diagonal matrix D, one can multiply successive rows of A by the successive diagonal entries of D, and to multiply A on the right by D one can multiply successive columns of A by the successive diagonal entries of D.*

Triangular Matrices
A square matrix in which all the entries above the main diagonal are zero is called ***lower triangular***, and a square matrix in which all the entries below the main diagonal are zero is called ***upper triangular***. A matrix that is either upper triangular or lower triangular is called ***triangular***.

EXAMPLE 2 Upper and Lower Triangular Matrices

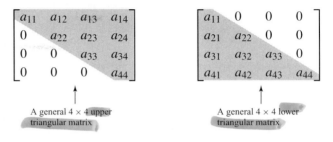

A general 4×4 upper triangular matrix

A general 4×4 lower triangular matrix

♦

REMARK. Observe that diagonal matrices are both upper triangular and lower triangular since they have zeros below and above the main diagonal. Observe also that a *square* matrix in row-echelon form is upper triangular since it has zeros below the main diagonal.

The following are four useful characterizations of triangular matrices. The reader will find it instructive to verify that the matrices in Example 2 have the stated properties.

- A square matrix $A = [a_{ij}]$ is upper triangular if and only if the ith row starts with at least $i - 1$ zeros.
- A square matrix $A = [a_{ij}]$ is lower triangular if and only if the jth column starts with at least $j - 1$ zeros.
- A square matrix $A = [a_{ij}]$ is upper triangular if and only if $a_{ij} = 0$ for $i > j$.
- A square matrix $A = [a_{ij}]$ is lower triangular if and only if $a_{ij} = 0$ for $i < j$.

The following theorem lists some of the basic properties of triangular matrices.

Theorem 1.7.1

(a) The transpose of a lower triangular matrix is upper triangular, and the transpose of an upper triangular matrix is lower triangular.

(b) The product of lower triangular matrices is lower triangular, and the product of upper triangular matrices is upper triangular.

(c) A triangular matrix is invertible if and only if its diagonal entries are all nonzero.

(d) The inverse of an invertible lower triangular matrix is lower triangular, and the inverse of an invertible upper triangular matrix is upper triangular.

Part (*a*) is evident from the fact that transposing a square matrix can be accomplished by reflecting the entries about the main diagonal; we omit the formal proof. We will prove (*b*), but we will defer the proofs of (*c*) and (*d*) to the next chapter, where we will have the tools to prove those results more efficiently.

Proof (b). We will prove the result for lower triangular matrices; the proof for upper triangular matrices is similar. Let $A = [a_{ij}]$ and $B = [b_{ij}]$ be lower triangular $n \times n$ matrices, and let $C = [c_{ij}]$ be the product $C = AB$. From the remark preceding this theorem, we can prove that C is lower triangular by showing that $c_{ij} = 0$ for $i < j$. But from the definition of matrix multiplication,

$$c_{ij} = a_{i1}b_{1j} + a_{i2}b_{2j} + \cdots + a_{in}b_{nj}$$

If we assume that $i < j$, then the terms in this expression can be grouped as follows:

$$c_{ij} = \underbrace{a_{i1}b_{1j} + a_{i2}b_{2j} + \cdots + a_{i(j-1)}b_{(j-1)j}}_{\substack{\text{Terms in which the row} \\ \text{number of } b \text{ is less than the} \\ \text{column number of } b}} + \underbrace{a_{ij}b_{jj} + \cdots + a_{in}b_{nj}}_{\substack{\text{Terms in which the row} \\ \text{number of } a \text{ is less than} \\ \text{the column number of } a}}$$

In the first grouping all of the b factors are zero since B is lower triangular, and in the second grouping all of the a factors are zero since A is lower triangular. Thus, $c_{ij} = 0$, which is what we wanted to prove. ∎

EXAMPLE 3 Upper Triangular Matrices

Consider the upper triangular matrices

$$A = \begin{bmatrix} 1 & 3 & -1 \\ 0 & 2 & 4 \\ 0 & 0 & 5 \end{bmatrix}, \quad B = \begin{bmatrix} 3 & -2 & 2 \\ 0 & 0 & -1 \\ 0 & 0 & 1 \end{bmatrix}$$

The matrix A is invertible, since its diagonal entries are nonzero, but the matrix B is not. We leave it for the reader to calculate the inverse of A by the method of Section 1.5 and show that

$$A^{-1} = \begin{bmatrix} 1 & -\frac{3}{2} & \frac{7}{5} \\ 0 & \frac{1}{2} & -\frac{2}{5} \\ 0 & 0 & \frac{1}{5} \end{bmatrix}$$

This inverse is upper triangular, as guaranteed by part (d) of Theorem 1.7.1. We also leave it for the reader to check that the product AB is

$$AB = \begin{bmatrix} 3 & -2 & -2 \\ 0 & 0 & 2 \\ 0 & 0 & 5 \end{bmatrix}$$

This product is upper triangular, as guaranteed by part (b) of Theorem 1.7.1. ◆

Symmetric Matrices

A square matrix A is called **symmetric** if $A = A^T$.

EXAMPLE 4 Symmetric Matrices

The following matrices are symmetric, since each is equal to its own transpose (verify).

$$\begin{bmatrix} 7 & -3 \\ -3 & 5 \end{bmatrix}, \quad \begin{bmatrix} 1 & 4 & 5 \\ 4 & -3 & 0 \\ 5 & 0 & 7 \end{bmatrix}, \quad \begin{bmatrix} d_1 & 0 & 0 & 0 \\ 0 & d_2 & 0 & 0 \\ 0 & 0 & d_3 & 0 \\ 0 & 0 & 0 & d_4 \end{bmatrix} \quad ◆$$

It is easy to recognize symmetric matrices by inspection: The entries on the main diagonal may be arbitrary, but as shown in (2), "mirror images" of entries across the main diagonal must be equal.

$$\begin{bmatrix} 1 & 4 & 5 \\ 4 & -3 & 0 \\ 5 & 0 & 7 \end{bmatrix} \tag{2}$$

This follows from the fact that transposing a square matrix can be accomplished by interchanging entries that are symmetrically positioned about the main diagonal. Expressed in terms of the individual entries, a matrix $A = [a_{ij}]$ is symmetric if and only if $a_{ij} = a_{ji}$ for all values of i and j. As illustrated in Example 4, all diagonal matrices are symmetric.

The following theorem lists the main algebraic properties of symmetric matrices. The proofs are direct consequences of Theorem 1.4.9 and are left for the reader.

Theorem 1.7.2

If A and B are symmetric matrices with the same size, and if k is any scalar, then:

(a) A^T is symmetric.
(b) $A + B$ and $A - B$ are symmetric.
(c) kA is symmetric.

REMARK. It is not true, in general, that the product of symmetric matrices is symmetric. To see why this is so, let A and B be symmetric matrices with the same size. Then from part (d) of Theorem 1.4.9 and the symmetry we have

$$(AB)^T = B^T A^T = BA$$

Since AB and BA are not usually equal, it follows that AB will not usually be symmetric. However, in the special case where $AB = BA$, the product AB will be symmetric. If A and B are matrices such that $AB = BA$, then we say that A and B ***commute***. In summary: *The product of two symmetric matrices is symmetric if and only if the matrices commute.*

EXAMPLE 5 Products of Symmetric Matrices

The first of the following equations shows a product of symmetric matrices that *is not* symmetric, and the second shows a product of symmetric matrices that *is* symmetric. We conclude that the factors in the first equation do not commute, but those in the second equation do. We leave it for the reader to verify that this is so.

$$\begin{bmatrix} 1 & 2 \\ 2 & 3 \end{bmatrix} \begin{bmatrix} -4 & 1 \\ 1 & 0 \end{bmatrix} = \begin{bmatrix} -2 & 1 \\ -5 & 2 \end{bmatrix}$$

$$\begin{bmatrix} 1 & 2 \\ 2 & 3 \end{bmatrix} \begin{bmatrix} -4 & 3 \\ 3 & -1 \end{bmatrix} = \begin{bmatrix} 2 & 1 \\ 1 & 3 \end{bmatrix} \qquad \blacklozenge$$

In general, a symmetric matrix need not be invertible; for example, a square zero matrix is symmetric, but not invertible. However, if a symmetric matrix is invertible, then that inverse is also symmetric.

Theorem 1.7.3

If A is an invertible symmetric matrix, then A^{-1} is symmetric.

Proof. Assume that A is symmetric and invertible. From Theorem 1.4.10 and the fact that $A = A^T$ we have

$$(A^{-1})^T = (A^T)^{-1} = A^{-1}$$

which proves that A^{-1} is symmetric. ∎

Products AA^T and A^TA Matrix products of the form AA^T and A^TA arise in a variety of applications. If A is an $m \times n$ matrix, then A^T is an $n \times m$ matrix, so the products AA^T and A^TA are both square matrices—the matrix AA^T has size $m \times m$ and the matrix A^TA has size $n \times n$. Such products are always symmetric since

$$(AA^T)^T = (A^T)^TA^T = AA^T \quad \text{and} \quad (A^TA)^T = A^T(A^T)^T = A^TA$$

EXAMPLE 6 The Product of a Matrix and Its Transpose Is Symmetric

Let A be the 2×3 matrix

$$A = \begin{bmatrix} 1 & -2 & 4 \\ 3 & 0 & -5 \end{bmatrix}$$

Then

$$A^TA = \begin{bmatrix} 1 & 3 \\ -2 & 0 \\ 4 & -5 \end{bmatrix} \begin{bmatrix} 1 & -2 & 4 \\ 3 & 0 & -5 \end{bmatrix} = \begin{bmatrix} 10 & -2 & -11 \\ -2 & 4 & -8 \\ -11 & -8 & 41 \end{bmatrix}$$

$$AA^T = \begin{bmatrix} 1 & -2 & 4 \\ 3 & 0 & -5 \end{bmatrix} \begin{bmatrix} 1 & 3 \\ -2 & 0 \\ 4 & -5 \end{bmatrix} = \begin{bmatrix} 21 & -17 \\ -17 & 34 \end{bmatrix}$$

Observe that A^TA and AA^T are symmetric as expected. ◆

Later in this text, we will obtain general conditions on A under which AA^T and A^TA are invertible. However, in the special case where A is *square* we have the following result.

Theorem 1.7.4

If A is an invertible matrix, then AA^T and A^TA are also invertible.

Proof. Since A is invertible, so is A^T by Theorem 1.4.10. Thus, AA^T and A^TA are invertible, since they are the products of invertible matrices. ∎

Exercise Set 1.7

1. Determine whether the matrix is invertible; if so, find the inverse by inspection.

(a) $\begin{bmatrix} 2 & 0 \\ 0 & -5 \end{bmatrix}$ (b) $\begin{bmatrix} 4 & 0 & 0 \\ 0 & 0 & 0 \\ 0 & 0 & 5 \end{bmatrix}$ (c) $\begin{bmatrix} -1 & 0 & 0 \\ 0 & 2 & 0 \\ 0 & 0 & \frac{1}{3} \end{bmatrix}$

2. Compute the product by inspection.

(a) $\begin{bmatrix} 3 & 0 & 0 \\ 0 & -1 & 0 \\ 0 & 0 & 2 \end{bmatrix} \begin{bmatrix} 2 & 1 \\ -4 & 1 \\ 2 & 5 \end{bmatrix}$ (b) $\begin{bmatrix} 2 & 0 & 0 \\ 0 & -1 & 0 \\ 0 & 0 & 4 \end{bmatrix} \begin{bmatrix} 4 & -1 & 3 \\ 1 & 2 & 0 \\ -5 & 1 & -2 \end{bmatrix} \begin{bmatrix} -3 & 0 & 0 \\ 0 & 5 & 0 \\ 0 & 0 & 2 \end{bmatrix}$

3. Find A^2, A^{-2}, and A^{-k} by inspection.

(a) $A = \begin{bmatrix} 1 & 0 \\ 0 & -2 \end{bmatrix}$ (b) $A = \begin{bmatrix} \frac{1}{2} & 0 & 0 \\ 0 & \frac{1}{3} & 0 \\ 0 & 0 & \frac{1}{4} \end{bmatrix}$

4. Which of the following matrices are symmetric?

(a) $\begin{bmatrix} 2 & -1 \\ 1 & 2 \end{bmatrix}$ (b) $\begin{bmatrix} 3 & 4 \\ 4 & 0 \end{bmatrix}$ (c) $\begin{bmatrix} 2 & -1 & 3 \\ -1 & 5 & 1 \\ 3 & 1 & 7 \end{bmatrix}$ (d) $\begin{bmatrix} 0 & 0 & 1 \\ 0 & 2 & 0 \\ 3 & 0 & 0 \end{bmatrix}$

5. By inspection, determine whether the given triangular matrix is invertible.

(a) $\begin{bmatrix} -1 & 2 & 4 \\ 0 & 3 & 0 \\ 0 & 0 & 5 \end{bmatrix}$ (b) $\begin{bmatrix} 0 & 1 & -2 & 5 \\ 0 & 1 & 5 & 6 \\ 0 & 0 & -3 & 1 \\ 0 & 0 & 0 & 5 \end{bmatrix}$

6. Find all values of a, b, and c for which A is symmetric.

$$A = \begin{bmatrix} 2 & a-2b+2c & 2a+b+c \\ 3 & 5 & a+c \\ 0 & -2 & 7 \end{bmatrix}$$

7. Find all values of a and b for which A and B are both not invertible.

$$A = \begin{bmatrix} a+b-1 & 0 \\ 0 & 3 \end{bmatrix}, \quad B = \begin{bmatrix} 5 & 0 \\ 0 & 2a-3b-7 \end{bmatrix}$$

8. Use the given equation to determine by inspection whether the matrices on the left commute.

(a) $\begin{bmatrix} 1 & -3 \\ -3 & 2 \end{bmatrix} \begin{bmatrix} 4 & 1 \\ 1 & 2 \end{bmatrix} = \begin{bmatrix} 1 & -5 \\ -10 & 1 \end{bmatrix}$ (b) $\begin{bmatrix} 2 & -1 \\ -1 & 3 \end{bmatrix} \begin{bmatrix} 3 & 2 \\ 2 & 1 \end{bmatrix} = \begin{bmatrix} 4 & 3 \\ 3 & 1 \end{bmatrix}$

9. Show that A and B commute if $a - d = 7b$.

$$A = \begin{bmatrix} 2 & 1 \\ 1 & -5 \end{bmatrix}, \quad B = \begin{bmatrix} a & b \\ b & d \end{bmatrix}$$

10. Find a diagonal matrix A that satisfies

(a) $A^5 = \begin{bmatrix} 1 & 0 & 0 \\ 0 & -1 & 0 \\ 0 & 0 & -1 \end{bmatrix}$ (b) $A^{-2} = \begin{bmatrix} 9 & 0 & 0 \\ 0 & 4 & 0 \\ 0 & 0 & 1 \end{bmatrix}$

11. (a) Factor A into the form $A = BD$, where D is a diagonal matrix.

$$A = \begin{bmatrix} 3a_{11} & 5a_{12} & 7a_{13} \\ 3a_{21} & 5a_{22} & 7a_{23} \\ 3a_{31} & 5a_{32} & 7a_{33} \end{bmatrix}$$

(b) Is your factorization the only one possible? Explain.

12. Verify Theorem 1.7.1b for the product AB, where

$$A = \begin{bmatrix} -1 & 2 & 5 \\ 0 & 1 & 3 \\ 0 & 0 & -4 \end{bmatrix}, \quad B = \begin{bmatrix} 2 & -8 & 0 \\ 0 & 2 & 1 \\ 0 & 0 & 3 \end{bmatrix}$$

13. Verify Theorem 1.7.1d for the matrices A and B in Exercise 12.

14. Verify Theorem 1.7.3 for the given matrix A.

(a) $A = \begin{bmatrix} 2 & -1 \\ -1 & 3 \end{bmatrix}$ (b) $A = \begin{bmatrix} 1 & -2 & 3 \\ -2 & 1 & -7 \\ 3 & -7 & 4 \end{bmatrix}$

15. Let A be a symmetric matrix.

(a) Show that A^2 is symmetric.

(b) Show that $2A^2 - 3A + I$ is symmetric.

16. Let A be a symmetric matrix.

(a) Show that A^k is symmetric if k is any nonnegative integer.

(b) If $p(x)$ is a polynomial, is $p(A)$ necessarily symmetric? Explain.

17. Let A be an upper triangular matrix and let $p(x)$ be a polynomial. Is $p(A)$ necessarily upper triangular? Explain.

18. Prove: If $A^T A = A$, then A is symmetric and $A = A^2$.

19. Find all 3×3 diagonal matrices A that satisfy $A^2 - 3A - 4I = 0$.

20. Let $A = [a_{ij}]$ be an $n \times n$ matrix. Determine whether A is symmetric.

(a) $a_{ij} = i^2 + j^2$ (b) $a_{ij} = i^2 - j^2$

(c) $a_{ij} = 2i + 2j$ (d) $a_{ij} = 2i^2 + 2j^3$

21. Based on your experience with Exercise 20, devise a general test that can be applied to a formula for a_{ij} to determine whether $A = [a_{ij}]$ is symmetric.

22. A square matrix A is called **skew-symmetric** if $A^T = -A$. Prove:

(a) If A is an invertible skew-symmetric matrix, then A^{-1} is skew-symmetric.

(b) If A and B are skew-symmetric, then so are A^T, $A + B$, $A - B$, and kA for any scalar k.

(c) Every square matrix A can be expressed as the sum of a symmetric matrix and a skew-symmetric matrix. [**Hint.** Note the identity $A = \frac{1}{2}(A + A^T) + \frac{1}{2}(A - A^T)$.]

23. We showed in the text that the product of symmetric matrices is symmetric if and only if the matrices commute. Is the product of commuting skew-symmetric matrices skew-symmetric? Explain. [**Note.** See Exercise 22 for terminology.]

24. If the $n \times n$ matrix A can be expressed as $A = LU$, where L is a lower triangular matrix and U is an upper triangular matrix, then the linear system $A\mathbf{x} = \mathbf{b}$ can be expressed as $LU\mathbf{x} = \mathbf{b}$ and can be solved in two steps:

Step 1. Let $U\mathbf{x} = \mathbf{y}$, so that $LU\mathbf{x} = \mathbf{b}$ can be expressed as $L\mathbf{y} = \mathbf{b}$. Solve this system.

Step 2. Solve the system $U\mathbf{x} = \mathbf{y}$ for $\mathbf{x}$.

In each part use this two-step method to solve the given system.

(a) $\begin{bmatrix} 1 & 0 & 0 \\ -2 & 3 & 0 \\ 2 & 4 & 1 \end{bmatrix} \begin{bmatrix} 2 & -1 & 3 \\ 0 & 1 & 2 \\ 0 & 0 & 4 \end{bmatrix} \begin{bmatrix} x_1 \\ x_2 \\ x_3 \end{bmatrix} = \begin{bmatrix} 1 \\ -2 \\ 0 \end{bmatrix}$

(b) $\begin{bmatrix} 2 & 0 & 0 \\ 4 & 1 & 0 \\ -3 & -2 & 3 \end{bmatrix} \begin{bmatrix} 3 & -5 & 2 \\ 0 & 4 & 1 \\ 0 & 0 & 2 \end{bmatrix} \begin{bmatrix} x_1 \\ x_2 \\ x_3 \end{bmatrix} = \begin{bmatrix} 4 \\ -5 \\ 2 \end{bmatrix}$

25. Find an upper triangular matrix that satisfies

$$A^3 = \begin{bmatrix} 1 & 30 \\ 0 & -8 \end{bmatrix}$$

Discussion and Discovery

26. What is the maximum number of distinct entries that an $n \times n$ symmetric matrix can have? Explain your reasoning.

27. Invent and prove a theorem that describes how to multiply two diagonal matrices.

28. Suppose that A is a square matrix and D is a diagonal matrix such that $AD = I$. What can you say about the matrix A? Explain your reasoning.

29. (a) Make up a consistent linear system of five equations in five unknowns that has a lower triangular coefficient matrix with no zeros on or below the main diagonal.
   (b) Devise an efficient procedure for solving your system by hand.
   (c) Invent an appropriate name for your procedure.

30. Indicate whether the statement is always true or sometimes false. Justify each answer.

   (a) If AA^T is singular, then so is A.
   (b) If $A + B$ is symmetric, then so are A and B.
   (c) If A is an $n \times n$ matrix and $A\mathbf{x} = \mathbf{0}$ has only the trivial solution, then so does $A^T\mathbf{x} = \mathbf{0}$.
   (d) If A^2 is symmetric, then so is A.

Chapter 1 Supplementary Exercises

1. Use Gauss–Jordan elimination to solve for x' and y' in terms of x and y.

$$x = \tfrac{3}{5}x' - \tfrac{4}{5}y'$$
$$y = \tfrac{4}{5}x' + \tfrac{3}{5}y'$$

2. Use Gauss–Jordan elimination to solve for x' and y' in terms of x and y.

$$x = x'\cos\theta - y'\sin\theta$$
$$y = x'\sin\theta + y'\cos\theta$$

3. Find a homogeneous linear system with two equations that are not multiples of one another and such that

$$x_1 = 1, \quad x_2 = -1, \quad x_3 = 1, \quad x_4 = 2$$

and

$$x_1 = 2, \quad x_2 = 0, \quad x_3 = 3, \quad x_4 = -1$$

are solutions of the system.

4. A box containing pennies, nickels, and dimes has 13 coins with a total value of 83 cents. How many coins of each type are in the box?

5. Find positive integers that satisfy

$$x + y + z = 9$$
$$x + 5y + 10z = 44$$

6. For which value(s) of a does the following system have zero, one, infinitely many solutions?

$$x_1 + x_2 + x_3 = 4$$
$$x_3 = 2$$
$$(a^2 - 4)x_3 = a - 2$$

7. Let

$$\begin{bmatrix} a & 0 & b & 2 \\ a & a & 4 & 4 \\ 0 & a & 2 & b \end{bmatrix}$$

be the augmented matrix for a linear system. For what values of a and b does the system have

   (a) a unique solution,   (b) a one-parameter solution,
   (c) a two-parameter solution,   (d) no solution?

8. Solve for x, y, and z.

$$xy - 2\sqrt{y} + 3zy = 8$$
$$2xy - 3\sqrt{y} + 2zy = 7$$
$$-xy + \sqrt{y} + 2zy = 4$$

9. Find a matrix K such that $AKB = C$ given that

$$A = \begin{bmatrix} 1 & 4 \\ -2 & 3 \\ 1 & -2 \end{bmatrix}, \quad B = \begin{bmatrix} 2 & 0 & 0 \\ 0 & 1 & -1 \end{bmatrix}, \quad C = \begin{bmatrix} 8 & 6 & -6 \\ 6 & -1 & 1 \\ -4 & 0 & 0 \end{bmatrix}$$

10. How should the coefficients a, b, and c be chosen so that the system

$$ax + by - 3z = -3$$
$$-2x - by + cz = -1$$
$$ax + 3y - cz = -3$$

has the solution $x = 1$, $y = -1$, and $z = 2$?

11. In each part solve the matrix equation for X.

(a) $X \begin{bmatrix} -1 & 0 & 1 \\ 1 & 1 & 0 \\ 3 & 1 & -1 \end{bmatrix} = \begin{bmatrix} 1 & 2 & 0 \\ -3 & 1 & 5 \end{bmatrix}$ (b) $X \begin{bmatrix} 1 & -1 & 2 \\ 3 & 0 & 1 \end{bmatrix} = \begin{bmatrix} -5 & -1 & 0 \\ 6 & -3 & 7 \end{bmatrix}$

(c) $\begin{bmatrix} 3 & 1 \\ -1 & 2 \end{bmatrix} X - X \begin{bmatrix} 1 & 4 \\ 2 & 0 \end{bmatrix} = \begin{bmatrix} 2 & -2 \\ 5 & 4 \end{bmatrix}$

12. (a) Express the equations

$$\begin{aligned} y_1 &= x_1 - x_2 + x_3 \\ y_2 &= 3x_1 + x_2 - 4x_3 \\ y_3 &= -2x_1 - 2x_2 + 3x_3 \end{aligned} \quad \text{and} \quad \begin{aligned} z_1 &= 4y_1 - y_2 + y_3 \\ z_2 &= -3y_1 + 5y_2 - y_3 \end{aligned}$$

in the matrix forms $Y = AX$ and $Z = BY$. Then use these to obtain a direct relationship $Z = CX$ between Z and X.

(b) Use the equation $Z = CX$ obtained in (a) to express z_1 and z_2 in terms of x_1, x_2, and x_3.

(c) Check the result in (b) by directly substituting the equations for y_1, y_2, and y_3 into the equations for z_1 and z_2 and then simplifying.

13. If A is $m \times n$ and B is $n \times p$, how many multiplication operations and how many addition operations are needed to calculate the matrix product AB?

14. Let A be a square matrix.

(a) Show that $(I - A)^{-1} = I + A + A^2 + A^3$ if $A^4 = 0$.

(b) Show that $(I - A)^{-1} = I + A + A^2 + \cdots + A^n$ if $A^{n+1} = 0$.

15. Find values of a, b, and c so that the graph of the polynomial $p(x) = ax^2 + bx + c$ passes through the points $(1, 2)$, $(-1, 6)$, and $(2, 3)$.

16. (*For readers who have studied calculus.*) Find values of a, b, and c so that the graph of the polynomial $p(x) = ax^2 + bx + c$ passes through the point $(-1, 0)$ and has a horizontal tangent at $(2, -9)$.

17. Let J_n be the $n \times n$ matrix each of whose entries is 1. Show that if $n > 1$, then

$$(I - J_n)^{-1} = I - \frac{1}{n-1} J_n$$

18. Show that if a square matrix A satisfies $A^3 + 4A^2 - 2A + 7I = 0$, then so does A^T.

19. Prove: If B is invertible, then $AB^{-1} = B^{-1}A$ if and only if $AB = BA$.

20. Prove: If A is invertible, then $A + B$ and $I + BA^{-1}$ are both invertible or both not invertible.

21. Prove that if A and B are $n \times n$ matrices, then

  (a) $\text{tr}(A + B) = \text{tr}(A) + \text{tr}(B)$ (b) $\text{tr}(kA) = k\,\text{tr}(A)$ (c) $\text{tr}(A^T) = \text{tr}(A)$ (d) $\text{tr}(AB) = \text{tr}(BA)$

22. Use Exercise 21 to show that there are no square matrices A and B such that

$$AB - BA = I$$

23. Prove: If A is an $m \times n$ matrix and B is the $n \times 1$ matrix each of whose entries is $1/n$, then

$$AB = \begin{bmatrix} \bar{r}_1 \\ \bar{r}_2 \\ \vdots \\ \bar{r}_m \end{bmatrix}$$

where $\bar{r}_i$ is the average of the entries in the ith row of A.

24. (*For readers who have studied calculus.*) If the entries of the matrix

$$C = \begin{bmatrix} c_{11}(x) & c_{12}(x) & \cdots & c_{1n}(x) \\ c_{21}(x) & c_{22}(x) & \cdots & c_{2n}(x) \\ \vdots & \vdots & & \vdots \\ c_{m1}(x) & c_{m2}(x) & \cdots & c_{mn}(x) \end{bmatrix}$$

are differentiable functions of x, then we define

$$\frac{dC}{dx} = \begin{bmatrix} c'_{11}(x) & c'_{12}(x) & \cdots & c'_{1n}(x) \\ c'_{21}(x) & c'_{22}(x) & \cdots & c'_{2n}(x) \\ \vdots & \vdots & & \vdots \\ c'_{m1}(x) & c'_{m2}(x) & \cdots & c'_{mn}(x) \end{bmatrix}$$

Show that if the entries in A and B are differentiable functions of x and the sizes of the matrices are such that the stated operations can be performed, then

  (a) $\dfrac{d}{dx}(kA) = k\dfrac{dA}{dx}$ (b) $\dfrac{d}{dx}(A+B) = \dfrac{dA}{dx} + \dfrac{dB}{dx}$ (c) $\dfrac{d}{dx}(AB) = \dfrac{dA}{dx}B + A\dfrac{dB}{dx}$

25. (*For readers who have studied calculus.*) Use part (c) of Exercise 24 to show that

$$\frac{dA^{-1}}{dx} = -A^{-1}\frac{dA}{dx}A^{-1}$$

State all the assumptions you make in obtaining this formula.

26. Find the values of A, B, and C that will make the equation

$$\frac{x^2 + x - 2}{(3x - 1)(x^2 + 1)} = \frac{A}{3x - 1} + \frac{Bx + C}{x^2 + 1}$$

an identity. [**Hint.** Multiply through by $(3x - 1)(x^2 + 1)$ and equate the corresponding coefficients of the polynomials on each side of the resulting equation.]

27. If P is an $n \times 1$ matrix such that $P^T P = 1$, then $H = I - 2PP^T$ is called the corresponding **Householder matrix** (named after the American mathematician A. S. Householder).

  (a) Verify that $P^T P = 1$ if $P^T = \left[\frac{3}{4}\ \frac{1}{6}\ \frac{1}{4}\ \frac{5}{12}\ \frac{5}{12}\right]$ and compute the corresponding Householder matrix.

  (b) Prove that if H is any Householder matrix, then $H = H^T$ and $H^T H = I$.

  (c) Verify that the Householder matrix found in part (a) satisfies the conditions proved in part (b).

28. Assuming that the stated inverses exist, prove the following equalities.

  (a) $(C^{-1} + D^{-1})^{-1} = C(C + D)^{-1}D$ (b) $(I + CD)^{-1}C = C(I + DC)^{-1}$
  (c) $(C + DD^T)^{-1}D = C^{-1}D(I + D^T C^{-1}D)^{-1}$

29. (a) Show that if $a \neq b$, then

$$a^n + a^{n-1}b + a^{n-2}b^2 + \cdots + ab^{n-1} + b^n = \frac{a^{n+1} - b^{n+1}}{a - b}$$

(b) Use the result in part (a) to find A^n if

$$A = \begin{bmatrix} a & 0 & 0 \\ 0 & b & 0 \\ 1 & 0 & c \end{bmatrix}$$

[**Note.** This exercise is based on a problem by John M. Johnson, *The Mathematics Teacher*, Vol. 85, No. 9, 1992.]

Chapter 1 Technology Exercises

The following exercises are designed to be solved using a technology utility. Typically, this will be MATLAB, *Mathematica*, Maple, Derive, or Mathcad, but it may also be some other type of linear algebra software or a scientific calculator with some linear algebra capabilities. For each exercise you will need to read the relevant documentation for the particular utility you are using. The goal of these exercises is to provide you with a basic proficiency with your technology utility. Once you have mastered the techniques in these exercises, you will be able to use your technology utility to solve many of the problems in the regular exercise sets.

Section 1.1

T1. (*Numbers and numerical operations*) Read your documentation on entering and displaying numbers and performing the basic arithmetic operations of addition, subtraction, multiplication, division, raising numbers to powers, and extraction of roots. Determine how to control the number of digits in the screen display of a decimal number. If you are using a CAS, in which case you can compute with exact numbers rather than decimal approximations, then learn how to enter such numbers as π, $\sqrt{2}$, and $\frac{1}{3}$ exactly and convert them to decimal form. Experiment with numbers of your own choosing until you feel you have mastered the procedures and operations.

Section 1.2

T1. (*Matrices and reduced row-echelon form*) Read your documentation on how to enter matrices and how to find the reduced row-echelon form of a matrix. Then use your utility to find the reduced row-echelon form of the augmented matrix in Example 4 of Section 1.2.

T2. (*Linear systems with a unique solution*) Read your documentation on how to solve a linear system, and then use your utility to solve the linear system in Example 3 of Section 1.1. Also, solve the system by reducing the augmented matrix to reduced row-echelon form.

T3. (*Linear systems with infinitely many solutions*) Technology utilities vary on how they handle linear systems with infinitely many solutions. See how your utility handles the system in Example 4 of Section 1.2.

T4. (*Inconsistent linear systems*) Technology utilities will often successfully identify inconsistent linear systems, but they can sometimes be fooled into reporting an inconsistent system as consistent or vice versa. This typically occurs when some of the numbers that occur in the computations are so small that roundoff error makes it difficult for the utility to determine whether or not they are equal to zero. Create some inconsistent linear systems and see how your utility handles them.

T5. A polynomial whose graph passes through a given set of points is called an ***interpolating polynomial*** for those points. Some technology utilities have specific commands for finding interpolating polynomials. If your utility has this capability, read the documentation and then use this feature to solve Exercise 25 of Section 1.2.

Section 1.3

T1. (*Matrix operations*) Read your documentation on how to perform the basic operations on matrices—addition, subtraction, multiplication by scalars, and multiplication of matrices. Then perform the computations in Examples 3, 4, and 5. See what happens when you try to perform an operation on matrices with inconsistent sizes.

T2. Evaluate the expression $A^5 - 3A^3 + 7A - 4I$ for the matrix

$$A = \begin{bmatrix} 1 & -2 & 3 \\ -4 & 5 & -6 \\ 7 & -8 & 9 \end{bmatrix}$$

T3. (*Extracting rows and columns*) Read your documentation on how to extract rows and columns from a matrix, and then use your utility to extract various rows and columns from a matrix of your choice.

T4. (*Transpose and trace*) Read your documentation on how to find the transpose and trace of a matrix, and then use your utility to find the transpose of the matrix A in Formula (12) and the trace of the matrix B in Example 11.

T5. (*Constructing an augmented matrix*) Read your documentation on how to create an augmented matrix $[A \mid \mathbf{b}]$ from matrices A and $\mathbf{b}$ that have previously been entered. Then use your utility to form the augmented matrix for the system $A\mathbf{x} = \mathbf{b}$ in Example 4 of Section 1.1 from the matrices A and $\mathbf{b}$.

Section 1.4

T1. (*Zero and identity matrices*) Typing in entries of a matrix can be tedious, so many technology utilities provide shortcuts for entering zero and identity matrices. Read your documentation on how to do this, and then enter some zero and identity matrices of various sizes.

T2. (*Inverse*) Read your documentation on how to find the inverse of a matrix, and then use your utility to perform the computations in Example 7.

T3. (*Formula for the inverse*) If you are working with a CAS, use it to confirm Theorem 1.4.5.

T4. (*Powers of a matrix*) Read your documentation on how to find powers of a matrix, and then use your utility to find various positive and negative powers of the matrix A in Example 8.

T5. Let

$$A = \begin{bmatrix} 1 & \frac{1}{2} & \frac{1}{3} \\ \frac{1}{4} & 1 & \frac{1}{5} \\ \frac{1}{6} & \frac{1}{7} & 1 \end{bmatrix}$$

Describe what happens to the matrix A^k when k is allowed to increase indefinitely (that is, as $k \to \infty$).

T6. By experimenting with different values of n, find an expression for the inverse of an $n \times n$ matrix of the form

$$A = \begin{bmatrix} 1 & 2 & 3 & 4 & \cdots & n-1 & n \\ 0 & 1 & 2 & 3 & \cdots & n-2 & n-1 \\ 0 & 0 & 1 & 2 & \cdots & n-3 & n-2 \\ \vdots & \vdots & \vdots & \vdots & & \vdots & \vdots \\ 0 & 0 & 0 & 0 & \cdots & 1 & 2 \\ 0 & 0 & 0 & 0 & \cdots & 0 & 1 \end{bmatrix}$$

Section 1.5

T1. (***Singular matrices***) Find the inverse of the matrix in Example 4, and then see what your utility does when you try to invert the matrix in Example 5.

Section 1.6

T1. (***Solving*** $A\mathbf{x} = \mathbf{b}$ ***by inversion***) Use the method of Example 4 to solve the system in Example 3 of Section 1.1.

T2. Solve the linear system $A\mathbf{x} = 2\mathbf{x}$, given that

$$A = \begin{bmatrix} 0 & 0 & -2 \\ 1 & 2 & 1 \\ 1 & 0 & 3 \end{bmatrix}$$

Section 1.7

T1. (***Diagonal, symmetric, and triangular matrices***) Many technology utilities provide shortcuts for entering diagonal, symmetric, and triangular matrices. Read your documentation on how to do this, and then experiment with entering various matrices of these types.

T2. (***Properties of triangular matrices***) Confirm the results in Theorem 1.7.1 using some triangular matrices of your choice.

Determinants

INTRODUCTION: We are all familiar with functions such as $f(x) = \sin x$ and $f(x) = x^2$, which associate a real number $f(x)$ with a real value of the variable x. Since both x and $f(x)$ assume only real values, such functions are described as real-valued functions of a real variable. In this section we shall study the "determinant function," which is a real-valued function of a matrix variable in the sense that it associates a real number $f(X)$ with a square matrix X. Our work on determinant functions will have important applications to the theory of systems of linear equations and will also lead us to an explicit formula for the inverse of an invertible matrix.

2.1 THE DETERMINANT FUNCTION

As noted in the introduction to this chapter, a "determinant" is a certain kind of function that associates a real number with a square matrix. In this section we will define this function, and we will apply it to 2×2 and 3×3 matrices.

Recall from Theorem 1.4.5 that the 2×2 matrix

$$A = \begin{bmatrix} a & b \\ c & d \end{bmatrix}$$

is invertible if $ad - bc \neq 0$. The expression $ad - bc$ occurs so frequently in mathematics that it has a name; it is called the ***determinant*** of the matrix A and is denoted by the symbol $\det(A)$. With this notation the formula for A^{-1} given in Theorem 1.4.5 is

$$A^{-1} = \frac{1}{\det(A)} \begin{bmatrix} d & -b \\ -c & a \end{bmatrix}$$

One of the goals of this chapter is to obtain analogs of this formula to square matrices of higher order. This will require that we extend the concept of a determinant to square matrices of all orders. To do this we will need some preliminary results about permutations.

Definition

A ***permutation*** of the set of integers $\{1, 2, \ldots, n\}$ is an arrangement of these integers in some order without omissions or repetitions.

EXAMPLE 1 Permutations of Three Integers

There are six different permutations of the set of integers $\{1, 2, 3\}$. These are

$$(1, 2, 3) \quad (2, 1, 3) \quad (3, 1, 2)$$
$$(1, 3, 2) \quad (2, 3, 1) \quad (3, 2, 1) \qquad \blacklozenge$$

One convenient method of systematically listing permutations is to use a *permutation tree*. This method is illustrated in our next example.

EXAMPLE 2 Permutations of Four Integers

List all permutations of the set of integers $\{1, 2, 3, 4\}$.

Solution.

Consider Figure 2.1.1. The four dots labeled 1, 2, 3, 4 at the top of the figure represent the possible choices for the first number in the permutation. The three branches emanating from these dots represent the possible choices for the second position in the permutation. Thus, if the permutation begins $(2, -, -, -)$, the three possibilities for the second position are 1, 3, and 4. The two branches emanating from each dot in the second position represent the possible choices for the third position. Thus, if the permutation begins $(2, 3, -, -)$, the two possible choices for the third position are 1 and 4. Finally,

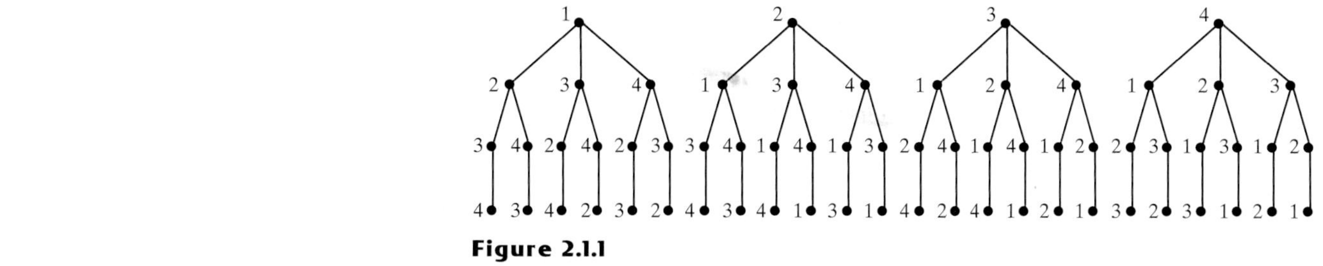

Figure 2.1.1

the single branch emanating from each dot in the third position represents the only possible choice for the fourth position. Thus, if the permutation begins with $(2, 3, 4, -)$, the only choice for the fourth position is 1. The different permutations can now be listed by tracing out all the possible paths through the "tree" from the first position to the last position. We obtain the following list by this process.

$(1, 2, 3, 4)$	$(2, 1, 3, 4)$	$(3, 1, 2, 4)$	$(4, 1, 2, 3)$
$(1, 2, 4, 3)$	$(2, 1, 4, 3)$	$(3, 1, 4, 2)$	$(4, 1, 3, 2)$
$(1, 3, 2, 4)$	$(2, 3, 1, 4)$	$(3, 2, 1, 4)$	$(4, 2, 1, 3)$
$(1, 3, 4, 2)$	$(2, 3, 4, 1)$	$(3, 2, 4, 1)$	$(4, 2, 3, 1)$
$(1, 4, 2, 3)$	$(2, 4, 1, 3)$	$(3, 4, 1, 2)$	$(4, 3, 1, 2)$
$(1, 4, 3, 2)$	$(2, 4, 3, 1)$	$(3, 4, 2, 1)$	$(4, 3, 2, 1)$

♦

From this example we see that there are 24 permutations of $\{1, 2, 3, 4\}$. This result could have been anticipated without actually listing the permutations by arguing as follows. Since the first position can be filled in four ways and then the second position in three ways, there are $4 \cdot 3$ ways of filling the first two positions. Since the third position can then be filled in two ways, there are $4 \cdot 3 \cdot 2$ ways of filling the first three positions. Finally, since the last position can then be filled in only one way, there are $4 \cdot 3 \cdot 2 \cdot 1 = 24$ ways of filling all four positions. In general, the set $\{1, 2, \ldots, n\}$ will have $n(n - 1)(n - 2) \cdots 2 \cdot 1 = n!$ different permutations.

We will denote a general permutation of the set $\{1, 2, \ldots, n\}$ by $(j_1, j_2, \ldots, j_n)$. Here, j_1 is the first integer in the permutation, j_2 is the second, and so on. An ***inversion*** is said to occur in a permutation $(j_1, j_2, \ldots, j_n)$ whenever a larger integer precedes a smaller one. The total number of inversions occurring in a permutation can be obtained as follows: (1) find the number of integers that are less than j_1 and that follow j_1 in the permutation; (2) find the number of integers that are less than j_2 and that follow j_2 in the permutation. Continue this counting process for $j_3, \ldots, j_{n-1}$. The sum of these numbers will be the total number of inversions in the permutation.

EXAMPLE 3 Counting Inversions

Determine the number of inversions in the following permutations:

$$\text{(a) } (6, 1, 3, 4, 5, 2) \quad \text{(b) } (2, 4, 1, 3) \quad \text{(c) } (1, 2, 3, 4)$$

Solution.

(a) The number of inversions is $5 + 0 + 1 + 1 + 1 = 8$.
(b) The number of inversions is $1 + 2 + 0 = 3$.
(c) There are no inversions in this permutation.

♦

Definition

A permutation is called *even* if the total number of inversions is an even integer and is called *odd* if the total number of inversions is an odd integer.

EXAMPLE 4 Classifying Permutations

The following table classifies the various permutations of $\{1, 2, 3\}$ as even or odd.

Permutation	Number of Inversions	Classification
$(1, 2, 3)$	0	even
$(1, 3, 2)$	1	odd
$(2, 1, 3)$	1	odd
$(2, 3, 1)$	2	even
$(3, 1, 2)$	2	even
$(3, 2, 1)$	3	odd

◆

Definition of a Determinant   By an *elementary product* from an $n \times n$ matrix A we shall mean any product of n entries from A, no two of which come from the same row or same column.

EXAMPLE 5 Elementary Products

List all elementary products from the matrices

$$\text{(a)} \begin{bmatrix} a_{11} & a_{12} \\ a_{21} & a_{22} \end{bmatrix} \quad \text{(b)} \begin{bmatrix} a_{11} & a_{12} & a_{13} \\ a_{21} & a_{22} & a_{23} \\ a_{31} & a_{32} & a_{33} \end{bmatrix}$$

Solution (a). Since each elementary product has two factors, and since each factor comes from a different row, an elementary product can be written in the form

$$a_{1\_}a_{2\_}$$

where the blanks designate column numbers. Since no two factors in the product come from the same column, the column numbers must be $1\,2$ or $2\,1$. Thus, the only elementary products are $a_{11}a_{22}$ and $a_{12}a_{21}$.

Solution (b). Since each elementary product has three factors, each of which comes from a different row, an elementary product can be written in the form

$$a_{1\_}a_{2\_}a_{3\_}$$

Since no two factors in the product come from the same column, the column numbers have no repetitions; consequently, they must form a permutation of the set $\{1, 2, 3\}$. These $3! = 6$ permutations yield the following list of elementary products.

$$a_{11}a_{22}a_{33} \qquad a_{12}a_{21}a_{33} \qquad a_{13}a_{21}a_{32}$$

$$a_{11}a_{23}a_{32} \qquad a_{12}a_{23}a_{31} \qquad a_{13}a_{22}a_{31}$$

◆

As this example points out, an $n \times n$ matrix A has $n!$ elementary products. They are the products of the form $a_{1j_1}a_{2j_2} \cdots a_{nj_n}$, where $(j_1, j_2, \ldots, j_n)$ is a permutation of the set $\{1, 2, \ldots, n\}$. By a ***signed elementary product from A*** we shall mean an elementary product $a_{1j_1}a_{2j_2} \cdots a_{nj_n}$ multiplied by $+1$ or -1. We use the $+$ if $(j_1, j_2, \ldots, j_n)$ is an even permutation and the $-$ if $(j_1, j_2, \ldots, j_n)$ is an odd permutation.

EXAMPLE 6 Signed Elementary Products

List all signed elementary products from the matrices

$$(a) \begin{bmatrix} a_{11} & a_{12} \\ a_{21} & a_{22} \end{bmatrix} \quad (b) \begin{bmatrix} a_{11} & a_{12} & a_{13} \\ a_{21} & a_{22} & a_{23} \\ a_{31} & a_{32} & a_{33} \end{bmatrix}$$

Solution.

(a)

Elementary Product	Associated Permutation	Even or Odd	Signed Elementary Product
$a_{11}a_{22}$	$(1, 2)$	even	$a_{11}a_{22}$
$a_{12}a_{21}$	$(2, 1)$	odd	$-a_{12}a_{21}$

(b)

Elementary Product	Associated Permutation	Even or Odd	Signed Elementary Product
$a_{11}a_{22}a_{33}$	$(1, 2, 3)$	even	$a_{11}a_{22}a_{33}$
$a_{11}a_{23}a_{32}$	$(1, 3, 2)$	odd	$-a_{11}a_{23}a_{32}$
$a_{12}a_{21}a_{33}$	$(2, 1, 3)$	odd	$-a_{12}a_{21}a_{33}$
$a_{12}a_{23}a_{31}$	$(2, 3, 1)$	even	$a_{12}a_{23}a_{31}$
$a_{13}a_{21}a_{32}$	$(3, 1, 2)$	even	$a_{13}a_{21}a_{32}$
$a_{13}a_{22}a_{31}$	$(3, 2, 1)$	odd	$-a_{13}a_{22}a_{31}$

We are now in a position to define the determinant function.

Definition

Let A be a square matrix. The ***determinant function*** is denoted by ***det***, and we define $\det(A)$ to be the sum of all signed elementary products from A. The number $\det(A)$ is called the ***determinant of A***.

EXAMPLE 7 Determinants of 2 × 2 and 3 × 3 Matrices

Referring to Example 6, we obtain

$$(a) \ \det \begin{bmatrix} a_{11} & a_{12} \\ a_{21} & a_{22} \end{bmatrix} = a_{11}a_{22} - a_{12}a_{21}$$

(b) $\det \begin{bmatrix} a_{11} & a_{12} & a_{13} \\ a_{21} & a_{22} & a_{23} \\ a_{31} & a_{32} & a_{33} \end{bmatrix} = a_{11}a_{22}a_{33} + a_{12}a_{23}a_{31} + a_{13}a_{21}a_{32}$
$$- a_{13}a_{22}a_{31} - a_{12}a_{21}a_{33} - a_{11}a_{23}a_{32} \qquad \blacklozenge$$

To avoid memorizing these unwieldy expressions, we suggest using the mnemonic devices given in Figure 2.1.2. The formula in part (a) of Example 7 is obtained from Figure 2.1.2*a* by multiplying the entries on the rightward arrow and subtracting the product of the entries on the leftward arrow. The formula in part (b) of Example 7 is obtained by recopying the first and second columns as shown in Figure 2.1.2*b*. The determinant is then computed by summing the products on the rightward arrows and subtracting the products on the leftward arrows.

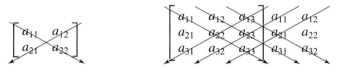

(*a*) Determinant of a 2 × 2 matrix   (*b*) Determinant of a 3 × 3 matrix

Figure 2.1.2

EXAMPLE 8 Evaluating Determinants

Evaluate the determinants of

$$A = \begin{bmatrix} 3 & 1 \\ 4 & -2 \end{bmatrix} \quad \text{and} \quad B = \begin{bmatrix} 1 & 2 & 3 \\ -4 & 5 & 6 \\ 7 & -8 & 9 \end{bmatrix}$$

Solution.

Using the method of Figure 2.1.2*a* gives

$$\det(A) = (3)(-2) - (1)(4) = -10$$

Using the method of Figure 2.1.2*b* gives

$$\det(B) = (45) + (84) + (96) - (105) - (-48) - (-72) = 240$$

$$\begin{bmatrix} 1 & 2 & 3 & 1 & 2 \\ -4 & 5 & 6 & -4 & 5 \\ 7 & -8 & 9 & 7 & -8 \end{bmatrix}$$

$\blacklozenge$

Warning. We emphasize that the methods shown in Figure 2.1.2 do not work for determinants of 4 × 4 matrices or higher.

Notation and Terminology   We conclude this section with some comments about terminology and notation. First, we note that the symbol $|A|$ is an alternative notation for $\det(A)$. For example, the determinant of a 3 × 3 matrix can be written as

$$\det \begin{bmatrix} a_{11} & a_{12} & a_{13} \\ a_{21} & a_{22} & a_{23} \\ a_{31} & a_{32} & a_{33} \end{bmatrix} \quad \text{or} \quad \begin{vmatrix} a_{11} & a_{12} & a_{13} \\ a_{21} & a_{22} & a_{23} \\ a_{31} & a_{32} & a_{33} \end{vmatrix}$$

In the latter notation the determinant of the matrix A in Example 8 would be written as

$$\begin{vmatrix} 3 & 1 \\ 4 & -2 \end{vmatrix} = -10$$

Strictly speaking, the determinant of a matrix is a number. However, it is common practice to "abuse" the terminology slightly and use the term "determinant" to refer to the *matrix* whose determinant is being computed. Thus, we might refer to

$$\begin{vmatrix} 3 & 1 \\ 4 & -2 \end{vmatrix}$$

as a 2×2 determinant and call 3 the entry in the first row and first column of the determinant.

Finally, we note that the determinant of A is often written symbolically as

$$\det(A) = \sum \pm a_{1j_1} a_{2j_2} \cdots a_{nj_n} \tag{1}$$

where $\sum$ indicates that the terms are to be summed over all permutations $(j_1, j_2, \ldots, j_n)$ and the $+$ or $-$ is selected in each term according to whether the permutation is even or odd. This notation is useful when the definition of a determinant needs to be emphasized.

REMARK. Evaluating determinants directly from the definition leads to computational difficulties. Indeed, evaluating a 4×4 determinant directly would involve computing $4! = 24$ signed elementary products, and a 10×10 determinant would require the computation of $10! = 3,628,800$ signed elementary products. Even the fastest of digital computers cannot handle the computation of a 25×25 determinant by this method in a practical amount of time. Much of the remainder of this chapter is devoted, therefore, to developing properties of determinants that will simplify their evaluation.

Exercise Set 2.1

1. Find the number of inversions in each of the following permutations of $\{1, 2, 3, 4, 5\}$.

(a) $(4\ 1\ 3\ 5\ 2)$ (b) $(5\ 3\ 4\ 2\ 1)$ (c) $(3\ 2\ 5\ 4\ 1)$ (d) $(5\ 4\ 3\ 2\ 1)$ (e) $(1\ 2\ 3\ 4\ 5)$ (f) $(1\ 4\ 2\ 3\ 5)$

2. Classify each of the permutations in Exercise 1 as even or odd.

In Exercises 3–12 evaluate the determinant.

3. $\begin{vmatrix} 3 & 5 \\ -2 & 4 \end{vmatrix}$
4. $\begin{vmatrix} 4 & 1 \\ 8 & 2 \end{vmatrix}$
5. $\begin{vmatrix} -5 & 6 \\ -7 & -2 \end{vmatrix}$
6. $\begin{vmatrix} \sqrt{2} & \sqrt{6} \\ 4 & \sqrt{3} \end{vmatrix}$
7. $\begin{vmatrix} a-3 & 5 \\ -3 & a-2 \end{vmatrix}$
8. $\begin{vmatrix} -2 & 7 & 6 \\ 5 & 1 & -2 \\ 3 & 8 & 4 \end{vmatrix}$

9. $\begin{vmatrix} -2 & 1 & 4 \\ 3 & 5 & -7 \\ 1 & 6 & 2 \end{vmatrix}$
10. $\begin{vmatrix} -1 & 1 & 2 \\ 3 & 0 & -5 \\ 1 & 7 & 2 \end{vmatrix}$
11. $\begin{vmatrix} 3 & 0 & 0 \\ 2 & -1 & 5 \\ 1 & 9 & -4 \end{vmatrix}$
12. $\begin{vmatrix} c & -4 & 3 \\ 2 & 1 & c^2 \\ 4 & c-1 & 2 \end{vmatrix}$

13. Find all values of λ for which $\det(A) = 0$.

(a) $\begin{bmatrix} \lambda-2 & 1 \\ -5 & \lambda+4 \end{bmatrix}$
(b) $\begin{bmatrix} \lambda-4 & 0 & 0 \\ 0 & \lambda & 2 \\ 0 & 3 & \lambda-1 \end{bmatrix}$

14. Classify each permutation of $\{1, 2, 3, 4\}$ as even or odd.

15. Use the results in Exercise 14 to construct a formula for the determinant of a 4×4 matrix.

16. Use the formula obtained in Exercise 15 to evaluate

$$\begin{vmatrix} 4 & -9 & 9 & 2 \\ -2 & 5 & 6 & 4 \\ 1 & 2 & -5 & -3 \\ 1 & -2 & 0 & -2 \end{vmatrix}$$

17. Use the determinant definition to evaluate

(a) $\begin{vmatrix} 0 & 0 & 0 & 0 & -3 \\ 0 & 0 & 0 & -4 & 0 \\ 0 & 0 & -1 & 0 & 0 \\ 0 & 2 & 0 & 0 & 0 \\ 5 & 0 & 0 & 0 & 0 \end{vmatrix}$ (b) $\begin{vmatrix} 5 & 0 & 0 & 0 & 0 \\ 0 & 0 & 0 & 0 & -4 \\ 0 & 0 & 3 & 0 & 0 \\ 0 & 0 & 0 & 1 & 0 \\ 0 & -2 & 0 & 0 & 0 \end{vmatrix}$

18. Solve for x.

$$\begin{vmatrix} x & -1 \\ 3 & 1-x \end{vmatrix} = \begin{vmatrix} 1 & 0 & -3 \\ 2 & x & -6 \\ 1 & 3 & x-5 \end{vmatrix}$$

19. Show that the value of the determinant

$$\begin{vmatrix} \sin\theta & \cos\theta & 0 \\ -\cos\theta & \sin\theta & 0 \\ \sin\theta - \cos\theta & \sin\theta + \cos\theta & 1 \end{vmatrix}$$

does not depend on θ.

20. Prove that the matrices

$$A = \begin{bmatrix} a & b \\ 0 & c \end{bmatrix} \quad \text{and} \quad B = \begin{bmatrix} d & e \\ 0 & f \end{bmatrix}$$

commute if and only if

$$\begin{vmatrix} b & a-c \\ e & d-f \end{vmatrix} = 0$$

Discussion and Discovery

21. Explain why the determinant of an $n \times n$ matrix with integer entries must be an integer.

22. What can you say about the determinant of an $n \times n$ matrix all of whose entries are 1? Explain your reasoning.

23. (a) Explain why the determinant of an $n \times n$ matrix with a row of zeros must have a zero determinant.
(b) Explain why the determinant of an $n \times n$ matrix with a column of zeros must have a zero determinant.

24. Use Formula (1) to discover a formula for the determinant of an $n \times n$ diagonal matrix. Express the formula in words.

25. Use Formula (1) to discover a formula for the determinant of an $n \times n$ upper triangular matrix. Express the formula in words. Do the same for a lower triangular matrix.

2.2 EVALUATING DETERMINANTS BY ROW REDUCTION

In this section we shall show that the determinant of a square matrix can be evaluated by reducing the matrix to row-echelon form. This method is important since it avoids the lengthy computations involved in directly applying the determinant definition.

A Basic Theorem   As discussed at the end of the last section, the definition of a determinant is helpful for proving theorems *about* determinants, but it does not provide a practical means for evaluating them, especially determinants of matrices larger than 3×3. Accordingly, we begin with a fundamental theorem that will lead us to an efficient procedure for evaluating the determinant of a matrix of any order n.

Theorem 2.2.1

Let A be a square matrix.

(a) If A has a row of zeros or a column of zeros, then $\det(A) = 0$.
(b) $\det(A) = \det(A^T)$.

Proof (a).   Since every signed elementary product from A has one factor from each row and one factor from each column, every signed elementary product would of necessity have a factor from a zero row or a factor from a zero column. In such cases, every signed elementary product is zero, and $\det(A)$, which is the sum of all the signed elementary products, is zero.   ■

We omit the proof of part (b), but recall that an elementary product has one factor from each row and each column, so it is evident that A and A^T have precisely the same set of elementary products. With the help of some theorems on permutations, which would take us too far afield to discuss, it can be shown that A and A^T actually have the same set of *signed* elementary products. This implies that $\det(A) = \det(A^T)$.

REMARK. Because of Theorem 2.2.1*b*, nearly every theorem about determinants that contains the word "row" in its statement is also true when the word "column" is substituted for "row." To prove a column statement one need only transpose the matrix in question to convert the column statement to a row statement, and then apply the corresponding known result for rows.

Triangular Matrices   The following theorem makes it easy to evaluate the determinant of a triangular matrix, regardless of its size.

Theorem 2.2.2

If A is an $n \times n$ *triangular matrix (upper triangular, lower triangular, or diagonal), then* $\det(A)$ *is the product of the entries on the main diagonal of the matrix; that is,* $\det(A) = a_{11}a_{22} \cdots a_{nn}$.

For simplicity of notation, we will prove the result for a 4×4 lower triangular matrix

$$A = \begin{bmatrix} a_{11} & 0 & 0 & 0 \\ a_{21} & a_{22} & 0 & 0 \\ a_{31} & a_{32} & a_{33} & 0 \\ a_{41} & a_{42} & a_{43} & a_{44} \end{bmatrix}$$

The argument in the $n \times n$ case is similar. A proof for upper triangular matrices can be obtained by applying Theorem 2.2.1b and observing that the transpose of an upper triangular matrix is a lower triangular matrix with the same diagonal entries.

***Proof of Theorem 2.2.2* (4×4 *lower triangular case*).** The only elementary product from A that can be nonzero is $a_{11}a_{22}a_{33}a_{44}$. To see that this is so, consider a typical elementary product $a_{1j_1}a_{2j_2}a_{3j_3}a_{4j_4}$. Since $a_{12} = a_{13} = a_{14} = 0$, we must have $j_1 = 1$ in order to have a nonzero elementary product. If $j_1 = 1$, we must have $j_2 \neq 1$, since no two factors come from the same column. Further, since $a_{23} = a_{24} = 0$, we must have $j_2 = 2$ in order to have a nonzero product. Continuing in this way, we obtain $j_3 = 3$ and $j_4 = 4$. Since $a_{11}a_{22}a_{33}a_{44}$ is multiplied by $+1$ in forming the signed elementary product, we obtain

$$\det(A) = a_{11}a_{22}a_{33}a_{44} \qquad ■$$

EXAMPLE 1 Determinant of an Upper Triangular Matrix

$$\begin{vmatrix} 2 & 7 & -3 & 8 & 3 \\ 0 & -3 & 7 & 5 & 1 \\ 0 & 0 & 6 & 7 & 6 \\ 0 & 0 & 0 & 9 & 8 \\ 0 & 0 & 0 & 0 & 4 \end{vmatrix} = (2)(-3)(6)(9)(4) = -1296 \qquad ◆$$

Elementary Row Operations The next theorem shows how an elementary row operation on a matrix affects the value of its determinant.

Theorem 2.2.3

Let A be an $n \times n$ matrix.

(a) *If B is the matrix that results when a single row or single column of A is multiplied by a scalar k, then $\det(B) = k \det(A)$.*

(b) *If B is the matrix that results when two rows or two columns of A are interchanged, then $\det(B) = -\det(A)$.*

(c) *If B is the matrix that results when a multiple of one row of A is added to another row or when a multiple of one column is added to another column, then $\det(B) = \det(A)$.*

A proof of this theorem can be obtained by using Formula (1) of Section 2.1 to compute the determinants involved, and then verifying the equalities. We omit the proof but give the following example that illustrates the theorem for 3×3 determinants.

EXAMPLE 2 Theorem 2.2.3 Applied to 3 × 3 Determinants

Relationship	Operation
$\begin{vmatrix} ka_{11} & ka_{12} & ka_{13} \\ a_{21} & a_{22} & a_{23} \\ a_{31} & a_{32} & a_{33} \end{vmatrix} = k \begin{vmatrix} a_{11} & a_{12} & a_{13} \\ a_{21} & a_{22} & a_{23} \\ a_{31} & a_{32} & a_{33} \end{vmatrix}$ $\det(B) = k \det(A)$	The first row of A is multiplied by k.
$\begin{vmatrix} a_{21} & a_{22} & a_{23} \\ a_{11} & a_{12} & a_{13} \\ a_{31} & a_{32} & a_{33} \end{vmatrix} = - \begin{vmatrix} a_{11} & a_{12} & a_{13} \\ a_{21} & a_{22} & a_{23} \\ a_{31} & a_{32} & a_{33} \end{vmatrix}$ $\det(B) = -\det(A)$	The first and second rows of A are interchanged.
$\begin{vmatrix} a_{11} + ka_{21} & a_{12} + ka_{22} & a_{13} + ka_{23} \\ a_{21} & a_{22} & a_{23} \\ a_{31} & a_{32} & a_{33} \end{vmatrix} = \begin{vmatrix} a_{11} & a_{12} & a_{13} \\ a_{21} & a_{22} & a_{23} \\ a_{31} & a_{32} & a_{33} \end{vmatrix}$ $\det(B) = \det(A)$	A multiple of the second row of A is added to the first row.

We will verify the equation in the last row of the table and leave the first two for the reader. With the help of Example 7 in Section 2.1 we obtain

$$\det(B) = (a_{11} + ka_{21})a_{22}a_{33} + (a_{12} + ka_{22})a_{23}a_{31} + (a_{13} + ka_{23})a_{21}a_{32}$$
$$- a_{31}a_{22}(a_{13} + ka_{23}) - a_{33}a_{21}(a_{12} + ka_{22}) - a_{32}a_{23}(a_{11} + ka_{21})$$
$$= \det(A) + k(a_{21}a_{22}a_{33} + a_{22}a_{23}a_{31} + a_{23}a_{21}a_{32}$$
$$- a_{31}a_{22}a_{23} - a_{33}a_{21}a_{22} - a_{32}a_{23}a_{21})$$
$$= \det(A) + 0 = \det(A) \qquad\qquad \blacklozenge$$

REMARK. As illustrated by the first equation in Example 2, part (*a*) of Theorem 2.2.3 allows us to bring a "common factor" from any row (or column) through the determinant sign.

Elementary Matrices   Recall that an elementary matrix results from performing a single elementary row operation on an identity matrix; thus, if we let $A = I_n$ in Theorem 2.2.3 [so that we have $\det(A) = \det(I_n) = 1$], then the matrix B is an elementary matrix, and the theorem yields the following result about determinants of elementary matrices.

Theorem 2.2.4

Let E be an $n \times n$ elementary matrix.

(*a*) *If E results from multiplying a row of I_n by k, then $\det(E) = k$.*

(*b*) *If E results from interchanging two rows of I_n, then $\det(E) = -1$.*

(*c*) *If E results from adding a multiple of one row of I_n to another, then $\det(E) = 1$.*

EXAMPLE 3 Determinants of Elementary Matrices

The following determinants of elementary matrices, which are evaluated by inspection, illustrate Theorem 2.2.4.

$$
\begin{vmatrix} 1 & 0 & 0 & 0 \\ 0 & 3 & 0 & 0 \\ 0 & 0 & 1 & 0 \\ 0 & 0 & 0 & 1 \end{vmatrix} = 3,
\qquad
\begin{vmatrix} 0 & 0 & 0 & 1 \\ 0 & 1 & 0 & 0 \\ 0 & 0 & 1 & 0 \\ 1 & 0 & 0 & 0 \end{vmatrix} = -1,
\qquad
\begin{vmatrix} 1 & 0 & 0 & 7 \\ 0 & 1 & 0 & 0 \\ 0 & 0 & 1 & 0 \\ 0 & 0 & 0 & 1 \end{vmatrix} = 1
$$

The second row of I_4 was multiplied by 3.

The first and last rows of I_4 were interchanged.

7 times the last row of I_4 was added to the first row. ◆

Matrices with Proportional Rows or Columns

If a square matrix A has two proportional rows, then a row of zeros can be introduced by adding a suitable multiple of one of the rows to the other. Similarly for columns. But adding a multiple of one row or column to another does not change the determinant, so from Theorem 2.2.1a, we must have $\det(A) = 0$. This proves the following theorem.

Theorem 2.2.5

If A is a square matrix with two proportional rows or two proportional columns, then $\det(A) = 0$.

EXAMPLE 4 Introducing Zero Rows

The following computation illustrates the introduction of a row of zeros when there are two proportional rows:

$$
\begin{vmatrix} 1 & 3 & -2 & 4 \\ 2 & 6 & -4 & 8 \\ 3 & 9 & 1 & 5 \\ 1 & 1 & 4 & 8 \end{vmatrix}
=
\begin{vmatrix} 1 & 3 & -2 & 4 \\ 0 & 0 & 0 & 0 \\ 3 & 9 & 1 & 5 \\ 1 & 1 & 4 & 8 \end{vmatrix} = 0
\longleftarrow
$$

The second row is 2 times the first, so we added -2 times the first row to the second to introduce a row of zeros.

Each of the following matrices has two proportional rows or columns; thus, each has a determinant of zero.

$$
\begin{bmatrix} -1 & 4 \\ -2 & 8 \end{bmatrix},
\qquad
\begin{bmatrix} 1 & -2 & 7 \\ -4 & 8 & 5 \\ 2 & -4 & 3 \end{bmatrix},
\qquad
\begin{bmatrix} 3 & -1 & 4 & -5 \\ 6 & -2 & 5 & 2 \\ 5 & 8 & 1 & 4 \\ -9 & 3 & -12 & 15 \end{bmatrix}
\qquad ◆
$$

Evaluating Determinants by Row Reduction

We shall now give a method for evaluating determinants that involves substantially less computation than applying the determinant definition directly. The idea of the method is to reduce the given matrix to upper triangular form by elementary row operations, then compute the determinant of the upper triangular matrix (an easy computation), then relate that determinant to that of the original matrix. Here is an example.

EXAMPLE 5 Using Row Reduction to Evaluate a Determinant

Evaluate $\det(A)$ where

$$A = \begin{bmatrix} 0 & 1 & 5 \\ 3 & -6 & 9 \\ 2 & 6 & 1 \end{bmatrix}$$

Solution.

We will reduce A to row-echelon form (which is upper triangular) and apply Theorem 2.2.3:

$$\det(A) = \begin{vmatrix} 0 & 1 & 5 \\ 3 & -6 & 9 \\ 2 & 6 & 1 \end{vmatrix} = - \begin{vmatrix} 3 & -6 & 9 \\ 0 & 1 & 5 \\ 2 & 6 & 1 \end{vmatrix}$$ ⟵ The first and second rows of A were interchanged.

$$= -3 \begin{vmatrix} 1 & -2 & 3 \\ 0 & 1 & 5 \\ 2 & 6 & 1 \end{vmatrix}$$ ⟵ A common factor of 3 from the first row was taken through the determinant sign.

$$= -3 \begin{vmatrix} 1 & -2 & 3 \\ 0 & 1 & 5 \\ 0 & 10 & -5 \end{vmatrix}$$ ⟵ -2 times the first row was added to the third row.

$$= -3 \begin{vmatrix} 1 & -2 & 3 \\ 0 & 1 & 5 \\ 0 & 0 & -55 \end{vmatrix}$$ ⟵ -10 times the second row was added to the third row.

$$= (-3)(-55) \begin{vmatrix} 1 & -2 & 3 \\ 0 & 1 & 5 \\ 0 & 0 & 1 \end{vmatrix}$$ ⟵ A common factor of -55 from the last row was taken through the determinant sign.

$$= (-3)(-55)(1) = 165$$ ◆

REMARK. The method of row reduction is well suited for computer evaluation of determinants because it is systematic and easily programmed. However, in subsequent sections we will develop methods that are often easier for hand computation.

EXAMPLE 6 Using Column Operations to Evaluate a Determinant

Compute the determinant of

$$A = \begin{bmatrix} 1 & 0 & 0 & 3 \\ 2 & 7 & 0 & 6 \\ 0 & 6 & 3 & 0 \\ 7 & 3 & 1 & -5 \end{bmatrix}$$

Solution.

This determinant could be computed as above by using elementary row operations to reduce A to row-echelon form, but we can put A in lower triangular form in one step by adding -3 times the first column to the fourth to obtain

$$\det(A) = \det \begin{bmatrix} 1 & 0 & 0 & 0 \\ 2 & 7 & 0 & 0 \\ 0 & 6 & 3 & 0 \\ 7 & 3 & 1 & -26 \end{bmatrix} = (1)(7)(3)(-26) = -546$$

This example points out the utility of keeping an eye open for column operations that can shorten computations. ◆

Exercise Set 2.2

1. Verify that $\det(A) = \det(A^T)$ for

(a) $A = \begin{bmatrix} -2 & 3 \\ 1 & 4 \end{bmatrix}$ (b) $A = \begin{bmatrix} 2 & -1 & 3 \\ 1 & 2 & 4 \\ 5 & -3 & 6 \end{bmatrix}$

2. Evaluate the following determinants by inspection.

(a) $\begin{vmatrix} 3 & -17 & 4 \\ 0 & 5 & 1 \\ 0 & 0 & -2 \end{vmatrix}$ (b) $\begin{vmatrix} \sqrt{2} & 0 & 0 & 0 \\ -8 & \sqrt{2} & 0 & 0 \\ 7 & 0 & -1 & 0 \\ 9 & 5 & 6 & 1 \end{vmatrix}$ (c) $\begin{vmatrix} -2 & 1 & 3 \\ 1 & -7 & 4 \\ -2 & 1 & 3 \end{vmatrix}$ (d) $\begin{vmatrix} 1 & -2 & 3 \\ 2 & -4 & 6 \\ 5 & -8 & 1 \end{vmatrix}$

3. Find the determinants of the following elementary matrices by inspection.

(a) $\begin{bmatrix} 1 & 0 & 0 & 0 \\ 0 & 1 & 0 & 0 \\ 0 & 0 & -5 & 0 \\ 0 & 0 & 0 & 1 \end{bmatrix}$ (b) $\begin{bmatrix} 1 & 0 & 0 & 0 \\ 0 & 0 & 1 & 0 \\ 0 & 1 & 0 & 0 \\ 0 & 0 & 0 & 1 \end{bmatrix}$ (c) $\begin{bmatrix} 1 & 0 & 0 & 0 \\ 0 & 1 & 0 & -9 \\ 0 & 0 & 1 & 0 \\ 0 & 0 & 0 & 1 \end{bmatrix}$

In Exercises 4–11 evaluate the determinant of the given matrix by reducing the matrix to row-echelon form.

4. $\begin{bmatrix} 3 & 6 & -9 \\ 0 & 0 & -2 \\ -2 & 1 & 5 \end{bmatrix}$ **5.** $\begin{bmatrix} 0 & 3 & 1 \\ 1 & 1 & 2 \\ 3 & 2 & 4 \end{bmatrix}$ **6.** $\begin{bmatrix} 1 & -3 & 0 \\ -2 & 4 & 1 \\ 5 & -2 & 2 \end{bmatrix}$ **7.** $\begin{bmatrix} 3 & -6 & 9 \\ -2 & 7 & -2 \\ 0 & 1 & 5 \end{bmatrix}$

8. $\begin{bmatrix} 1 & -2 & 3 & 1 \\ 5 & -9 & 6 & 3 \\ -1 & 2 & -6 & -2 \\ 2 & 8 & 6 & 1 \end{bmatrix}$ **9.** $\begin{bmatrix} 2 & 1 & 3 & 1 \\ 1 & 0 & 1 & 1 \\ 0 & 2 & 1 & 0 \\ 0 & 1 & 2 & 3 \end{bmatrix}$ **10.** $\begin{bmatrix} 0 & 1 & 1 & 1 \\ \frac{1}{2} & \frac{1}{2} & 1 & \frac{1}{2} \\ \frac{2}{3} & \frac{1}{3} & \frac{1}{3} & 0 \\ -\frac{1}{3} & \frac{2}{3} & 0 & 0 \end{bmatrix}$ **11.** $\begin{bmatrix} 1 & 3 & 1 & 5 & 3 \\ -2 & -7 & 0 & -4 & 2 \\ 0 & 0 & 1 & 0 & 1 \\ 0 & 0 & 2 & 1 & 1 \\ 0 & 0 & 0 & 1 & 1 \end{bmatrix}$

12. Given that $\begin{vmatrix} a & b & c \\ d & e & f \\ g & h & i \end{vmatrix} = -6$, find

(a) $\begin{vmatrix} d & e & f \\ g & h & i \\ a & b & c \end{vmatrix}$ (b) $\begin{vmatrix} 3a & 3b & 3c \\ -d & -e & -f \\ 4g & 4h & 4i \end{vmatrix}$ (c) $\begin{vmatrix} a+g & b+h & c+i \\ d & e & f \\ g & h & i \end{vmatrix}$ (d) $\begin{vmatrix} -3a & -3b & -3c \\ d & e & f \\ g-4d & h-4e & i-4f \end{vmatrix}$

13. Use row reduction to show that

$$\begin{vmatrix} 1 & 1 & 1 \\ a & b & c \\ a^2 & b^2 & c^2 \end{vmatrix} = (b-a)(c-a)(c-b)$$

14. Use an argument like that in the proof of Theorem 2.2.2 to show that

(a) $\det \begin{bmatrix} 0 & 0 & a_{13} \\ 0 & a_{22} & a_{23} \\ a_{31} & a_{32} & a_{33} \end{bmatrix} = -a_{13}a_{22}a_{31}$
(b) $\det \begin{bmatrix} 0 & 0 & 0 & a_{14} \\ 0 & 0 & a_{23} & a_{24} \\ 0 & a_{32} & a_{33} & a_{34} \\ a_{41} & a_{42} & a_{43} & a_{44} \end{bmatrix} = a_{14}a_{23}a_{32}a_{41}$

15. Prove the following special cases of Theorem 2.2.3.

(a) $\begin{vmatrix} ka_{11} & ka_{12} & ka_{13} \\ a_{21} & a_{22} & a_{23} \\ a_{31} & a_{32} & a_{33} \end{vmatrix} = k \begin{vmatrix} a_{11} & a_{12} & a_{13} \\ a_{21} & a_{22} & a_{23} \\ a_{31} & a_{32} & a_{33} \end{vmatrix}$
(b) $\begin{vmatrix} a_{21} & a_{22} & a_{23} \\ a_{11} & a_{12} & a_{13} \\ a_{31} & a_{32} & a_{33} \end{vmatrix} = - \begin{vmatrix} a_{11} & a_{12} & a_{13} \\ a_{21} & a_{22} & a_{23} \\ a_{31} & a_{32} & a_{33} \end{vmatrix}$

Discussion and Discovery

16. In each part, find $\det(A)$ by inspection, and explain your reasoning.

(a) $A = \begin{bmatrix} 0 & 0 & 1 \\ 0 & 1 & 0 \\ 1 & 0 & 0 \end{bmatrix}$
(b) $A = \begin{bmatrix} 0 & 0 & 0 & 1 \\ 0 & 0 & 1 & 0 \\ 0 & 1 & 0 & 0 \\ 1 & 0 & 0 & 0 \end{bmatrix}$

17. By inspection, solve the equation

$$\begin{vmatrix} x & 5 & 7 \\ 0 & x+1 & 6 \\ 0 & 0 & 2x-1 \end{vmatrix} = 0$$

Explain your reasoning.

18. (a) By inspection, find two solutions of the equation

$$\begin{vmatrix} 1 & x & x^2 \\ 1 & 1 & 1 \\ 1 & -3 & 9 \end{vmatrix} = 0$$

(b) Is it possible that there are other solutions? Justify your answer.

2.3 PROPERTIES OF THE DETERMINANT FUNCTION

In this section we shall develop some of the fundamental properties of the determinant function. Our work here will give us some further insight into the relationship between a square matrix and its determinant. One of the immediate consequences of this material will be an important determinant test for the invertibility of a matrix.

Basic Properties of Determinants   Suppose that A and B are $n \times n$ matrices and k is any scalar. We begin by considering possible relationships between $\det(A)$, $\det(B)$, and

$$\det(kA), \quad \det(A+B), \quad \text{and} \quad \det(AB)$$

Since a common factor of any row of a matrix can be moved through the det sign, and since each of the n rows in kA has a common factor of k, we obtain

$$\det(kA) = k^n \det(A) \tag{1}$$

For example,

$$\begin{vmatrix} ka_{11} & ka_{12} & ka_{13} \\ ka_{21} & ka_{22} & ka_{23} \\ ka_{31} & ka_{32} & ka_{33} \end{vmatrix} = k^3 \begin{vmatrix} a_{11} & a_{12} & a_{13} \\ a_{21} & a_{22} & a_{23} \\ a_{31} & a_{32} & a_{33} \end{vmatrix}$$

Unfortunately, no simple relationship exists between $\det(A)$, $\det(B)$, and $\det(A+B)$ in general. In particular, we emphasize that $\det(A+B)$ is usually *not* equal to $\det(A) + \det(B)$. The following example illustrates this fact.

EXAMPLE 1 $\det[A+B] \neq \det[A] + \det[B]$

Consider

$$A = \begin{bmatrix} 1 & 2 \\ 2 & 5 \end{bmatrix}, \quad B = \begin{bmatrix} 3 & 1 \\ 1 & 3 \end{bmatrix}, \quad A+B = \begin{bmatrix} 4 & 3 \\ 3 & 8 \end{bmatrix}$$

We have $\det(A) = 1$, $\det(B) = 8$, and $\det(A+B) = 23$; thus

$$\det(A+B) \neq \det(A) + \det(B) \qquad \blacklozenge$$

In spite of the negative tone of the preceding example, there is one important relationship concerning sums of determinants that is often useful. To obtain it, consider two 2×2 matrices that differ only in the second row:

$$A = \begin{bmatrix} a_{11} & a_{12} \\ a_{21} & a_{22} \end{bmatrix} \quad \text{and} \quad B = \begin{bmatrix} a_{11} & a_{12} \\ b_{21} & b_{22} \end{bmatrix}$$

We have

$$\det(A) + \det(B) = (a_{11}a_{22} - a_{12}a_{21}) + (a_{11}b_{22} - a_{12}b_{21})$$
$$= a_{11}(a_{22} + b_{22}) - a_{12}(a_{21} + b_{21})$$
$$= \det \begin{bmatrix} a_{11} & a_{12} \\ a_{21}+b_{21} & a_{22}+b_{22} \end{bmatrix}$$

Thus,

$$\det \begin{bmatrix} a_{11} & a_{12} \\ a_{21} & a_{22} \end{bmatrix} + \det \begin{bmatrix} a_{11} & a_{12} \\ b_{21} & b_{22} \end{bmatrix} = \det \begin{bmatrix} a_{11} & a_{12} \\ a_{21}+b_{21} & a_{22}+b_{22} \end{bmatrix}$$

This is a special case of the following general result.

Theorem 2.3.1

Let A, B, and C be $n \times n$ matrices that differ only in a single row, say the rth, and assume that the rth row of C can be obtained by adding corresponding entries in the rth rows of A and B. Then

$$\det(C) = \det(A) + \det(B)$$

The same result holds for columns.

EXAMPLE 2 Using Theorem 2.3.1

By evaluating the determinants, the reader can check that

$$\det \begin{bmatrix} 1 & 7 & 5 \\ 2 & 0 & 3 \\ 1+0 & 4+1 & 7+(-1) \end{bmatrix} = \det \begin{bmatrix} 1 & 7 & 5 \\ 2 & 0 & 3 \\ 1 & 4 & 7 \end{bmatrix} + \det \begin{bmatrix} 1 & 7 & 5 \\ 2 & 0 & 3 \\ 0 & 1 & -1 \end{bmatrix} \qquad \blacklozenge$$

Determinant of a Matrix Product   When one considers the complexity of the definitions of matrix multiplication and determinants, it would seem unlikely that any simple relationship should exist between them. This is what makes the elegant simplicity of the following result so surprising: We will show that if A and B are square matrices of the same size, then

$$\det(AB) = \det(A)\det(B) \tag{2}$$

The proof of this theorem is fairly intricate, so we will have to develop some preliminary results first. We begin with the special case of (2) in which A is an elementary matrix. Because this special case is only a prelude to (2), we call it a lemma.

Lemma 2.3.2

If B is an $n \times n$ matrix and E is an $n \times n$ elementary matrix, then

$$\det(EB) = \det(E)\det(B)$$

Proof. We shall consider three cases, each depending on the row operation that produces matrix E.

Case 1. If E results from multiplying a row of I_n by k, then by Theorem 1.5.1 EB results from B by multiplying a row by k; so from Theorem 2.2.3a we have

$$\det(EB) = k\det(B)$$

But from Theorem 2.2.4a we have $\det(E) = k$, so

$$\det(EB) = \det(E)\det(B)$$

Cases 2 and 3. The proofs of the cases where E results from interchanging two rows of I_n or from adding a multiple of one row to another follow the same pattern as Case 1 and are left as exercises.   ∎

REMARK. It follows by repeated applications of Lemma 2.3.2 that if B is an $n \times n$ matrix and $E_1, E_2, \ldots, E_r$ are $n \times n$ elementary matrices, then

$$\det(E_1 E_2 \cdots E_r B) = \det(E_1)\det(E_2)\cdots\det(E_r)\det(B) \tag{3}$$

For example,

$$\det(E_1 E_2 B) = \det(E_1)\det(E_2 B) = \det(E_1)\det(E_2)\det(B)$$

Determinant Test for Invertibility   The next theorem is one of the most fundamental in linear algebra; it provides an important criterion for invertibility in terms of determinants, and it will be used in proving (2).

Theorem 2.3.3

A square matrix A is invertible if and only if $\det(A) \neq 0$.

Proof. Let R be the reduced row-echelon form of A. As a preliminary step, we will show that $\det(A)$ and $\det(R)$ are both zero or both nonzero: Let $E_1, E_2, \ldots, E_r$ be the elementary matrices that correspond to the elementary row operations that produce R from A. Thus,

$$R = E_r \cdots E_2 E_1 A$$

and from (3)

$$\det(R) = \det(E_r) \cdots \det(E_2) \det(E_1) \det(A) \tag{4}$$

But from Theorem 2.2.4 the determinants of the elementary matrices are all nonzero. (Keep in mind that multiplying a row by zero is *not* an allowable elementary row operation, so $k \neq 0$ in this application of Theorem 2.2.4.) Thus, it follows from (4) that $\det(A)$ and $\det(R)$ are both zero or both nonzero. Now to the main body of the proof.

If A is invertible, then by Theorem 1.6.4 we have $R = I$, so $\det(R) = 1 \neq 0$ and consequently $\det(A) \neq 0$. Conversely, if $\det(A) \neq 0$, then $\det(R) \neq 0$, so R cannot have a row of zeros. It follows from Theorem 1.4.3 that $R = I$, so A is invertible by Theorem 1.6.4. ∎

It follows from Theorems 2.3.3 and 2.2.5 that a square matrix with two proportional rows or columns is not invertible.

EXAMPLE 3 Determinant Test for Invertibility

Since the first and third rows of

$$A = \begin{bmatrix} 1 & 2 & 3 \\ 1 & 0 & 1 \\ 2 & 4 & 6 \end{bmatrix}$$

are proportional, $\det(A) = 0$. Thus, A is not invertible. ◆

We are now ready for the main result in this section.

Theorem 2.3.4

If A and B are square matrices of the same size, then

$$\det(AB) = \det(A) \det(B)$$

Proof. We divide the proof into two cases that depend on whether or not A is invertible. If the matrix A is not invertible, then by Theorem 1.6.5 neither is the product AB. Thus, from Theorem 2.3.3, we have $\det(AB) = 0$ and $\det(A) = 0$, so it follows that $\det(AB) = \det(A) \det(B)$.

Now assume that A is invertible. By Theorem 1.6.4, the matrix A is expressible as a product of elementary matrices, say

$$A = E_1 E_2 \cdots E_r \tag{5}$$

so
$$AB = E_1 E_2 \cdots E_r B$$

Applying (3) to this equation yields
$$\det(AB) = \det(E_1)\det(E_2)\cdots\det(E_r)\det(B)$$

and applying (3) again yields
$$\det(AB) = \det(E_1 E_2 \cdots E_r)\det(B)$$

which, from (5), can be written as $\det(AB) = \det(A)\det(B)$. ■

EXAMPLE 4 Verifying That det[AB] = det[A] det[B]

Consider the matrices
$$A = \begin{bmatrix} 3 & 1 \\ 2 & 1 \end{bmatrix}, \qquad B = \begin{bmatrix} -1 & 3 \\ 5 & 8 \end{bmatrix}, \qquad AB = \begin{bmatrix} 2 & 17 \\ 3 & 14 \end{bmatrix}$$

We leave it for the reader to verify that
$$\det(A) = 1, \quad \det(B) = -23, \quad \text{and} \quad \det(AB) = -23$$

Thus, $\det(AB) = \det(A)\det(B)$ as guaranteed by Theorem 2.3.4. ◆

The following theorem gives a useful relationship between the determinant of an invertible matrix and the determinant of its inverse.

Theorem 2.3.5

If A is invertible, then
$$\det(A^{-1}) = \frac{1}{\det(A)}$$

Proof. Since $A^{-1}A = I$, it follows that $\det(A^{-1}A) = \det(I)$. Therefore, we must have $\det(A^{-1})\det(A) = 1$. Since $\det(A) \neq 0$, the proof can be completed by dividing through by $\det(A)$. ■

Linear Systems of the Form $A\mathbf{x} = \lambda\mathbf{x}$ Many applications of linear algebra are concerned with systems of n linear equations in n unknowns that are expressed in the form
$$A\mathbf{x} = \lambda\mathbf{x} \tag{6}$$

where λ is a scalar. Such systems are really homogeneous linear systems in disguise, since (6) can be rewritten as $\lambda\mathbf{x} - A\mathbf{x} = \mathbf{0}$ or, by inserting an identity matrix and factoring, as
$$(\lambda I - A)\mathbf{x} = \mathbf{0} \tag{7}$$

Here is an example.

EXAMPLE 5 Finding $\lambda I - A$

The linear system
$$x_1 + 3x_2 = \lambda x_1$$
$$4x_1 + 2x_2 = \lambda x_2$$

can be written in matrix form as

$$\begin{bmatrix} 1 & 3 \\ 4 & 2 \end{bmatrix} \begin{bmatrix} x_1 \\ x_2 \end{bmatrix} = \lambda \begin{bmatrix} x_1 \\ x_2 \end{bmatrix}$$

which is of form (6) with

$$A = \begin{bmatrix} 1 & 3 \\ 4 & 2 \end{bmatrix} \quad \text{and} \quad \mathbf{x} = \begin{bmatrix} x_1 \\ x_2 \end{bmatrix}$$

This system can be rewritten as

$$\lambda \begin{bmatrix} x_1 \\ x_2 \end{bmatrix} - \begin{bmatrix} 1 & 3 \\ 4 & 2 \end{bmatrix} \begin{bmatrix} x_1 \\ x_2 \end{bmatrix} = \begin{bmatrix} 0 \\ 0 \end{bmatrix}$$

or

$$\lambda \begin{bmatrix} 1 & 0 \\ 0 & 1 \end{bmatrix} \begin{bmatrix} x_1 \\ x_2 \end{bmatrix} - \begin{bmatrix} 1 & 3 \\ 4 & 2 \end{bmatrix} \begin{bmatrix} x_1 \\ x_2 \end{bmatrix} = \begin{bmatrix} 0 \\ 0 \end{bmatrix}$$

or

$$\begin{bmatrix} \lambda - 1 & -3 \\ -4 & \lambda - 2 \end{bmatrix} \begin{bmatrix} x_1 \\ x_2 \end{bmatrix} = \begin{bmatrix} 0 \\ 0 \end{bmatrix}$$

which is of form (7) with

$$\lambda I - A = \begin{bmatrix} \lambda - 1 & -3 \\ -4 & \lambda - 2 \end{bmatrix} \qquad \blacklozenge$$

The primary problem of interest for linear systems of the form (7) is to determine those values of λ for which the system has a nontrivial solution; such a value of λ is called a ***characteristic value*** or an ***eigenvalue***[†] of A. If λ is an eigenvalue of A, then the nontrivial solutions of (7) are called the ***eigenvectors*** of A corresponding to λ.

It follows from Theorem 2.3.3 that the system $(\lambda I - A)\mathbf{x} = \mathbf{0}$ has a nontrivial solution if and only if

$$\det(\lambda I - A) = 0 \tag{8}$$

This is called the ***characteristic equation*** of A; the eigenvalues of A can be found by solving this equation for λ.

Eigenvalues and eigenvectors will be studied again in subsequent chapters, where we will discuss their geometric interpretation and develop their properties in more depth.

EXAMPLE 6 Eigenvalues and Eigenvectors

Find the eigenvalues and corresponding eigenvectors of the matrix A in Example 5.

Solution.

The characteristic equation of A is

$$\det(\lambda I - A) = \begin{vmatrix} \lambda - 1 & -3 \\ -4 & \lambda - 2 \end{vmatrix} = 0 \quad \text{or} \quad \lambda^2 - 3\lambda - 10 = 0$$

[†]The word *eigenvalue* is a mixture of German and English. The German prefix *eigen* can be translated as "proper," which stems from the older literature where eigenvalues were known as *proper values*; they were also called *latent roots*.

The factored form of this equation is $(\lambda + 2)(\lambda - 5) = 0$, so the eigenvalues of A are $\lambda = -2$ and $\lambda = 5$.

By definition

$$\mathbf{x} = \begin{bmatrix} x_1 \\ x_2 \end{bmatrix}$$

is an eigenvector of A if and only if $\mathbf{x}$ is a nontrivial solution of $(\lambda I - A)\mathbf{x} = \mathbf{0}$; that is,

$$\begin{bmatrix} \lambda - 1 & -3 \\ -4 & \lambda - 2 \end{bmatrix} \begin{bmatrix} x_1 \\ x_2 \end{bmatrix} = \begin{bmatrix} 0 \\ 0 \end{bmatrix} \tag{9}$$

If $\lambda = -2$, then (9) becomes

$$\begin{bmatrix} -3 & -3 \\ -4 & -4 \end{bmatrix} \begin{bmatrix} x_1 \\ x_2 \end{bmatrix} = \begin{bmatrix} 0 \\ 0 \end{bmatrix}$$

Solving this system yields (verify) $x_1 = -t$, $x_2 = t$, so the eigenvectors corresponding to $\lambda = -2$ are the nonzero solutions of the form

$$\mathbf{x} = \begin{bmatrix} x_1 \\ x_2 \end{bmatrix} = \begin{bmatrix} -t \\ t \end{bmatrix}$$

Again from (9), the eigenvectors of A corresponding to $\lambda = 5$ are the nontrivial solutions of

$$\begin{bmatrix} 4 & -3 \\ -4 & 3 \end{bmatrix} \begin{bmatrix} x_1 \\ x_2 \end{bmatrix} = \begin{bmatrix} 0 \\ 0 \end{bmatrix}$$

We leave it for the reader to solve this system and show that the eigenvectors of A corresponding to $\lambda = 5$ are the nonzero solutions of the form

$$\mathbf{x} = \begin{bmatrix} \frac{3}{4}t \\ t \end{bmatrix} \qquad \blacklozenge$$

Summary   In Theorem 1.6.4 we listed five results that are equivalent to the invertibility of a matrix A. We conclude this section by merging Theorem 2.3.3 with that list to produce the following theorem that relates all of the major topics we have studied thus far.

Theorem 2.3.6   **Equivalent Statements**

If A is an $n \times n$ matrix, then the following are equivalent.

(a)   *A is invertible.*
(b) $A\mathbf{x} = \mathbf{0}$ *has only the trivial solution.*
(c)   *The reduced row-echelon form of A is I_n.*
(d)   *A is expressible as a product of elementary matrices.*
(e) $A\mathbf{x} = \mathbf{b}$ *is consistent for every $n \times 1$ matrix $\mathbf{b}$.*
(f) $A\mathbf{x} = \mathbf{b}$ *has exactly one solution for every $n \times 1$ matrix $\mathbf{b}$.*
(g) $\det(A) \neq 0$.

Exercise Set 2.3

1. Verify that $\det(kA) = k^n \det(A)$ for

(a) $A = \begin{bmatrix} -1 & 2 \\ 3 & 4 \end{bmatrix}$; $k = 2$ (b) $A = \begin{bmatrix} 2 & -1 & 3 \\ 3 & 2 & 1 \\ 1 & 4 & 5 \end{bmatrix}$; $k = -2$

2. Verify that $\det(AB) = \det(A)\det(B)$ for

$$A = \begin{bmatrix} 2 & 1 & 0 \\ 3 & 4 & 0 \\ 0 & 0 & 2 \end{bmatrix} \quad \text{and} \quad B = \begin{bmatrix} 1 & -1 & 3 \\ 7 & 1 & 2 \\ 5 & 0 & 1 \end{bmatrix}$$

3. By inspection, explain why $\det(A) = 0$.

$$A = \begin{bmatrix} -2 & 8 & 1 & 4 \\ 3 & 2 & 5 & 1 \\ 1 & 10 & 6 & 5 \\ 4 & -6 & 4 & -3 \end{bmatrix}$$

4. Use Theorem 2.3.3 to determine which of the following matrices are invertible.

(a) $\begin{bmatrix} 1 & 0 & -1 \\ 9 & -1 & 4 \\ 8 & 9 & -1 \end{bmatrix}$ (b) $\begin{bmatrix} 4 & 2 & 8 \\ -2 & 1 & -4 \\ 3 & 1 & 6 \end{bmatrix}$ (c) $\begin{bmatrix} \sqrt{2} & -\sqrt{7} & 0 \\ 3\sqrt{2} & -3\sqrt{7} & 0 \\ 5 & -9 & 0 \end{bmatrix}$ (d) $\begin{bmatrix} -3 & 0 & 1 \\ 5 & 0 & 6 \\ 8 & 0 & 3 \end{bmatrix}$

5. Let

$$A = \begin{bmatrix} a & b & c \\ d & e & f \\ g & h & i \end{bmatrix}$$

Assuming that $\det(A) = -7$, find

(a) $\det(3A)$ (b) $\det(A^{-1})$ (c) $\det(2A^{-1})$ (d) $\det((2A)^{-1})$ (e) $\det \begin{bmatrix} a & g & d \\ b & h & e \\ c & i & f \end{bmatrix}$

6. Without directly evaluating, show that $x = 0$ and $x = 2$ satisfy

$$\begin{vmatrix} x^2 & x & 2 \\ 2 & 1 & 1 \\ 0 & 0 & -5 \end{vmatrix} = 0$$

7. Without directly evaluating, show that

$$\det \begin{bmatrix} b+c & c+a & b+a \\ a & b & c \\ 1 & 1 & 1 \end{bmatrix} = 0$$

In Exercises 8–11 prove the identity without evaluating the determinants.

8. $\begin{vmatrix} a_1 & b_1 & a_1 + b_1 + c_1 \\ a_2 & b_2 & a_2 + b_2 + c_2 \\ a_3 & b_3 & a_3 + b_3 + c_3 \end{vmatrix} = \begin{vmatrix} a_1 & b_1 & c_1 \\ a_2 & b_2 & c_2 \\ a_3 & b_3 & c_3 \end{vmatrix}$

9. $\begin{vmatrix} a_1 + b_1 & a_1 - b_1 & c_1 \\ a_2 + b_2 & a_2 - b_2 & c_2 \\ a_3 + b_3 & a_3 - b_3 & c_3 \end{vmatrix} = -2 \begin{vmatrix} a_1 & b_1 & c_1 \\ a_2 & b_2 & c_2 \\ a_3 & b_3 & c_3 \end{vmatrix}$

10. $\begin{vmatrix} a_1 + b_1t & a_2 + b_2t & a_3 + b_3t \\ a_1t + b_1 & a_2t + b_2 & a_3t + b_3 \\ c_1 & c_2 & c_3 \end{vmatrix} = (1 - t^2) \begin{vmatrix} a_1 & a_2 & a_3 \\ b_1 & b_2 & b_3 \\ c_1 & c_2 & c_3 \end{vmatrix}$

11. $\begin{vmatrix} a_1 & b_1 + ta_1 & c_1 + rb_1 + sa_1 \\ a_2 & b_2 + ta_2 & c_2 + rb_2 + sa_2 \\ a_3 & b_3 + ta_3 & c_3 + rb_3 + sa_3 \end{vmatrix} = \begin{vmatrix} a_1 & a_2 & a_3 \\ b_1 & b_2 & b_3 \\ c_1 & c_2 & c_3 \end{vmatrix}$

12. For which value(s) of k does A fail to be invertible?

(a) $A = \begin{bmatrix} k - 3 & -2 \\ -2 & k - 2 \end{bmatrix}$ (b) $A = \begin{bmatrix} 1 & 2 & 4 \\ 3 & 1 & 6 \\ k & 3 & 2 \end{bmatrix}$

13. Use Theorem 2.3.3 to show that

$$\begin{bmatrix} \sin^2 \alpha & \sin^2 \beta & \sin^2 \gamma \\ \cos^2 \alpha & \cos^2 \beta & \cos^2 \gamma \\ 1 & 1 & 1 \end{bmatrix}$$

is not invertible for any values of α, β, and γ.

14. Express the following linear systems in the form $(\lambda I - A)\mathbf{x} = \mathbf{0}$.

(a) $\begin{aligned} x_1 + 2x_2 &= \lambda x_1 \\ 2x_1 + x_2 &= \lambda x_2 \end{aligned}$ (b) $\begin{aligned} 2x_1 + 3x_2 &= \lambda x_1 \\ 4x_1 + 3x_2 &= \lambda x_2 \end{aligned}$ (c) $\begin{aligned} 3x_1 + x_2 &= \lambda x_1 \\ -5x_1 - 3x_2 &= \lambda x_2 \end{aligned}$

15. For each of the systems in Exercise 14, find

(i) the characteristic equation;
(ii) the eigenvalues;
(iii) the eigenvectors corresponding to each of the eigenvalues.

16. Let A and B be $n \times n$ matrices. Show that if A is invertible, then $\det(B) = \det(A^{-1}BA)$.

17. (a) Express

$$\begin{vmatrix} a_1 + b_1 & c_1 + d_1 \\ a_2 + b_2 & c_2 + d_2 \end{vmatrix}$$

as a sum of four determinants whose entries contain no sums.

(b) Express

$$\begin{vmatrix} a_1 + b_1 & c_1 + d_1 & e_1 + f_1 \\ a_2 + b_2 & c_2 + d_2 & e_2 + f_2 \\ a_3 + b_3 & c_3 + d_3 & e_3 + f_3 \end{vmatrix}$$

as a sum of eight determinants whose entries contain no sums.

18. Prove that a square matrix A is invertible if and only if $A^T A$ is invertible.

19. Prove Cases 2 and 3 of Lemma 2.3.2.

Discussion and Discovery

20. Let A and B be $n \times n$ matrices. You know from earlier work that AB and BA need not be equal. Is the same true for $\det(AB)$ and $\det(BA)$? Explain your reasoning.

21. Let A and B be $n \times n$ matrices. You know from earlier work that AB is invertible if A and B are invertible. What can you say about the invertibility of AB if one or both of the factors are singular? Explain your reasoning.

22. Indicate whether the statement is always true or sometimes false. Justify each answer by giving a logical argument or a counterexample.

    (a) $\det(2A) = 2\det(A)$

    (b) $|A^2| = |A|^2$

    (c) $\det(I + A) = 1 + \det(A)$

    (d) If $\det(A) = 0$, then the homogeneous system $A\mathbf{x} = \mathbf{0}$ has infinitely many solutions.

23. Indicate whether the statement is always true or sometimes false. Justify your answer by giving a logical argument or a counterexample.

    (a) If $\det(A) = 0$, then A is not expressible as a product of elementary matrices.

    (b) If the reduced row-echelon form of A has a row of zeros, then $\det(A) = 0$.

    (c) The determinant of a matrix is unchanged if the columns are written in reverse order.

    (d) There is no square matrix A such that $\det(AA^T) = -1$.

2.4 COFACTOR EXPANSION; CRAMER'S RULE

In this section we shall consider a method for evaluating determinants that is useful for hand computations and is also important theoretically. As a consequence of our work here, we will obtain a formula for the inverse of an invertible matrix as well as a formula for the solution to certain systems of linear equations in terms of determinants.

Minors and Cofactors   In Example 7 of Section 2.1 we saw that the determinant of a 3×3 matrix

$$A = \begin{bmatrix} a_{11} & a_{12} & a_{13} \\ a_{21} & a_{22} & a_{23} \\ a_{33} & a_{32} & a_{33} \end{bmatrix}$$

is the number

$$\det(A) = a_{11}a_{22}a_{33} + a_{12}a_{23}a_{31} + a_{13}a_{21}a_{32}$$
$$- a_{13}a_{22}a_{31} - a_{12}a_{21}a_{33} - a_{11}a_{23}a_{32} \tag{1}$$

By rearranging terms and factoring, (1) can be rewritten as

$$\det(A) = a_{11}(a_{22}a_{33} - a_{23}a_{32}) - a_{12}(a_{21}a_{33} - a_{23}a_{31}) + a_{13}(a_{21}a_{32} - a_{22}a_{31}) \tag{2}$$

The expressions highlighted in color in (2) are themselves determinants:

$$M_{11} = \begin{vmatrix} a_{22} & a_{23} \\ a_{32} & a_{33} \end{vmatrix}, \qquad M_{12} = \begin{vmatrix} a_{21} & a_{23} \\ a_{31} & a_{33} \end{vmatrix}, \qquad M_{13} = \begin{vmatrix} a_{21} & a_{22} \\ a_{31} & a_{32} \end{vmatrix}$$

The submatrices of A that appear in these determinants are given a special name:

Definition

If A is a square matrix, then the **_minor of entry_** a_{ij} is denoted by M_{ij} and is defined to be the determinant of the submatrix that remains after the ith row and jth column are deleted from A. The number $(-1)^{i+j}M_{ij}$ is denoted by C_{ij} and is called the **_cofactor of entry_** a_{ij}.

EXAMPLE 1 Finding Minors and Cofactors

Let

$$A = \begin{bmatrix} 3 & 1 & -4 \\ 2 & 5 & 6 \\ 1 & 4 & 8 \end{bmatrix}$$

The minor of entry a_{11} is

$$M_{11} = \begin{vmatrix} 3 & 1 & -4 \\ 2 & 5 & 6 \\ 1 & 4 & 8 \end{vmatrix} = \begin{vmatrix} 5 & 6 \\ 4 & 8 \end{vmatrix} = 16$$

The cofactor of a_{11} is

$$C_{11} = (-1)^{1+1} M_{11} = M_{11} = 16$$

Similarly, the minor of entry a_{32} is

$$M_{32} = \begin{vmatrix} 3 & 1 & -4 \\ 2 & 5 & 6 \\ 1 & 4 & 8 \end{vmatrix} = \begin{vmatrix} 3 & -4 \\ 2 & 6 \end{vmatrix} = 26$$

The cofactor of a_{32} is

$$C_{32} = (-1)^{3+2} M_{32} = -M_{32} = -26 \qquad \blacklozenge$$

Notice that the cofactor and the minor of an element a_{ij} differ only in sign, that is, $C_{ij} = \pm M_{ij}$. A quick way for determining whether to use the $+$ or $-$ is to use the fact that the sign relating C_{ij} and M_{ij} is in the ith row and jth column of the "checkerboard" array

$$\begin{bmatrix} + & - & + & - & + & \cdots \\ - & + & - & + & - & \cdots \\ + & - & + & - & | & \cdots \\ - & + & - & + & - & \cdots \\ \vdots & \vdots & \vdots & \vdots & \vdots & \end{bmatrix}$$

For example, $C_{11} = M_{11}$, $C_{21} = -M_{21}$, $C_{12} = -M_{12}$, $C_{22} = M_{22}$, and so on.

Cofactor Expansions   In view of the definition on the preceding page, the expression in (2) can be written in terms of minors and cofactors as

$$\det(A) = a_{11}M_{11} + a_{12}(-M_{12}) + a_{13}M_{13}$$
$$= a_{11}C_{11} + a_{12}C_{12} + a_{13}C_{13} \qquad (3)$$

Equation (3) shows that the determinant of A can be computed by multiplying the entries in the first row of A by their corresponding cofactors and adding the resulting products. This method of evaluating $\det(A)$ is called ***cofactor expansion*** along the first row of A.

EXAMPLE 2 Cofactor Expansion Along the First Row

Let $A = \begin{bmatrix} 3 & 1 & 0 \\ -2 & -4 & 3 \\ 5 & 4 & -2 \end{bmatrix}$. Evaluate $\det(A)$ by cofactor expansion along the first row of A.

Solution.

From (3)

$$\det(A) = \begin{vmatrix} 3 & 1 & 0 \\ -2 & -4 & 3 \\ 5 & 4 & -2 \end{vmatrix} = 3 \begin{vmatrix} -4 & 3 \\ 4 & -2 \end{vmatrix} - 1 \begin{vmatrix} -2 & 3 \\ 5 & -2 \end{vmatrix} + 0 \begin{vmatrix} -2 & -4 \\ 5 & 4 \end{vmatrix}$$

$$= 3(-4) - (1)(-11) + 0 = -1 \qquad \blacklozenge$$

By rearranging the terms in (1) in various ways, it is possible to obtain other formulas like (3). There should be no trouble checking that all of the following are correct (see Exercise 28):

$$\det(A) = a_{11}C_{11} + a_{12}C_{12} + a_{13}C_{13}$$
$$= a_{11}C_{11} + a_{21}C_{21} + a_{31}C_{31}$$
$$= a_{21}C_{21} + a_{22}C_{22} + a_{23}C_{23}$$
$$= a_{12}C_{12} + a_{22}C_{22} + a_{32}C_{32}$$
$$= a_{31}C_{31} + a_{32}C_{32} + a_{33}C_{33}$$
$$= a_{13}C_{13} + a_{23}C_{23} + a_{33}C_{33} \qquad (4)$$

Notice that in each equation the entries and cofactors all come from the same row or column. These equations are called the ***cofactor expansions*** of $\det(A)$.

The results we have just given for 3×3 matrices form a special case of the following general theorem, which we state without proof.

Theorem 2.4.1   **Expansions by Cofactors**

The determinant of an $n \times n$ matrix A can be computed by multiplying the entries in any row (or column) by their cofactors and adding the resulting products; that is, for each $1 \le i \le n$ and $1 \le j \le n$,

$$\det(A) = a_{1j}C_{1j} + a_{2j}C_{2j} + \cdots + a_{nj}C_{nj}$$

(cofactor expansion along the jth column)

and

$$\det(A) = a_{i1}C_{i1} + a_{i2}C_{i2} + \cdots + a_{in}C_{in}$$

(cofactor expansion along the ith row)

EXAMPLE 3 **Cofactor Expansion Along the First Column**

Let A be the matrix in Example 2. Evaluate $\det(A)$ by cofactor expansion along the first column of A.

Solution.

From (4)

$$\det(A) = \begin{vmatrix} 3 & 1 & 0 \\ -2 & -4 & 3 \\ 5 & 4 & -2 \end{vmatrix} = 3 \begin{vmatrix} -4 & 3 \\ 4 & -2 \end{vmatrix} - (-2) \begin{vmatrix} 1 & 0 \\ 4 & -2 \end{vmatrix} + 5 \begin{vmatrix} 1 & 0 \\ -4 & 3 \end{vmatrix}$$

$$= 3(-4) - (-2)(-2) + 5(3) = -1$$

This agrees with the result obtained in Example 2. $\blacklozenge$

REMARK. In this example we had to compute three cofactors, but in Example 2 we only had to compute two of them, since the third was multiplied by zero. In general, the best strategy for evaluating a determinant by cofactor expansion is to expand along a row or column having the largest number of zeros.

Cofactor expansion and row or column operations can sometimes be used in combination to provide an effective method for evaluating determinants. The following example illustrates this idea.

EXAMPLE 4 Row Operations and Cofactor Expansion

Evaluate $\det(A)$ where

$$A = \begin{bmatrix} 3 & 5 & -2 & 6 \\ 1 & 2 & -1 & 1 \\ 2 & 4 & 1 & 5 \\ 3 & 7 & 5 & 3 \end{bmatrix}$$

Solution.

By adding suitable multiples of the second row to the remaining rows, we obtain

$$\det(A) = \begin{vmatrix} 0 & -1 & 1 & 3 \\ 1 & 2 & -1 & 1 \\ 0 & 0 & 3 & 3 \\ 0 & 1 & 8 & 0 \end{vmatrix}$$

$$= - \begin{vmatrix} -1 & 1 & 3 \\ 0 & 3 & 3 \\ 1 & 8 & 0 \end{vmatrix} \quad \longleftarrow \text{Cofactor expansion along the first column}$$

$$= - \begin{vmatrix} -1 & 1 & 3 \\ 0 & 3 & 3 \\ 0 & 9 & 3 \end{vmatrix} \quad \longleftarrow \text{We added the first row to the third row.}$$

$$= -(-1) \begin{vmatrix} 3 & 3 \\ 9 & 3 \end{vmatrix} \quad \longleftarrow \text{Cofactor expansion along the first column}$$

$$= -18 \qquad \blacklozenge$$

Adjoint of a Matrix   In a cofactor expansion we compute $\det(A)$ by multiplying the entries in a row or column by their cofactors and adding the resulting products. It turns out that if one multiplies the entries in any row by the corresponding cofactors from a *different* row, the sum of these products is always zero. (This result also holds for columns.) Although we omit the general proof, the next example illustrates the idea of the proof in a special case.

EXAMPLE 5 Entries and Cofactors from Different Rows

Let

$$A = \begin{bmatrix} a_{11} & a_{12} & a_{13} \\ a_{21} & a_{22} & a_{23} \\ a_{31} & a_{32} & a_{33} \end{bmatrix}$$

Consider the quantity

$$a_{11}C_{31} + a_{12}C_{32} + a_{13}C_{33}$$

that is formed by multiplying the entries in the first row by the cofactors of the corresponding entries in the third row and adding the resulting products. We now show that this quantity is equal to zero by the following trick. Construct a new matrix A' by replacing the third row of A with another copy of the first row. Thus,

$$A' = \begin{bmatrix} a_{11} & a_{12} & a_{13} \\ a_{21} & a_{22} & a_{23} \\ a_{11} & a_{12} & a_{13} \end{bmatrix}$$

Let C'_{31}, C'_{32}, C'_{33} be the cofactors of the entries in the third row of A'. Since the first two rows of A and A' are the same, and since the computations of C_{31}, C_{32}, C_{33}, C'_{31}, C'_{32}, and C'_{33} involve only entries from the first two rows of A and A', it follows that

$$C_{31} = C'_{31}, \qquad C_{32} = C'_{32}, \qquad C_{33} = C'_{33}$$

Since A' has two identical rows,

$$\det(A') = 0 \tag{5}$$

On the other hand, evaluating $\det(A')$ by cofactor expansion along the third row gives

$$\det(A') = a_{11}C'_{31} + a_{12}C'_{32} + a_{13}C'_{33} = a_{11}C_{31} + a_{12}C_{32} + a_{13}C_{33} \tag{6}$$

From (5) and (6) we obtain

$$a_{11}C_{31} + a_{12}C_{32} + a_{13}C_{33} = 0 \qquad\qquad\blacklozenge$$

Definition

If A is any $n \times n$ matrix and C_{ij} is the cofactor of a_{ij}, then the matrix

$$\begin{bmatrix} C_{11} & C_{12} & \cdots & C_{1n} \\ C_{21} & C_{22} & \cdots & C_{2n} \\ \vdots & \vdots & & \vdots \\ C_{n1} & C_{n2} & \cdots & C_{nn} \end{bmatrix}$$

is called the ***matrix of cofactors from A***. The transpose of this matrix is called the ***adjoint of A*** and is denoted by $\text{adj}(A)$.

EXAMPLE 6 Adjoint of a 3 × 3 Matrix

Let

$$A = \begin{bmatrix} 3 & 2 & -1 \\ 1 & 6 & 3 \\ 2 & -4 & 0 \end{bmatrix}$$

The cofactors of A are

$$\begin{array}{lll} C_{11} = 12 & C_{12} = 6 & C_{13} = -16 \\ C_{21} = 4 & C_{22} = 2 & C_{23} = 16 \\ C_{31} = 12 & C_{32} = -10 & C_{33} = 16 \end{array}$$

so that the matrix of cofactors is

$$\begin{bmatrix} 12 & 6 & -16 \\ 4 & 2 & 16 \\ 12 & -10 & 16 \end{bmatrix}$$

and the adjoint of A is

$$\text{adj}(A) = \begin{bmatrix} 12 & 4 & 12 \\ 6 & 2 & -10 \\ -16 & 16 & 16 \end{bmatrix}$$ ◆

We are now in a position to derive a formula for the inverse of an invertible matrix.

Theorem 2.4.2   **Inverse of a Matrix Using Its Adjoint**

If A is an invertible matrix, then

$$A^{-1} = \frac{1}{\det(A)} \text{adj}(A) \tag{7}$$

Proof. We show first that

$$A \, \text{adj}(A) = \det(A)I$$

Consider the product

$$A \, \text{adj}(A) = \begin{bmatrix} a_{11} & a_{12} & \cdots & a_{1n} \\ a_{21} & a_{22} & \cdots & a_{2n} \\ \vdots & \vdots & & \vdots \\ a_{i1} & a_{i2} & \cdots & a_{in} \\ \vdots & \vdots & & \vdots \\ a_{n1} & a_{n2} & \cdots & a_{nn} \end{bmatrix} \begin{bmatrix} C_{11} & C_{21} & \cdots & C_{j1} & \cdots & C_{n1} \\ C_{12} & C_{22} & \cdots & C_{j2} & \cdots & C_{n2} \\ \vdots & \vdots & & \vdots & & \vdots \\ C_{1n} & C_{2n} & \cdots & C_{jn} & \cdots & C_{nn} \end{bmatrix}$$

The entry in the ith row and jth column of the product $A \, \text{adj}(A)$ is

$$a_{i1}C_{j1} + a_{i2}C_{j2} + \cdots + a_{in}C_{jn} \tag{8}$$

(see the shaded lines above).

If $i = j$, then (8) is the cofactor expansion of $\det(A)$ along the ith row of A (Theorem 2.4.1), and if $i \neq j$, then the a's and the cofactors come from different rows of A, so the value of (8) is zero. Therefore,

$$A \, \text{adj}(A) = \begin{bmatrix} \det(A) & 0 & \cdots & 0 \\ 0 & \det(A) & \cdots & 0 \\ \vdots & \vdots & & \vdots \\ 0 & 0 & \cdots & \det(A) \end{bmatrix} = \det(A)I \tag{9}$$

Since A is invertible, $\det(A) \neq 0$. Therefore, Equation (9) can be rewritten as

$$\frac{1}{\det(A)}[A \, \text{adj}(A)] = I \quad \text{or} \quad A\left[\frac{1}{\det(A)} \text{adj}(A)\right] = I$$

Multiplying both sides on the left by A^{-1} yields

$$A^{-1} = \frac{1}{\det(A)} \text{adj}(A)$$ ∎

EXAMPLE 7 Using the Adjoint to Find an Inverse Matrix

Use (7) to find the inverse of the matrix A in Example 6.

Solution.

The reader can check that $\det(A) = 64$. Thus,

$$A^{-1} = \frac{1}{\det(A)}\mathrm{adj}(A) = \frac{1}{64}\begin{bmatrix} 12 & 4 & 12 \\ 6 & 2 & -10 \\ -16 & 16 & 16 \end{bmatrix} = \begin{bmatrix} \frac{12}{64} & \frac{4}{64} & \frac{12}{64} \\ \frac{6}{64} & \frac{2}{64} & -\frac{10}{64} \\ -\frac{16}{64} & \frac{16}{64} & \frac{16}{64} \end{bmatrix} \qquad \blacklozenge$$

Applications of Formula (7)

Although the method in the preceding example is reasonable for inverting 3×3 matrices by hand, the inversion algorithm discussed in Section 1.5 is more efficient for larger matrices. It should be kept in mind, however, that the method of Section 1.5 is just a computational procedure, whereas Formula (7) is an actual formula for the inverse. As we shall now see, this formula is useful for deriving properties of the inverse.

In Section 1.7 we stated two results about inverses without proof.

- **Theorem 1.7.1c:** A triangular matrix is invertible if and only if its diagonal entries are all nonzero.

- **Theorem 1.7.1d:** The inverse of an invertible lower triangular matrix is lower triangular, and the inverse of an invertible upper triangular matrix is upper triangular.

We will now prove these results using the adjoint formula for the inverse.

Proof of Theorem 1.7.1c. Let $A = [a_{ij}]$ be a triangular matrix, so that its diagonal entries are

$$a_{11}, a_{22}, \ldots, a_{nn}$$

From Theorems 2.2.2 and 2.3.3, the matrix A is invertible if and only if

$$\det(A) = a_{11}a_{22}\cdots a_{nn} \neq 0$$

which is true if and only if the diagonal entries are all nonzero. $\blacklozenge$

We leave it as an exercise for the reader to use the adjoint formula for A^{-1} to show that if $A = [a_{ij}]$ is an invertible triangular matrix, then the successive diagonal entries of A^{-1} are

$$\frac{1}{a_{11}}, \frac{1}{a_{22}}, \ldots, \frac{1}{a_{nn}}$$

(See Example 3 of Section 1.7.)

Proof of Theorem 1.7.1d. We will prove the result for upper triangular matrices and leave the lower triangular case as an exercise. Assume that A is upper triangular and invertible. Since

$$A^{-1} = \frac{1}{\det(A)}\mathrm{adj}(A)$$

we can prove that A^{-1} is upper triangular by showing that $\mathrm{adj}(A)$ is upper triangular, or equivalently, that the matrix of cofactors is lower triangular. We can do this by showing that every cofactor C_{ij} with $i < j$ (i.e., above the main diagonal) is zero. Since

$$C_{ij} = (-1)^{i+j}M_{ij}$$

it suffices to show that each minor M_{ij} with $i < j$ is zero. For this purpose let B_{ij} be the matrix that results when the ith row and jth column of A are deleted, so that

$$M_{ij} = \det(B_{ij}) \tag{10}$$

From the assumption that $i < j$ it follows that B_{ij} is upper triangular (Exercise 32). Since A is upper triangular, its $(i+1)$-st row begins with at least i zeros. But the ith row of B_{ij} is the $(i+1)$-st row of A with the entry in the jth column removed. Since $i < j$, none of the first i zeros is removed by deleting the jth column; thus, the ith row of B_{ij} starts with at least i zeros, which implies that this row has a zero on the main diagonal. It now follows from Theorem 2.2.2 that $\det(B_{ij}) = 0$ and from (10) that $M_{ij} = 0$. ∎

Cramer's Rule

The next theorem provides a formula for the solution of certain linear systems of n equations in n unknowns. This formula, known as *Cramer's rule* is of marginal interest for computational purposes, but it is useful for studying the mathematical properties of a solution without the need for solving the system.

Gabriel Cramer (1704–1752) was a Swiss mathematician. Although Cramer does not rank with the great mathematicians of his time, his contributions as a disseminator of mathematical ideas have earned him a well-deserved place in the history of mathematics. Cramer traveled extensively and met many of the leading mathematicians of his day.

Cramer's most widely known work, *Introduction à l'analyse des lignes courbes algébriques* (1750), was a study and classification of algebraic curves; Cramer's rule appeared in the appendix. Although the rule bears his name, variations of the idea were formulated earlier by various mathematicians. However, Cramer's superior notation helped clarify and popularize the technique.

Overwork combined with a fall from a carriage led to his death at the age of 48. Cramer was apparently a good-natured and pleasant person with broad interests. He wrote on philosophy of law and government and the history of mathematics. He served in public office, participated in artillery and fortifications activities for the government, instructed workers on techniques of cathedral repair, and undertook excavations of cathedral archives. Cramer received numerous honors for his activities.

> ### Theorem 2.4.3   Cramer's Rule
>
> *If $A\mathbf{x} = \mathbf{b}$ is a system of n linear equations in n unknowns such that $\det(A) \neq 0$, then the system has a unique solution. This solution is*
>
> $$x_1 = \frac{\det(A_1)}{\det(A)}, \quad x_2 = \frac{\det(A_2)}{\det(A)}, \ldots, \quad x_n = \frac{\det(A_n)}{\det(A)}$$
>
> *where A_j is the matrix obtained by replacing the entries in the jth column of A by the entries in the matrix*
>
> $$\mathbf{b} = \begin{bmatrix} b_1 \\ b_2 \\ \vdots \\ b_n \end{bmatrix}$$

Proof. If $\det(A) \neq 0$, then A is invertible and, by Theorem 1.6.2, $\mathbf{x} = A^{-1}\mathbf{b}$ is the unique solution of $A\mathbf{x} = \mathbf{b}$. Therefore, by Theorem 2.4.2 we have

$$\mathbf{x} = A^{-1}\mathbf{b} = \frac{1}{\det(A)}\text{adj}(A)\mathbf{b} = \frac{1}{\det(A)}\begin{bmatrix} C_{11} & C_{21} & \cdots & C_{n1} \\ C_{12} & C_{22} & \cdots & C_{n2} \\ \vdots & \vdots & & \vdots \\ C_{1n} & C_{2n} & \cdots & C_{nn} \end{bmatrix}\begin{bmatrix} b_1 \\ b_2 \\ \vdots \\ b_n \end{bmatrix}$$

Multiplying the matrices out gives

$$\mathbf{x} = \frac{1}{\det(A)}\begin{bmatrix} b_1 C_{11} + b_2 C_{21} + \cdots + b_n C_{n1} \\ b_1 C_{12} + b_2 C_{22} + \cdots + b_n C_{n2} \\ \vdots \qquad \vdots \qquad\qquad \vdots \\ b_1 C_{1n} + b_2 C_{2n} + \cdots + b_n C_{nn} \end{bmatrix}$$

The entry in the jth row of $\mathbf{x}$ is therefore

$$x_j = \frac{b_1 C_{1j} + b_2 C_{2j} + \cdots + b_n C_{nj}}{\det(A)} \tag{11}$$

Now let

$$A_j = \begin{bmatrix} a_{11} & a_{12} & \cdots & a_{1j-1} & b_1 & a_{1j+1} & \cdots & a_{1n} \\ a_{21} & a_{22} & \cdots & a_{2j-1} & b_2 & a_{2j+1} & \cdots & a_{2n} \\ \vdots & \vdots & & \vdots & \vdots & \vdots & & \vdots \\ a_{n1} & a_{n2} & \cdots & a_{nj-1} & b_n & a_{nj+1} & \cdots & a_{nn} \end{bmatrix}$$

Since A_j differs from A only in the jth column, it follows that the cofactors of entries $b_1, b_2, \ldots, b_n$ in A_j are the same as the cofactors of the corresponding entries in the jth column of A. The cofactor expansion of $\det(A_j)$ along the jth column is therefore

$$\det(A_j) = b_1 C_{1j} + b_2 C_{2j} + \cdots + b_n C_{nj}$$

Substituting this result in (11) gives

$$x_j = \frac{\det(A_j)}{\det(A)} \qquad \blacksquare$$

EXAMPLE 8 Using Cramer's Rule to Solve a Linear System

Use Cramer's rule to solve

$$\begin{aligned} x_1 + + 2x_3 &= 6 \\ -3x_1 + 4x_2 + 6x_3 &= 30 \\ -x_1 - 2x_2 + 3x_3 &= 8 \end{aligned}$$

Solution.

$$A = \begin{bmatrix} 1 & 0 & 2 \\ -3 & 4 & 6 \\ -1 & -2 & 3 \end{bmatrix}, \qquad A_1 = \begin{bmatrix} 6 & 0 & 2 \\ 30 & 4 & 6 \\ 8 & -2 & 3 \end{bmatrix},$$

$$A_2 = \begin{bmatrix} 1 & 6 & 2 \\ -3 & 30 & 6 \\ -1 & 8 & 3 \end{bmatrix}, \qquad A_3 = \begin{bmatrix} 1 & 0 & 6 \\ -3 & 4 & 30 \\ -1 & -2 & 8 \end{bmatrix}$$

Therefore,

$$x_1 = \frac{\det(A_1)}{\det(A)} = \frac{-40}{44} = \frac{-10}{11}, \qquad x_2 = \frac{\det(A_2)}{\det(A)} = \frac{72}{44} = \frac{18}{11},$$

$$x_3 = \frac{\det(A_3)}{\det(A)} = \frac{152}{44} = \frac{38}{11} \qquad \blacklozenge$$

REMARK. To solve a system of n equations in n unknowns by Cramer's rule, it is necessary to evaluate $n + 1$ determinants of $n \times n$ matrices. For systems with more than three equations, Gaussian elimination is far more efficient, since it is only necessary to reduce one $n \times (n + 1)$ augmented matrix. However, Cramer's rule does give a formula for the solution if the determinant of the coefficient matrix is nonzero.

Exercise Set 2.4

1. Let

$$A = \begin{bmatrix} 1 & -2 & 3 \\ 6 & 7 & -1 \\ -3 & 1 & 4 \end{bmatrix}$$

(a) Find all the minors of A.   (b) Find all the cofactors.

2. Let

$$A = \begin{bmatrix} 4 & -1 & 1 & 6 \\ 0 & 0 & -3 & 3 \\ 4 & 1 & 0 & 14 \\ 4 & 1 & 3 & 2 \end{bmatrix}$$

Find

(a) M_{13} and C_{13} (b) M_{23} and C_{23} (c) M_{22} and C_{22} (d) M_{21} and C_{21}

3. Evaluate the determinant of the matrix in Exercise 1 by a cofactor expansion along

(a) the first row (b) the first column (c) the second row
(d) the second column (e) the third row (f) the third column

4. For the matrix in Exercise 1, find

(a) adj(A) (b) A^{-1} using Theorem 2.4.2

In Exercises 5–10 evaluate det(A) by a cofactor expansion along a row or column of your choice.

5. $A = \begin{bmatrix} -3 & 0 & 7 \\ 2 & 5 & 1 \\ -1 & 0 & 5 \end{bmatrix}$
6. $A = \begin{bmatrix} 3 & 3 & 1 \\ 1 & 0 & -4 \\ 1 & -3 & 5 \end{bmatrix}$
7. $A = \begin{bmatrix} 1 & k & k^2 \\ 1 & k & k^2 \\ 1 & k & k^2 \end{bmatrix}$

8. $A = \begin{bmatrix} k+1 & k-1 & 7 \\ 2 & k-3 & 4 \\ 5 & k+1 & k \end{bmatrix}$
9. $A = \begin{bmatrix} 3 & 3 & 0 & 5 \\ 2 & 2 & 0 & -2 \\ 4 & 1 & -3 & 0 \\ 2 & 10 & 3 & 2 \end{bmatrix}$
10. $A = \begin{bmatrix} 4 & 0 & 0 & 1 & 0 \\ 3 & 3 & 3 & -1 & 0 \\ 1 & 2 & 4 & 2 & 3 \\ 9 & 4 & 6 & 2 & 3 \\ 2 & 2 & 4 & 2 & 3 \end{bmatrix}$

In Exercises 11–14 find A^{-1} using Theorem 2.4.2.

11. $A = \begin{bmatrix} 2 & 5 & 5 \\ -1 & -1 & 0 \\ 2 & 4 & 3 \end{bmatrix}$
12. $A = \begin{bmatrix} 2 & 0 & 3 \\ 0 & 3 & 2 \\ -2 & 0 & -4 \end{bmatrix}$

13. $A = \begin{bmatrix} 2 & -3 & 5 \\ 0 & 1 & -3 \\ 0 & 0 & 2 \end{bmatrix}$
14. $A = \begin{bmatrix} 2 & 0 & 0 \\ 8 & 1 & 0 \\ -5 & 3 & 6 \end{bmatrix}$

15. Let

$$A = \begin{bmatrix} 1 & 3 & 1 & 1 \\ 2 & 5 & 2 & 2 \\ 1 & 3 & 8 & 9 \\ 1 & 3 & 2 & 2 \end{bmatrix}$$

(a) Evaluate A^{-1} using Theorem 2.4.2.
(b) Evaluate A^{-1} using the method of Example 4 in Section 1.5.
(c) Which method involves less computation?

In Exercises 16–21 solve by Cramer's rule, where it applies.

16. $7x_1 - 2x_2 = 3$
$3x_1 + x_2 = 5$

17. $4x + 5y = 2$
$11x + y + 2z = 3$
$x + 5y + 2z = 1$

18. $x - 4y + z = 6$
$4x - y + 2z = -1$
$2x + 2y - 3z = -20$

19.
$x_1 - 3x_2 + x_3 = 4$
$2x_1 - x_2 = -2$
$4x_1 - 3x_3 = 0$

20. $-x_1 - 4x_2 + 2x_3 + x_4 = -32$
$2x_1 - x_2 + 7x_3 + 9x_4 = 14$
$-x_1 + x_2 + 3x_3 + x_4 = 11$
$x_1 - 2x_2 + x_3 - 4x_4 = -4$

21. $3x_1 - x_2 + x_3 = 4$
$-x_1 + 7x_2 - 2x_3 = 1$
$2x_1 + 6x_2 - x_3 = 5$

22. Show that the matrix

$$A = \begin{bmatrix} \cos\theta & \sin\theta & 0 \\ -\sin\theta & \cos\theta & 0 \\ 0 & 0 & 1 \end{bmatrix}$$

is invertible for all values of θ; then find A^{-1} using Theorem 2.4.2.

23. Use Cramer's rule to solve for y without solving for x, z, and w.

$$\begin{aligned} 4x + y + z + w &= 6 \\ 3x + 7y - z + w &= 1 \\ 7x + 3y - 5z + 8w &= -3 \\ x + y + \cdot z + 2w &= 3 \end{aligned}$$

24. Let $A\mathbf{x} = \mathbf{b}$ be the system in Exercise 23.

(a) Solve by Cramer's rule. (b) Solve by Gauss–Jordan elimination.
(c) Which method involves fewer computations?

25. Prove that if $\det(A) = 1$ and all the entries in A are integers, then all the entries in A^{-1} are integers.

26. Let $A\mathbf{x} = \mathbf{b}$ be a system of n linear equations in n unknowns with integer coefficients and integer constants. Prove that if $\det(A) = 1$, the solution $\mathbf{x}$ has integer entries.

27. Prove that if A is an invertible lower triangular matrix, then A^{-1} is lower triangular.

28. Derive the last cofactor expansion listed in Formula (4).

29. Prove: The equation of the line through the distinct points (a_1, b_1) and (a_2, b_2) can be written as

$$\begin{vmatrix} x & y & 1 \\ a_1 & b_1 & 1 \\ a_2 & b_2 & 1 \end{vmatrix} = 0$$

30. Prove: (x_1, y_1), (x_2, y_2), and (x_3, y_3) are collinear points if and only if

$$\begin{vmatrix} x_1 & y_1 & 1 \\ x_2 & y_2 & 1 \\ x_3 & y_3 & 1 \end{vmatrix} = 0$$

31. (a) If $A = \left[\begin{array}{c|c} A_{11} & A_{12} \\ \hline 0 & A_{22} \end{array} \right]$ is an "upper triangular" block matrix, where A_{11} and A_{22} are square

matrices, then $\det(A) = \det(A_{11})\det(A_{22})$. Use this result to evaluate $\det(A)$ for

$$\left[\begin{array}{cc|ccc} 2 & -1 & 2 & 5 & 6 \\ 4 & 3 & -1 & 3 & 4 \\ \hline 0 & 0 & 1 & 3 & 5 \\ 0 & 0 & -2 & 6 & 2 \\ 0 & 0 & 3 & 5 & 2 \end{array} \right]$$

(b) Verify your answer in part (a) by using a cofactor expansion to evaluate $\det(A)$.

32. Prove that if A is upper triangular and B_{ij} is the matrix that results when the ith row and jth column of A are deleted, then B_{ij} is upper triangular if $i < j$.

Discussion and Discovery

33. What is the maximum number of zeros that a 4×4 matrix can have without having a zero determinant? Explain your reasoning.

34. Let A be a matrix of the form

$$A = \begin{bmatrix} * & * & 0 & 0 & 0 \\ * & * & 0 & 0 & 0 \\ * & * & 0 & 0 & 0 \\ * & * & * & * & * \\ * & * & * & * & * \end{bmatrix}$$

How many different values can you obtain for $\det(A)$ by substituting numerical values (not necessarily all the same) for the $*$'s. Explain your reasoning.

35. Indicate whether the statement is always true or sometimes false. Justify your answer by giving a logical argument or a counterexample.

(a) $A \operatorname{adj}(A)$ is a diagonal matrix for every square matrix A.

(b) In theory, Cramer's rule can be used to solve any system of linear equations, though the amount of computation may be enormous.

(c) If A is invertible, then $\operatorname{adj}(A)$ must also be invertible.

(d) If A has a row of zeros, then so does $\operatorname{adj}(A)$.

Chapter 2 Supplementary Exercises

1. Use Cramer's rule to solve for x' and y' in terms of x and y.

$$x = \tfrac{3}{5}x' - \tfrac{4}{5}y'$$
$$y = \tfrac{4}{5}x' + \tfrac{3}{5}y'$$

2. Use Cramer's rule to solve for x' and y' in terms of x and y.

$$x = x'\cos\theta - y'\sin\theta$$
$$y = x'\sin\theta + y'\cos\theta$$

3. By examining the determinant of the coefficient matrix, show that the following system has a nontrivial solution if and only if $\alpha = \beta$.

$$\begin{aligned} x + \quad y + \alpha z &= 0 \\ x + \quad y + \beta z &= 0 \\ \alpha x + \beta y + \quad z &= 0 \end{aligned}$$

4. Let A be a 3×3 matrix, each of whose entries is 1 or 0. What is the largest possible value for $\det(A)$?

5. (a) For the triangle in the accompanying figure, use trigonometry to show that

$$b\cos\gamma + c\cos\beta = a$$
$$c\cos\alpha + a\cos\gamma = b$$
$$a\cos\beta + b\cos\alpha = c$$

and then apply Cramer's rule to show that

$$\cos\alpha = \frac{b^2 + c^2 - a^2}{2bc}$$

(b) Use Cramer's rule to obtain similar formulas for $\cos\beta$ and $\cos\gamma$.

Figure Ex-5

6. Use determinants to show that for all real values of λ the only solution of

$$x - 2y = \lambda x$$
$$x - \ \ y = \lambda y$$

is $x = 0$, $y = 0$.

7. Prove: If A is invertible, then $\mathrm{adj}(A)$ is invertible and

$$[\mathrm{adj}(A)]^{-1} = \frac{1}{\det(A)}A = \mathrm{adj}(A^{-1})$$

8. Prove: If A is an $n \times n$ matrix, then $\det[\mathrm{adj}(A)] = [\det(A)]^{n-1}$.

9. **(For readers who have studied calculus.)** Show that if $f_1(x)$, $f_2(x)$, $g_1(x)$, and $g_2(x)$ are differentiable functions, and if

$$W = \begin{vmatrix} f_1(x) & f_2(x) \\ g_1(x) & g_2(x) \end{vmatrix}, \quad \text{then} \quad \frac{dW}{dx} = \begin{vmatrix} f_1'(x) & f_2'(x) \\ g_1(x) & g_2(x) \end{vmatrix} + \begin{vmatrix} f_1(x) & f_2(x) \\ g_1'(x) & g_2'(x) \end{vmatrix}$$

10. (a) In the accompanying figure, the area of the triangle ABC is expressible as

$$\text{area } ABC = \text{area } ADEC + \text{area } CEFB - \text{area } ADFB$$

Use this and the fact that the area of a trapezoid equals $\frac{1}{2}$ the altitude times the sum of the parallel sides to show that

$$\text{area } ABC = \frac{1}{2}\begin{vmatrix} x_1 & y_1 & 1 \\ x_2 & y_2 & 1 \\ x_3 & y_3 & 1 \end{vmatrix}$$

[**Note.** In the derivation of this formula, the vertices are labeled so that the triangle is traced counterclockwise proceeding from (x_1, y_1) to (x_2, y_2) to (x_3, y_3). For a clockwise orientation, the determinant above yields the *negative* of the area.]

(b) Use the result in (a) to find the area of the triangle with vertices $(3, 3)$, $(4, 0)$, $(-2, -1)$.

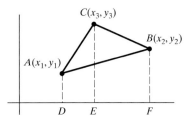

Figure Ex-10

11. Prove: If the entries in each row of an $n \times n$ matrix A add up to zero, then the determinant of A is zero. [**Hint.** Consider the product AX, where X is the $n \times 1$ matrix, each of whose entries is one.]

12. Let A be an $n \times n$ matrix and B the matrix that results when the rows of A are written in reverse order (last row becomes the first, and so forth). How are $\det(A)$ and $\det(B)$ related?

13. How will A^{-1} be affected if

(a) the ith and jth rows of A are interchanged;

(b) the ith row of A is multiplied by a nonzero scalar, c;

(c) c times the ith row of A is added to the jth row?

14. Let A be an $n \times n$ matrix. Suppose that B_1 is obtained by adding the same number t to each entry in the ith row of A and B_2 is obtained by subtracting t from each entry in the ith row of A. Show that $\det(A) = \frac{1}{2}[\det(B_1) + \det(B_2)]$.

15. Let

$$A = \begin{bmatrix} a_{11} & a_{12} & a_{13} \\ a_{21} & a_{22} & a_{23} \\ a_{31} & a_{32} & a_{33} \end{bmatrix}$$

(a) Express $\det(\lambda I - A)$ as a polynomial $p(\lambda) = \lambda^3 + b\lambda^2 + c\lambda + d$.

(b) Express the coefficients b and d in terms of determinants and traces.

16. Without directly evaluating the determinant, show that

$$\begin{vmatrix} \sin\alpha & \cos\alpha & \sin(\alpha + \delta) \\ \sin\beta & \cos\beta & \sin(\beta + \delta) \\ \sin\gamma & \cos\gamma & \sin(\gamma + \delta) \end{vmatrix} = 0$$

17. Use the fact that 21,375, 38,798, 34,162, 40,223, and 79,154 are all divisible by 19 to show that

$$\begin{vmatrix} 2 & 1 & 3 & 7 & 5 \\ 3 & 8 & 7 & 9 & 8 \\ 3 & 4 & 1 & 6 & 2 \\ 4 & 0 & 2 & 2 & 3 \\ 7 & 9 & 1 & 5 & 4 \end{vmatrix}$$

is divisible by 19 without directly evaluating the determinant.

18. Find the eigenvalues and corresponding eigenvectors for each of the following systems.

(a)
$$\begin{aligned} x_2 + 9x_3 &= \lambda x_1 \\ x_1 + 4x_2 - 7x_3 &= \lambda x_2 \\ x_1 \qquad\quad - 3x_3 &= \lambda x_3 \end{aligned}$$

(b)
$$\begin{aligned} x_2 + x_3 &= \lambda x_1 \\ x_1 \qquad - x_3 &= \lambda x_2 \\ x_1 + 5x_2 + 3x_3 &= \lambda x_3 \end{aligned}$$

Chapter 2 Technology Exercises

The following exercises are designed to be solved using a technology utility. Typically, this will be MATLAB, *Mathematica*, Maple, Derive, or Mathcad, but it may also be some other type of linear algebra software or a scientific calculator with some linear algebra capabilities. For each exercise you will need to read the relevant documentation for the particular utility you are using. The goal of these exercises is to provide you with a basic proficiency with your technology utility. Once you have mastered the techniques in these exercises, you will be able to use your technology utility to solve many of the problems in the regular exercise sets.

Section 2.1

T1. (*Determinants*) Read your documentation on how to compute determinants, and then compute the determinants in Example 8.

T2. (*Determinant formulas*) If you are working with a CAS, use it to confirm the formulas in Example 7. Also, use it to obtain the formula requested in Exercise 15 of Section 2.1.

T3. (*Simplifcation*) If you are working with a CAS, read the documentation on simplifying algebraic expressions and then use the determinant and simplification commands in combination to show that

$$\begin{vmatrix} a & b & c & d \\ -b & a & d & -c \\ -c & -d & a & b \\ -d & c & -b & a \end{vmatrix} = (a^2 + b^2 + c^2 + d^2)^2$$

T4. Use the method of Exercise T3 to find a simple formula for the determinant

$$\begin{vmatrix} (a+b)^2 & c^2 & c^2 \\ a^2 & (b+c)^2 & a^2 \\ b^2 & b^2 & (c+a)^2 \end{vmatrix}$$

Section 2.2

T1. (*Determinant of a transpose*) Confirm part (*b*) of Theorem 2.2.1 using some matrices of your choice.

Section 2.3

T1. (*Determinant of a product*) Confirm Theorem 2.3.4 for some matrices of your choice.

T2. (*Determinant of an inverse*) Confirm Theorem 2.3.5 for some matrices of your choice.

T3. (*Characteristic equation*) If you are working with a CAS, use it to find the characteristic equation of the matrix A in Example 6. Also, read your documentation on how to solve equations, and then solve the equation $\det(\lambda I - A) = 0$ for the eigenvalues of A.

Section 2.4

T1. (*Minors, cofactors, and adjoints*) Technology utilities vary widely on their treatment of minors, cofactors, and adjoints. For example, some utilities have commands for computing minors but not cofactors, and some provide direct commands for finding adjoints while others do not. Thus, depending on your utility, you may have to piece together commands or do some sign adjustment by hand to find cofactors and adjoints. Read your documentation, and then find the adjoint of the matrix A in Example 6.

Vectors in 2-Space and 3-Space

3

Chapter Contents

INTRODUCTION: Many physical quantities, such as area, length, mass, and temperature, are completely described once the magnitude of the quantity is given. Such quantities are called *scalars*. Other physical quantities are not completely determined until both a magnitude and a direction are specified. These quantities are called *vectors*. For example, wind movement is usually described by giving the speed and direction, say 20 mph northeast. The wind speed and wind direction form a vector called the wind *velocity*. Other examples of vectors are *force* and *displacement*. In this chapter our goal is to review some of the basic theory of vectors in two and three dimensions.

Note. Readers already familiar with the contents of this chapter can go to Chapter 4 with no loss of continuity.

3.1 INTRODUCTION TO VECTORS (GEOMETRIC)

In this section vectors in 2-space and 3-space will be introduced geometrically, arithmetic operations on vectors will be defined, and some basic properties of these arithmetic operations will be established.

(a) The vector $\overrightarrow{AB}$

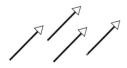

(b) Equivalent vectors

Figure 3.1.1

Geometric Vectors   Vectors can be represented geometrically as directed line segments or arrows in 2-space or 3-space. The direction of the arrow specifies the direction of the vector, and the length of the arrow describes its magnitude. The tail of the arrow is called the *initial point* of the vector, and the tip of the arrow the *terminal point*. Symbolically, we shall denote vectors in lowercase boldface type (for instance, **a**, **k**, **v**, **w**, and **x**). When discussing vectors, we shall refer to numbers as *scalars*. For now, all our scalars will be real numbers and will be denoted in lowercase italic type (for instance, a, k, v, w, and x).

If, as in Figure 3.1.1*a*, the initial point of a vector **v** is A and the terminal point is B, we write

$$\mathbf{v} = \overrightarrow{AB}$$

Vectors with the same length and same direction, such as those in Figure 3.1.1*b*, are called *equivalent*. Since we want a vector to be determined solely by its length and direction, equivalent vectors are regarded as *equal* even though they may be located in different positions. If **v** and **w** are equivalent, we write

$$\mathbf{v} = \mathbf{w}$$

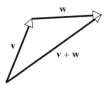

(a) The sum **v** + **w**

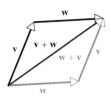

(b) **v** + **w** = **w** + **v**

Figure 3.1.2

> ### Definition
>
> If **v** and **w** are any two vectors, then the *sum* **v** + **w** is the vector determined as follows: Position the vector **w** so that its initial point coincides with the terminal point of **v**. The vector **v** + **w** is represented by the arrow from the initial point of **v** to the terminal point of **w** (Figure 3.1.2*a*).

In Figure 3.1.2*b* we have constructed two sums, **v** + **w** (color arrows) and **w** + **v** (gray arrows). It is evident that

$$\mathbf{v} + \mathbf{w} = \mathbf{w} + \mathbf{v}$$

and that the sum coincides with the diagonal of the parallelogram determined by **v** and **w** when these vectors are positioned so they have the same initial point.

The vector of length zero is called the *zero vector* and is denoted by **0**. We define

$$\mathbf{0} + \mathbf{v} = \mathbf{v} + \mathbf{0} = \mathbf{v}$$

Figure 3.1.3
The negative of
v has the same
length as **v**, but
is oppositely
directed.

for every vector **v**. Since there is no natural direction for the zero vector, we shall agree that it can be assigned any direction that is convenient for the problem being considered. If **v** is any nonzero vector, then −**v**, the *negative* of **v**, is defined to be the vector having the same magnitude as **v**, but oppositely directed (Figure 3.1.3). This vector has the property

$$\mathbf{v} + (-\mathbf{v}) = \mathbf{0}$$

(Why?) In addition, we define −**0** = **0**. Subtraction of vectors is defined as follows.

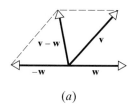

(a)

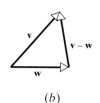

(b)

Figure 3.1.4

> ### Definition
>
> If **v** and **w** are any two vectors, then the ***difference*** of **w** from **v** is defined by
>
> $$\mathbf{v} - \mathbf{w} = \mathbf{v} + (-\mathbf{w})$$
>
> (Figure 3.1.4*a*).

To obtain the difference **v** − **w** without constructing −**w**, position **v** and **w** so their initial points coincide; the vector from the terminal point of **w** to the terminal point of **v** is then the vector **v** − **w** (Figure 3.1.4*b*).

> ### Definition
>
> If **v** is a nonzero vector and k is a nonzero real number (scalar), then the ***product*** $k\mathbf{v}$ is defined to be the vector whose length is $|k|$ times the length of **v** and whose direction is the same as that of **v** if $k > 0$ and opposite to that of **v** if $k < 0$. We define $k\mathbf{v} = \mathbf{0}$ if $k = 0$ or $\mathbf{v} = \mathbf{0}$.

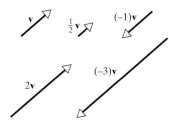

Figure 3.1.5

Figure 3.1.5 illustrates the relation between a vector **v** and the vectors $\frac{1}{2}\mathbf{v}$, $(-1)\mathbf{v}$, $2\mathbf{v}$, and $(-3)\mathbf{v}$. Note that the vector $(-1)\mathbf{v}$ has the same length as **v**, but is oppositely directed. Thus, $(-1)\mathbf{v}$ is just the negative of **v**; that is,

$$(-1)\mathbf{v} = -\mathbf{v}$$

A vector of the form $k\mathbf{v}$ is called a ***scalar multiple*** of **v**. As evidenced by Figure 3.1.5, vectors that are scalar multiples of each other are parallel. Conversely, it can be shown that nonzero parallel vectors are scalar multiples of each other. We omit the proof.

Vectors in Coordinate Systems

Problems involving vectors can often be simplified by introducing a rectangular coordinate system. For the moment we shall restrict the discussion to vectors in 2-space (the plane). Let **v** be any vector in the plane, and assume, as in Figure 3.1.6, that **v** has been positioned so its initial point is at the origin of a rectangular coordinate system. The coordinates (v_1, v_2) of the terminal point of **v** are called the ***components of*** **v**, and we write

$$\mathbf{v} = (v_1, v_2)$$

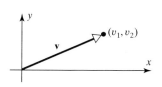

Figure 3.1.6

v_1 and v_2 are the components of **v**.

If equivalent vectors, **v** and **w**, are located so their initial points fall at the origin, then it is obvious that their terminal points must coincide (since the vectors have the same length and direction); thus, the vectors have the same components. Conversely, vectors with the same components are equivalent since they have the same length and same direction. In summary, two vectors

$$\mathbf{v} = (v_1, v_2) \quad \text{and} \quad \mathbf{w} = (w_1, w_2)$$

are equivalent if and only if

$$v_1 = w_1 \quad \text{and} \quad v_2 = w_2$$

The operations of vector addition and multiplication by scalars are easy to carry out in terms of components. As illustrated in Figure 3.1.7, if

$$\mathbf{v} = (v_1, v_2) \quad \text{and} \quad \mathbf{w} = (w_1, w_2)$$

then

$$\mathbf{v} + \mathbf{w} = (v_1 + w_1, v_2 + w_2) \tag{1}$$

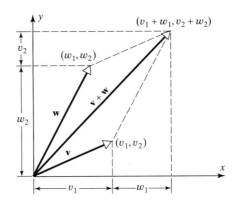

Figure 3.1.7

If $\mathbf{v} = (v_1, v_2)$ and k is any scalar, then by using a geometric argument involving similar triangles, it can be shown (Exercise 15) that

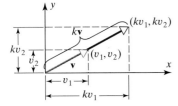

Figure 3.1.8

$$kv = (kv_1, kv_2) \qquad (2)$$

(Figure 3.1.8). Thus, for example, if $\mathbf{v} = (1, -2)$ and $\mathbf{w} = (7, 6)$, then

$$\mathbf{v} + \mathbf{w} = (1, -2) + (7, 6) = (1 + 7, -2 + 6) = (8, 4)$$

and

$$4\mathbf{v} = 4(1, -2) = (4(1), 4(-2)) = (4, -8)$$

Since $\mathbf{v} - \mathbf{w} = \mathbf{v} + (-1)\mathbf{w}$, it follows from Formulas (1) and (2) that

$$\mathbf{v} - \mathbf{w} = (v_1 - w_1, v_2 - w_2)$$

(Verify.)

Vectors in 3-Space   Just as vectors in the plane can be described by pairs of real numbers, vectors in 3-space can be described by triples of real numbers by introducing a *rectangular coordinate* system. To construct such a coordinate system, select a point O, called the *origin*, and choose three mutually perpendicular lines, called *coordinate axes*, passing through the origin. Label these axes x, y, and z, and select a positive direction for each coordinate axis as well as a unit of length for measuring distances (Figure 3.1.9a). Each pair of coordinate axes determines a plane called a *coordinate plane*. These are referred to as the *xy-plane*, the *xz-plane*, and the *yz-plane*. To each point P in 3-space we assign a triple of numbers (x, y, z), called the *coordinates of P*, as follows: Pass three planes through P parallel to the coordinate planes, and denote

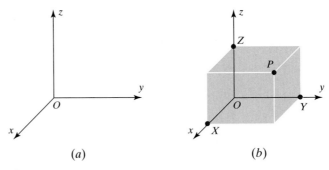

(a)   (b)

Figure 3.1.9

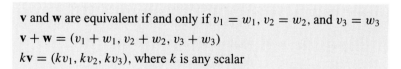

Figure 3.1.10

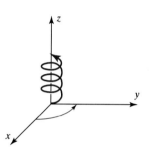

(a) Right-handed

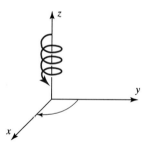

(b) Left-handed

Figure 3.1.11

the points of intersections of these planes with the three coordinate axes by X, Y, and Z (Figure 3.1.9b).

The coordinates of P are defined to be the signed lengths

$$x = OX, \qquad y = OY, \qquad z = OZ$$

In Figure 3.1.10a we have constructed the point whose coordinates are $(4, 5, 6)$ and in Figure 3.1.10b the point whose coordinates are $(-3, 2, -4)$.

Rectangular coordinate systems in 3-space fall into two categories, **_left-handed_** and **_right-handed_**. A right-handed system has the property that an ordinary screw pointed in the positive direction on the z-axis would be advanced if the positive x-axis is rotated $90°$ toward the positive y-axis (Figure 3.1.11a); the system is left-handed if the screw would be retracted (Figure 3.1.11b).

REMARK.   In this book we shall use only right-handed coordinate systems.

If, as in Figure 3.1.12, a vector $\mathbf{v}$ in 3-space is positioned so its initial point is at the origin of a rectangular coordinate system, then the coordinates of the terminal point are called the **_components_** of $\mathbf{v}$, and we write

$$\mathbf{v} = (v_1, v_2, v_3)$$

If $\mathbf{v} = (v_1, v_2, v_3)$ and $\mathbf{w} = (w_1, w_2, w_3)$ are two vectors in 3-space, then arguments similar to those used for vectors in a plane can be used to establish the following results.

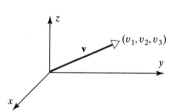

Figure 3.1.12

> $\mathbf{v}$ and $\mathbf{w}$ are equivalent if and only if $v_1 = w_1$, $v_2 = w_2$, and $v_3 = w_3$
>
> $\mathbf{v} + \mathbf{w} = (v_1 + w_1, v_2 + w_2, v_3 + w_3)$
>
> $k\mathbf{v} = (kv_1, kv_2, kv_3)$, where k is any scalar

EXAMPLE 1 Vector Computations with Components

If $\mathbf{v} = (1, -3, 2)$ and $\mathbf{w} = (4, 2, 1)$, then

$$\mathbf{v} + \mathbf{w} = (5, -1, 3), \qquad 2\mathbf{v} = (2, -6, 4), \qquad -\mathbf{w} = (-4, -2, -1),$$
$$\mathbf{v} - \mathbf{w} = \mathbf{v} + (-\mathbf{w}) = (-3, -5, 1) \qquad \blacklozenge$$

Sometimes a vector is positioned so that its initial point is not at the origin. If the vector $\overrightarrow{P_1 P_2}$ has initial point $P_1(x_1, y_1, z_1)$ and terminal point $P_2(x_2, y_2, z_2)$, then

$$\overrightarrow{P_1P_2} = (x_2 - x_1, y_2 - y_1, z_2 - z_1)$$

That is, the components of $\overrightarrow{P_1P_2}$ are obtained by subtracting the coordinates of the initial point from the coordinates of the terminal point. This may be seen using Figure 3.1.13: The vector $\overrightarrow{P_1P_2}$ is the difference of vectors $\overrightarrow{OP_2}$ and $\overrightarrow{OP_1}$, so

$$\overrightarrow{P_1P_2} = \overrightarrow{OP_2} - \overrightarrow{OP_1} = (x_2, y_2, z_2) - (x_1, y_1, z_1) = (x_2 - x_1, y_2 - y_1, z_2 - z_1)$$

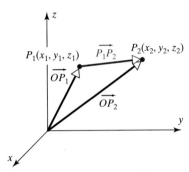

Figure 3.1.13

EXAMPLE 2 Finding the Components of a Vector

The components of the vector $\mathbf{v} = \overrightarrow{P_1P_2}$ with initial point $P_1(2, -1, 4)$ and terminal point $P_2(7, 5, -8)$ are

$$\mathbf{v} = (7 - 2, 5 - (-1), (-8) - 4) = (5, 6, -12) \qquad \blacklozenge$$

In 2-space the vector with initial point $P_1(x_1, y_1)$ and terminal point $P_2(x_2, y_2)$ is

$$\overrightarrow{P_1P_2} = (x_2 - x_1, y_2 - y_1)$$

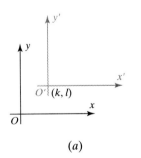

(a)

Translation of Axes The solutions to many problems can be simplified by translating the coordinate axes to obtain new axes parallel to the original ones.

In Figure 3.1.14a we have translated the axes of an xy-coordinate system to obtain an $x'y'$-coordinate system whose origin O' is at the point $(x, y) = (k, l)$. A point P in 2-space now has both (x, y) coordinates and (x', y') coordinates. To see how the two are related, consider the vector $\overrightarrow{O'P}$ (Figure 3.1.14b). In the xy-system its initial point is at (k, l) and its terminal point is at (x, y), so $\overrightarrow{O'P} = (x - k, y - l)$. In the $x'y'$-system its initial point is at $(0, 0)$ and its terminal point is at (x', y'), so $\overrightarrow{O'P} = (x', y')$. Therefore,

$$x' = x - k, \qquad y' = y - l$$

These formulas are called the ***translation equations***.

EXAMPLE 3 Using the Translation Equations

Suppose that an xy-coordinate system is translated to obtain an $x'y'$-coordinate system whose origin has xy-coordinates $(k, l) = (4, 1)$.

(a) Find the $x'y'$-coordinates of the point with the xy-coordinates $P(2, 0)$.
(b) Find the xy-coordinates of the point with $x'y'$-coordinates $Q(-1, 5)$.

(b)

Figure 3.1.14

Solution (a). The translation equations are

$$x' = x - 4, \qquad y' = y - 1$$

so the $x'y'$-coordinates of $P(2, 0)$ are $x' = 2 - 4 = -2$ and $y' = 0 - 1 = -1$.

Solution (b). The translation equations in (a) can be rewritten as

$$x = x' + 4, \qquad y = y' + 1$$

so the xy-coordinates of Q are $x = -1 + 4 = 3$ and $y = 5 + 1 = 6$. ◆

In 3-space the translation equations are

$$x' = x - k, \qquad y' = y - l, \qquad z' = z - m$$

where (k, l, m) are the xyz-coordinates of the $x'y'z'$-origin.

Exercise Set 3.1

1. Draw a right-handed coordinate system and locate the points whose coordinates are
- (a) $(3, 4, 5)$
- (b) $(-3, 4, 5)$
- (c) $(3, -4, 5)$
- (d) $(3, 4, -5)$
- (e) $(-3, -4, 5)$
- (f) $(-3, 4, -5)$
- (g) $(3, -4, -5)$
- (h) $(-3, -4, -5)$
- (i) $(-3, 0, 0)$
- (j) $(3, 0, 3)$
- (k) $(0, 0, -3)$
- (l) $(0, 3, 0)$

2. Sketch the following vectors with the initial points located at the origin:
- (a) $\mathbf{v}_1 = (3, 6)$
- (b) $\mathbf{v}_2 = (-4, -8)$
- (c) $\mathbf{v}_3 = (-4, -3)$
- (d) $\mathbf{v}_4 = (5, -4)$
- (e) $\mathbf{v}_5 = (3, 0)$
- (f) $\mathbf{v}_6 = (0, -7)$
- (g) $\mathbf{v}_7 = (3, 4, 5)$
- (h) $\mathbf{v}_8 = (3, 3, 0)$
- (i) $\mathbf{v}_9 = (0, 0, -3)$

3. Find the components of the vector having initial point P_1 and terminal point P_2.
- (a) $P_1(4, 8)$, $P_2(3, 7)$
- (b) $P_1(3, -5)$, $P_2(-4, -7)$
- (c) $P_1(-5, 0)$, $P_2(-3, 1)$
- (d) $P_1(0, 0)$, $P_2(a, b)$
- (e) $P_1(3, -7, 2)$, $P_2(-2, 5, -4)$
- (f) $P_1(-1, 0, 2)$, $P_2(0, -1, 0)$
- (g) $P_1(a, b, c)$, $P_2(0, 0, 0)$
- (h) $P_1(0, 0, 0)$, $P_2(a, b, c)$

4. Find a nonzero vector $\mathbf{u}$ with initial point $P(-1, 3, -5)$ such that
- (a) $\mathbf{u}$ has the same direction as $\mathbf{v} = (6, 7, -3)$
- (b) $\mathbf{u}$ is oppositely directed to $\mathbf{v} = (6, 7, -3)$

5. Find a nonzero vector $\mathbf{u}$ with terminal point $Q(3, 0, -5)$ such that
- (a) $\mathbf{u}$ has the same direction as $\mathbf{v} = (4, -2, -1)$
- (b) $\mathbf{u}$ is oppositely directed to $\mathbf{v} = (4, -2, -1)$

6. Let $\mathbf{u} = (-3, 1, 2)$, $\mathbf{v} = (4, 0, -8)$, and $\mathbf{w} = (6, -1, -4)$. Find the components of
- (a) $\mathbf{v} - \mathbf{w}$
- (b) $6\mathbf{u} + 2\mathbf{v}$
- (c) $-\mathbf{v} + \mathbf{u}$
- (d) $5(\mathbf{v} - 4\mathbf{u})$
- (e) $-3(\mathbf{v} - 8\mathbf{w})$
- (f) $(2\mathbf{u} - 7\mathbf{w}) - (8\mathbf{v} + \mathbf{u})$

7. Let $\mathbf{u}$, $\mathbf{v}$, and $\mathbf{w}$ be the vectors in Exercise 6. Find the components of the vector $\mathbf{x}$ that satisfies $2\mathbf{u} - \mathbf{v} + \mathbf{x} = 7\mathbf{x} + \mathbf{w}$.

8. Let $\mathbf{u}$, $\mathbf{v}$, and $\mathbf{w}$ be the vectors in Exercise 6. Find scalars c_1, c_2, and c_3 such that

$$c_1\mathbf{u} + c_2\mathbf{v} + c_3\mathbf{w} = (2, 0, 4)$$

9. Show that there do not exist scalars c_1, c_2, and c_3 such that

$$c_1(-2, 9, 6) + c_2(-3, 2, 1) + c_3(1, 7, 5) = (0, 5, 4)$$

10. Find all scalars c_1, c_2, and c_3 such that

$$c_1(1, 2, 0) + c_2(2, 1, 1) + c_3(0, 3, 1) = (0, 0, 0)$$

11. Let P be the point $(2, 3, -2)$ and Q the point $(7, -4, 1)$.
- (a) Find the midpoint of the line segment connecting P and Q.
- (b) Find the point on the line segment connecting P and Q that is $\frac{3}{4}$ of the way from P to Q.

12. Suppose an xy-coordinate system is translated to obtain an $x'y'$-coordinate system whose origin O' has xy-coordinates $(2, -3)$.

    (a) Find the $x'y'$-coordinates of the point P whose xy-coordinates are $(7, 5)$.

    (b) Find the xy-coordinates of the point Q whose $x'y'$-coordinates are $(-3, 6)$.

    (c) Draw the xy and $x'y'$-coordinate axes and locate the points P and Q.

13. Suppose that an xyz-coordinate system is translated to obtain an $x'y'z'$-coordinate system. Let $\mathbf{v}$ be a vector whose components are $\mathbf{v} = (v_1, v_2, v_3)$ in the xyz-system. Show that $\mathbf{v}$ has the same components in the $x'y'z'$-system.

14. Find the components of $\mathbf{u}, \mathbf{v}, \mathbf{u} + \mathbf{v}$, and $\mathbf{u} - \mathbf{v}$ for the vectors shown in the accompanying figure.

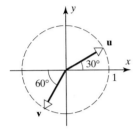

Figure Ex-14

15. Prove geometrically that if $\mathbf{v} = (v_1, v_2)$, then $k\mathbf{v} = (kv_1, kv_2)$. (Restrict the proof to the case $k > 0$ illustrated in Figure 3.1.8. The complete proof would involve various cases that depend on the sign of k and the quadrant in which the vector falls.)

Discussion and Discovery

16. Consider Figure 3.1.13. Discuss a geometric interpretation of the vector

$$\mathbf{u} = \overrightarrow{OP_1} + \tfrac{1}{2}(\overrightarrow{OP_2} - \overrightarrow{OP_1})$$

17. Draw a picture that shows four nonzero vectors whose sum is zero.

18. If you were given four nonzero vectors, how would you construct a fifth vector geometrically that is equal to the sum of the first four? Draw a picture to illustrate your method.

3.2 NORM OF A VECTOR; VECTOR ARITHMETIC

In this section we shall establish the basic rules of vector arithmetic.

Properties of Vector Operations   The following theorem lists the most important properties of vectors in 2-space and 3-space.

Theorem 3.2.1   **Properties of Vector Arithmetic**

If $\mathbf{u}, \mathbf{v}$, and $\mathbf{w}$ are vectors in 2- or 3-space and k and l are scalars, then the following relationships hold.

    (a) $\mathbf{u} + \mathbf{v} = \mathbf{v} + \mathbf{u}$ (b) $(\mathbf{u} + \mathbf{v}) + \mathbf{w} = \mathbf{u} + (\mathbf{v} + \mathbf{w})$

    (c) $\mathbf{u} + \mathbf{0} = \mathbf{0} + \mathbf{u} = \mathbf{u}$ (d) $\mathbf{u} + (-\mathbf{u}) = \mathbf{0}$

    (e) $k(l\mathbf{u}) = (kl)\mathbf{u}$ (f) $k(\mathbf{u} + \mathbf{v}) = k\mathbf{u} + k\mathbf{v}$

    (g) $(k + l)\mathbf{u} = k\mathbf{u} + l\mathbf{u}$ (h) $1\mathbf{u} = \mathbf{u}$

Before discussing the proof, we note that we have developed two approaches to vectors: *geometric*, in which vectors are represented by arrows or directed line segments, and *analytic*, in which vectors are represented by pairs or triples of numbers called components. As a consequence, the equations in Theorem 3.2.1 can be proved either geometrically or analytically. To illustrate, we shall prove part (*b*) both ways. The remaining proofs are left as exercises.

Proof of part (b) (analytic). We shall give the proof for vectors in 3-space; the proof for 2-space is similar. If $\mathbf{u} = (u_1, u_2, u_3)$, $\mathbf{v} = (v_1, v_2, v_3)$, and $\mathbf{w} = (w_1, w_2, w_3)$, then

$$
\begin{aligned}
(\mathbf{u} + \mathbf{v}) + \mathbf{w} &= [(u_1, u_2, u_3) + (v_1, v_2, v_3)] + (w_1, w_2, w_3) \\
&= (u_1 + v_1, u_2 + v_2, u_3 + v_3) + (w_1, w_2, w_3) \\
&= ([u_1 + v_1] + w_1, [u_2 + v_2] + w_2, [u_3 + v_3] + w_3) \\
&= (u_1 + [v_1 + w_1], u_2 + [v_2 + w_2], u_3 + [v_3 + w_3]) \\
&= (u_1, u_2, u_3) + (v_1 + w_1, v_2 + w_2, v_3 + w_3) \\
&= \mathbf{u} + (\mathbf{v} + \mathbf{w})
\end{aligned}
$$

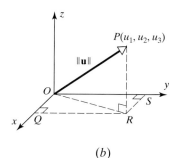

Figure 3.2.1
The vectors
$\mathbf{u} + (\mathbf{v} + \mathbf{w})$ and $(\mathbf{u} + \mathbf{v}) + \mathbf{w}$
are equal.

Proof of part (b) (geometric). Let $\mathbf{u}$, $\mathbf{v}$, and $\mathbf{w}$ be represented by $\overrightarrow{PQ}$, $\overrightarrow{QR}$, and $\overrightarrow{RS}$ as shown in Figure 3.2.1. Then

$$\mathbf{v} + \mathbf{w} = \overrightarrow{QS} \quad \text{and} \quad \mathbf{u} + (\mathbf{v} + \mathbf{w}) = \overrightarrow{PS}$$

Also,

$$\mathbf{u} + \mathbf{v} = \overrightarrow{PR} \quad \text{and} \quad (\mathbf{u} + \mathbf{v}) + \mathbf{w} = \overrightarrow{PS}$$

Therefore,

$$\mathbf{u} + (\mathbf{v} + \mathbf{w}) = (\mathbf{u} + \mathbf{v}) + \mathbf{w} \qquad \blacksquare$$

REMARK. In light of part (*b*) of this theorem, the symbol $\mathbf{u} + \mathbf{v} + \mathbf{w}$ is unambiguous since the same sum is obtained no matter where parentheses are inserted. Moreover, if the vectors $\mathbf{u}$, $\mathbf{v}$, and $\mathbf{w}$ are placed "tip to tail," then the sum $\mathbf{u} + \mathbf{v} + \mathbf{w}$ is the vector from the initial point of $\mathbf{u}$ to the terminal point of $\mathbf{w}$ (Figure 3.2.1).

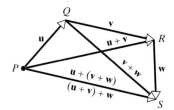

(a)

Norm of a Vector   The *length* of a vector $\mathbf{u}$ is often called the *norm* of $\mathbf{u}$ and is denoted by $\|\mathbf{u}\|$. It follows from the Theorem of Pythagoras that the norm of a vector $\mathbf{u} = (u_1, u_2)$ in 2-space is

$$\|\mathbf{u}\| = \sqrt{u_1^2 + u_2^2} \tag{1}$$

(Figure 3.2.2*a*). Let $\mathbf{u} = (u_1, u_2, u_3)$ be a vector in 3-space. Using Figure 3.2.2*b* and two applications of the Theorem of Pythagoras, we obtain

$$\|\mathbf{u}\|^2 = (OR)^2 + (RP)^2 = (OQ)^2 + (OS)^2 + (RP)^2 = u_1^2 + u_2^2 + u_3^2$$

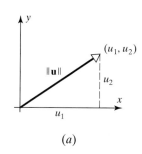

(b)

Figure 3.2.2

Thus,

$$\|\mathbf{u}\| = \sqrt{u_1^2 + u_2^2 + u_3^2} \tag{2}$$

A vector of norm 1 is called a *unit vector*.

If $P_1(x_1, y_1, z_1)$ and $P_2(x_2, y_2, z_2)$ are two points in 3-space, then the *distance* d between them is the norm of the vector $\overrightarrow{P_1 P_2}$ (Figure 3.2.3). Since

$$\overrightarrow{P_1 P_2} = (x_2 - x_1, y_2 - y_1, z_2 - z_1)$$

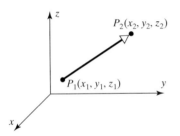

Figure 3.2.3
The distance between P_1 and P_2 is the norm of the vector $\overrightarrow{P_1 P_2}$.

it follows from (2) that

$$d = \sqrt{(x_2 - x_1)^2 + (y_2 - y_1)^2 + (z_2 - z_1)^2} \tag{3}$$

Similarly, if $P_1(x_1, y_1)$ and $P_2(x_2, y_2)$ are points in 2-space, then the distance between them is given by

$$d = \sqrt{(x_2 - x_1)^2 + (y_2 - y_1)^2} \tag{4}$$

EXAMPLE 1 Finding Norm and Distance

The norm of the vector $\mathbf{u} = (-3, 2, 1)$ is

$$\|\mathbf{u}\| = \sqrt{(-3)^2 + (2)^2 + (1)^2} = \sqrt{14}$$

The distance d between the points $P_1(2, -1, -5)$ and $P_2(4, -3, 1)$ is

$$d = \sqrt{(4 - 2)^2 + (-3 + 1)^2 + (1 + 5)^2} = \sqrt{44} = 2\sqrt{11} \qquad \blacklozenge$$

From the definition of the product $k\mathbf{u}$, the length of the vector $k\mathbf{u}$ is $|k|$ times the length of $\mathbf{u}$. Expressed as an equation, this statement says that

$$\|k\mathbf{u}\| = |k|\|\mathbf{u}\| \tag{5}$$

This useful formula is applicable in both 2-space and 3-space.

Exercise Set 3.2

1. Find the norm of **v**.
 (a) $\mathbf{v} = (4, -3)$ (b) $\mathbf{v} = (2, 3)$ (c) $\mathbf{v} = (-5, 0)$
 (d) $\mathbf{v} = (2, 2, 2)$ (e) $\mathbf{v} = (-7, 2, -1)$ (f) $\mathbf{v} = (0, 6, 0)$

2. Find the distance between P_1 and P_2.
 (a) $P_1(3, 4), P_2(5, 7)$ (b) $P_1(-3, 6), P_2(-1, -4)$
 (c) $P_1(7, -5, 1), P_2(-7, -2, -1)$ (d) $P_1(3, 3, 3), P_2(6, 0, 3)$

3. Let $\mathbf{u} = (2, -2, 3)$, $\mathbf{v} = (1, -3, 4)$, $\mathbf{w} = (3, 6, -4)$. In each part evaluate the expression.
 (a) $\|\mathbf{u} + \mathbf{v}\|$ (b) $\|\mathbf{u}\| + \|\mathbf{v}\|$ (c) $\|-2\mathbf{u}\| + 2\|\mathbf{u}\|$
 (d) $\|3\mathbf{u} - 5\mathbf{v} + \mathbf{w}\|$ (e) $\dfrac{1}{\|\mathbf{w}\|}\mathbf{w}$ (f) $\left\|\dfrac{1}{\|\mathbf{w}\|}\mathbf{w}\right\|$

4. Let $\mathbf{v} = (-1, 2, 5)$. Find all scalars k such that $\|k\mathbf{v}\| = 4$.

5. Let $\mathbf{u} = (7, -3, 1)$, $\mathbf{v} = (9, 6, 6)$, $\mathbf{w} = (2, 1, -8)$, $k = -2$, and $l = 5$. Verify that these vectors and scalars satisfy the stated equalities from Theorem 3.2.1.
 (a) part (b) (b) part (e) (c) part (f) (d) part (g)

6. (a) Show that if **v** is any nonzero vector, then $\dfrac{1}{\|\mathbf{v}\|}\mathbf{v}$ is a unit vector.
 (b) Use the result in part (a) to find a unit vector that has the same direction as the vector $\mathbf{v} = (3, 4)$.
 (c) Use the result in part (a) to find a unit vector that is oppositely directed to the vector $\mathbf{v} = (-2, 3, -6)$.

7. (a) Show that the components of the vector $\mathbf{v} = (v_1, v_2)$ in the accompanying figure are
   $v_1 = \|\mathbf{v}\| \cos \theta$ and $v_2 = \|\mathbf{v}\| \sin \theta$.

   (b) Let $\mathbf{u}$ and $\mathbf{v}$ be the vectors in the accompanying figure. Use the result in part (a) to find the components of $4\mathbf{u} - 5\mathbf{v}$.

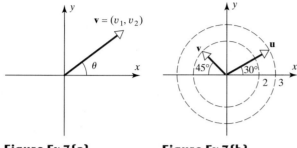

Figure Ex-7(a)   **Figure Ex-7(b)**

8. Let $\mathbf{p}_0 = (x_0, y_0, z_0)$ and $\mathbf{p} = (x, y, z)$. Describe the set of all points (x, y, z) for which $\|\mathbf{p} - \mathbf{p}_0\| = 1$.

9. Prove geometrically that if $\mathbf{u}$ and $\mathbf{v}$ are vectors in 2- or 3-space, then $\|\mathbf{u} + \mathbf{v}\| \le \|\mathbf{u}\| + \|\mathbf{v}\|$.

10. Prove parts (*a*), (*c*), and (*e*) of Theorem 3.2.1 analytically.

11. Prove parts (*d*), (*g*), and (*h*) of Theorem 3.2.1 analytically.

Discussion and Discovery

12. For the inequality stated in Exercise 9, is it possible to have $\|\mathbf{u} + \mathbf{v}\| = \|\mathbf{u}\| + \|\mathbf{v}\|$? Explain your reasoning.

13. (a) What relationship must hold for the point $\mathbf{p} = (a, b, c)$ to be equidistant from the origin and the xz-plane? Make sure that the relationship you state is valid for positive and negative values of a, b, and c.

    (b) What relationship must hold for the point $\mathbf{p} = (a, b, c)$ to be farther from the origin than from the xz-plane? Make sure that the relationship you state is valid for positive and negative values of a, b, and c.

14. (a) What does the inequality $\|\mathbf{x}\| < 1$ tell you about the point $\mathbf{x}$?

    (b) Write down an inequality that describes the set of points that lie outside the circle of radius 1, centered at the point $\mathbf{x}_0$.

15. The triangles in the accompanying figure should suggest a geometric proof of Theorem 3.2.1(*f*) for the case where $k > 0$. Give the proof.

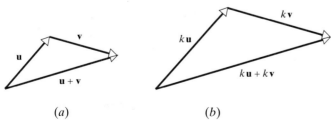

(*a*)   (*b*)

Figure Ex-15

3.3 DOT PRODUCT; PROJECTIONS

In this section we shall discuss an important way of multiplying vectors in 2-space or 3-space. We shall then give some applications of this multiplication to geometry.

Dot Product of Vectors   Let **u** and **v** be two nonzero vectors in 2-space or 3-space, and assume these vectors have been positioned so their initial points coincide. By the *angle between* **u** *and* **v**, we shall mean the angle θ determined by **u** and **v** that satisfies $0 \leq \theta \leq \pi$ (Figure 3.3.1).

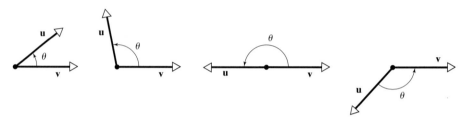

Figure 3.3.1 The angle θ between **u** and **v** satisfies $0 \leq \theta \leq \pi$.

> **Definition**
>
> If **u** and **v** are vectors in 2-space or 3-space and θ is the angle between **u** and **v**, then the *dot product* or *Euclidean inner product* **u** · **v** is defined by
>
> $$\mathbf{u} \cdot \mathbf{v} = \begin{cases} \|\mathbf{u}\| \|\mathbf{v}\| \cos\theta & \text{if } \mathbf{u} \neq \mathbf{0} \text{ and } \mathbf{v} \neq \mathbf{0} \\ 0 & \text{if } \mathbf{u} = \mathbf{0} \text{ or } \mathbf{v} = \mathbf{0} \end{cases} \tag{1}$$

EXAMPLE 1 Dot Product

As shown in Figure 3.3.2, the angle between the vectors **u** = (0, 0, 1) and **v** = (0, 2, 2) is 45°. Thus,

$$\mathbf{u} \cdot \mathbf{v} = \|\mathbf{u}\| \|\mathbf{v}\| \cos\theta = (\sqrt{0^2 + 0^2 + 1^2})(\sqrt{0^2 + 2^2 + 2^2})\left(\frac{1}{\sqrt{2}}\right) = 2 \quad \blacklozenge$$

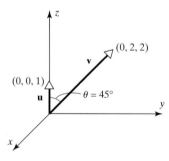

Figure 3.3.2

Component Form of the Dot Product   For purposes of computation, it is desirable to have a formula that expresses the dot product of two vectors in terms of the components of the vectors. We will derive such a formula for vectors in 3-space; the derivation for vectors in 2-space is similar.

Let **u** = (u_1, u_2, u_3) and **v** = (v_1, v_2, v_3) be two nonzero vectors. If, as shown in Figure 3.3.3, θ is the angle between **u** and **v**, then the law of cosines yields

$$\|\overrightarrow{PQ}\|^2 = \|\mathbf{u}\|^2 + \|\mathbf{v}\|^2 - 2\|\mathbf{u}\| \|\mathbf{v}\| \cos\theta \tag{2}$$

Since $\overrightarrow{PQ} = \mathbf{v} - \mathbf{u}$, we can rewrite (2) as

$$\|\mathbf{u}\| \|\mathbf{v}\| \cos\theta = \tfrac{1}{2}(\|\mathbf{u}\|^2 + \|\mathbf{v}\|^2 - \|\mathbf{v} - \mathbf{u}\|^2)$$

or

$$\mathbf{u} \cdot \mathbf{v} = \tfrac{1}{2}(\|\mathbf{u}\|^2 + \|\mathbf{v}\|^2 - \|\mathbf{v} - \mathbf{u}\|^2)$$

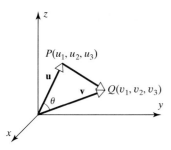

Figure 3.3.3

Substituting

$$\|\mathbf{u}\|^2 = u_1^2 + u_2^2 + u_3^2, \qquad \|\mathbf{v}\|^2 = v_1^2 + v_2^2 + v_3^2,$$

and

$$\|\mathbf{v} - \mathbf{u}\|^2 = (v_1 - u_1)^2 + (v_2 - u_2)^2 + (v_3 - u_3)^2$$

we obtain after simplifying

$$\mathbf{u} \cdot \mathbf{v} = u_1 v_1 + u_2 v_2 + u_3 v_3 \tag{3}$$

Although we derived this formula under the assumption that **u** and **v** are nonzero, the formula is also valid if $\mathbf{u} = \mathbf{0}$ or $\mathbf{v} = \mathbf{0}$ (verify).

If $\mathbf{u} = (u_1, u_2)$ and $\mathbf{v} = (v_1, v_2)$ are two vectors in 2-space, then the formula corresponding to (3) is

$$\mathbf{u} \cdot \mathbf{v} = u_1 v_1 + u_2 v_2 \tag{4}$$

Finding the Angle Between Vectors   If **u** and **v** are nonzero vectors, then Formula (1) can be written as

$$\cos\theta = \frac{\mathbf{u} \cdot \mathbf{v}}{\|\mathbf{u}\|\,\|\mathbf{v}\|} \tag{5}$$

EXAMPLE 2 Dot Product Using (3)

Consider the vectors $\mathbf{u} = (2, -1, 1)$ and $\mathbf{v} = (1, 1, 2)$. Find $\mathbf{u} \cdot \mathbf{v}$ and determine the angle θ between **u** and **v**.

Solution.

$$\mathbf{u} \cdot \mathbf{v} = u_1 v_1 + u_2 v_2 + u_3 v_3 = (2)(1) + (-1)(1) + (1)(2) = 3$$

For the given vectors we have $\|\mathbf{u}\| = \|\mathbf{v}\| = \sqrt{6}$, so that from (5)

$$\cos\theta = \frac{\mathbf{u} \cdot \mathbf{v}}{\|\mathbf{u}\|\,\|\mathbf{v}\|} = \frac{3}{\sqrt{6}\sqrt{6}} = \frac{1}{2}$$

Thus, $\theta = 60°$.   ◆

EXAMPLE 3 A Geometric Problem

Find the angle between a diagonal of a cube and one of its edges.

Solution.

Let k be the length of an edge and introduce a coordinate system as shown in Figure 3.3.4. If we let $\mathbf{u}_1 = (k, 0, 0)$, $\mathbf{u}_2 = (0, k, 0)$, and $\mathbf{u}_3 = (0, 0, k)$, then the vector

$$\mathbf{d} = (k, k, k) = \mathbf{u}_1 + \mathbf{u}_2 + \mathbf{u}_3$$

is a diagonal of the cube. The angle θ between **d** and the edge $\mathbf{u}_1$ satisfies

$$\cos\theta = \frac{\mathbf{u}_1 \cdot \mathbf{d}}{\|\mathbf{u}_1\|\,\|\mathbf{d}\|} = \frac{k^2}{(k)(\sqrt{3k^2})} = \frac{1}{\sqrt{3}}$$

Thus,

$$\theta = \cos^{-1}\left(\frac{1}{\sqrt{3}}\right) \approx 54.74° \qquad ◆$$

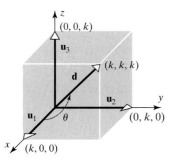

Figure 3.3.4

The following theorem shows how the dot product can be used to obtain information about the angle between two vectors; it also establishes an important relationship between the norm and the dot product.

Theorem 3.3.1

Let **u** *and* **v** *be vectors in 2- or 3-space.*

(*a*) $\mathbf{v} \cdot \mathbf{v} = \|\mathbf{v}\|^2$; *that is,* $\|\mathbf{v}\| = (\mathbf{v} \cdot \mathbf{v})^{1/2}$

(*b*) *If the vectors* **u** *and* **v** *are nonzero and* θ *is the angle between them, then*

θ *is acute*	*if and only if*	$\mathbf{u} \cdot \mathbf{v} > 0$
θ *is obtuse*	*if and only if*	$\mathbf{u} \cdot \mathbf{v} < 0$
$\theta = \pi/2$	*if and only if*	$\mathbf{u} \cdot \mathbf{v} = 0$

Proof (*a*). Since the angle θ between **v** and **v** is 0, we have

$$\mathbf{v} \cdot \mathbf{v} = \|\mathbf{v}\| \|\mathbf{v}\| \cos \theta = \|\mathbf{v}\|^2 \cos 0 = \|\mathbf{v}\|^2$$

Proof (*b*). Since θ satisfies $0 \leq \theta \leq \pi$, it follows that: θ is acute if and only if $\cos \theta > 0$; θ is obtuse if and only if $\cos \theta < 0$; and $\theta = \pi/2$ if and only if $\cos \theta = 0$. But $\cos \theta$ has the same sign as $\mathbf{u} \cdot \mathbf{v}$ since $\mathbf{u} \cdot \mathbf{v} = \|\mathbf{u}\| \|\mathbf{v}\| \cos \theta$, $\|\mathbf{u}\| > 0$, and $\|\mathbf{v}\| > 0$. Thus, the result follows. ∎

EXAMPLE 4 Finding Dot Products from Components

If $\mathbf{u} = (1, -2, 3)$, $\mathbf{v} = (-3, 4, 2)$, and $\mathbf{w} = (3, 6, 3)$, then

$$\mathbf{u} \cdot \mathbf{v} = (1)(-3) + (-2)(4) + (3)(2) = -5$$
$$\mathbf{v} \cdot \mathbf{w} = (-3)(3) + (4)(6) + (2)(3) = 21$$
$$\mathbf{u} \cdot \mathbf{w} = (1)(3) + (-2)(6) + (3)(3) = 0$$

Therefore, **u** and **v** make an obtuse angle, **v** and **w** make an acute angle, and **u** and **w** are perpendicular. ◆

Orthogonal Vectors Perpendicular vectors are also called *orthogonal* vectors. In light of Theorem 3.3.1*b*, two *nonzero* vectors are orthogonal if and only if their dot product is zero. If we agree to consider **u** and **v** to be perpendicular when either or both of these vectors is **0**, then we can state without exception that *two vectors* **u** *and* **v** *are orthogonal* (*perpendicular*) *if and only if* $\mathbf{u} \cdot \mathbf{v} = 0$. To indicate that **u** and **v** are orthogonal vectors we write $\mathbf{u} \perp \mathbf{v}$.

EXAMPLE 5 A Vector Perpendicular to a Line

Show that in 2-space the nonzero vector $\mathbf{n} = (a, b)$ is perpendicular to the line $ax + by + c = 0$.

Solution.

Let $P_1(x_1, y_1)$ and $P_2(x_2, y_2)$ be distinct points on the line, so that

$$ax_1 + by_1 + c = 0$$
$$ax_2 + by_2 + c = 0$$

(6)

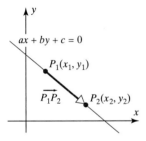

Figure 3.3.5

Since the vector $\overrightarrow{P_1 P_2} = (x_2 - x_1, y_2 - y_1)$ runs along the line (Figure 3.3.5), we need only show that **n** and $\overrightarrow{P_1 P_2}$ are perpendicular. But on subtracting the equations in (6) we obtain

$$a(x_2 - x_1) + b(y_2 - y_1) = 0$$

which can be expressed in the form

$$(a, b) \cdot (x_2 - x_1, y_2 - y_1) = 0 \quad \text{or} \quad \mathbf{n} \cdot \overrightarrow{P_1 P_2} = 0$$

Thus, **n** and $\overrightarrow{P_1 P_2}$ are perpendicular. ◆

The following theorem lists the most important properties of the dot product. They are useful in calculations involving vectors.

Theorem 3.3.2   **Properties of the Dot Product**

If **u**, **v**, *and* **w** *are vectors in 2- or 3-space and k is a scalar, then*:

(a) $\mathbf{u} \cdot \mathbf{v} = \mathbf{v} \cdot \mathbf{u}$
(b) $\mathbf{u} \cdot (\mathbf{v} + \mathbf{w}) = \mathbf{u} \cdot \mathbf{v} + \mathbf{u} \cdot \mathbf{w}$
(c) $k(\mathbf{u} \cdot \mathbf{v}) = (k\mathbf{u}) \cdot \mathbf{v} = \mathbf{u} \cdot (k\mathbf{v})$
(d) $\mathbf{v} \cdot \mathbf{v} > 0$ *if* $\mathbf{v} \neq \mathbf{0}$, *and* $\mathbf{v} \cdot \mathbf{v} = 0$ *if* $\mathbf{v} = \mathbf{0}$

Proof. We shall prove (*c*) for vectors in 3-space and leave the remaining proofs as exercises. Let $\mathbf{u} = (u_1, u_2, u_3)$ and $\mathbf{v} = (v_1, v_2, v_3)$; then

$$\begin{aligned} k(\mathbf{u} \cdot \mathbf{v}) &= k(u_1 v_1 + u_2 v_2 + u_3 v_3) \\ &= (ku_1)v_1 + (ku_2)v_2 + (ku_3)v_3 \\ &= (k\mathbf{u}) \cdot \mathbf{v} \end{aligned}$$

Similarly,

$$k(\mathbf{u} \cdot \mathbf{v}) = \mathbf{u} \cdot (k\mathbf{v}) \qquad ■$$

An Orthogonal Projection   In many applications it is of interest to "decompose" a vector **u** into a sum of two terms, one parallel to a specified nonzero vector **a** and the other perpendicular to **a**. If **u** and **a** are positioned so their initial points coincide at a point Q, we can decompose the vector **u** as follows (Figure 3.3.6): Drop a perpendicular from the tip of **u** to the line through **a**, and construct the vector $\mathbf{w}_1$ from Q to the foot of this perpendicular. Next form the difference

$$\mathbf{w}_2 = \mathbf{u} - \mathbf{w}_1$$

As indicated in Figure 3.3.6, the vector $\mathbf{w}_1$ is parallel to **a**, the vector $\mathbf{w}_2$ is perpendicular to **a**, and

$$\mathbf{w}_1 + \mathbf{w}_2 = \mathbf{w}_1 + (\mathbf{u} - \mathbf{w}_1) = \mathbf{u}$$

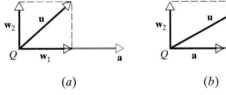 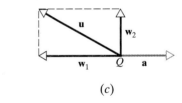

(a) $\qquad\qquad\qquad$ (b) $\qquad\qquad\qquad$ (c)

Figure 3.3.6 The vector **u** is the sum of $\mathbf{w}_1$ and $\mathbf{w}_2$, where $\mathbf{w}_1$ is parallel to **a** and $\mathbf{w}_2$ is perpendicular to **a**.

The vector $\mathbf{w}_1$ is called the ***orthogonal projection of*** $\mathbf{u}$ ***on*** $\mathbf{a}$ or sometimes the ***vector component of*** $\mathbf{u}$ ***along*** $\mathbf{a}$. It is denoted by

$$\text{proj}_{\mathbf{a}}\,\mathbf{u} \qquad\qquad (7)$$

The vector $\mathbf{w}_2$ is called the ***vector component of*** $\mathbf{u}$ ***orthogonal to*** $\mathbf{a}$. Since we have $\mathbf{w}_2 = \mathbf{u} - \mathbf{w}_1$, this vector can be written in notation (7) as

$$\mathbf{w}_2 = \mathbf{u} - \text{proj}_{\mathbf{a}}\,\mathbf{u}$$

The following theorem gives formulas for calculating $\text{proj}_{\mathbf{a}}\,\mathbf{u}$ and $\mathbf{u} - \text{proj}_{\mathbf{a}}\,\mathbf{u}$.

Theorem 3.3.3

If $\mathbf{u}$ *and* $\mathbf{a}$ *are vectors in 2-space or 3-space and if* $\mathbf{a} \neq \mathbf{0}$, *then*

$$\text{proj}_{\mathbf{a}}\,\mathbf{u} = \frac{\mathbf{u} \cdot \mathbf{a}}{\|\mathbf{a}\|^2}\mathbf{a} \quad \textit{(vector component of \textbf{u} along \textbf{a})}$$

$$\mathbf{u} - \text{proj}_{\mathbf{a}}\,\mathbf{u} = \mathbf{u} - \frac{\mathbf{u} \cdot \mathbf{a}}{\|\mathbf{a}\|^2}\mathbf{a} \quad \textit{(vector component of \textbf{u} orthogonal to \textbf{a})}$$

Proof. Let $\mathbf{w}_1 = \text{proj}_{\mathbf{a}}\,\mathbf{u}$ and $\mathbf{w}_2 = \mathbf{u} - \text{proj}_{\mathbf{a}}\,\mathbf{u}$. Since $\mathbf{w}_1$ is parallel to $\mathbf{a}$, it must be a scalar multiple of $\mathbf{a}$, so it can be written in the form $\mathbf{w}_1 = k\mathbf{a}$. Thus

$$\mathbf{u} = \mathbf{w}_1 + \mathbf{w}_2 = k\mathbf{a} + \mathbf{w}_2 \qquad\qquad (8)$$

Taking the dot product of both sides of (8) with $\mathbf{a}$ and using Theorems 3.3.1*a* and 3.3.2 yields

$$\mathbf{u} \cdot \mathbf{a} = (k\mathbf{a} + \mathbf{w}_2) \cdot \mathbf{a} = k\|\mathbf{a}\|^2 + \mathbf{w}_2 \cdot \mathbf{a} \qquad\qquad (9)$$

But $\mathbf{w}_2 \cdot \mathbf{a} = 0$ since $\mathbf{w}_2$ is perpendicular to $\mathbf{a}$; so (9) yields

$$k = \frac{\mathbf{u} \cdot \mathbf{a}}{\|\mathbf{a}\|^2}$$

Since $\text{proj}_{\mathbf{a}}\,\mathbf{u} = \mathbf{w}_1 = k\mathbf{a}$, we obtain

$$\text{proj}_{\mathbf{a}}\,\mathbf{u} = \frac{\mathbf{u} \cdot \mathbf{a}}{\|\mathbf{a}\|^2}\mathbf{a} \qquad\qquad \blacksquare$$

EXAMPLE 6 Vector Component of u Along a

Let $\mathbf{u} = (2, -1, 3)$ and $\mathbf{a} = (4, -1, 2)$. Find the vector component of $\mathbf{u}$ along $\mathbf{a}$ and the vector component of $\mathbf{u}$ orthogonal to $\mathbf{a}$.

Solution.

$$\mathbf{u} \cdot \mathbf{a} = (2)(4) + (-1)(-1) + (3)(2) = 15$$
$$\|\mathbf{a}\|^2 = 4^2 + (-1)^2 + 2^2 = 21$$

Thus, the vector component of $\mathbf{u}$ along $\mathbf{a}$ is

$$\text{proj}_{\mathbf{a}}\,\mathbf{u} = \frac{\mathbf{u} \cdot \mathbf{a}}{\|\mathbf{a}\|^2}\mathbf{a} = \tfrac{15}{21}(4, -1, 2) = \left(\tfrac{20}{7}, -\tfrac{5}{7}, \tfrac{10}{7}\right)$$

and the vector component of **u** orthogonal to **a** is

$$\mathbf{u} - \text{proj}_{\mathbf{a}}\, \mathbf{u} = (2, -1, 3) - \left(\tfrac{20}{7}, -\tfrac{5}{7}, \tfrac{10}{7}\right) = \left(-\tfrac{6}{7}, -\tfrac{2}{7}, \tfrac{11}{7}\right)$$

As a check, the reader may wish to verify that the vectors $\mathbf{u} - \text{proj}_{\mathbf{a}}\, \mathbf{u}$ and **a** are perpendicular by showing that their dot product is zero. ◆

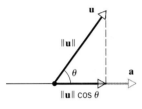

A formula for the length of the vector component of **u** along **a** can be obtained by writing

$$\|\text{proj}_{\mathbf{a}}\, \mathbf{u}\| = \left\| \frac{\mathbf{u} \cdot \mathbf{a}}{\|\mathbf{a}\|^2} \mathbf{a} \right\|$$

$$= \left| \frac{\mathbf{u} \cdot \mathbf{a}}{\|\mathbf{a}\|^2} \right| \|\mathbf{a}\| \qquad \longleftarrow \begin{array}{l}\text{Formula (5) of}\\ \text{Section 3.2}\end{array}$$

$$= \frac{|\mathbf{u} \cdot \mathbf{a}|}{\|\mathbf{a}\|^2} \|\mathbf{a}\| \qquad \longleftarrow \text{Since } \|\mathbf{a}\|^2 > 0$$

(a) $0 \le \theta < \dfrac{\pi}{2}$

which yields

$$\|\text{proj}_{\mathbf{a}}\, \mathbf{u}\| = \frac{|\mathbf{u} \cdot \mathbf{a}|}{\|\mathbf{a}\|} \tag{10}$$

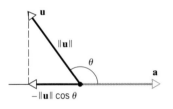

If θ denotes the angle between **u** and **a**, then $\mathbf{u} \cdot \mathbf{a} = \|\mathbf{u}\|\|\mathbf{a}\| \cos\theta$, so that (10) can also be written as

$$\|\text{proj}_{\mathbf{a}}\, \mathbf{u}\| = \|\mathbf{u}\|\,|\cos\theta| \tag{11}$$

(b) $\dfrac{\pi}{2} < \theta \le \pi$

Figure 3.3.7

(Verify.) A geometric interpretation of this result is given in Figure 3.3.7.

As an example, we will use vector methods to derive a formula for the distance from a point in the plane to a line.

EXAMPLE 7 Distance Between a Point and a Line

Find a formula for the distance D between point $P_0(x_0, y_0)$ and the line $ax + by + c = 0$.

Solution.

Let $Q(x_1, y_1)$ be any point on the line and position the vector $\mathbf{n} = (a, b)$ so that its initial point is at Q.

By virtue of Example 5, the vector **n** is perpendicular to the line (Figure 3.3.8). As indicated in the figure, the distance D is equal to the length of the orthogonal projection of $\overrightarrow{QP_0}$ on **n**; thus, from (10),

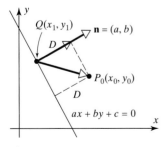

Figure 3.3.8

$$D = \|\text{proj}_{\mathbf{n}}\, \overrightarrow{QP_0}\| = \frac{|\overrightarrow{QP_0} \cdot \mathbf{n}|}{\|\mathbf{n}\|}$$

But

$$\overrightarrow{QP_0} = (x_0 - x_1, y_0 - y_1)$$

$$\overrightarrow{QP_0} \cdot \mathbf{n} = a(x_0 - x_1) + b(y_0 - y_1)$$

$$\|\mathbf{n}\| = \sqrt{a^2 + b^2}$$

so that

$$D = \frac{|a(x_0 - x_1) + b(y_0 - y_1)|}{\sqrt{a^2 + b^2}} \tag{12}$$

Since the point $Q(x_1, y_1)$ lies on the line, its coordinates satisfy the equation of the line, so

$$ax_1 + by_1 + c = 0 \quad \text{or} \quad c = -ax_1 - by_1$$

Substituting this expression in (12) yields the formula

$$D = \frac{|ax_0 + by_0 + c|}{\sqrt{a^2 + b^2}} \tag{13}$$

◆

EXAMPLE 8 Using the Distance Formula

It follows from Formula (13) that the distance D from the point $(1, -2)$ to the line $3x + 4y - 6 = 0$ is

$$D = \frac{|(3)(1) + 4(-2) - 6|}{\sqrt{3^2 + 4^2}} = \frac{|-11|}{\sqrt{25}} = \frac{11}{5}$$

◆

Exercise Set 3.3

1. Find $\mathbf{u} \cdot \mathbf{v}$.
   (a) $\mathbf{u} = (2, 3)$, $\mathbf{v} = (5, -7)$ (b) $\mathbf{u} = (-6, -2)$, $\mathbf{v} = (4, 0)$
   (c) $\mathbf{u} = (1, -5, 4)$, $\mathbf{v} = (3, 3, 3)$ (d) $\mathbf{u} = (-2, 2, 3)$, $\mathbf{v} = (1, 7, -4)$

2. In each part of Exercise 1, find the cosine of the angle θ between $\mathbf{u}$ and $\mathbf{v}$.

3. Determine whether $\mathbf{u}$ and $\mathbf{v}$ make an acute angle, make an obtuse angle, or are orthogonal.
   (a) $\mathbf{u} = (6, 1, 4)$, $\mathbf{v} = (2, 0, -3)$ (b) $\mathbf{u} = (0, 0, -1)$, $\mathbf{v} = (1, 1, 1)$
   (c) $\mathbf{u} = (-6, 0, 4)$, $\mathbf{v} = (3, 1, 6)$ (d) $\mathbf{u} = (2, 4, -8)$, $\mathbf{v} = (5, 3, 7)$

4. Find the orthogonal projection of $\mathbf{u}$ on $\mathbf{a}$.
   (a) $\mathbf{u} = (6, 2)$, $\mathbf{a} = (3, -9)$ (b) $\mathbf{u} = (-1, -2)$, $\mathbf{a} = (-2, 3)$
   (c) $\mathbf{u} = (3, 1, -7)$, $\mathbf{a} = (1, 0, 5)$ (d) $\mathbf{u} = (1, 0, 0)$, $\mathbf{a} = (4, 3, 8)$

5. In each part of Exercise 4, find the vector component of $\mathbf{u}$ orthogonal to $\mathbf{a}$.

6. In each part find $\|\text{proj}_{\mathbf{a}}\, \mathbf{u}\|$.
   (a) $\mathbf{u} = (1, -2)$, $\mathbf{a} = (-4, -3)$ (b) $\mathbf{u} = (5, 6)$, $\mathbf{a} = (2, -1)$
   (c) $\mathbf{u} = (3, 0, 4)$, $\mathbf{a} = (2, 3, 3)$ (d) $\mathbf{u} = (3, -2, 6)$, $\mathbf{a} = (1, 2, -7)$

7. Let $\mathbf{u} = (5, -2, 1)$, $\mathbf{v} = (1, 6, 3)$, and $k = -4$. Verify Theorem 3.3.2 for these quantities.

8. (a) Show that $\mathbf{v} = (a, b)$ and $\mathbf{w} = (-b, a)$ are orthogonal vectors.
   (b) Use the result in part (a) to find two vectors that are orthogonal to $\mathbf{v} = (2, -3)$.
   (c) Find two unit vectors that are orthogonal to $(-3, 4)$.

9. Let $\mathbf{u} = (3, 4)$, $\mathbf{v} = (5, -1)$, and $\mathbf{w} = (7, 1)$. Evaluate the expressions.
   (a) $\mathbf{u} \cdot (7\mathbf{v} + \mathbf{w})$ (b) $\|(\mathbf{u} \cdot \mathbf{w})\mathbf{w}\|$ (c) $\|\mathbf{u}\|(\mathbf{v} \cdot \mathbf{w})$ (d) $(\|\mathbf{u}\|\mathbf{v}) \cdot \mathbf{w}$

10. Find five different nonzero vectors that are orthogonal to $\mathbf{u} = (5, -2, 3)$.

11. Use vectors to find the cosines of the interior angles of the triangle with vertices $(0, -1)$, $(1, -2)$, and $(4, 1)$.

12. Show that $A(3, 0, 2)$, $B(4, 3, 0)$, and $C(8, 1, -1)$ are vertices of a right triangle. At which vertex is the right angle?

13. Find a unit vector that is orthogonal to both $\mathbf{u} = (1, 0, 1)$ and $\mathbf{v} = (0, 1, 1)$.

14. Let $\mathbf{p} = (2, k)$ and $\mathbf{q} = (3, 5)$. Find k such that

    (a) $\mathbf{p}$ and $\mathbf{q}$ are parallel

    (b) $\mathbf{p}$ and $\mathbf{q}$ are orthogonal

    (c) the angle between $\mathbf{p}$ and $\mathbf{q}$ is $\pi/3$

    (d) the angle between $\mathbf{p}$ and $\mathbf{q}$ is $\pi/4$

15. Use Formula (13) to calculate the distance between the point and the line.

    (a) $4x + 3y + 4 = 0$; $(-3, 1)$

    (b) $y = -4x + 2$; $(2, -5)$

    (c) $3x + y = 5$; $(1, 8)$

16. Establish the identity $\|\mathbf{u} + \mathbf{v}\|^2 + \|\mathbf{u} - \mathbf{v}\|^2 = 2\|\mathbf{u}\|^2 + 2\|\mathbf{v}\|^2$.

17. Establish the identity $\mathbf{u} \cdot \mathbf{v} = \frac{1}{4}\|\mathbf{u} + \mathbf{v}\|^2 - \frac{1}{4}\|\mathbf{u} - \mathbf{v}\|^2$.

18. Find the angle between a diagonal of a cube and one of its faces.

19. Let $\mathbf{i}$, $\mathbf{j}$, and $\mathbf{k}$ be unit vectors along the positive x, y, and z axes of a rectangular coordinate system in 3-space. If $\mathbf{v} = (a, b, c)$ is a nonzero vector, then the angles α, β, and γ between $\mathbf{v}$ and the vectors $\mathbf{i}$, $\mathbf{j}$, and $\mathbf{k}$, respectively, are called the ***direction angles*** of $\mathbf{v}$ (see accompanying figure), and the numbers $\cos\alpha$, $\cos\beta$, and $\cos\gamma$ are called the ***direction cosines*** of $\mathbf{v}$.

    (a) Show that $\cos\alpha = a/\|\mathbf{v}\|$.

    (b) Find $\cos\beta$ and $\cos\gamma$.

    (c) Show that $\mathbf{v}/\|\mathbf{v}\| = (\cos\alpha, \cos\beta, \cos\gamma)$.

    (d) Show that $\cos^2\alpha + \cos^2\beta + \cos^2\gamma = 1$.

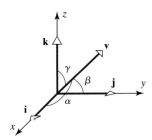

Figure Ex-19

20. Use the result in Exercise 19 to estimate, to the nearest degree, the angles that a diagonal of a box with dimensions 10 cm × 15 cm × 25 cm makes with edges of the box. [***Note.*** A calculator is needed.]

21. Referring to Exercise 19, show that two nonzero vectors, $\mathbf{v}_1$ and $\mathbf{v}_2$, in 3-space are perpendicular if and only if their direction cosines satisfy

$$\cos\alpha_1\cos\alpha_2 + \cos\beta_1\cos\beta_2 + \cos\gamma_1\cos\gamma_2 = 0$$

22. Show that if $\mathbf{v}$ is orthogonal to both $\mathbf{w}_1$ and $\mathbf{w}_2$, then $\mathbf{v}$ is orthogonal to $k_1\mathbf{w}_1 + k_2\mathbf{w}_2$ for all scalars k_1 and k_2.

23. Let $\mathbf{u}$ and $\mathbf{v}$ be nonzero vectors in 2- or 3-space, and let $k = \|\mathbf{u}\|$ and $l = \|\mathbf{v}\|$. Show that the vector $\mathbf{w} = l\mathbf{u} + k\mathbf{v}$ bisects the angle between $\mathbf{u}$ and $\mathbf{v}$.

Discussion and Discovery

24. In each part, something is wrong with the expression. What?

    (a) $\mathbf{u} \cdot (\mathbf{v} \cdot \mathbf{w})$ (b) $(\mathbf{u} \cdot \mathbf{v}) + \mathbf{w}$ (c) $\|\mathbf{u} \cdot \mathbf{v}\|$ (d) $k \cdot (\mathbf{u} + \mathbf{v})$

25. Is it possible to have $\text{proj}_{\mathbf{a}}\,\mathbf{u} = \text{proj}_{\mathbf{u}}\,\mathbf{a}$? Explain your reasoning.

26. If $\mathbf{u} \neq \mathbf{0}$, is it valid to cancel $\mathbf{u}$ from both sides of the equation $\mathbf{u} \cdot \mathbf{v} = \mathbf{u} \cdot \mathbf{w}$ and conclude that $\mathbf{v} = \mathbf{w}$? Explain your reasoning.

27. Suppose that **u**, **v**, and **w** are mutually orthogonal nonzero vectors in 3-space, and suppose that you know the dot products of these vectors with a vector **r** in 3-space. Find an expression for **r** in terms of **u**, **v**, **w**, and the dot products. [*Hint.* Look for an expression of the form **r** = c_1**u** + c_2**v** + c_3**w**.]

28. Suppose that **u** and **v** are orthogonal vectors in 2-space or 3-space. What famous theorem is described by the equation $\|\mathbf{u} + \mathbf{v}\|^2 = \|\mathbf{u}\|^2 + \|\mathbf{v}\|^2$? Draw a picture to support your answer.

3.4 CROSS PRODUCT

In many applications of vectors to problems in geometry, physics, and engineering, it is of interest to construct a vector in 3-space that is perpendicular to two given vectors. In this section we shall show how to do this.

Cross Product of Vectors   Recall from Section 3.3 that the dot product of two vectors in 2-space or 3-space produces a scalar. We will now define a type of vector multiplication that produces a vector as the product, but which is applicable only in 3-space.

> **Definition**
>
> If **u** = (u_1, u_2, u_3) and **v** = (v_1, v_2, v_3) are vectors in 3-space, then the ***cross product*** **u** × **v** is the vector defined by
>
> $$\mathbf{u} \times \mathbf{v} = (u_2 v_3 - u_3 v_2, \, u_3 v_1 - u_1 v_3, \, u_1 v_2 - u_2 v_1) \tag{1a}$$
>
> or in determinant notation
>
> $$\mathbf{u} \times \mathbf{v} = \left(\begin{vmatrix} u_2 & u_3 \\ v_2 & v_3 \end{vmatrix}, \, -\begin{vmatrix} u_1 & u_3 \\ v_1 & v_3 \end{vmatrix}, \, \begin{vmatrix} u_1 & u_2 \\ v_1 & v_2 \end{vmatrix} \right) \tag{1b}$$

REMARK.   Instead of memorizing (1b), you can obtain the components of **u** × **v** as follows:

- Form the 2 × 3 matrix $\begin{bmatrix} u_1 & u_2 & u_3 \\ v_1 & v_2 & v_3 \end{bmatrix}$ whose first row contains the components of **u** and whose second row contains the components of **v**.

- To find the first component of **u** × **v**, delete the first column and take the determinant; to find the second component, delete the second column and take the negative of the determinant; and to find the third component, delete the third column and take the determinant.

EXAMPLE 1   Calculating a Cross Product

Find **u** × **v**, where **u** = $(1, 2, -2)$ and **v** = $(3, 0, 1)$.

Solution.

From either (1) or the mnemonic in the preceding remark, we have

$$\mathbf{u} \times \mathbf{v} = \left(\begin{vmatrix} 2 & -2 \\ 0 & 1 \end{vmatrix}, \, -\begin{vmatrix} 1 & -2 \\ 3 & 1 \end{vmatrix}, \, \begin{vmatrix} 1 & 2 \\ 3 & 0 \end{vmatrix} \right)$$

$$= (2, -7, -6) \qquad \qquad \blacklozenge$$

There is an important difference between the dot product and cross product of two vectors—the dot product is a scalar and the cross product is a vector. The following theorem gives some important relationships between the dot product and cross product and also shows that $\mathbf{u} \times \mathbf{v}$ is orthogonal to both $\mathbf{u}$ and $\mathbf{v}$.

Theorem 3.4.1   **Relationships Involving Cross Product and Dot Product**

If $\mathbf{u}$, $\mathbf{v}$, *and* $\mathbf{w}$ *are vectors in 3-space, then*:

(*a*) $\mathbf{u} \cdot (\mathbf{u} \times \mathbf{v}) = 0$ (*$\mathbf{u} \times \mathbf{v}$ is orthogonal to $\mathbf{u}$*)

(*b*) $\mathbf{v} \cdot (\mathbf{u} \times \mathbf{v}) = 0$ (*$\mathbf{u} \times \mathbf{v}$ is orthogonal to $\mathbf{v}$*)

(*c*) $\|\mathbf{u} \times \mathbf{v}\|^2 = \|\mathbf{u}\|^2 \|\mathbf{v}\|^2 - (\mathbf{u} \cdot \mathbf{v})^2$ (*Lagrange's identity*)

(*d*) $\mathbf{u} \times (\mathbf{v} \times \mathbf{w}) = (\mathbf{u} \cdot \mathbf{w})\mathbf{v} - (\mathbf{u} \cdot \mathbf{v})\mathbf{w}$ (*relationship between cross and dot products*)

(*e*) $(\mathbf{u} \times \mathbf{v}) \times \mathbf{w} = (\mathbf{u} \cdot \mathbf{w})\mathbf{v} - (\mathbf{v} \cdot \mathbf{w})\mathbf{u}$ (*relationship between cross and dot products*)

Joseph Louis Lagrange (1736–1813) was a French-Italian mathematician and astronomer. Although his father wanted him to become a lawyer, Lagrange was attracted to mathematics and astronomy after reading a memoir by the astronomer Halley. At age 16 he began to study mathematics on his own and by age 19 was appointed to a professorship at the Royal Artillery School in Turin. The following year he solved some famous problems using new methods that eventually blossomed into a branch of mathematics called the *calculus of variations*. These methods and Lagrange's applications of them to problems in celestial mechanics were so monumental that by age 25 he was regarded by many of his contemporaries as the greatest living mathematician. One of Lagrange's most famous works is a memoir, *Mécanique Analytique*, in which he reduced the theory of mechanics to a few general formulas from which all other necessary equations could be derived.

Napoleon was a great admirer of Lagrange and showered him with many honors. In spite of his fame, Lagrange was a shy and modest man. On his death, he was buried with honor in the Pantheon.

Proof (a). Let $\mathbf{u} = (u_1, u_2, u_3)$ and $\mathbf{v} = (v_1, v_2, v_3)$. Then

$$\mathbf{u} \cdot (\mathbf{u} \times \mathbf{v}) = (u_1, u_2, u_3) \cdot (u_2 v_3 - u_3 v_2, u_3 v_1 - u_1 v_3, u_1 v_2 - u_2 v_1)$$
$$= u_1(u_2 v_3 - u_3 v_2) + u_2(u_3 v_1 - u_1 v_3) + u_3(u_1 v_2 - u_2 v_1) = 0$$

Proof (b). Similar to (*a*).

Proof (c). Since

$$\|\mathbf{u} \times \mathbf{v}\|^2 = (u_2 v_3 - u_3 v_2)^2 + (u_3 v_1 - u_1 v_3)^2 + (u_1 v_2 - u_2 v_1)^2 \qquad (2)$$

and

$$\|\mathbf{u}\|^2 \|\mathbf{v}\|^2 - (\mathbf{u} \cdot \mathbf{v})^2 = (u_1^2 + u_2^2 + u_3^2)(v_1^2 + v_2^2 + v_3^2) - (u_1 v_1 + u_2 v_2 + u_3 v_3)^2 \qquad (3)$$

the proof can be completed by "multiplying out" the right sides of (2) and (3) and verifying their equality.

Proof (d) and (e). See Exercises 26 and 27.   ■

EXAMPLE 2 $\mathbf{u} \times \mathbf{v}$ Is Perpendicular to $\mathbf{u}$ and to $\mathbf{v}$

Consider the vectors

$$\mathbf{u} = (1, 2, -2) \quad \text{and} \quad \mathbf{v} = (3, 0, 1)$$

In Example 1 we showed that

$$\mathbf{u} \times \mathbf{v} = (2, -7, -6)$$

Since

$$\mathbf{u} \cdot (\mathbf{u} \times \mathbf{v}) = (1)(2) + (2)(-7) + (-2)(-6) = 0$$

and

$$\mathbf{v} \cdot (\mathbf{u} \times \mathbf{v}) = (3)(2) + (0)(-7) + (1)(-6) = 0$$

$\mathbf{u} \times \mathbf{v}$ is orthogonal to both $\mathbf{u}$ and $\mathbf{v}$ as guaranteed by Theorem 3.4.1.   ◆

The main arithmetic properties of the cross product are listed in the next theorem.

> ### Theorem 3.4.2 — Properties of Cross Product
>
> *If* **u**, **v**, *and* **w** *are any vectors in 3-space and k is any scalar, then*:
>
> (*a*) $\mathbf{u} \times \mathbf{v} = -(\mathbf{v} \times \mathbf{u})$
> (*b*) $\mathbf{u} \times (\mathbf{v} + \mathbf{w}) = (\mathbf{u} \times \mathbf{v}) + (\mathbf{u} \times \mathbf{w})$
> (*c*) $(\mathbf{u} + \mathbf{v}) \times \mathbf{w} = (\mathbf{u} \times \mathbf{w}) + (\mathbf{v} \times \mathbf{w})$
> (*d*) $k(\mathbf{u} \times \mathbf{v}) = (k\mathbf{u}) \times \mathbf{v} = \mathbf{u} \times (k\mathbf{v})$
> (*e*) $\mathbf{u} \times \mathbf{0} = \mathbf{0} \times \mathbf{u} = \mathbf{0}$
> (*f*) $\mathbf{u} \times \mathbf{u} = \mathbf{0}$

The proofs follow immediately from Formula (1b) and properties of determinants; for example, (*a*) can be proved as follows:

Proof (*a*). Interchanging **u** and **v** in (1b) interchanges the rows of the three determinants on the right side of (1b) and hence changes the sign of each component in the cross product. Thus, $\mathbf{u} \times \mathbf{v} = -(\mathbf{v} \times \mathbf{u})$. ∎

The proofs of the remaining parts are left as exercises.

EXAMPLE 3 Standard Unit Vectors

Consider the vectors

$$\mathbf{i} = (1, 0, 0), \qquad \mathbf{j} = (0, 1, 0), \qquad \mathbf{k} = (0, 0, 1)$$

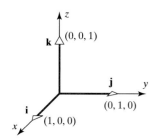

Figure 3.4.1
The standard unit vectors.

These vectors each have length 1 and lie along the coordinate axes (Figure 3.4.1). They are called the ***standard unit vectors*** in 3-space. Every vector $\mathbf{v} = (v_1, v_2, v_3)$ in 3-space is expressible in terms of **i**, **j**, and **k** since we can write

$$\mathbf{v} = (v_1, v_2, v_3) = v_1(1, 0, 0) + v_2(0, 1, 0) + v_3(0, 0, 1) = v_1\mathbf{i} + v_2\mathbf{j} + v_3\mathbf{k}$$

For example,

$$(2, -3, 4) = 2\mathbf{i} - 3\mathbf{j} + 4\mathbf{k}$$

From (1b) we obtain

$$\mathbf{i} \times \mathbf{j} = \left(\begin{vmatrix} 0 & 0 \\ 1 & 0 \end{vmatrix}, -\begin{vmatrix} 1 & 0 \\ 0 & 0 \end{vmatrix}, \begin{vmatrix} 1 & 0 \\ 0 & 1 \end{vmatrix} \right) = (0, 0, 1) = \mathbf{k} \qquad \blacklozenge$$

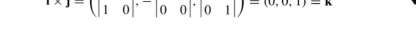

The reader should have no trouble obtaining the following results:

$$
\begin{array}{lll}
\mathbf{i} \times \mathbf{i} = \mathbf{0} & \mathbf{j} \times \mathbf{j} = \mathbf{0} & \mathbf{k} \times \mathbf{k} = \mathbf{0} \\
\mathbf{i} \times \mathbf{j} = \mathbf{k} & \mathbf{j} \times \mathbf{k} = \mathbf{i} & \mathbf{k} \times \mathbf{i} = \mathbf{j} \\
\mathbf{j} \times \mathbf{i} = -\mathbf{k} & \mathbf{k} \times \mathbf{j} = -\mathbf{i} & \mathbf{i} \times \mathbf{k} = -\mathbf{j}
\end{array}
$$

Figure 3.4.2

Figure 3.4.2 is helpful for remembering these results. Referring to this diagram, the cross product of two consecutive vectors going clockwise is the next vector around, and the cross product of two consecutive vectors going counterclockwise is the negative of the next vector around.

Determinant Form of Cross Product

It is also worth noting that a cross product can be represented symbolically in the form of a 3×3 determinant:

$$\mathbf{u} \times \mathbf{v} = \begin{vmatrix} \mathbf{i} & \mathbf{j} & \mathbf{k} \\ u_1 & u_2 & u_3 \\ v_1 & v_2 & v_3 \end{vmatrix} = \begin{vmatrix} u_2 & u_3 \\ v_2 & v_3 \end{vmatrix} \mathbf{i} - \begin{vmatrix} u_1 & u_3 \\ v_1 & v_3 \end{vmatrix} \mathbf{j} + \begin{vmatrix} u_1 & u_2 \\ v_1 & v_2 \end{vmatrix} \mathbf{k} \tag{4}$$

For example, if $\mathbf{u} = (1, 2, -2)$ and $\mathbf{v} = (3, 0, 1)$, then

$$\mathbf{u} \times \mathbf{v} = \begin{vmatrix} \mathbf{i} & \mathbf{j} & \mathbf{k} \\ 1 & 2 & -2 \\ 3 & 0 & 1 \end{vmatrix} = 2\mathbf{i} - 7\mathbf{j} - 6\mathbf{k}$$

which agrees with the result obtained in Example 1.

Warning. It is not true in general that $\mathbf{u} \times (\mathbf{v} \times \mathbf{w}) = (\mathbf{u} \times \mathbf{v}) \times \mathbf{w}$. For example,

$$\mathbf{i} \times (\mathbf{j} \times \mathbf{j}) = \mathbf{i} \times \mathbf{0} = \mathbf{0}$$

and

$$(\mathbf{i} \times \mathbf{j}) \times \mathbf{j} = \mathbf{k} \times \mathbf{j} = -\mathbf{i}$$

so that

$$\mathbf{i} \times (\mathbf{j} \times \mathbf{j}) \neq (\mathbf{i} \times \mathbf{j}) \times \mathbf{j}$$

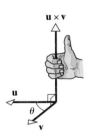

Figure 3.4.3

We know from Theorem 3.4.1 that $\mathbf{u} \times \mathbf{v}$ is orthogonal to both $\mathbf{u}$ and $\mathbf{v}$. If $\mathbf{u}$ and $\mathbf{v}$ are nonzero vectors, it can be shown that the direction of $\mathbf{u} \times \mathbf{v}$ can be determined using the following "right-hand rule"[†] (Figure 3.4.3): Let θ be the angle between $\mathbf{u}$ and $\mathbf{v}$, and suppose $\mathbf{u}$ is rotated through the angle θ until it coincides with $\mathbf{v}$. If the fingers of the right hand are cupped so they point in the direction of rotation, then the thumb indicates (roughly) the direction of $\mathbf{u} \times \mathbf{v}$.

The reader may find it instructive to practice this rule with the products

$$\mathbf{i} \times \mathbf{j} = \mathbf{k}, \qquad \mathbf{j} \times \mathbf{k} = \mathbf{i}, \qquad \mathbf{k} \times \mathbf{i} = \mathbf{j}$$

Geometric Interpretation of Cross Product

If $\mathbf{u}$ and $\mathbf{v}$ are vectors in 3-space, then the norm of $\mathbf{u} \times \mathbf{v}$ has a useful geometric interpretation. Lagrange's identity, given in Theorem 3.4.1, states that

$$\|\mathbf{u} \times \mathbf{v}\|^2 = \|\mathbf{u}\|^2 \|\mathbf{v}\|^2 - (\mathbf{u} \cdot \mathbf{v})^2 \tag{5}$$

If θ denotes the angle between $\mathbf{u}$ and $\mathbf{v}$, then $\mathbf{u} \cdot \mathbf{v} = \|\mathbf{u}\| \|\mathbf{v}\| \cos\theta$, so that (5) can be rewritten as

$$\begin{aligned} \|\mathbf{u} \times \mathbf{v}\|^2 &= \|\mathbf{u}\|^2 \|\mathbf{v}\|^2 - \|\mathbf{u}\|^2 \|\mathbf{v}\|^2 \cos^2\theta \\ &= \|\mathbf{u}\|^2 \|\mathbf{v}\|^2 (1 - \cos^2\theta) \\ &= \|\mathbf{u}\|^2 \|\mathbf{v}\|^2 \sin^2\theta \end{aligned}$$

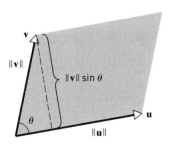

Figure 3.4.4

Since $0 \leq \theta \leq \pi$, it follows that $\sin\theta \geq 0$, so this can be rewritten as

$$\|\mathbf{u} \times \mathbf{v}\| = \|\mathbf{u}\| \|\mathbf{v}\| \sin\theta \tag{6}$$

But $\|\mathbf{v}\| \sin\theta$ is the altitude of the parallelogram determined by $\mathbf{u}$ and $\mathbf{v}$ (Figure 3.4.4). Thus, from (6), the area A of this parallelogram is given by

$$A = (\text{base})(\text{altitude}) = \|\mathbf{u}\| \|\mathbf{v}\| \sin\theta = \|\mathbf{u} \times \mathbf{v}\|$$

[†]Recall that we agreed to consider only right-handed coordinate systems in this text. Had we used left-handed systems instead, a "left-hand rule" would apply here.

This result is even correct if **u** and **v** are collinear, since the parallelogram determined by **u** and **v** has zero area and from (6) we have $\mathbf{u} \times \mathbf{v} = \mathbf{0}$ because $\theta = 0$ in this case. Thus we have the following theorem.

Theorem 3.4.3 | Area of a Parallelogram

*If **u** and **v** are vectors in 3-space, then $\|\mathbf{u} \times \mathbf{v}\|$ is equal to the area of the parallelogram determined by **u** and **v**.*

EXAMPLE 4 Area of a Triangle

Find the area of the triangle determined by the points $P_1(2, 2, 0)$, $P_2(-1, 0, 2)$, and $P_3(0, 4, 3)$.

Solution.

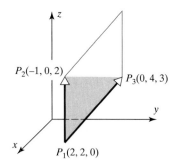

The area A of the triangle is $\frac{1}{2}$ the area of the parallelogram determined by the vectors $\overrightarrow{P_1 P_2}$ and $\overrightarrow{P_1 P_3}$ (Figure 3.4.5). Using the method discussed in Example 2 of Section 3.1, $\overrightarrow{P_1 P_2} = (-3, -2, 2)$ and $\overrightarrow{P_1 P_3} = (-2, 2, 3)$. It follows that

$$\overrightarrow{P_1 P_2} \times \overrightarrow{P_1 P_3} = (-10, 5, -10)$$

and consequently

$$A = \tfrac{1}{2}\|\overrightarrow{P_1 P_2} \times \overrightarrow{P_1 P_3}\| = \tfrac{1}{2}(15) = \tfrac{15}{2} \qquad \blacklozenge$$

Figure 3.4.5

Definition

If **u**, **v**, and **w** are vectors in 3-space, then

$$\mathbf{u} \cdot (\mathbf{v} \times \mathbf{w})$$

is called the ***scalar triple product*** of **u**, **v**, and **w**.

The scalar triple product of $\mathbf{u} = (u_1, u_2, u_3)$, $\mathbf{v} = (v_1, v_2, v_3)$, and $\mathbf{w} = (w_1, w_2, w_3)$ can be calculated from the formula

$$\mathbf{u} \cdot (\mathbf{v} \times \mathbf{w}) = \begin{vmatrix} u_1 & u_2 & u_3 \\ v_1 & v_2 & v_3 \\ w_1 & w_2 & w_3 \end{vmatrix} \tag{7}$$

This follows from Formula (4) since

$$\mathbf{u} \cdot (\mathbf{v} \times \mathbf{w}) = \mathbf{u} \cdot \left(\begin{vmatrix} v_2 & v_3 \\ w_2 & w_3 \end{vmatrix} \mathbf{i} - \begin{vmatrix} v_1 & v_3 \\ w_1 & w_3 \end{vmatrix} \mathbf{j} + \begin{vmatrix} v_1 & v_2 \\ w_1 & w_2 \end{vmatrix} \mathbf{k} \right)$$

$$= \begin{vmatrix} v_2 & v_3 \\ w_2 & w_3 \end{vmatrix} u_1 - \begin{vmatrix} v_1 & v_3 \\ w_1 & w_3 \end{vmatrix} u_2 + \begin{vmatrix} v_1 & v_2 \\ w_1 & w_2 \end{vmatrix} u_3$$

$$= \begin{vmatrix} u_1 & u_2 & u_3 \\ v_1 & v_2 & v_3 \\ w_1 & w_2 & w_3 \end{vmatrix}$$

EXAMPLE 5 Calculating a Scalar Triple Product

Calculate the scalar triple product $\mathbf{u} \cdot (\mathbf{v} \times \mathbf{w})$ of the vectors

$$\mathbf{u} = 3\mathbf{i} - 2\mathbf{j} - 5\mathbf{k}, \qquad \mathbf{v} = \mathbf{i} + 4\mathbf{j} - 4\mathbf{k}, \qquad \mathbf{w} = 3\mathbf{j} + 2\mathbf{k}$$

Solution.

From (7)

$$\mathbf{u} \cdot (\mathbf{v} \times \mathbf{w}) = \begin{vmatrix} 3 & -2 & -5 \\ 1 & 4 & -4 \\ 0 & 3 & 2 \end{vmatrix}$$

$$= 3 \begin{vmatrix} 4 & -4 \\ 3 & 2 \end{vmatrix} - (-2) \begin{vmatrix} 1 & -4 \\ 0 & 2 \end{vmatrix} + (-5) \begin{vmatrix} 1 & 4 \\ 0 & 3 \end{vmatrix}$$

$$= 60 + 4 - 15 = 49 \qquad ◆$$

REMARK. The symbol $(\mathbf{u} \cdot \mathbf{v}) \times \mathbf{w}$ makes no sense since we cannot form the cross product of a scalar and a vector. Thus, no ambiguity arises if we write $\mathbf{u} \cdot \mathbf{v} \times \mathbf{w}$ rather than $\mathbf{u} \cdot (\mathbf{v} \times \mathbf{w})$. However, for clarity we shall usually keep the parentheses.

It follows from (7) that

$$\mathbf{u} \cdot (\mathbf{v} \times \mathbf{w}) = \mathbf{w} \cdot (\mathbf{u} \times \mathbf{v}) = \mathbf{v} \cdot (\mathbf{w} \times \mathbf{u})$$

since the 3×3 determinants that represent these products can be obtained from one another by *two* row interchanges. (Verify.) These relationships can be remembered by moving the vectors $\mathbf{u}$, $\mathbf{v}$, and $\mathbf{w}$ clockwise around the vertices of the triangle in Figure 3.4.6.

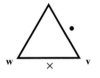

Figure 3.4.6

Geometric Interpretation of Determinants The next theorem provides a useful geometric interpretation of 2×2 and 3×3 determinants.

Theorem 3.4.4

(*a*) *The absolute value of the determinant*

$$\det \begin{bmatrix} u_1 & u_2 \\ v_1 & v_2 \end{bmatrix}$$

is equal to the area of the parallelogram in 2-space determined by the vectors $\mathbf{u} = (u_1, u_2)$ *and* $\mathbf{v} = (v_1, v_2)$. *(See Figure 3.4.7a.)*

(*b*) *The absolute value of the determinant*

$$\det \begin{bmatrix} u_1 & u_2 & u_3 \\ v_1 & v_2 & v_3 \\ w_1 & w_2 & w_3 \end{bmatrix}$$

is equal to the volume of the parallelepiped in 3-space determined by the vectors $\mathbf{u} = (u_1, u_2, u_3)$, $\mathbf{v} = (v_1, v_2, v_3)$, *and* $\mathbf{w} = (w_1, w_2, w_3)$. *(See Figure 3.4.7b.)*

Proof (a). The key to the proof is to use Theorem 3.4.3. However, that theorem applies to vectors in 3-space, whereas $\mathbf{u} = (u_1, u_2)$ and $\mathbf{v} = (v_1, v_2)$ are vectors in 2-space. To circumvent this "dimension problem," we shall view $\mathbf{u}$ and $\mathbf{v}$ as vectors in the xy-plane

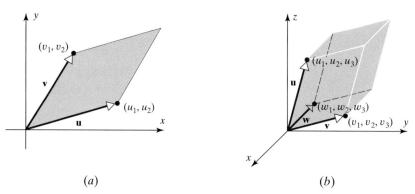

Figure 3.4.7

of an xyz-coordinate system (Figure 3.4.8a), in which case these vectors are expressed as $\mathbf{u} = (u_1, u_2, 0)$ and $\mathbf{v} = (v_1, v_2, 0)$. Thus,

$$\mathbf{u} \times \mathbf{v} = \begin{vmatrix} \mathbf{i} & \mathbf{j} & \mathbf{k} \\ u_1 & u_2 & 0 \\ v_1 & v_2 & 0 \end{vmatrix} = \begin{vmatrix} u_1 & u_2 \\ v_1 & v_2 \end{vmatrix} \mathbf{k} = \det \begin{bmatrix} u_1 & u_2 \\ v_1 & v_2 \end{bmatrix} \mathbf{k}$$

It now follows from Theorem 3.4.3 and the fact that $\|\mathbf{k}\| = 1$ that the area A of the parallelogram determined by $\mathbf{u}$ and $\mathbf{v}$ is

$$A = \|\mathbf{u} \times \mathbf{v}\| = \left\| \det \begin{bmatrix} u_1 & u_2 \\ v_1 & v_2 \end{bmatrix} \mathbf{k} \right\| = \left| \det \begin{bmatrix} u_1 & u_2 \\ v_1 & v_2 \end{bmatrix} \right| \|\mathbf{k}\| = \left| \det \begin{bmatrix} u_1 & u_2 \\ v_1 & v_2 \end{bmatrix} \right|$$

which completes the proof.

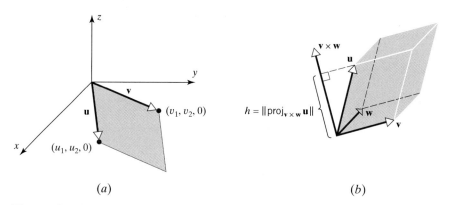

Figure 3.4.8

Proof (b). As shown in Figure 3.4.8b, take the base of the parallelepiped determined by $\mathbf{u}, \mathbf{v}$, and $\mathbf{w}$ to be the parallelogram determined by $\mathbf{v}$ and $\mathbf{w}$. It follows from Theorem 3.4.3 that the area of the base is $\|\mathbf{v} \times \mathbf{w}\|$ and, as illustrated in Figure 3.4.8b, the height h of the parallelepiped is the length of the orthogonal projection of $\mathbf{u}$ on $\mathbf{v} \times \mathbf{w}$. Therefore, by Formula (10) of Section 3.3,

$$h = \|\text{proj}_{\mathbf{v} \times \mathbf{w}} \, \mathbf{u}\| = \frac{|\mathbf{u} \cdot (\mathbf{v} \times \mathbf{w})|}{\|\mathbf{v} \times \mathbf{w}\|}$$

It follows that the volume V of the parallelepiped is

$$V = (\text{area of base}) \cdot \text{height} = \|\mathbf{v} \times \mathbf{w}\| \frac{|\mathbf{u} \cdot (\mathbf{v} \times \mathbf{w})|}{\|\mathbf{v} \times \mathbf{w}\|} = |\mathbf{u} \cdot (\mathbf{v} \times \mathbf{w})|$$

so from (7)

$$V = \left| \det \begin{bmatrix} u_1 & u_2 & u_3 \\ v_1 & v_2 & v_3 \\ w_1 & w_2 & w_3 \end{bmatrix} \right|$$

which completes the proof. ∎

REMARK. If V denotes the volume of the parallelepiped determined by vectors $\mathbf{u}$, $\mathbf{v}$, and $\mathbf{w}$, then it follows from Theorem 3.4.4 and Formula (7) that

$$V = \begin{bmatrix} \text{volume of parallelepiped} \\ \text{determined by } \mathbf{u}, \mathbf{v}, \text{ and } \mathbf{w} \end{bmatrix} = |\mathbf{u} \cdot (\mathbf{v} \times \mathbf{w})| \qquad (8)$$

From this and Theorem 3.3.1*b*, we can conclude that

$$\mathbf{u} \cdot (\mathbf{v} \times \mathbf{w}) = \pm V$$

where the $+$ or $-$ results depending on whether $\mathbf{u}$ makes an acute or obtuse angle with $\mathbf{v} \times \mathbf{w}$.

Formula (8) leads to a useful test for ascertaining whether three given vectors lie in the same plane. Since three vectors not in the same plane determine a parallelepiped of positive volume, it follows from (8) that $|\mathbf{u} \cdot (\mathbf{v} \times \mathbf{w})| = 0$ if and only if the vectors $\mathbf{u}$, $\mathbf{v}$, and $\mathbf{w}$ lie in the same plane. Thus, we have the following result.

Theorem 3.4.5

If the vectors $\mathbf{u} = (u_1, u_2, u_3)$, $\mathbf{v} = (v_1, v_2, v_3)$, *and* $\mathbf{w} = (w_1, w_2, w_3)$ *have the same initial point, then they lie in the same plane if and only if*

$$\mathbf{u} \cdot (\mathbf{v} \times \mathbf{w}) = \begin{vmatrix} u_1 & u_2 & u_3 \\ v_1 & v_2 & v_3 \\ w_1 & w_2 & w_3 \end{vmatrix} = 0$$

Independence of Cross Product and Coordinates

Initially, we defined a vector to be a directed line segment or arrow in 2-space or 3-space; coordinate systems and components were introduced later in order to simplify computations with vectors. Thus, a vector has a "mathematical existence" regardless of whether a coordinate system has been introduced. Further, the components of a vector are not determined by the vector alone; they depend as well on the coordinate system chosen. For example, in Figure 3.4.9 we have indicated a fixed vector $\mathbf{v}$ in the plane and two

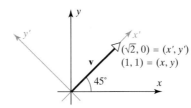

Figure 3.4.9

different coordinate systems. In the xy-coordinate system the components of **v** are $(1, 1)$, and in the $x'y'$-system they are $(\sqrt{2}, 0)$.

This raises an important question about our definition of cross product. Since we defined the cross product **u** × **v** in terms of the components of **u** and **v**, and since these components depend on the coordinate system chosen, it seems possible that two *fixed* vectors **u** and **v** might have different cross products in different coordinate systems. Fortunately, this is not the case. To see that this is so, we need only recall that

- **u** × **v** is perpendicular to both **u** and **v**.
- The orientation of **u** × **v** is determined by the right-hand rule.
- $\|\mathbf{u} \times \mathbf{v}\| = \|\mathbf{u}\|\|\mathbf{v}\| \sin \theta$.

These three properties completely determine the vector **u** × **v**: the first and second properties determine the direction, and the third property determines the length. Since these properties of **u** × **v** depend only on the lengths and relative positions of **u** and **v** and not on the particular right-hand coordinate system being used, the vector **u** × **v** will remain unchanged if a different right-hand coordinate system is introduced. Thus, we say that the definition of **u** × **v** is *coordinate free*. This result is of importance to physicists and engineers who often work with many coordinate systems in the same problem.

EXAMPLE 6 u × v Is Independent of the Coordinate System

Consider two perpendicular vectors **u** and **v**, each of length 1 (Figure 3.4.10*a*). If we introduce an xyz-coordinate system as shown in Figure 3.4.10*b*, then

$$\mathbf{u} = (1, 0, 0) = \mathbf{i} \quad \text{and} \quad \mathbf{v} = (0, 1, 0) = \mathbf{j}$$

so that

$$\mathbf{u} \times \mathbf{v} = \mathbf{i} \times \mathbf{j} = \mathbf{k} = (0, 0, 1)$$

However, if we introduce an $x'y'z'$-coordinate system as shown in Figure 3.4.10*c*, then

$$\mathbf{u} = (0, 0, 1) = \mathbf{k} \quad \text{and} \quad \mathbf{v} = (1, 0, 0) = \mathbf{i}$$

so that

$$\mathbf{u} \times \mathbf{v} = \mathbf{k} \times \mathbf{i} = \mathbf{j} = (0, 1, 0)$$

But it is clear from Figures 3.4.10*b* and 3.4.10*c* that the vector $(0, 0, 1)$ in the xyz-system is the same as the vector $(0, 1, 0)$ in the $x'y'z'$-system. Thus, we obtain the same vector **u** × **v** whether we compute with coordinates from the xyz-system or with coordinates from the $x'y'z'$-system. ◆

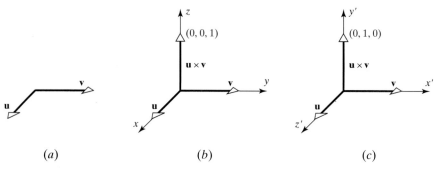

(a) (b) (c)

Figure 3.4.10

Exercise Set 3.4

1. Let $\mathbf{u} = (3, 2, -1)$, $\mathbf{v} = (0, 2, -3)$, and $\mathbf{w} = (2, 6, 7)$. Compute
   - (a) $\mathbf{v} \times \mathbf{w}$
   - (b) $\mathbf{u} \times (\mathbf{v} \times \mathbf{w})$
   - (c) $(\mathbf{u} \times \mathbf{v}) \times \mathbf{w}$
   - (d) $(\mathbf{u} \times \mathbf{v}) \times (\mathbf{v} \times \mathbf{w})$
   - (e) $\mathbf{u} \times (\mathbf{v} - 2\mathbf{w})$
   - (f) $(\mathbf{u} \times \mathbf{v}) - 2\mathbf{w}$

2. Find a vector that is orthogonal to both $\mathbf{u}$ and $\mathbf{v}$.
   - (a) $\mathbf{u} = (-6, 4, 2)$, $\mathbf{v} = (3, 1, 5)$
   - (b) $\mathbf{u} = (-2, 1, 5)$, $\mathbf{v} = (3, 0, -3)$

3. Find the area of the parallelogram determined by $\mathbf{u}$ and $\mathbf{v}$.
   - (a) $\mathbf{u} = (1, -1, 2)$, $\mathbf{v} = (0, 3, 1)$
   - (b) $\mathbf{u} = (2, 3, 0)$, $\mathbf{v} = (-1, 2, -2)$
   - (c) $\mathbf{u} = (3, -1, 4)$, $\mathbf{v} = (6, -2, 8)$

4. Find the area of the triangle having vertices P, Q, and R.
   - (a) $P(2, 6, -1)$, $Q(1, 1, 1)$, $R(4, 6, 2)$
   - (b) $P(1, -1, 2)$, $Q(0, 3, 4)$, $R(6, 1, 8)$

5. Verify parts (a), (b), and (c) of Theorem 3.4.1 for the vectors $\mathbf{u} = (4, 2, 1)$ and $\mathbf{v} = (-3, 2, 7)$.

6. Verify parts (a), (b), and (c) of Theorem 3.4.2 for $\mathbf{u} = (5, -1, 2)$, $\mathbf{v} = (6, 0, -2)$, and $\mathbf{w} = (1, 2, -1)$.

7. Find a vector $\mathbf{v}$ that is orthogonal to the vector $\mathbf{u} = (2, -3, 5)$.

8. Find the scalar triple product $\mathbf{u} \cdot (\mathbf{v} \times \mathbf{w})$.
   - (a) $\mathbf{u} = (-1, 2, 4)$, $\mathbf{v} = (3, 4, -2)$, $\mathbf{w} = (-1, 2, 5)$
   - (b) $\mathbf{u} = (3, -1, 6)$, $\mathbf{v} = (2, 4, 3)$, $\mathbf{w} = (5, -1, 2)$

9. Suppose that $\mathbf{u} \cdot (\mathbf{v} \times \mathbf{w}) = 3$. Find
   - (a) $\mathbf{u} \cdot (\mathbf{w} \times \mathbf{v})$
   - (b) $(\mathbf{v} \times \mathbf{w}) \cdot \mathbf{u}$
   - (c) $\mathbf{w} \cdot (\mathbf{u} \times \mathbf{v})$
   - (d) $\mathbf{v} \cdot (\mathbf{u} \times \mathbf{w})$
   - (e) $(\mathbf{u} \times \mathbf{w}) \cdot \mathbf{v}$
   - (f) $\mathbf{v} \cdot (\mathbf{w} \times \mathbf{w})$

10. Find the volume of the parallelepiped with sides $\mathbf{u}$, $\mathbf{v}$, and $\mathbf{w}$.
    - (a) $\mathbf{u} = (2, -6, 2)$, $\mathbf{v} = (0, 4, -2)$, $\mathbf{w} = (2, 2, -4)$
    - (b) $\mathbf{u} = (3, 1, 2)$, $\mathbf{v} = (4, 5, 1)$, $\mathbf{w} = (1, 2, 4)$

11. Determine whether $\mathbf{u}$, $\mathbf{v}$, and $\mathbf{w}$ lie in the same plane when positioned so that their initial points coincide.
    - (a) $\mathbf{u} = (-1, -2, 1)$, $\mathbf{v} = (3, 0, -2)$, $\mathbf{w} - (5, -4, 0)$
    - (b) $\mathbf{u} = (5, -2, 1)$, $\mathbf{v} = (4, -1, 1)$, $\mathbf{w} = (1, -1, 0)$
    - (c) $\mathbf{u} = (4, -8, 1)$, $\mathbf{v} = (2, 1, -2)$, $\mathbf{w} = (3, -4, 12)$

12. Find all unit vectors parallel to the yz-plane that are perpendicular to the vector $(3, -1, 2)$.

13. Find all unit vectors in the plane determined by $\mathbf{u} = (3, 0, 1)$ and $\mathbf{v} = (1, -1, 1)$ that are perpendicular to the vector $\mathbf{w} = (1, 2, 0)$.

14. Let $\mathbf{a} = (a_1, a_2, a_3)$, $\mathbf{b} = (b_1, b_2, b_3)$, $\mathbf{c} = (c_1, c_2, c_3)$, and $\mathbf{d} = (d_1, d_2, d_3)$. Show that
$$(\mathbf{a} + \mathbf{d}) \cdot (\mathbf{b} \times \mathbf{c}) = \mathbf{a} \cdot (\mathbf{b} \times \mathbf{c}) + \mathbf{d} \cdot (\mathbf{b} \times \mathbf{c})$$

15. Simplify $(\mathbf{u} + \mathbf{v}) \times (\mathbf{u} - \mathbf{v})$.

16. Use the cross product to find the sine of the angle between the vectors $\mathbf{u} = (2, 3, -6)$ and $\mathbf{v} = (2, 3, 6)$.

17. (a) Find the area of the triangle having vertices $A(1, 0, 1)$, $B(0, 2, 3)$, and $C(2, 1, 0)$.
    (b) Use the result of part (a) to find the length of the altitude from vertex C to side AB.

18. Show that if $\mathbf{u}$ is a vector from any point on a line to a point P not on the line, and $\mathbf{v}$ is a vector parallel to the line, then the distance between P and the line is given by $\|\mathbf{u} \times \mathbf{v}\|/\|\mathbf{v}\|$.

19. Use the result of Exercise 18 to find the distance between the point P and the line through the points A and B:
    - (a) $P(-3, 1, 2)$, $A(1, 1, 0)$, $B(-2, 3, -4)$
    - (b) $P(4, 3, 0)$, $A(2, 1, -3)$, $B(0, 2, -1)$

20. Prove: If θ is the angle between $\mathbf{u}$ and $\mathbf{v}$ and $\mathbf{u} \cdot \mathbf{v} \neq 0$, then $\tan\theta = \|\mathbf{u} \times \mathbf{v}\|/(\mathbf{u} \cdot \mathbf{v})$.

21. Consider the parallelepiped with sides $\mathbf{u} = (3, 2, 1)$, $\mathbf{v} = (1, 1, 2)$, and $\mathbf{w} = (1, 3, 3)$.

    (a) Find the area of the face determined by $\mathbf{u}$ and $\mathbf{w}$.
    (b) Find the angle between $\mathbf{u}$ and the plane containing the face determined by $\mathbf{v}$ and $\mathbf{w}$. [*Note.* The *angle between a vector and a plane* is defined to be the complement of the angle θ between the vector and that normal to the plane for which $0 \leq \theta \leq \pi/2$.]

22. Find a vector $\mathbf{n}$ perpendicular to the plane determined by the points $A(0, -2, 1)$, $B(1, -1, -2)$, and $C(-1, 1, 0)$. [See the note in Exercise 21.]

23. Let $\mathbf{m}$ and $\mathbf{n}$ be vectors whose components in the xyz-system of Figure 3.4.10 are $\mathbf{m} = (0, 0, 1)$ and $\mathbf{n} = (0, 1, 0)$.

    (a) Find the components of $\mathbf{m}$ and $\mathbf{n}$ in the $x'y'z'$-system of Figure 3.4.10.
    (b) Compute $\mathbf{m} \times \mathbf{n}$ using the components in the xyz-system.
    (c) Compute $\mathbf{m} \times \mathbf{n}$ using the components in the $xy'z'$-system.
    (d) Show that the vectors obtained in (b) and (c) are the same.

24. Prove the following identities.

    (a) $(\mathbf{u} + k\mathbf{v}) \times \mathbf{v} = \mathbf{u} \times \mathbf{v}$ (b) $\mathbf{u} \cdot (\mathbf{v} \times \mathbf{z}) = -(\mathbf{u} \times \mathbf{z}) \cdot \mathbf{v}$

25. Let $\mathbf{u}$, $\mathbf{v}$, and $\mathbf{w}$ be nonzero vectors in 3-space with the same initial point, but such that no two of them are collinear. Show that

    (a) $\mathbf{u} \times (\mathbf{v} \times \mathbf{w})$ lies in the plane determined by $\mathbf{v}$ and $\mathbf{w}$
    (b) $(\mathbf{u} \times \mathbf{v}) \times \mathbf{w}$ lies in the plane determined by $\mathbf{u}$ and $\mathbf{v}$

26. Prove part (*d*) of Theorem 3.4.1. [*Hint.* First prove the result in the case where $\mathbf{w} = \mathbf{i} = (1, 0, 0)$, then when $\mathbf{w} = \mathbf{j} = (0, 1, 0)$, and then when $\mathbf{w} = \mathbf{k} = (0, 0, 1)$. Finally prove it for an arbitrary vector $\mathbf{w} = (w_1, w_2, w_3)$ by writing $\mathbf{w} = w_1\mathbf{i} + w_2\mathbf{j} + w_3\mathbf{k}$.]

27. Prove part (*e*) of Theorem 3.4.1. [*Hint.* Apply part (*a*) of Theorem 3.4.2 to the result in part (*d*) of Theorem 3.4.1.]

28. Let $\mathbf{u} = (1, 3, -1)$, $\mathbf{v} = (1, 1, 2)$, and $\mathbf{w} = (3, -1, 2)$. Calculate $\mathbf{u} \times (\mathbf{v} \times \mathbf{w})$ using the technique of Exercise 26; then check your result by calculating directly.

29. Prove: If $\mathbf{a}$, $\mathbf{b}$, $\mathbf{c}$, and $\mathbf{d}$ lie in the same plane, then $(\mathbf{a} \times \mathbf{b}) \times (\mathbf{c} \times \mathbf{d}) = \mathbf{0}$.

30. It is a theorem of solid geometry that the volume of a tetrahedron is $\frac{1}{3}$(area of base) · (height). Use this result to prove that the volume of a tetrahedron whose sides are the vectors $\mathbf{a}$, $\mathbf{b}$, and $\mathbf{c}$ is $\frac{1}{6}|\mathbf{a} \cdot (\mathbf{b} \times \mathbf{c})|$ (see accompanying figure).

Figure Ex-30

31. Use the result of Exercise 30 to find the volume of the tetrahedron with vertices P, Q, R, S.

    (a) $P(-1, 2, 0)$, $Q(2, 1, -3)$, $R(1, 0, 1)$, $S(3, -2, 3)$
    (b) $P(0, 0, 0)$, $Q(1, 2, -1)$, $R(3, 4, 0)$, $S(-1, -3, 4)$

32. Prove part (*b*) of Theorem 3.4.2.

33. Prove parts (*c*) and (*d*) of Theorem 3.4.2.

34. Prove parts (*e*) and (*f*) of Theorem 3.4.2.

Discussion and Discovery

35. (a) Suppose that $\mathbf{u}$ and $\mathbf{v}$ are noncollinear vectors with their initial points at the origin in 3-space. Make a sketch that illustrates how $\mathbf{w} = \mathbf{v} \times (\mathbf{u} \times \mathbf{v})$ is oriented in relation to $\mathbf{u}$ and $\mathbf{v}$.

    (b) What can you say about the values of $\mathbf{u} \cdot \mathbf{w}$ and $\mathbf{v} \cdot \mathbf{w}$? Explain your reasoning.

36. If $\mathbf{u} \neq \mathbf{0}$, is it valid to cancel $\mathbf{u}$ from both sides of the equation $\mathbf{u} \times \mathbf{v} = \mathbf{u} \times \mathbf{w}$ and conclude that $\mathbf{v} = \mathbf{w}$? Explain your reasoning.

37. Something is wrong with one of the following expressions. Which one is it and what is wrong?

$$\mathbf{u} \cdot (\mathbf{v} \times \mathbf{w}), \quad \mathbf{u} \times \mathbf{v} \times \mathbf{w}, \quad \mathbf{u} \cdot \mathbf{v} \times \mathbf{w}$$

38. What can you say about the vectors $\mathbf{u}$ and $\mathbf{v}$ if $\mathbf{u} \times \mathbf{v} = \mathbf{0}$?

39. Give some examples of algebraic rules that hold for multiplication of real numbers but not for the cross product of vectors.

3.5 LINES AND PLANES IN 3-SPACE

In this section we shall use vectors to derive equations of lines and planes in 3-space. We shall then use these equations to solve some basic geometric problems.

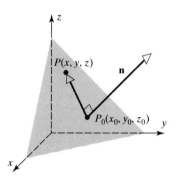

Figure 3.5.1
Plane with normal vector.

Planes in 3-Space In analytic geometry a line in 2-space can be specified by giving its slope and one of its points. Similarly, one can specify a plane in 3-space by giving its inclination and specifying one of its points. A convenient method for describing the inclination of a plane is to specify a nonzero vector, called a ***normal***, that is perpendicular to the plane.

Suppose that we want to find the equation of the plane passing through the point $P_0(x_0, y_0, z_0)$ and having the nonzero vector $\mathbf{n} = (a, b, c)$ as a normal. It is evident from Figure 3.5.1 that the plane consists precisely of those points $P(x, y, z)$ for which the vector $\overrightarrow{P_0 P}$ is orthogonal to $\mathbf{n}$, that is,

$$\mathbf{n} \cdot \overrightarrow{P_0 P} = 0 \tag{1}$$

Since $\overrightarrow{P_0 P} = (x - x_0, y - y_0, z - z_0)$, Equation (1) can be written as

$$a(x - x_0) + b(y - y_0) + c(z - z_0) = 0 \tag{2}$$

We call this the ***point-normal*** form of the equation of a plane.

EXAMPLE 1 Finding the Point-Normal Equation of a Plane

Find an equation of the plane passing through the point $(3, -1, 7)$ and perpendicular to the vector $\mathbf{n} = (4, 2, -5)$.

Solution.

From (2) a point-normal form is

$$4(x - 3) + 2(y + 1) - 5(z - 7) = 0 \qquad\qquad ◆$$

By multiplying out and collecting terms, (2) can be rewritten in the form

$$ax + by + cz + d = 0$$

where a, b, c, and d are constants, and a, b, and c are not all zero. For example, the equation in Example 1 can be rewritten as

$$4x + 2y - 5z + 25 = 0$$

As the next theorem shows, planes in 3-space are represented by equations of the form $ax + by + cz + d = 0$.

Theorem 3.5.1

If a, b, c, and d are constants and a, b, and c are not all zero, then the graph of the equation

$$ax + by + cz + d = 0 \tag{3}$$

is a plane having the vector $\mathbf{n} = (a, b, c)$ as a normal.

Equation (3) is a linear equation in x, y, and z; it is called the ***general form*** of the equation of a plane.

Proof. By hypothesis, the coefficients a, b, and c are not all zero. Assume, for the moment, that $a \neq 0$. Then the equation $ax + by + cz + d = 0$ can be rewritten in the form $a(x + (d/a)) + by + cz = 0$. But this is a point-normal form of the plane passing through the point $(-d/a, 0, 0)$ and having $\mathbf{n} = (a, b, c)$ as a normal.

If $a = 0$, then either $b \neq 0$ or $c \neq 0$. A straightforward modification of the above argument will handle these other cases. ∎

Just as the solutions of a system of linear equations

$$ax + by = k_1$$
$$cx + dy = k_2$$

correspond to points of intersection of the lines $ax + by = k_1$ and $cx + dy = k_2$ in the xy-plane, so the solutions of a system

$$ax + by + cz = k_1$$
$$dx + ey + fz = k_2 \tag{4}$$
$$gx + hy + iz = k_3$$

correspond to the points of intersection of the three planes $ax + by + cz = k_1$, $dx + ey + fz = k_2$, and $gx + hy + iz = k_3$.

In Figure 3.5.2 we have illustrated the geometric possibilities that occur when (4) has zero, one, or infinitely many solutions.

EXAMPLE 2 Equation of a Plane Through Three Points

Find the equation of the plane passing through the points $P_1(1, 2, -1)$, $P_2(2, 3, 1)$, and $P_3(3, -1, 2)$.

Solution.

Since the three points lie in the plane, their coordinates must satisfy the general equation $ax + by + cz + d = 0$ of the plane. Thus,

$$a + 2b - c + d = 0$$
$$2a + 3b + c + d = 0$$
$$3a - b + 2c + d = 0$$

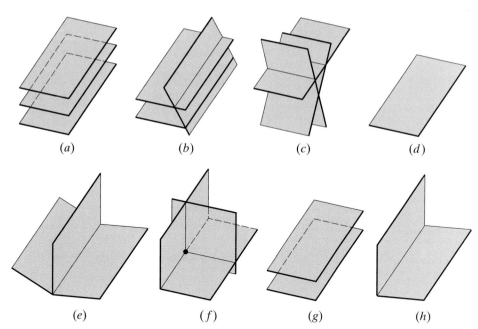

Figure 3.5.2 (*a*) No solutions (3 parallel planes). (*b*) No solutions (2 planes parallel). (*c*) No solutions (3 planes with no common intersection). (*d*) Infinitely many solutions (3 coincident planes). (*e*) Infinitely many solutions (3 planes intersecting in a line). (*f*) One solution (3 planes intersecting at a point). (*g*) No solutions (2 coincident planes parallel to a third plane). (*h*) Infinitely many solutions (2 coincident planes intersecting a third plane).

Solving this system gives $a = -\frac{9}{16}t$, $b = -\frac{1}{16}t$, $c = \frac{5}{16}t$, $d = t$. Letting $t = -16$, for example, yields the desired equation

$$9x + y - 5z - 16 = 0$$

We note that any other choice of t gives a multiple of this equation, so that any value of $t \neq 0$ would also give a valid equation of the plane.

Alternative Solution.

Since the points $P_1(1, 2, -1)$, $P_2(2, 3, 1)$, and $P_3(3, -1, 2)$ lie in the plane, the vectors $\overrightarrow{P_1 P_2} = (1, 1, 2)$ and $\overrightarrow{P_1 P_3} = (2, -3, 3)$ are parallel to the plane. Therefore, the equation $\overrightarrow{P_1 P_2} \times \overrightarrow{P_1 P_3} = (9, 1, -5)$ is normal to the plane, since it is perpendicular to both $\overrightarrow{P_1 P_2}$ and $\overrightarrow{P_1 P_3}$. From this and the fact that P_1 lies in the plane, a point-normal form for the equation of the plane is

$$9(x - 1) + (y - 2) - 5(z + 1) = 0 \quad \text{or} \quad 9x + y - 5z - 16 = 0 \qquad \blacklozenge$$

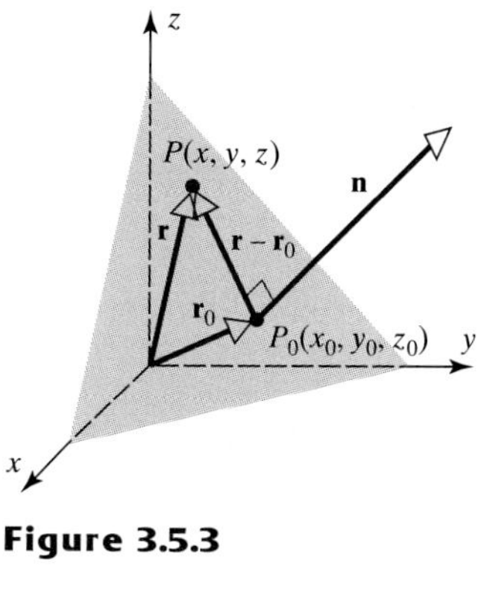

Vector Form of Equation of a Plane Vector notation provides a useful alternative way of writing the point-normal form of the equation of a plane: Referring to Figure 3.5.3, let $\mathbf{r} = (x, y, z)$ be the vector from the origin to the point $P(x, y, z)$, let $\mathbf{r}_0 = (x_0, y_0, z_0)$ be the vector from the origin to the point $P_0(x_0, y_0, z_0)$, and let $\mathbf{n} = (a, b, c)$ be a vector normal to the plane. Then $\overrightarrow{P_0 P} = \mathbf{r} - \mathbf{r}_0$, so Formula (1) can be rewritten as

$$\mathbf{n} \cdot (\mathbf{r} - \mathbf{r}_0) = 0 \tag{5}$$

This is called the ***vector form of the equation of a plane***.

Figure 3.5.3

EXAMPLE 3 Vector Equation of a Plane Using (5)

The equation

$$(-1, 2, 5) \cdot (x - 6, y - 3, z + 4) = 0$$

is the vector equation of the plane that passes through the point $(6, 3, -4)$ and is perpendicular to the vector $\mathbf{n} = (-1, 2, 5)$. ♦

Lines in 3-Space   We shall now show how to obtain equations for lines in 3-space. Suppose that l is the line in 3-space through the point $P_0(x_0, y_0, z_0)$ and parallel to the nonzero vector $\mathbf{v} = (a, b, c)$. It is clear (Figure 3.5.4) that l consists precisely of those points $P(x, y, z)$ for which the vector $\overrightarrow{P_0 P}$ is parallel to $\mathbf{v}$, that is, for which there is a scalar t such that

$$\overrightarrow{P_0 P} = t\mathbf{v} \tag{6}$$

In terms of components, (6) can be written as

$$(x - x_0, y - y_0, z - z_0) = (ta, tb, tc)$$

from which it follows that $x - x_0 = ta$, $y - y_0 = tb$, and $z - z_0 = tc$, so that

$$x = x_0 + ta, \qquad y = y_0 + tb, \qquad z = z_0 + tc$$

As the parameter t varies from $-\infty$ to $+\infty$, the point $P(x, y, z)$ traces out the line l. The equations

$$x = x_0 + ta, \quad y = y_0 + tb, \quad z = z_0 + tc \quad (-\infty < t < +\infty) \tag{7}$$

are called **parametric equations** for l.

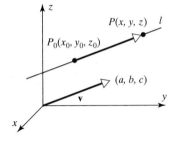

Figure 3.5.4
$\overrightarrow{P_0 P}$ is parallel to $\mathbf{v}$.

EXAMPLE 4 Parametric Equations of a Line

The line through the point $(1, 2, -3)$ and parallel to the vector $\mathbf{v} = (4, 5, -7)$ has parametric equations

$$x = 1 + 4t, \quad y = 2 + 5t, \quad z = -3 - 7t \quad (-\infty < t < +\infty) \qquad ♦$$

EXAMPLE 5 Intersection of a Line and the xy-Plane

(a) Find parametric equations for the line l passing through the points $P_1(2, 4, -1)$ and $P_2(5, 0, 7)$.

(b) Where does the line intersect the xy-plane?

Solution (a).   Since the vector $\overrightarrow{P_1 P_2} = (3, -4, 8)$ is parallel to l and $P_1(2, 4, -1)$ lies on l, the line l is given by

$$x = 2 + 3t, \quad y = 4 - 4t, \quad z = -1 + 8t \quad (-\infty < t < +\infty)$$

Solution (b).   The line intersects the xy-plane at the point where $z = -1 + 8t = 0$, that is, where $t = \frac{1}{8}$. Substituting this value of t in the parametric equations for l yields as the point of intersection

$$(x, y, z) = \left(\tfrac{19}{8}, \tfrac{7}{2}, 0 \right) \qquad ♦$$

EXAMPLE 6 Line of Intersection of Two Planes

Find parametric equations for the line of intersection of the planes

$$3x + 2y - 4z - 6 = 0 \quad \text{and} \quad x - 3y - 2z - 4 = 0$$

Solution.

The line of intersection consists of all points (x, y, z) that satisfy the two equations in the system

$$3x + 2y - 4z = 6$$
$$x - 3y - 2z = 4$$

Solving this system gives $x = \frac{26}{11} + \frac{16}{11}t$, $y = -\frac{6}{11} - \frac{2}{11}t$, $z = t$. Therefore, the line of intersection can be represented by the parametric equations

$$x = \tfrac{26}{11} + \tfrac{16}{11}t, \quad y = -\tfrac{6}{11} - \tfrac{2}{11}t, \quad z = t \quad (-\infty < t < +\infty) \qquad \blacklozenge$$

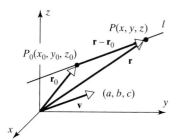

Figure 3.5.5
Vector interpretation of a line in 3-space.

Vector Form of Equation of a Line

Vector notation provides a useful alternative way of writing the parametric equations of a line: Referring to Figure 3.5.5, let $\mathbf{r} = (x, y, z)$ be the vector from the origin to the point $P(x, y, z)$, let $\mathbf{r}_0 = (x_0, y_0, z_0)$ be the vector from the origin to the point $P_0(x_0, y_0, z_0)$, and let $\mathbf{v} = (a, b, c)$ be a vector parallel to the line. Then $\overrightarrow{P_0 P} = \mathbf{r} - \mathbf{r}_0$, so Formula (6) can be rewritten as

$$\mathbf{r} - \mathbf{r}_0 = t\mathbf{v}$$

Taking into account the range of t-values, this can be rewritten as

$$\mathbf{r} = \mathbf{r}_0 + t\mathbf{v} \quad (-\infty < t < +\infty) \tag{8}$$

This is called the ***vector form of the equation of a line*** in 3-space.

EXAMPLE 7 A Line Parallel to a Given Vector

The equation

$$(x, y, z) = (-2, 0, 3) + t(4, -7, 1) \quad (-\infty < t < +\infty)$$

is the vector equation of the line through the point $(-2, 0, 3)$ that is parallel to the vector $\mathbf{v} = (4, -7, 1)$. $\qquad \blacklozenge$

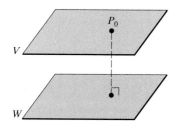

Figure 3.5.6
The distance between the parallel planes V and W is equal to the distance between P_0 and W.

Problems Involving Distance

We conclude this section by discussing two basic "distance problems" in 3-space:

Problems:
(a) Find the distance between a point and a plane.
(b) Find the distance between two parallel planes.

The two problems are related. If we can find the distance between a point and a plane, then we can find the distance between parallel planes by computing the distance between either one of the planes and an arbitrary point P_0 in the other (Figure 3.5.6).

Theorem 3.5.2 **Distance Between a Point and a Plane**

The distance D between a point $P_0(x_0, y_0, z_0)$ and the plane $ax + by + cz + d = 0$ is

$$D = \frac{|ax_0 + by_0 + cz_0 + d|}{\sqrt{a^2 + b^2 + c^2}} \qquad (9)$$

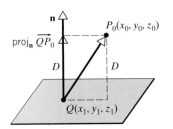

Figure 3.5.7
Distance from P_0 to plane.

Proof. Let $Q(x_1, y_1, z_1)$ be any point in the plane. Position the normal $\mathbf{n} = (a, b, c)$ so that its initial point is at Q. As illustrated in Figure 3.5.7, the distance D is equal to the length of the orthogonal projection of $\overrightarrow{QP_0}$ on $\mathbf{n}$. Thus, from (10) of Section 3.3,

$$D = \|\text{proj}_{\mathbf{n}} \overrightarrow{QP_0}\| = \frac{|\overrightarrow{QP_0} \cdot \mathbf{n}|}{\|\mathbf{n}\|}$$

But

$$\overrightarrow{QP_0} = (x_0 - x_1, y_0 - y_1, z_0 - z_1)$$

$$\overrightarrow{QP_0} \cdot \mathbf{n} = a(x_0 - x_1) + b(y_0 - y_1) + c(z_0 - z_1)$$

$$\|\mathbf{n}\| = \sqrt{a^2 + b^2 + c^2}$$

Thus,

$$D = \frac{|a(x_0 - x_1) + b(y_0 - y_1) + c(z_0 - z_1)|}{\sqrt{a^2 + b^2 + c^2}} \qquad (10)$$

Since the point $Q(x_1, y_1, z_1)$ lies in the plane, its coordinates satisfy the equation of the plane; thus

$$ax_1 + by_1 + cz_1 + d = 0$$

or

$$d = -ax_1 - by_1 - cz_1$$

Substituting this expression in (10) yields (9). ■

REMARK. Note the similarity between (9) and the formula for the distance between a point and a line in 2-space [(13) of Section 3.3].

EXAMPLE 8 Distance Between a Point and a Plane

Find the distance D between the point $(1, -4, -3)$ and the plane $2x - 3y + 6z = -1$.

Solution.

To apply (9), we first rewrite the equation of the plane in the form

$$2x - 3y + 6z + 1 = 0$$

Then

$$D = \frac{|2(1) + (-3)(-4) + 6(-3) + 1|}{\sqrt{2^2 + (-3)^2 + 6^2}} = \frac{|-3|}{7} = \frac{3}{7} \qquad ♦$$

Given two planes, either they intersect, in which case we can ask for their line of intersection, as in Example 6, or they are parallel, in which case we can ask for the distance between them. The following example illustrates the latter problem.

EXAMPLE 9 Distance Between Parallel Planes

The planes

$$x + 2y - 2z = 3 \quad \text{and} \quad 2x + 4y - 4z = 7$$

are parallel since their normals, $(1, 2, -2)$ and $(2, 4, -4)$, are parallel vectors. Find the distance between these planes.

Solution.

To find the distance D between the planes, we may select an arbitrary point in one of the planes and compute its distance to the other plane. By setting $y = z = 0$ in the equation $x + 2y - 2z = 3$, we obtain the point $P_0(3, 0, 0)$ in this plane. From (9), the distance between P_0 and the plane $2x + 4y - 4z = 7$ is

$$D = \frac{|2(3) + 4(0) + (-4)(0) - 7|}{\sqrt{2^2 + 4^2 + (-4)^2}} = \frac{1}{6} \qquad \blacklozenge$$

Exercise Set 3.5

1. Find a point-normal form of the equation of the plane passing through P and having $\mathbf{n}$ as a normal.
   (a) $P(-1, 3, -2)$; $\mathbf{n} = (-2, 1, -1)$ (b) $P(1, 1, 4)$; $\mathbf{n} = (1, 9, 8)$
   (c) $P(2, 0, 0)$; $\mathbf{n} = (0, 0, 2)$ (d) $P(0, 0, 0)$; $\mathbf{n} = (1, 2, 3)$

2. Write the equations of the planes in Exercise 1 in general form.

3. Find a point-normal form.
   (a) $-3x + 7y + 2z = 10$ (b) $x - 4z = 0$

4. Find an equation for the plane passing through the given points.
   (a) $P(-4, -1, -1)$, $Q(-2, 0, 1)$, $R(-1, -2, -3)$ (b) $P(5, 4, 3)$, $Q(4, 3, 1)$, $R(1, 5, 4)$

5. Determine whether the planes are parallel.
   (a) $4x - y + 2z = 5$ and $7x - 3y + 4z = 8$
   (b) $x - 4y - 3z - 2 = 0$ and $3x - 12y - 9z - 7 = 0$
   (c) $2y = 8x - 4z + 5$ and $x = \frac{1}{2}z + \frac{1}{4}y$

6. Determine whether the line and plane are parallel.
   (a) $x = -5 - 4t$, $y = 1 - t$, $z = 3 + 2t$; $x + 2y + 3z - 9 = 0$
   (b) $x = 3t$, $y = 1 + 2t$, $z = 2 - t$; $4x - y + 2z = 1$

7. Determine whether the planes are perpendicular.
   (a) $3x - y + z - 4 = 0$, $x + 2z = -1$ (b) $x - 2y + 3z = 4$, $-2x + 5y + 4z = -1$

8. Determine whether the line and plane are perpendicular.
   (a) $x = -2 - 4t$, $y = 3 - 2t$, $z = 1 + 2t$; $2x + y - z = 5$
   (b) $x = 2 + t$, $y = 1 - t$, $z = 5 + 3t$; $6x + 6y - 7 = 0$

9. Find parametric equations for the line passing through P and parallel to $\mathbf{n}$.
   (a) $P(3, -1, 2)$; $\mathbf{n} = (2, 1, 3)$ (b) $P(-2, 3, -3)$; $\mathbf{n} = (6, -6, -2)$
   (c) $P(2, 2, 6)$; $\mathbf{n} = (0, 1, 0)$ (d) $P(0, 0, 0)$; $\mathbf{n} = (1, -2, 3)$

10. Find parametric equations for the line passing through the given points.
    (a) $(5, -2, 4)$, $(7, 2, -4)$ (b) $(0, 0, 0)$, $(2, -1, -3)$

11. Find parametric equations for the line of intersection of the given planes.
    (a) $7x - 2y + 3z = -2$ and $-3x + y + 2z + 5 = 0$ (b) $2x + 3y - 5z = 0$ and $y = 0$

12. Find the vector form of the equation of the plane that passes through P_0 and has normal $\mathbf{n}$.
  (a) $P_0(-1, 2, 4)$; $\mathbf{n} = (-2, 4, 1)$ (b) $P_0(2, 0, -5)$; $\mathbf{n} = (-1, 4, 3)$
  (c) $P_0(5, -2, 1)$; $\mathbf{n} = (-1, 0, 0)$ (d) $P_0(0, 0, 0)$; $\mathbf{n} = (a, b, c)$

13. Determine whether the planes are parallel.
  (a) $(-1, 2, 4) \cdot (x - 5, y + 3, z - 7) = 0$; $(2, -4, -8) \cdot (x + 3, y + 5, z - 9) = 0$
  (b) $(3, 0, -1) \cdot (x + 1, y - 2, z - 3) = 0$; $(-1, 0, 3) \cdot (x + 1, y - z, z - 3) = 0$

14. Determine whether the planes are perpendicular.
  (a) $(-2, 1, 4) \cdot (x - 1, y, z + 3) = 0$; $(1, -2, 1) \cdot (x + 3, y - 5, z) = 0$
  (b) $(3, 0, -2) \cdot (x + 4, y - 7, z + 1) = 0$; $(1, 1, 1) \cdot (x, y, z) = 0$

15. Find the vector form of the equation of the line through P_0 and parallel to $\mathbf{v}$.
  (a) $P_0(-1, 2, 3)$; $\mathbf{v} = (7, -1, 5)$ (b) $P_0(2, 0, -1)$; $\mathbf{v} = (1, 1, 1)$
  (c) $P_0(2, -4, 1)$; $\mathbf{v} = (0, 0, -2)$ (d) $P_0(0, 0, 0)$; $\mathbf{v} = (a, b, c)$

16. Show that the line

$$x = 0, \quad y = t, \quad z = t \quad (-\infty < t < +\infty)$$

  (a) lies in the plane $6x + 4y - 4z = 0$
  (b) is parallel to and below the plane $5x - 3y + 3z = 1$
  (c) is parallel to and above the plane $6x + 2y - 2z = 3$

17. Find an equation for the plane through $(-2, 1, 7)$ that is perpendicular to the line $x - 4 = 2t$, $y + 2 = 3t, z = -5t$.

18. Find an equation of
  (a) the xy-plane   (b) the xz-plane   (c) the yz-plane

19. Find an equation of the plane that contains the point (x_0, y_0, z_0) and is
  (a) parallel to the xy-plane   (b) parallel to the yz-plane   (c) parallel to the xz-plane

20. Find an equation for the plane that passes through the origin and is parallel to the plane $7x + 4y - 2z + 3 = 0$.

21. Find an equation for the plane that passes through the point $(3, -6, 7)$ and is parallel to the plane $5x - 2y + z - 5 = 0$.

22. Find the point of intersection of the line

$$x - 9 = -5t, \quad y + 1 = -t, \quad z - 3 = t \quad (-\infty < t < +\infty)$$

  and the plane $2x - 3y + 4z + 7 = 0$.

23. Find an equation for the plane that contains the line $x = -1 + 3t, y = 5 + 2t, z = 2 - t$ and is perpendicular to the plane $2x - 4y + 2z = 9$.

24. Find an equation for the plane that passes through $(2, 4, -1)$ and contains the line of intersection of the planes $x - y - 4z = 2$ and $-2x + y + 2z = 3$.

25. Show that the points $(-1, -2, -3)$, $(-2, 0, 1)$, $(-4, -1, -1)$, and $(2, 0, 1)$ lie in the same plane.

26. Find parametric equations for the line through $(-2, 5, 0)$ that is parallel to the planes $2x + y - 4z = 0$ and $-x + 2y + 3z + 1 = 0$.

27. Find an equation for the plane through $(-2, 1, 5)$ that is perpendicular to the planes $4x - 2y + 2z = -1$ and $3x + 3y - 6z = 5$.

28. Find an equation for the plane through $(2, -1, 4)$ that is perpendicular to the line of intersection of the planes $4x + 2y + 2z = -1$ and $3x + 6y + 3z = 7$.

29. Find an equation for the plane that is perpendicular to the plane $8x - 2y + 6z = 1$ and passes through the points $P_1(-1, 2, 5)$ and $P_2(2, 1, 4)$.

30. Show that the lines

$$x = 3 - 2t, \quad y = 4 + t, \quad z = 1 - t \quad (-\infty < t < +\infty)$$

and

$$x = 5 + 2t, \quad y = 1 - t, \quad z = 7 + t \quad (-\infty < t < +\infty)$$

are parallel, and find an equation for the plane they determine.

31. Find an equation for the plane that contains the point $(1, -1, 2)$ and the line $x = t$, $y = t + 1$, $z = -3 + 2t$.

32. Find an equation for the plane that contains the line $x = 1 + t$, $y = 3t$, $z = 2t$ and is parallel to the line of intersection of the planes $-x + 2y + z = 0$ and $x + z + 1 = 0$.

33. Find an equation for the plane, each of whose points is equidistant from $(-1, -4, -2)$ and $(0, -2, 2)$.

34. Show that the line

$$x - 5 = -t, \quad y + 3 = 2t, \quad z + 1 = -5t \quad (-\infty < t < +\infty)$$

is parallel to the plane $-3x + y + z - 9 = 0$.

35. Show that the lines

$$x - 3 = 4t, \quad y - 4 = t, \quad z - 1 = 0 \quad (-\infty < t < +\infty)$$

and

$$x + 1 = 12t, \quad y - 7 = 6t, \quad z - 5 = 3t \quad (-\infty < t < +\infty)$$

intersect, and find the point of intersection.

36. Find an equation for the plane containing the lines in Exercise 35.

37. Find parametric equations for the line of intersection of the planes

(a) $-3x + 2y + z = -5$ and $7x + 3y - 2z = -2$
(b) $5x - 7y + 2z = 0$ and $y = 0$

38. Show that the plane whose intercepts with the coordinate axes are $x = a$, $y = b$, and $z = c$ has equation

$$\frac{x}{a} + \frac{y}{b} + \frac{z}{c} = 1$$

provided a, b, and c are nonzero.

39. Find the distance between the point and the plane.

(a) $(3, 1, -2)$; $x + 2y - 2z = 4$
(b) $(-1, 2, 1)$; $2x + 3y - 4z = 1$
(c) $(0, 3, -2)$; $x - y - z = 3$

40. Find the distance between the given parallel planes.

(a) $3x - 4y + z = 1$ and $6x - 8y + 2z = 3$
(b) $-4x + y - 3z = 0$ and $8x - 2y + 6z = 0$
(c) $2x - y + z = 1$ and $2x - y + z = -1$

41. Show that if a, b, and c are all nonzero, then the line

$$x = x_0 + at, \quad y = y_0 + bt, \quad z = z_0 + ct \quad (-\infty < t < +\infty)$$

consists of all points (x, y, z) that satisfy

$$\frac{x - x_0}{a} = \frac{y - y_0}{b} = \frac{z - z_0}{c}$$

These are called ***symmetric equations*** for the line.

42. Find symmetric equations for the lines in parts (a) and (b) of Exercise 9. [***Note.*** See Exercise 41 for terminology.]

43. In each part find equations for two planes whose intersection is the given line.

(a) $x = 7 - 4t, \quad y = -5 - 2t, \quad z = 5 + t \quad (-\infty < t < +\infty)$
(b) $x = 4t, \quad y = 2t, \quad z = 7t \quad (-\infty < t < +\infty)$
[***Hint.*** Each equality in the symmetric equations of a line represents a plane containing the line. See Exercise 41 for terminology.]

44. Two intersecting planes in 3-space determine two angles of intersection, an acute angle ($0 \leq \theta \leq 90°$) and its supplement $180° - \theta$ (see accompanying figure). If $\mathbf{n}_1$ and $\mathbf{n}_2$ are nonzero normals to the planes, then the angle between $\mathbf{n}_1$ and $\mathbf{n}_2$ is θ or $180° - \theta$, depending on the directions of the normals (see accompanying figure). In each part find the acute angle of intersection of the planes to the nearest degree.

   (a) $x = 0$ and $2x - y + z - 4 = 0$
   (b) $x + 2y - 2z = 5$ and $6x - 3y + 2z = 8$
   [*Note.* A calculator is needed.]

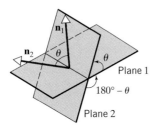

Figure Ex-44

45. Find the acute angle between the plane $x - y - 3z = 5$ and the line $x = 2 - t$, $y = 2t$, $z = 3t - 1$ to the nearest degree. [*Hint.* See Exercise 44.]

Discussion and Discovery

46. What do the lines $\mathbf{r} = \mathbf{r}_0 + t\mathbf{v}$ and $\mathbf{r} = \mathbf{r}_0 - t\mathbf{v}$ have in common? Explain.

47. What is the relationship between the line $x = x_0 + at$, $y = y_0 + bt$, $z = z_0 + ct$ and the plane $ax + by + cz = 0$? Explain your reasoning.

48. Let $\mathbf{r}_1$ and $\mathbf{r}_2$ be vectors from the origin to the points $P_1(x_1, y_1, z_1)$ and $P_2(x_2, y_2, z_2)$, respectively. What does the equation

$$\mathbf{r} = (1 - t)\mathbf{r}_1 + t\mathbf{r}_2 \quad (0 \leq t \leq 1)$$

represent geometrically? Explain your reasoning.

49. Write parametric equations for two perpendicular lines that pass through the point (x_0, y_0, z_0).

Chapter 3 Technology Exercises

The following exercises are designed to be solved using a technology utility. Typically, this will be MATLAB, *Mathematica*, Maple, Derive, or Mathcad, but it may also be some other type of linear algebra software or a scientific calculator with some linear algebra capabilities. For each exercise you will need to read the relevant documentation for the particular utility you are using. The goal of these exercises is to provide you with a basic proficiency with your technology utility. Once you have mastered the techniques in these exercises, you will be able to use your technology utility to solve many of the problems in the regular exercise sets.

Section 3.1

T1. (*Vectors*) Read your documentation on how to enter vectors and how to add, subtract, and multiply them by scalars. Then perform the computations in Example 1.

T2. (*Drawing vectors*) If you are using a technology utility that can draw line segments in two- or three-dimensional space, try drawing some line segments with initial and terminal points of your choice. You may also want to see if your utility allows you to create arrowheads, in which case you can make your line segments look like geometric vectors.

Section 3.3

T1. (*Dot product and norm*) Some technology utilities provide commands for calculating dot products and norms, while others only provide a command for the dot product. In the latter case, norms can be computed from the formula $\|\mathbf{v}\| = \sqrt{\mathbf{v} \cdot \mathbf{v}}$. Read your documentation on how to find dot products (and norms, if available), and then perform the computations in Example 2.

T2. (*Projections*) See if you can program your utility to calculate $\text{proj}_a \mathbf{u}$ when the user enters the vectors **a** and **u**. Check your work by having your program perform the computations in Example 6.

Section 3.4

T1. (*Cross product*) Read your documentation on how to find cross products, and then perform the computation in Example 1.

T2. (*Cross product formula*) If you are working with a CAS, then use it to confirm Formula (1a).

T3. (*Cross product properties*) If you are working with a CAS, use it to prove the results in Theorem 3.4.1.

T4. (*Area of a triangle*) See if you can program your technology utility to find the area of the triangle in 3-space determined by three points when the user enters their coordinates. Check your work by calculating the area of the triangle in Example 4.

T5. (*Triple scalar product formula*) If you are working with a CAS, use it to prove Formula (7) by showing that the difference between the two sides is zero.

T6. (*Volume of a parallelepiped*) See if you can program your technology utility to find the volume of the parallelepiped in 3-space determined by vectors **u**, **v**, and **w** when the user enters the vectors. Check your work by solving Exercise 10 in Exercise Set 3.4.

4

Euclidean Vector Spaces

INTRODUCTION: The idea of using pairs of numbers to locate points in the plane and triples of numbers to locate points in 3-space was first clearly spelled out in the mid-seventeenth century. By the latter part of the eighteenth century, mathematicians and physicists began to realize that there was no need to stop with triples. It was recognized that quadruples of numbers (a_1, a_2, a_3, a_4) could be regarded as points in "four-dimensional" space, quintuples $(a_1, a_2, a_3, a_4, a_5)$ as points in "five-dimensional" space, and so on, an n-tuple of numbers being a point in "n-dimensional" space. Our goal in this chapter is to study the properties of operations on vectors in this kind of space.

4.1 EUCLIDEAN *n*-SPACE

Although our geometric visualization does not extend beyond 3-space, it is nevertheless possible to extend many familiar ideas beyond 3-space by working with analytic or numerical properties of points and vectors rather than the geometric properties. In this section we shall make these ideas more precise.

Vectors in *n*-Space   We begin with a definition.

> ### Definition
>
> If n is a positive integer, then an ***ordered n-tuple*** is a sequence of n real numbers $(a_1, a_2, \ldots, a_n)$. The set of all ordered n-tuples is called ***n-space*** and is denoted by R^n.

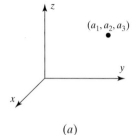

(a)

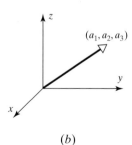

(b)

Figure 4.1.1
The ordered triple (a_1, a_2, a_3) can be interpreted geometrically as a point or a vector.

When $n = 2$ or 3, it is usual to use the terms ***ordered pair*** and ***ordered triple***, respectively, rather than ordered 2-tuple and 3-tuple. When $n = 1$, each ordered n-tuple consists of one real number, and so R^1 may be viewed as the set of real numbers. It is usual to write R rather than R^1 for this set.

It might have occurred to you in the study of 3-space that the symbol (a_1, a_2, a_3) has two different geometric interpretations: it can be interpreted as a point, in which case a_1, a_2, and a_3 are the coordinates (Figure 4.1.1a), or it can be interpreted as a vector, in which case a_1, a_2, and a_3 are the components (Figure 4.1.1b). It follows, therefore, that an ordered n-tuple $(a_1, a_2, \ldots, a_n)$ can be viewed either as a "generalized point" or a "generalized vector"—the distinction is mathematically unimportant. Thus, we can describe the 5-tuple $(-2, 4, 0, 1, 6)$ either as a point in R^5 or a vector in R^5.

> ### Definition
>
> Two vectors $\mathbf{u} = (u_1, u_2, \ldots, u_n)$ and $\mathbf{v} = (v_1, v_2, \ldots, v_n)$ in R^n are called ***equal*** if
>
> $$u_1 = v_1, \quad u_2 = v_2, \ldots, \quad u_n = v_n$$
>
> The ***sum*** $\mathbf{u} + \mathbf{v}$ is defined by
>
> $$\mathbf{u} + \mathbf{v} = (u_1 + v_1, u_2 + v_2, \ldots, u_n + v_n)$$
>
> and if k is any scalar, the ***scalar multiple*** $k\mathbf{u}$ is defined by
>
> $$k\mathbf{u} = (ku_1, ku_2, \ldots, ku_n)$$

The operations of addition and scalar multiplication in this definition are called the ***standard operations*** on R^n.

The ***zero vector*** in R^n is denoted by $\mathbf{0}$ and is defined to be the vector

$$\mathbf{0} = (0, 0, \ldots, 0)$$

If $\mathbf{u} = (u_1, u_2, \ldots, u_n)$ is any vector in R^n, then the **negative** (or **additive inverse**) of $\mathbf{u}$ is denoted by $-\mathbf{u}$ and is defined by

$$-\mathbf{u} = (-u_1, -u_2, \ldots, -u_n)$$

The **difference** of vectors in R^n is defined by

$$\mathbf{v} - \mathbf{u} = \mathbf{v} + (-\mathbf{u})$$

or in terms of components

$$\mathbf{v} - \mathbf{u} = (v_1 - u_1, v_2 - u_2, \ldots, v_n - u_n)$$

Properties of Vector Operations in *n*-Space   The most important arithmetic properties of addition and scalar multiplication of vectors in R^n are listed in the following theorem. The proofs are all easy and are left as exercises.

Theorem 4.1.1   **Properties of Vectors in R^n**

If $\mathbf{u} = (u_1, u_2, \ldots, u_n)$, $\mathbf{v} = (v_1, v_2, \ldots, v_n)$, *and* $\mathbf{w} = (w_1, w_2, \ldots, w_n)$ *are vectors in* R^n *and k and l are scalars, then:*

(a) $\mathbf{u} + \mathbf{v} = \mathbf{v} + \mathbf{u}$ (b) $\mathbf{u} + (\mathbf{v} + \mathbf{w}) = (\mathbf{u} + \mathbf{v}) + \mathbf{w}$

(c) $\mathbf{u} + \mathbf{0} = \mathbf{0} + \mathbf{u} = \mathbf{u}$ (d) $\mathbf{u} + (-\mathbf{u}) = \mathbf{0}$; *that is,* $\mathbf{u} - \mathbf{u} = \mathbf{0}$

(e) $k(l\mathbf{u}) = (kl)\mathbf{u}$ (f) $k(\mathbf{u} + \mathbf{v}) = k\mathbf{u} + k\mathbf{v}$

(g) $(k + l)\mathbf{u} = k\mathbf{u} + l\mathbf{u}$ (h) $1\mathbf{u} = \mathbf{u}$

This theorem enables us to manipulate vectors in R^n without expressing the vectors in terms of components. For example, to solve the vector equation $\mathbf{x} + \mathbf{u} = \mathbf{v}$ for $\mathbf{x}$, we can add $-\mathbf{u}$ to both sides and proceed as follows:

$$(\mathbf{x} + \mathbf{u}) + (-\mathbf{u}) = \mathbf{v} + (-\mathbf{u})$$
$$\mathbf{x} + (\mathbf{u} - \mathbf{u}) = \mathbf{v} - \mathbf{u}$$
$$\mathbf{x} + \mathbf{0} = \mathbf{v} - \mathbf{u}$$
$$\mathbf{x} = \mathbf{v} - \mathbf{u}$$

The reader will find it instructive to name the parts of Theorem 4.1.1 that justify the last three steps in this computation.

Euclidean *n*-Space   To extend the notions of distance, norm, and angle to R^n, we begin with the following generalization of the dot product on R^2 and R^3 [Formulas (3) and (4) of Section 3.3].

Definition

If $\mathbf{u} = (u_1, u_2, \ldots, u_n)$ and $\mathbf{v} = (v_1, v_2, \ldots, v_n)$ are any vectors in R^n, then the **Euclidean inner product** $\mathbf{u} \cdot \mathbf{v}$ is defined by

$$\mathbf{u} \cdot \mathbf{v} = u_1 v_1 + u_2 v_2 + \cdots + u_n v_n$$

Observe that when $n = 2$ or 3, the Euclidean inner product is the ordinary dot product.

EXAMPLE 1 Inner Product of Vectors in R^4

The Euclidean inner product of the vectors

$$\mathbf{u} = (-1, 3, 5, 7) \quad \text{and} \quad \mathbf{v} = (5, -4, 7, 0)$$

in R^4 is

$$\mathbf{u} \cdot \mathbf{v} = (-1)(5) + (3)(-4) + (5)(7) + (7)(0) = 18 \qquad \blacklozenge$$

Since so many of the familiar ideas from 2-space and 3-space carry over to n-space, it is common to refer to R^n with the operations of addition, scalar multiplication, and the Euclidean inner product as ***Euclidean n-space***.

The four main arithmetic properties of the Euclidean inner product are listed in the next theorem.

Theorem 4.1.2 **Properties of Euclidean Inner Product**

If $\mathbf{u}$, $\mathbf{v}$, *and* $\mathbf{w}$ *are vectors in* R^n *and k is any scalar, then*:

(*a*) $\mathbf{u} \cdot \mathbf{v} = \mathbf{v} \cdot \mathbf{u}$ (*b*) $(\mathbf{u} + \mathbf{v}) \cdot \mathbf{w} = \mathbf{u} \cdot \mathbf{w} + \mathbf{v} \cdot \mathbf{w}$

(*c*) $(k\mathbf{u}) \cdot \mathbf{v} = k(\mathbf{u} \cdot \mathbf{v})$ (*d*) $\mathbf{v} \cdot \mathbf{v} \geq 0$. *Further,* $\mathbf{v} \cdot \mathbf{v} = 0$ *if and only if* $\mathbf{v} = \mathbf{0}$.

We shall prove parts (*b*) and (*d*) and leave proofs of the rest as exercises.

Proof (b). Let $\mathbf{u} = (u_1, u_2, \ldots, u_n)$, $\mathbf{v} = (v_1, v_2, \ldots, v_n)$, and $\mathbf{w} = (w_1, w_2, \ldots, w_n)$. Then

$$(\mathbf{u} + \mathbf{v}) \cdot \mathbf{w} = (u_1 + v_1, u_2 + v_2, \ldots, u_n + v_n) \cdot (w_1, w_2, \ldots, w_n)$$

$$= (u_1 + v_1)w_1 + (u_2 + v_2)w_2 + \cdots + (u_n + v_n)w_n$$

$$= (u_1 w_1 + u_2 w_2 + \cdots + u_n w_n) + (v_1 w_1 + v_2 w_2 + \cdots + v_n w_n)$$

$$= \mathbf{u} \cdot \mathbf{w} + \mathbf{v} \cdot \mathbf{w}$$

Proof (d). We have $\mathbf{v} \cdot \mathbf{v} = v_1^2 + v_2^2 + \cdots + v_n^2 \geq 0$. Further, equality holds if and only if $v_1 = v_2 = \cdots = v_n = 0$, that is, if and only if $\mathbf{v} = \mathbf{0}$. ∎

EXAMPLE 2 Length and Distance in R^4

Theorem 4.1.2 allows us to perform computations with Euclidean inner products in much the same way that we perform them with ordinary arithmetic products. For example,

$$(3\mathbf{u} + 2\mathbf{v}) \cdot (4\mathbf{u} + \mathbf{v}) = (3\mathbf{u}) \cdot (4\mathbf{u} + \mathbf{v}) + (2\mathbf{v}) \cdot (4\mathbf{u} + \mathbf{v})$$

$$= (3\mathbf{u}) \cdot (4\mathbf{u}) + (3\mathbf{u}) \cdot \mathbf{v} + (2\mathbf{v}) \cdot (4\mathbf{u}) + (2\mathbf{v}) \cdot \mathbf{v}$$

$$= 12(\mathbf{u} \cdot \mathbf{u}) + 3(\mathbf{u} \cdot \mathbf{v}) + 8(\mathbf{v} \cdot \mathbf{u}) + 2(\mathbf{v} \cdot \mathbf{v})$$

$$= 12(\mathbf{u} \cdot \mathbf{u}) + 11(\mathbf{u} \cdot \mathbf{v}) + 2(\mathbf{v} \cdot \mathbf{v})$$

The reader should determine which parts of Theorem 4.1.2 were used in each step. $\blacklozenge$

Norm and Distance in Euclidean n-Space   By analogy with the familiar formulas in R^2 and R^3, we define the ***Euclidean norm*** (or ***Euclidean length***) of a vector $\mathbf{u} = (u_1, u_2, \ldots, u_n)$ in R^n by

$$\|\mathbf{u}\| = (\mathbf{u} \cdot \mathbf{u})^{1/2} = \sqrt{u_1^2 + u_2^2 + \cdots + u_n^2} \tag{1}$$

[Compare this formula to Formulas (1) and (2) in Section 3.2.]

Similarly, the ***Euclidean distance*** between the points $\mathbf{u} = (u_1, u_2, \dots, u_n)$ and $\mathbf{v} = (v_1, v_2, \dots, v_n)$ in R^n is defined by

$$d(\mathbf{u}, \mathbf{v}) = \|\mathbf{u} - \mathbf{v}\| = \sqrt{(u_1 - v_1)^2 + (u_2 - v_2)^2 + \cdots + (u_n - v_n)^2} \qquad (2)$$

[See Formulas (3) and (4) of Section 3.2.]

EXAMPLE 3 Finding Norm and Distance

If $\mathbf{u} = (1, 3, -2, 7)$ and $\mathbf{v} = (0, 7, 2, 2)$, then in the Euclidean space R^4

$$\|\mathbf{u}\| = \sqrt{(1)^2 + (3)^2 + (-2)^2 + (7)^2} = \sqrt{63} = 3\sqrt{7}$$

and

$$d(\mathbf{u}, \mathbf{v}) = \sqrt{(1 - 0)^2 + (3 - 7)^2 + (-2 - 2)^2 + (7 - 2)^2} = \sqrt{58} \qquad \blacklozenge$$

The following theorem provides one of the most important inequalities in linear algebra, the ***Cauchy–Schwarz*** inequality.

Augustin Louis (Baron de) Cauchy

Herman Amandus Schwarz

Augustin Louis (Baron de) Cauchy (1789–1857), French mathematician. Cauchy's early education was acquired from his father, a barrister and master of the classics. Cauchy entered L'Ecole Polytechnique in 1805 to study engineering, but because of poor health, was advised to concentrate on mathematics. His major mathematical work began in 1811 with a series of brilliant solutions to some difficult outstanding problems.

Cauchy's mathematical contributions for the next 35 years were brilliant and staggering in quantity, over 700 papers filling 26 modern volumes. Cauchy's work initiated the era of modern analysis; he brought to mathematics standards of precision and rigor undreamed of by earlier mathematicians.

Cauchy's life was inextricably tied to the political upheavals of the time. A strong partisan of the Bourbons, he left his wife and children in 1830 to follow the Bourbon king Charles X into exile. For his loyalty he was made a baron by the ex-king. Cauchy eventually returned to France, but refused to accept a university position until the government waived its requirement that he take a loyalty oath.

It is difficult to get a clear picture of the man. Devoutly Catholic, he sponsored charitable work for unwed mothers and criminals, and relief for Ireland. Yet other aspects of his life cast him in an unfavorable light. The Norwegian mathematician Abel described him as "mad, infinitely Catholic, and bigoted." Some writers praise his teaching, yet others say he rambled incoherently and, according to a report of the day, he once devoted an entire lecture to extracting the square root of seventeen to ten decimal places by a method well known to his students. In any event, Cauchy is undeniably one of the greatest minds in the history of science.

Herman Amandus Schwarz (1843–1921), German mathematician. Schwarz was the leading mathematician in Berlin in the first part of the twentieth century. Because of a devotion to his teaching duties at the University of Berlin and a propensity for treating both important and trivial facts with equal thoroughness, he did not publish in great volume. He tended to focus on narrow concrete problems, but his techniques were often extremely clever and influenced the work of other mathematicians. A version of the inequality that bears his name appeared in a paper about surfaces of minimal area published in 1885.

> ### Theorem 4.1.3   Cauchy–Schwarz Inequality in R^n
>
> *If* $\mathbf{u} = (u_1, u_2, \ldots, u_n)$ *and* $\mathbf{v} = (v_1, v_2, \ldots, v_n)$ *are vectors in* R^n, *then*
>
> $$|\mathbf{u} \cdot \mathbf{v}| \leq \|\mathbf{u}\|\|\mathbf{v}\| \qquad (3)$$

In terms of components, (3) is the same as

$$|u_1 v_1 + u_2 v_2 + \cdots + u_n v_n| \leq (u_1^2 + u_2^2 + \cdots + u_n^2)^{1/2}(v_1^2 + v_2^2 + \cdots + v_n^2)^{1/2} \quad (4)$$

We omit the proof at this time, since a more general version of this theorem will be proved later in the text. However, for vectors in R^2 and R^3, this result is a simple consequence of Formula (1) of Section 3.3: If $\mathbf{u}$ and $\mathbf{v}$ are nonzero vectors in R^2 or R^3, then

$$|\mathbf{u} \cdot \mathbf{v}| = |\|\mathbf{u}\|\|\mathbf{v}\| \cos\theta| = \|\mathbf{u}\|\|\mathbf{v}\| |\cos\theta| \leq \|\mathbf{u}\|\|\mathbf{v}\| \qquad (5)$$

and if either $\mathbf{u} = \mathbf{0}$ or $\mathbf{v} = \mathbf{0}$, then both sides of (3) are zero, so the inequality holds in this case as well.

The next two theorems list the basic properties of length and distance in Euclidean n-space.

> ### Theorem 4.1.4   Properties of Length in R^n
>
> *If* $\mathbf{u}$ *and* $\mathbf{v}$ *are vectors in* R^n *and* k *is any scalar, then*:
> (*a*) $\|\mathbf{u}\| \geq 0$ (*b*) $\|\mathbf{u}\| = 0$ *if and only if* $\mathbf{u} = \mathbf{0}$
> (*c*) $\|k\mathbf{u}\| = |k|\|\mathbf{u}\|$ (*d*) $\|\mathbf{u} + \mathbf{v}\| \leq \|\mathbf{u}\| + \|\mathbf{v}\|$ (*Triangle inequality*)

We shall prove (*c*) and (*d*) and leave (*a*) and (*b*) as exercises.

Proof (*c*). If $\mathbf{u} = (u_1, u_2, \ldots, u_n)$, then $k\mathbf{u} = (ku_1, ku_2, \ldots, ku_n)$, so

$$\|k\mathbf{u}\| = \sqrt{(ku_1)^2 + (ku_2)^2 + \cdots + (ku_n)^2}$$
$$= |k|\sqrt{u_1^2 + u_2^2 + \cdots + u_n^2}$$
$$= |k|\|\mathbf{u}\|$$

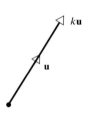

(*a*) $\|k\mathbf{u}\| = |k|\|\mathbf{u}\|$

Proof (*d*).

$$\|\mathbf{u} + \mathbf{v}\|^2 = (\mathbf{u} + \mathbf{v}) \cdot (\mathbf{u} + \mathbf{v}) = (\mathbf{u} \cdot \mathbf{u}) + 2(\mathbf{u} \cdot \mathbf{v}) + (\mathbf{v} \cdot \mathbf{v})$$
$$= \|\mathbf{u}\|^2 + 2(\mathbf{u} \cdot \mathbf{v}) + \|\mathbf{v}\|^2$$
$$\leq \|\mathbf{u}\|^2 + 2|\mathbf{u} \cdot \mathbf{v}| + \|\mathbf{v}\|^2 \quad \longleftarrow \text{ Property of absolute value}$$
$$\leq \|\mathbf{u}\|^2 + 2\|\mathbf{u}\|\|\mathbf{v}\| + \|\mathbf{v}\|^2 \quad \longleftarrow \text{ Cauchy–Schwarz inequality}$$
$$= (\|\mathbf{u}\| + \|\mathbf{v}\|)^2$$

The result now follows on taking square roots of both sides. ∎

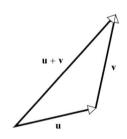

(*b*) $\|\mathbf{u} + \mathbf{v}\| \leq \|\mathbf{u}\| + \|\mathbf{v}\|$

Figure 4.1.2

Part (*c*) of this theorem states that multiplying a vector by a scalar k multiplies the length of that vector by a factor of $|k|$ (Figure 4.1.2*a*). Part (*d*) of this theorem is known as the ***triangle inequality*** because it generalizes the familiar result from Euclidean geometry which states that the sum of two sides of a triangle is at least as large as the third side (Figure 4.1.2*b*).

> **Theorem 4.1.5** **Properties of Distance in R^n**
>
> *If* **u**, **v**, *and* **w** *are vectors in R^n and k is any scalar, then*:
>
> (a) $d(\mathbf{u}, \mathbf{v}) \geq 0$ (b) $d(\mathbf{u}, \mathbf{v}) = 0$ *if and only if* $\mathbf{u} = \mathbf{v}$
>
> (c) $d(\mathbf{u}, \mathbf{v}) = d(\mathbf{v}, \mathbf{u})$ (d) $d(\mathbf{u}, \mathbf{v}) \leq d(\mathbf{u}, \mathbf{w}) + d(\mathbf{w}, \mathbf{v})$ (*Triangle inequality*)

The results in this theorem are immediate consequences of Theorem 4.1.4. We shall prove part (*d*) and leave the remaining parts as exercises.

Proof (*d*). From (2) and part (*d*) of Theorem 4.1.4 we have

$$d(\mathbf{u}, \mathbf{v}) = \|\mathbf{u} - \mathbf{v}\| = \|(\mathbf{u} - \mathbf{w}) + (\mathbf{w} - \mathbf{v})\|$$
$$\leq \|\mathbf{u} - \mathbf{w}\| + \|\mathbf{w} - \mathbf{v}\| = d(\mathbf{u}, \mathbf{w}) + d(\mathbf{w}, \mathbf{v}) \qquad ■$$

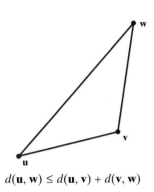

$$d(\mathbf{u}, \mathbf{w}) \leq d(\mathbf{u}, \mathbf{v}) + d(\mathbf{v}, \mathbf{w})$$

Figure 4.1.3

Part (*d*) of this theorem, which is also called the *triangle inequality*, generalizes the familiar result from Euclidean geometry which states that the shortest distance between two points is along a straight line (Figure 4.1.3).

Formula (1) expresses the norm of a vector in terms of a dot product. The following useful theorem expresses the dot product in terms of norms.

> **Theorem 4.1.6**
>
> *If* **u** *and* **v** *are vectors in R^n with the Euclidean inner product, then*
>
> $$\mathbf{u} \cdot \mathbf{v} = \tfrac{1}{4}\|\mathbf{u} + \mathbf{v}\|^2 - \tfrac{1}{4}\|\mathbf{u} - \mathbf{v}\|^2 \qquad (6)$$

Proof.
$$\|\mathbf{u} + \mathbf{v}\|^2 = (\mathbf{u} + \mathbf{v}) \cdot (\mathbf{u} + \mathbf{v}) = \|\mathbf{u}\|^2 + 2(\mathbf{u} \cdot \mathbf{v}) + \|\mathbf{v}\|^2$$
$$\|\mathbf{u} - \mathbf{v}\|^2 = (\mathbf{u} - \mathbf{v}) \cdot (\mathbf{u} - \mathbf{v}) = \|\mathbf{u}\|^2 - 2(\mathbf{u} \cdot \mathbf{v}) + \|\mathbf{v}\|^2$$

from which (6) follows by simple algebra. ■

Some problems that use this theorem are given in the exercises.

Orthogonality   Recall that in the Euclidean spaces R^2 and R^3 two vectors **u** and **v** are defined to be *orthogonal* (perpendicular) if $\mathbf{u} \cdot \mathbf{v} = 0$ (Section 3.3). Motivated by this, we make the following definition.

> **Definition**
>
> Two vectors **u** and **v** in R^n are called *orthogonal* if $\mathbf{u} \cdot \mathbf{v} = 0$.

EXAMPLE 4 Orthogonal Vectors in R^4

In the Euclidean space R^4 the vectors

$$\mathbf{u} = (-2, 3, 1, 4) \quad \text{and} \quad \mathbf{v} = (1, 2, 0, -1)$$

are orthogonal, since

$$\mathbf{u} \cdot \mathbf{v} = (-2)(1) + (3)(2) + (1)(0) + (4)(-1) = 0 \qquad ◆$$

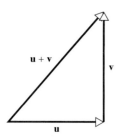

Figure 4.1.4

Properties of orthogonal vectors will be discussed in more detail later in the text, but we note at this point that many of the familiar properties of orthogonal vectors in the Euclidean spaces R^2 and R^3 continue to hold in the Euclidean space R^n. For example, if $\mathbf{u}$ and $\mathbf{v}$ are orthogonal vectors in R^2 or R^3, then $\mathbf{u}$, $\mathbf{v}$, and $\mathbf{u} + \mathbf{v}$ form the sides of a right triangle (Figure 4.1.4); thus, by the Theorem of Pythagoras

$$\|\mathbf{u} + \mathbf{v}\|^2 = \|\mathbf{u}\|^2 + \|\mathbf{v}\|^2$$

The following theorem shows that this result extends to R^n.

Theorem 4.1.7   **Pythagorean Theorem in R^n**

If $\mathbf{u}$ and $\mathbf{v}$ are orthogonal vectors in R^n with the Euclidean inner product, then

$$\|\mathbf{u} + \mathbf{v}\|^2 = \|\mathbf{u}\|^2 + \|\mathbf{v}\|^2$$

Proof.

$$\|\mathbf{u} + \mathbf{v}\|^2 = (\mathbf{u} + \mathbf{v}) \cdot (\mathbf{u} + \mathbf{v}) = \|\mathbf{u}\|^2 + 2(\mathbf{u} \cdot \mathbf{v}) + \|\mathbf{v}\|^2 = \|\mathbf{u}\|^2 + \|\mathbf{v}\|^2 \qquad \blacksquare$$

Alternative Notations for Vectors in R^n

It is often useful to write a vector $\mathbf{u} = (u_1, u_2, \ldots, u_n)$ in R^n in matrix notation as a row matrix or a column matrix:

$$\mathbf{u} = \begin{bmatrix} u_1 \\ u_2 \\ \vdots \\ u_n \end{bmatrix} \quad \text{or} \quad \mathbf{u} = [u_1 \quad u_2 \quad \cdots \quad u_n]$$

This is justified because the matrix operations

$$\mathbf{u} + \mathbf{v} = \begin{bmatrix} u_1 \\ u_2 \\ \vdots \\ u_n \end{bmatrix} + \begin{bmatrix} v_1 \\ v_2 \\ \vdots \\ v_n \end{bmatrix} = \begin{bmatrix} u_1 + v_1 \\ u_2 + v_2 \\ \vdots \\ u_n + v_n \end{bmatrix}, \qquad k\mathbf{u} = k\begin{bmatrix} u_1 \\ u_2 \\ \vdots \\ u_n \end{bmatrix} = \begin{bmatrix} ku_1 \\ ku_2 \\ \vdots \\ ku_n \end{bmatrix}$$

or

$$\mathbf{u} + \mathbf{v} = [u_1 \quad u_2 \quad \cdots \quad u_n] + [v_1 \quad v_2 \quad \cdots \quad v_n]$$
$$= [u_1 + v_1 \quad u_2 + v_2 \quad \cdots \quad u_n + v_n]$$
$$k\mathbf{u} = k[u_1 \quad u_2 \quad \cdots \quad u_n] = [ku_1 \quad ku_2 \quad \cdots \quad ku_n]$$

produce the same results as the vector operations

$$\mathbf{u} + \mathbf{v} = (u_1, u_2, \ldots, u_n) + (v_1, v_2, \ldots, v_n) = (u_1 + v_1, u_2 + v_2, \ldots, u_n + v_n)$$
$$k\mathbf{u} = k(u_1, u_2, \ldots, u_n) = (ku_1, ku_2, \ldots, ku_n)$$

The only difference is the form in which the vectors are written.

A Matrix Formula for the Dot Product

If we use column matrix notation for the vectors

$$\mathbf{u} = \begin{bmatrix} u_1 \\ u_2 \\ \vdots \\ u_n \end{bmatrix} \quad \text{and} \quad \mathbf{v} = \begin{bmatrix} v_1 \\ v_2 \\ \vdots \\ v_n \end{bmatrix}$$

and omit the brackets on 1×1 matrices, then it follows that

$$\mathbf{v}^T \mathbf{u} = [v_1 \quad v_2 \quad \cdots \quad v_n] \begin{bmatrix} u_1 \\ u_2 \\ \vdots \\ u_n \end{bmatrix} = [u_1 v_1 + u_2 v_2 + \cdots + u_n v_n] = [\mathbf{u} \cdot \mathbf{v}] = \mathbf{u} \cdot \mathbf{v}$$

Thus, for vectors in column matrix notation we have the following formula for the Euclidean inner product:

$$\mathbf{u} \cdot \mathbf{v} = \mathbf{v}^T \mathbf{u} \tag{7}$$

For example, if

$$\mathbf{u} = \begin{bmatrix} -1 \\ 3 \\ 5 \\ 7 \end{bmatrix} \quad \text{and} \quad \mathbf{v} = \begin{bmatrix} 5 \\ -4 \\ 7 \\ 0 \end{bmatrix}$$

then

$$\mathbf{u} \cdot \mathbf{v} = \mathbf{v}^T \mathbf{u} = [5 \quad -4 \quad 7 \quad 0] \begin{bmatrix} -1 \\ 3 \\ 5 \\ 7 \end{bmatrix} = [18] = 18$$

If A is an $n \times n$ matrix, then it follows from Formula (7) and properties of the transpose that

$$A\mathbf{u} \cdot \mathbf{v} = \mathbf{v}^T(A\mathbf{u}) = (\mathbf{v}^T A)\mathbf{u} = (A^T \mathbf{v})^T \mathbf{u} = \mathbf{u} \cdot A^T \mathbf{v}$$
$$\mathbf{u} \cdot A\mathbf{v} = (A\mathbf{v})^T \mathbf{u} = (\mathbf{v}^T A^T)\mathbf{u} = \mathbf{v}^T(A^T \mathbf{u}) = A^T \mathbf{u} \cdot \mathbf{v}$$

The resulting formulas

$$A\mathbf{u} \cdot \mathbf{v} = \mathbf{u} \cdot A^T \mathbf{v} \tag{8}$$

$$\mathbf{u} \cdot A\mathbf{v} = A^T \mathbf{u} \cdot \mathbf{v} \tag{9}$$

provide an important link between multiplication by an $n \times n$ matrix A and multiplication by A^T.

EXAMPLE 5 Verifying That $A\mathbf{u} \cdot \mathbf{v} = \mathbf{u} \cdot A^T \mathbf{v}$

Suppose that

$$A = \begin{bmatrix} 1 & -2 & 3 \\ 2 & 4 & 1 \\ -1 & 0 & 1 \end{bmatrix}, \quad \mathbf{u} = \begin{bmatrix} -1 \\ 2 \\ 4 \end{bmatrix}, \quad \mathbf{v} = \begin{bmatrix} -2 \\ 0 \\ 5 \end{bmatrix}$$

Then

$$A\mathbf{u} = \begin{bmatrix} 1 & -2 & 3 \\ 2 & 4 & 1 \\ -1 & 0 & 1 \end{bmatrix} \begin{bmatrix} -1 \\ 2 \\ 4 \end{bmatrix} = \begin{bmatrix} 7 \\ 10 \\ 5 \end{bmatrix}$$

$$A^T \mathbf{v} = \begin{bmatrix} 1 & 2 & -1 \\ -2 & 4 & 0 \\ 3 & 1 & 1 \end{bmatrix} \begin{bmatrix} -2 \\ 0 \\ 5 \end{bmatrix} = \begin{bmatrix} -7 \\ 4 \\ -1 \end{bmatrix}$$

from which we obtain

$$A\mathbf{u} \cdot \mathbf{v} = 7(-2) + 10(0) + 5(5) = 11$$
$$\mathbf{u} \cdot A^T\mathbf{v} = (-1)(-7) + 2(4) + 4(-1) = 11$$

Thus, $A\mathbf{u} \cdot \mathbf{v} = \mathbf{u} \cdot A^T\mathbf{v}$ as guaranteed by Formula (8). We leave it for the reader to verify that (9) also holds. ◆

A Dot Product View of Matrix Multiplication   Dot products provide another way of thinking about matrix multiplication. Recall that if $A = [a_{ij}]$ is an $m \times r$ matrix and $B = [b_{ij}]$ is an $r \times n$ matrix, then the ijth entry of AB is

$$a_{i1}b_{1j} + a_{i2}b_{2j} + \cdots + a_{ir}b_{rj}$$

which is the dot product of the ith row vector of A

$$[a_{i1} \quad a_{i2} \quad \cdots \quad a_{ir}]$$

and the jth column vector of B

$$\begin{bmatrix} b_{1j} \\ b_{2j} \\ \vdots \\ b_{rj} \end{bmatrix}$$

Thus, if the row vectors of A are $\mathbf{r}_1, \mathbf{r}_2, \ldots, \mathbf{r}_m$ and the column vectors of B are $\mathbf{c}_1, \mathbf{c}_2, \ldots, \mathbf{c}_n$, then the matrix product AB can be expressed as

$$AB = \begin{bmatrix} \mathbf{r}_1 \cdot \mathbf{c}_1 & \mathbf{r}_1 \cdot \mathbf{c}_2 & \cdots & \mathbf{r}_1 \cdot \mathbf{c}_n \\ \mathbf{r}_2 \cdot \mathbf{c}_1 & \mathbf{r}_2 \cdot \mathbf{c}_2 & \cdots & \mathbf{r}_2 \cdot \mathbf{c}_n \\ \vdots & \vdots & & \vdots \\ \mathbf{r}_m \cdot \mathbf{c}_1 & \mathbf{r}_m \cdot \mathbf{c}_2 & \cdots & \mathbf{r}_m \cdot \mathbf{c}_n \end{bmatrix} \tag{10}$$

In particular, a linear system $A\mathbf{x} = \mathbf{b}$ can be expressed in dot product form as

$$\begin{bmatrix} \mathbf{r}_1 \cdot \mathbf{x} \\ \mathbf{r}_2 \cdot \mathbf{x} \\ \vdots \\ \mathbf{r}_m \cdot \mathbf{x} \end{bmatrix} = \begin{bmatrix} b_1 \\ b_2 \\ \vdots \\ b_m \end{bmatrix} \tag{11}$$

where $\mathbf{r}_1, \mathbf{r}_2, \ldots, \mathbf{r}_m$ are the row vectors of A, and $b_1, b_2, \ldots, b_m$ are the entries of $\mathbf{b}$.

EXAMPLE 6 A Linear System Written in Dot Product Form

The following is an example of a linear system expressed in dot product form (11).

System	Dot Product Form
$3x_1 - 4x_2 + x_3 = 1$	$\begin{bmatrix} (3, -4, 1) \cdot (x_1, x_2, x_3) \\ (2, -7, -4) \cdot (x_1, x_2, x_3) \\ (1, 5, -8) \cdot (x_1, x_2, x_3) \end{bmatrix} = \begin{bmatrix} 1 \\ 5 \\ 0 \end{bmatrix}$
$2x_1 - 7x_2 - 4x_3 = 5$	
$x_1 + 5x_2 - 8x_3 = 0$	

◆

Exercise Set 4.1

1. Let $\mathbf{u} = (-3, 2, 1, 0)$, $\mathbf{v} = (4, 7, -3, 2)$, and $\mathbf{w} = (5, -2, 8, 1)$. Find

(a) $\mathbf{v} - \mathbf{w}$ (b) $2\mathbf{u} + 7\mathbf{v}$ (c) $-\mathbf{u} + (\mathbf{v} - 4\mathbf{w})$ (d) $6(\mathbf{u} - 3\mathbf{v})$ (e) $-\mathbf{v} - \mathbf{w}$ (f) $(6\mathbf{v} - \mathbf{w}) - (4\mathbf{u} + \mathbf{v})$

2. Let $\mathbf{u}$, $\mathbf{v}$, and $\mathbf{w}$ be the vectors in Exercise 1. Find the vector $\mathbf{x}$ that satisfies $5\mathbf{x} - 2\mathbf{v} = 2(\mathbf{w} - 5\mathbf{x})$.

3. Let $\mathbf{u}_1 = (-1, 3, 2, 0)$, $\mathbf{u}_2 = (2, 0, 4, -1)$, $\mathbf{u}_3 = (7, 1, 1, 4)$, and $\mathbf{u}_4 = (6, 3, 1, 2)$. Find scalars c_1, c_2, c_3, and c_4 such that $c_1\mathbf{u}_1 + c_2\mathbf{u}_2 + c_3\mathbf{u}_3 + c_4\mathbf{u}_4 = (0, 5, 6, -3)$.

4. Show that there do not exist scalars c_1, c_2, and c_3 such that
$$c_1(1, 0, 1, 0) + c_2(1, 0, -2, 1) + c_3(2, 0, 1, 2) = (1, -2, 2, 3)$$

5. In each part compute the Euclidean norm of the vector.

(a) $(-2, 5)$ (b) $(1, 2, -2)$ (c) $(3, 4, 0, -12)$ (d) $(-2, 1, 1, -3, 4)$

6. Let $\mathbf{u} = (4, 1, 2, 3)$, $\mathbf{v} = (0, 3, 8, -2)$, and $\mathbf{w} = (3, 1, 2, 2)$. Evaluate each expression.

(a) $\|\mathbf{u} + \mathbf{v}\|$ (b) $\|\mathbf{u}\| + \|\mathbf{v}\|$ (c) $\|-2\mathbf{u}\| + 2\|\mathbf{u}\|$

(d) $\|3\mathbf{u} - 5\mathbf{v} + \mathbf{w}\|$ (e) $\dfrac{1}{\|\mathbf{w}\|}\mathbf{w}$ (f) $\left\|\dfrac{1}{\|\mathbf{w}\|}\mathbf{w}\right\|$

7. Show that if $\mathbf{v}$ is a nonzero vector in R^n, then $(1/\|\mathbf{v}\|)\mathbf{v}$ has Euclidean norm 1.

8. Let $\mathbf{v} = (-2, 3, 0, 6)$. Find all scalars k such that $\|k\mathbf{v}\| = 5$.

9. Find the Euclidean inner product $\mathbf{u} \cdot \mathbf{v}$.

(a) $\mathbf{u} = (2, 5)$, $\mathbf{v} = (-4, 3)$ (b) $\mathbf{u} = (4, 8, 2)$, $\mathbf{v} = (0, 1, 3)$
(c) $\mathbf{u} = (3, 1, 4, -5)$, $\mathbf{v} = (2, 2, -4, -3)$ (d) $\mathbf{u} = (-1, 1, 0, 4, -3)$, $\mathbf{v} = (-2, -2, 0, 2, -1)$

10. (a) Find two vectors in R^2 with Euclidean norm 1 whose Euclidean inner product with $(3, -1)$ is zero.

(b) Show that there are infinitely many vectors in R^3 with Euclidean norm 1 whose Euclidean inner product with $(1, -3, 5)$ is zero.

11. Find the Euclidean distance between $\mathbf{u}$ and $\mathbf{v}$.

(a) $\mathbf{u} = (1, -2)$, $\mathbf{v} = (2, 1)$ (b) $\mathbf{u} = (2, -2, 2)$, $\mathbf{v} = (0, 4, -2)$
(c) $\mathbf{u} = (0, -2, -1, 1)$, $\mathbf{v} = (-3, 2, 4, 4)$ (d) $\mathbf{u} = (3, -3, -2, 0, -3)$, $\mathbf{v} = (-4, 1, -1, 5, 0)$

12. Verify parts (*b*), (*e*), (*f*), and (*g*) of Theorem 4.1.1 for $\mathbf{u} = (2, 0, -3, 1)$, $\mathbf{v} = (4, 0, 3, 5)$, $\mathbf{w} = (1, 6, 2, -1)$, $k = 5$, and $l = -3$.

13. Verify parts (*b*) and (*c*) of Theorem 4.1.2 for the values of $\mathbf{u}$, $\mathbf{v}$, $\mathbf{w}$, and k in Exercise 12.

14. In each part determine whether the given vectors are orthogonal.

(a) $\mathbf{u} = (-1, 3, 2)$, $\mathbf{v} = (4, 2, -1)$ (b) $\mathbf{u} = (-2, -2, -2)$, $\mathbf{v} = (1, 1, 1)$
(c) $\mathbf{u} = (u_1, u_2, u_3)$, $\mathbf{v} = (0, 0, 0)$ (d) $\mathbf{u} = (-4, 6, -10, 1)$, $\mathbf{v} = (2, 1, -2, 9)$
(e) $\mathbf{u} = (0, 3, -2, 1)$, $\mathbf{v} = (5, 2, -1, 0)$ (f) $\mathbf{u} = (a, b)$, $\mathbf{v} = (-b, a)$

15. For which values of k are $\mathbf{u}$ and $\mathbf{v}$ orthogonal?

(a) $\mathbf{u} = (2, 1, 3)$, $\mathbf{v} = (1, 7, k)$ (b) $\mathbf{u} = (k, k, 1)$, $\mathbf{v} = (k, 5, 6)$

16. Find two vectors of norm 1 that are orthogonal to the three vectors $\mathbf{u} = (2, 1, -4, 0)$, $\mathbf{v} = (-1, -1, 2, 2)$, and $\mathbf{w} = (3, 2, 5, 4)$.

17. In each part verify that the Cauchy–Schwarz inequality holds.

(a) $\mathbf{u} = (3, 2)$, $\mathbf{v} = (4, -1)$ (b) $\mathbf{u} = (-3, 1, 0)$, $\mathbf{v} = (2, -1, 3)$
(c) $\mathbf{u} = (-4, 2, 1)$, $\mathbf{v} = (8, -4, -2)$ (d) $\mathbf{u} = (0, -2, 2, 1)$, $\mathbf{v} = (-1, -1, 1, 1)$

18. In each part verify that Formulas (8) and (9) hold.

(a) $A = \begin{bmatrix} 2 & -1 \\ 3 & 4 \end{bmatrix}$, $\mathbf{u} = \begin{bmatrix} 3 \\ 1 \end{bmatrix}$, $\mathbf{v} = \begin{bmatrix} -2 \\ 6 \end{bmatrix}$ (b) $A = \begin{bmatrix} -1 & 2 & 4 \\ 3 & 1 & 0 \\ 5 & -2 & 3 \end{bmatrix}$, $\mathbf{u} = \begin{bmatrix} -1 \\ 2 \\ 5 \end{bmatrix}$, $\mathbf{v} = \begin{bmatrix} 0 \\ 2 \\ -4 \end{bmatrix}$

19. Solve the following linear system for x_1, x_2, and x_3.
$$(1, -1, 4) \cdot (x_1, x_2, x_3) = 10$$
$$(3, 2, 0) \cdot (x_1, x_2, x_3) = 1$$
$$(4, -5, -1) \cdot (x_1, x_2, x_3) = 7$$

20. Find $\mathbf{u} \cdot \mathbf{v}$ given that $\|\mathbf{u} + \mathbf{v}\| = 1$ and $\|\mathbf{u} - \mathbf{v}\| = 5$.

21. Use Theorem 4.1.6 to show that $\mathbf{u}$ and $\mathbf{v}$ are orthogonal vectors in R^n if $\|\mathbf{u} + \mathbf{v}\| = \|\mathbf{u} - \mathbf{v}\|$. Interpret this result geometrically in R^2.

22. The formulas for the vector components in Theorem 3.3.3 hold in R^n as well. Given that $\mathbf{a} = (-1, 1, 2, 3)$ and $\mathbf{u} = (2, 1, 4, -1)$, find the vector component of $\mathbf{u}$ along $\mathbf{a}$ and the vector component of $\mathbf{u}$ orthogonal to $\mathbf{a}$.

23. Determine whether the two lines

$$\mathbf{r} = (3, 2, 3, -1) + t(4, 6, 4, -2) \quad \text{and} \quad \mathbf{r} = (0, 3, 5, 4) + t(1, -3, -4, -2)$$

intersect in R^4.

24. Prove the following generalization of Theorem 4.1.7. If $\mathbf{v}_1, \mathbf{v}_2, \ldots, \mathbf{v}_r$ are pairwise orthogonal vectors in R^n, then

$$\|\mathbf{v}_1 + \mathbf{v}_2 + \cdots + \mathbf{v}_r\|^2 = \|\mathbf{v}_1\|^2 + \|\mathbf{v}_2\|^2 + \cdots + \|\mathbf{v}_r\|^2$$

25. Prove: If $\mathbf{u}$ and $\mathbf{v}$ are $n \times 1$ matrices and A is an $n \times n$ matrix, then

$$(\mathbf{v}^T A^T A \mathbf{u})^2 \leq (\mathbf{u}^T A^T A \mathbf{u})(\mathbf{v}^T A^T A \mathbf{v})$$

26. Use the Cauchy–Schwarz inequality to prove that for all real values of a, b, and θ,

$$(a \cos\theta + b \sin\theta)^2 \leq a^2 + b^2$$

27. Prove: If $\mathbf{u}$, $\mathbf{v}$, and $\mathbf{w}$ are vectors in R^n and k is any scalar, then

(a) $\mathbf{u} \cdot (k\mathbf{v}) = k(\mathbf{u} \cdot \mathbf{v})$ (b) $\mathbf{u} \cdot (\mathbf{v} + \mathbf{w}) = \mathbf{u} \cdot \mathbf{v} + \mathbf{u} \cdot \mathbf{w}$

28. Prove parts (*a*) through (*d*) of Theorem 4.1.1.

29. Prove parts (*e*) through (*h*) of Theorem 4.1.1.

30. Prove parts (*a*) and (*c*) of Theorem 4.1.2.

31. Prove parts (*a*) and (*b*) of Theorem 4.1.4.

32. Prove parts (*a*), (*b*), and (*c*) of Theorem 4.1.5.

33. Suppose that $a_1, a_2, \ldots, a_n$ are positive real numbers. In R^2, the vectors $\mathbf{v}_1 = (a_1, 0)$ and $\mathbf{v}_2 = (0, a_2)$ determine a rectangle of area $A = a_1 a_2$ (see accompanying figure), and in R^3 the vectors $\mathbf{v}_1 = (a_1, 0, 0)$, $\mathbf{v}_2 = (0, a_2, 0)$, and $\mathbf{v}_3 = (0, 0, a_3)$ determine a box of volume $V = a_1 a_2 a_3$ (see accompanying figure). The area A and the volume V are sometimes called the ***Euclidean measure*** of the rectangle and box, respectively.

(a) How would you define the Euclidean measure of the "box" in R^n that is determined by the vectors

$$\mathbf{v}_1 = (a_1, 0, 0, \ldots, 0), \quad \mathbf{v}_2 = (0, a_2, 0, \ldots, 0), \ldots, \quad \mathbf{v}_n = (0, 0, 0, \ldots, a_n)?$$

(b) How would you define the Euclidean length of the "diagonal" of the box in part (a)?

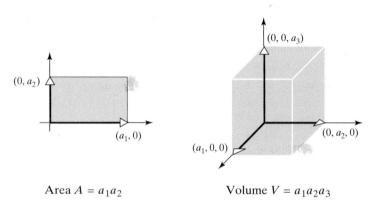

Area $A = a_1 a_2$ Volume $V = a_1 a_2 a_3$

Figure Ex-33

Discussion and Discovery

34. (a) Suppose that **u** and **v** are vectors in R^n. Show that

$$\|\mathbf{u} + \mathbf{v}\|^2 + \|\mathbf{u} - \mathbf{v}\|^2 = 2(\|\mathbf{u}\|^2 + \|\mathbf{v}\|^2)$$

(b) The result in part (a) states a theorem about parallelograms in R^2. What is the theorem?

35. (a) If **u** and **v** are orthogonal vectors in R^n such that $\|\mathbf{u}\| = 1$ and $\|\mathbf{v}\| = 1$, then $d(\mathbf{u}, \mathbf{v}) =$____.

(b) Draw a picture to illustrate this result.

36. In the accompanying figure the vectors **u**, **v**, and **u** − **v** form a triangle in R^2, and θ denotes the angle between **u** and **v**. It follows from the law of cosines in trigonometry that

$$\|\mathbf{u} - \mathbf{v}\|^2 = \|\mathbf{u}\|^2 + \|\mathbf{v}\|^2 - 2\|\mathbf{u}\|\|\mathbf{v}\|\cos\theta$$

Do you think that this formula still holds if **u** and **v** are vectors in R^n? Justify your answer.

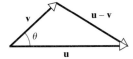

Figure Ex-36

37. Indicate whether the statement is always true or sometimes false. Justify your answer by giving a logical argument or a counterexample.

(a) If $\|\mathbf{u} + \mathbf{v}\|^2 = \|\mathbf{u}\|^2 + \|\mathbf{v}\|^2$, then **u** and **v** are orthogonal. T

(b) If **u** is orthogonal to **v** and **w**, then **u** is orthogonal to **v** + **w**. T

(c) If **u** is orthogonal to **v** + **w**, then **u** is orthogonal to **v** and **w**. F

(d) If $\|\mathbf{u} - \mathbf{v}\| = 0$, then **u** = **v**. T

(e) If $\|k\mathbf{u}\| = k\|\mathbf{u}\|$, then $k \geq 0$. T unless u = 0

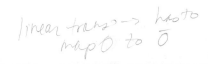

4.2 LINEAR TRANSFORMATIONS FROM R^n TO R^m

In this section we shall begin the study of functions of the form w $= F(x)$*, where the independent variable* x *is a vector in* R^n *and the dependent variable* w *is a vector in* R^m*. We shall concentrate on a special class of such functions called "linear transformations." Linear transformations are fundamental in the study of linear algebra and have many important applications in physics, engineering, social sciences, and various branches of mathematics.*

Functions from R^n to R Recall that a ***function*** is a rule f that associates with each element in a set A one and only one element in a set B. If f associates the element b with the element a, then we write $b = f(a)$ and say that b is the ***image*** of a under f or that $f(a)$ is the ***value*** of f at a. The set A is called the ***domain*** of f and the set B is called the ***codomain*** of f. The subset of B consisting of all possible values for f as a varies over A is called the ***range*** of f. For the most common functions, A and B are sets of real numbers, in which case f is called a ***real-valued function of a real variable***. Other common functions occur when B is a set of real numbers and A is a set of vectors in R^2, R^3, or more generally, in R^n. Some examples are shown in Table 1.

TABLE 1

Formula	Example	Classification	Description
$f(x)$	$f(x) = x^2$	Real-valued function of a real variable	Function from R to R
$f(x, y)$	$f(x, y) = x^2 + y^2$	Real-valued function of two real variables	Function from R^2 to R
$f(x, y, z)$	$f(x, y, z) = x^2 + y^2 + z^2$	Real-valued function of three real variables	Function from R^3 to R
$f(x_1, x_2, \ldots, x_n)$	$f(x_1, x_2, \ldots, x_n) = x_1^2 + x_2^2 + \cdots + x_n^2$	Real-valued function of n real variables	Function from R^n to R

Two functions f_1 and f_2 are regarded to be **equal**, written $f_1 = f_2$, if they have the same domain and $f_1(a) = f_2(a)$ for all a in the domain.

Functions from R^n to R^m

If the domain of a function f is R^n and the codomain is R^m (m and n possibly the same), then f is called a **map** or **transformation** from R^n to R^m, and we say that the function f **maps** R^n into R^m. We denote this by writing $f: R^n \rightarrow R^m$. The functions in Table 1 are transformations for which $m = 1$. In the case where $m = n$ the transformation $f: R^n \rightarrow R^n$ is called an **operator** on R^n. The first entry in Table 1 is an operator on R.

To illustrate one important way in which transformations can arise, suppose that $f_1, f_2, \ldots, f_m$ are real-valued functions of n real variables, say

$$
\begin{aligned}
w_1 &= f_1(x_1, x_2, \ldots, x_n) \\
w_2 &= f_2(x_1, x_2, \ldots, x_n) \\
&\;\;\vdots \qquad\qquad \vdots \\
w_m &= f_m(x_1, x_2, \ldots, x_n)
\end{aligned}
\tag{1}
$$

These m equations assign a unique point $(w_1, w_2, \ldots, w_m)$ in R^m to each point $(x_1, x_2, \ldots, x_n)$ in R^n and thus define a transformation from R^n to R^m. If we denote this transformation by T, then $T: R^n \rightarrow R^m$ and

$$T(x_1, x_2, \ldots, x_n) = (w_1, w_2, \ldots, w_m)$$

EXAMPLE 1 A Transformation from R^2 to R^3

The equations

$$
\begin{aligned}
w_1 &= x_1 + x_2 \\
w_2 &= 3x_1 x_2 \\
w_3 &= x_1^2 - x_2^2
\end{aligned}
$$

define a transformation $T: R^2 \rightarrow R^3$. With this transformation the image of the point (x_1, x_2) is

$$T(x_1, x_2) = (x_1 + x_2, 3x_1 x_2, x_1^2 - x_2^2)$$

Thus, for example, $T(1, -2) = (-1, -6, -3)$. ◆

Linear Transformations from R^n to R^m   In the special case where the equations in (1) are linear, the transformation $T: R^n \rightarrow R^m$ defined by those equations is called a ***linear transformation*** (or a ***linear operator*** if $m = n$). Thus, a linear transformation $T: R^n \rightarrow R^m$ is defined by equations of the form

$$
\begin{aligned}
w_1 &= a_{11}x_1 + a_{12}x_2 + \cdots + a_{1n}x_n \\
w_2 &= a_{21}x_1 + a_{22}x_2 + \cdots + a_{2n}x_n \\
&\ \ \vdots \qquad \vdots \qquad \vdots \qquad\qquad \vdots \\
w_m &= a_{m1}x_1 + a_{m2}x_2 + \cdots + a_{mn}x_n
\end{aligned}
\tag{2}
$$

or in matrix notation

$$
\begin{bmatrix} w_1 \\ w_2 \\ \vdots \\ w_m \end{bmatrix} = \begin{bmatrix} a_{11} & a_{12} & \cdots & a_{1n} \\ a_{21} & a_{22} & \cdots & a_{2n} \\ \vdots & \vdots & & \vdots \\ a_{m1} & a_{m2} & \cdots & a_{mn} \end{bmatrix} \begin{bmatrix} x_1 \\ x_2 \\ \vdots \\ x_n \end{bmatrix}
\tag{3}
$$

or more briefly by

$$
\mathbf{w} = A\mathbf{x} \tag{4}
$$

The matrix $A = [a_{ij}]$ is called the ***standard matrix*** for the linear transformation T, and T is called ***multiplication by A***.

EXAMPLE 2 A Linear Transformation from R^4 to R^3

The linear transformation $T: R^4 \rightarrow R^3$ defined by the equations

$$
\begin{aligned}
w_1 &= 2x_1 - 3x_2 + x_3 - 5x_4 \\
w_2 &= 4x_1 + x_2 - 2x_3 + x_4 \\
w_3 &= 5x_1 - x_2 + 4x_3
\end{aligned}
\tag{5}
$$

can be expressed in matrix form as

$$
\begin{bmatrix} w_1 \\ w_2 \\ w_3 \end{bmatrix} = \begin{bmatrix} 2 & -3 & 1 & -5 \\ 4 & 1 & -2 & 1 \\ 5 & -1 & 4 & 0 \end{bmatrix} \begin{bmatrix} x_1 \\ x_2 \\ x_3 \\ x_4 \end{bmatrix}
\tag{6}
$$

so the standard matrix for T is

$$
A = \begin{bmatrix} 2 & -3 & 1 & -5 \\ 4 & 1 & -2 & 1 \\ 5 & -1 & 4 & 0 \end{bmatrix}
$$

The image of a point (x_1, x_2, x_3, x_4) can be computed directly from the defining equations (5) or from (6) by matrix multiplication. For example, if $(x_1, x_2, x_3, x_4) = (1, -3, 0, 2)$, then substituting in (5) yields

$$
w_1 = 1, \qquad w_2 = 3, \qquad w_3 = 8
$$

(verify) or alternatively from (6)

$$
\begin{bmatrix} w_1 \\ w_2 \\ w_3 \end{bmatrix} = \begin{bmatrix} 2 & -3 & 1 & -5 \\ 4 & 1 & -2 & 1 \\ 5 & -1 & 4 & 0 \end{bmatrix} \begin{bmatrix} 1 \\ -3 \\ 0 \\ 2 \end{bmatrix} = \begin{bmatrix} 1 \\ 3 \\ 8 \end{bmatrix}
$$

◆

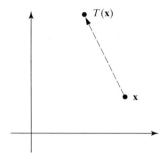

(*a*) *T* maps points to points

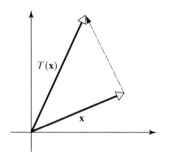

(*b*) *T* maps vectors to vectors

Figure 4.2.1

Some Notational Matters   If $T: R^n \to R^m$ is multiplication by A, and if it is important to emphasize that A is the standard matrix for T, we shall denote the linear transformation $T: R^n \to R^m$ by $T_A: R^n \to R^m$. Thus,

$$T_A(\mathbf{x}) = A\mathbf{x} \tag{7}$$

It is understood in this equation that the vector $\mathbf{x}$ in R^n is expressed as a column matrix.

Sometimes it is awkward to introduce a new letter to denote the standard matrix for a linear transformation $T: R^n \to R^m$. In such cases we will denote the standard matrix for T by the symbol $[T]$. With this notation equation (7) would take the form

$$T(\mathbf{x}) = [T]\mathbf{x} \tag{8}$$

Occasionally, the two notations for a standard matrix will be mixed, in which case we have the relationship

$$[T_A] = A \tag{9}$$

REMARK. Amidst all of this notation it is important to keep in mind that we have established a correspondence between $m \times n$ matrices and linear transformations from R^n to R^m: To each matrix A there corresponds a linear transformation T_A (multiplication by A), and to each linear transformation $T: R^n \to R^m$, there corresponds an $m \times n$ matrix $[T]$ (the standard matrix for T).

Geometry of Linear Transformations   Depending on whether n-tuples are regarded as points or vectors, the geometric effect of an operator $T: R^n \to R^n$ is to transform each point (or vector) in R^n into some new point (or vector) (Figure 4.2.1).

EXAMPLE 3 Zero Transformation from R^n to R^m

If 0 is the $m \times n$ zero matrix and $\mathbf{0}$ is the zero vector in R^n, then for every vector $\mathbf{x}$ in R^n

$$T_0(\mathbf{x}) = 0\mathbf{x} = \mathbf{0}$$

so multiplication by zero maps every vector in R^n into the zero vector in R^m. We call T_0 the ***zero transformation*** from R^n to R^m. Sometimes the zero transformation is denoted by 0. Although this is the same notation used for the zero matrix, the appropriate interpretation will usually be clear from the context.   ◆

EXAMPLE 4 Identity Operator on R^n

If I is the $n \times n$ identity matrix, then for every vector $\mathbf{x}$ in R^n

$$T_I(\mathbf{x}) = I\mathbf{x} = \mathbf{x}$$

so multiplication by I maps every vector in R^n into itself. We call T_I the ***identity operator*** on R^n. Sometimes the identity operator is denoted by I. Although this is the same notation used for the identity matrix, the appropriate interpretation will usually be clear from the context.   ◆

Among the most important linear operators on R^2 and R^3 are those that produce reflections, projections, and rotations. We shall now discuss such operators.

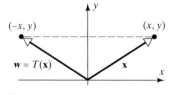

Figure 4.2.2

Reflection Operators   Consider the operator $T: R^2 \to R^2$ that maps each vector into its symmetric image about the *y*-axis (Figure 4.2.2).

If we let $\mathbf{w} = T(\mathbf{x})$, then the equations relating the components of $\mathbf{x}$ and $\mathbf{w}$ are

$$\begin{aligned} w_1 &= -x = -x + 0y \\ w_2 &= \quad y = 0x + \quad y \end{aligned} \tag{10}$$

or in matrix form,

$$\begin{bmatrix} w_1 \\ w_2 \end{bmatrix} = \begin{bmatrix} -1 & 0 \\ 0 & 1 \end{bmatrix} \begin{bmatrix} x \\ y \end{bmatrix} \tag{11}$$

Since the equations in (10) are linear, T is a linear operator and from (11) the standard matrix for T is

$$[T] = \begin{bmatrix} -1 & 0 \\ 0 & 1 \end{bmatrix}$$

In general, operators on R^2 and R^3 that map each vector into its symmetric image about some line or plane are called ***reflection operators***. Such operators are linear. Tables 2 and 3 list some of the common reflection operators.

Projection Operators   Consider the operator $T: R^2 \to R^2$ that maps each vector into its orthogonal projection on the x-axis (Figure 4.2.3). The equations relating the components of $\mathbf{x}$ and $\mathbf{w} = T(\mathbf{x})$ are

$$\begin{aligned} w_1 &= x = \quad x + 0y \\ w_2 &= 0 = 0x + 0y \end{aligned} \tag{12}$$

or in matrix form,

$$\begin{bmatrix} w_1 \\ w_2 \end{bmatrix} = \begin{bmatrix} 1 & 0 \\ 0 & 0 \end{bmatrix} \begin{bmatrix} x \\ y \end{bmatrix} \tag{13}$$

The equations in (12) are linear, so T is a linear operator and from (13) the standard

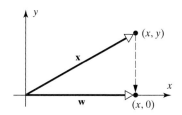

Figure 4.2.3

TABLE 2

Operator	Illustration	Equations	Standard Matrix
Reflection about the y-axis		$\begin{aligned} w_1 &= -x \\ w_2 &= \quad y \end{aligned}$	$\begin{bmatrix} -1 & 0 \\ 0 & 1 \end{bmatrix}$
Reflection about the x-axis		$\begin{aligned} w_1 &= \quad x \\ w_2 &= -y \end{aligned}$	$\begin{bmatrix} 1 & 0 \\ 0 & -1 \end{bmatrix}$
Reflection about the line $y = x$		$\begin{aligned} w_1 &= y \\ w_2 &= x \end{aligned}$	$\begin{bmatrix} 0 & 1 \\ 1 & 0 \end{bmatrix}$

TABLE 3

Operator	Illustration	Equations	Standard Matrix
Reflection about the xy-plane		$w_1 = x$ $w_2 = y$ $w_3 = -z$	$\begin{bmatrix} 1 & 0 & 0 \\ 0 & 1 & 0 \\ 0 & 0 & -1 \end{bmatrix}$
Reflection about the xz-plane		$w_1 = x$ $w_2 = -y$ $w_3 = z$	$\begin{bmatrix} 1 & 0 & 0 \\ 0 & -1 & 0 \\ 0 & 0 & 1 \end{bmatrix}$
Reflection about the yz-plane		$w_1 = -x$ $w_2 = y$ $w_3 = z$	$\begin{bmatrix} -1 & 0 & 0 \\ 0 & 1 & 0 \\ 0 & 0 & 1 \end{bmatrix}$

matrix for T is

$$[T] = \begin{bmatrix} 1 & 0 \\ 0 & 0 \end{bmatrix}$$

In general, a **_projection operator_** (or more precisely an **_orthogonal projection operator_**) on R^2 or R^3 is any operator that maps each vector into its orthogonal projection on a line or plane through the origin. It can be shown that such operators are linear. Some of the basic projection operators on R^2 and R^3 are listed in Tables 4 and 5.

TABLE 4

Operator	Illustration	Equations	Standard Matrix
Orthogonal projection on the x-axis		$w_1 = x$ $w_2 = 0$	$\begin{bmatrix} 1 & 0 \\ 0 & 0 \end{bmatrix}$
Orthogonal projection on the y-axis		$w_1 = 0$ $w_2 = y$	$\begin{bmatrix} 0 & 0 \\ 0 & 1 \end{bmatrix}$

TABLE 5

Operator	Illustration	Equations	Standard Matrix
Orthogonal projection on the xy-plane		$w_1 = x$ $w_2 = y$ $w_3 = 0$	$\begin{bmatrix} 1 & 0 & 0 \\ 0 & 1 & 0 \\ 0 & 0 & 0 \end{bmatrix}$
Orthogonal projection on the xz-plane		$w_1 = x$ $w_2 = 0$ $w_3 = z$	$\begin{bmatrix} 1 & 0 & 0 \\ 0 & 0 & 0 \\ 0 & 0 & 1 \end{bmatrix}$
Orthogonal projection on the yz-plane		$w_1 = 0$ $w_2 = y$ $w_3 = z$	$\begin{bmatrix} 0 & 0 & 0 \\ 0 & 1 & 0 \\ 0 & 0 & 1 \end{bmatrix}$

Rotation Operators   An operator that rotates each vector in R^2 through a fixed angle θ is called a ***rotation operator*** on R^2. Table 6 gives formulas for the rotation operators on R^2. To show how these results are derived, consider the rotation operator that rotates each vector counterclockwise through a fixed positive angle θ. To find equations relating $\mathbf{x}$ and $\mathbf{w} = T(\mathbf{x})$, let ϕ be the angle from the positive x-axis to $\mathbf{x}$, and let r be the common length of $\mathbf{x}$ and $\mathbf{w}$ (Figure 4.2.4).

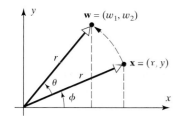

Figure 4.2.4

Then from basic trigonometry

$$x = r\cos\phi, \qquad y = r\sin\phi \tag{14}$$

and

$$w_1 = r\cos(\theta + \phi), \qquad w_2 = r\sin(\theta + \phi) \tag{15}$$

Using trigonometric identities on (15) yields

$$w_1 = r\cos\theta\cos\phi - r\sin\theta\sin\phi$$
$$w_2 = r\sin\theta\cos\phi + r\cos\theta\sin\phi$$

and substituting (14) yields

TABLE 6

Operator	Illustration	Equations	Standard Matrix
Rotation through an angle θ		$w_1 = x\cos\theta - y\sin\theta$ $w_2 = x\sin\theta + y\cos\theta$	$\begin{bmatrix} \cos\theta & -\sin\theta \\ \sin\theta & \cos\theta \end{bmatrix}$

$$w_1 = x \cos \theta - y \sin \theta$$
$$w_2 = x \sin \theta + y \cos \theta \tag{16}$$

The equations in (16) are linear, so T is a linear operator; moreover, it follows from these equations that the standard matrix for T is

$$[T] = \begin{bmatrix} \cos \theta & -\sin \theta \\ \sin \theta & \cos \theta \end{bmatrix}$$

EXAMPLE 5 Rotation

If each vector in R^2 is rotated through an angle of $\pi/6 \, (= 30°)$, then the image $\mathbf{w}$ of a vector

$$\mathbf{x} = \begin{bmatrix} x \\ y \end{bmatrix}$$

is

$$\mathbf{w} = \begin{bmatrix} \cos \pi/6 & -\sin \pi/6 \\ \sin \pi/6 & \cos \pi/6 \end{bmatrix} \begin{bmatrix} x \\ y \end{bmatrix} = \begin{bmatrix} \sqrt{3}/2 & -1/2 \\ 1/2 & \sqrt{3}/2 \end{bmatrix} \begin{bmatrix} x \\ y \end{bmatrix} = \begin{bmatrix} \dfrac{\sqrt{3}}{2}x - \dfrac{1}{2}y \\ \dfrac{1}{2}x + \dfrac{\sqrt{3}}{2}y \end{bmatrix}$$

For example, the image of the vector

$$\mathbf{x} = \begin{bmatrix} 1 \\ 1 \end{bmatrix} \quad \text{is} \quad \mathbf{w} = \begin{bmatrix} \dfrac{\sqrt{3}-1}{2} \\ \dfrac{1+\sqrt{3}}{2} \end{bmatrix} \qquad \blacklozenge$$

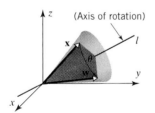

(*a*) Angle of rotation

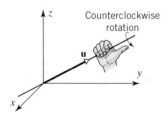

(*b*) Right-hand rule

Figure 4.2.5

A rotation of vectors in R^3 is usually described in relation to a ray emanating from the origin, called the **axis of rotation**. As a vector revolves around the axis of rotation it sweeps out some portion of a cone (Figure 4.2.5*a*). The **angle of rotation**, which is measured in the base of the cone, is described as "clockwise" or "counterclockwise" in relation to a viewpoint that is along the axis of rotation *looking toward the origin*. For example, in Figure 4.2.5*a* the vector $\mathbf{w}$ results from rotating the vector $\mathbf{x}$ counterclockwise around the axis l through an angle θ. As in R^2, angles are *positive* if they are generated by counterclockwise rotations and *negative* if they are generated by clockwise rotations.

The most common way of describing a general axis of rotation is to specify a nonzero vector $\mathbf{u}$ that runs along the axis of rotation and has its initial point at the origin. The counterclockwise direction for a rotation about the axis can then be determined by a "right-hand rule" (Figure 4.2.5*b*): If the thumb of the right hand points in the direction of $\mathbf{u}$, then the cupped fingers point in a counterclockwise direction.

A **rotation operator** on R^3 is a linear operator that rotates each vector in R^3 about some rotation axis through a fixed angle θ. In Table 7 we have described the rotation operators on R^3 whose axes of rotation are the positive coordinate axes. For each of these rotations one of the components is unchanged by the rotation, and the relationships between the other components can be derived by the same procedure used to derive (16). For example, in the rotation about the z-axis, the z-components of $\mathbf{x}$ and $\mathbf{w} = T(\mathbf{x})$ are the same, and the x- and y-components are related as in (16). This yields the rotation equations shown in the last row of Table 7.

TABLE 7

Operator	Illustration	Equations	Standard Matrix
Counterclockwise rotation about the positive x-axis through an angle θ		$w_1 = x$ $w_2 = y \cos\theta - z \sin\theta$ $w_3 = y \sin\theta + z \cos\theta$	$\begin{bmatrix} 1 & 0 & 0 \\ 0 & \cos\theta & -\sin\theta \\ 0 & \sin\theta & \cos\theta \end{bmatrix}$
Counterclockwise rotation about the positive y-axis through an angle θ		$w_1 = x \cos\theta + z \sin\theta$ $w_2 = y$ $w_3 = -x \sin\theta + z \cos\theta$	$\begin{bmatrix} \cos\theta & 0 & \sin\theta \\ 0 & 1 & 0 \\ -\sin\theta & 0 & \cos\theta \end{bmatrix}$
Counterclockwise rotation about the positive z-axis through an angle θ		$w_1 = x \cos\theta - y \sin\theta$ $w_2 = x \sin\theta + y \cos\theta$ $w_3 = z$	$\begin{bmatrix} \cos\theta & -\sin\theta & 0 \\ \sin\theta & \cos\theta & 0 \\ 0 & 0 & 1 \end{bmatrix}$

For completeness, we note that the standard matrix for a counterclockwise rotation through an angle θ about an axis in R^3, which is determined by an arbitrary *unit vector* $\mathbf{u} = (a, b, c)$ that has its initial point at the origin, is

$$\begin{bmatrix} a^2(1 - \cos\theta) + \cos\theta & ab(1 - \cos\theta) - c\sin\theta & ac(1 - \cos\theta) + b\sin\theta \\ ab(1 - \cos\theta) + c\sin\theta & b^2(1 - \cos\theta) + \cos\theta & bc(1 - \cos\theta) - a\sin\theta \\ ac(1 - \cos\theta) - b\sin\theta & bc(1 - \cos\theta) + a\sin\theta & c^2(1 - \cos\theta) + \cos\theta \end{bmatrix} \quad (17)$$

The derivation can be found in the book *Principles of Interactive Computer Graphics*, by W. M. Newman and R. F. Sproull, New York, McGraw-Hill, 1979. The reader may find it instructive to derive the results in Table 7 as special cases of this more general result.

Dilation and Contraction Operators   If k is a nonnegative scalar, then the operator $T(\mathbf{x}) = k\mathbf{x}$ on R^2 or R^3 is called a ***contraction with factor k*** if $0 \le k \le 1$ and a ***dilation with factor k*** if $k \ge 1$. The geometric effect of a contraction is to compress each vector by a factor of k (Figure 4.2.6a), and the effect of a dilation is to stretch each vector by a factor of k (Figure 4.2.6b). A contraction compresses R^2 or R^3 uniformly toward the origin from all directions, and a dilation stretches R^2 or R^3 uniformly away from the origin in all directions.

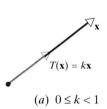

(a) $0 \le k < 1$

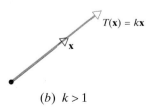

(b) $k > 1$

Figure 4.2.6

TABLE 8

Operator	Illustration	Equations	Standard Matrix
Contraction with factor k on R^2 $(0 \le k \le 1)$		$w_1 = kx$ $w_2 = ky$	$\begin{bmatrix} k & 0 \\ 0 & k \end{bmatrix}$
Dilation with factor k on R^2 $(k \ge 1)$		$w_1 = kx$ $w_2 = ky$	

TABLE 9

Operator	Illustration	Equations	Standard Matrix
Contraction with factor k on R^3 $(0 \le k \le 1)$		$w_1 = kx$ $w_2 = ky$ $w_3 = kz$	$\begin{bmatrix} k & 0 & 0 \\ 0 & k & 0 \\ 0 & 0 & k \end{bmatrix}$
Dilation with factor k on R^3 $(k \ge 1)$		$w_1 = kx$ $w_2 = ky$ $w_3 = kz$	

The most extreme contraction occurs when $k = 0$, in which case $T(\mathbf{x}) = k\mathbf{x}$ reduces to the zero operator $T(\mathbf{x}) = \mathbf{0}$, which compresses every vector into a single point (the origin). If $k = 1$, then $T(\mathbf{x}) = k\mathbf{x}$ reduces to the identity operator $T(\mathbf{x}) = \mathbf{x}$, which leaves each vector unchanged; this can be regarded as either a contraction or a dilation. Tables 8 and 9 list the dilation and contraction operators on R^2 and R^3.

Compositions of Linear Transformations   If $T_A \colon R^n \to R^k$ and $T_B \colon R^k \to R^m$ are linear transformations, then for each $\mathbf{x}$ in R^n one can first compute $T_A(\mathbf{x})$, which is a vector in R^k, and then one can compute $T_B(T_A(\mathbf{x}))$, which is a vector in R^m. Thus, the application of T_A followed by T_B produces a transformation from R^n to R^m. This transformation is called the ***composition of T_B with T_A*** and is denoted by $T_B \circ T_A$ (read "T_B circle T_A"). Thus,

$$(T_B \circ T_A)(\mathbf{x}) = T_B(T_A(\mathbf{x})) \tag{18}$$

The composition $T_B \circ T_A$ is linear since

$$(T_B \circ T_A)(\mathbf{x}) = T_B(T_A(\mathbf{x})) = B(A\mathbf{x}) = (BA)\mathbf{x} \tag{19}$$

so $T_B \circ T_A$ is multiplication by BA, which is a linear transformation. Formula (19) also tells us that the standard matrix for $T_B \circ T_A$ is BA. This is expressed by the formula

$$T_B \circ T_A = T_{BA} \tag{20}$$

REMARK. Formula (20) captures an important idea: *Multiplying matrices is equivalent to composing the corresponding linear transformations in the right-to-left order of the factors.*

There is an alternative form of Formula (20): If $T_1: R^n \rightarrow R^k$ and $T_2: R^k \rightarrow R^m$ are linear transformations, then because the standard matrix for the composition $T_2 \circ T_1$ is the product of the standard matrices of T_2 and T_1, we have

$$[T_2 \circ T_1] = [T_2][T_1] \tag{21}$$

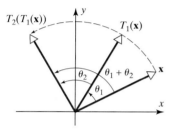

Figure 4.2.7

EXAMPLE 6 Composition of Two Rotations

Let $T_1: R^2 \rightarrow R^2$ and $T_2: R^2 \rightarrow R^2$ be the linear operators that rotate vectors through the angles θ_1 and θ_2, respectively. Thus, the operation

$$(T_2 \circ T_1)(\mathbf{x}) = T_2(T_1(\mathbf{x}))$$

first rotates $\mathbf{x}$ through the angle θ_1, then rotates $T_1(\mathbf{x})$ through the angle θ_2. It follows that the net effect of $T_2 \circ T_1$ is to rotate each vector in R^2 through the angle $\theta_1 + \theta_2$ (Figure 4.2.7).

Thus, the standard matrices for these linear operators are

$$[T_1] = \begin{bmatrix} \cos\theta_1 & -\sin\theta_1 \\ \sin\theta_1 & \cos\theta_1 \end{bmatrix}, \qquad [T_2] = \begin{bmatrix} \cos\theta_2 & -\sin\theta_2 \\ \sin\theta_2 & \cos\theta_2 \end{bmatrix},$$

$$[T_2 \circ T_1] = \begin{bmatrix} \cos(\theta_1 + \theta_2) & -\sin(\theta_1 + \theta_2) \\ \sin(\theta_1 + \theta_2) & \cos(\theta_1 + \theta_2) \end{bmatrix}$$

These matrices should satisfy (21). With the help of some basic trigonometric identities, we can show that this is so as follows:

$$[T_2][T_1] = \begin{bmatrix} \cos\theta_2 & -\sin\theta_2 \\ \sin\theta_2 & \cos\theta_2 \end{bmatrix} \begin{bmatrix} \cos\theta_1 & -\sin\theta_1 \\ \sin\theta_1 & \cos\theta_1 \end{bmatrix}$$

$$= \begin{bmatrix} \cos\theta_2 \cos\theta_1 - \sin\theta_2 \sin\theta_1 & -(\cos\theta_2 \sin\theta_1 + \sin\theta_2 \cos\theta_1) \\ \sin\theta_2 \cos\theta_1 + \cos\theta_2 \sin\theta_1 & -\sin\theta_2 \sin\theta_1 + \cos\theta_2 \cos\theta_1 \end{bmatrix}$$

$$= \begin{bmatrix} \cos(\theta_1 + \theta_2) & -\sin(\theta_1 + \theta_2) \\ \sin(\theta_1 + \theta_2) & \cos(\theta_1 + \theta_2) \end{bmatrix}$$

$$= [T_2 \circ T_1] \qquad \blacklozenge$$

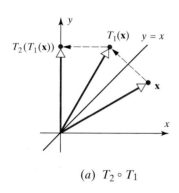

(*a*) $T_2 \circ T_1$

REMARK. In general, the order in which linear transformations are composed matters. This is to be expected, since the composition of two linear transformations corresponds to the multiplication of their standard matrices, and we know that the order in which matrices are multiplied makes a difference.

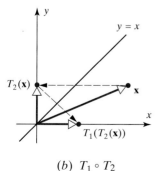

(*b*) $T_1 \circ T_2$

Figure 4.2.8

EXAMPLE 7 Composition Is Not Commutative

Let $T_1: R^2 \rightarrow R^2$ be the reflection operator about the line $y = x$, and let $T_2: R^2 \rightarrow R^2$ be the orthogonal projection on the y-axis. Figure 4.2.8 illustrates graphically that $T_1 \circ T_2$ and $T_2 \circ T_1$ have different effects on a vector $\mathbf{x}$. This same conclusion can be reached by showing that the standard matrices for T_1 and T_2 do not commute:

$$[T_1 \circ T_2] = [T_1][T_2] = \begin{bmatrix} 0 & 1 \\ 1 & 0 \end{bmatrix} \begin{bmatrix} 0 & 0 \\ 0 & 1 \end{bmatrix} = \begin{bmatrix} 0 & 1 \\ 0 & 0 \end{bmatrix}$$

$$[T_2 \circ T_1] = [T_2][T_1] = \begin{bmatrix} 0 & 0 \\ 0 & 1 \end{bmatrix} \begin{bmatrix} 0 & 1 \\ 1 & 0 \end{bmatrix} = \begin{bmatrix} 0 & 0 \\ 1 & 0 \end{bmatrix}$$

so $[T_2 \circ T_1] \neq [T_1 \circ T_2]$. ◆

EXAMPLE 8 Composition of Two Reflections

Let $T_1: R^2 \rightarrow R^2$ be the reflection about the y-axis, and let $T_2: R^2 \rightarrow R^2$ be the reflection about the x-axis. In this case $T_1 \circ T_2$ and $T_2 \circ T_1$ are the same; both map each vector $\mathbf{x} = (x, y)$ into its negative $-\mathbf{x} = (-x, -y)$ (Figure 4.2.9):

$$(T_1 \circ T_2)(x, y) = T_1(x, -y) = (-x, -y)$$
$$(T_2 \circ T_1)(x, y) = T_2(-x, y) = (-x, -y)$$

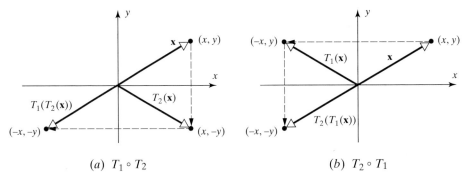

(a) $T_1 \circ T_2$ (b) $T_2 \circ T_1$

Figure 4.2.9

The equality of $T_1 \circ T_2$ and $T_2 \circ T_1$ can also be deduced by showing that the standard matrices for T_1 and T_2 commute:

$$[T_1 \circ T_2] = [T_1][T_2] = \begin{bmatrix} -1 & 0 \\ 0 & 1 \end{bmatrix} \begin{bmatrix} 1 & 0 \\ 0 & -1 \end{bmatrix} = \begin{bmatrix} -1 & 0 \\ 0 & -1 \end{bmatrix}$$

$$[T_2 \circ T_1] = [T_2][T_1] = \begin{bmatrix} 1 & 0 \\ 0 & -1 \end{bmatrix} \begin{bmatrix} -1 & 0 \\ 0 & 1 \end{bmatrix} = \begin{bmatrix} -1 & 0 \\ 0 & -1 \end{bmatrix}$$

The operator $T(\mathbf{x}) = -\mathbf{x}$ on R^2 or R^3 is called the ***reflection about the origin***. As the computations above show, the standard matrix for this operator on R^2 is

$$[T] = \begin{bmatrix} -1 & 0 \\ 0 & -1 \end{bmatrix}$$ ◆

Compositions of Three or More Linear Transformations

Compositions can be defined for three or more linear transformations. For example, consider the linear transformations

$$T_1: R^n \rightarrow R^k, \qquad T_2: R^k \rightarrow R^l, \qquad T_3: R^l \rightarrow R^m$$

We define the composition $(T_3 \circ T_2 \circ T_1): R^n \rightarrow R^m$ by

$$(T_3 \circ T_2 \circ T_1)(\mathbf{x}) = T_3(T_2(T_1(\mathbf{x})))$$

It can be shown that this composition is a linear transformation and that the standard matrix for $T_3 \circ T_2 \circ T_1$ is related to the standard matrices for T_1, T_2, and T_3 by

$$[T_3 \circ T_2 \circ T_1] = [T_3][T_2][T_1] \tag{22}$$

which is a generalization of (21). If the standard matrices for T_1, T_2, and T_3 are denoted by A, B, and C, respectively, then we also have the following generalization of (20):

$$T_C \circ T_B \circ T_A = T_{CBA} \tag{23}$$

EXAMPLE 9 Composition of Three Transformations

Find the standard matrix for the linear operator $T: R^3 \rightarrow R^3$ that first rotates a vector counterclockwise about the z-axis through an angle θ, then reflects the resulting vector about the yz-plane, and then projects that vector orthogonally onto the xy-plane.

Solution.

The linear transformation T can be expressed as the composition

$$T = T_3 \circ T_2 \circ T_1$$

where T_1 is the rotation about the z-axis, T_2 is the reflection about the yz-plane, and T_3 is the orthogonal projection on the xy-plane. From Tables 3, 5, and 7 the standard matrices for these linear transformations are

$$[T_1] = \begin{bmatrix} \cos\theta & -\sin\theta & 0 \\ \sin\theta & \cos\theta & 0 \\ 0 & 0 & 1 \end{bmatrix}, \quad [T_2] = \begin{bmatrix} -1 & 0 & 0 \\ 0 & 1 & 0 \\ 0 & 0 & 1 \end{bmatrix}, \quad [T_3] = \begin{bmatrix} 1 & 0 & 0 \\ 0 & 1 & 0 \\ 0 & 0 & 0 \end{bmatrix}$$

Thus, from (22) the standard matrix for T is $[T] = [T_3][T_2][T_1]$; that is,

$$[T] = \begin{bmatrix} 1 & 0 & 0 \\ 0 & 1 & 0 \\ 0 & 0 & 0 \end{bmatrix} \begin{bmatrix} -1 & 0 & 0 \\ 0 & 1 & 0 \\ 0 & 0 & 1 \end{bmatrix} \begin{bmatrix} \cos\theta & -\sin\theta & 0 \\ \sin\theta & \cos\theta & 0 \\ 0 & 0 & 1 \end{bmatrix}$$

$$= \begin{bmatrix} -\cos\theta & \sin\theta & 0 \\ \sin\theta & \cos\theta & 0 \\ 0 & 0 & 0 \end{bmatrix}$$

◆

Exercise Set 4.2

1. Find the domain and codomain of the transformation defined by the equations, and determine whether the transformation is linear.

*linear
* must map zero to 0*

(a) $w_1 = 3x_1 - 2x_2 + 4x_3$
    $w_2 = 5x_1 - 8x_2 + x_3$

(b) $w_1 = 2x_1x_2 - x_2$
    $w_2 = x_1 + 3x_1x_2$
    $w_3 = x_1 + x_2$

(c) $w_1 = 5x_1 - x_2 + x_3$
    $w_2 = -x_1 + x_2 + 7x_3$
    $w_3 = 2x_1 - 4x_2 - x_3$

(d) $w_1 = x_1^2 - 3x_2 + x_3 - 2x_4$
    $w_2 = 3x_1 - 4x_2 - x_3^2 + x_4$

2. Find the standard matrix for the linear transformation defined by the equations.

(a) $w_1 = 2x_1 - 3x_2 + x_4$
    $w_2 = 3x_1 + 5x_2 - x_4$

(b) $w_1 = 7x_1 + 2x_2 - 8x_3$
    $w_2 = -x_2 + 5x_3$
    $w_3 = 4x_1 + 7x_2 - x_3$

(c) $w_1 = -x_1 + x_2$
    $w_2 = 3x_1 - 2x_2$
    $w_3 = 5x_1 - 7x_2$

(d) $w_1 = x_1$
    $w_2 = x_1 + x_2$
    $w_3 = x_1 + x_2 + x_3$
    $w_4 = x_1 + x_2 + x_3 + x_4$

3. Find the standard matrix for the linear transformation $T: R^3 \to R^3$ given by

$$w_1 = 3x_1 + 5x_2 - x_3$$
$$w_2 = 4x_1 - x_2 + x_3$$
$$w_3 = 3x_1 + 2x_2 - x_3$$

and then calculate $T(-1, 2, 4)$ by directly substituting in the equations and also by matrix multiplication.

4. Find the standard matrix for the linear operator T defined by the formula.

   (a) $T(x_1, x_2) = (2x_1 - x_2, x_1 + x_2)$ (b) $T(x_1, x_2) = (x_1, x_2)$
   (c) $T(x_1, x_2, x_3) = (x_1 + 2x_2 + x_3, x_1 + 5x_2, x_3)$ (d) $T(x_1, x_2, x_3) = (4x_1, 7x_2, -8x_3)$

5. Find the standard matrix for the linear transformation T defined by the formula.

   (a) $T(x_1, x_2) = (x_2, -x_1, x_1 + 3x_2, x_1 - x_2)$
   (b) $T(x_1, x_2, x_3, x_4) = (7x_1 + 2x_2 - x_3 + x_4, x_2 + x_3, -x_1)$
   (c) $T(x_1, x_2, x_3) = (0, 0, 0, 0, 0)$
   (d) $T(x_1, x_2, x_3, x_4) = (x_4, x_1, x_3, x_2, x_1 - x_3)$

6. In each part the standard matrix $[T]$ of a linear transformation T is given. Use it to find $T(x)$. [Express the answers in matrix form.]

   (a) $[T] = \begin{bmatrix} 1 & 2 \\ 3 & 4 \end{bmatrix}$; $\mathbf{x} = \begin{bmatrix} 3 \\ -2 \end{bmatrix}$ (b) $[T] = \begin{bmatrix} -1 & 2 & 0 \\ 3 & 1 & 5 \end{bmatrix}$; $\mathbf{x} = \begin{bmatrix} -1 \\ 1 \\ 3 \end{bmatrix}$

   (c) $[T] = \begin{bmatrix} -2 & 1 & 4 \\ 3 & 5 & 7 \\ 6 & 0 & -1 \end{bmatrix}$; $\mathbf{x} = \begin{bmatrix} x_1 \\ x_2 \\ x_3 \end{bmatrix}$ (d) $[T] = \begin{bmatrix} -1 & 1 \\ 2 & 4 \\ 7 & 8 \end{bmatrix}$; $\mathbf{x} = \begin{bmatrix} x_1 \\ x_2 \end{bmatrix}$

7. In each part use the standard matrix for T to find $T(\mathbf{x})$; then check the result by calculating $T(\mathbf{x})$ directly.

   (a) $T(x_1, x_2) = (-x_1 + x_2, x_2)$; $\mathbf{x} = (-1, 4)$
   (b) $T(x_1, x_2, x_3) = (2x_1 - x_2 + x_3, x_2 + x_3, 0)$; $\mathbf{x} = (2, 1, -3)$

8. Use matrix multiplication to find the reflection of $(-1, 2)$ about

   (a) the x-axis   (b) the y-axis   (c) the line $y = x$

9. Use matrix multiplication to find the reflection of $(2, -5, 3)$ about

   (a) the xy-plane   (b) the xz-plane   (c) the yz-plane

10. Use matrix multiplication to find the orthogonal projection of $(2, -5)$ on

   (a) the x-axis   (b) the y-axis

11. Use matrix multiplication to find the orthogonal projection of $(-2, 1, 3)$ on

   (a) the xy-plane   (b) the xz-plane   (c) the yz-plane

12. Use matrix multiplication to find the image of the vector $(3, -4)$ when it is rotated through an angle of

   (a) $\theta = 30°$ (b) $\theta = -60°$ (c) $\theta = 45°$ (d) $\theta = 90°$

13. Use matrix multiplication to find the image of the vector $(-2, 1, 2)$ if it is rotated

   (a) $30°$ about the x-axis   (b) $45°$ about the y-axis
   (c) $90°$ about the z-axis

14. Find the standard matrix for the linear operator that rotates a vector in R^3 through an angle of $-60°$ about

   (a) the x-axis   (b) the y-axis   (c) the z-axis

15. Use matrix multiplication to find the image of the vector $(-2, 1, 2)$ if it is rotated

   (a) $-30°$ about the x-axis   (b) $-45°$ about the y-axis
   (c) $-90°$ about the z-axis

16. Find the standard matrix for the stated composition of linear operators on R^2.

   (a) A rotation of $90°$, followed by a reflection about the line $y = x$.
   (b) An orthogonal projection on the y-axis, followed by a contraction with factor $k = \frac{1}{2}$.
   (c) A reflection about the x-axis, followed by a dilation with factor $k = 3$.

17. Find the standard matrix for the stated composition of linear operators on R^2.

   (a) A rotation of $60°$, followed by an orthogonal projection on the x-axis, followed by a reflection about the line $y = x$.
   (b) A dilation with factor $k = 2$, followed by a rotation of $45°$, followed by a reflection about the y-axis.
   (c) A rotation of $15°$, followed by a rotation of $105°$, followed by a rotation of $60°$.

18. Find the standard matrix for the stated composition of linear operators on R^3.

   (a) A reflection about the yz-plane, followed by an orthogonal projection on the xz-plane.
   (b) A rotation of $45°$ about the y-axis, followed by a dilation with factor $k = \sqrt{2}$.
   (c) An orthogonal projection on the xy-plane, followed by a reflection about the yz-plane.

19. Find the standard matrix for the stated composition of linear operators on R^3.

   (a) A rotation of $30°$ about the x-axis, followed by a rotation of $30°$ about the z-axis, followed by a contraction with factor $k = \frac{1}{4}$.
   (b) A reflection about the xy-plane, followed by a reflection about the xz-plane, followed by an orthogonal projection on the yz-plane.
   (c) A rotation of $270°$ about the x-axis, followed by a rotation of $90°$ about the y-axis, followed by a rotation of $180°$ about the z-axis.

20. Determine whether $T_1 \circ T_2 = T_2 \circ T_1$.

   (a) $T_1 : R^2 \to R^2$ is the orthogonal projection on the x-axis and $T_2 : R^2 \to R^2$ is the orthogonal projection on the y-axis.
   (b) $T_1 : R^2 \to R^2$ is the rotation through an angle θ_1 and $T_2 : R^2 \to R^2$ is the rotation through an angle θ_2.
   (c) $T_1 : R^2 \to R^2$ is the orthogonal projection on the x-axis and $T_2 : R^2 \to R^2$ is the rotation through an angle θ.

21. Determine whether $T_1 \circ T_2 = T_2 \circ T_1$.

   (a) $T_1 : R^3 \to R^3$ is a dilation by a factor k and $T_2 : R^3 \to R^3$ is the rotation about the z-axis through an angle θ.
   (b) $T_1 : R^3 \to R^3$ is the rotation about the x-axis through an angle θ_1 and $T_2 : R^3 \to R^3$ is the rotation about the z-axis through an angle θ_2.

22. In R^3 the ***orthogonal projections*** on the x-axis, y-axis, and z-axis are defined by

$$T_1(x, y, z) = (x, 0, 0), \qquad T_2(x, y, z) = (0, y, 0), \qquad T_3(x, y, z) = (0, 0, z)$$

respectively.

   (a) Show that the orthogonal projections on the coordinate axes are linear operators and find their standard matrices.
   (b) Show that if $T : R^3 \to R^3$ is an orthogonal projection on one of the coordinate axes, then for every vector $\mathbf{x}$ in R^3, the vectors $T(\mathbf{x})$ and $\mathbf{x} - T(\mathbf{x})$ are orthogonal vectors.
   (c) Make a sketch showing $\mathbf{x}$ and $\mathbf{x} - T(\mathbf{x})$ in the case where T is the orthogonal projection on the x-axis.

23. Derive the standard matrices for the rotations about the x-axis, y-axis, and z-axis in R^3 from Formula (17).

24. Use Formula (17) to find the standard matrix for a rotation $90°$ about the axis determined by the vector $\mathbf{v} = (1, 1, 1)$. [***Note.*** Formula (17) requires that the vector defining the axis of rotation have length 1.]

25. Verify Formula (21) for the given linear transformations.

(a) $T_1(x_1, x_2) = (x_1 + x_2, x_1 - x_2)$ and $T_2(x_1, x_2) = (3x_1, 2x_1 + 4x_2)$

(b) $T_1(x_1, x_2) = (4x_1, -2x_1 + x_2, -x_1 - 3x_2)$ and $T_2(x_1, x_2, x_3) = (x_1 + 2x_2 - x_3, 4x_1 - x_3)$

(c) $T_1(x_1, x_2, x_3) = (-x_1 + x_2, -x_2 + x_3, -x_3 + x_1)$ and $T_2(x_1, x_2, x_3) = (-2x_1, 3x_3, -4x_2)$

26. It can be proved that if A is a 2×2 matrix with $\det(A) = 1$ and such that the column vectors of A are orthogonal and have length 1, then multiplication by A is a rotation through some angle θ. Verify that

$$A = \begin{bmatrix} -1/\sqrt{2} & -1/\sqrt{2} \\ 1/\sqrt{2} & -1/\sqrt{2} \end{bmatrix}$$

satisfies the stated conditions and find the angle of rotation.

27. The result stated in Exercise 26 is also true in R^3: It can be proved that if A is a 3×3 matrix with $\det(A) = 1$ and such that the column vectors of A are pairwise orthogonal and have length 1, then multiplication by A is a rotation about some axis of rotation through some angle θ. Use Formula (17) to show that if A satisfies the stated conditions, then the angle of rotation satisfies the equation

$$\cos\theta = \frac{\operatorname{tr}(A) - 1}{2}$$

28. Let A be a 3×3 matrix (other than the identity matrix) satisfying the conditions stated in Exercise 27. It can be shown that if $\mathbf{x}$ is any nonzero vector in R^3, then the vector $\mathbf{u} = A\mathbf{x} + A^T\mathbf{x} + [1 - \operatorname{tr}(A)]\mathbf{x}$ determines an axis of rotation when $\mathbf{u}$ is positioned with its initial point at the origin. [See *The Axis of Rotation: Analysis, Algebra, Geometry*, by Dan Kalman, *Mathematics Magazine*, Vol. 62, No. 4, Oct. 1989.]

(a) Show that multiplication by

$$A = \begin{bmatrix} \frac{1}{9} & -\frac{4}{9} & \frac{8}{9} \\ \frac{8}{9} & \frac{4}{9} & \frac{1}{9} \\ -\frac{4}{9} & \frac{7}{9} & \frac{4}{9} \end{bmatrix}$$

is a rotation.

(b) Find a vector of length 1 that defines an axis for the rotation.

(c) Use the result in Exercise 27 to find the angle of rotation about the axis obtained in part (b).

Discussion and Discovery

29. In words, describe the geometric effect of multiplying a vector $\mathbf{x}$ by the matrix A.

(a) $A = \begin{bmatrix} 2 & 0 \\ 0 & 0 \end{bmatrix}$ (b) $A = \begin{bmatrix} 2 & 0 \\ 0 & -2 \end{bmatrix}$

30. In words, describe the geometric effect of multiplying a vector $\mathbf{x}$ by the matrix A.

(a) $A = \begin{bmatrix} 2 & 0 \\ 0 & 3 \end{bmatrix}$ (b) $A = \begin{bmatrix} \sqrt{3}/2 & -1/2 \\ 1/2 & \sqrt{3}/2 \end{bmatrix}$

31. In words, describe the geometric effect of multiplying a vector $\mathbf{x}$ by the matrix

$$A = \begin{bmatrix} \cos^2\theta - \sin^2\theta & -2\sin\theta\cos\theta \\ 2\sin\theta\cos\theta & \cos^2\theta - \sin^2\theta \end{bmatrix}$$

32. If multiplication by A rotates a vector $\mathbf{x}$ in the xy-plane through an angle θ, what is the effect of multiplying $\mathbf{x}$ by A^T? Explain your reasoning.

4.3 PROPERTIES OF LINEAR TRANSFORMATIONS FROM R^n TO R^m

In this section we shall investigate the relationship between the invertibility of a matrix and properties of the corresponding matrix transformation. We shall also obtain a characterization of linear transformations from R^n to R^m that will form the basis for more general linear transformations that will be discussed in subsequent sections, and we shall discuss some geometric properties of eigenvectors.

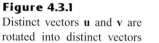

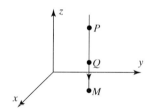

Figure 4.3.1
Distinct vectors **u** and **v** are rotated into distinct vectors $T(\mathbf{u})$ and $T(\mathbf{v})$.

One-to-One Linear Transformations   Linear transformations that map distinct vectors (or points) into distinct vectors (or points) are of special importance. One example of such a transformation is the linear operator $T: R^2 \rightarrow R^2$ that rotates each vector through an angle θ. It is obvious geometrically that if **u** and **v** are distinct vectors in R^2, then so are the rotated vectors $T(\mathbf{u})$ and $T(\mathbf{v})$ (Figure 4.3.1).

In contrast, if $T: R^3 \rightarrow R^3$ is the orthogonal projection of R^3 on the xy-plane, then distinct points on the same vertical line get mapped into the same point in the xy-plane (Figure 4.3.2).

Definition

A linear transformation $T: R^n \rightarrow R^m$ is said to be ***one-to-one*** if T maps distinct vectors (points) in R^n into distinct vectors (points) in R^m.

REMARK. It follows from this definition that for each vector **w** in the range of a one-to-one linear transformation T, there is exactly one vector **x** such that $T(\mathbf{x}) = \mathbf{w}$.

Figure 4.3.2
The distinct points P and Q are mapped into the same point M.

EXAMPLE 1   One-to-One Linear Transformations

In the terminology of the preceding definition, the rotation operator of Figure 4.3.1 is one-to-one, but the orthogonal projection operator of Figure 4.3.2 is not.   ◆

Let A be an $n \times n$ matrix, and let $T_A: R^n \rightarrow R^n$ be multiplication by A. We shall now investigate relationships between the invertibility of A and properties of T_A.

Recall from Theorem 2.3.6 (with **w** in place of **b**) that the following are equivalent:

- A is invertible.
- $A\mathbf{x} = \mathbf{w}$ is consistent for every $n \times 1$ matrix **w**.
- $A\mathbf{x} = \mathbf{w}$ has exactly one solution for every $n \times 1$ matrix **w**.

However, the last of these statements is actually stronger than necessary. One can show that the following are equivalent (Exercise 24):

- A is invertible.
- $A\mathbf{x} = \mathbf{w}$ is consistent for every $n \times 1$ matrix **w**.
- $A\mathbf{x} = \mathbf{w}$ has exactly one solution when the system is consistent.

Translating these into the corresponding statements about the linear operator T_A, we deduce that the following are equivalent:

- A is invertible.
- For every vector $\mathbf{w}$ in R^n, there is some vector $\mathbf{x}$ in R^n such that $T_A(\mathbf{x}) = \mathbf{w}$. Stated another way, the range of T_A is all of R^n.
- For every vector $\mathbf{w}$ in the range of T_A, there is exactly one vector $\mathbf{x}$ in R^n such that $T_A(\mathbf{x}) = \mathbf{w}$. Stated another way, T_A is one-to-one.

In summary, we have established the following theorem about linear operators on R^n.

Theorem 4.3.1   Equivalent Statements

If A is an $n \times n$ matrix and $T_A: R^n \rightarrow R^n$ is multiplication by A, then the following statements are equivalent.

(a) A *is invertible.*   *(b)* *The range of T_A is R^n.*   *(c)* T_A *is one-to-one.*

EXAMPLE 2 Applying Theorem 4.3.1

In Example 1 we observed that the rotation operator $T: R^2 \rightarrow R^2$ illustrated in Figure 4.3.1 is one-to-one. It follows from Theorem 4.3.1 that the range of T must be all of R^2 and that the standard matrix for T must be invertible. To show that the range of T is all of R^2, we must show that every vector $\mathbf{w}$ in R^2 is the image of some vector $\mathbf{x}$ under T. But this is clearly so, since the vector $\mathbf{x}$ obtained by rotating $\mathbf{w}$ through the angle $-\theta$ maps into $\mathbf{w}$ when rotated through the angle θ. Moreover, from Table 6 of Section 4.2, the standard matrix for T is

$$[T] = \begin{bmatrix} \cos\theta & -\sin\theta \\ \sin\theta & \cos\theta \end{bmatrix}$$

which is invertible, since

$$\det[T] = \begin{vmatrix} \cos\theta & -\sin\theta \\ \sin\theta & \cos\theta \end{vmatrix} = \cos^2\theta + \sin^2\theta = 1 \neq 0 \qquad \blacklozenge$$

EXAMPLE 3 Applying Theorem 4.3.1

In Example 1 we observed that the projection operator $T: R^3 \rightarrow R^3$ illustrated in Figure 4.3.2 is *not* one-to-one. It follows from Theorem 4.3.1 that the range of T is *not* all of R^3 and the standard matrix for T is not invertible. To show that the range of T is not all of R^3, we must find a vector $\mathbf{w}$ in R^3 that is not the image of any vector $\mathbf{x}$ under T. But any vector $\mathbf{w}$ outside of the xy-plane has this property, since all images under T lie in the xy-plane. Moreover, from Table 5 of Section 4.2, the standard matrix for T is

$$[T] = \begin{bmatrix} 1 & 0 & 0 \\ 0 & 1 & 0 \\ 0 & 0 & 0 \end{bmatrix}$$

which is not invertible, since $\det[T] = 0$. $\qquad \blacklozenge$

Inverse of a One-to-One Linear Operator   If $T_A: R^n \rightarrow R^n$ is

a one-to-one linear operator, then from Theorem 4.3.1 the matrix A is invertible. Thus, $T_{A^{-1}}: R^n \rightarrow R^n$ is itself a linear operator; it is called the ***inverse of T_A***. The linear operators T_A and $T_{A^{-1}}$ cancel the effect of one another in the sense that for all $\mathbf{x}$ in R^n

$$T_A(T_{A^{-1}}(\mathbf{x})) = AA^{-1}\mathbf{x} = I\mathbf{x} = \mathbf{x}$$
$$T_{A^{-1}}(T_A(\mathbf{x})) = A^{-1}A\mathbf{x} = I\mathbf{x} = \mathbf{x}$$

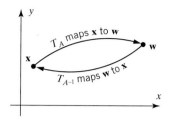

Figure 4.3.3

or equivalently,

$$T_A \circ T_{A^{-1}} = T_{AA^{-1}} = T_I$$
$$T_{A^{-1}} \circ T_A = T_{A^{-1}A} = T_I$$

From a more geometric viewpoint, if $\mathbf{w}$ is the image of $\mathbf{x}$ under T_A, then $T_{A^{-1}}$ maps $\mathbf{w}$ back into $\mathbf{x}$, since

$$T_{A^{-1}}(\mathbf{w}) = T_{A^{-1}}(T_A(\mathbf{x})) = \mathbf{x}$$

(Figure 4.3.3).

Before turning to an example, it will be helpful to touch on a notational matter. When a one-to-one linear operator on R^n is written as $T: R^n \rightarrow R^n$ (rather than $T_A: R^n \rightarrow R^n$), then the inverse of the operator T is denoted by T^{-1} (rather than $T_{A^{-1}}$). Since the standard matrix for T^{-1} is the inverse of the standard matrix for T, we have

$$[T^{-1}] = [T]^{-1} \tag{1}$$

EXAMPLE 4 Standard Matrix for T^{-1}

Let $T: R^2 \rightarrow R^2$ be the operator that rotates each vector in R^2 through the angle θ; so from Table 6 of Section 4.2

$$[T] = \begin{bmatrix} \cos\theta & -\sin\theta \\ \sin\theta & \cos\theta \end{bmatrix} \tag{2}$$

It is evident geometrically that to undo the effect of T one must rotate each vector in R^2 through the angle $-\theta$. But this is exactly what the operator T^{-1} does, since the standard matrix for T^{-1} is

$$[T^{-1}] = [T]^{-1} = \begin{bmatrix} \cos\theta & \sin\theta \\ -\sin\theta & \cos\theta \end{bmatrix} = \begin{bmatrix} \cos(-\theta) & -\sin(-\theta) \\ \sin(-\theta) & \cos(-\theta) \end{bmatrix}$$

(verify), which is identical to (2) except that θ is replaced by $-\theta$.   ◆

EXAMPLE 5 Finding T^{-1}

Show that the linear operator $T: R^2 \rightarrow R^2$ defined by the equations

$$w_1 = 2x_1 + x_2$$
$$w_2 = 3x_1 + 4x_2$$

is one-to-one, and find $T^{-1}(w_1, w_2)$.

Solution.

The matrix form of these equations is

$$\begin{bmatrix} w_1 \\ w_2 \end{bmatrix} = \begin{bmatrix} 2 & 1 \\ 3 & 4 \end{bmatrix} \begin{bmatrix} x_1 \\ x_2 \end{bmatrix}$$

so the standard matrix for T is

$$[T] = \begin{bmatrix} 2 & 1 \\ 3 & 4 \end{bmatrix}$$

This matrix is invertible (so T is one-to-one) and the standard matrix for T^{-1} is

$$[T^{-1}] = [T]^{-1} = \begin{bmatrix} \frac{4}{5} & -\frac{1}{5} \\ -\frac{3}{5} & \frac{2}{5} \end{bmatrix}$$

Thus,

$$[T^{-1}]\begin{bmatrix} w_1 \\ w_2 \end{bmatrix} = \begin{bmatrix} \frac{4}{5} & -\frac{1}{5} \\ -\frac{3}{5} & \frac{2}{5} \end{bmatrix} \begin{bmatrix} w_1 \\ w_2 \end{bmatrix} = \begin{bmatrix} \frac{4}{5}w_1 - \frac{1}{5}w_2 \\ -\frac{3}{5}w_1 + \frac{2}{5}w_2 \end{bmatrix}$$

from which we conclude that

$$T^{-1}(w_1, w_2) = (\tfrac{4}{5}w_1 - \tfrac{1}{5}w_2, -\tfrac{3}{5}w_1 + \tfrac{2}{5}w_2) \qquad \blacklozenge$$

Linearity Properties   In the preceding section we defined a transformation $T: R^n \to R^m$ to be linear if the equations relating $\mathbf{x}$ and $\mathbf{w} = T(\mathbf{x})$ are linear equations. The following theorem provides an alternative characterization of linearity. This theorem is fundamental and will be the basis for extending the concept of a linear transformation to more general settings later in this text.

Theorem 4.3.2   Properties of Linear Transformations

A transformation $T: R^n \to R^m$ is linear if and only if the following relationships hold for all vectors $\mathbf{u}$ and $\mathbf{v}$ in R^n and every scalar c.

(a) $T(\mathbf{u} + \mathbf{v}) = T(\mathbf{u}) + T(\mathbf{v})$ (b) $T(c\mathbf{u}) = cT(\mathbf{u})$

Proof. Assume first that T is a linear transformation, and let A be the standard matrix for T. It follows from the basic arithmetic properties of matrices that

$$T(\mathbf{u} + \mathbf{v}) = A(\mathbf{u} + \mathbf{v}) = A\mathbf{u} + A\mathbf{v} = T(\mathbf{u}) + T(\mathbf{v})$$

and

$$T(c\mathbf{u}) = A(c\mathbf{u}) = c(A\mathbf{u}) = cT(\mathbf{u})$$

Conversely, assume that properties (a) and (b) hold for the transformation T. We can prove that T is linear by finding a matrix A with the property that

$$T(\mathbf{x}) = A\mathbf{x} \tag{3}$$

for all vectors $\mathbf{x}$ in R^n. This will show that T is multiplication by A and therefore linear. But before we can produce this matrix we need to observe that property (a) can be extended to three or more terms; for example, if $\mathbf{u}$, $\mathbf{v}$, and $\mathbf{w}$ are any vectors in R^n, then by first grouping $\mathbf{v}$ and $\mathbf{w}$ and applying property (a) we obtain

$$T(\mathbf{u} + \mathbf{v} + \mathbf{w}) = T(\mathbf{u} + (\mathbf{v} + \mathbf{w})) = T(\mathbf{u}) + T(\mathbf{v} + \mathbf{w}) = T(\mathbf{u}) + T(\mathbf{v}) + T(\mathbf{w})$$

More generally, for any vectors $\mathbf{v}_1, \mathbf{v}_2, \ldots, \mathbf{v}_k$ in R^n we have

$$T(\mathbf{v}_1 + \mathbf{v}_2 + \cdots + \mathbf{v}_k) = T(\mathbf{v}_1) + T(\mathbf{v}_2) + \cdots + T(\mathbf{v}_k)$$

Now, to find the matrix A let $\mathbf{e}_1, \mathbf{e}_2, \ldots, \mathbf{e}_n$ be the vectors

$$\mathbf{e}_1 = \begin{bmatrix} 1 \\ 0 \\ 0 \\ \vdots \\ 0 \end{bmatrix}, \quad \mathbf{e}_2 = \begin{bmatrix} 0 \\ 1 \\ 0 \\ \vdots \\ 0 \end{bmatrix}, \ldots, \quad \mathbf{e}_n = \begin{bmatrix} 0 \\ 0 \\ 0 \\ \vdots \\ 1 \end{bmatrix} \tag{4}$$

and let A be the matrix whose successive column vectors are $T(\mathbf{e}_1), T(\mathbf{e}_2), \ldots, T(\mathbf{e}_n)$; that is,

$$A = [T(\mathbf{e}_1) \mid T(\mathbf{e}_2) \mid \cdots \mid T(\mathbf{e}_n)] \tag{5}$$

If

$$\mathbf{x} = \begin{bmatrix} x_1 \\ x_2 \\ \vdots \\ x_n \end{bmatrix}$$

is any vector in R^n, then as discussed in Section 1.3, the product $A\mathbf{x}$ is a linear combination of the column vectors of A with coefficients from $\mathbf{x}$, so that

$$\begin{aligned}
A\mathbf{x} &= x_1 T(\mathbf{e}_1) + x_2 T(\mathbf{e}_2) + \cdots + x_n T(\mathbf{e}_n) \\
&= T(x_1 \mathbf{e}_1) + T(x_2 \mathbf{e}_2) + \cdots + T(x_n \mathbf{e}_n) \qquad \longleftarrow \text{Property } (b)\\
&= T(x_1 \mathbf{e}_1 + x_2 \mathbf{e}_2 + \cdots + x_n \mathbf{e}_n) \qquad \longleftarrow \text{Property } (a) \text{ for } n \text{ terms}\\
&= T(\mathbf{x})
\end{aligned}$$

which completes the proof. ∎

Expression (5) is important in its own right, since it provides an explicit formula for the standard matrix of a linear operator $T: R^n \to R^m$ in terms of the images of the vectors $\mathbf{e}_1, \mathbf{e}_2, \ldots, \mathbf{e}_n$ under T. For reasons that will be discussed later, the vectors $\mathbf{e}_1, \mathbf{e}_2, \ldots, \mathbf{e}_n$ in (4) are called the **standard basis** vectors for R^n. In R^2 and R^3 these are the vectors of length 1 along the coordinate axes (Figure 4.3.4).

Because of its importance we shall state (5) as a theorem for future reference.

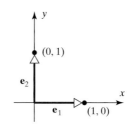

(*a*) Standard basis for R^2

Theorem 4.3.3

If $T: R^n \to R^m$ is a linear transformation, and $\mathbf{e}_1, \mathbf{e}_2, \ldots, \mathbf{e}_n$ are the standard basis vectors for R^n, then the standard matrix for T is

$$[T] = [T(\mathbf{e}_1) \mid T(\mathbf{e}_2) \mid \cdots \mid T(\mathbf{e}_n)] \tag{6}$$

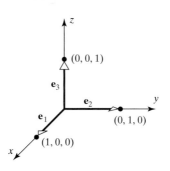

(*b*) Standard basis for R^3

Figure 4.3.4

Formula (6) is a powerful tool for finding standard matrices and analyzing the geometric effect of a linear transformation. For example, suppose that $T: R^3 \to R^3$ is the orthogonal projection on the xy-plane. Referring to Figure 4.3.4, it is evident geometrically that

$$T(\mathbf{e}_1) = \mathbf{e}_1 = \begin{bmatrix} 1 \\ 0 \\ 0 \end{bmatrix}, \qquad T(\mathbf{e}_2) = \mathbf{e}_2 = \begin{bmatrix} 0 \\ 1 \\ 0 \end{bmatrix}, \qquad T(\mathbf{e}_3) = \mathbf{0} = \begin{bmatrix} 0 \\ 0 \\ 0 \end{bmatrix}$$

so by (6)

$$[T] = \begin{bmatrix} 1 & 0 & 0 \\ 0 & 1 & 0 \\ 0 & 0 & 0 \end{bmatrix}$$

which agrees with the result in Table 5 of Section 4.2.

Using (6) another way, suppose that $T_A: R^3 \to R^2$ is multiplication by

$$A = \begin{bmatrix} -1 & 2 & 1 \\ 3 & 0 & 6 \end{bmatrix}$$

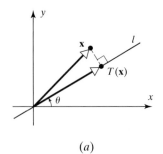

(a)

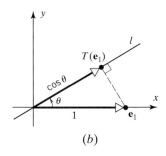

(b)

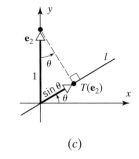

(c)

Figure 4.3.5

The images of the standard basis vectors can be read directly from the columns of the matrix A:

$$T_A\left(\begin{bmatrix} 1 \\ 0 \\ 0 \end{bmatrix}\right) = \begin{bmatrix} -1 \\ 3 \end{bmatrix}, \qquad T_A\left(\begin{bmatrix} 0 \\ 1 \\ 0 \end{bmatrix}\right) = \begin{bmatrix} 2 \\ 0 \end{bmatrix}, \qquad T_A\left(\begin{bmatrix} 0 \\ 0 \\ 1 \end{bmatrix}\right) = \begin{bmatrix} 1 \\ 6 \end{bmatrix}$$

EXAMPLE 6 Standard Matrix for a Projection Operator

Let l be the line in the xy-plane that passes through the origin and makes an angle θ with the positive x-axis, where $0 \leq \theta < \pi$. As illustrated in Figure 4.3.5a, let $T: R^2 \to R^2$ be a linear operator that maps each vector into its orthogonal projection on l.

(a) Find the standard matrix for T.
(b) Find the orthogonal projection of the vector $\mathbf{x} = (1, 5)$ onto the line through the origin that makes an angle of $\theta = \pi/6$ with the positive x-axis.

Solution (a). From (6)

$$[T] = [T(\mathbf{e}_1) \mid T(\mathbf{e}_2)]$$

where $\mathbf{e}_1$ and $\mathbf{e}_2$ are the standard basis vectors for R^2. We consider the case where $0 \leq \theta \leq \pi/2$; the case where $\pi/2 < \theta < \pi$ is similar. Referring to Figure 4.3.5b, we have $\|T(\mathbf{e}_1)\| = \cos\theta$, so

$$T(\mathbf{e}_1) = \begin{bmatrix} \|T(\mathbf{e}_1)\| \cos\theta \\ \|T(\mathbf{e}_1)\| \sin\theta \end{bmatrix} = \begin{bmatrix} \cos^2\theta \\ \sin\theta\cos\theta \end{bmatrix}$$

and referring to Figure 4.3.5c, we have $\|T(\mathbf{e}_2)\| = \sin\theta$, so

$$T(\mathbf{e}_2) = \begin{bmatrix} \|T(\mathbf{e}_2)\| \cos\theta \\ \|T(\mathbf{e}_2)\| \sin\theta \end{bmatrix} = \begin{bmatrix} \sin\theta\cos\theta \\ \sin^2\theta \end{bmatrix}$$

Thus, the standard matrix for T is

$$[T] = \begin{bmatrix} \cos^2\theta & \sin\theta\cos\theta \\ \sin\theta\cos\theta & \sin^2\theta \end{bmatrix}$$

Solution (b). Since $\sin\pi/6 = 1/2$ and $\cos\pi/6 = \sqrt{3}/2$, it follows from part (a) that the standard matrix for this projection operator is

$$[T] = \begin{bmatrix} 3/4 & \sqrt{3}/4 \\ \sqrt{3}/4 & 1/4 \end{bmatrix}$$

Thus,

$$T\left(\begin{bmatrix} 1 \\ 5 \end{bmatrix}\right) = \begin{bmatrix} 3/4 & \sqrt{3}/4 \\ \sqrt{3}/4 & 1/4 \end{bmatrix} \begin{bmatrix} 1 \\ 5 \end{bmatrix} = \begin{bmatrix} \dfrac{3+5\sqrt{3}}{4} \\ \dfrac{\sqrt{3}+5}{4} \end{bmatrix}$$

or in horizontal notation

$$T(1, 5) = \left(\frac{3+5\sqrt{3}}{4}, \frac{\sqrt{3}+5}{4} \right) \qquad \blacklozenge$$

Geometric Interpretation of Eigenvectors

Recall from Section 2.3 that if A is an $n \times n$ matrix, then λ is called an *eigenvalue* of A if there is a nonzero vector $\mathbf{x}$ such that

$$A\mathbf{x} = \lambda\mathbf{x} \quad \text{or equivalently,} \quad (\lambda I - A)\mathbf{x} = \mathbf{0}$$

The nonzero vectors **x** satisfying this equation are called the *eigenvectors* of A corresponding to λ.

Eigenvalues and eigenvectors can also be defined for linear operators on R^n; the definitions parallel those for matrices.

Definition

If $T: R^n \to R^n$ is a linear operator, then a scalar λ is called an ***eigenvalue of T*** if there is a nonzero **x** in R^n such that

$$T(\mathbf{x}) = \lambda \mathbf{x} \tag{7}$$

Those nonzero vectors **x** that satisfy this equation are called the ***eigenvectors of T corresponding to*** λ.

Observe that if A is the standard matrix for T, then (7) can be written as

$$A\mathbf{x} = \lambda\mathbf{x}$$

from which it follows that

- The eigenvalues of T are precisely the eigenvalues of its standard matrix A.
- **x** is an eigenvector of T corresponding to λ if and only if **x** is an eigenvector of A corresponding to λ.

If λ is an eigenvalue of A and **x** is a corresponding eigenvector, then $A\mathbf{x} = \lambda\mathbf{x}$, so multiplication by A maps **x** into a scalar multiple of itself. In R^2 and R^3, this means that *multiplication by A maps each eigenvector **x** into a vector that lies on the same line as* **x** (Figure 4.3.6).

Recall from Section 4.2 that if $\lambda \geq 0$, then the linear operator $A\mathbf{x} = \lambda\mathbf{x}$ compresses **x** by a factor of λ if $0 \leq \lambda \leq 1$ or stretches **x** by a factor of λ if $\lambda \geq 1$. If $\lambda < 0$, then $A\mathbf{x} = \lambda\mathbf{x}$ reverses the direction of **x**, and compresses the reversed vector by a factor of $|\lambda|$ if $0 \leq |\lambda| \leq 1$ or stretches the reversed vector by a factor of $|\lambda|$ if $|\lambda| \geq 1$ (Figure 4.3.7).

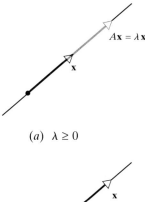

(a) $\lambda \geq 0$

(b) $\lambda \leq 0$

Figure 4.3.6

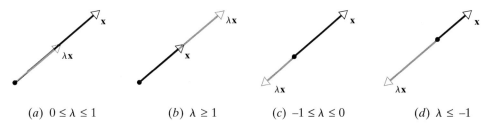

(a) $0 \leq \lambda \leq 1$ (b) $\lambda \geq 1$ (c) $-1 \leq \lambda \leq 0$ (d) $\lambda \leq -1$

Figure 4.3.7

EXAMPLE 7 Eigenvalues of a Linear Operator

Let $T: R^2 \to R^2$ be the linear operator that rotates each vector through an angle θ. It is evident geometrically that unless θ is a multiple of π, then T does not map any nonzero vector **x** onto the same line as **x**; consequently, T has no real eigenvalues. But if θ *is* a multiple of π, then every nonzero vector **x** is mapped onto the same line as **x**, so *every* nonzero vector is an eigenvector of T. Let us verify these geometric observations algebraically. The standard matrix for T is

$$A = \begin{bmatrix} \cos\theta & -\sin\theta \\ \sin\theta & \cos\theta \end{bmatrix}$$

As discussed in Section 2.3, the eigenvalues of this matrix are the solutions of the characteristic equation

$$\det(\lambda I - A) = \begin{vmatrix} \lambda - \cos\theta & \sin\theta \\ -\sin\theta & \lambda - \cos\theta \end{vmatrix} = 0$$

that is,

$$(\lambda - \cos\theta)^2 + \sin^2\theta = 0 \tag{8}$$

But if θ is not a multiple of π, then $\sin^2\theta > 0$, so this equation has no real solution for λ and consequently A has no real eigenvectors.[†] If θ *is* a multiple of π, then $\sin\theta = 0$ and either $\cos\theta = 1$ or $\cos\theta = -1$, depending on the particular multiple of π. In the case where $\sin\theta = 0$ and $\cos\theta = 1$, the characteristic equation (8) becomes $(\lambda - 1)^2 = 0$, so $\lambda = 1$ is the only eigenvalue of A. In this case the matrix A is

$$A = \begin{bmatrix} 1 & 0 \\ 0 & 1 \end{bmatrix} = I$$

Thus, for all $\mathbf{x}$ in R^2,

$$T(\mathbf{x}) = A\mathbf{x} = I\mathbf{x} = \mathbf{x}$$

so T maps every vector to itself, and hence to the same line. In the case where $\sin\theta = 0$ and $\cos\theta = -1$, the characteristic equation (8) becomes $(\lambda + 1)^2 = 0$, so that $\lambda = -1$ is the only eigenvalue of A. In this case the matrix A is

$$A = \begin{bmatrix} -1 & 0 \\ 0 & -1 \end{bmatrix} = -I$$

Thus, for all $\mathbf{x}$ in R^2,

$$T(\mathbf{x}) = A\mathbf{x} = -I\mathbf{x} = -\mathbf{x}$$

so T maps every vector to its negative, and hence to the same line as $\mathbf{x}$. ◆

EXAMPLE 8 Eigenvalues of a Linear Operator

Let $T: R^3 \rightarrow R^3$ be the orthogonal projection on the xy-plane. Vectors in the xy-plane are mapped into themselves under T, so each nonzero vector in the xy-plane is an eigenvector corresponding to the eigenvalue $\lambda = 1$. Every vector $\mathbf{x}$ along the z-axis is mapped into $\mathbf{0}$ under T, which is on the same line as $\mathbf{x}$, so every nonzero vector on the z-axis is an eigenvector corresponding to the eigenvalue $\lambda = 0$. Vectors not in the xy-plane or along the z-axis are not mapped into scalar multiples of themselves, so there are no other eigenvectors or eigenvalues.

To verify these geometric observations algebraically, recall from Table 5 of Section 4.2 that the standard matrix for T is

$$A = \begin{bmatrix} 1 & 0 & 0 \\ 0 & 1 & 0 \\ 0 & 0 & 0 \end{bmatrix}$$

The characteristic equation of A is

$$\det(\lambda I - A) = \begin{vmatrix} \lambda - 1 & 0 & 0 \\ 0 & \lambda - 1 & 0 \\ 0 & 0 & \lambda \end{vmatrix} = 0 \quad \text{or} \quad (\lambda - 1)^2\lambda = 0$$

which has the solutions $\lambda = 0$ and $\lambda = 1$ anticipated above.

[†]There are applications that require complex scalars and vectors with complex components. In such cases, complex eigenvalues and eigenvectors with complex components are allowed. However, these have no direct geometric significance here. In later chapters we will discuss such eigenvalues and eigenvectors, but until explicitly stated otherwise, it will be assumed that only real eigenvalues and eigenvectors with real components are to be considered.

As discussed in Section 2.3, the eigenvectors of the matrix A corresponding to an eigenvalue λ are the nonzero solutions of

$$\begin{bmatrix} \lambda - 1 & 0 & 0 \\ 0 & \lambda - 1 & 0 \\ 0 & 0 & \lambda \end{bmatrix} \begin{bmatrix} x_1 \\ x_2 \\ x_3 \end{bmatrix} = \begin{bmatrix} 0 \\ 0 \\ 0 \end{bmatrix} \tag{9}$$

If $\lambda = 0$, this system is

$$\begin{bmatrix} -1 & 0 & 0 \\ 0 & -1 & 0 \\ 0 & 0 & 0 \end{bmatrix} \begin{bmatrix} x_1 \\ x_2 \\ x_3 \end{bmatrix} = \begin{bmatrix} 0 \\ 0 \\ 0 \end{bmatrix}$$

which has the solutions $x_1 = 0$, $x_2 = 0$, $x_3 = t$ (verify), or in matrix form

$$\begin{bmatrix} x_1 \\ x_2 \\ x_3 \end{bmatrix} = \begin{bmatrix} 0 \\ 0 \\ t \end{bmatrix}$$

As anticipated, these are the vectors along the z-axis. If $\lambda = 1$, then system (9) is

$$\begin{bmatrix} 0 & 0 & 0 \\ 0 & 0 & 0 \\ 0 & 0 & 1 \end{bmatrix} \begin{bmatrix} x_1 \\ x_2 \\ x_3 \end{bmatrix} = \begin{bmatrix} 0 \\ 0 \\ 0 \end{bmatrix}$$

which has the solutions $x_1 = s$, $x_2 = t$, $x_3 = 0$ (verify), or in matrix form,

$$\begin{bmatrix} x_1 \\ x_2 \\ x_3 \end{bmatrix} = \begin{bmatrix} s \\ t \\ 0 \end{bmatrix}$$

As anticipated, these are the vectors in the xy-plane. ◆

Summary In Theorem 2.3.6 we listed six results that are equivalent to the invertibility of a matrix A. We conclude this section by merging Theorem 4.3.1 with that list to produce the following theorem that relates all of the major topics we have studied thus far.

Theorem 4.3.4   **Equivalent Statements**

If A is an $n \times n$ matrix, and if $T_A \colon R^n \to R^n$ is multiplication by A, then the following are equivalent.

(a) *A is invertible.*
(b) $A\mathbf{x} = \mathbf{0}$ *has only the trivial solution.*
(c) *The reduced row-echelon form of A is I_n.*
(d) *A is expressible as a product of elementary matrices.*
(e) $A\mathbf{x} = \mathbf{b}$ *is consistent for every $n \times 1$ matrix* $\mathbf{b}$.
(f) $A\mathbf{x} = \mathbf{b}$ *has exactly one solution for every $n \times 1$ matrix* $\mathbf{b}$.
(g) $\det(A) \neq 0$.
(h) *The range of T_A is R^n.*
(i) T_A *is one-to-one.*

Exercise Set 4.3

1. By inspection, determine whether the linear operator is one-to-one.

(a) the orthogonal projection on the x-axis in R^2 (b) the reflection about the y-axis in R^2

(c) the reflection about the line $y = x$ in R^2 (d) a contraction with factor $k > 0$ in R^2

(e) a rotation about the z-axis in R^3 (f) a reflection about the xy-plane in R^3

(g) a dilation with factor $k > 0$ in R^3

2. Find the standard matrix for the linear operator defined by the equations and use Theorem 4.3.4 to determine whether the operator is one-to-one.

(a) $w_1 = 8x_1 + 4x_2$ (b) $w_1 = 2x_1 - 3x_2$ (c) $w_1 = -x_1 + 3x_2 + 2x_3$ (d) $w_1 = x_1 + 2x_2 + 3x_3$
$\quad\ \ w_2 = 2x_1 + x_2$ $\quad\ \ w_2 = 5x_1 + x_2$ $\quad\ \ w_2 = 2x_1 \qquad\quad + 4x_3$ $\quad\ \ w_2 = 2x_1 + 5x_2 + 3x_3$
$\qquad\qquad\qquad\qquad\qquad\qquad\qquad\qquad\qquad\quad w_3 = x_1 + 3x_2 + 6x_3$ $\quad\ \ w_3 = x_1 \qquad\qquad + 8x_3$

3. Show that the range of the linear operator defined by the equations

$$w_1 = 4x_1 - 2x_2$$
$$w_2 = 2x_1 - x_2$$

is not all of R^2, and find a vector that is not in the range.

4. Show that the range of the linear operator defined by the equations

$$w_1 = x_1 - 2x_2 + x_3$$
$$w_2 = 5x_1 - x_2 + 3x_3$$
$$w_3 = 4x_1 + x_2 + 2x_3$$

is not R^3, and find a vector that is not in the range.

5. Determine whether the linear operator $T: R^2 \to R^2$ defined by the equations is one-to-one; if so, find the standard matrix for the inverse operator, and find $T^{-1}(w_1, w_2)$.

(a) $w_1 = x_1 + 2x_2$ (b) $w_1 = 4x_1 - 6x_2$ (c) $w_1 = -x_2$ (d) $w_1 = 3x_1$
$\quad\ \ w_2 = -x_1 + x_2$ $\quad\ \ w_2 = -2x_1 + 3x_2$ $\quad\ \ w_2 = -x_1$ $\quad\ \ w_2 = -5x_1$

6. Determine whether the linear operator $T: R^3 \to R^3$ defined by the equations is one-to-one; if so, find the standard matrix for the inverse operator, and find $T^{-1}(w_1, w_2, w_3)$.

(a) $w_1 = x_1 - 2x_2 + 2x_3$ (b) $w_1 = x_1 - 3x_2 + 4x_3$ (c) $w_1 = x_1 + 4x_2 - x_3$ (d) $w_1 = x_1 + 2x_2 + x_3$
$\quad\ \ w_2 = 2x_1 + x_2 + x_3$ $\quad\ \ w_2 = -x_1 + x_2 + x_3$ $\quad\ \ w_2 = 2x_1 + 7x_2 + x_3$ $\quad\ \ w_2 = -2x_1 + x_2 + 4x_3$
$\quad\ \ w_3 = x_1 + x_2$ $\quad\ \ w_3 = -2x_2 + 5x_3$ $\quad\ \ w_3 = x_1 + 3x_2$ $\quad\ \ w_3 = 7x_1 + 4x_2 - 5x_3$

7. By inspection, determine the inverse of the given one-to-one linear operator.

(a) the reflection about the x-axis in R^2 (b) the rotation through an angle of $\pi/4$ in R^2

(c) the dilation by a factor of 3 in R^2 (d) the reflection about the yz-plane in R^3

(e) the contraction by a factor of $\frac{1}{5}$ in R^3

In Exercises 8 and 9 use Theorem 4.3.2 to determine whether $T: R^2 \to R^2$ is a linear operator.

8. (a) $T(x, y) = (2x, y)$ (b) $T(x, y) = (x^2, y)$ (c) $T(x, y) = (-y, x)$ (d) $T(x, y) = (x, 0)$

9. (a) $T(x, y) = (2x + y, x - y)$ (b) $T(x, y) = (x + 1, y)$ (c) $T(x, y) = (y, y)$ (d) $T(x, y) = (\sqrt[3]{x}, \sqrt[3]{y})$

In Exercises 10 and 11 use Theorem 4.3.2 to determine whether $T: R^3 \to R^2$ is a linear transformation.

10. (a) $T(x, y, z) = (x, x + y + z)$ (b) $T(x, y, z) = (1, 1)$

11. (a) $T(x, y, z) = (0, 0)$ (b) $T(x, y, z) = (3x - 4y, 2x - 5z)$

12. In each part use Theorem 4.3.3 to find the standard matrix for the linear operator from the images of the standard basis vectors.

(a) the reflection operators on R^2 in Table 2 of Section 4.2

(b) the reflection operators on R^3 in Table 3 of Section 4.2

(c) the projection operators on R^2 in Table 4 of Section 4.2

(d) the projection operators on R^3 in Table 5 of Section 4.2

(e) the rotation operators on R^2 in Table 6 of Section 4.2

(f) the dilation and contraction operators on R^3 in Table 9 of Section 4.2

13. Use Theorem 4.3.3 to find the standard matrix for $T: R^2 \rightarrow R^2$ from the images of the standard basis vectors.

   (a) $T: R^2 \rightarrow R^2$ projects a vector orthogonally onto the x-axis and then reflects that vector about the y-axis.

   (b) $T: R^2 \rightarrow R^2$ reflects a vector about the line $y = x$ and then reflects that vector about the x-axis.

   (c) $T: R^2 \rightarrow R^2$ dilates a vector by a factor of 3, then reflects that vector about the line $y = x$, and then projects that vector orthogonally onto the y-axis.

14. Use Theorem 4.3.3 to find the standard matrix for $T: R^3 \rightarrow R^3$ from the images of the standard basis vectors.

   (a) $T: R^3 \rightarrow R^3$ reflects a vector about the xz-plane and then contracts that vector by a factor of $\frac{1}{5}$.

   (b) $T: R^3 \rightarrow R^3$ projects a vector orthogonally onto the xz-plane and then projects that vector orthogonally onto the xy-plane.

   (c) $T: R^3 \rightarrow R^3$ reflects a vector about the xy-plane, then reflects that vector about the xz-plane, and then reflects that vector about the yz-plane.

15. Let $T_A: R^3 \rightarrow R^3$ be multiplication by

$$A = \begin{bmatrix} -1 & 3 & 0 \\ 2 & 1 & 2 \\ 4 & 5 & -3 \end{bmatrix}$$

   and let e_1, e_2, and e_3 be the standard basis vectors for R^3. Find the following vectors by inspection.

   (a) $T_A(e_1)$, $T_A(e_2)$, and $T_A(e_3)$ (b) $T_A(e_1 + e_2 + e_3)$ (c) $T_A(7e_3)$

16. Determine whether multiplication by A is a one-to-one linear transformation.

   (a) $A = \begin{bmatrix} 1 & -1 \\ 2 & 0 \\ 3 & -4 \end{bmatrix}$ (b) $A = \begin{bmatrix} 1 & 2 & 3 \\ -1 & 0 & -4 \end{bmatrix}$

17. Use the result in Example 6 to find the orthogonal projection of x onto the line through the origin that makes an angle θ with the positive x-axis.

   (a) $x = (-1, 2)$; $\theta = 45°$ (b) $x = (1, 0)$; $\theta = 30°$ (c) $x = (1, 5)$; $\theta = 120°$

18. Use the type of argument given in Example 8 to find the eigenvalues and corresponding eigenvectors of T. Check your conclusions by calculating the eigenvalues and corresponding eigenvectors from the standard matrix for T.

   (a) $T: R^2 \rightarrow R^2$ is the reflection about the x-axis.

   (b) $T: R^2 \rightarrow R^2$ is the reflection about the line $y = x$.

   (c) $T: R^2 \rightarrow R^2$ is the orthogonal projection on the x-axis.

   (d) $T: R^2 \rightarrow R^2$ is the contraction by a factor of $\frac{1}{2}$.

19. Follow the directions of Exercise 18.

   (a) $T: R^3 \rightarrow R^3$ is the reflection about the yz-plane.

   (b) $T: R^3 \rightarrow R^3$ is the orthogonal projection on the xz-plane.

   (c) $T: R^3 \rightarrow R^3$ is the dilation by a factor of 2.

   (d) $T: R^3 \rightarrow R^3$ is a rotation of $45°$ about the z-axis.

20. (a) Is a composition of one-to-one linear transformations one-to-one? Justify your conclusion.

   (b) Can the composition of a one-to-one linear transformation and a linear transformation that is not one-to-one be one-to-one? Account for both possible orders of composition and justify your conclusion.

21. Show that $T(x, y) = (0, 0)$ defines a linear operator on R^2 but $T(x, y) = (1, 1)$ does not.

22. Prove that if $T: R^n \rightarrow R^m$ is a linear transformation, then $T(\mathbf{0}) = \mathbf{0}$; that is, T maps the zero vector in R^n into the zero vector in R^m.

23. Let l be the line in the xy-plane that passes through the origin and makes an angle θ with the positive x-axis, where $0 \leq \theta < \pi$. Let $T: R^2 \rightarrow R^2$ be the linear operator that reflects each vector about l (see accompanying figure).

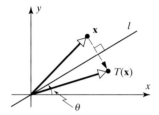

Figure Ex-23

(a) Use the method of Example 6 to find the standard matrix for T.

(b) Find the reflection of the vector $\mathbf{x} = (1, 5)$ about the line l through the origin that makes an angle of $\theta = 30°$ with the positive x-axis.

24. Prove: An $n \times n$ matrix A is invertible if and only if the linear system $A\mathbf{x} = \mathbf{w}$ has exactly one solution for every vector $\mathbf{w}$ in R^n for which the system is consistent.

Discussion and Discovery

25. Indicate whether the statement is always true or sometimes false. Justify your answer by giving a logical argument or a counterexample.

(a) If T maps R^n into R^m, and $T(\mathbf{0}) = \mathbf{0}$, then T is linear.

(b) If $T: R^n \rightarrow R^m$ is a one-to-one linear transformation, then there are no distinct vectors $\mathbf{u}$ and $\mathbf{v}$ in R^n such that $T(\mathbf{u} - \mathbf{v}) = \mathbf{0}$.

(c) If $T: R^n \rightarrow R^n$ is a linear operator, and if $T(\mathbf{x}) = 2\mathbf{x}$ for some vector $\mathbf{x}$, then $\lambda = 2$ is an eigenvalue of T.

(d) If T maps R^n into R^m, and if $T(c_1\mathbf{u} + c_2\mathbf{v}) = c_1 T(\mathbf{u}) + c_2 T(\mathbf{v})$ for all scalars c_1 and c_2 and for all vectors $\mathbf{u}$ and $\mathbf{v}$ in R^n, then T is linear.

26. Let A be an $n \times n$ matrix such that $\det(A) = 0$, and let $T: R^n \rightarrow R^n$ be multiplication by A.

(a) What can you say about the range of the linear operator T? Give an example that illustrates your conclusion.

(b) What can you say about the number of vectors that T maps into $\mathbf{0}$?

Chapter 4 Technology Exercises

The following exercises are designed to be solved using a technology utility. Typically, this will be MATLAB, *Mathematica*, Maple, Derive, or Mathcad, but it may also be some other type of linear algebra software or a scientific calculator with some linear algebra capabilities. For each exercise you will need to read the relevant documentation for the particular utility you are using. The goal of these exercises is to provide you with a basic proficiency with your technology utility. Once you have mastered the techniques in these exercises, you will be able to use your technology utility to solve many of the problems in the regular exercise sets.

Section 4.1

T1. (***Vector operations in*** R^n) With most technology utilities, the commands for operating on vectors in R^n are the same as those for vectors in R^2 and R^3, and the command for computing a dot product produces the Euclidean inner product in R^n. Use your utility to perform computations in Exercises 1, 3, and 9 of Section 4.1.

Section 4.2

T1. (***Rotations***) Find the standard matrix for the linear operator on R^3 that performs a counterclockwise rotation of 45° about the x-axis, followed by a counterclockwise rotation of 60° about the y-axis, followed by a counterclockwise rotation of 30° about the z-axis. Then find the image of the point $(1, 1, 1)$ under this operator.

General Vector Spaces

INTRODUCTION: In the last chapter we generalized vectors from 2- and 3-space to vectors in n-space. In this chapter we shall generalize the concept of vector still further. We shall state a set of axioms that, if satisfied by a class of objects, will entitle those objects to be called "vectors." These generalized vectors will include, among other things, various kinds of matrices and functions. Our work in this chapter is not an idle exercise in theoretical mathematics; it will provide a powerful tool for extending our geometric visualization to a wide variety of important mathematical problems where geometric intuition would not otherwise be available. We can visualize vectors in R^2 and R^3 as arrows, which enables us to draw or form mental pictures to help solve problems. Because the axioms we give to define our new kinds of vectors will be based on properties of vectors in R^2 and R^3, the new vectors will have many familiar properties. Consequently, when we want to solve a problem involving our new kinds of vectors, say matrices or functions, we may be able to get a foothold on the problem by visualizing what the corresponding problem would be like in R^2 and R^3.

5.1 REAL VECTOR SPACES

In this section we shall extend the concept of a vector by extracting the most important properties of familiar vectors and turning them into axioms. Thus, when a set of objects satisfies these axioms, they will automatically have the most important properties of familiar vectors, thereby making it reasonable to regard these objects as new kinds of vectors.

Vector Space Axioms   The following definition consists of ten axioms. As you read each axiom, keep in mind that you have already seen each of them as parts of various definitions and theorems in the preceding two chapters (for instance, see Theorem 4.1.1). Remember too, you do not prove axioms; they are simply the "rules of the game."

Definition

Let V be an arbitrary nonempty set of objects on which two operations are defined, addition and multiplication by scalars (numbers). By **addition** we mean a rule for associating with each pair of objects $\mathbf{u}$ and $\mathbf{v}$ in V an object $\mathbf{u} + \mathbf{v}$, called the **sum** of $\mathbf{u}$ and $\mathbf{v}$; by **scalar multiplication** we mean a rule for associating with each scalar k and each object $\mathbf{u}$ in V an object $k\mathbf{u}$, called the **scalar multiple** of $\mathbf{u}$ by k. If the following axioms are satisfied by all objects $\mathbf{u}$, $\mathbf{v}$, $\mathbf{w}$ in V and all scalars k and l, then we call V a **vector space** and we call the objects in V **vectors**.

(1) If $\mathbf{u}$ and $\mathbf{v}$ are objects in V, then $\mathbf{u} + \mathbf{v}$ is in V.

(2) $\mathbf{u} + \mathbf{v} = \mathbf{v} + \mathbf{u}$

(3) $\mathbf{u} + (\mathbf{v} + \mathbf{w}) = (\mathbf{u} + \mathbf{v}) + \mathbf{w}$

(4) There is an object $\mathbf{0}$ in V, called a **zero vector** for V, such that $\mathbf{0} + \mathbf{u} = \mathbf{u} + \mathbf{0} = \mathbf{u}$ for all $\mathbf{u}$ in V.

(5) For each $\mathbf{u}$ in V, there is an object $-\mathbf{u}$ in V, called a **negative** of $\mathbf{u}$, such that $\mathbf{u} + (-\mathbf{u}) = (-\mathbf{u}) + \mathbf{u} = \mathbf{0}$.

(6) If k is any scalar and $\mathbf{u}$ is any object in V, then $k\mathbf{u}$ is in V.

(7) $k(\mathbf{u} + \mathbf{v}) = k\mathbf{u} + k\mathbf{v}$

(8) $(k + l)\mathbf{u} = k\mathbf{u} + l\mathbf{u}$

(9) $k(l\mathbf{u}) = (kl)(\mathbf{u})$

(10) $1\mathbf{u} = \mathbf{u}$

REMARK. Depending on the application, scalars may be real numbers or complex numbers. Vector spaces in which the scalars are complex numbers are called **complex vector spaces**, and those in which the scalars must be real are called **real vector spaces**. In Chapter 10 we shall discuss complex vector spaces; until then, *all of our scalars will be real numbers.*

The reader should keep in mind that the definition of a vector space specifies neither the nature of the vectors nor the operations. Any kind of object can be a vector, and the operations of addition and scalar multiplication may not have any relationship or similarity to the standard vector operations on R^n. The only requirement is that the ten vector space axioms be satisfied. Some authors use the notations $\oplus$ and $\odot$ for vector addition and scalar multiplication to distinguish these operations from addition and multiplication of real numbers; we will not use this convention, however.

Examples of Vector Spaces The following examples will illustrate the variety of possible vector spaces. In each example we will specify a nonempty set V and two operations: addition and scalar multiplication; then we shall verify that the ten vector space axioms are satisfied, thereby entitling V, with the specified operations, to be called a vector space.

EXAMPLE 1 R^n Is a Vector Space

The set $V = R^n$ with the standard operations of addition and scalar multiplication defined in Section 4.1 is a vector space. Axioms 1 and 6 follow from the definitions of the standard operations on R^n; the remaining axioms follow from Theorem 4.1.1. ◆

The three most important special cases of R^n are R (the real numbers), R^2 (the vectors in the plane), and R^3 (the vectors in 3-space).

EXAMPLE 2 A Vector Space of 2×2 Matrices

Show that the set V of all 2×2 matrices with real entries is a vector space if vector addition is defined to be matrix addition and vector scalar multiplication is defined to be matrix scalar multiplication.

Solution.

In this example we will find it convenient to verify the axioms in the following order: 1, 6, 2, 3, 7, 8, 9, 4, 5, and 10. Let

$$\mathbf{u} = \begin{bmatrix} u_{11} & u_{12} \\ u_{21} & u_{22} \end{bmatrix} \quad \text{and} \quad \mathbf{v} = \begin{bmatrix} v_{11} & v_{12} \\ v_{21} & v_{22} \end{bmatrix}$$

To prove Axiom 1, we must show that $\mathbf{u} + \mathbf{v}$ is an object in V; that is, we must show that $\mathbf{u} + \mathbf{v}$ is a 2×2 matrix. But this follows from the definition of matrix addition, since

$$\mathbf{u} + \mathbf{v} = \begin{bmatrix} u_{11} & u_{12} \\ u_{21} & u_{22} \end{bmatrix} + \begin{bmatrix} v_{11} & v_{12} \\ v_{21} & v_{22} \end{bmatrix} = \begin{bmatrix} u_{11} + v_{11} & u_{12} + v_{12} \\ u_{21} + v_{21} & u_{22} + v_{22} \end{bmatrix}$$

Similarly, Axiom 6 holds because for any real number k we have

$$k\mathbf{u} = k\begin{bmatrix} u_{11} & u_{12} \\ u_{21} & u_{22} \end{bmatrix} = \begin{bmatrix} ku_{11} & ku_{12} \\ ku_{21} & ku_{22} \end{bmatrix}$$

so that $k\mathbf{u}$ is a 2×2 matrix and consequently is an object in V.

Axiom 2 follows from Theorem 1.4.1*a* since

$$\mathbf{u} + \mathbf{v} = \begin{bmatrix} u_{11} & u_{12} \\ u_{21} & u_{22} \end{bmatrix} + \begin{bmatrix} v_{11} & v_{12} \\ v_{21} & v_{22} \end{bmatrix} = \begin{bmatrix} v_{11} & v_{12} \\ v_{21} & v_{22} \end{bmatrix} + \begin{bmatrix} u_{11} & u_{12} \\ u_{21} & u_{22} \end{bmatrix} = \mathbf{v} + \mathbf{u}$$

Similarly, Axiom 3 follows from part (*b*) of that theorem; and Axioms 7, 8, and 9 follow from parts (*h*), (*j*), and (*l*), respectively.

To prove Axiom 4, we must find an object **0** in V such that $\mathbf{0} + \mathbf{u} = \mathbf{u} + \mathbf{0} = \mathbf{u}$ for all **u** in V. This can be done by defining **0** to be

$$\mathbf{0} = \begin{bmatrix} 0 & 0 \\ 0 & 0 \end{bmatrix}$$

With this definition

$$\mathbf{0} + \mathbf{u} = \begin{bmatrix} 0 & 0 \\ 0 & 0 \end{bmatrix} + \begin{bmatrix} u_{11} & u_{12} \\ u_{21} & u_{22} \end{bmatrix} = \begin{bmatrix} u_{11} & u_{12} \\ u_{21} & u_{22} \end{bmatrix} = \mathbf{u}$$

and similarly $\mathbf{u} + \mathbf{0} = \mathbf{u}$. To prove Axiom 5, we must show that each object **u** in V has a negative $-\mathbf{u}$ such that $\mathbf{u} + (-\mathbf{u}) = \mathbf{0}$ and $(-\mathbf{u}) + \mathbf{u} = \mathbf{0}$. This can be done by defining the negative of **u** to be

$$-\mathbf{u} = \begin{bmatrix} -u_{11} & -u_{12} \\ -u_{21} & -u_{22} \end{bmatrix}$$

With this definition

$$\mathbf{u} + (-\mathbf{u}) = \begin{bmatrix} u_{11} & u_{12} \\ u_{21} & u_{22} \end{bmatrix} + \begin{bmatrix} -u_{11} & -u_{12} \\ -u_{21} & -u_{22} \end{bmatrix} = \begin{bmatrix} 0 & 0 \\ 0 & 0 \end{bmatrix} = \mathbf{0}$$

and similarly $(-\mathbf{u}) + \mathbf{u} = \mathbf{0}$. Finally, Axiom 10 is a simple computation:

$$1\mathbf{u} = 1\begin{bmatrix} u_{11} & u_{12} \\ u_{21} & u_{22} \end{bmatrix} = \begin{bmatrix} u_{11} & u_{12} \\ u_{21} & u_{22} \end{bmatrix} = \mathbf{u} \qquad \blacklozenge$$

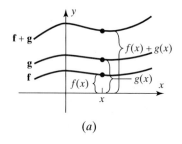

(a)

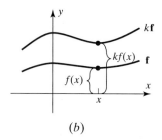

(b)

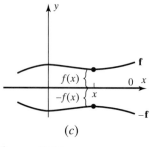

(c)

Figure 5.1.1

EXAMPLE 3 A Vector Space of $m \times n$ Matrices

Example 2 is a special case of a more general class of vector spaces. The arguments in that example can be adapted to show that the set V of all $m \times n$ matrices with real entries, together with the operations of matrix addition and scalar multiplication, is a vector space. The $m \times n$ zero matrix is the zero vector **0**, and if **u** is the $m \times n$ matrix U, then the matrix $-U$ is the negative $-\mathbf{u}$ of the vector **u**. We shall denote this vector space by the symbol M_{mn}. $\qquad \blacklozenge$

EXAMPLE 4 A Vector Space of Real-Valued Functions

Let V be the set of real-valued functions defined on the entire real line $(-\infty, \infty)$. If $\mathbf{f} = f(x)$ and $\mathbf{g} = g(x)$ are two such functions and k is any real number, define the sum function $\mathbf{f} + \mathbf{g}$ and the scalar multiple $k\mathbf{f}$, respectively, by

$$(\mathbf{f} + \mathbf{g})(x) = f(x) + g(x) \quad \text{and} \quad (k\mathbf{f})(x) = kf(x)$$

In other words, the value of the function $\mathbf{f} + \mathbf{g}$ at x is obtained by adding together the values of **f** and **g** at x (Figure 5.1.1a). Similarly, the value of $k\mathbf{f}$ at x is k times the value of **f** at x (Figure 5.1.1b). In the exercises we shall ask you to show that V is a vector space with respect to these operations. This vector space is denoted by $F(-\infty, \infty)$. If **f** and **g** are vectors in this space, then to say that $\mathbf{f} = \mathbf{g}$ is equivalent to saying that $f(x) = g(x)$ for all x in the interval $(-\infty, \infty)$.

The vector **0** in $F(-\infty, \infty)$ is the constant function that is identically zero for all values of x. The graph of this function is the line that coincides with the x-axis. The negative of a vector **f** is the function $-\mathbf{f} = -f(x)$. Geometrically, the graph of $-\mathbf{f}$ is the reflection of the graph of **f** across the x-axis (Figure 5.1.1c). $\qquad \blacklozenge$

REMARK. In the preceding example we focused attention on the interval $(-\infty, \infty)$. Had we restricted our attention to some closed interval $[a, b]$ or some open interval (a, b), the functions defined on those intervals with the operations stated in the example would also have produced vector spaces. Those vector spaces are denoted by $F[a, b]$ and $F(a, b)$, respectively.

EXAMPLE 5 A Set That Is Not a Vector Space

Let $V = R^2$ and define addition and scalar multiplication operations as follows: If $\mathbf{u} = (u_1, u_2)$ and $\mathbf{v} = (v_1, v_2)$, then define

$$\mathbf{u} + \mathbf{v} = (u_1 + v_1, u_2 + v_2)$$

and if k is any real number, then define

$$k\mathbf{u} = (ku_1, 0)$$

For example, if $\mathbf{u} = (2, 4)$, $\mathbf{v} = (-3, 5)$, and $k = 7$, then

$$\mathbf{u} + \mathbf{v} = (2 + (-3), 4 + 5) = (-1, 9)$$
$$k\mathbf{u} = 7\mathbf{u} = (7 \cdot 2, 0) = (14, 0)$$

The addition operation is the standard addition operation on R^2, but the scalar multiplication operation is not the standard scalar multiplication. In the exercises we will ask you to show that the first nine vector space axioms are satisfied; however, there are values of $\mathbf{u}$ for which Axiom 10 fails to hold. For example, if $\mathbf{u} = (u_1, u_2)$ is such that $u_2 \neq 0$, then

$$1\mathbf{u} = 1(u_1, u_2) = (1 \cdot u_1, 0) = (u_1, 0) \neq \mathbf{u}$$

Thus, V is not a vector space with the stated operations. ◆

EXAMPLE 6 Every Plane Through the Origin Is a Vector Space

Let V be any plane through the origin in R^3. We shall show that the points in V form a vector space under the standard addition and scalar multiplication operations for vectors in R^3. From Example 1, we know that R^3 itself is a vector space under these operations. Thus, Axioms 2, 3, 7, 8, 9, and 10 hold for all points in R^3 and consequently for all points in the plane V. We therefore need only show that Axioms 1, 4, 5, and 6 are satisfied.

Since the plane V passes through the origin, it has an equation of the form

$$ax + by + cz = 0 \tag{1}$$

(Theorem 3.5.1). Thus, if $\mathbf{u} = (u_1, u_2, u_3)$ and $\mathbf{v} = (v_1, v_2, v_3)$ are points in V, then $au_1 + bu_2 + cu_3 = 0$ and $av_1 + bv_2 + cv_3 = 0$. Adding these equations gives

$$a(u_1 + v_1) + b(u_2 + v_2) + c(u_3 + v_3) = 0$$

This equality tells us that the coordinates of the point

$$\mathbf{u} + \mathbf{v} = (u_1 + v_1, u_2 + v_2, u_3 + v_3)$$

satisfy (1); thus, $\mathbf{u} + \mathbf{v}$ lies in the plane V. This proves that Axiom 1 is satisfied. The verifications of Axioms 4 and 6 are left as exercises; however, we shall prove that Axiom 5 is satisfied. Multiplying $au_1 + bu_2 + cu_3 = 0$ through by -1 gives

$$a(-u_1) + b(-u_2) + c(-u_3) = 0$$

Thus, $-\mathbf{u} = (-u_1, -u_2, -u_3)$ lies in V. This establishes Axiom 5. ◆

EXAMPLE 7 The Zero Vector Space

Let V consist of a single object, which we denote by **0**, and define

$$\mathbf{0} + \mathbf{0} = \mathbf{0} \quad \text{and} \quad k\mathbf{0} = \mathbf{0}$$

for all scalars k. It is easy to check that all the vector space axioms are satisfied. We call this the *zero vector space*. ◆

Some Properties of Vectors As we progress, we shall add more examples of vector spaces to our list. We conclude this section with a theorem that gives a useful list of vector properties.

Theorem 5.1.1

*Let V be a vector space, **u** a vector in V, and k a scalar; then*:

(a) $0\mathbf{u} = \mathbf{0}$
(b) $k\mathbf{0} = \mathbf{0}$
(c) $(-1)\mathbf{u} = -\mathbf{u}$
(d) *If $k\mathbf{u} = \mathbf{0}$, then $k = 0$ or $\mathbf{u} = \mathbf{0}$.*

We shall prove parts (a) and (c) and leave proofs of the remaining parts as exercises.

Proof (a). We can write

$$0\mathbf{u} + 0\mathbf{u} = (0 + 0)\mathbf{u} \quad \text{[Axiom 8]}$$
$$= 0\mathbf{u} \qquad \text{[Property of the number 0]}$$

By Axiom 5 the vector $0\mathbf{u}$ has a negative, $-0\mathbf{u}$. Adding this negative to both sides above yields

$$[0\mathbf{u} + 0\mathbf{u}] + (-0\mathbf{u}) = 0\mathbf{u} + (-0\mathbf{u})$$

or

$$0\mathbf{u} + [0\mathbf{u} + (-0\mathbf{u})] = 0\mathbf{u} + (-0\mathbf{u}) \quad \text{[Axiom 3]}$$
$$0\mathbf{u} + \mathbf{0} = \mathbf{0} \qquad \text{[Axiom 5]}$$
$$0\mathbf{u} = \mathbf{0} \qquad \text{[Axiom 4]}$$

Proof (c). To show $(-1)\mathbf{u} = -\mathbf{u}$, we must demonstrate that $\mathbf{u} + (-1)\mathbf{u} = \mathbf{0}$. To see this, observe that

$$\mathbf{u} + (-1)\mathbf{u} = 1\mathbf{u} + (-1)\mathbf{u} \quad \text{[Axiom 10]}$$
$$= (1 + (-1))\mathbf{u} \quad \text{[Axiom 8]}$$
$$= 0\mathbf{u} \qquad \text{[Property of numbers]}$$
$$= \mathbf{0} \qquad \text{[Part (a) above]} \quad ■$$

Exercise Set 5.1

In Exercises 1–16 a set of objects is given together with operations of addition and scalar multiplication. Determine which sets are vector spaces under the given operations. For those that are not, list all axioms that fail to hold.

1. The set of all triples of real numbers (x, y, z) with the operations

   $$(x, y, z) + (x', y', z') = (x + x', y + y', z + z') \quad \text{and} \quad k(x, y, z) = (kx, y, z)$$

2. The set of all triples of real numbers (x, y, z) with the operations

   $$(x, y, z) + (x', y', z') = (x + x', y + y', z + z') \quad \text{and} \quad k(x, y, z) = (0, 0, 0)$$

3. The set of all pairs of real numbers (x, y) with the operations

   $$(x, y) + (x', y') = (x + x', y + y') \quad \text{and} \quad k(x, y) = (2kx, 2ky)$$

4. The set of all real numbers x with the standard operations of addition and multiplication.

5. The set of all pairs of real numbers of the form $(x, 0)$ with the standard operations on R^2.

6. The set of all pairs of real numbers of the form (x, y), where $x \geq 0$, with the standard operations on R^2.

7. The set of all n-tuples of real numbers of the form $(x, x, \ldots, x)$ with the standard operations on R^n.

8. The set of all pairs of real numbers (x, y) with the operations

   $$(x, y) + (x', y') = (x + x' + 1, y + y' + 1) \quad \text{and} \quad k(x, y) = (kx, ky)$$

9. The set of all 2×2 matrices of the form

   $$\begin{bmatrix} a & 1 \\ 1 & b \end{bmatrix}$$

   with matrix addition and scalar multiplication.

10. The set of all 2×2 matrices of the form

    $$\begin{bmatrix} a & 0 \\ 0 & b \end{bmatrix}$$

    with matrix addition and scalar multiplication.

11. The set of all real-valued functions f defined everywhere on the real line and such that $f(1) = 0$, with the operations defined in Example 4.

12. The set of all 2×2 matrices of the form

    $$\begin{bmatrix} a & a + b \\ a + b & b \end{bmatrix}$$

    with matrix addition and scalar multiplication.

13. The set of all pairs of real numbers of the form $(1, x)$ with the operations

    $$(1, y) + (1, y') = (1, y + y') \quad \text{and} \quad k(1, y) = (1, ky)$$

14. The set of polynomials of the form $a + bx$ with the operations

    $$(a_0 + a_1 x) + (b_0 + b_1 x) = (a_0 + b_0) + (a_1 + b_1)x \quad \text{and} \quad k(a_0 + a_1 x) = (ka_0) + (ka_1)x$$

15. The set of all positive real numbers with operations

    $$x + y = xy \quad \text{and} \quad kx = x^k$$

16. The set of all pairs of real numbers (x, y) with operations

    $$(x, y) + (x', y') = (xx', yy') \quad \text{and} \quad k(x, y) = (kx, ky)$$

17. Show that the first nine vector space axioms are satisfied if $V = R^2$ has the addition and scalar multiplication operations defined in Example 5.

18. Prove that a line passing through the origin in R^3 is a vector space under the standard operations on R^3.

19. Complete the unfinished details of Example 4.

20. Complete the unfinished details of Example 6.

Discussion and Discovery

21. We showed in Example 6 that every plane in R^3 that passes through the origin is a vector space under the standard operations on R^3. Is the same true for planes that do not pass through the origin? Explain your reasoning.

22. It was shown in Exercise 14 above that the set of polynomials of degree 1 or less is a vector space under the operations stated in that exercise. Is the set of polynomials whose degree is exactly 1 a vector space under those operations? Explain your reasoning.

23. Consider the set whose only element is the moon. Is this set a vector space under the operations moon + moon = moon and k(moon) = moon for every real number k? Explain your reasoning.

24. Do you think that it is possible to have a vector space with exactly two distinct vectors in it? Explain your reasoning.

25. The following is a proof of part (b) of Theorem 5.1.1. Justify each step by filling in the blank line with the word "hypothesis" or by specifying the number of one of the vector space axioms given in this section.

Hypothesis: Let $\mathbf{u}$ be any vector in a vector space V, $\mathbf{0}$ the zero vector in V, and k a scalar.

Conclusion: Then $k\mathbf{0} = \mathbf{0}$.

Proof: (1) First, $k\mathbf{0} + k\mathbf{u} = k(\mathbf{0} + \mathbf{u})$. _____

  (2) $= k\mathbf{u}$ _____

  (3) Since $k\mathbf{u}$ is in V, $-k\mathbf{u}$ is in V. _____

  (4) Therefore, $(k\mathbf{0} + k\mathbf{u}) + (-k\mathbf{u}) = k\mathbf{u} + (-k\mathbf{u})$. _____

  (5) $k\mathbf{0} + (k\mathbf{u} + (-k\mathbf{u})) = k\mathbf{u} + (-k\mathbf{u})$ _____

  (6) $k\mathbf{0} + \mathbf{0} = \mathbf{0}$ _____

  (7) Finally, $k\mathbf{0} = \mathbf{0}$. _____

26. Prove part (d) of Theorem 5.1.1.

27. The following is a proof that the cancellation law for addition holds in a vector space. Justify each step by filling in the blank line with the word "hypothesis" or by specifying the number of one of the vector space axioms given in this section.

Hypothesis: Let $\mathbf{u}$, $\mathbf{v}$, and $\mathbf{w}$ be vectors in a vector space V and suppose that $\mathbf{u} + \mathbf{w} = \mathbf{v} + \mathbf{w}$.

Conclusion: Then $\mathbf{u} = \mathbf{v}$.

Proof: (1) First, $(\mathbf{u} + \mathbf{w}) + (-\mathbf{w})$ and $(\mathbf{v} + \mathbf{w}) + (-\mathbf{w})$ are vectors in V. _____

  (2) Then $(\mathbf{u} + \mathbf{w}) + (-\mathbf{w}) = (\mathbf{v} + \mathbf{w}) + (-\mathbf{w})$. _____

  (3) The left side of the equality in step (2) is

  $(\mathbf{u} + \mathbf{w}) + (-\mathbf{w}) = \mathbf{u} + (\mathbf{w} + (-\mathbf{w}))$ _____

  $= \mathbf{u}$ _____

  (4) The right side of the equality in step (2) is

  $(\mathbf{v} + \mathbf{w}) + (-\mathbf{w}) = \mathbf{v} + (\mathbf{w} + (-\mathbf{w}))$ _____

  $= \mathbf{v}$ _____

  From the equality in step (2), it follows from steps (3) and (4) that $\mathbf{u} = \mathbf{v}$.

28. Do you think it is possible for a vector space to have two different zero vectors? That is, is it possible to have two *different* vectors $\mathbf{0}_1$ and $\mathbf{0}_2$ such that these vectors both satisfy Axiom 4? Explain your reasoning.

29. Do you think that it is possible for a vector **u** in a vector space to have two different negatives? That is, is it possible to have two *different* vectors $(-\mathbf{u})_1$ and $(-\mathbf{u})_2$, both of which satisfy Axiom 5? Explain your reasoning.

30. The set of ten axioms of a vector space is not an independent set because Axiom 2 can be deduced from other axioms in the set. Using the expression

$$(\mathbf{u} + \mathbf{v}) - (\mathbf{v} + \mathbf{u})$$

and Axiom 7 as a starting point, prove that $\mathbf{u} + \mathbf{v} = \mathbf{v} + \mathbf{u}$. [***Hint.*** You can use Theorem 5.1.1 since the proof of each part of that theorem does not use Axiom 2.]

5.2 SUBSPACES

It is possible for one vector space to be contained within another vector space. For example, we showed in the preceding section that planes through the origin are vector spaces that are contained in the vector space R^3. In this section we shall study this important concept in detail.

A subset of a vector space V that is itself a vector space with respect to the operations of vector addition and scalar multiplication defined on V is given a special name.

> ### Definition
>
> A subset W of a vector space V is called a ***subspace*** of V if W is itself a vector space under the addition and scalar multiplication defined on V.

In general, one must verify the ten vector space axioms to show that a set W with addition and scalar multiplication forms a vector space. However, if W is part of a larger set V that is already known to be a vector space, then certain axioms need not be verified for W because they are "inherited" from V. For example, there is no need to check that $\mathbf{u} + \mathbf{v} = \mathbf{v} + \mathbf{u}$ (Axiom 2) for W because this holds for all vectors in V and consequently for all vectors in W. Other axioms inherited by W from V are 3, 7, 8, 9, and 10. Thus, to show that a set W is a subspace of a vector space V, we need only verify Axioms 1, 4, 5, and 6. The following theorem shows that even Axioms 4 and 5 can be omitted.

> ### Theorem 5.2.1
>
> *If W is a set of one or more vectors from a vector space V, then W is a subspace of V if and only if the following conditions hold.*
>
> *(a) If $\mathbf{u}$ and $\mathbf{v}$ are vectors in W, then $\mathbf{u} + \mathbf{v}$ is in W.*
> *(b) If k is any scalar and $\mathbf{u}$ is any vector in W, then $k\mathbf{u}$ is in W.*

Proof. If W is a subspace of V, then all the vector space axioms are satisfied; in particular, Axioms 1 and 6 hold. But these are precisely conditions (*a*) and (*b*).

Conversely, assume conditions (*a*) and (*b*) hold. Since these conditions are vector space Axioms 1 and 6, we need only show that W satisfies the remaining eight axioms.

Axioms 2, 3, 7, 8, 9, and 10 are automatically satisfied by the vectors in W since they are satisfied by all vectors in V. Therefore, to complete the proof, we need only verify that Axioms 4 and 5 are satisfied by vectors in W.

Let $\mathbf{u}$ be any vector in W. By condition (b), $k\mathbf{u}$ is in W for every scalar k. Setting $k = 0$, it follows from Theorem 5.1.1 that $0\mathbf{u} = \mathbf{0}$ is in W, and setting $k = -1$, it follows that $(-1)\mathbf{u} = -\mathbf{u}$ is in W. ∎

REMARK. A set W of one or more vectors from a vector space V is said to be ***closed under addition*** if condition (a) in Theorem 5.2.1 holds and ***closed under scalar multiplication*** if condition (b) holds. Thus, Theorem 5.2.1 states that W *is a subspace of V if and only if W is closed under addition and closed under scalar multiplication.*

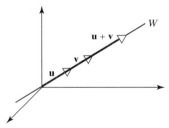

Figure 5.2.1

The vectors $\mathbf{u} + \mathbf{v}$ and $k\mathbf{u}$ both lie in the same plane as $\mathbf{u}$ and $\mathbf{v}$.

EXAMPLE 1 Testing for a Subspace

In Example 6 of Section 5.1 we verified the ten vector space axioms to show that the points in a plane through the origin of R^3 form a subspace of R^3. In light of Theorem 5.2.1 we can see that much of that work was unnecessary; it would have been sufficient to verify that the plane is closed under addition and scalar multiplication (Axioms 1 and 6). In Section 5.1 we verified those two axioms algebraically; however, they can also be proved geometrically as follows: Let W be any plane through the origin and let $\mathbf{u}$ and $\mathbf{v}$ be any vectors in W. Then $\mathbf{u} + \mathbf{v}$ must lie in W because it is the diagonal of the parallelogram determined by $\mathbf{u}$ and $\mathbf{v}$ (Figure 5.2.1), and $k\mathbf{u}$ must lie in W for any scalar k because $k\mathbf{u}$ lies on a line through $\mathbf{u}$. Thus, W is closed under addition and scalar multiplication, so it is a subspace of R^3. ◆

EXAMPLE 2 Lines Through the Origin Are Subspaces

Show that a line through the origin of R^3 is a subspace of R^3.

Solution.

Let W be a line through the origin of R^3. It is evident geometrically that the sum of two vectors on this line also lies on the line and that a scalar multiple of a vector on the line is on the line as well (Figure 5.2.2). Thus, W is closed under addition and scalar multiplication, so it is a subspace of R^3. In the exercises we will ask you to prove this result algebraically using parametric equations for the line. ◆

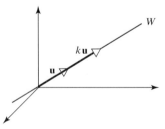

(a) W is closed under addition.

(b) W is closed under scalar multiplication.

Figure 5.2.2

EXAMPLE 3 A Subset of R^2 That Is Not a Subspace

Let W be the set of all points (x, y) in R^2 such that $x \geq 0$ and $y \geq 0$. These are the points in the first quadrant. The set W is *not* a subspace of R^2 since it is not closed under scalar multiplication. For example, $\mathbf{v} = (1, 1)$ lies in W, but its negative $(-1)\mathbf{v} = -\mathbf{v} = (-1, -1)$ does not (Figure 5.2.3). ◆

Every nonzero vector space V has at least two subspaces: V itself is a subspace, and the set $\{\mathbf{0}\}$ consisting of just the zero vector in V is a subspace called the ***zero subspace***.

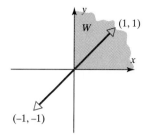

Figure 5.2.3
W is not closed under scalar multiplication.

Combining this with Examples 1 and 2, we obtain the following list of subspaces of R^2 and R^3:

Subspaces of R^2	Subspaces of R^3
• $\{0\}$	• $\{0\}$
• Lines through the origin	• Lines through the origin
• R^2	• Planes through the origin
	• R^3

Later, we will show that these are the only subspaces of R^2 and R^3.

EXAMPLE 4 Subspaces of M_{nn}

From Theorem 1.7.2, the sum of two symmetric matrices is symmetric, and a scalar multiple of a symmetric matrix is symmetric. Thus, the set of $n \times n$ symmetric matrices is a subspace of the vector space M_{nn} of all $n \times n$ matrices. Similarly, the set of $n \times n$ upper triangular matrices, the set of $n \times n$ lower triangular matrices, and the set of $n \times n$ diagonal matrices all form subspaces of M_{nn}, since each of these sets is closed under addition and scalar multiplication. ◆

EXAMPLE 5 A Subspace of Polynomials of Degree $\leq n$

Let n be a nonnegative integer, and let W consist of all functions expressible in the form

$$p(x) = a_0 + a_1 x + \cdots + a_n x^n \tag{1}$$

where $a_0, \ldots, a_n$ are real numbers. Thus, W consists of all real polynomials of degree n or less. The set W is a subspace of the vector space of all real-valued functions discussed in Example 4 of the preceding section. To see this, let $\mathbf{p}$ and $\mathbf{q}$ be the polynomials

$$p(x) = a_0 + a_1 x + \cdots + a_n x^n \quad \text{and} \quad q(x) = b_0 + b_1 x + \cdots + b_n x^n$$

Then

$$(\mathbf{p} + \mathbf{q})(x) = p(x) + q(x) = (a_0 + b_0) + (a_1 + b_1)x + \cdots + (a_n + b_n)x^n$$

and

$$(k\mathbf{p})(x) = kp(x) = (ka_0) + (ka_1)x + \cdots + (ka_n)x^n$$

These functions have the form given in (1), so $\mathbf{p} + \mathbf{q}$ and $k\mathbf{p}$ lie in W. We shall denote the vector space W in this example by the symbol P_n. ◆

Calculus Required

EXAMPLE 6 Subspaces of Functions Continuous on $(-\infty, \infty)$

Recall from calculus that if $\mathbf{f}$ and $\mathbf{g}$ are continuous functions on the interval $(-\infty, \infty)$ and k is a constant, then $\mathbf{f} + \mathbf{g}$ and $k\mathbf{f}$ are also continuous. Thus, the continuous functions on the interval $(-\infty, \infty)$ form a subspace of $F(-\infty, \infty)$, since they are closed under addition and scalar multiplication. We denote this subspace by $C(-\infty, \infty)$. Similarly, if $\mathbf{f}$ and $\mathbf{g}$ have continuous first derivatives on $(-\infty, \infty)$, then so do $\mathbf{f} + \mathbf{g}$ and $k\mathbf{f}$. Thus, the functions with continuous first derivatives on $(-\infty, \infty)$ form a subspace of $F(-\infty, \infty)$. We denote this subspace by $C^1(-\infty, \infty)$, where the superscript 1 is used to emphasize the *first* derivative. However, it is a theorem of calculus that every differentiable function is continuous, so $C^1(-\infty, \infty)$ is actually a subspace of $C(-\infty, \infty)$.

To take this a step further, for each positive integer m, the functions with continuous mth derivatives on $(-\infty, \infty)$ form a subspace of $C^1(-\infty, \infty)$ as do the functions that have continuous derivatives of all orders. We denote the subspace of functions with continuous mth derivatives on $(-\infty, \infty)$ by $C^m(-\infty, \infty)$, and we denote the subspace of functions that have continuous derivatives of all orders on $(-\infty, \infty)$ by $C^\infty(-\infty, \infty)$. Finally, it is a theorem of calculus that polynomials have continuous derivatives of all orders, so P_n is a subspace of $C^\infty(-\infty, \infty)$. The hierarchy of subspaces discussed in this example is pictured in Figure 5.2.4. ◆

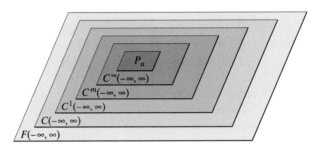

Figure 5.2.4

REMARK. In the preceding example we focused on the interval $(-\infty, \infty)$. Had we focused on a closed interval $[a, b]$, then the subspaces corresponding to those defined in the example would be denoted by $C[a, b]$, $C^m[a, b]$, and $C^\infty[a, b]$. Similarly, on an open interval (a, b) they would be denoted by $C(a, b)$, $C^m(a, b)$, and $C^\infty(a, b)$.

Solution Spaces of Homogeneous Systems

If $Ax = b$ is a system of linear equations, then each vector $\mathbf{x}$ that satisfies this equation is called a *solution vector* of the system. The following theorem shows that the solution vectors of a *homogeneous* linear system form a vector space, which we shall call the *solution space* of the system.

Theorem 5.2.2

If $Ax = 0$ is a homogeneous linear system of m equations in n unknowns, then the set of solution vectors is a subspace of R^n.

Proof. Let W be the set of solution vectors. There is at least one vector in W, namely $\mathbf{0}$. To show that W is closed under addition and scalar multiplication, we must show that if $\mathbf{x}$ and $\mathbf{x}'$ are any solution vectors and k is any scalar, then $\mathbf{x} + \mathbf{x}'$ and $k\mathbf{x}$ are also solution vectors. But if $\mathbf{x}$ and $\mathbf{x}'$ are solution vectors, then

$$A\mathbf{x} = \mathbf{0} \quad \text{and} \quad A\mathbf{x}' = \mathbf{0}$$

from which it follows that

$$A(\mathbf{x} + \mathbf{x}') = A\mathbf{x} + A\mathbf{x}' = \mathbf{0} + \mathbf{0} = \mathbf{0}$$

and

$$A(k\mathbf{x}) = kA\mathbf{x} = k\mathbf{0} = \mathbf{0}$$

which proves that $\mathbf{x} + \mathbf{x}'$ and $k\mathbf{x}$ are solution vectors. ∎

5.2 Subspaces ••• 215

EXAMPLE 7 Solution Spaces That Are Subspaces of R^3

Consider the linear systems

(a) $\begin{bmatrix} 1 & -2 & 3 \\ 2 & -4 & 6 \\ 3 & -6 & 9 \end{bmatrix} \begin{bmatrix} x \\ y \\ z \end{bmatrix} = \begin{bmatrix} 0 \\ 0 \\ 0 \end{bmatrix}$
(b) $\begin{bmatrix} 1 & -2 & 3 \\ -3 & 7 & -8 \\ -2 & 4 & -6 \end{bmatrix} \begin{bmatrix} x \\ y \\ z \end{bmatrix} = \begin{bmatrix} 0 \\ 0 \\ 0 \end{bmatrix}$

(c) $\begin{bmatrix} 1 & -2 & 3 \\ -3 & 7 & -8 \\ 4 & 1 & 2 \end{bmatrix} \begin{bmatrix} x \\ y \\ z \end{bmatrix} = \begin{bmatrix} 0 \\ 0 \\ 0 \end{bmatrix}$
(d) $\begin{bmatrix} 0 & 0 & 0 \\ 0 & 0 & 0 \\ 0 & 0 & 0 \end{bmatrix} \begin{bmatrix} x \\ y \\ z \end{bmatrix} = \begin{bmatrix} 0 \\ 0 \\ 0 \end{bmatrix}$

Each of these systems has three unknowns, so the solutions form subspaces of R^3. Geometrically, this means that each solution space must be a line through the origin, a plane through the origin, the origin only, or all of R^3. We shall now verify that this is so (leaving it to the reader to solve the systems).

Solution.

(*a*) The solutions are

$$x = 2s - 3t, \qquad y = s, \qquad z = t$$

from which it follows that

$$x = 2y - 3z \quad \text{or} \quad x - 2y + 3z = 0$$

This is the equation of the plane through the origin with $\mathbf{n} = (1, -2, 3)$ as a normal vector.

(*b*) The solutions are

$$x = -5t, \qquad y = -t, \qquad z = t$$

which are parametric equations for the line through the origin parallel to the vector $\mathbf{v} - (-5, -1, 1)$.

(*c*) The solution is $x = 0$, $y = 0$, $z = 0$, so the solution space is the origin only, that is, $\{\mathbf{0}\}$.

(*d*) The solutions are

$$x = r, \qquad y = s, \qquad z = t$$

where r, s, and t have arbitrary values, so the solution space is all of R^3. ◆

In Section 1.3 we introduced the concept of a linear combination of column vectors. The following definition extends this idea to more general vectors.

Definition

A vector $\mathbf{w}$ is called a ***linear combination*** of the vectors $\mathbf{v}_1, \mathbf{v}_2, \ldots, \mathbf{v}_r$ if it can be expressed in the form

$$\mathbf{w} = k_1\mathbf{v}_1 + k_2\mathbf{v}_2 + \cdots + k_r\mathbf{v}_r$$

where $k_1, k_2, \ldots, k_r$ are scalars.

REMARK. If $r = 1$, then the equation in the preceding definition reduces to $\mathbf{w} = k_1\mathbf{v}_1$; that is, $\mathbf{w}$ is a linear combination of a single vector $\mathbf{v}_1$ if it is a scalar multiple of $\mathbf{v}_1$.

EXAMPLE 8 Vectors in R^3 Are Linear Combinations of i, j, and k

Every vector $\mathbf{v} = (a, b, c)$ in R^3 is expressible as a linear combination of the standard basis vectors

$$\mathbf{i} = (1, 0, 0), \quad \mathbf{j} = (0, 1, 0), \quad \mathbf{k} = (0, 0, 1)$$

since

$$\mathbf{v} = (a, b, c) = a(1, 0, 0) + b(0, 1, 0) + c(0, 0, 1) = a\mathbf{i} + b\mathbf{j} + c\mathbf{k} \qquad \blacklozenge$$

EXAMPLE 9 Checking a Linear Combination

Consider the vectors $\mathbf{u} = (1, 2, -1)$ and $\mathbf{v} = (6, 4, 2)$ in R^3. Show that $\mathbf{w} = (9, 2, 7)$ is a linear combination of $\mathbf{u}$ and $\mathbf{v}$ and that $\mathbf{w}' = (4, -1, 8)$ is *not* a linear combination of $\mathbf{u}$ and $\mathbf{v}$.

Solution.

In order for $\mathbf{w}$ to be a linear combination of $\mathbf{u}$ and $\mathbf{v}$, there must be scalars k_1 and k_2 such that $\mathbf{w} = k_1\mathbf{u} + k_2\mathbf{v}$; that is,

$$(9, 2, 7) = k_1(1, 2, -1) + k_2(6, 4, 2)$$

or

$$(9, 2, 7) = (k_1 + 6k_2, 2k_1 + 4k_2, -k_1 + 2k_2)$$

Equating corresponding components gives

$$k_1 + 6k_2 = 9$$
$$2k_1 + 4k_2 = 2$$
$$-k_1 + 2k_2 = 7$$

Solving this system yields $k_1 = -3$, $k_2 = 2$, so

$$\mathbf{w} = -3\mathbf{u} + 2\mathbf{v}$$

Similarly, for $\mathbf{w}'$ to be a linear combination of $\mathbf{u}$ and $\mathbf{v}$, there must be scalars k_1 and k_2 such that $\mathbf{w}' = k_1\mathbf{u} + k_2\mathbf{v}$; that is,

$$(4, -1, 8) = k_1(1, 2, -1) + k_2(6, 4, 2)$$

or

$$(4, -1, 8) = (k_1 + 6k_2, 2k_1 + 4k_2, -k_1 + 2k_2)$$

Equating corresponding components gives

$$k_1 + 6k_2 = 4$$
$$2k_1 + 4k_2 = -1$$
$$-k_1 + 2k_2 = 8$$

This system of equations is inconsistent (verify), so no such scalars k_1 and k_2 exist. Consequently, $\mathbf{w}'$ is not a linear combination of $\mathbf{u}$ and $\mathbf{v}$. $\qquad \blacklozenge$

Spanning If $\mathbf{v}_1, \mathbf{v}_2, \ldots, \mathbf{v}_r$ are vectors in a vector space V, then generally some vectors in V may be linear combinations of $\mathbf{v}_1, \mathbf{v}_2, \ldots, \mathbf{v}_r$ and others may not. The following theorem shows that if we construct a set W consisting of all those vectors that are expressible as linear combinations of $\mathbf{v}_1, \mathbf{v}_2, \ldots, \mathbf{v}_r$, then W forms a subspace of V.

Theorem 5.2.3

If $\mathbf{v}_1, \mathbf{v}_2, \ldots, \mathbf{v}_r$ are vectors in a vector space V, then:

(a) *The set W of all linear combinations of $\mathbf{v}_1, \mathbf{v}_2, \ldots, \mathbf{v}_r$ is a subspace of V.*

(b) *W is the smallest subspace of V that contains $\mathbf{v}_1, \mathbf{v}_2, \ldots, \mathbf{v}_r$ in the sense that every other subspace of V that contains $\mathbf{v}_1, \mathbf{v}_2, \ldots, \mathbf{v}_r$ must contain W.*

Proof (a). To show that W is a subspace of V, we must prove that it is closed under addition and scalar multiplication. There is at least one vector in W, namely, $\mathbf{0}$, since $\mathbf{0} = 0\mathbf{v}_1 + 0\mathbf{v}_2 + \cdots + 0\mathbf{v}_r$. If $\mathbf{u}$ and $\mathbf{v}$ are vectors in W, then

$$\mathbf{u} = c_1\mathbf{v}_1 + c_2\mathbf{v}_2 + \cdots + c_r\mathbf{v}_r$$

and

$$\mathbf{v} = k_1\mathbf{v}_1 + k_2\mathbf{v}_2 + \cdots + k_r\mathbf{v}_r$$

where $c_1, c_2, \ldots, c_r, k_1, k_2, \ldots, k_r$ are scalars. Therefore,

$$\mathbf{u} + \mathbf{v} = (c_1 + k_1)\mathbf{v}_1 + (c_2 + k_2)\mathbf{v}_2 + \cdots + (c_r + k_r)\mathbf{v}_r$$

and, for any scalar k,

$$k\mathbf{u} = (kc_1)\mathbf{v}_1 + (kc_2)\mathbf{v}_2 + \cdots + (kc_r)\mathbf{v}_r$$

Thus, $\mathbf{u} + \mathbf{v}$ and $k\mathbf{u}$ are linear combinations of $\mathbf{v}_1, \mathbf{v}_2, \ldots, \mathbf{v}_r$ and consequently lie in W. Therefore, W is closed under addition and scalar multiplication.

Proof (b). Each vector $\mathbf{v}_i$ is a linear combination of $\mathbf{v}_1, \mathbf{v}_2, \ldots, \mathbf{v}_r$ since we can write

$$\mathbf{v}_i = 0\mathbf{v}_1 + 0\mathbf{v}_2 + \cdots + 1\mathbf{v}_i + \cdots + 0\mathbf{v}_r$$

Therefore, the subspace W contains each of the vectors $\mathbf{v}_1, \mathbf{v}_2, \ldots, \mathbf{v}_r$. Let W' be any other subspace that contains $\mathbf{v}_1, \mathbf{v}_2, \ldots, \mathbf{v}_r$. Since W' is closed under addition and scalar multiplication, it must contain all linear combinations of $\mathbf{v}_1, \mathbf{v}_2, \ldots, \mathbf{v}_r$. Thus, W' contains cach vector of W. ∎

We make the following definition.

Definition

If $S = \{\mathbf{v}_1, \mathbf{v}_2, \ldots, \mathbf{v}_r\}$ is a set of vectors in a vector space V, then the subspace W of V consisting of all linear combinations of the vectors in S is called the ***space spanned*** by $\mathbf{v}_1, \mathbf{v}_2, \ldots, \mathbf{v}_r$, and we say that the vectors $\mathbf{v}_1, \mathbf{v}_2, \ldots, \mathbf{v}_r$ ***span*** W. To indicate that W is the space spanned by the vectors in the set $S = \{\mathbf{v}_1, \mathbf{v}_2, \ldots, \mathbf{v}_r\}$, we write

$$W = \text{span}(S) \quad \text{or} \quad W = \text{span}\{\mathbf{v}_1, \mathbf{v}_2, \ldots, \mathbf{v}_r\}$$

EXAMPLE 10 Spaces Spanned by One or Two Vectors

If $\mathbf{v}_1$ and $\mathbf{v}_2$ are noncollinear vectors in R^3 with their initial points at the origin, then span$\{\mathbf{v}_1, \mathbf{v}_2\}$, which consists of all linear combinations $k_1\mathbf{v}_1 + k_2\mathbf{v}_2$, is the plane determined by $\mathbf{v}_1$ and $\mathbf{v}_2$ (see Figure 5.2.5a). Similarly, if $\mathbf{v}$ is a nonzero vector in R^2 or R^3, then span$\{\mathbf{v}\}$, which is the set of all scalar multiples $k\mathbf{v}$, is the line determined by $\mathbf{v}$ (see Figure 5.2.5b). ◆

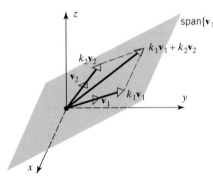

(a) Span$\{\mathbf{v}_1, \mathbf{v}_2\}$ is the plane through the origin determined by $\mathbf{v}_1$ and $\mathbf{v}_2$.

(b) Span$\{\mathbf{v}\}$ is the line through the origin determined by $\mathbf{v}$.

Figure 5.2.5

EXAMPLE 11 Spanning Set for P_n

The polynomials $1, x, x^2, \ldots, x^n$ span the vector space P_n defined in Example 5 since each polynomial $\mathbf{p}$ in P_n can be written as

$$\mathbf{p} = a_0 + a_1 x + \cdots + a_n x^n$$

which is a linear combination of $1, x, x^2, \ldots, x^n$. We can denote this by writing

$$P_n = \text{span}\{1, x, x^2, \ldots, x^n\} \qquad \blacklozenge$$

EXAMPLE 12 Three Vectors That Do Not Span R^3

Determine whether $\mathbf{v}_1 = (1, 1, 2)$, $\mathbf{v}_2 = (1, 0, 1)$, and $\mathbf{v}_3 = (2, 1, 3)$ span the vector space R^3.

Solution.

We must determine whether an arbitrary vector $\mathbf{b} = (b_1, b_2, b_3)$ in R^3 can be expressed as a linear combination

$$\mathbf{b} = k_1 \mathbf{v}_1 + k_2 \mathbf{v}_2 + k_3 \mathbf{v}_3$$

of the vectors $\mathbf{v}_1$, $\mathbf{v}_2$, and $\mathbf{v}_3$. Expressing this equation in terms of components gives

$$(b_1, b_2, b_3) = k_1(1, 1, 2) + k_2(1, 0, 1) + k_3(2, 1, 3)$$

or

$$(b_1, b_2, b_3) = (k_1 + k_2 + 2k_3, k_1 + k_3, 2k_1 + k_2 + 3k_3)$$

or

$$
\begin{aligned}
k_1 + k_2 + 2k_3 &= b_1 \\
k_1 \quad\;\; + \; k_3 &= b_2 \\
2k_1 + k_2 + 3k_3 &= b_3
\end{aligned}
$$

The problem thus reduces to determining whether this system is consistent for all values of b_1, b_2, and b_3. By parts (e) and (g) of Theorem 4.3.4, this system is consistent for all b_1, b_2, and b_3 if and only if the coefficient matrix

$$A = \begin{bmatrix} 1 & 1 & 2 \\ 1 & 0 & 1 \\ 2 & 1 & 3 \end{bmatrix}$$

has a nonzero determinant. However, $\det(A) = 0$ (verify), so that $\mathbf{v}_1$, $\mathbf{v}_2$, and $\mathbf{v}_3$ do not span R^3. ◆

Spanning sets are not unique. For example, any two noncollinear vectors that lie in the plane shown in Figure 5.2.5 will span that same plane, and any nonzero vector on the line in that figure will span the same line. We leave the proof of the following useful theorem as an exercise.

Theorem 5.2.4

If $S = \{\mathbf{v}_1, \mathbf{v}_2, \ldots, \mathbf{v}_r\}$ and $S' = \{\mathbf{w}_1, \mathbf{w}_2, \ldots, \mathbf{w}_k\}$ are two sets of vectors in a vector space V, then

$$span\{\mathbf{v}_1, \mathbf{v}_2, \ldots, \mathbf{v}_r\} = span\{\mathbf{w}_1, \mathbf{w}_2, \ldots, \mathbf{w}_k\}$$

if and only if each vector in S is a linear combination of those in S' and each vector in S' is a linear combination of those in S.

Exercise Set 5.2

1. Use Theorem 5.2.1 to determine which of the following are subspaces of R^3.

   (a) all vectors of the form $(a, 0, 0)$
   (b) all vectors of the form $(a, 1, 1)$
   (c) all vectors of the form (a, b, c), where $b = a + c$
   (d) all vectors of the form (a, b, c), where $b = a + c + 1$

2. Use Theorem 5.2.1 to determine which of the following are subspaces of M_{22}.

   (a) all 2×2 matrices with integer entries
   (b) all matrices

   $$\begin{bmatrix} a & b \\ c & d \end{bmatrix}$$

   where $a + b + c + d = 0$
   (c) all 2×2 matrices A such that $\det(A) = 0$
   (d) all matrices of the form

   $$\begin{bmatrix} a & b \\ 0 & c \end{bmatrix}$$

3. Use Theorem 5.2.1 to determine which of the following are subspaces of P_3.

   (a) all polynomials $a_0 + a_1 x + a_2 x^2 + a_3 x^3$ for which $a_0 = 0$
   (b) all polynomials $a_0 + a_1 x + a_2 x^2 + a_3 x^3$ for which $a_0 + a_1 + a_2 + a_3 = 0$
   (c) all polynomials $a_0 + a_1 x + a_2 x^2 + a_3 x^3$ for which a_0, a_1, a_2, and a_3 are integers
   (d) all polynomials of the form $a_0 + a_1 x$, where a_0 and a_1 are real numbers

4. Use Theorem 5.2.1 to determine which of the following are subspaces of the space $F(-\infty, \infty)$.

   (a) all f such that $f(x) \leq 0$ for all x (b) all f such that $f(0) = 0$
   (c) all f such that $f(0) = 2$ (d) all constant functions
   (e) all f of the form $k_1 + k_2 \sin x$, where k_1 and k_2 are real numbers

5. Use Theorem 5.2.1 to determine which of the following are subspaces of M_{nn}.

   (a) all $n \times n$ matrices A such that $\text{tr}(A) = 0$

   (b) all $n \times n$ matrices A such that $A^T = -A$

   (c) all $n \times n$ matrices A such that the linear system $A\mathbf{x} = \mathbf{0}$ has only the trivial solution

   (d) all $n \times n$ matrices A such that $AB = BA$ for a fixed $n \times n$ matrix B

6. Determine whether the solution space of the system $A\mathbf{x} = \mathbf{0}$ is a line through the origin, a plane through the origin, or the origin only. If it is a plane, find an equation for it; if it is a line, find parametric equations for it.

   (a) $A = \begin{bmatrix} -1 & 1 & 1 \\ 3 & -1 & 0 \\ 2 & -4 & -5 \end{bmatrix}$ (b) $A = \begin{bmatrix} 1 & -2 & 3 \\ -3 & 6 & 9 \\ -2 & 4 & -6 \end{bmatrix}$ (c) $A = \begin{bmatrix} 1 & 2 & 3 \\ 2 & 5 & 3 \\ 1 & 0 & 8 \end{bmatrix}$

   (d) $A = \begin{bmatrix} 1 & 2 & -6 \\ 1 & 4 & 4 \\ 3 & 10 & 6 \end{bmatrix}$ (e) $A = \begin{bmatrix} 1 & -1 & 1 \\ 2 & -1 & 4 \\ 3 & 1 & 11 \end{bmatrix}$ (f) $A = \begin{bmatrix} 1 & -3 & 1 \\ 2 & -6 & 2 \\ 3 & -9 & 3 \end{bmatrix}$

7. Which of the following are linear combinations of $\mathbf{u} = (0, -2, 2)$ and $\mathbf{v} = (1, 3, -1)$?

   (a) $(2, 2, 2)$ (b) $(3, 1, 5)$ (c) $(0, 4, 5)$ (d) $(0, 0, 0)$

8. Express the following as linear combinations of $\mathbf{u} = (2, 1, 4)$, $\mathbf{v} = (1, -1, 3)$, and $\mathbf{w} = (3, 2, 5)$.

   (a) $(-9, -7, -15)$ (b) $(6, 11, 6)$ (c) $(0, 0, 0)$ (d) $(7, 8, 9)$

9. Express the following as linear combinations of $\mathbf{p}_1 = 2 + x + 4x^2$, $\mathbf{p}_2 = 1 - x + 3x^2$, and $\mathbf{p}_3 = 3 + 2x + 5x^2$.

   (a) $-9 - 7x - 15x^2$ (b) $6 + 11x + 6x^2$ (c) 0 (d) $7 + 8x + 9x^2$

10. Which of the following are linear combinations of

$$A = \begin{bmatrix} 4 & 0 \\ -2 & -2 \end{bmatrix}, \quad B = \begin{bmatrix} 1 & -1 \\ 2 & 3 \end{bmatrix}, \quad C = \begin{bmatrix} 0 & 2 \\ 1 & 4 \end{bmatrix}?$$

   (a) $\begin{bmatrix} 6 & -8 \\ -1 & -8 \end{bmatrix}$ (b) $\begin{bmatrix} 0 & 0 \\ 0 & 0 \end{bmatrix}$ (c) $\begin{bmatrix} 6 & 0 \\ 3 & 8 \end{bmatrix}$ (d) $\begin{bmatrix} -1 & 5 \\ 7 & 1 \end{bmatrix}$

11. In each part determine whether the given vectors span R^3.

   (a) $\mathbf{v}_1 = (2, 2, 2)$, $\mathbf{v}_2 = (0, 0, 3)$, $\mathbf{v}_3 = (0, 1, 1)$

   (b) $\mathbf{v}_1 = (2, -1, 3)$, $\mathbf{v}_2 = (4, 1, 2)$, $\mathbf{v}_3 = (8, -1, 8)$

   (c) $\mathbf{v}_1 = (3, 1, 4)$, $\mathbf{v}_2 = (2, -3, 5)$, $\mathbf{v}_3 = (5, -2, 9)$, $\mathbf{v}_4 = (1, 4, -1)$

   (d) $\mathbf{v}_1 = (1, 2, 6)$, $\mathbf{v}_2 = (3, 4, 1)$, $\mathbf{v}_3 = (4, 3, 1)$, $\mathbf{v}_4 = (3, 3, 1)$

12. Let $\mathbf{f} = \cos^2 x$ and $\mathbf{g} = \sin^2 x$. Which of the following lie in the space spanned by $\mathbf{f}$ and $\mathbf{g}$?

   (a) $\cos 2x$ (b) $3 + x^2$ (c) 1 (d) $\sin x$ (e) 0

13. Determine whether the following polynomials span P_2.

$$\mathbf{p}_1 = 1 - x + 2x^2, \qquad \mathbf{p}_2 = 3 + x, \qquad \mathbf{p}_3 = 5 - x + 4x^2, \qquad \mathbf{p}_4 = -2 - 2x + 2x^2$$

14. Let $\mathbf{v}_1 = (2, 1, 0, 3)$, $\mathbf{v}_2 = (3, -1, 5, 2)$, and $\mathbf{v}_3 = (-1, 0, 2, 1)$. Which of the following vectors are in span$\{\mathbf{v}_1, \mathbf{v}_2, \mathbf{v}_3\}$?

   (a) $(2, 3, -7, 3)$ (b) $(0, 0, 0, 0)$ (c) $(1, 1, 1, 1)$ (d) $(-4, 6, -13, 4)$

15. Find an equation for the plane spanned by the vectors $\mathbf{u} = (-1, 1, 1)$ and $\mathbf{v} = (3, 4, 4)$.

16. Find parametric equations for the line spanned by the vector $\mathbf{u} = (3, -2, 5)$.

17. Show that the solution vectors of a consistent nonhomogeneous system of m linear equations in n unknowns do not form a subspace of R^n.

18. Prove Theorem 5.2.4.

19. Use Theorem 5.2.4 to show that $\mathbf{v}_1 = (1, 6, 4)$, $\mathbf{v}_2 = (2, 4, -1)$, $\mathbf{v}_3 = (-1, 2, 5)$ and $\mathbf{w}_1 = (1, -2, -5)$, $\mathbf{w}_2 = (0, 8, 9)$ span the same subspace of R^3.

20. A line L through the origin in R^3 can be represented by parametric equations of the form $x = at$, $y = bt$, and $z = ct$. Use these equations to show that L is a subspace of R^3; that is, if $\mathbf{v}_1 = (x_1, y_1, z_1)$ and $\mathbf{v}_2 = (x_2, y_2, z_2)$ are points on L and k is any real number, then $k\mathbf{v}_1$ and $\mathbf{v}_1 + \mathbf{v}_2$ are also points on L.

21. (*For readers who have studied calculus.*) Show that the following sets of functions are subspaces of $F(-\infty, \infty)$.

(a) all everywhere continuous functions   (b) all everywhere differentiable functions
(c) all everywhere differentiable functions that satisfy $\mathbf{f}' + 2\mathbf{f} = \mathbf{0}$

22. (*For readers who have studied calculus.*) Show that the set of continuous functions $\mathbf{f} = f(x)$ on $[a, b]$ such that

$$\int_a^b f(x)\, dx = 0$$

is a subspace of $C[a, b]$.

Discussion and Discovery

23. Indicate whether the statement is always true or sometimes false. Justify your answer by giving a logical argument or a counterexample.

(a) If $A\mathbf{x} = \mathbf{b}$ is any consistent linear system of m equations in n unknowns, then the solution set is a subspace of R^n.
(b) If W is a set of one or more vectors from a vector space V, and if $k\mathbf{u} + \mathbf{v}$ is a vector in W for all vectors $\mathbf{u}$ and $\mathbf{v}$ in W and for all scalars k, then W is a subspace of V.
(c) If S is a finite set of vectors in a vector space V, then span(S) must be closed under addition and scalar multiplication.
(d) The intersection of two subspaces of a vector space V is also a subspace of V.
(e) If span(S_1) = span(S_2), then $S_1 = S_2$.

24. (a) Under what conditions will two vectors in R^3 span a plane? A line?
(b) Under what conditions will it be true that span$\{\mathbf{u}\}$ = span$\{\mathbf{v}\}$? Explain.
(c) If $A\mathbf{x} = \mathbf{b}$ is a consistent system of m equations in n unknowns, under what conditions will it be true that the solution set is a subspace of R^n? Explain.

25. Recall that lines through the origin are subspaces of R^2. If W_1 is the line $y = x$ and W_2 is the line $y = -x$, is the union $W_1 \cup W_2$ a subspace of R^2? Explain your reasoning.

26. (a) Let M_{22} be the vector space of 2×2 matrices. Find four matrices that span M_{22}.
(b) In words, describe a set of matrices that spans M_{nn}.

27. We showed in Example 8 that the vectors $\mathbf{i}, \mathbf{j}, \mathbf{k}$ span R^3. In general, what geometric property must a set of three vectors in R^3 have if they are to span R^3?

5.3 LINEAR INDEPENDENCE

In the preceding section we learned that a set of vectors $S = \{\mathbf{v}_1, \mathbf{v}_2, \ldots, \mathbf{v}_r\}$ spans a given vector space V if every vector in V is expressible as a linear combination of the vectors in S. In general, there may be more than one way to express a vector in V as a linear combination of vectors in a spanning set. In this section we shall study conditions under which each vector in V is expressible as a linear combination of the spanning vectors in exactly one way. Spanning sets with this property play a fundamental role in the study of vector spaces.

> ### Definition
>
> If $S = \{\mathbf{v}_1, \mathbf{v}_2, \ldots, \mathbf{v}_r\}$ is a nonempty set of vectors, then the vector equation
>
> $$k_1\mathbf{v}_1 + k_2\mathbf{v}_2 + \cdots + k_r\mathbf{v}_r = \mathbf{0}$$
>
> has at least one solution, namely
>
> $$k_1 = 0, \quad k_2 = 0, \ldots, \quad k_r = 0$$
>
> If this is the only solution, then S is called a ***linearly independent*** set. If there are other solutions, then S is called a ***linearly dependent*** set.

EXAMPLE 1 A Linearly Dependent Set

If $\mathbf{v}_1 = (2, -1, 0, 3)$, $\mathbf{v}_2 = (1, 2, 5, -1)$, and $\mathbf{v}_3 = (7, -1, 5, 8)$, then the set of vectors $S = \{\mathbf{v}_1, \mathbf{v}_2, \mathbf{v}_3\}$ is linearly dependent, since $3\mathbf{v}_1 + \mathbf{v}_2 - \mathbf{v}_3 = \mathbf{0}$. ◆

EXAMPLE 2 A Linearly Dependent Set

The polynomials

$$\mathbf{p}_1 = 1 - x, \quad \mathbf{p}_2 = 5 + 3x - 2x^2, \quad \text{and} \quad \mathbf{p}_3 = 1 + 3x - x^2$$

form a linearly dependent set in P_2 since $3\mathbf{p}_1 - \mathbf{p}_2 + 2\mathbf{p}_3 = \mathbf{0}$. ◆

EXAMPLE 3 Linearly Independent Sets

Consider the vectors $\mathbf{i} = (1, 0, 0)$, $\mathbf{j} = (0, 1, 0)$, and $\mathbf{k} = (0, 0, 1)$ in R^3. In terms of components the vector equation

$$k_1\mathbf{i} + k_2\mathbf{j} + k_3\mathbf{k} = \mathbf{0}$$

becomes

$$k_1(1, 0, 0) + k_2(0, 1, 0) + k_3(0, 0, 1) = (0, 0, 0)$$

or equivalently,

$$(k_1, k_2, k_3) = (0, 0, 0)$$

This implies that $k_1 = 0$, $k_2 = 0$, and $k_3 = 0$, so the set $S = \{\mathbf{i}, \mathbf{j}, \mathbf{k}\}$ is linearly independent. A similar argument can be used to show that the vectors

$$\mathbf{e}_1 = (1, 0, 0, \ldots, 0), \quad \mathbf{e}_2 = (0, 1, 0, \ldots, 0), \ldots, \quad \mathbf{e}_n = (0, 0, 0, \ldots, 1)$$

form a linearly independent set in R^n. ◆

EXAMPLE 4 Determining Linear Independence/Dependence

Determine whether the vectors

$$\mathbf{v}_1 = (1, -2, 3), \quad \mathbf{v}_2 = (5, 6, -1), \quad \mathbf{v}_3 = (3, 2, 1)$$

form a linearly dependent set or a linearly independent set.

Solution.

In terms of components the vector equation

$$k_1\mathbf{v}_1 + k_2\mathbf{v}_2 + k_3\mathbf{v}_3 = \mathbf{0}$$

becomes

$$k_1(1, -2, 3) + k_2(5, 6, -1) + k_3(3, 2, 1) = (0, 0, 0)$$

or equivalently,

$$(k_1 + 5k_2 + 3k_3, -2k_1 + 6k_2 + 2k_3, 3k_1 - k_2 + k_3) = (0, 0, 0)$$

Equating corresponding components gives

$$\begin{aligned} k_1 + 5k_2 + 3k_3 &= 0 \\ -2k_1 + 6k_2 + 2k_3 &= 0 \\ 3k_1 - k_2 + k_3 &= 0 \end{aligned}$$

Thus, $\mathbf{v}_1$, $\mathbf{v}_2$, and $\mathbf{v}_3$ form a linearly dependent set if this system has a nontrivial solution, or a linearly independent set if it has only the trivial solution. Solving this system yields

$$k_1 = -\tfrac{1}{2}t, \qquad k_2 = -\tfrac{1}{2}t, \qquad k_3 = t$$

Thus, the system has nontrivial solutions and $\mathbf{v}_1$, $\mathbf{v}_2$, and $\mathbf{v}_3$ form a linearly dependent set. Alternatively, we could show the existence of nontrivial solutions without solving the system by showing that the coefficient matrix has determinant zero and consequently is not invertible (verify). ◆

EXAMPLE 5   Linearly Independent Set in P_n

Show that the polynomials

$$1, x, x^2, \ldots, x^n$$

form a linearly independent set of vectors in P_n.

Solution.

Let $\mathbf{p}_0 = 1$, $\mathbf{p}_1 = x$, $\mathbf{p}_2 = x^2$, ..., $\mathbf{p}_n = x^n$ and assume that some linear combination of these polynomials is zero, say

$$a_0\mathbf{p}_0 + a_1\mathbf{p}_1 + a_2\mathbf{p}_2 + \cdots + a_n\mathbf{p}_n = \mathbf{0}$$

or equivalently,

$$a_0 + a_1 x + a_2 x^2 + \cdots + a_n x^n = 0 \quad \text{for all } x \text{ in } (-\infty, \infty) \tag{1}$$

We must show that

$$a_0 = a_1 = a_2 = \cdots = a_n = 0.$$

To see that this is so, recall from algebra that a *nonzero* polynomial of degree n has at most n distinct roots. But this implies that $a_0 = a_1 = a_2 = \cdots = a_n = 0$; otherwise, it would follow from (1) that $a_0 + a_1 x + a_2 x^2 + \cdots + a_n x^n$ is a nonzero polynomial with infinitely many roots. ◆

The term "linearly dependent" suggests that the vectors "depend" on each other in some way. The following theorem shows that this is in fact the case.

Theorem 5.3.1

A set S with two or more vectors is:

(a) *Linearly dependent if and only if at least one of the vectors in S is expressible as a linear combination of the other vectors in S.*

(b) *Linearly independent if and only if no vector in S is expressible as a linear combination of the other vectors in S.*

We shall prove part (a) and leave the proof of part (b) as an exercise.

Proof (a). Let $S = \{\mathbf{v}_1, \mathbf{v}_2, \ldots, \mathbf{v}_r\}$ be a set with two or more vectors. If we assume that S is linearly dependent, then there are scalars $k_1, k_2, \ldots, k_r$, not all zero, such that

$$k_1\mathbf{v}_1 + k_2\mathbf{v}_2 + \cdots + k_r\mathbf{v}_r = \mathbf{0} \qquad (2)$$

To be specific, suppose that $k_1 \neq 0$. Then (2) can be rewritten as

$$\mathbf{v}_1 = \left(-\frac{k_2}{k_1}\right)\mathbf{v}_2 + \cdots + \left(-\frac{k_r}{k_1}\right)\mathbf{v}_r$$

which expresses $\mathbf{v}_1$ as a linear combination of the other vectors in S. Similarly, if $k_j \neq 0$ in (2) for some $j = 2, 3, \ldots, r$, then $\mathbf{v}_j$ is expressible as a linear combination of the other vectors in S.

Conversely, let us assume that at least one of the vectors in S is expressible as a linear combination of the other vectors. To be specific, suppose that

$$\mathbf{v}_1 = c_2\mathbf{v}_2 + c_3\mathbf{v}_3 + \cdots + c_r\mathbf{v}_r$$

so

$$\mathbf{v}_1 - c_2\mathbf{v}_2 - c_3\mathbf{v}_3 - \cdots - c_r\mathbf{v}_r = \mathbf{0}$$

It follows that S is linearly dependent since the equation

$$k_1\mathbf{v}_1 + k_2\mathbf{v}_2 + \cdots + k_r\mathbf{v}_r = \mathbf{0}$$

is satisfied by

$$k_1 = 1, \quad k_2 = -c_2, \ldots, \quad k_r = -c_r$$

which are not all zero. The proof in the case where some vector other than $\mathbf{v}_1$ is expressible as a linear combination of the other vectors in S is similar. ∎

EXAMPLE 6 Example 1 Revisited

In Example 1 we saw that the vectors

$$\mathbf{v}_1 = (2, -1, 0, 3), \quad \mathbf{v}_2 = (1, 2, 5, -1), \quad \text{and} \quad \mathbf{v}_3 = (7, -1, 5, 8)$$

form a linearly dependent set. It follows from Theorem 5.3.1 that at least one of these vectors is expressible as a linear combination of the other two. In this example each vector is expressible as a linear combination of the other two since it follows from the equation $3\mathbf{v}_1 + \mathbf{v}_2 - \mathbf{v}_3 = \mathbf{0}$ (see Example 1) that

$$\mathbf{v}_1 = -\tfrac{1}{3}\mathbf{v}_2 + \tfrac{1}{3}\mathbf{v}_3, \quad \mathbf{v}_2 = -3\mathbf{v}_1 + \mathbf{v}_3, \quad \text{and} \quad \mathbf{v}_3 = 3\mathbf{v}_1 + \mathbf{v}_2 \qquad \blacklozenge$$

EXAMPLE 7 Example 3 Revisited

In Example 3 we saw that the vectors $\mathbf{i} = (1, 0, 0)$, $\mathbf{j} = (0, 1, 0)$, and $\mathbf{k} = (0, 0, 1)$ form a linearly independent set. Thus, it follows from Theorem 5.3.1 that none of these vectors is expressible as a linear combination of the other two. To see directly that this is so, suppose that **k** is expressible as

$$\mathbf{k} = k_1 \mathbf{i} + k_2 \mathbf{j}$$

Then, in terms of components,

$$(0, 0, 1) = k_1(1, 0, 0) + k_2(0, 1, 0) \quad \text{or} \quad (0, 0, 1) = (k_1, k_2, 0)$$

But the last equation is not satisfied by any values of k_1 and k_2, so **k** cannot be expressed as a linear combination of **i** and **j**. Similarly, **i** is not expressible as a linear combination of **j** and **k**, and **j** is not expressible as a linear combination of **i** and **k**. ◆

The following theorem gives two simple facts about linear independence that are important to know.

Theorem 5.3.2

(a) *A finite set of vectors that contains the zero vector is linearly dependent.*

(b) *A set with exactly two vectors is linearly independent if and only if neither vector is a scalar multiple of the other.*

We shall prove part (a) and leave the proof of part (b) as an exercise.

Proof (a). For any vectors $\mathbf{v}_1, \mathbf{v}_2, \ldots, \mathbf{v}_r$, the set $S = \{\mathbf{v}_1, \mathbf{v}_2, \ldots, \mathbf{v}_r, \mathbf{0}\}$ is linearly dependent since the equation

$$0\mathbf{v}_1 + 0\mathbf{v}_2 + \cdots + 0\mathbf{v}_r + 1(\mathbf{0}) = \mathbf{0}$$

expresses **0** as a linear combination of the vectors in S with coefficients that are not all zero. ■

EXAMPLE 8 Using Theorem 5.3.2*b*

The functions $\mathbf{f}_1 = x$ and $\mathbf{f}_2 = \sin x$ form a linearly independent set of vectors in $F(-\infty, \infty)$, since neither function is a constant multiple of the other. ◆

Geometric Interpretation of Linear Independence

Linear independence has some useful geometric interpretations in R^2 and R^3:

- In R^2 or R^3, a set of two vectors is linearly independent if and only if the vectors do not lie on the same line when they are placed with their initial points at the origin (Figure 5.3.1).

- In R^3, a set of three vectors is linearly independent if and only if the vectors do not lie in the same plane when they are placed with their initial points at the origin (Figure 5.3.2).

The first result follows from the fact that two vectors are linearly independent if and only if neither vector is a scalar multiple of the other. Geometrically, this is equivalent to stating that the vectors do not lie on the same line when they are positioned with their initial points at the origin.

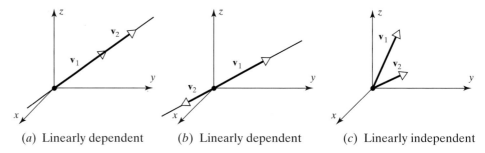

(*a*) Linearly dependent   (*b*) Linearly dependent   (*c*) Linearly independent

Figure 5.3.1

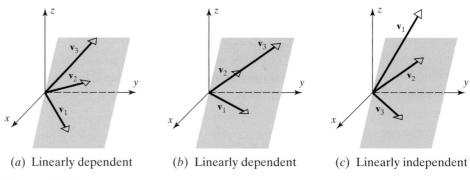

(*a*) Linearly dependent   (*b*) Linearly dependent   (*c*) Linearly independent

Figure 5.3.2

The second result follows from the fact that three vectors are linearly independent if and only if none of the vectors is a linear combination of the other two. Geometrically, this is equivalent to stating that none of the vectors lies in the same plane as the other two, or alternatively, that the three vectors do not lie in a common plane when they are positioned with their initial points at the origin (why?).

The next theorem shows that a linearly independent set in R^n can contain at most n vectors.

Theorem 5.3.3

Let $S = \{\mathbf{v}_1, \mathbf{v}_2, \ldots, \mathbf{v}_r\}$ be a set of vectors in R^n. If $r > n$, then S is linearly dependent.

Proof. Suppose that

$$\mathbf{v}_1 = (v_{11}, v_{12}, \ldots, v_{1n})$$
$$\mathbf{v}_2 = (v_{21}, v_{22}, \ldots, v_{2n})$$
$$\vdots \qquad\qquad \vdots$$
$$\mathbf{v}_r = (v_{r1}, v_{r2}, \ldots, v_{rn})$$

Consider the equation

$$k_1\mathbf{v}_1 + k_2\mathbf{v}_2 + \cdots + k_r\mathbf{v}_r = \mathbf{0}$$

If, as illustrated in Example 4, we express both sides of this equation in terms of components and then equate corresponding components, we obtain the system

$$v_{11}k_1 + v_{21}k_2 + \cdots + v_{r1}k_r = 0$$
$$v_{12}k_1 + v_{22}k_2 + \cdots + v_{r2}k_r = 0$$
$$\vdots \qquad \vdots \qquad\qquad \vdots \qquad \vdots$$
$$v_{1n}k_1 + v_{2n}k_2 + \cdots + v_{rn}k_r = 0$$

This is a homogeneous system of n equations in the r unknowns $k_1, \ldots, k_r$. Since $r > n$, it follows from Theorem 1.2.1 that the system has nontrivial solutions. Therefore, $S = \{\mathbf{v}_1, \mathbf{v}_2, \ldots, \mathbf{v}_r\}$ is a linearly dependent set. ∎

REMARK. The preceding theorem tells us that a set in R^2 with more than two vectors is linearly dependent, and a set in R^3 with more than three vectors is linearly dependent.

Calculus Required

Linear Independence of Functions
Sometimes linear dependence of functions can be deduced from known identities. For example, the functions

$$\mathbf{f}_1 = \sin^2 x, \quad \mathbf{f}_2 = \cos^2 x, \quad \text{and} \quad \mathbf{f}_3 = 5$$

form a linearly dependent set in $F(-\infty, \infty)$, since the equation

$$5\mathbf{f}_1 + 5\mathbf{f}_2 - \mathbf{f}_3 = 5\sin^2 x + 5\cos^2 x - 5 = 5(\sin^2 x + \cos^2 x) - 5 = \mathbf{0}$$

expresses $\mathbf{0}$ as a linear combination of $\mathbf{f}_1$, $\mathbf{f}_2$, and $\mathbf{f}_3$ with coefficients that are not all zero. However, it is only in special situations that such identities can be applied. Although there is no general method that can be used to establish linear independence or linear dependence of functions in $F(-\infty, \infty)$, we shall now develop a theorem that can sometimes be used to show that a given set of functions is linearly independent.

If $\mathbf{f}_1 = f_1(x)$, $\mathbf{f}_2 = f_2(x), \ldots, \mathbf{f}_n = f_n(x)$ are $n - 1$ times differentiable functions on the interval $(-\infty, \infty)$, then the determinant of

$$W(x) = \begin{bmatrix} f_1(x) & f_2(x) & \cdots & f_n(x) \\ f_1'(x) & f_2'(x) & \cdots & f_n'(x) \\ \vdots & \vdots & & \vdots \\ f_1^{(n-1)}(x) & f_2^{(n-1)}(x) & \cdots & f_n^{(n-1)}(x) \end{bmatrix}$$

is called the **Wronskian** of $f_1, f_2, \ldots, f_n$. As we shall now show, this determinant is useful for ascertaining whether the functions $\mathbf{f}_1, \mathbf{f}_2, \ldots, \mathbf{f}_n$ form a linearly independent set of vectors in the vector space $C^{(n-1)}(-\infty, \infty)$.

Suppose, for the moment, that $\mathbf{f}_1, \mathbf{f}_2, \ldots, \mathbf{f}_n$ are linearly dependent vectors in $C^{(n-1)}(-\infty, \infty)$. Then there exist scalars $k_1, k_2, \ldots, k_n$, *not all zero*, such that

$$k_1 f_1(x) + k_2 f_2(x) + \cdots + k_n f_n(x) = 0$$

for all x in the interval $(-\infty, \infty)$. Combining this equation with the equations obtained by $n - 1$ successive differentiations yields

$$k_1 f_1(x) + k_2 f_2(x) + \cdots + k_n f_n(x) = 0$$
$$k_1 f_1'(x) + k_2 f_2'(x) + \cdots + k_n f_n'(x) = 0$$
$$\vdots \qquad\qquad \vdots \qquad\qquad \vdots \qquad\qquad \vdots$$
$$k_1 f_1^{(n-1)}(x) + k_2 f_2^{(n-1)}(x) + \cdots + k_n f_n^{(n-1)}(x) = 0$$

Thus, the linear dependence of $\mathbf{f}_1, \mathbf{f}_2, \ldots, \mathbf{f}_n$ implies that the linear system

$$
\begin{bmatrix}
f_1(x) & f_2(x) & \cdots & f_n(x) \\
f_1'(x) & f_2'(x) & \cdots & f_n'(x) \\
\vdots & \vdots & & \vdots \\
f_1^{(n-1)}(x) & f_2^{(n-1)}(x) & \cdots & f_n^{(n-1)}(x)
\end{bmatrix}
\begin{bmatrix}
k_1 \\ k_2 \\ \vdots \\ k_n
\end{bmatrix}
=
\begin{bmatrix}
0 \\ 0 \\ \vdots \\ 0
\end{bmatrix}
$$

has a nontrivial solution for *every* x in the interval $(-\infty, \infty)$. This implies in turn that for every x in $(-\infty, \infty)$ the coefficient matrix is not invertible, or equivalently, that its determinant (the Wronskian) is zero for every x in $(-\infty, \infty)$. Thus, if the Wronskian is *not* identically zero on $(-\infty, \infty)$, then the functions $\mathbf{f}_1, \mathbf{f}_2, \ldots, \mathbf{f}_n$ must be linearly independent vectors in $C^{(n-1)}(-\infty, \infty)$. This is the content of the following theorem.

Theorem 5.3.4

If the functions $\mathbf{f}_1, \mathbf{f}_2, \ldots, \mathbf{f}_n$ have $n - 1$ continuous derivatives on the interval $(-\infty, \infty)$, and if the Wronskian of these functions is not identically zero on $(-\infty, \infty)$, then these functions form a linearly independent set of vectors in $C^{(n-1)}(-\infty, \infty)$.

EXAMPLE 9 Linearly Independent Set in $C^1(-\infty, \infty)$

Show that the functions $\mathbf{f}_1 = x$ and $\mathbf{f}_2 = \sin x$ form a linearly independent set of vectors in $C^1(-\infty, \infty)$.

Solution.

In Example 8 we showed that these vectors form a linearly independent set by noting that neither vector is a scalar multiple of the other. However, for illustrative purposes, we shall obtain this same result using Theorem 5.3.4. The Wronskian is

$$
W(x) = \begin{vmatrix} x & \sin x \\ 1 & \cos x \end{vmatrix} = x \cos x - \sin x
$$

This function does not have value zero for all x in the interval $(-\infty, \infty)$ (verify), so $\mathbf{f}_1$ and $\mathbf{f}_2$ form a linearly independent set. ◆

EXAMPLE 10 Linearly Independent Set in $C^2(-\infty, \infty)$

Show that $\mathbf{f}_1 = 1$, $\mathbf{f}_2 = e^x$, and $\mathbf{f}_3 = e^{2x}$ form a linearly independent set of vectors in $C^2(-\infty, \infty)$.

Solution.

The Wronskian is

$$
W(x) = \begin{vmatrix} 1 & e^x & e^{2x} \\ 0 & e^x & 2e^{2x} \\ 0 & e^x & 4e^{2x} \end{vmatrix} = 2e^{3x}
$$

This function does not have value zero for all x (in fact, for any x) in the interval $(-\infty, \infty)$, so $\mathbf{f}_1$, $\mathbf{f}_2$, and $\mathbf{f}_3$ form a linearly independent set. ◆

Józef Maria Hoëne-Wroński (1776–1853) was a Polish-French mathematician and philosopher. Wroński received his early education in Poznán and Warsaw. He served as an artillery officer in the Prussian army in a national uprising in 1794, was taken prisoner by the Russian army, and on his release studied philosophy at various German universities. He became a French citizen in 1800 and eventually settled in Paris, where he did research in analysis leading to some controversial mathematical papers and relatedly to a famous court trial over financial matters. Several years thereafter, his proposed research on the determination of longitude at sea was rebuffed by the British Board of Longitude and Wrónski turned to studies in Messianic philosophy. In the 1830s he investigated the feasibility of caterpillar vehicles to compete with trains, with no luck, and spent his last years in poverty. Much of his mathematical work was fraught with errors and imprecision, but it often contained valuable isolated results and ideas. Some writers attribute this lifelong pattern of argumentation to psychopathic tendencies and an exaggeration of the importance of his own work.

REMARK.   The converse of Theorem 5.3.4 is false. If the Wronskian of $\mathbf{f}_1, \mathbf{f}_2, \ldots, \mathbf{f}_n$ is identically zero on $(-\infty, \infty)$, then no conclusion can be reached about the linear independence of $\{\mathbf{f}_1, \mathbf{f}_2, \ldots, \mathbf{f}_n\}$; this set of vectors may be linearly independent or linearly dependent. (We omit the details.)

Exercise Set 5.3

1. Explain why the following are linearly dependent sets of vectors. (Solve this problem by inspection.)

   (a) $\mathbf{u}_1 = (-1, 2, 4)$ and $\mathbf{u}_2 = (5, -10, -20)$ in R^3 (b) $\mathbf{u}_1 = (3, -1)$, $\mathbf{u}_2 = (4, 5)$, $\mathbf{u}_3 = (-4, 7)$ in R^2

   (c) $\mathbf{p}_1 = 3 - 2x + x^2$ and $\mathbf{p}_2 = 6 - 4x + 2x^2$ in P_2 (d) $A = \begin{bmatrix} -3 & 4 \\ 2 & 0 \end{bmatrix}$ and $B = \begin{bmatrix} 3 & -4 \\ -2 & 0 \end{bmatrix}$ in M_{22}

2. Which of the following sets of vectors in R^3 are linearly dependent?

   (a) $(4, -1, 2)$, $(-4, 10, 2)$ (b) $(-3, 0, 4)$, $(5, -1, 2)$, $(1, 1, 3)$
   (c) $(8, -1, 3)$, $(4, 0, 1)$ (d) $(-2, 0, 1)$, $(3, 2, 5)$, $(6, -1, 1)$, $(7, 0, -2)$

3. Which of the following sets of vectors in R^4 are linearly dependent?

   (a) $(3, 8, 7, -3)$, $(1, 5, 3, -1)$, $(2, -1, 2, 6)$, $(1, 4, 0, 3)$
   (b) $(0, 0, 2, 2)$, $(3, 3, 0, 0)$, $(1, 1, 0, -1)$
   (c) $(0, 3, -3, -6)$, $(-2, 0, 0, -6)$, $(0, -4, -2, -2)$, $(0, -8, 4, -4)$
   (d) $(3, 0, -3, 6)$, $(0, 2, 3, 1)$, $(0, -2, -2, 0)$, $(-2, 1, 2, 1)$

4. Which of the following sets of vectors in P_2 are linearly dependent?

   (a) $2 - x + 4x^2$, $3 + 6x + 2x^2$, $2 + 10x - 4x^2$ (b) $3 + x + x^2$, $2 - x + 5x^2$, $4 - 3x^2$
   (c) $6 - x^2$, $1 + x + 4x^2$ (d) $1 + 3x + 3x^2$, $x + 4x^2$, $5 + 6x + 3x^2$, $7 + 2x - x^2$

5. Assume that $\mathbf{v}_1$, $\mathbf{v}_2$, and $\mathbf{v}_3$ are vectors in R^3 that have their initial points at the origin. In each part determine whether the three vectors lie in a plane.

   (a) $\mathbf{v}_1 = (2, -2, 0)$, $\mathbf{v}_2 = (6, 1, 4)$, $\mathbf{v}_3 = (2, 0, -4)$ (b) $\mathbf{v}_1 = (-6, 7, 2)$, $\mathbf{v}_2 = (3, 2, 4)$, $\mathbf{v}_3 = (4, -1, 2)$

6. Assume that $\mathbf{v}_1$, $\mathbf{v}_2$, and $\mathbf{v}_3$ are vectors in R^3 that have their initial points at the origin. In each part determine whether the three vectors lie on the same line.

   (a) $\mathbf{v}_1 = (-1, 2, 3)$, $\mathbf{v}_2 = (2, -4, -6)$, $\mathbf{v}_3 = (-3, 6, 0)$ (b) $\mathbf{v}_1 = (2, -1, 4)$, $\mathbf{v}_2 = (4, 2, 3)$, $\mathbf{v}_3 = (2, 7, -6)$
   (c) $\mathbf{v}_1 = (4, 6, 8)$, $\mathbf{v}_2 = (2, 3, 4)$, $\mathbf{v}_3 = (-2, -3, -4)$

7. (a) Show that the vectors $\mathbf{v}_1 = (0, 3, 1, -1)$, $\mathbf{v}_2 = (6, 0, 5, 1)$, and $\mathbf{v}_3 = (4, -7, 1, 3)$ form a linearly dependent set in R^4.
   (b) Express each vector as a linear combination of the other two.

8. For which real values of λ do the following vectors form a linearly dependent set in R^3?

   $$\mathbf{v}_1 = \left(\lambda, -\tfrac{1}{2}, -\tfrac{1}{2}\right), \qquad \mathbf{v}_2 = \left(-\tfrac{1}{2}, \lambda, -\tfrac{1}{2}\right), \qquad \mathbf{v}_3 = \left(-\tfrac{1}{2}, -\tfrac{1}{2}, \lambda\right)$$

9. Show that if $\{\mathbf{v}_1, \mathbf{v}_2, \mathbf{v}_3\}$ is a linearly independent set of vectors, then so are $\{\mathbf{v}_1, \mathbf{v}_2\}$, $\{\mathbf{v}_1, \mathbf{v}_3\}$, $\{\mathbf{v}_2, \mathbf{v}_3\}$, $\{\mathbf{v}_1\}$, $\{\mathbf{v}_2\}$, and $\{\mathbf{v}_3\}$.

10. Show that if $S = \{\mathbf{v}_1, \mathbf{v}_2, \ldots, \mathbf{v}_r\}$ is a linearly independent set of vectors, then so is every nonempty subset of S.

11. Show that if $\{\mathbf{v}_1, \mathbf{v}_2, \mathbf{v}_3\}$ is a linearly dependent set of vectors in a vector space V, and $\mathbf{v}_4$ is any vector in V, then $\{\mathbf{v}_1, \mathbf{v}_2, \mathbf{v}_3, \mathbf{v}_4\}$ is also linearly dependent.

12. Show that if $\{\mathbf{v}_1, \mathbf{v}_2, \ldots, \mathbf{v}_r\}$ is a linearly dependent set of vectors in a vector space V, and if $\mathbf{v}_{r+1}, \ldots, \mathbf{v}_n$ are any vectors in V, then $\{\mathbf{v}_1, \mathbf{v}_2, \ldots, \mathbf{v}_{r+1}, \ldots, \mathbf{v}_n\}$ is also linearly dependent.

13. Show that every set with more than three vectors from P_2 is linearly dependent.

14. Show that if $\{v_1, v_2\}$ is linearly independent and v_3 does not lie in span $\{v_1, v_2\}$, then $\{v_1, v_2, v_3\}$ is linearly independent.

15. Prove: For any vectors u, v, and w, the vectors $u - v$, $v - w$, and $w - u$ form a linearly dependent set.

16. Prove: The space spanned by two vectors in R^3 is either a line through the origin, a plane through the origin, or the origin itself.

17. Under what conditions is a set with one vector linearly independent?

18. Are the vectors v_1, v_2, and v_3 in part (a) of the accompanying figure linearly independent? What about those in part (b)? Explain.

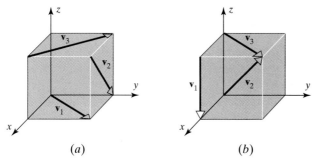

(a)   (b)

Figure Ex-18

19. Use appropriate identities, where required, to determine which of the following sets of vectors in $F(-\infty, \infty)$ are linearly dependent.

(a) 6, $3\sin^2 x$, $2\cos^2 x$ (b) x, $\cos x$ (c) 1, $\sin x$, $\sin 2x$

(d) $\cos 2x$, $\sin^2 x$, $\cos^2 x$ (e) $(3 - x)^2$, $x^2 - 6x$, 5 (f) 0, $\cos^3 \pi x$, $\sin^5 3\pi x$

20. (*For readers who have studied calculus.*) Use the Wronskian to show that the following sets of vectors are linearly independent.

(a) 1, x, e^x (b) $\sin x$, $\cos x$, $x \sin x$ (c) e^x, xe^x, $x^2 e^x$ (d) 1, x, x^2

21. Use part (a) of Theorem 5.3.1 to prove part (b).

22. Prove part (b) of Theorem 5.3.2.

Discussion and Discovery

23. Indicate whether the statement is always true or sometimes false. Justify your answer by giving a logical argument or a counterexample.

(a) The set of 2×2 matrices that contain exactly two ones and two zeros is a linearly independent set in M_{22}.

(b) If $\{v_1, v_2\}$ is a linearly dependent set, then each vector is a scalar multiple of the other.

(c) If $\{v_1, v_2, v_3\}$ is a linearly independent set, then so is the set $\{kv_1, kv_2, kv_3\}$ for every nonzero scalar k.

(d) The converse of Theorem 5.3.2a is also true.

24. If $\{v_1, v_2, v_3\}$ is a linearly dependent set with nonzero vectors, then each vector in the set is expressible as a linear combination of the other two.

25. Theorem 5.3.3 implies that four nonzero vectors in R^3 must be linearly independent. Give an informal geometric argument to explain this result.

26. (a) In Example 3 we showed that the mutually orthogonal vectors i, j, and k form a linearly independent set of vectors in R^3. Do you think that every set of three nonzero mutually orthogonal vectors in R^3 is linearly independent? Justify your conclusion with a geometric argument.

(b) Justify your conclusion with an algebraic argument. [***Hint.*** Use dot products.]

5.4 BASIS AND DIMENSION

We usually think of a line as being one-dimensional, a plane as two-dimensional, and the space around us as three-dimensional. It is the primary purpose of this section to make this intuitive notion of "dimension" more precise.

Nonrectangular Coordinate Systems   In plane analytic geometry we learned to associate a point P in the plane with a pair of coordinates (a, b) by projecting P onto a pair of perpendicular coordinate axes (Figure 5.4.1a). By this process, each point in the plane is assigned a unique set of coordinates and, conversely, each pair of coordinates is associated with a unique point in the plane. We describe this by saying that the coordinate system establishes a ***one-to-one correspondence*** between points in the plane and ordered pairs of real numbers. Although perpendicular coordinate axes are the most common, any two nonparallel lines can be used to define a coordinate system in the plane. For example, in Figure 5.4.1b, we have attached a pair of coordinates (a, b) to the point P by projecting P parallel to the nonperpendicular coordinate axes. Similarly, in 3-space any three noncoplanar coordinate axes can be used to define a coordinate system (Figure 5.4.1c).

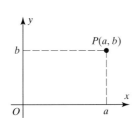

(*a*) Coordinates of P in a rectangular coordinate system in 2-space

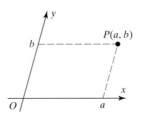

(*b*) Coordinates of P in a nonrectangular coordinate system in 2-space

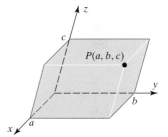

(*c*) Coordinates of P in a nonrectangular coordinate system in 3-space

Figure 5.4.1

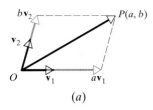

(*a*)

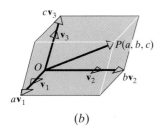

(*b*)

Figure 5.4.2

Our first objective in this section is to extend the concept of a coordinate system to general vector spaces. As a start, it will be helpful to reformulate the notion of a coordinate system in 2-space or 3-space using vectors rather than coordinate axes to specify the coordinate system. This can be done by replacing each coordinate axis with a vector of length 1 that points in the positive direction of the axis. In Figure 5.4.2a, for example, $\mathbf{v}_1$ and $\mathbf{v}_2$ are such vectors. As illustrated in that figure, if P is any point in the plane, the vector $\overrightarrow{OP}$ can be written as a linear combination of $\mathbf{v}_1$ and $\mathbf{v}_2$ by projecting P parallel to $\mathbf{v}_1$ and $\mathbf{v}_2$ to make $\overrightarrow{OP}$ the diagonal of a parallelogram determined by vectors $a\mathbf{v}_1$ and $b\mathbf{v}_2$:

$$\overrightarrow{OP} = a\mathbf{v}_1 + b\mathbf{v}_2$$

It is evident that the numbers a and b in this vector formula are precisely the coordinates of P in the coordinate system of Figure 5.4.1b. Similarly, the coordinates (a, b, c) of the point P in Figure 5.4.1c can be obtained by expressing $\overrightarrow{OP}$ as a linear combination of the vectors shown in Figure 5.4.2b.

Informally stated, vectors that specify a coordinate system are called "basis vectors" for that system. Although we used basis vectors of length 1 in the preceding discussion, we shall see in a moment that this is not essential—nonzero vectors of any length will suffice.

The scales of measurement along the coordinate axes are essential ingredients of any coordinate system. Usually, one tries to use the same scale on each axis and have the integer points on the axes spaced 1 unit of distance apart. However, this is not always practical or appropriate: Unequal scales, or scales in which the integral points are more or less than 1 unit apart, may be required to fit a particular graph on a printed page or to represent physical quantities with diverse units in the same coordinate system (time in seconds on one axis and temperature in hundreds of degrees on another, for example). When a coordinate system is specified by a set of basis vectors, then the lengths of those vectors correspond to the distances between successive integer points on the coordinate axes (Figure 5.4.3). Thus, it is the directions of the basis vectors that define the positive directions of the coordinate axes and the lengths of the basis vectors that establish the scales of measurement.

The following key definition will make the preceding ideas more precise and enable us to extend the concept of a coordinate system to general vector spaces.

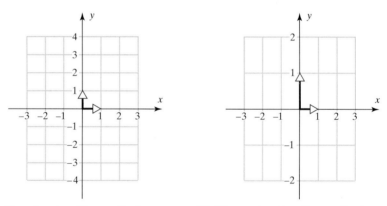

(a) Equal scales. Perpendicular axes.   (b) Unequal scales. Perpendicular axes.

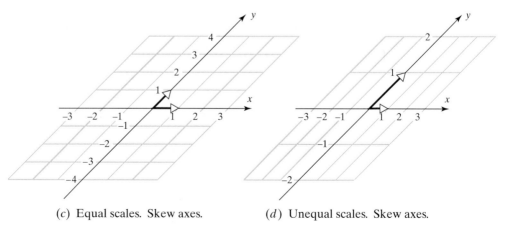

(c) Equal scales. Skew axes.   (d) Unequal scales. Skew axes.

Figure 5.4.3

> ## Definition
>
> If V is any vector space and $S = \{\mathbf{v}_1, \mathbf{v}_2, \ldots, \mathbf{v}_n\}$ is a set of vectors in V, then S is called a **basis** for V if the following two conditions hold:
>
> (a) S is linearly independent.
> (b) S spans V.

A basis is the vector space generalization of a coordinate system in 2-space and 3-space. The following theorem will help us to see why this is so.

> ## Theorem 5.4.1   **Uniqueness of Basis Representation**
>
> *If $S = \{\mathbf{v}_1, \mathbf{v}_2, \ldots, \mathbf{v}_n\}$ is a basis for a vector space V, then every vector $\mathbf{v}$ in V can be expressed in the form $\mathbf{v} = c_1\mathbf{v}_1 + c_2\mathbf{v}_2 + \cdots + c_n\mathbf{v}_n$ in exactly one way.*

Proof. Since S spans V, it follows from the definition of a spanning set that every vector in V is expressible as a linear combination of the vectors in S. To see that there is only *one* way to express a vector as a linear combination of the vectors in S, suppose that some vector $\mathbf{v}$ can be written as

$$\mathbf{v} = c_1\mathbf{v}_1 + c_2\mathbf{v}_2 + \cdots + c_n\mathbf{v}_n$$

and also as

$$\mathbf{v} = k_1\mathbf{v}_1 + k_2\mathbf{v}_2 + \cdots + k_n\mathbf{v}_n$$

Subtracting the second equation from the first gives

$$\mathbf{0} = (c_1 - k_1)\mathbf{v}_1 + (c_2 - k_2)\mathbf{v}_2 + \cdots + (c_n - k_n)\mathbf{v}_n$$

Since the right side of this equation is a linear combination of vectors in S, the linear independence of S implies that

$$c_1 - k_1 = 0, \quad c_2 - k_2 = 0, \ldots, \quad c_n - k_n = 0$$

that is,

$$c_1 = k_1, \quad c_2 = k_2, \ldots, \quad c_n = k_n$$

Thus, the two expressions for $\mathbf{v}$ are the same.   ∎

Coordinates Relative to a Basis

If $S = \{\mathbf{v}_1, \mathbf{v}_2, \ldots, \mathbf{v}_n\}$ is a basis for a vector space V, and

$$\mathbf{v} = c_1\mathbf{v}_1 + c_2\mathbf{v}_2 + \cdots + c_n\mathbf{v}_n$$

is the expression for a vector $\mathbf{v}$ in terms of the basis S, then the scalars $c_1, c_2, \ldots, c_n$ are called the ***coordinates*** of $\mathbf{v}$ relative to the basis S. The vector $(c_1, c_2, \ldots, c_n)$ in R^n constructed from these coordinates is called the ***coordinate vector of $\mathbf{v}$ relative to S***; it is denoted by

$$(\mathbf{v})_S = (c_1, c_2, \ldots, c_n)$$

REMARK. It should be noted that coordinate vectors depend not only on the basis S but also on the order in which the basis vectors are written; a change in the order of the basis vectors results in a corresponding change of order for the entries in the coordinate vectors.

EXAMPLE 1 Standard Basis for R^3

In Example 3 of the preceding section we showed that if

$$\mathbf{i} = (1, 0, 0), \quad \mathbf{j} = (0, 1, 0), \quad \text{and} \quad \mathbf{k} = (0, 0, 1)$$

then $S = \{\mathbf{i}, \mathbf{j}, \mathbf{k}\}$ is a linearly independent set in R^3. This set also spans R^3 since any vector $\mathbf{v} = (a, b, c)$ in R^3 can be written as

$$\mathbf{v} = (a, b, c) = a(1, 0, 0) + b(0, 1, 0) + c(0, 0, 1) = a\mathbf{i} + b\mathbf{j} + c\mathbf{k} \tag{1}$$

Thus, S is a basis for R^3; it is called the ***standard basis*** for R^3. Looking at the coefficients of $\mathbf{i}$, $\mathbf{j}$, and $\mathbf{k}$ in (1), it follows that the coordinates of $\mathbf{v}$ relative to the standard basis are a, b, and c, so

$$(\mathbf{v})_S = (a, b, c)$$

Comparing this result to (1) we see that

$$\mathbf{v} = (\mathbf{v})_S$$

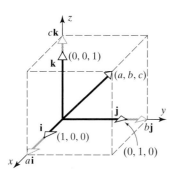

Figure 5.4.4

This equation states that the components of a vector $\mathbf{v}$ relative to a rectangular xyz-coordinate system and the coordinates of $\mathbf{v}$ relative to the standard basis are the same; thus, the coordinate system and the basis produce precisely the same one-to-one correspondence between points in 3-space and ordered triples of real numbers (Figure 5.4.4). ◆

The results in the preceding example are a special case of those in the next example.

EXAMPLE 2 Standard Basis for R^n

In Example 3 of the preceding section we showed that if

$$\mathbf{e}_1 = (1, 0, 0, \ldots, 0), \quad \mathbf{e}_2 = (0, 1, 0, \ldots, 0), \ldots, \quad \mathbf{e}_n = (0, 0, 0, \ldots, 1)$$

then

$$S = \{\mathbf{e}_1, \mathbf{e}_2, \ldots, \mathbf{e}_n\}$$

is a linearly independent set in R^n. Moreover, this set also spans R^n since any vector $\mathbf{v} = (v_1, v_2, \ldots, v_n)$ in R^n can be written as

$$\mathbf{v} = v_1\mathbf{e}_1 + v_2\mathbf{e}_2 + \cdots + v_n\mathbf{e}_n \tag{2}$$

Thus, S is a basis for R^n; it is called the ***standard basis for R^n***. It follows from (2) that the coordinates of $\mathbf{v} = (v_1, v_2, \ldots, v_n)$ relative to the standard basis are $v_1, v_2, \ldots, v_n$, so

$$(\mathbf{v})_S = (v_1, v_2, \ldots, v_n)$$

As in Example 1, we have $\mathbf{v} = (\mathbf{v})_S$, so a vector $\mathbf{v}$ and its coordinate vector relative to the standard basis for R^n are the same. ◆

REMARK. We will see in a subsequent example that a vector and its coordinate vector will not be the same; the equality that we observed in the two preceding examples is a special situation that occurs only with the standard basis for R^n.

REMARK. In R^2 and R^3, the standard basis vectors are commonly denoted by $\mathbf{i}$, $\mathbf{j}$, and $\mathbf{k}$, rather than by $\mathbf{e}_1$, $\mathbf{e}_2$, and $\mathbf{e}_3$. We shall use both notations, depending on the particular situation.

EXAMPLE 3 Demonstrating That a Set of Vectors Is a Basis

Let $v_1 = (1, 2, 1)$, $v_2 = (2, 9, 0)$, and $v_3 = (3, 3, 4)$. Show that the set $S = \{v_1, v_2, v_3\}$ is a basis for R^3.

Solution.

To show that the set S spans R^3, we must show that an arbitrary vector $\mathbf{b} = (b_1, b_2, b_3)$ can be expressed as a linear combination

$$\mathbf{b} = c_1 \mathbf{v}_1 + c_2 \mathbf{v}_2 + c_3 \mathbf{v}_3$$

of the vectors in S. Expressing this equation in terms of components gives

$$(b_1, b_2, b_3) = c_1(1, 2, 1) + c_2(2, 9, 0) + c_3(3, 3, 4)$$

or

$$(b_1, b_2, b_3) = (c_1 + 2c_2 + 3c_3, \, 2c_1 + 9c_2 + 3c_3, \, c_1 + 4c_3)$$

or, on equating corresponding components,

$$\begin{aligned}
c_1 + 2c_2 + 3c_3 &= b_1 \\
2c_1 + 9c_2 + 3c_3 &= b_2 \\
c_1 \phantom{{}+ 9c_2} + 4c_3 &= b_3
\end{aligned} \tag{3}$$

Thus, to show that S spans R^3, we must demonstrate that system (3) has a solution for all choices of $\mathbf{b} = (b_1, b_2, b_3)$.

To prove that S is linearly independent, we must show that the only solution of

$$c_1 \mathbf{v}_1 + c_2 \mathbf{v}_2 + c_3 \mathbf{v}_3 = \mathbf{0} \tag{4}$$

is $c_1 = c_2 = c_3 = 0$. As above, if (4) is expressed in terms of components, the verification of independence reduces to showing that the homogeneous system

$$\begin{aligned}
c_1 + 2c_2 + 3c_3 &= 0 \\
2c_1 + 9c_2 + 3c_3 &= 0 \\
c_1 \phantom{{}+ 9c_2} + 4c_3 &= 0
\end{aligned} \tag{5}$$

has only the trivial solution. Observe that systems (3) and (5) have the same coefficient matrix. Thus, by parts (b), (e), and (g) of Theorem 4.3.4, we can simultaneously prove that S is linearly independent and spans R^3 by demonstrating that in systems (3) and (5) the matrix of coefficients has a nonzero determinant. From

$$A = \begin{bmatrix} 1 & 2 & 3 \\ 2 & 9 & 3 \\ 1 & 0 & 4 \end{bmatrix} \quad \text{we find} \quad \det(A) = \begin{vmatrix} 1 & 2 & 3 \\ 2 & 9 & 3 \\ 1 & 0 & 4 \end{vmatrix} = -1$$

and so S is a basis for R^3.   ◆

EXAMPLE 4 Representing a Vector Using Two Bases

Let $S = \{v_1, v_2, v_3\}$ be the basis for R^3 in the preceding example.

(a) Find the coordinate vector of $\mathbf{v} = (5, -1, 9)$ with respect to S.
(b) Find the vector $\mathbf{v}$ in R^3 whose coordinate vector with respect to the basis S is $(\mathbf{v})_S = (-1, 3, 2)$.

Solution (a). We must find scalars c_1, c_2, c_3 such that

$$\mathbf{v} = c_1\mathbf{v}_1 + c_2\mathbf{v}_2 + c_3\mathbf{v}_3$$

or, in terms of components,

$$(5, -1, 9) = c_1(1, 2, 1) + c_2(2, 9, 0) + c_3(3, 3, 4)$$

Equating corresponding components gives

$$
\begin{aligned}
c_1 + 2c_2 + 3c_3 &= 5 \\
2c_1 + 9c_2 + 3c_3 &= -1 \\
c_1 + 4c_3 &= 9
\end{aligned}
$$

Solving this system, we obtain $c_1 = 1$, $c_2 = -1$, $c_3 = 2$ (verify). Therefore,

$$(\mathbf{v})_S = (1, -1, 2)$$

Solution (b). Using the definition of the coordinate vector $(\mathbf{v})_S$, we obtain

$$
\begin{aligned}
\mathbf{v} &= (-1)\mathbf{v}_1 + 3\mathbf{v}_2 + 2\mathbf{v}_3 \\
&= (-1)(1, 2, 1) + 3(2, 9, 0) + 2(3, 3, 4) = (11, 31, 7) \qquad \blacklozenge
\end{aligned}
$$

EXAMPLE 5 Standard Basis for P_n

(a) Show that $S = \{1, x, x^2, \ldots, x^n\}$ is a basis for the vector space P_n of polynomials of the form $a_0 + a_1x + \cdots + a_nx^n$.
(b) Find the coordinate vector of the polynomial $\mathbf{p} = a_0 + a_1x + a_2x^2$ relative to the basis $S = \{1, x, x^2\}$ for P_2.

Solution (a). We showed that S spans P_n in Example 11 of Section 5.2, and we showed that S is a linearly independent set in Example 5 of Section 5.3. Thus, S is a basis for P_n; it is called the ***standard basis for P_n***.

Solution (b). The coordinates of $\mathbf{p} = a_0 + a_1x + a_2x^2$ are the scalar coefficients of the basis vectors 1, x, and x^2, so $(\mathbf{p})_S = (a_0, a_1, a_2)$. $\qquad \blacklozenge$

EXAMPLE 6 Standard Basis for M_{mn}

Let

$$
M_1 = \begin{bmatrix} 1 & 0 \\ 0 & 0 \end{bmatrix}, \qquad
M_2 = \begin{bmatrix} 0 & 1 \\ 0 & 0 \end{bmatrix}, \qquad
M_3 = \begin{bmatrix} 0 & 0 \\ 1 & 0 \end{bmatrix}, \qquad
M_4 = \begin{bmatrix} 0 & 0 \\ 0 & 1 \end{bmatrix}
$$

The set $S = \{M_1, M_2, M_3, M_4\}$ is a basis for the vector space M_{22} of 2×2 matrices. To see that S spans M_{22}, note that an arbitrary vector (matrix)

$$\begin{bmatrix} a & b \\ c & d \end{bmatrix}$$

can be written as

$$
\begin{bmatrix} a & b \\ c & d \end{bmatrix} = a\begin{bmatrix} 1 & 0 \\ 0 & 0 \end{bmatrix} + b\begin{bmatrix} 0 & 1 \\ 0 & 0 \end{bmatrix} + c\begin{bmatrix} 0 & 0 \\ 1 & 0 \end{bmatrix} + d\begin{bmatrix} 0 & 0 \\ 0 & 1 \end{bmatrix}
$$

$$= aM_1 + bM_2 + cM_3 + dM_4$$

To see that S is linearly independent, assume that

$$aM_1 + bM_2 + cM_3 + dM_4 = 0$$

That is,

$$a \begin{bmatrix} 1 & 0 \\ 0 & 0 \end{bmatrix} + b \begin{bmatrix} 0 & 1 \\ 0 & 0 \end{bmatrix} + c \begin{bmatrix} 0 & 0 \\ 1 & 0 \end{bmatrix} + d \begin{bmatrix} 0 & 0 \\ 0 & 1 \end{bmatrix} = \begin{bmatrix} 0 & 0 \\ 0 & 0 \end{bmatrix}$$

It follows that

$$\begin{bmatrix} a & b \\ c & d \end{bmatrix} = \begin{bmatrix} 0 & 0 \\ 0 & 0 \end{bmatrix}$$

Thus, $a = b = c = d = 0$, so S is linearly independent. The basis S in this example is called the *standard basis for* M_{22}. More generally, the **standard basis for** M_{mn} consists of the mn different matrices with a single 1 and zeros for the remaining entries. ◆

EXAMPLE 7 Basis for the Subspace span(S)

If $S = \{\mathbf{v}_1, \mathbf{v}_2, \dots, \mathbf{v}_r\}$ is a *linearly independent* set in a vector space V, then S is a basis for the subspace span(S) since the set S spans span(S) by definition of span(S). ◆

Definition

A nonzero vector space V is called **finite-dimensional** if it contains a finite set of vectors $\{\mathbf{v}_1, \mathbf{v}_2, \dots, \mathbf{v}_n\}$ that forms a basis. If no such set exists, V is called **infinite-dimensional**. In addition, we shall regard the zero vector space to be finite-dimensional.

EXAMPLE 8 Some Finite- and Infinite-Dimensional Spaces

By Examples 2, 5, and 6, the vector spaces R^n, P_n, and M_{mn} are finite-dimensional. The vector spaces $F(-\infty, \infty)$, $C(-\infty, \infty)$, $C^m(-\infty, \infty)$, and $C^\infty(-\infty, \infty)$ are infinite-dimensional (Exercise 23). ◆

The next theorem will provide the key to the concept of dimension.

Theorem 5.4.2

Let V be a finite-dimensional vector space and $\{\mathbf{v}_1, \mathbf{v}_2, \dots, \mathbf{v}_n\}$ any basis.

(a) If a set has more than n vectors, then it is linearly dependent.

(b) If a set has fewer than n vectors, then it does not span V.

Proof (a). Let $S' = \{\mathbf{w}_1, \mathbf{w}_2, \dots, \mathbf{w}_m\}$ be any set of m vectors in V, where $m > n$. We want to show that S' is linearly dependent. Since $S = \{\mathbf{v}_1, \mathbf{v}_2, \dots, \mathbf{v}_n\}$ is a basis, each $\mathbf{w}_i$ can be expressed as a linear combination of the vectors in S, say

$$\begin{aligned}
\mathbf{w}_1 &= a_{11}\mathbf{v}_1 + a_{21}\mathbf{v}_2 + \cdots + a_{n1}\mathbf{v}_n \\
\mathbf{w}_2 &= a_{12}\mathbf{v}_1 + a_{22}\mathbf{v}_2 + \cdots + a_{n2}\mathbf{v}_n \\
&\vdots \qquad \vdots \qquad \vdots \qquad\qquad \vdots \\
\mathbf{w}_m &= a_{1m}\mathbf{v}_1 + a_{2m}\mathbf{v}_2 + \cdots + a_{nm}\mathbf{v}_n
\end{aligned} \tag{6}$$

To show that S' is linearly dependent, we must find scalars $k_1, k_2, \dots, k_m$, not all zero, such that

$$k_1\mathbf{w}_1 + k_2\mathbf{w}_2 + \cdots + k_m\mathbf{w}_m = \mathbf{0} \tag{7}$$

Using the equations in (6), we can rewrite (7) as

$$(k_1 a_{11} + k_2 a_{12} + \cdots + k_m a_{1m})\mathbf{v}_1$$
$$+ (k_1 a_{21} + k_2 a_{22} + \cdots + k_m a_{2m})\mathbf{v}_2$$
$$\ddots$$
$$+ (k_1 a_{n1} + k_2 a_{n2} + \cdots + k_m a_{nm})\mathbf{v}_n = \mathbf{0}$$

Thus, from the linear independence of S, the problem of proving that S' is a linearly dependent set reduces to showing there are scalars $k_1, k_2, \ldots, k_m$, not all zero, that satisfy

$$
\begin{aligned}
a_{11}k_1 + a_{12}k_2 + \cdots + a_{1m}k_m &= 0 \\
a_{21}k_1 + a_{22}k_2 + \cdots + a_{2m}k_m &= 0 \\
\vdots \qquad \vdots \qquad\quad \vdots \qquad \vdots & \\
a_{n1}k_1 + a_{n2}k_2 + \cdots + a_{nm}k_m &= 0
\end{aligned}
\tag{8}
$$

But (8) has more unknowns than equations, so the proof is complete since Theorem 1.2.1 guarantees the existence of nontrivial solutions.

Proof (b). Let $S' = \{\mathbf{w}_1, \mathbf{w}_2, \ldots, \mathbf{w}_m\}$ be any set of m vectors in V, where $m < n$. We want to show that S' does not span V. The proof will be by contradiction: We will show that assuming S' spans V leads to a contradiction of the linear independence of $\{\mathbf{v}_1, \mathbf{v}_2, \ldots, \mathbf{v}_n\}$.

If S' spans V, then every vector in V is a linear combination of the vectors is S'. In particular, each basis vector $\mathbf{v}_i$ is a linear combination of the vectors in S', say

$$
\begin{aligned}
\mathbf{v}_1 &= a_{11}\mathbf{w}_1 + a_{21}\mathbf{w}_2 + \cdots + a_{m1}\mathbf{w}_m \\
\mathbf{v}_2 &= a_{12}\mathbf{w}_1 + a_{22}\mathbf{w}_2 + \cdots + a_{m2}\mathbf{w}_m \\
\vdots \qquad &\vdots \qquad\quad \vdots \qquad\qquad \vdots \\
\mathbf{v}_n &= a_{1n}\mathbf{w}_1 + a_{2n}\mathbf{w}_2 + \cdots + a_{mn}\mathbf{w}_m
\end{aligned}
\tag{9}
$$

To obtain our contradiction we will show that there are scalars $k_1, k_2, \ldots, k_n$, not all zero, such that

$$k_1\mathbf{v}_1 + k_2\mathbf{v}_2 + \cdots + k_n\mathbf{v}_n = \mathbf{0} \tag{10}$$

But observe that (9) and (10) have the same form as (6) and (7) except that m and n are interchanged and the $\mathbf{w}$'s and $\mathbf{v}$'s are interchanged. Thus, the computations that led to (8) now yield

$$
\begin{aligned}
a_{11}k_1 + a_{12}k_2 + \cdots + a_{1n}k_n &= 0 \\
a_{21}k_1 + a_{22}k_2 + \cdots + a_{2n}k_n &= 0 \\
\vdots \qquad\quad \vdots \qquad\qquad \vdots \qquad \vdots & \\
a_{m1}k_1 + a_{m2}k_2 + \cdots + a_{mn}k_n &= 0
\end{aligned}
$$

This linear system has more unknowns than equations, and hence has nontrivial solutions by Theorem 1.2.1. ∎

It follows from the preceding theorem that if $S = \{\mathbf{v}_1, \mathbf{v}_2, \ldots, \mathbf{v}_n\}$ is any basis for a vector space V, then all sets in V that simultaneously span V and are linearly independent must have precisely n vectors. Thus, all bases for V must have the same number of vectors as the arbitrary basis S. This yields the following result, which is one of the most important in linear algebra.

Theorem 5.4.3

All bases for a finite-dimensional vector space have the same number of vectors.

To see how this theorem relates to the concept of "dimension," recall that the standard basis for R^n has n vectors (Example 2). Thus, Theorem 5.4.3 implies that all bases for R^n have n vectors. In particular, every basis for R^3 has three vectors, every basis for R^2 has two vectors, and every basis for R^1 $(=R)$ has one vector. Intuitively, R^3 is three-dimensional, R^2 (a plane) is two-dimensional, and R (a line) is one-dimensional. Thus, for familiar vector spaces, the number of vectors in a basis is the same as the dimension. This suggests the following definition.

Definition

The *dimension* of a finite-dimensional vector space V, denoted by $\dim(V)$, is defined to be the number of vectors in a basis for V. In addition, we define the zero vector space to have dimension zero.

REMARK.   From here on we shall follow a common convention of regarding the empty set to be a basis for the zero vector space. This is consistent with the preceding definition, since the empty set has no vectors and the zero vector space has dimension zero.

EXAMPLE 9 Dimensions of Some Vector Spaces

$$\dim(R^n) = n \qquad \text{[The standard basis has } n \text{ vectors (Example 2).]}$$
$$\dim(P_n) = n + 1 \quad \text{[The standard basis has } n + 1 \text{ vectors (Example 5).]}$$
$$\dim(M_{mn}) = mn \quad \text{[The standard basis has } mn \text{ vectors (Example 6).]} \qquad \blacklozenge$$

EXAMPLE 10 Dimension of a Solution Space

Determine a basis for and the dimension of the solution space of the homogeneous system

$$
\begin{aligned}
2x_1 + 2x_2 - x_3 \quad\;\;\; + x_5 &= 0 \\
-x_1 - x_2 + 2x_3 - 3x_4 + x_5 &= 0 \\
x_1 + x_2 - 2x_3 \quad\;\;\; - x_5 &= 0 \\
x_3 + x_4 + x_5 &= 0
\end{aligned}
$$

Solution.

In Example 7 of Section 1.2 it was shown that the general solution of the given system is

$$x_1 = -s - t, \qquad x_2 = s, \qquad x_3 = -t, \qquad x_4 = 0, \qquad x_5 = t$$

Therefore, the solution vectors can be written as

$$
\begin{bmatrix} x_1 \\ x_2 \\ x_3 \\ x_4 \\ x_5 \end{bmatrix} =
\begin{bmatrix} -s - t \\ s \\ -t \\ 0 \\ t \end{bmatrix} =
\begin{bmatrix} -s \\ s \\ 0 \\ 0 \\ 0 \end{bmatrix} +
\begin{bmatrix} -t \\ 0 \\ -t \\ 0 \\ t \end{bmatrix} = s
\begin{bmatrix} -1 \\ 1 \\ 0 \\ 0 \\ 0 \end{bmatrix} + t
\begin{bmatrix} -1 \\ 0 \\ -1 \\ 0 \\ 1 \end{bmatrix}
$$

which shows that the vectors

$$\mathbf{v}_1 = \begin{bmatrix} -1 \\ 1 \\ 0 \\ 0 \\ 0 \end{bmatrix} \quad \text{and} \quad \mathbf{v}_2 = \begin{bmatrix} -1 \\ 0 \\ -1 \\ 0 \\ 1 \end{bmatrix}$$

span the solution space. Since they are also linearly independent (verify), $\{\mathbf{v}_1, \mathbf{v}_2\}$ is a basis, and the solution space is two-dimensional. ◆

Some Fundamental Theorems   We shall devote the remainder of this section to a series of theorems that reveal the subtle interrelationships among the concepts of spanning, linear independence, basis, and dimension. These theorems are not idle exercises in mathematical theory—they are essential to the understanding of vector spaces, and many practical applications of linear algebra build on them.

The following theorem, which we call the *Plus/Minus Theorem* (our own name), establishes two basic principles on which most of the theorems to follow will rely.

Theorem 5.4.4   **Plus/Minus Theorem**

Let S be a nonempty set of vectors in a vector space V.

(a) If S is a linearly independent set, and if **v** *is a vector in V that is outside of span(S), then the set S ∪ {**v**} that results by inserting* **v** *into S is still linearly independent.*

(b) If **v** *is a vector in S that is expressible as a linear combination of other vectors in S, and if S − {**v**} denotes the set obtained by removing* **v** *from S, then S and S − {**v**} span the same space; that is,*

$$span(S) = span(S - \{\mathbf{v}\})$$

We shall defer the proof to the end of the section, so that we may move more immediately to the consequences of the theorem. However, the theorem can be visualized in R^3 as follows:

(a) A set S of two linearly independent vectors in R^3 spans a plane through the origin. If we enlarge S by inserting any vector **v** outside of this plane (Figure 5.4.5a), then the resulting set of three vectors is still linearly independent since none of the three vectors lies in the same plane as the other two.

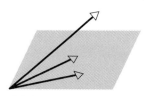  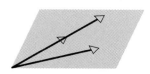

(a) None of the three vectors lies in the same plane as the other two.

(b) Any of the vectors can be removed, and the remaining two will still span the plane.

(c) Either of the collinear vectors can be removed, and the remaining two will still span the plane.

Figure 5.4.5

(b) If S is a set of three noncollinear vectors in R^3 that lie in a common plane through the origin (Figure 5.4.5b, c), then the three vectors span the plane. However, if we remove from S any vector $\mathbf{v}$ that is a linear combination of the other two, the remaining set of two vectors still spans the plane.

In general, to show that a set of vectors $\{\mathbf{v}_1, \mathbf{v}_2, \ldots, \mathbf{v}_n\}$ is a basis for a vector space V, we must show that the vectors are linearly independent and span V. However, if we happen to know that V has dimension n (so that $\{\mathbf{v}_1, \mathbf{v}_2, \ldots, \mathbf{v}_n\}$ contains the right number of vectors for a basis), then it suffices to check *either* linear independence *or* spanning—the remaining condition will hold automatically. This is the content of the following theorem.

Theorem 5.4.5

If V is an n-dimensional vector space, and if S is a set in V with exactly n vectors, then S is a basis for V if either S spans V or S is linearly independent.

Proof. Assume that S has exactly n vectors and spans V. To prove that S is a basis we must show that S is a linearly independent set. But if this is not so, then some vector $\mathbf{v}$ in S is a linear combination of the remaining vectors. If we remove this vector from S, then it follows from the Plus/Minus Theorem (5.4.4b) that the remaining set of $n - 1$ vectors still spans V. But this is impossible, since it follows from Theorem 5.4.2b that no set with fewer than n vectors can span an n-dimensional vector space. Thus, S is linearly independent.

Assume that S has exactly n vectors and is a linearly independent set. To prove that S is a basis we must show that S spans V. But if this is not so, then there is some vector $\mathbf{v}$ in V that is not in span(S). If we insert this vector into S, then it follows from the Plus/Minus Theorem (5.4.4a) that this set of $n + 1$ vectors is still linearly independent. But this is impossible, since it follows from Theorem 5.4.2a that no set with more than n vectors in an n-dimensional vector space can be linearly independent. Thus, S spans V. ∎

EXAMPLE 11 Checking for a Basis

(a) Show that $\mathbf{v}_1 = (-3, 7)$ and $\mathbf{v}_2 = (5, 5)$ form a basis for R^2 by inspection.
(b) Show that $\mathbf{v}_1 = (2, 0, -1)$, $\mathbf{v}_2 = (4, 0, 7)$, $\mathbf{v}_3 = (-1, 1, 4)$ form a basis for R^3 by inspection.

Solution (a). Since neither vector is a scalar multiple of the other, the two vectors form a linearly independent set in the two-dimensional space R^2, and hence form a basis by Theorem 5.4.5.

Solution (b). The vectors $\mathbf{v}_1$ and $\mathbf{v}_2$ form a linearly independent set in the xz-plane (why?). The vector $\mathbf{v}_3$ is outside of the xz-plane, so the set $\{\mathbf{v}_1, \mathbf{v}_2, \mathbf{v}_3\}$ is also linearly independent. Since R^3 is three-dimensional, Theorem 5.4.5 implies that $\{\mathbf{v}_1, \mathbf{v}_2, \mathbf{v}_3\}$ is a basis for R^3. ◆

The following theorem shows that for a finite-dimensional vector space V every set that spans V contains a basis for V within it, and every linearly independent set in V is part of some basis for V.

Theorem 5.4.6

Let S be a finite set of vectors in a finite-dimensional vector space V.

(*a*) *If S spans V but is not a basis for V, then S can be reduced to a basis for V by removing appropriate vectors from S.*

(*b*) *If S is a linearly independent set that is not already a basis for V, then S can be enlarged to a basis for V by inserting appropriate vectors into S.*

Proof (a). If S is a set of vectors that spans V but is not a basis for V, then S is a linearly dependent set. Thus, some vector **v** in S is expressible as a linear combination of the other vectors in S. By the Plus/Minus Theorem (5.4.4*b*), we can remove **v** from S, and the resulting set S' will still span V. If S' is linearly independent, then S' is a basis for V, and we are done. If S' is linearly dependent, then we can remove some appropriate vector from S' to produce a set S'' that still spans V. We can continue removing vectors in this way until we finally arrive at a set of vectors in S that is linearly independent and spans V. This subset of S is a basis for V.

Proof (b). Suppose that $\dim(V) = n$. If S is a linearly independent set that is not already a basis for V, then S fails to span V, and there is some vector **v** in V that is not in span(S). By the Plus/Minus Theorem (5.4.4*a*), we can insert **v** into S, and the resulting set S' will still be linearly independent. If S' spans V, then S' is a basis for V, and we are done. If S' does not span V, then we can insert an appropriate vector into S' to produce a set S'' that is still linearly independent. We can continue inserting vectors in this way until we reach a set with n linearly independent vectors in V. This set will be a basis for V by Theorem 5.4.5. ∎

In the next section we shall give numerical examples that illustrate this theorem.

It can be proved (Exercise 29) that any subspace of a finite-dimensional vector space is finite-dimensional. We conclude this section with a theorem which shows that the dimension of a subspace of a finite-dimensional vector space V cannot exceed the dimension of V itself and that the only way a subspace can have the same dimension as V is if the subspace is the entire vector space V. Figure 5.4.6 illustrates this idea in R^3. In that figure observe that successively larger subspaces increase in dimension.

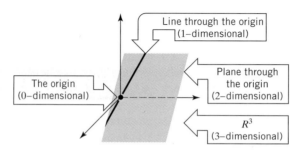

Figure 5.4.6

Theorem 5.4.7

If W is a subspace of a finite-dimensional vector space V, then $\dim(W) \le \dim(V)$; *moreover, if* $\dim(W) = \dim(V)$, *then* $W = V$.

Proof. Since V is finite-dimensional, so is W by Exercise 29. Accordingly, suppose that $S = \{\mathbf{w}_1, \mathbf{w}_2, \ldots, \mathbf{w}_m\}$ is a basis for W. Either S is also a basis for V or it is not. If it is, then $\dim(W) = \dim(V) = m$. If it is not, then by Theorem 5.4.6*b*, vectors can be added to the linearly independent set S to make it into a basis for V, so $\dim(W) < \dim(V)$. Thus, $\dim(W) \leq \dim(V)$ in all cases. If $\dim(W) = \dim(V)$, then S is a set of m linearly independent vectors in the m-dimensional vector space V; hence, S is a basis for V by Theorem 5.4.5. This implies that $W = V$ (why?). ■

Additional Proofs

Proof of Theorem 5.4.4a. Assume that $S = \{\mathbf{v}_1, \mathbf{v}_2, \ldots, \mathbf{v}_r\}$ is a linearly independent set of vectors in V, and $\mathbf{v}$ is a vector in V outside of span(S). To show that $S' = \{\mathbf{v}_1, \mathbf{v}_2, \ldots, \mathbf{v}_r, \mathbf{v}\}$ is a linearly independent set, we must show that the only scalars that satisfy

$$k_1\mathbf{v}_1 + k_2\mathbf{v}_2 + \cdots + k_r\mathbf{v}_r + k_{r+1}\mathbf{v} = \mathbf{0} \tag{11}$$

are $k_1 = k_2 = \cdots = k_r = k_{r+1} = 0$. But we must have $k_{r+1} = 0$; otherwise, we could solve (11) for $\mathbf{v}$ as a linear combination of $\mathbf{v}_1, \mathbf{v}_2, \ldots, \mathbf{v}_r$, contradicting the assumption that $\mathbf{v}$ is outside of span(S). Thus, (11) simplifies to

$$k_1\mathbf{v}_1 + k_2\mathbf{v}_2 + \cdots + k_r\mathbf{v}_r = \mathbf{0} \tag{12}$$

which, by the linear independence of $\{\mathbf{v}_1, \mathbf{v}_2, \ldots, \mathbf{v}_r\}$, implies that

$$k_1 = k_2 = \cdots = k_r = 0$$

Proof of Theorem 5.4.4b. Assume that $S = \{\mathbf{v}_1, \mathbf{v}_2, \ldots, \mathbf{v}_r\}$ is a set of vectors in V, and to be specific suppose that $\mathbf{v}_r$ is a linear combination of $\mathbf{v}_1, \mathbf{v}_2, \ldots, \mathbf{v}_{r-1}$, say

$$\mathbf{v}_r = c_1\mathbf{v}_1 + c_2\mathbf{v}_2 + \cdots + c_{r-1}\mathbf{v}_{r-1} \tag{13}$$

We want to show that if $\mathbf{v}_r$ is removed from S, then the remaining set of vectors $\{\mathbf{v}_1, \mathbf{v}_2, \ldots, \mathbf{v}_{r-1}\}$ still spans span(S); that is, we must show that every vector $\mathbf{w}$ in span(S) is expressible as a linear combination of $\{\mathbf{v}_1, \mathbf{v}_2, \ldots, \mathbf{v}_{r-1}\}$. But if $\mathbf{w}$ is in span(S), then $\mathbf{w}$ is expressible in the form

$$\mathbf{w} = k_1\mathbf{v}_1 + k_2\mathbf{v}_2 + \cdots + k_{r-1}\mathbf{v}_{r-1} + k_r\mathbf{v}_r$$

or on substituting (13),

$$\mathbf{w} = k_1\mathbf{v}_1 + k_2\mathbf{v}_2 + \cdots + k_{r-1}\mathbf{v}_{r-1} + k_r(c_1\mathbf{v}_1 + c_2\mathbf{v}_2 + \cdots + c_{r-1}\mathbf{v}_{r-1})$$

which expresses $\mathbf{w}$ as a linear combination of $\mathbf{v}_1, \mathbf{v}_2, \ldots, \mathbf{v}_{r-1}$. ■

Exercise Set 5.4

1. Explain why the following sets of vectors are *not* bases for the indicated vector spaces. (Solve this problem by inspection.)

(a) $\mathbf{u}_1 = (1, 2)$, $\mathbf{u}_2 = (0, 3)$, $\mathbf{u}_3 = (2, 7)$ for R^2

(b) $\mathbf{u}_1 = (-1, 3, 2)$, $\mathbf{u}_2 = (6, 1, 1)$ for R^3

(c) $\mathbf{p}_1 = 1 + x + x^2$, $\mathbf{p}_2 = x - 1$ for P_2

(d) $A = \begin{bmatrix} 1 & 1 \\ 2 & 3 \end{bmatrix}$, $B = \begin{bmatrix} 6 & 0 \\ -1 & 4 \end{bmatrix}$, $C = \begin{bmatrix} 3 & 0 \\ 1 & 7 \end{bmatrix}$, $D = \begin{bmatrix} 5 & 1 \\ 4 & 2 \end{bmatrix}$, $E = \begin{bmatrix} 7 & 1 \\ 2 & 9 \end{bmatrix}$ for M_{22}

2. Which of the following sets of vectors are bases for R^2?

(a) $(2, 1), (3, 0)$ (b) $(4, 1), (-7, -8)$ (c) $(0, 0), (1, 3)$ (d) $(3, 9), (-4, -12)$

3. Which of the following sets of vectors are bases for R^3?
   - (a) $(1, 0, 0), (2, 2, 0), (3, 3, 3)$
   - (b) $(3, 1, -4), (2, 5, 6), (1, 4, 8)$
   - (c) $(2, -3, 1), (4, 1, 1), (0, -7, 1)$
   - (d) $(1, 6, 4), (2, 4, -1), (-1, 2, 5)$

4. Which of the following sets of vectors are bases for P_2?
   - (a) $1 - 3x + 2x^2, 1 + x + 4x^2, 1 - 7x$
   - (b) $4 + 6x + x^2, -1 + 4x + 2x^2, 5 + 2x - x^2$
   - (c) $1 + x + x^2, x + x^2, x^2$
   - (d) $-4 + x + 3x^2, 6 + 5x + 2x^2, 8 + 4x + x^2$

5. Show that the following set of vectors is a basis for M_{22}.

   $$\begin{bmatrix} 3 & 6 \\ 3 & -6 \end{bmatrix}, \quad \begin{bmatrix} 0 & -1 \\ -1 & 0 \end{bmatrix}, \quad \begin{bmatrix} 0 & -8 \\ -12 & -4 \end{bmatrix}, \quad \begin{bmatrix} 1 & 0 \\ -1 & 2 \end{bmatrix}$$

6. Let V be the space spanned by $v_1 = \cos^2 x$, $v_2 = \sin^2 x$, $v_3 = \cos 2x$.
   - (a) Show that $S = \{v_1, v_2, v_3\}$ is not a basis for V.
   - (b) Find a basis for V.

7. Find the coordinate vector of w relative to the basis $S = \{u_1, u_2\}$ for R^2.
   - (a) $u_1 = (1, 0)$, $u_2 = (0, 1)$; $w = (3, -7)$
   - (b) $u_1 = (2, -4)$, $u_2 = (3, 8)$; $w = (1, 1)$
   - (c) $u_1 = (1, 1)$, $u_2 = (0, 2)$; $w = (a, b)$

8. Find the coordinate vector of v relative to the basis $S = \{v_1, v_2, v_3\}$.
   - (a) $v = (2, -1, 3)$; $v_1 = (1, 0, 0)$, $v_2 = (2, 2, 0)$, $v_3 = (3, 3, 3)$
   - (b) $v = (5, -12, 3)$; $v_1 = (1, 2, 3)$, $v_2 = (-4, 5, 6)$, $v_3 = (7, -8, 9)$

9. Find the coordinate vector of p relative to the basis $S = \{p_1, p_2, p_3\}$.
   - (a) $p = 4 - 3x + x^2$; $p_1 = 1$, $p_2 = x$, $p_3 = x^2$
   - (b) $p = 2 - x + x^2$; $p_1 = 1 + x$, $p_2 = 1 + x^2$, $p_3 = x + x^2$

10. Find the coordinate vector of A relative to the basis $S = \{A_1, A_2, A_3, A_4\}$.

    $$A = \begin{bmatrix} 2 & 0 \\ -1 & 3 \end{bmatrix}; \quad A_1 = \begin{bmatrix} -1 & 1 \\ 0 & 0 \end{bmatrix}, \quad A_2 = \begin{bmatrix} 1 & 1 \\ 0 & 0 \end{bmatrix}, \quad A_3 = \begin{bmatrix} 0 & 0 \\ 1 & 0 \end{bmatrix}, \quad A_4 = \begin{bmatrix} 0 & 0 \\ 0 & 1 \end{bmatrix}$$

In Exercises 11–16 determine the dimension of and a basis for the solution space of the system.

11. $\begin{aligned} x_1 + x_2 - x_3 &= 0 \\ -2x_1 - x_2 + 2x_3 &= 0 \\ -x_1 \qquad + x_3 &= 0 \end{aligned}$

12. $\begin{aligned} 3x_1 + x_2 + x_3 + x_4 &= 0 \\ 5x_1 - x_2 + x_3 - x_4 &= 0 \end{aligned}$

13. $\begin{aligned} x_1 - 4x_2 + 3x_3 - x_4 &= 0 \\ 2x_1 - 8x_2 + 6x_3 - 2x_4 &= 0 \end{aligned}$

14. $\begin{aligned} x_1 - 3x_2 + x_3 &= 0 \\ 2x_1 - 6x_2 + 2x_3 &= 0 \\ 3x_1 - 9x_2 + 3x_3 &= 0 \end{aligned}$

15. $\begin{aligned} 2x_1 + x_2 + 3x_3 &= 0 \\ x_1 \qquad + 5x_3 &= 0 \\ x_2 + x_3 &= 0 \end{aligned}$

16. $\begin{aligned} x + y + z &= 0 \\ 3x + 2y - 2z &= 0 \\ 4x + 3y - z &= 0 \\ 6x + 5y + z &= 0 \end{aligned}$

17. Determine bases for the following subspaces of R^3.
    - (a) the plane $3x - 2y + 5z = 0$
    - (b) the plane $x - y = 0$
    - (c) the line $x = 2t, y = -t, z = 4t$
    - (d) all vectors of the form (a, b, c), where $b = a + c$

18. Determine the dimensions of the following subspaces of R^4.

    - (a) all vectors of the form $(a, b, c, 0)$
    - (b) all vectors of the form (a, b, c, d), where $d = a + b$ and $c = a - b$
    - (c) all vectors of the form (a, b, c, d), where $a = b = c = d$

19. Determine the dimension of the subspace of P_3 consisting of all polynomials $a_0 + a_1x + a_2x^2 + a_3x^3$ for which $a_0 = 0$.

20. Find a standard basis vector that can be added to the set $\{v_1, v_2\}$ to produce a basis for R^3.
    - (a) $v_1 = (-1, 2, 3)$, $v_2 = (1, -2, -2)$
    - (b) $v_1 = (1, -1, 0)$, $v_2 = (3, 1, -2)$

21. Find standard basis vectors that can be added to the set $\{v_1, v_2\}$ to produce a basis for R^4.

    $$v_1 = (1, -4, 2, -3), \qquad v_2 = (-3, 8, -4, 6)$$

22. Let $\{\mathbf{v}_1, \mathbf{v}_2, \mathbf{v}_3\}$ be a basis for a vector space V. Show that $\{\mathbf{u}_1, \mathbf{u}_2, \mathbf{u}_3\}$ is also a basis, where $\mathbf{u}_1 = \mathbf{v}_1$, $\mathbf{u}_2 = \mathbf{v}_1 + \mathbf{v}_2$, and $\mathbf{u}_3 = \mathbf{v}_1 + \mathbf{v}_2 + \mathbf{v}_3$.

23. (a) Show that for every positive integer n, one can find $n + 1$ linearly independent vectors in $F(-\infty, \infty)$. [***Hint.*** Look for polynomials.]
   (b) Use the result in part (a) to prove that $F(-\infty, \infty)$ is infinite-dimensional.
   (c) Prove that $C(-\infty, \infty)$, $C^m(-\infty, \infty)$, and $C^\infty(-\infty, \infty)$ are infinite-dimensional vector spaces.

24. Let S be a basis for an n-dimensional vector space V. Show that if $\mathbf{v}_1, \mathbf{v}_2, \ldots, \mathbf{v}_r$ form a linearly independent set of vectors in V, then the coordinate vectors $(\mathbf{v}_1)_S, (\mathbf{v}_2)_S, \ldots, (\mathbf{v}_r)_S$ form a linearly independent set in R^n, and conversely.

25. Using the notation from Exercise 24, show that if $\mathbf{v}_1, \mathbf{v}_2, \ldots, \mathbf{v}_r$ span V, then the coordinate vectors $(\mathbf{v}_1)_S, (\mathbf{v}_2)_S, \ldots, (\mathbf{v}_r)_S$ span R^n, and conversely.

26. Find a basis for the subspace of P_2 spanned by the given vectors.
   (a) $-1 + x - 2x^2$, $3 + 3x + 6x^2$, 9 (b) $1 + x$, x^2, $-2 + 2x^2$, $-3x$
   (c) $1 + x - 3x^2$, $2 + 2x - 6x^2$, $3 + 3x - 9x^2$
   [***Hint.*** Let S be the standard basis for P_2 and work with the coordinate vectors relative to S; note Exercises 24, 25.]

27. The accompanying figure shows a rectangular xy-coordinate system and an $x'y'$-coordinate system with skewed axes. Assuming that 1-unit scales are used on all the axes, find the $x'y'$-coordinates of the points whose xy-coordinates are given.
   (a) $(1, 1)$ (b) $(1, 0)$ (c) $(0, 1)$ (d) (a, b)

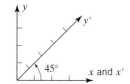

Figure Ex-27

28. The accompanying figure shows a rectangular xy-coordinate system determined by the unit basis vectors $\mathbf{i}$ and $\mathbf{j}$ and an $x'y'$-coordinate system determined by unit basis vectors $\mathbf{u}_1$ and $\mathbf{u}_2$. Find the $x'y'$-coordinates of the points whose xy-coordinates are given.
   (a) $(\sqrt{3}, 1)$ (b) $(1, 0)$ (c) $(0, 1)$ (d) (a, b)

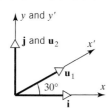

Figure Ex-28

29. Prove: Any subspace of a finite-dimensional vector space is finite-dimensional.

Discussion and Discovery

30. The basis that we gave for M_{22} in Example 6 consisted of noninvertible matrices. Do you think that there is a basis for M_{22} consisting of invertible matrices? Justify your answer.

31. (a) The vector space of all diagonal $n \times n$ matrices has dimension _____.
   (b) The vector space of all symmetric $n \times n$ matrices has dimension _____.
   (c) The vector space of all upper triangular $n \times n$ matrices has dimension _____.

32. (a) For a 3×3 matrix A, explain in words why the set $I_3, A, A^2, \ldots, A^9$ must be linearly dependent if the ten matrices are distinct.
   (b) State a corresponding result for an $n \times n$ matrix A.

33. State the two parts of Theorem 5.4.2 in contrapositive form. [See Exercise 34 of Section 1.4.]

34. (a) The equation $x_1 + x_2 + \cdots + x_n = 0$ can be viewed as a linear system of one equation in n unknowns. Make a conjecture about the dimension of its solution space.

   (b) Confirm your conjecture by finding a basis.

35. (a) Show that the set W of polynomials in P_2 such that $p(1) = 0$ is a subspace of P_2.

   (b) Make a conjecture about the dimension of W.

   (c) Confirm your conjecture by finding a basis for W.

5.5 ROW SPACE, COLUMN, SPACE, AND NULLSPACE

In this section we shall study three important vector spaces that are associated with matrices. Our work here will provide us with a deeper understanding of the relationships between the solutions of a linear system of equations and properties of its coefficient matrix.

We begin with some definitions.

Definition

For an $m \times n$ matrix

$$A = \begin{bmatrix} a_{11} & a_{12} & \cdots & a_{1n} \\ a_{21} & a_{22} & \cdots & a_{2n} \\ \vdots & \vdots & & \vdots \\ a_{m1} & a_{m2} & \cdots & a_{mn} \end{bmatrix}$$

the vectors

$$\mathbf{r}_1 = [a_{11} \quad a_{12} \quad \cdots \quad a_{1n}]$$
$$\mathbf{r}_2 = [a_{21} \quad a_{22} \quad \cdots \quad a_{2n}]$$
$$\vdots$$
$$\mathbf{r}_m = [a_{m1} \quad a_{m2} \quad \cdots \quad a_{mn}]$$

in R^n formed from the rows of A are called the **row vectors** of A, and the vectors

$$\mathbf{c}_1 = \begin{bmatrix} a_{11} \\ a_{21} \\ \vdots \\ a_{m1} \end{bmatrix}, \quad \mathbf{c}_2 = \begin{bmatrix} a_{12} \\ a_{22} \\ \vdots \\ a_{m2} \end{bmatrix}, \dots, \quad \mathbf{c}_n = \begin{bmatrix} a_{1n} \\ a_{2n} \\ \vdots \\ a_{mn} \end{bmatrix}$$

in R^m formed from the columns of A are called the **column vectors** of A.

EXAMPLE 1 Row and Column Vectors in a 2 × 3 Matrix

Let

$$A = \begin{bmatrix} 2 & 1 & 0 \\ 3 & -1 & 4 \end{bmatrix}$$

The row vectors of A are

$$\mathbf{r}_1 = [2 \quad 1 \quad 0] \quad \text{and} \quad \mathbf{r}_2 = [3 \quad -1 \quad 4]$$

and the column vectors of A are

$$\mathbf{c}_1 = \begin{bmatrix} 2 \\ 3 \end{bmatrix}, \quad \mathbf{c}_2 = \begin{bmatrix} 1 \\ -1 \end{bmatrix}, \quad \text{and} \quad \mathbf{c}_3 = \begin{bmatrix} 0 \\ 4 \end{bmatrix} \qquad \blacklozenge$$

The following definition defines three important vector spaces associated with a matrix.

Definition

If A is an $m \times n$ matrix, then the subspace of R^n spanned by the row vectors of A is called the *row space* of A, and the subspace of R^m spanned by the column vectors is called the *column space* of A. The solution space of the homogeneous system of equations $A\mathbf{x} = \mathbf{0}$, which is a subspace of R^n, is called the *nullspace* of A.

In this section and the next we shall be concerned with the following two general questions:

• What relationships exist between the solutions of a linear system $A\mathbf{x} = \mathbf{b}$ and the row space, column space, and nullspace of the coefficient matrix A?

• What relationships exist between the row space, column space, and nullspace of a matrix?

To investigate the first of these questions, suppose that

$$A = \begin{bmatrix} a_{11} & a_{12} & \cdots & a_{1n} \\ a_{21} & a_{22} & \cdots & a_{2n} \\ \vdots & \vdots & & \vdots \\ a_{m1} & a_{m2} & \cdots & a_{mn} \end{bmatrix} \quad \text{and} \quad \mathbf{x} = \begin{bmatrix} x_1 \\ x_2 \\ \vdots \\ x_n \end{bmatrix}$$

It follows from Formula (10) of Section 1.3 that if $\mathbf{c}_1, \mathbf{c}_2, \ldots, \mathbf{c}_n$ denote the column vectors of A, then the product $A\mathbf{x}$ can be expressed as a linear combination of these column vectors with coefficients from $\mathbf{x}$; that is,

$$A\mathbf{x} = x_1\mathbf{c}_1 + x_2\mathbf{c}_2 + \cdots + x_n\mathbf{c}_n \tag{1}$$

Thus, a linear system, $A\mathbf{x} = \mathbf{b}$, of m equations in n unknowns can be written as

$$x_1\mathbf{c}_1 + x_2\mathbf{c}_2 + \cdots + x_n\mathbf{c}_n = \mathbf{b} \tag{2}$$

from which we conclude that $A\mathbf{x} = \mathbf{b}$ is consistent if and only if $\mathbf{b}$ is expressible as a linear combination of the column vectors of A or, equivalently, if and only if $\mathbf{b}$ is in the column space of A. This yields the following theorem.

Theorem 5.5.1

A system of linear equations $A\mathbf{x} = \mathbf{b}$ is consistent if and only if $\mathbf{b}$ is in the column space of A.

EXAMPLE 2 A Vector b in the Column Space of A

Let $A\mathbf{x} = \mathbf{b}$ be the linear system

$$\begin{bmatrix} -1 & 3 & 2 \\ 1 & 2 & -3 \\ 2 & 1 & -2 \end{bmatrix} \begin{bmatrix} x_1 \\ x_2 \\ x_3 \end{bmatrix} = \begin{bmatrix} 1 \\ -9 \\ -3 \end{bmatrix}$$

Show that $\mathbf{b}$ is in the column space of A, and express $\mathbf{b}$ as a linear combination of the column vectors of A.

Solution.

Solving the system by Gaussian elimination yields (verify)

$$x_1 = 2, \qquad x_2 = -1, \qquad x_3 = 3$$

Since the system is consistent, $\mathbf{b}$ is in the column space of A. Moreover, from (2) and the solution obtained, it follows that

$$2 \begin{bmatrix} -1 \\ 1 \\ 2 \end{bmatrix} - \begin{bmatrix} 3 \\ 2 \\ 1 \end{bmatrix} + 3 \begin{bmatrix} 2 \\ -3 \\ -2 \end{bmatrix} = \begin{bmatrix} 1 \\ -9 \\ -3 \end{bmatrix} \qquad \blacklozenge$$

The next theorem establishes a fundamental relationship between the solutions of a nonhomogeneous linear system $A\mathbf{x} = \mathbf{b}$ and those of the corresponding homogeneous linear system $A\mathbf{x} = \mathbf{0}$ with the same coefficient matrix.

Theorem 5.5.2

If $\mathbf{x}_0$ denotes any single solution of a consistent linear system $A\mathbf{x} = \mathbf{b}$, and if $\mathbf{v}_1, \mathbf{v}_2, \ldots, \mathbf{v}_k$ form a basis for the nullspace of A, that is, the solution space of the homogeneous system $A\mathbf{x} = \mathbf{0}$, then every solution of $A\mathbf{x} = \mathbf{b}$ can be expressed in the form

$$\mathbf{x} = \mathbf{x}_0 + c_1\mathbf{v}_1 + c_2\mathbf{v}_2 + \cdots + c_k\mathbf{v}_k \tag{3}$$

and, conversely, for all choices of scalars $c_1, c_2, \ldots, c_k$, the vector $\mathbf{x}$ in this formula is a solution of $A\mathbf{x} = \mathbf{b}$.

Proof. Assume that $\mathbf{x}_0$ is any fixed solution of $A\mathbf{x} = \mathbf{b}$ and that $\mathbf{x}$ is an arbitrary solution. Then

$$A\mathbf{x}_0 = \mathbf{b} \quad \text{and} \quad A\mathbf{x} = \mathbf{b}$$

Subtracting these equations yields

$$A\mathbf{x} - A\mathbf{x}_0 = \mathbf{0} \quad \text{or} \quad A(\mathbf{x} - \mathbf{x}_0) = \mathbf{0}$$

which shows that $\mathbf{x} - \mathbf{x}_0$ is a solution of the homogeneous system $A\mathbf{x} = \mathbf{0}$. Since $\mathbf{v}_1, \mathbf{v}_2, \ldots, \mathbf{v}_k$ is a basis for the solution space of this system, we can express $\mathbf{x} - \mathbf{x}_0$ as a linear combination of these vectors, say

$$\mathbf{x} - \mathbf{x}_0 = c_1\mathbf{v}_1 + c_2\mathbf{v}_2 + \cdots + c_k\mathbf{v}_k$$

Thus,

$$\mathbf{x} = \mathbf{x}_0 + c_1\mathbf{v}_1 + c_2\mathbf{v}_2 + \cdots + c_k\mathbf{v}_k$$

which proves the first part of the theorem. Conversely, for all choices of the scalars $c_1, c_2, \ldots, c_k$ in (3) we have

$$A\mathbf{x} = A(\mathbf{x}_0 + c_1\mathbf{v}_1 + c_2\mathbf{v}_2 + \cdots + c_k\mathbf{v}_k)$$

or

$$A\mathbf{x} = A\mathbf{x}_0 + c_1(A\mathbf{v}_1) + c_2(A\mathbf{v}_2) + \cdots + c_k(A\mathbf{v}_k)$$

But $\mathbf{x}_0$ is a solution of the nonhomogeneous system and $\mathbf{v}_1, \mathbf{v}_2, \ldots, \mathbf{v}_k$ are solutions of the homogeneous system, so the last equation implies that

$$A\mathbf{x} = \mathbf{b} + \mathbf{0} + \mathbf{0} + \cdots + \mathbf{0} = \mathbf{b}$$

which shows that $\mathbf{x}$ is a solution of $A\mathbf{x} = \mathbf{b}$. ■

General and Particular Solutions

There is some terminology associated with Formula (3). The vector $\mathbf{x}_0$ is called a ***particular solution*** of $A\mathbf{x} = \mathbf{b}$. The expression $\mathbf{x}_0 + c_1\mathbf{v}_1 + c_2\mathbf{v}_2 + \cdots + c_k\mathbf{v}_k$ is called the ***general solution*** of $A\mathbf{x} = \mathbf{b}$, and the expression $c_1\mathbf{v}_1 + c_2\mathbf{v}_2 + \cdots + c_k\mathbf{v}_k$ is called the ***general solution*** of $A\mathbf{x} = \mathbf{0}$. With this terminology, Formula (3) states, *the general solution of $A\mathbf{x} = \mathbf{b}$ is the sum of any particular solution of $A\mathbf{x} = \mathbf{b}$ and the general solution of $A\mathbf{x} = \mathbf{0}$.*

For linear systems with two or three unknowns, Theorem 5.5.2 has a nice geometric interpretation in R^2 and R^3. For example, consider the case where $A\mathbf{x} = \mathbf{0}$ and $A\mathbf{x} = \mathbf{b}$ are linear systems with two unknowns. The solutions of $A\mathbf{x} = \mathbf{0}$ form a subspace of R^2, and hence constitute a line through the origin, the origin only, or all of R^2. From Theorem 5.5.2, the solutions of $A\mathbf{x} = \mathbf{b}$ can be obtained by adding any particular solution of $A\mathbf{x} = \mathbf{b}$, say $\mathbf{x}_0$, to the solutions of $A\mathbf{x} = \mathbf{0}$. Assuming that $\mathbf{x}_0$ is positioned with its initial point at the origin, this has the geometric effect of translating the solution space of $A\mathbf{x} = \mathbf{0}$ so that the point at the origin is moved to the tip of $\mathbf{x}_0$ (Figure 5.5.1). This means that the solution vectors of $A\mathbf{x} = \mathbf{b}$ form a line through the tip of $\mathbf{x}_0$, the point at the tip of $\mathbf{x}_0$, or all of R^2. (Can you visualize the last case?) Similarly, for linear systems with three unknowns, the solutions of $A\mathbf{x} = \mathbf{b}$ constitute a plane through the tip of any particular solution $\mathbf{x}_0$, a line through the tip of $\mathbf{x}_0$, the point at the tip of $\mathbf{x}_0$, or all of R^3.

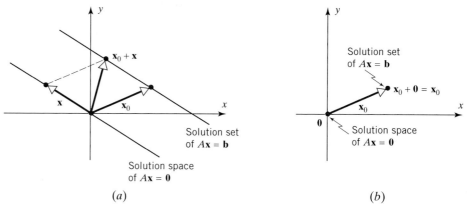

(*a*)   (*b*)

Figure 5.5.1 Adding $\mathbf{x}_0$ to each vector $\mathbf{x}$ in the solution space of $A\mathbf{x} = \mathbf{0}$ translates the solution space.

EXAMPLE 3 **General Solution of a Linear System $A\mathbf{x} = \mathbf{b}$**

In Example 4 of Section 1.2 we solved the nonhomogeneous linear system

$$
\begin{array}{rcl}
x_1 + 3x_2 - 2x_3 \phantom{{}-5x_3} + 2x_5 \phantom{{}+18x_6} &=& 0 \\
2x_1 + 6x_2 - 5x_3 - 2x_4 + 4x_5 - 3x_6 &=& -1 \\
5x_3 + 10x_4 \phantom{{}+4x_5} + 15x_6 &=& 5 \\
2x_1 + 6x_2 \phantom{{}-5x_3} + 8x_4 + 4x_5 + 18x_6 &=& 6
\end{array}
\tag{4}
$$

and obtained

$$
x_1 = -3r - 4s - 2t, \quad x_2 = r, \quad x_3 = -2s, \quad x_4 = s, \quad x_5 = t, \quad x_6 = \tfrac{1}{3}
$$

This result can be written in vector form as

$$
\begin{bmatrix} x_1 \\ x_2 \\ x_3 \\ x_4 \\ x_5 \\ x_6 \end{bmatrix}
=
\begin{bmatrix} -3r - 4s - 2t \\ r \\ -2s \\ s \\ t \\ \tfrac{1}{3} \end{bmatrix}
=
\underbrace{\begin{bmatrix} 0 \\ 0 \\ 0 \\ 0 \\ 0 \\ \tfrac{1}{3} \end{bmatrix}}_{\mathbf{x}_0}
+
\underbrace{r\begin{bmatrix} -3 \\ 1 \\ 0 \\ 0 \\ 0 \\ 0 \end{bmatrix} + s\begin{bmatrix} -4 \\ 0 \\ -2 \\ 1 \\ 0 \\ 0 \end{bmatrix} + t\begin{bmatrix} -2 \\ 0 \\ 0 \\ 0 \\ 1 \\ 0 \end{bmatrix}}_{\mathbf{x}}
\tag{5}
$$

which is the general solution of (4). The vector $\mathbf{x}_0$ in (5) is a particular solution of (4); the linear combination $\mathbf{x}$ in (5) is the general solution of the homogeneous system

$$
\begin{array}{rcl}
x_1 + 3x_2 - 2x_3 \phantom{{}-5x_3} + 2x_5 \phantom{{}+18x_6} &=& 0 \\
2x_1 + 6x_2 - 5x_3 - 2x_4 + 4x_5 - 3x_6 &=& 0 \\
5x_3 + 10x_4 \phantom{{}+4x_5} + 15x_6 &=& 0 \\
2x_1 + 6x_2 \phantom{{}-5x_3} + 8x_4 + 4x_5 + 18x_6 &=& 0
\end{array}
$$

(verify). ◆

Bases for Row Spaces, Column Spaces, and Nullspaces

We first developed elementary row operations for the purpose of solving linear systems, and we know from that work that performing an elementary row operation on an augmented matrix does not change the solution set of the corresponding linear system. It follows that applying an elementary row operation to a matrix A does not change the solution set of the corresponding linear system $A\mathbf{x} = \mathbf{0}$, or, stated another way, it does not change the nullspace of A. Thus, we have the following theorem.

Theorem 5.5.3

Elementary row operations do not change the nullspace of a matrix.

EXAMPLE 4 **Basis for Nullspace**

Find a basis for the nullspace of

$$
A = \begin{bmatrix} 2 & 2 & -1 & 0 & 1 \\ -1 & -1 & 2 & -3 & 1 \\ 1 & 1 & -2 & 0 & -1 \\ 0 & 0 & 1 & 1 & 1 \end{bmatrix}
$$

Solution.

The nullspace of A is the solution space of the homogeneous system

$$
\begin{aligned}
2x_1 + 2x_2 - \ x_3 \qquad\quad + x_5 &= 0 \\
-x_1 - \ x_2 + 2x_3 - 3x_4 + x_5 &= 0 \\
x_1 + \ x_2 - 2x_3 \qquad\quad - x_5 &= 0 \\
x_3 + \ x_4 + x_5 &= 0
\end{aligned}
$$

In Example 10 of Section 5.4 we showed that the vectors

$$
\mathbf{v}_1 = \begin{bmatrix} -1 \\ 1 \\ 0 \\ 0 \\ 0 \end{bmatrix} \quad \text{and} \quad \mathbf{v}_2 = \begin{bmatrix} -1 \\ 0 \\ -1 \\ 0 \\ 1 \end{bmatrix}
$$

form a basis for this space.   ◆

The following theorem is a companion to Theorem 5.5.3.

Theorem 5.5.4

Elementary row operations do not change the row space of a matrix.

Proof. Suppose that the row vectors of a matrix A are $\mathbf{r}_1, \mathbf{r}_2, \ldots, \mathbf{r}_m$, and let B be obtained from A by performing an elementary row operation. We shall show that every vector in the row space of B is also in the row space of A, and conversely that every vector in the row space of A is in the row space of B. We can then conclude that A and B have the same row space.

Consider the possibilities: If the row operation is a row interchange, then B and A have the same row vectors and consequently have the same row space. If the row operation is multiplication of a row by a nonzero scalar or the addition of a multiple of one row to another, then the row vectors $\mathbf{r}'_1, \mathbf{r}'_2, \ldots, \mathbf{r}'_m$ of B are linear combinations of $\mathbf{r}_1, \mathbf{r}_2, \ldots, \mathbf{r}_m$; thus, they lie in the row space of A. Since a vector space is closed under addition and scalar multiplication, all linear combinations of $\mathbf{r}'_1, \mathbf{r}'_2, \ldots, \mathbf{r}'_m$ will also lie in the row space of A. Therefore, each vector in the row space of B is in the row space of A.

Since B is obtained from A by performing a row operation, A can be obtained from B by performing the inverse operation (Section 1.5). Thus, the argument above shows that the row space of A is contained in the row space of B.   ∎

In light of Theorems 5.5.3 and 5.5.4, one might anticipate that elementary row operations should not change the column space of a matrix. However, this is *not* so—elementary row operations can change the column space. For example, consider the matrix

$$
A = \begin{bmatrix} 1 & 3 \\ 2 & 6 \end{bmatrix}
$$

The second column is a scalar multiple of the first, so the column space of A consists of all scalar multiples of the first column vector. However, if we add -2 times the first row of A to the second row, we obtain

$$B = \begin{bmatrix} 1 & 3 \\ 0 & 0 \end{bmatrix}$$

Here again the second column is a scalar multiple of the first, so the column space of B consists of all scalar multiples of the first column vector. This is not the same as the column space of A.

Although elementary row operations can change the column space of a matrix, we shall show that whatever relationships of linear independence or linear dependence exist among the column vectors prior to a row operation will also hold for the corresponding columns of the matrix that results from that operation. To make this more precise, suppose that a matrix B results by performing an elementary row operation on an $m \times n$ matrix A. By Theorem 5.5.3 the two homogeneous linear systems

$$A\mathbf{x} = \mathbf{0} \quad \text{and} \quad B\mathbf{x} = \mathbf{0}$$

have the same solution set. Thus, the first system has a nontrivial solution if and only if the same is true of the second. But if the column vectors of A and B, respectively, are

$$\mathbf{c}_1, \mathbf{c}_2, \ldots, \mathbf{c}_n \quad \text{and} \quad \mathbf{c}'_1, \mathbf{c}'_2, \ldots, \mathbf{c}'_n$$

then from (2) the two systems can be rewritten as

$$x_1\mathbf{c}_1 + x_2\mathbf{c}_2 + \cdots + x_n\mathbf{c}_n = \mathbf{0} \tag{6}$$

and

$$x_1\mathbf{c}'_1 + x_2\mathbf{c}'_2 + \cdots + x_n\mathbf{c}'_n = \mathbf{0} \tag{7}$$

Thus, (6) has a nontrivial solution for $x_1, x_2, \ldots, x_n$ if and only if the same is true of (7). This implies that the column vectors of A are linearly independent if and only if the same is true of B. Although we shall omit the proof, this conclusion also applies to any subset of the column vectors. Thus, we have the following result.

Theorem 5.5.5

If A and B are row equivalent matrices, then:

(a) *A given set of column vectors of A is linearly independent if and only if the corresponding column vectors of B are linearly independent.*

(b) *A given set of column vectors of A forms a basis for the column space of A if and only if the corresponding column vectors of B form a basis for the column space of B.*

The following theorem makes it possible to find bases for the row and column spaces of a matrix in row-echelon form by inspection.

Theorem 5.5.6

If a matrix R is in row-echelon form, then the row vectors with the leading 1's (i.e., the nonzero row vectors) form a basis for the row space of R, and the column vectors with the leading 1's of the row vectors form a basis for the column space of R.

Since this result is virtually self-evident when one looks at numerical examples, we shall omit the proof; the proof involves little more than an analysis of the positions of the 0's and 1's of R.

EXAMPLE 5 Bases for Row and Column Spaces

The matrix

$$R = \begin{bmatrix} 1 & -2 & 5 & 0 & 3 \\ 0 & 1 & 3 & 0 & 0 \\ 0 & 0 & 0 & 1 & 0 \\ 0 & 0 & 0 & 0 & 0 \end{bmatrix}$$

is in row-echelon form. From Theorem 5.5.6 the vectors

$$\mathbf{r}_1 = \begin{bmatrix} 1 & -2 & 5 & 0 & 3 \end{bmatrix}$$
$$\mathbf{r}_2 = \begin{bmatrix} 0 & 1 & 3 & 0 & 0 \end{bmatrix}$$
$$\mathbf{r}_3 = \begin{bmatrix} 0 & 0 & 0 & 1 & 0 \end{bmatrix}$$

form a basis for the row space of R, and the vectors

$$\mathbf{c}_1 = \begin{bmatrix} 1 \\ 0 \\ 0 \\ 0 \end{bmatrix}, \qquad \mathbf{c}_2 = \begin{bmatrix} -2 \\ 1 \\ 0 \\ 0 \end{bmatrix}, \qquad \mathbf{c}_4 = \begin{bmatrix} 0 \\ 0 \\ 1 \\ 0 \end{bmatrix}$$

form a basis for the column space of R. ◆

EXAMPLE 6 Bases for Row and Column Spaces

Find bases for the row and column spaces of

$$A = \begin{bmatrix} 1 & -3 & 4 & -2 & 5 & 4 \\ 2 & -6 & 9 & -1 & 8 & 2 \\ 2 & -6 & 9 & -1 & 9 & 7 \\ -1 & 3 & -4 & 2 & -5 & -4 \end{bmatrix}$$

Solution.

Since elementary row operations do not change the row space of a matrix, we can find a basis for the row space of A by finding a basis for the row space of any row-echelon form of A. Reducing A to row-echelon form we obtain (verify)

$$R = \begin{bmatrix} 1 & -3 & 4 & -2 & 5 & 4 \\ 0 & 0 & 1 & 3 & -2 & -6 \\ 0 & 0 & 0 & 0 & 1 & 5 \\ 0 & 0 & 0 & 0 & 0 & 0 \end{bmatrix}$$

By Theorem 5.5.6 the nonzero row vectors of R form a basis for the row space of R, and hence form a basis for the row space of A. These basis vectors are

$$\mathbf{r}_1 = \begin{bmatrix} 1 & -3 & 4 & -2 & 5 & 4 \end{bmatrix}$$
$$\mathbf{r}_2 = \begin{bmatrix} 0 & 0 & 1 & 3 & -2 & -6 \end{bmatrix}$$
$$\mathbf{r}_3 = \begin{bmatrix} 0 & 0 & 0 & 0 & 1 & 5 \end{bmatrix}$$

Keeping in mind that A and R may have different column spaces, we cannot find a basis for the column space of A *directly* from the column vectors of R. However, it follows from Theorem 5.5.5*b* that if we can find a set of column vectors of R that forms a basis for the column space of R, then the *corresponding* column vectors of A will form a basis for the column space of A.

The first, third, and fifth columns of R contain the leading 1's of the row vectors, so

$$\mathbf{c}_1' = \begin{bmatrix} 1 \\ 0 \\ 0 \\ 0 \end{bmatrix}, \quad \mathbf{c}_3' = \begin{bmatrix} 4 \\ 1 \\ 0 \\ 0 \end{bmatrix}, \quad \mathbf{c}_5' = \begin{bmatrix} 5 \\ -2 \\ 1 \\ 0 \end{bmatrix}$$

form a basis for the column space of R; thus the corresponding column vectors of A, namely,

$$\mathbf{c}_1 = \begin{bmatrix} 1 \\ 2 \\ 2 \\ -1 \end{bmatrix}, \quad \mathbf{c}_3 = \begin{bmatrix} 4 \\ 9 \\ 9 \\ -4 \end{bmatrix}, \quad \mathbf{c}_5 = \begin{bmatrix} 5 \\ 8 \\ 9 \\ -5 \end{bmatrix}$$

form a basis for the column space of A. ◆

EXAMPLE 7 Basis for a Vector Space Using Row Operations

Find a basis for the space spanned by the vectors

$$\mathbf{v}_1 = (1, -2, 0, 0, 3), \quad \mathbf{v}_2 = (2, -5, -3, -2, 6), \quad \mathbf{v}_3 = (0, 5, 15, 10, 0),$$
$$\mathbf{v}_4 = (2, 6, 18, 8, 6)$$

Solution.

Except for a variation in notation, the space spanned by these vectors is the row space of the matrix

$$\begin{bmatrix} 1 & -2 & 0 & 0 & 3 \\ 2 & -5 & -3 & -2 & 6 \\ 0 & 5 & 15 & 10 & 0 \\ 2 & 6 & 18 & 8 & 6 \end{bmatrix}$$

Reducing this matrix to row-echelon form we obtain

$$\begin{bmatrix} 1 & -2 & 0 & 0 & 3 \\ 0 & 1 & 3 & 2 & 0 \\ 0 & 0 & 1 & 1 & 0 \\ 0 & 0 & 0 & 0 & 0 \end{bmatrix}$$

The nonzero row vectors in this matrix are

$$\mathbf{w}_1 = (1, -2, 0, 0, 3), \quad \mathbf{w}_2 = (0, 1, 3, 2, 0), \quad \mathbf{w}_3 = (0, 0, 1, 1, 0)$$

These vectors form a basis for the row space and consequently form a basis for the subspace of R^5 spanned by $\mathbf{v}_1$, $\mathbf{v}_2$, $\mathbf{v}_3$, and $\mathbf{v}_4$. ◆

Observe that in Example 6 the basis vectors obtained for the column space of A consisted of column vectors of A, but the basis vectors obtained for the row space of A were not all row vectors of A. The following example illustrates a procedure for finding a basis for the row space of a matrix A that consists entirely of row vectors of A.

EXAMPLE 8 Basis for the Row Space of a Matrix

Find a basis for the row space of

$$A = \begin{bmatrix} 1 & -2 & 0 & 0 & 3 \\ 2 & -5 & -3 & -2 & 6 \\ 0 & 5 & 15 & 10 & 0 \\ 2 & 6 & 18 & 8 & 6 \end{bmatrix}$$

consisting entirely of row vectors from A.

Solution.

We will transpose A, thereby converting the row space of A into the column space of A^T; then we will use the method of Example 6 to find a basis for the column space of A^T; and then we will transpose again to convert column vectors back to row vectors. Transposing A yields

$$A^T = \begin{bmatrix} 1 & 2 & 0 & 2 \\ -2 & -5 & 5 & 6 \\ 0 & -3 & 15 & 18 \\ 0 & -2 & 10 & 8 \\ 3 & 6 & 0 & 6 \end{bmatrix}$$

Reducing this matrix to row-echelon form yields

$$\begin{bmatrix} 1 & 2 & 0 & 2 \\ 0 & 1 & -5 & -10 \\ 0 & 0 & 0 & 1 \\ 0 & 0 & 0 & 0 \\ 0 & 0 & 0 & 0 \end{bmatrix}$$

The first, second, and fourth columns contain the leading 1's, so the corresponding column vectors in A^T form a basis for the column space of A^T; these are

$$\mathbf{c}_1 = \begin{bmatrix} 1 \\ -2 \\ 0 \\ 0 \\ 3 \end{bmatrix}, \quad \mathbf{c}_2 = \begin{bmatrix} 2 \\ -5 \\ -3 \\ -2 \\ 6 \end{bmatrix}, \quad \text{and} \quad \mathbf{c}_4 = \begin{bmatrix} 2 \\ 6 \\ 18 \\ 8 \\ 6 \end{bmatrix}$$

Transposing again and adjusting the notation appropriately yields the basis vectors

$$\mathbf{r}_1 = [1 \quad -2 \quad 0 \quad 0 \quad 3], \qquad \mathbf{r}_2 = [2 \quad -5 \quad -3 \quad -2 \quad 6],$$

and

$$\mathbf{r}_4 = [2 \quad 6 \quad 18 \quad 8 \quad 6]$$

for the row space of A. ◆

We know from Theorem 5.5.5 that elementary row operations do not alter relationships of linear independence and linear dependence among the column vectors; however, Formulas (6) and (7) imply an even deeper result. Because these formulas actually have *the same scalar coefficients* $x_1, x_2, \ldots, x_n$, it follows that elementary row operations do not alter the *formulas* (linear combinations) that relate linearly dependent column vectors. We omit the formal proof.

EXAMPLE 9 Basis and Linear Combinations

(a) Find a subset of the vectors
$$\mathbf{v}_1 = (1, -2, 0, 3), \qquad \mathbf{v}_2 = (2, -5, -3, 6),$$
$$\mathbf{v}_3 = (0, 1, 3, 0), \qquad \mathbf{v}_4 = (2, -1, 4, -7), \qquad \mathbf{v}_5 = (5, -8, 1, 2)$$

that forms a basis for the space spanned by these vectors.

(b) Express each vector not in the basis as a linear combination of the basis vectors.

Solution (*a*). We begin by constructing a matrix that has $\mathbf{v}_1, \mathbf{v}_2, \ldots, \mathbf{v}_5$ as its column vectors:

$$\begin{bmatrix} 1 & 2 & 0 & 2 & 5 \\ -2 & -5 & 1 & -1 & -8 \\ 0 & -3 & 3 & 4 & 1 \\ 3 & 6 & 0 & -7 & 2 \end{bmatrix} \qquad (8)$$
$$\begin{array}{ccccc} \uparrow & \uparrow & \uparrow & \uparrow & \uparrow \\ \mathbf{v}_1 & \mathbf{v}_2 & \mathbf{v}_3 & \mathbf{v}_4 & \mathbf{v}_5 \end{array}$$

The first part of our problem can be solved by finding a basis for the column space of this matrix. Reducing the matrix to *reduced* row-echelon form and denoting the column vectors of the resulting matrix by $\mathbf{w}_1, \mathbf{w}_2, \mathbf{w}_3, \mathbf{w}_4$, and $\mathbf{w}_5$ yields

$$\begin{bmatrix} 1 & 0 & 2 & 0 & 1 \\ 0 & 1 & -1 & 0 & 1 \\ 0 & 0 & 0 & 1 & 1 \\ 0 & 0 & 0 & 0 & 0 \end{bmatrix} \qquad (9)$$
$$\begin{array}{ccccc} \uparrow & \uparrow & \uparrow & \uparrow & \uparrow \\ \mathbf{w}_1 & \mathbf{w}_2 & \mathbf{w}_3 & \mathbf{w}_4 & \mathbf{w}_5 \end{array}$$

The leading 1's occur in columns 1, 2, and 4, so that by Theorem 5.5.6
$$\{\mathbf{w}_1, \mathbf{w}_2, \mathbf{w}_4\}$$
is a basis for the column space of (9) and consequently
$$\{\mathbf{v}_1, \mathbf{v}_2, \mathbf{v}_4\}$$
is a basis for the column space of (9).

Solution (*b*). We shall start by expressing $\mathbf{w}_3$ and $\mathbf{w}_5$ as linear combinations of the basis vectors $\mathbf{w}_1, \mathbf{w}_2, \mathbf{w}_4$. The simplest way of doing this is to express $\mathbf{w}_3$ and $\mathbf{w}_5$ in terms of basis vectors with smaller subscripts. Thus, we shall express $\mathbf{w}_3$ as a linear combination of $\mathbf{w}_1$ and $\mathbf{w}_2$, and we shall express $\mathbf{w}_5$ as a linear combination of $\mathbf{w}_1, \mathbf{w}_2$, and $\mathbf{w}_4$. By inspection of (9), these linear combinations are
$$\mathbf{w}_3 = 2\mathbf{w}_1 - \mathbf{w}_2$$
$$\mathbf{w}_5 = \mathbf{w}_1 + \mathbf{w}_2 + \mathbf{w}_4$$

We call these the ***dependency equations***. The corresponding relationships in (8) are
$$\mathbf{v}_3 = 2\mathbf{v}_1 - \mathbf{v}_2$$
$$\mathbf{v}_5 = \mathbf{v}_1 + \mathbf{v}_2 + \mathbf{v}_4 \qquad \blacklozenge$$

The procedure illustrated in the preceding example is sufficiently important that we shall summarize the steps:

Given a set of vectors $S = \{\mathbf{v}_1, \mathbf{v}_2, \ldots, \mathbf{v}_k\}$ in R^n, the following procedure produces a subset of these vectors that forms a basis for span(S) and expresses those vectors of S that are not in the basis as linear combinations of the basis vectors.

Step 1. Form the matrix A having $\mathbf{v}_1, \mathbf{v}_2, \ldots, \mathbf{v}_k$ as its column vectors.

Step 2. Reduce the matrix A to its reduced row-echelon form R, and let $\mathbf{w}_1, \mathbf{w}_2, \ldots, \mathbf{w}_k$ be the column vectors of R.

Step 3. Identify the columns that contain the leading 1's in R. The corresponding column vectors of A are the basis vectors for span(S).

Step 4. Express each column vector of R that does *not* contain a leading 1 as a linear combination of preceding column vectors that do contain leading 1's. (You will be able to do this by inspection.) This yields a set of dependency equations involving the column vectors of R. The corresponding equations for the column vectors of A express the vectors not in the basis as linear combinations of the basis vectors.

Exercise Set 5.5

1. List the row vectors and column vectors of the matrix

$$\begin{bmatrix} 2 & -1 & 0 & 1 \\ 3 & 5 & 7 & -1 \\ 1 & 4 & 2 & 7 \end{bmatrix}$$

2. Express the product $A\mathbf{x}$ as a linear combination of the column vectors of A.

(a) $\begin{bmatrix} 2 & 3 \\ -1 & 4 \end{bmatrix} \begin{bmatrix} 1 \\ 2 \end{bmatrix}$

(b) $\begin{bmatrix} 4 & 0 & -1 \\ 3 & 6 & 2 \\ 0 & -1 & 4 \end{bmatrix} \begin{bmatrix} -2 \\ 3 \\ 5 \end{bmatrix}$

(c) $\begin{bmatrix} -3 & 6 & 2 \\ 5 & -4 & 0 \\ 2 & 3 & -1 \\ 1 & 8 & 3 \end{bmatrix} \begin{bmatrix} -1 \\ 2 \\ 5 \end{bmatrix}$

(d) $\begin{bmatrix} 2 & 1 & 5 \\ 6 & 3 & -8 \end{bmatrix} \begin{bmatrix} 3 \\ 0 \\ -5 \end{bmatrix}$

3. Determine whether $\mathbf{b}$ is in the column space of A, and if so, express $\mathbf{b}$ as a linear combination of the column vectors of A.

(a) $A = \begin{bmatrix} 1 & 3 \\ 4 & -6 \end{bmatrix}$; $\mathbf{b} = \begin{bmatrix} -2 \\ 10 \end{bmatrix}$

(b) $A = \begin{bmatrix} 1 & 1 & 2 \\ 1 & 0 & 1 \\ 2 & 1 & 3 \end{bmatrix}$; $\mathbf{b} = \begin{bmatrix} -1 \\ 0 \\ 2 \end{bmatrix}$

(c) $A = \begin{bmatrix} 1 & -1 & 1 \\ 9 & 3 & 1 \\ 1 & 1 & 1 \end{bmatrix}$; $\mathbf{b} = \begin{bmatrix} 5 \\ 1 \\ -1 \end{bmatrix}$

(d) $A = \begin{bmatrix} 1 & -1 & 1 \\ 1 & 1 & -1 \\ -1 & -1 & 1 \end{bmatrix}$; $\mathbf{b} = \begin{bmatrix} 2 \\ 0 \\ 0 \end{bmatrix}$

(e) $A = \begin{bmatrix} 1 & 2 & 0 & 1 \\ 0 & 1 & 2 & 1 \\ 1 & 2 & 1 & 3 \\ 0 & 1 & 2 & 2 \end{bmatrix}$; $\mathbf{b} = \begin{bmatrix} 4 \\ 3 \\ 5 \\ 7 \end{bmatrix}$

4. Suppose that $x_1 = -1$, $x_2 = 2$, $x_3 = 4$, $x_4 = -3$ is a solution of a nonhomogeneous linear system $A\mathbf{x} = \mathbf{b}$ and that the solution set of the homogeneous system $A\mathbf{x} = \mathbf{0}$ is given by the formulas

$$x_1 = -3r + 4s, \quad x_2 = r - s, \quad x_3 = r, \quad x_4 = s$$

(a) Find the vector form of the general solution of $A\mathbf{x} = \mathbf{0}$.

(b) Find the vector form of the general solution of $A\mathbf{x} = \mathbf{b}$.

5. Find the vector form of the general solution of the given linear system $A\mathbf{x} = \mathbf{b}$; then use that result to find the vector form of the general solution of $A\mathbf{x} = \mathbf{0}$.

(a) $\quad x_1 - 3x_2 = 1$
$\quad\quad 2x_1 - 6x_2 = 2$

(b) $\quad x_1 + x_2 + 2x_3 = 5$
$\quad\quad x_1 + x_3 = -2$
$\quad\quad 2x_1 + x_2 + 3x_3 = 3$

(c) $\quad x_1 - 2x_2 + x_3 + 2x_4 = -1$
$\quad\quad 2x_1 - 4x_2 + 2x_3 + 4x_4 = -2$
$\quad\quad -x_1 + 2x_2 - x_3 - 2x_4 = 1$
$\quad\quad 3x_1 - 6x_2 + 3x_3 + 6x_4 = -3$

(d) $\quad x_1 + 2x_2 - 3x_3 + x_4 = 4$
$\quad\quad -2x_1 + x_2 + 2x_3 + x_4 = -1$
$\quad\quad -x_1 + 3x_2 - x_3 + 2x_4 = 3$
$\quad\quad 4x_1 - 7x_2 - 5x_4 = -5$

6. Find a basis for the nullspace of A.

(a) $A = \begin{bmatrix} 1 & -1 & 3 \\ 5 & -4 & -4 \\ 7 & -6 & 2 \end{bmatrix}$

(b) $A = \begin{bmatrix} 2 & 0 & -1 \\ 4 & 0 & -2 \\ 0 & 0 & 0 \end{bmatrix}$

(c) $A = \begin{bmatrix} 1 & 4 & 5 & 2 \\ 2 & 1 & 3 & 0 \\ -1 & 3 & 2 & 2 \end{bmatrix}$

(d) $A = \begin{bmatrix} 1 & 4 & 5 & 6 & 9 \\ 3 & -2 & 1 & 4 & -1 \\ -1 & 0 & -1 & -2 & -1 \\ 2 & 3 & 5 & 7 & 8 \end{bmatrix}$

(e) $A = \begin{bmatrix} 1 & -3 & 2 & 2 & 1 \\ 0 & 3 & 6 & 0 & -3 \\ 2 & -3 & -2 & 4 & 4 \\ 3 & -6 & 0 & 6 & 5 \\ -2 & 9 & 2 & -4 & -5 \end{bmatrix}$

7. In each part, a matrix in row-echelon form is given. By inspection, find bases for the row and column spaces of A.

(a) $\begin{bmatrix} 1 & 0 & 2 \\ 0 & 0 & 1 \\ 0 & 0 & 0 \end{bmatrix}$

(b) $\begin{bmatrix} 1 & -3 & 0 & 0 \\ 0 & 1 & 0 & 0 \\ 0 & 0 & 0 & 0 \\ 0 & 0 & 0 & 0 \end{bmatrix}$

(c) $\begin{bmatrix} 1 & 2 & 4 & 5 \\ 0 & 1 & -3 & 0 \\ 0 & 0 & 1 & -3 \\ 0 & 0 & 0 & 1 \\ 0 & 0 & 0 & 0 \end{bmatrix}$

(d) $\begin{bmatrix} 1 & 2 & -1 & 5 \\ 0 & 1 & 4 & 3 \\ 0 & 0 & 1 & -7 \\ 0 & 0 & 0 & 1 \end{bmatrix}$

8. For the matrices in Exercise 6, find a basis for the row space of A by reducing the matrix to row-echelon form.

9. For the matrices in Exercise 6, find a basis for the column space of A.

10. For the matrices in Exercise 6, find a basis for the row space of A consisting entirely of row vectors of A.

11. Find a basis for the subspace of R^4 spanned by the given vectors.

(a) $(1, 1, -4, -3)$, $(2, 0, 2, -2)$, $(2, -1, 3, 2)$ (b) $(-1, 1, -2, 0)$, $(3, 3, 6, 0)$, $(9, 0, 0, 3)$
(c) $(1, 1, 0, 0)$, $(0, 0, 1, 1)$, $(-2, 0, 2, 2)$, $(0, -3, 0, 3)$

12. Find a subset of the vectors that forms a basis for the space spanned by the vectors; then express each vector that is not in the basis as a linear combination of the basis vectors.

(a) $\mathbf{v}_1 = (1, 0, 1, 1)$, $\mathbf{v}_2 = (-3, 3, 7, 1)$, $\mathbf{v}_3 = (-1, 3, 9, 3)$, $\mathbf{v}_4 = (-5, 3, 5, -1)$
(b) $\mathbf{v}_1 = (1, -2, 0, 3)$, $\mathbf{v}_2 = (2, -4, 0, 6)$, $\mathbf{v}_3 = (-1, 1, 2, 0)$, $\mathbf{v}_4 = (0, -1, 2, 3)$
(c) $\mathbf{v}_1 = (1, -1, 5, 2)$, $\mathbf{v}_2 = (-2, 3, 1, 0)$, $\mathbf{v}_3 = (4, -5, 9, 4)$, $\mathbf{v}_4 = (0, 4, 2, -3)$,
$\quad \mathbf{v}_5 = (-7, 18, 2, -8)$

13. Prove that the row vectors of an $n \times n$ invertible matrix A form a basis for R^n.

14. (a) Let

$$A = \begin{bmatrix} 0 & 1 & 0 \\ 1 & 0 & 0 \\ 0 & 0 & 0 \end{bmatrix}$$

and consider a rectangular xyz-coordinate system in 3-space. Show that the nullspace of A consists of all points on the z-axis and the column space consists of all points in the xy-plane (see accompanying figure).

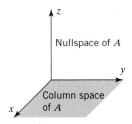

Figure Ex-14

(b) Find a 3×3 matrix whose nullspace is the x-axis and whose column space is the yz-plane.

Discussion and Discovery

15. Indicate whether the statement is always true or sometimes false. Justify your answer by giving a logical argument or a counterexample.

(a) If E is an elementary matrix, then A and EA must have the same nullspace.

(b) If E is an elementary matrix, then A and EA must have the same row space.

(c) If E is an elementary matrix, then A and EA must have the same column space.

(d) If $A\mathbf{x} = \mathbf{b}$ does not have any solutions, then $\mathbf{b}$ is not in the column space of A.

(e) The row space and nullspace of an invertible matrix are the same.

16. (a) Find all 2×2 matrices whose nullspace is the line $3x - 5y = 0$.

(b) Sketch the nullspaces of the following matrices:

$$A = \begin{bmatrix} 1 & 4 \\ 0 & 5 \end{bmatrix}, \qquad B = \begin{bmatrix} 1 & 0 \\ 0 & 5 \end{bmatrix}, \qquad C = \begin{bmatrix} 6 & 2 \\ 3 & 1 \end{bmatrix}, \qquad D = \begin{bmatrix} 0 & 0 \\ 0 & 0 \end{bmatrix}$$

17. The equation $x_1 + x_2 + x_3 = 1$ can be viewed as a linear system of one equation in three unknowns. Express its general solution as a particular solution plus the general solution of the corresponding homogeneous system. [Write the vectors in column form.]

18. Suppose that A and B are $n \times n$ matrices and A is invertible. Invent and prove a theorem that describes how the row spaces of AB and B are related.

5.6 RANK AND NULLITY

In the preceding section we investigated the relationships between systems of linear equations and the row space, column space, and nullspace of the coefficient matrix. In this section we shall be concerned with relationships between the dimensions of the row space, column space, and nullspace of a matrix and its transpose. The results we will obtain are fundamental and will provide deeper insights into linear systems and linear transformations.

Four Fundamental Matrix Spaces   If we consider a matrix A and its transpose A^T together, then there are six vector spaces of interest:

> row space of A row space of A^T
>
> column space of A column space of A^T
>
> nullspace of A nullspace of A^T

However, transposing a matrix converts row vectors into column vectors and column vectors into row vectors, so that, except for a difference in notation, the row space of A^T is the same as the column space of A, and the column space of A^T is the same as the row space of A. This leaves four vector spaces of interest:

> row space of A column space of A
>
> nullspace of A nullspace of A^T

These are known as the ***fundamental matrix spaces*** associated with A. If A is an $m \times n$ matrix, then the row space of A and nullspace of A are subspaces of R^n and the column space of A and the nullspace of A^T are subspaces of R^m. Our primary goal in this section is to establish relationships between the dimensions of these four vector spaces.

Row and Column Spaces Have Equal Dimensions

In Example 6 of Section 5.5, we found that the row and column spaces of the matrix

$$A = \begin{bmatrix} 1 & -3 & 4 & -2 & 5 & 4 \\ 2 & -6 & 9 & -1 & 8 & 2 \\ 2 & -6 & 9 & -1 & 9 & 7 \\ -1 & 3 & -4 & 2 & -5 & -4 \end{bmatrix}$$

each have three basis vectors; that is, both are three-dimensional. It is not accidental that these dimensions are the same; it is a consequence of the following general result.

Theorem 5.6.1

If A is any matrix, then the row space and column space of A have the same dimension.

Proof. Let R be any row-echelon form of A. It follows from Theorem 5.5.4 that

$$\dim(\text{row space of } A) = \dim(\text{row space of } R)$$

and it follows from Theorem 5.5.5b that

$$\dim(\text{column space of } A) = \dim(\text{column space of } R)$$

Thus, the proof will be complete if we can show that the row space and column space of R have the same dimension. But the dimension of the row space of R is the number of nonzero rows and the dimension of the column space of R is the number of columns that contain leading 1's (Theorem 5.5.6). However, the nonzero rows are precisely the rows in which the leading 1's occur, so the number of leading 1's and the number of nonzero rows is the same. This shows that the row space and column space of R have the same dimension. ■

The dimensions of the row space, column space, and nullspace of a matrix are such important numbers that there is some notation and terminology associated with them.

> ### Definition
>
> The common dimension of the row space and column space of a matrix A is called the **rank** of A and is denoted by rank(A); the dimension of the nullspace of A is called the **nullity** of A and is denoted by nullity(A).

EXAMPLE 1   Rank and Nullity of a 4 × 6 Matrix

Find the rank and nullity of the matrix

$$A = \begin{bmatrix} -1 & 2 & 0 & 4 & 5 & -3 \\ 3 & -7 & 2 & 0 & 1 & 4 \\ 2 & -5 & 2 & 4 & 6 & 1 \\ 4 & -9 & 2 & -4 & -4 & 7 \end{bmatrix}$$

Solution.

The reduced row-echelon form of A is

$$\begin{bmatrix} 1 & 0 & -4 & -28 & -37 & 13 \\ 0 & 1 & -2 & -12 & -16 & 5 \\ 0 & 0 & 0 & 0 & 0 & 0 \\ 0 & 0 & 0 & 0 & 0 & 0 \end{bmatrix} \tag{1}$$

(verify). Since there are two nonzero rows (or equivalently, two leading 1's), the row space and column space are both two-dimensional, so rank (A) $= 2$. To find the nullity of A, we must find the dimension of the solution space of the linear system $A\mathbf{x} = \mathbf{0}$. This system can be solved by reducing the augmented matrix to reduced row-echelon form. The resulting matrix will be identical to (1), except with an additional last column of zeros, and the corresponding system of equations will be

$$\begin{aligned} x_1 - 4x_3 - 28x_4 - 37x_5 + 13x_6 &= 0 \\ x_2 - 2x_3 - 12x_4 - 16x_5 + 5x_6 &= 0 \end{aligned}$$

or, on solving for the leading variables,

$$\begin{aligned} x_1 &= 4x_3 + 28x_4 + 37x_5 - 13x_6 \\ x_2 &= 2x_3 + 12x_4 + 16x_5 - 5x_6 \end{aligned} \tag{2}$$

It follows that the general solution of the system is

$$\begin{aligned} x_1 &= 4r + 28s + 37t - 13u \\ x_2 &= 2r + 12s + 16t - 5u \\ x_3 &= r \\ x_4 &= s \\ x_5 &= t \\ x_6 &= u \end{aligned}$$

or equivalently,

$$
\begin{bmatrix} x_1 \\ x_2 \\ x_3 \\ x_4 \\ x_5 \\ x_6 \end{bmatrix} = r \begin{bmatrix} 4 \\ 2 \\ 1 \\ 0 \\ 0 \\ 0 \end{bmatrix} + s \begin{bmatrix} 28 \\ 12 \\ 0 \\ 1 \\ 0 \\ 0 \end{bmatrix} + t \begin{bmatrix} 37 \\ 16 \\ 0 \\ 0 \\ 1 \\ 0 \end{bmatrix} + u \begin{bmatrix} -13 \\ -5 \\ 0 \\ 0 \\ 0 \\ 1 \end{bmatrix} \tag{3}
$$

The four vectors on the right side of (3) form a basis for the solution space, so that $\text{nullity}(A) = 4$. ◆

The following theorem shows that a matrix and its transpose have the same rank.

Theorem 5.6.2

If A is any matrix, then $rank(A) = rank(A^T)$.

Proof.

$$\text{rank}(A) = \dim(\text{row space of } A) = \dim(\text{column space of } A^T) = \text{rank}(A^T). \quad \blacksquare$$

The following theorem establishes an important relationship between the rank and nullity of a matrix.

Theorem 5.6.3   **Dimension Theorem for Matrices**

If A is a matrix with n columns, then

$$rank(A) + nullity(A) = n \tag{4}$$

Proof. Since A has n columns, the homogeneous linear system $A\mathbf{x} = \mathbf{0}$ has n unknowns (variables). These fall into two categories: the leading variables and the free variables. Thus,

$$\begin{bmatrix} \text{number of leading} \\ \text{variables} \end{bmatrix} + \begin{bmatrix} \text{number of free} \\ \text{variables} \end{bmatrix} = n$$

But the number of leading variables is the same as the number of leading 1's in the reduced row-echelon form of A, and this is the rank of A. Thus,

$$\text{rank}(A) + \begin{bmatrix} \text{number of free} \\ \text{variables} \end{bmatrix} = n$$

The number of free variables is equal to the nullity of A. This is so because the nullity of A is the dimension of the solution space of $A\mathbf{x} = \mathbf{0}$, which is the same as the number of parameters in the general solution [see (3), for example], which is the same as the number of free variables. Thus,

$$\text{rank}(A) + \text{nullity}(A) = n \quad \blacksquare$$

The proof of the preceding theorem contains two results that are of importance in their own right.

> ## Theorem 5.6.4
>
> *If A is an m × n matrix, then*:
> (*a*) *rank*(*A*) = *the number of leading variables in the solution of* A**x** = **0**.
> (*b*) *nullity*(*A*) = *the number of parameters in the general solution of* A**x** = **0**.

EXAMPLE 2   The Sum of Rank and Nullity

The matrix

$$A = \begin{bmatrix} -1 & 2 & 0 & 4 & 5 & -3 \\ 3 & -7 & 2 & 0 & 1 & 4 \\ 2 & -5 & 2 & 4 & 6 & 1 \\ 4 & -9 & 2 & -4 & -4 & 7 \end{bmatrix}$$

has 6 columns, so

$$\text{rank}(A) + \text{nullity}(A) = 6$$

This is consistent with Example 1, where we showed that

$$\text{rank}(A) = 2 \quad \text{and} \quad \text{nullity}(A) = 4 \qquad \blacklozenge$$

EXAMPLE 3   Number of Parameters in a General Solution

Find the number of parameters in the general solution of A**x** = **0** if A is a 5 × 7 matrix of rank 3.

Solution.

From (4),

$$\text{nullity}(A) = n - \text{rank}(A) = 7 - 3 = 4$$

Thus, there are four parameters. ♦

Suppose now that *A* is an *m* × *n* matrix of rank *r*; it follows from Theorem 5.6.2 that A^T is an *n* × *m* matrix of rank *r*. Applying Theorem 5.6.3 to *A* and A^T yields

$$\text{nullity}(A) = n - r, \qquad \text{nullity}(A^T) = m - r$$

from which we deduce the following table relating the dimensions of the four fundamental spaces of an *m* × *n* matrix *A* of rank *r*.

Fundamental Space	Dimension
Row space of *A*	*r*
Column space of *A*	*r*
Nullspace of *A*	*n* − *r*
Nullspace of A^T	*m* − *r*

Maximum Value for Rank If A is an $m \times n$ matrix, then the row vectors lie in R^n and the column vectors lie in R^m. This implies that the row space of A is at most n-dimensional and that the column space is at most m-dimensional. Since the row and column spaces have the same dimension (the rank of A), we must conclude that if $m \neq n$, then the rank of A is at most the smaller of the values of m and n. We denote this by writing

$$\text{rank}(A) \leq \min(m, n) \tag{5}$$

where $\min(m, n)$ denotes the smaller of the numbers m and n if $m \neq n$ or their common value if $m = n$.

EXAMPLE 4 Maximum Value of Rank for a 7 × 4 Matrix

If A is a 7×4 matrix, then the rank of A is at most 4 and, consequently, the seven row vectors must be linearly dependent. If A is a 4×7 matrix, then again the rank of A is at most 4 and, consequently, the seven column vectors must be linearly dependent. ◆

Linear Systems of m Equations in n Unknowns In earlier sections we obtained a wide range of theorems concerning linear systems of n equations in n unknowns. (See Theorem 4.3.4.) We shall now turn our attention to linear systems of m equations in n unknowns in which m and n need not be the same.

The following theorem provides conditions under which a linear system of m equations in n unknowns is guaranteed to be consistent.

Theorem 5.6.5 **The Consistency Theorem**

If $A\mathbf{x} = \mathbf{b}$ is a linear system of m equations in n unknowns, then the following are equivalent.

(a) $A\mathbf{x} = \mathbf{b}$ is consistent.

(b) $\mathbf{b}$ is in the column space of A.

(c) The coefficient matrix A and the augmented matrix $[A \mid \mathbf{b}]$ have the same rank.

Proof. It suffices to prove the two equivalences $(a) \Leftrightarrow (b)$ and $(b) \Leftrightarrow (c)$, since it will then follow as a matter of logic that $(a) \Leftrightarrow (c)$.

$(a) \Leftrightarrow (b)$. See Theorem 5.5.1.

$(b) \Rightarrow (c)$. We will show that if $\mathbf{b}$ is in the column space of A, then the column spaces of A and $[A \mid \mathbf{b}]$ are actually the same, from which it will follow that these two matrices have the same rank.

By definition, the column space of a matrix is the space spanned by its column vectors, so the column spaces of A and $[A \mid \mathbf{b}]$ can be expressed as

$$\text{span}\{\mathbf{c}_1, \mathbf{c}_2, \ldots, \mathbf{c}_n\} \quad \text{and} \quad \text{span}\{\mathbf{c}_1, \mathbf{c}_2, \ldots, \mathbf{c}_n, \mathbf{b}\}$$

respectively. If $\mathbf{b}$ is in the column space of A, then each vector in the set $\{\mathbf{c}_1, \mathbf{c}_2, \ldots, \mathbf{c}_n, \mathbf{b}\}$ is a linear combination of the vectors in $\{\mathbf{c}_1, \mathbf{c}_2, \ldots, \mathbf{c}_n\}$ and conversely (why?). Thus, from Theorem 5.2.4 the column spaces of A and $[A \mid \mathbf{b}]$ are the same.

$(c) \Rightarrow (b)$. Assume that A and $[A \mid \mathbf{b}]$ have the same rank r. By Theorem 5.4.6a, there is some subset of the column vectors of A that forms a basis for the column space of A. Suppose that those column vectors are

$$\mathbf{c}_1', \mathbf{c}_2', \ldots, \mathbf{c}_r'$$

These r basis vectors also belong to the r-dimensional column space of $[A \mid \mathbf{b}]$; hence they also form a basis for the column space of $[A \mid \mathbf{b}]$ by Theorem 5.4.6a. This means that $\mathbf{b}$ is expressible as a linear combination of $\mathbf{c}_1', \mathbf{c}_2', \ldots, \mathbf{c}_r'$, and consequently $\mathbf{b}$ lies in the column space of A. ∎

It is not hard to visualize why this theorem is true if one views the rank of a matrix as the number of nonzero rows in its reduced row-echelon form. For example, the augmented matrix for the system

$$
\begin{aligned}
x_1 - 2x_2 - 3x_3 + 2x_4 &= -4 \\
-3x_1 + 7x_2 - x_3 + x_4 &= -3 \\
2x_1 - 5x_2 + 4x_3 - 3x_4 &= 7 \\
-3x_1 + 6x_2 + 9x_3 - 6x_4 &= -1
\end{aligned}
\quad \text{is} \quad
\begin{bmatrix}
1 & -2 & -3 & 2 & -4 \\
-3 & 7 & -1 & 1 & -3 \\
2 & -5 & 4 & -3 & 7 \\
-3 & 6 & 9 & -6 & -1
\end{bmatrix}
$$

which has the following reduced row-echelon form (verify):

$$
\begin{bmatrix}
1 & 0 & -23 & 16 & 0 \\
0 & 1 & -10 & 7 & 0 \\
0 & 0 & 0 & 0 & 1 \\
0 & 0 & 0 & 0 & 0
\end{bmatrix}
$$

We see from the third row in this matrix that the system is inconsistent. However, it is also because of this row that the reduced row-echelon form of the augmented matrix has fewer zero rows than the reduced row-echelon form of the coefficient matrix. This forces the coefficient matrix and the augmented matrix for the system to have different ranks.

The Consistency Theorem is concerned with conditions under which a linear system $A\mathbf{x} = \mathbf{b}$ is consistent for a *specific* vector $\mathbf{b}$. The following theorem is concerned with conditions under which a linear system is consistent for *all possible* choices of $\mathbf{b}$.

Theorem 5.6.6

If $A\mathbf{x} = \mathbf{b}$ is a linear system of m equations in n unknowns, then the following are equivalent.

(a) $A\mathbf{x} = \mathbf{b}$ *is consistent for every $m \times 1$ matrix $\mathbf{b}$.*
(b) *The column vectors of A span R^m.*
(c) $rank(A) = m$.

Proof. It suffices to prove the two equivalences (a) ⟺ (b) and (a) ⟺ (c), since it will then follow as a matter of logic that (b) ⟺ (c).

(a) ⟺ (b). From Formula (2) of Section 5.5, the system $A\mathbf{x} = \mathbf{b}$ can be expressed as

$$
x_1\mathbf{c}_1 + x_2\mathbf{c}_2 + \cdots + x_n\mathbf{c}_n = \mathbf{b}
$$

from which we can conclude that $A\mathbf{x} = \mathbf{b}$ is consistent for every $m \times 1$ matrix $\mathbf{b}$ if and only if every such $\mathbf{b}$ is expressible as a linear combination of the column vectors $\mathbf{c}_1, \mathbf{c}_2, \ldots, \mathbf{c}_n$, or equivalently, if and only if these column vectors span R^m.

(a) ⟹ (c). From the assumption that $A\mathbf{x} = \mathbf{b}$ is consistent for every $m \times 1$ matrix $\mathbf{b}$, and from parts (a) and (b) of the Consistency Theorem (5.6.5), it follows that every vector $\mathbf{b}$ in R^m lies in the column space of A; that is, the column space of A is all of R^m. Thus, $\mathrm{rank}(A) = \dim(R^m) = m$.

(c) ⇒ (a). From the assumption that $\text{rank}(A) = m$, it follows that the column space of A is a subspace of R^m of dimension m, and hence must be all of R^m by Theorem 5.4.7. It now follows from parts (a) and (b) of the Consistency Theorem (5.6.5) that $A\mathbf{x} = \mathbf{b}$ is consistent for every vector $\mathbf{b}$ in R^m, since every such $\mathbf{b}$ is in the column space of A. ∎

A linear system with more equations than unknowns is called an ***overdetermined linear system***. If $A\mathbf{x} = \mathbf{b}$ is an overdetermined linear system of m equations in n unknowns (so that $m > n$), then the column vectors of A cannot span R^m; it follows from the last theorem that *for a fixed $m \times n$ matrix A with $m > n$, the overdetermined linear system $A\mathbf{x} = \mathbf{b}$ cannot be consistent for every possible $\mathbf{b}$.*

EXAMPLE 5 An Overdetermined System

The linear system

$$
\begin{aligned}
x_1 - 2x_2 &= b_1 \\
x_1 - x_2 &= b_2 \\
x_1 + x_2 &= b_3 \\
x_1 + 2x_2 &= b_4 \\
x_1 + 3x_2 &= b_5
\end{aligned}
$$

is overdetermined, so it cannot be consistent for all possible values of b_1, b_2, b_3, b_4, and b_5. Exact conditions under which the system is consistent can be obtained by solving the linear system by Gauss–Jordan elimination. We leave it for the reader to show that the augmented matrix is row equivalent to

$$
\begin{bmatrix}
1 & 0 & 2b_2 - b_1 \\
0 & 1 & b_2 - b_1 \\
0 & 0 & b_3 - 3b_2 + 2b_1 \\
0 & 0 & b_4 - 4b_2 + 3b_1 \\
0 & 0 & b_5 - 5b_2 + 4b_1
\end{bmatrix}
$$

Thus, the system is consistent if and only if b_1, b_2, b_3, b_4, and b_5 satisfy the conditions

$$
\begin{aligned}
2b_1 - 3b_2 + b_3 \quad\quad\quad &= 0 \\
3b_1 - 4b_2 \quad\quad + b_4 \quad &= 0 \\
4b_1 - 5b_2 \quad\quad\quad\quad + b_5 &= 0
\end{aligned}
$$

or, on solving this homogeneous linear system,

$$
b_1 = 5r - 4s, \quad b_2 = 4r - 3s, \quad b_3 = 2r - s, \quad b_4 = r, \quad b_5 = s
$$

where r and s are arbitrary. ◆

In Formula (3) of Theorem 5.5.2, the scalars $c_1, c_2, \ldots, c_k$ are the arbitrary parameters in the general solutions of both $A\mathbf{x} = \mathbf{b}$ and $A\mathbf{x} = \mathbf{0}$. Thus, these two systems have the same number of parameters in their general solutions. Moreover, it follows from part (b) of Theorem 5.6.4 that the number of such parameters is $\text{nullity}(A)$. This fact and the Dimension Theorem for Matrices (5.6.3) yields the following theorem.

Theorem 5.6.7

If $A\mathbf{x} = \mathbf{b}$ is a consistent linear system of m equations in n unknowns, and if A has rank r, then the general solution of the system contains $n - r$ parameters.

EXAMPLE 6   Number of Parameters in a General Solution

If A is a 5×7 matrix with rank 4, and if $A\mathbf{x} = \mathbf{b}$ is a consistent linear system, then the general solution of the system contains $7 - 4 = 3$ parameters. ◆

In earlier sections we obtained a wide range of conditions under which a homogeneous linear system $A\mathbf{x} = \mathbf{0}$ of n equations in n unknowns is guaranteed to have only the trivial solution. (See Theorem 4.3.4.) The following theorem obtains some corresponding results for systems of m equations in n unknowns, where m and n may differ.

Theorem 5.6.8

If A is an $m \times n$ matrix, then the following are equivalent.

(*a*) *$A\mathbf{x} = \mathbf{0}$ has only the trivial solution.*
(*b*) *The column vectors of A are linearly independent.*
(*c*) *$A\mathbf{x} = \mathbf{b}$ has at most one solution (none or one) for every $m \times 1$ matrix $\mathbf{b}$.*

Proof. It suffices to prove the two equivalences (*a*) ⇔ (*b*) and (*a*) ⇔ (*c*), since it will then follow as a matter of logic that (*b*) ⇔ (*c*).

(***a***) ⇔ (***b***). If $\mathbf{c}_1, \mathbf{c}_2, \ldots, \mathbf{c}_n$ are the column vectors of A, then the linear system $A\mathbf{x} = \mathbf{0}$ can be written as

$$x_1\mathbf{c}_1 + x_2\mathbf{c}_2 + \cdots + x_n\mathbf{c}_n = \mathbf{0} \tag{6}$$

If $\mathbf{c}_1, \mathbf{c}_2, \ldots, \mathbf{c}_n$ are linearly independent vectors, then this equation is satisfied only by $x_1 = x_2 = \cdots = x_n = 0$, which means that $A\mathbf{x} = \mathbf{0}$ has only the trivial solution. Conversely, if $A\mathbf{x} = \mathbf{0}$ has only the trivial solution, then Equation (6) is satisfied only by $x_1 = x_2 = \cdots = x_n = 0$, which means that $\mathbf{c}_1, \mathbf{c}_2, \ldots, \mathbf{c}_n$ are linearly independent.

(***a***) ⇒ (***c***). Assume that $A\mathbf{x} = \mathbf{0}$ has only the trivial solution. Either $A\mathbf{x} = \mathbf{b}$ is consistent or it is not. If it is not consistent, then there are no solutions of $A\mathbf{x} = \mathbf{b}$, and we are done. If $A\mathbf{x} = \mathbf{b}$ is consistent, let $\mathbf{x}_0$ be any solution. From the discussion following Theorem 5.5.2 and the fact that $A\mathbf{x} = \mathbf{0}$ has only the trivial solution, we conclude that the general solution of $A\mathbf{x} = \mathbf{b}$ is $\mathbf{x}_0 + \mathbf{0} = \mathbf{x}_0$. Thus, the only solution of $A\mathbf{x} = \mathbf{b}$ is $\mathbf{x}_0$.

(***c***) ⇒ (***a***). Assume that $A\mathbf{x} = \mathbf{b}$ has at most one solution for every $m \times 1$ matrix $\mathbf{b}$. Then, in particular, $A\mathbf{x} = \mathbf{0}$ has at most one solution. Thus, $A\mathbf{x} = \mathbf{0}$ has only the trivial solution. ∎

A linear system with more unknowns than equations is called an ***underdetermined linear system***. If $A\mathbf{x} = \mathbf{b}$ is a consistent underdetermined linear system of m equations in n unknowns (so that $m < n$), then it follows from Theorem 5.6.7 that the general solution has at least one parameter (why?); hence *a consistent underdetermined linear system* *must have infinitely many solutions.* In particular, an underdetermined homogeneous linear system has infinitely many solutions; but this was already proved in Chapter 1 (Theorem 1.2.1).

EXAMPLE 7 An Underdetermined System

If A is a 5×7 matrix, then for every 7×1 matrix $\mathbf{b}$, the linear system $A\mathbf{x} = \mathbf{b}$ is underdetermined. Thus, $A\mathbf{x} = \mathbf{b}$ must be consistent for some $\mathbf{b}$, and for each such $\mathbf{b}$ the general solution must have $7 - r$ parameters, where r is the rank of A. ◆

SUMMARY   In Theorem 4.3.4 we listed eight results that are equivalent to the invertibility of a matrix A. We conclude this section by adding eight more results to that list to produce the following theorem that relates all of the major topics we have studied thus far.

Theorem 5.6.9   Equivalent Statements

If A is an $n \times n$ matrix, and if $T_A: R^n \rightarrow R^n$ is multiplication by A, then the following are equivalent.

(a) A is invertible.
(b) $A\mathbf{x} = \mathbf{0}$ has only the trivial solution.
(c)   The reduced row-echelon form of A is I_n.
(d) A is expressible as a product of elementary matrices.
(e) $A\mathbf{x} = \mathbf{b}$ is consistent for every $n \times 1$ matrix $\mathbf{b}$.
(f) $A\mathbf{x} = \mathbf{b}$ has exactly one solution for every $n \times 1$ matrix $\mathbf{b}$.
(g) $\det(A) \neq 0$.
(h)   The range of T_A is R^n.
(i) T_A is one-to-one.
(j)   The column vectors of A are linearly independent.
(k)   The row vectors of A are linearly independent.
(l)   The column vectors of A span R^n.
(m)   The row vectors of A span R^n.
(n)   The column vectors of A form a basis for R^n.
(o)   The row vectors of A form a basis for R^n.
(p) A has rank n.
(q) A has nullity 0.

Proof. We already know from Theorem 4.3.4 that statements (a) through (i) are equivalent. To complete the proof we will show that (j) through (q) are equivalent to (b) by proving the sequence of implications (b) $\Rightarrow$ (j) $\Rightarrow$ (k) $\Rightarrow$ (l) $\Rightarrow$ (m) $\Rightarrow$ (n) $\Rightarrow$ (o) $\Rightarrow$ (p) $\Rightarrow$ (q) $\Rightarrow$ (b).

(b) $\Rightarrow$ (j). If $A\mathbf{x} = \mathbf{0}$ has only the trivial solution, then by Theorem 5.6.8 the column vectors of A are linearly independent.

(j) $\Rightarrow$ (k) $\Rightarrow$ (l) $\Rightarrow$ (m) $\Rightarrow$ (n) $\Rightarrow$ (o). This follows from Theorem 5.4.5 and the fact that R^n is an n-dimensional vector space. (The details are omitted.)

$(o) \Rightarrow (p)$. If the n row vectors of A form a basis for R^n, then the row space of A is n-dimensional and A has rank n.

$(p) \Rightarrow (q)$. This follows from the Dimension Theorem (5.6.3).

$(q) \Rightarrow (b)$. If A has nullity 0, then the solution space of $A\mathbf{x} = \mathbf{0}$ has dimension 0, which means that it contains only the zero vector. Hence, $A\mathbf{x} = \mathbf{0}$ has only the trivial solution. ∎

Exercise Set 5.6

1. Verify that $\text{rank}(A) = \text{rank}(A^T)$.

$$A = \begin{bmatrix} 1 & 2 & 4 & 0 \\ -3 & 1 & 5 & 2 \\ -2 & 3 & 9 & 2 \end{bmatrix}$$

2. Find the rank and nullity of the matrix; then verify that the values obtained satisfy Formula (4) of the Dimension Theorem.

(a) $A = \begin{bmatrix} 1 & -1 & 3 \\ 5 & -4 & -4 \\ 7 & -6 & 2 \end{bmatrix}$ (b) $A = \begin{bmatrix} 2 & 0 & -1 \\ 4 & 0 & -2 \\ 0 & 0 & 0 \end{bmatrix}$ (c) $A = \begin{bmatrix} 1 & 4 & 5 & 2 \\ 2 & 1 & 3 & 0 \\ -1 & 3 & 2 & 2 \end{bmatrix}$

(d) $A = \begin{bmatrix} 1 & 4 & 5 & 6 & 9 \\ 3 & -2 & 1 & 4 & -1 \\ -1 & 0 & -1 & -2 & -1 \\ 2 & 3 & 5 & 7 & 8 \end{bmatrix}$ (e) $A = \begin{bmatrix} 1 & -3 & 2 & 2 & 1 \\ 0 & 3 & 6 & 0 & -3 \\ 2 & -3 & -2 & 4 & 4 \\ 3 & -6 & 0 & 6 & 5 \\ -2 & 9 & 2 & -4 & -5 \end{bmatrix}$

3. In each part of Exercise 2 use the results obtained to find the number of leading variables and the number of parameters in the solution of $A\mathbf{x} = \mathbf{0}$ without solving the system.

4. In each part use the information in the table to find the dimension of the row space of A, the column space of A, the nullspace of A, and the nullspace of A^T.

	(a)	(b)	(c)	(d)	(e)	(f)	(g)
Size of A	3×3	3×3	3×3	5×9	9×5	4×4	6×2
Rank(A)	3	2	1	2	2	0	2

5. In each part, find the largest possible value for the rank of A and the smallest possible value for the nullity of A.

(a) A is 4×4 (b) A is 3×5 (c) A is 5×3

6. If A is an $m \times n$ matrix, what is the largest possible value for its rank and the smallest possible value for its nullity? [**Hint.** See Exercise 5.]

7. In each part, use the information in the table to determine whether the linear system $A\mathbf{x} = \mathbf{b}$ is consistent. If so, state the number of parameters in its general solution.

	(a)	(b)	(c)	(d)	(e)	(f)	(g)
Size of A	3×3	3×3	3×3	5×9	5×9	4×4	6×2
Rank(A)	3	2	1	2	2	0	2
Rank[$A \mid \mathbf{b}$]	3	3	1	2	3	0	2

8. For each of the matrices in Exercise 7, find the nullity of A, and determine the number of parameters in the general solution of the homogeneous linear system $A\mathbf{x} = \mathbf{0}$.

9. What conditions must be satisfied by b_1, b_2, b_3, b_4, and b_5 for the overdetermined linear system

$$x_1 - 3x_2 = b_1$$
$$x_1 - 2x_2 = b_2$$
$$x_1 + x_2 = b_3$$
$$x_1 - 4x_2 = b_4$$
$$x_1 + 5x_2 = b_5$$

to be consistent?

10. Let

$$A = \begin{bmatrix} a_{11} & a_{12} & a_{13} \\ a_{21} & a_{22} & a_{23} \end{bmatrix}$$

Show that A has rank 2 if and only if one or more of the determinants

$$\begin{vmatrix} a_{11} & a_{12} \\ a_{21} & a_{22} \end{vmatrix}, \quad \begin{vmatrix} a_{11} & a_{13} \\ a_{21} & a_{23} \end{vmatrix}, \quad \begin{vmatrix} a_{12} & a_{13} \\ a_{22} & a_{23} \end{vmatrix}$$

is nonzero.

11. Suppose that A is a 3×3 matrix whose nullspace is a line through the origin in 3-space. Can the row or column space of A also be a line through the origin? Explain.

12. Discuss how the rank of A varies with t.

(a) $A = \begin{bmatrix} 1 & 1 & t \\ 1 & t & 1 \\ t & 1 & 1 \end{bmatrix}$ (b) $A = \begin{bmatrix} t & 3 & -1 \\ 3 & 6 & -2 \\ -1 & -3 & t \end{bmatrix}$

13. Are there values of r and s for which the rank of

$$\begin{bmatrix} 1 & 0 & 0 \\ 0 & r-2 & 2 \\ 0 & s-1 & r+2 \\ 0 & 0 & 3 \end{bmatrix}$$

is one or two? If so, find those values.

14. Use the result in Exercise 10 to show that the set of points (x, y, z) in R^3 for which the matrix

$$\begin{bmatrix} x & y & z \\ 1 & x & y \end{bmatrix}$$

has rank 1 is the curve with parametric equations $x = t$, $y = t^2$, $z = t^3$.

15. Prove: If $k \neq 0$, then A and kA have the same rank.

Discussion and Discovery

16. (a) Give an example of a 3×3 matrix whose column space is a plane through the origin in 3-space.
    (b) What kind of geometric object is the nullspace of your matrix?
    (c) What kind of geometric object is the row space of your matrix?
    (d) In general, if the column space of a 3×3 matrix is a plane through the origin in 3-space, what can you say about the geometric properties of the nullspace and row space? Explain your reasoning.

17. Indicate whether the statement is always true or sometimes false. Justify your answer by giving a logical argument or a counterexample.

    (a) If A is not square, then the row vectors of A must be linearly dependent.

(b) If A is square, then either the row vectors or the column vectors of A must be linearly independent.

(c) If the row vectors and the column vectors of A are linearly independent, then A must be square.

(d) Adding one additional column to a matrix A increases its rank by one.

18. (a) If A is a 3×5 matrix, then the number of leading 1's in the reduced row-echelon form of A is at most ————. Why?

(b) If A is a 3×5 matrix, then the number of parameters in the general solution of $A\mathbf{x} = \mathbf{0}$ is at most ————. Why?

(c) If A is a 5×3 matrix, then the number of leading 1's in the reduced row-echelon form of A is at most ————. Why?

(d) If A is a 5×3 matrix, then the number of parameters in the general solution of $A\mathbf{x} = \mathbf{0}$ is at most ————. Why?

19. (a) If A is a 3×5 matrix, then the rank of A is at most ————. Why?

(b) If A is a 3×5 matrix, then the nullity of A is at most ————. Why?

(c) If A is a 3×5 matrix, then the rank of A^T is at most ————. Why?

(d) If A is a 3×5 matrix, then the nullity of A^T is at most ————. Why?

Chapter 5 Supplementary Exercises

1. In each part, the solution space is a subspace of R^3 and so must be a line through the origin, a plane through the origin, all of R^3, or the origin only. For each system, determine which is the case. If the subspace is a plane, find an equation for it, and if it is a line, find parametric equations.

(a) $0x + 0y + 0z = 0$

(b) $\begin{aligned} 2x - 3y + z &= 0 \\ 6x - 9y + 3z &= 0 \\ -4x + 6y - 2z &= 0 \end{aligned}$

(c) $\begin{aligned} x - 2y + 7z &= 0 \\ -4x + 8y + 5z &= 0 \\ 2x - 4y + 3z &= 0 \end{aligned}$

(d) $\begin{aligned} x + 4y + 8z &= 0 \\ 2x + 5y + 6z &= 0 \\ 3x + y - 4z &= 0 \end{aligned}$

2. For what values of s is the solution space of

$$\begin{aligned} x_1 + x_2 + sx_3 &= 0 \\ x_1 + sx_2 + x_3 &= 0 \\ sx_1 + x_2 + x_3 &= 0 \end{aligned}$$

a line through the origin, a plane through the origin, the origin only, or all of R^3?

3. (a) Express $(4a, a - b, a + 2b)$ as a linear combination of $(4, 1, 1)$ and $(0, -1, 2)$.

(b) Express $(3a + b + 3c, -a + 4b - c, 2a + b + 2c)$ as a linear combination of $(3, -1, 2)$ and $(1, 4, 1)$.

(c) Express $(2a - b + 4c, 3a - c, 4b + c)$ as a linear combination of three nonzero vectors.

4. Let W be the space spanned by $\mathbf{f} = \sin x$ and $\mathbf{g} = \cos x$.

(a) Show that for any value of θ, $\mathbf{f}_1 = \sin(x + \theta)$ and $\mathbf{g}_1 = \cos(x + \theta)$ are vectors in W.

(b) Show that $\mathbf{f}_1$ and $\mathbf{g}_1$ form a basis for W.

5. (a) Express $\mathbf{v} = (1, 1)$ as a linear combination of $\mathbf{v}_1 = (1, -1)$, $\mathbf{v}_2 = (3, 0)$, $\mathbf{v}_3 = (2, 1)$ in two different ways.

(b) Show that this does not violate Theorem 5.4.1.

6. Let A be an $n \times n$ matrix, and let $\mathbf{v}_1, \mathbf{v}_2, \dots, \mathbf{v}_n$ be linearly independent vectors in R^n expressed as $n \times 1$ matrices. What must be true about A for $A\mathbf{v}_1, A\mathbf{v}_2, \dots, A\mathbf{v}_n$ to be linearly independent?

7. Must a basis for P_n contain a polynomial of degree k for each $k = 0, 1, 2, \dots, n$? Justify your answer.

8. For purposes of this problem, let us define a "checkerboard matrix" to be a square matrix $A = [a_{ij}]$ such that

$$a_{ij} = \begin{cases} 1 & \text{if } i + j \text{ is even} \\ 0 & \text{if } i + j \text{ is odd} \end{cases}$$

Find the rank and nullity of the following checkerboard matrices:

(a) the 3×3 checkerboard matrix   (b) the 4×4 checkerboard matrix   (c) the $n \times n$ checkerboard matrix

9. For purposes of this exercise, let us define an "X-matrix" to be a square matrix with an odd number of rows and columns that has 0's everywhere except on the two diagonals, where it has 1's. Find the rank and nullity of the following X-matrices:

(a) $\begin{bmatrix} 1 & 0 & 1 \\ 0 & 1 & 0 \\ 1 & 0 & 1 \end{bmatrix}$ (b) $\begin{bmatrix} 1 & 0 & 0 & 0 & 1 \\ 0 & 1 & 0 & 1 & 0 \\ 0 & 0 & 1 & 0 & 0 \\ 0 & 1 & 0 & 1 & 0 \\ 1 & 0 & 0 & 0 & 1 \end{bmatrix}$ (c) the X-matrix of size $(2n + 1) \times (2n + 1)$

10. In each part, show that the set of polynomials is a subspace of P_n and find a basis for it.

(a) all polynomials in P_n such that $p(-x) = p(x)$ (b) all polynomials in P_n such that $p(0) = 0$

11. (*For readers who have studied calculus.*) Show that the set of all polynomials in P_n that have a horizontal tangent at $x = 0$ is a subspace of P_n. Find a basis for this subspace.

12. In advanced linear algebra one proves the following determinant criterion for rank: *The rank of a matrix A is r if and only if A has some $r \times r$ submatrix with a nonzero determinant, and all square submatrices of larger size have determinant zero.* (A submatrix of A is any matrix obtained by deleting rows or columns of A. The matrix A itself is also considered to be a submatrix of A.) In each part use this criterion to find the rank of the matrix.

(a) $\begin{bmatrix} 1 & 2 & 0 \\ 2 & 4 & -1 \end{bmatrix}$ (b) $\begin{bmatrix} 1 & 2 & 3 \\ 2 & 4 & 6 \end{bmatrix}$ (c) $\begin{bmatrix} 1 & 0 & 1 \\ 2 & -1 & 3 \\ 3 & -1 & 4 \end{bmatrix}$ (d) $\begin{bmatrix} 1 & -1 & 2 & 0 \\ 3 & 1 & 0 & 0 \\ -1 & 2 & 4 & 0 \end{bmatrix}$

13. Use the result in Exercise 12 to find the possible ranks for matrices of the form

$$\begin{bmatrix} 0 & 0 & 0 & 0 & 0 & a_{16} \\ 0 & 0 & 0 & 0 & 0 & a_{26} \\ 0 & 0 & 0 & 0 & 0 & a_{36} \\ 0 & 0 & 0 & 0 & 0 & a_{46} \\ a_{51} & a_{52} & a_{53} & a_{54} & a_{55} & a_{56} \end{bmatrix}$$

14. Prove: If S is a basis for a vector space V, then for any vectors $\mathbf{u}$ and $\mathbf{v}$ in V and any scalar k the following relationships hold:

(a) $(\mathbf{u} + \mathbf{v})_S = (\mathbf{u})_S + (\mathbf{v})_S$ (b) $(k\mathbf{u})_S = k(\mathbf{u})_S$

Chapter 5 Technology Exercises

The following exercises are designed to be solved using a technology utility. Typically, this will be MATLAB, *Mathematica*, Maple, Derive, or Mathcad, but it may also be some other type of linear algebra software or a scientific calculator with some linear algebra capabilities. For each exercise you will need to read the relevant documentation for the particular utility you are using. The goal of these exercises is to provide you with a basic proficiency with your technology utility. Once you have mastered the techniques in these exercises, you will be able to use your technology utility to solve many of the problems in the regular exercise sets.

Section 5.2

T1. (a) Some technology utilities do not have direct commands for finding linear combinations of vectors in R^n. However, you can use matrix multiplication to calculate a linear combination by creating a matrix A with the vectors as columns and a column vector $\mathbf{x}$ with the coefficients as entries. Use this method to compute the vector

$$\mathbf{v} = 6(8, -2, 1, -4) + 17(-3, 9, 11, 6) - 9(0, -1, 2, 4)$$

Check your work by hand.

(b) Use your technology utility to determine whether the vector $(9, 1, 0)$ is a linear combination of the vectors $(1, 2, 3)$, $(1, 4, 6)$, and $(2, -3, -5)$.

Section 5.4

T1. (***Linear independence***) Devise three different procedures for using your technology utility to determine whether a set of n vectors in R^n is linearly independent, and use all of your procedures to determine whether the vectors

$$\mathbf{v}_1 = (4, -5, 2, 6), \quad \mathbf{v}_2 = (2, -2, 1, 3), \quad \mathbf{v}_3 = (6, -3, 3, 9), \quad \mathbf{v}_4 = (4, -1, 5, 6)$$

are linearly independent.

T2. (***Dimension***) Devise three different procedures for using your technology utility to determine the dimension of the subspace spanned by a set of vectors in R^n, and use all of your procedures to determine the dimension of the subspace of R^5 spanned by the vectors

$$\mathbf{v}_1 = (2, 2, -1, 0, 1), \qquad \mathbf{v}_2 = (-1, -1, 2, -3, 1),$$
$$\mathbf{v}_3 = (1, 1, -2, 0, -1), \qquad \mathbf{v}_4 = (0, 0, 1, 1, 1)$$

Section 5.5

T1. (***Basis for row space***) Some technology utilities provide a command for finding a basis for the row space of a matrix. If your utility has this capability, read the documentation and then use your utility to find a basis for the row space of the matrix in Example 6.

T2. (***Basis for column space***) Some technology utilities provide a command for finding a basis for the column space of a matrix. If your utility has this capability, read the documentation and then use your utility to find a basis for the column space of the matrix in Example 6.

T3. (***Nullspace***) Some technology utilities provide a command for finding a basis for the nullspace of a matrix. If your utility has this capability, read the documentation and then check your understanding of the procedure by finding a basis for the nullspace of the matrix A in Example 4. Use this result to find the general solution of the homogeneous system $A\mathbf{x} = \mathbf{0}$.

Section 5.6

T1. (***Rank and nullity***) Read your documentation on finding the rank of a matrix, and then use your utility to find the rank of the matrix A in Example 1. Find the nullity of the matrix using Theorem 5.6.3 and the rank.

T2. There is a result, called ***Sylvester's inequality***, which states that if A and B are $n \times n$ matrices with rank r_A and r_B, respectively, then the rank r_{AB} of AB satisfies the inequality $r_A + r_B - n \leq r_{AB} \leq \min(r_A, r_B)$, where $\min(r_A, r_B)$ denotes the smaller of r_A and r_B or their common value if the two ranks are the same. Use your technology utility to confirm this result for some matrices of your choice.

Inner Product Spaces

Chapter Contents

INTRODUCTION: In Section 3.3 we defined the Euclidean inner product on the spaces R^2 and R^3. Then, in Section 4.1 we extended that concept to R^n and used it to define notions of length, distance, and angle in R^n. In this section we shall extend the concept of an inner product still further by extracting the most important properties of the Euclidean inner product on R^n and turning them into axioms that are applicable in general vector spaces. Thus, when these axioms are satisfied they will produce generalized inner products that automatically have the most important properties of Euclidean inner products. It will then be reasonable to use these generalized inner products to define notions of length, distance, and angle in general vector spaces.

6.1 INNER PRODUCTS

In this section we shall use the most important properties of the Euclidean inner product as axioms for defining the general concept of an inner product. We will then show how an inner product can be used to define notions of length and distance in vector spaces other than R^n.

General Inner Products   In Section 4.1 we denoted the Euclidean inner product of two vectors in R^n by the notation $\mathbf{u} \cdot \mathbf{v}$. It will be convenient in this section to introduce the alternative notation $\langle \mathbf{u}, \mathbf{v} \rangle$ for the general inner product. With this new notation, the fundamental properties of the Euclidean inner product that were listed in Theorem 4.1.2 are precisely the axioms in the following definition.

> **Definition**
>
> An **inner product** on a real vector space V is a function that associates a real number $\langle \mathbf{u}, \mathbf{v} \rangle$ with each pair of vectors $\mathbf{u}$ and $\mathbf{v}$ in V in such a way that the following axioms are satisfied for all vectors $\mathbf{u}$, $\mathbf{v}$, and $\mathbf{w}$ in V and all scalars k.
>
> (1) $\langle \mathbf{u}, \mathbf{v} \rangle = \langle \mathbf{v}, \mathbf{u} \rangle$ [Symmetry axiom]
> (2) $\langle \mathbf{u} + \mathbf{v}, \mathbf{w} \rangle = \langle \mathbf{u}, \mathbf{w} \rangle + \langle \mathbf{v}, \mathbf{w} \rangle$ [Additivity axiom]
> (3) $\langle k\mathbf{u}, \mathbf{v} \rangle = k\langle \mathbf{u}, \mathbf{v} \rangle$ [Homogeneity axiom]
> (4) $\langle \mathbf{v}, \mathbf{v} \rangle \geq 0$ [Positivity axiom]
> and $\langle \mathbf{v}, \mathbf{v} \rangle = 0$
> if and only if $\mathbf{v} = \mathbf{0}$
>
> A real vector space with an inner product is called a **real inner product space**.

REMARK.   In Chapter 10 we shall study inner products over complex vector spaces. However, until that time we shall use the term "inner product space" to mean "real inner product space."

Because the inner product axioms are based on properties of the Euclidean inner product, the Euclidean inner product automatically satisfies these axioms; this is the content of the following example.

EXAMPLE 1   Euclidean Inner Product on R^n

If $\mathbf{u} = (u_1, u_2, \ldots, u_n)$ and $\mathbf{v} = (v_1, v_2, \ldots, v_n)$ are vectors in R^n, then the formula

$$\langle \mathbf{u}, \mathbf{v} \rangle = \mathbf{u} \cdot \mathbf{v} = u_1 v_1 + u_2 v_2 + \cdots + u_n v_n$$

defines $\langle \mathbf{u}, \mathbf{v} \rangle$ to be the Euclidean inner product on R^n. The four inner product axioms hold by Theorem 4.1.2. ◆

The Euclidean inner product is the most important inner product on R^n. However, there are various applications in which it is desirable to modify the Euclidean inner product by *weighting* its terms differently. More precisely, if

$$w_1, w_2, \ldots, w_n$$

are *positive* real numbers, which we shall call **weights**, and if $\mathbf{u} = (u_1, u_2, \ldots, u_n)$ and $\mathbf{v} = (v_1, v_2, \ldots, v_n)$ are vectors in R^n, then it can be shown (Exercise 26) that the formula

$$\langle \mathbf{u}, \mathbf{v} \rangle = w_1 u_1 v_1 + w_2 u_2 v_2 + \cdots + w_n u_n v_n \tag{1}$$

defines an inner product on R^n; it is called the **weighted Euclidean inner product with weights $w_1, w_2, \ldots, w_n$**.

To illustrate one way in which a weighted Euclidean inner product can arise, suppose that some physical experiment can produce any of n possible numerical values

$$x_1, x_2, \ldots, x_n$$

and that m repetitions of the experiment yields these values with various frequencies; that is, x_1 occurs f_1 times, x_2 occurs f_2 times, and so forth. Since there are a total of m repetitions of the experiment,

$$f_1 + f_2 + \cdots + f_n = m$$

Thus, the **arithmetic average** or **mean** of the observed numerical values (denoted by $\bar{x}$) is

$$\bar{x} = \frac{f_1 x_1 + f_2 x_2 + \cdots + f_n x_n}{f_1 + f_2 + \cdots + f_n} = \frac{1}{m}(f_1 x_1 + f_2 x_2 + \cdots + f_n x_n) \tag{2}$$

If we let

$$\mathbf{f} = (f_1, f_2, \ldots, f_n)$$
$$\mathbf{x} = (x_1, x_2, \ldots, x_n)$$
$$w_1 = w_2 = \cdots = w_n = 1/m$$

then (2) can be expressed as the weighted inner product

$$\bar{x} = \langle \mathbf{f}, \mathbf{x} \rangle = w_1 f_1 x_1 + w_2 f_2 x_2 + \cdots + w_n f_n x_n$$

REMARK.   It will always be assumed that R^n has the Euclidean inner product unless some other inner product is specifically specified. As defined in Section 4.1, we refer to R^n with the Euclidean inner product as *Euclidean n-space*.

EXAMPLE 2   Weighted Euclidean Inner Product

Let $\mathbf{u} = (u_1, u_2)$ and $\mathbf{v} = (v_1, v_2)$ be vectors in R^2. Verify that the weighted Euclidean inner product

$$\langle \mathbf{u}, \mathbf{v} \rangle = 3u_1 v_1 + 2u_2 v_2$$

satisfies the four inner product axioms.

Solution.

Note first that if $\mathbf{u}$ and $\mathbf{v}$ are interchanged in this equation, the right side remains the same. Therefore,

$$\langle \mathbf{u}, \mathbf{v} \rangle = \langle \mathbf{v}, \mathbf{u} \rangle$$

If $\mathbf{w} = (w_1, w_2)$, then

$$\langle \mathbf{u} + \mathbf{v}, \mathbf{w} \rangle = 3(u_1 + v_1)w_1 + 2(u_2 + v_2)w_2$$
$$= (3u_1 w_1 + 2u_2 w_2) + (3v_1 w_1 + 2v_2 w_2)$$
$$= \langle \mathbf{u}, \mathbf{w} \rangle + \langle \mathbf{v}, \mathbf{w} \rangle$$

which establishes the second axiom.

Next,
$$\langle k\mathbf{u}, \mathbf{v} \rangle = 3(ku_1)v_1 + 2(ku_2)v_2 = k(3u_1v_1 + 2u_2v_2) = k\langle \mathbf{u}, \mathbf{v} \rangle$$
which establishes the third axiom.

Finally,
$$\langle \mathbf{v}, \mathbf{v} \rangle = 3v_1v_1 + 2v_2v_2 = 3v_1^2 + 2v_2^2$$

Obviously, $\langle \mathbf{v}, \mathbf{v} \rangle = 3v_1^2 + 2v_2^2 \geq 0$. Further, $\langle \mathbf{v}, \mathbf{v} \rangle = 3v_1^2 + 2v_2^2 = 0$ if and only if $v_1 = v_2 = 0$, that is, if and only if $\mathbf{v} = (v_1, v_2) = \mathbf{0}$. Thus, the fourth axiom is satisfied. ◆

Length and Distance in Inner Product Spaces

Before discussing more examples of inner products, we shall pause to explain how inner products are used to introduce notions of length and distance in inner product spaces. Recall that in Euclidean n-space the Euclidean length of a vector $\mathbf{u} = (u_1, u_2, \ldots, u_n)$ can be expressed in terms of the Euclidean inner product as

$$\|\mathbf{u}\| = (\mathbf{u} \cdot \mathbf{u})^{1/2}$$

and the Euclidean distance between two arbitrary points $\mathbf{u} = (u_1, u_2, \ldots, u_n)$ and $\mathbf{v} = (v_1, v_2, \ldots, v_n)$ can be expressed as

$$d(\mathbf{u}, \mathbf{v}) = \|\mathbf{u} - \mathbf{v}\| = [(\mathbf{u} - \mathbf{v}) \cdot (\mathbf{u} - \mathbf{v})]^{1/2}$$

[see Formulas (1) and (2) of Section 4.1]. Motivated by these formulas, we make the following definition.

Definition

If V is an inner product space, then the **norm** (or **length**) of a vector $\mathbf{u}$ in V is denoted by $\|\mathbf{u}\|$ and is defined by

$$\|\mathbf{u}\| = \langle \mathbf{u}, \mathbf{u} \rangle^{1/2}$$

The **distance** between two points (vectors) $\mathbf{u}$ and $\mathbf{v}$ is denoted by $d(\mathbf{u}, \mathbf{v})$ and is defined by

$$d(\mathbf{u}, \mathbf{v}) = \|\mathbf{u} - \mathbf{v}\|$$

EXAMPLE 3 Norm and Distance in R^n

If $\mathbf{u} = (u_1, u_2, \ldots, u_n)$ and $\mathbf{v} = (v_1, v_2, \ldots, v_n)$ are vectors in R^n with the Euclidean inner product, then

$$\|\mathbf{u}\| = \langle \mathbf{u}, \mathbf{u} \rangle^{1/2} = (\mathbf{u} \cdot \mathbf{u})^{1/2} = \sqrt{u_1^2 + u_2^2 + \cdots + u_n^2}$$

and

$$d(\mathbf{u}, \mathbf{v}) = \|\mathbf{u} - \mathbf{v}\| = \langle \mathbf{u} - \mathbf{v}, \mathbf{u} - \mathbf{v} \rangle^{1/2} = [(\mathbf{u} - \mathbf{v}) \cdot (\mathbf{u} - \mathbf{v})]^{1/2}$$
$$= \sqrt{(u_1 - v_1)^2 + (u_2 - v_2)^2 + \cdots + (u_n - v_n)^2}$$

Observe that these are simply the standard formulas for the Euclidean norm and distance discussed in Section 4.1 [see Formulas (1) and (2) in that section]. ◆

EXAMPLE 4 Using a Weighted Euclidean Inner Product

It is important to keep in mind that norm and distance depend on the inner product being used. If the inner product is changed, then the norms and distances between vectors also change. For example, for the vectors $\mathbf{u} = (1, 0)$ and $\mathbf{v} = (0, 1)$ in R^2 with the Euclidean inner product, we have

$$\|\mathbf{u}\| = \sqrt{1^2 + 0^2} = 1$$

and

$$d(\mathbf{u}, \mathbf{v}) = \|\mathbf{u} - \mathbf{v}\| = \|(1, -1)\| = \sqrt{1^2 + (-1)^2} = \sqrt{2}$$

However, if we change to the weighted Euclidean inner product

$$\langle \mathbf{u}, \mathbf{v} \rangle = 3u_1 v_1 + 2u_2 v_2$$

then we obtain

$$\|\mathbf{u}\| = \langle \mathbf{u}, \mathbf{u} \rangle^{1/2} = [3(1)(1) + 2(0)(0)]^{1/2} = \sqrt{3}$$

and

$$d(\mathbf{u}, \mathbf{v}) = \|\mathbf{u} - \mathbf{v}\| = \langle (1, -1), (1, -1) \rangle^{1/2}$$
$$= [3(1)(1) + 2(-1)(-1)]^{1/2} = \sqrt{5} \qquad \blacklozenge$$

Unit Circles and Spheres in Inner Product Spaces   If *V* is
an inner product space, then the set of points in *V* that satisfy

$$\|\mathbf{u}\| = 1$$

is called the ***unit sphere*** or sometimes the ***unit circle*** in *V*. In R^2 and R^3 these are the
points that lie 1 unit away from the origin.

EXAMPLE 5 Unusual Unit Circles in R^2

(a) Sketch the unit circle in an *xy*-coordinate system in R^2 using the Euclidean inner
     product $\langle \mathbf{u}, \mathbf{v} \rangle = u_1 v_1 + u_2 v_2$.
(b) Sketch the unit circle in an *xy*-coordinate system in R^2 using the weighted Euclidean
     inner product $\langle \mathbf{u}, \mathbf{v} \rangle = \frac{1}{9} u_1 v_1 + \frac{1}{4} u_2 v_2$.

Solution (*a*). If $\mathbf{u} = (x, y)$, then $\|\mathbf{u}\| = \langle \mathbf{u}, \mathbf{u} \rangle^{1/2} = \sqrt{x^2 + y^2}$, so the equation of the
unit circle is $\sqrt{x^2 + y^2} = 1$, or on squaring both sides,

$$x^2 + y^2 = 1$$

As expected, the graph of this equation is a circle of radius 1 centered at the origin
(Figure 6.1.1*a*).

Solution (*b*). If $\mathbf{u} = (x, y)$, then $\|\mathbf{u}\| = \langle \mathbf{u}, \mathbf{u} \rangle^{1/2} = \sqrt{\frac{1}{9} x^2 + \frac{1}{4} y^2}$, so the equation of
the unit circle is $\sqrt{\frac{1}{9} x^2 + \frac{1}{4} y^2} = 1$, or on squaring both sides,

$$\frac{x^2}{9} + \frac{y^2}{4} = 1$$

The graph of this equation is the ellipse shown in Figure 6.1.1*b*. $\blacklozenge$

It would be reasonable for you to feel uncomfortable with the results in the last
example: For even though our definitions of length and distance reduce to the standard
definitions when applied to R^2 with the Euclidean inner product, it does require a stretch
of the imagination to think of the unit "circle" as having an elliptical shape. However,
even though nonstandard inner products distort familiar spaces and lead to strange values

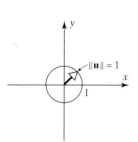

(*a*) The unit circle with
    the Euclidean norm
    $\|\mathbf{u}\| = \sqrt{x^2 + y^2}$

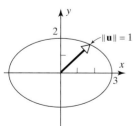

(*b*) The unit circle with
    the Euclidean norm
    $\|\mathbf{u}\| = \sqrt{\frac{1}{9} x^2 + \frac{1}{4} y^2}$

Figure 6.1.1

for lengths and distances, many of the basic theorems of Euclidean geometry continue to apply in these unusual spaces. For example, it is a basic fact in Euclidean geometry that the sum of the lengths of two sides of a triangle is at least as large as the length of the third side (Figure 6.1.2*a*). We shall see later that this familiar result holds in all inner product spaces, regardless of how unusual the inner product might be. As another example, recall the theorem from Euclidean geometry which states that the sum of the squares of the diagonals of a parallelogram is equal to the sum of the squares of the four sides (Figure 6.1.2*b*). This result also holds in all inner product spaces, regardless of the inner product (Exercise 20).

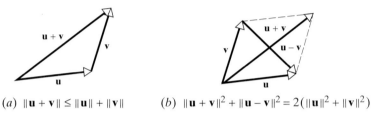

(*a*) $\|\mathbf{u} + \mathbf{v}\| \leq \|\mathbf{u}\| + \|\mathbf{v}\|$ (*b*) $\|\mathbf{u} + \mathbf{v}\|^2 + \|\mathbf{u} - \mathbf{v}\|^2 = 2(\|\mathbf{u}\|^2 + \|\mathbf{v}\|^2)$

Figure 6.1.2

Inner Products Generated by Matrices

The Euclidean inner product and the weighted Euclidean inner products are special cases of a general class of inner products on R^n, which we shall now describe. Let

$$\mathbf{u} = \begin{bmatrix} u_1 \\ u_2 \\ \vdots \\ u_n \end{bmatrix} \quad \text{and} \quad \mathbf{v} = \begin{bmatrix} v_1 \\ v_2 \\ \vdots \\ v_n \end{bmatrix}$$

be vectors in R^n (expressed as $n \times 1$ matrices), and let A be an *invertible* $n \times n$ matrix. It can be shown (Exercise 30) that if $\mathbf{u} \cdot \mathbf{v}$ is the Euclidean inner product on R^n, then the formula

$$\langle \mathbf{u}, \mathbf{v} \rangle = A\mathbf{u} \cdot A\mathbf{v} \tag{3}$$

defines an inner product; it is called the ***inner product on R^n generated by A***.

Recalling that the Euclidean inner product $\mathbf{u} \cdot \mathbf{v}$ can be written as the matrix product $\mathbf{v}^T\mathbf{u}$ [see (7) in Section 4.1], it follows that (3) can be written in the alternative form

$$\langle \mathbf{u}, \mathbf{v} \rangle = (A\mathbf{v})^T A\mathbf{u}$$

or equivalently,

$$\langle \mathbf{u}, \mathbf{v} \rangle = \mathbf{v}^T A^T A\mathbf{u} \tag{4}$$

EXAMPLE 6 Inner Product Generated by the Identity Matrix

The inner product on R^n generated by the $n \times n$ identity matrix is the Euclidean inner product, since substituting $A = I$ in (3) yields

$$\langle \mathbf{u}, \mathbf{v} \rangle = I\mathbf{u} \cdot I\mathbf{v} = \mathbf{u} \cdot \mathbf{v}$$

The weighted Euclidean inner product $\langle \mathbf{u}, \mathbf{v} \rangle = 3u_1v_1 + 2u_2v_2$ discussed in Example 2 is the inner product on R^2 generated by

$$A = \begin{bmatrix} \sqrt{3} & 0 \\ 0 & \sqrt{2} \end{bmatrix}$$

because substituting this matrix in (4) yields

$$\langle \mathbf{u}, \mathbf{v} \rangle = [v_1 \ \ v_2] \begin{bmatrix} \sqrt{3} & 0 \\ 0 & \sqrt{2} \end{bmatrix} \begin{bmatrix} \sqrt{3} & 0 \\ 0 & \sqrt{2} \end{bmatrix} \begin{bmatrix} u_1 \\ u_2 \end{bmatrix}$$

$$= [v_1 \ \ v_2] \begin{bmatrix} 3 & 0 \\ 0 & 2 \end{bmatrix} \begin{bmatrix} u_1 \\ u_2 \end{bmatrix}$$

$$= 3u_1 v_1 + 2u_2 v_2$$

In general, the weighted Euclidean inner product

$$\langle \mathbf{u}, \mathbf{v} \rangle = w_1 u_1 v_1 + w_2 u_2 v_2 + \cdots + w_n u_n v_n$$

is the inner product on R^n generated by

$$A = \begin{bmatrix} \sqrt{w_1} & 0 & 0 & \cdots & 0 \\ 0 & \sqrt{w_2} & 0 & \cdots & 0 \\ \vdots & \vdots & \vdots & & \vdots \\ 0 & 0 & 0 & \cdots & \sqrt{w_n} \end{bmatrix} \tag{5}$$

(verify). ◆

In the following examples we shall describe some inner products on vector spaces other than R^n.

EXAMPLE 7 An Inner Product on M_{22}

If

$$U = \begin{bmatrix} u_1 & u_2 \\ u_3 & u_4 \end{bmatrix} \quad \text{and} \quad V = \begin{bmatrix} v_1 & v_2 \\ v_3 & v_4 \end{bmatrix}$$

are any two 2×2 matrices, then the following formula defines an inner product on M_{22} (verify):

$$\langle U, V \rangle = \text{tr}(U^T V) = \text{tr}(V^T U) = u_1 v_1 + u_2 v_2 + u_3 v_3 + u_4 v_4$$

For example, if

$$U = \begin{bmatrix} 1 & 2 \\ 3 & 4 \end{bmatrix} \quad \text{and} \quad V = \begin{bmatrix} -1 & 0 \\ 3 & 2 \end{bmatrix}$$

then

$$\langle U, V \rangle = 1(-1) + 2(0) + 3(3) + 4(2) = 16$$

The norm of a matrix U relative to this inner product is

$$\|U\| = \langle U, U \rangle^{1/2} = \sqrt{u_1^2 + u_2^2 + u_3^2 + u_4^2}$$

and the unit sphere in this space consists of all 2×2 matrices U whose entries satisfy the equation $\|U\| = 1$, which on squaring yields

$$u_1^2 + u_2^2 + u_3^2 + u_4^2 = 1$$ ◆

EXAMPLE 8 An Inner Product on P_2

If

$$\mathbf{p} = a_0 + a_1 x + a_2 x^2 \quad \text{and} \quad \mathbf{q} = b_0 + b_1 x + b_2 x^2$$

are any two vectors in P_2, then the following formula defines an inner product on P_2 (verify):

$$\langle \mathbf{p}, \mathbf{q} \rangle = a_0 b_0 + a_1 b_1 + a_2 b_2$$

The norm of the polynomial **p** relative to this inner product is

$$\|\mathbf{p}\| = \langle \mathbf{p}, \mathbf{p} \rangle^{1/2} = \sqrt{a_0^2 + a_1^2 + a_2^2}$$

and the unit sphere in this space consists of all polynomials **p** in P_2 whose coefficients satisfy the equation $\|\mathbf{p}\| = 1$, which on squaring yields

$$a_0^2 + a_1^2 + a_2^2 = 1 \qquad \blacklozenge$$

Calculus Required

EXAMPLE 9 An Inner Product on $C[a, b]$

Let $\mathbf{f} = f(x)$ and $\mathbf{g} = g(x)$ be two continuous functions in $C[a, b]$ and define

$$\langle \mathbf{f}, \mathbf{g} \rangle = \int_a^b f(x)g(x)\, dx \qquad (6)$$

We shall show that this formula defines an inner product on $C[a, b]$ by verifying the four inner product axioms for functions $\mathbf{f} = f(x)$, $\mathbf{g} = g(x)$, and $\mathbf{s} = s(x)$ in $C[a, b]$:

(1) $\langle \mathbf{f}, \mathbf{g} \rangle = \displaystyle\int_a^b f(x)g(x)\, dx = \int_a^b g(x)f(x)\, dx = \langle \mathbf{g}, \mathbf{f} \rangle$

which proves that Axiom 1 holds.

$$
\begin{aligned}
(2)\ \langle \mathbf{f} + \mathbf{g}, \mathbf{s} \rangle &= \int_a^b (f(x) + g(x))s(x)\, dx \\
&= \int_a^b f(x)s(x)\, dx + \int_a^b g(x)s(x)\, dx \\
&= \langle \mathbf{f}, \mathbf{s} \rangle + \langle \mathbf{g}, \mathbf{s} \rangle
\end{aligned}
$$

which proves that Axiom 2 holds.

(3) $\langle k\mathbf{f}, \mathbf{g} \rangle = \displaystyle\int_a^b kf(x)g(x)\, dx = k \int_a^b f(x)g(x)\, dx = k\langle \mathbf{f}, \mathbf{g} \rangle$

which proves that Axiom 3 holds.

(4) If $\mathbf{f} = f(x)$ is any function in $C[a, b]$, then $f^2(x) \geq 0$ for all x in $[a, b]$; therefore,

$$\langle \mathbf{f}, \mathbf{f} \rangle = \int_a^b f^2(x)\, dx \geq 0$$

Further, because $f^2(x) \geq 0$ and $\mathbf{f} = f(x)$ is continuous on $[a, b]$, it follows that $\int_a^b f^2(x)\, dx = 0$ if and only if $f(x) = 0$ for all x in $[a, b]$. Therefore, we have $\langle \mathbf{f}, \mathbf{f} \rangle = \int_a^b f^2(x)\, dx = 0$ if and only if $\mathbf{f} = \mathbf{0}$. This proves that Axiom 4 holds. $\blacklozenge$

Calculus Required

EXAMPLE 10 Norm of a Vector in $C[a, b]$

If $C[a, b]$ has the inner product defined in the preceding example, then the norm of a function $\mathbf{f} = f(x)$ relative to this inner product is

$$\|\mathbf{f}\| = \langle \mathbf{f}, \mathbf{f} \rangle^{1/2} = \sqrt{\int_a^b f^2(x)\, dx} \qquad (7)$$

and the unit sphere in this space consists of all functions **f** in $C[a, b]$ that satisfy the equation $\|\mathbf{f}\| = 1$, which on squaring yields

$$\int_a^b f^2(x)\, dx = 1 \qquad \blacklozenge$$

Calculus Required REMARK. Since polynomials are continuous functions on $(-\infty, \infty)$, they are continuous on any closed interval $[a, b]$. Thus, for all such intervals the vector space P_n is a subspace of $C[a, b]$, and Formula (6) defines an inner product on P_n.

Calculus Required REMARK. Recall from calculus that the arc length of a curve $y = f(x)$ over an interval $[a, b]$ is given by the formula

$$L = \int_a^b \sqrt{1 + [f'(x)]^2}\, dx \tag{8}$$

Do not confuse this concept of arc length with $\|\mathbf{f}\|$, which is the length (norm) of $\mathbf{f}$ when $\mathbf{f}$ is viewed as a vector in $C[a, b]$. Formulas (7) and (8) are quite different.

The following theorem lists some basic algebraic properties of inner products.

Theorem 6.1.1 **Properties of Inner Products**

If $\mathbf{u}$, $\mathbf{v}$, and $\mathbf{w}$ are vectors in a real inner product space, and k is any scalar, then:

(a) $\langle \mathbf{0}, \mathbf{v} \rangle = \langle \mathbf{v}, \mathbf{0} \rangle = 0$ (b) $\langle \mathbf{u}, \mathbf{v} + \mathbf{w} \rangle = \langle \mathbf{u}, \mathbf{v} \rangle + \langle \mathbf{u}, \mathbf{w} \rangle$

(c) $\langle \mathbf{u}, k\mathbf{v} \rangle = k\langle \mathbf{u}, \mathbf{v} \rangle$ (d) $\langle \mathbf{u} - \mathbf{v}, \mathbf{w} \rangle = \langle \mathbf{u}, \mathbf{w} \rangle - \langle \mathbf{v}, \mathbf{w} \rangle$

(e) $\langle \mathbf{u}, \mathbf{v} - \mathbf{w} \rangle = \langle \mathbf{u}, \mathbf{v} \rangle - \langle \mathbf{u}, \mathbf{w} \rangle$

Proof. We shall prove part (b) and leave the proofs of the remaining parts as exercises.

$$\langle \mathbf{u}, \mathbf{v} + \mathbf{w} \rangle = \langle \mathbf{v} + \mathbf{w}, \mathbf{u} \rangle \quad \textbf{[By symmetry]}$$
$$= \langle \mathbf{v}, \mathbf{u} \rangle + \langle \mathbf{w}, \mathbf{u} \rangle \quad \textbf{[By additivity]}$$
$$= \langle \mathbf{u}, \mathbf{v} \rangle + \langle \mathbf{u}, \mathbf{w} \rangle \quad \textbf{[By symmetry]} \qquad \blacksquare$$

The following example illustrates how Theorem 6.1.1 and the defining properties of inner products can be used to perform algebraic computations with inner products. As you read through the example, you will find it instructive to justify the steps.

EXAMPLE 11 Calculating with Inner Products

$$\langle \mathbf{u} - 2\mathbf{v}, 3\mathbf{u} + 4\mathbf{v} \rangle = \langle \mathbf{u}, 3\mathbf{u} + 4\mathbf{v} \rangle - \langle 2\mathbf{v}, 3\mathbf{u} + 4\mathbf{v} \rangle$$
$$= \langle \mathbf{u}, 3\mathbf{u} \rangle + \langle \mathbf{u}, 4\mathbf{v} \rangle - \langle 2\mathbf{v}, 3\mathbf{u} \rangle - \langle 2\mathbf{v}, 4\mathbf{v} \rangle$$
$$= 3\langle \mathbf{u}, \mathbf{u} \rangle + 4\langle \mathbf{u}, \mathbf{v} \rangle - 6\langle \mathbf{v}, \mathbf{u} \rangle - 8\langle \mathbf{v}, \mathbf{v} \rangle$$
$$= 3\|\mathbf{u}\|^2 + 4\langle \mathbf{u}, \mathbf{v} \rangle - 6\langle \mathbf{u}, \mathbf{v} \rangle - 8\|\mathbf{v}\|^2$$
$$= 3\|\mathbf{u}\|^2 - 2\langle \mathbf{u}, \mathbf{v} \rangle - 8\|\mathbf{v}\|^2 \qquad \blacklozenge$$

Since Theorem 6.1.1 is a general result, it is guaranteed to hold for *all* real inner product spaces. This is the real power of the axiomatic development of vector spaces and inner products—a single theorem proves a multitude of results at once. For example, we are guaranteed without any further proof that the five properties given in Theorem 6.1.1 are true for the inner product on R^n generated by any matrix A [Formula (3)]. For example, let us check part (b) of Theorem 6.1.1 for this inner product:

$$\langle \mathbf{u}, \mathbf{v} + \mathbf{w} \rangle = (\mathbf{v} + \mathbf{w})^T A^T A \mathbf{u}$$
$$= (\mathbf{v}^T + \mathbf{w}^T) A^T A \mathbf{u} \qquad \textbf{[Property of transpose]}$$
$$= (\mathbf{v}^T A^T A \mathbf{u}) + (\mathbf{w}^T A^T A \mathbf{u}) \qquad \textbf{[Property of matrix multiplication]}$$
$$= \langle \mathbf{u}, \mathbf{v} \rangle + \langle \mathbf{u}, \mathbf{w} \rangle$$

The reader will find it instructive to check the remaining parts of Theorem 6.1.1 for this inner product.

Exercise Set 6.1

1. Let $\langle \mathbf{u}, \mathbf{v} \rangle$ be the Euclidean inner product on R^2, and let $\mathbf{u} = (3, -2)$, $\mathbf{v} = (4, 5)$, $\mathbf{w} = (-1, 6)$, and $k = -4$. Verify that

   (a) $\langle \mathbf{u}, \mathbf{v} \rangle = \langle \mathbf{v}, \mathbf{u} \rangle$ (b) $\langle \mathbf{u} + \mathbf{v}, \mathbf{w} \rangle = \langle \mathbf{u}, \mathbf{w} \rangle + \langle \mathbf{v}, \mathbf{w} \rangle$ (c) $\langle \mathbf{u}, \mathbf{v} + \mathbf{w} \rangle = \langle \mathbf{u}, \mathbf{v} \rangle + \langle \mathbf{u}, \mathbf{w} \rangle$

   (d) $\langle k\mathbf{u}, \mathbf{v} \rangle = k\langle \mathbf{u}, \mathbf{v} \rangle = \langle \mathbf{u}, k\mathbf{v} \rangle$ (e) $\langle \mathbf{0}, \mathbf{v} \rangle = \langle \mathbf{v}, \mathbf{0} \rangle = 0$

2. Repeat Exercise 1 for the weighted Euclidean inner product $\langle \mathbf{u}, \mathbf{v} \rangle = 4u_1v_1 + 5u_2v_2$.

3. Compute $\langle \mathbf{u}, \mathbf{v} \rangle$ using the inner product in Example 7.

   (a) $\mathbf{u} = \begin{bmatrix} 3 & -2 \\ 4 & 8 \end{bmatrix}$, $\mathbf{v} = \begin{bmatrix} -1 & 3 \\ 1 & 1 \end{bmatrix}$ (b) $\mathbf{u} = \begin{bmatrix} 1 & 2 \\ -3 & 5 \end{bmatrix}$, $\mathbf{v} = \begin{bmatrix} 4 & 6 \\ 0 & 8 \end{bmatrix}$

4. Compute $\langle \mathbf{p}, \mathbf{q} \rangle$ using the inner product in Example 8.

   (a) $\mathbf{p} = -2 + x + 3x^2$, $\mathbf{q} = 4 - 7x^2$ (b) $\mathbf{p} = -5 + 2x + x^2$, $\mathbf{q} = 3 + 2x - 4x^2$

5. (a) Use Formula (3) to show that $\langle \mathbf{u}, \mathbf{v} \rangle = 9u_1v_1 + 4u_2v_2$ is the inner product on R^2 generated by

   $$A = \begin{bmatrix} 3 & 0 \\ 0 & 2 \end{bmatrix}$$

   (b) Use the inner product in part (a) to compute $\langle \mathbf{u}, \mathbf{v} \rangle$ if $\mathbf{u} = (-3, 2)$ and $\mathbf{v} = (1, 7)$.

6. (a) Use Formula (3) to show that $\langle \mathbf{u}, \mathbf{v} \rangle = 5u_1v_1 - u_1v_2 - u_2v_1 + 10u_2v_2$ is the inner product on R^2 generated by

   $$A = \begin{bmatrix} 2 & 1 \\ -1 & 3 \end{bmatrix}$$

   (b) Use the inner product in part (a) to compute $\langle \mathbf{u}, \mathbf{v} \rangle$ if $\mathbf{u} = (0, -3)$ and $\mathbf{v} = (6, 2)$.

7. Let $\mathbf{u} = (u_1, u_2)$ and $\mathbf{v} = (v_1, v_2)$. In each part, the given expression is an inner product on R^2. Find a matrix that generates it.

   (a) $\langle \mathbf{u}, \mathbf{v} \rangle = 3u_1v_1 + 5u_2v_2$ (b) $\langle \mathbf{u}, \mathbf{v} \rangle = 4u_1v_1 + 6u_2v_2$

8. Let $\mathbf{u} = (u_1, u_2)$ and $\mathbf{v} = (v_1, v_2)$. Show that the following are inner products on R^2 by verifying that the inner product axioms hold.

   (a) $\langle \mathbf{u}, \mathbf{v} \rangle = 3u_1v_1 + 5u_2v_2$ (b) $\langle \mathbf{u}, \mathbf{v} \rangle = 4u_1v_1 + u_2v_1 + u_1v_2 + 4u_2v_2$

9. Let $\mathbf{u} = (u_1, u_2, u_3)$ and $\mathbf{v} = (v_1, v_2, v_3)$. Determine which of the following are inner products on R^3. For those that are not, list the axioms that do not hold.

   (a) $\langle \mathbf{u}, \mathbf{v} \rangle = u_1v_1 + u_3v_3$ (b) $\langle \mathbf{u}, \mathbf{v} \rangle = u_1^2v_1^2 + u_2^2v_2^2 + u_3^2v_3^2$

   (c) $\langle \mathbf{u}, \mathbf{v} \rangle = 2u_1v_1 + u_2v_2 + 4u_3v_3$ (d) $\langle \mathbf{u}, \mathbf{v} \rangle = u_1v_1 - u_2v_2 + u_3v_3$

10. In each part use the given inner product on R^2 to find $\|\mathbf{w}\|$, where $\mathbf{w} = (-1, 3)$.

    (a) the Euclidean inner product

    (b) the weighted Euclidean inner product $\langle \mathbf{u}, \mathbf{v} \rangle = 3u_1v_1 + 2u_2v_2$, where $\mathbf{u} = (u_1, u_2)$ and $\mathbf{v} = (v_1, v_2)$

    (c) the inner product generated by the matrix

    $$A = \begin{bmatrix} 1 & 2 \\ -1 & 3 \end{bmatrix}$$

11. Use the inner products in Exercise 10 to find $d(\mathbf{u}, \mathbf{v})$ for $\mathbf{u} = (-1, 2)$ and $\mathbf{v} = (2, 5)$.

12. Let P_2 have the inner product in Example 8. In each part find $\|\mathbf{p}\|$. $= \langle \overline{p}, \overline{p} \rangle^{1/2} = \sqrt{a_0^2 + a_1^2 + a_2^2}$

    (a) $\mathbf{p} = -2 + 3x + 2x^2$ (b) $\mathbf{p} = 4 - 3x^2$

13. Let M_{22} have the inner product in Example 7. In each part find $\|A\|$. $\|A\| = \langle A, A \rangle^{1/2} = \sqrt{u_1^2 + u_2^2 + u_3^2 + u_4^2}$

    (a) $A = \begin{bmatrix} -2 & 5 \\ 3 & 6 \end{bmatrix}$ (b) $A = \begin{bmatrix} 0 & 0 \\ 0 & 0 \end{bmatrix}$

14. Let P_2 have the inner product in Example 8. Find $d(\mathbf{p}, \mathbf{q})$.

    $$\mathbf{p} = 3 - x + x^2, \qquad \mathbf{q} = 2 + 5x^2$$

15. Let M_{22} have the inner product in Example 7. Find $d(A, B)$.

    (a) $A = \begin{bmatrix} 2 & 6 \\ 9 & 4 \end{bmatrix}$, $B = \begin{bmatrix} -4 & 7 \\ 1 & 6 \end{bmatrix}$ (b) $A = \begin{bmatrix} -2 & 4 \\ 1 & 0 \end{bmatrix}$, $B = \begin{bmatrix} -5 & 1 \\ 6 & 2 \end{bmatrix}$

16. Suppose that $\mathbf{u}$, $\mathbf{v}$, and $\mathbf{w}$ are vectors such that

    $$\langle \mathbf{u}, \mathbf{v} \rangle = 2, \quad \langle \mathbf{v}, \mathbf{w} \rangle = -3, \quad \langle \mathbf{u}, \mathbf{w} \rangle = 5, \quad \|\mathbf{u}\| = 1, \quad \|\mathbf{v}\| = 2, \quad \|\mathbf{w}\| = 7$$

    Evaluate the given expression.

    (a) $\langle \mathbf{u} + \mathbf{v}, \mathbf{v} + \mathbf{w} \rangle$ (b) $\langle 2\mathbf{v} - \mathbf{w}, 3\mathbf{u} + 2\mathbf{w} \rangle$ (c) $\langle \mathbf{u} - \mathbf{v} - 2\mathbf{w}, 4\mathbf{u} + \mathbf{v} \rangle$
    (d) $\|\mathbf{u} + \mathbf{v}\|$ (e) $\|2\mathbf{w} - \mathbf{v}\|$ (f) $\|\mathbf{u} - 2\mathbf{v} + 4\mathbf{w}\|$

17. (*For readers who have studied calculus.*) Let the vector space P_2 have the inner product

    $$\langle \mathbf{p}, \mathbf{q} \rangle = \int_{-1}^{1} p(x)q(x)\,dx$$

    (a) Find $\|\mathbf{p}\|$ for $\mathbf{p} = 1$, $\mathbf{p} = x$, and $\mathbf{p} = x^2$.   (b) Find $d(\mathbf{p}, \mathbf{q})$ if $\mathbf{p} = 1$ and $\mathbf{q} = x$.

18. Sketch the unit circle in R^2 using the given inner product.

    (a) $\langle \mathbf{u}, \mathbf{v} \rangle = \frac{1}{4}u_1 v_1 + \frac{1}{16}u_2 v_2$ (b) $\langle \mathbf{u}, \mathbf{v} \rangle = 2u_1 v_1 + u_2 v_2$

19. Find a weighted Euclidean inner product on R^2 for which the unit circle is the ellipse shown in the accompanying figure.

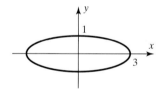

Figure Ex-19

20. Show that the following identity holds for vectors in any inner product space.

    $$\|\mathbf{u} + \mathbf{v}\|^2 + \|\mathbf{u} - \mathbf{v}\|^2 = 2\|\mathbf{u}\|^2 + 2\|\mathbf{v}\|^2$$

21. Show that the following identity holds for vectors in any inner product space.

    $$\langle \mathbf{u}, \mathbf{v} \rangle = \tfrac{1}{4}\|\mathbf{u} + \mathbf{v}\|^2 - \tfrac{1}{4}\|\mathbf{u} - \mathbf{v}\|^2$$

22. Let $U = \begin{bmatrix} u_1 & u_2 \\ u_3 & u_4 \end{bmatrix}$ and $V = \begin{bmatrix} v_1 & v_2 \\ v_3 & v_4 \end{bmatrix}$.

    Show that $\langle U, V \rangle = u_1 v_1 + u_2 v_3 + u_3 v_2 + u_4 v_4$ is *not* an inner product on M_{22}.

23. Let $\mathbf{p} = p(x)$ and $\mathbf{q} = q(x)$ be polynomials in P_2. Show that

    $$\langle \mathbf{p}, \mathbf{q} \rangle = p(0)q(0) + p\left(\tfrac{1}{2}\right) q\left(\tfrac{1}{2}\right) + p(1)q(1)$$

    is an inner product on P_2. Is this an inner product on P_3? Explain.

24. Prove: If $\langle \mathbf{u}, \mathbf{v} \rangle$ is the Euclidean inner product on R^n, and if A is an $n \times n$ matrix, then

    $$\langle \mathbf{u}, A\mathbf{v} \rangle = \langle A^T \mathbf{u}, \mathbf{v} \rangle$$

    [***Hint.*** Use the fact that $\langle \mathbf{u}, \mathbf{v} \rangle = \mathbf{u} \cdot \mathbf{v} = \mathbf{v}^T \mathbf{u}$.]

25. Verify the result in Exercise 24 for the Euclidean inner product on R^3 and

$$\mathbf{u} = \begin{bmatrix} -1 \\ 2 \\ 4 \end{bmatrix}, \quad \mathbf{v} = \begin{bmatrix} 3 \\ 0 \\ -2 \end{bmatrix}, \quad A = \begin{bmatrix} 1 & -2 & 1 \\ 3 & 4 & 0 \\ 5 & -1 & 2 \end{bmatrix}$$

26. Let $\mathbf{u} = (u_1, u_2, \ldots, u_n)$ and $\mathbf{v} = (v_1, v_2, \ldots, v_n)$. Show that

$$\langle \mathbf{u}, \mathbf{v} \rangle = w_1 u_1 v_1 + w_2 u_2 v_2 + \cdots + w_n u_n v_n$$

is an inner product on R^n if $w_1, w_2, \ldots, w_n$ are positive real numbers.

27. (*For readers who have studied calculus.*) Use the inner product

$$\langle \mathbf{p}, \mathbf{q} \rangle = \int_{-1}^{1} p(x)q(x)\,dx$$

to compute $\langle \mathbf{p}, \mathbf{q} \rangle$ for the vectors $\mathbf{p} = p(x)$ and $\mathbf{q} = q(x)$ in P_3.

(a) $\mathbf{p} = 1 - x + x^2 + 5x^3$ $\mathbf{q} = x - 3x^2$
(b) $\mathbf{p} = x - 5x^3$ $\mathbf{q} = 2 + 8x^2$

28. (*For readers who have studied calculus.*) In each part use the inner product

$$\langle \mathbf{f}, \mathbf{g} \rangle = \int_{0}^{1} f(x)g(x)\,dx$$

to compute $\langle \mathbf{f}, \mathbf{g} \rangle$ for the vectors $\mathbf{f} = f(x)$ and $\mathbf{g} = g(x)$ in $C[0, 1]$.

(a) $\mathbf{f} = \cos 2\pi x$, $\mathbf{g} = \sin 2\pi x$ (b) $\mathbf{f} = x$, $\mathbf{g} = e^x$ (c) $\mathbf{f} = \tan \dfrac{\pi}{4} x$, $\mathbf{g} = 1$

29. Show that the inner product in Example 7 can be written as $\langle U, V \rangle = \text{tr}(U^T V)$.

30. Prove that Formula (3) defines an inner product on R^n. [**Hint.** Use the alternative version of Formula (3) given by (4).]

31. Show that matrix (5) generates the weighted Euclidean inner product $\langle \mathbf{u}, \mathbf{v} \rangle = w_1 u_1 v_1 + w_2 u_2 v_2 + \cdots + w_n u_n v_n$ on R^n.

Discussion and Discovery

32. The following is a proof of part (*c*) of Theorem 6.1.1. Fill in each blank line with the name of an inner product axiom that justifies the step.

Hypothesis: Let $\mathbf{u}$ and $\mathbf{v}$ be vectors in a real inner product space.

Conclusion: $\langle \mathbf{u}, k\mathbf{v} \rangle = k\langle \mathbf{u}, \mathbf{v} \rangle$.

Proof: (1) $\langle \mathbf{u}, k\mathbf{v} \rangle = \langle k\mathbf{v}, \mathbf{u} \rangle$ _____

   (2) $= k\langle \mathbf{v}, \mathbf{u} \rangle$ _____

   (3) $= k\langle \mathbf{u}, \mathbf{v} \rangle$ _____

33. Prove parts (*a*), (*d*), and (*e*) of Theorem 6.1.1, justifying each step with the name of a vector space axiom or by referring to previously established results.

34. Create a weighted Euclidean inner product $\langle \mathbf{u}, \mathbf{v} \rangle = au_1 v_1 + bu_2 v_2$ on R^2 for which the unit circle in an xy-coordinate system is the ellipse shown in the accompanying figure.

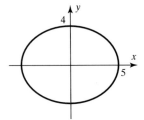

Figure Ex-34

6.2 ANGLE AND ORTHOGONALITY IN INNER PRODUCT SPACES

In this section we shall define the notion of an angle between two vectors in an inner product space, and we shall use this concept to obtain some basic relations between vectors in an inner product, including a fundamental geometric relationship between the nullspace and column space of a matrix.

Cauchy–Schwarz Inequality

Recall from Formula (1) of Section 3.3 that if **u** and **v** are nonzero vectors in R^2 or R^3 and θ is the angle between them, then

$$\mathbf{u} \cdot \mathbf{v} = \|\mathbf{u}\| \|\mathbf{v}\| \cos \theta \tag{1}$$

or alternatively,

$$\cos \theta = \frac{\mathbf{u} \cdot \mathbf{v}}{\|\mathbf{u}\| \|\mathbf{v}\|} \tag{2}$$

Our first goal in this section is to define the concept of an angle between two vectors in a general inner product space. For such a definition to be reasonable, we would want it to be consistent with Formula (2) when it is applied to the special case of R^2 and R^3 with the Euclidean inner product. Thus, we will want our definition of the angle θ between two nonzero vectors in an inner product space to satisfy the relationship

$$\cos \theta = \frac{\langle \mathbf{u}, \mathbf{v} \rangle}{\|\mathbf{u}\| \|\mathbf{v}\|} \tag{3}$$

However, because $|\cos \theta| \leq 1$, there would be no hope of satisfying (3) unless we were assured that every pair of nonzero vectors in an inner product space satisfies the inequality

$$\left| \frac{\langle \mathbf{u}, \mathbf{v} \rangle}{\|\mathbf{u}\| \|\mathbf{v}\|} \right| \leq 1$$

Fortunately, we will be able to prove that this is the case by using the following generalization of the Cauchy–Schwarz inequality (see Theorem 4.1.3).

Theorem 6.2.1 **Cauchy–Schwarz Inequality**

If **u** *and* **v** *are vectors in a real inner product space, then*

$$|\langle \mathbf{u}, \mathbf{v} \rangle| \leq \|\mathbf{u}\| \|\mathbf{v}\| \tag{4}$$

Proof. We warn the reader in advance that the proof presented here depends on a clever trick that is not easy to motivate. If $\mathbf{u} = \mathbf{0}$, then $\langle \mathbf{u}, \mathbf{v} \rangle = \langle \mathbf{u}, \mathbf{u} \rangle = 0$, so that the two sides of (4) are equal. Assume now that $\mathbf{u} \neq \mathbf{0}$. Let $a = \langle \mathbf{u}, \mathbf{u} \rangle$, $b = 2\langle \mathbf{u}, \mathbf{v} \rangle$, $c = \langle \mathbf{v}, \mathbf{v} \rangle$, and let t be any real number. By the positivity axiom, the inner product of any vector with itself is always nonnegative. Therefore,

$$0 \leq \langle t\mathbf{u} + \mathbf{v}, t\mathbf{u} + \mathbf{v} \rangle = \langle \mathbf{u}, \mathbf{u} \rangle t^2 + 2\langle \mathbf{u}, \mathbf{v} \rangle t + \langle \mathbf{v}, \mathbf{v} \rangle$$
$$= at^2 + bt + c$$

This inequality implies that the quadratic polynomial $at^2 + bt + c$ has either no real roots or a repeated real root. Therefore, its discriminant must satisfy the inequality

$b^2 - 4ac \leq 0$. Expressing the coefficients a, b, and c in terms of the vectors $\mathbf{u}$ and $\mathbf{v}$ gives $4\langle \mathbf{u}, \mathbf{v} \rangle^2 - 4\langle \mathbf{u}, \mathbf{u} \rangle \langle \mathbf{v}, \mathbf{v} \rangle \leq 0$, or equivalently,

$$\langle \mathbf{u}, \mathbf{v} \rangle^2 \leq \langle \mathbf{u}, \mathbf{u} \rangle \langle \mathbf{v}, \mathbf{v} \rangle$$

Taking square roots of both sides and using the fact that $\langle \mathbf{u}, \mathbf{u} \rangle$ and $\langle \mathbf{v}, \mathbf{v} \rangle$ are nonnegative yields

$$|\langle \mathbf{u}, \mathbf{v} \rangle| \leq \langle \mathbf{u}, \mathbf{u} \rangle^{1/2} \langle \mathbf{v}, \mathbf{v} \rangle^{1/2} \quad \text{or equivalently,} \quad |\langle \mathbf{u}, \mathbf{v} \rangle| \leq \|\mathbf{u}\|\|\mathbf{v}\|$$

which completes the proof. ∎

For reference, we note that the Cauchy–Schwarz inequality can be written in the following two alternative forms:

$$\langle \mathbf{u}, \mathbf{v} \rangle^2 \leq \langle \mathbf{u}, \mathbf{u} \rangle \langle \mathbf{v}, \mathbf{v} \rangle \tag{5}$$

$$\langle \mathbf{u}, \mathbf{v} \rangle^2 \leq \|\mathbf{u}\|^2 \|\mathbf{v}\|^2 \tag{6}$$

The first of these formulas was obtained in the proof of Theorem 6.2.1, and the second is derived from the first using the fact that $\|\mathbf{u}\|^2 = \langle \mathbf{u}, \mathbf{u} \rangle$ and $\|\mathbf{v}\|^2 = \langle \mathbf{v}, \mathbf{v} \rangle$.

EXAMPLE 1 Cauchy–Schwarz Inequality in R^n

The Cauchy–Schwarz inequality for R^n (Theorem 4.1.3) follows as a special case of Theorem 6.2.1 by taking $\langle \mathbf{u}, \mathbf{v} \rangle$ to be the Euclidean inner product $\mathbf{u} \cdot \mathbf{v}$. ◆

The next two theorems show that the basic properties of length and distance that were established in Theorems 4.1.4 and 4.1.5 for vectors in Euclidean n-space continue to hold in general inner product spaces. This is strong evidence that our definitions of inner product, length, and distance are well chosen.

Theorem 6.2.2 **Properties of Length**

If $\mathbf{u}$ and $\mathbf{v}$ are vectors in an inner product space V, and if k is any scalar, then:

(a) $\|\mathbf{u}\| \geq 0$ (b) $\|\mathbf{u}\| = 0$ *if and only if* $\mathbf{u} = \mathbf{0}$
(c) $\|k\mathbf{u}\| = |k|\|\mathbf{u}\|$ (d) $\|\mathbf{u} + \mathbf{v}\| \leq \|\mathbf{u}\| + \|\mathbf{v}\|$ (*Triangle inequality*)

Theorem 6.2.3 **Properties of Distance**

If $\mathbf{u}$, $\mathbf{v}$, and $\mathbf{w}$ are vectors in an inner product space V, and if k is any scalar, then:

(a) $d(\mathbf{u}, \mathbf{v}) \geq 0$ (b) $d(\mathbf{u}, \mathbf{v}) = 0$ *if and only if* $\mathbf{u} = \mathbf{v}$
(c) $d(\mathbf{u}, \mathbf{v}) = d(\mathbf{v}, \mathbf{u})$ (d) $d(\mathbf{u}, \mathbf{v}) \leq d(\mathbf{u}, \mathbf{w}) + d(\mathbf{w}, \mathbf{v})$ (*Triangle inequality*)

We shall prove part (*d*) of Theorem 6.2.2 and leave the remaining parts of Theorems 6.2.2 and 6.2.3 as exercises.

Proof of Theorem 6.2.2d. By definition

$$\|\mathbf{u} + \mathbf{v}\|^2 = \langle \mathbf{u} + \mathbf{v}, \mathbf{u} + \mathbf{v} \rangle$$
$$= \langle \mathbf{u}, \mathbf{u} \rangle + 2\langle \mathbf{u}, \mathbf{v} \rangle + \langle \mathbf{v}, \mathbf{v} \rangle$$
$$\leq \langle \mathbf{u}, \mathbf{u} \rangle + 2|\langle \mathbf{u}, \mathbf{v} \rangle| + \langle \mathbf{v}, \mathbf{v} \rangle \quad \textbf{[Property of absolute value]}$$
$$\leq \langle \mathbf{u}, \mathbf{u} \rangle + 2\|\mathbf{u}\|\,\|\mathbf{v}\| + \langle \mathbf{v}, \mathbf{v} \rangle \quad \textbf{[By (4)]}$$
$$= \|\mathbf{u}\|^2 + 2\|\mathbf{u}\|\,\|\mathbf{v}\| + \|\mathbf{v}\|^2$$
$$= (\|\mathbf{u}\| + \|\mathbf{v}\|)^2$$

Taking square roots gives $\|\mathbf{u} + \mathbf{v}\| \leq \|\mathbf{u}\| + \|\mathbf{v}\|$. ∎ ■

Angle Between Vectors

We shall now show how the Cauchy–Schwarz inequality can be used to define angles in general inner product spaces. Suppose that $\mathbf{u}$ and $\mathbf{v}$ are nonzero vectors in an inner product space V. If we divide both sides of Formula (6) by $\|\mathbf{u}\|^2\|\mathbf{v}\|^2$, we obtain

$$\left[\frac{\langle \mathbf{u}, \mathbf{v} \rangle}{\|\mathbf{u}\|\,\|\mathbf{v}\|} \right]^2 \leq 1$$

or equivalently,

$$-1 \leq \frac{\langle \mathbf{u}, \mathbf{v} \rangle}{\|\mathbf{u}\|\,\|\mathbf{v}\|} \leq 1 \tag{7}$$

Now if θ is an angle whose radian measure varies from 0 to π, then $\cos\theta$ assumes every value between -1 and 1 inclusive exactly once (Figure 6.2.1).

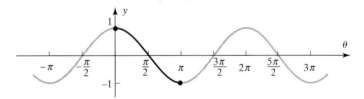

Figure 6.2.1

Thus, from (7) there is a unique angle θ such that

$$\cos\theta = \frac{\langle \mathbf{u}, \mathbf{v} \rangle}{\|\mathbf{u}\|\,\|\mathbf{v}\|} \quad \text{and} \quad 0 \leq \theta \leq \pi \tag{8}$$

We define θ to be the ***angle between*** $\mathbf{u}$ *and* $\mathbf{v}$. Observe that in R^2 or R^3 with the Euclidean inner product, (8) agrees with the usual formula for the cosine of the angle between two nonzero vectors [Formula (2)].

EXAMPLE 2 Cosine of an Angle Between Two Vectors in R^4

Let R^4 have the Euclidean inner product. Find the cosine of the angle θ between the vectors $\mathbf{u} = (4, 3, 1, -2)$ and $\mathbf{v} = (-2, 1, 2, 3)$.

Solution.

We leave it for the reader to verify that

$$\|\mathbf{u}\| = \sqrt{30}, \quad \|\mathbf{v}\| = \sqrt{18}, \quad \text{and} \quad \langle \mathbf{u}, \mathbf{v} \rangle = -9$$

so that

$$\cos\theta = \frac{\langle \mathbf{u}, \mathbf{v}\rangle}{\|\mathbf{u}\|\,\|\mathbf{v}\|} = -\frac{9}{\sqrt{30}\sqrt{18}} = -\frac{3}{2\sqrt{15}}$$ ◆

Orthogonality   Example 2 is primarily a mathematical exercise, for there is relatively little need to find angles between vectors, except in R^2 and R^3 with the Euclidean inner product. However, a problem of major importance in all inner product spaces is to determine whether two vectors are *orthogonal*—that is, whether the angle between them is $\theta = \pi/2$.

It follows from (8) that if $\mathbf{u}$ and $\mathbf{v}$ are *nonzero* vectors in an inner product space and θ is the angle between them, then $\cos\theta = 0$ if and only if $\langle \mathbf{u}, \mathbf{v}\rangle = 0$. Equivalently, for nonzero vectors we have $\theta = \pi/2$ if and only if $\langle \mathbf{u}, \mathbf{v}\rangle = 0$. If we agree to consider the angle between $\mathbf{u}$ and $\mathbf{v}$ to be $\pi/2$ when either or both of these vectors is $\mathbf{0}$, then we can state without exception that the angle between $\mathbf{u}$ and $\mathbf{v}$ is $\pi/2$ if and only if $\langle \mathbf{u}, \mathbf{v}\rangle = 0$. This suggests the following definition.

> **Definition**
>
> Two vectors $\mathbf{u}$ and $\mathbf{v}$ in an inner product space are called ***orthogonal*** if $\langle \mathbf{u}, \mathbf{v}\rangle = 0$.

Observe that in the special case where $\langle \mathbf{u}, \mathbf{v}\rangle = \mathbf{u} \cdot \mathbf{v}$ is the Euclidean inner product on R^n, this definition reduces to the definition of orthogonality in Euclidean n-space given in Section 4.1. We also emphasize that orthogonality depends on the inner product; two vectors can be orthogonal with respect to one inner product but not another.

EXAMPLE 3 Orthogonal Vectors in M_{22}

If M_{22} has the inner product of Example 7 in the preceding section, then the matrices

$$U = \begin{bmatrix} 1 & 0 \\ 1 & 1 \end{bmatrix} \quad \text{and} \quad V = \begin{bmatrix} 0 & 2 \\ 0 & 0 \end{bmatrix}$$

are orthogonal, since

$$\langle U, V\rangle = 1(0) + 0(2) + 1(0) + 1(0) = 0$$ ◆

Calculus Required

EXAMPLE 4 Orthogonal Vectors in P_2

Let P_2 have the inner product

$$\langle \mathbf{p}, \mathbf{q}\rangle = \int_{-1}^{1} p(x)q(x)\,dx$$

and let $\mathbf{p} = x$ and $\mathbf{q} = x^2$. Then

$$\|\mathbf{p}\| = \langle \mathbf{p}, \mathbf{p}\rangle^{1/2} = \left[\int_{-1}^{1} xx\,dx\right]^{1/2} = \left[\int_{-1}^{1} x^2\,dx\right]^{1/2} = \sqrt{\frac{2}{3}}$$

$$\|\mathbf{q}\| = \langle \mathbf{q}, \mathbf{q}\rangle^{1/2} = \left[\int_{-1}^{1} x^2 x^2\,dx\right]^{1/2} = \left[\int_{-1}^{1} x^4\,dx\right]^{1/2} = \sqrt{\frac{2}{5}}$$

$$\langle \mathbf{p}, \mathbf{q}\rangle = \int_{-1}^{1} xx^2\,dx = \int_{-1}^{1} x^3\,dx = 0$$

Because $\langle \mathbf{p}, \mathbf{q}\rangle = 0$, the vectors $\mathbf{p} = x$ and $\mathbf{q} = x^2$ are orthogonal relative to the given inner product. ◆

In Section 4.1 we proved the Theorem of Pythagoras for vectors in Euclidean n-space. The following theorem extends this result to vectors in any inner product space.

Theorem 6.2.4 **Generalized Theorem of Pythagoras**

*If **u** and **v** are orthogonal vectors in an inner product space, then*

$$\|\mathbf{u} + \mathbf{v}\|^2 = \|\mathbf{u}\|^2 + \|\mathbf{v}\|^2$$

Proof. The orthogonality of **u** and **v** implies that $\langle \mathbf{u}, \mathbf{v} \rangle = 0$, so

$$\|\mathbf{u} + \mathbf{v}\|^2 = \langle \mathbf{u} + \mathbf{v}, \mathbf{u} + \mathbf{v} \rangle = \|\mathbf{u}\|^2 + 2\langle \mathbf{u}, \mathbf{v} \rangle + \|\mathbf{v}\|^2$$
$$= \|\mathbf{u}\|^2 + \|\mathbf{v}\|^2 \qquad \blacksquare$$

Calculus Required

EXAMPLE 5 Theorem of Pythagoras in P_2

In Example 4 we showed that $\mathbf{p} = x$ and $\mathbf{q} = x^2$ are orthogonal relative to the inner product

$$\langle \mathbf{p}, \mathbf{q} \rangle = \int_{-1}^{1} p(x)q(x)\,dx$$

on P_2. It follows from the Theorem of Pythagoras that

$$\|\mathbf{p} + \mathbf{q}\|^2 = \|\mathbf{p}\|^2 + \|\mathbf{q}\|^2$$

Thus, from the computations in Example 4 we have

$$\|\mathbf{p} + \mathbf{q}\|^2 = \left(\sqrt{\frac{2}{3}}\right)^2 + \left(\sqrt{\frac{2}{5}}\right)^2 = \frac{2}{3} + \frac{2}{5} = \frac{16}{15}$$

We can check this result by direct integration:

$$\|\mathbf{p} + \mathbf{q}\|^2 = \langle \mathbf{p} + \mathbf{q}, \mathbf{p} + \mathbf{q} \rangle = \int_{-1}^{1} (x + x^2)(x + x^2)\,dx$$
$$= \int_{-1}^{1} x^2\,dx + 2\int_{-1}^{1} x^3\,dx + \int_{-1}^{1} x^4\,dx = \frac{2}{3} + 0 + \frac{2}{5} = \frac{16}{15} \qquad \blacklozenge$$

Orthogonal Complements If V is a plane through the origin of R^3 with the Euclidean inner product, then the set of all vectors that are orthogonal to every vector in V forms the line L through the origin that is perpendicular to V (Figure 6.2.2). In the language of linear algebra we say that the line and the plane are *orthogonal complements* of one another. The following definition extends this concept to general inner product spaces.

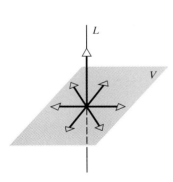

Figure 6.2.2
Every vector in L is orthogonal to every vector in V.

Definition

Let W be a subspace of an inner product space V. A vector **u** in V is said to be **orthogonal to W** if it is orthogonal to every vector in W, and the set of all vectors in V that are orthogonal to W is called the **orthogonal complement of W**.

Recall from geometry that the symbol $\perp$ is used to indicate perpendicularity. In linear algebra the orthogonal complement of a subspace W is denoted by $W^{\perp}$ (read "W perp"). The following theorem lists the basic properties of orthogonal complements.

> **Theorem 6.2.5** **Properties of Orthogonal Complements**
>
> *If W is a subspace of a finite-dimensional inner product space V, then:*
>
> (a) $W^\perp$ *is a subspace of V.*
> (b) *The only vector common to W and $W^\perp$ is* **0**.
> (c) *The orthogonal complement of $W^\perp$ is W; that is,* $(W^\perp)^\perp = W$.

We shall prove parts (*a*) and (*b*). The proof of (*c*) requires results covered later in this chapter, so its proof is left for the exercises at the end of the chapter.

Proof (a). Note first that $\langle \mathbf{0}, \mathbf{w} \rangle = 0$ for every vector $\mathbf{w}$ in W, so $W^\perp$ contains at least the zero vector. We want to show that $W^\perp$ is closed under addition and scalar multiplication; that is, we want to show that the sum of two vectors in $W^\perp$ is orthogonal to every vector in W and that any scalar multiple of a vector in $W^\perp$ is orthogonal to every vector in W. Let $\mathbf{u}$ and $\mathbf{v}$ be any vectors in $W^\perp$, let k be any scalar, and let $\mathbf{w}$ be any vector in W. Then from the definition of $W^\perp$ we have $\langle \mathbf{u}, \mathbf{w} \rangle = 0$ and $\langle \mathbf{v}, \mathbf{w} \rangle = 0$. Using basic properties of the inner product we have

$$\langle \mathbf{u} + \mathbf{v}, \mathbf{w} \rangle = \langle \mathbf{u}, \mathbf{w} \rangle + \langle \mathbf{v}, \mathbf{w} \rangle = 0 + 0 = 0$$
$$\langle k\mathbf{u}, \mathbf{w} \rangle = k\langle \mathbf{u}, \mathbf{w} \rangle = k(0) = 0$$

which proves that $\mathbf{u} + \mathbf{v}$ and $k\mathbf{u}$ are in $W^\perp$.

Proof (b). If $\mathbf{v}$ is common to W and $W^\perp$, then $\langle \mathbf{v}, \mathbf{v} \rangle = 0$, which implies that $\mathbf{v} = \mathbf{0}$ by Axiom 4 for inner products. ■

REMARK. Because W and $W^\perp$ are orthogonal complements of one another by part (*c*) of the preceding theorem, we shall say that W and $W^\perp$ are ***orthogonal complements***.

A Geometric Link Between Nullspace and Row Space

The following fundamental theorem provides a geometric link between the nullspace and row space of a matrix.

> **Theorem 6.2.6**
>
> *If A is an m × n matrix, then:*
>
> (a) *The nullspace of A and the row space of A are orthogonal complements in R^n with respect to the Euclidean inner product.*
> (b) *The nullspace of A^T and the column space of A are orthogonal complements in R^m with respect to the Euclidean inner product.*

Proof (a). We want to show that the orthogonal complement of the row space of A is the nullspace of A. To do this we must show that if a vector $\mathbf{v}$ is orthogonal to every vector in the row space, then $A\mathbf{v} = \mathbf{0}$, and conversely, if $A\mathbf{v} = \mathbf{0}$, then $\mathbf{v}$ is orthogonal to every vector in the row space.

Assume first that $\mathbf{v}$ is orthogonal to every vector in the row space of A. Then in particular $\mathbf{v}$ is orthogonal to the row vectors $\mathbf{r}_1, \mathbf{r}_2, \ldots, \mathbf{r}_n$ of A; that is,

$$\mathbf{r}_1 \cdot \mathbf{v} = \mathbf{r}_2 \cdot \mathbf{v} = \cdots = \mathbf{r}_n \cdot \mathbf{v} = 0 \qquad (9)$$

But by Formula (11) of Section 4.1, the linear system $A\mathbf{x} = \mathbf{0}$ can be expressed in dot product notation as

$$\begin{bmatrix} \mathbf{r}_1 \cdot \mathbf{x} \\ \mathbf{r}_2 \cdot \mathbf{x} \\ \vdots \\ \mathbf{r}_n \cdot \mathbf{x} \end{bmatrix} = \begin{bmatrix} 0 \\ 0 \\ \vdots \\ 0 \end{bmatrix} \tag{10}$$

so it follows from (9) that $\mathbf{v}$ is a solution of this system and hence lies in the nullspace of A.

Conversely, assume that $\mathbf{v}$ is a vector in the nullspace of A, so that $A\mathbf{v} = \mathbf{0}$. It follows from (10) that

$$\mathbf{r}_1 \cdot \mathbf{v} = \mathbf{r}_2 \cdot \mathbf{v} = \cdots = \mathbf{r}_n \cdot \mathbf{v} = 0$$

But if $\mathbf{r}$ is any vector in the row space of A, then $\mathbf{r}$ is expressible as a linear combination of the row vectors of A, say

$$\mathbf{r} = c_1 \mathbf{r}_1 + c_2 \mathbf{r}_2 + \cdots + c_n \mathbf{r}_n$$

Thus,

$$\begin{aligned} \mathbf{r} \cdot \mathbf{v} &= (c_1 \mathbf{r}_1 + c_2 \mathbf{r}_2 + \cdots + c_n \mathbf{r}_n) \cdot \mathbf{v} \\ &= c_1 (\mathbf{r}_1 \cdot \mathbf{v}) + c_2 (\mathbf{r}_2 \cdot \mathbf{v}) + \cdots + c_n (\mathbf{r}_n \cdot \mathbf{v}) \\ &= 0 + 0 + \cdots + 0 = 0 \end{aligned}$$

which proves that $\mathbf{v}$ is orthogonal to every vector in the row space of A.

Proof (b). Since the column space of A is the row space of A^T (except for a difference in notation), the proof follows by applying the result in part (*a*) to A^T. ∎

The following example shows how Theorem 6.2.6 can be used to find a basis for the orthogonal complement of a subspace of Euclidean n-space.

EXAMPLE 6 Basis for an Orthogonal Complement

Let W be the subspace of R^5 spanned by the vectors

$$\mathbf{w}_1 = (2, 2, -1, 0, 1), \qquad \mathbf{w}_2 = (-1, -1, 2, -3, 1),$$
$$\mathbf{w}_3 = (1, 1, -2, 0, -1), \qquad \mathbf{w}_4 = (0, 0, 1, 1, 1)$$

Find a basis for the orthogonal complement of W.

Solution.

The space W spanned by $\mathbf{w}_1$, $\mathbf{w}_2$, $\mathbf{w}_3$, and $\mathbf{w}_4$ is the same as the row space of the matrix

$$A = \begin{bmatrix} 2 & 2 & -1 & 0 & 1 \\ -1 & -1 & 2 & -3 & 1 \\ 1 & 1 & -2 & 0 & -1 \\ 0 & 0 & 1 & 1 & 1 \end{bmatrix}$$

and by part (*a*) of Theorem 6.2.6 the nullspace of A is the orthogonal complement of W. In Example 4 of Section 5.5 we showed that

$$\mathbf{v}_1 = \begin{bmatrix} -1 \\ 1 \\ 0 \\ 0 \\ 0 \end{bmatrix} \quad \text{and} \quad \mathbf{v}_2 = \begin{bmatrix} -1 \\ 0 \\ -1 \\ 0 \\ 1 \end{bmatrix}$$

form a basis for this nullspace. Expressing these vectors in the same notation as $\mathbf{w}_1$, $\mathbf{w}_2$, $\mathbf{w}_3$, and $\mathbf{w}_4$, we conclude that the vectors

$$\mathbf{v}_1 = (-1, 1, 0, 0, 0) \quad \text{and} \quad \mathbf{v}_2 = (-1, 0, -1, 0, 1)$$

form a basis for the orthogonal complement of W. As a check, the reader may want to verify that $\mathbf{v}_1$ and $\mathbf{v}_2$ are orthogonal to $\mathbf{w}_1$, $\mathbf{w}_2$, $\mathbf{w}_3$, and $\mathbf{w}_4$ by calculating the necessary dot products. ◆

Summary We leave it for the reader to show that in any inner product space V, the zero space $\{\mathbf{0}\}$ and the entire space V are orthogonal complements. Thus, if A is an $n \times n$ matrix, to say that $A\mathbf{x} = \mathbf{0}$ has only the trivial solution is equivalent to saying that the orthogonal complement of the nullspace of A is all of R^n, or equivalently, that the rowspace of A is all of R^n. This enables us to add two new results to the seventeen listed in Theorem 5.6.9.

Theorem 6.2.7 Equivalent Statements

If A is an $n \times n$ matrix, and if $T_A : R^n \to R^n$ is multiplication by A, then the following are equivalent.

(a) *A is invertible.*
(b) *$A\mathbf{x} = \mathbf{0}$ has only the trivial solution.*
(c) *The reduced row-echelon form of A is I_n.*
(d) *A is expressible as a product of elementary matrices.*
(e) *$A\mathbf{x} = \mathbf{b}$ is consistent for every $n \times 1$ matrix $\mathbf{b}$.*
(f) *$A\mathbf{x} = \mathbf{b}$ has exactly one solution for every $n \times 1$ matrix $\mathbf{b}$.*
(g) *$\det(A) \neq 0$.*
(h) *The range of T_A is R^n.*
(i) *T_A is one-to-one.*
(j) *The column vectors of A are linearly independent.*
(k) *The row vectors of A are linearly independent.*
(l) *The column vectors of A span R^n.*
(m) *The row vectors of A span R^n.*
(n) *The column vectors of A form a basis for R^n.*
(o) *The row vectors of A form a basis for R^n.*
(p) *A has rank n.*
(q) *A has nullity 0.*
(r) *The orthogonal complement of the nullspace of A is R^n.*
(s) *The orthogonal complement of the row space of A is $\{\mathbf{0}\}$.*

This theorem relates all of the major topics we have studied thus far.

Exercise Set 6.2

1. In each part determine whether the given vectors are orthogonal with respect to the Euclidean inner product.

   (a) $\mathbf{u} = (-1, 3, 2)$, $\mathbf{v} = (4, 2, -1)$ (b) $\mathbf{u} = (-2, -2, -2)$, $\mathbf{v} = (1, 1, 1)$ (c) $\mathbf{u} = (u_1, u_2, u_3)$, $\mathbf{v} = (0, 0, 0)$
   (d) $\mathbf{u} = (-4, 6, -10, 1)$, $\mathbf{v} = (2, 1, -2, 9)$ (e) $\mathbf{u} = (0, 3, -2, 1)$, $\mathbf{v} = (5, 2, -1, 0)$ (f) $\mathbf{u} = (a, b)$, $\mathbf{v} = (-b, a)$

2. Let R^4 have the Euclidean inner product, and let $\mathbf{u} = (-1, 1, 0, 2)$. Determine whether the vector $\mathbf{u}$ is orthogonal to the subspace spanned by the vectors $\mathbf{w}_1 = (0, 0, 0, 0)$, $\mathbf{w}_2 = (1, -1, 3, 0)$, and $\mathbf{w}_3 = (4, 0, 9, 2)$.

3. Let R^2, R^3, and R^4 have the Euclidean inner product. In each part find the cosine of the angle between $\mathbf{u}$ and $\mathbf{v}$.

(a) $\mathbf{u} = (1, -3)$, $\mathbf{v} = (2, 4)$ (b) $\mathbf{u} = (-1, 0)$, $\mathbf{v} = (3, 8)$

(c) $\mathbf{u} = (-1, 5, 2)$, $\mathbf{v} = (2, 4, -9)$ (d) $\mathbf{u} = (4, 1, 8)$, $\mathbf{v} = (1, 0, -3)$

(e) $\mathbf{u} = (1, 0, 1, 0)$, $\mathbf{v} = (-3, -3, -3, -3)$ (f) $\mathbf{u} = (2, 1, 7, -1)$, $\mathbf{v} = (4, 0, 0, 0)$

4. Let P_2 have the inner product in Example 8 of Section 6.1. Find the cosine of the angle between $\mathbf{p}$ and $\mathbf{q}$.

(a) $\mathbf{p} = -1 + 5x + 2x^2$, $\mathbf{q} = 2 + 4x - 9x^2$ (b) $\mathbf{p} = x - x^2$, $\mathbf{q} = 7 + 3x + 3x^2$

5. Show that $\mathbf{p} = 1 - x + 2x^2$ and $\mathbf{q} = 2x + x^2$ are orthogonal with respect to the inner product in Exercise 4.

6. Let M_{22} have the inner product in Example 7 of Section 6.1. Find the cosine of the angle between A and B.

(a) $A = \begin{bmatrix} 2 & 6 \\ 1 & -3 \end{bmatrix}$, $B = \begin{bmatrix} 3 & 2 \\ 1 & 0 \end{bmatrix}$ (b) $A = \begin{bmatrix} 2 & 4 \\ -1 & 3 \end{bmatrix}$, $B = \begin{bmatrix} -3 & 1 \\ 4 & 2 \end{bmatrix}$

7. Let

$$A = \begin{bmatrix} 2 & 1 \\ -1 & 3 \end{bmatrix}$$

Which of the following matrices are orthogonal to A with respect to the inner product in Exercise 6?

(a) $\begin{bmatrix} -3 & 0 \\ 0 & 2 \end{bmatrix}$ (b) $\begin{bmatrix} 1 & 1 \\ 0 & -1 \end{bmatrix}$ (c) $\begin{bmatrix} 0 & 0 \\ 0 & 0 \end{bmatrix}$ (d) $\begin{bmatrix} 2 & 1 \\ 5 & 2 \end{bmatrix}$

8. Let R^3 have the Euclidean inner product. For which values of k are $\mathbf{u}$ and $\mathbf{v}$ orthogonal?

(a) $\mathbf{u} = (2, 1, 3)$, $\mathbf{v} = (1, 7, k)$ (b) $\mathbf{u} = (k, k, 1)$, $\mathbf{v} = (k, 5, 6)$

9. Let R^4 have the Euclidean inner product. Find two vectors of norm 1 that are orthogonal to the three vectors $\mathbf{u} = (2, 1, -4, 0)$, $\mathbf{v} = (-1, -1, 2, 2)$, and $\mathbf{w} = (3, 2, 5, 4)$.

10. In each part verify that the Cauchy–Schwarz inequality holds for the given vectors using the Euclidean inner product.

(a) $\mathbf{u} = (3, 2)$, $\mathbf{v} = (4, -1)$ (b) $\mathbf{u} = (-3, 1, 0)$, $\mathbf{v} = (2, -1, 3)$

(c) $\mathbf{u} = (-4, 2, 1)$, $\mathbf{v} = (8, -4, -2)$ (d) $\mathbf{u} = (0, -2, 2, 1)$, $\mathbf{v} = (-1, -1, 1, 1)$

11. In each part verify that the Cauchy–Schwarz inequality holds for the given vectors.

(a) $\mathbf{u} = (-2, 1)$ and $\mathbf{v} = (1, 0)$ using the inner product of Example 2 of Section 6.1

(b) $U = \begin{bmatrix} -1 & 2 \\ 6 & 1 \end{bmatrix}$ and $V = \begin{bmatrix} 1 & 0 \\ 3 & 3 \end{bmatrix}$

using the inner product in Example 7 of Section 6.1

(c) $\mathbf{p} = -1 + 2x + x^2$ and $\mathbf{q} = 2 - 4x^2$ using the inner product given in Example 8 of Section 6.1

12. Let W be the line in R^2 with equation $y = 2x$. Find an equation for $W^\perp$.

13. (a) Let W be the plane in R^3 with equation $x - 2y - 3z = 0$. Find parametric equations for $W^\perp$.

(b) Let W be the line in R^3 with parametric equations

$$x = 2t, \quad y = -5t, \quad z = 4t \quad (-\infty < t < \infty)$$

Find an equation for $W^\perp$.

(c) Let W be the intersection of the two planes

$$x + y + z = 0 \quad \text{and} \quad x - y + z = 0$$

in R^3. Find an equation for $W^\perp$.

14. Let

$$A = \begin{bmatrix} 1 & 2 & -1 & 2 \\ 3 & 5 & 0 & 4 \\ 1 & 1 & 2 & 0 \end{bmatrix}$$

(a) Find bases for the row space and nullspace of A.

(b) Verify that every vector in the row space is orthogonal to every vector in the nullspace (as guaranteed by Theorem 6.2.6a).

15. Let A be the matrix in Exercise 14.

(a) Find bases for the column space of A and nullspace of A^T.

(b) Verify that every vector in the column space of A is orthogonal to every vector in the nullspace of A^T (as guaranteed by Theorem 6.2.6b).

16. Find a basis for the orthogonal complement of the subspace of R^n spanned by the vectors.

(a) $\mathbf{v}_1 = (1, -1, 3)$, $\mathbf{v}_2 = (5, -4, -4)$, $\mathbf{v}_3 = (7, -6, 2)$

(b) $\mathbf{v}_1 = (2, 0, -1)$, $\mathbf{v}_2 = (4, 0, -2)$

(c) $\mathbf{v}_1 = (1, 4, 5, 2)$, $\mathbf{v}_2 = (2, 1, 3, 0)$, $\mathbf{v}_3 = (-1, 3, 2, 2)$

(d) $\mathbf{v}_1 = (1, 4, 5, 6, 9)$, $\mathbf{v}_2 = (3, -2, 1, 4, -1)$, $\mathbf{v}_3 = (-1, 0, -1, -2, -1)$, $\mathbf{v}_4 = (2, 3, 5, 7, 8)$

17. Let V be an inner product space. Show that if $\mathbf{u}$ and $\mathbf{v}$ are orthogonal vectors in V such that $\|\mathbf{u}\| = \|\mathbf{v}\| = 1$, then $\|\mathbf{u} - \mathbf{v}\| = \sqrt{2}$.

18. Let V be an inner product space. Show that if $\mathbf{w}$ is orthogonal to both $\mathbf{u}_1$ and $\mathbf{u}_2$, it is orthogonal to $k_1\mathbf{u}_1 + k_2\mathbf{u}_2$ for all scalars k_1 and k_2. Interpret this result geometrically in the case where V is R^3 with the Euclidean inner product.

19. Let V be an inner product space. Show that if $\mathbf{w}$ is orthogonal to each of the vectors $\mathbf{u}_1, \mathbf{u}_2, \ldots, \mathbf{u}_r$, then it is orthogonal to every vector in span$\{\mathbf{u}_1, \mathbf{u}_2, \ldots, \mathbf{u}_r\}$.

20. Let $\{\mathbf{v}_1, \mathbf{v}_2, \ldots, \mathbf{v}_r\}$ be a basis for an inner product space V. Show that the zero vector is the only vector in V that is orthogonal to all of the basis vectors.

21. Let $\{\mathbf{w}_1, \mathbf{w}_2, \ldots, \mathbf{w}_k\}$ be a basis for a subspace W of V. Show that $W^\perp$ consists of all vectors in V that are orthogonal to every basis vector.

22. Prove the following generalization of Theorem 6.2.4. If $\mathbf{v}_1, \mathbf{v}_2, \ldots, \mathbf{v}_r$ are pairwise orthogonal vectors in an inner product space V, then

$$\|\mathbf{v}_1 + \mathbf{v}_2 + \cdots + \mathbf{v}_r\|^2 = \|\mathbf{v}_1\|^2 + \|\mathbf{v}_2\|^2 + \cdots + \|\mathbf{v}_r\|^2$$

23. Prove the following parts of Theorem 6.2.2:

(a) part (a)   (b) part (b)   (c) part (c)

24. Prove the following parts of Theorem 6.2.3:

(a) part (a)   (b) part (b)   (c) part (c)   (d) part (d)

25. Prove: If $\mathbf{u}$ and $\mathbf{v}$ are $n \times 1$ matrices and A is an $n \times n$ matrix, then

$$(\mathbf{v}^T A^T A \mathbf{u})^2 \leq (\mathbf{u}^T A^T A \mathbf{u})(\mathbf{v}^T A^T A \mathbf{v})$$

26. Use the Cauchy–Schwarz inequality to prove that for all real values of a, b, and θ,

$$(a \cos \theta + b \sin \theta)^2 \leq a^2 + b^2$$

27. Prove: If $w_1, w_2, \ldots, w_n$ are positive real numbers and if $\mathbf{u} = (u_1, u_2, \ldots, u_n)$ and $\mathbf{v} = (v_1, v_2, \ldots, v_n)$ are any two vectors in R^n, then

$$|w_1 u_1 v_1 + w_2 u_2 v_2 + \cdots + w_n u_n v_n| \leq (w_1 u_1^2 + w_2 u_2^2 + \cdots + w_n u_n^2)^{1/2} (w_1 v_1^2 + w_2 v_2^2 + \cdots + w_n v_n^2)^{1/2}$$

28. Show that equality holds in the Cauchy–Schwarz inequality if and only if $\mathbf{u}$ and $\mathbf{v}$ are linearly dependent.

29. Use vector methods to prove that a triangle that is inscribed in a circle so that it has a diameter for a side must be a right triangle. [**Hint.** Express the vectors $\overrightarrow{AB}$ and $\overrightarrow{BC}$ in the accompanying figure in terms of **u** and **v**.]

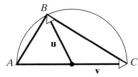

Figure Ex-29

30. With respect to the Euclidean inner product, the vectors $\mathbf{u} = (1, \sqrt{3}\,)$ and $\mathbf{v} = (-1, \sqrt{3}\,)$ have norm 2, and the angle between them is 60° (see accompanying figure). Find a weighted Euclidean inner product with respect to which **u** and **v** are orthogonal unit vectors.

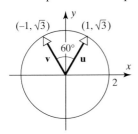

Figure Ex-30

31. (*For readers who have studied calculus.*) Let $f(x)$ and $g(x)$ be continuous functions on $[0, 1]$. Prove:

(a) $\left[\int_0^1 f(x)g(x)\,dx \right]^2 \le \left[\int_0^1 f^2(x)\,dx \right] \left[\int_0^1 g^2(x)\,dx \right]$

(b) $\left[\int_0^1 [f(x) + g(x)]^2\,dx \right]^{1/2} \le \left[\int_0^1 f^2(x)\,dx \right]^{1/2} + \left[\int_0^1 g^2(x)\,dx \right]^{1/2}$

[**Hint.** Use the Cauchy–Schwarz inequality.]

32. (*For readers who have studied calculus.*) Let $C[0, \pi]$ have the inner product

$$\langle \mathbf{f}, \mathbf{g} \rangle = \int_0^\pi f(x)g(x)\,dx$$

and let $\mathbf{f}_n = \cos nx$ $(n = 0, 1, 2, \ldots)$. Show that if $k \ne l$, then $\mathbf{f}_k$ and $\mathbf{f}_l$ are orthogonal with respect to the given inner product.

Discussion and Discovery

33. (a) Let W be the line $y = x$ in an xy-coordinate system in R^2. Describe the subspace $W^\perp$.

(b) Let W be the y-axis in an xyz-coordinate system in R^3. Describe the subspace $W^\perp$.

(c) Let W be the yz-plane of an xyz-coordinate system in R^3. Describe the subspace $W^\perp$.

34. Let $A\mathbf{x} = \mathbf{0}$ be a homogeneous system of three equations in the unknowns x, y, and z.

(a) If the solution space is a line through the origin in R^3, what kind of geometric object is the row space of A? Explain your reasoning.

(b) If the column space of A is a line through the origin, what kind of geometric object is the solution space of the homogeneous system $A^T\mathbf{x} = \mathbf{0}$? Explain your reasoning.

(c) If the homogeneous system $A^T\mathbf{x} = \mathbf{0}$ has a unique solution, what can you say about the row space and column space of A? Explain your reasoning.

35. Indicate whether the statement is always true or sometimes false. Justify your answer by giving a logical argument or a counterexample.

(a) If V is a subspace of R^n and W is a subspace of V, then $W^\perp$ is a subspace of $V^\perp$.

(b) $\|\mathbf{u} + \mathbf{v} + \mathbf{w}\| \le \|\mathbf{u}\| + \|\mathbf{v}\| + \|\mathbf{w}\|$ for all vectors **u**, **v**, and **w** in an inner product space.

(c) If **u** is in the row space and the nullspace of a square matrix A, then $\mathbf{u} = \mathbf{0}$.

(d) If **u** is in the row space and the column space of an $n \times n$ matrix A, then $\mathbf{u} = \mathbf{0}$.

36. Let M_{22} have the inner product $\langle U, V \rangle = \text{tr}(U^T V) = \text{tr}(V^T U)$ that was defined in Example 7 of Section 6.1. Describe the orthogonal complement of

(a) the subspace of all diagonal matrices   (b) the subspace of symmetric matrices

6.3 ORTHONORMAL BASES; GRAM–SCHMIDT PROCESS; QR-DECOMPOSITION

In many problems involving vector spaces, the problem solver is free to choose any basis for the vector space that seems appropriate. In inner product spaces the solution of a problem is often greatly simplified by choosing a basis in which the vectors are orthogonal to one another. In this section we shall show how such bases can be obtained.

Definition

A set of vectors in an inner product space is called an ***orthogonal set*** if all pairs of distinct vectors in the set are orthogonal. An orthogonal set in which each vector has norm 1 is called ***orthonormal***.

EXAMPLE 1   An Orthogonal Set in R^3

Let

$$\mathbf{u}_1 = (0, 1, 0), \qquad \mathbf{u}_2 = (1, 0, 1), \qquad \mathbf{u}_3 = (1, 0, -1)$$

and assume that R^3 has the Euclidean inner product. It follows that the set of vectors $S = \{\mathbf{u}_1, \mathbf{u}_2, \mathbf{u}_3\}$ is orthogonal since $\langle \mathbf{u}_1, \mathbf{u}_2 \rangle = \langle \mathbf{u}_1, \mathbf{u}_3 \rangle = \langle \mathbf{u}_2, \mathbf{u}_3 \rangle = 0$.   ◆

If $\mathbf{v}$ is a nonzero vector in an inner product space, then by part (c) of Theorem 6.2.2 the vector

$$\frac{1}{\|\mathbf{v}\|} \mathbf{v}$$

has norm 1, since

$$\left\| \frac{1}{\|\mathbf{v}\|} \mathbf{v} \right\| = \left| \frac{1}{\|\mathbf{v}\|} \right| \|\mathbf{v}\| = \frac{1}{\|\mathbf{v}\|} \|\mathbf{v}\| = 1$$

The process of multiplying a nonzero vector $\mathbf{v}$ by the reciprocal of its length to obtain a vector of norm 1 is called ***normalizing*** $\mathbf{v}$. An orthogonal set of *nonzero* vectors can always be converted to an orthonormal set by normalizing each of its vectors.

EXAMPLE 2   Constructing an Orthonormal Set

The Euclidean norms of the vectors in Example 1 are

$$\|\mathbf{u}_1\| = 1, \qquad \|\mathbf{u}_2\| = \sqrt{2}, \qquad \|\mathbf{u}_3\| = \sqrt{2}$$

Consequently, normalizing $\mathbf{u}_1$, $\mathbf{u}_2$, and $\mathbf{u}_3$ yields

$$\mathbf{v}_1 = \frac{\mathbf{u}_1}{\|\mathbf{u}_1\|} = (0, 1, 0), \quad \mathbf{v}_2 = \frac{\mathbf{u}_2}{\|\mathbf{u}_2\|} = \left(\frac{1}{\sqrt{2}}, 0, \frac{1}{\sqrt{2}}\right),$$

$$\mathbf{v}_3 = \frac{\mathbf{u}_3}{\|\mathbf{u}_3\|} = \left(\frac{1}{\sqrt{2}}, 0, -\frac{1}{\sqrt{2}}\right)$$

We leave it for you to verify that the set $S = \{\mathbf{v}_1, \mathbf{v}_2, \mathbf{v}_3\}$ is orthonormal by showing that

$$\langle \mathbf{v}_1, \mathbf{v}_2 \rangle = \langle \mathbf{v}_1, \mathbf{v}_3 \rangle = \langle \mathbf{v}_2, \mathbf{v}_3 \rangle = 0 \quad \text{and} \quad \|\mathbf{v}_1\| = \|\mathbf{v}_2\| = \|\mathbf{v}_3\| = 1 \qquad \blacklozenge$$

In an inner product space, a basis consisting of orthonormal vectors is called an **orthonormal basis**, and a basis consisting of orthogonal vectors is called an **orthogonal basis**. A familiar example of an orthonormal basis is the standard basis for R^3 with the Euclidean inner product:

$$\mathbf{i} = (1, 0, 0), \quad \mathbf{j} = (0, 1, 0), \quad \mathbf{k} = (0, 0, 1)$$

This is the basis that is associated with rectangular coordinate systems (see Figure 5.4.4). More generally, in R^n with the Euclidean inner product, the standard basis

$$\mathbf{e}_1 = (1, 0, 0, \ldots, 0), \quad \mathbf{e}_2 = (0, 1, 0, \ldots, 0), \ldots, \quad \mathbf{e}_n = (0, 0, 0, \ldots, 1)$$

is orthonormal.

Coordinates Relative to Orthonormal Bases
The interest in finding orthonormal bases for inner product spaces is motivated in part by the following theorem, which shows that it is exceptionally simple to express a vector in terms of an orthonormal basis.

Theorem 6.3.1

If $S = \{\mathbf{v}_1, \mathbf{v}_2, \ldots, \mathbf{v}_n\}$ is an orthonormal basis for an inner product space V, and $\mathbf{u}$ is any vector in V, then

$$\mathbf{u} = \langle \mathbf{u}, \mathbf{v}_1 \rangle \mathbf{v}_1 + \langle \mathbf{u}, \mathbf{v}_2 \rangle \mathbf{v}_2 + \cdots + \langle \mathbf{u}, \mathbf{v}_n \rangle \mathbf{v}_n$$

Proof. Since $S = \{\mathbf{v}_1, \mathbf{v}_2, \ldots, \mathbf{v}_n\}$ is a basis, a vector $\mathbf{u}$ can be expressed in the form

$$\mathbf{u} = k_1 \mathbf{v}_1 + k_2 \mathbf{v}_2 + \cdots + k_n \mathbf{v}_n$$

We shall complete the proof by showing that $k_i = \langle \mathbf{u}, \mathbf{v}_i \rangle$ for $i = 1, 2, \ldots, n$. For each vector $\mathbf{v}_i$ in S we have

$$\langle \mathbf{u}, \mathbf{v}_i \rangle = \langle k_1 \mathbf{v}_1 + k_2 \mathbf{v}_2 + \cdots + k_n \mathbf{v}_n, \mathbf{v}_i \rangle$$
$$= k_1 \langle \mathbf{v}_1, \mathbf{v}_i \rangle + k_2 \langle \mathbf{v}_2, \mathbf{v}_i \rangle + \cdots + k_n \langle \mathbf{v}_n, \mathbf{v}_i \rangle$$

Since $S = \{\mathbf{v}_1, \mathbf{v}_2, \ldots, \mathbf{v}_n\}$ is an orthonormal set, we have

$$\langle \mathbf{v}_i, \mathbf{v}_i \rangle = \|\mathbf{v}_i\|^2 = 1 \quad \text{and} \quad \langle \mathbf{v}_j, \mathbf{v}_i \rangle = 0 \quad \text{if } j \neq i$$

Therefore, the above expression for $\langle \mathbf{u}, \mathbf{v}_i \rangle$ simplifies to

$$\langle \mathbf{u}, \mathbf{v}_i \rangle = k_i \qquad \blacksquare$$

Using the terminology and notation introduced in Section 5.4, the scalars

$$\langle \mathbf{u}, \mathbf{v}_1 \rangle, \langle \mathbf{u}, \mathbf{v}_2 \rangle, \ldots, \langle \mathbf{u}, \mathbf{v}_n \rangle$$

in Theorem 6.3.1 are the coordinates of the vector $\mathbf{u}$ relative to the orthonormal basis $S = \{\mathbf{v}_1, \mathbf{v}_2, \ldots, \mathbf{v}_n\}$ and

$$(\mathbf{u})_S = (\langle \mathbf{u}, \mathbf{v}_1 \rangle, \langle \mathbf{u}, \mathbf{v}_2 \rangle, \ldots, \langle \mathbf{u}, \mathbf{v}_n \rangle)$$

is the coordinate vector of $\mathbf{u}$ relative to this basis.

EXAMPLE 3 Coordinate Vector Relative to an Orthonormal Basis

Let

$$\mathbf{v}_1 = (0, 1, 0), \qquad \mathbf{v}_2 = \left(-\tfrac{4}{5}, 0, \tfrac{3}{5}\right), \qquad \mathbf{v}_3 = \left(\tfrac{3}{5}, 0, \tfrac{4}{5}\right)$$

It is easy to check that $S = \{\mathbf{v}_1, \mathbf{v}_2, \mathbf{v}_3\}$ is an orthonormal basis for R^3 with the Euclidean inner product. Express the vector $\mathbf{u} = (1, 1, 1)$ as a linear combination of the vectors in S, and find the coordinate vector $(\mathbf{u})_S$.

Solution.

$$\langle \mathbf{u}, \mathbf{v}_1 \rangle = 1, \quad \langle \mathbf{u}, \mathbf{v}_2 \rangle = -\tfrac{1}{5}, \quad \text{and} \quad \langle \mathbf{u}, \mathbf{v}_3 \rangle = \tfrac{7}{5}$$

Therefore, by Theorem 6.3.1 we have

$$\mathbf{u} = \mathbf{v}_1 - \tfrac{1}{5}\mathbf{v}_2 + \tfrac{7}{5}\mathbf{v}_3$$

that is,

$$(1, 1, 1) = (0, 1, 0) - \tfrac{1}{5}\left(-\tfrac{4}{5}, 0, \tfrac{3}{5}\right) + \tfrac{7}{5}\left(\tfrac{3}{5}, 0, \tfrac{4}{5}\right)$$

The coordinate vector of $\mathbf{u}$ relative to S is

$$(\mathbf{u})_S = (\langle \mathbf{u}, \mathbf{v}_1 \rangle, \langle \mathbf{u}, \mathbf{v}_2 \rangle, \langle \mathbf{u}, \mathbf{v}_3 \rangle) = \left(1, -\tfrac{1}{5}, \tfrac{7}{5}\right) \qquad \blacklozenge$$

REMARK. The usefulness of Theorem 6.3.1 should be evident from this example if it is kept in mind that for nonorthonormal bases, it is usually necessary to solve a system of equations in order to express a vector in terms of the basis.

Orthonormal bases for inner product spaces are convenient because, as the following theorem shows, many familiar formulas hold for such bases.

Theorem 6.3.2

If S is an orthonormal basis for an n-dimensional inner product space, and if

$$(\mathbf{u})_S = (u_1, u_2, \ldots, u_n) \quad and \quad (\mathbf{v})_S = (v_1, v_2, \ldots, v_n)$$

then:

(a) $\|\mathbf{u}\| = \sqrt{u_1^2 + u_2^2 + \cdots + u_n^2}$

(b) $d(\mathbf{u}, \mathbf{v}) = \sqrt{(u_1 - v_1)^2 + (u_2 - v_2)^2 + \cdots + (u_n - v_n)^2}$

(c) $\langle \mathbf{u}, \mathbf{v} \rangle = u_1 v_1 + u_2 v_2 + \cdots + u_n v_n$

The proof is left for the exercises.

REMARK. Observe that the right side of the equality in part (a) is the norm of the coordinate vector $(\mathbf{u})_S$ with respect to the Euclidean inner product on R^n, and the right side of the equality in part (c) is the Euclidean inner product of $(\mathbf{u})_S$ and $(\mathbf{v})_S$. Thus, by working with orthonormal bases, the computation of general norms and inner products can be reduced to the computation of Euclidean norms and inner products of the coordinate vectors.

EXAMPLE 4 Calculating Norms Using Orthonormal Bases

If R^3 has the Euclidean inner product, then the norm of the vector $\mathbf{u} = (1, 1, 1)$ is

$$\|\mathbf{u}\| = (\mathbf{u} \cdot \mathbf{u})^{1/2} = \sqrt{1^2 + 1^2 + 1^2} = \sqrt{3}$$

However, if we let R^3 have the orthonormal basis S in the last example, then we know from that example that the coordinate vector of $\mathbf{u}$ relative to S is

$$(\mathbf{u})_S = \left(1, -\tfrac{1}{5}, \tfrac{7}{5}\right)$$

The norm of $\mathbf{u}$ can also be calculated from this vector using part (a) of Theorem 6.3.2. This yields

$$\|\mathbf{u}\| = \sqrt{1^2 + \left(-\tfrac{1}{5}\right)^2 + \left(\tfrac{7}{5}\right)^2} = \sqrt{\tfrac{75}{25}} = \sqrt{3} \qquad \blacklozenge$$

Coordinates Relative to Orthogonal Bases

If $S = \{\mathbf{v}_1, \mathbf{v}_2, \ldots, \mathbf{v}_n\}$ is an *orthogonal* basis for a vector space V, then normalizing each of these vectors yields the orthonormal basis

$$S' = \left\{ \frac{\mathbf{v}_1}{\|\mathbf{v}_1\|}, \frac{\mathbf{v}_2}{\|\mathbf{v}_2\|}, \ldots, \frac{\mathbf{v}_n}{\|\mathbf{v}_n\|} \right\}$$

Thus, if $\mathbf{u}$ is any vector in V, it follows from Theorem 6.3.1 that

$$\mathbf{u} = \left\langle \mathbf{u}, \frac{\mathbf{v}_1}{\|\mathbf{v}_1\|} \right\rangle \frac{\mathbf{v}_1}{\|\mathbf{v}_1\|} + \left\langle \mathbf{u}, \frac{\mathbf{v}_2}{\|\mathbf{v}_2\|} \right\rangle \frac{\mathbf{v}_2}{\|\mathbf{v}_2\|} + \cdots + \left\langle \mathbf{u}, \frac{\mathbf{v}_n}{\|\mathbf{v}_n\|} \right\rangle \frac{\mathbf{v}_n}{\|\mathbf{v}_n\|}$$

which by part (c) of Theorem 6.1.1 can be rewritten as

$$\mathbf{u} - \frac{\langle \mathbf{u}, \mathbf{v}_1 \rangle}{\|\mathbf{v}_1\|^2} \mathbf{v}_1 + \frac{\langle \mathbf{u}, \mathbf{v}_2 \rangle}{\|\mathbf{v}_2\|^2} \mathbf{v}_2 + \cdots + \frac{\langle \mathbf{u}, \mathbf{v}_n \rangle}{\|\mathbf{v}_n\|^2} \mathbf{v}_n \tag{1}$$

This formula expresses $\mathbf{u}$ as a linear combination of the vectors in the orthogonal basis S. Some problems requiring the use of this formula are given in the exercises.

It is self-evident that if $\mathbf{v}_1$, $\mathbf{v}_2$, and $\mathbf{v}_3$ are three nonzero mutually perpendicular vectors in R^3, then none of these vectors lies in the same plane as the other two; that is, the vectors are linearly independent. The following theorem generalizes this result.

Theorem 6.3.3

If $S = \{\mathbf{v}_1, \mathbf{v}_2, \ldots, \mathbf{v}_n\}$ is an orthogonal set of nonzero vectors in an inner product space, then S is linearly independent.

Proof. Assume that

$$k_1 \mathbf{v}_1 + k_2 \mathbf{v}_2 + \cdots + k_n \mathbf{v}_n = \mathbf{0} \tag{2}$$

To demonstrate that $S = \{\mathbf{v}_1, \mathbf{v}_2, \ldots, \mathbf{v}_n\}$ is linearly independent, we must prove that $k_1 = k_2 = \cdots = k_n = 0$.

For each $\mathbf{v}_i$ in S, it follows from (2) that

$$\langle k_1 \mathbf{v}_1 + k_2 \mathbf{v}_2 + \cdots + k_n \mathbf{v}_n, \mathbf{v}_i \rangle = \langle \mathbf{0}, \mathbf{v}_i \rangle = 0$$

or equivalently,

$$k_1 \langle \mathbf{v}_1, \mathbf{v}_i \rangle + k_2 \langle \mathbf{v}_2, \mathbf{v}_i \rangle + \cdots + k_n \langle \mathbf{v}_n, \mathbf{v}_i \rangle = 0$$

From the orthogonality of S it follows that $\langle \mathbf{v}_j, \mathbf{v}_i \rangle = 0$ when $j \neq i$, so that this equation reduces to

$$k_i \langle \mathbf{v}_i, \mathbf{v}_i \rangle = 0$$

Since the vectors in S are assumed to be nonzero, $\langle \mathbf{v}_i, \mathbf{v}_i \rangle \neq 0$ by the positivity axiom for inner products. Therefore, $k_i = 0$. Since the subscript i is arbitrary, we have $k_1 = k_2 = \cdots = k_n = 0$; thus, S is linearly independent. ∎

EXAMPLE 5 Using Theorem 6.3.3

In Example 2 we showed that the vectors

$$\mathbf{v}_1 = (0, 1, 0), \quad \mathbf{v}_2 = \left(\frac{1}{\sqrt{2}}, 0, \frac{1}{\sqrt{2}} \right), \quad \text{and} \quad \mathbf{v}_3 = \left(\frac{1}{\sqrt{2}}, 0, -\frac{1}{\sqrt{2}} \right)$$

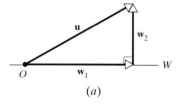

(a)

form an orthonormal set with respect to the Euclidean inner product on R^3. By Theorem 6.3.3, these vectors form a linearly independent set, and since R^3 is three-dimensional, $S = \{\mathbf{v}_1, \mathbf{v}_2, \mathbf{v}_3\}$ is an orthonormal basis for R^3 by Theorem 5.4.5. ◆

Orthogonal Projections   We shall now develop some results that will help us to construct orthogonal and orthonormal bases for inner product spaces.

In R^2 or R^3 with the Euclidean inner product, it is evident geometrically that if W is a line or a plane through the origin, then each vector $\mathbf{u}$ in the space can be expressed as a sum

$$\mathbf{u} = \mathbf{w}_1 + \mathbf{w}_2$$

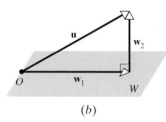

(b)

Figure 6.3.1

where $\mathbf{w}_1$ is in W and $\mathbf{w}_2$ is perpendicular to W (Figure 6.3.1). This result is a special case of the following general theorem whose proof is given at the end of this section.

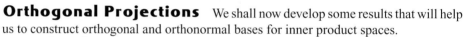

Theorem 6.3.4   **Projection Theorem**

If W is a finite-dimensional subspace of an inner product space V, then every vector $\mathbf{u}$ in V can be expressed in exactly one way as

$$\mathbf{u} = \mathbf{w}_1 + \mathbf{w}_2 \tag{3}$$

where $\mathbf{w}_1$ is in W and $\mathbf{w}_2$ is in $W^\perp$.

The vector $\mathbf{w}_1$ in the preceding theorem is called the ***orthogonal projection of* u *on*** W and is denoted by $\operatorname{proj}_W \mathbf{u}$. The vector $\mathbf{w}_2$ is called the ***component of* u *orthogonal to* W** and is denoted by $\operatorname{proj}_{W^\perp} \mathbf{u}$. Thus, Formula (3) in the Projection Theorem can be expressed as

$$\mathbf{u} = \operatorname{proj}_W \mathbf{u} + \operatorname{proj}_{W^\perp} \mathbf{u} \tag{4}$$

Since $\mathbf{w}_2 = \mathbf{u} - \mathbf{w}_1$, it follows that

$$\operatorname{proj}_{W^\perp} \mathbf{u} = \mathbf{u} - \operatorname{proj}_W \mathbf{u}$$

so Formula (4) can also be written as

$$\mathbf{u} = \operatorname{proj}_W \mathbf{u} + (\mathbf{u} - \operatorname{proj}_W \mathbf{u}) \tag{5}$$

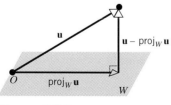

Figure 6.3.2

(Figure 6.3.2).

The following theorem, whose proof is requested in the exercises, provides formulas for calculating orthogonal projections.

Theorem 6.3.5

Let W be a finite-dimensional subspace of an inner product space V.

(a) If $\{\mathbf{v}_1, \mathbf{v}_2, \ldots, \mathbf{v}_r\}$ *is an orthonormal basis for W, and* $\mathbf{u}$ *is any vector in V, then*

$$\text{proj}_W \mathbf{u} = \langle \mathbf{u}, \mathbf{v}_1 \rangle \mathbf{v}_1 + \langle \mathbf{u}, \mathbf{v}_2 \rangle \mathbf{v}_2 + \cdots + \langle \mathbf{u}, \mathbf{v}_r \rangle \mathbf{v}_r \tag{6}$$

(b) If $\{\mathbf{v}_1, \mathbf{v}_2, \ldots, \mathbf{v}_r\}$ *is an orthogonal basis for W, and* $\mathbf{u}$ *is any vector in V, then*

$$\text{proj}_W \mathbf{u} = \frac{\langle \mathbf{u}, \mathbf{v}_1 \rangle}{\|\mathbf{v}_1\|^2} \mathbf{v}_1 + \frac{\langle \mathbf{u}, \mathbf{v}_2 \rangle}{\|\mathbf{v}_2\|^2} \mathbf{v}_2 + \cdots + \frac{\langle \mathbf{u}, \mathbf{v}_r \rangle}{\|\mathbf{v}_r\|^2} \mathbf{v}_r \tag{7}$$

EXAMPLE 6   Calculating Projections

Let R^3 have the Euclidean inner product, and let W be the subspace spanned by the orthonormal vectors $\mathbf{v}_1 = (0, 1, 0)$ and $\mathbf{v}_2 = \left(-\frac{4}{5}, 0, \frac{3}{5}\right)$. From (6) the orthogonal projection of $\mathbf{u} = (1, 1, 1)$ on W is

$$\text{proj}_W \mathbf{u} = \langle \mathbf{u}, \mathbf{v}_1 \rangle \mathbf{v}_1 + \langle \mathbf{u}, \mathbf{v}_2 \rangle \mathbf{v}_2$$
$$= (1)(0, 1, 0) + \left(-\tfrac{1}{5}\right)\left(-\tfrac{4}{5}, 0, \tfrac{3}{5}\right)$$
$$= \left(\tfrac{4}{25}, 1, -\tfrac{3}{25}\right)$$

The component of $\mathbf{u}$ orthogonal to W is

$$\text{proj}_{W^\perp} \mathbf{u} = \mathbf{u} - \text{proj}_W \mathbf{u} = (1, 1, 1) - \left(\tfrac{4}{25}, 1, -\tfrac{3}{25}\right) = \left(\tfrac{21}{25}, 0, \tfrac{28}{25}\right)$$

Observe that $\text{proj}_{W^\perp} \mathbf{u}$ is orthogonal to both $\mathbf{v}_1$ and $\mathbf{v}_2$ so that this vector is orthogonal to each vector in the space W spanned by $\mathbf{v}_1$ and $\mathbf{v}_2$ as it should be.   ◆

Finding Orthogonal and Orthonormal Bases

We have seen that orthonormal bases enjoy a variety of useful properties. Our next theorem, which is the main result in this section, shows that every nonzero finite-dimensional vector space has an orthonormal basis. The proof of this result is extremely important, since it provides an algorithm, or method, for converting an arbitrary basis into an orthonormal basis.

Theorem 6.3.6

Every nonzero finite-dimensional inner product space has an orthonormal basis.

Proof. Let V be any nonzero finite-dimensional inner product space, and suppose that $\{\mathbf{u}_1, \mathbf{u}_2, \ldots, \mathbf{u}_n\}$ is any basis for V. It suffices to show that V has an orthogonal basis, since the vectors in the orthogonal basis can be normalized to produce an orthonormal basis for V. The following sequence of steps will produce an orthogonal basis $\{\mathbf{v}_1, \mathbf{v}_2, \ldots, \mathbf{v}_n\}$ for V.

Step 1. Let $\mathbf{v}_1 = \mathbf{u}_1$.

Step 2. As illustrated in Figure 6.3.3, we can obtain a vector $\mathbf{v}_2$ that is orthogonal to $\mathbf{v}_1$ by computing the component of $\mathbf{u}_2$ that is orthogonal to the space W_1 spanned by $\mathbf{v}_1$. We use Formula (7):

$$\mathbf{v}_2 = \mathbf{u}_2 - \text{proj}_{W_1} \mathbf{u}_2 = \mathbf{u}_2 - \frac{\langle \mathbf{u}_2, \mathbf{v}_1 \rangle}{\|\mathbf{v}_1\|^2} \mathbf{v}_1$$

Of course, if $\mathbf{v}_2 = \mathbf{0}$, then $\mathbf{v}_2$ is not a basis vector. But this cannot happen, since it would then follow from the preceding formula for $\mathbf{v}_2$ that

$$\mathbf{u}_2 = \frac{\langle \mathbf{u}_2, \mathbf{v}_1 \rangle}{\|\mathbf{v}_1\|^2} \mathbf{v}_1 = \frac{\langle \mathbf{u}_2, \mathbf{v}_1 \rangle}{\|\mathbf{u}_1\|^2} \mathbf{u}_1$$

which says that $\mathbf{u}_2$ is a multiple of $\mathbf{u}_1$, contradicting the linear independence of the basis $S = \{\mathbf{u}_1, \mathbf{u}_2, \ldots, \mathbf{u}_n\}$.

Step 3. To construct a vector $\mathbf{v}_3$ that is orthogonal to both $\mathbf{v}_1$ and $\mathbf{v}_2$, we compute the component of $\mathbf{u}_3$ orthogonal to the space W_2 spanned by $\mathbf{v}_1$ and $\mathbf{v}_2$ (Figure 6.3.4). From (7)

$$\mathbf{v}_3 = \mathbf{u}_3 - \text{proj}_{W_2} \mathbf{u}_3 = \mathbf{u}_3 - \frac{\langle \mathbf{u}_3, \mathbf{v}_1 \rangle}{\|\mathbf{v}_1\|^2} \mathbf{v}_1 - \frac{\langle \mathbf{u}_3, \mathbf{v}_2 \rangle}{\|\mathbf{v}_2\|^2} \mathbf{v}_2$$

As in Step (2), the linear independence of $\{\mathbf{u}_1, \mathbf{u}_2, \ldots, \mathbf{u}_n\}$ ensures that $\mathbf{v}_3 \neq \mathbf{0}$. We leave the details as an exercise.

Step 4. To determine a vector $\mathbf{v}_4$ that is orthogonal to $\mathbf{v}_1$, $\mathbf{v}_2$, and $\mathbf{v}_3$, we compute the component of $\mathbf{u}_4$ orthogonal to the space W_3 spanned by $\mathbf{v}_1$, $\mathbf{v}_2$, and $\mathbf{v}_3$. From (7)

$$\mathbf{v}_4 = \mathbf{u}_4 - \text{proj}_{W_3} \mathbf{u}_4 = \mathbf{u}_4 - \frac{\langle \mathbf{u}_4, \mathbf{v}_1 \rangle}{\|\mathbf{v}_1\|^2} \mathbf{v}_1 - \frac{\langle \mathbf{u}_4, \mathbf{v}_2 \rangle}{\|\mathbf{v}_2\|^2} \mathbf{v}_2 - \frac{\langle \mathbf{u}_4, \mathbf{v}_3 \rangle}{\|\mathbf{v}_3\|^2} \mathbf{v}_3$$

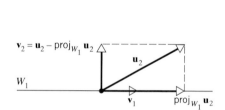

Figure 6.3.3

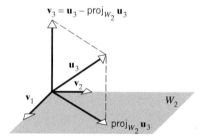

Figure 6.3.4

Continuing in this way, we will obtain, after n steps, an orthogonal set of vectors, $\{\mathbf{v}_1, \mathbf{v}_2, \ldots, \mathbf{v}_n\}$. Since V is n-dimensional and every orthogonal set is linearly independent, the set $\{\mathbf{v}_1, \mathbf{v}_2, \ldots, \mathbf{v}_n\}$ is an orthogonal basis for V. ■

The preceding step-by-step construction for converting an arbitrary basis into an orthogonal basis is called the ***Gram–Schmidt process***.

EXAMPLE 7 Using the Gram–Schmidt Process

Consider the vector space R^3 with the Euclidean inner product. Apply the Gram–Schmidt process to transform the basis vectors $\mathbf{u}_1 = (1, 1, 1)$, $\mathbf{u}_2 = (0, 1, 1)$, $\mathbf{u}_3 = (0, 0, 1)$ into an orthogonal basis $\{\mathbf{v}_1, \mathbf{v}_2, \mathbf{v}_3\}$; then normalize the orthogonal basis vectors to obtain an orthonormal basis $\{\mathbf{q}_1, \mathbf{q}_2, \mathbf{q}_3\}$.

Solution.

Step 1. $\mathbf{v}_1 = \mathbf{u}_1 = (1, 1, 1)$

Jörgen Pederson Gram (1850–1916) was a Danish actuary. Gram's early education was at village schools supplemented by private tutoring. After graduating from high school he obtained a master's degree in mathematics with specialization in the newly developing modern algebra. Gram then took a position as an actuary for the Hafnia Life Insurance Company, where he developed mathematical foundations of accident insurance for the company Skjold. He served on the Board of Directors of Hafnia and directed Skjold until 1910, at which time he became director of the Danish Insurance Board. During his employ as an actuary, he earned a Ph.D. based on his dissertation "On Series Development Utilizing the Least Squares Method." It was in this thesis that his contributions to the Gram–Schmidt process were first formulated. Gram eventually became interested in abstract number theory and won a gold medal from the Royal Danish Society of Sciences and Letters for his contributions to that field. However, he also had a lifelong interest in the interplay between theoretical and applied mathematics that led to four treatises on Danish forest management. Gram was killed one evening in a bicycle collision on the way to a meeting of the Royal Danish Society.

Step 2.
$$\mathbf{v}_2 = \mathbf{u}_2 - \text{proj}_{W_1}\mathbf{u}_2 = \mathbf{u}_2 - \frac{\langle \mathbf{u}_2, \mathbf{v}_1 \rangle}{\|\mathbf{v}_1\|^2}\mathbf{v}_1$$

$$= (0, 1, 1) - \frac{2}{3}(1, 1, 1) = \left(-\frac{2}{3}, \frac{1}{3}, \frac{1}{3}\right)$$

Step 3.
$$\mathbf{v}_3 = \mathbf{u}_3 - \text{proj}_{W_2}\mathbf{u}_3 = \mathbf{u}_3 - \frac{\langle \mathbf{u}_3, \mathbf{v}_1 \rangle}{\|\mathbf{v}_1\|^2}\mathbf{v}_1 - \frac{\langle \mathbf{u}_3, \mathbf{v}_2 \rangle}{\|\mathbf{v}_2\|^2}\mathbf{v}_2$$

$$= (0, 0, 1) - \frac{1}{3}(1, 1, 1) - \frac{1/3}{2/3}\left(-\frac{2}{3}, \frac{1}{3}, \frac{1}{3}\right)$$

$$= \left(0, -\frac{1}{2}, \frac{1}{2}\right)$$

Erhardt Schmidt (1876–1959) was a German mathematician. Schmidt received his doctoral degree from Göttingen University in 1905, where he studied under one of the giants of mathematics, David Hilbert. He eventually went to teach at Berlin University in 1917, where he stayed for the rest of his life. Schmidt made important contributions to a variety of mathematical fields, but is most noteworthy for fashioning many of Hilbert's diverse ideas into a general concept (called a Hilbert space), which is fundamental in the study of infinite-dimensional vector spaces. Schmidt first described the process that bears his name in a paper on integral equations published in 1907.

Thus,

$$\mathbf{v}_1 = (1, 1, 1), \qquad \mathbf{v}_2 = \left(-\frac{2}{3}, \frac{1}{3}, \frac{1}{3}\right), \qquad \mathbf{v}_3 = \left(0, -\frac{1}{2}, \frac{1}{2}\right)$$

form an orthogonal basis for R^3. The norms of these vectors are

$$\|\mathbf{v}_1\| = \sqrt{3}, \qquad \|\mathbf{v}_2\| = \frac{\sqrt{6}}{3}, \qquad \|\mathbf{v}_3\| = \frac{1}{\sqrt{2}}$$

so an orthonormal basis for R^3 is

$$\mathbf{q}_1 = \frac{\mathbf{v}_1}{\|\mathbf{v}_1\|} = \left(\frac{1}{\sqrt{3}}, \frac{1}{\sqrt{3}}, \frac{1}{\sqrt{3}}\right), \qquad \mathbf{q}_2 = \frac{\mathbf{v}_2}{\|\mathbf{v}_2\|} = \left(-\frac{2}{\sqrt{6}}, \frac{1}{\sqrt{6}}, \frac{1}{\sqrt{6}}\right),$$

$$\mathbf{q}_3 = \frac{\mathbf{v}_3}{\|\mathbf{v}_3\|} = \left(0, -\frac{1}{\sqrt{2}}, \frac{1}{\sqrt{2}}\right) \qquad \blacklozenge$$

REMARK. In the preceding example we used the Gram–Schmidt process to produce an orthogonal basis; then, after the entire orthogonal basis was obtained, we normalized to obtain an orthonormal basis. Alternatively, one can normalize each orthogonal basis vector as soon as it is obtained, thereby generating the orthonormal basis step by step. However, this method has the slight disadvantage of producing more square roots to manipulate.

The Gram–Schmidt process with subsequent normalization not only converts an arbitrary basis $\{\mathbf{u}_1, \mathbf{u}_2, \ldots, \mathbf{u}_n\}$ into an orthonormal basis $\{\mathbf{q}_1, \mathbf{q}_2, \ldots, \mathbf{q}_n\}$, but it does it in such a way that for $k \geq 2$ the following relationships hold:

- $\{\mathbf{q}_1, \mathbf{q}_2, \ldots, \mathbf{q}_k\}$ is an orthonormal basis for the space spanned by $\{\mathbf{u}_1, \mathbf{u}_2, \ldots, \mathbf{u}_k\}$.
- $\mathbf{q}_k$ is orthogonal to the space spanned by $\{\mathbf{u}_1, \mathbf{u}_2, \ldots, \mathbf{u}_{k-1}\}$.

We omit the proofs, but these facts should become evident after some thoughtful examination of the proof of Theorem 6.3.6.

QR-Decomposition

We pose the following problem.

Problem. If A is an $m \times n$ matrix with linearly independent column vectors, and if Q is the matrix with orthonormal column vectors that results from applying the Gram–Schmidt process to the column vectors of A, what relationship, if any, exists between A and Q?

To solve this problem, suppose that the column vectors of A are $\mathbf{u}_1, \mathbf{u}_2, \ldots, \mathbf{u}_n$ and the orthonormal column vectors of Q are $\mathbf{q}_1, \mathbf{q}_2, \ldots, \mathbf{q}_n$; thus,

$$A = [\mathbf{u}_1 \mid \mathbf{u}_2 \mid \cdots \mid \mathbf{u}_n] \quad \text{and} \quad Q = [\mathbf{q}_1 \mid \mathbf{q}_2 \mid \cdots \mid \mathbf{q}_n]$$

It follows from Theorem 6.3.1 that $\mathbf{u}_1, \mathbf{u}_2, \ldots, \mathbf{u}_n$ are expressible in terms of the vectors $\mathbf{q}_1, \mathbf{q}_2, \ldots, \mathbf{q}_n$ as

$$\mathbf{u}_1 = \langle \mathbf{u}_1, \mathbf{q}_1 \rangle \mathbf{q}_1 + \langle \mathbf{u}_1, \mathbf{q}_2 \rangle \mathbf{q}_2 + \cdots + \langle \mathbf{u}_1, \mathbf{q}_n \rangle \mathbf{q}_n$$
$$\mathbf{u}_2 = \langle \mathbf{u}_2, \mathbf{q}_1 \rangle \mathbf{q}_1 + \langle \mathbf{u}_2, \mathbf{q}_2 \rangle \mathbf{q}_2 + \cdots + \langle \mathbf{u}_2, \mathbf{q}_n \rangle \mathbf{q}_n$$
$$\vdots \qquad \vdots \qquad \vdots \qquad \qquad \vdots$$
$$\mathbf{u}_n = \langle \mathbf{u}_n, \mathbf{q}_1 \rangle \mathbf{q}_1 + \langle \mathbf{u}_n, \mathbf{q}_2 \rangle \mathbf{q}_2 + \cdots + \langle \mathbf{u}_n, \mathbf{q}_n \rangle \mathbf{q}_n$$

Recalling from Section 1.3 that the jth column vector of a matrix product is a linear combination of the column vectors of the first factor with coefficients coming from the jth column of the second factor, it follows that these relationships can be expressed in matrix form as

$$[\mathbf{u}_1 \mid \mathbf{u}_2 \mid \cdots \mid \mathbf{u}_n] = [\mathbf{q}_1 \mid \mathbf{q}_2 \mid \cdots \mid \mathbf{q}_n] \begin{bmatrix} \langle \mathbf{u}_1, \mathbf{q}_1 \rangle & \langle \mathbf{u}_2, \mathbf{q}_1 \rangle & \cdots & \langle \mathbf{u}_n, \mathbf{q}_1 \rangle \\ \langle \mathbf{u}_1, \mathbf{q}_2 \rangle & \langle \mathbf{u}_2, \mathbf{q}_2 \rangle & \cdots & \langle \mathbf{u}_n, \mathbf{q}_2 \rangle \\ \vdots & \vdots & & \vdots \\ \langle \mathbf{u}_1, \mathbf{q}_n \rangle & \langle \mathbf{u}_2, \mathbf{q}_n \rangle & \cdots & \langle \mathbf{u}_n, \mathbf{q}_n \rangle \end{bmatrix}$$

or more briefly as

$$A = QR \tag{8}$$

However, it is a property of the Gram–Schmidt process that for $j \geq 2$, the vector $\mathbf{q}_j$ is orthogonal to $\mathbf{u}_1, \mathbf{u}_2, \ldots, \mathbf{u}_{j-1}$; thus, all entries below the main diagonal of R are zero,

$$R = \begin{bmatrix} \langle \mathbf{u}_1, \mathbf{q}_1 \rangle & \langle \mathbf{u}_2, \mathbf{q}_1 \rangle & \cdots & \langle \mathbf{u}_n, \mathbf{q}_1 \rangle \\ 0 & \langle \mathbf{u}_2, \mathbf{q}_2 \rangle & \cdots & \langle \mathbf{u}_n, \mathbf{q}_2 \rangle \\ \vdots & \vdots & & \vdots \\ 0 & 0 & \cdots & \langle \mathbf{u}_n, \mathbf{q}_n \rangle \end{bmatrix} \tag{9}$$

We leave it as an exercise to show that the diagonal entries of R are nonzero, so that R is invertible. Thus, (8) is a factorization of A into the product of a matrix Q with orthonormal column vectors and an invertible upper triangular matrix R. We call (8) the **QR-decomposition of A**. In summary, we have the following theorem.

Theorem 6.3.7 *QR-Decomposition*

If A is an $m \times n$ matrix with linearly independent column vectors, then A can be factored as

$$A = QR$$

where Q is an $m \times n$ matrix with orthonormal column vectors, and R is an $n \times n$ invertible upper triangular matrix.

REMARK. Recall from Theorem 6.2.7 that if A is an $n \times n$ matrix, then the invertibility of A is equivalent to linear independence of the column vectors; thus, every invertible matrix has a *QR*-decomposition.

EXAMPLE 8 *QR*-**Decomposition of a 3 × 3 Matrix**

Find the *QR*-decomposition of

$$A = \begin{bmatrix} 1 & 0 & 0 \\ 1 & 1 & 0 \\ 1 & 1 & 1 \end{bmatrix}$$

Solution.

The column vectors of *A* are

$$\mathbf{u}_1 = \begin{bmatrix} 1 \\ 1 \\ 1 \end{bmatrix}, \qquad \mathbf{u}_2 = \begin{bmatrix} 0 \\ 1 \\ 1 \end{bmatrix}, \qquad \mathbf{u}_3 = \begin{bmatrix} 0 \\ 0 \\ 1 \end{bmatrix}$$

Applying the Gram–Schmidt process with subsequent normalization to these column vectors yields the orthonormal vectors (see Example 7)

$$\mathbf{q}_1 = \begin{bmatrix} 1/\sqrt{3} \\ 1/\sqrt{3} \\ 1/\sqrt{3} \end{bmatrix}, \qquad \mathbf{q}_2 = \begin{bmatrix} -2/\sqrt{6} \\ 1/\sqrt{6} \\ 1/\sqrt{6} \end{bmatrix}, \qquad \mathbf{q}_3 = \begin{bmatrix} 0 \\ -1/\sqrt{2} \\ 1/\sqrt{2} \end{bmatrix}$$

and from (9) the matrix *R* is

$$R = \begin{bmatrix} \langle \mathbf{u}_1, \mathbf{q}_1 \rangle & \langle \mathbf{u}_2, \mathbf{q}_1 \rangle & \langle \mathbf{u}_3, \mathbf{q}_1 \rangle \\ 0 & \langle \mathbf{u}_2, \mathbf{q}_2 \rangle & \langle \mathbf{u}_3, \mathbf{q}_2 \rangle \\ 0 & 0 & \langle \mathbf{u}_3, \mathbf{q}_3 \rangle \end{bmatrix} = \begin{bmatrix} 3/\sqrt{3} & 2/\sqrt{3} & 1/\sqrt{3} \\ 0 & 2/\sqrt{6} & 1/\sqrt{6} \\ 0 & 0 & 1/\sqrt{2} \end{bmatrix}$$

Thus, the *QR*-decomposition of *A* is

$$\underbrace{\begin{bmatrix} 1 & 0 & 0 \\ 1 & 1 & 0 \\ 1 & 1 & 1 \end{bmatrix}}_{A} = \underbrace{\begin{bmatrix} 1/\sqrt{3} & -2/\sqrt{6} & 0 \\ 1/\sqrt{3} & 1/\sqrt{6} & -1/\sqrt{2} \\ 1/\sqrt{3} & 1/\sqrt{6} & 1/\sqrt{2} \end{bmatrix}}_{Q} \underbrace{\begin{bmatrix} 3/\sqrt{3} & 2/\sqrt{3} & 1/\sqrt{3} \\ 0 & 2/\sqrt{6} & 1/\sqrt{6} \\ 0 & 0 & 1/\sqrt{2} \end{bmatrix}}_{R} \qquad \blacklozenge$$

The Role of the *QR*-Decomposition in Linear Algebra

In recent years the *QR*-decomposition has assumed growing importance as the mathematical foundation for a wide variety of practical numerical algorithms, including a widely used algorithm for computing eigenvalues of large matrices. Such algorithms are discussed in textbooks that deal with advanced numerical methods of linear algebra.

Additional Proof

Proof of Theorem 6.3.4. There are two parts to the proof. First we must find vectors $\mathbf{w}_1$ and $\mathbf{w}_2$ with the stated properties, and then we must show that these are the only such vectors.

By the Gram–Schmidt process there is an orthonormal basis $\{\mathbf{v}_1, \mathbf{v}_2, \dots, \mathbf{v}_n\}$ for *W*. Let

$$\mathbf{w}_1 = \langle \mathbf{u}, \mathbf{v}_1 \rangle \mathbf{v}_1 + \langle \mathbf{u}, \mathbf{v}_2 \rangle \mathbf{v}_2 + \cdots + \langle \mathbf{u}, \mathbf{v}_n \rangle \mathbf{v}_n \tag{10}$$

and

$$\mathbf{w}_2 = \mathbf{u} - \mathbf{w}_1 \tag{11}$$

It follows that $\mathbf{w}_1 + \mathbf{w}_2 = \mathbf{w}_1 + (\mathbf{u} - \mathbf{w}_1) = \mathbf{u}$, so it remains to show that $\mathbf{w}_1$ is in W and $\mathbf{w}_2$ is orthogonal to W. But $\mathbf{w}_1$ lies in W because it is a linear combination of the basis vectors for W. To show that $\mathbf{w}_2$ is orthogonal to W, we must show that $\langle \mathbf{w}_2, \mathbf{w} \rangle = 0$ for every vector $\mathbf{w}$ in W. But if $\mathbf{w}$ is any vector in W, it can be expressed as a linear combination

$$\mathbf{w} = k_1 \mathbf{v}_1 + k_2 \mathbf{v}_2 + \cdots + k_n \mathbf{v}_n$$

of the basis vectors $\mathbf{v}_1, \mathbf{v}_2, \ldots, \mathbf{v}_n$. Thus,

$$\langle \mathbf{w}_2, \mathbf{w} \rangle = \langle \mathbf{u} - \mathbf{w}_1, \mathbf{w} \rangle = \langle \mathbf{u}, \mathbf{w} \rangle - \langle \mathbf{w}_1, \mathbf{w} \rangle \tag{12}$$

But

$$\langle \mathbf{u}, \mathbf{w} \rangle = \langle \mathbf{u}, k_1 \mathbf{v}_1 + k_2 \mathbf{v}_2 + \cdots + k_n \mathbf{v}_n \rangle$$
$$= k_1 \langle \mathbf{u}, \mathbf{v}_1 \rangle + k_2 \langle \mathbf{u}, \mathbf{v}_2 \rangle + \cdots + k_n \langle \mathbf{u}, \mathbf{v}_n \rangle$$

and by part (c) of Theorem 6.3.2

$$\langle \mathbf{w}_1, \mathbf{w} \rangle = \langle \mathbf{u}, \mathbf{v}_1 \rangle k_1 + \langle \mathbf{u}, \mathbf{v}_2 \rangle k_2 + \cdots + \langle \mathbf{u}, \mathbf{v}_n \rangle k_n$$

Thus, $\langle \mathbf{u}, \mathbf{w} \rangle$ and $\langle \mathbf{w}_1, \mathbf{w} \rangle$ are equal, so (12) yields $\langle \mathbf{w}_2, \mathbf{w} \rangle = 0$, which is what we want to show.

To see that (10) and (11) are the only vectors with the properties stated in the theorem, suppose that we can also write

$$\mathbf{u} = \mathbf{w}_1' + \mathbf{w}_2' \tag{13}$$

where $\mathbf{w}_1'$ is in W and $\mathbf{w}_2'$ is orthogonal to W. If we subtract from (13) the equation

$$\mathbf{u} = \mathbf{w}_1 + \mathbf{w}_2$$

we obtain

$$\mathbf{0} = (\mathbf{w}_1' - \mathbf{w}_1) + (\mathbf{w}_2' - \mathbf{w}_2)$$

or

$$\mathbf{w}_1 - \mathbf{w}_1' = \mathbf{w}_2' - \mathbf{w}_2 \tag{14}$$

Since $\mathbf{w}_2$ and $\mathbf{w}_2'$ are orthogonal to W, their difference is also orthogonal to W, since for any vector $\mathbf{w}$ in W we can write

$$\langle \mathbf{w}, \mathbf{w}_2' - \mathbf{w}_2 \rangle = \langle \mathbf{w}, \mathbf{w}_2' \rangle - \langle \mathbf{w}, \mathbf{w}_2 \rangle = 0 - 0 = 0$$

But $\mathbf{w}_2' - \mathbf{w}_2$ is itself a vector in W, since from (14) it is the difference of the two vectors $\mathbf{w}_1$ and $\mathbf{w}_1'$ which lie in the subspace W. Thus, $\mathbf{w}_2' - \mathbf{w}_2$ must be orthogonal to itself; that is,

$$\langle \mathbf{w}_2' - \mathbf{w}_2, \mathbf{w}_2' - \mathbf{w}_2 \rangle = 0$$

But this implies that $\mathbf{w}_2' - \mathbf{w}_2 = 0$ by Axiom 4 for inner products. Thus, $\mathbf{w}_2' = \mathbf{w}_2$, and by (14), $\mathbf{w}_1' = \mathbf{w}_1$. ∎

Exercise Set 6.3

1. Which of the following sets of vectors are orthogonal with respect to the Euclidean inner product on R^2?

   (a) $(0, 1)$, $(2, 0)$ (b) $(-1/\sqrt{2}, 1/\sqrt{2})$, $(1/\sqrt{2}, 1/\sqrt{2})$ (c) $(-1/\sqrt{2}, -1/\sqrt{2})$, $(1/\sqrt{2}, 1/\sqrt{2})$ (d) $(0, 0)$, $(0, 1)$

2. Which of the sets in Exercise 1 are orthonormal with respect to the Euclidean inner product on R^2?

3. Which of the following sets of vectors are orthogonal with respect to the Euclidean inner product on R^3?

(a) $\left(\frac{1}{\sqrt{2}}, 0, \frac{1}{\sqrt{2}}\right), \left(\frac{1}{\sqrt{3}}, \frac{1}{\sqrt{3}}, -\frac{1}{\sqrt{3}}\right), \left(-\frac{1}{\sqrt{2}}, 0, \frac{1}{\sqrt{2}}\right)$

(b) $\left(\frac{2}{3}, -\frac{2}{3}, \frac{1}{3}\right), \left(\frac{2}{3}, \frac{1}{3}, -\frac{2}{3}\right), \left(\frac{1}{3}, \frac{2}{3}, \frac{2}{3}\right)$

(c) $(1, 0, 0), \left(0, \frac{1}{\sqrt{2}}, \frac{1}{\sqrt{2}}\right), (0, 0, 1)$

(d) $\left(\frac{1}{\sqrt{6}}, \frac{1}{\sqrt{6}}, -\frac{2}{\sqrt{6}}\right), \left(\frac{1}{\sqrt{2}}, -\frac{1}{\sqrt{2}}, 0\right)$

4. Which of the sets in Exercise 3 are orthonormal with respect to the Euclidean inner product on R^3?

5. Which of the following sets of polynomials are orthonormal with respect to the inner product on P_2 discussed in Example 8 of Section 6.1?

(a) $\frac{2}{3} - \frac{2}{3}x + \frac{1}{3}x^2, \; \frac{2}{3} + \frac{1}{3}x - \frac{2}{3}x^2, \; \frac{1}{3} + \frac{2}{3}x + \frac{2}{3}x^2$

(b) $1, \; \frac{1}{\sqrt{2}}x + \frac{1}{\sqrt{2}}x^2, \; x^2$

6. Which of the following sets of matrices are orthonormal with respect to the inner product on M_{22} discussed in Example 7 of Section 6.1?

(a) $\begin{bmatrix} 1 & 0 \\ 0 & 0 \end{bmatrix}, \begin{bmatrix} 0 & \frac{2}{3} \\ \frac{1}{3} & -\frac{2}{3} \end{bmatrix}, \begin{bmatrix} 0 & \frac{2}{3} \\ -\frac{2}{3} & \frac{1}{3} \end{bmatrix}, \begin{bmatrix} 0 & \frac{1}{3} \\ \frac{2}{3} & \frac{2}{3} \end{bmatrix}$

(b) $\begin{bmatrix} 1 & 0 \\ 0 & 0 \end{bmatrix}, \begin{bmatrix} 0 & 1 \\ 0 & 0 \end{bmatrix}, \begin{bmatrix} 0 & 0 \\ 1 & 1 \end{bmatrix}, \begin{bmatrix} 0 & 0 \\ 1 & -1 \end{bmatrix}$

7. Verify that the given set of vectors is orthogonal with respect to the Euclidean inner product; then convert it to an orthonormal set by normalizing the vectors.

(a) $(-1, 2), (6, 3)$ (b) $(1, 0, -1), (2, 0, 2), (0, 5, 0)$ (c) $\left(\frac{1}{5}, \frac{1}{5}, \frac{1}{5}\right), \left(-\frac{1}{2}, \frac{1}{2}, 0\right), \left(\frac{1}{3}, \frac{1}{3}, -\frac{2}{3}\right)$

8. Verify that the set of vectors $\{(1, 0), (0, 1)\}$ is orthogonal with respect to the inner product $\langle \mathbf{u}, \mathbf{v} \rangle = 4u_1 v_1 + u_2 v_2$ on R^2; then convert it to an orthonormal set by normalizing the vectors.

9. Verify that the vectors $\mathbf{v}_1 = \left(-\frac{3}{5}, \frac{4}{5}, 0\right)$, $\mathbf{v}_2 = \left(\frac{4}{5}, \frac{3}{5}, 0\right)$, $\mathbf{v}_3 = (0, 0, 1)$ form an orthonormal basis for R^3 with the Euclidean inner product; then use Theorem 6.3.1 to express each of the following as linear combinations of $\mathbf{v}_1$, $\mathbf{v}_2$, and $\mathbf{v}_3$.

(a) $(1, -1, 2)$ (b) $(3, -7, 4)$ (c) $\left(\frac{1}{7}, -\frac{3}{7}, \frac{5}{7}\right)$

10. Verify that the vectors

$$\mathbf{v}_1 = (1, -1, 2, -1), \quad \mathbf{v}_2 = (-2, 2, 3, 2), \quad \mathbf{v}_3 = (1, 2, 0, -1), \quad \mathbf{v}_4 = (1, 0, 0, 1)$$

form an orthogonal basis for R^4 with the Euclidean inner product; then use Formula (1) to express each of the following as linear combinations of $\mathbf{v}_1$, $\mathbf{v}_2$, $\mathbf{v}_3$, and $\mathbf{v}_4$.

(a) $(1, 1, 1, 1)$ (b) $(\sqrt{2}, -3\sqrt{2}, 5\sqrt{2}, -\sqrt{2})$ (c) $\left(-\frac{1}{3}, \frac{2}{3}, -\frac{1}{3}, \frac{4}{3}\right)$

11. In each part an orthonormal basis relative to the Euclidean inner product is given. Use Theorem 6.3.1 to find the coordinate vector of $\mathbf{w}$ with respect to that basis.

(a) $\mathbf{w} = (3, 7)$; $\mathbf{u}_1 = \left(\frac{1}{\sqrt{2}}, -\frac{1}{\sqrt{2}}\right)$, $\mathbf{u}_2 = \left(\frac{1}{\sqrt{2}}, \frac{1}{\sqrt{2}}\right)$

(b) $\mathbf{w} = (-1, 0, 2)$; $\mathbf{u}_1 = \left(\frac{2}{3}, -\frac{2}{3}, \frac{1}{3}\right)$, $\mathbf{u}_2 = \left(\frac{2}{3}, \frac{1}{3}, -\frac{2}{3}\right)$, $\mathbf{u}_3 = \left(\frac{1}{3}, \frac{2}{3}, \frac{2}{3}\right)$

12. Let R^2 have the Euclidean inner product, and let $S = \{\mathbf{w}_1, \mathbf{w}_2\}$ be the orthonormal basis with $\mathbf{w}_1 = \left(\frac{3}{5}, -\frac{4}{5}\right)$, $\mathbf{w}_2 = \left(\frac{4}{5}, \frac{3}{5}\right)$.

(a) Find the vectors $\mathbf{u}$ and $\mathbf{v}$ that have coordinate vectors $(\mathbf{u})_S = (1, 1)$ and $(\mathbf{v})_S = (-1, 4)$.

(b) Compute $\|\mathbf{u}\|$, $d(\mathbf{u}, \mathbf{v})$, and $\langle \mathbf{u}, \mathbf{v} \rangle$ by applying Theorem 6.3.2 to the coordinate vectors $(\mathbf{u})_S$ and $(\mathbf{v})_S$; then check the results by performing the computations directly on $\mathbf{u}$ and $\mathbf{v}$.

13. Let R^3 have the Euclidean inner product, and let $S = \{\mathbf{w}_1, \mathbf{w}_2, \mathbf{w}_3\}$ be the orthonormal basis with $\mathbf{w}_1 = \left(0, -\frac{3}{5}, \frac{4}{5}\right)$, $\mathbf{w}_2 = (1, 0, 0)$, and $\mathbf{w}_3 = \left(0, \frac{4}{5}, \frac{3}{5}\right)$.

(a) Find the vectors $\mathbf{u}$, $\mathbf{v}$, and $\mathbf{w}$ that have the coordinate vectors $(\mathbf{u})_S = (-2, 1, 2)$, $(\mathbf{v})_S = (3, 0, -2)$, and $(\mathbf{w})_S = (5, -4, 1)$.

(b) Compute $\|\mathbf{v}\|$, $d(\mathbf{u}, \mathbf{w})$, and $\langle \mathbf{w}, \mathbf{v} \rangle$ by applying Theorem 6.3.2 to the coordinate vectors $(\mathbf{u})_S$, $(\mathbf{v})_S$, and $(\mathbf{w})_S$; then check the results by performing the computations directly on $\mathbf{u}$, $\mathbf{v}$, and $\mathbf{w}$.

14. In each part, S represents some orthonormal basis for a four-dimensional inner product space. Use the given information to find $\|\mathbf{u}\|$, $\|\mathbf{v} - \mathbf{w}\|$, $\|\mathbf{v} + \mathbf{w}\|$, and $\langle \mathbf{v}, \mathbf{w} \rangle$.

   (a) $(\mathbf{u})_S = (-1, 2, 1, 3)$, $(\mathbf{v})_S = (0, -3, 1, 5)$, $(\mathbf{w})_S = (-2, -4, 3, 1)$
   (b) $(\mathbf{u})_S = (0, 0, -1, -1)$, $(\mathbf{v})_S = (5, 5, -2, -2)$, $(\mathbf{w})_S = (3, 0, -3, 0)$

15. (a) Show that the vectors $\mathbf{v}_1 = (1, -2, 3, -4)$, $\mathbf{v}_2 = (2, 1, -4, -3)$, $\mathbf{v}_3 = (-3, 4, 1, -2)$, and $\mathbf{v}_4 = (4, 3, 2, 1)$ form an orthogonal basis for R^4 with the Euclidean inner product.
   (b) Use (1) to express $\mathbf{u} = (-1, 2, 3, 7)$ as a linear combination of the vectors in (a).

16. Let R^2 have the Euclidean inner product. Use the Gram–Schmidt process to transform the basis $\{\mathbf{u}_1, \mathbf{u}_2\}$ into an orthonormal basis. Draw both sets of basis vectors in the xy-plane.

   (a) $\mathbf{u}_1 = (1, -3)$, $\mathbf{u}_2 = (2, 2)$ (b) $\mathbf{u}_1 = (1, 0)$, $\mathbf{u}_2 = (3, -5)$

17. Let R^3 have the Euclidean inner product. Use the Gram–Schmidt process to transform the basis $\{\mathbf{u}_1, \mathbf{u}_2, \mathbf{u}_3\}$ into an orthonormal basis.

   (a) $\mathbf{u}_1 = (1, 1, 1)$, $\mathbf{u}_2 = (-1, 1, 0)$, $\mathbf{u}_3 = (1, 2, 1)$ (b) $\mathbf{u}_1 = (1, 0, 0)$, $\mathbf{u}_2 = (3, 7, -2)$, $\mathbf{u}_3 = (0, 4, 1)$

18. Let R^4 have the Euclidean inner product. Use the Gram–Schmidt process to transform the basis $\{\mathbf{u}_1, \mathbf{u}_2, \mathbf{u}_3, \mathbf{u}_4\}$ into an orthonormal basis.

$$\mathbf{u}_1 = (0, 2, 1, 0), \quad \mathbf{u}_2 = (1, -1, 0, 0), \quad \mathbf{u}_3 = (1, 2, 0, -1), \quad \mathbf{u}_4 = (1, 0, 0, 1)$$

19. Let R^3 have the Euclidean inner product. Find an orthonormal basis for the subspace spanned by $(0, 1, 2)$, $(-1, 0, 1)$, $(-1, 1, 3)$.

20. Let R^3 have the inner product $\langle \mathbf{u}, \mathbf{v} \rangle = u_1 v_1 + 2u_2 v_2 + 3u_3 v_3$. Use the Gram–Schmidt process to transform $\mathbf{u}_1 = (1, 1, 1)$, $\mathbf{u}_2 = (1, 1, 0)$, $\mathbf{u}_3 = (1, 0, 0)$ into an orthonormal basis.

21. The subspace of R^3 spanned by the vectors $\mathbf{u}_1 = \left(\frac{4}{5}, 0, -\frac{3}{5}\right)$ and $\mathbf{u}_2 = (0, 1, 0)$ is a plane passing through the origin. Express $\mathbf{w} = (1, 2, 3)$ in the form $\mathbf{w} = \mathbf{w}_1 + \mathbf{w}_2$, where $\mathbf{w}_1$ lies in the plane and $\mathbf{w}_2$ is perpendicular to the plane.

22. Repeat Exercise 21 with $\mathbf{u}_1 = (1, 1, 1)$ and $\mathbf{u}_2 = (2, 0, -1)$.

23. Let R^4 have the Euclidean inner product. Express $\mathbf{w} = (-1, 2, 6, 0)$ in the form $\mathbf{w} = \mathbf{w}_1 + \mathbf{w}_2$, where $\mathbf{w}_1$ is in the space W spanned by $\mathbf{u}_1 = (-1, 0, 1, 2)$ and $\mathbf{u}_2 = (0, 1, 0, 1)$, and $\mathbf{w}_2$ is orthogonal to W.

24. Find the QR-decomposition of the matrix, where possible.

(a) $\begin{bmatrix} 1 & -1 \\ 2 & 3 \end{bmatrix}$ (b) $\begin{bmatrix} 1 & 2 \\ 0 & 1 \\ 1 & 4 \end{bmatrix}$ (c) $\begin{bmatrix} 1 & 1 \\ -2 & 1 \\ 2 & 1 \end{bmatrix}$ (d) $\begin{bmatrix} 1 & 0 & 2 \\ 0 & 1 & 1 \\ 1 & 2 & 0 \end{bmatrix}$ (e) $\begin{bmatrix} 1 & 2 & 1 \\ 1 & 1 & 1 \\ 0 & 3 & 1 \end{bmatrix}$ (f) $\begin{bmatrix} 1 & 0 & 1 \\ -1 & 1 & 1 \\ 1 & 0 & 1 \\ -1 & 1 & 1 \end{bmatrix}$

25. Let $\{\mathbf{v}_1, \mathbf{v}_2, \mathbf{v}_3\}$ be an orthonormal basis for an inner product space V. Show that if $\mathbf{w}$ is a vector in V, then $\|\mathbf{w}\|^2 = \langle \mathbf{w}, \mathbf{v}_1 \rangle^2 + \langle \mathbf{w}, \mathbf{v}_2 \rangle^2 + \langle \mathbf{w}, \mathbf{v}_3 \rangle^2$.

26. Let $\{\mathbf{v}_1, \mathbf{v}_2, \ldots, \mathbf{v}_n\}$ be an orthonormal basis for an inner product space V. Show that if $\mathbf{w}$ is a vector in V, then $\|\mathbf{w}\|^2 = \langle \mathbf{w}, \mathbf{v}_1 \rangle^2 + \langle \mathbf{w}, \mathbf{v}_2 \rangle^2 + \cdots + \langle \mathbf{w}, \mathbf{v}_n \rangle^2$.

27. In Step 3 of the proof of Theorem 6.3.6, it was stated that "the linear independence of $\{\mathbf{u}_1, \mathbf{u}_2, \ldots, \mathbf{u}_n\}$ ensures that $\mathbf{v}_3 \neq \mathbf{0}$." Prove this statement.

28. Prove that the diagonal entries of R in Formula (9) are nonzero.

29. (*For readers who have studied calculus.*) Let the vector space P_2 have the inner product

$$\langle \mathbf{p}, \mathbf{q} \rangle = \int_{-1}^{1} p(x)q(x)\,dx$$

   Apply the Gram–Schmidt process to transform the standard basis $S = \{1, x, x^2\}$ into an orthonormal basis. (The polynomials in the resulting basis are called the first three *normalized Legendre polynomials*.)

30. (*For readers who have studied calculus.*) Use Theorem 6.3.1 to express the following as linear combinations of the first three normalized Legendre polynomials (Exercise 29).

   (a) $1 + x + 4x^2$ (b) $2 - 7x^2$ (c) $4 + 3x$

31. (*For readers who have studied calculus.*) Let P_2 have the inner product

$$\langle \mathbf{p}, \mathbf{q} \rangle = \int_0^1 p(x)q(x)\, dx$$

Apply the Gram–Schmidt process to transform the standard basis $S = \{1, x, x^2\}$ into an orthonormal basis.

32. Prove Theorem 6.3.2.

33. Prove Theorem 6.3.5.

Discussion and Discovery

34. (a) It follows from Theorem 6.3.6 that every plane through the origin in R^3 must have an orthonormal basis with respect to the Euclidean inner product. In words, explain how you would go about finding an orthonormal basis for a plane if you knew its equation.

   (b) Use your method to find an orthonormal basis for the plane $x + 2y - z = 0$.

35. Find vectors $\mathbf{x}$ and $\mathbf{y}$ in R^2 that are orthonormal with respect to the inner product $\langle \mathbf{u}, \mathbf{v} \rangle = 3u_1 v_1 + 2u_2 v_2$ but are not orthonormal with respect to the Euclidean inner product.

36. If W is a line through the origin of R^3 with the Euclidean inner product, and if $\mathbf{u}$ is a vector in R^3, then Theorem 6.3.4 implies that $\mathbf{u}$ can be expressed uniquely as $\mathbf{u} = \mathbf{w}_1 + \mathbf{w}_2$, where $\mathbf{w}_1$ is a vector in W and $\mathbf{w}_2$ is a vector in $W^\perp$. Draw a picture that illustrates this.

37. Indicate whether the statement is always true or sometimes false. Justify your answer by giving a logical argument or a counterexample.

   (a) A linearly dependent set of vectors in an inner product space cannot be orthonormal.

   (b) Every finite-dimensional vector space has an orthonormal basis.

   (c) $\text{proj}_W \mathbf{u}$ is orthogonal to $\text{proj}_{W^\perp} \mathbf{u}$ in any inner product space.

   (d) Every matrix with a nonzero determinant has a QR-decomposition.

6.4 BEST APPROXIMATION; LEAST SQUARES

In this section we shall show how orthogonal projections can be used to solve certain approximation problems. The results obtained in this section have a wide variety of applications in both mathematics and science.

Orthogonal Projections Viewed as Approximations
If P is a point in ordinary 3-space and W is a plane through the origin, then the point Q in W closest to P is obtained by dropping a perpendicular from P to W (Figure 6.4.1a).

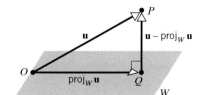

(*a*) Q is the point in W closest to P.   (*b*) $\|\mathbf{u} - \mathbf{w}\|$ is minimized by $\mathbf{w} = \text{proj}_W \mathbf{u}$.

Figure 6.4.1

Therefore, if we let $\mathbf{u} = \overrightarrow{OP}$, the distance between P and W is given by

$$\|\mathbf{u} - \text{proj}_W \mathbf{u}\|$$

In other words, among all vectors $\mathbf{w}$ in W the vector $\mathbf{w} = \text{proj}_W \mathbf{u}$ minimizes the distance $\|\mathbf{u} - \mathbf{w}\|$ (Figure 6.4.1b).

There is another way of thinking about this idea. View $\mathbf{u}$ as a fixed vector that we would like to approximate by a vector in W. Any such approximation $\mathbf{w}$ will result in an "error vector,"

$$\mathbf{u} - \mathbf{w}$$

which, unless $\mathbf{u}$ is in W, cannot be made equal to $\mathbf{0}$. However, by choosing

$$\mathbf{w} = \text{proj}_W \mathbf{u}$$

we can make the length of the error vector

$$\|\mathbf{u} - \mathbf{w}\| = \|\mathbf{u} - \text{proj}_W \mathbf{u}\|$$

as small as possible. Thus, we can describe $\text{proj}_W \mathbf{u}$ as the "best approximation" to $\mathbf{u}$ by vectors in W. The following theorem will make these intuitive ideas precise.

Theorem 6.4.1 **Best Approximation Theorem**

*If W is a finite-dimensional subspace of an inner product space V, and if $\mathbf{u}$ is a vector in V, then $\text{proj}_W \mathbf{u}$ is the **best approximation** to $\mathbf{u}$ from W in the sense that*

$$\|\mathbf{u} - \text{proj}_W \mathbf{u}\| < \|\mathbf{u} - \mathbf{w}\|$$

for every vector $\mathbf{w}$ in W that is different from $\text{proj}_W \mathbf{u}$.

Proof. For every vector $\mathbf{w}$ in W we can write

$$\mathbf{u} - \mathbf{w} = (\mathbf{u} - \text{proj}_W \mathbf{u}) + (\text{proj}_W \mathbf{u} - \mathbf{w}) \tag{1}$$

But $\text{proj}_W \mathbf{u} - \mathbf{w}$, being a difference of vectors in W, is in W; and $\mathbf{u} - \text{proj}_W \mathbf{u}$ is orthogonal to W, so that the two terms on the right side of (1) are orthogonal. Thus, by the Theorem of Pythagoras (6.2.4),

$$\|\mathbf{u} - \mathbf{w}\|^2 = \|\mathbf{u} - \text{proj}_W \mathbf{u}\|^2 + \|\text{proj}_W \mathbf{u} - \mathbf{w}\|^2$$

If $\mathbf{w} \neq \text{proj}_W \mathbf{u}$, then the second term in this sum will be positive, so that

$$\|\mathbf{u} - \mathbf{w}\|^2 > \|\mathbf{u} - \text{proj}_W \mathbf{u}\|^2$$

or equivalently,

$$\|\mathbf{u} - \mathbf{w}\| > \|\mathbf{u} - \text{proj}_W \mathbf{u}\| \qquad \blacksquare$$

Applications of this theorem will be given later in the text.

Least Squares Solutions of Linear Systems Up to now we have been concerned primarily with consistent systems of linear equations. However, inconsistent linear systems are also important in physical applications. It is a common situation that some physical problem leads to a linear system $A\mathbf{x} = \mathbf{b}$ that should be consistent on theoretical grounds but fails to be so because "measurement errors" in the entries of A and $\mathbf{b}$ perturb the system enough to cause inconsistency. In such situations one looks for a value of $\mathbf{x}$ that comes "as close as possible" to being a solution in the sense that it minimizes the value of $\|A\mathbf{x} - \mathbf{b}\|$ with respect to the Euclidean inner product. The quantity $\|A\mathbf{x} - \mathbf{b}\|$ can be viewed as a measure of the "error" that results from regarding $\mathbf{x}$ as an approximate solution of the linear system $A\mathbf{x} = \mathbf{b}$. If the system is consistent and $\mathbf{x}$ is an exact solution, then the error is zero, since $\|A\mathbf{x} - \mathbf{b}\| = \|\mathbf{0}\| = 0$. In general, the larger the value of $\|A\mathbf{x} - \mathbf{b}\|$, the more poorly $\mathbf{x}$ approximates a solution of the system.

Least Squares Problem.   Given a linear system $A\mathbf{x} = \mathbf{b}$ of m equations in n unknowns, find a vector $\mathbf{x}$, if possible, that minimizes $\|A\mathbf{x} - \mathbf{b}\|$ with respect to the Euclidean inner product on R^m. Such a vector is called a ***least squares solution*** of $A\mathbf{x} = \mathbf{b}$.

REMARK. To understand the origin of the term *least squares*, let $\mathbf{e} = A\mathbf{x} - \mathbf{b}$, which we can view as the error vector that results from the approximation $\mathbf{x}$. If $\mathbf{e} = (e_1, e_2, \ldots, e_m)$, then a least squares solution minimizes $\|\mathbf{e}\| = (e_1^2 + e_2^2 + \cdots + e_m^2)^{1/2}$; hence it also minimizes $\|\mathbf{e}\|^2 = e_1^2 + e_2^2 + \cdots + e_m^2$—thus, the term *least squares*.

To solve the least squares problem, let W be the column space of A. For each $n \times 1$ matrix $\mathbf{x}$, the product $A\mathbf{x}$ is a linear combination of the column vectors of A. Thus, as $\mathbf{x}$ varies over R^n the vector $A\mathbf{x}$ varies over all possible linear combinations of the column vectors of A; that is, $A\mathbf{x}$ varies over the entire column space W. Geometrically, solving the least squares problem amounts to finding a vector $\mathbf{x}$ in R^n such that $A\mathbf{x}$ is the closest vector in W to $\mathbf{b}$ (Figure 6.4.2).

It follows from the Best Approximation Theorem (6.4.1) that the closest vector in W to $\mathbf{b}$ is the orthogonal projection of $\mathbf{b}$ on W. Thus, for a vector $\mathbf{x}$ to be a least squares solution of $A\mathbf{x} = \mathbf{b}$, this vector must satisfy

$$A\mathbf{x} = \text{proj}_W \mathbf{b} \tag{2}$$

One could attempt to find least squares solutions of $A\mathbf{x} = \mathbf{b}$ by first calculating the vector $\text{proj}_W \mathbf{b}$ and then solving (2); however, there is a better approach: It follows from the Projection Theorem (6.3.4) and Formula (5) of Section 6.3 that

$$\mathbf{b} - A\mathbf{x} = \mathbf{b} - \text{proj}_W \mathbf{b}$$

is orthogonal to W. But W is the column space of A, so it follows from Theorem 6.2.6 that $\mathbf{b} - A\mathbf{x}$ lies in the nullspace of A^T. Therefore, a least squares solution of $A\mathbf{x} = \mathbf{b}$ must satisfy

$$A^T(\mathbf{b} - A\mathbf{x}) = 0$$

or equivalently,

$$A^T A\mathbf{x} = A^T \mathbf{b} \tag{3}$$

This is called the ***normal system*** associated with $A\mathbf{x} = \mathbf{b}$, and the individual equations are called the ***normal equations*** associated with $A\mathbf{x} = \mathbf{b}$. Thus, the problem of finding a least squares solution of $A\mathbf{x} = \mathbf{b}$ has been reduced to the problem of finding an exact solution of the associated normal system.

Note the following observations about the normal system:

- The normal system involves n equations in n unknowns (verify).
- The normal system is consistent, since it is satisfied by a least squares solution of $A\mathbf{x} = \mathbf{b}$.
- The normal system may have infinitely many solutions, in which case all of its solutions are least squares solutions of $A\mathbf{x} = \mathbf{b}$.

From these observations and Formula (2) we have the following theorem.

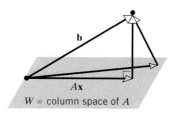

Figure 6.4.2

A least squares solution $\mathbf{x}$ produces the vector $A\mathbf{x}$ in W closest to $\mathbf{b}$.

W = column space of A

Theorem 6.4.2

For any linear system $A\mathbf{x} = \mathbf{b}$, *the associated normal system*

$$A^TA\mathbf{x} = A^T\mathbf{b}$$

is consistent, and all solutions of the normal system are least squares solutions of $A\mathbf{x} = \mathbf{b}$. *Moreover, if W is the column space of A, and* $\mathbf{x}$ *is any least squares solution of* $A\mathbf{x} = \mathbf{b}$, *then the orthogonal projection of* $\mathbf{b}$ *on W is*

$$\text{proj}_W \mathbf{b} = A\mathbf{x}$$

Uniqueness of Least Squares Solutions   Before we examine some numerical examples, we shall establish conditions under which a linear system is guaranteed to have a unique least squares solution. We shall need the following theorem.

Theorem 6.4.3

If A is an $m \times n$ *matrix, then the following are equivalent.*

(a) A has linearly independent column vectors.   *(b)* A^TA *is invertible.*

Proof. We shall prove that $(a) \Rightarrow (b)$ and leave the proof that $(b) \Rightarrow (a)$ as an exercise.

(a) $\Rightarrow$ (b). Assume that A has linearly independent column vectors. The matrix A^TA has size $n \times n$, so we can prove that this matrix is invertible by showing that the linear system $A^TA\mathbf{x} = \mathbf{0}$ has only the trivial solution. But if $\mathbf{x}$ is any solution of this system, then $A\mathbf{x}$ is in the nullspace of A^T and also in the column space of A. By Theorem 6.2.6 these spaces are orthogonal complements, so part (b) of Theorem 6.2.5 implies that $A\mathbf{x} = \mathbf{0}$. But A has linearly independent column vectors, so $\mathbf{x} = \mathbf{0}$ by Theorem 5.6.8.   ∎

The next theorem is a direct consequence of Theorems 6.4.2 and 6.4.3. We omit the details.

Theorem 6.4.4

If A is an $m \times n$ *matrix with linearly independent column vectors, then for every* $m \times 1$ *matrix* $\mathbf{b}$, *the linear system* $A\mathbf{x} = \mathbf{b}$ *has a unique least squares solution. This solution is given by*

$$\mathbf{x} = (A^TA)^{-1}A^T\mathbf{b} \tag{4}$$

Moreover, if W is the column space of A, then the orthogonal projection of $\mathbf{b}$ *on W is*

$$\text{proj}_W \mathbf{b} = A\mathbf{x} = A(A^TA)^{-1}A^T\mathbf{b} \tag{5}$$

REMARK. Formulas (4) and (5) have various theoretical applications, but they are not efficient for numerical calculations. Least squares solutions of $A\mathbf{x} = \mathbf{b}$ are best computed by using Gaussian elimination or Gauss–Jordan elimination to solve the normal equations, and the orthogonal projection of $\mathbf{b}$ on the column space of A is best obtained by computing $A\mathbf{x}$, where $\mathbf{x}$ is the least squares solution of $A\mathbf{x} = \mathbf{b}$.

EXAMPLE 1 Least Squares Solution

Find the least squares solution of the linear system $A\mathbf{x} = \mathbf{b}$ given by

$$\begin{aligned} x_1 - \quad x_2 &= 4 \\ 3x_1 + 2x_2 &= 1 \\ -2x_1 + 4x_2 &= 3 \end{aligned}$$

and find the orthogonal projection of $\mathbf{b}$ on the column space of A.

Solution.

Here

$$A = \begin{bmatrix} 1 & -1 \\ 3 & 2 \\ -2 & 4 \end{bmatrix} \quad \text{and} \quad \mathbf{b} = \begin{bmatrix} 4 \\ 1 \\ 3 \end{bmatrix}$$

Observe that A has linearly independent column vectors, so we know in advance that there is a unique least squares solution. We have

$$A^T A = \begin{bmatrix} 1 & 3 & -2 \\ -1 & 2 & 4 \end{bmatrix} \begin{bmatrix} 1 & -1 \\ 3 & 2 \\ -2 & 4 \end{bmatrix} = \begin{bmatrix} 14 & -3 \\ -3 & 21 \end{bmatrix}$$

$$A^T \mathbf{b} = \begin{bmatrix} 1 & 3 & -2 \\ -1 & 2 & 4 \end{bmatrix} \begin{bmatrix} 4 \\ 1 \\ 3 \end{bmatrix} = \begin{bmatrix} 1 \\ 10 \end{bmatrix}$$

so the normal system $A^T A \mathbf{x} = A^T \mathbf{b}$ in this case is

$$\begin{bmatrix} 14 & -3 \\ -3 & 21 \end{bmatrix} \begin{bmatrix} x_1 \\ x_2 \end{bmatrix} = \begin{bmatrix} 1 \\ 10 \end{bmatrix}$$

Solving this system yields the least squares solution

$$x_1 = \tfrac{17}{95}, \qquad x_2 = \tfrac{143}{285}$$

From (5), the orthogonal projection of $\mathbf{b}$ on the column space of A is

$$A\mathbf{x} = \begin{bmatrix} 1 & -1 \\ 3 & 2 \\ -2 & 4 \end{bmatrix} \begin{bmatrix} \tfrac{17}{95} \\ \tfrac{143}{285} \end{bmatrix} = \begin{bmatrix} -\tfrac{92}{285} \\ \tfrac{439}{285} \\ \tfrac{94}{57} \end{bmatrix} \qquad \blacklozenge$$

EXAMPLE 2 Orthogonal Projection on a Subspace

Find the orthogonal projection of the vector $\mathbf{u} = (-3, -3, 8, 9)$ on the subspace of R^4 spanned by the vectors

$$\mathbf{u}_1 = (3, 1, 0, 1), \qquad \mathbf{u}_2 = (1, 2, 1, 1), \qquad \mathbf{u}_3 = (-1, 0, 2, -1)$$

Solution.

One could solve this problem by first using the Gram–Schmidt process to convert $\{\mathbf{u}_1, \mathbf{u}_2, \mathbf{u}_3\}$ into an orthonormal basis, and then applying the method used in Example 6 of Section 6.3. However, the following method is more efficient.

The subspace W of R^4 spanned by $\mathbf{u}_1$, $\mathbf{u}_2$, and $\mathbf{u}_3$ is the column space of the matrix

$$A = \begin{bmatrix} 3 & 1 & -1 \\ 1 & 2 & 0 \\ 0 & 1 & 2 \\ 1 & 1 & -1 \end{bmatrix}$$

Thus, if $\mathbf{u}$ is expressed as a column vector, we can find the orthogonal projection of $\mathbf{u}$ on W by finding a least squares solution of the system $A\mathbf{x} = \mathbf{u}$ and then calculating $\mathrm{proj}_W \mathbf{u} = A\mathbf{x}$ from the least squares solution. The computations are as follows: The system $A\mathbf{x} = \mathbf{u}$ is

$$\begin{bmatrix} 3 & 1 & -1 \\ 1 & 2 & 0 \\ 0 & 1 & 2 \\ 1 & 1 & -1 \end{bmatrix} \begin{bmatrix} x_1 \\ x_2 \\ x_3 \end{bmatrix} = \begin{bmatrix} -3 \\ -3 \\ 8 \\ 9 \end{bmatrix}$$

so

$$A^T A = \begin{bmatrix} 3 & 1 & 0 & 1 \\ 1 & 2 & 1 & 1 \\ -1 & 0 & 2 & -1 \end{bmatrix} \begin{bmatrix} 3 & 1 & -1 \\ 1 & 2 & 0 \\ 0 & 1 & 2 \\ 1 & 1 & -1 \end{bmatrix} = \begin{bmatrix} 11 & 6 & -4 \\ 6 & 7 & 0 \\ -4 & 0 & 6 \end{bmatrix}$$

$$A^T \mathbf{u} = \begin{bmatrix} 3 & 1 & 0 & 1 \\ 1 & 2 & 1 & 1 \\ -1 & 0 & 2 & -1 \end{bmatrix} \begin{bmatrix} -3 \\ -3 \\ 8 \\ 9 \end{bmatrix} = \begin{bmatrix} -3 \\ 8 \\ 10 \end{bmatrix}$$

The normal system $A^T A\mathbf{x} = A^T \mathbf{u}$ in this case is

$$\begin{bmatrix} 11 & 6 & -4 \\ 6 & 7 & 0 \\ -4 & 0 & 6 \end{bmatrix} \begin{bmatrix} x_1 \\ x_2 \\ x_3 \end{bmatrix} = \begin{bmatrix} -3 \\ 8 \\ 10 \end{bmatrix}$$

Solving this system yields as the least squares solution of $A\mathbf{x} = \mathbf{u}$

$$\mathbf{x} = \begin{bmatrix} x_1 \\ x_2 \\ x_3 \end{bmatrix} = \begin{bmatrix} -1 \\ 2 \\ 1 \end{bmatrix}$$

(verify), so

$$\mathrm{proj}_W \mathbf{u} = A\mathbf{x} = \begin{bmatrix} 3 & 1 & -1 \\ 1 & 2 & 0 \\ 0 & 1 & 2 \\ 1 & 1 & -1 \end{bmatrix} \begin{bmatrix} -1 \\ 2 \\ 1 \end{bmatrix} = \begin{bmatrix} -2 \\ 3 \\ 4 \\ 0 \end{bmatrix}$$

or in horizontal notation (which is consistent with the original phrasing of the problem), $\mathrm{proj}_W \mathbf{u} = (-2, 3, 4, 0)$. ◆

In Section 4.2 we discussed some basic orthogonal projection operators on R^2 and R^3 (Tables 4 and 5). The concept of an orthogonal projection operator can be extended to higher dimensional Euclidean spaces as follows.

Definition

If W is a subspace of R^m, then the transformation $P: R^m \rightarrow W$ that maps each vector $\mathbf{x}$ in R^m into its orthogonal projection $\text{proj}_W \mathbf{x}$ in W is called the ***orthogonal projection of R^m on W.***

We leave it as an exercise to show that orthogonal projections are linear operators. It follows from Formula (5) that the standard matrix for the orthogonal projection of R^m on W is

$$[P] = A(A^T A)^{-1} A^T \tag{6}$$

where A is constructed using any basis for W as its column vectors.

EXAMPLE 3 Verifying Formula (6)

In Table 5 of Section 4.2 we showed that the standard matrix for the orthogonal projection of R^3 on the xy-plane is

$$[P] = \begin{bmatrix} 1 & 0 & 0 \\ 0 & 1 & 0 \\ 0 & 0 & 0 \end{bmatrix} \tag{7}$$

To see that this is consistent with Formula (6), take the unit vectors along the positive x and y axes as a basis for the xy-plane, so that

$$A = \begin{bmatrix} 1 & 0 \\ 0 & 1 \\ 0 & 0 \end{bmatrix}$$

We leave it for the reader to verify that $A^T A$ is the 2×2 identity matrix; thus, (6) simplifies to

$$[P] = AA^T = \begin{bmatrix} 1 & 0 \\ 0 & 1 \\ 0 & 0 \end{bmatrix} \begin{bmatrix} 1 & 0 & 0 \\ 0 & 1 & 0 \end{bmatrix} = \begin{bmatrix} 1 & 0 & 0 \\ 0 & 1 & 0 \\ 0 & 0 & 0 \end{bmatrix}$$

which agrees with (7). ◆

EXAMPLE 4 Standard Matrix for an Orthogonal Projection

Find the standard matrix for the orthogonal projection P of R^2 on the line l that passes through the origin and makes an angle θ with the positive x-axis.

Solution.

The line l is a one-dimensional subspace of R^2. As illustrated in Figure 6.4.3, we can take $\mathbf{v} = (\cos\theta, \sin\theta)$ as a basis for this subspace, so

$$A = \begin{bmatrix} \cos\theta \\ \sin\theta \end{bmatrix}$$

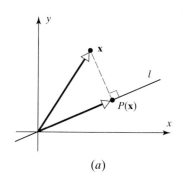

(a)

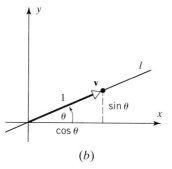

(b)

Figure 6.4.3

We leave it for the reader to show that $A^T A$ is the 1×1 identity matrix; thus, Formula (6) simplifies to

$$[P] = AA^T = \begin{bmatrix} \cos\theta \\ \sin\theta \end{bmatrix} [\cos\theta \quad \sin\theta] = \begin{bmatrix} \cos^2\theta & \sin\theta\cos\theta \\ \sin\theta\cos\theta & \sin^2\theta \end{bmatrix}$$

Note that this agrees with Example 6 of Section 4.3. ◆

Summary   Theorem 6.4.3 enables us to add an additional result to Theorem 6.2.7.

Theorem 6.4.5   **Equivalent Statements**

If A is an $n \times n$ matrix, and if $T_A: R^n \to R^n$ is multiplication by A, then the following are equivalent.

(*a*) *A is invertible.*
(*b*) *$A\mathbf{x} = \mathbf{0}$ has only the trivial solution.*
(*c*) *The reduced row-echelon form of A is I_n.*
(*d*) *A is expressible as a product of elementary matrices.*
(*e*) *$A\mathbf{x} = \mathbf{b}$ is consistent for every $n \times 1$ matrix $\mathbf{b}$.*
(*f*) *$A\mathbf{x} = \mathbf{b}$ has exactly one solution for every $n \times 1$ matrix $\mathbf{b}$.*
(*g*) *$\det(A) \neq 0$.*
(*h*) *The range of T_A is R^n.*
(*i*) *T_A is one-to-one.*
(*j*) *The column vectors of A are linearly independent.*
(*k*) *The row vectors of A are linearly independent.*
(*l*) *The column vectors of A span R^n.*
(*m*) *The row vectors of A span R^n.*
(*n*) *The column vectors of A form a basis for R^n.*
(*o*) *The row vectors of A form a basis for R^n.*
(*p*) *A has rank n.*
(*q*) *A has nullity 0.*
(*r*) *The orthogonal complement of the nullspace of A is R^n.*
(*s*) *The orthogonal complement of the row space of A is $\{\mathbf{0}\}$.*
(*t*) *$A^T A$ is invertible.*

This theorem relates all of the major topics we have studied thus far.

Exercise Set 6.4

1. Find the normal system associated with the given linear system.

(a) $\begin{bmatrix} 1 & -1 \\ 2 & 3 \\ 4 & 5 \end{bmatrix} \begin{bmatrix} x_1 \\ x_2 \end{bmatrix} = \begin{bmatrix} 2 \\ -1 \\ 5 \end{bmatrix}$ (b) $\begin{bmatrix} 2 & -1 & 0 \\ 3 & 1 & 2 \\ -1 & 4 & 5 \\ 1 & 2 & 4 \end{bmatrix} \begin{bmatrix} x_1 \\ x_2 \\ x_3 \end{bmatrix} = \begin{bmatrix} -1 \\ 0 \\ 1 \\ 2 \end{bmatrix}$

2. In each part find $\det(A^T A)$, and apply Theorem 6.4.3 to determine whether A has linearly independent column vectors.

(a) $A = \begin{bmatrix} -1 & 3 & 2 \\ 2 & 1 & 3 \\ 0 & 1 & 1 \end{bmatrix}$ (b) $A = \begin{bmatrix} 2 & -1 & 3 \\ 0 & 1 & 1 \\ -1 & 0 & -2 \\ 4 & -5 & 3 \end{bmatrix}$

3. Find the least squares solution of the linear system $A\mathbf{x} = \mathbf{b}$, and find the orthogonal projection of $\mathbf{b}$ on the column space of A.

(a) $A = \begin{bmatrix} 1 & 1 \\ -1 & 1 \\ -1 & 2 \end{bmatrix}$, $\mathbf{b} = \begin{bmatrix} 7 \\ 0 \\ -7 \end{bmatrix}$ (b) $A = \begin{bmatrix} 2 & -2 \\ 1 & 1 \\ 3 & 1 \end{bmatrix}$, $\mathbf{b} = \begin{bmatrix} 2 \\ -1 \\ 1 \end{bmatrix}$

(c) $A = \begin{bmatrix} 1 & 0 & -1 \\ 2 & 1 & -2 \\ 1 & 1 & 0 \\ 1 & 1 & -1 \end{bmatrix}$, $\mathbf{b} = \begin{bmatrix} 6 \\ 0 \\ 9 \\ 3 \end{bmatrix}$ (d) $A = \begin{bmatrix} 2 & 0 & -1 \\ 1 & -2 & 2 \\ 2 & -1 & 0 \\ 0 & 1 & -1 \end{bmatrix}$, $\mathbf{b} = \begin{bmatrix} 0 \\ 6 \\ 0 \\ 6 \end{bmatrix}$

4. Find the orthogonal projection of $\mathbf{u}$ on the subspace of R^3 spanned by the vectors $\mathbf{v}_1$ and $\mathbf{v}_2$.

(a) $\mathbf{u} = (2, 1, 3)$; $\mathbf{v}_1 = (1, 1, 0)$, $\mathbf{v}_2 = (1, 2, 1)$

(b) $\mathbf{u} = (1, -6, 1)$; $\mathbf{v}_1 = (-1, 2, 1)$, $\mathbf{v}_2 = (2, 2, 4)$

5. Find the orthogonal projection of $\mathbf{u}$ on the subspace of R^4 spanned by the vectors $\mathbf{v}_1$, $\mathbf{v}_2$, and $\mathbf{v}_3$.

(a) $\mathbf{u} = (6, 3, 9, 6)$; $\mathbf{v}_1 = (2, 1, 1, 1)$, $\mathbf{v}_2 = (1, 0, 1, 1)$, $\mathbf{v}_3 = (-2, -1, 0, -1)$

(b) $\mathbf{u} = (-2, 0, 2, 4)$; $\mathbf{v}_1 = (1, 1, 3, 0)$, $\mathbf{v}_2 = (-2, -1, -2, 1)$, $\mathbf{v}_3 = (-3, -1, 1, 3)$

6. Find the orthogonal projection of $\mathbf{u} = (5, 6, 7, 2)$ on the solution space of the homogeneous linear system

$$\begin{aligned} x_1 + x_2 + x_3 \phantom{{}+ x_4} &= 0 \\ 2x_2 + x_3 + x_4 &= 0 \end{aligned}$$

7. Use Formula (6) and the method of Example 3 to find the standard matrix for the orthogonal projection $P: R^2 \to R^2$ onto

(a) the x-axis   (b) the y-axis

[*Note.* Compare your results to Table 4 of Section 4.2.]

8. Use Formula (6) and the method of Example 3 to find the standard matrix for the orthogonal projection $P: R^3 \to R^3$ onto

(a) the xz-plane   (b) the yz-plane

[*Note.* Compare your results to Table 5 of Section 4.2.]

9. Let W be the plane with equation $5x - 3y + z = 0$.

(a) Find a basis for W.

(b) Use Formula (6) to find the standard matrix for the orthogonal projection onto W.

(c) Use the matrix obtained in (b) to find the orthogonal projection of a point $P_0(x_0, y_0, z_0)$ on W.

(d) Find the distance between the point $P_0(1, -2, 4)$ and the plane W, and check your result using Theorem 3.5.2.

10. Let W be the line with parametric equations

$$x = 2t, \quad y = -t, \quad z = 4t \quad (-\infty < t < \infty)$$

(a) Find a basis for W.

(b) Use Formula (6) to find the standard matrix for the orthogonal projection onto W.

(c) Use the matrix obtained in (b) to find the orthogonal projection of a point $P_0(x_0, y_0, z_0)$ on W.

(d) Find the distance between the point $P_0(2, 1, -3)$ and the line W.

11. For the linear systems in Exercise 3, verify that the *error vector* $A\bar{\mathbf{x}} - \mathbf{b}$ resulting from the least squares solution $\bar{\mathbf{x}}$ is orthogonal to the column space of A.

12. Prove: If A has linearly independent column vectors, and if $A\mathbf{x} = \mathbf{b}$ is consistent, then the least squares solution of $A\mathbf{x} = \mathbf{b}$ and the exact solution of $A\mathbf{x} = \mathbf{b}$ are the same.

13. Prove: If A has linearly independent column vectors, and if $\mathbf{b}$ is orthogonal to the column space of A, then the least squares solution of $A\mathbf{x} = \mathbf{b}$ is $\mathbf{x} = \mathbf{0}$.

14. Let $P: R^m \to W$ be the orthogonal projection of R^m onto a subspace W.

   (a) Prove that $[P]^2 = [P]$.
   (b) What does the result in part (a) imply about the composition $P \circ P$?
   (c) Show that $[P]$ is symmetric.
   (d) Verify that the matrices in Tables 4 and 5 of Section 4.2 have the properties in parts (a) and (c).

15. Let A be an $m \times n$ matrix with linearly independent row vectors. Find a standard matrix for the orthogonal projection of R^n onto the row space of A. [**Hint.** Start with Formula (6).]

Discussion and Discovery

16. The following is the proof that $(b) \Rightarrow (a)$ in Theorem 6.4.3. Justify each line by filling in the blank appropriately.

   Hypothesis: Suppose that A is an $m \times n$ matrix and $A^T A$ is invertible.

   Conclusion: A has linearly independent column vectors.

   Proof: (1) If $\mathbf{x}$ is a solution of $A\mathbf{x} = \mathbf{0}$, then $A^T A\mathbf{x} = \mathbf{0}$. _____

       (2) Thus, $\mathbf{x} = \mathbf{0}$. _____

       (3) Thus, the column vectors of A are linearly independent. _____

17. Let A be an $m \times n$ matrix with linearly independent column vectors, and let $\mathbf{b}$ be an $m \times 1$ matrix. Give a formula in terms of A and A^T for

   (a) the vector in the column space of A that is closest to $\mathbf{b}$ relative to the Euclidean inner product;
   (b) the least squares solution of $A\mathbf{x} = \mathbf{b}$ relative to the Euclidean inner product;
   (c) the error in the least squares solution of $A\mathbf{x} = \mathbf{b}$ relative to the Euclidean inner product;
   (d) the standard matrix for the orthogonal projection of R^m onto the column space of A relative to the Euclidean inner product.

6.5 ORTHOGONAL MATRICES; CHANGE OF BASIS

A basis that is suitable for one problem may not be suitable for another, so it is a common process in the study of vector spaces to change from one basis to another. Because a basis is the vector space generalization of a coordinate system, changing bases is akin to changing coordinate axes in R^2 and R^3. In this section we shall study various problems relating to change of basis. We shall also develop properties of square matrices with orthonormal column vectors. Such matrices arise in many contexts, including problems involving a change from one orthonormal basis to another.

Matrices whose inverses can be obtained by transposition are sufficiently important that there is some terminology associated with them.

> **Definition**
>
> A square matrix A with the property
>
> $$A^{-1} = A^T$$
>
> is said to be an ***orthogonal matrix***.

It follows from this definition that *a square matrix A is orthogonal if and only if*

$$AA^T = A^T A = I \tag{1}$$

In fact, it follows from Theorem 1.6.3 that a square matrix A is orthogonal if *either* $AA^T = I$ or $A^T A = I$.

EXAMPLE 1 A 3 × 3 Orthogonal Matrix

The matrix

$$A = \begin{bmatrix} \frac{3}{7} & \frac{2}{7} & \frac{6}{7} \\ -\frac{6}{7} & \frac{3}{7} & \frac{2}{7} \\ \frac{2}{7} & \frac{6}{7} & -\frac{3}{7} \end{bmatrix}$$

is orthogonal, since

$$A^T A = \begin{bmatrix} \frac{3}{7} & -\frac{6}{7} & \frac{2}{7} \\ \frac{2}{7} & \frac{3}{7} & \frac{6}{7} \\ \frac{6}{7} & \frac{2}{7} & -\frac{3}{7} \end{bmatrix} \begin{bmatrix} \frac{3}{7} & \frac{2}{7} & \frac{6}{7} \\ -\frac{6}{7} & \frac{3}{7} & \frac{2}{7} \\ \frac{2}{7} & \frac{6}{7} & -\frac{3}{7} \end{bmatrix} = \begin{bmatrix} 1 & 0 & 0 \\ 0 & 1 & 0 \\ 0 & 0 & 1 \end{bmatrix} \qquad ◆$$

EXAMPLE 2 A Rotation Matrix Is Orthogonal

Recall from Table 6 of Section 4.2 that the standard matrix for the counterclockwise rotation of R^2 through an angle θ is

$$A = \begin{bmatrix} \cos\theta & -\sin\theta \\ \sin\theta & \cos\theta \end{bmatrix}$$

This matrix is orthogonal for all choices of θ, since

$$A^T A = \begin{bmatrix} \cos\theta & \sin\theta \\ -\sin\theta & \cos\theta \end{bmatrix} \begin{bmatrix} \cos\theta & -\sin\theta \\ \sin\theta & \cos\theta \end{bmatrix} = \begin{bmatrix} 1 & 0 \\ 0 & 1 \end{bmatrix}$$

In fact, it is a simple matter to check that all of the "reflection matrices" in Tables 2 and 3 and all of the "rotation matrices" in Tables 6 and 7 of Section 4.2 are orthogonal matrices. ◆

Observe that for the orthogonal matrices in Examples 1 and 2, both the row vectors and the column vectors form orthonormal sets with respect to the Euclidean inner product (verify). This is not accidental; it is a consequence of the following theorem.

Theorem 6.5.1

The following are equivalent for an n × n matrix A.

(*a*) *A is orthogonal.*
(*b*) *The row vectors of A form an orthonormal set in R^n with the Euclidean inner product.*
(*c*) *The column vectors of A form an orthonormal set in R^n with the Euclidean inner product.*

Proof. We shall prove the equivalence of (*a*) and (*b*) and leave the equivalence of (*a*) and (*c*) as an exercise.

(*a*) ⇔ (*b*). The entry in the ith row and jth column of the matrix product AA^T is the dot product of the ith row vector of A and the jth column vector of A^T. But, except for a difference in notation, the jth column vector of A^T is the jth row vector of A. Thus, if the row vectors of A are $\mathbf{r}_1, \mathbf{r}_2, \ldots, \mathbf{r}_n$, then the matrix product AA^T can be expressed as

$$AA^T = \begin{bmatrix} \mathbf{r}_1 \cdot \mathbf{r}_1 & \mathbf{r}_1 \cdot \mathbf{r}_2 & \cdots & \mathbf{r}_1 \cdot \mathbf{r}_n \\ \mathbf{r}_2 \cdot \mathbf{r}_1 & \mathbf{r}_2 \cdot \mathbf{r}_2 & \cdots & \mathbf{r}_2 \cdot \mathbf{r}_n \\ \vdots & \vdots & & \vdots \\ \mathbf{r}_n \cdot \mathbf{r}_1 & \mathbf{r}_n \cdot \mathbf{r}_2 & \cdots & \mathbf{r}_n \cdot \mathbf{r}_n \end{bmatrix}$$

Thus, $AA^T = I$ if and only if

$$\mathbf{r}_1 \cdot \mathbf{r}_1 = \mathbf{r}_2 \cdot \mathbf{r}_2 = \cdots = \mathbf{r}_n \cdot \mathbf{r}_n = 1$$

and

$$\mathbf{r}_i \cdot \mathbf{r}_j = 0 \quad \text{when } i \neq j$$

which are true if and only if $\{\mathbf{r}_1, \mathbf{r}_2, \ldots, \mathbf{r}_n\}$ is an orthonormal set in R^n. ∎

REMARK. In light of Theorem 6.5.1, it seems more appropriate to call orthogonal matrices *orthonormal matrices*. However, we will not do so in deference to historical tradition.

The following theorem lists some additional fundamental properties of orthogonal matrices. The proofs are all straightforward and are left for the reader.

Theorem 6.5.2

(*a*) *The inverse of an orthogonal matrix is orthogonal.*
(*b*) *A product of orthogonal matrices is orthogonal.*
(*c*) *If A is orthogonal, then $\det(A) = 1$ or $\det(A) = -1$.*

EXAMPLE 3 det[A] = ±1 for an Orthogonal Matrix A

The matrix

$$A = \begin{bmatrix} 1/\sqrt{2} & 1/\sqrt{2} \\ -1/\sqrt{2} & 1/\sqrt{2} \end{bmatrix}$$

is orthogonal since its row (and column) vectors form orthonormal sets in R^2. We leave it for the reader to check that $\det(A) = 1$. Interchanging the rows produces an orthogonal matrix for which $\det(A) = -1$. ◆

Orthogonal Matrices as Linear Operators

We observed in Example 2 that the standard matrices for the basic reflection and rotation operators on R^2 and R^3 are orthogonal. The next theorem will help explain why this is so.

Theorem 6.5.3

If A is an $n \times n$ matrix, then the following are equivalent.

(*a*) *A is orthogonal*
(*b*) *$\|A\mathbf{x}\| = \|\mathbf{x}\|$ for all $\mathbf{x}$ in R^n.*
(*c*) *$A\mathbf{x} \cdot A\mathbf{y} = \mathbf{x} \cdot \mathbf{y}$ for all $\mathbf{x}$ and $\mathbf{y}$ in R^n.*

Proof. We shall prove the sequence of implications $(a) \Rightarrow (b) \Rightarrow (c) \Rightarrow (a)$.

$(a) \Rightarrow (b)$. Assume that A is orthogonal, so that $A^T A = I$. Then from Formula (8) of Section 4.1,

$$\|A\mathbf{x}\| = (A\mathbf{x} \cdot A\mathbf{x})^{1/2} = (\mathbf{x} \cdot A^T A\mathbf{x})^{1/2} = (\mathbf{x} \cdot \mathbf{x})^{1/2} = \|\mathbf{x}\|$$

$(b) \Rightarrow (c)$. Assume that $\|A\mathbf{x}\| = \|\mathbf{x}\|$ for all $\mathbf{x}$ in R^n. From Theorem 4.1.6 we have

$$A\mathbf{x} \cdot A\mathbf{y} = \tfrac{1}{4}\|A\mathbf{x} + A\mathbf{y}\|^2 - \tfrac{1}{4}\|A\mathbf{x} - A\mathbf{y}\|^2 = \tfrac{1}{4}\|A(\mathbf{x} + \mathbf{y})\|^2 - \tfrac{1}{4}\|A(\mathbf{x} - \mathbf{y})\|^2$$

$$= \tfrac{1}{4}\|\mathbf{x} + \mathbf{y}\|^2 - \tfrac{1}{4}\|\mathbf{x} - \mathbf{y}\|^2 = \mathbf{x} \cdot \mathbf{y}$$

$(c) \Rightarrow (a)$. Assume that $A\mathbf{x} \cdot A\mathbf{y} = \mathbf{x} \cdot \mathbf{y}$ for all $\mathbf{x}$ and $\mathbf{y}$ in R^n. Then from Formula (8) of Section 4.1 we have

$$\mathbf{x} \cdot \mathbf{y} = \mathbf{x} \cdot A^T A\mathbf{y}$$

which can be rewritten as

$$\mathbf{x} \cdot (A^T A\mathbf{y} - \mathbf{y}) = 0 \quad \text{or} \quad \mathbf{x} \cdot (A^T A - I)\mathbf{y} = 0$$

Since this holds for all $\mathbf{x}$ in R^n, it holds in particular if

$$\mathbf{x} = (A^T A - I)\mathbf{y}, \quad \text{so} \quad (A^T A - I)\mathbf{y} \cdot (A^T A - I)\mathbf{y} = 0$$

from which we can conclude that

$$(A^T A - I)\mathbf{y} = \mathbf{0} \tag{2}$$

(why?). Thus, (2) is a homogeneous system of linear equations that is satisfied by every $\mathbf{y}$ in R^n. But this implies that the coefficient matrix must be zero (why?), so $A^T A = I$ and, consequently, A is orthogonal. ∎

If $T: R^n \rightarrow R^n$ is multiplication by an orthogonal matrix A, then T is called an ***orthogonal operator*** on R^n. It follows from parts (a) and (b) of the preceding theorem that the orthogonal operators on R^n are precisely those operators that leave the lengths of all vectors unchanged. Since reflections and rotations of R^2 and R^3 have this property, this explains our observation in Example 2 that the standard matrices for the basic reflections and rotations of R^2 and R^3 are orthogonal.

Coordinate Matrices

Recall from Theorem 5.4.1 that if $S = \{\mathbf{v}_1, \mathbf{v}_2, \ldots, \mathbf{v}_n\}$ is a basis for a vector space V, then each vector $\mathbf{v}$ in V can be expressed uniquely as a linear combination of the basis vectors, say

$$\mathbf{v} = k_1\mathbf{v}_1 + k_2\mathbf{v}_2 + \cdots + k_n\mathbf{v}_n$$

The scalars $k_1, k_2, \ldots, k_n$ are the coordinates of $\mathbf{v}$ relative to S, and the vector

$$(\mathbf{v})_S = (k_1, k_2, \ldots, k_n)$$

is the coordinate vector of $\mathbf{v}$ relative to S. In this section it will be convenient to list the coordinates as entries of an $n \times 1$ matrix. Thus, we define

$$[\mathbf{v}]_S = \begin{bmatrix} k_1 \\ k_2 \\ \vdots \\ k_n \end{bmatrix}$$

to be the ***coordinate matrix*** of $\mathbf{v}$ relative to S.

Change of Basis In applications it is common to work with more than one coordinate system, and in such cases it is usually necessary to know the relationships between the coordinates of a fixed point or vector in the various coordinate systems. Since a basis is the vector space generalization of a coordinate system, we are led to consider the following problem.

Change of Basis Problem. If we change the basis for a vector space V from some old basis B to some new basis B', how is the old coordinate matrix $[\mathbf{v}]_B$ of a vector $\mathbf{v}$ related to the new coordinate matrix $[\mathbf{v}]_{B'}$?

For simplicity, we will solve this problem for two-dimensional spaces. The solution for n-dimensional spaces is similar and is left for the reader. Let

$$B = \{\mathbf{u}_1, \mathbf{u}_2\} \quad \text{and} \quad B' = \{\mathbf{u}_1', \mathbf{u}_2'\}$$

be the old and new bases, respectively. We will need the coordinate matrices for the new basis vectors relative to the old basis. Suppose they are

$$[\mathbf{u}_1']_B = \begin{bmatrix} a \\ b \end{bmatrix} \quad \text{and} \quad [\mathbf{u}_2']_B = \begin{bmatrix} c \\ d \end{bmatrix} \tag{3}$$

That is,

$$\begin{aligned} \mathbf{u}_1' &= a\mathbf{u}_1 + b\mathbf{u}_2 \\ \mathbf{u}_2' &= c\mathbf{u}_1 + d\mathbf{u}_2 \end{aligned} \tag{4}$$

Now let $\mathbf{v}$ be any vector in V and let

$$[\mathbf{v}]_{B'} = \begin{bmatrix} k_1 \\ k_2 \end{bmatrix} \tag{5}$$

be the new coordinate matrix, so that

$$\mathbf{v} = k_1\mathbf{u}_1' + k_2\mathbf{u}_2' \tag{6}$$

In order to find the old coordinates of $\mathbf{v}$ we must express $\mathbf{v}$ in terms of the old basis B. To do this, we substitute (4) into (6). This yields

$$\mathbf{v} = k_1(a\mathbf{u}_1 + b\mathbf{u}_2) + k_2(c\mathbf{u}_1 + d\mathbf{u}_2)$$

or

$$\mathbf{v} = (k_1 a + k_2 c)\mathbf{u}_1 + (k_1 b + k_2 d)\mathbf{u}_2$$

Thus, the old coordinate matrix for $\mathbf{v}$ is

$$[\mathbf{v}]_B = \begin{bmatrix} k_1 a + k_2 c \\ k_1 b + k_2 d \end{bmatrix}$$

which can be written as

$$[\mathbf{v}]_B = \begin{bmatrix} a & c \\ b & d \end{bmatrix} \begin{bmatrix} k_1 \\ k_2 \end{bmatrix} \quad \text{or, from (5),} \quad [\mathbf{v}]_B = \begin{bmatrix} a & c \\ b & d \end{bmatrix} [\mathbf{v}]_{B'}$$

This equation states that the old coordinate matrix $[\mathbf{v}]_B$ results when we multiply the new coordinate matrix $[\mathbf{v}]_{B'}$ on the left by the matrix

$$P = \begin{bmatrix} a & c \\ b & d \end{bmatrix}$$

The columns of this matrix are the coordinates of the new basis vectors relative to the old basis [see (3)]. Thus, we have the following solution of the change of basis problem.

Solution of the Change of Basis Problem. If we change the basis for a vector space V from some old basis $B = \{\mathbf{u}_1, \mathbf{u}_2, \ldots, \mathbf{u}_n\}$ to some new basis $B' = \{\mathbf{u}'_1, \mathbf{u}'_2, \ldots, \mathbf{u}'_n\}$, then the old coordinate matrix $[\mathbf{v}]_B$ of a vector $\mathbf{v}$ is related to the new coordinate matrix $[\mathbf{v}]_{B'}$ of the same vector $\mathbf{v}$ by the equation

$$[\mathbf{v}]_B = P[\mathbf{v}]_{B'} \tag{7}$$

where the columns of P are the coordinate matrices of the new basis vectors relative to the old basis; that is, the column vectors of P are

$$[\mathbf{u}'_1]_B, [\mathbf{u}'_2]_B, \ldots, [\mathbf{u}'_n]_B$$

Transition Matrices The matrix P is called the ***transition matrix*** from B' to B; it can be expressed in terms of its column vectors as

$$P = \big[[\mathbf{u}'_1]_B \mid [\mathbf{u}'_2]_B \mid \cdots \mid [\mathbf{u}'_n]_B\big] \tag{8}$$

EXAMPLE 4 Finding a Transition Matrix

Consider the bases $B = \{\mathbf{u}_1, \mathbf{u}_2\}$ and $B' = \{\mathbf{u}'_1, \mathbf{u}'_2\}$ for R^2, where

$$\mathbf{u}_1 = (1, 0); \quad \mathbf{u}_2 = (0, 1); \quad \mathbf{u}'_1 = (1, 1); \quad \mathbf{u}'_2 = (2, 1)$$

(a) Find the transition matrix from B' to B.

(b) Use (7) to find $[\mathbf{v}]_B$ if

$$[\mathbf{v}]_{B'} = \begin{bmatrix} -3 \\ 5 \end{bmatrix}$$

*Solution (**a**).* First we must find the coordinate matrices for the new basis vectors $\mathbf{u}'_1$ and $\mathbf{u}'_2$ relative to the old basis B. By inspection

$$\mathbf{u}'_1 = \mathbf{u}_1 + \mathbf{u}_2$$
$$\mathbf{u}'_2 = 2\mathbf{u}_1 + \mathbf{u}_2$$

so that

$$[\mathbf{u}'_1]_B = \begin{bmatrix} 1 \\ 1 \end{bmatrix} \quad \text{and} \quad [\mathbf{u}'_2]_B = \begin{bmatrix} 2 \\ 1 \end{bmatrix}$$

Thus, the transition matrix from B' to B is

$$P = \begin{bmatrix} 1 & 2 \\ 1 & 1 \end{bmatrix}$$

*Solution (**b**).* Using (7) and the transition matrix in part (a),

$$[\mathbf{v}]_B = \begin{bmatrix} 1 & 2 \\ 1 & 1 \end{bmatrix}\begin{bmatrix} -3 \\ 5 \end{bmatrix} = \begin{bmatrix} 7 \\ 2 \end{bmatrix}$$

As a check, we should be able to recover the vector $\mathbf{v}$ either from $[\mathbf{v}]_B$ or $[\mathbf{v}]_{B'}$. We leave it for the reader to show that $-3\mathbf{u}'_1 + 5\mathbf{u}'_2 = 7\mathbf{u}_1 + 2\mathbf{u}_2 = \mathbf{v} = (7, 2)$. ◆

EXAMPLE 5 A Different Viewpoint on Example 4

Consider the vectors $\mathbf{u}_1 = (1, 0)$, $\mathbf{u}_2 = (0, 1)$, $\mathbf{u}'_1 = (1, 1)$, $\mathbf{u}'_2 = (2, 1)$. In Example 4 we found the transition matrix from the basis $B' = \{\mathbf{u}'_1, \mathbf{u}'_2\}$ for R^2 to the basis $B = \{\mathbf{u}_1, \mathbf{u}_2\}$.

However, we can just as well ask for the transition matrix from B to B'. To obtain this matrix, we simply change our point of view and regard B' as the old basis and B as the new basis. As usual, the columns of the transition matrix will be the coordinates of the new basis vectors relative to the old basis.

By equating corresponding components and solving the resulting linear system, the reader should be able to show that

$$\mathbf{u}_1 = -\mathbf{u}_1' + \mathbf{u}_2'$$
$$\mathbf{u}_2 = 2\mathbf{u}_1' - \mathbf{u}_2'$$

so that

$$[\mathbf{u}_1]_{B'} = \begin{bmatrix} -1 \\ 1 \end{bmatrix} \quad \text{and} \quad [\mathbf{u}_2]_{B'} = \begin{bmatrix} 2 \\ -1 \end{bmatrix}$$

Thus, the transition matrix from B to B' is

$$Q = \begin{bmatrix} -1 & 2 \\ 1 & -1 \end{bmatrix} \qquad \blacklozenge$$

If we multiply the transition matrix from B' to B obtained in Example 4 and the transition matrix from B to B' obtained in Example 5, we find

$$PQ = \begin{bmatrix} 1 & 2 \\ 1 & 1 \end{bmatrix} \begin{bmatrix} -1 & 2 \\ 1 & -1 \end{bmatrix} = \begin{bmatrix} 1 & 0 \\ 0 & 1 \end{bmatrix} = I$$

which shows that $Q = P^{-1}$. The following theorem shows that this is not accidental.

Theorem 6.5.4

If P is the transition matrix from a basis B' to a basis B for a finite-dimensional vector space V, then:

(a) P is invertible.   *(b) P^{-1} is the transition matrix from B to B'.*

Proof. Let Q be the transition matrix from B to B'. We shall show that $PQ = I$ and thus conclude that $Q = P^{-1}$ to complete the proof.

Assume that $B = \{\mathbf{u}_1, \mathbf{u}_2, \ldots, \mathbf{u}_n\}$ and suppose that

$$PQ = \begin{bmatrix} c_{11} & c_{12} & \cdots & c_{1n} \\ c_{21} & c_{22} & \cdots & c_{2n} \\ \vdots & \vdots & & \vdots \\ c_{n1} & c_{n2} & \cdots & c_{nn} \end{bmatrix}$$

From (7)

$$[\mathbf{x}]_B = P[\mathbf{x}]_{B'} \quad \text{and} \quad [\mathbf{x}]_{B'} = Q[\mathbf{x}]_B$$

for all $\mathbf{x}$ in V. Multiplying the second equation through on the left by P and substituting the first gives

$$[\mathbf{x}]_B = PQ[\mathbf{x}]_B \qquad (9)$$

for all $\mathbf{x}$ in V. Letting $\mathbf{x} = \mathbf{u}_1$ in (9) gives

$$\begin{bmatrix} 1 \\ 0 \\ \vdots \\ 0 \end{bmatrix} = \begin{bmatrix} c_{11} & c_{12} & \cdots & c_{1n} \\ c_{21} & c_{22} & \cdots & c_{2n} \\ \vdots & \vdots & & \vdots \\ c_{n1} & c_{n2} & \cdots & c_{nn} \end{bmatrix} \begin{bmatrix} 1 \\ 0 \\ \vdots \\ 0 \end{bmatrix} \quad \text{or} \quad \begin{bmatrix} 1 \\ 0 \\ \vdots \\ 0 \end{bmatrix} = \begin{bmatrix} c_{11} \\ c_{21} \\ \vdots \\ c_{n1} \end{bmatrix}$$

Similarly, successively substituting $\mathbf{x} = \mathbf{u}_2, \ldots, \mathbf{u}_n$ in (9) yields

$$\begin{bmatrix} c_{12} \\ c_{22} \\ \vdots \\ c_{n2} \end{bmatrix} = \begin{bmatrix} 0 \\ 1 \\ \vdots \\ 0 \end{bmatrix}, \ldots, \begin{bmatrix} c_{1n} \\ c_{2n} \\ \vdots \\ c_{nn} \end{bmatrix} = \begin{bmatrix} 0 \\ 0 \\ \vdots \\ 1 \end{bmatrix}$$

Therefore, $PQ = I$. ∎

To summarize, if P is the transition matrix from a basis B' to a basis B, then for every vector $\mathbf{v}$ the following relationships hold:

$$[\mathbf{v}]_B = P[\mathbf{v}]_{B'} \tag{10}$$

$$[\mathbf{v}]_{B'} = P^{-1}[\mathbf{v}]_B \tag{11}$$

Change of Orthonormal Basis The following theorem shows that in an inner product space, the transition matrix from one orthonormal basis to another is orthogonal.

Theorem 6.5.5

If P is the transition matrix from one orthonormal basis to another orthonormal basis for an inner product space, then P is an orthogonal matrix; that is,
$$P^{-1} = P^T$$

Proof. Assume that V is an n-dimensional inner product space and that P is the transition matrix from an orthonormal basis B' to an orthonormal basis B. To prove that P is orthogonal we shall use Theorem 6.5.3 and show that $\|P\mathbf{x}\| = \|\mathbf{x}\|$ for every vector $\mathbf{x}$ in R^n.

Recall from Theorem 6.3.2a that for any orthonormal basis for V, the norm of any vector $\mathbf{u}$ in V is the same as the norm of its coordinate vector in R^n with respect to the Euclidean inner product. Thus, for any vector $\mathbf{u}$ in V we have

$$\|\mathbf{u}\| = \|[\mathbf{u}]_{B'}\| = \|[\mathbf{u}]_B\|$$

or from (10)

$$\|\mathbf{u}\| = \|[\mathbf{u}]_{B'}\| = \|P[\mathbf{u}]_{B'}\| \tag{12}$$

where the first norm is with respect to the inner product on V and the second and third are with respect to the Euclidean inner product on R^n.

Now let $\mathbf{x}$ be any vector in R^n, and let $\mathbf{u}$ be the vector in V whose coordinate matrix with respect to the basis B' is $\mathbf{x}$; that is, $[\mathbf{u}]_{B'} = \mathbf{x}$. Thus, from (12)

$$\|\mathbf{u}\| = \|\mathbf{x}\| = \|P\mathbf{x}\|$$

which proves that P is orthogonal. ∎

EXAMPLE 6 Application to Rotation of Axes in 2-Space

In many problems a rectangular xy-coordinate system is given and a new $x'y'$-coordinate system is obtained by rotating the xy-system counterclockwise about the origin through an angle θ. When this is done, each point Q in the plane has two sets of coordinates:

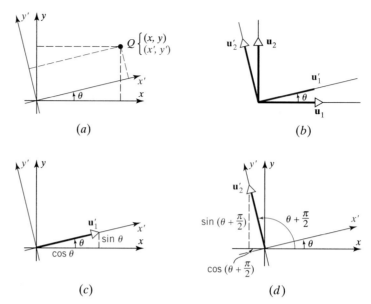

(a) *(b)*

(c) *(d)*

Figure 6.5.1

coordinates (x, y) relative to the xy-system and coordinates (x', y') relative to the $x'y'$-system (Figure 6.5.1a).

By introducing unit vectors $\mathbf{u}_1$ and $\mathbf{u}_2$ along the positive x and y axes and unit vectors $\mathbf{u}'_1$ and $\mathbf{u}'_2$ along the positive x' and y' axes, we can regard this rotation as a change from an old basis $B = \{\mathbf{u}_1, \mathbf{u}_2\}$ to a new basis $B' = \{\mathbf{u}'_1, \mathbf{u}'_2\}$ (Figure 6.5.1b). Thus, the new coordinates (x', y') and the old coordinates (x, y) of a point Q will be related by

$$\begin{bmatrix} x' \\ y' \end{bmatrix} = P^{-1} \begin{bmatrix} x \\ y \end{bmatrix} \tag{13}$$

where P is the transition from B' to B. To find P we must determine the coordinate matrices of the new basis vectors $\mathbf{u}'_1$ and $\mathbf{u}'_2$ relative to the old basis. As indicated in Figure 6.5.1c, the components of $\mathbf{u}'_1$ in the old basis are $\cos \theta$ and $\sin \theta$ so that

$$[\mathbf{u}'_1]_B = \begin{bmatrix} \cos \theta \\ \sin \theta \end{bmatrix}$$

Similarly, from Figure 6.5.1d, we see that the components of $\mathbf{u}'_2$ in the old basis are $\cos(\theta + \pi/2) = -\sin \theta$ and $\sin(\theta + \pi/2) = \cos \theta$, so that

$$[\mathbf{u}'_2]_B = \begin{bmatrix} -\sin \theta \\ \cos \theta \end{bmatrix}$$

Thus, the transition matrix from B' to B is

$$P = \begin{bmatrix} \cos \theta & -\sin \theta \\ \sin \theta & \cos \theta \end{bmatrix}$$

Observe that P is an orthogonal matrix, as expected, since B and B' are orthonormal bases. Thus,

$$P^{-1} = P^T = \begin{bmatrix} \cos \theta & \sin \theta \\ -\sin \theta & \cos \theta \end{bmatrix}$$

so (13) yields

$$\begin{bmatrix} x' \\ y' \end{bmatrix} = \begin{bmatrix} \cos\theta & \sin\theta \\ -\sin\theta & \cos\theta \end{bmatrix} \begin{bmatrix} x \\ y \end{bmatrix} \tag{14}$$

or equivalently,

$$\begin{aligned} x' &= x\cos\theta + y\sin\theta \\ y' &= -x\sin\theta + y\cos\theta \end{aligned}$$

For example, if the axes are rotated $\theta = \pi/4$, then since

$$\sin\frac{\pi}{4} = \cos\frac{\pi}{4} = \frac{1}{\sqrt{2}}$$

Equation (14) becomes

$$\begin{bmatrix} x' \\ y' \end{bmatrix} = \begin{bmatrix} \dfrac{1}{\sqrt{2}} & \dfrac{1}{\sqrt{2}} \\ -\dfrac{1}{\sqrt{2}} & \dfrac{1}{\sqrt{2}} \end{bmatrix} \begin{bmatrix} x \\ y \end{bmatrix}$$

Thus, if the old coordinates of a point Q are $(x, y) = (2, -1)$, then

$$\begin{bmatrix} x' \\ y' \end{bmatrix} = \begin{bmatrix} \dfrac{1}{\sqrt{2}} & \dfrac{1}{\sqrt{2}} \\ -\dfrac{1}{\sqrt{2}} & \dfrac{1}{\sqrt{2}} \end{bmatrix} \begin{bmatrix} 2 \\ -1 \end{bmatrix} = \begin{bmatrix} \dfrac{1}{\sqrt{2}} \\ -\dfrac{3}{\sqrt{2}} \end{bmatrix}$$

so the new coordinates of Q are $(x', y') = (1/\sqrt{2}, -3/\sqrt{2})$. ◆

REMARK. Observe that the coefficient matrix in (14) is the same as the standard matrix for the linear operator that rotates the vectors of R^2 through the angle $-\theta$ (Table 6 of Section 4.2). This is to be expected since rotating the coordinate axes through the angle θ with the vectors of R^2 kept fixed has the same effect as rotating the vectors through the angle $-\theta$ with the axes kept fixed.

EXAMPLE 7 Application to Rotation of Axes in 3-Space

Suppose that a rectangular xyz-coordinate system is rotated around its z-axis counterclockwise (looking down the positive z-axis) through an angle θ (Figure 6.5.2). If we introduce unit vectors $\mathbf{u}_1$, $\mathbf{u}_2$, and $\mathbf{u}_3$ along the positive x, y, and z axes and unit vectors $\mathbf{u}'_1$, $\mathbf{u}'_2$, and $\mathbf{u}'_3$ along the positive x', y', and z' axes, we can regard the rotation as a change from the old basis $B = \{\mathbf{u}_1, \mathbf{u}_2, \mathbf{u}_3\}$ to the new basis $B' = \{\mathbf{u}'_1, \mathbf{u}'_2, \mathbf{u}'_3\}$. In light of Example 6 it should be evident that

$$[\mathbf{u}'_1]_B = \begin{bmatrix} \cos\theta \\ \sin\theta \\ 0 \end{bmatrix} \quad \text{and} \quad [\mathbf{u}'_2]_B = \begin{bmatrix} -\sin\theta \\ \cos\theta \\ 0 \end{bmatrix}$$

Moreover, since $\mathbf{u}'_3$ extends 1 unit up the positive z'-axis,

$$[\mathbf{u}'_3]_B = \begin{bmatrix} 0 \\ 0 \\ 1 \end{bmatrix}$$

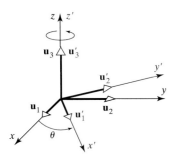

Figure 6.5.2

Thus, the transition matrix from B' to B is

$$P = \begin{bmatrix} \cos\theta & -\sin\theta & 0 \\ \sin\theta & \cos\theta & 0 \\ 0 & 0 & 1 \end{bmatrix}$$

and the transition matrix from B to B' is

$$P^{-1} = \begin{bmatrix} \cos\theta & \sin\theta & 0 \\ -\sin\theta & \cos\theta & 0 \\ 0 & 0 & 1 \end{bmatrix}$$

(verify). Thus, the new coordinates (x', y', z') of a point Q can be computed from its old coordinates (x, y, z) by

$$\begin{bmatrix} x' \\ y' \\ z' \end{bmatrix} = \begin{bmatrix} \cos\theta & \sin\theta & 0 \\ -\sin\theta & \cos\theta & 0 \\ 0 & 0 & 1 \end{bmatrix} \begin{bmatrix} x \\ y \\ z \end{bmatrix} \qquad \blacklozenge$$

Exercise Set 6.5

1. (a) Show that the matrix

$$A = \begin{bmatrix} \frac{4}{5} & 0 & -\frac{3}{5} \\ -\frac{9}{25} & \frac{4}{5} & -\frac{12}{25} \\ \frac{12}{25} & \frac{3}{5} & \frac{16}{25} \end{bmatrix}$$

is orthogonal in three ways: by calculating $A^T A$, by using part (b) of Theorem 6.5.1, and by using part (c) of Theorem 6.5.1.

   (b) Find the inverse of the matrix A in part (a).

2. (a) Show that the matrix

$$A = \begin{bmatrix} \frac{1}{3} & \frac{2}{3} & \frac{2}{3} \\ \frac{2}{3} & -\frac{2}{3} & \frac{1}{3} \\ -\frac{2}{3} & -\frac{1}{3} & \frac{2}{3} \end{bmatrix}$$

is orthogonal.

   (b) Let $T: R^3 \to R^3$ be multiplication by the matrix A in part (a). Find $T(\mathbf{x})$ for the vector $\mathbf{x} = (-2, 3, 5)$. Using the Euclidean inner product on R^3, verify that $\|T(\mathbf{x})\| = \|\mathbf{x}\|$.

3. Determine which of the following matrices are orthogonal. For those that are orthogonal, find the inverse.

   (a) $\begin{bmatrix} 1 & 0 \\ 0 & 1 \end{bmatrix}$ (b) $\begin{bmatrix} 1/\sqrt{2} & -1/\sqrt{2} \\ 1/\sqrt{2} & 1/\sqrt{2} \end{bmatrix}$ (c) $\begin{bmatrix} 0 & 1 & 1/\sqrt{2} \\ 1 & 0 & 0 \\ 0 & 0 & 1/\sqrt{2} \end{bmatrix}$

   (d) $\begin{bmatrix} -1/\sqrt{2} & 1/\sqrt{6} & 1/\sqrt{3} \\ 0 & -2/\sqrt{6} & 1/\sqrt{3} \\ 1/\sqrt{2} & 1/\sqrt{6} & 1/\sqrt{3} \end{bmatrix}$ (e) $\begin{bmatrix} \frac{1}{2} & \frac{1}{2} & \frac{1}{2} & \frac{1}{2} \\ \frac{1}{2} & -\frac{5}{6} & \frac{1}{6} & \frac{1}{6} \\ \frac{1}{2} & \frac{1}{6} & \frac{1}{6} & -\frac{5}{6} \\ \frac{1}{2} & \frac{1}{6} & -\frac{5}{6} & \frac{1}{6} \end{bmatrix}$ (f) $\begin{bmatrix} 1 & 0 & 0 & 0 \\ 0 & 1/\sqrt{3} & -1/2 & 0 \\ 0 & 1/\sqrt{3} & 0 & 1 \\ 0 & 1/\sqrt{3} & 1/2 & 0 \end{bmatrix}$

4. Verify that the reflection matrices in Tables 2 and 3 of Section 4.2 are orthogonal.

5. Find the coordinate matrix for **w** relative to the basis $S = \{\mathbf{u}_1, \mathbf{u}_2\}$ for R^2.

(a) $\mathbf{u}_1 = (1, 0)$, $\mathbf{u}_2 = (0, 1)$; $\mathbf{w} = (3, -7)$ (b) $\mathbf{u}_1 = (2, -4)$, $\mathbf{u}_2 = (3, 8)$; $\mathbf{w} = (1, 1)$

(c) $\mathbf{u}_1 = (1, 1)$, $\mathbf{u}_2 = (0, 2)$; $\mathbf{w} = (a, b)$

6. Find the coordinate matrix for **v** relative to $S = \{\mathbf{v}_1, \mathbf{v}_2, \mathbf{v}_3\}$.

(a) $\mathbf{v} = (2, -1, 3)$; $\mathbf{v}_1 = (1, 0, 0)$, $\mathbf{v}_2 = (2, 2, 0)$, $\mathbf{v}_3 = (3, 3, 3)$

(b) $\mathbf{v} = (5, -12, 3)$; $\mathbf{v}_1 = (1, 2, 3)$, $\mathbf{v}_2 = (-4, 5, 6)$, $\mathbf{v}_3 = (7, -8, 9)$

7. Find the coordinate matrix for **p** relative to $S = \{\mathbf{p}_1, \mathbf{p}_2, \mathbf{p}_3\}$.

(a) $\mathbf{p} = 4 - 3x + x^2$; $\mathbf{p}_1 = 1$, $\mathbf{p}_2 = x$, $\mathbf{p}_3 = x^2$

(b) $\mathbf{p} = 2 - x + x^2$; $\mathbf{p}_1 = 1 + x$, $\mathbf{p}_2 = 1 + x^2$, $\mathbf{p}_3 = x + x^2$

8. Find the coordinate matrix for A relative to $S = \{A_1, A_2, A_3, A_4\}$.

$$A = \begin{bmatrix} 2 & 0 \\ -1 & 3 \end{bmatrix}, \quad A_1 = \begin{bmatrix} -1 & 1 \\ 0 & 0 \end{bmatrix}, \quad A_2 = \begin{bmatrix} 1 & 1 \\ 0 & 0 \end{bmatrix}, \quad A_3 = \begin{bmatrix} 0 & 0 \\ 1 & 0 \end{bmatrix}, \quad A_4 = \begin{bmatrix} 0 & 0 \\ 0 & 1 \end{bmatrix}$$

9. Consider the coordinate matrices

$$[\mathbf{w}]_S = \begin{bmatrix} 6 \\ -1 \\ 4 \end{bmatrix}, \quad [\mathbf{q}]_S = \begin{bmatrix} 3 \\ 0 \\ 4 \end{bmatrix}, \quad [B]_S = \begin{bmatrix} -8 \\ 7 \\ 6 \\ 3 \end{bmatrix}$$

(a) Find **w** if S is the basis in Exercise 6(a).

(b) Find **q** if S is the basis in Exercise 7(a).

(c) Find B if S is the basis in Exercise 8.

10. Consider the bases $B = \{\mathbf{u}_1, \mathbf{u}_2\}$ and $B' = \{\mathbf{v}_1, \mathbf{v}_2\}$ for R^2, where

$$\mathbf{u}_1 = \begin{bmatrix} 1 \\ 0 \end{bmatrix}, \quad \mathbf{u}_2 = \begin{bmatrix} 0 \\ 1 \end{bmatrix}, \quad \mathbf{v}_1 = \begin{bmatrix} 2 \\ 1 \end{bmatrix}, \quad \text{and} \quad \mathbf{v}_2 = \begin{bmatrix} -3 \\ 4 \end{bmatrix}$$

(a) Find the transition matrix from B' to B.

(b) Find the transition matrix from B to B'.

(c) Compute the coordinate matrix $[\mathbf{w}]_B$, where

$$\mathbf{w} = \begin{bmatrix} 3 \\ -5 \end{bmatrix}$$

and use (11) to compute $[\mathbf{w}]_{B'}$.

(d) Check your work by computing $[\mathbf{w}]_{B'}$ directly.

11. Repeat the directions of Exercise 10 with the same vector **w** but with

$$\mathbf{u}_1 = \begin{bmatrix} 2 \\ 2 \end{bmatrix}, \quad \mathbf{u}_2 = \begin{bmatrix} 4 \\ -1 \end{bmatrix}, \quad \mathbf{v}_1 = \begin{bmatrix} 1 \\ 3 \end{bmatrix}, \quad \mathbf{v}_2 = \begin{bmatrix} -1 \\ -1 \end{bmatrix}$$

12. Consider the bases $B = \{\mathbf{u}_1, \mathbf{u}_2, \mathbf{u}_3\}$ and $B' = \{\mathbf{v}_1, \mathbf{v}_2, \mathbf{v}_3\}$ for R^3, where

$$\mathbf{u}_1 = \begin{bmatrix} -3 \\ 0 \\ -3 \end{bmatrix}, \quad \mathbf{u}_2 = \begin{bmatrix} -3 \\ 2 \\ -1 \end{bmatrix}, \quad \mathbf{u}_3 = \begin{bmatrix} 1 \\ 6 \\ -1 \end{bmatrix}, \quad \mathbf{v}_1 = \begin{bmatrix} -6 \\ -6 \\ 0 \end{bmatrix}, \quad \mathbf{v}_2 = \begin{bmatrix} -2 \\ -6 \\ 4 \end{bmatrix}, \quad \mathbf{v}_3 = \begin{bmatrix} -2 \\ -3 \\ 7 \end{bmatrix}$$

(a) Find the transition matrix from B to B'.

(b) Compute the coordinate matrix $[\mathbf{w}]_B$, where

$$\mathbf{w} = \begin{bmatrix} -5 \\ 8 \\ -5 \end{bmatrix}$$

and use (11) to compute $[\mathbf{w}]_{B'}$.

(c) Check your work by computing $[\mathbf{w}]_{B'}$ directly.

13. Repeat the directions of Exercise 12 with the same vector **w**, but with

$$\mathbf{u}_1 = \begin{bmatrix} 2 \\ 1 \\ 1 \end{bmatrix}, \quad \mathbf{u}_2 = \begin{bmatrix} 2 \\ -1 \\ 1 \end{bmatrix}, \quad \mathbf{u}_3 = \begin{bmatrix} 1 \\ 2 \\ 1 \end{bmatrix}, \quad \mathbf{v}_1 = \begin{bmatrix} 3 \\ 1 \\ -5 \end{bmatrix}, \quad \mathbf{v}_2 = \begin{bmatrix} 1 \\ 1 \\ -3 \end{bmatrix}, \quad \mathbf{v}_3 = \begin{bmatrix} -1 \\ 0 \\ 2 \end{bmatrix}$$

14. Consider the bases $B = \{\mathbf{p}_1, \mathbf{p}_2\}$ and $B' = \{\mathbf{q}_1, \mathbf{q}_2\}$ for P_1, where

$$\mathbf{p}_1 = 6 + 3x, \quad \mathbf{p}_2 = 10 + 2x, \quad \mathbf{q}_1 = 2, \quad \mathbf{q}_2 = 3 + 2x$$

(a) Find the transition matrix from B' to B.

(b) Find the transition matrix from B to B'.

(c) Compute the coordinate matrix $[\mathbf{p}]_B$, where $\mathbf{p} = -4 + x$, and use (11) to compute $[\mathbf{p}]_{B'}$.

(d) Check your work by computing $[\mathbf{p}]_{B'}$ directly.

15. Let V be the space spanned by $\mathbf{f}_1 = \sin x$ and $\mathbf{f}_2 = \cos x$.

(a) Show that $\mathbf{g}_1 = 2 \sin x + \cos x$ and $\mathbf{g}_2 = 3 \cos x$ form a basis for V.

(b) Find the transition matrix from $B' = \{\mathbf{g}_1, \mathbf{g}_2\}$ to $B = \{\mathbf{f}_1, \mathbf{f}_2\}$.

(c) Find the transition matrix from B to B'.

(d) Compute the coordinate matrix $[\mathbf{h}]_B$, where $\mathbf{h} = 2 \sin x - 5 \cos x$, and use (11) to obtain $[\mathbf{h}]_{B'}$.

(e) Check your work by computing $[\mathbf{h}]_{B'}$ directly.

16. Let a rectangular $x'y'$-coordinate system be obtained by rotating a rectangular xy-coordinate system counterclockwise through the angle $\theta = 3\pi/4$.

(a) Find the $x'y'$-coordinates of the point whose xy-coordinates are $(-2, 6)$.

(b) Find the xy-coordinates of the point whose $x'y'$-coordinates are $(5, 2)$.

17. Repeat Exercise 16 with $\theta = \pi/3$.

18. Let a rectangular $x'y'z'$-coordinate system be obtained by rotating a rectangular xyz-coordinate system counterclockwise about the z-axis (looking down the z-axis) through the angle $\theta = \pi/4$.

(a) Find the $x'y'z'$-coordinates of the point whose xyz-coordinates are $(-1, 2, 5)$.

(b) Find the xyz-coordinates of the point whose $x'y'z'$-coordinates are $(1, 6, -3)$.

19. Repeat Exercise 18 for a rotation of $\theta = \pi/3$ counterclockwise about the y-axis (looking along the positive y-axis toward the origin).

20. Repeat Exercise 18 for a rotation of $\theta = 3\pi/4$ counterclockwise about the x-axis (looking along the positive x-axis toward the origin).

21. (a) A rectangular $x'y'z'$-coordinate system is obtained by rotating an xyz-coordinate system counterclockwise about the y-axis through an angle θ (looking along the positive y-axis toward the origin). Find a matrix A such that

$$\begin{bmatrix} x' \\ y' \\ z' \end{bmatrix} = A \begin{bmatrix} x \\ y \\ z \end{bmatrix}$$

where (x, y, z) and (x', y', z') are the coordinates of the same point in the xyz- and $x'y'z'$-systems, respectively.

(b) Repeat part (a) for a rotation about the x-axis.

22. A rectangular $x''y''z''$-coordinate system is obtained by first rotating a rectangular xyz-coordinate system $60°$ counterclockwise about the z-axis (looking down the positive z-axis) to obtain an $x'y'z'$-coordinate system, and then rotating the $x'y'z'$-coordinate system $45°$ counterclockwise about the y'-axis (looking along the positive y'-axis toward the origin). Find a matrix A such that

$$\begin{bmatrix} x'' \\ y'' \\ z'' \end{bmatrix} = A \begin{bmatrix} x \\ y \\ z \end{bmatrix}$$

where (x, y, z) and (x'', y'', z'') are the xyz- and $x''y''z''$-coordinates of the same point.

23. What conditions must a and b satisfy for the matrix

$$\begin{bmatrix} a+b & b-a \\ a-b & b+a \end{bmatrix}$$

to be orthogonal?

24. Prove that a 2×2 orthogonal matrix A has one of two possible forms:

$$A = \begin{bmatrix} \cos\theta & -\sin\theta \\ \sin\theta & \cos\theta \end{bmatrix} \quad \text{or} \quad A = \begin{bmatrix} \cos\theta & \sin\theta \\ \sin\theta & -\cos\theta \end{bmatrix}$$

where $0 \leq \theta < 2\pi$. [***Hint.*** Start with a general 2×2 matrix $A = (a_{ij})$, and use the fact that the column vectors form an orthonormal set in R^2.]

25. (a) Use the result in Exercise 24 to prove that multiplication by a 2×2 orthogonal matrix is either a rotation or a rotation followed by a reflection about the x-axis.

   (b) Show that multiplication by A is a rotation if $\det(A) = 1$ and a rotation followed by a reflection if $\det(A) = -1$.

26. Use the result in Exercise 25 to determine whether multiplication by A is a rotation or a rotation followed by a reflection about the x-axis. Find the angle of rotation in either case.

(a) $A = \begin{bmatrix} -1/\sqrt{2} & 1/\sqrt{2} \\ -1/\sqrt{2} & -1/\sqrt{2} \end{bmatrix}$ (b) $A = \begin{bmatrix} -1/2 & \sqrt{3}/2 \\ \sqrt{3}/2 & 1/2 \end{bmatrix}$

27. The result in Exercise 25 has an analog for 3×3 orthogonal matrices: It can be proved that multiplication by a 3×3 orthogonal matrix A is a rotation about some axis if $\det(A) = 1$ and is a rotation about some axis followed by a reflection about some coordinate plane if $\det(A) = -1$. Determine whether multiplication by A is a rotation or a rotation followed by a reflection.

(a) $A = \begin{bmatrix} \frac{3}{7} & \frac{2}{7} & \frac{6}{7} \\ -\frac{6}{7} & \frac{3}{7} & \frac{2}{7} \\ \frac{2}{7} & \frac{6}{7} & -\frac{3}{7} \end{bmatrix}$ (b) $A = \begin{bmatrix} \frac{2}{7} & \frac{3}{7} & \frac{6}{7} \\ \frac{3}{7} & -\frac{6}{7} & \frac{2}{7} \\ \frac{6}{7} & \frac{2}{7} & -\frac{3}{7} \end{bmatrix}$

28. Use the result in Exercise 27 and part (*b*) of Theorem 6.5.2 to show that a composition of rotations can always be accomplished by a single rotation about some appropriate axis.

29. Prove the equivalence of statements (*a*) and (*c*) in Theorem 6.5.1.

Discussion and Discovery

30. A linear operator on R^2 is called ***rigid*** if it does not change the lengths of vectors, and it is called ***angle preserving*** if it does not change the angle between nonzero vectors.

   (a) Name two different types of linear operators that are rigid.
   (b) Name two different types of linear operators that are angle preserving.
   (c) Are there any linear operators on R^2 that are rigid and not angle preserving? Angle preserving and not rigid? Justify your answer.

31. Referring to Exercise 30, what can you say about $\det(A)$ if A is the standard matrix for a rigid linear operator on R^2.

32. Find a, b, and c such that the matrix

$$A = \begin{bmatrix} a & 1/\sqrt{2} & -1/\sqrt{2} \\ b & 1/\sqrt{6} & 1/\sqrt{6} \\ c & 1/\sqrt{3} & 1/\sqrt{3} \end{bmatrix}$$

is orthogonal. Are the values of a, b, and c unique? Explain.

Chapter 6 Supplementary Exercises

1. Let R^4 have the Euclidean inner product.

   (a) Find a vector in R^4 that is orthogonal to $\mathbf{u}_1 = (1, 0, 0, 0)$ and $\mathbf{u}_4 = (0, 0, 0, 1)$ and makes equal angles with $\mathbf{u}_2 = (0, 1, 0, 0)$ and $\mathbf{u}_3 = (0, 0, 1, 0)$.

   (b) Find a vector $\mathbf{x} = (x_1, x_2, x_3, x_4)$ of length 1 that is orthogonal to $\mathbf{u}_1$ and $\mathbf{u}_4$ above and such that the cosine of the angle between $\mathbf{x}$ and $\mathbf{u}_2$ is twice the cosine of the angle between $\mathbf{x}$ and $\mathbf{u}_3$.

2. Show that if $\mathbf{x}$ is a nonzero column vector in R^n, then the $n \times n$ matrix

$$A = I_n - \frac{2}{\|\mathbf{x}\|^2} \mathbf{x}\mathbf{x}^T$$

   is both orthogonal and symmetric.

3. Let $A\mathbf{x} = \mathbf{0}$ be a system of m equations in n unknowns. Show that

$$\mathbf{x} = \begin{bmatrix} x_1 \\ x_2 \\ \vdots \\ x_n \end{bmatrix}$$

   is a solution of the system if and only if the vector $\mathbf{x} = (x_1, x_2, \ldots, x_n)$ is orthogonal to every row vector of A in the Euclidean inner product on R^n.

4. Use the Cauchy–Schwarz inequality to show that if $a_1, a_2, \ldots, a_n$ are positive real numbers, then

$$(a_1 + a_2 + \cdots + a_n)\left(\frac{1}{a_1} + \frac{1}{a_2} + \cdots + \frac{1}{a_n}\right) \geq n^2$$

5. Show that if $\mathbf{x}$ and $\mathbf{y}$ are vectors in an inner product space and c is any scalar, then

$$\|c\mathbf{x} + \mathbf{y}\|^2 = c^2\|\mathbf{x}\|^2 + 2c\langle\mathbf{x}, \mathbf{y}\rangle + \|\mathbf{y}\|^2$$

6. Let R^3 have the Euclidean inner product. Find two vectors of length 1 that are orthogonal to all three of the vectors $\mathbf{u}_1 = (1, 1, -1)$, $\mathbf{u}_2 = (-2, -1, 2)$, and $\mathbf{u}_3 = (-1, 0, 1)$.

7. Find a weighted Euclidean inner product on R^n such that the vectors

$$\begin{aligned} \mathbf{v}_1 &= (1, 0, 0, \ldots, 0) \\ \mathbf{v}_2 &= (0, \sqrt{2}, 0, \ldots, 0) \\ \mathbf{v}_3 &= (0, 0, \sqrt{3}, \ldots, 0) \\ &\vdots \\ \mathbf{v}_n &= (0, 0, 0, \ldots, \sqrt{n}) \end{aligned}$$

   form an orthonormal set.

8. Is there a weighted Euclidean inner product on R^2 for which the vectors $(1, 2)$ and $(3, -1)$ form an orthonormal set? Justify your answer.

9. Prove: If Q is an orthogonal matrix, then each entry of Q is the same as its cofactor if $\det(Q) = 1$ and is the negative of its cofactor if $\det(Q) = -1$.

10. If $\mathbf{u}$ and $\mathbf{v}$ are vectors in an inner product space V, then $\mathbf{u}$, $\mathbf{v}$, and $\mathbf{u} - \mathbf{v}$ can be regarded as sides of a "triangle" in V (see accompanying figure). Prove that the law of cosines holds for any such triangle; that is, $\|\mathbf{u} - \mathbf{v}\|^2 = \|\mathbf{u}\|^2 + \|\mathbf{v}\|^2 - 2\|\mathbf{u}\|\|\mathbf{v}\|\cos\theta$, where θ is the angle between $\mathbf{u}$ and $\mathbf{v}$.

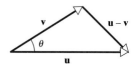

Figure Ex-10

11. (a) In R^3 the vectors $(k, 0, 0)$, $(0, k, 0)$, and $(0, 0, k)$ form the edges of a cube with diagonal (k, k, k) (Figure 3.3.4). Similarly, in R^n the vectors

$$(k, 0, 0, \ldots, 0), \quad (0, k, 0, \ldots, 0), \ldots, \quad (0, 0, 0, \ldots, k)$$

can be regarded as edges of a "cube" with diagonal $(k, k, k, \ldots, k)$. Show that each of the above edges makes an angle of θ with the diagonal, where $\cos \theta = 1/\sqrt{n}$.

(b) **(For readers who have studied calculus.)** What happens to the angle θ in part (a) as the dimension of R^n approaches $+\infty$?

12. Let **u** and **v** be vectors in an inner product space.

(a) Prove that $\|\mathbf{u}\| = \|\mathbf{v}\|$ if and only if $\mathbf{u} + \mathbf{v}$ and $\mathbf{u} - \mathbf{v}$ are orthogonal.

(b) Give a geometric interpretation of this result in R^2 with the Euclidean inner product.

13. Let **u** be a vector in an inner product space V, and let $\{\mathbf{v}_1, \mathbf{v}_2, \ldots, \mathbf{v}_n\}$ be an orthonormal basis for V. Show that if α_i is the angle between **u** and $\mathbf{v}_i$, then

$$\cos^2 \alpha_1 + \cos^2 \alpha_2 + \cdots + \cos^2 \alpha_n = 1$$

14. Prove: If $\langle \mathbf{u}, \mathbf{v} \rangle_1$ and $\langle \mathbf{u}, \mathbf{v} \rangle_2$ are two inner products on a vector space V, then the quantity $\langle \mathbf{u}, \mathbf{v} \rangle = \langle \mathbf{u}, \mathbf{v} \rangle_1 + \langle \mathbf{u}, \mathbf{v} \rangle_2$ is also an inner product.

15. Show that the inner product on R^n generated by any orthogonal matrix is the Euclidean inner product.

16. Prove part (c) of Theorem 6.2.5.

Chapter 6 Technology Exercises

The following exercises are designed to be solved using a technology utility. Typically, this will be MATLAB, *Mathematica*, Maple, Derive, or Mathcad, but it may also be some other type of linear algebra software or a scientific calculator with some linear algebra capabilities. For each exercise you will need to read the relevant documentation for the particular utility you are using. The goal of these exercises is to provide you with a basic proficiency with your technology utility. Once you have mastered the techniques in these exercises, you will be able to use your technology utility to solve many of the problems in the regular exercise sets.

Section 6.1

T1. (*Weighted Euclidean inner products*) See if you can program your utility so it produces the value of a weighted Euclidean inner product when the user enters n, the weights, and the vectors. Check your work by having the program do some specific computations.

T2. (*Inner product on M_{22}*) See if you can program your utility to produce the inner product in Example 7 when the user enters the matrices U and V. Check your work by having the program do some specific computations.

T3. (*Inner product on $C[a, b]$*) If you are using a CAS or a technology utility that can do numerical integration, see if you can program the utility to compute the inner product given in Example 9 when the user enters a, b, and the functions $f(x)$ and $g(x)$. Check your work by having the program do some specific calculations.

Section 6.3

T1. (*Normalizing a vector*) See if you can create a program that will normalize a nonzero vector **v** in R^n when the user enters **v**.

T2. (*Gram–Schmidt process*) Read your documentation on performing the Gram–Schmidt process, and then use your utility to perform the computations in Example 7.

T3. (*QR-decomposition*) Read your documentation on performing the Gram–Schmidt process, and then use your utility to perform the computations in Example 8.

Section 6.4

T1. (*Least squares*) Read your documentation on finding least squares solutions of linear systems, and then use your utility to find the least squares solution of the system in Example 1.

T2. (*Orthogonal projection on a subspace*) Use the least squares capability of your technology utility to find the least squares solution **x** of the normal system in Example 2, and then complete the computations in the example by computing $A\mathbf{x}$. If you are successful, then see if you can create a program that will produce the orthogonal projection of a vector **u** in R^4 on a subspace W when the user enters **u** and a set of vectors that spans W. [*Suggestion.* As the first step, have the program create the matrix A that has the spanning vectors as columns.] Check your work by having your program find the orthogonal projection in Example 2.

Eigenvalues, Eigenvectors

INTRODUCTION: If A is an $n \times n$ matrix and $\mathbf{x}$ is a vector in R^n, then $A\mathbf{x}$ is also a vector in R^n, but usually there is no simple geometric relationship between $\mathbf{x}$ and $A\mathbf{x}$. However, in the special case where $\mathbf{x}$ is a nonzero vector and $A\mathbf{x}$ is a scalar multiple of $\mathbf{x}$, a simple geometric relationship occurs. For example, if A is a 2×2 matrix, and if $\mathbf{x}$ is a nonzero vector such that $A\mathbf{x}$ is a scalar multiple of $\mathbf{x}$, say $A\mathbf{x} = \lambda\mathbf{x}$, then each vector on the line through the origin determined by $\mathbf{x}$ gets mapped back onto the same line under multiplication by A.

Nonzero vectors that get mapped into scalar multiples of themselves under a linear operator arise naturally in the study of vibrations, genetics, population dynamics, quantum mechanics, and economics, as well as in geometry. In this chapter we will study such vectors and their applications.

7.1 EIGENVALUES AND EIGENVECTORS

In Section 2.3 we introduced the concepts of eigenvalue and eigenvector. In this section we will study those ideas in more detail to set the stage for applications of them in later sections.

Review We begin with a review of some concepts that were developed in Sections 2.3 and 4.3.

Definition

If A is an $n \times n$ matrix, then a nonzero vector $\mathbf{x}$ in R^n is called an **eigenvector** of A if $A\mathbf{x}$ is a scalar multiple of $\mathbf{x}$; that is,

$$A\mathbf{x} = \lambda\mathbf{x}$$

for some scalar λ. The scalar λ is called an **eigenvalue** of A, and $\mathbf{x}$ is said to be an eigenvector of A **corresponding** to λ.

In R^2 and R^3 multiplication by A maps each eigenvector $\mathbf{x}$ of A (if any) onto the same line through the origin as $\mathbf{x}$. Depending on the sign and the magnitude of the eigenvalue λ corresponding to $\mathbf{x}$, the linear operator $A\mathbf{x} = \lambda\mathbf{x}$ compresses or stretches $\mathbf{x}$ by a factor of λ, with a reversal of direction in the case where λ is negative (Figure 7.1.1).

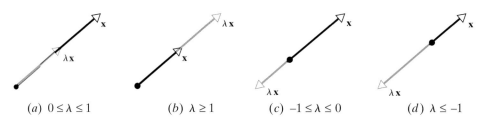

$(a)\ 0 \le \lambda \le 1$ $(b)\ \lambda \ge 1$ $(c)\ -1 \le \lambda \le 0$ $(d)\ \lambda \le -1$

Figure 7.1.1

EXAMPLE 1 Eigenvector of a 2 × 2 Matrix

The vector $\mathbf{x} = \begin{bmatrix} 1 \\ 2 \end{bmatrix}$ is an eigenvector of

$$A = \begin{bmatrix} 3 & 0 \\ 8 & -1 \end{bmatrix}$$

corresponding to the eigenvalue $\lambda = 3$, since

$$A\mathbf{x} = \begin{bmatrix} 3 & 0 \\ 8 & -1 \end{bmatrix} \begin{bmatrix} 1 \\ 2 \end{bmatrix} = \begin{bmatrix} 3 \\ 6 \end{bmatrix} = 3\mathbf{x}$$

       ◆

To find the eigenvalues of an $n \times n$ matrix A we rewrite $A\mathbf{x} = \lambda\mathbf{x}$ as

$$A\mathbf{x} = \lambda I\mathbf{x}$$

or equivalently,

$$(\lambda I - A)\mathbf{x} = \mathbf{0} \tag{1}$$

For λ to be an eigenvalue, there must be a nonzero solution of this equation. However, by Theorem 6.4.5, Equation (1) has a nonzero solution if and only if

$$\det(\lambda I - A) = 0$$

This is called the ***characteristic equation*** of A; the scalars satisfying this equation are the eigenvalues of A. When expanded, the determinant $\det(\lambda I - A)$ is a polynomial p in λ called the ***characteristic polynomial*** of A.

It can be shown (Exercise 15) that if A is an $n \times n$ matrix, then the characteristic polynomial of A has degree n and the coefficient of λ^n is 1; that is, the characteristic polynomial $p(x)$ of an $n \times n$ matrix has the form

$$p(\lambda) = \det(\lambda I - A) = \lambda^n + c_1\lambda^{n-1} + \cdots + c_n$$

It follows from the Fundamental Theorem of Algebra that the characteristic equation

$$\lambda^n + c_1\lambda^{n-1} + \cdots + c_n = 0$$

has at most n distinct solutions, so an $n \times n$ matrix has at most n distinct eigenvalues.

The reader may wish to review Example 6 of Section 2.3, where we found the eigenvalues of a 2×2 matrix by solving the characteristic equation. The following example involves a 3×3 matrix.

EXAMPLE 2 Eigenvalues of a 3 × 3 Matrix

Find the eigenvalues of

$$A = \begin{bmatrix} 0 & 1 & 0 \\ 0 & 0 & 1 \\ 4 & -17 & 8 \end{bmatrix}$$

Solution.

The characteristic polynomial of A is

$$\det(\lambda I - A) = \det \begin{bmatrix} \lambda & -1 & 0 \\ 0 & \lambda & -1 \\ -4 & 17 & \lambda - 8 \end{bmatrix} = \lambda^3 - 8\lambda^2 + 17\lambda - 4$$

The eigenvalues of A must therefore satisfy the cubic equation

$$\lambda^3 - 8\lambda^2 + 17\lambda - 4 = 0 \tag{2}$$

To solve this equation, we shall begin by searching for integer solutions. This task can be greatly simplified by exploiting the fact that all integer solutions (if there are any) to a polynomial equation with integer coefficients

$$\lambda^n + c_1\lambda^{n-1} + \cdots + c_n = 0$$

must be divisors of the constant term, c_n. Thus, the only possible integer solutions of (2) are the divisors of -4, that is, $\pm 1, \pm 2, \pm 4$. Successively substituting these values in (2) shows that $\lambda = 4$ is an integer solution. As a consequence, $\lambda - 4$ must be a factor of the left side of (2). Dividing $\lambda - 4$ into $\lambda^3 - 8\lambda^2 + 17\lambda - 4$ shows that (2) can be rewritten as

$$(\lambda - 4)(\lambda^2 - 4\lambda + 1) = 0$$

Thus, the remaining solutions of (2) satisfy the quadratic equation

$$\lambda^2 - 4\lambda + 1 = 0$$

which can be solved by the quadratic formula. Thus, the eigenvalues of A are

$$\lambda = 4, \quad \lambda = 2 + \sqrt{3}, \quad \text{and} \quad \lambda = 2 - \sqrt{3} \qquad \blacklozenge$$

EXAMPLE 3 Eigenvalues of an Upper Triangular Matrix

Find the eigenvalues of the upper triangular matrix

$$A = \begin{bmatrix} a_{11} & a_{12} & a_{13} & a_{14} \\ 0 & a_{22} & a_{23} & a_{24} \\ 0 & 0 & a_{33} & a_{34} \\ 0 & 0 & 0 & a_{44} \end{bmatrix}$$

Solution.

Recalling that the determinant of a triangular matrix is the product of the entries on the main diagonal (Theorem 2.2.2), we obtain

$$\det(\lambda I - A) = \det \begin{bmatrix} \lambda - a_{11} & -a_{12} & -a_{13} & -a_{14} \\ 0 & \lambda - a_{22} & -a_{23} & -a_{24} \\ 0 & 0 & \lambda - a_{33} & -a_{34} \\ 0 & 0 & 0 & \lambda - a_{44} \end{bmatrix}$$

$$= (\lambda - a_{11})(\lambda - a_{22})(\lambda - a_{33})(\lambda - a_{44})$$

Thus, the characteristic equation is

$$(\lambda - a_{11})(\lambda - a_{22})(\lambda - a_{33})(\lambda - a_{44}) = 0$$

and the eigenvalues are

$$\lambda = a_{11}, \quad \lambda = a_{22}, \quad \lambda = a_{33}, \quad \lambda = a_{44}$$

which are precisely the diagonal entries of A. $\qquad \blacklozenge$

The following general theorem should be evident from the computations in the preceding example.

Theorem 7.1.1

If A is an $n \times n$ triangular matrix (upper triangular, lower triangular, or diagonal), then the eigenvalues of A are the entries on the main diagonal of A.

EXAMPLE 4 Eigenvalues of a Lower Triangular Matrix

By inspection, the eigenvalues of the lower triangular matrix

$$A = \begin{bmatrix} \frac{1}{2} & 0 & 0 \\ -1 & \frac{2}{3} & 0 \\ 5 & -8 & -\frac{1}{4} \end{bmatrix}$$

are $\lambda = \frac{1}{2}$, $\lambda = \frac{2}{3}$, and $\lambda = -\frac{1}{4}$. $\qquad \blacklozenge$

REMARK. In practical problems, the matrix A is often so large that computing the characteristic equation is not practical. As a result, various approximation methods are used to obtain eigenvalues.

Complex Eigenvalues   It is possible for the characteristic equation of a matrix with real entries to have complex solutions. For example, the characteristic polynomial of the matrix

$$A = \begin{bmatrix} -2 & -1 \\ 5 & 2 \end{bmatrix}$$

is

$$\det(\lambda I - A) = \det \begin{bmatrix} \lambda + 2 & 1 \\ -5 & \lambda - 2 \end{bmatrix} = \lambda^2 + 1$$

so the characteristic equation is $\lambda^2 + 1 = 0$, the solutions of which are the imaginary numbers $\lambda = i$ and $\lambda = -i$. Thus, we are forced to consider complex eigenvalues, even for real matrices. This, in turn, leads us to consider the possibility of complex vector spaces, that is, vector spaces in which scalars are allowed to have complex values. Such vector spaces will be considered in Chapter 10. For now, we will allow complex eigenvalues, but our discussion of eigenvectors will be limited to matrices with real eigenvalues.

The following theorem summarizes our discussion thus far.

Theorem 7.1.2   **Equivalent Statements**

If A is an $n \times n$ matrix and λ is a real number, then the following are equivalent.

(a) λ *is an eigenvalue of A.*
(b) *The system of equations $(\lambda I - A)\mathbf{x} = \mathbf{0}$ has nontrivial solutions.*
(c) *There is a nonzero vector $\mathbf{x}$ in R^n such that $A\mathbf{x} = \lambda \mathbf{x}$.*
(d) λ *is a solution of the characteristic equation $\det(\lambda I - A) = 0$.*

Finding Bases for Eigenspaces   Now that we know how to find eigenvalues, we turn to the problem of finding eigenvectors. The eigenvectors of A corresponding to an eigenvalue λ are the nonzero vectors $\mathbf{x}$ that satisfy $A\mathbf{x} = \lambda \mathbf{x}$. Equivalently, the eigenvectors corresponding to λ are the nonzero vectors in the solution space of $(\lambda I - A)\mathbf{x} = \mathbf{0}$. We call this solution space the *eigenspace* of A corresponding to λ.

EXAMPLE 5 **Bases for Eigenspaces**

Find bases for the eigenspaces of

$$A = \begin{bmatrix} 0 & 0 & -2 \\ 1 & 2 & 1 \\ 1 & 0 & 3 \end{bmatrix}$$

Solution.

The characteristic equation of matrix A is $\lambda^3 - 5\lambda^2 + 8\lambda - 4 = 0$, or in factored form, $(\lambda - 1)(\lambda - 2)^2 = 0$ (verify); thus, the eigenvalues of A are $\lambda = 1$ and $\lambda = 2$, so there are two eigenspaces of A.

By definition,

$$\mathbf{x} = \begin{bmatrix} x_1 \\ x_2 \\ x_3 \end{bmatrix}$$

is an eigenvector of A corresponding to λ if and only if $\mathbf{x}$ is a nontrivial solution of $(\lambda I - A)\mathbf{x} = \mathbf{0}$, that is, of

$$\begin{bmatrix} \lambda & 0 & 2 \\ -1 & \lambda - 2 & -1 \\ -1 & 0 & \lambda - 3 \end{bmatrix} \begin{bmatrix} x_1 \\ x_2 \\ x_3 \end{bmatrix} = \begin{bmatrix} 0 \\ 0 \\ 0 \end{bmatrix} \tag{3}$$

If $\lambda = 2$, then (3) becomes

$$\begin{bmatrix} 2 & 0 & 2 \\ -1 & 0 & -1 \\ -1 & 0 & -1 \end{bmatrix} \begin{bmatrix} x_1 \\ x_2 \\ x_3 \end{bmatrix} = \begin{bmatrix} 0 \\ 0 \\ 0 \end{bmatrix}$$

Solving this system yields (verify)

$$x_1 = -s, \qquad x_2 = t, \qquad x_3 = s$$

Thus, the eigenvectors of A corresponding to $\lambda = 2$ are the nonzero vectors of the form

$$\mathbf{x} = \begin{bmatrix} -s \\ t \\ s \end{bmatrix} = \begin{bmatrix} -s \\ 0 \\ s \end{bmatrix} + \begin{bmatrix} 0 \\ t \\ 0 \end{bmatrix} = s \begin{bmatrix} -1 \\ 0 \\ 1 \end{bmatrix} + t \begin{bmatrix} 0 \\ 1 \\ 0 \end{bmatrix}$$

Since

$$\begin{bmatrix} -1 \\ 0 \\ 1 \end{bmatrix} \quad \text{and} \quad \begin{bmatrix} 0 \\ 1 \\ 0 \end{bmatrix}$$

are linearly independent, these vectors form a basis for the eigenspace corresponding to $\lambda = 2$.

If $\lambda = 1$, then (3) becomes

$$\begin{bmatrix} 1 & 0 & 2 \\ -1 & -1 & -1 \\ -1 & 0 & -2 \end{bmatrix} \begin{bmatrix} x_1 \\ x_2 \\ x_3 \end{bmatrix} = \begin{bmatrix} 0 \\ 0 \\ 0 \end{bmatrix}$$

Solving this system yields (verify)

$$x_1 = -2s, \qquad x_2 = s, \qquad x_3 = s$$

Thus, the eigenvectors corresponding to $\lambda = 1$ are the nonzero vectors of the form

$$\begin{bmatrix} -2s \\ s \\ s \end{bmatrix} = s \begin{bmatrix} -2 \\ 1 \\ 1 \end{bmatrix} \quad \text{so that} \quad \begin{bmatrix} -2 \\ 1 \\ 1 \end{bmatrix}$$

is a basis for the eigenspace corresponding to $\lambda = 1$. ◆

Powers of a Matrix

Once the eigenvalues and eigenvectors of a matrix A are found, it is a simple matter to find the eigenvalues and eigenvectors of any positive integer power of A; for example, if λ is an eigenvalue of A and $\mathbf{x}$ is a corresponding eigenvector, then

$$A^2\mathbf{x} = A(A\mathbf{x}) = A(\lambda\mathbf{x}) = \lambda(A\mathbf{x}) = \lambda(\lambda\mathbf{x}) = \lambda^2\mathbf{x}$$

which shows that λ^2 is an eigenvalue of A^2 and $\mathbf{x}$ is a corresponding eigenvector. In general, we have the following result.

Theorem 7.1.3

If k is a positive integer, λ is an eigenvalue of a matrix A, and $\mathbf{x}$ is a corresponding eigenvector, then λ^k is an eigenvalue of A^k and $\mathbf{x}$ is a corresponding eigenvector.

EXAMPLE 6 Using Theorem 7.1.3

In Example 5 we showed that the eigenvalues of

$$A = \begin{bmatrix} 0 & 0 & -2 \\ 1 & 2 & 1 \\ 1 & 0 & 3 \end{bmatrix}$$

are $\lambda = 2$ and $\lambda = 1$, so from Theorem 7.1.3 both $\lambda = 2^7 = 128$ and $\lambda = 1^7 = 1$ are eigenvalues of A^7. We also showed that

$$\begin{bmatrix} -1 \\ 0 \\ 1 \end{bmatrix} \quad \text{and} \quad \begin{bmatrix} 0 \\ 1 \\ 0 \end{bmatrix}$$

are eigenvectors of A corresponding to the eigenvalue $\lambda = 2$, so from Theorem 7.1.3 they are also eigenvectors of A^7 corresponding to $\lambda = 2^7 = 128$. Similarly, the eigenvector

$$\begin{bmatrix} -2 \\ 1 \\ 1 \end{bmatrix}$$

of A corresponding to the eigenvalue $\lambda = 1$ is also an eigenvector of A^7 corresponding to $\lambda = 1^7 = 1$. ◆

Eigenvalues and Invertibility   The next theorem establishes a relationship between the eigenvalues and the invertibility of a matrix.

Theorem 7.1.4

A square matrix A is invertible if and only if $\lambda = 0$ is not an eigenvalue of A.

Proof. Assume that A is an $n \times n$ matrix and observe first that $\lambda = 0$ is a solution of the characteristic equation

$$\lambda^n + c_1\lambda^{n-1} + \cdots + c_n = 0$$

if and only if the constant term c_n is zero. Thus, it suffices to prove that A is invertible if and only if $c_n \neq 0$. But

$$\det(\lambda I - A) = \lambda^n + c_1\lambda^{n-1} + \cdots + c_n$$

or on setting $\lambda = 0$,

$$\det(-A) = c_n \quad \text{or} \quad (-1)^n \det(A) = c_n$$

It follows from the last equation that $\det(A) = 0$ if and only if $c_n = 0$, and this in turn implies that A is invertible if and only if $c_n \neq 0$. ∎

EXAMPLE 7 Using Theorem 7.1.4

The matrix A in Example 5 is invertible since it has eigenvalues $\lambda = 1$ and $\lambda = 2$, neither of which is zero. We leave it for the reader to check this conclusion by showing that $\det(A) \neq 0$. ◆

Summary   Theorem 7.1.4 enables us to add an additional result to Theorem 6.4.5.

Theorem 7.1.5   Equivalent Statements

If A is an $n \times n$ matrix, and if $T_A: R^n \rightarrow R^n$ is multiplication by A, then the following are equivalent.

(a) *A is invertible.*
(b) *$A\mathbf{x} = \mathbf{0}$ has only the trivial solution.*
(c) *The reduced row-echelon form of A is I_n.*
(d) *A is expressible as a product of elementary matrices.*
(e) *$A\mathbf{x} = \mathbf{b}$ is consistent for every $n \times 1$ matrix $\mathbf{b}$.*
(f) *$A\mathbf{x} = \mathbf{b}$ has exactly one solution for every $n \times 1$ matrix $\mathbf{b}$.*
(g) *$\det(A) \neq 0$.*
(h) *The range of T_A is R^n.*
(i) *T_A is one-to-one.*
(j) *The column vectors of A are linearly independent.*
(k) *The row vectors of A are linearly independent.*
(l) *The column vectors of A span R^n.*
(m) *The row vectors of A span R^n.*
(n) *The column vectors of A form a basis for R^n.*
(o) *The row vectors of A form a basis for R^n.*
(p) *A has rank n.*
(q) *A has nullity 0.*
(r) *The orthogonal complement of the nullspace of A is R^n.*
(s) *The orthogonal complement of the row space of A is $\{\mathbf{0}\}$.*
(t) *$A^T A$ is invertible.*
(u) *$\lambda = 0$ is not an eigenvalue of A.*

This theorem relates all of the major topics we have studied thus far.

Exercise Set 7.1

1. Find the characteristic equations of the following matrices:

(a) $\begin{bmatrix} 3 & 0 \\ 8 & -1 \end{bmatrix}$ (b) $\begin{bmatrix} 10 & -9 \\ 4 & -2 \end{bmatrix}$ (c) $\begin{bmatrix} 0 & 3 \\ 4 & 0 \end{bmatrix}$ (d) $\begin{bmatrix} -2 & -7 \\ 1 & 2 \end{bmatrix}$ (e) $\begin{bmatrix} 0 & 0 \\ 0 & 0 \end{bmatrix}$ (f) $\begin{bmatrix} 1 & 0 \\ 0 & 1 \end{bmatrix}$

2. Find the eigenvalues of the matrices in Exercise 1.

3. Find bases for the eigenspaces of the matrices in Exercise 1.

4. Find the characteristic equations of the following matrices:

(a) $\begin{bmatrix} 4 & 0 & 1 \\ -2 & 1 & 0 \\ -2 & 0 & 1 \end{bmatrix}$ (b) $\begin{bmatrix} 3 & 0 & -5 \\ \frac{1}{5} & -1 & 0 \\ 1 & 1 & -2 \end{bmatrix}$ (c) $\begin{bmatrix} -2 & 0 & 1 \\ -6 & -2 & 0 \\ 19 & 5 & -4 \end{bmatrix}$

(d) $\begin{bmatrix} -1 & 0 & 1 \\ -1 & 3 & 0 \\ -4 & 13 & -1 \end{bmatrix}$ (e) $\begin{bmatrix} 5 & 0 & 1 \\ 1 & 1 & 0 \\ -7 & 1 & 0 \end{bmatrix}$ (f) $\begin{bmatrix} 5 & 6 & 2 \\ 0 & -1 & -8 \\ 1 & 0 & -2 \end{bmatrix}$

5. Find the eigenvalues of the matrices in Exercise 4.

6. Find bases for the eigenspaces of the matrices in Exercise 4.

7. Find the characteristic equations of the following matrices:

(a) $\begin{bmatrix} 0 & 0 & 2 & 0 \\ 1 & 0 & 1 & 0 \\ 0 & 1 & -2 & 0 \\ 0 & 0 & 0 & 1 \end{bmatrix}$ (b) $\begin{bmatrix} 10 & -9 & 0 & 0 \\ 4 & -2 & 0 & 0 \\ 0 & 0 & -2 & -7 \\ 0 & 0 & 1 & 2 \end{bmatrix}$

8. Find the eigenvalues of the matrices in Exercise 7.

9. Find bases for the eigenspaces of the matrices in Exercise 7.

10. By inspection, find the eigenvalues of the following matrices:

(a) $\begin{bmatrix} -1 & 6 \\ 0 & 5 \end{bmatrix}$ (b) $\begin{bmatrix} 3 & 0 & 0 \\ -2 & 7 & 0 \\ 4 & 8 & 1 \end{bmatrix}$ (c) $\begin{bmatrix} -\frac{1}{3} & 0 & 0 & 0 \\ 0 & -\frac{1}{3} & 0 & 0 \\ 0 & 0 & 1 & 0 \\ 0 & 0 & 0 & \frac{1}{2} \end{bmatrix}$

11. Find the eigenvalues of A^9 for

$$A = \begin{bmatrix} 1 & 3 & 7 & 11 \\ 0 & \frac{1}{2} & 3 & 8 \\ 0 & 0 & 0 & 4 \\ 0 & 0 & 0 & 2 \end{bmatrix}$$

12. Find the eigenvalues and bases for the eigenspaces of A^{25} for

$$A = \begin{bmatrix} -1 & -2 & -2 \\ 1 & 2 & 1 \\ -1 & -1 & 0 \end{bmatrix}$$

13. Let A be a 2×2 matrix, and call a line through the origin of R^2 *invariant* under A if $A\mathbf{x}$ lies on the line when $\mathbf{x}$ does. Find equations for all lines in R^2, if any, that are invariant under the given matrix.

(a) $A = \begin{bmatrix} 4 & -1 \\ 2 & 1 \end{bmatrix}$ (b) $A = \begin{bmatrix} 0 & 1 \\ -1 & 0 \end{bmatrix}$ (c) $A = \begin{bmatrix} 2 & 3 \\ 0 & 2 \end{bmatrix}$

14. Find $\det(A)$ given that A has $p(\lambda)$ as its characteristic polynomial.

(a) $p(\lambda) = \lambda^3 - 2\lambda^2 + \lambda + 5$ (b) $p(\lambda) = \lambda^4 - \lambda^3 + 7$

[**Hint.** See the proof of Theorem 7.1.4.]

15. Let A be an $n \times n$ matrix.

(a) Prove that the characteristic polynomial of A has degree n.

(b) Prove that the coefficient of λ^n in the characteristic polynomial is 1.

16. Show that the characteristic equation of a 2×2 matrix A can be expressed as $\lambda^2 - \text{tr}(A)\lambda + \det(A) = 0$, where $\text{tr}(A)$ is the trace of A.

17. Use the result in Exercise 16 to show that if

$$A = \begin{bmatrix} a & b \\ c & d \end{bmatrix}$$

then the solutions of the characteristic equation of A are

$$\lambda = \tfrac{1}{2}\left[(a+d) \pm \sqrt{(a-d)^2 + 4bc}\,\right]$$

Use this result to show that A has

(a) two distinct real eigenvalues if $(a-d)^2 + 4bc > 0$ (b) one real eigenvalue if $(a-d)^2 + 4bc = 0$
(c) no real eigenvalues if $(a-d)^2 + 4bc < 0$

18. Let A be the matrix in Exercise 17. Show that if $(a-d)^2 + 4bc > 0$ and $b \neq 0$, then eigenvectors of A corresponding to the eigenvalues

$$\lambda_1 = \tfrac{1}{2}\left[(a+d) + \sqrt{(a-d)^2 + 4bc}\,\right] \quad \text{and} \quad \lambda_2 = \tfrac{1}{2}\left[(a+d) - \sqrt{(a-d)^2 + 4bc}\,\right]$$

are

$$\begin{bmatrix} -b \\ a - \lambda_1 \end{bmatrix} \quad \text{and} \quad \begin{bmatrix} -b \\ a - \lambda_2 \end{bmatrix}$$

respectively.

19. Prove: If $a, b, c,$ and d are integers such that $a + b = c + d$, then

$$A = \begin{bmatrix} a & b \\ c & d \end{bmatrix}$$

has integer eigenvalues, namely, $\lambda_1 = a + b$ and $\lambda_2 = a - c$. [**Hint.** See Exercise 17.]

20. Prove: If λ is an eigenvalue of an invertible matrix A and $\mathbf{x}$ is a corresponding eigenvector, then $1/\lambda$ is an eigenvalue of A^{-1}, and $\mathbf{x}$ is a corresponding eigenvector.

21. Prove: If λ is an eigenvalue of A, $\mathbf{x}$ is a corresponding eigenvector, and s is a scalar, then $\lambda - s$ is an eigenvalue of $A - sI$, and $\mathbf{x}$ is a corresponding eigenvector.

22. Find the eigenvalues and bases for the eigenspaces of

$$A = \begin{bmatrix} -2 & 2 & 3 \\ -2 & 3 & 2 \\ -4 & 2 & 5 \end{bmatrix}$$

Then use Exercises 20 and 21 to find the eigenvalues and bases for the eigenspaces of

(a) A^{-1} (b) $A - 3I$ (c) $A + 2I$

23. (a) Prove that if A is a square matrix, then A and A^T have the same eigenvalues. [**Hint.** Look at the characteristic equation $\det(\lambda I - A) = 0$.]

(b) Show that A and A^T need not have the same eigenspaces. [**Hint.** Use the result in Exercise 18 to find a 2×2 matrix for which A and A^T have different eigenspaces.]

Discussion and Discovery

24. Indicate whether the statement is always true or sometimes false. Justify your answer by giving a logical argument or a counterexample. In each part, A is an $n \times n$ matrix.

(a) If $A\mathbf{x} = \lambda\mathbf{x}$ for some nonzero scalar λ, then $\mathbf{x}$ is an eigenvector of A.
(b) If λ is not an eigenvalue of A, then the linear system $(\lambda I - A)\mathbf{x} = \mathbf{0}$ has only the trivial solution.
(c) If $\lambda = 0$ is an eigenvalue of A, then A^2 is singular.
(d) If the characteristic polynomial of A is $p(\lambda) = \lambda^n + 1$, then A is invertible.

25. Suppose that the characteristic polynomial of some matrix A is $p(\lambda) = (\lambda-1)(\lambda-3)^2(\lambda-4)^3$.
In each part answer the question and explain your reasoning.

(a) What is the size of A?

(b) Is A invertible?

(c) How many eigenspaces does A have?

7.2 DIAGONALIZATION

In this section we shall be concerned with the problem of finding a basis for R^n that consists of eigenvectors of a given $n \times n$ matrix A. Such bases can be used to study geometric properties of A and to simplify various numerical computations involving A. These bases are also of physical significance in a wide variety of applications, some of which will be considered later in this text.

The Matrix Diagonalization Problem Our first objective in this section is to show that the following two problems, which on the surface seem quite different, are actually equivalent.

The Eigenvector Problem. Given an $n \times n$ matrix A, does there exist a basis for R^n consisting of eigenvectors of A?

The Diagonalization Problem (Matrix Form). Given an $n \times n$ matrix A, does there exist an invertible matrix P such that $P^{-1}AP$ is a diagonal matrix?

The latter problem suggests the following terminology.

Definition

A square matrix A is called ***diagonalizable*** if there is an invertible matrix P such that $P^{-1}AP$ is a diagonal matrix; the matrix P is said to ***diagonalize*** A.

The following theorem shows that the eigenvector problem and the diagonalization problem are equivalent.

Theorem 7.2.1

If A is an $n \times n$ matrix, then the following are equivalent.

(a) A is diagonalizable.

(b) A has n linearly independent eigenvectors.

Proof $(a) \Rightarrow (b)$. Since A is assumed diagonalizable, there is an invertible matrix

$$P = \begin{bmatrix} p_{11} & p_{12} & \cdots & p_{1n} \\ p_{21} & p_{22} & \cdots & p_{2n} \\ \vdots & \vdots & & \vdots \\ p_{n1} & p_{n2} & \cdots & p_{nn} \end{bmatrix}$$

such that $P^{-1}AP$ is diagonal, say $P^{-1}AP = D$, where

$$D = \begin{bmatrix} \lambda_1 & 0 & \cdots & 0 \\ 0 & \lambda_2 & \cdots & 0 \\ \vdots & \vdots & & \vdots \\ 0 & 0 & \cdots & \lambda_n \end{bmatrix}$$

It follows from the formula $P^{-1}AP = D$ that $AP = PD$; that is,

$$AP = \begin{bmatrix} p_{11} & p_{12} & \cdots & p_{1n} \\ p_{21} & p_{22} & \cdots & p_{2n} \\ \vdots & \vdots & & \vdots \\ p_{n1} & p_{n2} & \cdots & p_{nn} \end{bmatrix} \begin{bmatrix} \lambda_1 & 0 & \cdots & 0 \\ 0 & \lambda_2 & \cdots & 0 \\ \vdots & \vdots & & \vdots \\ 0 & 0 & \cdots & \lambda_n \end{bmatrix} = \begin{bmatrix} \lambda_1 p_{11} & \lambda_2 p_{12} & \cdots & \lambda_n p_{1n} \\ \lambda_1 p_{21} & \lambda_2 p_{22} & \cdots & \lambda_n p_{2n} \\ \vdots & \vdots & & \vdots \\ \lambda_1 p_{n1} & \lambda_2 p_{n2} & \cdots & \lambda_n p_{nn} \end{bmatrix} \tag{1}$$

If we now let $\mathbf{p}_1, \mathbf{p}_2, \ldots, \mathbf{p}_n$ denote the column vectors of P, then from (1) the successive columns of AP are $\lambda_1\mathbf{p}_1, \lambda_2\mathbf{p}_2, \ldots, \lambda_n\mathbf{p}_n$. However, from Formula (6) of Section 1.3, the successive columns of AP are $A\mathbf{p}_1, A\mathbf{p}_2, \ldots, A\mathbf{p}_n$. Thus, we must have

$$A\mathbf{p}_1 = \lambda_1\mathbf{p}_1, \quad A\mathbf{p}_2 = \lambda_2\mathbf{p}_2, \ldots, \quad A\mathbf{p}_n = \lambda_n\mathbf{p}_n \tag{2}$$

Since P is invertible, its column vectors are all nonzero; thus, it follows from (2) that $\lambda_1, \lambda_2, \ldots, \lambda_n$ are eigenvalues of A, and $\mathbf{p}_1, \mathbf{p}_2, \ldots, \mathbf{p}_n$ are corresponding eigenvectors. Since P is invertible, it follows from Theorem 7.1.5 that $\mathbf{p}_1, \mathbf{p}_2, \ldots, \mathbf{p}_n$ are linearly independent. Thus, A has n linearly independent eigenvectors.

(b) $\Rightarrow$ **(a).** Assume that A has n linearly independent eigenvectors, $\mathbf{p}_1, \mathbf{p}_2, \ldots, \mathbf{p}_n$, with corresponding eigenvalues $\lambda_1, \lambda_2, \ldots, \lambda_n$, and let

$$P = \begin{bmatrix} p_{11} & p_{12} & \cdots & p_{1n} \\ p_{21} & p_{22} & \cdots & p_{2n} \\ \vdots & \vdots & & \vdots \\ p_{n1} & p_{n2} & \cdots & p_{nn} \end{bmatrix}$$

be the matrix whose column vectors are $\mathbf{p}_1, \mathbf{p}_2, \ldots, \mathbf{p}_n$. By Formula (6) of Section 1.3, the column vectors of the product AP are

$$A\mathbf{p}_1, A\mathbf{p}_2, \ldots, A\mathbf{p}_n$$

But

$$A\mathbf{p}_1 = \lambda_1\mathbf{p}_1, \quad A\mathbf{p}_2 = \lambda_2\mathbf{p}_2, \ldots, \quad A\mathbf{p}_n = \lambda_n\mathbf{p}_n$$

so that

$$\begin{aligned} AP &= \begin{bmatrix} \lambda_1 p_{11} & \lambda_2 p_{12} & \cdots & \lambda_n p_{1n} \\ \lambda_1 p_{21} & \lambda_2 p_{22} & \cdots & \lambda_n p_{2n} \\ \vdots & \vdots & & \vdots \\ \lambda_1 p_{n1} & \lambda_2 p_{n2} & \cdots & \lambda_n p_{nn} \end{bmatrix} \\ &= \begin{bmatrix} p_{11} & p_{12} & \cdots & p_{1n} \\ p_{21} & p_{22} & \cdots & p_{2n} \\ \vdots & \vdots & & \vdots \\ p_{n1} & p_{n2} & \cdots & p_{nn} \end{bmatrix} \begin{bmatrix} \lambda_1 & 0 & \cdots & 0 \\ 0 & \lambda_2 & \cdots & 0 \\ \vdots & \vdots & & \vdots \\ 0 & 0 & \cdots & \lambda_n \end{bmatrix} = PD \end{aligned} \tag{3}$$

where D is the diagonal matrix having the eigenvalues $\lambda_1, \lambda_2, \ldots, \lambda_n$ on the main diagonal. Since the column vectors of P are linearly independent, P is invertible; thus, (3) can be rewritten as $P^{-1}AP = D$; that is, A is diagonalizable. ∎

Procedure for Diagonalizing a Matrix   The preceding theorem guarantees that an $n \times n$ matrix A with n linearly independent eigenvectors is diagonalizable, and the proof provides the following method for diagonalizing A.

Step 1. Find n linearly independent eigenvectors of A, say, $\mathbf{p}_1, \mathbf{p}_2, \ldots, \mathbf{p}_n$.

Step 2. Form the matrix P having $\mathbf{p}_1, \mathbf{p}_2, \ldots, \mathbf{p}_n$ as its column vectors.

Step 3. The matrix $P^{-1}AP$ will then be diagonal with $\lambda_1, \lambda_2, \ldots, \lambda_n$ as its successive diagonal entries, where λ_i is the eigenvalue corresponding to $\mathbf{p}_i$, for $i = 1, 2, \ldots, n$.

In order to carry out Step 1 of this procedure, one first needs a way of determining whether a given $n \times n$ matrix A has n linearly independent eigenvectors, and then one needs a method for finding them. One can address both problems at once by finding bases for the eigenspaces of A. Later in this section we will show that those basis vectors, as a combined set, are linearly independent, so that if there is a total of n such vectors, then A is diagonalizable, and the n basis vectors can be used as the column vectors of the diagonalizing matrix P. If there are fewer than n basis vectors, then A is not diagonalizable.

EXAMPLE 1   Finding a Matrix P That Diagonalizes a Matrix A

Find a matrix P that diagonalizes

$$A = \begin{bmatrix} 0 & 0 & -2 \\ 1 & 2 & 1 \\ 1 & 0 & 3 \end{bmatrix}$$

Solution.

From Example 5 of the preceding section we found the characteristic equation of A to be

$$(\lambda - 1)(\lambda - 2)^2 = 0$$

and we found the following bases for the eigenspaces:

$$\lambda = 2: \quad \mathbf{p}_1 = \begin{bmatrix} -1 \\ 0 \\ 1 \end{bmatrix}, \quad \mathbf{p}_2 = \begin{bmatrix} 0 \\ 1 \\ 0 \end{bmatrix} \qquad \lambda = 1: \quad \mathbf{p}_3 = \begin{bmatrix} -2 \\ 1 \\ 1 \end{bmatrix}$$

There are three basis vectors in total, so the matrix A is diagonalizable and

$$P = \begin{bmatrix} -1 & 0 & -2 \\ 0 & 1 & 1 \\ 1 & 0 & 1 \end{bmatrix}$$

diagonalizes A. As a check, the reader should verify that

$$P^{-1}AP = \begin{bmatrix} 1 & 0 & 2 \\ 1 & 1 & 1 \\ -1 & 0 & -1 \end{bmatrix} \begin{bmatrix} 0 & 0 & -2 \\ 1 & 2 & 1 \\ 1 & 0 & 3 \end{bmatrix} \begin{bmatrix} -1 & 0 & -2 \\ 0 & 1 & 1 \\ 1 & 0 & 1 \end{bmatrix} = \begin{bmatrix} 2 & 0 & 0 \\ 0 & 2 & 0 \\ 0 & 0 & 1 \end{bmatrix} \quad \blacklozenge$$

There is no preferred order for the columns of P. Since the ith diagonal entry of $P^{-1}AP$ is an eigenvalue for the ith column vector of P, changing the order of the columns of P just changes the order of the eigenvalues on the diagonal of $P^{-1}AP$. Thus, had we written

$$P = \begin{bmatrix} -1 & -2 & 0 \\ 0 & 1 & 1 \\ 1 & 1 & 0 \end{bmatrix}$$

in Example 1, we would have obtained

$$P^{-1}AP = \begin{bmatrix} 2 & 0 & 0 \\ 0 & 1 & 0 \\ 0 & 0 & 2 \end{bmatrix}$$

EXAMPLE 2 A Matrix That Is Not Diagonalizable

Find a matrix P that diagonalizes

$$A = \begin{bmatrix} 1 & 0 & 0 \\ 1 & 2 & 0 \\ -3 & 5 & 2 \end{bmatrix}$$

Solution.

The characteristic polynomial of A is

$$\det(\lambda I - A) = \begin{vmatrix} \lambda - 1 & 0 & 0 \\ -1 & \lambda - 2 & 0 \\ 3 & -5 & \lambda - 2 \end{vmatrix} = (\lambda - 1)(\lambda - 2)^2$$

so the characteristic equation is

$$(\lambda - 1)(\lambda - 2)^2 = 0$$

Thus, the eigenvalues of A are $\lambda = 1$ and $\lambda = 2$. We leave it for the reader to show that bases for the eigenspaces are

$$\lambda = 1: \quad \mathbf{p}_1 = \begin{bmatrix} \frac{1}{8} \\ -\frac{1}{8} \\ 1 \end{bmatrix} \qquad \lambda = 2: \quad \mathbf{p}_2 = \begin{bmatrix} 0 \\ 0 \\ 1 \end{bmatrix}$$

Since A is a 3×3 matrix and there are only two basis vectors in total, A is not diagonalizable.

Alternative Solution.

If one is interested only in determining whether a matrix is diagonalizable and is not concerned with actually finding a diagonalizing matrix P, then it is not necessary to compute bases for the eigenspaces; it suffices to find the dimensions of the eigenspaces. For this example, the eigenspace corresponding to $\lambda = 1$ is the solution space of the system

$$\begin{bmatrix} 0 & 0 & 0 \\ -1 & -1 & 0 \\ 3 & -5 & -1 \end{bmatrix} \begin{bmatrix} x_1 \\ x_2 \\ x_3 \end{bmatrix} = \begin{bmatrix} 0 \\ 0 \\ 0 \end{bmatrix}$$

The coefficient matrix has rank 2 (verify). Thus, the nullity of this matrix is 1 by Theorem 5.6.3, and hence the solution space is one-dimensional.

The eigenspace corresponding to $\lambda = 2$ is the solution space of the system

$$\begin{bmatrix} 1 & 0 & 0 \\ -1 & 0 & 0 \\ 3 & -5 & 0 \end{bmatrix} \begin{bmatrix} x_1 \\ x_2 \\ x_3 \end{bmatrix} = \begin{bmatrix} 0 \\ 0 \\ 0 \end{bmatrix}$$

This coefficient matrix also has rank 2 and nullity 1 (verify), so the eigenspace corresponding to $\lambda = 2$ is also one-dimensional. Since the eigenspaces produce a total of two basis vectors, the matrix A is not diagonalizable. ◆

There is an assumption in Example 1 that the column vectors of P, which are made up of basis vectors from the various eigenspaces of A, are linearly independent. The following theorem addresses this issue.

Theorem 7.2.2

If $\mathbf{v}_1, \mathbf{v}_2, \ldots, \mathbf{v}_k$ are eigenvectors of A corresponding to distinct eigenvalues $\lambda_1, \lambda_2, \ldots, \lambda_k$, then $\{\mathbf{v}_1, \mathbf{v}_2, \ldots, \mathbf{v}_k\}$ is a linearly independent set.

Proof. Let $\mathbf{v}_1, \mathbf{v}_2, \ldots, \mathbf{v}_k$ be eigenvectors of A corresponding to distinct eigenvalues $\lambda_1, \lambda_2, \ldots, \lambda_k$. We shall assume that $\mathbf{v}_1, \mathbf{v}_2, \ldots, \mathbf{v}_k$ are linearly dependent and obtain a contradiction. We can then conclude that $\mathbf{v}_1, \mathbf{v}_2, \ldots, \mathbf{v}_k$ are linearly independent.

Since an eigenvector is nonzero by definition, $\{\mathbf{v}_1\}$ is linearly independent. Let r be the largest integer such that $\{\mathbf{v}_1, \mathbf{v}_2, \ldots, \mathbf{v}_r\}$ is linearly independent. Since we are assuming that $\{\mathbf{v}_1, \mathbf{v}_2, \ldots, \mathbf{v}_k\}$ is linearly dependent, r satisfies $1 \leq r < k$. Moreover, by definition of r, $\{\mathbf{v}_1, \mathbf{v}_2, \ldots, \mathbf{v}_{r+1}\}$ is linearly dependent. Thus, there are scalars $c_1, c_2, \ldots, c_{r+1}$, not all zero, such that

$$c_1\mathbf{v}_1 + c_2\mathbf{v}_2 + \cdots + c_{r+1}\mathbf{v}_{r+1} = \mathbf{0} \tag{4}$$

Multiplying both sides of (4) by A and using

$$A\mathbf{v}_1 = \lambda_1\mathbf{v}_1, \quad A\mathbf{v}_2 = \lambda_2\mathbf{v}_2, \ldots, \quad A\mathbf{v}_{r+1} = \lambda_{r+1}\mathbf{v}_{r+1}$$

we obtain

$$c_1\lambda_1\mathbf{v}_1 + c_2\lambda_2\mathbf{v}_2 + \cdots + c_{r+1}\lambda_{r+1}\mathbf{v}_{r+1} = \mathbf{0} \tag{5}$$

Multiplying both sides of (4) by λ_{r+1} and subtracting the resulting equation from (5) yields

$$c_1(\lambda_1 - \lambda_{r+1})\mathbf{v}_1 + c_2(\lambda_2 - \lambda_{r+1})\mathbf{v}_2 + \cdots + c_r(\lambda_r - \lambda_{r+1})\mathbf{v}_r = \mathbf{0}$$

Since $\{\mathbf{v}_1, \mathbf{v}_2, \ldots, \mathbf{v}_r\}$ is a linearly independent set, this equation implies that

$$c_1(\lambda_1 - \lambda_{r+1}) = c_2(\lambda_2 - \lambda_{r+1}) = \cdots = c_r(\lambda_r - \lambda_{r+1}) = 0$$

and since $\lambda_1, \lambda_2, \ldots, \lambda_{r+1}$ are distinct, it follows that

$$c_1 = c_2 = \cdots = c_r = 0 \tag{6}$$

Substituting these values in (4) yields

$$c_{r+1}\mathbf{v}_{r+1} = \mathbf{0}$$

Since the eigenvector $\mathbf{v}_{r+1}$ is nonzero, it follows that

$$c_{r+1} = 0 \tag{7}$$

Equations 6 and 7 contradict the fact that $c_1, c_2, \ldots, c_{r+1}$ are not all zero; this completes the proof. ∎

REMARK. Theorem 7.2.2 is a special case of a more general result: Suppose that $\lambda_1, \lambda_2, \ldots, \lambda_k$ are distinct eigenvalues and that we choose a linearly independent set in each of the corresponding eigenspaces. If we then merge all these vectors into a single set, the result will still be a linearly independent set. For example, if we choose three linearly independent vectors from one eigenspace and two linearly independent vectors from another eigenspace, then the five vectors together form a linearly independent set. We omit the proof.

As a consequence of Theorem 7.2.2, we obtain the following important result.

Theorem 7.2.3

If an $n \times n$ matrix A has n distinct eigenvalues, then A is diagonalizable.

Proof. If $\mathbf{v}_1, \mathbf{v}_2, \ldots, \mathbf{v}_n$ are eigenvectors corresponding to the distinct eigenvalues $\lambda_1, \lambda_2, \ldots, \lambda_n$, then by Theorem 7.2.2, $\mathbf{v}_1, \mathbf{v}_2, \ldots, \mathbf{v}_n$ are linearly independent. Thus, A is diagonalizable by Theorem 7.2.1. ∎

EXAMPLE 3 Using Theorem 7.2.3

We saw in Example 2 of the preceding section that

$$A = \begin{bmatrix} 0 & 1 & 0 \\ 0 & 0 & 1 \\ 4 & -17 & 8 \end{bmatrix}$$

has three distinct eigenvalues, $\lambda = 4$, $\lambda = 2 + \sqrt{3}$, $\lambda = 2 - \sqrt{3}$. Therefore, A is diagonalizable. Further,

$$P^{-1}AP = \begin{bmatrix} 4 & 0 & 0 \\ 0 & 2 + \sqrt{3} & 0 \\ 0 & 0 & 2 - \sqrt{3} \end{bmatrix}$$

for some invertible matrix P. If desired, the matrix P can be found using the method shown in Example 1 of this section. ◆

EXAMPLE 4 A Diagonalizable Matrix

From Theorem 7.1.1 the eigenvalues of a triangular matrix are the entries on its main diagonal. Thus, a triangular matrix with distinct entries on the main diagonal is diagonalizable. For example,

$$A = \begin{bmatrix} -1 & 2 & 4 & 0 \\ 0 & 3 & 1 & 7 \\ 0 & 0 & 5 & 8 \\ 0 & 0 & 0 & -2 \end{bmatrix}$$

is a diagonalizable matrix. ◆

Geometric and Algebraic Multiplicity Theorem 7.2.3 does not completely settle the diagonalization problem, since it is possible for an $n \times n$ matrix A to be diagonalizable without having n distinct eigenvalues. We saw this in Example 1, where the given 3×3 matrix had only two distinct eigenvalues, yet was diagonalizable. What really matters for diagonalizability are the dimensions of the eigenspaces—those dimensions must add up to n in order for an $n \times n$ matrix to be diagonalizable. Examples 1 and 2 illustrate this; the matrices in those examples have the same characteristic equation and the same eigenvalues, but the matrix in Example 1 is diagonalizable because the dimensions of the eigenspaces add to 3 and the matrix in Example 2 is not diagonalizable because the dimensions only add to 2.

A full excursion into the study of diagonalizability is left for more advanced courses, but we shall touch on one theorem that is important to a fuller understanding of diagonalizability. It can be proved that if λ_0 is an eigenvalue of A, then the dimension of the eigenspace corresponding to λ_0 cannot exceed the number of times that $\lambda - \lambda_0$ appears as a factor in the characteristic polynomial of A. For example, in Examples 1 and 2 the characteristic polynomial is

$$(\lambda - 1)(\lambda - 2)^2$$

Thus, the eigenspace corresponding to $\lambda = 1$ is at most (hence exactly) one-dimensional and the eigenspace corresponding to $\lambda = 2$ is at most two-dimensional. In Example 1, the eigenspace corresponding to $\lambda = 2$ actually had dimension 2, resulting in diagonalizability, but in Example 2 that eigenspace had only dimension 1, resulting in the failure of diagonalizability.

There is some terminology that relates to these ideas. If λ_0 is an eigenvalue of an $n \times n$ matrix A, then the dimension of the eigenspace corresponding to λ_0 is called the ***geometric multiplicity*** of λ_0, and the number of times that $\lambda - \lambda_0$ appears as a factor in the characteristic polynomial of A is called the ***algebraic multiplicity*** of A. The following theorem, which we state without proof, summarizes the preceding discussion.

Theorem 7.2.4 **Geometric and Algebraic Multiplicity**

If A is a square matrix, then:

(a) *For every eigenvalue of A the geometric multiplicity is less than or equal to the algebraic multiplicity.*

(b) *A is diagonalizable if and only if the geometric multiplicity is equal to the algebraic multiplicity for every eigenvalue.*

Computing Powers of a Matrix There are numerous problems in applied mathematics that require the computation of high powers of a square matrix. We shall conclude this section by showing how diagonalization can be used to simplify such computations for diagonalizable matrices.

If A is an $n \times n$ matrix and P is an invertible matrix, then

$$(P^{-1}AP)^2 = P^{-1}APP^{-1}AP = P^{-1}AIAP = P^{-1}A^2P$$

More generally, for any positive integer k

$$(P^{-1}AP)^k = P^{-1}A^kP \tag{8}$$

It follows from this equation that if A is diagonalizable, and $P^{-1}AP = D$ is a diagonal matrix, then

$$P^{-1}A^kP = (P^{-1}AP)^k = D^k \tag{9}$$

Solving this equation for A^k yields

$$A^k = PD^k P^{-1} \tag{10}$$

This last equation expresses the kth power of A in terms of the kth power of the diagonal matrix D. But D^k is easy to compute; for example, if

$$D = \begin{bmatrix} d_1 & 0 & \cdots & 0 \\ 0 & d_2 & \cdots & 0 \\ \vdots & \vdots & & \vdots \\ 0 & 0 & \cdots & d_n \end{bmatrix}, \quad \text{then} \quad D^k = \begin{bmatrix} d_1^k & 0 & \cdots & 0 \\ 0 & d_2^k & \cdots & 0 \\ \vdots & \vdots & & \vdots \\ 0 & 0 & \cdots & d_n^k \end{bmatrix}$$

EXAMPLE 5 Power of a Matrix

Use (10) to find A^{13}, where

$$A = \begin{bmatrix} 0 & 0 & -2 \\ 1 & 2 & 1 \\ 1 & 0 & 3 \end{bmatrix}$$

Solution.

We showed in Example 1 that the matrix A is diagonalized by

$$P = \begin{bmatrix} -1 & 0 & -2 \\ 0 & 1 & 1 \\ 1 & 0 & 1 \end{bmatrix}$$

and that

$$D = P^{-1}AP = \begin{bmatrix} 2 & 0 & 0 \\ 0 & 2 & 0 \\ 0 & 0 & 1 \end{bmatrix}$$

Thus, from (10)

$$\begin{aligned}
A^{13} = PD^{13}P^{-1} &= \begin{bmatrix} -1 & 0 & -2 \\ 0 & 1 & 1 \\ 1 & 0 & 1 \end{bmatrix} \begin{bmatrix} 2^{13} & 0 & 0 \\ 0 & 2^{13} & 0 \\ 0 & 0 & 1^{13} \end{bmatrix} \begin{bmatrix} 1 & 0 & 2 \\ 1 & 1 & 1 \\ -1 & 0 & -1 \end{bmatrix} \\
&= \begin{bmatrix} -8190 & 0 & -16382 \\ 8191 & 8192 & 8191 \\ 8191 & 0 & 16383 \end{bmatrix}
\end{aligned} \tag{11}$$

◆

REMARK. With the method in the preceding example, most of the work is in diagonalizing A. Once that work is done, it can be used to compute any power of A. Thus, to compute A^{1000} we need only change the exponents from 13 to 1000 in (11).

Exercise Set 7.2

1. Let A be a 6×6 matrix with characteristic equation $\lambda^2(\lambda - 1)(\lambda - 2)^3 = 0$. What are the possible dimensions for eigenspaces of A?

2. Let

$$A = \begin{bmatrix} 4 & 0 & 1 \\ 2 & 3 & 2 \\ 1 & 0 & 4 \end{bmatrix}$$

(a) Find the eigenvalues of A.

(b) For each eigenvalue λ find the rank of the matrix $\lambda I - A$.

(c) Is A diagonalizable? Justify your conclusion.

In Exercises 3–7 use the method of Exercise 2 to determine whether the matrix is diagonalizable.

3. $\begin{bmatrix} 2 & 0 \\ 1 & 2 \end{bmatrix}$ **4.** $\begin{bmatrix} 2 & -3 \\ 1 & -1 \end{bmatrix}$ **5.** $\begin{bmatrix} 3 & 0 & 0 \\ 0 & 2 & 0 \\ 0 & 1 & 2 \end{bmatrix}$ **6.** $\begin{bmatrix} -1 & 0 & 1 \\ -1 & 3 & 0 \\ -4 & 13 & -1 \end{bmatrix}$ **7.** $\begin{bmatrix} 2 & -1 & 0 & 1 \\ 0 & 2 & 1 & -1 \\ 0 & 0 & 3 & 2 \\ 0 & 0 & 0 & 3 \end{bmatrix}$

In Exercises 8–11 find a matrix P that diagonalizes A, and determine $P^{-1}AP$.

8. $A = \begin{bmatrix} -14 & 12 \\ -20 & 17 \end{bmatrix}$ **9.** $A = \begin{bmatrix} 1 & 0 \\ 6 & -1 \end{bmatrix}$ **10.** $A = \begin{bmatrix} 1 & 0 & 0 \\ 0 & 1 & 1 \\ 0 & 1 & 1 \end{bmatrix}$ **11.** $A = \begin{bmatrix} 2 & 0 & -2 \\ 0 & 3 & 0 \\ 0 & 0 & 3 \end{bmatrix}$

In Exercises 12–17 determine whether A is diagonalizable. If so, find a matrix P that diagonalizes A, and determine $P^{-1}AP$.

12. $A = \begin{bmatrix} 19 & -9 & -6 \\ 25 & -11 & -9 \\ 17 & -9 & -4 \end{bmatrix}$ **13.** $A = \begin{bmatrix} -1 & 4 & -2 \\ -3 & 4 & 0 \\ -3 & 1 & 3 \end{bmatrix}$ **14.** $A = \begin{bmatrix} 5 & 0 & 0 \\ 1 & 5 & 0 \\ 0 & 1 & 5 \end{bmatrix}$

15. $A = \begin{bmatrix} 0 & 0 & 0 \\ 0 & 0 & 0 \\ 3 & 0 & 1 \end{bmatrix}$ **16.** $A = \begin{bmatrix} -2 & 0 & 0 & 0 \\ 0 & -2 & 0 & 0 \\ 0 & 0 & 3 & 0 \\ 0 & 0 & 1 & 3 \end{bmatrix}$ **17.** $A = \begin{bmatrix} -2 & 0 & 0 & 0 \\ 0 & -2 & 5 & -5 \\ 0 & 0 & 3 & 0 \\ 0 & 0 & 0 & 3 \end{bmatrix}$

18. Use the method of Example 5 to compute A^{10}, where

$$A = \begin{bmatrix} 1 & 0 \\ -1 & 2 \end{bmatrix}$$

19. Use the method of Example 5 to compute A^{11}, where

$$A = \begin{bmatrix} -1 & 7 & -1 \\ 0 & 1 & 0 \\ 0 & 15 & -2 \end{bmatrix}$$

20. In each part compute the stated power of

$$A = \begin{bmatrix} 1 & -2 & 8 \\ 0 & -1 & 0 \\ 0 & 0 & -1 \end{bmatrix}$$

(a) A^{1000} (b) A^{-1000} (c) A^{2301} (d) A^{-2301}

21. Find A^n if n is a positive integer and

$$A = \begin{bmatrix} 3 & -1 & 0 \\ -1 & 2 & -1 \\ 0 & -1 & 3 \end{bmatrix}$$

22. Let

$$A = \begin{bmatrix} a & b \\ c & d \end{bmatrix}$$

Show:

(a) A is diagonalizable if $(a - d)^2 + 4bc > 0$.

(b) A is not diagonalizable if $(a - d)^2 + 4bc < 0$.
[**Hint.** See Exercise 17 of Section 7.1.]

23. In the case where the matrix A in Exercise 22 is diagonalizable, find a matrix P that diagonalizes A. [**Hint.** See Exercise 18 of Section 7.1.]

24. Prove that if A is a diagonalizable matrix, then the rank of A is the number of nonzero eigenvalues of A.

Discussion and Discovery

25. Indicate whether the statement is always true or sometimes false. Justify your answer by giving a logical argument or a counterexample.

(a) A square matrix with linearly independent column vectors is diagonalizable.

(b) If A is diagonalizable, then there is a unique matrix P such that $P^{-1}AP$ is a diagonal matrix.

(c) If $\mathbf{v}_1$, $\mathbf{v}_2$, and $\mathbf{v}_3$ come from different eigenspaces of A, then it is impossible to express $\mathbf{v}_3$ as a linear combination of $\mathbf{v}_1$ and $\mathbf{v}_2$.

(d) If A is diagonalizable and invertible, then A^{-1} is diagonalizable.

26. Suppose that the characteristic polynomial of some matrix A is $p(\lambda) = (\lambda - 1)(\lambda - 3)^2(\lambda - 4)^3$. In each part answer the question and explain your reasoning.

(a) What can you say about the dimensions of the eigenspaces of A?

(b) What can you say about the dimensions of the eigenspaces if you know that A is diagonalizable?

(c) If $\{\mathbf{v}_1, \mathbf{v}_2, \mathbf{v}_3\}$ is a linearly independent set of eigenvectors of A all of which correspond to the same eigenvalue of A, what can you say about the eigenvalue?

27. (*For readers who have studied calculus.*) If $A_1, A_2, \ldots, A_k, \ldots$ is an infinite sequence of $n \times n$ matrices, then the sequence is said to **converge** to the $n \times n$ matrix A if the entries in the ith row and jth column of the sequence converge to the entry in the ith row and jth column of A for all i and j. In that case we call A the **limit** of the sequence and write $\lim_{k \to +\infty} A_k = A$. The algebraic properties of such limits mirror those of numerical limits. Thus, for example, if P is an invertible $n \times n$ matrix whose entries do not depend on k, then $\lim_{k \to +\infty} A^k = A$ if and only if $\lim_{k \to +\infty} P^{-1}A^kP = P^{-1}AP$.

(a) Suppose that A is an $n \times n$ diagonalizable matrix. Under what conditions on the eigenvalues of A will the sequence $A, A^2, \ldots, A^k, \ldots$ converge? Explain your reasoning.

(b) What is the limit when your conditions are satisfied?

28. (*For readers who have studied calculus.*) If $A_1 + A_2 + \cdots + A_k + \cdots$ is an infinite series of $n \times n$ matrices, then the series is said to **converge** if its sequence of partial sums converges to some limit A in the sense defined in Exercise 27. In that case we call A the **sum** of the series and write $A = A_1 + A_2 + \cdots + A_k + \cdots$.

(a) From calculus, under what conditions on x does the geometric series

$$1 + x + x^2 + \cdots + x^k + \cdots$$

converge? What is the sum?

(b) Based on Exercise 27, under what conditions on the eigenvalues of A would you expect the geometric matrix series $I + A + A^2 + \cdots + A^k + \cdots$ to converge? Explain your reasoning.

(c) What is the sum of the series when it converges?

7.3 ORTHOGONAL DIAGONALIZATION

In this section we shall be concerned with the problem of finding an orthonormal basis for R^n with the Euclidean inner product consisting of eigenvectors of a given $n \times n$ matrix A. Our earlier work on symmetric matrices and orthogonal matrices will play an important role in the discussion that follows.

Orthogonal Diagonalization Problem As in the preceding section, we begin by stating two problems. Our goal is to show that the problems are equivalent.

The Orthonormal Eigenvector Problem. Given an $n \times n$ matrix A, does there exist an orthonormal basis for R^n with the Euclidean inner product consisting of eigenvectors of the matrix A?

The Orthogonal Diagonalization Problem (*Matrix Form*). Given an $n \times n$ matrix A, does there exist an orthogonal matrix P such that the matrix $P^{-1}AP = P^TAP$ is diagonal? If there is such a matrix, then A is said to be ***orthogonally diagonalizable*** and P is said to ***orthogonally diagonalize*** A.

For the latter problem we have two questions to consider:

- Which matrices are orthogonally diagonalizable?
- How do we find an orthogonal matrix to carry out the diagonalization?

With regard to the first question, we note that there is no hope of orthogonally diagonalizing a matrix A unless A is symmetric (i.e., $A = A^T$). To see why this is so, suppose that

$$P^TAP = D \tag{1}$$

where P is an orthogonal matrix and D is a diagonal matrix. Since P is orthogonal, $PP^T = P^TP = I$, so it follows that (1) can be written as

$$A = PDP^T \tag{2}$$

Since D is a diagonal matrix, we have $D = D^T$, so transposing both sides of (2) yields

$$A^T = (PDP^T)^T = (P^T)^T D^T P^T = PDP^T = A$$

so A must be symmetric.

Conditions for Orthogonal Diagonalizability The following theorem shows that every symmetric matrix is, in fact, orthogonally diagonalizable. In this theorem, and for the remainder of this section, *orthogonal* will mean orthogonal with respect to the Euclidean inner product on R^n.

Theorem 7.3.1

If A is an $n \times n$ matrix, then the following are equivalent.

(*a*) *A is orthogonally diagonalizable.*

(*b*) *A has an orthonormal set of n eigenvectors.*

(*c*) *A is symmetric.*

Proof (a) ⇒ (b). Since A is orthogonally diagonalizable, there is an orthogonal matrix P such that $P^{-1}AP$ is diagonal. As shown in the proof of Theorem 7.2.1, the n column vectors of P are eigenvectors of A. Since P is orthogonal, these column vectors are orthonormal (see Theorem 6.5.1), so that A has n orthonormal eigenvectors.

(b) ⇒ (a). Assume that A has an orthonormal set of n eigenvectors $\{\mathbf{p}_1, \mathbf{p}_2, \ldots, \mathbf{p}_n\}$. As shown in the proof of Theorem 7.2.1, the matrix P with these eigenvectors as columns diagonalizes A. Since these eigenvectors are orthonormal, P is orthogonal and thus orthogonally diagonalizes A.

(a) ⇒ (c). In the proof that $(a) ⇒ (b)$ we showed that an orthogonally diagonalizable $n \times n$ matrix A is orthogonally diagonalized by an $n \times n$ matrix P whose columns form an orthonormal set of eigenvectors of A. Let D be the diagonal matrix

$$D = P^{-1}AP$$

Thus,

$$A = PDP^{-1} \quad \text{or} \quad A = PDP^T$$

since P is orthogonal. Therefore,

$$A^T = (PDP^T)^T = PD^TP^T = PDP^T = A$$

which shows that A is symmetric.

(c) ⇒ (a). The proof of this part is beyond the scope of this text and will be omitted.

∎

Symmetric Matrices
Our next goal is to devise a procedure for orthogonally diagonalizing a symmetric matrix, but before we can do so, we need a critical theorem about eigenvalues and eigenvectors of symmetric matrices.

Theorem 7.3.2

If A is a symmetric matrix, then:

(a) *The eigenvalues of A are all real numbers.*
(b) *Eigenvectors from different eigenspaces are orthogonal.*

Proof (a). The proof of part (a), which requires results about complex vector spaces, is discussed in Section 10.6.

Proof (b). Let $\mathbf{v}_1$ and $\mathbf{v}_2$ be eigenvectors corresponding to distinct eigenvalues λ_1 and λ_2 of the matrix A. We want to show that $\mathbf{v}_1 \cdot \mathbf{v}_2 = 0$. The proof of this involves the trick of starting with the expression $A\mathbf{v}_1 \cdot \mathbf{v}_2$. It follows from Formula (8) of Section 4.1 and the symmetry of A that

$$A\mathbf{v}_1 \cdot \mathbf{v}_2 = \mathbf{v}_1 \cdot A^T\mathbf{v}_2 = \mathbf{v}_1 \cdot A\mathbf{v}_2 \tag{3}$$

But $\mathbf{v}_1$ is an eigenvector of A corresponding to λ_1 and $\mathbf{v}_2$ is an eigenvector of A corresponding to λ_2, so (3) yields the relationship

$$\lambda_1\mathbf{v}_1 \cdot \mathbf{v}_2 = \mathbf{v}_1 \cdot \lambda_2\mathbf{v}_2$$

which can be rewritten as

$$(\lambda_1 - \lambda_2)(\mathbf{v}_1 \cdot \mathbf{v}_2) = 0 \tag{4}$$

But $\lambda_1 - \lambda_2 \neq 0$, since λ_1 and λ_2 were assumed distinct. Thus, it follows from (4) that $\mathbf{v}_1 \cdot \mathbf{v}_2 = 0$. ∎

REMARK. We remind the reader that we have assumed to this point that all of our matrices have real entries. Indeed, we shall see in Chapter 10 that part (*a*) of Theorem 7.3.2 is false for matrices with complex entries.

Diagonalization of Symmetric Matrices

As a consequence of the preceding theorem we obtain the following procedure for orthogonally diagonalizing a symmetric matrix.

Step 1. Find a basis for each eigenspace of A.

Step 2. Apply the Gram–Schmidt process to each of these bases to obtain an orthonormal basis for each eigenspace.

Step 3. Form the matrix P whose columns are the basis vectors constructed in Step 2; this matrix orthogonally diagonalizes A.

The justification of this procedure should be clear: Theorem 7.3.2 ensures that eigenvectors from *different* eigenspaces are orthogonal, while the application of the Gram–Schmidt process ensures that the eigenvectors obtained within the *same* eigenspace are orthonormal. Therefore, the *entire* set of eigenvectors obtained by this procedure is orthonormal.

EXAMPLE 1 An Orthogonal Matrix P That Diagonalizes a Matrix A

Find an orthogonal matrix P that diagonalizes

$$A = \begin{bmatrix} 4 & 2 & 2 \\ 2 & 4 & 2 \\ 2 & 2 & 4 \end{bmatrix}$$

Solution.

The characteristic equation of A is

$$\det(\lambda I - A) = \det \begin{bmatrix} \lambda - 4 & -2 & -2 \\ -2 & \lambda - 4 & -2 \\ -2 & -2 & \lambda - 4 \end{bmatrix} = (\lambda - 2)^2(\lambda - 8) = 0$$

Thus, the eigenvalues of A are $\lambda = 2$ and $\lambda = 8$. By the method used in Example 5 of Section 7.1, it can be shown that

$$\mathbf{u}_1 = \begin{bmatrix} -1 \\ 1 \\ 0 \end{bmatrix} \quad \text{and} \quad \mathbf{u}_2 = \begin{bmatrix} -1 \\ 0 \\ 1 \end{bmatrix} \tag{5}$$

form a basis for the eigenspace corresponding to $\lambda = 2$. Applying the Gram–Schmidt process to $\{\mathbf{u}_1, \mathbf{u}_2\}$ yields the following orthonormal eigenvectors (verify):

$$\mathbf{v}_1 = \begin{bmatrix} -1/\sqrt{2} \\ 1/\sqrt{2} \\ 0 \end{bmatrix} \quad \text{and} \quad \mathbf{v}_2 = \begin{bmatrix} -1/\sqrt{6} \\ -1/\sqrt{6} \\ 2/\sqrt{6} \end{bmatrix} \tag{6}$$

The eigenspace corresponding to $\lambda = 8$ has

$$\mathbf{u}_3 = \begin{bmatrix} 1 \\ 1 \\ 1 \end{bmatrix}$$

as a basis. Applying the Gram–Schmidt process to $\{\mathbf{u}_3\}$ yields

$$\mathbf{v}_3 = \begin{bmatrix} 1/\sqrt{3} \\ 1/\sqrt{3} \\ 1/\sqrt{3} \end{bmatrix}$$

Finally, using $\mathbf{v}_1$, $\mathbf{v}_2$, and $\mathbf{v}_3$ as column vectors we obtain

$$P = \begin{bmatrix} -1/\sqrt{2} & -1/\sqrt{6} & 1/\sqrt{3} \\ 1/\sqrt{2} & -1/\sqrt{6} & 1/\sqrt{3} \\ 0 & 2/\sqrt{6} & 1/\sqrt{3} \end{bmatrix}$$

which orthogonally diagonalizes A. (As a check, the reader may wish to verify that $P^T AP$ is a diagonal matrix.) ◆

Exercise Set 7.3

1. Find the characteristic equation of the given symmetric matrix, and then by inspection determine the dimensions of the eigenspaces.

(a) $\begin{bmatrix} 1 & 2 \\ 2 & 4 \end{bmatrix}$

(b) $\begin{bmatrix} 1 & -4 & 2 \\ -4 & 1 & -2 \\ 2 & -2 & -2 \end{bmatrix}$

(c) $\begin{bmatrix} 1 & 1 & 1 \\ 1 & 1 & 1 \\ 1 & 1 & 1 \end{bmatrix}$

(d) $\begin{bmatrix} 4 & 2 & 2 \\ 2 & 4 & 2 \\ 2 & 2 & 4 \end{bmatrix}$

(e) $\begin{bmatrix} 4 & 4 & 0 & 0 \\ 4 & 4 & 0 & 0 \\ 0 & 0 & 0 & 0 \\ 0 & 0 & 0 & 0 \end{bmatrix}$

(f) $\begin{bmatrix} 2 & -1 & 0 & 0 \\ -1 & 2 & 0 & 0 \\ 0 & 0 & 2 & -1 \\ 0 & 0 & -1 & 2 \end{bmatrix}$

In Exercises 2–9 find a matrix P that orthogonally diagonalizes A, and determine $P^{-1}AP$.

2. $A = \begin{bmatrix} 3 & 1 \\ 1 & 3 \end{bmatrix}$

3. $A = \begin{bmatrix} 6 & 2\sqrt{3} \\ 2\sqrt{3} & 7 \end{bmatrix}$

4. $A = \begin{bmatrix} 6 & -2 \\ -2 & 3 \end{bmatrix}$

5. $A = \begin{bmatrix} -2 & 0 & -36 \\ 0 & -3 & 0 \\ -36 & 0 & -23 \end{bmatrix}$

6. $A = \begin{bmatrix} 1 & 1 & 0 \\ 1 & 1 & 0 \\ 0 & 0 & 0 \end{bmatrix}$

7. $A = \begin{bmatrix} 2 & -1 & -1 \\ -1 & 2 & -1 \\ -1 & -1 & 2 \end{bmatrix}$

8. $A = \begin{bmatrix} 3 & 1 & 0 & 0 \\ 1 & 3 & 0 & 0 \\ 0 & 0 & 0 & 0 \\ 0 & 0 & 0 & 0 \end{bmatrix}$

9. $A = \begin{bmatrix} -7 & 24 & 0 & 0 \\ 24 & 7 & 0 & 0 \\ 0 & 0 & -7 & 24 \\ 0 & 0 & 24 & 7 \end{bmatrix}$

10. Assuming that $b \neq 0$, find a matrix that orthogonally diagonalizes

$$\begin{bmatrix} a & b \\ b & a \end{bmatrix}$$

11. Prove that if A is any $m \times n$ matrix, then $A^T A$ has an orthonormal set of n eigenvectors.

12. (a) Show that if $\mathbf{v}$ is any $n \times 1$ matrix and I is the $n \times n$ identity matrix, then $I - \mathbf{v}\mathbf{v}^T$ is orthogonally diagonalizable.

(b) Find a matrix P that orthogonally diagonalizes $I - \mathbf{v}\mathbf{v}^T$ if

$$\mathbf{v} = \begin{bmatrix} 1 \\ 0 \\ 1 \end{bmatrix}$$

13. Use the result in Exercise 17 of Section 7.1 to prove Theorem 7.3.2a for 2×2 symmetric matrices.

Discussion and Discovery

14. Indicate whether the statement is always true or sometimes false. Justify your answer by giving a logical argument or a counterexample.

(a) If A is a square matrix, then $A A^T$ and $A^T A$ are orthogonally diagonalizable.

(b) If $\mathbf{v}_1$ and $\mathbf{v}_2$ are eigenvectors from distinct eigenspaces of a symmetric matrix, then $\|\mathbf{v}_1 + \mathbf{v}_2\|^2 = \|\mathbf{v}_1\|^2 + \|\mathbf{v}_2\|^2$.

(c) An orthogonal matrix is orthogonally diagonalizable.

(d) If A is an invertible orthogonally diagonalizable matrix, then A^{-1} is orthogonally diagonalizable.

15. Does there exist a 3×3 symmetric matrix with eigenvalues $\lambda_1 = -1$, $\lambda_2 = 3$, $\lambda_3 = 7$ and corresponding eigenvectors

$$\begin{bmatrix} 0 \\ 1 \\ -1 \end{bmatrix}, \quad \begin{bmatrix} 1 \\ 0 \\ 0 \end{bmatrix}, \quad \begin{bmatrix} 0 \\ 1 \\ 1 \end{bmatrix}?$$

If so, find such a matrix, and if not, explain why.

Chapter 7 Supplementary Exercises

1. (a) Show that if $0 < \theta < \pi$, then

$$A = \begin{bmatrix} \cos\theta & -\sin\theta \\ \sin\theta & \cos\theta \end{bmatrix}$$

has no eigenvalues and consequently no eigenvectors.

(b) Give a geometric explanation of the result in (a).

2. Find the eigenvalues of

$$A = \begin{bmatrix} 0 & 1 & 0 \\ 0 & 0 & 1 \\ k^3 & -3k^2 & 3k \end{bmatrix}$$

3. (a) Show that if D is a diagonal matrix with nonnegative entries on the main diagonal, then there is a matrix S such that $S^2 = D$.

(b) Show that if A is a diagonalizable matrix with nonnegative eigenvalues, then there is a matrix S such that $S^2 = A$.

(c) Find a matrix S such that $S^2 = A$, if

$$A = \begin{bmatrix} 1 & 3 & 1 \\ 0 & 4 & 5 \\ 0 & 0 & 9 \end{bmatrix}$$

4. (a) Prove: If A is a square matrix, then A and A^T have the same eigenvalues.

(b) Show that A and A^T need not have the same eigenvectors. [**Hint.** Use Exercise 18 of Section 7.1 to find a 2×2 matrix A in which A and A^T have different eigenvectors.]

5. Prove: If A is a square matrix and $p(\lambda) = \det(\lambda I - A)$ is the characteristic polynomial of A, then the coefficient of λ^{n-1} in $p(\lambda)$ is the negative of the trace of A.

6. Prove: If $b \neq 0$, then

$$A = \begin{bmatrix} a & b \\ 0 & a \end{bmatrix}$$

is not diagonalizable.

7. In advanced linear algebra, one proves the **Cayley–Hamilton Theorem**, which states that a square matrix A satisfies its characteristic equation; that is, if

$$c_0 + c_1\lambda + c_2\lambda^2 + \cdots + c_{n-1}\lambda^{n-1} + \lambda^n = 0$$

is the characteristic equation of A, then

$$c_0 I + c_1 A + c_2 A^2 + \cdots + c_{n-1}A^{n-1} + A^n = 0$$

Verify this result for

(a) $A = \begin{bmatrix} 3 & 6 \\ 1 & 2 \end{bmatrix}$ (b) $A = \begin{bmatrix} 0 & 1 & 0 \\ 0 & 0 & 1 \\ 1 & -3 & 3 \end{bmatrix}$

Exercises 8–10 use the Cayley–Hamilton Theorem, stated in Exercise 7.

8. Use Exercise 16 of Section 7.1 to prove the Cayley–Hamilton Theorem for 2×2 matrices.

9. The Cayley–Hamilton Theorem provides an efficient method for calculating powers of a matrix. For example, if A is a 2×2 matrix with characteristic equation

$$c_0 + c_1\lambda + \lambda^2 = 0$$

then $c_0 I + c_1 A + A^2 = 0$, so

$$A^2 = -c_1 A - c_0 I$$

Multiplying through by A yields $A^3 = -c_1 A^2 - c_0 A$, which expresses A^3 in terms of A^2 and A, and multiplying through by A^2 yields $A^4 = -c_1 A^3 - c_0 A^2$, which expresses A^4 in terms of A^3 and A^2. Continuing in this way, we can calculate successive powers of A simply by expressing them in terms of lower powers. Use this procedure to calculate A^2, A^3, A^4, and A^5 for

$$A = \begin{bmatrix} 3 & 6 \\ 1 & 2 \end{bmatrix}$$

10. Use the method of the preceding exercise to calculate A^3 and A^4 for

$$A = \begin{bmatrix} 0 & 1 & 0 \\ 0 & 0 & 1 \\ 1 & -3 & 3 \end{bmatrix}$$

11. Find the eigenvalues of the matrix

$$A = \begin{bmatrix} c_1 & c_2 & \cdots & c_n \\ c_1 & c_2 & \cdots & c_n \\ \vdots & \vdots & & \vdots \\ c_1 & c_2 & \cdots & c_n \end{bmatrix}$$

12. (a) It was shown in Exercise 15 of Section 7.1 that if A is an $n \times n$ matrix, then the coefficient of λ^n in the characteristic polynomial of A is 1. (A polynomial with this property is called *monic*.) Show that the matrix

$$\begin{bmatrix} 0 & 0 & 0 & \cdots & 0 & -c_0 \\ 1 & 0 & 0 & \cdots & 0 & -c_1 \\ 0 & 1 & 0 & \cdots & 0 & -c_2 \\ \vdots & \vdots & \vdots & & \vdots & \vdots \\ 0 & 0 & 0 & \cdots & 1 & -c_{n-1} \end{bmatrix}$$

has characteristic polynomial $p(\lambda) = c_0 + c_1\lambda + \cdots + c_{n-1}\lambda^{n-1} + \lambda^n$. This shows that every monic polynomial is the characteristic polynomial of some matrix. The matrix in this example is called the ***companion matrix*** of $p(\lambda)$. [***Hint.*** Evaluate all determinants in the problem by adding a multiple of the second row to the first to introduce a zero at the top of the first column, and then expanding by cofactors along the first column.]

(b) Find a matrix with characteristic polynomial $p(\lambda) = 1 - 2\lambda + \lambda^2 + 3\lambda^3 + \lambda^4$.

13. A square matrix A is called ***nilpotent*** if $A^n = 0$ for some positive integer n. What can you say about the eigenvalues of a nilpotent matrix?

14. Prove: If A is an $n \times n$ matrix and n is odd, then A has at least one real eigenvalue.

15. Find a 3×3 matrix A that has eigenvalues $\lambda = 0, 1$, and -1 with corresponding eigenvectors

$$\begin{bmatrix} 0 \\ 1 \\ -1 \end{bmatrix}, \quad \begin{bmatrix} 1 \\ -1 \\ 1 \end{bmatrix}, \quad \begin{bmatrix} 0 \\ 1 \\ 1 \end{bmatrix}$$

respectively.

16. Suppose that a 4×4 matrix A has eigenvalues $\lambda_1 = 1, \lambda_2 = -2, \lambda_3 = 3$, and $\lambda_4 = -3$.

(a) Use the method of Exercise 14 of Section 7.1 to find $\det(A)$.

(b) Use Exercise 5 above to find $\operatorname{tr}(A)$.

17. Let A be a square matrix such that $A^3 = A$. What can you say about the eigenvalues of A?

Chapter 7 Technology Exercises

The following exercises are designed to be solved using a technology utility. Typically, this will be MATLAB, *Mathematica*, Maple, Derive, or Mathcad, but it may also be some other type of linear algebra software or a scientific calculator with some linear algebra capabilities. For each exercise you will need to read the relevant documentation for the particular utility you are using. The goal of these exercises is to provide you with a basic proficiency with your technology utility. Once you have mastered the techniques in these exercises, you will be able to use your technology utility to solve many of the problems in the regular exercise sets.

Section 7.1

T1. (***Characteristic polynomial***) Some technology utilities have a specific command for finding characteristic polynomials, and in others you must use the determinant function to compute $\det(\lambda I - A)$. Read your documentation to determine which method you must use, and then use your utility to find $p(\lambda)$ for the matrix in Example 2.

T2. (***Solving the characteristic equation***) Depending on the particular characteristic polynomial, your technology utility may or may not be successful in solving the characteristic equation for the eigenvalues. See if your utility can find the eigenvalues in Example 2 by solving the characteristic equation $p(\lambda) = 0$.

T3. (a) Read the statement of the Cayley–Hamilton Theorem in Supplementary Exercise 7 of this chapter and then use your technology utility to do that exercise.

    (b) If you are working with a CAS, use it to prove the Cayley–Hamilton Theorem for 3×3 matrices.

T4. (*Eigenvalues*) Some technology utilities have specific commands for finding the eigenvalues of a matrix directly (though the procedure may not be successful in all cases). If your utility has this capability, read the documentation and then compute the eigenvalues in Example 2 directly.

T5. (*Eigenvectors*) One way to use a technology utility to find eigenvectors corresponding to an eigenvalue λ is to solve the linear system $(\lambda I - A)\mathbf{x} = 0$. Another way is to use a command for finding a basis for the nullspace of $\lambda I - A$ (if available). However, some utilities have specific commands for finding eigenvectors. Read your documentation, and then explore various procedures for finding the eigenvectors in Examples 5 and 6.

Section 7.2

T1. (*Diagonalization*) Some technology utilities have specific commands for diagonalizing a matrix. If your utility has this capability, you may find it described as a "Jordan decomposition" or by some variation of this that involves the word "Jordan." If your utility has this capability, read the documentation and then use your utility to perform the computations in Example 2. [*Note.* Your software may or may not produce the eigenvalues of A and the columns of P in the same order as the example.]

Section 7.3

T1. (*Orthogonal diagonalization*) Use your technology utility to check the computations in Example 1.

8

Linear Transformations

NTRODUCTION: In Sections 4.2 and 4.3 we studied linear transformations from R^n to R^m. In this chapter we shall define and study linear transformations from an arbitrary vector space V to another arbitrary vector space W. The results we obtain here have important applications in physics, engineering, and various branches of mathematics.

8.1 General Linear Transformations

In Section 4.2 we defined linear transformations from R^n to R^m. In this section we shall extend this idea by defining the more general concept of a linear transformation from one vector space to another.

Definitions and Terminology   Recall that a linear transformation from R^n to R^m was first defined as a function

$$T(x_1, x_2, \ldots, x_n) = (w_1, w_2, \ldots, w_m)$$

for which the equations relating $w_1, w_2, \ldots, w_m$ and $x_1, x_2, \ldots, x_n$ are linear. Subsequently, we showed that a transformation $T: R^n \rightarrow R^m$ is linear if and only if the two relationships

$$T(\mathbf{u} + \mathbf{v}) = T(\mathbf{u}) + T(\mathbf{v}) \quad \text{and} \quad T(c\mathbf{u}) = cT(\mathbf{u})$$

hold for all vectors $\mathbf{u}$ and $\mathbf{v}$ in R^n and every scalar c (see Theorem 4.3.2). We shall use these properties as the starting point for general linear transformations.

Definition

If $T: V \rightarrow W$ is a function from a vector space V into a vector space W, then T is called a ***linear transformation*** from V to W if for all vectors $\mathbf{u}$ and $\mathbf{v}$ in V and all scalars c

(a) $T(\mathbf{u} + \mathbf{v}) = T(\mathbf{u}) + T(\mathbf{v})$ (b) $T(c\mathbf{u}) = cT(\mathbf{u})$

In the special case where $V = W$, the linear transformation $T: V \rightarrow V$ is called a ***linear operator*** on V.

EXAMPLE 1 Matrix Transformations

Because the preceding definition of a linear transformation was based on Theorem 4.3.2, linear transformations from R^n to R^m, as defined in Section 4.2, are linear transformations under this more general definition as well. We shall call linear transformations from R^n to R^m ***matrix transformations***, since they can be carried out by matrix multiplication. ◆

EXAMPLE 2 The Zero Transformation

Let V and W be any two vector spaces. The mapping $T: V \rightarrow W$ such that $T(\mathbf{v}) = \mathbf{0}$ for every $\mathbf{v}$ in V is a linear transformation called the ***zero transformation***. To see that T is linear, observe that

$$T(\mathbf{u} + \mathbf{v}) = \mathbf{0}, \quad T(\mathbf{u}) = \mathbf{0}, \quad T(\mathbf{v}) = \mathbf{0}, \quad \text{and} \quad T(k\mathbf{u}) = \mathbf{0}$$

Therefore,

$$T(\mathbf{u} + \mathbf{v}) = T(\mathbf{u}) + T(\mathbf{v}) \quad \text{and} \quad T(k\mathbf{u}) = kT(\mathbf{u}) \qquad ◆$$

EXAMPLE 3 The Identity Operator

Let V be any vector space. The mapping $I: V \rightarrow V$ defined by $I(\mathbf{v}) = \mathbf{v}$ is called the ***identity operator*** on V. The verification that I is linear is left for the reader. ◆

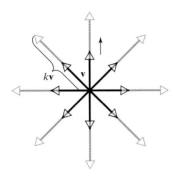

(a) Dilation of V

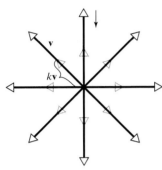

(b) Contraction of V

Figure 8.1.1

EXAMPLE 4 Dilation and Contraction Operators

Let V be any vector space and k any fixed scalar. We leave it as an exercise to check that the function $T: V \to V$ defined by

$$T(\mathbf{v}) = k\mathbf{v}$$

is a linear operator on V. This linear operator is called a ***dilation*** of V with factor k if $k > 1$ and is called a ***contraction*** of V with factor k if $0 < k < 1$. Geometrically, the dilation "stretches" each vector in V by a factor of k, and the contraction of V "compresses" each vector by a factor of k (Figure 8.1.1). ◆

EXAMPLE 5 Orthogonal Projections

In Section 6.4 we defined the orthogonal projection of R^m onto a subspace W. [See Formula (6) and the definition preceding it in that section.] Orthogonal projections can also be defined in general inner product spaces as follows: Suppose that W is a finite-dimensional subspace of an inner product space V; then the ***orthogonal projection of V onto W*** is the transformation defined by

$$T(\mathbf{v}) = \text{proj}_W \mathbf{v}$$

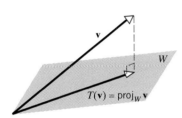

Figure 8.1.2
The orthogonal projection of V onto W.

(Figure 8.1.2). It follows from Theorem 6.3.5 that if

$$S = \{\mathbf{w}_1, \mathbf{w}_2, \ldots, \mathbf{w}_r\}$$

is any orthonormal basis for W, then $T(\mathbf{v})$ is given by the formula

$$T(\mathbf{v}) = \text{proj}_W \mathbf{v} = \langle \mathbf{v}, \mathbf{w}_1 \rangle \mathbf{w}_1 + \langle \mathbf{v}, \mathbf{w}_2 \rangle \mathbf{w}_2 + \cdots + \langle \mathbf{v}, \mathbf{w}_r \rangle \mathbf{w}_r$$

The proof that T is a linear transformation follows from properties of the inner product. For example,

$$T(\mathbf{u} + \mathbf{v}) = \langle \mathbf{u} + \mathbf{v}, \mathbf{w}_1 \rangle \mathbf{w}_1 + \langle \mathbf{u} + \mathbf{v}, \mathbf{w}_2 \rangle \mathbf{w}_2 + \cdots + \langle \mathbf{u} + \mathbf{v}, \mathbf{w}_r \rangle \mathbf{w}_r$$
$$= \langle \mathbf{u}, \mathbf{w}_1 \rangle \mathbf{w}_1 + \langle \mathbf{u}, \mathbf{w}_2 \rangle \mathbf{w}_2 + \cdots + \langle \mathbf{u}, \mathbf{w}_r \rangle \mathbf{w}_r$$
$$+ \langle \mathbf{v}, \mathbf{w}_1 \rangle \mathbf{w}_1 + \langle \mathbf{v}, \mathbf{w}_2 \rangle \mathbf{w}_2 + \cdots + \langle \mathbf{v}, \mathbf{w}_r \rangle \mathbf{w}_r$$
$$= T(\mathbf{u}) + T(\mathbf{v})$$

Similarly, $T(k\mathbf{u}) = kT(\mathbf{u})$. ◆

EXAMPLE 6 Computing an Orthogonal Projection

As a special case of the preceding example, let $V = R^3$ have the Euclidean inner product. The vectors $\mathbf{w}_1 = (1, 0, 0)$ and $\mathbf{w}_2 = (0, 1, 0)$ form an orthonormal basis for the xy-plane. Thus, if $\mathbf{v} = (x, y, z)$ is any vector in R^3, the orthogonal projection of R^3 onto the xy-plane is given by

$$T(\mathbf{v}) = \langle \mathbf{v}, \mathbf{w}_1 \rangle \mathbf{w}_1 + \langle \mathbf{v}, \mathbf{w}_2 \rangle \mathbf{w}_2$$
$$= x(1, 0, 0) + y(0, 1, 0)$$
$$= (x, y, 0)$$

(See Figure 8.1.3.) ◆

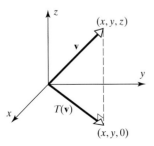

Figure 8.1.3
The orthogonal projection of R^3 onto the xy-plane.

EXAMPLE 7 A Linear Transformation from a Space V to R^n

Let $S = \{\mathbf{w}_1, \mathbf{w}_2, \ldots, \mathbf{w}_n\}$ be a basis for an n-dimensional vector space V, and let

$$(\mathbf{v})_S = (k_1, k_2, \ldots, k_n)$$

be the coordinate vector relative to S of a vector $\mathbf{v}$ in V; thus

$$\mathbf{v} = k_1\mathbf{w}_1 + k_2\mathbf{w}_2 + \cdots + k_n\mathbf{w}_n$$

Define $T: V \to R^n$ to be the function that maps $\mathbf{v}$ into its coordinate vector relative to S; that is,

$$T(\mathbf{v}) = (\mathbf{v})_S = (k_1, k_2, \ldots, k_n)$$

The function T is a linear transformation. To see that this is so, suppose that $\mathbf{u}$ and $\mathbf{v}$ are vectors in V and that

$$\mathbf{u} = c_1\mathbf{w}_1 + c_2\mathbf{w}_2 + \cdots + c_n\mathbf{w}_n \quad \text{and} \quad \mathbf{v} = d_1\mathbf{w}_1 + d_2\mathbf{w}_2 + \cdots + d_n\mathbf{w}_n$$

Thus,

$$(\mathbf{u})_S = (c_1, c_2, \ldots, c_n) \quad \text{and} \quad (\mathbf{v})_S = (d_1, d_2, \ldots, d_n)$$

But

$$\mathbf{u} + \mathbf{v} = (c_1 + d_1)\mathbf{w}_1 + (c_2 + d_2)\mathbf{w}_2 + \cdots + (c_n + d_n)\mathbf{w}_n$$
$$k\mathbf{u} = (kc_1)\mathbf{w}_1 + (kc_2)\mathbf{w}_2 + \cdots + (kc_n)\mathbf{w}_n$$

so that

$$(\mathbf{u} + \mathbf{v})_S = (c_1 + d_1, c_2 + d_2, \ldots, c_n + d_n)$$
$$(k\mathbf{u})_S = (kc_1, kc_2, \ldots, kc_n)$$

Therefore,

$$(\mathbf{u} + \mathbf{v})_S = (\mathbf{u})_S + (\mathbf{v})_S \quad \text{and} \quad (k\mathbf{u})_S = k(\mathbf{u})_S$$

Expressing these equations in terms of T, we obtain

$$T(\mathbf{u} + \mathbf{v}) = T(\mathbf{u}) + T(\mathbf{v}) \quad \text{and} \quad T(k\mathbf{u}) = kT(\mathbf{u})$$

which shows that T is a linear transformation. ◆

REMARK. The computations in the preceding example could just as well have been performed using coordinate matrices rather than coordinate vectors; that is,

$$[\mathbf{u} + \mathbf{v}]_S = [\mathbf{u}]_S + [\mathbf{v}]_S \quad \text{and} \quad [k\mathbf{u}]_S = k[\mathbf{u}]_S$$

EXAMPLE 8 A Linear Transformation from P_n to P_{n+1}

Let $\mathbf{p} = p(x) = c_0 + c_1x + \cdots + c_nx^n$ be a polynomial in P_n, and define the function $T: P_n \to P_{n+1}$ by

$$T(\mathbf{p}) = T(p(x)) = xp(x) = c_0x + c_1x^2 + \cdots + c_nx^{n+1}$$

The function T is a linear transformation, since for any scalar k and any polynomials $\mathbf{p}_1$ and $\mathbf{p}_2$ in P_n we have

$$T(\mathbf{p}_1 + \mathbf{p}_2) = T(p_1(x) + p_2(x)) = x(p_1(x) + p_2(x))$$
$$= xp_1(x) + xp_2(x) = T(\mathbf{p}_1) + T(\mathbf{p}_2)$$

and

$$T(k\mathbf{p}) = T(kp(x)) = x(kp(x)) = k(xp(x)) = kT(\mathbf{p}) \quad ◆$$

EXAMPLE 9 A Linear Operator on P_n

Let $\mathbf{p} = p(x) = c_0 + c_1x + \cdots + c_nx^n$ be a polynomial in P_n, and let a and b be any scalars. We leave it as an exercise to show that the function T defined by

$$T(\mathbf{p}) = T(p(x)) = p(ax + b) = c_0 + c_1(ax + b) + \cdots + c_n(ax + b)^n$$

is a linear operator. For example, if $ax + b = 3x - 5$, then $T: P_2 \rightarrow P_2$ would be the linear operator given by the formula

$$T(c_0 + c_1 x + c_2 x^2) = c_0 + c_1(3x - 5) + c_2(3x - 5)^2$$ ♦

EXAMPLE 10   A Linear Transformation Using an Inner Product

Let V be an inner product space and let $\mathbf{v}_0$ be any fixed vector in V. Let $T: V \rightarrow R$ be the transformation that maps a vector $\mathbf{v}$ into its inner product with $\mathbf{v}_0$; that is,

$$T(\mathbf{v}) = \langle \mathbf{v}, \mathbf{v}_0 \rangle$$

From the properties of an inner product

$$T(\mathbf{u} + \mathbf{v}) = \langle \mathbf{u} + \mathbf{v}, \mathbf{v}_0 \rangle = \langle \mathbf{u}, \mathbf{v}_0 \rangle + \langle \mathbf{v}, \mathbf{v}_0 \rangle = T(\mathbf{u}) + T(\mathbf{v})$$

and

$$T(k\mathbf{u}) = \langle k\mathbf{u}, \mathbf{v}_0 \rangle = k \langle \mathbf{u}, \mathbf{v}_0 \rangle = kT(\mathbf{u})$$

so that T is a linear transformation.   ♦

Calculus Required

EXAMPLE 11   A Linear Transformation from $C^1(-\infty, \infty)$ to $F(-\infty, \infty)$

Let $V = C^1(-\infty, \infty)$ be the vector space of functions with continuous first derivatives on $(-\infty, \infty)$ and let $W = F(-\infty, \infty)$ be the vector space of all real-valued functions defined on $(-\infty, \infty)$. Let $D: V \rightarrow W$ be the transformation that maps a function $\mathbf{f} = f(x)$ into its derivative; that is,

$$D(\mathbf{f}) = f'(x)$$

From the properties of differentiation, we have

$$D(\mathbf{f} + \mathbf{g}) = D(\mathbf{f}) + D(\mathbf{g}) \quad \text{and} \quad D(k\mathbf{f}) = kD(\mathbf{f})$$

Thus, D is a linear transformation.   ♦

Calculus Required

EXAMPLE 12   A Linear Transformation from $C(-\infty, \infty)$ to $C^1(-\infty, \infty)$

Let $V = C(-\infty, \infty)$ be the vector space of continuous functions on $(-\infty, \infty)$, and let $W = C^1(-\infty, \infty)$ be the vector space of functions with continuous first derivatives on $(-\infty, \infty)$. Let $J: V \rightarrow W$ be the transformation that maps $\mathbf{f} = f(x)$ into the integral $\int_0^x f(t)\,dt$. For example, if $\mathbf{f} = x^2$, then

$$J(\mathbf{f}) = \int_0^x t^2\,dt = \left.\frac{t^3}{3}\right|_0^x = \frac{x^3}{3}$$

From the properties of integration, we have

$$J(\mathbf{f} + \mathbf{g}) = \int_0^x (f(t) + g(t))\,dt = \int_0^x f(t)\,dt + \int_0^x g(t)\,dt = J(\mathbf{f}) + J(\mathbf{g})$$

$$J(c\mathbf{f}) = \int_0^x cf(t)\,dt = c\int_0^x f(t)\,dt = cJ(\mathbf{f})$$

So J is a linear transformation.   ♦

EXAMPLE 13 A Transformation That Is Not Linear

Let $T: M_{nn} \rightarrow R$ be the transformation that maps an $n \times n$ matrix into its determinant; that is,

$$T(A) = \det(A)$$

If $n > 1$, then this transformation does not satisfy either of the properties required of a linear transformation. For example, we saw in Example 1 of Section 2.3 that

$$\det(A_1 + A_2) \neq \det(A_1) + \det(A_2)$$

in general. Moreover, $\det(cA) = c^n \det(A)$, so

$$\det(cA) \neq c \det(A)$$

in general. Thus, T is *not* a linear transformation. ◆

Properties of Linear Transformations

If $T: V \rightarrow W$ is a linear transformation, then for any vectors $\mathbf{v}_1$ and $\mathbf{v}_2$ in V and any scalars c_1 and c_2, we have

$$T(c_1\mathbf{v}_1 + c_2\mathbf{v}_2) = T(c_1\mathbf{v}_1) + T(c_2\mathbf{v}_2) = c_1 T(\mathbf{v}_1) + c_2 T(\mathbf{v}_2)$$

and more generally, if $\mathbf{v}_1, \mathbf{v}_2, \ldots, \mathbf{v}_n$ are vectors in V and $c_1, c_2, \ldots, c_n$ are scalars, then

$$T(c_1\mathbf{v}_1 + c_2\mathbf{v}_2 + \cdots + c_n\mathbf{v}_n) = c_1 T(\mathbf{v}_1) + c_2 T(\mathbf{v}_2) + \cdots + c_n T(\mathbf{v}_n) \qquad (1)$$

Formula (1) is sometimes described by saying that *linear transformations preserve linear combinations*.

The following theorem lists three basic properties that are common to all linear transformations.

Theorem 8.1.1

If $T: V \rightarrow W$ is a linear transformation, then:

(a) $T(\mathbf{0}) = \mathbf{0}$
(b) $T(-\mathbf{v}) = -T(\mathbf{v})$ *for all* $\mathbf{v}$ *in* V
(c) $T(\mathbf{v} - \mathbf{w}) = T(\mathbf{v}) - T(\mathbf{w})$ *for all* $\mathbf{v}$ *and* $\mathbf{w}$ *in* V

Proof. Let $\mathbf{v}$ be any vector in V. Since $0\mathbf{v} = \mathbf{0}$, we have

$$T(\mathbf{0}) = T(0\mathbf{v}) = 0T(\mathbf{v}) = \mathbf{0}$$

which proves (a). Also,

$$T(-\mathbf{v}) = T((-1)\mathbf{v}) = (-1)T(\mathbf{v}) = -T(\mathbf{v})$$

which proves (b). Finally, $\mathbf{v} - \mathbf{w} = \mathbf{v} + (-1)\mathbf{w}$; thus,

$$T(\mathbf{v} - \mathbf{w}) = T(\mathbf{v} + (-1)\mathbf{w})$$
$$= T(\mathbf{v}) + (-1)T(\mathbf{w})$$
$$= T(\mathbf{v}) - T(\mathbf{w})$$

which proves (c). ∎

In words, part (a) of the preceding theorem states that a linear transformation maps $\mathbf{0}$ into $\mathbf{0}$. This property is useful for identifying transformations that are *not* linear. For example, if $\mathbf{x}_0$ is a fixed nonzero vector in R^2, then the transformation

$$T(\mathbf{x}) = \mathbf{x} + \mathbf{x}_0$$

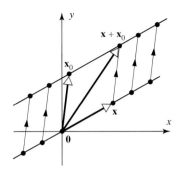

Figure 8.1.4

$T(\mathbf{x}) = \mathbf{x} + \mathbf{x}_0$ translates each point $\mathbf{x}$ along a line parallel to $\mathbf{x}_0$ through a distance $\|\mathbf{x}_0\|$.

has the geometric effect of translating each point $\mathbf{x}$ in a direction parallel to $\mathbf{x}_0$ through a distance of $\|\mathbf{x}_0\|$ (Figure 8.1.4). This is not a linear transformation, since $T(\mathbf{0}) = \mathbf{x}_0$, so T does not map $\mathbf{0}$ to $\mathbf{0}$.

Finding Linear Transformations from Images of Basis Vectors

Theorem 4.3.3 shows that if T is a matrix transformation, then the standard matrix for T can be obtained from the images of the standard basis vectors. Stated another way, *a matrix transformation is completely determined by its images of the standard basis vectors.* This is a special case of a more general result: If $T: V \to W$ is a linear transformation, and if $\{\mathbf{v}_1, \mathbf{v}_2, \dots, \mathbf{v}_n\}$ is any basis for V, then the image $T(\mathbf{v})$ of any vector $\mathbf{v}$ in V can be calculated from the images

$$T(\mathbf{v}_1), T(\mathbf{v}_2), \dots, T(\mathbf{v}_n)$$

of the basis vectors. This can be done by first expressing $\mathbf{v}$ as a linear combination of the basis vectors, say

$$\mathbf{v} = c_1\mathbf{v}_1 + c_2\mathbf{v}_2 + \cdots + c_n\mathbf{v}_n$$

and then using Formula (1) to write

$$T(\mathbf{v}) = c_1 T(\mathbf{v}_1) + c_2 T(\mathbf{v}_2) + \cdots + c_n T(\mathbf{v}_n)$$

In words, *a linear transformation is completely determined by its images of any basis vectors.*

EXAMPLE 14 Computing with Images of Basis Vectors

Consider the basis $S = \{\mathbf{v}_1, \mathbf{v}_2, \mathbf{v}_3\}$ for R^3, where $\mathbf{v}_1 = (1, 1, 1)$, $\mathbf{v}_2 = (1, 1, 0)$, and $\mathbf{v}_3 = (1, 0, 0)$. Let $T: R^3 \to R^2$ be the linear transformation such that

$$T(\mathbf{v}_1) = (1, 0), \qquad T(\mathbf{v}_2) = (2, -1), \qquad T(\mathbf{v}_3) = (4, 3)$$

Find a formula for $T(x_1, x_2, x_3)$; then use this formula to compute $T(2, -3, 5)$.

Solution.

We first express $\mathbf{x} = (x_1, x_2, x_3)$ as a linear combination of $\mathbf{v}_1 = (1, 1, 1)$, $\mathbf{v}_2 = (1, 1, 0)$, and $\mathbf{v}_3 = (1, 0, 0)$. If we write

$$(x_1, x_2, x_3) = c_1(1, 1, 1) + c_2(1, 1, 0) + c_3(1, 0, 0)$$

then on equating corresponding components we obtain

$$c_1 + c_2 + c_3 = x_1$$
$$c_1 + c_2 \qquad = x_2$$
$$c_1 \qquad\qquad = x_3$$

which yields $c_1 = x_3$, $c_2 = x_2 - x_3$, $c_3 = x_1 - x_2$, so that

$$(x_1, x_2, x_3) = x_3(1, 1, 1) + (x_2 - x_3)(1, 1, 0) + (x_1 - x_2)(1, 0, 0)$$
$$= x_3\mathbf{v}_1 + (x_2 - x_3)\mathbf{v}_2 + (x_1 - x_2)\mathbf{v}_3$$

Thus,

$$T(x_1, x_2, x_3) = x_3 T(\mathbf{v}_1) + (x_2 - x_3)T(\mathbf{v}_2) + (x_1 - x_2)T(\mathbf{v}_3)$$
$$= x_3(1, 0) + (x_2 - x_3)(2, -1) + (x_1 - x_2)(4, 3)$$
$$= (4x_1 - 2x_2 - x_3, 3x_1 - 4x_2 + x_3)$$

From this formula we obtain

$$T(2, -3, 5) = (9, 23) \qquad \blacklozenge$$

In Section 4.2 we defined the composition of matrix transformations. The following definition extends that concept to general linear transformations.

Definition

If $T_1: U \rightarrow V$ and $T_2: V \rightarrow W$ are linear transformations, the ***composition of T_2 with T_1***, denoted by $T_2 \circ T_1$ (read "T_2 circle T_1"), is the function defined by the formula

$$(T_2 \circ T_1)(\mathbf{u}) = T_2(T_1(\mathbf{u})) \tag{2}$$

where $\mathbf{u}$ is a vector in U.

REMARK. Observe that this definition requires that the domain of T_2 (which is V) contain the range of T_1; this is essential for the formula $T_2(T_1(\mathbf{u}))$ to make sense (Figure 8.1.5). The reader should compare (2) to Formula (18) in Section 4.2.

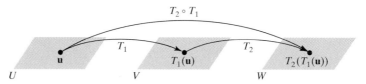

Figure 8.1.5 The composition of T_2 with T_1.

The next result shows that the composition of two linear transformations is itself a linear transformation.

Theorem 8.1.2

If $T_1: U \rightarrow V$ and $T_2: V \rightarrow W$ are linear transformations, then $(T_2 \circ T_1): U \rightarrow W$ is also a linear transformation.

Proof. If $\mathbf{u}$ and $\mathbf{v}$ are vectors in U and c is a scalar, then it follows from (2) and the linearity of T_1 and T_2 that

$$(T_2 \circ T_1)(\mathbf{u} + \mathbf{v}) = T_2(T_1(\mathbf{u} + \mathbf{v})) = T_2(T_1(\mathbf{u}) + T_1(\mathbf{v}))$$
$$= T_2(T_1(\mathbf{u})) + T_2(T_1(\mathbf{v}))$$
$$= (T_2 \circ T_1)(\mathbf{u}) + (T_2 \circ T_1)(\mathbf{v})$$

and

$$(T_2 \circ T_1)(c\mathbf{u}) = T_2(T_1(c\mathbf{u})) = T_2(cT_1(\mathbf{u}))$$
$$= cT_2(T_1(\mathbf{u})) = c(T_2 \circ T_1)(\mathbf{u})$$

Thus, $T_2 \circ T_1$ satisfies the two requirements of a linear transformation. ∎

EXAMPLE 15 Composition of Linear Transformations

Let $T_1: P_1 \rightarrow P_2$ and $T_2: P_2 \rightarrow P_2$ be the linear transformations given by the formulas

$$T_1(p(x)) = xp(x) \quad \text{and} \quad T_2(p(x)) = p(2x + 4)$$

Then the composition $(T_2 \circ T_1): P_1 \rightarrow P_2$ is given by the formula

$$(T_2 \circ T_1)(p(x)) = T_2(T_1(p(x))) = T_2(xp(x)) = (2x + 4)p(2x + 4)$$

In particular, if $p(x) = c_0 + c_1 x$, then

$$(T_2 \circ T_1)(p(x)) = (T_2 \circ T_1)(c_0 + c_1 x) = (2x + 4)(c_0 + c_1(2x + 4))$$
$$= c_0(2x + 4) + c_1(2x + 4)^2 \qquad \blacklozenge$$

EXAMPLE 16 Composition with the Identity Operator

If $T: V \to V$ is any linear operator, and if $I: V \to V$ is the identity operator (Example 3), then for all vectors **v** in V we have

$$(T \circ I)(\mathbf{v}) = T(I(\mathbf{v})) = T(\mathbf{v})$$
$$(I \circ T)(\mathbf{v}) = I(T(\mathbf{v})) = T(\mathbf{v})$$

It follows that $T \circ I$ and $I \circ T$ are the same as T; that is,

$$T \circ I = T \qquad \text{and} \qquad I \circ T = T \qquad (3)$$

We conclude this section by noting that compositions can be defined for more than two linear transformations. For example, if

$$T_1: U \to V, \quad T_2: V \to W, \quad \text{and} \quad T_3: W \to Y$$

are linear transformations, then the composition $T_3 \circ T_2 \circ T_1$ is defined by

$$(T_3 \circ T_2 \circ T_1)(\mathbf{u}) = T_3(T_2(T_1(\mathbf{u}))) \qquad (4)$$

(Figure 8.1.6).

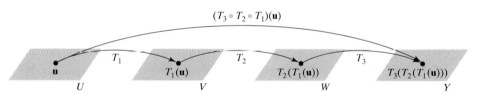

Figure 8.1.6 The composition of three linear transformations.

Exercise Set 8.1

1. Use the definition of a linear operator that was given in this section to show that the function $T: R^2 \to R^2$ given by the formula $T(x_1, x_2) = (x_1 + 2x_2, 3x_1 - x_2)$ is a linear operator.

2. Use the definition of a linear transformation given in this section to show that the function $T: R^3 \to R^2$ given by the formula $T(x_1, x_2, x_3) = (2x_1 - x_2 + x_3, x_2 - 4x_3)$ is a linear transformation.

In Exercises 3–10 determine whether the function is a linear transformation. Justify your answer.

3. $T: V \to R$, where V is an inner product space, and $T(\mathbf{u}) = \|\mathbf{u}\|$.

4. $T: R^3 \to R^3$, where $\mathbf{v}_0$ is a fixed vector in R^3 and $T(\mathbf{u}) = \mathbf{u} \times \mathbf{v}_0$.

5. $T: M_{22} \to M_{23}$, where B is a fixed 2×3 matrix and $T(A) = AB$.

6. $T: M_{nn} \to R$, where $T(A) = \text{tr}(A)$.

7. $F: M_{mn} \to M_{nm}$, where $F(A) = A^T$.

8. $T: M_{22} \rightarrow R$, where

(a) $T\left(\begin{bmatrix} a & b \\ c & d \end{bmatrix}\right) = 3a - 4b + c - d$ (b) $T\left(\begin{bmatrix} a & b \\ c & d \end{bmatrix}\right) = a^2 + b^2$

9. $T: P_2 \rightarrow P_2$, where

(a) $T(a_0 + a_1 x + a_2 x^2) = a_0 + a_1(x + 1) + a_2(x + 1)^2$
(b) $T(a_0 + a_1 x + a_2 x^2) = (a_0 + 1) + (a_1 + 1)x + (a_2 + 1)x^2$

10. $T: F(-\infty, \infty) \rightarrow F(-\infty, \infty)$, where

(a) $T(f(x)) = 1 + f(x)$ (b) $T(f(x)) = f(x + 1)$

11. Show that the function T in Example 9 is a linear operator.

12. Consider the basis $S = \{v_1, v_2\}$ for R^2, where $v_1 = (1, 1)$ and $v_2 = (1, 0)$, and let $T: R^2 \rightarrow R^2$ be the linear operator such that

$$T(v_1) = (1, -2) \quad \text{and} \quad T(v_2) = (-4, 1)$$

Find a formula for $T(x_1, x_2)$, and use that formula to find $T(5, -3)$.

13. Consider the basis $S = \{v_1, v_2\}$ for R^2, where $v_1 = (-2, 1)$ and $v_2 = (1, 3)$, and let $T: R^2 \rightarrow R^3$ be the linear transformation such that

$$T(v_1) = (-1, 2, 0) \quad \text{and} \quad T(v_2) = (0, -3, 5)$$

Find a formula for $T(x_1, x_2)$, and use that formula to find $T(2, -3)$.

14. Consider the basis $S = \{v_1, v_2, v_3\}$ for R^3, where $v_1 = (1, 1, 1)$, $v_2 = (1, 1, 0)$, and $v_3 = (1, 0, 0)$, and let $T: R^3 \rightarrow R^3$ be the linear operator such that

$$T(v_1) = (2, -1, 4), \quad T(v_2) = (3, 0, 1), \quad T(v_3) = (-1, 5, 1)$$

Find a formula for $T(x_1, x_2, x_3)$, and use that formula to find $T(2, 4, -1)$.

15. Consider the basis $S = \{v_1, v_2, v_3\}$ for R^3, where $v_1 = (1, 2, 1)$, $v_2 = (2, 9, 0)$, and $v_3 = (3, 3, 4)$, and let $T: R^3 \rightarrow R^2$ be the linear transformation such that

$$T(v_1) = (1, 0), \quad T(v_2) = (-1, 1), \quad T(v_3) = (0, 1)$$

Find a formula for $T(x_1, x_2, x_3)$, and use that formula to find $T(7, 13, 7)$.

16. Let v_1, v_2, and v_3 be vectors in a vector space V and $T: V \rightarrow R^3$ a linear transformation for which

$$T(v_1) = (1, -1, 2), \quad T(v_2) = (0, 3, 2), \quad T(v_3) = (-3, 1, 2)$$

Find $T(2v_1 - 3v_2 + 4v_3)$.

17. Find the domain and codomain of $T_2 \circ T_1$, and find $(T_2 \circ T_1)(x, y)$.

(a) $T_1(x, y) = (2x, 3y)$, $T_2(x, y) = (x - y, x + y)$
(b) $T_1(x, y) = (x - 3y, 0)$, $T_2(x, y) = (4x - 5y, 3x - 6y)$
(c) $T_1(x, y) = (2x, -3y, x + y)$, $T_2(x, y, z) = (x - y, y + z)$
(d) $T_1(x, y) = (x - y, y + z, x - z)$, $T_2(x, y, z) = (0, x + y + z)$

18. Find the domain and codomain of $T_3 \circ T_2 \circ T_1$, and find $(T_3 \circ T_2 \circ T_1)(x, y)$.

(a) $T_1(x, y) = (-2y, 3x, x - 2y)$, $T_2(x, y, z) = (y, z, x)$, $T_3(x, y, z) = (x + z, y - z)$
(b) $T_1(x, y) = (x + y, y, -x)$, $T_2(x, y, z) = (0, x + y + z, 3y)$, $T_3(x, y, z) = (3x + 2y, 4z - x - 3y)$

19. Let $T_1: M_{22} \rightarrow R$ and $T_2: M_{22} \rightarrow M_{22}$ be the linear transformations given by $T_1(A) = \text{tr}(A)$ and $T_2(A) = A^T$.

(a) Find $(T_1 \circ T_2)(A)$, where $A = \begin{bmatrix} a & b \\ c & d \end{bmatrix}$.   (b) Can you find $(T_2 \circ T_1)(A)$? Explain.

20. Let $T_1: P_n \rightarrow P_n$ and $T_2: P_n \rightarrow P_n$ be the linear operators given by $T_1(p(x)) = p(x - 1)$ and $T_2(p(x)) = p(x + 1)$. Find $(T_1 \circ T_2)(p(x))$ and $(T_2 \circ T_1)(p(x))$.

21. Let $T_1: V \rightarrow V$ be the dilation $T_1(v) = 4v$. Find a linear operator $T_2: V \rightarrow V$ such that $T_1 \circ T_2 = I$ and $T_2 \circ T_1 = I$.

22. Suppose that the linear transformations $T_1: P_2 \rightarrow P_2$ and $T_2: P_2 \rightarrow P_3$ are given by the formulas $T_1(p(x)) = p(x + 1)$ and $T_2(p(x)) = xp(x)$. Find $(T_2 \circ T_1)(a_0 + a_1 x + a_2 x^2)$.

23. Let $q_0(x)$ be a fixed polynomial of degree m, and define a function T with domain P_n by the formula $T(p(x)) = p(q_0(x))$.

    (a) Show that T is a linear transformation.   (b) What is the codomain of T?

24. Use the definition of $T_3 \circ T_2 \circ T_1$ given by Formula (4) to prove that

    (a) $T_3 \circ T_2 \circ T_1$ is a linear transformation   (b) $T_3 \circ T_2 \circ T_1 = (T_3 \circ T_2) \circ T_1$
    (c) $T_3 \circ T_2 \circ T_1 = T_3 \circ (T_2 \circ T_1)$

25. Let $T: R^3 \rightarrow R^3$ be the orthogonal projection of R^3 onto the xy-plane. Show that $T \circ T = T$.

26. (a) Let $T: V \rightarrow W$ be a linear transformation and let k be a scalar. Define the function $(kT): V \rightarrow W$ by $(kT)(\mathbf{v}) = k(T(\mathbf{v}))$. Show that kT is a linear transformation.
    (b) Find $(3T)(x_1, x_2)$ if $T: R^2 \rightarrow R^2$ is given by the formula $T(x_1, x_2) = (2x_1 - x_2, x_2 + x_1)$.

27. (a) Let $T_1: V \rightarrow W$ and $T_2: V \rightarrow W$ be linear transformations. Define the functions $(T_1 + T_2): V \rightarrow W$ and $(T_1 - T_2): V \rightarrow W$ by

    $$(T_1 + T_2)(\mathbf{v}) = T_1(\mathbf{v}) + T_2(\mathbf{v})$$
    $$(T_1 - T_2)(\mathbf{v}) = T_1(\mathbf{v}) - T_2(\mathbf{v})$$

    Show that $T_1 + T_2$ and $T_1 - T_2$ are linear transformations.
    (b) Find $(T_1 + T_2)(x, y)$ and $(T_1 - T_2)(x, y)$ if $T_1: R^2 \rightarrow R^2$ and $T_2: R^2 \rightarrow R^2$ are given by the formulas $T_1(x, y) = (2y, 3x)$ and $T_2(x, y) = (y, x)$.

28. (a) Prove that if a_1, a_2, b_1, and b_2 are any scalars, then the formula

    $$F(x, y) = (a_1 x + b_1 y, a_2 x + b_2 y)$$

    defines a linear operator on R^2.
    (b) Does the formula $F(x, y) = (a_1 x^2 + b_1 y^2, a_2 x^2 + b_2 y^2)$ define a linear operator on R^2? Explain.

29. Let $\{\mathbf{v}_1, \mathbf{v}_2, \ldots, \mathbf{v}_n\}$ be a basis for a vector space V, and let $T: V \rightarrow W$ be a linear transformation. Show that if $T(\mathbf{v}_1) = T(\mathbf{v}_2) = \cdots = T(\mathbf{v}_n) = \mathbf{0}$, then T is the zero transformation.

30. Let $\{\mathbf{v}_1, \mathbf{v}_2, \ldots, \mathbf{v}_n\}$ be a basis for a vector space V, and let $T: V \rightarrow V$ be a linear operator. Show that if $T(\mathbf{v}_1) = \mathbf{v}_1, T(\mathbf{v}_2) = \mathbf{v}_2, \ldots, T(\mathbf{v}_n) = \mathbf{v}_n$, then T is the identity transformation on V.

31. (**For readers who have studied calculus.**) Let

    $$D(\mathbf{f}) = f'(x) \quad \text{and} \quad J(\mathbf{f}) = \int_0^x f(t)\, dt$$

    be the linear transformations in Examples 11 and 12. Find $(J \circ D)(\mathbf{f})$ for

    (a) $\mathbf{f}(x) = x^2 + 3x + 2$ (b) $\mathbf{f}(x) = \sin x$ (c) $\mathbf{f}(x) = e^x + 3$

32. (**For readers who have studied calculus.**) Let $V = C[a, b]$ be the vector space of functions continuous on $[a, b]$, and let $T: V \rightarrow V$ be the transformation defined by

    $$T(\mathbf{f}) = 5f(x) + 3 \int_a^x f(t)\, dt$$

    Is T a linear operator?

Discussion and Discovery

33. Indicate whether the statement is always true or sometimes false. Justify your answer by giving a logical argument or a counterexample. In each part V and W are vector spaces.

    (a) If $T(c_1 \mathbf{v}_1 + c_2 \mathbf{v}_2) = c_1 T(\mathbf{v}_1) + c_2 T(\mathbf{v}_2)$ for all vectors $\mathbf{v}_1$ and $\mathbf{v}_2$ in V and all scalars c_1 and c_2, then T is a linear transformation.
    (b) If $\mathbf{v}$ is a nonzero vector in V, then there is exactly one linear transformation $T: V \rightarrow W$ such that $T(-\mathbf{v}) = -T(\mathbf{v})$.

(c) There is exactly one linear transformation $T : V \rightarrow W$ for which $T(\mathbf{u} + \mathbf{v}) = T(\mathbf{u} - \mathbf{v})$ for all vectors $\mathbf{u}$ and $\mathbf{v}$ in V.

(d) If $\mathbf{v}_0$ is a nonzero vector in V, then the formula $T(\mathbf{v}) = \mathbf{v}_0 + \mathbf{v}$ defines a linear operator on V.

34. If $B = \{\mathbf{v}_1, \mathbf{v}_2, \dots, \mathbf{v}_n\}$ is a basis for a vector space V, how many different linear operators can be created that map each vector in B back into B? Explain your reasoning.

8.2 KERNEL AND RANGE

In this section we shall develop some basic properties of linear transformations that generalize properties of matrix transformations obtained earlier in the text.

Kernel and Range   Recall that if A is an $m \times n$ matrix, then the nullspace of A consists of all vectors $\mathbf{x}$ in R^n such that $A\mathbf{x} = \mathbf{0}$, and by Theorem 5.5.1 the column space of A consists of all vectors $\mathbf{b}$ in R^m for which there is at least one vector $\mathbf{x}$ in R^n such that $A\mathbf{x} = \mathbf{b}$. From the viewpoint of matrix transformations, the nullspace of A consists of all vectors in R^n that multiplication by A maps into $\mathbf{0}$, and the column space of A consists of all vectors in R^m that are images of at least one vector in R^n under multiplication by A. The following definition extends these ideas to general linear transformations.

> **Definition**
>
> If $T : V \rightarrow W$ is a linear transformation, then the set of vectors in V that T maps into $\mathbf{0}$ is called the **kernel** of T; it is denoted by $\ker(T)$. The set of all vectors in W that are images under T of at least one vector in V is called the **range** of T; it is denoted by $R(T)$.

EXAMPLE 1   Kernel and Range of a Matrix Transformation

If $T_A : R^n \rightarrow R^m$ is multiplication by the $m \times n$ matrix A, then from the discussion preceding the definition above, the kernel of T_A is the nullspace of A, and the range of T_A is the column space of A.   ◆

EXAMPLE 2   Kernel and Range of the Zero Transformation

Let $T : V \rightarrow W$ be the zero transformation (Example 2 of Section 8.1). Since T maps *every* vector in V into $\mathbf{0}$, it follows that $\ker(T) = V$. Moreover, since $\mathbf{0}$ is the *only* image under T of vectors in V, we have $R(T) = \{\mathbf{0}\}$.   ◆

EXAMPLE 3   Kernel and Range of the Identity Operator

Let $I : V \rightarrow V$ be the identity operator (Example 3 of Section 8.1). Since $I(\mathbf{v}) = \mathbf{v}$ for all vectors in V, *every* vector in V is the image of some vector (namely, itself); thus, $R(I) = V$. Since the *only* vector that I maps into $\mathbf{0}$ is $\mathbf{0}$, it follows that $\ker(I) = \{\mathbf{0}\}$.   ◆

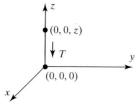

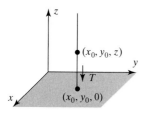

(a) ker(*T*) is the *z*-axis.

EXAMPLE 4 Kernel and Range of an Orthogonal Projection

Let $T: R^3 \to R^3$ be the orthogonal projection on the *xy*-plane. The kernel of *T* is the set of points that *T* maps into $\mathbf{0} = (0, 0, 0)$; these are the points on the *z*-axis (Figure 8.2.1*a*). Since *T* maps every point in R^3 into the *xy*-plane, the range of *T* must be some subset of this plane. But every point $(x_0, y_0, 0)$ in the *xy*-plane is the image under *T* of some point; in fact, it is the image of all points on the vertical line that passes through $(x_0, y_0, 0)$ (Figure 8.2.1*b*). Thus $R(T)$ is the entire *xy*-plane. ◆

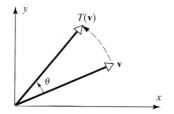

(b) $R(T)$ is the entire *xy*-plane.

Figure 8.2.1

EXAMPLE 5 Kernel and Range of a Rotation

Let $T: R^2 \to R^2$ be the linear operator that rotates each vector in the *xy*-plane through the angle θ (Figure 8.2.2). Since *every* vector in the *xy*-plane can be obtained by rotating some vector through the angle θ (why?), we have $R(T) = R^2$. Moreover, the only vector that rotates into $\mathbf{0}$ is $\mathbf{0}$, so ker(*T*) = {$\mathbf{0}$}. ◆

Calculus Required

EXAMPLE 6 Kernel of a Differentiation Transformation

Let $V = C^1(-\infty, \infty)$ be the vector space of functions with continuous first derivatives on $(-\infty, \infty)$, let $W = F(-\infty, \infty)$ be the vector space of all real-valued functions defined on $(-\infty, \infty)$, and let $D: V \to W$ be the differentiation transformation $D(\mathbf{f}) = f'(x)$. The kernel of *D* is the set of functions in *V* with derivative zero. From calculus, this is the set of constant functions on $(-\infty, \infty)$. ◆

Figure 8.2.2

Properties of Kernel and Range

In all of the preceding examples, ker(*T*) and $R(T)$ turned out to be *subspaces*. In Examples 2, 3, and 5 they were either the zero subspace or the entire vector space. In Example 4 the kernel was a line through the origin and the range was a plane through the origin, both of which are subspaces of R^3. All of this is not accidental; it is a consequence of the following general result.

Theorem 8.2.1

If $T: V \to W$ *is linear transformation, then*:

(a) The kernel of T is a subspace of V.   *(b) The range of T is a subspace of W.*

Proof (a). To show that ker(*T*) is a subspace, we must show that it contains at least one vector and is closed under addition and scalar multiplication. By part (*a*) of Theorem 8.1.1, the vector $\mathbf{0}$ is in ker(*T*), so this set contains at least one vector. Let $\mathbf{v}_1$ and $\mathbf{v}_2$ be vectors in ker(*T*), and let *k* be any scalar. Then

$$T(\mathbf{v}_1 + \mathbf{v}_2) = T(\mathbf{v}_1) + T(\mathbf{v}_2) = \mathbf{0} + \mathbf{0} = \mathbf{0}$$

so that $\mathbf{v}_1 + \mathbf{v}_2$ is in ker(*T*). Also,

$$T(k\mathbf{v}_1) = kT(\mathbf{v}_1) = k\mathbf{0} = \mathbf{0}$$

so that $k\mathbf{v}_1$ is in ker(*T*).

Proof (b). Since $T(\mathbf{0}) = \mathbf{0}$, there is at least one vector in $R(T)$. Let $\mathbf{w}_1$ and $\mathbf{w}_2$ be vectors in the range of *T*, and let *k* be any scalar. To prove this part we must show that $\mathbf{w}_1 + \mathbf{w}_2$ and $k\mathbf{w}_1$ are in the range of *T*; that is, we must find vectors $\mathbf{a}$ and $\mathbf{b}$ in *V* such that $T(\mathbf{a}) = \mathbf{w}_1 + \mathbf{w}_2$ and $T(\mathbf{b}) = k\mathbf{w}_1$.

Since $\mathbf{w}_1$ and $\mathbf{w}_2$ are in the range of T, there are vectors $\mathbf{a}_1$ and $\mathbf{a}_2$ in V such that $T(\mathbf{a}_1) = \mathbf{w}_1$ and $T(\mathbf{a}_2) = \mathbf{w}_2$. Let $\mathbf{a} = \mathbf{a}_1 + \mathbf{a}_2$ and $\mathbf{b} = k\mathbf{a}_1$. Then

$$T(\mathbf{a}) = T(\mathbf{a}_1 + \mathbf{a}_2) = T(\mathbf{a}_1) + T(\mathbf{a}_2) = \mathbf{w}_1 + \mathbf{w}_2$$

and

$$T(\mathbf{b}) = T(k\mathbf{a}_1) = kT(\mathbf{a}_1) = k\mathbf{w}_1$$

which completes the proof. ∎

In Section 5.6 we defined the rank of a matrix to be the dimension of its column (or row) space and the nullity to be the dimension of its nullspace. The following definition extends these definitions to general linear transformations.

Definition

If $T: V \rightarrow W$ is a linear transformation, then the dimension of the range of T is called the *rank of T* and is denoted by rank(T); the dimension of the kernel is called the *nullity of T* and is denoted by nullity(T).

If A is an $m \times n$ matrix and $T_A: R^n \rightarrow R^m$ is multiplication by A, then we know from Example 1 that the kernel of T_A is the nullspace of A and the range of T_A is the column space of A. Thus, we have the following relationship between the rank and nullity of a matrix and the rank and nullity of the corresponding matrix transformation.

Theorem 8.2.2

If A is an $m \times n$ matrix and $T_A: R^n \rightarrow R^m$ is multiplication by A, then:

(a) nullity(T_A) = nullity(A)   (b) rank(T_A) = rank(A)

EXAMPLE 7 Finding Rank and Nullity

Let $T_A: R^6 \rightarrow R^4$ be multiplication by

$$A = \begin{bmatrix} -1 & 2 & 0 & 4 & 5 & -3 \\ 3 & -7 & 2 & 0 & 1 & 4 \\ 2 & -5 & 2 & 4 & 6 & 1 \\ 4 & -9 & 2 & -4 & -4 & 7 \end{bmatrix}$$

Find the rank and nullity of T_A.

Solution.

In Example 1 of Section 5.6 we showed that rank(A) = 2 and nullity(A) = 4. Thus, from Theorem 8.2.2 we have rank(T_A) = 2 and nullity(T_A) = 4. ◆

EXAMPLE 8 Finding Rank and Nullity

Let $T: R^3 \rightarrow R^3$ be the orthogonal projection on the xy-plane. From Example 4, the kernel of T is the z-axis, which is one-dimensional; and the range of T is the xy-plane, which is two-dimensional. Thus,

$$\text{nullity}(T) = 1 \quad \text{and} \quad \text{rank}(T) = 2$$

◆

Dimension Theorem for Linear Transformations   Recall from the Dimension Theorem for Matrices (Theorem 5.6.3) that if A is a matrix with n columns, then

$$\text{rank}(A) + \text{nullity}(A) = n$$

The following theorem, whose proof is deferred to the end of the section, extends this result to general linear transformations.

Theorem 8.2.3   **Dimension Theorem for Linear Transformations**

If $T: V \to W$ is a linear transformation from an n-dimensional vector space V to a vector space W, then

$$\text{rank}(T) + \text{nullity}(T) = n \qquad (1)$$

In words, this theorem states that *for linear transformations the rank plus the nullity is equal to the dimension of the domain.*

REMARK.   If A is an $m \times n$ matrix and $T_A: R^n \to R^m$ is multiplication by A, then the domain of T_A has dimension n, so Theorem 8.2.3 agrees with Theorem 5.6.3 in this case.

EXAMPLE 9   Using the Dimension Theorem

Let $T: R^2 \to R^2$ be the linear operator that rotates each vector in the xy-plane through an angle θ. We showed in Example 5 that $\ker(T) = \{\mathbf{0}\}$ and $R(T) = R^2$. Thus,

$$\text{rank}(T) + \text{nullity}(T) = 2 + 0 = 2$$

which is consistent with the fact that the domain of T is two-dimensional.   ◆

Additional Proof

Proof of Theorem 8.2.3.   We must show that

$$\dim(R(T)) + \dim(\ker(T)) = n$$

We shall give the proof for the case where $1 \leq \dim(\ker(T)) < n$. The cases where $\dim(\ker(T)) = 0$ and $\dim(\ker(T)) = n$ are left as exercises. Assume $\dim(\ker(T)) = r$, and let $\mathbf{v}_1, \ldots, \mathbf{v}_r$ be a basis for the kernel. Since $\{\mathbf{v}_1, \ldots, \mathbf{v}_r\}$ is linearly independent, Theorem 5.4.6b states that there are $n - r$ vectors, $\mathbf{v}_{r+1}, \ldots, \mathbf{v}_n$, such that the extended set $\{\mathbf{v}_1, \ldots, \mathbf{v}_r, \mathbf{v}_{r+1}, \ldots, \mathbf{v}_n\}$ is a basis for V. To complete the proof, we shall show that the $n - r$ vectors in the set $S = \{T(\mathbf{v}_{r+1}), \ldots, T(\mathbf{v}_n)\}$ form a basis for the range of T. It will then follow that

$$\dim(R(T)) + \dim(\ker(T)) = (n - r) + r = n$$

First we show that S spans the range of T. If $\mathbf{b}$ is any vector in the range of T, then $\mathbf{b} = T(\mathbf{v})$ for some vector $\mathbf{v}$ in V. Since $\{\mathbf{v}_1, \ldots, \mathbf{v}_r, \mathbf{v}_{r+1}, \ldots, \mathbf{v}_n\}$ is a basis for V, the vector $\mathbf{v}$ can be written in the form

$$\mathbf{v} = c_1\mathbf{v}_1 + \cdots + c_r\mathbf{v}_r + c_{r+1}\mathbf{v}_{r+1} + \cdots + c_n\mathbf{v}_n$$

Since $\mathbf{v}_1, \ldots, \mathbf{v}_r$ lie in the kernel of T, we have $T(\mathbf{v}_1) = \cdots = T(\mathbf{v}_r) = \mathbf{0}$, so that

$$\mathbf{b} = T(\mathbf{v}) = c_{r+1}T(\mathbf{v}_{r+1}) + \cdots + c_nT(\mathbf{v}_n)$$

Thus, S spans the range of T.

Finally, we show that S is a linearly independent set and consequently forms a basis for the range of T. Suppose that some linear combination of the vectors in S is zero; that is,

$$k_{r+1}T(\mathbf{v}_{r+1}) + \cdots + k_n T(\mathbf{v}_n) = \mathbf{0} \tag{2}$$

We must show $k_{r+1} = \cdots = k_n = 0$. Since T is linear, (2) can be rewritten as

$$T(k_{r+1}\mathbf{v}_{r+1} + \cdots + k_n\mathbf{v}_n) = \mathbf{0}$$

which says that $k_{r+1}\mathbf{v}_{r+1} + \cdots + k_n\mathbf{v}_n$ is in the kernel of T. This vector can therefore be written as a linear combination of the basis vectors $\{\mathbf{v}_1, \ldots, \mathbf{v}_r\}$, say

$$k_{r+1}\mathbf{v}_{r+1} + \cdots + k_n\mathbf{v}_n = k_1\mathbf{v}_1 + \cdots + k_r\mathbf{v}_r$$

Thus,

$$k_1\mathbf{v}_1 + \cdots + k_r\mathbf{v}_r - k_{r+1}\mathbf{v}_{r+1} - \cdots - k_n\mathbf{v}_n = \mathbf{0}$$

Since $\{\mathbf{v}_1, \ldots, \mathbf{v}_n\}$ is linearly independent, all of the k's are zero; in particular, $k_{r+1} = \cdots = k_n = 0$, which completes the proof. ∎

Exercise Set 8.2

1. Let $T: R^2 \to R^2$ be the linear operator given by the formula

$$T(x, y) = (2x - y, -8x + 4y)$$

Which of the following vectors are in $R(T)$?

(a) $(1, -4)$ (b) $(5, 0)$ (c) $(-3, 12)$

2. Let $T: R^2 \to R^2$ be the linear operator in Exercise 1. Which of the following vectors are in $\ker(T)$?

(a) $(5, 10)$ (b) $(3, 2)$ (c) $(1, 1)$

3. Let $T: R^4 \to R^3$ be the linear transformation given by the formula

$$T(x_1, x_2, x_3, x_4) = (4x_1 + x_2 - 2x_3 - 3x_4, 2x_1 + x_2 + x_3 - 4x_4, 6x_1 - 9x_3 + 9x_4)$$

Which of the following are in $R(T)$?

(a) $(0, 0, 6)$ (b) $(1, 3, 0)$ (c) $(2, 4, 1)$

4. Let $T: R^4 \to R^3$ be the linear transformation in Exercise 3. Which of the following are in $\ker(T)$?

(a) $(3, -8, 2, 0)$ (b) $(0, 0, 0, 1)$ (c) $(0, -4, 1, 0)$

5. Let $T: P_2 \to P_3$ be the linear transformation defined by $T(p(x)) = xp(x)$. Which of the following are in $\ker(T)$?

(a) x^2 (b) 0 (c) $1 + x$

6. Let $T: P_2 \to P_3$ be the linear transformation in Exercise 5. Which of the following are in $R(T)$?

(a) $x + x^2$ (b) $1 + x$ (c) $3 - x^2$

7. Find a basis for the kernel of

(a) the linear operator in Exercise 1   (b) the linear transformation in Exercise 3
(c) the linear transformation in Exercise 5

8. Find a basis for the range of

(a) the linear operator in Exercise 1   (b) the linear transformation in Exercise 3
(c) the linear transformation in Exercise 5

9. Verify Formula (1) of the dimension theorem for

(a) the linear operator in Exercise 1   (b) the linear transformation in Exercise 3
(c) the linear transformation in Exercise 5.

In Exercises 10–13 let T be multiplication by the matrix A. Find

(a) a basis for the range of T (b) a basis for the kernel of T
(c) the rank and nullity of T (d) the rank and nullity of A

10. $A = \begin{bmatrix} 1 & -1 & 3 \\ 5 & 6 & -4 \\ 7 & 4 & 2 \end{bmatrix}$ **11.** $A = \begin{bmatrix} 2 & 0 & -1 \\ 4 & 0 & -2 \\ 0 & 0 & 0 \end{bmatrix}$

12. $A = \begin{bmatrix} 4 & 1 & 5 & 2 \\ 1 & 2 & 3 & 0 \end{bmatrix}$ **13.** $A = \begin{bmatrix} 1 & 4 & 5 & 0 & 9 \\ 3 & -2 & 1 & 0 & -1 \\ -1 & 0 & -1 & 0 & -1 \\ 2 & 3 & 5 & 1 & 8 \end{bmatrix}$

14. Describe the kernel and range of

(a) the orthogonal projection on the xz-plane
(b) the orthogonal projection on the yz-plane
(c) the orthogonal projection on the plane with the equation $y = x$

15. Let V be any vector space, and let $T: V \to V$ be defined by $T(\mathbf{v}) = 3\mathbf{v}$.

(a) What is the kernel of T?   (b) What is the range of T?

16. In each part use the given information to find the nullity of T.

(a) $T: R^5 \to R^7$ has rank 3.   (b) $T: P_4 \to P_3$ has rank 1.
(c) The range of $T: R^6 \to R^3$ is R^3.   (d) $T: M_{22} \to M_{22}$ has rank 3.

17. Let A be a 7×6 matrix such that $A\mathbf{x} = \mathbf{0}$ has only the trivial solution, and let $T: R^6 \to R^7$ be multiplication by A. Find the rank and nullity of T.

18. Let A be a 5×7 matrix with rank 4.

(a) What is the dimension of the solution space of $A\mathbf{x} = \mathbf{0}$?
(b) Is $A\mathbf{x} = \mathbf{b}$ consistent for all vectors $\mathbf{b}$ in R^5? Explain.

19. Let $T: R^3 \to V$ be a linear transformation from R^3 to any vector space. Show that the kernel of T is a line through the origin, a plane through the origin, the origin only, or all of R^3.

20. Let $T: V \to R^3$ be a linear transformation from any vector space to R^3. Show that the range of T is a line through the origin, a plane through the origin, the origin only, or all of R^3.

21. Let $T: R^3 \to R^3$ be multiplication by

$$\begin{bmatrix} 1 & 3 & 4 \\ 3 & 4 & 7 \\ -2 & 2 & 0 \end{bmatrix}$$

(a) Show that the kernel of T is a line through the origin, and find parametric equations for it.
(b) Show that the range of T is a plane through the origin, and find an equation for it.

22. Prove: If $\{\mathbf{v}_1, \mathbf{v}_2, \ldots, \mathbf{v}_n\}$ is a basis for V and $\mathbf{w}_1, \mathbf{w}_2, \ldots, \mathbf{w}_n$ are vectors in W, not necessarily distinct, then there exists a linear transformation $T: V \to W$ such that

$$T(\mathbf{v}_1) = \mathbf{w}_1, \quad T(\mathbf{v}_2) = \mathbf{w}_2, \ldots, \quad T(\mathbf{v}_n) = \mathbf{w}_n$$

23. Prove the dimension theorem in the cases

(a) $\dim(\ker(T)) = 0$ (b) $\dim(\ker(T)) = n$

24. (*For readers who have studied calculus.*) Let $D: P_3 \to P_2$ be the differentiation transformation $D(\mathbf{p}) = p'(x)$. Describe the kernel of D.

25. (*For readers who have studied calculus.*) Let $J: P_1 \to R$ be the integration transformation $J(\mathbf{p}) = \int_{-1}^{1} p(x)\, dx$. Describe the kernel of J.

26. (*For readers who have studied calculus.*) Let $D: V \to W$ be the differentiation transformation $D(\mathbf{f}) = f'(x)$, where $V = C^3(-\infty, \infty)$ and $W = F(-\infty, \infty)$. Describe the kernels of $D \circ D$ and $D \circ D \circ D$.

Discussion and Discovery

27. Fill in the blanks.
   (a) If $T_A: R^n \to R^m$ is multiplication by A, then the nullspace of A corresponds to the _____ of T_A, and the column space of A corresponds to the _____ of T_A.
   (b) If $T: R^3 \to R^3$ is the orthogonal projection on the plane $x + y + z = 0$, then the kernel of T is the line through the origin that is parallel to the vector _____.
   (c) If V is a finite-dimensional vector space and $T: V \to W$ is a linear transformation, then the dimension of the range of T plus the dimension of the kernel of T is _____.
   (d) If $T_A: R^5 \to R^3$ is multiplication by A, and if $\text{rank}(T_A) = 2$, then the general solution of $A\mathbf{x} = \mathbf{0}$ has _____ (how many?) parameters.

28. (a) If $T: R^3 \to R^3$ is a linear operator, and if the kernel of T is a line through the origin, then what kind of geometric object is the range of T? Explain your reasoning.
   (b) If $T: R^3 \to R^3$ is a linear operator, and if the range of T is a plane through the origin, then what kind of geometric object is the kernel of T? Explain your reasoning.

29. (*For readers who have studied calculus.*) Let V be the vector space of real-valued functions with continuous derivatives of all orders on the interval $(-\infty, \infty)$, and let $W = F(-\infty, \infty)$ be the vector space of real-valued functions defined on $(-\infty, \infty)$.
   (a) Find a linear transformation $T: V \to W$ whose kernel is P_3.
   (b) Find a linear transformation $T: V \to W$ whose kernel is P_n.

8.3 INVERSE LINEAR TRANSFORMATIONS

In Section 4.3 we discussed properties of one-to-one linear transformations from R^n to R^m. In this section we shall extend those ideas to more general kinds of linear transformations.

Recall from Section 4.3 that a linear transformation from R^n to R^m is called *one-to-one* if it maps distinct vectors in R^n into distinct vectors in R^m. The following definition generalizes that idea.

> **Definition**
>
> A linear transformation $T: V \to W$ is said to be ***one-to-one*** if T maps distinct vectors in V into distinct vectors in W.

EXAMPLE 1 A One-to-One Linear Transformation

Recall from Theorem 4.3.1 that if A is an $n \times n$ matrix and $T_A: R^n \to R^n$ is multiplication by A, then T_A is one-to-one if and only if A is an invertible matrix. $\blacklozenge$

EXAMPLE 2   A One-to-One Linear Transformation

Let $T: P_n \to P_{n+1}$ be the linear transformation

$$T(\mathbf{p}) = T(p(x)) = xp(x)$$

discussed in Example 8 of Section 8.1. If

$$\mathbf{p} = p(x) = c_0 + c_1 x + \cdots + c_n x^n \quad \text{and} \quad \mathbf{q} = q(x) = d_0 + d_1 x + \cdots + d_n x^n$$

are distinct polynomials, then they differ in at least one coefficient. Thus,

$$T(\mathbf{p}) = c_0 x + c_1 x^2 + \cdots + c_n x^{n+1} \quad \text{and} \quad T(\mathbf{q}) = d_0 x + d_1 x^2 + \cdots + d_n x^{n+1}$$

also differ in at least one coefficient. Thus, T is one-to-one, since it maps distinct polynomials $\mathbf{p}$ and $\mathbf{q}$ into distinct polynomials $T(\mathbf{p})$ and $T(\mathbf{q})$.   ◆

Calculus Required

EXAMPLE 3   A Transformation That Is Not One-to-One

Let

$$D: C^1(-\infty, \infty) \to F(-\infty, \infty)$$

be the differentiation transformation discussed in Example 11 of Section 8.1. This linear transformation is *not* one-to-one because it maps functions that differ by a constant into the same function. For example,

$$D(x^2) = D(x^2 + 1) = 2x \qquad ◆$$

The following theorem establishes a relationship between a one-to-one linear transformation and its kernel.

Theorem 8.3.1   **Equivalent Statements**

If $T: V \to W$ is a linear transformation, then the following are equivalent.

(*a*) *T is one-to-one.*
(*b*) *The kernel of T contains only the zero vector; that is, $\ker(T) = \{\mathbf{0}\}$.*
(*c*) *nullity$(T) = 0$.*

Proof. We leave it as a simple exercise to show the equivalence of (*b*) and (*c*); we shall complete the proof by proving the equivalence of (*a*) and (*b*).

(***a***) $\Rightarrow$ (***b***). Assume that T is one-to-one, and let $\mathbf{v}$ be any vector in $\ker(T)$. Since $\mathbf{v}$ and $\mathbf{0}$ both lie in $\ker(T)$, we have $T(\mathbf{v}) = \mathbf{0}$ and $T(\mathbf{0}) = \mathbf{0}$, so $T(\mathbf{v}) = T(\mathbf{0})$. But this implies that $\mathbf{v} = \mathbf{0}$, since T is one-to-one; thus, $\ker(T)$ contains only the zero vector.

(***b***) $\Rightarrow$ (***a***). Assume that $\ker(T) = \{\mathbf{0}\}$ and that $\mathbf{v}$ and $\mathbf{w}$ are distinct vectors in V; that is,

$$\mathbf{v} - \mathbf{w} \neq \mathbf{0} \tag{1}$$

To prove that T is one-to-one we must show that $T(\mathbf{v})$ and $T(\mathbf{w})$ are distinct vectors. But if this were not so, then we would have $T(\mathbf{v}) = T(\mathbf{w})$. Therefore,

$$T(\mathbf{v}) - T(\mathbf{w}) = \mathbf{0} \quad \text{or} \quad T(\mathbf{v} - \mathbf{w}) = \mathbf{0}$$

and so $\mathbf{v} - \mathbf{w}$ is in the kernel of T. Since $\ker(T) = \{\mathbf{0}\}$, this implies that $\mathbf{v} - \mathbf{w} = \mathbf{0}$, which contradicts (1). Thus, $T(\mathbf{v})$ and $T(\mathbf{w})$ must be distinct. ∎

EXAMPLE 4 Using Theorem 8.3.1

In each part determine whether the linear transformation is one-to-one by finding the kernel or the nullity and applying Theorem 8.3.1.

(a) $T: R^2 \to R^2$ rotates each vector through the angle θ.
(b) $T: R^3 \to R^3$ is the orthogonal projection on the xy-plane.
(c) $T: R^6 \to R^4$ is multiplication by the matrix

$$A = \begin{bmatrix} -1 & 2 & 0 & 4 & 5 & -3 \\ 3 & -7 & 2 & 0 & 1 & 4 \\ 2 & -5 & 2 & 4 & 6 & 1 \\ 4 & -9 & 2 & -4 & -4 & 7 \end{bmatrix}$$

Solution (a). From Example 5 of Section 8.2, $\ker(T) = \{\mathbf{0}\}$, so T is one-to-one.

Solution (b). From Example 4 of Section 8.2, $\ker(T)$ contains nonzero vectors, so T is not one-to-one.

Solution (c). From Example 7 of Section 8.2, $\text{nullity}(T) = 4$, so T is not one-to-one. ◆

In the special case where T is a *linear operator* on a *finite-dimensional* vector space, a fourth equivalent statement can be added to those in Theorem 8.3.1.

Theorem 8.3.2

If V is a finite-dimensional vector space, and $T: V \to V$ is a linear operator, then the following are equivalent.

(a) *T is one-to-one.*   (b) *$\ker(T) = \{\mathbf{0}\}$.*
(c) *$\text{nullity}(T) = 0$.*   (d) *The range of T is V; that is, $R(T) = V$.*

Proof. We already know that (a), (b), and (c) are equivalent, so we can complete the proof by proving the equivalence of (c) and (d).

(c) ⇒ (d). Suppose that $\dim(V) = n$ and $\text{nullity}(T) = 0$. It follows from the dimension theorem (Theorem 8.2.3) that

$$\text{rank}(T) = n - \text{nullity}(T) = n$$

By definition, $\text{rank}(T)$ is the dimension of the range of T, so the range of T has dimension n. It now follows from Theorem 5.4.7 that the range of T is V, since the two spaces have the same dimension.

(d) ⇒ (c). Suppose that $\dim(V) = n$ and $R(T) = V$. It follows from these relationships that $\dim(R(T)) = n$, or equivalently, $\text{rank}(T) = n$. Thus, it follows from the dimension theorem (Theorem 8.2.3) that

$$\text{nullity}(T) = n - \text{rank}(T) = n - n = 0$$ ∎

EXAMPLE 5   A Transformation That Is Not One-To-One

Let $T_A: R^4 \to R^4$ be multiplication by

$$A = \begin{bmatrix} 1 & 3 & -2 & 4 \\ 2 & 6 & -4 & 8 \\ 3 & 9 & 1 & 5 \\ 1 & 1 & 4 & 8 \end{bmatrix}$$

Determine whether T_A is one-to-one.

Solution.

As noted in Example 1, the given problem is equivalent to determining whether A is invertible. But $\det(A) = 0$, since the first two rows of A are proportional and consequently A is not invertible. Thus, T_A is not one-to-one.   ◆

Inverse Linear Transformations

In Section 4.3 we defined the *inverse* of a one-to-one matrix operator $T_A: R^n \to R^n$ to be $T_{A^{-1}}: R^n \to R^n$, and we showed that if $\mathbf{w}$ is the image of a vector $\mathbf{x}$ under T_A, then T_A^{-1} maps $\mathbf{w}$ back into $\mathbf{x}$. We shall now extend these ideas to general linear transformations.

Recall that if $T: V \to W$ is a linear transformation, then the range of T, denoted by $R(T)$, is the subspace of W consisting of all images under T of vectors in V. If T is one-to-one, then each vector $\mathbf{w}$ in $R(T)$ is the image of a unique vector $\mathbf{v}$ in V. This uniqueness allows us to define a new function, called the ***inverse of*** T, and denoted by T^{-1}, which maps $\mathbf{w}$ back into $\mathbf{v}$ (Figure 8.3.1).

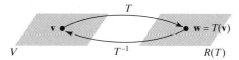

Figure 8.3.1 The inverse of T maps $T(\mathbf{v})$ back into $\mathbf{v}$.

It can be proved (Exercise 19) that $T^{-1}: R(T) \to V$ is a linear transformation. Moreover, it follows from the definition of T^{-1} that

$$T^{-1}(T(\mathbf{v})) = T^{-1}(\mathbf{w}) = \mathbf{v} \tag{2a}$$

$$T(T^{-1}(\mathbf{w})) = T(\mathbf{v}) = \mathbf{w} \tag{2b}$$

so that T and T^{-1}, when applied in succession in either order, cancel the effect of one another.

REMARK.   It is important to note that if $T: V \to W$ is a one-to-one linear transformation, then the domain of T^{-1} is the *range* of T. The range may or may not be all of W. However, in the special case where $T: V \to V$ is a one-to-one linear operator it follows from Theorem 8.3.2 that $R(T) = V$; that is, the domain of T^{-1} is *all* of V.

EXAMPLE 6 An Inverse Transformation

In Example 2 we showed that the linear transformation $T: P_n \to P_{n+1}$ given by

$$T(\mathbf{p}) = T(p(x)) = xp(x)$$

is one-to-one; thus, T has an inverse. Here, the range of T is *not* all of P_{n+1}; rather, $R(T)$ is the subspace of P_{n+1} consisting of polynomials with a zero constant term. This is evident from the formula for T:

$$T(c_0 + c_1 x + \cdots + c_n x^n) = c_0 x + c_1 x^2 + \cdots + c_n x^{n+1}$$

It follows that $T^{-1}: R(T) \to P_n$ is given by the formula

$$T^{-1}(c_0 x + c_1 x^2 + \cdots + c_n x^{n+1}) = c_0 + c_1 x + \cdots + c_n x^n$$

For example, in the case where $n \geq 3$,

$$T^{-1}(2x - x^2 + 5x^3 + 3x^4) = 2 - x + 5x^2 + 3x^3 \qquad \blacklozenge$$

EXAMPLE 7 An Inverse Transformation

Let $T: R^3 \to R^3$ be the linear operator defined by the formula

$$T(x_1, x_2, x_3) = (3x_1 + x_2, -2x_1 - 4x_2 + 3x_3, 5x_1 + 4x_2 - 2x_3)$$

Determine whether T is one-to-one; if so, find $T^{-1}(x_1, x_2, x_3)$.

Solution.

From Theorem 4.3.3 the standard matrix for T is

$$[T] = \begin{bmatrix} 3 & 1 & 0 \\ -2 & -4 & 3 \\ 5 & 4 & -2 \end{bmatrix}$$

(verify). This matrix is invertible and from Formula (1) of Section 4.3 the standard matrix for T^{-1} is

$$[T^{-1}] = [T]^{-1} = \begin{bmatrix} 4 & -2 & -3 \\ -11 & 6 & 9 \\ -12 & 7 & 10 \end{bmatrix}$$

It follows that

$$T^{-1}\left(\begin{bmatrix} x_1 \\ x_2 \\ x_3 \end{bmatrix}\right) = [T^{-1}]\begin{bmatrix} x_1 \\ x_2 \\ x_3 \end{bmatrix} = \begin{bmatrix} 4 & -2 & -3 \\ -11 & 6 & 9 \\ -12 & 7 & 10 \end{bmatrix}\begin{bmatrix} x_1 \\ x_2 \\ x_3 \end{bmatrix} = \begin{bmatrix} 4x_1 - 2x_2 - 3x_3 \\ -11x_1 + 6x_2 + 9x_3 \\ -12x_1 + 7x_2 + 10x_3 \end{bmatrix}$$

Expressing this result in horizontal notation yields

$$T^{-1}(x_1, x_2, x_3) = (4x_1 - 2x_2 - 3x_3, -11x_1 + 6x_2 + 9x_3, -12x_1 + 7x_2 + 10x_3) \qquad \blacklozenge$$

The following theorem shows that a composition of one-to-one linear transformations is one-to-one, and it relates the inverse of the composition to the inverses of the individual linear transformations.

Theorem 8.3.3

If $T_1: U \to V$ and $T_2: V \to W$ are one-to-one linear transformations, then:

(a) $T_2 \circ T_1$ is one-to-one.   *(b) $(T_2 \circ T_1)^{-1} = T_1^{-1} \circ T_2^{-1}$.*

Proof (a). We want to show that $T_2 \circ T_1$ maps distinct vectors in U into distinct vectors in W. But if **u** and **v** are distinct vectors in U, then $T_1(\mathbf{u})$ and $T_1(\mathbf{v})$ are distinct vectors in V since T_1 is one-to-one. This and the fact that T_2 is one-to-one implies that

$$T_2(T_1(\mathbf{u})) \quad \text{and} \quad T_2(T_1(\mathbf{v}))$$

are also distinct vectors. But these expressions can also be written as

$$(T_2 \circ T_1)(\mathbf{u}) \quad \text{and} \quad (T_2 \circ T_1)(\mathbf{v})$$

so $T_2 \circ T_1$ maps **u** and **v** into distinct vectors in W.

Proof (b). We want to show that

$$(T_2 \circ T_1)^{-1}(\mathbf{w}) = (T_1^{-1} \circ T_2^{-1})(\mathbf{w})$$

for every vector **w** in the range of $T_2 \circ T_1$. For this purpose let

$$\mathbf{u} = (T_2 \circ T_1)^{-1}(\mathbf{w}) \tag{3}$$

so our goal is to show that

$$\mathbf{u} = (T_1^{-1} \circ T_2^{-1})(\mathbf{w})$$

But it follows from (3) that

$$(T_2 \circ T_1)(\mathbf{u}) = \mathbf{w}$$

or equivalently,

$$T_2(T_1(\mathbf{u})) = \mathbf{w}$$

Now, taking T_2^{-1} of each side of this equation and then T_1^{-1} of each side of the result yields (verify)

$$\mathbf{u} = T_1^{-1}(T_2^{-1}(\mathbf{w}))$$

or equivalently,

$$\mathbf{u} = (T_1^{-1} \circ T_2^{-1})(\mathbf{w}) \qquad\qquad \blacksquare$$

In words, part (*b*) of Theorem 8.3.3 states that *the inverse of a composition is the composition of the inverse in the reverse order.* This result can be extended to compositions of three or more linear transformations; for example,

$$(T_3 \circ T_2 \circ T_1)^{-1} = T_1^{-1} \circ T_2^{-1} \circ T_3^{-1} \tag{4}$$

In the special case where T_A, T_B, and T_C are matrix operators on R^n, Formula (4) can be written as

$$(T_C \circ T_B \circ T_A)^{-1} = T_A^{-1} \circ T_B^{-1} \circ T_C^{-1}$$

or equivalently,

$$(T_{CBA})^{-1} = T_{A^{-1}B^{-1}C^{-1}} \tag{5}$$

In words, this formula states that *the standard matrix for the inverse of a composition is the product of the inverses of the standard matrices of the individual operators in the reverse order.*

Some problems that use Formula (5) are given in the exercises.

Exercise Set 8.3

1. In each part find $\ker(T)$, and determine whether the linear transformation T is one-to-one.

   (a) $T: R^2 \to R^2$, where $T(x, y) = (y, x)$
   (b) $T: R^2 \to R^2$, where $T(x, y) = (0, 2x + 3y)$
   (c) $T: R^2 \to R^2$, where $T(x, y) = (x + y, x - y)$
   (d) $T: R^2 \to R^3$, where $T(x, y) = (x, y, x + y)$
   (e) $T: R^2 \to R^3$, where $T(x, y) = (x - y, y - x, 2x - 2y)$
   (f) $T: R^3 \to R^2$, where $T(x, y, z) = (x + y + z, x - y - z)$

2. In each part let $T: R^2 \to R^2$ be multiplication by A. Determine whether T has an inverse; if so, find

$$T^{-1}\left(\begin{bmatrix} x_1 \\ x_2 \end{bmatrix}\right)$$

   (a) $A = \begin{bmatrix} 5 & 2 \\ 2 & 1 \end{bmatrix}$
   (b) $A = \begin{bmatrix} 6 & -3 \\ 4 & -2 \end{bmatrix}$
   (c) $A = \begin{bmatrix} 4 & 7 \\ -1 & 3 \end{bmatrix}$

3. In each part let $T: R^3 \to R^3$ be multiplication by A. Determine whether T has an inverse; if so, find

$$T^{-1}\left(\begin{bmatrix} x_1 \\ x_2 \\ x_3 \end{bmatrix}\right)$$

   (a) $A = \begin{bmatrix} 1 & 5 & 2 \\ 1 & 2 & 1 \\ -1 & 1 & 0 \end{bmatrix}$
   (b) $A = \begin{bmatrix} 1 & 4 & -1 \\ 1 & 2 & 1 \\ -1 & 1 & 0 \end{bmatrix}$
   (c) $A = \begin{bmatrix} 1 & 0 & 1 \\ 0 & 1 & 1 \\ 1 & 1 & 0 \end{bmatrix}$
   (d) $A = \begin{bmatrix} 1 & -1 & 1 \\ 0 & 2 & -1 \\ 2 & 3 & 0 \end{bmatrix}$

4. In each part determine whether multiplication by A is a one-to-one linear transformation.

   (a) $A = \begin{bmatrix} 1 & -2 \\ 2 & -4 \\ -3 & 6 \end{bmatrix}$
   (b) $A = \begin{bmatrix} 1 & 3 & 5 & 7 \\ 2 & -1 & 2 & 4 \\ -1 & 3 & 0 & 0 \end{bmatrix}$
   (c) $A = \begin{bmatrix} 4 & -2 \\ 1 & 5 \\ 5 & 3 \end{bmatrix}$

5. As indicated in the accompanying figure, let $T: R^2 \to R^2$ be the orthogonal projection on the line $y = x$.

   (a) Find the kernel of T.
   (b) Is T one-to-one? Justify your conclusion.

6. As indicated in the accompanying figure, let $T: R^2 \to R^2$ be the linear operator that reflects each point about the y-axis.

   (a) Find the kernel of T.
   (b) Is T one-to-one? Justify your conclusion.

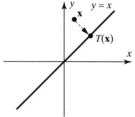

Figure Ex-5

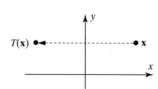

Figure Ex-6

7. In each part use the given information to determine whether the linear transformation T is one-to-one.

   (a) $T: R^m \to R^m$; $\text{nullity}(T) = 0$
   (b) $T: R^n \to R^n$; $\text{rank}(T) = n - 1$
   (c) $T: R^m \to R^n$; $n < m$
   (d) $T: R^n \to R^n$; $R(T) = R^n$

8. In each part determine whether the linear transformation T is one-to-one.

   (a) $T: P_2 \to P_3$, where $T(a_0 + a_1 x + a_2 x^2) = x(a_0 + a_1 x + a_2 x^2)$
   (b) $T: P_2 \to P_2$, where $T(p(x)) = p(x + 1)$

9. Let A be a square matrix such that $\det(A) = 0$. Is multiplication by A a one-to-one linear transformation? Justify your conclusion.

10. In each part determine whether the linear operator $T: R^n \to R^n$ is one-to-one; if so, find $T^{-1}(x_1, x_2, \ldots, x_n)$.

    (a) $T(x_1, x_2, \ldots, x_n) = (0, x_1, x_2, \ldots, x_{n-1})$ (b) $T(x_1, x_2, \ldots, x_n) = (x_n, x_{n-1}, \ldots, x_2, x_1)$

    (c) $T(x_1, x_2, \ldots, x_n) = (x_2, x_3, \ldots, x_n, x_1)$

11. Let $T: R^n \to R^n$ be the linear operator defined by the formula

$$T(x_1, x_2, \ldots, x_n) = (a_1 x_1, a_2 x_2, \ldots, a_n x_n)$$

    (a) Under what conditions will T have an inverse?

    (b) Assuming that the conditions determined in part (a) are satisfied, find a formula for $T^{-1}(x_1, x_2, \ldots, x_n)$.

12. Let $T_1: R^2 \to R^2$ and $T_2: R^2 \to R^2$ be the linear operators given by the formulas

$$T_1(x, y) = (x + y, x - y) \quad \text{and} \quad T_2(x, y) = (2x + y, x - 2y)$$

    (a) Show that T_1 and T_2 are one-to-one.

    (b) Find formulas for $T_1^{-1}(x, y)$, and $T_2^{-1}(x, y)$, and $(T_2 \circ T_1)^{-1}(x, y)$.

    (c) Verify that $(T_2 \circ T_1)^{-1} = T_1^{-1} \circ T_2^{-1}$.

13. Let $T_1: P_2 \to P_3$ and $T_2: P_3 \to P_3$ be the linear transformations given by the formulas

$$T_1(p(x)) = xp(x) \quad \text{and} \quad T_2(p(x)) = p(x + 1)$$

    (a) Find formulas for $T_1^{-1}(p(x))$, $T_2^{-1}(p(x))$, and $(T_2 \circ T_1)^{-1}(p(x))$.

    (b) Verify that $(T_2 \circ T_1)^{-1} = T_1^{-1} \circ T_2^{-1}$.

14. Let $T_A: R^3 \to R^3$, $T_B: R^3 \to R^3$, and $T_C: R^3 \to R^3$ be the reflections about the xy-plane, the xz-plane, and the yz-plane, respectively. Verify Formula (5) for these linear operators.

15. Let $T: P_1 \to R^2$ be the function defined by the formula

$$T(p(x)) = (p(0), p(1))$$

    (a) Find $T(1 - 2x)$.   (b) Show that T is a linear transformation.

    (c) Show that T is one-to-one.   (d) Find $T^{-1}(2, 3)$, and sketch its graph.

16. Prove: If V and W are finite-dimensional vector spaces such that $\dim W < \dim V$, then there is no one-to-one linear transformation $T: V \to W$.

17. In each part determine whether the linear operator $T: M_{22} \to M_{22}$ is one-to-one. If so, find

$$T^{-1}\left(\begin{bmatrix} a & b \\ c & d \end{bmatrix}\right)$$

    (a) $T\left(\begin{bmatrix} a & b \\ c & d \end{bmatrix}\right) = \begin{bmatrix} a & 0 \\ 0 & d \end{bmatrix}$ (b) $T\left(\begin{bmatrix} a & b \\ c & d \end{bmatrix}\right) = \begin{bmatrix} a & c \\ b & d \end{bmatrix}$ (c) $T\left(\begin{bmatrix} a & b \\ c & d \end{bmatrix}\right) = \begin{bmatrix} d & -b \\ -c & a \end{bmatrix}$

18. Let $T: R^2 \to R^2$ be the linear operator given by the formula $T(x, y) = (x + ky, -y)$. Show that T is one-to-one for every real value of k and that $T^{-1} = T$.

19. Prove that if $T: V \to W$ is a one-to-one linear transformation, then $T^{-1}: R(T) \to V$ is a linear transformation.

20. (*For readers who have studied calculus.*) Let $J: P_1 \to R$ be the integration transformation $J(\mathbf{p}) = \int_{-1}^{1} p(x)\, dx$. Determine whether J is one-to-one. Justify your conclusion.

21. (*For readers who have studied calculus.*) Let V be the vector space $C^1[0, 1]$ and let $T: V \to R$ be defined by $T(\mathbf{f}) = f(0) + 2f'(0) + 3f'(1)$. Verify that T is a linear transformation. Determine whether T is one-to-one. Justify your conclusion.

Discussion and Discovery

22. Indicate whether the statement is always true or sometimes false. Justify your answer by giving a logical argument or a counterexample.

(a) If $T: R^2 \rightarrow R^2$ is the orthogonal projection onto the x-axis, then $T^{-1}: R^2 \rightarrow R^2$ maps each point on the x-axis onto a line that is perpendicular to the x-axis.

(b) If $T_1: U \rightarrow V$ and $T_2: V \rightarrow W$ are linear transformations and if T_1 is not one-to-one, then neither is $T_2 \circ T_1$.

(c) In the xy-plane, a rotation about the origin followed by a reflection about a coordinate axis is one-to-one.

23. Do you think that the formula $T(a, b, c) = ax^2 + bx + c$ defines a one-to-one linear transformation from R^3 to P_2? Explain your reasoning.

24. Let E be a fixed 2×2 elementary matrix. Do you think that the formula $T(A) = EA$ defines a one-to-one linear operator on M_{22}? Explain your reasoning.

25. Let $\mathbf{a}$ be a fixed vector in R^3. Do you think that the formula $T(\mathbf{v}) = \mathbf{a} \times \mathbf{v}$ defines a one-to-one linear operator on R^3? Explain your reasoning.

8.4 MATRICES OF GENERAL LINEAR TRANSFORMATIONS

In this section we shall show that if V and W are finite-dimensional vector spaces (not necessarily R^n and R^m), then with a little ingenuity any linear transformation $T: V \rightarrow W$ can be regarded as a matrix transformation. The basic idea is to work with coordinate matrices of the vectors rather than with the vectors themselves.

Matrices of Linear Transformations   Suppose that V is an n-dimensional vector space and W an m-dimensional vector space. If we choose bases B and B' for V and W, respectively, then for each $\mathbf{x}$ in V, the coordinate matrix $[\mathbf{x}]_B$ will be a vector in R^n, and the coordinate matrix $[T(\mathbf{x})]_{B'}$ will be a vector in R^m (Figure 8.4.1).

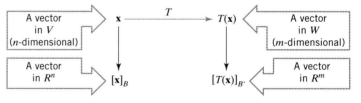

Figure 8.4.1

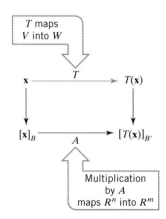

Figure 8.4.2

If, as illustrated in Figure 8.4.2, we complete the rectangle suggested by Figure 8.4.1, we obtain a mapping from R^n to R^m, which can be shown to be a linear transformation. If we let A be the standard matrix for this transformation, then

$$A[\mathbf{x}]_B = [T(\mathbf{x})]_{B'} \tag{1}$$

The matrix A in (1) is called the ***matrix for T with respect to the bases B and B'***.

Later in this section, we shall give some of the uses of the matrix A in (1), but first, let us show how it can be computed. For this purpose, let $B = \{\mathbf{u}_1, \mathbf{u}_2, \ldots, \mathbf{u}_n\}$ be a basis for the n-dimensional space V and $B' = \{\mathbf{v}_1, \mathbf{v}_2, \ldots, \mathbf{v}_m\}$ a basis for the m-dimensional space W. We are looking for an $m \times n$ matrix

$$
A = \begin{bmatrix}
a_{11} & a_{12} & \cdots & a_{1n} \\
a_{21} & a_{22} & \cdots & a_{2n} \\
\vdots & \vdots & & \vdots \\
a_{m1} & a_{m2} & \cdots & a_{mn}
\end{bmatrix}
$$

such that (1) holds for all vectors $\mathbf{x}$ in V. In particular, we want this equation to hold for the basis vectors $\mathbf{u}_1, \mathbf{u}_2, \ldots, \mathbf{u}_n$; that is,

$$
A[\mathbf{u}_1]_B = [T(\mathbf{u}_1)]_{B'}, \quad A[\mathbf{u}_2]_B = [T(\mathbf{u}_2)]_{B'}, \ldots, \quad A[\mathbf{u}_n]_B = [T(\mathbf{u}_n)]_{B'} \tag{2}
$$

But

$$
[\mathbf{u}_1]_B = \begin{bmatrix} 1 \\ 0 \\ 0 \\ \vdots \\ 0 \end{bmatrix}, \quad [\mathbf{u}_2]_B = \begin{bmatrix} 0 \\ 1 \\ 0 \\ \vdots \\ 0 \end{bmatrix}, \ldots, \quad [\mathbf{u}_n]_B = \begin{bmatrix} 0 \\ 0 \\ 0 \\ \vdots \\ 1 \end{bmatrix}
$$

so

$$
A[\mathbf{u}_1]_B = \begin{bmatrix}
a_{11} & a_{12} & \cdots & a_{1n} \\
a_{21} & a_{22} & \cdots & a_{2n} \\
\vdots & \vdots & & \vdots \\
a_{m1} & a_{m2} & \cdots & a_{mn}
\end{bmatrix} \begin{bmatrix} 1 \\ 0 \\ 0 \\ \vdots \\ 0 \end{bmatrix} = \begin{bmatrix} a_{11} \\ a_{21} \\ \vdots \\ a_{m1} \end{bmatrix}
$$

$$
A[\mathbf{u}_2]_B = \begin{bmatrix}
a_{11} & a_{12} & \cdots & a_{1n} \\
a_{21} & a_{22} & \cdots & a_{2n} \\
\vdots & \vdots & & \vdots \\
a_{m1} & a_{m2} & \cdots & a_{mn}
\end{bmatrix} \begin{bmatrix} 0 \\ 1 \\ 0 \\ \vdots \\ 0 \end{bmatrix} = \begin{bmatrix} a_{12} \\ a_{22} \\ \vdots \\ a_{m2} \end{bmatrix}
$$

$$
\vdots
$$

$$
A[\mathbf{u}_n]_B = \begin{bmatrix}
a_{11} & a_{12} & \cdots & a_{1n} \\
a_{21} & a_{22} & \cdots & a_{2n} \\
\vdots & \vdots & & \vdots \\
a_{m1} & a_{m2} & \cdots & a_{mn}
\end{bmatrix} \begin{bmatrix} 0 \\ 0 \\ 0 \\ \vdots \\ 1 \end{bmatrix} = \begin{bmatrix} a_{1n} \\ a_{2n} \\ \vdots \\ a_{mn} \end{bmatrix}
$$

Substituting these results into (2) yields

$$
\begin{bmatrix} a_{11} \\ a_{21} \\ \vdots \\ a_{m1} \end{bmatrix} = [T(\mathbf{u}_1)]_{B'}, \quad \begin{bmatrix} a_{12} \\ a_{22} \\ \vdots \\ a_{m2} \end{bmatrix} = [T(\mathbf{u}_2)]_{B'}, \ldots, \quad \begin{bmatrix} a_{1n} \\ a_{2n} \\ \vdots \\ a_{mn} \end{bmatrix} = [T(\mathbf{u}_n)]_{B'}
$$

which shows that the successive columns of A are the coordinate matrices of

$$
T(\mathbf{u}_1), T(\mathbf{u}_2), \ldots, T(\mathbf{u}_n)
$$

with respect to the basis B'. Thus, the matrix for T with respect to the bases B and B' is

$$
A = \left[[T(\mathbf{u}_1)]_{B'} \mid [T(\mathbf{u}_2)]_{B'} \mid \cdots \mid [T(\mathbf{u}_n)]_{B'} \right] \tag{3}
$$

This matrix is commonly denoted by the symbol

$$
[T]_{B',B}
$$

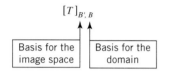

Figure 8.4.3

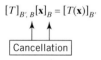

Figure 8.4.4

so that the preceding formula can also be written as

$$[T]_{B',B} = \left[[T(\mathbf{u}_1)]_{B'} \mid [T(\mathbf{u}_2)]_{B'} \mid \cdots \mid [T(\mathbf{u}_n)]_{B'} \right] \tag{4}$$

and from (1) this matrix has the property

$$[T]_{B',B}[\mathbf{x}]_B = [T(\mathbf{x})]_{B'} \tag{4a}$$

REMARK. Observe that in the notation $[T]_{B',B}$ the right subscript is a basis for the domain of T, and the left subscript is a basis for the image space of T (Figure 8.4.3). Moreover, observe how the subscript B seems to "cancel out" in Formula (4a) (Figure 8.4.4).

Matrices of Linear Operators

In the special case where $V = W$ (so that $T: V \to V$ is a linear operator) it is usual to take $B = B'$ when constructing a matrix for T. In this case the resulting matrix is called the **matrix for T with respect to the basis B** and is usually denoted by $[T]_B$ rather than $[T]_{B,B}$. If $B = \{\mathbf{u}_1, \mathbf{u}_2, \ldots, \mathbf{u}_n\}$, then in this case Formulas (4) and (4a) become

$$[T]_B = \left[[T(\mathbf{u}_1)]_B \mid [T(\mathbf{u}_2)]_B \mid \cdots \mid [T(\mathbf{u}_n)]_B \right] \tag{5}$$

and

$$[T]_B[\mathbf{x}]_B = [T(\mathbf{x})]_B \tag{5a}$$

Phrased informally, (4a) and (5a) state that *the matrix for T times the coordinate matrix for $\mathbf{x}$ is the coordinate matrix for $T(\mathbf{x})$.*

EXAMPLE 1 Matrix for a Linear Transformation

Let $T: P_1 \to P_2$ be the linear transformation defined by

$$T(p(x)) = xp(x)$$

Find the matrix for T with respect to the standard bases

$$B = \{\mathbf{u}_1, \mathbf{u}_2\} \quad \text{and} \quad B' = \{\mathbf{v}_1, \mathbf{v}_2, \mathbf{v}_3\}$$

where

$$\mathbf{u}_1 = 1, \quad \mathbf{u}_2 = x; \quad \mathbf{v}_1 = 1, \quad \mathbf{v}_2 = x, \quad \mathbf{v}_3 = x^2$$

Solution.

From the given formula for T we obtain

$$T(\mathbf{u}_1) = T(1) = (x)(1) = x$$
$$T(\mathbf{u}_2) = T(x) = (x)(x) = x^2$$

By inspection, we can determine the coordinate matrices for $T(\mathbf{u}_1)$ and $T(\mathbf{u}_2)$ relative to B'; they are

$$[T(\mathbf{u}_1)]_{B'} = \begin{bmatrix} 0 \\ 1 \\ 0 \end{bmatrix}, \qquad [T(\mathbf{u}_2)]_{B'} = \begin{bmatrix} 0 \\ 0 \\ 1 \end{bmatrix}$$

Thus, the matrix for T with respect to B and B' is

$$[T]_{B',B} = \left[[T(\mathbf{u}_1)]_{B'} \mid [T(\mathbf{u}_2)]_{B'} \right] = \begin{bmatrix} 0 & 0 \\ 1 & 0 \\ 0 & 1 \end{bmatrix} \qquad \blacklozenge$$

EXAMPLE 2 Verifying Formula (4a)

Let $T: P_1 \rightarrow P_2$ be the linear transformation in Example 1. Show that the matrix

$$[T]_{B',B} = \begin{bmatrix} 0 & 0 \\ 1 & 0 \\ 0 & 1 \end{bmatrix}$$

(obtained in Example 1) satisfies (4a) for every vector $\mathbf{x} = a + bx$ in P_1.

Solution.

Since $\mathbf{x} = p(x) = a + bx$, we have

$$T(\mathbf{x}) = xp(x) = ax + bx^2$$

For the bases B and B' in Example 1, it follows by inspection that

$$[\mathbf{x}]_B = [ax + b]_B = \begin{bmatrix} a \\ b \end{bmatrix} \quad \text{and} \quad [T(\mathbf{x})]_{B'} = [ax + bx^2] = \begin{bmatrix} 0 \\ a \\ b \end{bmatrix}$$

Thus,

$$[T]_{B',B}[\mathbf{x}]_B = \begin{bmatrix} 0 & 0 \\ 1 & 0 \\ 0 & 1 \end{bmatrix} \begin{bmatrix} a \\ b \end{bmatrix} = \begin{bmatrix} 0 \\ a \\ b \end{bmatrix} = [T(\mathbf{x})]_{B'}$$

so (4a) holds. ◆

EXAMPLE 3 Matrix for a Linear Transformation

Let $T: R^2 \rightarrow R^3$ be the linear transformation defined by

$$T\left(\begin{bmatrix} x_1 \\ x_2 \end{bmatrix}\right) = \begin{bmatrix} x_2 \\ -5x_1 + 13x_2 \\ -7x_1 + 16x_2 \end{bmatrix}$$

Find the matrix for the transformation T with respect to the bases $B = \{\mathbf{u}_1, \mathbf{u}_2\}$ for R^2 and $B' = \{\mathbf{v}_1, \mathbf{v}_2, \mathbf{v}_3\}$ for R^3, where

$$\mathbf{u}_1 = \begin{bmatrix} 3 \\ 1 \end{bmatrix}, \quad \mathbf{u}_2 = \begin{bmatrix} 5 \\ 2 \end{bmatrix}; \quad \mathbf{v}_1 = \begin{bmatrix} 1 \\ 0 \\ -1 \end{bmatrix}, \quad \mathbf{v}_2 = \begin{bmatrix} -1 \\ 2 \\ 2 \end{bmatrix}, \quad \mathbf{v}_3 = \begin{bmatrix} 0 \\ 1 \\ 2 \end{bmatrix}$$

Solution.

From the formula for T,

$$T(\mathbf{u}_1) = \begin{bmatrix} 1 \\ -2 \\ -5 \end{bmatrix}, \qquad T(\mathbf{u}_2) = \begin{bmatrix} 2 \\ 1 \\ -3 \end{bmatrix}$$

Expressing these vectors as linear combinations of $\mathbf{v}_1$, $\mathbf{v}_2$, and $\mathbf{v}_3$ we obtain (verify)

$$T(\mathbf{u}_1) = \mathbf{v}_1 - 2\mathbf{v}_3, \qquad T(\mathbf{u}_2) = 3\mathbf{v}_1 + \mathbf{v}_2 - \mathbf{v}_3$$

Thus,

$$[T(\mathbf{u}_1)]_{B'} = \begin{bmatrix} 1 \\ 0 \\ -2 \end{bmatrix}, \qquad [T(\mathbf{u}_2)]_{B'} = \begin{bmatrix} 3 \\ 1 \\ -1 \end{bmatrix}$$

so

$$[T]_{B',B} = \big[[T(\mathbf{u}_1)]_{B'} \mid [T(\mathbf{u}_2)]_{B'}\big] = \begin{bmatrix} 1 & 3 \\ 0 & 1 \\ -2 & -1 \end{bmatrix} \qquad \blacklozenge$$

EXAMPLE 4 Verifying Formula (5a)

Let $T: R^2 \to R^2$ be the linear operator defined by

$$T\left(\begin{bmatrix} x_1 \\ x_2 \end{bmatrix}\right) = \begin{bmatrix} x_1 + x_2 \\ -2x_1 + 4x_2 \end{bmatrix}$$

and let $B = \{\mathbf{u}_1, \mathbf{u}_2\}$ be the basis, where

$$\mathbf{u}_1 = \begin{bmatrix} 1 \\ 1 \end{bmatrix}, \qquad \mathbf{u}_2 = \begin{bmatrix} 1 \\ 2 \end{bmatrix}$$

(a) Find $[T]_B$.
(b) Verify that (5a) holds for every vector $\mathbf{x}$ in R^2.

Solution (*a*). From the given formula for T,

$$T(\mathbf{u}_1) = \begin{bmatrix} 2 \\ 2 \end{bmatrix} = 2\mathbf{u}_1, \qquad T(\mathbf{u}_2) = \begin{bmatrix} 3 \\ 6 \end{bmatrix} = 3\mathbf{u}_2$$

Therefore,

$$[T(\mathbf{u}_1)]_B = \begin{bmatrix} 2 \\ 0 \end{bmatrix} \quad \text{and} \quad [T(\mathbf{u}_2)]_B = \begin{bmatrix} 0 \\ 3 \end{bmatrix}$$

Consequently,

$$[T]_B = \big[[T(\mathbf{u}_1)]_B \mid [T(\mathbf{u}_2)]_B\big] = \begin{bmatrix} 2 & 0 \\ 0 & 3 \end{bmatrix}$$

Solution (*b*). If

$$\mathbf{x} = \begin{bmatrix} x_1 \\ x_2 \end{bmatrix} \tag{6}$$

is any vector in R^2, then from the given formula for T

$$T(\mathbf{x}) = \begin{bmatrix} x_1 + x_2 \\ -2x_1 + 4x_2 \end{bmatrix} \tag{7}$$

To find $[\mathbf{x}]_B$ and $[T(\mathbf{x})]_B$, we must express (6) and (7) as linear combinations of $\mathbf{u}_1$ and $\mathbf{u}_2$. This yields the vector equations

$$\begin{bmatrix} x_1 \\ x_2 \end{bmatrix} = k_1 \begin{bmatrix} 1 \\ 1 \end{bmatrix} + k_2 \begin{bmatrix} 1 \\ 2 \end{bmatrix} \tag{8}$$

$$\begin{bmatrix} x_1 + x_2 \\ -2x_1 + 4x_2 \end{bmatrix} = c_1 \begin{bmatrix} 1 \\ 1 \end{bmatrix} + c_2 \begin{bmatrix} 1 \\ 2 \end{bmatrix} \tag{9}$$

Equating corresponding entries yields the linear systems

$$\begin{aligned} k_1 + \ k_2 &= x_1 \\ k_1 + 2k_2 &= x_2 \end{aligned}$$ (10)

and

$$\begin{aligned} c_1 + \ c_2 &= \ x_1 + \ x_2 \\ c_1 + 2c_2 &= -2x_1 + 4x_2 \end{aligned}$$ (11)

Solving (10) for k_1 and k_2 yields

$$k_1 = 2x_1 - x_2, \qquad k_2 = -x_1 + x_2$$

so that

$$[\mathbf{x}]_B = \begin{bmatrix} 2x_1 - x_2 \\ -x_1 + x_2 \end{bmatrix}$$

and solving (11) for c_1 and c_2 yields

$$c_1 = 4x_1 - 2x_2, \qquad c_2 = -3x_1 + 3x_2$$

so that

$$[T(\mathbf{x})]_B = \begin{bmatrix} 4x_1 - 2x_2 \\ -3x_1 + 3x_2 \end{bmatrix}$$

Thus,

$$[T]_B[\mathbf{x}]_B = \begin{bmatrix} 2 & 0 \\ 0 & 3 \end{bmatrix} \begin{bmatrix} 2x_1 - x_2 \\ -x_1 + x_2 \end{bmatrix} = \begin{bmatrix} 4x_1 - 2x_2 \\ -3x_1 + 3x_2 \end{bmatrix} = [T(\mathbf{x})]_B$$

so (5a) holds. ◆

Matrices of Identity Operators

EXAMPLE 5 Matrices of Identity Operators

If $B = \{\mathbf{u}_1, \mathbf{u}_2, \ldots, \mathbf{u}_n\}$ is a basis for a finite-dimensional vector space V and $I : V \to V$ is the identity operator on V, then

$$I(\mathbf{u}_1) = \mathbf{u}_1, \quad I(\mathbf{u}_2) = \mathbf{u}_2, \ldots, \quad I(\mathbf{u}_n) = \mathbf{u}_n$$

Therefore,

$$[I(\mathbf{u}_1)]_B = \begin{bmatrix} 1 \\ 0 \\ 0 \\ \vdots \\ 0 \end{bmatrix}, \quad [I(\mathbf{u}_2)]_B = \begin{bmatrix} 0 \\ 1 \\ 0 \\ \vdots \\ 0 \end{bmatrix}, \ldots, \quad [I(\mathbf{u}_n)]_B = \begin{bmatrix} 0 \\ 0 \\ 0 \\ \vdots \\ 1 \end{bmatrix}$$

Thus,

$$[I]_B = \begin{bmatrix} 1 & 0 & \cdots & 0 \\ 0 & 1 & \cdots & 0 \\ 0 & 0 & \cdots & 0 \\ \vdots & \vdots & & \vdots \\ 0 & 0 & \cdots & 1 \end{bmatrix} = I$$

Consequently, the matrix of the identity operator with respect to any basis is the $n \times n$ identity matrix. This result could have been anticipated from Formula (5a), since the formula yields

$$[I]_B[\mathbf{x}]_B = [I(\mathbf{x})]_B = [\mathbf{x}]_B$$

which is consistent with the fact that $[I]_B = I$. ◆

We leave it as an exercise to prove the following result.

Theorem 8.4.1

If $T: R^n \to R^m$ is a linear transformation and if B and B' are the standard bases for R^n and R^m, respectively, then

$$[T]_{B', B} = [T] \tag{12}$$

This theorem tells us that in the special case where T maps R^n into R^m, the matrix for T with respect to the standard bases is the standard matrix for T. In this special case Formula (4a) of this section reduces to

$$[T]\mathbf{x} = T(\mathbf{x})$$

Why Matrices of Linear Transformations Are Important

There are two primary reasons for studying matrices for general linear transformations, one theoretical and the other quite practical:

- Answers to theoretical questions about the structure of general linear transformations on finite-dimensional vector spaces can often be obtained by studying just the matrix transformations. Such matters are considered in detail in more advanced linear algebra courses, but will be touched on in later sections.

- These matrices make it possible to compute images of vectors using matrix multiplication. Such computations can be performed rapidly on computers.

To focus on the latter idea, let $T: V \to W$ be a linear transformation. As shown in Figure 8.4.5, the matrix $[T]_{B', B}$ can be used to calculate $T(\mathbf{x})$ in three steps by the following *indirect* procedure:

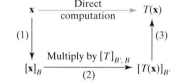

Figure 8.4.5

(1) Compute the coordinate matrix $[\mathbf{x}]_B$.

(2) Multiply $[\mathbf{x}]_B$ on the left by $[T]_{B', B}$ to produce $[T(\mathbf{x})]_{B'}$.

(3) Reconstruct $T(\mathbf{x})$ from its coordinate matrix $[T(\mathbf{x})]_{B'}$.

EXAMPLE 6 Linear Operator on P_2

Let $T: P_2 \to P_2$ be the linear operator defined by

$$T(p(x)) = p(3x - 5)$$

that is, $T(c_0 + c_1 x + c_2 x^2) = c_0 + c_1(3x - 5) + c_2(3x - 5)^2$.

(a) Find $[T]_B$ with respect to the basis $B = \{1, x, x^2\}$.
(b) Use the indirect procedure to compute $T(1 + 2x + 3x^2)$.
(c) Check the result in (b) by computing $T(1 + 2x + 3x^2)$ directly.

Solution (a). From the formula for T,

$$T(1) = 1, \quad T(x) = 3x - 5, \quad T(x^2) = (3x - 5)^2 = 9x^2 - 30x + 25$$

so that

$$[T(1)]_B = \begin{bmatrix} 1 \\ 0 \\ 0 \end{bmatrix}, \quad [T(x)]_B = \begin{bmatrix} -5 \\ 3 \\ 0 \end{bmatrix}, \quad [T(x^2)]_B = \begin{bmatrix} 25 \\ -30 \\ 9 \end{bmatrix}$$

Thus,

$$[T]_B = \begin{bmatrix} 1 & -5 & 25 \\ 0 & 3 & -30 \\ 0 & 0 & 9 \end{bmatrix}$$

Solution (b). The coordinate matrix relative to B for the vector $\mathbf{p} = 1 + 2x + 3x^2$ is

$$[\mathbf{p}]_B = \begin{bmatrix} 1 \\ 2 \\ 3 \end{bmatrix}$$

Thus, from (5a)

$$[T(1 + 2x + 3x^2)]_B = [T(\mathbf{p})]_B = [T]_B[\mathbf{p}]_B$$

$$= \begin{bmatrix} 1 & -5 & 25 \\ 0 & 3 & -30 \\ 0 & 0 & 9 \end{bmatrix} \begin{bmatrix} 1 \\ 2 \\ 3 \end{bmatrix} = \begin{bmatrix} 66 \\ -84 \\ 27 \end{bmatrix}$$

from which it follows that

$$T(1 + 2x + 3x^2) = 66 - 84x + 27x^2$$

Solution (c). By direct computation

$$\begin{aligned} T(1 + 2x + 3x^2) &= 1 + 2(3x - 5) + 3(3x - 5)^2 \\ &= 1 + 6x - 10 + 27x^2 - 90x + 75 \\ &= 66 - 84x + 27x^2 \end{aligned}$$

which agrees with the result in (b).   ◆

Matrices of Compositions and Inverse Transformations

We shall now mention two theorems that are generalizations of Formula (21) of Section 4.2 and Formula (1) of Section 4.3. The proofs are omitted.

Theorem 8.4.2

If $T_1: U \rightarrow V$ and $T_2: V \rightarrow W$ are linear transformations, and if B, B'', and B' are bases for U, V, and W, respectively, then

$$[T_2 \circ T_1]_{B',B} = [T_2]_{B',B''}[T_1]_{B'',B} \tag{13}$$

Theorem 8.4.3

If $T: V \rightarrow V$ is a linear operator, and if B is a basis for V, then the following are equivalent.

(a) T is one-to-one.   *(b) $[T]_B$ is invertible.*

Moreover, when these equivalent conditions hold

$$[T^{-1}]_B = [T]_B^{-1} \tag{14}$$

$[T_2 \circ T_1]_{B', B} = [T_2]_{B', B''} \quad [T_1]_{B'', B}$

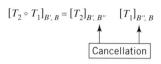

Figure 8.4.6

REMARK. In (13), observe how the interior subscript B'' (the basis for the intermediate space V) seems to "cancel out," leaving only the bases for the domain and image space of the composition as subscripts (Figure 8.4.6). This cancellation of interior subscripts suggests the following extension of Formula (13) to compositions of three linear transformations (Figure 8.4.7).

$$[T_3 \circ T_2 \circ T_1]_{B', B} = [T_3]_{B', B'''}[T_2]_{B''', B''}[T_1]_{B'', B} \tag{15}$$

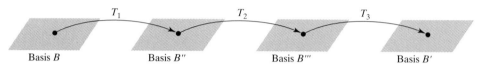

Basis B Basis B'' Basis B''' Basis B'

Figure 8.4.7

The following example illustrates Theorem 8.4.2.

EXAMPLE 7 Using Theorem 8.4.2

Let $T_1: P_1 \to P_2$ be the linear transformation defined by

$$T_1(p(x)) = xp(x)$$

and let $T_2: P_2 \to P_2$ be the linear operator defined by

$$T_2(p(x)) = p(3x - 5)$$

Then the composition $(T_2 \circ T_1): P_1 \to P_2$ is given by

$$(T_2 \circ T_1)(p(x)) = T_2(T_1(p(x))) = T_2(xp(x)) = (3x - 5)p(3x - 5)$$

Thus, if $p(x) = c_0 + c_1 x$, then

$$(T_2 \circ T_1)(c_0 + c_1 x) = (3x - 5)(c_0 + c_1(3x - 5))$$
$$= c_0(3x - 5) + c_1(3x - 5)^2 \tag{16}$$

In this example, P_1 plays the role of U in Theorem 8.4.2 and P_2 plays the roles of both V and W; thus we can take $B' = B''$ in (13) so that the formula simplifies to

$$[T_2 \circ T_1]_{B', B} = [T_2]_{B'}[T_1]_{B', B} \tag{17}$$

Let us choose $B = \{1, x\}$ to be the basis for P_1 and $B' = \{1, x, x^2\}$ to be the basis for P_2. We showed in Examples 1 and 6 that

$$[T_1]_{B', B} = \begin{bmatrix} 0 & 0 \\ 1 & 0 \\ 0 & 1 \end{bmatrix} \quad \text{and} \quad [T_2]_{B'} = \begin{bmatrix} 1 & -5 & 25 \\ 0 & 3 & -30 \\ 0 & 0 & 9 \end{bmatrix}$$

Thus, it follows from (17) that

$$[T_2 \circ T_1]_{B', B} = \begin{bmatrix} 1 & -5 & 25 \\ 0 & 3 & -30 \\ 0 & 0 & 9 \end{bmatrix}\begin{bmatrix} 0 & 0 \\ 1 & 0 \\ 0 & 1 \end{bmatrix} = \begin{bmatrix} -5 & 25 \\ 3 & -30 \\ 0 & 9 \end{bmatrix} \tag{18}$$

As a check, we will calculate $[T_2 \circ T_1]_{B', B}$ directly from Formula (4). Since $B = \{1, x\}$, it follows from Formula (4) with $\mathbf{u}_1 = 1$ and $\mathbf{u}_2 = x$ that

$$[T_2 \circ T_1]_{B'B} = \left[[(T_2 \circ T_1)(1)]_{B'} \mid [(T_2 \circ T_1)(x)]_{B'} \right] \tag{19}$$

Using (16) yields

$$(T_2 \circ T_1)(1) = 3x - 5 \quad \text{and} \quad (T_2 \circ T_1)(x) = (3x - 5)^2 = 9x^2 - 30x + 25$$

Since $B' = \{1, x, x^2\}$, it follows from this that

$$[(T_2 \circ T_1)(1)]_{B'} = \begin{bmatrix} -5 \\ 3 \\ 0 \end{bmatrix} \quad \text{and} \quad [(T_2 \circ T_1)(x)]_{B'} = \begin{bmatrix} 25 \\ -30 \\ 9 \end{bmatrix}$$

Substituting in (19) yields

$$[T_2 \circ T_1]_{B',B} = \begin{bmatrix} -5 & 25 \\ 3 & -30 \\ 0 & 9 \end{bmatrix}$$

which agrees with (18). ◆

Exercise Set 8.4

1. Let $T: P_2 \to P_3$ be the linear transformation defined by $T(p(x)) = xp(x)$.
   (a) Find the matrix for T with respect to the standard bases

   $$B = \{\mathbf{u}_1, \mathbf{u}_2, \mathbf{u}_3\} \quad \text{and} \quad B' = \{\mathbf{v}_1, \mathbf{v}_2, \mathbf{v}_3, \mathbf{v}_4\}$$

   where

   $$\begin{aligned} \mathbf{u}_1 &= 1, & \mathbf{u}_2 &= x, & \mathbf{u}_3 &= x^2 \\ \mathbf{v}_1 &= 1, & \mathbf{v}_2 &= x, & \mathbf{v}_3 &= x^2, & \mathbf{v}_4 &= x^3 \end{aligned}$$

   (b) Verify that the matrix $[T]_{B',B}$ obtained in part (a) satisfies Formula (4a) for every vector $\mathbf{x} = c_0 + c_1 x + c_2 x^2$ in P_2.

2. Let $T: P_2 \to P_1$ be the linear transformation defined by

   $$T(a_0 + a_1 x + a_2 x^2) = (a_0 + a_1) - (2a_1 + 3a_2)x$$

   (a) Find the matrix for T with respect to the standard bases $B = \{1, x, x^2\}$ and $B' = \{1, x\}$ for P_2 and P_1.
   (b) Verify that the matrix $[T]_{B',B}$ obtained in part (a) satisfies Formula (4a) for every vector $\mathbf{x} = c_0 + c_1 x + c_2 x^2$ in P_2.

3. Let $T: P_2 \to P_2$ be the linear operator defined by

   $$T(a_0 + a_1 x + a_2 x^2) = a_0 + a_1(x - 1) + a_2(x - 1)^2$$

   (a) Find the matrix for T with respect to the standard basis $B = \{1, x, x^2\}$ for P_2.
   (b) Verify that the matrix $[T]_B$ obtained in part (a) satisfies Formula (5a) for every vector $\mathbf{x} = a_0 + a_1 x + a_2 x^2$ in P_2.

4. Let $T: R^2 \to R^2$ be the linear operator defined by

   $$T\left(\begin{bmatrix} x_1 \\ x_2 \end{bmatrix}\right) = \begin{bmatrix} x_1 - x_2 \\ x_1 + x_2 \end{bmatrix}$$

   and let $B = \{\mathbf{u}_1, \mathbf{u}_2\}$ be the basis for which

   $$\mathbf{u}_1 = \begin{bmatrix} 1 \\ 1 \end{bmatrix} \quad \text{and} \quad \mathbf{u}_2 = \begin{bmatrix} -1 \\ 0 \end{bmatrix}$$

   (a) Find $[T]_B$.
   (b) Verify that Formula (5a) holds for every vector $\mathbf{x}$ in R^2.

5. Let $T: R^2 \to R^3$ be defined by

   $$T\left(\begin{bmatrix} x_1 \\ x_2 \end{bmatrix}\right) = \begin{bmatrix} x_1 + 2x_2 \\ -x_1 \\ 0 \end{bmatrix}$$

(a) Find the matrix $[T]_{B',B}$ with respect to the bases $B = \{\mathbf{u}_1, \mathbf{u}_2\}$ and $B' = \{\mathbf{v}_1, \mathbf{v}_2, \mathbf{v}_3\}$, where

$$\mathbf{u}_1 = \begin{bmatrix} 1 \\ 3 \end{bmatrix}, \quad \mathbf{u}_2 = \begin{bmatrix} -2 \\ 4 \end{bmatrix}, \quad \mathbf{v}_1 = \begin{bmatrix} 1 \\ 1 \\ 1 \end{bmatrix}, \quad \mathbf{v}_2 = \begin{bmatrix} 2 \\ 2 \\ 0 \end{bmatrix}, \quad \mathbf{v}_3 = \begin{bmatrix} 3 \\ 0 \\ 0 \end{bmatrix}$$

(b) Verify that Formula (4a) holds for every vector

$$\mathbf{x} = \begin{bmatrix} x_1 \\ x_2 \end{bmatrix}$$

in R^2.

6. Let $T: R^3 \to R^3$ be defined by $T(x_1, x_2, x_3) = (x_1 - x_2, x_2 - x_1, x_1 - x_3)$.

(a) Find the matrix for T with respect to the basis $B = \{\mathbf{v}_1, \mathbf{v}_2, \mathbf{v}_3\}$, where

$$\mathbf{v}_1 = (1, 0, 1), \quad \mathbf{v}_2 = (0, 1, 1), \quad \mathbf{v}_3 = (1, 1, 0)$$

(b) Verify that Formula (5a) holds for every vector $\mathbf{x} = (x_1, x_2, x_3)$ in R^3.

7. Let $T: P_2 \to P_2$ be the linear operator defined by $T(p(x)) = p(2x + 1)$; that is,

$$T(c_0 + c_1 x + c_2 x^2) = c_0 + c_1(2x + 1) + c_2(2x + 1)^2$$

(a) Find $[T]_B$ with respect to the basis $B = \{1, x, x^2\}$.

(b) Use the indirect procedure illustrated in Figure 8.4.5 to compute $T(2 - 3x + 4x^2)$.

(c) Check the result obtained in part (b) by computing $T(2 - 3x + 4x^2)$ directly.

8. Let $T: P_2 \to P_3$ be the linear transformation defined by $T(p(x)) = xp(x - 3)$; that is,

$$T(c_0 + c_1 x + c_2 x^2) = x(c_0 + c_1(x - 3) + c_2(x - 3)^2)$$

(a) Find $[T]_{B',B}$ with respect to the bases $B = \{1, x, x^2\}$ and $B' = \{1, x, x^2, x^3\}$.

(b) Use the indirect procedure illustrated in Figure 8.4.5 to compute $T(1 + x - x^2)$.

(c) Check the result obtained in part (b) by computing $T(1 + x - x^2)$ directly.

9. Let $\mathbf{v}_1 = \begin{bmatrix} 1 \\ 3 \end{bmatrix}$ and $\mathbf{v}_2 = \begin{bmatrix} -1 \\ 4 \end{bmatrix}$, and let

$$A = \begin{bmatrix} 1 & 3 \\ -2 & 5 \end{bmatrix}$$

be the matrix for $T: R^2 \to R^2$ with respect to the basis $B = \{\mathbf{v}_1, \mathbf{v}_2\}$.

(a) Find $[T(\mathbf{v}_1)]_B$ and $[T(\mathbf{v}_2)]_B$. (b) Find $T(\mathbf{v}_1)$ and $T(\mathbf{v}_2)$.

(c) Find a formula for $T\left(\begin{bmatrix} x_1 \\ x_2 \end{bmatrix}\right)$. (d) Use the formula obtained in (c) to compute $T\left(\begin{bmatrix} 1 \\ 1 \end{bmatrix}\right)$.

10. Let $A = \begin{bmatrix} 3 & -2 & 1 & 0 \\ 1 & 6 & 2 & 1 \\ -3 & 0 & 7 & 1 \end{bmatrix}$ be the matrix of $T: R^4 \to R^3$ with respect to the bases

$B = \{\mathbf{v}_1, \mathbf{v}_2, \mathbf{v}_3, \mathbf{v}_4\}$ and $B' = \{\mathbf{w}_1, \mathbf{w}_2, \mathbf{w}_3\}$, where

$$\mathbf{v}_1 = \begin{bmatrix} 0 \\ 1 \\ 1 \\ 1 \end{bmatrix}, \quad \mathbf{v}_2 = \begin{bmatrix} 2 \\ 1 \\ -1 \\ -1 \end{bmatrix}, \quad \mathbf{v}_3 = \begin{bmatrix} 1 \\ 4 \\ -1 \\ 2 \end{bmatrix}, \quad \mathbf{v}_4 = \begin{bmatrix} 6 \\ 9 \\ 4 \\ 2 \end{bmatrix}$$

$$\mathbf{w}_1 = \begin{bmatrix} 0 \\ 8 \\ 8 \end{bmatrix}, \quad \mathbf{w}_2 = \begin{bmatrix} -7 \\ 8 \\ 1 \end{bmatrix}, \quad \mathbf{w}_3 = \begin{bmatrix} -6 \\ 9 \\ 1 \end{bmatrix}$$

(a) Find $[T(\mathbf{v}_1)]_{B'}$, $[T(\mathbf{v}_2)]_{B'}$, $[T(\mathbf{v}_3)]_{B'}$, and $[T(\mathbf{v}_4)]_{B'}$.   (b) Find $T(\mathbf{v}_1)$, $T(\mathbf{v}_2)$, $T(\mathbf{v}_3)$, and $T(\mathbf{v}_4)$.

(c) Find a formula for $T\left(\begin{bmatrix} x_1 \\ x_2 \\ x_3 \\ x_4 \end{bmatrix}\right)$.   (d) Use the formula obtained in (c) to compute $T\left(\begin{bmatrix} 2 \\ 2 \\ 0 \\ 0 \end{bmatrix}\right)$.

11. Let $A = \begin{bmatrix} 1 & 3 & -1 \\ 2 & 0 & 5 \\ 6 & -2 & 4 \end{bmatrix}$ be the matrix of $T: P_2 \to P_2$ with respect to the basis

$B = \{\mathbf{v}_1, \mathbf{v}_2, \mathbf{v}_3\}$, where $\mathbf{v}_1 = 3x + 3x^2$, $\mathbf{v}_2 = -1 + 3x + 2x^2$, $\mathbf{v}_3 = 3 + 7x + 2x^2$.

(a) Find $[T(\mathbf{v}_1)]_B$, $[T(\mathbf{v}_2)]_B$, and $[T(\mathbf{v}_3)]_B$.   (b) Find $T(\mathbf{v}_1)$, $T(\mathbf{v}_2)$, and $T(\mathbf{v}_3)$.

(c) Find a formula for $T(a_0 + a_1x + a_2x^2)$.   (d) Use the formula obtained in (c) to compute $T(1 + x^2)$.

12. Let $T_1: P_1 \to P_2$ be the linear transformation defined by

$$T_1(p(x)) = xp(x)$$

and let $T_2: P_2 \to P_2$ be the linear operator defined by

$$T_2(p(x)) = p(2x + 1)$$

Let $B = \{1, x\}$ and $B' = \{1, x, x^2\}$ be the standard bases for P_1 and P_2.

(a) Find $[T_2 \circ T_1]_{B',B}$, $[T_2]_{B'}$, and $[T_1]_{B',B}$.

(b) State a formula relating the matrices in part (a).

(c) Verify that the matrices in part (a) satisfy the formula you stated in part (b).

13. Let $T_1: P_1 \to P_2$ be the linear transformation defined by

$$T_1(c_0 + c_1x) = 2c_0 - 3c_1x$$

and let $T_2: P_2 \to P_3$ be the linear transformation defined by

$$T_1(c_0 + c_1x + c_2x^2) = 3c_0x + 3c_1x^2 + 3c_2x^3$$

Let $B = \{1, x\}$, $B'' = \{1, x, x^2\}$, and $B' = \{1, x, x^2, x^3\}$.

(a) Find $[T_2 \circ T_1]_{B',B}$, $[T_2]_{B',B''}$, and $[T_1]_{B'',B}$.

(b) State a formula relating the matrices in part (a).

(c) Verify that the matrices in part (a) satisfy the formula you stated in part (b).

14. Show that if $T: V \to W$ is the zero transformation, then the matrix of T with respect to any bases for V and W is a zero matrix.

15. Show that if $T: V \to V$ is a contraction or a dilation of V (Example 4 of Section 8.1), then the matrix of T with respect to any basis for V is a positive scalar multiple of the identity matrix.

16. Let $B = \{\mathbf{v}_1, \mathbf{v}_2, \mathbf{v}_3, \mathbf{v}_4\}$ be a basis for a vector space V. Find the matrix with respect to B of the linear operator $T: V \to V$ defined by $T(\mathbf{v}_1) = \mathbf{v}_2$, $T(\mathbf{v}_2) = \mathbf{v}_3$, $T(\mathbf{v}_3) = \mathbf{v}_4$, $T(\mathbf{v}_4) = \mathbf{v}_1$.

17. Prove that if B and B' are the standard bases for R^n and R^m, respectively, then the matrix for a linear transformation $T: R^n \to R^m$ with respect to the bases B and B' is the standard matrix for T.

18. (*For readers who have studied calculus.*) Let $D: P_2 \to P_2$ be the differentiation operator $D(\mathbf{p}) = p'(x)$. In parts (a) and (b) find the matrix of D with respect to the basis $B = \{\mathbf{p}_1, \mathbf{p}_2, \mathbf{p}_3\}$.

(a) $\mathbf{p}_1 = 1$, $\mathbf{p}_2 = x$, $\mathbf{p}_3 = x^2$

(b) $\mathbf{p}_1 = 2$, $\mathbf{p}_2 = 2 - 3x$, $\mathbf{p}_3 = 2 - 3x + 8x^2$

(c) Use the matrix in part (a) to compute $D(6 - 6x + 24x^2)$.

(d) Repeat the directions for part (c) for the matrix in part (b).

19. (*For readers who have studied calculus.*) In each part, $B = \{f_1, f_2, f_3\}$ is a basis for a subspace V of the vector space of real-valued functions defined on the real line. Find the matrix with respect to B of the differentiation operator $D: V \to V$.

(a) $f_1 = 1$, $f_2 = \sin x$, $f_3 = \cos x$ (b) $f_1 = 1$, $f_2 = e^x$, $f_3 = e^{2x}$

(c) $f_1 = e^{2x}$, $f_2 = xe^{2x}$, $f_3 = x^2 e^{2x}$ (d) Use the matrix in part (c) to compute $D(4e^{2x} + 6xe^{2x} - 10x^2 e^{2x})$.

Discussion and Discovery

20. Let V be a four-dimensional vector space with basis B, let W be a seven-dimensional vector space with basis B', and let $T: V \to W$ be a linear transformation. Identify the four vector spaces that contain the vectors at the corners of the accompanying diagram.

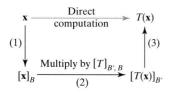

21. In each part, fill in the missing part of the equation.

(a) $[T_2 \circ T_1]_{B',B} = [T_2]\_\,?\,\_[T_1]_{B'',B}$ (b) $[T_3 \circ T_2 \circ T_1]_{B',B} = [T_3]\_\,?\,\_[T_2]_{B''',B''}[T_1]_{B'',B}$

22. Give two reasons why matrices for general linear transformations are important.

8.5 SIMILARITY

The matrix of a linear operator $T: V \to V$ depends on the basis selected for V. One of the fundamental problems of linear algebra is to choose a basis for V that makes the matrix for T as simple as possible—a diagonal or triangular matrix, for example. In this section we shall study this problem.

Simple Matrices for Linear Operators   Standard bases do not necessarily produce the simplest matrices for linear operators. For example, consider the linear operator $T: R^2 \to R^2$ defined by

$$T\left(\begin{bmatrix} x_1 \\ x_2 \end{bmatrix}\right) = \begin{bmatrix} x_1 + x_2 \\ -2x_1 + 4x_2 \end{bmatrix} \tag{1}$$

and the standard basis $B = \{e_1, e_2\}$ for R^2, where

$$e_1 = \begin{bmatrix} 1 \\ 0 \end{bmatrix}, \qquad e_2 = \begin{bmatrix} 0 \\ 1 \end{bmatrix}$$

By Theorem 8.4.1, the matrix for T with respect to this basis is the standard matrix for T; that is,

$$[T]_B = [T] = [T(e_1) \mid T(e_2)]$$

From (1),

$$T(e_1) = \begin{bmatrix} 1 \\ -2 \end{bmatrix}, \qquad T(e_2) = \begin{bmatrix} 1 \\ 4 \end{bmatrix}$$

so

$$[T]_B = \begin{bmatrix} 1 & 1 \\ -2 & 4 \end{bmatrix} \tag{2}$$

In comparison, we showed in Example 4 of Section 8.4 that if

$$\mathbf{u}_1 = \begin{bmatrix} 1 \\ 1 \end{bmatrix}, \qquad \mathbf{u}_2 = \begin{bmatrix} 1 \\ 2 \end{bmatrix} \tag{3}$$

then the matrix for T with respect to the basis $B' = \{\mathbf{u}_1, \mathbf{u}_2\}$ is the diagonal matrix

$$[T]_{B'} = \begin{bmatrix} 2 & 0 \\ 0 & 3 \end{bmatrix} \tag{4}$$

This matrix is "simpler" than (2) in the sense that diagonal matrices enjoy special properties that more general matrices do not.

One of the major themes in more advanced linear algebra courses is to determine the "simplest possible form" that can be obtained for the matrix of a linear operator by choosing the basis appropriately. Sometimes it is possible to obtain a diagonal matrix (as above, for example); other times one must settle for a triangular matrix or some other form. We will only be able to touch on this important topic in this text.

The problem of finding a basis that produces the simplest possible matrix for a linear operator $T: V \to V$ can be attacked by first finding a matrix for T relative to *any* basis, say a standard basis, where applicable, then changing the basis in a manner that simplifies the matrix. Before pursuing this idea, it will be helpful to review some concepts about changing bases.

Recall from Formula (8) in Section 6.5 that if the sets $B = \{\mathbf{u}_1, \mathbf{u}_2, \ldots, \mathbf{u}_n\}$ and $B' = \{\mathbf{u}'_1, \mathbf{u}'_2, \ldots, \mathbf{u}'_n\}$ are bases for a vector space V, then the *transition matrix* from B' to B is given by the formula

$$P = \left[[\mathbf{u}'_1]_B \mid [\mathbf{u}'_2]_B \mid \cdots \mid [\mathbf{u}'_n]_B \right] \tag{5}$$

This matrix has the property that for every vector $\mathbf{v}$ in V

$$P[\mathbf{v}]_{B'} = [\mathbf{v}]_B \tag{6}$$

that is, multiplication by P maps the coordinate matrix for $\mathbf{v}$ relative to B' into the coordinate matrix for $\mathbf{v}$ relative to B [see Formula (7) in Section 6.5]. We showed in Theorem 6.5.4 that P is invertible and P^{-1} is the transition matrix from B to B'.

The following theorem gives a useful alternative viewpoint about transition matrices; it shows that the transition matrix from a basis B' to a basis B can be regarded as the matrix of an identity operator.

Theorem 8.5.1

If B and B' are bases for a finite-dimensional vector space V, and if $I: V \to V$ is the identity operator, then $[I]_{B,B'}$ is the transition matrix from B' to B.

Proof. Suppose that $B = \{\mathbf{u}_1, \mathbf{u}_2, \ldots, \mathbf{u}_n\}$ and $B' = \{\mathbf{u}'_1, \mathbf{u}'_2, \ldots, \mathbf{u}'_n\}$ are bases for V. Using the fact that $I(\mathbf{v}) = \mathbf{v}$ for all $\mathbf{v}$ in V, it follows from Formula (4) of Section 8.4 with B and B' *reversed* that

$$[I]_{B,B'} = \left[[I(\mathbf{u}'_1)]_B \mid [I(\mathbf{u}'_2)]_B \mid \cdots \mid [I(\mathbf{u}'_n)]_B \right]$$

$$= \left[[\mathbf{u}'_1]_B \mid [\mathbf{u}'_2]_B \mid \cdots \mid [\mathbf{u}'_n]_B \right]$$

Thus, from (5), we have $[I]_{B,B'} = P$, which shows that $[I]_{B,B'}$ is the transition matrix from B' to B. ■

The result in this theorem is illustrated in Figure 8.5.1.

Figure 8.5.1 $[I]_{B,B'}$ is the transition matrix from B' to B.

Effect of Changing Bases on Matrices of Linear Operators
We are now ready to consider the main problem in this section.

Problem. If B and B' are two bases for a finite-dimensional vector space V, and if $T: V \rightarrow V$ is a linear operator, what relationship, if any, exists between the matrices $[T]_B$ and $[T]_{B'}$?

The answer to this question can be obtained by considering the composition of the three linear operators on V pictured in Figure 8.5.2.

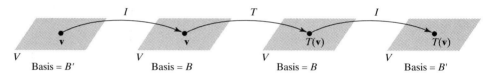

Figure 8.5.2

In this figure $\mathbf{v}$ is first mapped into itself by the identity operator, then $\mathbf{v}$ is mapped into $T(\mathbf{v})$ by T, then $T(\mathbf{v})$ is mapped into itself by the identity operator. All four vector spaces involved in the composition are the same (namely, V); however, the bases for the spaces vary. Since the starting vector is $\mathbf{v}$ and the final vector is $T(\mathbf{v})$, the composition is the same as T; that is,

$$T = I \circ T \circ I \tag{7}$$

If, as illustrated in Figure 8.5.2, the first and last vector spaces are assigned the basis B' and the middle two spaces are assigned the basis B, then it follows from (7) and Formula (15) of Section 8.4 (with an appropriate adjustment in the names of the bases) that

$$[T]_{B',B'} = [I \circ T \circ I]_{B',B'} = [I]_{B',B}[T]_{B,B}[I]_{B,B'} \tag{8}$$

or in simpler notation,

$$[T]_{B'} = [I]_{B',B}[T]_B[I]_{B,B'} \tag{9}$$

But it follows from Theorem 8.5.1 that $[I]_{B,B'}$ is the transition matrix from B' to B and consequently $[I]_{B',B}$ is the transition matrix from B to B'. Thus, if we let $P = [I]_{B,B'}$, then $P^{-1} = [I]_{B',B}$, so (9) can be written as

$$[T]_{B'} = P^{-1}[T]_B P$$

In summary, we have the following theorem.

Theorem 8.5.2

Let $T: V \rightarrow V$ be a linear operator on a finite-dimensional vector space V, and let B and B' be bases for V. Then

$$[T]_{B'} = P^{-1}[T]_B P \tag{10}$$

where P is the transition matrix from B' to B.

$$[T]_{B'} = [I]_{B',B}[T]_B[I]_{B,B'}$$

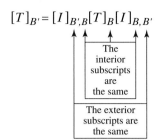

The interior subscripts are the same

The exterior subscripts are the same

Figure 8.5.3

Warning. When applying Theorem 8.5.2, it is easy to forget whether P is the transition matrix from B to B' (incorrect) or from B' to B (correct). As indicated in Figure 8.5.3, it may help to write (10) in form (9), keeping in mind that the three "interior" subscripts are the same, and the two exterior subscripts are the same. Once you master the pattern shown in this figure, you need only remember that $P = [I]_{B,B'}$ is the transition matrix from B' to B and $P^{-1} = [I]_{B',B}$ is its inverse.

EXAMPLE 1 Using Theorem 8.5.2

Let $T: R^2 \rightarrow R^2$ be defined by

$$T\left(\begin{bmatrix} x_1 \\ x_2 \end{bmatrix}\right) = \begin{bmatrix} x_1 + x_2 \\ -2x_1 + 4x_2 \end{bmatrix}$$

Find the matrix of T with respect to the standard basis $B = \{\mathbf{e}_1, \mathbf{e}_2\}$ for R^2, then use Theorem 8.5.2 to find the matrix of T with respect to the basis $B' = \{\mathbf{u}'_1, \mathbf{u}'_2\}$, where

$$\mathbf{u}'_1 = \begin{bmatrix} 1 \\ 1 \end{bmatrix} \quad \text{and} \quad \mathbf{u}'_2 = \begin{bmatrix} 1 \\ 2 \end{bmatrix}$$

Solution.

We showed earlier in this section [see (2)] that

$$[T]_B = \begin{bmatrix} 1 & 1 \\ -2 & 4 \end{bmatrix}$$

To find $[T]_{B'}$ from (10), we will need to find the transition matrix

$$P = [I]_{B,B'} = \begin{bmatrix} [\mathbf{u}'_1]_B \mid [\mathbf{u}'_2]_B \end{bmatrix}$$

[see (5)]. By inspection

$$\mathbf{u}'_1 = \mathbf{e}_1 + \mathbf{e}_2$$
$$\mathbf{u}'_2 = \mathbf{e}_1 + 2\mathbf{e}_2$$

so that

$$[\mathbf{u}'_1]_B = \begin{bmatrix} 1 \\ 1 \end{bmatrix} \quad \text{and} \quad [\mathbf{u}'_2]_B = \begin{bmatrix} 1 \\ 2 \end{bmatrix}$$

Thus, the transition matrix from B' to B is

$$P = \begin{bmatrix} 1 & 1 \\ 1 & 2 \end{bmatrix}$$

The reader can check that

$$P^{-1} = \begin{bmatrix} 2 & -1 \\ -1 & 1 \end{bmatrix}$$

so that by Theorem 8.5.2 the matrix of T relative to the basis B' is

$$[T]_{B'} = P^{-1}[T]_B P = \begin{bmatrix} 2 & -1 \\ -1 & 1 \end{bmatrix} \begin{bmatrix} 1 & 1 \\ -2 & 4 \end{bmatrix} \begin{bmatrix} 1 & 1 \\ 1 & 2 \end{bmatrix} = \begin{bmatrix} 2 & 0 \\ 0 & 3 \end{bmatrix}$$

which agrees with (4). ◆

Similarity   The relationship in Formula (10) is of such importance that there is some terminology associated with it.

> **Definition**
>
> If A and B are square matrices, we say that **B is similar to A** if there is an invertible matrix P such that $B = P^{-1}AP$.

REMARK. It is left as an exercise to show that if a matrix B is similar to a matrix A, then necessarily A is similar to B. Therefore, we shall usually simply say that **A and B are similar**.

Similarity Invariants   Similar matrices often have properties in common; for example, if A and B are similar matrices, then A and B have the same determinant. To see that this is so, suppose that

$$B = P^{-1}AP$$

Thus,

$$\det(B) = \det(P^{-1}AP) = \det(P^{-1})\det(A)\det(P)$$

$$= \frac{1}{\det(P)}\det(A)\det(P) = \det(A)$$

We make the following definition.

> **Definition**
>
> A property of square matrices is said to be a **similarity invariant** or **invariant under similarity** if that property is shared by any two similar matrices.

In the terminology of this definition, the determinant of a square matrix is a similarity invariant. Table 1 lists some other important similarity invariants. The proofs of some of the results in Table 1 are given in the exercises.

TABLE 1 *Similarity Invariants*

Property	Description
Determinant	A and $P^{-1}AP$ have the same determinant.
Invertibility	A is invertible if and only if $P^{-1}AP$ is invertible.
Rank	A and $P^{-1}AP$ have the same rank.
Nullity	A and $P^{-1}AP$ have the same nullity.
Trace	A and $P^{-1}AP$ have the same trace.
Characteristic polynomial	A and $P^{-1}AP$ have the same characteristic polynomial.
Eigenvalues	A and $P^{-1}AP$ have the same eigenvalues.
Eigenspace dimension	If λ is an eigenvalue of A and $P^{-1}AP$, then the eigenspace of A corresponding to λ and the eigenspace of $P^{-1}AP$ corresponding to λ have the same dimension.

It follows from Theorem 8.5.2 that *two matrices representing the same linear operator $T: V \rightarrow V$ with respect to different bases are similar.* Thus, if B is a basis for V, and the matrix $[T]_B$ has some property that is invariant under similarity, then for every basis B', the matrix $[T]_{B'}$ has that same property. For example, for any two bases B and B' we must have

$$\det([T]_B) = \det([T]_{B'})$$

It follows from this equation that the value of the determinant depends on T, but not on the particular basis that is used to obtain the matrix for T. Thus, the determinant can be regarded as a property of the linear operator T; indeed, if V is a finite-dimensional vector space, then we can *define* the ***determinant of the linear operator*** T to be

$$\det(T) = \det([T]_B) \tag{11}$$

where B is any basis for V.

EXAMPLE 2 Determinant of a Linear Operator

Let $T: R^2 \rightarrow R^2$ be defined by

$$T\left(\begin{bmatrix} x_1 \\ x_2 \end{bmatrix}\right) = \begin{bmatrix} x_1 + x_2 \\ -2x_1 + 4x_2 \end{bmatrix}$$

Find $\det(T)$.

Solution.

We can choose any basis B and calculate $\det([T]_B)$. If we take the standard basis, then from Example 1

$$[T]_B = \begin{bmatrix} 1 & 1 \\ -2 & 4 \end{bmatrix}, \quad \text{so} \quad \det(T) = \begin{vmatrix} 1 & 1 \\ -2 & 4 \end{vmatrix} = 6$$

Had we chosen the basis $B' = \{\mathbf{u}_1, \mathbf{u}_2\}$ of Example 1, then we would have obtained

$$[T]_{B'} = \begin{bmatrix} 2 & 0 \\ 0 & 3 \end{bmatrix}, \quad \text{so} \quad \det(T) = \begin{vmatrix} 2 & 0 \\ 0 & 3 \end{vmatrix} = 6$$

which agrees with the preceding computation. ◆

EXAMPLE 3 Reflection About a Line

Let l be the line in the xy-plane that passes through the origin and makes an angle θ with the positive x-axis, where $0 \leq \theta < \pi$. As illustrated in Figure 8.5.4, let $T: R^2 \rightarrow R^2$ be the linear operator that maps each vector into its reflection about the line l.

(a) Find the standard matrix for T.

(b) Find the reflection of the vector $\mathbf{x} = (1, 2)$ about the line l through the origin that makes an angle of $\theta = \pi/6$ with the positive x-axis.

Solution (*a*). We could proceed as in Example (6) of Section 4.3 and try to construct the standard matrix from the formula

$$[T]_B = [T] = [T(\mathbf{e}_1) \mid T(\mathbf{e}_2)]$$

where $B = \{\mathbf{e}_1, \mathbf{e}_2\}$ is the standard basis for R^2. However, it is easier to use a different strategy: Instead of finding $[T]_B$ directly, we shall first find the matrix $[T]_{B'}$, where

$$B' = \{\mathbf{u}_1', \mathbf{u}_2'\}$$

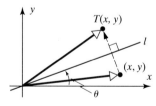

Figure 8.5.4

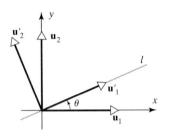

Figure 8.5.5

is the basis consisting of a unit vector $\mathbf{u}_1'$ along l and a unit vector $\mathbf{u}_2'$ perpendicular to l (Figure 8.5.5).

Once we have found $[T]_{B'}$, we shall perform a change of basis to find $[T]_B$. The computations are as follows:

$$T(\mathbf{u}_1') = \mathbf{u}_1' \quad \text{and} \quad T(\mathbf{u}_2') = -\mathbf{u}_2'$$

so

$$[T(\mathbf{u}_1')]_{B'} = \begin{bmatrix} 1 \\ 0 \end{bmatrix} \quad \text{and} \quad T[(\mathbf{u}_2')]_{B'} = \begin{bmatrix} 0 \\ -1 \end{bmatrix}$$

Thus,

$$[T]_{B'} = \begin{bmatrix} 1 & 0 \\ 0 & -1 \end{bmatrix}$$

From the computations in Example 6 of Section 6.5, the transition matrix from B' to B is

$$P = \begin{bmatrix} [\mathbf{u}_1']_B & | & [\mathbf{u}_2']_B \end{bmatrix} = \begin{bmatrix} \cos\theta & -\sin\theta \\ \sin\theta & \cos\theta \end{bmatrix} \tag{12}$$

It follows from Formula (10) that

$$[T]_B = P[T]_{B'}P^{-1}$$

Thus, from (12) the standard matrix for T is

$$[T] = P[T]_{B'}P^{-1} = \begin{bmatrix} \cos\theta & -\sin\theta \\ \sin\theta & \cos\theta \end{bmatrix} \begin{bmatrix} 1 & 0 \\ 0 & -1 \end{bmatrix} \begin{bmatrix} \cos\theta & \sin\theta \\ -\sin\theta & \cos\theta \end{bmatrix}$$

$$= \begin{bmatrix} \cos^2\theta - \sin^2\theta & 2\sin\theta\cos\theta \\ 2\sin\theta\cos\theta & \sin^2\theta - \cos^2\theta \end{bmatrix}$$

$$= \begin{bmatrix} \cos 2\theta & \sin 2\theta \\ \sin 2\theta & -\cos 2\theta \end{bmatrix}$$

Solution (b). It follows from part (a) that the formula for T in matrix notation is

$$T\left(\begin{bmatrix} x \\ y \end{bmatrix} \right) = \begin{bmatrix} \cos 2\theta & \sin 2\theta \\ \sin 2\theta & -\cos 2\theta \end{bmatrix} \begin{bmatrix} x \\ y \end{bmatrix}$$

Substituting $\theta = \pi/6$ in this formula yields

$$T\left(\begin{bmatrix} x \\ y \end{bmatrix} \right) = \begin{bmatrix} \frac{1}{2} & \frac{\sqrt{3}}{2} \\ \frac{\sqrt{3}}{2} & -\frac{1}{2} \end{bmatrix} \begin{bmatrix} x \\ y \end{bmatrix}$$

so

$$T\left(\begin{bmatrix} 1 \\ 2 \end{bmatrix} \right) = \begin{bmatrix} \frac{1}{2} & \frac{\sqrt{3}}{2} \\ \frac{\sqrt{3}}{2} & -\frac{1}{2} \end{bmatrix} \begin{bmatrix} 1 \\ 2 \end{bmatrix} = \begin{bmatrix} \frac{1}{2} + \sqrt{3} \\ \frac{\sqrt{3}}{2} - 1 \end{bmatrix}$$

Thus, $T(1, 2) = (\frac{1}{2} + \sqrt{3}, \frac{\sqrt{3}}{2} - 1)$. ◆

Eigenvalues of a Linear Operator

Eigenvectors and eigenvalues can be defined for linear operators as well as matrices. A scalar λ is called an *eigenvalue* of a linear operator $T: V \to V$ if there is a nonzero vector $\mathbf{x}$ in V such that $T\mathbf{x} = \lambda\mathbf{x}$. The vector $\mathbf{x}$ is called an *eigenvector* of T corresponding to λ. Equivalently, the eigenvectors of T corresponding to λ are the nonzero vectors in the kernel of $\lambda I - T$ (Exercise 15). This kernel is called the *eigenspace* of T corresponding to λ.

It can be shown that if V is a finite-dimensional vector space, and B is *any* basis for V, then

1. The eigenvalues of T are the same as the eigenvalues of $[T]_B$.

2. A vector $\mathbf{x}$ is an eigenvector of T corresponding to λ if and only if its coordinate matrix $[\mathbf{x}]_B$ is an eigenvector of $[T]_B$ corresponding to λ.

We omit the proofs.

EXAMPLE 4 Eigenvalues and Bases for Eigenspaces

Find the eigenvalues and bases for the eigenspaces of the linear operator $T: P_2 \rightarrow P_2$ defined by

$$T(a + bx + cx^2) = -2c + (a + 2b + c)x + (a + 3c)x^2$$

Solution.

The matrix for T with respect to the standard basis $B = \{1, x, x^2\}$ is

$$[T]_B = \begin{bmatrix} 0 & 0 & -2 \\ 1 & 2 & 1 \\ 1 & 0 & 3 \end{bmatrix}$$

(verify). The eigenvalues of T are $\lambda = 1$ and $\lambda = 2$ (Example 5 of Section 7.1). Also from that example, the eigenspace of $[T]_B$ corresponding to $\lambda = 2$ has the basis $\{\mathbf{u}_1, \mathbf{u}_2\}$, where

$$\mathbf{u}_1 = \begin{bmatrix} -1 \\ 0 \\ 1 \end{bmatrix}, \qquad \mathbf{u}_2 = \begin{bmatrix} 0 \\ 1 \\ 0 \end{bmatrix}$$

and the eigenspace of $[T]_B$ corresponding to $\lambda = 1$ has the basis $\{\mathbf{u}_3\}$, where

$$\mathbf{u}_3 = \begin{bmatrix} -2 \\ 1 \\ 1 \end{bmatrix}$$

The matrices $\mathbf{u}_1$, $\mathbf{u}_2$, and $\mathbf{u}_3$ are the coordinate matrices relative to B of

$$\mathbf{p}_1 = -1 + x^2, \qquad \mathbf{p}_2 = x, \qquad \mathbf{p}_3 = -2 + x + x^2$$

Thus, the eigenspace of T corresponding to $\lambda = 2$ has the basis

$$\{\mathbf{p}_1, \mathbf{p}_2\} = \{-1 + x^2, x\}$$

and that corresponding to $\lambda = 1$ has the basis

$$\{\mathbf{p}_3\} = \{-2 + x + x^2\}$$

As a check, the reader should use the given formula for T to verify that $T(\mathbf{p}_1) = 2\mathbf{p}_1$, $T(\mathbf{p}_2) = 2\mathbf{p}_2$, and $T(\mathbf{p}_3) = \mathbf{p}_3$. ◆

EXAMPLE 5 Diagonal Matrix for a Linear Operator

Let $T: R^3 \rightarrow R^3$ be the linear operator given by

$$T\left(\begin{bmatrix} x_1 \\ x_2 \\ x_3 \end{bmatrix}\right) = \begin{bmatrix} -2x_3 \\ x_1 + 2x_2 + x_3 \\ x_1 + 3x_3 \end{bmatrix}$$

Find a basis for R^3 relative to which the matrix for T is diagonal.

Solution.

First we will find the standard matrix for T; then we will look for a change of basis that diagonalizes the standard matrix.

If $B = \{\mathbf{e}_1, \mathbf{e}_2, \mathbf{e}_3\}$ denotes the standard basis for R^3, then

$$T(\mathbf{e}_1) = T\left(\begin{bmatrix} 1 \\ 0 \\ 0 \end{bmatrix}\right) = \begin{bmatrix} 0 \\ 1 \\ 1 \end{bmatrix}, \quad T(\mathbf{e}_2) = T\left(\begin{bmatrix} 0 \\ 1 \\ 0 \end{bmatrix}\right) = \begin{bmatrix} 0 \\ 2 \\ 0 \end{bmatrix}, \quad T(\mathbf{e}_3) = T\left(\begin{bmatrix} 0 \\ 0 \\ 1 \end{bmatrix}\right) = \begin{bmatrix} -2 \\ 1 \\ 3 \end{bmatrix}$$

so that the standard matrix for T is

$$[T] = \begin{bmatrix} 0 & 0 & -2 \\ 1 & 2 & 1 \\ 1 & 0 & 3 \end{bmatrix} \tag{13}$$

We now want to change from the standard basis B to a new basis $B' = \{\mathbf{u}_1', \mathbf{u}_2', \mathbf{u}_3'\}$ in order to obtain a diagonal matrix for T. If we let P be the transition matrix from the unknown basis B' to the standard basis B, then by Theorem 8.5.2 the matrices $[T]$ and $[T]_{B'}$ will be related by

$$[T]_{B'} = P^{-1}[T]P \tag{14}$$

In Example 1 of Section 7.2 we found that the matrix in (13) is diagonalized by

$$P = \begin{bmatrix} -1 & 0 & -2 \\ 0 & 1 & 1 \\ 1 & 0 & 1 \end{bmatrix}$$

Since P represents the transition matrix from the basis $B' = \{\mathbf{u}_1', \mathbf{u}_2', \mathbf{u}_3'\}$ to the standard basis $B = \{\mathbf{e}_1, \mathbf{e}_2, \mathbf{e}_3\}$, the columns of P are $[\mathbf{u}_1']_B$, $[\mathbf{u}_2']_B$, and $[\mathbf{u}_3']_B$, so that

$$[\mathbf{u}_1']_B = \begin{bmatrix} -1 \\ 0 \\ 1 \end{bmatrix}, \quad [\mathbf{u}_2']_B = \begin{bmatrix} 0 \\ 1 \\ 0 \end{bmatrix}, \quad [\mathbf{u}_3']_B = \begin{bmatrix} -2 \\ 1 \\ 1 \end{bmatrix}$$

Thus,

$$\mathbf{u}_1' = (-1)\mathbf{e}_1 + (0)\mathbf{e}_2 + (1)\mathbf{e}_3 = \begin{bmatrix} -1 \\ 0 \\ 1 \end{bmatrix}$$

$$\mathbf{u}_2' = (0)\mathbf{e}_1 + (1)\mathbf{e}_2 + (0)\mathbf{e}_3 = \begin{bmatrix} 0 \\ 1 \\ 0 \end{bmatrix}$$

$$\mathbf{u}_3' = (-2)\mathbf{e}_1 + (1)\mathbf{e}_2 + (1)\mathbf{e}_3 = \begin{bmatrix} -2 \\ 1 \\ 1 \end{bmatrix}$$

are basis vectors that produce a diagonal matrix for $[T]_{B'}$. As a check, let us compute $[T]_{B'}$ directly. From the given formula for T we have

$$T(\mathbf{u}_1') = \begin{bmatrix} -2 \\ 0 \\ 2 \end{bmatrix} = 2\mathbf{u}_1', \quad T(\mathbf{u}_2') = \begin{bmatrix} 0 \\ 2 \\ 0 \end{bmatrix} = 2\mathbf{u}_2', \quad T(\mathbf{u}_3') = \begin{bmatrix} -2 \\ 1 \\ 1 \end{bmatrix} = \mathbf{u}_3'$$

so that

$$[T(\mathbf{u}_1')]_{B'} = \begin{bmatrix} 2 \\ 0 \\ 0 \end{bmatrix}, \quad [T(\mathbf{u}_2')]_{B'} = \begin{bmatrix} 0 \\ 2 \\ 0 \end{bmatrix}, \quad [T(\mathbf{u}_3')]_{B'} = \begin{bmatrix} 0 \\ 0 \\ 1 \end{bmatrix}$$

Thus,

$$[T]_{B'} = \left[[T(\mathbf{u}'_1)]_{B'} \mid [T(\mathbf{u}'_2)]_{B'} \mid [T(\mathbf{u}'_3)]_{B'} \right] = \begin{bmatrix} 2 & 0 & 0 \\ 0 & 2 & 0 \\ 0 & 0 & 1 \end{bmatrix}$$

This is consistent with (14) since

$$P^{-1}[T]P = \begin{bmatrix} 1 & 0 & 2 \\ 1 & 1 & 1 \\ -1 & 0 & -1 \end{bmatrix} \begin{bmatrix} 0 & 0 & -2 \\ 1 & 2 & 1 \\ 1 & 0 & 3 \end{bmatrix} \begin{bmatrix} -1 & 0 & -2 \\ 0 & 1 & 1 \\ 1 & 0 & 1 \end{bmatrix}$$

$$= \begin{bmatrix} 2 & 0 & 0 \\ 0 & 2 & 0 \\ 0 & 0 & 1 \end{bmatrix}$$

♦

Exercise Set 8.5

In Exercises 1–7 find the matrix of T with respect to B, and use Theorem 8.5.2 to compute the matrix of T with respect to B'.

1. $T: R^2 \rightarrow R^2$ is defined by

$$T\left(\begin{bmatrix} x_1 \\ x_2 \end{bmatrix}\right) = \begin{bmatrix} x_1 - 2x_2 \\ -x_2 \end{bmatrix}$$

$B = \{\mathbf{u}_1, \mathbf{u}_2\}$ and $B' = \{\mathbf{v}_1, \mathbf{v}_2\}$, where

$$\mathbf{u}_1 = \begin{bmatrix} 1 \\ 0 \end{bmatrix}, \quad \mathbf{u}_2 = \begin{bmatrix} 0 \\ 1 \end{bmatrix}, \quad \mathbf{v}_1 = \begin{bmatrix} 2 \\ 1 \end{bmatrix}, \quad \mathbf{v}_2 = \begin{bmatrix} -3 \\ 4 \end{bmatrix}$$

2. $T: R^2 \rightarrow R^2$ is defined by

$$T\left(\begin{bmatrix} x_1 \\ x_2 \end{bmatrix}\right) = \begin{bmatrix} x_1 + 7x_2 \\ 3x_1 - 4x_2 \end{bmatrix}$$

$B = \{\mathbf{u}_1, \mathbf{u}_2\}$ and $B' = \{\mathbf{v}_1, \mathbf{v}_2\}$, where

$$\mathbf{u}_1 = \begin{bmatrix} 2 \\ 2 \end{bmatrix}, \quad \mathbf{u}_2 = \begin{bmatrix} 4 \\ -1 \end{bmatrix}, \quad \mathbf{v}_1 = \begin{bmatrix} 1 \\ 3 \end{bmatrix}, \quad \mathbf{v}_2 = \begin{bmatrix} -1 \\ -1 \end{bmatrix}$$

3. $T: R^2 \rightarrow R^2$ is the rotation about the origin through $45°$; B and B' are the bases in Exercise 1.

4. $T: R^3 \rightarrow R^3$ is defined by

$$T\left(\begin{bmatrix} x_1 \\ x_2 \\ x_3 \end{bmatrix}\right) = \begin{bmatrix} x_1 + 2x_2 - x_3 \\ -x_2 \\ x_1 + 7x_3 \end{bmatrix}$$

B is the standard basis for R^3 and $B' = \{\mathbf{v}_1, \mathbf{v}_2, \mathbf{v}_3\}$, where

$$\mathbf{v}_1 = \begin{bmatrix} 1 \\ 0 \\ 0 \end{bmatrix}, \quad \mathbf{v}_2 = \begin{bmatrix} 1 \\ 1 \\ 0 \end{bmatrix}, \quad \mathbf{v}_3 = \begin{bmatrix} 1 \\ 1 \\ 1 \end{bmatrix}$$

5. $T: R^3 \rightarrow R^3$ is the orthogonal projection on the xy-plane; B and B' are as in Exercise 4.

6. $T: R^2 \rightarrow R^2$ is defined by $T(\mathbf{x}) = 5\mathbf{x}$; B and B' are the bases in Exercise 2.

7. $T: P_1 \rightarrow P_1$ is defined by $T(a_0 + a_1x) = a_0 + a_1(x + 1)$; $B = \{\mathbf{p}_1, \mathbf{p}_2\}$ and $B' = \{\mathbf{q}_1, \mathbf{q}_2\}$, where $\mathbf{p}_1 = 6 + 3x$, $\mathbf{p}_2 = 10 + 2x$, $\mathbf{q}_1 = 2$, $\mathbf{q}_2 = 3 + 2x$.

8. Find $\det(T)$.

(a) $T: R^2 \to R^2$, where $T(x_1, x_2) = (3x_1 - 4x_2, -x_1 + 7x_2)$

(b) $T: R^3 \to R^3$, where $T(x_1, x_2, x_3) = (x_1 - x_2, x_2 - x_3, x_3 - x_1)$

(c) $T: P_2 \to P_2$, where $T(p(x)) = p(x - 1)$

9. Prove that the following are similarity invariants:

(a) rank   (b) nullity   (c) invertibility

10. Let $T: P_4 \to P_4$ be the linear operator given by the formula $T(p(x)) = p(2x + 1)$.

(a) Find a matrix for T with respect to some convenient basis; then use Theorem 8.2.2 to find the rank and nullity of T.

(b) Use the result in part (a) to determine whether T is one-to-one.

11. In each part find a basis for R^2 relative to which the matrix for T is diagonal.

(a) $T\left(\begin{bmatrix} x_1 \\ x_2 \end{bmatrix}\right) = \begin{bmatrix} x_1 - x_2 \\ 2x_1 + 4x_2 \end{bmatrix}$ (b) $T\left(\begin{bmatrix} x_1 \\ x_2 \end{bmatrix}\right) = \begin{bmatrix} 4x_1 - x_2 \\ -3x_1 + x_2 \end{bmatrix}$

12. In each part find a basis for R^3 relative to which the matrix for T is diagonal.

(a) $T\left(\begin{bmatrix} x_1 \\ x_2 \\ x_3 \end{bmatrix}\right) = \begin{bmatrix} -2x_1 + x_2 - x_3 \\ x_1 - 2x_2 - x_3 \\ -x_1 - x_2 - 2x_3 \end{bmatrix}$ (b) $T\left(\begin{bmatrix} x_1 \\ x_2 \\ x_3 \end{bmatrix}\right) = \begin{bmatrix} -x_2 + x_3 \\ -x_1 + x_3 \\ x_1 + x_2 \end{bmatrix}$

(c) $T\left(\begin{bmatrix} x_1 \\ x_2 \\ x_3 \end{bmatrix}\right) = \begin{bmatrix} 4x_1 + x_3 \\ 2x_1 + 3x_2 + 2x_3 \\ x_1 + 4x_3 \end{bmatrix}$

13. Let $T: P_2 \to P_2$ be defined by

$$T(a_0 + a_1 x + a_2 x^2) = (5a_0 + 6a_1 + 2a_2) - (a_1 + 8a_2)x + (a_0 - 2a_2)x^2$$

(a) Find the eigenvalues of T.   (b) Find bases for the eigenspaces of T.

14. Let $T: M_{22} \to M_{22}$ be defined by

$$T\left(\begin{bmatrix} a & b \\ c & d \end{bmatrix}\right) = \begin{bmatrix} 2c & a + c \\ b - 2c & d \end{bmatrix}$$

(a) Find the eigenvalues of T.   (b) Find bases for the eigenspaces of T.

15. Let λ be an eigenvalue of a linear operator $T: V \to V$. Prove that the eigenvectors of T corresponding to λ are the nonzero vectors in the kernel of $\lambda I - T$.

16. Prove that if A and B are similar matrices, then A^2 and B^2 are also similar. More generally, prove that A^k and B^k are similar, where k is any positive integer.

17. Let C and D be $m \times n$ matrices, and let $B = \{v_1, v_2, \ldots, v_n\}$ be a basis for a vector space V. Show that if $C[x]_B = D[x]_B$ for all x in V, then $C = D$.

18. Let l be a line in the xy-plane that passes through the origin and makes an angle θ with the positive x-axis. As illustrated in the accompanying figure, let $T: R^2 \to R^2$ be the orthogonal projection of R^2 onto l. Use the method of Example 3 to show that

$$T\left(\begin{bmatrix} x \\ y \end{bmatrix}\right) = \begin{bmatrix} \cos^2 \theta & \sin \theta \cos \theta \\ \sin \theta \cos \theta & \sin^2 \theta \end{bmatrix} \begin{bmatrix} x \\ y \end{bmatrix}$$

[**Note.** See Example 6 of Section 4.3.]

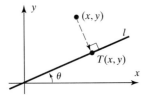

Figure Ex-18

Discussion and Discovery

19. Indicate whether the statement is always true or sometimes false. Justify your answer by giving a logical argument or a counterexample.

(a) A matrix cannot be similar to itself.

(b) If A is similar to B, and B is similar to C, then A is similar to C.

(c) If A and B are similar and B is singular, then A is singular.

(d) If A and B are invertible and similar, then A^{-1} and B^{-1} are similar.

20. Find two nonzero 2×2 matrices that are not similar, and explain why they are not.

21. Complete the proof by filling in the blanks with an appropriate justification.

Hypothesis: A and B are similar matrices.

Conclusion: A and B have the same characteristic polynomial (and hence the same eigenvalues).

Proof: (1) $\det(\lambda I - B) = \det(\lambda I - P^{-1}AP)$ _____

(2) $\qquad = \det(\lambda P^{-1}P - P^{-1}AP)$ _____

(3) $\qquad = \det(P^{-1}(\lambda I - A)P)$ _____

(4) $\qquad = \det(P^{-1})\det(\lambda I - A)\det(P)$ _____

(5) $\qquad = \det(P^{-1})\det(P)\det(\lambda I - A)$ _____

(6) $\qquad = \det(\lambda I - A)$ _____

22. If A and B are similar matrices, say $B = P^{-1}AP$, then Exercise 21 shows that A and B have the same eigenvalues. Suppose that λ is one of the common eigenvalues and $\mathbf{x}$ is a corresponding eigenvector for A. See if you can find an eigenvector of B corresponding to λ, expressed in terms of λ, $\mathbf{x}$, and P.

23. Since the standard basis for R^n is so simple, why would one want to represent a linear operator on R^n in another basis?

Chapter 8 Supplementary Exercises

1. Let A be an $n \times n$ matrix, B a nonzero $n \times 1$ matrix, and $\mathbf{x}$ a vector in R^n expressed in matrix notation. Is $T(\mathbf{x}) = A\mathbf{x} + B$ a linear operator on R^n? Justify your answer.

2. Let

$$A = \begin{bmatrix} \cos\theta & -\sin\theta \\ \sin\theta & \cos\theta \end{bmatrix}$$

(a) Show that

$$A^2 = \begin{bmatrix} \cos 2\theta & -\sin 2\theta \\ \sin 2\theta & \cos 2\theta \end{bmatrix} \quad \text{and} \quad A^3 = \begin{bmatrix} \cos 3\theta & -\sin 3\theta \\ \sin 3\theta & \cos 3\theta \end{bmatrix}$$

(b) Guess the form of the matrix A^n for any positive integer n.

(c) By considering the geometric effect of $T: R^2 \to R^2$, where T is multiplication by A, obtain the result in (b) geometrically.

3. Let $\mathbf{v}_0$ be a fixed vector in an inner product space V, and let $T: V \to V$ be defined by $T(\mathbf{v}) = \langle \mathbf{v}, \mathbf{v}_0 \rangle \mathbf{v}_0$. Show that T is a linear operator on V.

4. Let $\mathbf{v}_1, \mathbf{v}_2, \ldots, \mathbf{v}_m$ be fixed vectors in R^n, and let $T: R^n \to R^m$ be the function defined by $T(\mathbf{x}) = (\mathbf{x} \cdot \mathbf{v}_1, \mathbf{x} \cdot \mathbf{v}_2, \ldots, \mathbf{x} \cdot \mathbf{v}_m)$, where $\mathbf{x} \cdot \mathbf{v}_i$ is the Euclidean inner product on R^n.

(a) Show that T is a linear transformation.

(b) Show that the matrix with row vectors $\mathbf{v}_1, \mathbf{v}_2, \ldots, \mathbf{v}_m$ is the standard matrix for T.

5. Let $\{e_1, e_2, e_3, e_4\}$ be the standard basis for R^4 and $T: R^4 \to R^3$ the linear transformation for which

$$T(e_1) = (1, 2, 1), \qquad T(e_2) = (0, 1, 0),$$
$$T(e_3) = (1, 3, 0), \qquad T(e_4) = (1, 1, 1)$$

   (a) Find bases for the range and kernel of T.   (b) Find the rank and nullity of T.

6. Suppose that vectors in R^3 are denoted by 1×3 matrices, and define $T: R^3 \to R^3$ by

$$T([x_1 \ \ x_2 \ \ x_3]) = [x_1 \ \ x_2 \ \ x_3] \begin{bmatrix} -1 & 2 & 4 \\ 3 & 0 & 1 \\ 2 & 2 & 5 \end{bmatrix}$$

   (a) Find a basis for the kernel of T.   (b) Find a basis for the range of T.

7. Let $B = \{v_1, v_2, v_3, v_4\}$ be a basis for a vector space V and $T: V \to V$ the linear operator for which

$$T(v_1) = v_1 + v_2 + v_3 + 3v_4$$
$$T(v_2) = v_1 - v_2 + 2v_3 + 2v_4$$
$$T(v_3) = 2v_1 - 4v_2 + 5v_3 + 3v_4$$
$$T(v_4) = -2v_1 + 6v_2 - 6v_3 - 2v_4$$

   (a) Find the rank and nullity of T.   (b) Determine whether T is one-to-one.

8. Let V and W be vector spaces, T, T_1, and T_2 linear transformations from V to W, and k a scalar. Define new transformations, $T_1 + T_2$ and kT, by the formulas

$$(T_1 + T_2)(x) = T_1(x) + T_2(x)$$
$$(kT)(x) = k(T(x))$$

   (a) Show that $(T_1 + T_2): V \to W$ and $kT: V \to W$ are linear transformations.
   (b) Show that the set of all linear transformations from V to W with the operations in (a) forms a vector space.

9. Let A and B be similar matrices. Prove:

   (a) A^T and B^T are similar.
   (b) If A and B are invertible, then A^{-1} and B^{-1} are similar.

10. (**Fredholm Alternative Theorem.**) Let $T: V \to V$ be a linear operator on an n-dimensional vector space. Prove that exactly one of the following statements holds:

   (i) The equation $T(x) = b$ has a solution for all vectors b in V.
   (ii) Nullity of $T > 0$.

11. Let $T: M_{22} \to M_{22}$ be the linear operator defined by

$$T(X) = \begin{bmatrix} 1 & 1 \\ 0 & 0 \end{bmatrix} X + X \begin{bmatrix} 0 & 0 \\ 1 & 1 \end{bmatrix}$$

   Find the rank and nullity of T.

12. Prove: If A and B are similar matrices, and if B and C are similar matrices, then A and C are similar matrices.

13. Let $L: M_{22} \to M_{22}$ be the linear operator defined by $L(M) = M^T$. Find the matrix for L with respect to the standard basis for M_{22}.

14. Let $B = \{u_1, u_2, u_3\}$ and $B' = \{v_1, v_2, v_3\}$ be bases for a vector space V, and let

$$P = \begin{bmatrix} 2 & -1 & 3 \\ 1 & 1 & 4 \\ 0 & 1 & 2 \end{bmatrix}$$

   be the transition matrix from B' to B.

   (a) Express v_1, v_2, v_3 as linear combinations of u_1, u_2, u_3.
   (b) Express u_1, u_2, u_3 as linear combinations of v_1, v_2, v_3.

15. Let $B = \{\mathbf{u}_1, \mathbf{u}_2, \mathbf{u}_3\}$ be a basis for a vector space V and $T: V \rightarrow V$ a linear operator such that

$$[T]_B = \begin{bmatrix} -3 & 4 & 7 \\ 1 & 0 & -2 \\ 0 & 1 & 0 \end{bmatrix}$$

Find $[T]_{B'}$, where $B' = \{\mathbf{v}_1, \mathbf{v}_2, \mathbf{v}_3\}$ is the basis for V defined by

$$\mathbf{v}_1 = \mathbf{u}_1, \qquad \mathbf{v}_2 = \mathbf{u}_1 + \mathbf{u}_2, \qquad \mathbf{v}_3 = \mathbf{u}_1 + \mathbf{u}_2 + \mathbf{u}_3$$

16. Show that the matrices

$$\begin{bmatrix} 1 & 1 \\ -1 & 4 \end{bmatrix} \quad \text{and} \quad \begin{bmatrix} 2 & 1 \\ 1 & 3 \end{bmatrix}$$

are similar, but that

$$\begin{bmatrix} 3 & 1 \\ -6 & -2 \end{bmatrix} \quad \text{and} \quad \begin{bmatrix} -1 & 2 \\ 1 & 0 \end{bmatrix}$$

are not.

17. Suppose that $T: V \rightarrow V$ is a linear operator and B is a basis for V such that for any vector $\mathbf{x}$ in V

$$[T(\mathbf{x})]_B = \begin{bmatrix} x_1 - x_2 + x_3 \\ x_2 \\ x_1 - x_3 \end{bmatrix} \quad \text{if} \quad [\mathbf{x}]_B = \begin{bmatrix} x_1 \\ x_2 \\ x_3 \end{bmatrix}$$

Find $[T]_B$.

18. Let $T: V \rightarrow V$ be a linear operator. Prove that T is one-to-one if and only if $\det(T) \neq 0$.

19. (*For readers who have studied calculus.*)

(a) Show that if $\mathbf{f} = f(x)$, then the function $D: C^2(-\infty, \infty) \rightarrow F(-\infty, \infty)$ defined by $D(\mathbf{f}) = f''(x)$ is a linear transformation.

(b) Find a basis for the kernel of D.

(c) Show that the functions satisfying the equation $D(\mathbf{f}) = f(x)$ form a two-dimensional subspace of $C^2(-\infty, \infty)$, and find a basis for this subspace.

20. Let $T: P_2 \rightarrow R^3$ be the function defined by the formula

$$T(p(x)) = \begin{bmatrix} p(-1) \\ p(0) \\ p(1) \end{bmatrix}$$

(a) Find $T(x^2 + 5x + 6)$.

(b) Show that T is a linear transformation.

(c) Show that T is one-to-one.

(d) Find

$$T^{-1}\left(\begin{bmatrix} 0 \\ 3 \\ 0 \end{bmatrix} \right)$$

(e) Sketch the graph of the polynomial in part (d).

21. Let x_1, x_2, and x_3 be distinct real numbers such that $x_1 < x_2 < x_3$, and let $T: P_2 \rightarrow R^3$ be the function defined by the formula

$$T(p(x)) = \begin{bmatrix} p(x_1) \\ p(x_2) \\ p(x_3) \end{bmatrix}$$

(a) Show that T is a linear transformation.

(b) Show that T is one-to-one.

(c) Verify that if a_1, a_2, and a_3 are any real numbers, then

$$T^{-1}\left(\begin{bmatrix} a_1 \\ a_2 \\ a_3 \end{bmatrix}\right) = a_1 P_1(x) + a_2 P_2(x) + a_3 P_3(x)$$

where

$$P_1(x) = \frac{(x - x_2)(x - x_3)}{(x_1 - x_2)(x_1 - x_3)}, \qquad P_2(x) = \frac{(x - x_1)(x - x_3)}{(x_2 - x_1)(x_2 - x_3)}, \qquad P_3(x) = \frac{(x - x_1)(x - x_2)}{(x_3 - x_1)(x_3 - x_2)}$$

(d) What relationship exists between the graph of the function

$$a_1 P_1(x) + a_2 P_2(x) + a_3 P_3(x)$$

and the points (x_1, a_1), (x_2, a_2), and (x_3, a_3)?

22. (*For readers who have studied calculus.*) Let $p(x)$ and $q(x)$ be continuous functions, and let V be the subspace of $C(-\infty, +\infty)$ consisting of all twice differentiable functions. Define $L: V \rightarrow V$ by

$$L(y(x)) = y''(x) + p(x)y'(x) + q(x)y(x)$$

(a) Show that L is a linear transformation.
(b) Consider the special case where $p(x) = 0$ and $q(x) = 1$. Show that the function $\phi(x) = c_1 \sin x + c_2 \cos x$ is in the nullspace of L for all real values of c_1 and c_2.

23. (*For readers who have studied calculus.*) Let $D: P_n \rightarrow P_n$ be the differentiation operator $D(\mathbf{p}) = \mathbf{p}'$. Show that the matrix for D with respect to the basis $B = \{1, x, x^2, \ldots, x^n\}$ is

$$\begin{bmatrix} 0 & 1 & 0 & 0 & \cdots & 0 \\ 0 & 0 & 2 & 0 & \cdots & 0 \\ 0 & 0 & 0 & 3 & \cdots & 0 \\ \vdots & \vdots & \vdots & \vdots & & \vdots \\ 0 & 0 & 0 & 0 & \cdots & n \\ 0 & 0 & 0 & 0 & \cdots & 0 \end{bmatrix}$$

24. (*For readers who have studied calculus.*) It can be shown that for any real number c, the vectors

$$1, x - c, \frac{(x - c)^2}{2!}, \ldots, \frac{(x - c)^n}{n!}$$

form a basis for P_n. Find the matrix for the differentiation operator of Exercise 23 with respect to this basis.

25. (*For readers who have studied calculus.*) Let $J: P_n \rightarrow P_{n+1}$ be the integration transformation defined by

$$J(\mathbf{p}) = \int (a_0 + a_1 x + \cdots + a_n x^n)\, dx = a_0 x + \frac{a_1}{2}x^2 + \cdots + \frac{a_n}{n+1}x^{n+1}$$

where $\mathbf{p} = a_0 + a_1 x + \cdots + a_n x^n$. Find the matrix for T with respect to the standard bases for P_n and P_{n+1}.

Chapter 8 Technology Exercises

The following exercise is designed to be solved using a technology utility. Typically, this will be MATLAB, *Mathematica*, Maple, Derive, or Mathcad, but it may also be some other type of linear algebra software or a scientific calculator with some linear algebra capabilities. For this exercise you will need to read the relevant documentation for the particular utility you are using. The goal of this exercise is to provide you with a basic proficiency with your technology utility. Once you have mastered the techniques in this exercise, you will be able to use your technology utility to solve many of the problems in the regular exercise sets.

Section 8.5

T1. (*Similarity invariants*) Choose a nonzero 3×3 matrix A and an invertible 3×3 matrix P. Compute $P^{-1}AP$ and confirm the statements in Table 1.

Additional Topics

9

INTRODUCTION: In this chapter we shall see how some of the topics that we have studied in earlier chapters can be applied to other areas of mathematics such as differential equations, analytic geometry, curve fitting, and Fourier series. The chapter concludes by returning once again to the fundamental problem of solving systems of linear equations $A\mathbf{x} = \mathbf{b}$. This time we solve a system, not by another elimination procedure, but by factoring the coefficient matrix into two different triangular matrices. This is the method that is generally used in computer programs for solving large-scale linear systems in real-world applications.

9.1 APPLICATION TO DIFFERENTIAL EQUATIONS

Many laws of physics, chemistry, biology, engineering, and economics are described in terms of differential equations, that is, equations involving functions and their derivatives. The purpose of this section is to illustrate one way in which linear algebra can be applied to certain systems of differential equations. The scope of this section is narrow, but it illustrates an important area of application of linear algebra.

Terminology   One of the simplest differential equations is

$$y' = ay \tag{1}$$

where $y = f(x)$ is an unknown function to be determined, $y' = dy/dx$ is its derivative, and a is a constant. Like most differential equations, (1) has infinitely many solutions; they are the functions of the form

$$y = ce^{ax} \tag{2}$$

where c is an arbitrary constant. Each function of this form is a solution of $y' = ay$ since

$$y' = cae^{ax} = ay$$

Conversely, every solution of $y' = ay$ must be a function of the form ce^{ax} (Exercise 5), so that (2) describes all solutions of $y' = ay$. We call (2) the ***general solution*** of $y' = ay$.

Sometimes the physical problem that generates a differential equation imposes some added conditions that enable us to isolate one ***particular solution*** from the general solution. For example, if we require that the solution of $y' = ay$ satisfy the added condition

$$y(0) = 3 \tag{3}$$

that is, $y = 3$ when $x = 0$, then on substituting these values in the general solution $y = ce^{ax}$ we obtain a value for c, namely, $3 = ce^0 = c$. Thus,

$$y = 3e^{ax}$$

is the only solution of $y' = ay$ that satisfies the added condition. A condition such as (3), which specifies the value of the solution at a point, is called an ***initial condition***, and the problem of solving a differential equation subject to an initial condition is called an ***initial-value problem***.

Linear Systems of First-Order Equations   In this section we will be concerned with solving systems of differential equations having the form

$$
\begin{aligned}
y_1' &= a_{11}y_1 + a_{12}y_2 + \cdots + a_{1n}y_n \\
y_2' &= a_{21}y_1 + a_{22}y_2 + \cdots + a_{2n}y_n \\
&\ \ \vdots \qquad \vdots \qquad \vdots \qquad\qquad \vdots \\
y_n' &= a_{n1}y_1 + a_{n2}y_2 + \cdots + a_{nn}y_n
\end{aligned}
\tag{4}
$$

where $y_1 = f_1(x)$, $y_2 = f_2(x)$, ..., $y_n = f_n(x)$ are functions to be determined, and the a_{ij}'s are constants. In matrix notation (4) can be written as

$$
\begin{bmatrix} y_1' \\ y_2' \\ \vdots \\ y_n' \end{bmatrix} =
\begin{bmatrix}
a_{11} & a_{12} & \cdots & a_{1n} \\
a_{21} & a_{22} & \cdots & a_{2n} \\
\vdots & \vdots & & \vdots \\
a_{n1} & a_{n2} & \cdots & a_{nn}
\end{bmatrix}
\begin{bmatrix} y_1 \\ y_2 \\ \vdots \\ y_n \end{bmatrix}
$$

or more briefly,

$$\mathbf{y}' = A\mathbf{y}$$

EXAMPLE 1 Solution of a System with Initial Conditions

(a) Write the following system in matrix form:

$$
\begin{aligned}
y_1' &= 3y_1 \\
y_2' &= -2y_2 \\
y_3' &= 5y_3
\end{aligned}
$$

(b) Solve the system.

(c) Find a solution of the system that satisfies the initial conditions $y_1(0) = 1$, $y_2(0) = 4$, and $y_3(0) = -2$.

Solution (a).

$$
\begin{bmatrix} y_1' \\ y_2' \\ y_3' \end{bmatrix} =
\begin{bmatrix} 3 & 0 & 0 \\ 0 & -2 & 0 \\ 0 & 0 & 5 \end{bmatrix}
\begin{bmatrix} y_1 \\ y_2 \\ y_3 \end{bmatrix}
\tag{5}
$$

or

$$
\mathbf{y}' = \begin{bmatrix} 3 & 0 & 0 \\ 0 & -2 & 0 \\ 0 & 0 & 5 \end{bmatrix} \mathbf{y}
$$

Solution (b). Because each equation involves only one unknown function, we can solve the equations individually. From (2), we obtain

$$
\begin{aligned}
y_1 &- c_1 e^{3x} \\
y_2 &= c_2 e^{-2x} \\
y_3 &= c_3 e^{5x}
\end{aligned}
$$

or in matrix notation,

$$
\mathbf{y} = \begin{bmatrix} y_1 \\ y_2 \\ y_3 \end{bmatrix} =
\begin{bmatrix} c_1 e^{3x} \\ c_2 e^{-2x} \\ c_3 e^{5x} \end{bmatrix}
\tag{6}
$$

Solution (c). From the given initial conditions, we obtain

$$
\begin{aligned}
1 &= y_1(0) = c_1 e^0 = c_1 \\
4 &= y_2(0) = c_2 e^0 = c_2 \\
-2 &= y_3(0) = c_3 e^0 = c_3
\end{aligned}
$$

so that the solution satisfying the initial conditions is

$$
y_1 = e^{3x}, \qquad y_2 = 4e^{-2x}, \qquad y_3 = -2e^{5x}
$$

or in matrix notation,

$$
\mathbf{y} = \begin{bmatrix} y_1 \\ y_2 \\ y_3 \end{bmatrix} =
\begin{bmatrix} e^{3x} \\ 4e^{-2x} \\ -2e^{5x} \end{bmatrix}
\qquad\qquad\blacklozenge
$$

The system in the preceding example is easy to solve because each equation involves only one unknown function, and this is the case because the matrix of coefficients for the system in (5) is diagonal. But how do we handle a system $\mathbf{y}' = A\mathbf{y}$ in which the matrix A is not diagonal? The idea is simple: Try to make a substitution for $\mathbf{y}$ that will yield a new system with a diagonal coefficient matrix; solve this new simpler system, and then use this solution to determine the solution of the original system.

The kind of substitution we have in mind is

$$
\begin{aligned}
y_1 &= p_{11}u_1 + p_{12}u_2 + \cdots + p_{1n}u_n \\
y_2 &= p_{21}u_1 + p_{22}u_2 + \cdots + p_{2n}u_n \\
&\vdots \qquad \vdots \qquad \vdots \qquad \qquad \vdots \\
y_n &= p_{n1}u_1 + p_{n2}u_2 + \cdots + p_{nn}u_n
\end{aligned} \tag{7}
$$

or in matrix notation,

$$
\begin{bmatrix} y_1 \\ y_2 \\ \vdots \\ y_n \end{bmatrix} = \begin{bmatrix} p_{11} & p_{12} & \cdots & p_{1n} \\ p_{21} & p_{22} & \cdots & p_{2n} \\ \vdots & \vdots & & \vdots \\ p_{n1} & p_{n2} & \cdots & p_{nn} \end{bmatrix} \begin{bmatrix} u_1 \\ u_2 \\ \vdots \\ u_n \end{bmatrix} \quad \text{or more briefly,} \quad \mathbf{y} = P\mathbf{u}
$$

In this substitution the p_{ij}'s are constants to be determined in such a way that the new system involving the unknown functions $u_1, u_2, \ldots, u_n$ has a diagonal coefficient matrix. We leave it for the reader to differentiate each equation in (7) and deduce

$$\mathbf{y}' = P\mathbf{u}'$$

If we make the substitutions $\mathbf{y} = P\mathbf{u}$ and $\mathbf{y}' = P\mathbf{u}'$ in the original system

$$\mathbf{y}' = A\mathbf{y}$$

and if we assume P to be invertible, we obtain

$$P\mathbf{u}' = A(P\mathbf{u})$$

or

$$\mathbf{u}' = (P^{-1}AP)\mathbf{u} \quad \text{or} \quad \mathbf{u}' = D\mathbf{u}$$

where $D = P^{-1}AP$. The choice for P is now clear; if we want the new coefficient matrix D to be diagonal, we must choose P to be a matrix that diagonalizes A.

Solution by Diagonalization The preceding discussion suggests the following procedure for solving a system $\mathbf{y}' = A\mathbf{y}$ with a diagonalizable coefficient matrix A.

Step 1. Find a matrix P that diagonalizes A.

Step 2. Make the substitutions $\mathbf{y} = P\mathbf{u}$ and $\mathbf{y}' = P\mathbf{u}'$ to obtain a new "diagonal system" $\mathbf{u}' = D\mathbf{u}$, where $D = P^{-1}AP$.

Step 3. Solve $\mathbf{u}' = D\mathbf{u}$.

Step 4. Determine $\mathbf{y}$ from the equation $\mathbf{y} = P\mathbf{u}$.

EXAMPLE 2 Solution Using Diagonalization

(a) Solve the system

$$
\begin{aligned}
y_1' &= y_1 + y_2 \\
y_2' &= 4y_1 - 2y_2
\end{aligned}
$$

(b) Find the solution that satisfies the initial conditions $y_1(0) = 1$, $y_2(0) = 6$.

Solution (a). The coefficient matrix for the system is

$$A = \begin{bmatrix} 1 & 1 \\ 4 & -2 \end{bmatrix}$$

As discussed in Section 7.2, A will be diagonalized by any matrix P whose columns are linearly independent eigenvectors of A. Since

$$\det(\lambda I - A) = \begin{vmatrix} \lambda - 1 & -1 \\ -4 & \lambda + 2 \end{vmatrix} = \lambda^2 + \lambda - 6 = (\lambda + 3)(\lambda - 2)$$

the eigenvalues of A are $\lambda = 2$, $\lambda = -3$. By definition,

$$\mathbf{x} = \begin{bmatrix} x_1 \\ x_2 \end{bmatrix}$$

is an eigenvector of A corresponding to λ if and only if $\mathbf{x}$ is a nontrivial solution of $(\lambda I - A)\mathbf{x} = \mathbf{0}$, that is, of

$$\begin{bmatrix} \lambda - 1 & -1 \\ -4 & \lambda + 2 \end{bmatrix} \begin{bmatrix} x_1 \\ x_2 \end{bmatrix} = \begin{bmatrix} 0 \\ 0 \end{bmatrix}$$

If $\lambda = 2$, this system becomes

$$\begin{bmatrix} 1 & -1 \\ -4 & 4 \end{bmatrix} \begin{bmatrix} x_1 \\ x_2 \end{bmatrix} = \begin{bmatrix} 0 \\ 0 \end{bmatrix}$$

Solving this system yields $x_1 = t$, $x_2 = t$, so that

$$\begin{bmatrix} x_1 \\ x_2 \end{bmatrix} = \begin{bmatrix} t \\ t \end{bmatrix} = t \begin{bmatrix} 1 \\ 1 \end{bmatrix}$$

Thus,

$$\mathbf{p}_1 = \begin{bmatrix} 1 \\ 1 \end{bmatrix}$$

is a basis for the eigenspace corresponding to $\lambda = 2$. Similarly, the reader can show that

$$\mathbf{p}_2 = \begin{bmatrix} -\frac{1}{4} \\ 1 \end{bmatrix}$$

is a basis for the eigenspace corresponding to $\lambda = -3$. Thus,

$$P = \begin{bmatrix} 1 & -\frac{1}{4} \\ 1 & 1 \end{bmatrix}$$

diagonalizes A, and

$$D = P^{-1}AP = \begin{bmatrix} 2 & 0 \\ 0 & -3 \end{bmatrix}$$

Therefore, the substitution

$$\mathbf{y} = P\mathbf{u} \quad \text{and} \quad \mathbf{y}' = P\mathbf{u}'$$

yields the new "diagonal system"

$$\mathbf{u}' = D\mathbf{u} = \begin{bmatrix} 2 & 0 \\ 0 & -3 \end{bmatrix} \mathbf{u} \quad \text{or} \quad \begin{matrix} u_1' = 2u_1 \\ u_2' = -3u_2 \end{matrix}$$

From (2) the solution of this system is

$$u_1 = c_1 e^{2x} \qquad \text{or} \quad \mathbf{u} = \begin{bmatrix} c_1 e^{2x} \\ c_2 e^{-3x} \end{bmatrix}$$
$$u_2 = c_2 e^{-3x}$$

so that the equation $\mathbf{y} = P\mathbf{u}$ yields as the solution for $\mathbf{y}$

$$\mathbf{y} = \begin{bmatrix} y_1 \\ y_2 \end{bmatrix} = \begin{bmatrix} 1 & -\frac{1}{4} \\ 1 & 1 \end{bmatrix} \begin{bmatrix} c_1 e^{2x} \\ c_2 e^{-3x} \end{bmatrix} = \begin{bmatrix} c_1 e^{2x} - \frac{1}{4} c_2 e^{-3x} \\ c_1 e^{2x} + c_2 e^{-3x} \end{bmatrix}$$

or

$$y_1 = c_1 e^{2x} - \tfrac{1}{4} c_2 e^{-3x}$$
$$y_2 = c_1 e^{2x} + c_2 e^{-3x} \tag{8}$$

Solution (b). If we substitute the given initial conditions in (8), we obtain

$$c_1 - \tfrac{1}{4} c_2 = 1$$
$$c_1 + c_2 = 6$$

Solving this system we obtain $c_1 = 2$, $c_2 = 4$, so that from (8) the solution satisfying the initial conditions is

$$y_1 = 2e^{2x} - e^{-3x}$$
$$y_2 = 2e^{2x} + 4e^{-3x} \qquad \blacklozenge$$

We have assumed in this section that the coefficient matrix of $\mathbf{y}' = A\mathbf{y}$ is diagonalizable. If this is not the case, other methods must be used to solve the system. Such methods are discussed in more advanced texts.

Exercise Set 9.1

1. (a) Solve the system

$$y_1' = y_1 + 4y_2$$
$$y_2' = 2y_1 + 3y_2$$

(b) Find the solution that satisfies the initial conditions $y_1(0) = 0$, $y_2(0) = 0$.

2. (a) Solve the system

$$y_1' = y_1 + 3y_2$$
$$y_2' = 4y_1 + 5y_2$$

(b) Find the solution that satisfies the conditions $y_1(0) = 2$, $y_2'(0) = 1$.

3. (a) Solve the system

$$y_1' = 4y_1 + y_3$$
$$y_2' = -2y_1 + y_2$$
$$y_3' = -2y_1 + y_3$$

(b) Find the solution that satisfies the initial conditions $y_1(0) = -1$, $y_2(0) = 1$, $y_3(0) = 0$.

4. Solve the system

$$y_1' = 4y_1 + 2y_2 + 2y_3$$
$$y_2' = 2y_1 + 4y_2 + 2y_3$$
$$y_3' = 2y_1 + 2y_2 + 4y_3$$

5. Show that every solution of $y' = ay$ has the form $y = ce^{ax}$. [***Hint.*** Let $y = f(x)$ be a solution of the equation and show that $f(x)e^{-ax}$ is constant.]

6. Show that if A is diagonalizable and

$$\mathbf{y} = \begin{bmatrix} y_1 \\ y_2 \\ \vdots \\ y_n \end{bmatrix}$$

satisfies $\mathbf{y}' = A\mathbf{y}$, then each y_i is a linear combination of $e^{\lambda_1 x}, e^{\lambda_2 x}, \ldots, e^{\lambda_n x}$, where $\lambda_1, \lambda_2, \ldots, \lambda_n$ are the eigenvalues of A.

7. It is possible to solve a single differential equation by expressing the equation as a system and then using the method of this section. For the differential equation $y'' - y' - 6y = 0$, show that the substitutions $y_1 = y$ and $y_2 = y'$ lead to the system

$$y_1' = y_2$$
$$y_2' = 6y_1 + y_2$$

Solve this system and then solve the original differential equation.

8. Discuss: How can the procedure in Exercise 7 by used to solve $y''' - 6y'' + 11y' - 6y = 0$? Carry out your ideas.

Discussion and Discovery

9. (a) By rewriting (8) in matrix form, show that the solution of the system in Example 2 can be expressed as

$$\mathbf{y} = c_1 e^{2x} \begin{bmatrix} 1 \\ 1 \end{bmatrix} + c_2 e^{-3x} \begin{bmatrix} -\frac{1}{4} \\ 1 \end{bmatrix}$$

This is called the *general solution* of the system.

(b) Note that in part (a) the vector in the first term is an eigenvector corresponding to the eigenvalue $\lambda_1 = 2$ and the vector in the second term is an eigenvector corresponding to the eigenvalue $\lambda_2 = -3$. This is a special case of the following general result:

> ### Theorem
>
> *If the coefficient matrix A of the system $\mathbf{y}' = A\mathbf{y}$ in (4) is diagonalizable, then the general solution of the system can be expressed as*
>
> $$\mathbf{y} = c_1 e^{\lambda_1 x} \mathbf{x}_1 + c_2 e^{\lambda_2 x} \mathbf{x}_2 + \cdots + c_n e^{\lambda_n x} \mathbf{x}_n$$
>
> *where $\lambda_1, \lambda_2, \ldots, \lambda_n$ are the eigenvalues of A, and $\mathbf{x}_i$ is an eigenvector of A corresponding to λ_i.*

Prove this result by tracing through the four-step procedure discussed in the section with

$$D = \begin{bmatrix} \lambda_1 & 0 & \cdots & 0 \\ 0 & \lambda_2 & \cdots & 0 \\ \vdots & \vdots & & \vdots \\ 0 & 0 & \cdots & \lambda_n \end{bmatrix} \quad \text{and} \quad P = [\mathbf{x}_1 \mid \mathbf{x}_2 \mid \cdots \mid \mathbf{x}_n].$$

9.2 GEOMETRY OF LINEAR OPERATORS ON R^2

In Section 4.2 we studied some of the geometric properties of linear operators on R^2 and R^3. In this section we shall study linear operators on R^2 in a little more depth. Some of the ideas that will be developed here have important applications to the field of computer graphics.

Vectors or Points   If $T: R^2 \to R^2$ is the matrix operator whose standard matrix is

$$A = \begin{bmatrix} a & b \\ c & d \end{bmatrix}$$

then

$$T\left(\begin{bmatrix} x \\ y \end{bmatrix}\right) = \begin{bmatrix} a & b \\ c & d \end{bmatrix} \begin{bmatrix} x \\ y \end{bmatrix} = \begin{bmatrix} ax + by \\ cx + dy \end{bmatrix} \tag{1}$$

There are two equally good geometric interpretations of this formula. We may view the entries in the matrices

$$\begin{bmatrix} x \\ y \end{bmatrix} \quad \text{and} \quad \begin{bmatrix} ax + by \\ cx + dy \end{bmatrix}$$

either as components of vectors or coordinates of points. With the first interpretation, T maps arrows to arrows, and with the second, points to points (Figure 9.2.1). The choice is a matter of taste.

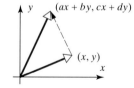
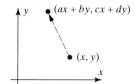

(*a*) T maps vectors to vectors.   (*b*) T maps points to points.

Figure 9.2.1

In this section we shall view linear operators on R^2 as mapping points to points. One useful device for visualizing the behavior of a linear operator is to observe its effect on the points of simple figures in the plane. For example, Table 1 shows the effect of some basic linear operators on a unit square that has been partially colored.

In Section 4.2 we discussed reflections, projections, rotations, contractions, and dilations of R^2. We shall now consider some other basic linear operators on R^2.

Expansions and Compressions   If the x-coordinate of each point in the plane is multiplied by a positive constant k, then the effect is to expand or compress each plane figure in the x-direction. If $0 < k < 1$, the result is a compression, and if $k > 1$, an expansion (Figure 9.2.2). We call such an operator an ***expansion*** (or ***compression***) ***in the x-direction with factor k***. Similarly, if the y-coordinate of each point is multiplied by a positive constant k, we obtain an ***expansion*** (or ***compression***) ***in the y-direction with factor k***. It can be shown that expansions and compressions along the coordinate axes are linear transformations.

TABLE 1

Operator	Standard Matrix	Effect on the Unit Square
Reflection about the y-axis	$\begin{bmatrix} -1 & 0 \\ 0 & 1 \end{bmatrix}$	
Reflection about the x-axis	$\begin{bmatrix} 1 & 0 \\ 0 & -1 \end{bmatrix}$	
Reflection about the line $y = x$	$\begin{bmatrix} 0 & 1 \\ 1 & 0 \end{bmatrix}$	
Counterclockwise rotation through an angle θ	$\begin{bmatrix} \cos\theta & -\sin\theta \\ \sin\theta & \cos\theta \end{bmatrix}$	

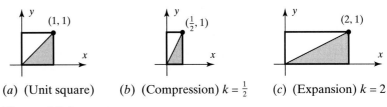

(a) (Unit square)   (b) (Compression) $k = \frac{1}{2}$ (c) (Expansion) $k = 2$

Figure 9.2.2

If $T: R^2 \rightarrow R^2$ is an expansion or compression in the x-direction with factor k, then

$$T(\mathbf{e}_1) = T\left(\begin{bmatrix} 1 \\ 0 \end{bmatrix}\right) = \begin{bmatrix} k \\ 0 \end{bmatrix}, \qquad T(\mathbf{e}_2) = T\left(\begin{bmatrix} 0 \\ 1 \end{bmatrix}\right) = \begin{bmatrix} 0 \\ 1 \end{bmatrix}$$

so the standard matrix for T is

$$\begin{bmatrix} k & 0 \\ 0 & 1 \end{bmatrix}$$

Similarly, the standard matrix for an expansion or compression in the y-direction is

$$\begin{bmatrix} 1 & 0 \\ 0 & k \end{bmatrix}$$

EXAMPLE 1 Operating with Diagonal Matrices

Suppose that the xy-plane is first expanded or compressed by a factor of k_1 in the x-direction and then is expanded or compressed by a factor of k_2 in the y-direction. Find a single matrix operator that performs both operations.

Solution.

The standard matrices for the two operations are

$$\begin{bmatrix} k_1 & 0 \\ 0 & 1 \end{bmatrix} \qquad \begin{bmatrix} 1 & 0 \\ 0 & k_2 \end{bmatrix}$$

x-expansion (compression) y-expansion (compression)

Thus, the standard matrix for the composition of the x-operation followed by the y-operation is

$$A = \begin{bmatrix} 1 & 0 \\ 0 & k_2 \end{bmatrix} \begin{bmatrix} k_1 & 0 \\ 0 & 1 \end{bmatrix} = \begin{bmatrix} k_1 & 0 \\ 0 & k_2 \end{bmatrix} \tag{2}$$

This shows that multiplication by a diagonal 2×2 matrix expands or compresses the plane in the x-direction and also in the y-direction. In the special case where k_1 and k_2 are the same, say $k_1 = k_2 = k$, note that (2) simplifies to

$$A = \begin{bmatrix} k & 0 \\ 0 & k \end{bmatrix}$$

which is a dilation or a contraction (Table 8 of Section 4.2). ◆

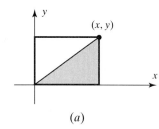

(a)

(b) Shear in x-direction with factor $k > 0$

(c) Shear in x-direction with factor $k < 0$

Figure 9.2.3

Shears

A *shear in the x-direction with factor k* is a transformation that moves each point (x, y) parallel to the x-axis by an amount ky to the new position $(x + ky, y)$. Under such a transformation, points on the x-axis are unmoved since $y = 0$. However, as we progress away from the x-axis, the magnitude of y increases, so that points farther from the x-axis move a greater distance than those closer (Figure 9.2.3).

A *shear in the y-direction with factor k* is a transformation that moves each point (x, y) parallel to the y-axis by an amount kx to the new position $(x, y + kx)$. Under such a transformation, points on the y-axis remain fixed, and points farther from the y-axis move a greater distance than do those closer.

It can be shown that shears are linear transformations. If $T: R^2 \rightarrow R^2$ is a shear with factor k in the x-direction, then

$$T(\mathbf{e}_1) = T\left(\begin{bmatrix} 1 \\ 0 \end{bmatrix}\right) = \begin{bmatrix} 1 \\ 0 \end{bmatrix}, \qquad T(\mathbf{e}_2) = T\left(\begin{bmatrix} 0 \\ 1 \end{bmatrix}\right) = \begin{bmatrix} k \\ 1 \end{bmatrix}$$

so the standard matrix for T is

$$\begin{bmatrix} 1 & k \\ 0 & 1 \end{bmatrix}$$

Similarly, the standard matrix for a shear in the y-direction with factor k is

$$\begin{bmatrix} 1 & 0 \\ k & 1 \end{bmatrix}$$

REMARK. Multiplication by the 2×2 identity matrix is the identity operator on R^2. This operator can be viewed as a rotation through $0°$, or as a shear along either axis with $k = 0$, or as a compression or expansion along either axis with factor $k = 1$.

EXAMPLE 2 Finding Matrix Transformations

(a) Find a matrix transformation from R^2 to R^2 that first shears by a factor of 2 in the x-direction and then reflects about $y = x$.

(b) Find a matrix transformation from R^2 to R^2 that first reflects about $y = x$ and then shears by a factor of 2 in the x-direction.

Solution (a). The standard matrix for the shear is

$$A_1 = \begin{bmatrix} 1 & 2 \\ 0 & 1 \end{bmatrix}$$

and for the reflection is

$$A_2 = \begin{bmatrix} 0 & 1 \\ 1 & 0 \end{bmatrix}$$

Thus, the standard matrix for the shear followed by the reflection is

$$A_2 A_1 = \begin{bmatrix} 0 & 1 \\ 1 & 0 \end{bmatrix} \begin{bmatrix} 1 & 2 \\ 0 & 1 \end{bmatrix} = \begin{bmatrix} 0 & 1 \\ 1 & 2 \end{bmatrix}$$

Solution (b). The reflection followed by the shear is represented by

$$A_1 A_2 = \begin{bmatrix} 1 & 2 \\ 0 & 1 \end{bmatrix} \begin{bmatrix} 0 & 1 \\ 1 & 0 \end{bmatrix} = \begin{bmatrix} 2 & 1 \\ 1 & 0 \end{bmatrix} \qquad \blacklozenge$$

In the last example, note that $A_1 A_2 \neq A_2 A_1$, so that the effect of shearing and then reflecting is different from reflecting and then shearing. This is illustrated geometrically in Figure 9.2.4, where we show the effects of the transformations on a unit square.

EXAMPLE 3 Transformations Using Elementary Matrices

Show that if $T: R^2 \rightarrow R^2$ is multiplication by an *elementary matrix*, then the transformation is one of the following:

(a) a shear along a coordinate axis
(b) a reflection about $y = x$
(c) a compression along a coordinate axis
(d) an expansion along a coordinate axis
(e) a reflection about a coordinate axis
(f) a compression or expansion along a coordinate axis followed by a reflection about a coordinate axis.

Solution.

Because a 2×2 elementary matrix results from performing a single elementary row operation on the 2×2 identity matrix, it must have one of the following forms (verify):

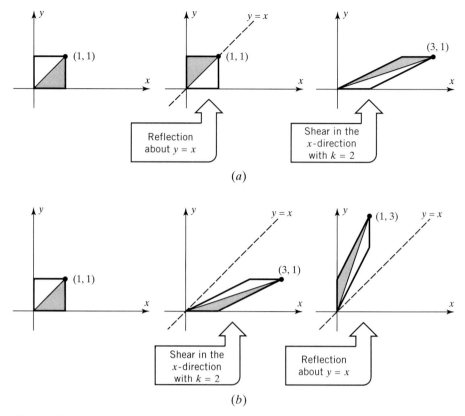

Figure 9.2.4

$$\begin{bmatrix} 1 & 0 \\ k & 1 \end{bmatrix}, \quad \begin{bmatrix} 1 & k \\ 0 & 1 \end{bmatrix}, \quad \begin{bmatrix} 0 & 1 \\ 1 & 0 \end{bmatrix}, \quad \begin{bmatrix} k & 0 \\ 0 & 1 \end{bmatrix}, \quad \begin{bmatrix} 1 & 0 \\ 0 & k \end{bmatrix}$$

The first two matrices represent shears along coordinate axes, and the third, a reflection about $y = x$. If $k > 0$, the last two matrices represent compressions or expansions along coordinate axes depending on whether $0 \leq k \leq 1$ or $k \geq 1$. If $k < 0$, and if we express k in the form $k = -k_1$, where $k_1 > 0$, then the last two matrices can be written as

$$\begin{bmatrix} k & 0 \\ 0 & 1 \end{bmatrix} = \begin{bmatrix} -k_1 & 0 \\ 0 & 1 \end{bmatrix} = \begin{bmatrix} -1 & 0 \\ 0 & 1 \end{bmatrix} \begin{bmatrix} k_1 & 0 \\ 0 & 1 \end{bmatrix} \tag{3}$$

$$\begin{bmatrix} 1 & 0 \\ 0 & k \end{bmatrix} = \begin{bmatrix} 1 & 0 \\ 0 & -k_1 \end{bmatrix} = \begin{bmatrix} 1 & 0 \\ 0 & -1 \end{bmatrix} \begin{bmatrix} 1 & 0 \\ 0 & k_1 \end{bmatrix} \tag{4}$$

Since $k_1 > 0$, the product in (3) represents a compression or expansion along the x-axis followed by a reflection about the y-axis, and (4) represents a compression or expansion along the y-axis followed by a reflection about the x-axis. In the case where $k = -1$, transformations (3) and (4) are simply reflections about the y-axis and x-axis, respectively. ◆

Reflections, rotations, expansions, compressions, and shears are all one-to-one linear operators. This is evident geometrically, since all of those operators map distinct points into distinct points. This can also be checked algebraically by verifying that the standard matrices for those operators are invertible.

EXAMPLE 4 A Transformation and Its Inverse

It is intuitively clear that if we compress the xy-plane by a factor of $\frac{1}{2}$ in the y-direction, then we must expand the xy-plane by a factor of 2 in the y-direction to move each point back to its original position. This is indeed the case, since

$$A = \begin{bmatrix} 1 & 0 \\ 0 & \frac{1}{2} \end{bmatrix}$$

represents a compression of factor $\frac{1}{2}$ in the y-direction, and

$$A^{-1} = \begin{bmatrix} 1 & 0 \\ 0 & 2 \end{bmatrix}$$

is an expansion of factor 2 in the y-direction. ◆

Geometric Properties of Linear Operators on R^2

We conclude this section with two theorems that provide some insight into the geometric properties of linear operators on R^2.

Theorem 9.2.1

If $T: R^2 \to R^2$ is multiplication by an invertible matrix A, then the geometric effect of T is the same as an appropriate succession of shears, compressions, expansions, and reflections.

Proof. Since A is invertible, it can be reduced to the identity by a finite sequence of elementary row operations. An elementary row operation can be performed by multiplying on the left by an elementary matrix, and so there exist elementary matrices $E_1, E_2, \ldots, E_k$ such that

$$E_k \cdots E_2 E_1 A = I$$

Solving for A yields

$$A = E_1^{-1} E_2^{-1} \cdots E_k^{-1} I$$

or equivalently,

$$A = E_1^{-1} E_2^{-1} \cdots E_k^{-1} \tag{5}$$

This equation expresses A as a product of elementary matrices (since the inverse of an elementary matrix is also elementary by Theorem 1.5.2). The result now follows from Example 3. ∎

EXAMPLE 5 Geometric Effect of Multiplication by a Matrix

Assuming that k_1 and k_2 are positive, express the diagonal matrix

$$A = \begin{bmatrix} k_1 & 0 \\ 0 & k_2 \end{bmatrix}$$

as a product of elementary matrices and describe the geometric effect of multiplication by A in terms of expansions and compressions.

Solution.

From Example 1 we have

$$A = \begin{bmatrix} k_1 & 0 \\ 0 & k_2 \end{bmatrix} = \begin{bmatrix} 1 & 0 \\ 0 & k_2 \end{bmatrix} \begin{bmatrix} k_1 & 0 \\ 0 & 1 \end{bmatrix}$$

which shows that multiplication by A has the geometric effect of expanding or compressing by a factor of k_1 in the x-direction and then expanding or compressing by a factor of k_2 in the y-direction. ◆

EXAMPLE 6 Analyzing the Geometric Effect of a Matrix Operator

Express

$$A = \begin{bmatrix} 1 & 2 \\ 3 & 4 \end{bmatrix}$$

as a product of elementary matrices, and then describe the geometric effect of multiplication by A in terms of shears, compressions, expansions, and reflections.

Solution.

A can be reduced to I as follows:

$$\begin{bmatrix} 1 & 2 \\ 3 & 4 \end{bmatrix} \longrightarrow \begin{bmatrix} 1 & 2 \\ 0 & -2 \end{bmatrix} \longrightarrow \begin{bmatrix} 1 & 2 \\ 0 & 1 \end{bmatrix} \longrightarrow \begin{bmatrix} 1 & 0 \\ 0 & 1 \end{bmatrix}$$

| Add -3 times the first row to the second. | Multiply the second row by $-\frac{1}{2}$. | Add -2 times the second row to the first. |

The three successive row operations can be performed by multiplying on the left successively by

$$E_1 = \begin{bmatrix} 1 & 0 \\ -3 & 1 \end{bmatrix}, \qquad E_2 = \begin{bmatrix} 1 & 0 \\ 0 & -\frac{1}{2} \end{bmatrix}, \qquad E_3 = \begin{bmatrix} 1 & -2 \\ 0 & 1 \end{bmatrix}$$

Inverting these matrices and using (5) yields

$$A = E_1^{-1} E_2^{-1} E_3^{-1} = \begin{bmatrix} 1 & 0 \\ 3 & 1 \end{bmatrix} \begin{bmatrix} 1 & 0 \\ 0 & -2 \end{bmatrix} \begin{bmatrix} 1 & 2 \\ 0 & 1 \end{bmatrix}$$

Reading from right to left and noting that

$$\begin{bmatrix} 1 & 0 \\ 0 & -2 \end{bmatrix} = \begin{bmatrix} 1 & 0 \\ 0 & -1 \end{bmatrix} \begin{bmatrix} 1 & 0 \\ 0 & 2 \end{bmatrix}$$

it follows that the effect of multiplying by A is equivalent to

(1) shearing by a factor of 2 in the x-direction,
(2) then expanding by a factor of 2 in the y-direction,
(3) then reflecting about the x-axis,
(4) then shearing by a factor of 3 in the y-direction. ◆

The proofs for parts of the following theorem are discussed in the exercises.

Theorem 9.2.2 **Images of Lines**

If $T: R^2 \to R^2$ is multiplication by an invertible matrix, then:

(a) *The image of a straight line is a straight line.*

(b) *The image of a straight line through the origin is a straight line through the origin.*

(c) *The images of parallel straight lines are parallel straight lines.*

(d) *The image of the line segment joining points P and Q is the line segment joining the images of P and Q.*

(e) *The images of three points lie on a line if and only if the points themselves lie on some line.*

REMARK.   It follows from parts (c), (d), and (e) that multiplication by an invertible 2×2 matrix A maps triangles into triangles and parallelograms into parallelograms.

EXAMPLE 7 Image of a Square

The square with vertices $P_1(0, 0)$, $P_2(1, 0)$, $P_3(1, 1)$, and $P_4(0, 1)$ is called the **unit square**. Sketch the image of the unit square under multiplication by

$$A = \begin{bmatrix} -1 & 2 \\ 2 & -1 \end{bmatrix}$$

Solution.

Since

$$\begin{bmatrix} -1 & 2 \\ 2 & -1 \end{bmatrix}\begin{bmatrix} 0 \\ 0 \end{bmatrix} = \begin{bmatrix} 0 \\ 0 \end{bmatrix} \qquad \begin{bmatrix} -1 & 2 \\ 2 & -1 \end{bmatrix}\begin{bmatrix} 1 \\ 0 \end{bmatrix} = \begin{bmatrix} -1 \\ 2 \end{bmatrix}$$

$$\begin{bmatrix} -1 & 2 \\ 2 & -1 \end{bmatrix}\begin{bmatrix} 0 \\ 1 \end{bmatrix} = \begin{bmatrix} 2 \\ -1 \end{bmatrix} \qquad \begin{bmatrix} -1 & 2 \\ 2 & -1 \end{bmatrix}\begin{bmatrix} 1 \\ 1 \end{bmatrix} = \begin{bmatrix} 1 \\ 1 \end{bmatrix}$$

the image of the square is a parallelogram with vertices $(0, 0)$, $(-1, 2)$, $(2, -1)$, and $(1, 1)$ (Figure 9.2.5). ◆

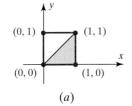

(a)

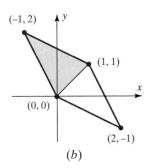

(b)

Figure 9.2.5

EXAMPLE 8 Image of a Line

According to Theorem 9.2.2, the invertible matrix

$$A = \begin{bmatrix} 3 & 1 \\ 2 & 1 \end{bmatrix}$$

maps the line $y = 2x + 1$ into another line. Find its equation.

Solution.

Let (x, y) be a point on the line $y = 2x + 1$ and let (x', y') be its image under multiplication by A. Then

$$\begin{bmatrix} x' \\ y' \end{bmatrix} = \begin{bmatrix} 3 & 1 \\ 2 & 1 \end{bmatrix}\begin{bmatrix} x \\ y \end{bmatrix} \quad \text{and} \quad \begin{bmatrix} x \\ y \end{bmatrix} = \begin{bmatrix} 3 & 1 \\ 2 & 1 \end{bmatrix}^{-1}\begin{bmatrix} x' \\ y' \end{bmatrix} = \begin{bmatrix} 1 & -1 \\ -2 & 3 \end{bmatrix}\begin{bmatrix} x' \\ y' \end{bmatrix}$$

so

$$x = x' - y'$$
$$y = -2x' + 3y'$$

Substituting in $y = 2x + 1$ yields

$$-2x' + 3y' = 2(x' - y') + 1 \quad \text{or equivalently,} \quad y' = \tfrac{4}{5}x' + \tfrac{1}{5}$$

Thus, (x', y') satisfies

$$y = \tfrac{4}{5}x + \tfrac{1}{5}$$

which is the equation we want.   ◆

Exercise Set 9.2

1. Find the standard matrix for the plane linear transformation $T : R^2 \to R^2$ that maps a point (x, y) into (see accompanying figure)

(a) its reflection about the line $y = -x$ (b) its reflection through the origin

(c) its orthogonal projection on the x-axis   (d) its orthogonal projection on the y-axis

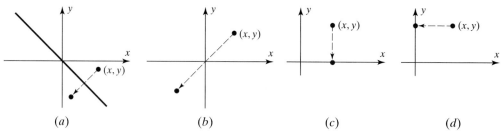

(a)   (b)   (c)   (d)

Figure Ex-1

2. For each part of Exercise 1, use the matrix you have obtained to compute $T(2, 1)$. Check your answers geometrically by plotting the points $(2, 1)$ and $T(2, 1)$.

3. Find the standard matrix for the linear operator $T : R^3 \to R^3$ that maps a point (x, y, z) into

(a) its reflection through the xy-plane   (b) its reflection through the xz-plane

(c) its reflection through the yz-plane

4. For each part of Exercise 3, use the matrix you have obtained to compute $T(1, 1, 1)$. Check your answers geometrically by sketching the vectors $(1, 1, 1)$ and $T(1, 1, 1)$.

5. Find the standard matrix for the linear operator $T : R^3 \to R^3$ that

(a) rotates each vector 90° counterclockwise about the z-axis (looking along the positive z-axis toward the origin)

(b) rotates each vector 90° counterclockwise about the x-axis (looking along the positive x-axis toward the origin)

(c) rotates each vector 90° counterclockwise about the y-axis (looking along the positive y-axis toward the origin)

6. Sketch the image of the rectangle with vertices $(0, 0)$, $(1, 0)$, $(1, 2)$, and $(0, 2)$ under

(a) a reflection about the x-axis

(b) a reflection about the y-axis

(c) a compression of factor $k = \tfrac{1}{4}$ in the y-direction

(d) an expansion of factor $k = 2$ in the x-direction

(e) a shear of factor $k = 3$ in the x-direction

(f) a shear of factor $k = 2$ in the y-direction

7. Sketch the image of the square with vertices $(0, 0)$, $(1, 0)$, $(0, 1)$, and $(1, 1)$ under multiplication by

$$A = \begin{bmatrix} -3 & 0 \\ 0 & 1 \end{bmatrix}$$

8. Find the matrix that rotates a point (x, y) about the origin through
   (a) $45°$ (b) $90°$ (c) $180°$ (d) $270°$ (e) $-30°$

9. Find the matrix that shears by
   (a) a factor of $k = 4$ in the y-direction   (b) a factor of $k = -2$ in the x-direction

10. Find the matrix that compresses or expands by
    (a) a factor of $\frac{1}{3}$ in the y-direction   (b) a factor of 6 in the x-direction

11. In each part, describe the geometric effect of multiplication by the given matrix.

    (a) $\begin{bmatrix} 3 & 0 \\ 0 & 1 \end{bmatrix}$ (b) $\begin{bmatrix} 1 & 0 \\ 0 & -5 \end{bmatrix}$ (c) $\begin{bmatrix} 1 & 4 \\ 0 & 1 \end{bmatrix}$

12. Express the matrix as a product of elementary matrices, and then describe the effect of multiplication by the given matrix in terms of compressions, expansions, reflections, and shears.

    (a) $\begin{bmatrix} 2 & 0 \\ 0 & 3 \end{bmatrix}$ (b) $\begin{bmatrix} 1 & 4 \\ 2 & 9 \end{bmatrix}$ (c) $\begin{bmatrix} 0 & -2 \\ 4 & 0 \end{bmatrix}$ (d) $\begin{bmatrix} 1 & -3 \\ 4 & 6 \end{bmatrix}$

13. In each part, find a single matrix that performs the indicated succession of operations:

    (a) compresses by a factor of $\frac{1}{2}$ in the x-direction, then expands by a factor of 5 in the y-direction
    (b) expands by a factor of 5 in the y-direction, then shears by a factor of 2 in the y-direction
    (c) reflects about $y = x$, then rotates through an angle of $180°$ about the origin

14. In each part, find a single matrix that performs the indicated succession of operations:

    (a) reflects about the y-axis, then expands by a factor of 5 in the x-direction, and then reflects about $y = x$
    (b) rotates through $30°$ about the origin, then shears by a factor of -2 in the y-direction, and then expands by a factor of 3 in the y-direction

15. By matrix inversion, show the following:

    (a) The inverse transformation for a reflection about $y = x$ is a reflection about $y = x$.
    (b) The inverse transformation for a compression along an axis is an expansion along that axis.
    (c) The inverse transformation for a reflection about a coordinate axis is a reflection about that axis.
    (d) The inverse transformation for a shear along a coordinate axis is a shear along that axis.

16. Find the equation of the image of the line $y = -4x + 3$ under multiplication by

$$A = \begin{bmatrix} 4 & -3 \\ 3 & -2 \end{bmatrix}$$

17. In parts (a) through (e) find the equation of the image of the line $y = 2x$ under
    (a) a shear of factor 3 in the x-direction   (b) a compression of factor $\frac{1}{2}$ in the y-direction
    (c) a reflection about $y = x$ (d) a reflection about the y-axis
    (e) a rotation of $60°$ about the origin

18. Find the matrix for a shear in the x-direction that transforms the triangle with vertices $(0, 0)$, $(2, 1)$, and $(3, 0)$ into a right triangle with the right angle at the origin.

19. (a) Show that multiplication by

$$A = \begin{bmatrix} 3 & 1 \\ 6 & 2 \end{bmatrix}$$

maps every point in the plane onto the line $y = 2x$.

(b) It follows from (a) that the noncollinear points $(1, 0)$, $(0, 1)$, $(-1, 0)$ are mapped on a line. Does this violate part (e) of Theorem 9.2.2?

20. Prove part (a) of Theorem 9.2.2. [*Hint.* A line in the plane has an equation of the form $Ax + By + C = 0$, where A and B are not both zero. Use the method of Example 8 to show that the image of this line under multiplication by the invertible matrix

$$\begin{bmatrix} a & b \\ c & d \end{bmatrix}$$

has the equation $A'x + B'y + C = 0$, where

$$A' = (dA - cB)/(ad - bc) \quad \text{and} \quad B' = (-bA + aB)/(ad - bc)$$

Then show that A' and B' are not both zero to conclude that the image is a line.]

21. Use the hint in Exercise 20 to prove parts (b) and (c) of Theorem 9.2.2.

22. In each part find the standard matrix for the linear operator $T: R^3 \rightarrow R^3$ described by the accompanying figure.

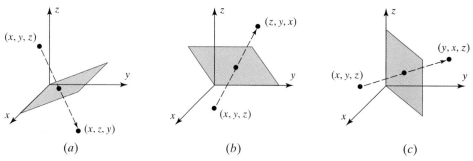

(a)   (b)   (c)

Figure Ex-22

23. In R^3 the **shear in the xy-direction with factor k** is the linear transformation that moves each point (x, y, z) parallel to the xy-plane to the new position $(x + kz, y + kz, z)$. (See accompanying figure.)

(a) Find the standard matrix for the shear in the xy-direction with factor k.

(b) How would you define the shear in the xz-direction with factor k and the shear in the yz-direction with factor k? Find the standard matrices for these linear transformations.

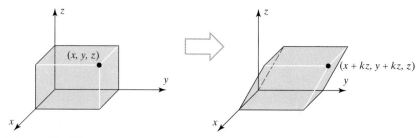

Figure Ex-23

24. In each part find as many linearly independent eigenvectors as you can by inspection (by visualizing the geometric effect of the transformation on R^2). For each of your eigenvectors find the corresponding eigenvalue by inspection, then check your results by computing the eigenvalues and bases for the eigenspaces from the standard matrix for the transformation.

(a) reflection about the x-axis

(b) reflection about the y-axis

(c) reflection about $y = x$

(d) shear in the x-direction with factor k

(e) shear in the y-direction with factor k

(f) rotation through the angle θ

9.3 LEAST SQUARES FITTING TO DATA

In this section we shall use results about orthogonal projections in inner product spaces to obtain a technique for fitting a line or other polynomial curve to a set of experimentally determined points in the plane.

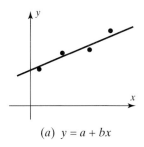

(a) $y = a + bx$

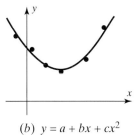

(b) $y = a + bx + cx^2$

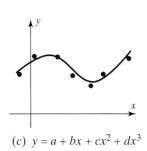

(c) $y = a + bx + cx^2 + dx^3$

Figure 9.3.1

Fitting a Curve to Data   A common problem in experimental work is to obtain a mathematical relationship $y = f(x)$ between two variables x and y by "fitting" a curve to points in the plane corresponding to various experimentally determined values of x and y, say

$$(x_1, y_1), (x_2, y_2), \ldots, (x_n, y_n)$$

On the basis of theoretical considerations or simply by the pattern of the points one decides on the general form of the curve $y = f(x)$ to be fitted. Some possibilities are (Figure 9.3.1)

(a) A straight line: $y = a + bx$

(b) A quadratic polynomial: $y = a + bx + cx^2$

(c) A cubic polynomial: $y = a + bx + cx^2 + dx^3$

Because the points are obtained experimentally, there is usually some measurement "error" in the data, making it impossible to find a curve of the desired form that passes through all the points. Thus, the idea is to choose the curve (by determining its coefficients) that "best" fits the data. We begin with the simplest case: fitting a straight line to the data points.

Least Squares Fit of a Straight Line   Suppose we want to fit a straight line $y = a + bx$ to the experimentally determined points

$$(x_1, y_1), (x_2, y_2), \ldots, (x_n, y_n)$$

If the data points are collinear, the line would pass through all n points, and so the unknown coefficients a and b would satisfy

$$y_1 = a + bx_1$$
$$y_2 = a + bx_2$$
$$\vdots$$
$$y_n = a + bx_n$$

We can write this system in matrix form as

$$\begin{bmatrix} 1 & x_1 \\ 1 & x_2 \\ \vdots & \vdots \\ 1 & x_n \end{bmatrix} \begin{bmatrix} a \\ b \end{bmatrix} = \begin{bmatrix} y_1 \\ y_2 \\ \vdots \\ y_n \end{bmatrix}$$

or, more compactly, as

$$M\mathbf{v} = \mathbf{y} \tag{1}$$

where

$$\mathbf{y} = \begin{bmatrix} y_1 \\ y_2 \\ \vdots \\ y_n \end{bmatrix}, \qquad M = \begin{bmatrix} 1 & x_1 \\ 1 & x_2 \\ \vdots & \vdots \\ 1 & x_n \end{bmatrix}, \qquad \mathbf{v} = \begin{bmatrix} a \\ b \end{bmatrix} \tag{2}$$

If the data points are not collinear, then it is impossible to find coefficients a and b that satisfy system (1) exactly; that is, the system is inconsistent. In this case we shall look for a least squares solution

$$\mathbf{v} = \mathbf{v}^* = \begin{bmatrix} a^* \\ b^* \end{bmatrix}$$

We call a line $y = a^* + b^*x$ whose coefficients come from a least squares solution a **regression line** or a **least squares straight line fit** to the data. To explain this terminology, recall that a least squares solution of (1) minimizes

$$\|\mathbf{y} - M\mathbf{v}\| \tag{3}$$

If we express the square of (3) in terms of components, we obtain

$$\|\mathbf{y} - M\mathbf{v}\|^2 = (y_1 - a - bx_1)^2 + (y_2 - a - bx_2)^2 + \cdots + (y_n - a - bx_n)^2 \tag{4}$$

If we now let

$$d_1 = |y_1 - a - bx_1|, \quad d_2 = |y_2 - a - bx_2|, \ldots, \quad d_n = |y_n - a - bx_n|$$

then (4) can be written as

$$\|\mathbf{y} - M\mathbf{v}\|^2 = d_1^2 + d_2^2 + \cdots + d_n^2 \tag{5}$$

As illustrated in Figure 9.3.2, d_i can be interpreted as the vertical distance between the line $y = a + bx$ and the data point (x_i, y_i). This distance is a measure of the "error" at the point (x_i, y_i) resulting from the inexact fit of $y = a + bx$ to the data points. Since (3) and (5) are minimized by the same vector $\mathbf{v}^*$, the least squares straight line fit minimizes the sum of the squares of these errors, hence the name *least squares straight line fit*.

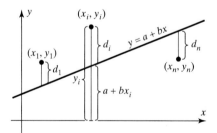

Figure 9.3.2 d_i measures the vertical error in the least squares straight line.

Normal Equations   Recall from Theorem 6.4.2 that the least squares solutions of (1) can be obtained by solving the associated normal system

$$M^T M\mathbf{v} = M^T\mathbf{y}$$

the equations of which are called the **normal equations**.

In the exercises it will be shown that the column vectors of M are linearly independent if and only if the n data points do not lie on a vertical line in the xy-plane. In this case it follows from Theorem 6.4.4 that the least squares solution is unique and is given by

$$\mathbf{v}^* = (M^T M)^{-1} M^T\mathbf{y}$$

In summary, we have the following theorem.

Theorem 9.3.1 Least Squares Solution

Let $(x_1, y_1), (x_2, y_2), \ldots, (x_n, y_n)$ be a set of two or more data points, not all lying on a vertical line, and let

$$M = \begin{bmatrix} 1 & x_1 \\ 1 & x_2 \\ \vdots & \vdots \\ 1 & x_n \end{bmatrix} \quad and \quad \mathbf{y} = \begin{bmatrix} y_1 \\ y_2 \\ \vdots \\ y_n \end{bmatrix}.$$

Then there is a unique least squares straight line fit

$$y = a^* + b^* x$$

to the data points. Moreover,

$$\mathbf{v}^* = \begin{bmatrix} a^* \\ b^* \end{bmatrix}$$

is given by the formula

$$\mathbf{v}^* = (M^T M)^{-1} M^T \mathbf{y} \tag{6}$$

which expresses the fact that $\mathbf{v} = \mathbf{v}^*$ *is the unique solution of the normal equations*

$$M^T M \mathbf{v} = M^T \mathbf{y} \tag{7}$$

EXAMPLE 1 Least Squares Line: Using Formula [6]

Find the least squares straight line fit to the four points $(0, 1)$, $(1, 3)$, $(2, 4)$, and $(3, 4)$. (See Figure 9.3.3.)

Solution.

We have

$$M = \begin{bmatrix} 1 & 0 \\ 1 & 1 \\ 1 & 2 \\ 1 & 3 \end{bmatrix}, \quad M^T M = \begin{bmatrix} 4 & 6 \\ 6 & 14 \end{bmatrix}, \quad and \quad (M^T M)^{-1} = \frac{1}{10} \begin{bmatrix} 7 & -3 \\ -3 & 2 \end{bmatrix}$$

$$\mathbf{v}^* = (M^T M)^{-1} M^T \mathbf{y} = \frac{1}{10} \begin{bmatrix} 7 & -3 \\ -3 & 2 \end{bmatrix} \begin{bmatrix} 1 & 1 & 1 & 1 \\ 0 & 1 & 2 & 3 \end{bmatrix} \begin{bmatrix} 1 \\ 3 \\ 4 \\ 4 \end{bmatrix} = \begin{bmatrix} 1.5 \\ 1 \end{bmatrix}$$

And so the desired line is $y = 1.5 + x$. ◆

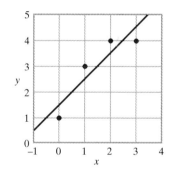

Figure 9.3.3

EXAMPLE 2 Spring Constant

Hooke's law in physics states that the length x of a uniform spring is a linear function of the force y applied to it. If we write $y = a + bx$, then the coefficient b is called the spring constant. Suppose a particular unstretched spring has a measured length of 6.1 inches (i.e., $x = 6.1$ when $y = 0$). Forces of 2 pounds, 4 pounds, and 6 pounds

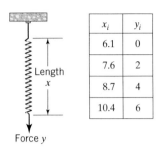

x_i	y_i
6.1	0
7.6	2
8.7	4
10.4	6

Figure 9.3.4

are then applied to the spring, and the corresponding lengths are found to be 7.6 inches, 8.7 inches, and 10.4 inches (see Figure 9.3.4). Find the spring constant of this spring.

Solution.

We have

$$M = \begin{bmatrix} 1 & 6.1 \\ 1 & 7.6 \\ 1 & 8.7 \\ 1 & 10.4 \end{bmatrix}, \qquad \mathbf{y} = \begin{bmatrix} 0 \\ 2 \\ 4 \\ 6 \end{bmatrix},$$

and

$$\mathbf{v}^* = \begin{bmatrix} a^* \\ b^* \end{bmatrix} = (M^T M)^{-1} M^T \mathbf{y} \simeq \begin{bmatrix} -8.6 \\ 1.4 \end{bmatrix}$$

where the numerical values have been rounded to one decimal place. Thus, the estimated value of the spring constant is $b^* \simeq 1.4$ pounds/inch. ◆

Least Squares Fit of a Polynomial

The technique described for fitting a straight line to data points generalizes easily to fitting a polynomial of any specified degree to data points. Let us attempt to fit a polynomial of fixed degree m

$$y = a_0 + a_1 x + \cdots + a_m x^m \tag{8}$$

to n points

$$(x_1, y_1), (x_2, y_2), \ldots, (x_n, y_n)$$

Substituting these n values of x and y into (8) yields the n equations

$$y_1 = a_0 + a_1 x_1 + \cdots + a_m x_1^m$$
$$y_2 = a_0 + a_1 x_2 + \cdots + a_m x_2^m$$
$$\vdots \qquad \vdots \qquad \vdots \qquad \vdots$$
$$y_n = a_0 + a_1 x_n + \cdots + a_m x_n^m$$

or in matrix form,

$$M\mathbf{v} = \mathbf{y} \tag{9}$$

where

$$\mathbf{y} = \begin{bmatrix} y_1 \\ y_2 \\ \vdots \\ y_n \end{bmatrix}, \qquad M = \begin{bmatrix} 1 & x_1 & x_1^2 & \cdots & x_1^m \\ 1 & x_2 & x_2^2 & \cdots & x_2^m \\ \vdots & \vdots & \vdots & & \vdots \\ 1 & x_n & x_n^2 & \cdots & x_n^m \end{bmatrix}, \qquad \mathbf{v} = \begin{bmatrix} a_0 \\ a_1 \\ \vdots \\ a_m \end{bmatrix} \tag{10}$$

As before, the solutions of the normal equations

$$M^T M \mathbf{v} = M^T \mathbf{y}$$

determine coefficients of polynomials that minimize

$$\|\mathbf{y} - M\mathbf{v}\|$$

Conditions that guarantee the invertibility of $M^T M$ are discussed in the exercises. If $M^T M$ is invertible, then the normal equations have a unique solution $\mathbf{v} = \mathbf{v}^*$, which is given by

$$\mathbf{v}^* = (M^T M)^{-1} M^T \mathbf{y} \tag{11}$$

EXAMPLE 3 Fitting a Quadratic Curve to Data

According to Newton's second law of motion, a body near the earth's surface falls vertically downward according to the equation

$$s = s_0 + v_0 t + \tfrac{1}{2} g t^2 \tag{12}$$

where

$s = $ vertical displacement downward relative to some fixed point

$s_0 = $ initial displacement at time $t = 0$

$v_0 = $ initial velocity at time $t = 0$

$g = $ acceleration of gravity at the earth's surface

Suppose that a laboratory experiment is performed to evaluate g using this equation. A weight is released with unknown initial displacement and velocity, and at certain times the distances fallen relative to some fixed reference point are measured. In particular, suppose that at times $t = .1, .2, .3, .4,$ and $.5$ seconds it is found that the weight has fallen $s = -0.18, 0.31, 1.03, 2.48,$ and 3.73 feet, respectively, from the reference point. Find an approximate value of g using these data.

Solution.

The mathematical problem is to fit a quadratic curve

$$s = a_0 + a_1 t + a_2 t^2 \tag{13}$$

to the five data points:

$$(.1, -0.18), \quad (.2, 0.31), \quad (.3, 1.03), \quad (.4, 2.48), \quad (.5, 3.73)$$

With the appropriate adjustments in notation, the matrices M and $\mathbf{y}$ in (10) are

$$M = \begin{bmatrix} 1 & t_1 & t_1^2 \\ 1 & t_2 & t_2^2 \\ 1 & t_3 & t_3^2 \\ 1 & t_4 & t_4^2 \\ 1 & t_5 & t_5^2 \end{bmatrix} = \begin{bmatrix} 1 & .1 & .01 \\ 1 & .2 & .04 \\ 1 & .3 & .09 \\ 1 & .4 & .16 \\ 1 & .5 & .25 \end{bmatrix}, \quad \mathbf{y} = \begin{bmatrix} s_1 \\ s_2 \\ s_3 \\ s_4 \\ s_5 \end{bmatrix} = \begin{bmatrix} -0.18 \\ 0.31 \\ 1.03 \\ 2.48 \\ 3.73 \end{bmatrix}$$

Thus, from (11)

$$\mathbf{v}^* = \begin{bmatrix} a_0^* \\ a_1^* \\ a_2^* \end{bmatrix} = (M^T M)^{-1} M^T \mathbf{y} \simeq \begin{bmatrix} -0.40 \\ 0.35 \\ 16.1 \end{bmatrix}$$

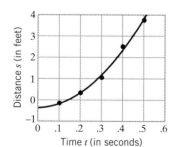

Figure 9.3.5

From (12) and (13) we have $a_2 = \tfrac{1}{2} g$, and so the estimated value of g is

$$g = 2a_2^* = 2(16.1) = 32.2 \text{ feet/second}^2$$

If desired, we can also estimate the initial displacement and initial velocity of the weight:

$$s_0 = a_0^* = -0.40 \text{ feet}$$
$$v_0 = a_1^* = 0.35 \text{ feet/second}$$

In Figure 9.3.5 we have plotted the five data points and the approximating polynomial.
◆

Exercise Set 9.3

1. Find the least squares straight line fit to the three points $(0, 0)$, $(1, 2)$, and $(2, 7)$.

2. Find the least squares straight line fit to the four points $(0, 1)$, $(2, 0)$, $(3, 1)$, and $(3, 2)$.

3. Find the quadratic polynomial that best fits the four points $(2, 0)$, $(3, -10)$, $(5, -48)$, and $(6, -76)$.

4. Find the cubic polynomial that best fits the five points $(-1, -14)$, $(0, -5)$, $(1, -4)$, $(2, 1)$, and $(3, 22)$.

5. Show that the matrix M in Equation (2) has linearly independent columns if and only if at least two of the numbers $x_1, x_2, \ldots, x_n$ are distinct.

6. Show that the columns of the $n \times (m + 1)$ matrix M in Equation (10) are linearly independent if $n > m$ and at least $m + 1$ of the numbers $x_1, x_2, \ldots, x_n$ are distinct. [**Hint.** A nonzero polynomial of degree m has at most m distinct roots.]

7. Let M be the matrix in Equation (10). Using Exercise 6, show that a sufficient condition for the matrix $M^T M$ to be invertible is that $n > m$ and at least $m + 1$ of the numbers $x_1, x_2, \ldots, x_n$ are distinct.

8. The owner of a rapidly expanding business finds that for the first five months of the year the sales (in thousands) are $4.0, $4.4, $5.2, $6.4, and $8.0. The owner plots these figures on a graph and conjectures that for the rest of the year the sales curve can be approximated by a quadratic polynomial. Find the least squares quadratic polynomial fit to the sales curve, and use it to project the sales for the twelfth month of the year.

9.4 APPROXIMATION PROBLEMS; FOURIER SERIES

In this section we shall use results about orthogonal projections in inner product spaces to solve problems that involve approximating a given function by simpler functions. Such problems arise in a variety of engineering and scientific applications.

Best Approximations All of the problems that we will study in this section will be special cases of the following general problem.

Approximation Problem. Given a function f that is continuous on an interval $[a, b]$, find the "best possible approximation" to f using only functions from a specified subspace W of $C[a, b]$.

Here are some examples of such problems:

(a) Find the best possible approximation to e^x over $[0, 1]$ by a polynomial of the form $a_0 + a_1 x + a_2 x^2$.

(b) Find the best possible approximation to $\sin \pi x$ over $[-1, 1]$ by a function of the form $a_0 + a_1 e^x + a_2 e^{2x} + a_3 e^{3x}$.

(c) Find the best possible approximation to x over $[0, 2\pi]$ by a function of the form $a_0 + a_1 \sin x + a_2 \sin 2x + b_1 \cos x + b_2 \cos 2x$.

In the first example W is the subspace of $C[0, 1]$ spanned by 1, x, and x^2; in the second example W is the subspace of $C[-1, 1]$ spanned by 1, e^x, e^{2x}, and e^{3x}; and in the third example W is the subspace of $C[0, 2\pi]$ spanned by 1, $\sin x$, $\sin 2x$, $\cos x$, and $\cos 2x$.

Measurements of Error To solve approximation problems of the preceding types, we must make the phrase "best approximation over $[a, b]$" mathematically precise; to do this we need a precise way of measuring the error that results when one continuous function is approximated by another over $[a, b]$. If we were concerned only with approximating $f(x)$ at a single point x_0, then the error at x_0 by an approximation $g(x)$ would be simply

$$\text{error} = |f(x_0) - g(x_0)|$$

sometimes called the **deviation** between f and g at x_0 (Figure 9.4.1). However, we are concerned with approximation over the entire interval $[a, b]$, not at a single point. Consequently, in one part of the interval an approximation g_1 to f may have smaller deviations from f than an approximation g_2 to f, and in another part of the interval it might be the other way around. How do we decide which is the better overall approximation? What we need is some way of measuring the overall error in an approximation $g(x)$. One possible measure of overall error is obtained by integrating the deviation $|f(x) - g(x)|$ over the entire interval $[a, b]$; that is,

$$\text{error} = \int_a^b |f(x) - g(x)|\, dx \tag{1}$$

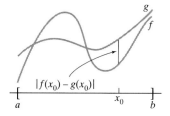

Figure 9.4.1
The deviation between f and g at x_0.

Geometrically (1) is the area between the graphs of $f(x)$ and $g(x)$ over the interval $[a, b]$ (Figure 9.4.2); the greater area, the greater the overall error.

While (1) is natural and appealing geometrically, most mathematicians and scientists generally favor the following alternative measure of error, called the **mean square error**.

$$\boxed{\text{mean square error} = \int_a^b [f(x) - g(x)]^2\, dx}$$

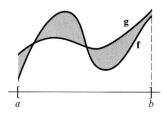

Figure 9.4.2
The area between the graphs of **f** and **g** over $[a, b]$ measures the error in approximating f by g over $[a, b]$.

Mean square error emphasizes the effect of larger errors because of the squaring and has the added advantage that it allows us to bring to bear the theory of inner product spaces. To see how, suppose that **f** is a continuous function on $[a, b]$ that we want to approximate by a function **g** from a subspace W of $C[a, b]$, and suppose that $C[a, b]$ is given the inner product

$$\langle \mathbf{f}, \mathbf{g} \rangle = \int_a^b f(x)g(x)\, dx$$

It follows that

$$\|\mathbf{f} - \mathbf{g}\|^2 = \langle \mathbf{f} - \mathbf{g}, \mathbf{f} - \mathbf{g} \rangle = \int_a^b [f(x) - g(x)]^2\, dx = \text{mean square error}$$

so that minimizing the mean square error is the same as minimizing $\|\mathbf{f} - \mathbf{g}\|^2$. Thus, the approximation problem posed informally at the beginning of this section can be restated more precisely as follows:

Least Squares Approximation

Least Squares Approximation Problem. Let **f** be a function that is continuous on an interval $[a, b]$, let $C[a, b]$ have the inner product

$$\langle \mathbf{f}, \mathbf{g} \rangle = \int_a^b f(x)g(x)\, dx$$

and let W be a finite-dimensional subspace of $C[a, b]$. Find a function **g** in W that minimizes

$$\|\mathbf{f} - \mathbf{g}\|^2 = \int_a^b [f(x) - g(x)]^2\, dx$$

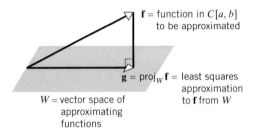

f = function in $C[a, b]$ to be approximated

g = proj$_W$ **f** = least squares approximation to **f** from W

W = vector space of approximating functions

Figure 9.4.3

Since $\|\mathbf{f} - \mathbf{g}\|^2$ and $\|\mathbf{f} - \mathbf{g}\|$ are minimized by the same function **g**, the preceding problem is equivalent to looking for a function **g** in W that is closest to **f**. But we know from Theorem 6.4.1 that $\mathbf{g} = \text{proj}_W \mathbf{f}$ is such a function (Figure 9.4.3). Thus, we have the following result.

Solution of the Least Squares Approximation Problem. If **f** is a continuous function on $[a, b]$, and W is a finite-dimensional subspace of $C[a, b]$, then the function **g** in W that minimizes the mean square error

$$\int_a^b [f(x) - g(x)]^2 \, dx$$

is $\mathbf{g} = \text{proj}_W \mathbf{f}$, where the orthogonal projection is relative to the inner product

$$\langle \mathbf{f}, \mathbf{g} \rangle = \int_a^b f(x) g(x) \, dx$$

The function $\mathbf{g} = \text{proj}_W \mathbf{f}$ is called the ***least squares approximation*** to **f** from W.

Fourier Series   A function of the form

$$t(x) = c_0 + c_1 \cos x + c_2 \cos 2x + \cdots + c_n \cos nx$$
$$+ d_1 \sin x + d_2 \sin 2x + \cdots + d_n \sin nx \tag{2}$$

is called a ***trigonometric polynomial***; if c_n and d_n are not both zero, then $t(x)$ is said to have ***order n***. For example,

$$t(x) = 2 + \cos x - 3 \cos 2x + 7 \sin 4x$$

is a trigonometric polynomial with

$$c_0 = 2, \quad c_1 = 1, \quad c_2 = -3, \quad c_3 = 0, \quad c_4 = 0, \quad d_1 = 0, \quad d_2 = 0, \quad d_3 = 0, \quad d_4 = 7$$

The order of $t(x)$ is 4.

It is evident from (2) that the trigonometric polynomials of order n or less are the various possible linear combinations of

$$1, \quad \cos x, \quad \cos 2x, \ldots, \quad \cos nx, \quad \sin x, \quad \sin 2x, \ldots, \quad \sin nx \tag{3}$$

It can be shown that these $2n + 1$ functions are linearly independent and that consequently for any interval $[a, b]$ they form a basis for a $(2n + 1)$-dimensional subspace of $C[a, b]$.

Let us now consider the problem of finding the least squares approximation of a continuous function $f(x)$ over the interval $[0, 2\pi]$ by a trigonometric polynomial of order n or less. As noted above, the least squares approximation to **f** from W is the orthogonal projection of **f** on W. To find this orthogonal projection, we must find an orthonormal basis $\mathbf{g}_0, \mathbf{g}_1, \ldots, \mathbf{g}_{2n}$ for W, after which we can compute the orthogonal projection on W from the formula

$$\text{proj}_W \mathbf{f} = \langle \mathbf{f}, \mathbf{g}_0 \rangle \mathbf{g}_0 + \langle \mathbf{f}, \mathbf{g}_1 \rangle \mathbf{g}_1 + \cdots + \langle \mathbf{f}, \mathbf{g}_{2n} \rangle \mathbf{g}_{2n} \tag{4}$$

[see Theorem 6.3.5]. An orthonormal basis for W can be obtained by applying the Gram–Schmidt process to the basis (3), using the inner product

$$\langle \mathbf{f}, \mathbf{g} \rangle = \int_0^{2\pi} f(x) g(x)\, dx$$

This yields (Exercise 6) the orthonormal basis

$$\mathbf{g}_0 = \frac{1}{\sqrt{2\pi}}, \quad \mathbf{g}_1 = \frac{1}{\sqrt{\pi}} \cos x, \dots, \quad \mathbf{g}_n = \frac{1}{\sqrt{\pi}} \cos nx,$$

$$\mathbf{g}_{n+1} = \frac{1}{\sqrt{\pi}} \sin x, \dots, \quad \mathbf{g}_{2n} = \frac{1}{\sqrt{\pi}} \sin nx \tag{5}$$

If we introduce the notation

$$a_0 = \frac{2}{\sqrt{2\pi}} \langle \mathbf{f}, \mathbf{g}_0 \rangle, \quad a_1 = \frac{1}{\sqrt{\pi}} \langle \mathbf{f}, \mathbf{g}_1 \rangle, \dots, \quad a_n = \frac{1}{\sqrt{\pi}} \langle \mathbf{f}, \mathbf{g}_n \rangle$$

$$b_1 = \frac{1}{\sqrt{\pi}} \langle \mathbf{f}, \mathbf{g}_{n+1} \rangle, \dots, \quad b_n = \frac{1}{\sqrt{\pi}} \langle \mathbf{f}, \mathbf{g}_{2n} \rangle \tag{6}$$

then on substituting (5) in (4) we obtain

$$\text{proj}_W \mathbf{f} = \frac{a_0}{2} + [a_1 \cos x + \cdots + a_n \cos nx] + [b_1 \sin x + \cdots + b_n \sin nx] \tag{7}$$

where

$$a_0 = \frac{2}{\sqrt{2\pi}} \langle \mathbf{f}, \mathbf{g}_0 \rangle = \frac{2}{\sqrt{2\pi}} \int_0^{2\pi} f(x) \frac{1}{\sqrt{2\pi}}\, dx = \frac{1}{\pi} \int_0^{2\pi} f(x)\, dx$$

$$a_1 = \frac{1}{\sqrt{\pi}} \langle \mathbf{f}, \mathbf{g}_1 \rangle = \frac{1}{\sqrt{\pi}} \int_0^{2\pi} f(x) \frac{1}{\sqrt{\pi}} \cos x\, dx = \frac{1}{\pi} \int_0^{2\pi} f(x) \cos x\, dx$$

$$\vdots$$

$$a_n = \frac{1}{\sqrt{\pi}} \langle \mathbf{f}, \mathbf{g}_n \rangle = \frac{1}{\sqrt{\pi}} \int_0^{2\pi} f(x) \frac{1}{\sqrt{\pi}} \cos nx\, dx = \frac{1}{\pi} \int_0^{2\pi} f(x) \cos nx\, dx$$

$$b_1 = \frac{1}{\sqrt{\pi}} \langle \mathbf{f}, \mathbf{g}_{n+1} \rangle = \frac{1}{\sqrt{\pi}} \int_0^{2\pi} f(x) \frac{1}{\sqrt{\pi}} \sin x\, dx = \frac{1}{\pi} \int_0^{2\pi} f(x) \sin x\, dx$$

$$\vdots$$

$$b_n = \frac{1}{\sqrt{\pi}} \langle \mathbf{f}, \mathbf{g}_{2n} \rangle = \frac{1}{\sqrt{\pi}} \int_0^{2\pi} f(x) \frac{1}{\sqrt{\pi}} \sin nx\, dx = \frac{1}{\pi} \int_0^{2\pi} f(x) \sin nx\, dx$$

In short,

$$a_k = \frac{1}{\pi} \int_0^{2\pi} f(x) \cos kx\, dx, \quad b_k = \frac{1}{\pi} \int_0^{2\pi} f(x) \sin kx\, dx \tag{8}$$

The numbers $a_0, a_1, \dots, a_n, b_1, \dots, b_n$ are called the ***Fourier coefficients*** of $\mathbf{f}$.

EXAMPLE 1 Least Squares Approximations

Find the least squares approximation of $f(x) = x$ on $[0, 2\pi]$ by

(a) a trigonometric polynomial of order 2 or less;

(b) a trigonometric polynomial of order n or less.

Jean Baptiste Joseph Fourier (1768–1830) was a French mathematician and physicist who discovered the Fourier series and related ideas while working on problems of heat diffusion. This discovery is one of the most influential in the history of mathematics; it is the cornerstone of many fields of mathematical research and a basic tool in many branches of engineering. Fourier, a political activist during the French revolution, spent time in jail for his defense of many victims during the Terror. He later became a favorite of Napoleon and was named both a baron and a count.

Solution (a).

$$a_0 = \frac{1}{\pi} \int_0^{2\pi} f(x)\,dx = \frac{1}{\pi} \int_0^{2\pi} x\,dx = 2\pi \qquad (9a)$$

For $k = 1, 2, \ldots$ integration by parts yields (verify)

$$a_k = \frac{1}{\pi} \int_0^{2\pi} f(x) \cos kx\,dx = \frac{1}{\pi} \int_0^{2\pi} x \cos kx\,dx = 0 \qquad (9b)$$

$$b_k = \frac{1}{\pi} \int_0^{2\pi} f(x) \sin kx\,dx = \frac{1}{\pi} \int_0^{2\pi} x \sin kx\,dx = -\frac{2}{k} \qquad (9c)$$

Thus, the least squares approximation to x on $[0, 2\pi]$ by a trigonometric polynomial of order 2 or less is

$$x \simeq \frac{a_0}{2} + a_1 \cos x + a_2 \cos 2x + b_1 \sin x + b_2 \sin 2x$$

or from (9a), (9b), and (9c)

$$x \simeq \pi - 2 \sin x - \sin 2x$$

Solution (b). The least squares approximation to x on $[0, 2\pi]$ by a trigonometric polynomial of order n or less is

$$x \simeq \frac{a_0}{2} + [a_1 \cos x + \cdots + a_n \cos nx] + [b_1 \sin x + \cdots + b_n \sin nx]$$

or from (9a), (9b), and (9c)

$$x \simeq \pi - 2\left(\sin x + \frac{\sin 2x}{2} + \frac{\sin 3x}{3} + \cdots + \frac{\sin nx}{n} \right)$$

The graphs of $y = x$ and some of these approximations are shown in Figure 9.4.4.

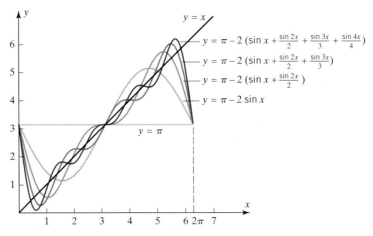

Figure 9.4.4

It is natural to expect that the mean square error will diminish as the number of terms in the least squares approximation

$$f(x) \simeq \frac{a_0}{2} + \sum_{k=1}^n (a_k \cos kx + b_k \sin kx)$$

increases. It can be proved that for functions f in $C[0, 2\pi]$ the mean square error approaches zero as $n \to +\infty$; this is denoted by writing

$$f(x) = \frac{a_0}{2} + \sum_{k=1}^{\infty}(a_k \cos kx + b_k \sin kx)$$

The right side of this equation is called the **_Fourier series_** for f over the interval $[0, 2\pi]$. Such series are of major importance in engineering, science, and mathematics. ♦

Exercise Set 9.4

1. Find the least squares approximation of $f(x) = 1 + x$ over the interval $[0, 2\pi]$ by
   (a) a trigonometric polynomial of order 2 or less
   (b) a trigonometric polynomial of order n or less

2. Find the least squares approximation of $f(x) = x^2$ over the interval $[0, 2\pi]$ by
   (a) a trigonometric polynomial of order 3 or less
   (b) a trigonometric polynomial of order n or less

3. (a) Find the least squares approximation of x over the interval $[0, 1]$ by a function of the form $a + be^x$.
   (b) Find the mean square error of the approximation.

4. (a) Find the least squares approximation of e^x over the interval $[0, 1]$ by a polynomial of the form $a_0 + a_1 x$.
   (b) Find the mean square error of the approximation.

5. (a) Find the least squares approximation of $\sin \pi x$ over the interval $[-1, 1]$ by a polynomial of the form $a_0 + a_1 x + a_2 x^2$.
   (b) Find the mean square error of the approximation.

6. Use the Gram–Schmidt process to obtain the orthonormal basis (5) from the basis (3).

7. Carry out the integrations in (9a), (9b), and (9c).

8. Find the Fourier series of $f(x) = \pi - x$ over the interval $[0, 2\pi]$.

9.5 QUADRATIC FORMS

In this section we shall study functions in which the terms are squares of variables or products of two variables. Such functions arise in a variety of applications, including geometry, vibrations of mechanical systems, statistics, and electrical engineering.

Quadratic Forms   Up to now, we have been interested primarily in linear equations; that is, equations of the form

$$a_1 x_1 + a_2 x_2 + \cdots + a_n x_n = b$$

The expression on the left side of this equation, namely,

$$a_1 x_1 + a_2 x_2 + \cdots + a_n x_n$$

is a function of n variables, called a **_linear form_**. In a linear form, all variables occur to the first power, and there are no products of variables in the expression. Here, we will be concerned with **_quadratic forms_**, which are functions of the form

$$a_1 x_1^2 + a_2 x_2^2 + \cdots + a_n x_n^2 + \text{(all possible terms of the form } a_k x_i x_j \text{ for } i < j) \quad (1)$$

For example, a quadratic form in the variables x_1 and x_2 is

$$a_1 x_1^2 + a_2 x_2^2 + a_3 x_1 x_2 \tag{2}$$

and a quadratic form in the variables x_1, x_2, and x_3 is

$$a_1 x_1^2 + a_2 x_2^2 + a_3 x_3^2 + a_4 x_1 x_2 + a_5 x_1 x_3 + a_6 x_2 x_3 \tag{3}$$

The terms in a quadratic form that involve products of different variables are called the *cross-product terms*. Thus, in (2) the last term is a cross-product term, and in (3) the last three terms are cross-product terms.

If we follow the convention of omitting brackets on 1×1 matrices, then (2) can be written in matrix form as

$$\begin{bmatrix} x_1 & x_2 \end{bmatrix} \begin{bmatrix} a_1 & a_3/2 \\ a_3/2 & a_2 \end{bmatrix} \begin{bmatrix} x_1 \\ x_2 \end{bmatrix} \tag{4}$$

and (3) can be written as

$$\begin{bmatrix} x_1 & x_2 & x_3 \end{bmatrix} \begin{bmatrix} a_1 & a_4/2 & a_5/2 \\ a_4/2 & a_2 & a_6/2 \\ a_5/2 & a_6/2 & a_3 \end{bmatrix} \begin{bmatrix} x_1 \\ x_2 \\ x_3 \end{bmatrix} \tag{5}$$

(verify by multiplying out). Note that the products in (4) and (5) are both of the form $\mathbf{x}^T A \mathbf{x}$, where $\mathbf{x}$ is the column vector of variables, and A is a symmetric matrix whose diagonal entries are the coefficients of the squared terms and whose entries off the main diagonal are half the coefficients of the cross-product terms. More precisely, the diagonal entry in row i and column i is the coefficient of x_i^2, and the off-diagonal entry in row i and column j is half the coefficient of the product $x_i x_j$. Here are some examples.

EXAMPLE 1 Matrix Representation of Quadratic Forms

$$2x^2 + 6xy - 7y^2 = \begin{bmatrix} x & y \end{bmatrix} \begin{bmatrix} 2 & 3 \\ 3 & -7 \end{bmatrix} \begin{bmatrix} x \\ y \end{bmatrix}$$

$$4x^2 - 5y^2 = \begin{bmatrix} x & y \end{bmatrix} \begin{bmatrix} 4 & 0 \\ 0 & -5 \end{bmatrix} \begin{bmatrix} x \\ y \end{bmatrix}$$

$$xy = \begin{bmatrix} x & y \end{bmatrix} \begin{bmatrix} 0 & \frac{1}{2} \\ \frac{1}{2} & 0 \end{bmatrix} \begin{bmatrix} x \\ y \end{bmatrix}$$

$$x_1^2 + 7x_2^2 - 3x_3^2 + 4x_1 x_2 - 2x_1 x_3 + 6x_2 x_3 = \begin{bmatrix} x_1 & x_2 & x_3 \end{bmatrix} \begin{bmatrix} 1 & 2 & -1 \\ 2 & 7 & 3 \\ -1 & 3 & -3 \end{bmatrix} \begin{bmatrix} x_1 \\ x_2 \\ x_3 \end{bmatrix}$$

◆

Symmetric matrices are useful, but not essential, for representing quadratic forms. For example, the quadratic form $2x^2 + 6xy - 7y^2$, which we represented in Example 1 as $\mathbf{x}^T A \mathbf{x}$ with a symmetric matrix A, can also be written as

$$2x^2 + 6xy - 7y^2 = \begin{bmatrix} x & y \end{bmatrix} \begin{bmatrix} 2 & 5 \\ 1 & -7 \end{bmatrix} \begin{bmatrix} x \\ y \end{bmatrix}$$

where the coefficient 6 of the cross-product term has been split as $5 + 1$ rather than $3 + 3$, as in the symmetric representation. However, symmetric matrices are usually more convenient to work with, so it will always be understood that A is symmetric when we write a quadratic form as $\mathbf{x}^T A \mathbf{x}$, even if not stated explicitly. When convenient, we can use Formula (7) of Section 4.1 to express a quadratic form $\mathbf{x}^T A \mathbf{x}$ in terms of the Euclidean inner product as

$$\mathbf{x}^T A \mathbf{x} = A\mathbf{x} \cdot \mathbf{x} \quad \text{or by symmetry of the dot product,} \quad \mathbf{x}^T A \mathbf{x} = \mathbf{x} \cdot A\mathbf{x}$$

If preferred, we can use the notation $\mathbf{u} \cdot \mathbf{v} = \langle \mathbf{u}, \mathbf{v} \rangle$ for the dot product and write these expressions as

$$\mathbf{x}^T A \mathbf{x} = \mathbf{x}^T (A\mathbf{x}) = \langle A\mathbf{x}, \mathbf{x} \rangle = \langle \mathbf{x}, A\mathbf{x} \rangle \tag{6}$$

Problems Involving Quadratic Forms

The study of quadratic forms is an extensive topic that we can only touch on in this section. The following are some of the important mathematical problems relating to quadratic forms.

- Find the maximum and minimum values of the quadratic form $\mathbf{x}^T A \mathbf{x}$ if $\mathbf{x}$ is constrained so that
$$\|\mathbf{x}\| = (x_1^2 + x_2^2 + \cdots + x_n^2)^{1/2} = 1$$

- What conditions must A satisfy in order for a quadratic form to satisfy the inequality $\mathbf{x}^T A \mathbf{x} > 0$ for all $\mathbf{x} \neq \mathbf{0}$?

- If $\mathbf{x}^T A \mathbf{x}$ is a quadratic form in two or three variables and c is a constant, what does the graph of the equation $\mathbf{x}^T A \mathbf{x} = c$ look like?

- If P is an orthogonal matrix, the change of variables $\mathbf{x} = P\mathbf{y}$ converts the quadratic form $\mathbf{x}^T A \mathbf{x}$ to $(P\mathbf{y})^T A (P\mathbf{y}) = \mathbf{y}^T (P^T A P)\mathbf{y}$. But $P^T A P$ is a symmetric matrix if A is (verify), so $\mathbf{y}^T (P^T A P)\mathbf{y}$ is a new quadratic form in the variables of $\mathbf{y}$. It is important to know if P can be chosen so that this new quadratic form has no cross-product terms.

In this section we shall study the first two problems, and in the following sections we shall study the last two. The following theorem provides a solution to the first problem. The proof is deferred to the end of this section.

Theorem 9.5.1

Let A be a symmetric $n \times n$ matrix whose eigenvalues in decreasing size order are $\lambda_1 \geq \lambda_2 \geq \cdots \geq \lambda_n$. If $\mathbf{x}$ is constrained so that $\|\mathbf{x}\| = 1$ relative to the Euclidean inner product on R^n, then:

(a) $\lambda_1 \geq \mathbf{x}^T A \mathbf{x} \geq \lambda_n$.

(b) $\mathbf{x}^T A \mathbf{x} = \lambda_n$ *if $\mathbf{x}$ is an eigenvector of A corresponding to λ_n and $\mathbf{x}^T A \mathbf{x} = \lambda_1$ if $\mathbf{x}$ is an eigenvector of A corresponding to λ_1.*

It follows from this theorem that subject to the constraint

$$\|\mathbf{x}\| = (x_1^2 + x_2^2 + \cdots + x_n^2)^{1/2} = 1$$

the quadratic form $\mathbf{x}^T A \mathbf{x}$ has a maximum value of λ_1 (the largest eigenvalue) and a minimum value of λ_n (the smallest eigenvalue).

EXAMPLE 2 Consequences of Theorem 9.5.1

Find the maximum and minimum values of the quadratic form

$$x_1^2 + x_2^2 + 4x_1 x_2$$

subject to the constraint $x_1^2 + x_2^2 = 1$, and determine values of x_1 and x_2 at which the maximum and minimum occur.

Solution.

The quadratic form can be written as

$$x_1^2 + x_2^2 + 4x_1x_2 = \mathbf{x}^T A \mathbf{x} = [x_1 \quad x_2] \begin{bmatrix} 1 & 2 \\ 2 & 1 \end{bmatrix} \begin{bmatrix} x_1 \\ x_2 \end{bmatrix}$$

The characteristic equation of A is

$$\det(\lambda I - A) = \det \begin{bmatrix} \lambda - 1 & -2 \\ -2 & \lambda - 1 \end{bmatrix} = \lambda^2 - 2\lambda - 3 = (\lambda - 3)(\lambda + 1) = 0$$

Thus, the eigenvalues of A are $\lambda = 3$ and $\lambda = -1$, which are the maximum and minimum values, respectively, of the quadratic form subject to the constraint. To find values of x_1 and x_2 at which these extreme values occur, we must find eigenvectors corresponding to these eigenvalues and then normalize these eigenvectors to satisfy the condition $x_1^2 + x_2^2 = 1$.

We leave it for the reader to show that bases for the eigenspaces are

$$\lambda = 3: \quad \begin{bmatrix} 1 \\ 1 \end{bmatrix}, \qquad \lambda = -1: \quad \begin{bmatrix} 1 \\ -1 \end{bmatrix}$$

Normalizing these eigenvectors yields

$$\begin{bmatrix} 1/\sqrt{2} \\ 1/\sqrt{2} \end{bmatrix}, \qquad \begin{bmatrix} 1/\sqrt{2} \\ -1/\sqrt{2} \end{bmatrix}$$

Thus, subject to the constraint $x_1^2 + x_2^2 = 1$, the maximum value of the quadratic form is $\lambda = 3$, which occurs if $x_1 = 1/\sqrt{2}$, $x_2 = 1/\sqrt{2}$; and the minimum value is $\lambda = -1$, which occurs if $x_1 = 1/\sqrt{2}$, $x_2 = -1/\sqrt{2}$. Moreover, alternative bases for the eigenspaces can be obtained by multiplying the basis vectors above by -1. Thus, the maximum value, $\lambda = 3$, also occurs if $x_1 = -1/\sqrt{2}$, $x_2 = -1/\sqrt{2}$, and the minimum value, $\lambda = -1$, also occurs if $x_1 = -1/\sqrt{2}$, $x_2 = 1/\sqrt{2}$. ◆

Definition

A quadratic form $\mathbf{x}^T A \mathbf{x}$ is called *positive definite* if $\mathbf{x}^T A \mathbf{x} > 0$ for all $\mathbf{x} \neq \mathbf{0}$, and a symmetric matrix A is called a *positive definite matrix* if $\mathbf{x}^T A \mathbf{x}$ is a positive definite quadratic form.

The following theorem is the main result about positive definite matrices.

Theorem 9.5.2

A symmetric matrix A is positive definite if and only if all the eigenvalues of A are positive.

Proof. Assume that A is positive definite, and let λ be any eigenvalue of A. If $\mathbf{x}$ is an eigenvector of A corresponding to λ, then $\mathbf{x} \neq \mathbf{0}$ and $A\mathbf{x} = \lambda\mathbf{x}$, so

$$0 < \mathbf{x}^T A \mathbf{x} = \mathbf{x}^T \lambda \mathbf{x} = \lambda \mathbf{x}^T \mathbf{x} = \lambda \|\mathbf{x}\|^2 \tag{7}$$

where $\|\mathbf{x}\|$ is the Euclidean norm of $\mathbf{x}$. Since $\|\mathbf{x}\|^2 > 0$ it follows that $\lambda > 0$, which is what we wanted to show.

Conversely, assume that all eigenvalues of A are positive. We must show that $\mathbf{x}^T A\mathbf{x} > 0$ for all $\mathbf{x} \neq \mathbf{0}$. But if $\mathbf{x} \neq \mathbf{0}$, we can normalize $\mathbf{x}$ to obtain the vector $\mathbf{y} = \mathbf{x}/\|\mathbf{x}\|$ with the property $\|\mathbf{y}\| = 1$. It now follows from Theorem 9.5.1 that

$$\mathbf{y}^T A\mathbf{y} \geq \lambda_n > 0$$

where λ_n is the smallest eigenvalue of A. Thus,

$$\mathbf{y}^T A\mathbf{y} = \left(\frac{\mathbf{x}}{\|\mathbf{x}\|}\right)^T A\left(\frac{\mathbf{x}}{\|\mathbf{x}\|}\right) = \frac{1}{\|\mathbf{x}\|^2}\mathbf{x}^T A\mathbf{x} > 0$$

Multiplying through by $\|\mathbf{x}\|^2$ yields

$$\mathbf{x}^T A\mathbf{x} > 0$$

which is what we wanted to show. ∎

EXAMPLE 3 Showing That a Matrix Is Positive Definite

In Example 1 of Section 7.3 we showed that the symmetric matrix

$$A = \begin{bmatrix} 4 & 2 & 2 \\ 2 & 4 & 2 \\ 2 & 2 & 4 \end{bmatrix}$$

has eigenvalues $\lambda = 2$ and $\lambda = 8$. Since these are positive, the matrix A is positive definite, and for all $\mathbf{x} \neq \mathbf{0}$

$$\mathbf{x}^T A\mathbf{x} = 4x_1^2 + 4x_2^2 + 4x_3^2 + 4x_1x_2 + 4x_1x_3 + 4x_2x_3 > 0 \qquad ◆$$

Our next objective is to give a criterion that can be used to determine whether a symmetric matrix is positive definite without finding its eigenvalues. To do this it will be helpful to introduce some terminology. If

$$A = \begin{bmatrix} a_{11} & a_{12} & \cdots & a_{1n} \\ a_{21} & a_{22} & \cdots & a_{2n} \\ \vdots & \vdots & & \vdots \\ a_{n1} & a_{n2} & \cdots & a_{nn} \end{bmatrix}$$

is a square matrix, then the ***principal submatrices*** of A are the submatrices formed from the first r rows and r columns of A for $r = 1, 2, \ldots, n$. These submatrices are

$$A_1 = [a_{11}], \quad A_2 = \begin{bmatrix} a_{11} & a_{12} \\ a_{21} & a_{22} \end{bmatrix}, \quad A_3 = \begin{bmatrix} a_{11} & a_{12} & a_{13} \\ a_{21} & a_{22} & a_{23} \\ a_{31} & a_{32} & a_{33} \end{bmatrix}, \ldots, \quad A_n = A = \begin{bmatrix} a_{11} & a_{12} & \cdots & a_{1n} \\ a_{21} & a_{22} & \cdots & a_{2n} \\ \vdots & \vdots & & \vdots \\ a_{n1} & a_{n2} & \cdots & a_{nn} \end{bmatrix}$$

Theorem 9.5.3

A symmetric matrix A is positive definite if and only if the determinant of every principal submatrix is positive.

We omit the proof.

EXAMPLE 4 Working with Principal Submatrices

The matrix

$$A = \begin{bmatrix} 2 & -1 & -3 \\ -1 & 2 & 4 \\ -3 & 4 & 9 \end{bmatrix}$$

is positive definite since

$$|2| = 2, \quad \begin{vmatrix} 2 & -1 \\ -1 & 2 \end{vmatrix} = 3, \quad \begin{vmatrix} 2 & -1 & -3 \\ -1 & 2 & 4 \\ -3 & 4 & 9 \end{vmatrix} = 1$$

all of which are positive. Thus, we are guaranteed that all eigenvalues of A are positive and $\mathbf{x}^T A \mathbf{x} > 0$ for all $\mathbf{x} \neq \mathbf{0}$. ◆

REMARK. A symmetric matrix A and the quadratic form $\mathbf{x}^T A \mathbf{x}$ are called

positive semidefinite if $\mathbf{x}^T A \mathbf{x} \geq 0$ for all $\mathbf{x}$
negative definite if $\mathbf{x}^T A \mathbf{x} < 0$ for $\mathbf{x} \neq \mathbf{0}$
negative semidefinite if $\mathbf{x}^T A \mathbf{x} \leq 0$ for all $\mathbf{x}$
indefinite if $\mathbf{x}^T A \mathbf{x}$ has both positive and negative values

Theorems 9.5.2 and 9.5.3 can be modified in an obvious way to apply to matrices of the first three types. For example, a symmetric matrix A is positive semidefinite if and only if all of its eigenvalues are *nonnegative*. Also, A is positive semidefinite if and only if all its principal submatrices have *nonnegative* determinants.

Optional

Proof of Theorem 9.5.1a. Since A is symmetric, it follows from Theorem 7.3.1 that there is an orthonormal basis for R^n consisting of eigenvectors of A. Suppose that $S = \{\mathbf{v}_1, \mathbf{v}_2, \ldots, \mathbf{v}_n\}$ is such a basis, where $\mathbf{v}_i$ is the eigenvector corresponding to the eigenvalue λ_i. If $\langle \ , \ \rangle$ denotes the Euclidean inner product, then it follows from Theorem 6.3.1 that for any $\mathbf{x}$ in R^n

$$\mathbf{x} = \langle \mathbf{x}, \mathbf{v}_1 \rangle \mathbf{v}_1 + \langle \mathbf{x}, \mathbf{v}_2 \rangle \mathbf{v}_2 + \cdots + \langle \mathbf{x}, \mathbf{v}_n \rangle \mathbf{v}_n$$

Thus,

$$Ax = \langle \mathbf{x}, \mathbf{v}_1 \rangle A\mathbf{v}_1 + \langle \mathbf{x}, \mathbf{v}_2 \rangle A\mathbf{v}_2 + \cdots + \langle \mathbf{x}, \mathbf{v}_n \rangle A\mathbf{v}_n$$
$$= \langle \mathbf{x}, \mathbf{v}_1 \rangle \lambda_1 \mathbf{v}_1 + \langle \mathbf{x}, \mathbf{v}_2 \rangle \lambda_2 \mathbf{v}_2 + \cdots + \langle \mathbf{x}, \mathbf{v}_n \rangle \lambda_n \mathbf{v}_n$$
$$= \lambda_1 \langle \mathbf{x}, \mathbf{v}_1 \rangle \mathbf{v}_1 + \lambda_2 \langle \mathbf{x}, \mathbf{v}_2 \rangle \mathbf{v}_2 + \cdots + \lambda_n \langle \mathbf{x}, \mathbf{v}_n \rangle \mathbf{v}_n$$

It follows that the coordinate vectors for $\mathbf{x}$ and $A\mathbf{x}$ relative to the basis S are

$$(\mathbf{x})_S = (\langle \mathbf{x}, \mathbf{v}_1 \rangle, \langle \mathbf{x}, \mathbf{v}_2 \rangle, \ldots, \langle \mathbf{x}, \mathbf{v}_n \rangle)$$
$$(A\mathbf{x})_S = (\lambda_1 \langle \mathbf{x}, \mathbf{v}_1 \rangle, \lambda_2 \langle \mathbf{x}, \mathbf{v}_2 \rangle, \ldots, \lambda_n \langle \mathbf{x}, \mathbf{v}_n \rangle)$$

Thus, from Theorem 6.3.2c and the fact that $\|\mathbf{x}\| = 1$, we obtain

$$\|\mathbf{x}\|^2 = \langle \mathbf{x}, \mathbf{v}_1 \rangle^2 + \langle \mathbf{x}, \mathbf{v}_2 \rangle^2 + \cdots + \langle \mathbf{x}, \mathbf{v}_n \rangle^2 = 1$$
$$\langle \mathbf{x}, A\mathbf{x} \rangle = \lambda_1 \langle \mathbf{x}, \mathbf{v}_1 \rangle^2 + \lambda_2 \langle \mathbf{x}, \mathbf{v}_2 \rangle^2 + \cdots + \lambda_n \langle \mathbf{x}, \mathbf{v}_n \rangle^2$$

Using these two equations and Formula (6), we can prove that $\mathbf{x}^T A \mathbf{x} \leq \lambda_1$ as follows:

$$\mathbf{x}^T A \mathbf{x} = \langle \mathbf{x}, A\mathbf{x} \rangle = \lambda_1 \langle \mathbf{x}, \mathbf{v}_1 \rangle^2 + \lambda_2 \langle \mathbf{x}, \mathbf{v}_2 \rangle^2 + \cdots + \lambda_n \langle \mathbf{x}, \mathbf{v}_n \rangle^2$$
$$\leq \lambda_1 \langle \mathbf{x}, \mathbf{v}_1 \rangle^2 + \lambda_1 \langle \mathbf{x}, \mathbf{v}_2 \rangle^2 + \cdots + \lambda_1 \langle \mathbf{x}, \mathbf{v}_n \rangle^2$$
$$= \lambda_1 (\langle \mathbf{x}, \mathbf{v}_1 \rangle^2 + \langle \mathbf{x}, \mathbf{v}_2 \rangle^2 + \cdots + \langle \mathbf{x}, \mathbf{v}_n \rangle^2)$$
$$= \lambda_1$$

The proof that $\lambda_n \leq \mathbf{x}^T A \mathbf{x}$ is similar and is left as an exercise.

Proof of Theorem 9.5.1b. If $\mathbf{x}$ is an eigenvector of A corresponding to λ_1 and $\|\mathbf{x}\| = 1$, then

$$\mathbf{x}^T A \mathbf{x} = \langle \mathbf{x}, A\mathbf{x} \rangle = \langle \mathbf{x}, \lambda_1 \mathbf{x} \rangle = \lambda_1 \langle \mathbf{x}, \mathbf{x} \rangle = \lambda_1 \|\mathbf{x}\|^2 = \lambda_1$$

Similarly, $\mathbf{x}^T A \mathbf{x} = \lambda_n$ if $\|\mathbf{x}\| = 1$ and $\mathbf{x}$ is an eigenvector of A corresponding to λ_n. ∎

Exercise Set 9.5

1. Which of the following are quadratic forms?

(a) $x^2 - \sqrt{2}xy$ (b) $5x_1^2 - 2x_2^3 + 4x_1 x_2$ (c) $4x_1^2 - 3x_2^2 + x_3^2 - 5x_1 x_3$

(d) $x_1^2 - 7x_2^2 + x_3^2 + 4x_1 x_2 x_3$ (e) $x_1 x_2 - 3x_1 x_3 + 2x_2 x_3$ (f) $x_1^2 - 6x_2^2 + x_1 - 5x_2$

(g) $(x_1 - 3x_2)^2$ (h) $(x_1 - x_3)^2 + 2(x_1 + 4x_2)^2$

2. Express the following quadratic forms in the matrix notation $\mathbf{x}^T A \mathbf{x}$, where A is a symmetric matrix.

(a) $3x_1^2 + 7x_2^2$ (b) $4x_1^2 - 9x_2^2 - 6x_1 x_2$ (c) $5x_1^2 + 5x_1 x_2$ (d) $-7x_1 x_2$

3. Express the following quadratic forms in the matrix notation $\mathbf{x}^T A \mathbf{x}$, where A is a symmetric matrix.

(a) $9x_1^2 - x_2^2 + 4x_3^2 + 6x_1 x_2 - 8x_1 x_3 + x_2 x_3$ (b) $x_1^2 + x_2^2 - 3x_3^2 - 5x_1 x_2 + 9x_1 x_3$

(c) $x_1 x_2 + x_1 x_3 + x_2 x_3$ (d) $\sqrt{2}x_1^2 - \sqrt{3}x_3^2 + 2\sqrt{2}x_1 x_2 - 8\sqrt{3}x_1 x_3$

(e) $x_1^2 + x_2^2 - x_3^2 - x_4^2 + 2x_1 x_2 - 10x_1 x_4 + 4x_3 x_4$

4. In each part find a formula for the quadratic form that does not use matrices.

(a) $\begin{bmatrix} x & y \end{bmatrix} \begin{bmatrix} 2 & -3 \\ -3 & 5 \end{bmatrix} \begin{bmatrix} x \\ y \end{bmatrix}$ (b) $\begin{bmatrix} x_1 & x_2 \end{bmatrix} \begin{bmatrix} 7 & \frac{5}{2} \\ \frac{5}{2} & 0 \end{bmatrix} \begin{bmatrix} x_1 \\ x_2 \end{bmatrix}$ (c) $\begin{bmatrix} x & y & z \end{bmatrix} \begin{bmatrix} 1 & 0 & 0 \\ 0 & -3 & 0 \\ 0 & 0 & 5 \end{bmatrix} \begin{bmatrix} x \\ y \\ z \end{bmatrix}$

(d) $\begin{bmatrix} x_1 & x_2 & x_3 \end{bmatrix} \begin{bmatrix} -2 & \frac{7}{2} & \frac{1}{2} \\ \frac{7}{2} & 0 & 6 \\ \frac{1}{2} & 6 & 3 \end{bmatrix} \begin{bmatrix} x_1 \\ x_2 \\ x_3 \end{bmatrix}$ (e) $\begin{bmatrix} x_1 & x_2 & x_3 & x_4 \end{bmatrix} \begin{bmatrix} 0 & 1 & 1 & 1 \\ 1 & 0 & 1 & 1 \\ 1 & 1 & 0 & 1 \\ 1 & 1 & 1 & 0 \end{bmatrix} \begin{bmatrix} x_1 \\ x_2 \\ x_3 \\ x_4 \end{bmatrix}$

5. In each part find the maximum and minimum values of the quadratic form subject to the constraint $x_1^2 + x_2^2 = 1$, and determine the values of x_1 and x_2 at which the maximum and minimum occur.

(a) $5x_1^2 - x_2^2$ (b) $7x_1^2 + 4x_2^2 + x_1 x_2$ (c) $5x_1^2 + 2x_2^2 - x_1 x_2$ (d) $2x_1^2 + x_2^2 + 3x_1 x_2$

6. In each part find the maximum and minimum values of the quadratic form subject to the constraint $x_1^2 + x_2^2 + x_3^2 = 1$, and determine the values of x_1, x_2, and x_3 at which the maximum and minimum occur.

(a) $x_1^2 + x_2^2 + 2x_3^2 - 2x_1 x_2 + 4x_1 x_3 + 4x_2 x_3$ (b) $2x_1^2 + x_2^2 + x_3^2 + 2x_1 x_3 + 2x_1 x_2$

(c) $3x_1^2 + 2x_2^2 + 3x_3^2 + 2x_1 x_3$

7. Use Theorem 9.5.2 to determine which of the following matrices are positive definite.

(a) $\begin{bmatrix} 2 & 3 \\ 3 & 2 \end{bmatrix}$ (b) $\begin{bmatrix} 5 & -1 \\ -1 & 5 \end{bmatrix}$ (c) $\begin{bmatrix} 2 & -2 \\ -2 & -1 \end{bmatrix}$

8. Use Theorem 9.5.3 to determine which of the matrices in Exercise 7 are positive definite.

9. Use Theorem 9.5.2 to determine which of the following matrices are positive definite.

(a) $\begin{bmatrix} 3 & -1 & 0 \\ -1 & 2 & -1 \\ 0 & -1 & 3 \end{bmatrix}$ (b) $\begin{bmatrix} 0 & 1 & 1 \\ 1 & 0 & 1 \\ 1 & 1 & 0 \end{bmatrix}$ (c) $\begin{bmatrix} 1 & 2 & 1 \\ 2 & 1 & 1 \\ 1 & 1 & 3 \end{bmatrix}$

10. Use Theorem 9.5.3 to determine which of the matrices in Exercise 9 are positive definite.

11. In each part classify the quadratic form as positive definite, positive semidefinite, negative definite, negative semidefinite, or indefinite.

(a) $x_1^2 + x_2^2$ (b) $-x_1^2 - 3x_2^2$ (c) $(x_1 - x_2)^2$
(d) $-(x_1 - x_2)^2$ (e) $x_1^2 - x_2^2$ (f) $x_1 x_2$

12. In each part classify the matrix as positive definite, positive semidefinite, negative definite, negative semidefinite, or indefinite.

(a) $\begin{bmatrix} 3 & 0 & 0 \\ 0 & -2 & 0 \\ 0 & 0 & 1 \end{bmatrix}$ (b) $\begin{bmatrix} -5 & 0 & 0 \\ 0 & 0 & 0 \\ 0 & 0 & 1 \end{bmatrix}$ (c) $\begin{bmatrix} 6 & 7 & 1 \\ 7 & 9 & 2 \\ 1 & 2 & 1 \end{bmatrix}$

(d) $\begin{bmatrix} -4 & 7 & 8 \\ 7 & -3 & 9 \\ 8 & 9 & -1 \end{bmatrix}$ (e) $\begin{bmatrix} 0 & 0 & 0 \\ 0 & 0 & 0 \\ 0 & 0 & 0 \end{bmatrix}$ (f) $\begin{bmatrix} 1 & 0 & 0 \\ 0 & 1 & 0 \\ 0 & 0 & 1 \end{bmatrix}$

13. Let $\mathbf{x}^T A \mathbf{x}$ be a quadratic form in $x_1, x_2, \ldots, x_n$ and define $T : R^n \to R$ by $T(\mathbf{x}) = \mathbf{x}^T A \mathbf{x}$.
    (a) Show that $T(\mathbf{x} + \mathbf{y}) = T(\mathbf{x}) + 2\mathbf{x}^T A \mathbf{y} + T(\mathbf{y})$. (b) Show that $T(k\mathbf{x}) = k^2 T(\mathbf{x})$.
    (c) Is T a linear transformation? Explain.

14. In each part find all values of k for which the quadratic form is positive definite.
    (a) $x_1^2 + kx_2^2 - 4x_1 x_2$ (b) $5x_1^2 + x_2^2 + kx_3^2 + 4x_1 x_2 - 2x_1 x_3 - 2x_2 x_3$
    (c) $3x_1^2 + x_2^2 + 2x_3^2 + 2x_1 x_3 + 2kx_2 x_3$

15. Express the quadratic form $(c_1 x_1 + c_2 x_2 + \cdots + c_n x_n)^2$ in the matrix notation $\mathbf{x}^T A \mathbf{x}$, where A is symmetric.

16. Let $\mathbf{x} = (x_1, x_2, \ldots, x_n)$. In statistics the quantity

$$\bar{x} = \frac{1}{n}(x_1 + x_2 + \cdots + x_n)$$

is called the **sample mean** of $x_1, x_2, \ldots, x_n$, and

$$s_{\mathbf{x}}^2 = \frac{1}{n-1}[(x_1 - \bar{x})^2 + (x_2 - \bar{x})^2 + \cdots + (x_n - \bar{x})^2]$$

is called the **sample variance**.
    (a) Express the quadratic form $s_{\mathbf{x}}^2$ in the matrix notation $\mathbf{x}^T A \mathbf{x}$, where A is symmetric.
    (b) Is $s_{\mathbf{x}}^2$ a positive definite quadratic form? Explain.

17. Complete the proof of Theorem 9.5.1 by showing that $\lambda_n \leq \mathbf{x}^T A \mathbf{x}$ if $\|\mathbf{x}\| = 1$ and $\lambda_n = \mathbf{x}^T A \mathbf{x}$ if $\mathbf{x}$ is an eigenvector of A corresponding to λ_n.

9.6 DIAGONALIZING QUADRATIC FORMS; CONIC SECTIONS

In this section we shall show how to remove the cross-product terms from a quadratic form by changing variables, and we shall use our results to study the graphs of the conic sections.

Diagonalization of Quadratic Forms   Let

$$\mathbf{x}^T A \mathbf{x} = [x_1 \quad x_2 \quad \cdots \quad x_n] \begin{bmatrix} a_{11} & a_{12} & \cdots & a_{1n} \\ a_{21} & a_{22} & \cdots & a_{2n} \\ \vdots & \vdots & & \vdots \\ a_{n1} & a_{n2} & \cdots & a_{nn} \end{bmatrix} \begin{bmatrix} x_1 \\ x_2 \\ \vdots \\ x_n \end{bmatrix} \tag{1}$$

be a quadratic form, where A is a symmetric matrix. We know from Theorem 7.3.1 that there is an orthogonal matrix P that diagonalizes A; that is,

$$P^T A P = D = \begin{bmatrix} \lambda_1 & 0 & \cdots & 0 \\ 0 & \lambda_2 & \cdots & 0 \\ \vdots & \vdots & & \vdots \\ 0 & 0 & \cdots & \lambda_n \end{bmatrix}$$

where $\lambda_1, \lambda_2, \ldots, \lambda_n$ are the eigenvalues of A. If we let

$$\mathbf{y} = \begin{bmatrix} y_1 \\ y_2 \\ \vdots \\ y_n \end{bmatrix}$$

where $y_1, y_2, \ldots, y_n$ are new variables, and if we make the substitution $\mathbf{x} = P\mathbf{y}$ in (1), then we obtain

$$\mathbf{x}^T A \mathbf{x} = (P\mathbf{y})^T A P\mathbf{y} = \mathbf{y}^T P^T A P\mathbf{y} = \mathbf{y}^T D\mathbf{y}$$

But

$$\mathbf{y}^T D\mathbf{y} = [y_1 \quad y_2 \quad \cdots \quad y_n] \begin{bmatrix} \lambda_1 & 0 & \cdots & 0 \\ 0 & \lambda_2 & \cdots & 0 \\ \vdots & \vdots & & \vdots \\ 0 & 0 & \cdots & \lambda_n \end{bmatrix} \begin{bmatrix} y_1 \\ y_2 \\ \vdots \\ y_n \end{bmatrix}$$

$$= \lambda_1 y_1^2 + \lambda_2 y_2^2 + \cdots + \lambda_n y_n^2$$

which is a quadratic form with no cross-product terms.

In summary, we have the following result.

Theorem 9.6.1

Let $\mathbf{x}^T A \mathbf{x}$ be a quadratic form in the variables $x_1, x_2, \ldots, x_n$, where A is symmetric. If P orthogonally diagonalizes A, and if the new variables $y_1, y_2, \ldots, y_n$ are defined by the equation $\mathbf{x} = P\mathbf{y}$, then substituting this equation in $\mathbf{x}^T A \mathbf{x}$ yields

$$\mathbf{x}^T A \mathbf{x} = \mathbf{y}^T D\mathbf{y} = \lambda_1 y_1^2 + \lambda_2 y_2^2 + \cdots + \lambda_n y_n^2$$

where $\lambda_1, \lambda_2, \ldots, \lambda_n$ are the eigenvalues of A and

$$D = P^T A P = \begin{bmatrix} \lambda_1 & 0 & \cdots & 0 \\ 0 & \lambda_2 & \cdots & 0 \\ \vdots & \vdots & & \vdots \\ 0 & 0 & \cdots & \lambda_n \end{bmatrix}$$

The matrix P in this theorem is said to ***orthogonally diagonalize*** the quadratic form or ***reduce the quadratic form to a sum of squares***.

EXAMPLE 1 Reducing a Quadratic Form to a Sum of Squares

Find a change of variables that will reduce the quadratic form $x_1^2 - x_3^2 - 4x_1x_2 + 4x_2x_3$ to a sum of squares, and express the quadratic form in terms of the new variables.

Solution.

The quadratic form can be written as

$$[x_1 \quad x_2 \quad x_3] \begin{bmatrix} 1 & -2 & 0 \\ -2 & 0 & 2 \\ 0 & 2 & -1 \end{bmatrix} \begin{bmatrix} x_1 \\ x_2 \\ x_3 \end{bmatrix}$$

The characteristic equation of the 3×3 matrix is

$$\begin{vmatrix} \lambda - 1 & 2 & 0 \\ 2 & \lambda & -2 \\ 0 & -2 & \lambda + 1 \end{vmatrix} = \lambda^3 - 9\lambda = \lambda(\lambda + 3)(\lambda - 3) = 0$$

so the eigenvalues are $\lambda = 0$, $\lambda = -3$, $\lambda = 3$. We leave it for the reader to show that orthonormal bases for the three eigenspaces are

$$\lambda = 0: \begin{bmatrix} \frac{2}{3} \\ \frac{1}{3} \\ \frac{2}{3} \end{bmatrix}, \quad \lambda = -3: \begin{bmatrix} -\frac{1}{3} \\ -\frac{2}{3} \\ \frac{2}{3} \end{bmatrix}, \quad \lambda = 3: \begin{bmatrix} -\frac{2}{3} \\ \frac{2}{3} \\ \frac{1}{3} \end{bmatrix}$$

Thus, a substitution $\mathbf{x} = P\mathbf{y}$ that eliminates cross-product terms is

$$\begin{bmatrix} x_1 \\ x_2 \\ x_3 \end{bmatrix} = \begin{bmatrix} \frac{2}{3} & -\frac{1}{3} & -\frac{2}{3} \\ \frac{1}{3} & -\frac{2}{3} & \frac{2}{3} \\ \frac{2}{3} & \frac{2}{3} & \frac{1}{3} \end{bmatrix} \begin{bmatrix} y_1 \\ y_2 \\ y_3 \end{bmatrix}$$

or equivalently,

$$x_1 = \tfrac{2}{3}y_1 - \tfrac{1}{3}y_2 - \tfrac{2}{3}y_3$$
$$x_2 = \tfrac{1}{3}y_1 - \tfrac{2}{3}y_2 + \tfrac{2}{3}y_3$$
$$x_3 = \tfrac{2}{3}y_1 + \tfrac{2}{3}y_2 + \tfrac{1}{3}y_3$$

The new quadratic form is

$$[y_1 \quad y_2 \quad y_3] \begin{bmatrix} 0 & 0 & 0 \\ 0 & -3 & 0 \\ 0 & 0 & 3 \end{bmatrix} \begin{bmatrix} y_1 \\ y_2 \\ y_3 \end{bmatrix}$$

or equivalently,

$$-3y_2^2 + 3y_3^2 \qquad \blacklozenge$$

REMARK. There are other methods for eliminating the cross-product terms from a quadratic form, which we shall not discuss here. Two such methods, *Lagrange's reduction* and *Kronecker's reduction*, are discussed in more advanced books.

Conic Sections
We shall now apply our work on quadratic forms to the study of equations of the form

$$ax^2 + 2bxy + cy^2 + dx + ey + f = 0 \qquad (2)$$

where $a, b, \ldots, f$ are real numbers, and at least one of the numbers a, b, c is not zero. An equation of this type is called a *quadratic equation* in x and y, and

$$ax^2 + 2bxy + cy^2$$

is called the *associated quadratic form*.

EXAMPLE 2 Coefficients in a Quadratic Equation

In the quadratic equation

$$3x^2 + 5xy - 7y^2 + 2x + 7 = 0$$

the constants in (2) are

$$a = 3, \quad b = \tfrac{5}{2}, \quad c = -7, \quad d = 2, \quad e = 0, \quad f = 7 \qquad \blacklozenge$$

EXAMPLE 3 Examples of Associated Quadratic Forms

Quadratic Equation	Associated Quadratic Form
$3x^2 + 5xy - 7y^2 + 2x + 7 = 0$	$3x^2 + 5xy - 7y^2$
$4x^2 - 5y^2 + 8y + 9 = 0$	$4x^2 - 5y^2$
$xy + y = 0$	xy

$\blacklozenge$

Graphs of quadratic equations in x and y are called **conics** or **conic sections**. The most important conics are ellipses, circles, hyperbolas, and parabolas; these are called the **nondegenerate** conics. The remaining conics are called **degenerate** and include single points and pairs of lines (see Exercise 15).

A nondegenerate conic is said to be in **standard position** relative to the coordinate axes if its equation can be expressed in one of the forms given in Figure 9.6.1 (p. 458).

EXAMPLE 4 Three Conics

From Figure 9.6.1, the equation

$$\frac{x^2}{4} + \frac{y^2}{9} = 1$$

matches the form of an ellipse with $k = 2$ and $l = 3$. Thus, the ellipse is in standard position, intersecting the x-axis at $(-2, 0)$ and $(2, 0)$ and intersecting the y-axis at $(0, -3)$ and $(0, 3)$.

The equation $x^2 - 8y^2 = -16$ can be rewritten as $y^2/2 - x^2/16 = 1$, which is of the form $y^2/k^2 - x^2/l^2 = 1$ with $k = \sqrt{2}, l = 4$. Its graph is thus a hyperbola in standard position intersecting the y-axis at $(0, -\sqrt{2})$ and $(0, \sqrt{2})$.

The equation $5x^2 + 2y = 0$ can be rewritten as $x^2 = -\tfrac{2}{5}y$, which is of the form $x^2 = ky$ with $k = -\tfrac{2}{5}$. Since $k < 0$, its graph is a parabola in standard position opening downward. $\blacklozenge$

Significance of the Cross-Product Term
Observe that no conic in standard position has an xy-term (i.e., a cross-product term) in its equation; the presence of an xy term in the equation of a nondegenerate conic indicates that the conic is rotated out of standard position (Figure 9.6.2*a* on page 459). Also, no conic in standard position has both an x^2 and x term or both a y^2 and y term. If there is no cross-product term, the occurrence of either of these pairs in the equation of a nondegenerate conic indicates that the conic is translated out of standard position (Figure 9.6.2*b*). The occurrence of either of these pairs and a cross-product term usually indicates that the conic is both rotated and translated out of standard position (Figure 9.6.2*c*).

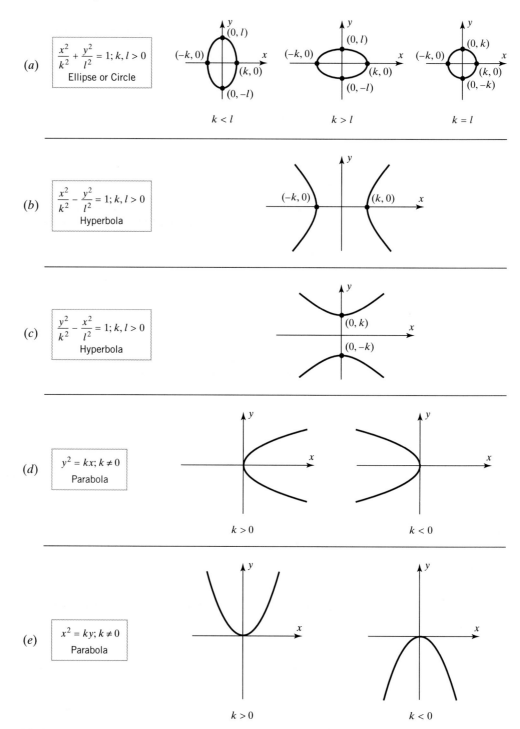

(a) $\dfrac{x^2}{k^2} + \dfrac{y^2}{l^2} = 1; k, l > 0$
Ellipse or Circle

(b) $\dfrac{x^2}{k^2} - \dfrac{y^2}{l^2} = 1; k, l > 0$
Hyperbola

(c) $\dfrac{y^2}{k^2} - \dfrac{x^2}{l^2} = 1; k, l > 0$
Hyperbola

(d) $y^2 = kx; k \neq 0$
Parabola

(e) $x^2 = ky; k \neq 0$
Parabola

Figure 9.6.1

One technique for identifying the graph of a nondegenerate conic that is not in standard position consists of rotating and translating the xy-coordinate axes to obtain an $x'y'$-coordinate system relative to which the conic is in standard position. Once this is done, the equation of the conic in the $x'y'$-system will have one of the forms given in Figure 9.6.1 and can then easily be identified.

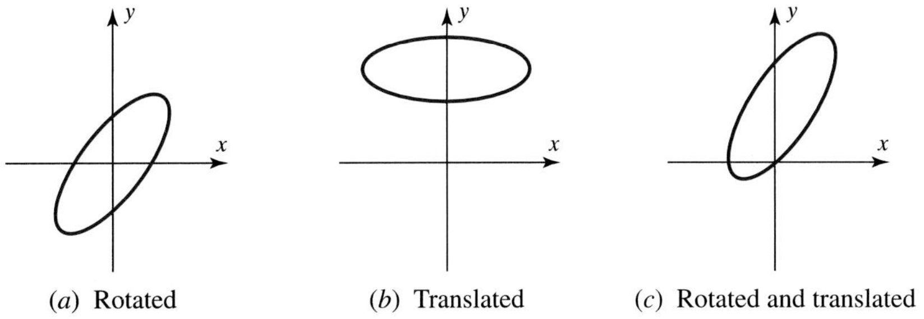

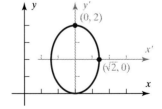

(a) Rotated   (b) Translated   (c) Rotated and translated

Figure 9.6.2

EXAMPLE 5 Completing the Square and Translating

Since the quadratic equation

$$2x^2 + y^2 - 12x - 4y + 18 = 0$$

contains x^2, x, y^2, and y terms but no cross-product term, its graph is a conic that is translated out of standard position but not rotated. This conic can be brought into standard position by suitably translating coordinate axes. To do this, first collect x and y terms. This yields

$$(2x^2 - 12x) + (y^2 - 4y) + 18 = 0 \quad \text{or} \quad 2(x^2 - 6x) + (y^2 - 4y) = -18$$

By completing the squares[†] on the two expressions in parentheses, we obtain

$$2(x^2 - 6x + 9) + (y^2 - 4y + 4) = -18 + 18 + 4$$

or

$$2(x - 3)^2 + (y - 2)^2 = 4 \tag{3}$$

If we translate the coordinate axes by means of the translation equations

$$x' = x - 3, \qquad y' = y - 2$$

then (3) becomes

$$2x'^2 + y'^2 = 4 \quad \text{or} \quad \frac{x'^2}{2} + \frac{y'^2}{4} = 1$$

which is the equation of an ellipse in standard position in the $x'y'$-system. This ellipse is sketched in Figure 9.6.3. ◆

Eliminating the Cross-Product Term

We shall now show how to identify conics that are rotated out of standard position. If we omit the brackets on 1×1 matrices, then (2) can be written in the matrix form

$$[x \quad y] \begin{bmatrix} a & b \\ b & c \end{bmatrix} \begin{bmatrix} x \\ y \end{bmatrix} + [d \quad e] \begin{bmatrix} x \\ y \end{bmatrix} + f = 0$$

or

$$\mathbf{x}^T A \mathbf{x} + K \mathbf{x} + f = 0$$

where

$$\mathbf{x} = \begin{bmatrix} x \\ y \end{bmatrix}, \qquad A = \begin{bmatrix} a & b \\ b & c \end{bmatrix}, \qquad K = [d \quad e]$$

Figure 9.6.3

$$\frac{x'^2}{2} + \frac{y'^2}{4} = 1$$

[†]To complete the square on an expression of the form $x^2 + px$, add and subtract the constant $(p/2)^2$ to obtain

$$x^2 + px = x^2 + px + \left(\frac{p}{2}\right)^2 - \left(\frac{p}{2}\right)^2 = \left(x + \frac{p}{2}\right)^2 - \left(\frac{p}{2}\right)^2$$

Now consider a conic C whose equation in xy-coordinates is

$$\mathbf{x}^T A \mathbf{x} + K \mathbf{x} + f = 0 \tag{4}$$

We would like to rotate the xy-coordinate axes so that the equation of the conic in the new $x'y'$-coordinate system has no cross-product term. This can be done as follows.

Step 1. Find a matrix

$$P = \begin{bmatrix} p_{11} & p_{12} \\ p_{21} & p_{22} \end{bmatrix}$$

that orthogonally diagonalizes the matrix A.

Step 2. Interchange the columns of P, if necessary, to make $\det(P) = 1$. This assures that the orthogonal coordinate transformation

$$\mathbf{x} = P\mathbf{x}', \quad \text{that is,} \quad \begin{bmatrix} x \\ y \end{bmatrix} = P \begin{bmatrix} x' \\ y' \end{bmatrix} \tag{5}$$

is a rotation.

Step 3. To obtain the equation for C in the $x'y'$-system, substitute (5) into (4). This yields

$$(P\mathbf{x}')^T A (P\mathbf{x}') + K(P\mathbf{x}') + f = 0$$

or

$$(\mathbf{x}')^T (P^T A P)\mathbf{x}' + (KP)\mathbf{x}' + f = 0 \tag{6}$$

Since P orthogonally diagonalizes A,

$$P^T A P = \begin{bmatrix} \lambda_1 & 0 \\ 0 & \lambda_2 \end{bmatrix}$$

where λ_1 and λ_2 are eigenvalues of A. Thus, (6) can be rewritten as

$$\begin{bmatrix} x' & y' \end{bmatrix} \begin{bmatrix} \lambda_1 & 0 \\ 0 & \lambda_2 \end{bmatrix} \begin{bmatrix} x' \\ y' \end{bmatrix} + \begin{bmatrix} d & e \end{bmatrix} \begin{bmatrix} p_{11} & p_{12} \\ p_{21} & p_{22} \end{bmatrix} \begin{bmatrix} x' \\ y' \end{bmatrix} + f = 0$$

or

$$\lambda_1 x'^2 + \lambda_2 y'^2 + d' x' + e' y' + f = 0$$

(where $d' = dp_{11} + ep_{21}$ and $e' = dp_{12} + ep_{22}$). This equation has no cross-product term.

The following theorem summarizes this discussion.

Theorem 9.6.2 **Principal Axes Theorem for R^2**

Let

$$ax^2 + 2bxy + cy^2 + dx + ey + f = 0$$

be the equation of a conic C, and let

$$\mathbf{x}^T A \mathbf{x} = ax^2 + 2bxy + cy^2$$

be the associated quadratic form. Then the coordinate axes can be rotated so that the equation for C in the new $x'y'$-coordinate system has the form

$$\lambda_1 x'^2 + \lambda_2 y'^2 + d' x' + e' y' + f = 0$$

where λ_1 and λ_2 are the eigenvalues of A. The rotation can be accomplished by the substitution

$$\mathbf{x} = P\mathbf{x}'$$

where P orthogonally diagonalizes A and $\det(P) = 1$.

EXAMPLE 6 Eliminating the Cross-Product Term

Describe the conic C whose equation is $5x^2 - 4xy + 8y^2 - 36 = 0$.

Solution.

The matrix form of this equation is

$$\mathbf{x}^T A\mathbf{x} - 36 = 0 \tag{7}$$

where

$$A = \begin{bmatrix} 5 & -2 \\ -2 & 8 \end{bmatrix}$$

The characteristic equation of A is

$$\det(\lambda I - A) = \det\begin{bmatrix} \lambda - 5 & 2 \\ 2 & \lambda - 8 \end{bmatrix} = (\lambda - 9)(\lambda - 4) = 0$$

so the eigenvalues of A are $\lambda = 4$ and $\lambda = 9$. We leave it for the reader to show that orthonormal bases for the eigenspaces are

$$\lambda = 4: \quad \mathbf{v}_1 = \begin{bmatrix} 2/\sqrt{5} \\ 1/\sqrt{5} \end{bmatrix}, \qquad \lambda = 9: \quad \mathbf{v}_2 = \begin{bmatrix} -1/\sqrt{5} \\ 2/\sqrt{5} \end{bmatrix}$$

Thus,

$$P = \begin{bmatrix} 2/\sqrt{5} & -1/\sqrt{5} \\ 1/\sqrt{5} & 2/\sqrt{5} \end{bmatrix}$$

orthogonally diagonalizes A. Moreover, $\det(P) = 1$ so that the orthogonal coordinate transformation

$$\mathbf{x} = P\mathbf{x}' \tag{8}$$

is a rotation. Substituting (8) into (7) yields

$$(P\mathbf{x}')^T A(P\mathbf{x}') - 36 = 0 \quad \text{or} \quad (\mathbf{x}')^T(P^T AP)\mathbf{x}' - 36 = 0$$

Since

$$P^T AP = \begin{bmatrix} 4 & 0 \\ 0 & 9 \end{bmatrix}$$

this equation can be written as

$$[x' \quad y']\begin{bmatrix} 4 & 0 \\ 0 & 9 \end{bmatrix}\begin{bmatrix} x' \\ y' \end{bmatrix} - 36 = 0$$

or

$$4x'^2 + 9y'^2 - 36 = 0 \quad \text{or} \quad \frac{x'^2}{9} + \frac{y'^2}{4} = 1$$

which is the equation of the ellipse sketched in Figure 9.6.4. In that figure, the vectors $\mathbf{v}_1$ and $\mathbf{v}_2$ are the column vectors of P. ◆

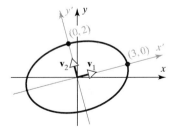

Figure 9.6.4

$$\frac{x'^2}{9} + \frac{y'^2}{4} = 1$$

EXAMPLE 7 Eliminating the Cross-Product Term Plus Translation

Describe the conic C whose equation is

$$5x^2 - 4xy + 8y^2 + 4\sqrt{5}x - 16\sqrt{5}y + 4 = 0$$

Solution.

The matrix form of this equation is

$$\mathbf{x}^T A \mathbf{x} + K \mathbf{x} + 4 = 0 \tag{9}$$

where

$$A = \begin{bmatrix} 5 & -2 \\ -2 & 8 \end{bmatrix} \quad \text{and} \quad K = [4\sqrt{5} \;\; -16\sqrt{5}\,]$$

As shown in Example 6,

$$P = \begin{bmatrix} \dfrac{2}{\sqrt{5}} & -\dfrac{1}{\sqrt{5}} \\[2mm] \dfrac{1}{\sqrt{5}} & \dfrac{2}{\sqrt{5}} \end{bmatrix}$$

orthogonally diagonalizes A and has determinant 1. Substituting $\mathbf{x} = P\mathbf{x}'$ into (9) gives

$$(P\mathbf{x}')^T A (P\mathbf{x}') + K(P\mathbf{x}') + 4 = 0$$

or

$$(\mathbf{x}')^T (P^T A P)\mathbf{x}' + (KP)\mathbf{x}' + 4 = 0 \tag{10}$$

Since

$$P^T A P = \begin{bmatrix} 4 & 0 \\ 0 & 9 \end{bmatrix} \quad \text{and} \quad KP = \begin{bmatrix} \dfrac{20}{\sqrt{5}} & -\dfrac{80}{\sqrt{5}} \end{bmatrix} \begin{bmatrix} \dfrac{2}{\sqrt{5}} & -\dfrac{1}{\sqrt{5}} \\[2mm] \dfrac{1}{\sqrt{5}} & \dfrac{2}{\sqrt{5}} \end{bmatrix} = \begin{bmatrix} -8 & -36 \end{bmatrix}$$

(10) can be written as

$$4x'^2 + 9y'^2 - 8x' - 36y' + 4 = 0 \tag{11}$$

To bring the conic into standard position, the $x'y'$ axes must be translated. Proceeding as in Example 5, we rewrite (11) as

$$4(x'^2 - 2x') + 9(y'^2 - 4y') = -4$$

Completing the squares yields

$$4(x'^2 - 2x' + 1) + 9(y'^2 - 4y' + 4) = -4 + 4 + 36$$

or

$$4(x' - 1)^2 + 9(y' - 2)^2 = 36 \tag{12}$$

If we translate the coordinate axes by means of the translation equations

$$x'' = x' - 1, \quad y'' = y' - 2$$

then (12) becomes

$$4x''^2 + 9y''^2 = 36 \quad \text{or} \quad \frac{x''^2}{9} + \frac{y''^2}{4} = 1$$

which is the equation of the ellipse sketched in Figure 9.6.5. In that figure, the vectors $\mathbf{v}_1$ and $\mathbf{v}_2$ are the column vectors of P. ◆

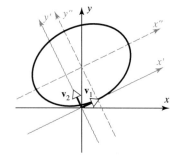

Figure 9.6.5

$$\frac{x''^2}{9} + \frac{y''^2}{4} = 1$$

Exercise Set 9.6

1. In each part find a change of variables that reduces the quadratic form to a sum or difference of squares, and express the quadratic form in terms of the new variables.

(a) $2x_1^2 + 2x_2^2 - 2x_1 x_2$ (b) $5x_1^2 + 2x_2^2 + 4x_1 x_2$ (c) $2x_1 x_2$ (d) $-3x_1^2 + 5x_2^2 + 2x_1 x_2$

2. In each part find a change of variables that reduces the quadratic form to a sum or difference of squares, and express the quadratic form in terms of the new variables.

   (a) $3x_1^2 + 4x_2^2 + 5x_3^2 + 4x_1x_2 - 4x_2x_3$ (b) $2x_1^2 + 5x_2^2 + 5x_3^2 + 4x_1x_2 - 4x_1x_3 - 8x_2x_3$

   (c) $-5x_1^2 + x_2^2 - x_3^2 + 6x_1x_3 + 4x_1x_2$ (d) $2x_1x_3 + 6x_2x_3$

3. Find the quadratic forms associated with the following quadratic equations.

   (a) $2x^2 - 3xy + 4y^2 - 7x + 2y + 7 = 0$ (b) $x^2 - xy + 5x + 8y - 3 = 0$ (c) $5xy = 8$

   (d) $4x^2 - 2y^2 = 7$ (e) $y^2 + 7x - 8y - 5 = 0$

4. Find the matrices of the quadratic forms in Exercise 3.

5. Express each of the quadratic equations in Exercise 3 in the matrix form $\mathbf{x}^T A \mathbf{x} + K\mathbf{x} + f = 0$.

6. Name the following conics.

   (a) $2x^2 + 5y^2 = 20$ (b) $4x^2 + 9y^2 = 1$ (c) $x^2 - y^2 - 8 = 0$ (d) $4y^2 - 5x^2 = 20$

   (e) $x^2 + y^2 - 25 = 0$ (f) $7y^2 - 2x = 0$ (g) $-x^2 = 2y$ (h) $3x - 11y^2 = 0$

   (i) $y - x^2 = 0$ (j) $x^2 - 3 = -y^2$

7. In each part a translation will put the conic in standard position. Name the conic and give its equation in the translated coordinate system.

   (a) $9x^2 + 4y^2 - 36x - 24y + 36 = 0$ (b) $x^2 - 16y^2 + 8x + 128y = 256$

   (c) $y^2 - 8x - 14y + 49 = 0$ (d) $x^2 + y^2 + 6x - 10y + 18 = 0$

   (e) $2x^2 - 3y^2 + 6x + 20y = -41$ (f) $x^2 + 10x + 7y = -32$

8. The following nondegenerate conics are rotated out of standard position. In each part rotate the coordinate axes to remove the xy term. Name the conic and give its equation in the rotated coordinate system.

   (a) $2x^2 - 4xy - y^2 + 8 = 0$ (b) $5x^2 + 4xy + 5y^2 = 9$

   (c) $11x^2 + 24xy + 4y^2 - 15 = 0$

In Exercises 9–14 translate and rotate the coordinate axes, if necessary, to put the conic in standard position. Name the conic and give its equation in the final coordinate system.

9. $9x^2 - 4xy + 6y^2 - 10x - 20y = 5$ 10. $3x^2 - 8xy - 12y^2 - 30x - 64y = 0$

11. $2x^2 - 4xy - y^2 - 4x - 8y = -14$ 12. $21x^2 + 6xy + 13y^2 - 114x + 34y + 73 = 0$

13. $x^2 - 6xy - 7y^2 + 10x + 2y + 9 = 0$ 14. $4x^2 - 20xy + 25y^2 - 15x - 6y = 0$

15. The graph of a quadratic equation in x and y can, in certain cases, be a point, a line, or a pair of lines. These are called **degenerate** conics. It is also possible that the equation is not satisfied by any real values of x and y. In such cases the equation has no graph; it is said to represent an **imaginary conic**. Each of the following represents a degenerate or imaginary conic. Where possible, sketch the graph.

   (a) $x^2 - y^2 = 0$ (b) $x^2 + 3y^2 + 7 = 0$ (c) $8x^2 + 7y^2 = 0$

   (d) $x^2 - 2xy + y^2 = 0$ (e) $9x^2 + 12xy + 4y^2 - 52 = 0$ (f) $x^2 + y^2 - 2x - 4y = -5$

9.7 QUADRIC SURFACES

In this section we shall apply the diagonalization techniques developed in the preceding section to quadratic equations in three variables, and we shall use our results to study quadric surfaces.

Quadric Surfaces   An equation of the form

$$ax^2 + by^2 + cz^2 + 2dxy + 2exz + 2fyz + gx + hy + iz + j = 0 \qquad (1)$$

where $a, b, \ldots, f$ are not all zero, is called a **quadratic equation in x, y, and z**; the expression

$$ax^2 + by^2 + cz^2 + 2dxy + 2exz + 2fyz$$

is called the ***associated quadratic form***.

Equation (1) can be written in the matrix form

$$[x \quad y \quad z] \begin{bmatrix} a & d & e \\ d & b & f \\ e & f & c \end{bmatrix} \begin{bmatrix} x \\ y \\ z \end{bmatrix} + [g \quad h \quad i] \begin{bmatrix} x \\ y \\ z \end{bmatrix} + j = 0$$

or

$$\mathbf{x}^T A \mathbf{x} + K \mathbf{x} + j = 0$$

where

$$\mathbf{x} = \begin{bmatrix} x \\ y \\ z \end{bmatrix}, \qquad A = \begin{bmatrix} a & d & e \\ d & b & f \\ e & f & c \end{bmatrix}, \qquad K = [g \quad h \quad i]$$

EXAMPLE 1 Associated Quadratic Form

The quadratic form associated with the quadratic equation

$$3x^2 + 2y^2 - z^2 + 4xy + 3xz - 8yz + 7x + 2y + 3z - 7 = 0$$

is

$$3x^2 + 2y^2 - z^2 + 4xy + 3xz - 8yz \qquad\qquad \blacklozenge$$

Graphs of quadratic equations in x, y, and z are called ***quadrics*** or ***quadric surfaces***. The simplest equations for quadric surfaces occur when those surfaces are placed in certain ***standard positions*** relative to the coordinate axes. Figure 9.7.1 shows the six basic quadric surfaces and the equations for those surfaces when the surfaces are in the standard positions shown in the figure. If a quadric surface is cut by a plane, then the curve of intersection is called the ***trace*** of the plane on the surface. To help visualize the quadric surfaces in Figure 9.7.1, we have shown and described the traces made by planes parallel to the coordinate planes. The presence of one or more of the cross-product terms xy, xz, and yz in the equation of a quadric indicates that the quadric is rotated out of standard position; the presence of both x^2 and x terms, y^2 and y terms, or z^2 and z terms in a quadric with no cross-product term indicates the quadric is translated out of standard position.

EXAMPLE 2 Identifying a Quadric Surface

Describe the quadric surface whose equation is

$$4x^2 + 36y^2 - 9z^2 - 16x - 216y + 304 = 0$$

Solution.

Rearranging terms gives

$$4(x^2 - 4x) + 36(y^2 - 6y) - 9z^2 = -304$$

Completing the squares yields

$$4(x^2 - 4x + 4) + 36(y^2 - 6y + 9) - 9z^2 = -304 + 16 + 324$$

or

$$4(x - 2)^2 + 36(y - 3)^2 - 9z^2 = 36$$

Surface	Equation	Surface	Equation
Ellipsoid 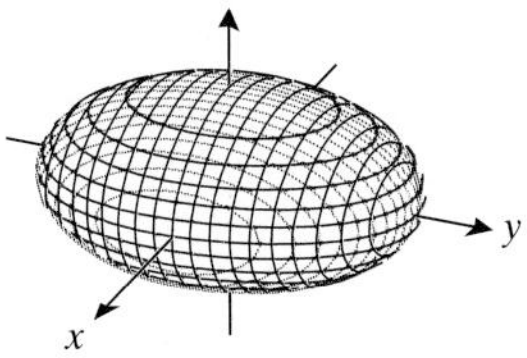	$\dfrac{x^2}{l^2} + \dfrac{y^2}{m^2} + \dfrac{z^2}{n^2} = 1$ The traces in the coordinate planes are ellipses, as are the traces in those planes that are parallel to the coordinate planes and intersect the surface in more than one point.	Elliptic cone 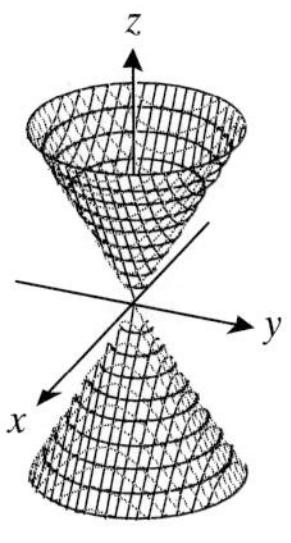	$z^2 = \dfrac{x^2}{l^2} + \dfrac{y^2}{m^2}$ The trace in the xy-plane is a point (the origin), and the traces in planes parallel to the xy-plane are ellipses. The traces in the yz- and xz-planes are pairs of lines intersecting at the origin. The traces in planes parallel to these are hyperbolas.
Hyperboloid of one sheet 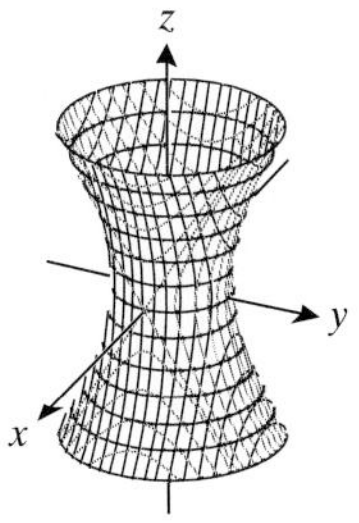	$\dfrac{x^2}{l^2} + \dfrac{y^2}{m^2} - \dfrac{z^2}{n^2} = 1$ The trace in the xy-plane is an ellipse, as are the traces in planes parallel to the xy-plane. The traces in the yz-plane and xz-plane are hyperbolas, as are the traces in those planes that are parallel to these and do not pass through the x- or y-intercepts. At these intercepts the traces are pairs of intersecting lines.	Elliptic paraboloid 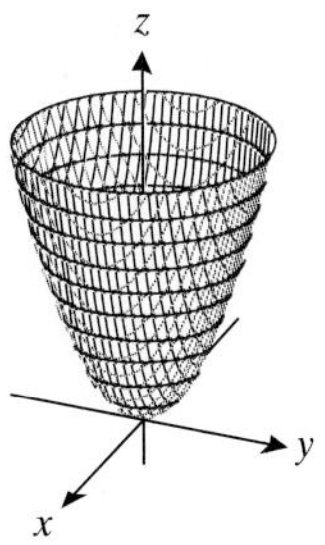	$z = \dfrac{x^2}{l^2} + \dfrac{y^2}{m^2}$ The trace in the xy-plane is a point (the origin), and the traces in planes parallel to and above the xy-plane are ellipses. The traces in the yz- and xz-planes are parabolas, as are the traces in planes parallel to these.
Hyperboloid of two sheets 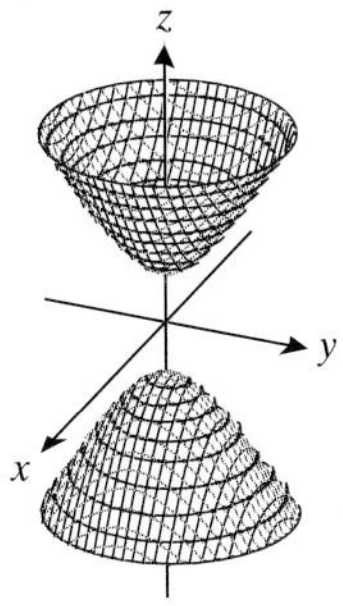	$\dfrac{z^2}{l^2} - \dfrac{x^2}{m^2} - \dfrac{y^2}{n^2} = 1$ There is no trace in the xy-plane. In planes parallel to the xy-plane that intersect the surface in more than one point the traces are ellipses. In the yz- and xz-planes, the traces are hyperbolas, as are the traces in those planes that are parallel to these and intersect the surface in more than one point.	Hyperbolic paraboloid 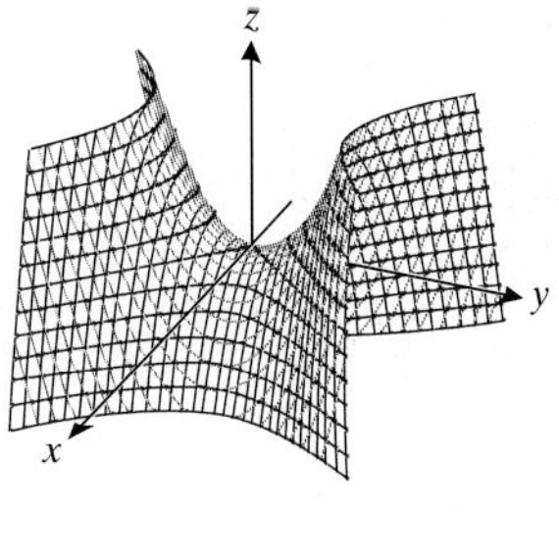	$z = \dfrac{y^2}{m^2} - \dfrac{x^2}{l^2}$ The trace in the xy-plane is a pair of lines intersecting at the origin. The traces in planes parallel to the xy-plane are hyperbolas. The hyperbolas above the xy-plane open in the y-direction, and those below in the x-direction. The traces in the yz- and xz-planes are parabolas, as are the traces in planes parallel to these.

Figure 9.7.1

or

$$\frac{(x-2)^2}{9} + (y-3)^2 - \frac{z^2}{4} = 1$$

Translating the axes by means of the translation equations

$$x' = x - 2, \qquad y' = y - 3, \qquad z' = z$$

yields

$$\frac{x'^2}{9} + y'^2 - \frac{z'^2}{4} = 1$$

which is the equation of a hyperboloid of one sheet. ◆

Eliminating Cross-Product Terms The procedure for identifying quadrics that are rotated out of standard position is similar to the procedure for conics. Let Q be a quadric surface whose equation in xyz-coordinates is

$$\mathbf{x}^T A \mathbf{x} + K \mathbf{x} + j = 0 \qquad (2)$$

We want to rotate the xyz-coordinate axes so that the equation of the quadric in the new $x'y'z'$-coordinate system has no cross-product terms. This can be done as follows:

Step 1. Find a matrix P that orthogonally diagonalizes $\mathbf{x}^T A \mathbf{x}$.

Step 2. Interchange two columns of P, if necessary, to make $\det(P) = 1$. This assures that the orthogonal coordinate transformation

$$\mathbf{x} = P \mathbf{x}', \quad \text{that is,} \quad \begin{bmatrix} x \\ y \\ z \end{bmatrix} = P \begin{bmatrix} x' \\ y' \\ z' \end{bmatrix}, \qquad (3)$$

is a rotation.

Step 3. Substitute (3) into (2). This will produce an equation for the quadric in $x'y'z'$-coordinates with no cross-product terms. (The proof is similar to that for conics and is left as an exercise.)

The following theorem summarizes this discussion.

Theorem 9.7.1   **Principal Axes Theorem for R^3**

Let

$$ax^2 + by^2 + cz^2 + 2dxy + 2exz + 2fyz + gx + hy + iz + j = 0$$

be the equation of a quadric Q, and let

$$\mathbf{x}^T A \mathbf{x} = ax^2 + by^2 + cz^2 + 2dxy + 2exz + 2fyz$$

be the associated quadratic form. The coordinate axes can be rotated so that the equation of Q in the $x'y'z'$-coordinate system has the form

$$\lambda_1 x'^2 + \lambda_2 y'^2 + \lambda_3 z'^2 + g'x' + h'y' + i'z' + j = 0$$

where λ_1, λ_2, and λ_3 are the eigenvalues of A. The rotation can be accomplished by the substitution

$$\mathbf{x} = P \mathbf{x}'$$

where P orthogonally diagonalizes A and $\det(P) = 1$.

EXAMPLE 3 Eliminating Cross-Product Terms

Describe the quadric surface whose equation is

$$4x^2 + 4y^2 + 4z^2 + 4xy + 4xz + 4yz - 3 = 0$$

Solution.

The matrix form of the above quadratic equation is

$$\mathbf{x}^T A \mathbf{x} - 3 = 0 \qquad (4)$$

where

$$A = \begin{bmatrix} 4 & 2 & 2 \\ 2 & 4 & 2 \\ 2 & 2 & 4 \end{bmatrix}$$

As shown in Example 1 of Section 7.3, the eigenvalues of A are $\lambda = 2$ and $\lambda = 8$, and A is orthogonally diagonalized by the matrix

$$P = \begin{bmatrix} -1/\sqrt{2} & -1/\sqrt{6} & 1/\sqrt{3} \\ 1/\sqrt{2} & -1/\sqrt{6} & 1/\sqrt{3} \\ 0 & 2/\sqrt{6} & 1/\sqrt{3} \end{bmatrix}$$

where the first two column vectors in P are eigenvectors corresponding to $\lambda = 2$, and the third column vector is an eigenvector corresponding to $\lambda = 8$.

Since $\det(P) = 1$ (verify), the orthogonal coordinate transformation $\mathbf{x} = P\mathbf{x}'$ is a rotation. Substituting this expression in (4) yields

$$(P\mathbf{x}')^T A (P\mathbf{x}') - 3 = 0$$

or equivalently,

$$(\mathbf{x}')^T (P^T A P) \mathbf{x}' - 3 = 0 \tag{5}$$

But

$$P^T A P = \begin{bmatrix} 2 & 0 & 0 \\ 0 & 2 & 0 \\ 0 & 0 & 8 \end{bmatrix}$$

so (5) becomes

$$[x' \ \ y' \ \ z'] \begin{bmatrix} 2 & 0 & 0 \\ 0 & 2 & 0 \\ 0 & 0 & 8 \end{bmatrix} \begin{bmatrix} x' \\ y' \\ z' \end{bmatrix} - 3 = 0$$

or

$$2x'^2 + 2y'^2 + 8z'^2 = 3$$

This can be rewritten as

$$\frac{x'^2}{3/2} + \frac{y'^2}{3/2} + \frac{z'^2}{3/8} = 1$$

which is the equation of an ellipsoid. ◆

Exercise Set 9.7

1. Find the quadratic forms associated with the following quadratic equations.
   (a) $x^2 + 2y^2 - z^2 + 4xy - 5yz + 7x + 2z = 3$ (b) $3x^2 + 7z^2 + 2xy - 3xz + 4yz - 3x = 4$
   (c) $xy + xz + yz = 1$ (d) $x^2 + y^2 - z^2 = 7$
   (e) $3z^2 + 3xz - 14y + 9 = 0$ (f) $2z^2 + 2xz + y^2 + 2x - y + 3z = 0$

2. Find the matrices of the quadratic forms in Exercise 1.

3. Express each of the quadratic equations given in Exercise 1 in the matrix form
   $\mathbf{x}^T A \mathbf{x} + K\mathbf{x} + j = 0$.

4. Name the following quadrics.
   (a) $36x^2 + 9y^2 + 4z^2 - 36 = 0$ (b) $2x^2 + 6y^2 - 3z^2 = 18$ (c) $6x^2 - 3y^2 - 2z^2 - 6 = 0$
   (d) $9x^2 + 4y^2 - z^2 = 0$ (e) $16x^2 + y^2 = 16z$ (f) $7x^2 - 3y^2 + z = 0$
   (g) $x^2 + y^2 + z^2 = 25$

5. In each part determine the translation equations that will put the quadric in standard position, and find the equation of the quadric in the translated coordinate system. Name the quadric.

(a) $9x^2 + 36y^2 + 4z^2 - 18x - 144y - 24z + 153 = 0$ (b) $6x^2 + 3y^2 - 2z^2 + 12x - 18y - 8z = -7$

(c) $3x^2 - 3y^2 - z^2 + 42x + 144 = 0$ (d) $4x^2 + 9y^2 - z^2 - 54y - 50z = 544$

(e) $x^2 + 16y^2 + 2x - 32y - 16z - 15 = 0$ (f) $7x^2 - 3y^2 + 126x + 72y + z + 135 = 0$

(g) $x^2 + y^2 + z^2 - 2x + 4y - 6z = 11$

6. In each part find a rotation $\mathbf{x} = P\mathbf{x}'$ that removes the cross-product terms, and give its equation in the $x'y'z'$-system. Name the quadric.

(a) $2x^2 + 3y^2 + 23z^2 + 72xz + 150 = 0$ (b) $4x^2 + 4y^2 + 4z^2 + 4xy + 4xz + 4yz - 5 = 0$

(c) $144x^2 + 100y^2 + 81z^2 - 216xz - 540x - 720z = 0$ (d) $2xy + z = 0$

In Exercises 7–10 translate and rotate the coordinate axes to put the quadric in standard position. Name the quadric and give its equation in the final coordinate system.

7. $2xy + 2xz + 2yz - 6x - 6y - 4z = -9$

8. $7x^2 + 7y^2 + 10z^2 - 2xy - 4xz + 4yz - 12x + 12y + 60z = 24$

9. $2xy - 6x + 10y + z - 31 = 0$

10. $2x^2 + 2y^2 + 5z^2 - 4xy - 2xz + 2yz + 10x - 26y - 2z = 0$

11. Prove Theorem 9.7.1.

9.8 COMPARISON OF PROCEDURES FOR SOLVING LINEAR SYSTEMS

In this section we shall discuss some practical aspects of solving systems of linear equations, inverting matrices, and finding eigenvalues. Although we have previously discussed methods for performing these computations, those methods are not directly applicable to the computer solution of the large-scale problems that arise in real-world applications.

Counting Operations Since computers are limited in the number of decimal places they can carry, they round off or truncate most numerical quantities. For example, a computer designed to store eight decimal places might record $\frac{2}{3}$ either as .66666667 (rounded off) or .66666666 (truncated). In either case, an error is introduced that we shall call **roundoff error** or **rounding error**.

The main practical considerations in solving linear algebra problems on digital computers are minimizing the computer time (and thus cost) needed to obtain the solution, and minimizing inaccuracies due to roundoff errors. Thus, a good computer algorithm uses as few operations as possible and performs those operations in a way that minimizes the effect of roundoff errors.

In this text we have studied four methods for solving a linear system, $A\mathbf{x} = \mathbf{b}$, of n equations in n unknowns:

1. Gaussian elimination with back-substitution

2. Gauss–Jordan elimination

3. Computing A^{-1}, then $\mathbf{x} = A^{-1}\mathbf{b}$

4. Cramer's rule

To determine how these methods compare as computational tools, we need to know how many arithmetic operations each requires. On a large modern computer, typical execution times in microseconds (1 microsecond $= 10^{-6}$ second) for the basic arithmetic operations are

$$\text{Multiplication} = 1.0 \text{ microsecond}$$
$$\text{Division} = 3.0 \text{ microseconds}$$
$$\text{Addition} = 0.5 \text{ microsecond}$$
$$\text{Subtraction} = 0.5 \text{ microsecond}$$

In our analysis we shall group divisions and multiplications together (average execution time $= 2.0$ microseconds), and we shall also group additions and subtractions together (average execution time $= 0.5$ microsecond). We shall refer to either multiplications or divisions as "multiplications" and to additions or subtractions as "additions."

In Table 1 we list the number of operations required to solve a linear system $A\mathbf{x} = \mathbf{b}$ of n equations in n unknowns by each of the four methods discussed in the text, as well as the number of operations required to invert A or to compute its determinant by row reduction.

TABLE 1 *Operation Counts for an Invertible $n \times n$ Matrix A*

Method	Number of Additions	Number of Multiplications
Solve $A\mathbf{x} = \mathbf{b}$ by Gauss–Jordan elimination	$\frac{1}{3}n^3 + \frac{1}{2}n^2 - \frac{5}{6}n$	$\frac{1}{3}n^3 + n^2 - \frac{1}{3}n$
Solve $A\mathbf{x} = \mathbf{b}$ by Gaussian elimination	$\frac{1}{3}n^3 + \frac{1}{2}n^2 - \frac{5}{6}n$	$\frac{1}{3}n^3 + n^2 - \frac{1}{3}n$
Find A^{-1} by reducing $[A \mid I]$ to $[I \mid A^{-1}]$	$n^3 - 2n^2 + n$	n^3
Solve $A\mathbf{x} = \mathbf{b}$ as $\mathbf{x} = A^{-1}\mathbf{b}$	$n^3 - n^2$	$n^3 + n^2$
Find $\det(A)$ by row reduction	$\frac{1}{3}n^3 - \frac{1}{2}n^2 + \frac{1}{6}n$	$\frac{1}{3}n^3 + \frac{2}{3}n - 1$
Solve $A\mathbf{x} = \mathbf{b}$ by Cramer's rule	$\frac{1}{3}n^4 - \frac{1}{6}n^3 - \frac{1}{3}n^2 + \frac{1}{6}n$	$\frac{1}{3}n^4 + \frac{1}{3}n^3 + \frac{2}{3}n^2 + \frac{2}{3}n - 1$

Note that the text methods of Gauss–Jordan elimination and Gaussian elimination have the same operation counts. It is not hard to see why this is so. Both methods begin by reducing the augmented matrix to row-echelon form. This is called the ***forward phase*** or ***forward pass***. Then the solution is completed by back-substitution in Gaussian elimination and by continued reduction to reduced row-echelon form in Gauss–Jordan elimination. This is called the ***backward phase*** or ***backward pass***. It turns out that the number of operations required for the backward phase is the same whether one uses back-substitution or continued reduction to reduced row-echelon form. Thus, the text methods of Gaussian elimination and Gauss–Jordan elimination have the same operation counts.

REMARK. There is a common variation of Gauss–Jordan elimination that is less efficient than the one presented in this text. In our method the augmented matrix is first reduced to reduced row-echelon form by introducing zeros *below* the leading 1's; then the reduction is completed by introducing zeros above the leading 1's. An alternative procedure is to introduce zeros *above* and *below* a leading 1 as soon as it is obtained. This method requires

$$\frac{n^3}{2} - \frac{n}{2} \quad \text{additions} \quad \text{and} \quad \frac{n^3}{2} + \frac{n^2}{2} \quad \text{multiplications}$$

both of which are larger than our values for all $n \geq 3$.

To illustrate how the results in Table 1 are computed, we shall derive the operation counts for Gauss–Jordan elimination. For this discussion we need the following formulas for the sum of the first n positive integers and the sum of the squares of the first n positive integers:

$$1 + 2 + 3 + \cdots + n = \frac{n(n+1)}{2} \tag{1}$$

$$1^2 + 2^2 + 3^2 + \cdots + n^2 = \frac{n(n+1)(2n+1)}{6} \tag{2}$$

Derivations of these formulas are discussed in the exercises. We also need formulas for the sum of the first $n-1$ positive integers and the sum of the squares of the first $n-1$ positive integers. These can be obtained by substituting $n-1$ for n in (1) and (2).

$$1 + 2 + 3 + \cdots + (n-1) = \frac{(n-1)n}{2} \tag{3}$$

$$1^2 + 2^2 + 3^2 + \cdots + (n-1)^2 = \frac{(n-1)n(2n-1)}{6} \tag{4}$$

Operation Count for Gauss–Jordan Elimination   Let $A\mathbf{x} = \mathbf{b}$ be a system of n linear equations in n unknowns, and assume that A is invertible, so that the system has a unique solution. Also assume, for simplicity, that no row interchanges are required to put the augmented matrix $[A \mid \mathbf{b}]$ in reduced row-echelon form. This assumption is justified by the fact that row interchanges are performed as bookkeeping operations on a computer and require much less time than arithmetic operations.

Since no row interchanges are required, the first step in the Gauss–Jordan elimination process is to introduce a leading 1 in the first row by multiplying the elements in that row by the reciprocal of the leftmost entry in the row. We shall represent this step schematically as follows:

← × denotes a quantity to be computed.
• denotes a quantity that is not computed.
The matrix size is $n \times (n+1)$.

Note that the leading 1 is simply recorded and requires no computation; only the remaining n entries in the first row must be computed.

The following is a schematic description of the steps and the number of operations required to reduce $[A \mid \mathbf{b}]$ to row-echelon form.

Step 1
$$\begin{bmatrix} 1 & \times & \times & \cdots & \times & \times & \big| & \times \\ \bullet & \bullet & \bullet & \cdots & \bullet & \bullet & \big| & \bullet \\ \bullet & \bullet & \bullet & \cdots & \bullet & \bullet & \big| & \bullet \\ \vdots & \vdots & \vdots & & \vdots & \vdots & \big| & \vdots \\ \bullet & \bullet & \bullet & \cdots & \bullet & \bullet & \big| & \bullet \\ \bullet & \bullet & \bullet & \cdots & \bullet & \bullet & \big| & \bullet \end{bmatrix}$$

⟵ n multiplications
0 additions

Step 1a
$$\begin{bmatrix} 1 & \bullet & \bullet & \cdots & \bullet & \bullet & \big| & \bullet \\ 0 & \times & \times & \cdots & \times & \times & \big| & \times \\ 0 & \times & \times & \cdots & \times & \times & \big| & \times \\ \vdots & \vdots & \vdots & & \vdots & \vdots & \big| & \vdots \\ 0 & \times & \times & \cdots & \times & \times & \big| & \times \\ 0 & \times & \times & \cdots & \times & \times & \big| & \times \end{bmatrix}$$

n multiplications/row
n additions/row
$n - 1$ rows requiring computations

⟵ $n(n-1)$ multiplications
$n(n-1)$ additions

Step 2
$$\begin{bmatrix} 1 & \bullet & \bullet & \cdots & \bullet & \bullet & \big| & \bullet \\ 0 & 1 & \times & \cdots & \times & \times & \big| & \times \\ 0 & \bullet & \bullet & \cdots & \bullet & \bullet & \big| & \bullet \\ \vdots & \vdots & \vdots & & \vdots & \vdots & \big| & \vdots \\ 0 & \bullet & \bullet & \cdots & \bullet & \bullet & \big| & \bullet \\ 0 & \bullet & \bullet & \cdots & \bullet & \bullet & \big| & \bullet \end{bmatrix}$$

⟵ $n - 1$ multiplications
0 additions

Step 2a
$$\begin{bmatrix} 1 & \bullet & \bullet & \cdots & \bullet & \bullet & \big| & \bullet \\ 0 & 1 & \bullet & \cdots & \bullet & \bullet & \big| & \bullet \\ 0 & 0 & \times & \cdots & \times & \times & \big| & \times \\ \vdots & \vdots & \vdots & & \vdots & \vdots & \big| & \vdots \\ 0 & 0 & \times & \cdots & \times & \times & \big| & \times \\ 0 & 0 & \times & \cdots & \times & \times & \big| & \times \end{bmatrix}$$

$n - 1$ multiplications/row
$n - 1$ additions/row
$n - 2$ rows requiring computations

⟵ $(n-1)(n-2)$ multiplications
$(n-1)(n-2)$ additions

Step 3
$$\begin{bmatrix} 1 & \bullet & \bullet & \cdots & \bullet & \bullet & \big| & \bullet \\ 0 & 1 & \bullet & \cdots & \bullet & \bullet & \big| & \bullet \\ 0 & 0 & 1 & \cdots & \times & \times & \big| & \times \\ \vdots & \vdots & \vdots & & \vdots & \vdots & \big| & \vdots \\ 0 & 0 & \bullet & \cdots & \bullet & \bullet & \big| & \bullet \\ 0 & 0 & \bullet & \cdots & \bullet & \bullet & \big| & \bullet \end{bmatrix}$$

⟵ $n - 2$ multiplications
0 additions

Step 3a
$$\begin{bmatrix} 1 & \bullet & \bullet & \cdots & \bullet & \bullet & \big| & \bullet \\ 0 & 1 & \bullet & \cdots & \bullet & \bullet & \big| & \bullet \\ 0 & 0 & 1 & \cdots & \bullet & \bullet & \big| & \bullet \\ \vdots & \vdots & \vdots & & \vdots & \vdots & \big| & \vdots \\ 0 & 0 & 0 & \cdots & \times & \times & \big| & \times \\ 0 & 0 & 0 & \cdots & \times & \times & \big| & \times \end{bmatrix}$$

$n - 2$ multiplications/row
$n - 2$ additions/row
$n - 3$ rows requiring computations

⟵ $(n-2)(n-3)$ multiplications
$(n-2)(n-3)$ additions

Step $(n-1)$

$$\left[\begin{array}{ccccccc|c} 1 & \bullet & \bullet & \cdots & \bullet & \bullet & & \bullet \\ 0 & 1 & \bullet & \cdots & \bullet & \bullet & & \bullet \\ 0 & 0 & 1 & \cdots & \bullet & \bullet & & \bullet \\ \vdots & \vdots & \vdots & & \vdots & \vdots & & \vdots \\ 0 & 0 & 0 & \cdots & 1 & \times & & \times \\ 0 & 0 & 0 & \cdots & \bullet & \bullet & & \bullet \end{array}\right]$$

$\longleftarrow$ 2 multiplications
0 additions

Step $(n-1)$a

$$\left[\begin{array}{ccccccc|c} 1 & \bullet & \bullet & \cdots & \bullet & \bullet & & \bullet \\ 0 & 1 & \bullet & \cdots & \bullet & \bullet & & \bullet \\ 0 & 0 & 1 & \cdots & \bullet & \bullet & & \bullet \\ \vdots & \vdots & \vdots & & \vdots & \vdots & & \vdots \\ 0 & 0 & 0 & \cdots & 1 & \bullet & & \bullet \\ 0 & 0 & 0 & \cdots & 0 & \times & & \times \end{array}\right]$$

2 multiplications/row
2 additions/row
1 row requiring computations

$\longleftarrow$ 2 multiplications
2 additions

Step n

$$\left[\begin{array}{ccccccc|c} 1 & \bullet & \bullet & \cdots & \bullet & \bullet & & \bullet \\ 0 & 1 & \bullet & \cdots & \bullet & \bullet & & \bullet \\ 0 & 0 & 1 & \cdots & \bullet & \bullet & & \bullet \\ \vdots & \vdots & \vdots & & \vdots & \vdots & & \vdots \\ 0 & 0 & 0 & \cdots & 1 & \bullet & & \bullet \\ 0 & 0 & 0 & \cdots & 0 & 1 & & \times \end{array}\right]$$

$\longleftarrow$ 1 multiplication
0 additions

Thus, the number of operations required to complete successive steps is as follows:

Steps 1 and 1a
Multiplications: $\quad n + n(n-1) = n^2$
Additions: $\quad n(n-1) = n^2 - n$

Steps 2 and 2a
Multiplications: $\quad (n-1) + (n-1)(n-2) = (n-1)^2$
Additions: $\quad (n-1)(n-2) = (n-1)^2 - (n-1)$

Steps 3 and 3a
Multiplications: $\quad (n-2) + (n-2)(n-3) = (n-2)^2$
Additions: $\quad (n-2)(n-3) = (n-2)^2 - (n-2)$

$\vdots$

Steps $(n-1)$ and $(n-1)$a
Multiplications: $\quad 4 \;(= 2^2)$
Additions: $\quad 2 \;(= 2^2 - 2)$

Step n
Multiplications: $\quad 1 \;(= 1^2)$
Additions: $\quad 0 \;(= 1^2 - 1)$

Therefore, the total number of operations required to reduce $[A \mid \mathbf{b}]$ to row-echelon form is

Multiplications: $\quad n^2 + (n-1)^2 + (n-2)^2 + \cdots + 1^2$

Additions: $\qquad [n^2 + (n-1)^2 + (n-2)^2 + \cdots + 1^2]$
$$- [n + (n-1) + (n-2) + \cdots + 1]$$

or, on applying Formulas (1) and (2),

$$\text{Multiplications:} \quad \frac{n(n+1)(2n+1)}{6} = \frac{n^3}{3} + \frac{n^2}{2} + \frac{n}{6} \tag{5}$$

$$\text{Additions:} \quad \frac{n(n+1)(2n+1)}{6} - \frac{n(n+1)}{2} = \frac{n^3}{3} - \frac{n}{3} \tag{6}$$

This completes the operation count for the forward phase. For the backward phase we must put the row-echelon form of $[A \mid \mathbf{b}]$ into reduced row-echelon form by introducing zeros above the leading 1's. The operations are as follows:

Step 1
$$\begin{bmatrix}
1 & \bullet & \bullet & \cdots & \bullet & 0 & \times \\
0 & 1 & \bullet & \cdots & \bullet & 0 & \times \\
0 & 0 & 1 & \cdots & \bullet & 0 & \times \\
\vdots & \vdots & \vdots & & \vdots & \vdots & \vdots \\
0 & 0 & 0 & \cdots & 1 & 0 & \times \\
0 & 0 & 0 & \cdots & 0 & 1 & \bullet
\end{bmatrix}$$
$\longleftarrow$ $n - 1$ multiplications $n - 1$ additions

Step 2
$$\begin{bmatrix}
1 & \bullet & \bullet & \cdots & 0 & 0 & \times \\
0 & 1 & \bullet & \cdots & 0 & 0 & \times \\
0 & 0 & 1 & \cdots & 0 & 0 & \times \\
\vdots & \vdots & \vdots & & \vdots & \vdots & \vdots \\
0 & 0 & 0 & \cdots & 1 & 0 & \bullet \\
0 & 0 & 0 & \cdots & 0 & 1 & \bullet
\end{bmatrix}$$
$\longleftarrow$ $n - 2$ multiplications $n - 2$ additions

Step $(n - 2)$
$$\begin{bmatrix}
1 & \bullet & 0 & \cdots & 0 & 0 & \times \\
0 & 1 & 0 & \cdots & 0 & 0 & \times \\
0 & 0 & 1 & \cdots & 0 & 0 & \bullet \\
\vdots & \vdots & \vdots & & \vdots & \vdots & \vdots \\
0 & 0 & 0 & \cdots & 1 & 0 & \bullet \\
0 & 0 & 0 & \cdots & 0 & 1 & \bullet
\end{bmatrix}$$
$\longleftarrow$ 2 multiplications 2 additions

Step $(n - 1)$
$$\begin{bmatrix}
1 & 0 & 0 & \cdots & 0 & 0 & \times \\
0 & 1 & 0 & \cdots & 0 & 0 & \bullet \\
0 & 0 & 1 & \cdots & 0 & 0 & \bullet \\
\vdots & \vdots & \vdots & & \vdots & \vdots & \vdots \\
0 & 0 & 0 & \cdots & 1 & 0 & \bullet \\
0 & 0 & 0 & \cdots & 0 & 1 & \bullet
\end{bmatrix}$$
$\longleftarrow$ 1 multiplication 1 addition

Thus, the number of operations required for the backward phase is

$$\text{Multiplications:} \quad (n - 1) + (n - 2) + \cdots + 2 + 1$$
$$\text{Additions:} \quad (n - 1) + (n - 2) + \cdots + 2 + 1$$

or, on applying Formula (3),

$$\text{Multiplications:} \quad \frac{(n-1)n}{2} = \frac{n^2}{2} - \frac{n}{2} \tag{7}$$

$$\text{Additions:} \quad \frac{(n-1)n}{2} = \frac{n^2}{2} - \frac{n}{2} \tag{8}$$

Thus, from (5), (6), (7), and (8) the total operation count for Gauss–Jordan elimination is

$$\text{Multiplications:} \quad \left(\frac{n^3}{3} + \frac{n^2}{2} + \frac{n}{6}\right) + \left(\frac{n^2}{2} - \frac{n}{2}\right) = \frac{n^3}{3} + n^2 - \frac{n}{3} \qquad (9)$$

$$\text{Additions:} \quad \left(\frac{n^3}{3} - \frac{n}{3}\right) + \left(\frac{n^2}{2} - \frac{n}{2}\right) = \frac{n^3}{3} + \frac{n^2}{2} - \frac{5n}{6} \qquad (10)$$

Comparison of Methods for Solving Linear Systems In practical applications it is not uncommon to encounter linear systems with thousands of equations in thousands of unknowns. Thus, we shall be interested in Table 1 for large values of n. It is a fact about polynomials that for large values of the variable, a polynomial can be approximated well by its term of highest degree; that is, if $a_k \neq 0$, then

$$a_0 + a_1 x + \cdots + a_k x^k \approx a_k x^k \quad \text{for large } x$$

(Exercise 12). Thus, for large values of n the operation counts in Table 1 can be approximated as shown in Table 2.

TABLE 2 *Approximate Operation Counts for an Invertible $n \times n$ Matrix A with Large n*

Method	Number of Additions	Number of Multiplications
Solve $A\mathbf{x} = \mathbf{b}$ by Gauss–Jordan elimination	$\approx \dfrac{n^3}{3}$	$\approx \dfrac{n^3}{3}$
Solve $A\mathbf{x} = \mathbf{b}$ by Gaussian elimination	$\approx \dfrac{n^3}{3}$	$\approx \dfrac{n^3}{3}$
Find A^{-1} by reducing $[A \mid I]$ to $[I \mid A^{-1}]$	$\approx n^3$	$\approx n^3$
Solve $A\mathbf{x} = \mathbf{b}$ as $\mathbf{x} = A^{-1}\mathbf{b}$	$\approx n^3$	$\approx n^3$
Find $\det(A)$ by row reduction	$\approx \dfrac{n^3}{3}$	$\approx \dfrac{n^3}{3}$
Solve $A\mathbf{x} = \mathbf{b}$ by Cramer's rule	$\approx \dfrac{n^4}{3}$	$\approx \dfrac{n^4}{3}$

It follows from Table 2 that for large n the best methods for solving $A\mathbf{x} = \mathbf{b}$ are Gaussian elimination and Gauss–Jordan elimination. The method of multiplying by A^{-1} is much worse than these (it requires three times as many operations), and the poorest of the four methods is Cramer's rule.

REMARK. We observed in the remark following Table 1 that if Gauss–Jordan elimination is performed by introducing zeros above and below leading 1's as soon as they are obtained, then the operation count is

$$\frac{n^3}{2} - \frac{n}{2} \quad \text{additions} \quad \text{and} \quad \frac{n^3}{2} + \frac{n^2}{2} \quad \text{multiplications}$$

Thus, for large n this procedure requires $\approx n^3/2$ multiplications, which is 50% greater than the $n^3/3$ multiplications required by the text method. Similarly for additions.

It is reasonable to ask if it is possible to devise other methods for solving linear systems that might require significantly fewer than the $\approx n^3/3$ additions and multiplications needed in Gaussian elimination and Gauss–Jordan elimination. The answer is a qualified "yes." In recent years methods have been devised that require $\approx Cn^q$ multiplications, where q is slightly larger than 2.5. However, these methods have little practical value because the programming is complicated, the constant C is very large, and the number of additions required is excessive. In short, there is currently no practical method for solving general linear systems that significantly improves on the operation counts for Gaussian elimination and the text method of Gauss–Jordan elimination.

Exercise Set 9.8

1. Find the number of additions and multiplications required to compute AB if A is an $m \times n$ matrix and B is an $n \times p$ matrix.

2. Use the result in Exercise 1 to find the number of additions and multiplications required to compute A^k by direct multiplication if A is an $n \times n$ matrix.

3. Assuming A to be an $n \times n$ matrix, use the formulas in Table 1 to determine the number of operations required for the procedures in Table 3.

TABLE 3

	$n = 5$		$n = 10$		$n = 100$		$n = 1000$	
	+	×	+	×	+	×	+	×
Solve $Ax = b$ by Gauss–Jordan elimination								
Solve $Ax = b$ by Gaussian elimination								
Find A^{-1} by reducing $[A \mid I]$ to $[I \mid A^{-1}]$								
Solve $Ax = b$ as $x = A^{-1}b$								
Find $\det(A)$ by row reduction								
Solve $Ax = b$ by Cramer's rule								

4. Assuming a computer execution time of 2.0 microseconds for multiplications and 0.5 microsecond for additions, use the results in Exercise 3 to fill in the execution times in seconds for the procedures in Table 4 on the following page.

5. Derive the formula

$$1 + 2 + 3 + \cdots + n = \frac{n(n+1)}{2}$$

[**Hint.** Let $S_n = 1 + 2 + 3 + \cdots + n$. Write the terms of S_n in reverse order and add the two expressions for S_n.]

6. Use the result in Exercise 5 to show that

$$1 + 2 + 3 + \cdots + (n-1) = \frac{(n-1)n}{2}$$

TABLE 4

	$n = 5$	$n = 10$	$n = 100$	$n = 1000$
	Execution Time (sec)	Execution Time (sec)	Execution Time (sec)	Execution Time (sec)
Solve $A\mathbf{x} = \mathbf{b}$ by Gauss–Jordan elimination				
Solve $A\mathbf{x} = \mathbf{b}$ by Gaussian elimination				
Find A^{-1} by reducing $[A \mid I]$ to $[I \mid A^{-1}]$				
Solve $A\mathbf{x} = \mathbf{b}$ as $\mathbf{x} = A^{-1}\mathbf{b}$				
Find $\det(A)$ by row reduction				
Solve $A\mathbf{x} = \mathbf{b}$ by Cramer's rule				

7. Derive the formula

$$1^2 + 2^2 + 3^2 + \cdots + n^2 = \frac{n(n+1)(2n+1)}{6}$$

using the following steps.

   (a) Show that $(k+1)^3 - k^3 = 3k^2 + 3k + 1$.
   (b) Show that

$$[2^3 - 1^3] + [3^3 - 2^3] + [4^3 - 3^3] + \cdots + [(n+1)^3 - n^3] = (n+1)^3 - 1$$

   (c) Apply (a) to each term on the left side of (b) to show that

$$(n+1)^3 - 1 = 3[1^2 + 2^2 + 3^2 + \cdots + n^2] + 3[1 + 2 + 3 + \cdots + n] + n$$

   (d) Solve the equation in (c) for $1^2 + 2^2 + 3^2 + \cdots + n^2$, use the result of Exercise 5, and then simplify.

8. Use the result in Exercise 7 to show that

$$1^2 + 2^2 + 3^2 + \cdots + (n-1)^2 = \frac{(n-1)n(2n-1)}{6}$$

9. Let R be a row-echelon form of an invertible $n \times n$ matrix. Show that solving the linear system $R\mathbf{x} = \mathbf{b}$ by back-substitution requires

$$\frac{n^2}{2} - \frac{n}{2} \quad \text{multiplications}$$

$$\frac{n^2}{2} - \frac{n}{2} \quad \text{additions}$$

10. Show that to reduce an invertible $n \times n$ matrix to I_n by the text method requires

$$\frac{n^3}{3} - \frac{n}{3} \quad \text{multiplications}$$

$$\frac{n^3}{3} - \frac{n^2}{2} + \frac{n}{6} \quad \text{additions}$$

   [**Note.** Assume that no row interchanges are required.]

11. Consider the variation of Gauss–Jordan elimination in which zeros are introduced above and below a leading 1 as soon as it is obtained, and let A be an invertible $n \times n$ matrix. Show

that to solve a linear system $A\mathbf{x} = \mathbf{b}$ using this version of Gauss–Jordan elimination requires

$$\frac{n^3}{2} + \frac{n^2}{2} \quad \text{multiplications}$$

$$\frac{n^3}{2} - \frac{n}{2} \quad \text{additions}$$

[*Note.* Assume that no row interchanges are required.]

12. (*For readers who have studied calculus.*) Show that if $p(x) = a_0 + a_1 x + \cdots + a_k x^k$, where $a_k \neq 0$, then

$$\lim_{x \to +\infty} \frac{p(x)}{a_k x^k} = 1$$

This result justifies the approximation $a_0 + a_1 x + \cdots + a_k x^k \approx a_k x^k$ for large values of x.

9.9 *LU*-DECOMPOSITIONS

With Gaussian elimination and Gauss–Jordan elimination, a linear system is solved by operating systematically on an augmented matrix. In this section we shall discuss a different approach, one based on factoring the coefficient matrix into a product of lower and upper triangular matrices. This method is well suited for computers and is the basis for many practical computer programs.[†]

Solving Linear Systems by Factoring   We shall proceed in two stages. First, we shall show how a linear system $A\mathbf{x} = \mathbf{b}$ can be solved very easily once the coefficient matrix A is factored into a product of lower and upper triangular matrices. Second, we shall show how to construct such factorizations.

If an $n \times n$ matrix A can be factored into a product of $n \times n$ matrices as

$$A = LU$$

where L is lower triangular and U is upper triangular, then the linear system $A\mathbf{x} = \mathbf{b}$ can be solved as follows:

Step 1. Rewrite the system $A\mathbf{x} = \mathbf{b}$ as

$$LU\mathbf{x} = \mathbf{b} \tag{1}$$

Step 2. Define a new $n \times 1$ matrix $\mathbf{y}$ by

$$U\mathbf{x} = \mathbf{y} \tag{2}$$

Step 3. Use (2) to rewrite (1) as $L\mathbf{y} = \mathbf{b}$ and solve this system for $\mathbf{y}$.

Step 4. Substitute $\mathbf{y}$ in (2) and solve for $\mathbf{x}$.

Although this procedure replaces the problem of solving the single system $A\mathbf{x} = \mathbf{b}$ by the problem of solving the two systems, $L\mathbf{y} = \mathbf{b}$ and $U\mathbf{x} = \mathbf{y}$, the latter systems are

---

[†]In 1979 an important library of machine-independent linear algebra programs called LINPAK was developed at Argonne National Laboratories. Many of the programs in that library use the factorization methods that we will study in this section. Variations of the LINPAK routines are used in many computer programs, including MATLAB, *Mathematica*, and Maple.

easy to solve because the coefficient matrices are triangular. The following example illustrates this procedure.

EXAMPLE 1 Solving a System by Factorization

Later in this section we will derive the factorization

$$\begin{bmatrix} 2 & 6 & 2 \\ -3 & -8 & 0 \\ 4 & 9 & 2 \end{bmatrix} = \begin{bmatrix} 2 & 0 & 0 \\ -3 & 1 & 0 \\ 4 & -3 & 7 \end{bmatrix}\begin{bmatrix} 1 & 3 & 1 \\ 0 & 1 & 3 \\ 0 & 0 & 1 \end{bmatrix}$$

Use this result and the method described above to solve the system

$$\begin{bmatrix} 2 & 6 & 2 \\ -3 & -8 & 0 \\ 4 & 9 & 2 \end{bmatrix}\begin{bmatrix} x_1 \\ x_2 \\ x_3 \end{bmatrix} = \begin{bmatrix} 2 \\ 2 \\ 3 \end{bmatrix} \tag{3}$$

Solution.

Rewrite (3) as

$$\begin{bmatrix} 2 & 0 & 0 \\ -3 & 1 & 0 \\ 4 & -3 & 7 \end{bmatrix}\begin{bmatrix} 1 & 3 & 1 \\ 0 & 1 & 3 \\ 0 & 0 & 1 \end{bmatrix}\begin{bmatrix} x_1 \\ x_2 \\ x_3 \end{bmatrix} = \begin{bmatrix} 2 \\ 2 \\ 3 \end{bmatrix} \tag{4}$$

As specified in Step 2 above, define y_1, y_2, and y_3 by the equation

$$\begin{bmatrix} 1 & 3 & 1 \\ 0 & 1 & 3 \\ 0 & 0 & 1 \end{bmatrix}\begin{bmatrix} x_1 \\ x_2 \\ x_3 \end{bmatrix} = \begin{bmatrix} y_1 \\ y_2 \\ y_3 \end{bmatrix} \tag{5}$$

so (4) can be rewritten as

$$\begin{bmatrix} 2 & 0 & 0 \\ -3 & 1 & 0 \\ 4 & -3 & 7 \end{bmatrix}\begin{bmatrix} y_1 \\ y_2 \\ y_3 \end{bmatrix} = \begin{bmatrix} 2 \\ 2 \\ 3 \end{bmatrix}$$

or equivalently,

$$\begin{aligned} 2y_1 &= 2 \\ -3y_1 + y_2 &= 2 \\ 4y_1 - 3y_2 + 7y_3 &= 3 \end{aligned}$$

The procedure for solving this system is similar to back-substitution except that the equations are solved from the top down instead of from the bottom up. This procedure, called *forward-substitution*, yields

$$y_1 = 1, \qquad y_2 = 5, \qquad y_3 = 2$$

(Verify.) Substituting these values in (5) yields the linear system

$$\begin{bmatrix} 1 & 3 & 1 \\ 0 & 1 & 3 \\ 0 & 0 & 1 \end{bmatrix}\begin{bmatrix} x_1 \\ x_2 \\ x_3 \end{bmatrix} = \begin{bmatrix} 1 \\ 5 \\ 2 \end{bmatrix}$$

or equivalently,

$$\begin{aligned} x_1 + 3x_2 + x_3 &= 1 \\ x_2 + 3x_3 &= 5 \\ x_3 &= 2 \end{aligned}$$

Solving this system by back-substitution yields the solution

$$x_1 = 2, \qquad x_2 = -1, \qquad x_3 = 2$$

(Verify.) ◆

_LU_-Decompositions Now that we have seen how a linear system of n equations in n unknowns can be solved by factoring the coefficient matrix, we shall turn to the problem of constructing such factorizations. To motivate the method, suppose that an $n \times n$ matrix A has been reduced to a row-echelon form U by a sequence of elementary row operations. By Theorem 1.5.1 each of these operations can be accomplished by multiplying on the left by an appropriate elementary matrix. Thus, we can find elementary matrices $E_1, E_2, \ldots, E_k$ such that

$$E_k \cdots E_2 E_1 A = U \tag{6}$$

By Theorem 1.5.2, $E_1, E_2, \ldots, E_k$ are invertible, so we can multiply both sides of Equation (6) on the left successively by

$$E_k^{-1}, \ldots, E_2^{-1}, E_1^{-1}$$

to obtain

$$A = E_1^{-1} E_2^{-1} \cdots E_k^{-1} U \tag{7}$$

In Exercise 15 we will help the reader to show that the matrix L defined by

$$L = E_1^{-1} E_2^{-1} \cdots E_k^{-1} \tag{8}$$

is lower triangular provided that *no row interchanges are used in reducing A to U.* Assuming this to be so, substituting (8) into (7) yields

$$A = LU$$

which is a factorization of A into a product of a lower triangular matrix and an upper triangular matrix.

The following theorem summarizes the above result.

Theorem 9.9.1

If A is a square matrix that can be reduced to a row-echelon form U by Gaussian elimination without row interchanges, then A can be factored as $A = LU$, where L is a lower triangular matrix.

Definition

A factorization of a square matrix A as $A = LU$, where L is lower triangular and U is upper triangular, is called an **_LU_-decomposition** or **triangular decomposition** of the matrix A.

EXAMPLE 2 An _LU_-Decomposition

Find an LU-decomposition of

$$A = \begin{bmatrix} 2 & 6 & 2 \\ -3 & -8 & 0 \\ 4 & 9 & 2 \end{bmatrix}$$

Solution.

To obtain an *LU*-decomposition, $A = LU$, we shall reduce A to a row-echelon form U, then calculate L from (8). The steps are as follows:

	Reduction to Row-Echelon Form	Elementary Matrix Corresponding to the Row Operation	Inverse of the Elementary Matrix
	$\begin{bmatrix} 2 & 6 & 2 \\ -3 & -8 & 0 \\ 4 & 9 & 2 \end{bmatrix}$		
Step 1		$E_1 = \begin{bmatrix} \frac{1}{2} & 0 & 0 \\ 0 & 1 & 0 \\ 0 & 0 & 1 \end{bmatrix}$	$E_1^{-1} = \begin{bmatrix} 2 & 0 & 0 \\ 0 & 1 & 0 \\ 0 & 0 & 1 \end{bmatrix}$
	$\begin{bmatrix} 1 & 3 & 1 \\ -3 & -8 & 0 \\ 4 & 9 & 2 \end{bmatrix}$		
Step 2		$E_2 = \begin{bmatrix} 1 & 0 & 0 \\ 3 & 1 & 0 \\ 0 & 0 & 1 \end{bmatrix}$	$E_2^{-1} = \begin{bmatrix} 1 & 0 & 0 \\ -3 & 1 & 0 \\ 0 & 0 & 1 \end{bmatrix}$
	$\begin{bmatrix} 1 & 3 & 1 \\ 0 & 1 & 3 \\ 4 & 9 & 2 \end{bmatrix}$		
Step 3		$E_3 = \begin{bmatrix} 1 & 0 & 0 \\ 0 & 1 & 0 \\ -4 & 0 & 1 \end{bmatrix}$	$E_3^{-1} = \begin{bmatrix} 1 & 0 & 0 \\ 0 & 1 & 0 \\ 4 & 0 & 1 \end{bmatrix}$
	$\begin{bmatrix} 1 & 3 & 1 \\ 0 & 1 & 3 \\ 0 & -3 & -2 \end{bmatrix}$		
Step 4		$E_4 = \begin{bmatrix} 1 & 0 & 0 \\ 0 & 1 & 0 \\ 0 & 3 & 1 \end{bmatrix}$	$E_4^{-1} = \begin{bmatrix} 1 & 0 & 0 \\ 0 & 1 & 0 \\ 0 & -3 & 1 \end{bmatrix}$
	$\begin{bmatrix} 1 & 3 & 1 \\ 0 & 1 & 3 \\ 0 & 0 & 7 \end{bmatrix}$		
Step 5		$E_5 = \begin{bmatrix} 1 & 0 & 0 \\ 0 & 1 & 0 \\ 0 & 0 & \frac{1}{7} \end{bmatrix}$	$E_5^{-1} = \begin{bmatrix} 1 & 0 & 0 \\ 0 & 1 & 0 \\ 0 & 0 & 7 \end{bmatrix}$
	$\begin{bmatrix} 1 & 3 & 1 \\ 0 & 1 & 3 \\ 0 & 0 & 1 \end{bmatrix}$		

Thus,

$$U = \begin{bmatrix} 1 & 3 & 1 \\ 0 & 1 & 3 \\ 0 & 0 & 1 \end{bmatrix}$$

and, from (8),

$$L = \begin{bmatrix} 2 & 0 & 0 \\ 0 & 1 & 0 \\ 0 & 0 & 1 \end{bmatrix} \begin{bmatrix} 1 & 0 & 0 \\ -3 & 1 & 0 \\ 0 & 0 & 1 \end{bmatrix} \begin{bmatrix} 1 & 0 & 0 \\ 0 & 1 & 0 \\ 4 & 0 & 1 \end{bmatrix} \begin{bmatrix} 1 & 0 & 0 \\ 0 & 1 & 0 \\ 0 & -3 & 1 \end{bmatrix} \begin{bmatrix} 1 & 0 & 0 \\ 0 & 1 & 0 \\ 0 & 0 & 7 \end{bmatrix}$$

$$= \begin{bmatrix} 2 & 0 & 0 \\ -3 & 1 & 0 \\ 4 & -3 & 7 \end{bmatrix}$$

so

$$\begin{bmatrix} 2 & 6 & 2 \\ -3 & -8 & 0 \\ 4 & 9 & 2 \end{bmatrix} = \begin{bmatrix} 2 & 0 & 0 \\ -3 & 1 & 0 \\ 4 & -3 & 7 \end{bmatrix} \begin{bmatrix} 1 & 3 & 1 \\ 0 & 1 & 3 \\ 0 & 0 & 1 \end{bmatrix}$$

is an *LU*-decomposition of *A*.   ◆

Procedure for Finding *LU*-Decompositions   As this example shows, most of the work in constructing an *LU*-decomposition is expended in the calculation of *L*. However, *all* this work can be eliminated by some careful bookkeeping of the operations used to reduce *A* to *U*. Because we are assuming that no row interchanges are required to reduce *A* to *U*, there are only two types of operations involved: multiplying a row by a nonzero constant, and adding a multiple of one row to another. The first operation is used to introduce the leading 1's and the second to introduce zeros below the leading 1's.

In Example 2, the multipliers needed to introduce the leading 1's in successive rows were as follows:

$$\tfrac{1}{2} \text{ for the first row}$$

$$1 \text{ for the second row}$$

$$\tfrac{1}{7} \text{ for the third row}$$

Note that in (9) the successive diagonal entries in *L* were precisely the reciprocals of these multipliers:

$$L = \begin{bmatrix} ② & 0 & 0 \\ -3 & ① & 0 \\ 4 & -3 & ⑦ \end{bmatrix} \tag{9}$$

Next, observe that to introduce zeros below the leading 1 in the first row we used the operations

add 3 times the first row to the second

add −4 times the first row to the third

and to introduce the zero below the leading 1 in the second row, we used the operation

add 3 times the second row to the third

Now note in (10) that in each position below the main diagonal of L the entry is the *negative* of the multiplier in the operation that introduced the zero in that position in U:

$$L = \begin{bmatrix} 2 & 0 & 0 \\ -3 & 1 & 0 \\ 4 & -3 & 7 \end{bmatrix} \qquad (10)$$

In summary, we have the following procedure for constructing an LU-decomposition of a square matrix A provided that A can be reduced to row-echelon form without row interchanges.

Step 1. Reduce A to a row-echelon form U by Gaussian elimination without row interchanges, keeping track of the multipliers used to introduce the leading 1's and the multipliers used to introduce the zeros below the leading 1's.

Step 2. In each position along the main diagonal of L, place the reciprocal of the multiplier that introduced the leading 1 in that position in U.

Step 3. In each position below the main diagonal of L, place the negative of the multiplier used to introduce the zero in that position in U.

Step 4. Form the decomposition $A = LU$.

EXAMPLE 3 Finding an *LU*-Decomposition

Find an LU-decomposition of

$$L = \begin{bmatrix} 6 & -2 & 0 \\ 9 & -1 & 1 \\ 3 & 7 & 5 \end{bmatrix}$$

Solution.

We begin by reducing A to row-echelon form, keeping track of all multipliers.

$$\begin{bmatrix} 6 & -2 & 0 \\ 9 & -1 & 1 \\ 3 & 7 & 5 \end{bmatrix}$$

$$\begin{bmatrix} ① & -\frac{1}{3} & 0 \\ 9 & -1 & 1 \\ 3 & 7 & 5 \end{bmatrix} \longleftarrow \text{multiplier} = \tfrac{1}{6}$$

$$\begin{bmatrix} 1 & -\frac{1}{3} & 0 \\ ⓪ & 2 & 1 \\ ⓪ & 8 & 5 \end{bmatrix} \begin{array}{l} \longleftarrow \text{multiplier} = -9 \\ \longleftarrow \text{multiplier} = -3 \end{array}$$

$$\begin{bmatrix} 1 & -\frac{1}{3} & 0 \\ 0 & ① & \frac{1}{2} \\ 0 & 8 & 5 \end{bmatrix} \longleftarrow \text{multiplier} = \tfrac{1}{2}$$

$$\begin{bmatrix} 1 & -\frac{1}{3} & 0 \\ 0 & 1 & \frac{1}{2} \\ 0 & ⓪ & 1 \end{bmatrix} \longleftarrow \text{multiplier} = -8$$

$$\begin{bmatrix} 1 & -\frac{1}{3} & 0 \\ 0 & 1 & \frac{1}{2} \\ 0 & 0 & ① \end{bmatrix} \longleftarrow \text{multiplier} = 1$$

$\longleftarrow$ No actual operation is performed here, since there is already a leading 1 in the third row.

Constructing L from the multipliers yields the LU-decomposition.

$$A = LU = \begin{bmatrix} 6 & 0 & 0 \\ 9 & 2 & 0 \\ 3 & 8 & 1 \end{bmatrix} \begin{bmatrix} 1 & -\frac{1}{3} & 0 \\ 0 & 1 & \frac{1}{2} \\ 0 & 0 & 1 \end{bmatrix} \qquad \blacklozenge$$

We conclude this section by briefly discussing two fundamental questions about LU-decompositions:

1. Does every square matrix have an LU-decomposition?

2. Can a square matrix have more than one LU-decomposition?

We already know that if a square matrix A can be reduced to row-echelon form by Gaussian elimination without row interchanges, then A has an LU-decomposition. In general, if row interchanges are required to reduce matrix A to row-echelon form, then there is no LU-decomposition of A. However, in such cases it is possible to factor A in the form

$$A = PLU$$

where L is lower triangular, U is upper triangular, and P is a matrix obtained by interchanging the rows of I_n appropriately (see Exercise 17).

In the absence of additional restrictions, LU-decompositions are not unique. For example, if

$$A = LU = \begin{bmatrix} l_{11} & 0 & 0 \\ l_{21} & l_{22} & 0 \\ l_{31} & l_{32} & l_{33} \end{bmatrix} \begin{bmatrix} 1 & u_{12} & u_{13} \\ 0 & 1 & u_{23} \\ 0 & 0 & 1 \end{bmatrix}$$

and L has nonzero diagonal entries, then we can shift the diagonal entries from the left factor to the right factor by writing

$$A = \begin{bmatrix} 1 & 0 & 0 \\ \dfrac{l_{21}}{l_{11}} & 1 & 0 \\ \dfrac{l_{31}}{l_{11}} & \dfrac{l_{32}}{l_{22}} & 1 \end{bmatrix} \begin{bmatrix} l_{11} & 0 & 0 \\ 0 & l_{22} & 0 \\ 0 & 0 & l_{33} \end{bmatrix} \begin{bmatrix} 1 & u_{12} & u_{13} \\ 0 & 1 & u_{23} \\ 0 & 0 & 1 \end{bmatrix}$$

$$= \begin{bmatrix} 1 & 0 & 0 \\ \dfrac{l_{21}}{l_{11}} & 1 & 0 \\ \dfrac{l_{31}}{l_{11}} & \dfrac{l_{32}}{l_{22}} & 1 \end{bmatrix} \begin{bmatrix} l_{11} & l_{11}u_{12} & l_{11}u_{13} \\ 0 & l_{22} & l_{22}u_{23} \\ 0 & 0 & l_{33} \end{bmatrix}$$

which is another triangular decomposition of A.

1. Use the method of Example 1 and the *LU*-decomposition

$$\begin{bmatrix} 3 & -6 \\ -2 & 5 \end{bmatrix} = \begin{bmatrix} 3 & 0 \\ -2 & 1 \end{bmatrix} \begin{bmatrix} 1 & -2 \\ 0 & 1 \end{bmatrix}$$

to solve the system

$$3x_1 - 6x_2 = 0$$
$$-2x_1 + 5x_2 = 1$$

2. Use the method of Example 1 and the *LU*-decomposition

$$\begin{bmatrix} 3 & -6 & -3 \\ 2 & 0 & 6 \\ -4 & 7 & 4 \end{bmatrix} = \begin{bmatrix} 3 & 0 & 0 \\ 2 & 4 & 0 \\ -4 & -1 & 2 \end{bmatrix} \begin{bmatrix} 1 & -2 & -1 \\ 0 & 1 & 2 \\ 0 & 0 & 1 \end{bmatrix}$$

to solve the system

$$3x_1 - 6x_2 - 3x_3 = -3$$
$$2x_1 \qquad + 6x_3 = -22$$
$$-4x_1 + 7x_2 + 4x_3 = \quad 3$$

In Exercises 3–10 find an *LU*-decomposition of the coefficient matrix; then use the method of Example 1 to solve the system.

3. $\begin{bmatrix} 2 & 8 \\ -1 & -1 \end{bmatrix} \begin{bmatrix} x_1 \\ x_2 \end{bmatrix} = \begin{bmatrix} -2 \\ -2 \end{bmatrix}$

4. $\begin{bmatrix} -5 & -10 \\ 6 & 5 \end{bmatrix} \begin{bmatrix} x_1 \\ x_2 \end{bmatrix} = \begin{bmatrix} -10 \\ 19 \end{bmatrix}$

5. $\begin{bmatrix} 2 & -2 & -2 \\ 0 & -2 & 2 \\ -1 & 5 & 2 \end{bmatrix} \begin{bmatrix} x_1 \\ x_2 \\ x_3 \end{bmatrix} = \begin{bmatrix} -4 \\ -2 \\ 6 \end{bmatrix}$

6. $\begin{bmatrix} -3 & 12 & -6 \\ 1 & -2 & 2 \\ 0 & 1 & 1 \end{bmatrix} \begin{bmatrix} x_1 \\ x_2 \\ x_3 \end{bmatrix} = \begin{bmatrix} -33 \\ 7 \\ -1 \end{bmatrix}$

7. $\begin{bmatrix} 5 & 5 & 10 \\ -8 & -7 & -9 \\ 0 & 4 & 26 \end{bmatrix} \begin{bmatrix} x_1 \\ x_2 \\ x_3 \end{bmatrix} = \begin{bmatrix} 0 \\ 1 \\ 4 \end{bmatrix}$

8. $\begin{bmatrix} -1 & -3 & -4 \\ 3 & 10 & -10 \\ -2 & -4 & 11 \end{bmatrix} \begin{bmatrix} x_1 \\ x_2 \\ x_3 \end{bmatrix} = \begin{bmatrix} -6 \\ -3 \\ 9 \end{bmatrix}$

9. $\begin{bmatrix} -1 & 0 & 1 & 0 \\ 2 & 3 & -2 & 6 \\ 0 & -1 & 2 & 0 \\ 0 & 0 & 1 & 5 \end{bmatrix} \begin{bmatrix} x_1 \\ x_2 \\ x_3 \\ x_4 \end{bmatrix} = \begin{bmatrix} 5 \\ -1 \\ 3 \\ 7 \end{bmatrix}$

10. $\begin{bmatrix} 2 & -4 & 0 & 0 \\ 1 & 2 & 12 & 0 \\ 0 & -1 & -4 & -5 \\ 0 & 0 & 2 & 11 \end{bmatrix} \begin{bmatrix} x_1 \\ x_2 \\ x_3 \\ x_4 \end{bmatrix} = \begin{bmatrix} 8 \\ 0 \\ 1 \\ 0 \end{bmatrix}$

11. Let

$$A = \begin{bmatrix} 2 & 1 & -1 \\ -2 & -1 & 2 \\ 2 & 1 & 0 \end{bmatrix}$$

   (a) Find an *LU*-decomposition of *A*.
   (b) Express *A* in the form $A = L_1 D U_1$, where L_1 is lower triangular with 1's along the main diagonal, U_1 is upper triangular, and *D* is a diagonal matrix.
   (c) Express *A* in the form $A = L_2 U_2$, where L_2 is lower triangular with 1's along the main diagonal and U_2 is upper triangular.

12. Show that the matrix

$$\begin{bmatrix} 0 & 1 \\ 1 & 0 \end{bmatrix}$$

has no *LU*-decomposition.

13. Let

$$A = \begin{bmatrix} a & b \\ c & d \end{bmatrix}$$

   (a) Prove: If $a \neq 0$, then A has a unique LU-decomposition with 1's along the main diagonal of L.

   (b) Find the LU-decomposition described in (a).

14. Let $A\mathbf{x} = \mathbf{b}$ be a linear system of n equations in n unknowns, and assume that A is an invertible matrix that can be reduced to row-echelon form without row interchanges. How many additions and multiplications are required to solve the system by the method of Example 1? [*Note.* Count subtractions as additions and divisions as multiplications.]

15. Recall from Theorem 1.7.1b that a product of lower triangular matrices is lower triangular. Use this fact to prove that the matrix L in (8) is lower triangular.

16. Use the result in Exercise 15 to prove that a product of finitely many upper triangular matrices is upper triangular. [*Hint.* Take transposes.]

17. Prove: If A is any $n \times n$ matrix, then A can be factored as $A = PLU$, where L is lower triangular, U is upper triangular, and P can be obtained by interchanging the rows of I_n appropriately. [*Hint.* Let U be a row-echelon form of A, and let all row interchanges required in the reduction of A to U be performed first.]

18. Factor

$$A = \begin{bmatrix} 3 & -1 & 0 \\ 3 & -1 & 1 \\ 0 & 2 & 1 \end{bmatrix}$$

as $A = PLU$, where P is obtained from I_3 by interchanging rows appropriately, L is lower triangular, and U is upper triangular.

Chapter 9 Technology Exercises

The following exercises are designed to be solved using a technology utility. Typically, this will be MATLAB, *Mathematica*, Maple, Derive, or Mathcad, but it may also be some other type of linear algebra software or a scientific calculator with some linear algebra capabilities. For each exercise you will need to read the relevant documentation for the particular utility you are using. The goal of these exercises is to provide you with a basic proficiency with your technology utility. Once you have mastered the techniques in these exercises, you will be able to use your technology utility to solve many of the problems in the regular exercise sets.

Section 9.3

T1. (*Least squares straight line fit*) Read your documentation on finding the least squares straight line fit to a set of data points, and then use your utility to find the line of best fit to the data in Example 1. Do not imitate the method in the example; rather, use the command provided by your utility.

T2. (*Least squares polynomial fit*) Read your documentation on fitting polynomials to a set of data points by least squares, and then use your utility to find the polynomial fit to the data in Example 3. Do not imitate the method in the example; rather, use the command provided by your utility.

Section 9.7

T1. (***Quadric surfaces***) Use your technology utility to perform the computations in Example 3.

Section 9.9

T1. (***LU-decomposition***) Technology utilities vary widely on how they handle LU-decompositions. For example, some programs perform row interchanges to reduce roundoff error and hence produce a PLU-decomposition, even when asked for an LU-decomposition. Read your documentation, and then see what happens when you use your utility to find an LU-decomposition of the matrix A in Example 3.

10

Complex Vector Spaces

INTRODUCTION: Up to now we have considered only vector spaces for which the scalars are real numbers. However, for many important applications of vectors it is desirable to allow the scalars to be complex numbers. One advantage of allowing complex scalars is that all matrices with scalar entries have eigenvalues, which is not true if only real scalars are allowed.

In the first three sections of this chapter we will review some of the basic properties of complex numbers, and in subsequent sections we will discuss vector spaces in which scalars can be complex numbers.

10.1 COMPLEX NUMBERS

In this section we shall review the definition of a complex number and discuss the operations of addition, subtraction, and multiplication of such numbers. We will also consider matrices with complex entries and explain how addition and subtraction of complex numbers can be viewed as operations on vectors.

Complex Numbers   Since $x^2 \geq 0$ for every real number x, the equation $x^2 = -1$ has no real solutions. To deal with this problem, mathematicians of the eighteenth century introduced the "imaginary" number,

$$i = \sqrt{-1}$$

which they assumed had the property

$$i^2 = (\sqrt{-1})^2 = -1$$

but which otherwise could be treated like an ordinary number. Expressions of the form

$$a + bi \tag{1}$$

where a and b are real numbers were called "complex numbers," and these were manipulated according to the standard rules of arithmetic with the added property that $i^2 = -1$.

By the beginning of the nineteenth century it was recognized that a complex number (1) could be regarded as an alternative symbol for the ordered pair

$$(a, b)$$

of real numbers, and that operations of addition, subtraction, multiplication, and division could be defined on these ordered pairs so that the familiar laws of arithmetic hold and $i^2 = -1$. This is the approach we will follow.

> **Definition**
>
> A **complex number** is an ordered pair of real numbers, denoted either by (a, b) or $a + bi$, where $i^2 = -1$.

EXAMPLE 1 Two Notations for a Complex Number

Some examples of complex numbers in both notations are as follows:

Ordered Pair	Equivalent Notation
$(3, 4)$	$3 + 4i$
$(-1, 2)$	$-1 + 2i$
$(0, 1)$	$0 + i$
$(2, 0)$	$2 + 0i$
$(4, -2)$	$4 + (-2)i$

For simplicity, the last three complex numbers would usually be abbreviated as

$$0 + i = i, \quad 2 + 0i = 2, \quad 4 + (-2)i = 4 - 2i \qquad \blacklozenge$$

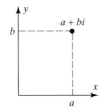

(*a*) Complex number as a point

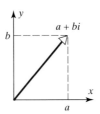

(*b*) Complex number as a vector

Figure 10.1.1

Geometrically, a complex number can be viewed either as a point or a vector in the *xy*-plane (Figure 10.1.1).

EXAMPLE 2 Complex Numbers as Points and as Vectors

Some complex numbers are shown as points in Figure 10.1.2*a* and as vectors in Figure 10.1.2*b*.

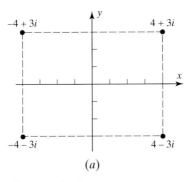

(*a*)

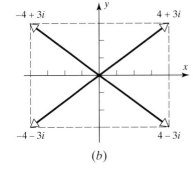

(*b*)

Figure 10.1.2

◆

The Complex Plane

Sometimes it is convenient to use a single letter, such as *z*, to denote a complex number. Thus, we might write

$$z = a + bi$$

The real number *a* is called the ***real part of*** *z* and the real number *b* the ***imaginary part of*** *z*. These numbers are denoted by Re(*z*) and Im(*z*), respectively. Thus,

$$\text{Re}(4 - 3i) = 4 \quad \text{and} \quad \text{Im}(4 - 3i) = -3$$

When complex numbers are represented geometrically in an *xy*-coordinate system, the *x*-axis is called the ***real axis***, the *y*-axis the ***imaginary axis***, and the plane is called the ***complex plane*** (Figure 10.1.3).

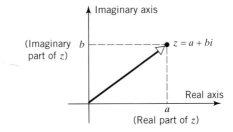

Figure 10.1.3 Complex plane.

Operations on Complex Numbers

Just as two vectors in R^2 are defined to be equal if they have the same components, so we define two complex numbers to be equal if their real parts are equal and their imaginary parts are equal:

Definition

Two complex numbers, $a + bi$ and $c + di$, are defined to be ***equal***, written

$$a + bi = c + di$$

if $a = c$ and $b = d$.

If $b = 0$, then the complex number $a + bi$ reduces to $a + 0i$, which we write simply as a. Thus, for any real number a,

$$a = a + 0i$$

so that the real numbers can be regarded as complex numbers with an imaginary part of zero. Geometrically, the real numbers correspond to points on the real axis. If we have $a = 0$, then $a + bi$ reduces to $0 + bi$, which we usually write as bi. These complex numbers, which correspond to points on the imaginary axis, are called ***pure imaginary numbers***.

Just as vectors in R^2 are added by adding corresponding components, so complex numbers are added by adding their real parts and adding their imaginary parts:

$$(a + bi) + (c + di) = (a + c) + (b + d)i \tag{2}$$

The operations of subtraction and multiplication by a *real* number are also similar to the corresponding vector operations in R^2:

$$(a + bi) - (c + di) = (a - c) + (b - d)i \tag{3}$$

$$k(a + bi) = (ka) + (kb)i, \quad k \text{ real} \tag{4}$$

Because the operations of addition, subtraction, and multiplication of a complex number by a real number parallel the corresponding operations for vectors in R^2, the familiar geometric interpretations of these operations hold for complex numbers (see Figure 10.1.4).

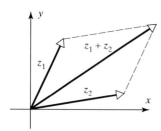

(*a*) The sum of two complex numbers

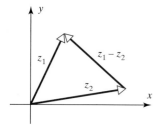

(*b*) The difference of two complex numbers

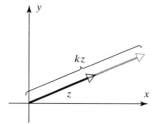

(*c*) The product of a complex number z and a positive real number k

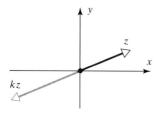

(*d*) The product of a complex number z and a negative real number k

Figure 10.1.4

It follows from (4) that $(-1)z + z = 0$ (verify), so we denote $(-1)z$ as $-z$ and call it the ***negative of z***.

EXAMPLE 3 Adding, Subtracting, and Multiplying by Real Numbers

If $z_1 = 4 - 5i$ and $z_2 = -1 + 6i$, find $z_1 + z_2$, $z_1 - z_2$, $3z_1$, and $-z_2$.

Solution.

$$z_1 + z_2 = (4 - 5i) + (-1 + 6i) = (4 - 1) + (-5 + 6)i = 3 + i$$
$$z_1 - z_2 = (4 - 5i) - (-1 + 6i) = (4 + 1) + (-5 - 6)i = 5 - 11i$$
$$3z_1 = 3(4 - 5i) = 12 - 15i$$
$$-z_2 = (-1)z_2 = (-1)(-1 + 6i) = 1 - 6i \qquad \blacklozenge$$

So far, there has been a parallel between complex numbers and vectors in R^2. However, we now define multiplication of complex numbers, an operation with no vector analog in R^2. To motivate the definition, we expand the product

$$(a + bi)(c + di)$$

following the usual rules of algebra, but treating i^2 as -1. This yields

$$(a + bi)(c + di) = ac + bdi^2 + adi + bci$$
$$= (ac - bd) + (ad + bc)i$$

which suggests the following *definition*:

$$(a + bi)(c + di) = (ac - bd) + (ad + bc)i \qquad (5)$$

EXAMPLE 4 Multiplying Complex Numbers

$$(3 + 2i)(4 + 5i) = (3 \cdot 4 - 2 \cdot 5) + (3 \cdot 5 + 2 \cdot 4)i$$
$$= 2 + 23i$$
$$(4 - i)(2 - 3i) = [4 \cdot 2 - (-1)(-3)] + [(4)(-3) + (-1)(2)]i$$
$$= 5 - 14i$$
$$i^2 = (0 + i)(0 + i) = (0 \cdot 0 - 1 \cdot 1) + (0 \cdot 1 + 1 \cdot 0)i = -1 \qquad \blacklozenge$$

We leave it as an exercise to verify the following rules of complex arithmetic:

$$z_1 + z_2 = z_2 + z_1$$
$$z_1 z_2 = z_2 z_1$$
$$z_1 + (z_2 + z_3) = (z_1 + z_2) + z_3$$
$$z_1(z_2 z_3) = (z_1 z_2)z_3$$
$$z_1(z_2 + z_3) = z_1 z_2 + z_1 z_3$$
$$0 + z = z$$
$$z + (-z) = 0$$
$$1 \cdot z = z$$

These rules make it possible to multiply complex numbers without using Formula (5) directly. Following the procedure used to motivate this formula, we can simply multiply each term of $a + bi$ by each term of $c + di$, set $i^2 = -1$, and simplify.

EXAMPLE 5 Multiplication of Complex Numbers

$$(3 + 2i)(4 + i) = 12 + 3i + 8i + 2i^2 = 12 + 11i - 2 = 10 + 11i$$
$$(5 - \tfrac{1}{2}i)(2 + 3i) = 10 + 15i - i - \tfrac{3}{2}i^2 = 10 + 14i + \tfrac{3}{2} = \tfrac{23}{2} + 14i$$
$$i(1 + i)(1 - 2i) = i(1 - 2i + i - 2i^2) = i(3 - i) = 3i - i^2 = 1 + 3i \qquad \blacklozenge$$

REMARK. Unlike the real numbers, there is no size ordering for the complex numbers. Thus, the order symbols $<$, $\leq$, $>$, and $\geq$ are not used with complex numbers.

Now that we have defined addition, subtraction, and multiplication of complex numbers, it is possible to add, subtract, and multiply matrices with complex entries and multiply a matrix by a complex number. Without going into detail, we note that the matrix operations and terminology discussed in Chapter 1 carry over without change to matrices with complex entries.

EXAMPLE 6 Matrices with Complex Entries

If

$$A = \begin{bmatrix} 1 & -i \\ 1+i & 4-i \end{bmatrix} \quad \text{and} \quad B = \begin{bmatrix} i & 1-i \\ 2-3i & 4 \end{bmatrix}$$

then

$$A + B = \begin{bmatrix} 1+i & 1-2i \\ 3-2i & 8-i \end{bmatrix}, \qquad A - B = \begin{bmatrix} 1-i & -1 \\ -1+4i & -i \end{bmatrix}$$

$$iA = \begin{bmatrix} i & -i^2 \\ i+i^2 & 4i-i^2 \end{bmatrix} = \begin{bmatrix} i & 1 \\ -1+i & 1+4i \end{bmatrix}$$

$$AB = \begin{bmatrix} 1 & -i \\ 1+i & 4-i \end{bmatrix} \begin{bmatrix} i & 1-i \\ 2-3i & 4 \end{bmatrix}$$

$$= \begin{bmatrix} 1 \cdot i + (-i) \cdot (2-3i) & 1 \cdot (1-i) + (-i) \cdot 4 \\ (1+i) \cdot i + (4-i) \cdot (2-3i) & (1+i) \cdot (1-i) + (4-i) \cdot 4 \end{bmatrix}$$

$$= \begin{bmatrix} -3-i & 1-5i \\ 4-13i & 18-4i \end{bmatrix} \qquad \blacklozenge$$

Exercise Set 10.1

1. In each part plot the point and sketch the vector that corresponds to the given complex number.

   (a) $2 + 3i$ (b) -4 (c) $-3 - 2i$ (d) $-5i$

2. Express each complex number in Exercise 1 as an ordered pair of real numbers.

3. In each part use the given information to find the real numbers x and y.

   (a) $x - iy = -2 + 3i$ (b) $(x + y) + (x - y)i = 3 + i$

4. Given that $z_1 = 1 - 2i$ and $z_2 = 4 + 5i$, find

   (a) $z_1 + z_2$ (b) $z_1 - z_2$ (c) $4z_1$ (d) $-z_2$ (e) $3z_1 + 4z_2$ (f) $\tfrac{1}{2}z_1 - \tfrac{3}{2}z_2$

5. In each part solve for z.

   (a) $z + (1 - i) = 3 + 2i$ (b) $-5z = 5 + 10i$ (c) $(i - z) + (2z - 3i) = -2 + 7i$

6. In each part sketch the vectors z_1, z_2, $z_1 + z_2$, and $z_1 - z_2$.

   (a) $z_1 = 3 + i$, $z_2 = 1 + 4i$ (b) $z_1 = -2 + 2i$, $z_2 = 4 + 5i$

7. In each part sketch the vectors z and kz.

   (a) $z = 1 + i$, $k = 2$ (b) $z = -3 - 4i$, $k = -2$ (c) $z = 4 + 6i$, $k = \frac{1}{2}$

8. In each part find real numbers k_1 and k_2 that satisfy the equation.

   (a) $k_1 i + k_2(1 + i) = 3 - 2i$ (b) $k_1(2 + 3i) + k_2(1 - 4i) = 7 + 5i$

9. In each part find $z_1 z_2$, z_1^2, and z_2^2.

   (a) $z_1 = 3i$, $z_2 = 1 - i$ (b) $z_1 = 4 + 6i$, $z_2 = 2 - 3i$ (c) $z_1 = \frac{1}{3}(2 + 4i)$, $z_2 = \frac{1}{2}(1 - 5i)$

10. Given that $z_1 = 2 - 5i$ and $z_2 = -1 - i$, find

   (a) $z_1 - z_1 z_2$ (b) $(z_1 + 3z_2)^2$ (c) $[z_1 + (1 + z_2)]^2$ (d) $iz_2 - z_1^2$

In Exercises 11–18 perform the calculations and express the result in the form $a + bi$.

11. $(1 + 2i)(4 - 6i)^2$

12. $(2 - i)(3 + i)(4 - 2i)$

13. $(1 - 3i)^3$

14. $i(1 + 7i) - 3i(4 + 2i)$

15. $[(2 + i)(\frac{1}{2} + \frac{3}{4}i)]^2$

16. $(\sqrt{2} + i) - i\sqrt{2}(1 + \sqrt{2}i)$

17. $(1 + i + i^2 + i^3)^{100}$

18. $(3 - 2i)^2 - (3 + 2i)^2$

19. Let

$$A = \begin{bmatrix} 1 & i \\ -i & 3 \end{bmatrix}, \qquad B = \begin{bmatrix} 2 & 2+i \\ 3-i & 4 \end{bmatrix}$$

   Find

   (a) $A + 3iB$ (b) BA (c) AB (d) $B^2 - A^2$

20. Let

$$A = \begin{bmatrix} 3+2i & 0 \\ -i & 2 \\ 1+i & 1-i \end{bmatrix}, \qquad B = \begin{bmatrix} -i & 2 \\ 0 & i \end{bmatrix}, \qquad C = \begin{bmatrix} -1-i & 0 & -i \\ 3 & 2i & -5 \end{bmatrix}$$

   Find

   (a) $A(BC)$ (b) $(BC)A$ (c) $(CA)B^2$ (d) $(1 + i)(AB) + (3 - 4i)A$

21. Show that

   (a) $\text{Im}(iz) = \text{Re}(z)$ (b) $\text{Re}(iz) = -\text{Im}(z)$

22. In each part solve the equation by the quadratic formula and check your results by substituting the solutions into the given equation.

   (a) $z^2 + 2z + 2 = 0$ (b) $z^2 - z + 1 = 0$

23. (a) Show that if n is a positive integer, then the only possible values for i^n are 1, -1, i, and $-i$.

   (b) Find i^{2509}. [**Hint.** The value of i^n can be determined from the remainder when n is divided by 4.]

24. Prove: If $z_1 z_2 = 0$, then $z_1 = 0$ or $z_2 = 0$.

25. Use the result of Exercise 24 to prove: If $zz_1 = zz_2$ and $z \neq 0$, then $z_1 = z_2$.

26. Prove that for all complex numbers z_1, z_2, and z_3

   (a) $z_1 + z_2 = z_2 + z_1$ (b) $z_1 + (z_2 + z_3) = (z_1 + z_2) + z_3$

27. Prove that for all complex numbers z_1, z_2, and z_3

   (a) $z_1 z_2 = z_2 z_1$ (b) $z_1(z_2 z_3) = (z_1 z_2)z_3$

28. Prove that $z_1(z_2 + z_3) = z_1 z_2 + z_1 z_3$ for all complex numbers z_1, z_2, and z_3.

29. In quantum mechanics the **Dirac matrices** are

$$\beta = \begin{bmatrix} 1 & 0 & 0 & 0 \\ 0 & 1 & 0 & 0 \\ 0 & 0 & -1 & 0 \\ 0 & 0 & 0 & -1 \end{bmatrix}, \quad \alpha_x = \begin{bmatrix} 0 & 0 & 0 & 1 \\ 0 & 0 & 1 & 0 \\ 0 & 1 & 0 & 0 \\ 1 & 0 & 0 & 0 \end{bmatrix},$$

$$\alpha_y = \begin{bmatrix} 0 & 0 & 0 & -i \\ 0 & 0 & i & 0 \\ 0 & -i & 0 & 0 \\ i & 0 & 0 & 0 \end{bmatrix}, \quad \alpha_z = \begin{bmatrix} 0 & 0 & 1 & 0 \\ 0 & 0 & 0 & -1 \\ 1 & 0 & 0 & 0 \\ 0 & -1 & 0 & 0 \end{bmatrix}$$

(a) Prove that $\beta^2 = \alpha_x^2 = \alpha_y^2 = \alpha_z^2 = I$.

(b) Two matrices A and B are called **anticommutative** if $AB = -BA$. Prove that any two distinct Dirac matrices are anticommutative.

10.2 DIVISION OF COMPLEX NUMBERS

In the last section we defined multiplication of complex numbers. In this section we shall define division of complex numbers as the inverse of multiplication.

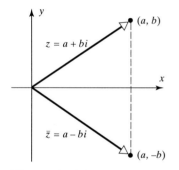

Figure 10.2.1
The conjugate of a complex number.

We begin with some preliminary ideas.

Complex Conjugates   If $z = a + bi$ is any complex number, then the **complex conjugate** of z (also called the **conjugate** of z) is denoted by the symbol $\bar{z}$ (read "z bar") and is defined by

$$\bar{z} = a - bi$$

In words, $\bar{z}$ is obtained by reversing the sign of the imaginary part of z. Geometrically, $\bar{z}$ is the reflection of z about the real axis (Figure 10.2.1).

EXAMPLE 1 Examples of Conjugates

$$
\begin{array}{ll}
z = 3 + 2i & \bar{z} = 3 - 2i \\
z = -4 - 2i & \bar{z} = -4 + 2i \\
z = i & \bar{z} = -i \\
z = 4 & z = 4
\end{array}
$$

◆

REMARK. The last line in Example 1 illustrates the fact that a real number is the same as its conjugate. More precisely, it can be shown (Exercise 22) that $z = \bar{z}$ if and only if z is a real number.

If a complex number z is viewed as a vector in R^2, then the norm or length of the vector is called the modulus (or *absolute value*) of z. More precisely:

> **Definition**
>
> The **modulus** of a complex number $z = a + bi$, denoted by $|z|$, is defined by
>
> $$|z| = \sqrt{a^2 + b^2} \tag{1}$$

If $b = 0$, then $z = a$ is a real number, and

$$|z| = \sqrt{a^2 + 0^2} = \sqrt{a^2} = |a|$$

so the modulus of a real number is simply its absolute value. Thus, the modulus of z is also called the **absolute value** of z.

EXAMPLE 2 Modulus of a Complex Number

Find $|z|$ if $z = 3 - 4i$.

Solution.

From (1) with $a = 3$ and $b = -4$, $|z| = \sqrt{(3)^2 + (-4)^2} = \sqrt{25} = 5$.   ◆

Paul Adrien Maurice Dirac (1902–1984) was a British theoretical physicist who devised a new form of quantum mechanics and a theory that predicted electron spin and the existence of a fundamental atomic particle called a positron. In 1933 he received the Nobel prize for physics and in 1939, the medal of the Royal Society.

The following theorem establishes a basic relationship between $\bar{z}$ and $|z|$.

Theorem 10.2.1

For any complex number z,

$$z\bar{z} = |z|^2$$

Proof. If $z = a + bi$, then

$$z\bar{z} = (a + bi)(a - bi) = a^2 - abi + bai - b^2 i^2 = a^2 + b^2 = |z|^2 \qquad ■$$

Division of Complex Numbers

We now turn to the division of complex numbers. Our objective is to define division as the inverse of multiplication. Thus, if $z_2 \neq 0$, then our definition of $z = z_1/z_2$ should be such that

$$z_1 = z_2 z \qquad (2)$$

Our procedure will be to prove that (2) has a unique solution for z if $z_2 \neq 0$, and then define z_1/z_2 to be this value of z. As with real numbers, division by zero is not allowed.

Theorem 10.2.2

If $z_2 \neq 0$, then Equation (2) has a unique solution, which is

$$z = \frac{1}{|z_2|^2} z_1 \bar{z}_2 \qquad (3)$$

Proof. Let $z = x + iy$, $z_1 = x_1 + iy_1$, and $z_2 = x_2 + iy_2$. Then (2) can be written as

$$x_1 + iy_1 = (x_2 + iy_2)(x + iy)$$

or

$$x_1 + iy_1 = (x_2 x - y_2 y) + i(y_2 x + x_2 y)$$

or, on equating real and imaginary parts,

$$x_2 x - y_2 y = x_1$$
$$y_2 x + x_2 y = y_1$$

or

$$\begin{bmatrix} x_2 & -y_2 \\ y_2 & x_2 \end{bmatrix} \begin{bmatrix} x \\ y \end{bmatrix} = \begin{bmatrix} x_1 \\ y_1 \end{bmatrix} \qquad (4)$$

Since $z_2 = x_2 + i y_2 \neq 0$, it follows that x_2 and y_2 are not both zero, so

$$\begin{vmatrix} x_2 & -y_2 \\ y_2 & x_2 \end{vmatrix} = x_2^2 + y_2^2 \neq 0$$

Thus, by Cramer's rule (Theorem 2.4.3), system (4) has the unique solution

$$x = \frac{\begin{vmatrix} x_1 & -y_2 \\ y_1 & x_2 \end{vmatrix}}{\begin{vmatrix} x_2 & -y_2 \\ y_2 & x_2 \end{vmatrix}} = \frac{x_1 x_2 + y_1 y_2}{x_2^2 + y_2^2} = \frac{x_1 x_2 + y_1 y_2}{|z_2|^2}$$

$$y = \frac{\begin{vmatrix} x_2 & x_1 \\ y_2 & y_1 \end{vmatrix}}{\begin{vmatrix} x_2 & -y_2 \\ y_2 & x_2 \end{vmatrix}} = \frac{y_1 x_2 - x_1 y_2}{x_2^2 + y_2^2} = \frac{y_1 x_2 - x_1 y_2}{|z_2|^2}$$

Thus,

$$z = x + iy = \frac{1}{|z_2|^2} \left[(x_1 x_2 + y_1 y_2) + i(y_1 x_2 - x_1 y_2) \right]$$

$$= \frac{1}{|z_2|^2} (x_1 + i y_1)(x_2 - i y_2) = \frac{1}{|z_2|^2} z_1 \bar{z}_2 \quad \blacksquare$$

Thus, for $z_2 \neq 0$ we define

$$\frac{z_1}{z_2} = \frac{1}{|z_2|^2} z_1 \bar{z}_2 \tag{5}$$

REMARK. To remember this formula, multiply numerator and denominator of z_1/z_2 by $\bar{z}_2$:

$$\frac{z_1}{z_2} = \frac{z_1 \bar{z}_2}{z_2 \bar{z}_2} = \frac{z_1 \bar{z}_2}{|z_2|^2} = \frac{1}{|z_2|^2} z_1 \bar{z}_2$$

EXAMPLE 3 Quotient in the Form $a + bi$

Express

$$\frac{3 + 4i}{1 - 2i}$$

in the form $a + bi$.

Solution.

From (5) with $z_1 = 3 + 4i$ and $z_2 = 1 - 2i$,

$$\frac{3 + 4i}{1 - 2i} = \frac{1}{|1 - 2i|^2} (3 + 4i)(\overline{1 - 2i}) = \frac{1}{5}(3 + 4i)(1 + 2i)$$

$$= \frac{1}{5}(-5 + 10i) = -1 + 2i$$

Alternative Solution. As in the remark above, multiply numerator and denominator by the conjugate of the denominator:

$$\frac{3 + 4i}{1 - 2i} = \frac{3 + 4i}{1 - 2i} \cdot \frac{1 + 2i}{1 + 2i} = \frac{-5 + 10i}{5} = -1 + 2i \qquad \blacklozenge$$

Systems of linear equations with complex coefficients arise in various applications. Without going into detail we note that all the results about linear systems studied in Chapters 1 and 2 carry over without change to systems with complex coefficients.

EXAMPLE 4 A Linear System with Complex Coefficients

Use Cramer's rule to solve

$$ix + 2y = 1 - 2i$$
$$4x - iy = -1 + 3i$$

Solution.

$$x = \frac{\begin{vmatrix} 1 - 2i & 2 \\ -1 + 3i & -i \end{vmatrix}}{\begin{vmatrix} i & 2 \\ 4 & -i \end{vmatrix}} = \frac{(-i)(1 - 2i) - 2(-1 + 3i)}{i(-i) - 2(4)} = \frac{-7i}{-7} = i$$

$$y = \frac{\begin{vmatrix} i & 1 - 2i \\ 4 & -1 + 3i \end{vmatrix}}{\begin{vmatrix} i & 2 \\ 4 & -i \end{vmatrix}} = \frac{(i)(-1 + 3i) - 4(1 - 2i)}{i(-i) - 2(4)} = \frac{-7 + 7i}{-7} = 1 - i$$

Thus, the solution is $x = i$, $y = 1 - i$.   ◆

We conclude this section by listing some properties of the complex conjugate that will be useful in later sections.

Theorem 10.2.3 **Properties of the Conjugate**

For any complex numbers z, z_1, and z_2

(a) $\overline{z_1 + z_2} = \bar{z}_1 + \bar{z}_2$ (b) $\overline{z_1 - z_2} = \bar{z}_1 - \bar{z}_2$

(c) $\overline{z_1 z_2} = \bar{z}_1 \bar{z}_2$ (d) $\overline{(z_1/z_2)} = \bar{z}_1/\bar{z}_2$

(e) $\bar{\bar{z}} = z$

We prove (*a*) and leave the rest as exercises.

Proof (a). Let $z_1 = a_1 + b_1 i$ and $z_2 = a_2 + b_2 i$; then

$$\overline{z_1 + z_2} = \overline{(a_1 + a_2) + (b_1 + b_2)i}$$
$$= (a_1 + a_2) - (b_1 + b_2)i$$
$$= (a_1 - b_1 i) + (a_2 - b_2 i)$$
$$= \bar{z}_1 + \bar{z}_2$$ ■

REMARK. It is possible to extend part (*a*) of Theorem 10.2.3 to n terms and part (*c*) to n factors. More precisely,

$$\overline{z_1 + z_2 + \cdots + z_n} = \bar{z}_1 + \bar{z}_2 + \cdots + \bar{z}_n$$
$$\overline{z_1 z_2 \cdots z_n} = \bar{z}_1 \bar{z}_2 \cdots \bar{z}_n$$

Exercise Set 10.2

1. In each part find $\bar{z}$.

   (a) $z = 2 + 7i$ (b) $z = -3 - 5i$ (c) $z = 5i$ (d) $z = -i$ (e) $z = -9$ (f) $z = 0$

2. In each part find $|z|$.

   (a) $z = i$ (b) $z = -7i$ (c) $z = -3 - 4i$ (d) $z = 1 + i$ (e) $z = -8$ (f) $z = 0$

3. Verify that $z\bar{z} = |z|^2$ for

   (a) $z = 2 - 4i$ (b) $z = -3 + 5i$ (c) $z = \sqrt{2} - \sqrt{2}i$

4. Given that $z_1 = 1 - 5i$ and $z_2 = 3 + 4i$, find

   (a) z_1/z_2 (b) $\bar{z}_1/z_2$ (c) $z_1/\bar{z}_2$ (d) $\overline{(z_1/z_2)}$ (e) $z_1/|z_2|$ (f) $|z_1/z_2|$

5. In each part find $1/z$.

   (a) $z = i$ (b) $z = 1 - 5i$ (c) $z = \dfrac{-i}{7}$

6. Given that $z_1 = 1 + i$ and $z_2 = 1 - 2i$, find

   (a) $z_1 - \left(\dfrac{z_1}{z_2}\right)$ (b) $\dfrac{z_1 - 1}{z_2}$ (c) $z_1^2 - \left(\dfrac{iz_1}{z_2}\right)$ (d) $\dfrac{z_1}{iz_2}$

In Exercises 7–14 perform the calculations and express the result in the form $a + bi$.

7. $\dfrac{i}{1+i}$

8. $\dfrac{2}{(1-i)(3+i)}$

9. $\dfrac{1}{(3+4i)^2}$

10. $\dfrac{2+i}{i(-3+4i)}$

11. $\dfrac{\sqrt{3}+i}{(1-i)(\sqrt{3}-i)}$

12. $\dfrac{1}{i(3-2i)(1+i)}$

13. $\dfrac{i}{(1-i)(1-2i)(1+2i)}$

14. $\dfrac{1-2i}{3+4i} - \dfrac{2+i}{5i}$

15. In each part solve for z.

    (a) $iz = 2 - i$ (b) $(4 - 3i)\bar{z} = i$

16. Use Theorem 10.2.3 to prove the following identities:

    (a) $\overline{\bar{z} + 5i} = z - 5i$ (b) $\overline{iz} = -i\bar{z}$ (c) $\dfrac{\overline{i + \bar{z}}}{i - z} = -1$

17. In each part sketch the set of points in the complex plane that satisfies the equation.

    (a) $|z| = 2$ (b) $|z - (1+i)| = 1$ (c) $|z - i| = |z + i|$ (d) $\operatorname{Im}(\bar{z} + i) = 3$

18. In each part sketch the set of points in the complex plane that satisfies the given condition(s).

    (a) $|z + i| \le 1$ (b) $1 < |z| < 2$ (c) $|2z - 4i| < 1$ (d) $|z| \le |z + i|$

19. Given that $z = x + iy$, find

    (a) $\operatorname{Re}(\overline{iz})$ (b) $\operatorname{Im}(\overline{iz})$ (c) $\operatorname{Re}(i\bar{z})$ (d) $\operatorname{Im}(i\bar{z})$

20. (a) Show that if n is a positive integer, then the only possible values for $(1/i)^n$ are $1, -1, i$, and $-i$.

    (b) Find $(1/i)^{2509}$. [**Hint.** See Exercise 23(b) of Section 10.1.]

21. Prove:

    (a) $\dfrac{1}{2}(z + \bar{z}) = \operatorname{Re}(z)$ (b) $\dfrac{1}{2i}(z - \bar{z}) = \operatorname{Im}(z)$

22. Prove: $z = \bar{z}$ if and only if z is a real number.

23. Given that $z_1 = x_1 + iy_1$ and $z_2 = x_2 + iy_2 \ne 0$, find

    (a) $\operatorname{Re}\left(\dfrac{z_1}{z_2}\right)$ (b) $\operatorname{Im}\left(\dfrac{z_1}{z_2}\right)$

24. Prove: If $(\bar{z})^2 = z^2$, then z is either real or pure imaginary.

25. Prove that $|z| = |\bar{z}|$.

26. Prove:

(a) $\overline{z_1 - z_2} = \bar{z}_1 - \bar{z}_2$ (b) $\overline{z_1 z_2} = \bar{z}_1 \bar{z}_2$ (c) $\overline{(z_1/z_2)} = \bar{z}_1/\bar{z}_2$ (d) $\bar{\bar{z}} = z$

27. (a) Prove that $\overline{z^2} = (\bar{z})^2$.

(b) Prove that if n is a positive integer, then $\overline{z^n} = (\bar{z})^n$.

(c) Is the result in (b) true if n is a negative integer? Explain.

In Exercises 28–31 solve the system of linear equations by Cramer's rule.

28. $ix_1 - ix_2 = -2$
$2x_1 + x_2 = i$

29. $x_1 + x_2 = 2$
$x_1 - x_2 = 2i$

30. $x_1 + x_2 + x_3 = 3$
$x_1 + x_2 - x_3 = 2 + 2i$
$x_1 - x_2 + x_3 = -1$

31. $ix_1 + 3x_2 + (1 + i)x_3 = -i$
$x_1 + ix_2 + \quad 3x_3 = -2i$
$x_1 + \quad x_2 + \quad x_3 = 0$

In Exercises 32 and 33 solve the system of linear equations by Gauss–Jordan elimination.

32. $\begin{bmatrix} -1 & -1-i \\ -1+i & -2 \end{bmatrix} \begin{bmatrix} x_1 \\ x_2 \end{bmatrix} = \begin{bmatrix} 0 \\ 0 \end{bmatrix}$ **33.** $\begin{bmatrix} 2 & -1-i \\ -1+i & 1 \end{bmatrix} \begin{bmatrix} x_1 \\ x_2 \end{bmatrix} = \begin{bmatrix} 0 \\ 0 \end{bmatrix}$

34. Solve the following system of linear equations by Gauss–Jordan elimination.

$x_1 + \quad ix_2 - \quad ix_3 = 0$
$-x_1 + \quad (1 - i)x_2 + 2ix_3 = 0$
$2x_1 + (-1 + 2i)x_2 - 3ix_3 = 0$

35. In each part use the formula in Theorem 1.4.5 to compute the inverse of the matrix, and check your result by showing that $AA^{-1} = A^{-1}A = I$.

(a) $A = \begin{bmatrix} i & -2 \\ 1 & i \end{bmatrix}$ (b) $A = \begin{bmatrix} 2 & i \\ 1 & 0 \end{bmatrix}$

36. Let $p(x) = a_0 + a_1 x + a_2 x^2 + \cdots + a_n x^n$ be a polynomial for which the coefficients $a_0, a_1, a_2, \ldots, a_n$ are real. Prove that if z is a solution of the equation $p(z) = 0$, then so is $\bar{z}$.

37. Prove: For any complex number z, $|\mathrm{Re}(z)| \leq |z|$ and $|\mathrm{Im}(z)| \leq |z|$.

38. Prove that

$$\frac{|\mathrm{Re}(z)| + |\mathrm{Im}(z)|}{\sqrt{2}} \leq |z|$$

[**Hint.** Let $z = x + iy$ and use the fact that $(|x| - |y|)^2 \geq 0$.]

39. In each part use the method of Example 4 in Section 1.5 to find A^{-1}, and check your result by showing that $AA^{-1} = A^{-1}A = I$.

(a) $A = \begin{bmatrix} 1 & 1+i & 0 \\ 0 & 1 & i \\ -i & 1-2i & 2 \end{bmatrix}$ (b) $A = \begin{bmatrix} i & 0 & -i \\ 0 & 1 & -1-4i \\ 2-i & i & 3 \end{bmatrix}$

40. Show that $|z - 1| = |\bar{z} - 1|$. Discuss the geometric interpretation of the result.

41. (a) If $z_1 = a_1 + b_1 i$ and $z_2 = a_2 + b_2 i$, find $|z_1 - z_2|$ and interpret the result geometrically.

(b) Use part (a) to show that the complex numbers 12, $6 + 2i$, and $8 + 8i$ are vertices of a right triangle.

42. Use Theorem 10.2.3 to show that if the coefficients a, b, and c in a quadratic polynomial are real, then the solutions of the equation $az^2 + bz + c = 0$ are complex conjugates. What can you conclude if a, b, and c are complex?

10.3 POLAR FORM OF A COMPLEX NUMBER

In this section we shall discuss a way to represent complex numbers using trigonometric properties. Our work will lead to an important formula for powers of complex numbers and a method for finding nth roots of complex numbers.

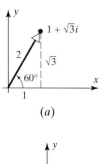

Figure 10.3.1

Polar Form   If $z = x + iy$ is a nonzero complex number, $r = |z|$, and θ measures the angle from the positive real axis to the vector z, then, as suggested by Figure 10.3.1,

$$x = r\cos\theta, \qquad y = r\sin\theta \tag{1}$$

so that $z = x + iy$ can be written as $z = r\cos\theta + ir\sin\theta$ or

$$z = r(\cos\theta + i\sin\theta) \tag{2}$$

This is called a ***polar form of z***.

Argument of a Complex Number   The angle θ is called an ***argument of z*** and is denoted by

$$\theta = \arg\ z$$

The argument of z is not uniquely determined because we can add or subtract any multiple of 2π from θ to produce another value of the argument. However, there is only one value of the argument in radians that satisfies

$$-\pi < \theta \le \pi$$

This is called the ***principal argument of z*** and is denoted by

$$\theta = \text{Arg}\ z$$

EXAMPLE 1   Polar Forms

Express the following complex numbers in polar form using their principal arguments:

$$\text{(a) } z = 1 + \sqrt{3}i \qquad \text{(b) } z = -1 - i$$

Solution (a). The value of r is

$$r = |z| = \sqrt{1^2 + (\sqrt{3})^2} = \sqrt{4} = 2$$

and since $x = 1$ and $y = \sqrt{3}$, it follows from (1) that

$$1 = 2\cos\theta \quad \text{and} \quad \sqrt{3} = 2\sin\theta$$

so $\cos\theta = 1/2$ and $\sin\theta = \sqrt{3}/2$. The only value of θ that satisfies these relations and meets the requirement $-\pi < \theta \le \pi$ is $\theta = \pi/3\,(= 60°)$ (see Figure 10.3.2a). Thus, a polar form of z is

$$z = 2\left(\cos\frac{\pi}{3} + i\sin\frac{\pi}{3}\right)$$

Solution (b). The value of r is

$$r = |z| = \sqrt{(-1)^2 + (-1)^2} = \sqrt{2}$$

and since $x = -1$, $y = -1$, it follows from (1) that

$$-1 = \sqrt{2}\cos\theta \quad \text{and} \quad -1 = \sqrt{2}\sin\theta$$

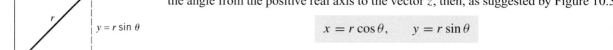

(a)

(b)

Figure 10.3.2

so $\cos\theta = -1/\sqrt{2}$ and $\sin\theta = -1/\sqrt{2}$. The only value of θ that satisfies these relations and meets the requirement $-\pi < \theta \leq \pi$ is $\theta = -3\pi/4\,(= -135°)$ (Figure 10.3.2b). Thus, a polar form of z is

$$z = \sqrt{2}\left(\cos\frac{-3\pi}{4} + i\sin\frac{-3\pi}{4}\right)$$ ◆

Multiplication and Division Interpreted Geometrically

We now show how polar forms can be used to give geometric interpretations of multiplication and division of complex numbers. Let

$$z_1 = r_1(\cos\theta_1 + i\sin\theta_1) \quad \text{and} \quad z_2 = r_2(\cos\theta_2 + i\sin\theta_2)$$

Multiplying, we obtain

$$z_1 z_2 = r_1 r_2[(\cos\theta_1\cos\theta_2 - \sin\theta_1\sin\theta_2) + i(\sin\theta_1\cos\theta_2 + \cos\theta_1\sin\theta_2)]$$

Recalling the trigonometric identities

$$\cos(\theta_1 + \theta_2) = \cos\theta_1\cos\theta_2 - \sin\theta_1\sin\theta_2$$
$$\sin(\theta_1 + \theta_2) = \sin\theta_1\cos\theta_2 + \cos\theta_1\sin\theta_2$$

we obtain

$$z_1 z_2 = r_1 r_2[\cos(\theta_1 + \theta_2) + i\sin(\theta_1 + \theta_2)] \tag{3}$$

which is a polar form of the complex number with modulus $r_1 r_2$ and argument $\theta_1 + \theta_2$. Thus, we have shown that

$$|z_1 z_2| = |z_1||z_2| \tag{4}$$

and

$$\arg(z_1 z_2) = \arg z_1 + \arg z_2$$

(Why?) In words, *the product of two complex numbers is obtained by multiplying their moduli and adding their arguments* (Figure 10.3.3).

We leave it as an exercise to show that if $z_2 \neq 0$, then

$$\frac{z_1}{z_2} = \frac{r_1}{r_2}[\cos(\theta_1 - \theta_2) + i\sin(\theta_1 - \theta_2)] \tag{5}$$

from which it follows that

$$\left|\frac{z_1}{z_2}\right| = \frac{|z_1|}{|z_2|} \quad \text{if } z_2 \neq 0$$

and

$$\arg\left(\frac{z_1}{z_2}\right) = \arg z_1 - \arg z_2$$

In words, *the quotient of two complex numbers is obtained by dividing their moduli and subtracting their arguments* (*in the appropriate order*).

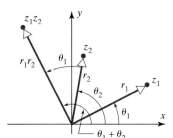

Figure 10.3.3
The product of two complex numbers.

EXAMPLE 2 A Quotient Using Polar Forms

Let

$$z_1 = 1 + \sqrt{3}i \quad \text{and} \quad z_2 = \sqrt{3} + i$$

Polar forms of these complex numbers are

$$z_1 = 2\left(\cos\frac{\pi}{3} + i\sin\frac{\pi}{3}\right) \quad \text{and} \quad z_2 = 2\left(\cos\frac{\pi}{6} + i\sin\frac{\pi}{6}\right)$$

(verify) so that from (3)

$$z_1 z_2 = 4\left[\cos\left(\frac{\pi}{3}+\frac{\pi}{6}\right)+i\sin\left(\frac{\pi}{3}+\frac{\pi}{6}\right)\right]$$

$$= 4\left[\cos\frac{\pi}{2}+i\sin\frac{\pi}{2}\right] = 4[0+i] = 4i$$

and from (5)

$$\frac{z_1}{z_2} = 1\cdot\left[\cos\left(\frac{\pi}{3}-\frac{\pi}{6}\right)+i\sin\left(\frac{\pi}{3}-\frac{\pi}{6}\right)\right]$$

$$= \cos\frac{\pi}{6}+i\sin\frac{\pi}{6} = \frac{\sqrt{3}}{2}+\frac{1}{2}i$$

As a check, we calculate $z_1 z_2$ and z_1/z_2 directly without using polar forms for z_1 and z_2:

$$z_1 z_2 = (1+\sqrt{3}i)(\sqrt{3}+i) = (\sqrt{3}-\sqrt{3})+(3+1)i = 4i$$

$$\frac{z_1}{z_2} = \frac{1+\sqrt{3}i}{\sqrt{3}+i}\cdot\frac{\sqrt{3}-i}{\sqrt{3}-i} = \frac{(\sqrt{3}+\sqrt{3})+(-i+3i)}{4} = \frac{\sqrt{3}}{2}+\frac{1}{2}i$$

which agrees with our previous results. ◆

The complex number i has a modulus of 1 and an argument of $\pi/2\ (=90°)$, so the product iz has the same modulus as z, but its argument is $90°$ greater than that of z. In short, *multiplying z by i rotates z counterclockwise by $90°$* (Figure 10.3.4).

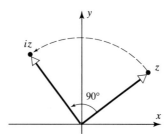

Figure 10.3.4 Multiplying by i rotates z counterclockwise by $90°$.

Demoivre's Formula   If n is a positive integer and $z = r(\cos\theta + i\sin\theta)$, then from Formula (3),

$$z^n = \underbrace{z\cdot z\cdot z\cdots z}_{n\text{-factors}} = r^n[\cos\underbrace{(\theta+\theta+\cdots+\theta)}_{n\text{-terms}}+i\sin\underbrace{(\theta+\theta+\cdots+\theta)}_{n\text{-terms}}]$$

or

$$z^n = r^n(\cos n\theta + i\sin n\theta) \tag{6}$$

Moreover, (6) also holds for negative integers if $z\neq 0$ (see Exercise 23).

In the special case where $r = 1$, we have $z = \cos\theta + i\sin\theta$, so (6) becomes

$$(\cos\theta + i\sin\theta)^n = \cos n\theta + i\sin n\theta \tag{7}$$

which is called ***DeMoivre's formula***. Although we derived (7) assuming n to be a positive integer, it will be shown in the exercises that this formula is valid for all integers n.

Finding nth Roots   We now show how DeMoivre's formula can be used to obtain roots of complex numbers. If n is a positive integer, and z is any complex number, then we define an ***nth root of z*** to be any complex number w that satisfies the equation

$$w^n = z \tag{8}$$

We denote an nth root of z by $z^{1/n}$. If $z \neq 0$, then we can derive formulas for the nth roots of z as follows. Let

$$w = \rho(\cos\alpha + i\sin\alpha) \quad \text{and} \quad z = r(\cos\theta + i\sin\theta)$$

If we assume that w satisfies (8), then it follows from (6) that

$$\rho^n(\cos n\alpha + i\sin n\alpha) = r(\cos\theta + i\sin\theta) \qquad (9)$$

Comparing the moduli of the two sides, we see that $\rho^n = r$ or

$$\rho = \sqrt[n]{r}$$

where $\sqrt[n]{r}$ denotes the real positive nth root of r. Moreover, in order to have the equalities $\cos n\alpha = \cos\theta$ and $\sin n\alpha = \sin\theta$ in (9), the angles $n\alpha$ and θ must either be equal or differ by a multiple of 2π. That is,

$$n\alpha = \theta + 2k\pi \quad \text{or} \quad \alpha = \frac{\theta}{n} + \frac{2k\pi}{n}, \qquad k = 0, \pm 1, \pm 2, \ldots$$

Thus, the values of $w = \rho(\cos\alpha + i\sin\alpha)$ that satisfy (8) are given by

$$w = \sqrt[n]{r}\left[\cos\left(\frac{\theta}{n} + \frac{2k\pi}{n}\right) + i\sin\left(\frac{\theta}{n} + \frac{2k\pi}{n}\right)\right], \qquad k = 0, \pm 1, \pm 2, \ldots$$

Although there are infinitely many values of k, it can be shown (see Exercise 16) that $k = 0, 1, 2, \ldots, n-1$ produce distinct values of w satisfying (8), but all other choices of k yield duplicates of these. Therefore, there are exactly n different nth roots of $z = r(\cos\theta + i\sin\theta)$, and these are given by

$$z^{1/n} = \sqrt[n]{r}\left[\cos\left(\frac{\theta}{n} + \frac{2k\pi}{n}\right) + i\sin\left(\frac{\theta}{n} + \frac{2k\pi}{n}\right)\right], \qquad k = 0, 1, 2, \ldots, n-1 \qquad (10)$$

Abraham DeMoivre (1667–1754) was a French mathematician who made important contributions to probability, statistics, and trigonometry. He developed the concept of statistically independent events, wrote a major and influential treatise on probability, and helped transform trigonometry from a branch of geometry into a branch of analysis through his use of complex numbers. In spite of his important work, he barely managed to eke out a living as a tutor and a consultant on gambling and insurance.

EXAMPLE 3 Cube Roots of a Complex Number

Find all cube roots of -8.

Solution.

Since -8 lies on the negative real axis, we can use $\theta = \pi$ as an argument. Moreover, $r = |z| = |-8| = 8$, so a polar form of -8 is

$$-8 = 8(\cos\pi + i\sin\pi)$$

From (10) with $n = 3$ it follows that

$$(-8)^{1/3} = \sqrt[3]{8}\left[\cos\left(\frac{\pi}{3} + \frac{2k\pi}{3}\right) + i\sin\left(\frac{\pi}{3} + \frac{2k\pi}{3}\right)\right], \qquad k = 0, 1, 2$$

Thus, the cube roots of -8 are

$$2\left(\cos\frac{\pi}{3} + i\sin\frac{\pi}{3}\right) = 2\left(\frac{1}{2} + \frac{\sqrt{3}}{2}i\right) = 1 + \sqrt{3}i$$

$$2(\cos\pi + i\sin\pi) = 2(-1) = -2$$

$$2\cos\left(\frac{5\pi}{3} + i\sin\frac{5\pi}{3}\right) = 2\left(\frac{1}{2} - \frac{\sqrt{3}}{2}i\right) = 1 - \sqrt{3}i \qquad \blacklozenge$$

As shown in Figure 10.3.5, the three cube roots of -8 obtained in Example 3 are equally spaced $\pi/3$ radians ($= 120°$) apart around the circle of radius 2 centered at the

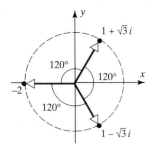

Figure 10.3.5
The cube roots of -8.

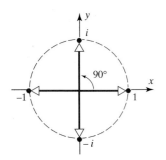

Figure 10.3.6
The fourth roots of 1.

origin. This is not accidental. In general, it follows from Formula (10) that the nth roots of z lie on the circle of radius $\sqrt[n]{r}$ ($= \sqrt[n]{|z|}$), and are equally spaced $2\pi/n$ radians apart. (Can you see why?) Thus, once one nth root of z is found, the remaining $n-1$ roots can be generated by rotating this root successively through increments of $2\pi/n$ radians.

EXAMPLE 4 Fourth Roots of a Complex Number

Find all fourth roots of 1.

Solution.

We could apply Formula (10). Instead, we observe that $w = 1$ is one fourth root of 1, so that the remaining three roots can be generated by rotating this root through increments of $2\pi/4 = \pi/2$ radians ($= 90°$). From Figure 10.3.6 we see that the fourth roots of 1 are

$$1, \quad i, \quad -1, \quad -i$$ ◆

Complex Exponents

We conclude this section with some comments on notation.

In more detailed studies of complex numbers, complex exponents are defined, and it is proved that

$$\cos\theta + i\sin\theta = e^{i\theta} \tag{11}$$

where e is an irrational real number given approximately by $e \approx 2.71828\ldots$. (For readers who have studied calculus, a proof of this result is given in Exercise 18.)

It follows from (11) that the polar form

$$z = r(\cos\theta + i\sin\theta)$$

can be written more briefly as

$$z = re^{i\theta} \tag{12}$$

EXAMPLE 5 Expressing a Complex Number in Form [12]

In Example 1 it was shown that

$$1 + \sqrt{3}i = 2\left(\cos\frac{\pi}{3} + i\sin\frac{\pi}{3}\right)$$

From (12) this can also be written as

$$1 + \sqrt{3}i = 2e^{i\pi/3}$$ ◆

It can be proved that complex exponents follow the same laws as real exponents, so that if

$$z_1 = r_1 e^{i\theta_1} \quad \text{and} \quad z_2 = r_2 e^{i\theta_2}$$

are nonzero complex numbers, then

$$z_1 z_2 = r_1 r_2 e^{i\theta_1 + i\theta_2} = r_1 r_2 e^{i(\theta_1 + \theta_2)}$$

$$\frac{z_1}{z_2} = \frac{r_1}{r_2} e^{i\theta_1 - i\theta_2} = \frac{r_1}{r_2} e^{i(\theta_1 - \theta_2)}$$

But these are just Formulas (3) and (5) in a different notation.

We conclude this section with a useful formula for $\bar{z}$ in polar notation. If

$$z = re^{i\theta} = r(\cos\theta + i\sin\theta)$$

then

$$\bar{z} = r(\cos\theta - i\sin\theta) \tag{13}$$

Recalling the trigonometric identities

$$\sin(-\theta) = -\sin\theta \quad \text{and} \quad \cos(-\theta) = \cos\theta$$

we can rewrite (13) as

$$\bar{z} = r[\cos(-\theta) + i\sin(-\theta)] = re^{i(-\theta)}$$

or equivalently,

$$\bar{z} = re^{-i\theta} \tag{14}$$

In the special case where $r = 1$, the polar form of z is $z = e^{i\theta}$, and (14) yields the formula

$$\overline{e^{i\theta}} = e^{-i\theta} \tag{15}$$

Exercise Set 10.3

1. In each part find the principal argument of z.

    (a) $z = 1$ (b) $z = i$ (c) $z = -i$ (d) $z = 1 + i$ (e) $z = -1 + \sqrt{3}i$ (f) $z = 1 - i$

2. In each part find the value of $\theta = \arg(1 - \sqrt{3}i)$ that satisfies the given condition.

    (a) $0 < \theta \leq 2\pi$ (b) $-\pi < \theta \leq \pi$ (c) $-\dfrac{\pi}{6} \leq \theta < \dfrac{11\pi}{6}$

3. In each part express the complex number in polar form using its principal argument.

    (a) $2i$ (b) -4 (c) $5 + 5i$ (d) $-6 + 6\sqrt{3}i$ (e) $-3 - 3i$ (f) $2\sqrt{3} - 2i$

4. Given that $z_1 = 2(\cos\pi/4 + i\sin\pi/4)$ and $z_2 = 3(\cos\pi/6 + i\sin\pi/6)$, find a polar form of

    (a) $z_1 z_2$ (b) $\dfrac{z_1}{z_2}$ (c) $\dfrac{z_2}{z_1}$ (d) $\dfrac{z_1^5}{z_2^2}$

5. Express $z_1 = i$, $z_2 = 1 - \sqrt{3}i$, and $z_3 = \sqrt{3} + i$ in polar form, and use your results to find $z_1 z_2 / z_3$. Check your results by performing the calculations without using polar forms.

6. Use Formula (6) to find

    (a) $(1 + i)^{12}$ (b) $\left(\dfrac{1}{\sqrt{2}} - \dfrac{1}{\sqrt{2}}i\right)^{-6}$ (c) $(\sqrt{3} + i)^7$ (d) $(1 - i\sqrt{3})^{-10}$

7. In each part find all the roots and sketch them as vectors in the complex plane.

    (a) $(-i)^{1/2}$ (b) $(1 + \sqrt{3}i)^{1/2}$ (c) $(-27)^{1/3}$ (d) $(i)^{1/3}$ (e) $(-1)^{1/4}$ (f) $(-8 + 8\sqrt{3}i)^{1/4}$

8. Use the method of Example 4 to find all cube roots of 1.

9. Use the method of Example 4 to find all sixth roots of 1.

10. Find all square roots of $1 + i$ and express your results in polar form.

11. Find all solutions of the equation $z^4 - 16 = 0$.

12. Find four solutions of the equation $z^4 + 8 = 0$ and use your results to factor $z^4 + 8$ into two quadratic factors with real coefficients.

13. It was shown in the text that multiplying z by i rotates z counterclockwise by $90°$. What is the geometric effect of dividing z by i?

14. In each part use (6) to calculate the given power.

(a) $(1+i)^8$ (b) $(-2\sqrt{3}+2i)^{-9}$

15. In each part find $\text{Re}(z)$ and $\text{Im}(z)$.

(a) $z = 3e^{i\pi}$ (b) $z = 3e^{-i\pi}$ (c) $\bar{z} = \sqrt{2}e^{\pi i/2}$ (d) $\bar{z} = -3e^{-2\pi i}$

16. (a) Show that the values of $z^{1/n}$ in Formula (10) are all different.

(b) Show that integer values of k other than $k = 0, 1, 2, \ldots, n-1$ produce values of $z^{1/n}$ that are duplicates of those in Formula (10).

17. Show that Formula (7) is valid if $n = 0$ or n is a negative integer.

18. (*For readers who have studied calculus.*) To prove Formula (11) recall that the Maclaurin series for e^x is

$$e^x = 1 + x + \frac{x^2}{2!} + \cdots + \frac{x^n}{n!} + \cdots$$

(a) By substituting $x = i\theta$ in this series and simplifying, show that

$$e^{i\theta} = \left(1 - \frac{\theta^2}{2!} + \frac{\theta^4}{4!} - \frac{\theta^6}{6!} + \cdots\right) + i\left(\theta - \frac{\theta^3}{3!} + \frac{\theta^5}{5!} - \frac{\theta^7}{7!} + \cdots\right)$$

(b) Use the result in part (a) to obtain (11).

19. Derive Formula (5).

20. When $n = 2$ and $n = 3$, Equation (7) gives

$$(\cos\theta + i\sin\theta)^2 = \cos 2\theta + i\sin 2\theta$$
$$(\cos\theta + i\sin\theta)^3 = \cos 3\theta + i\sin 3\theta$$

Use these two equations to obtain trigonometric identities for $\cos 2\theta$, $\sin 2\theta$, $\cos 3\theta$, and $\sin 3\theta$.

21. Use (11) to show that

$$\cos\theta = \frac{e^{i\theta} + e^{-i\theta}}{2} \quad \text{and} \quad \sin\theta = \frac{e^{i\theta} - e^{-i\theta}}{2i}$$

22. Show that if $(a+bi)^3 = 8$, then $a^2 + b^2 = 4$.

23. Show that Formula (6) is valid for negative integer exponents if $z \neq 0$.

10.4 COMPLEX VECTOR SPACES

In this section we shall develop the basic properties of vector spaces with complex scalars and discuss some of the ways in which they differ from real vector spaces. However, before going further the reader should review the vector space axioms given in Section 5.1.

Basic Properties   Recall that a vector space in which the scalars are allowed to be complex numbers is called a ***complex vector space***. Linear combinations of vectors in a complex vector space are defined exactly as in a real vector space except that the scalars are allowed to be complex numbers. More precisely, a vector **w** is called a ***linear combination*** of the vectors of $\mathbf{v}_1, \mathbf{v}_2, \ldots, \mathbf{v}_r$ if **w** can be expressed in the form

$$\mathbf{w} = k_1\mathbf{v}_1 + k_2\mathbf{v}_2 + \cdots + k_r\mathbf{v}_r$$

where $k_1, k_2, \ldots, k_r$ are complex numbers.

The notions of ***linear independence, spanning, basis, dimension,*** and ***subspace*** carry over without change to complex vector spaces, and the theorems developed in Chapter 5 continue to hold with R^n changed to C^n.

Among the real vector spaces the most important one is R^n, the space of n-tuples of real numbers, with addition and scalar multiplication performed coordinatewise. Among the complex vector spaces the most important one is C^n, the space of n-tuples of complex numbers, with addition and scalar multiplication performed coordinatewise. A vector **u** in C^n can be written either in vector notation

$$\mathbf{u} = (u_1, u_2, \ldots, u_n)$$

or in matrix notation

$$\mathbf{u} = \begin{bmatrix} u_1 \\ u_2 \\ \vdots \\ u_n \end{bmatrix}$$

where

$$u_1 = a_1 + b_1 i, \quad u_2 = a_2 + b_2 i, \ldots, \quad u_n = a_n + b_n i$$

EXAMPLE 1   Vector Addition and Scalar Multiplication

If

$$\mathbf{u} = (i, 1 + i, -2) \quad \text{and} \quad \mathbf{v} = (2 + i, 1 - i, 3 + 2i)$$

then

$$\mathbf{u} + \mathbf{v} = (i, 1 + i, -2) + (2 + i, 1 - i, 3 + 2i) = (2 + 2i, 2, 1 + 2i)$$

and

$$i\mathbf{u} = i(i, 1 + i, -2) = (i^2, i + i^2, -2i) = (-1, -1 + i, -2i) \qquad \blacklozenge$$

In C^n as in R^n, the vectors

$$\mathbf{e}_1 = (1, 0, 0, \ldots, 0), \quad \mathbf{e}_2 = (0, 1, 0, \ldots, 0), \ldots, \quad \mathbf{e}_n = (0, 0, 0, \ldots, 1)$$

form a basis. It is called the ***standard basis*** for C^n. Since there are n vectors in this basis, C^n is an n-dimensional vector space.

REMARK. Do not confuse the complex number $i = \sqrt{-1}$ with the vector $\mathbf{i} = (1, 0, 0)$ from the standard basis for R^3 (see Example 3, Section 3.4). The complex number i will always be set in lightface type and the vector **i** in boldface.

EXAMPLE 2   Complex M_{mn}

In Example 3 of Section 5.1 we defined the vector space M_{mn} of $m \times n$ matrices with real entries. The complex analog of this space is the vector space of $m \times n$ matrices with complex entries and the operations of matrix addition and scalar multiplication. We refer to this space as complex M_{mn}. ◆

EXAMPLE 3 Complex-Valued Function of a Real Variable

If $f_1(x)$ and $f_2(x)$ are real-valued functions of the real variable x, then the expression

$$f(x) = f_1(x) + if_2(x)$$

is called a *complex-valued function of the real variable x*. Some examples are

$$f(x) = 2x + ix^3 \quad \text{and} \quad g(x) = 2\sin x + i\cos x \qquad (1)$$

Let V be the set of all complex-valued functions that are defined on the entire line. If $\mathbf{f} = f_1(x) + if_2(x)$ and $\mathbf{g} = g_1(x) + ig_2(x)$ are two such functions and k is any complex number, then we define the *sum* function $\mathbf{f} + \mathbf{g}$ and the *scalar multiple* $k\mathbf{f}$ by

$$(\mathbf{f} + \mathbf{g})(x) = [f_1(x) + g_1(x)] + i[f_2(x) + g_2(x)]$$
$$(k\mathbf{f})(x) = kf_1(x) + ikf_2(x)$$

For example, if $\mathbf{f} = f(x)$ and $\mathbf{g} = g(x)$ are the functions in (1), then

$$(\mathbf{f} + \mathbf{g})(x) = (2x + 2\sin x) + i(x^3 + \cos x)$$
$$(i\mathbf{f})(x) = 2xi + i^2x^3 = -x^3 + 2xi$$
$$((1+i)\mathbf{g})(x) = (1+i)(2\sin x + i\cos x) = (2\sin x - \cos x) + i(2\sin x + \cos x)$$

It can be shown that V together with the stated operations is a complex vector space. It is the complex analog of the vector space $F(-\infty, \infty)$ of real-valued functions discussed in Example 4 of Section 5.1. ◆

Calculus Required

EXAMPLE 4 Complex $C(-\infty, \infty)$

If $f(x) = f_1(x) + if_2(x)$ is a complex-valued function of the real variable x, then f is said to be *continuous* if $f_1(x)$ and $f_2(x)$ are continuous. We leave it as an exercise to show that the set of all continuous complex-valued functions of a real variable x is a subspace of the vector space of all complex-valued functions of x. This space is the complex analog of the vector space $C(-\infty, \infty)$ discussed in Example 6 of Section 5.2 and is called complex $C(-\infty, \infty)$. A closely related example is complex $C[a, b]$, the vector space of all complex-valued functions that are continuous on the closed interval $[a, b]$. ◆

Recall that in R^n the Euclidean inner product of two vectors

$$\mathbf{u} = (u_1, u_2, \ldots, u_n) \quad \text{and} \quad \mathbf{v} = (v_1, v_2, \ldots, v_n)$$

was defined as

$$\mathbf{u} \cdot \mathbf{v} = u_1v_1 + u_2v_2 + \cdots + u_nv_n \qquad (2)$$

and the Euclidean norm (or length) of $\mathbf{u}$ as

$$\|\mathbf{u}\| = (\mathbf{u} \cdot \mathbf{u})^{1/2} = \sqrt{u_1^2 + u_2^2 + \cdots + u_n^2} \qquad (3)$$

Unfortunately, these definitions are not appropriate for vectors in C^n. For example, if (3) were applied to the vector $\mathbf{u} = (i, 1)$ in C^2, we would obtain

$$\|\mathbf{u}\| = \sqrt{i^2 + 1} = \sqrt{0} = 0$$

so $\mathbf{u}$ would be a *nonzero* vector with zero length—a situation that is clearly unsatisfactory.

To extend the notions of norm, distance, and angle to C^n properly, we must modify the inner product slightly.

Definition

If $\mathbf{u} = (u_1, u_2, \ldots, u_n)$ and $\mathbf{v} = (v_1, v_2, \ldots, v_n)$ are vectors in C^n, then their *complex Euclidean inner product* $\mathbf{u} \cdot \mathbf{v}$ is defined by

$$\mathbf{u} \cdot \mathbf{v} = u_1 \overline{v}_1 + u_2 \overline{v}_2 + \cdots + u_n \overline{v}_n$$

where $\overline{v}_1, \overline{v}_2, \ldots, \overline{v}_n$ are the conjugates of $v_1, v_2, \ldots, v_n$.

REMARK. Observe that the Euclidean inner product of vectors in C^n is a complex number, whereas the Euclidean inner product of vectors in R^n is a real number.

EXAMPLE 5 Complex $C(-\infty, \infty)$

The complex Euclidean inner product of the vectors

$$\mathbf{u} = (-i, 2, 1 + 3i) \quad \text{and} \quad \mathbf{v} = (1 - i, 0, 1 + 3i)$$

is

$$
\begin{aligned}
\mathbf{u} \cdot \mathbf{v} &= (-i)\overline{(1 - i)} + (2)(\overline{0}) + (1 + 3i)\overline{(1 + 3i)} \\
&= (-i)(1 + i) + (2)(0) + (1 + 3i)(1 - 3i) \\
&= -i - i^2 + 1 - 9i^2 = 11 - i
\end{aligned}
$$

♦

Theorem 4.1.2 listed the four main properties of the Euclidean inner product on R^n. The following theorem is the corresponding result for the complex Euclidean inner product on C^n.

Theorem 10.4.1 Properties of the Complex Inner Product

If $\mathbf{u}$, $\mathbf{v}$, *and* $\mathbf{w}$ *are vectors in* C^n, *and k is any complex number, then*:

(a) $\mathbf{u} \cdot \mathbf{v} = \overline{\mathbf{v} \cdot \mathbf{u}}$
(b) $(\mathbf{u} + \mathbf{v}) \cdot \mathbf{w} = \mathbf{u} \cdot \mathbf{w} + \mathbf{v} \cdot \mathbf{w}$
(c) $(k\mathbf{u}) \cdot \mathbf{v} = k(\mathbf{u} \cdot \mathbf{v})$
(d) $\mathbf{v} \cdot \mathbf{v} \geq 0$. *Further,* $\mathbf{v} \cdot \mathbf{v} = 0$ *if and only if* $\mathbf{v} = \mathbf{0}$.

Note the difference between part (*a*) of this theorem and part (*a*) of Theorem 4.1.2. We will prove parts (*a*) and (*d*) and leave the rest as exercises.

***Proof* (*a*).** Let $\mathbf{u} = (u_1, u_2, \ldots, u_n)$ and $\mathbf{v} = (v_1, v_2, \ldots, v_n)$. Then

$$\mathbf{u} \cdot \mathbf{v} = u_1 \overline{v}_1 + u_2 \overline{v}_2 + \cdots + u_n \overline{v}_n$$

and

$$\mathbf{v} \cdot \mathbf{u} = v_1 \overline{u}_1 + v_2 \overline{u}_2 + \cdots + v_n \overline{u}_n$$

so

$$
\begin{aligned}
\overline{\mathbf{v} \cdot \mathbf{u}} &= \overline{v_1 \overline{u}_1 + v_2 \overline{u}_2 + \cdots + v_n \overline{u}_n} \\
&= \overline{v}_1 \overline{\overline{u}}_1 + \overline{v}_2 \overline{\overline{u}}_2 + \cdots + \overline{v}_n \overline{\overline{u}}_n && \text{[Theorem 10.2.3, parts (}a\text{) and (}c\text{)]} \\
&= \overline{v}_1 u_1 + \overline{v}_2 u_2 + \cdots + \overline{v}_n u_n && \text{[Theorem 10.2.3, part (}e\text{)]} \\
&= u_1 \overline{v}_1 + u_2 \overline{v}_2 + \cdots + u_n \overline{v}_n \\
&= \mathbf{u} \cdot \mathbf{v}
\end{aligned}
$$

Proof (d).

$$\mathbf{v} \cdot \mathbf{v} = v_1 \bar{v}_1 + v_2 \bar{v}_2 + \cdots + v_n \bar{v}_n = |v_1|^2 + |v_2|^2 + \cdots + |v_n|^2 \geq 0$$

Moreover, equality holds if and only if $|v_1| = |v_2| = \cdots = |v_n| = 0$. But this is true if and only if $v_1 = v_2 = \cdots = v_n = 0$, that is, if and only if $\mathbf{v} = \mathbf{0}$. ∎

REMARK. We leave it as an exercise to prove that

$$\mathbf{u} \cdot (k\mathbf{v}) = \bar{k}(\mathbf{u} \cdot \mathbf{v})$$

for vectors in C^n. Compare this to the corresponding formula

$$\mathbf{u} \cdot (k\mathbf{v}) = k(\mathbf{u} \cdot \mathbf{v})$$

for vectors in R^n.

Norm and Distance in C^n

By analogy with (3), we define the *Euclidean norm* (or *Euclidean length*) of a vector $\mathbf{u} = (u_1, u_2, \ldots, u_n)$ in C^n by

$$\|\mathbf{u}\| = (\mathbf{u} \cdot \mathbf{u})^{1/2} = \sqrt{|u_1|^2 + |u_2|^2 + \cdots + |u_n|^2}$$

and we define the *Euclidean distance* between the points $\mathbf{u} = (u_1, u_2, \ldots, u_n)$ and $\mathbf{v} = (v_1, v_2, \ldots, v_n)$ by

$$d(\mathbf{u}, \mathbf{v}) = \|\mathbf{u} - \mathbf{v}\| = \sqrt{|u_1 - v_1|^2 + |u_2 - v_2|^2 + \cdots + |u_n - v_n|^2}$$

EXAMPLE 6 Norm and Distance

If $\mathbf{u} = (i, 1 + i, 3)$ and $\mathbf{v} = (1 - i, 2, 4i)$, then

$$\|\mathbf{u}\| = \sqrt{|i|^2 + |1 + i|^2 + |3|^2} = \sqrt{1 + 2 + 9} = \sqrt{12} = 2\sqrt{3}$$

and

$$d(\mathbf{u}, \mathbf{v}) = \sqrt{|i - (1 - i)|^2 + |(1 + i) - 2|^2 + |3 - 4i|^2}$$

$$= \sqrt{|-1 + 2i|^2 + |-1 + i|^2 + |3 - 4i|^2} = \sqrt{5 + 2 + 25} = \sqrt{32} = 4\sqrt{2}$$

◆

The complex vector space C^n with norm and inner product defined above is called *complex Euclidean n-space*.

Exercise Set 10.4

1. Let $\mathbf{u} = (2i, 0, -1, 3)$, $\mathbf{v} = (-i, i, 1 + i, -1)$, and $\mathbf{w} = (1 + i, -i, -1 + 2i, 0)$. Find

(a) $\mathbf{u} - \mathbf{v}$ (b) $i\mathbf{v} + 2\mathbf{w}$ (c) $-\mathbf{w} + \mathbf{v}$ (d) $3(\mathbf{u} - (1 + i)\mathbf{v})$ (e) $-i\mathbf{v} + 2i\mathbf{w}$ (f) $2\mathbf{v} - (\mathbf{u} + \mathbf{w})$

2. Let $\mathbf{u}$, $\mathbf{v}$, and $\mathbf{w}$ be the vectors in Exercise 1. Find the vector $\mathbf{x}$ that satisfies
$\mathbf{u} - \mathbf{v} + i\mathbf{x} = 2i\mathbf{x} + \mathbf{w}$.

3. Let $\mathbf{u}_1 = (1 - i, i, 0)$, $\mathbf{u}_2 = (2i, 1 + i, 1)$, and $\mathbf{u}_3 = (0, 2i, 2 - i)$. Find scalars c_1, c_2, and c_3 such that $c_1\mathbf{u}_1 + c_2\mathbf{u}_2 + c_3\mathbf{u}_3 = (-3 + i, 3 + 2i, 3 - 4i)$.

4. Show that there do not exist scalars c_1, c_2, and c_3 such that

$$c_1(i, 2 - i, 2 + i) + c_2(1 + i, -2i, 2) + c_3(3, i, 6 + i) = (i, i, i)$$

5. Find the Euclidean norm of $\mathbf{v}$ if

    (a) $\mathbf{v} = (1, i)$ (b) $\mathbf{v} = (1 + i, 3i, 1)$ (c) $\mathbf{v} = (2i, 0, 2i + 1, -1)$ (d) $\mathbf{v} = (-i, i, i, 3, 3 + 4i)$

6. Let $\mathbf{u} = (3i, 0, -i)$, $\mathbf{v} = (0, 3 + 4i, -2i)$, and $\mathbf{w} = (1 + i, 2i, 0)$. Find

    (a) $\|\mathbf{u} + \mathbf{v}\|$ (b) $\|\mathbf{u}\| + \|\mathbf{v}\|$ (c) $\|-i\mathbf{u}\| + i\|\mathbf{u}\|$ (d) $\|3\mathbf{u} - 5\mathbf{v} + \mathbf{w}\|$ (e) $\dfrac{1}{\|\mathbf{w}\|}\mathbf{w}$ (f) $\left\|\dfrac{1}{\|\mathbf{w}\|}\mathbf{w}\right\|$

7. Show that if $\mathbf{v}$ is a nonzero vector in C^n, then $(1/\|\mathbf{v}\|)\mathbf{v}$ has Euclidean norm 1.

8. Find all scalars k such that $\|k\mathbf{v}\| = 1$, where $\mathbf{v} = (3i, 4i)$.

9. Find the Euclidean inner product $\mathbf{u} \cdot \mathbf{v}$ if

    (a) $\mathbf{u} = (-i, 3i)$, $\mathbf{v} = (3i, 2i)$
    (b) $\mathbf{u} = (3 - 4i, 2 + i, -6i)$, $\mathbf{v} = (1 + i, 2 - i, 4)$
    (c) $\mathbf{u} = (1 - i, 1 + i, 2i, 3)$, $\mathbf{v} = (4 + 6i, -5i, -1 + i, i)$

In Exercises 10 and 11 a set of objects is given together with operations of addition and scalar multiplication. Determine which sets are complex vector spaces under the given operations. For those that are not, list all axioms that fail to hold.

10. The set of all triples of complex numbers (z_1, z_2, z_3) with the operations

$$(z_1, z_2, z_3) + (z_1', z_2', z_3') = (z_1 + z_1', z_2 + z_2', z_3 + z_3')$$

and

$$k(z_1, z_2, z_3) = (\bar{k}z_1, \bar{k}z_2, \bar{k}z_3)$$

11. The set of all complex 2×2 matrices of the form

$$\begin{bmatrix} z & 0 \\ 0 & \bar{z} \end{bmatrix}$$

    with the standard matrix operations of addition and scalar multiplication.

12. Use Theorem 5.2.1 to determine which of the following sets are subspaces of C^3:

    (a) all vectors of the form $(z, 0, 0)$
    (b) all vectors of the form (z, i, i)
    (c) all vectors of the form (z_1, z_2, z_3), where $z_3 = \bar{z}_1 + \bar{z}_2$
    (d) all vectors of the form (z_1, z_2, z_3), where $z_3 = z_1 + z_2 + i$

13. Let $T: C^3 \to C^3$ be a linear operator defined by $T(\mathbf{x}) = A\mathbf{x}$, where

$$A = \begin{bmatrix} i & -i & -1 \\ 1 & -i & 1 + i \\ 0 & 1 - i & 1 \end{bmatrix}$$

    Find the kernel and nullity of T.

14. Use Theorem 5.2.1 to determine which of the following are subspaces of M_{22}:

    (a) All complex matrices of the form   (b) All complex matrices of the form

$$\begin{bmatrix} z_1 & z_2 \\ z_3 & z_4 \end{bmatrix} \qquad\qquad \begin{bmatrix} z_1 & z_2 \\ z_3 & z_4 \end{bmatrix}$$

    where z_1 and z_2 are real.   where $z_1 + z_4 = 0$.

    (c) All 2×2 complex matrices A such that $(\bar{A})^T = A$, where $\bar{A}$ is the matrix whose entries are the conjugates of the corresponding entries in A.

15. Use Theorem 5.2.1 to determine which of the following are subspaces of the vector space of complex-valued functions of the real variable x:

    (a) all f such that $f(1) = 0$
    (b) all f such that $f(0) = i$
    (c) all f such that $f(-x) = \overline{f(x)}$
    (d) all f of the form $k_1 + k_2 e^{ix}$, where k_1 and k_2 are complex numbers

16. Which of the following are linear combinations of $\mathbf{u} = (i, -i, 3i)$ and $\mathbf{v} = (2i, 4i, 0)$?

   (a) $(3i, 3i, 3i)$ (b) $(4i, 2i, 6i)$ (c) $(i, 5i, 6i)$ (d) $(0, 0, 0)$

17. Express the following as linear combinations of $\mathbf{u} = (1, 0, -i)$, $\mathbf{v} = (1+i, 1, 1-2i)$, and $\mathbf{w} = (0, i, 2)$.

   (a) $(1, 1, 1)$ (b) $(i, 0, -i)$ (c) $(0, 0, 0)$ (d) $(2-i, 1, 1+i)$

18. In each part determine whether the given vectors span C^3.

   (a) $\mathbf{v}_1 = (i, i, i)$, $\mathbf{v}_2 = (2i, 2i, 0)$, $\mathbf{v}_3 = (3i, 0, 0)$
   (b) $\mathbf{v}_1 = (1+i, 2-i, 3+i)$, $\mathbf{v}_2 = (2+3i, 0, 1-i)$
   (c) $\mathbf{v}_1 = (1, 0, -i)$, $\mathbf{v}_2 = (1+i, 1, 1-2i)$, $\mathbf{v}_3 = (0, i, 2)$
   (d) $\mathbf{v}_1 = (1, i, 0)$, $\mathbf{v}_2 = (0, -i, 1)$, $\mathbf{v}_3 = (1, 0, 1)$

19. Determine which of the following lie in the space spanned by

$$\mathbf{f} = e^{ix} \quad \text{and} \quad \mathbf{g} = e^{-ix}$$

   (a) $\cos x$ (b) $\sin x$ (c) $\cos x + 3i \sin x$

20. Explain why the following are linearly dependent sets of vectors. (Solve this problem by inspection.)

   (a) $\mathbf{u}_1 = (1-i, i)$ and $\mathbf{u}_2 = (1+i, -1)$ in C^2
   (b) $\mathbf{u}_1 = (1, -i)$, $\mathbf{u}_2 = (2+i, -1)$, $\mathbf{u}_3 = (4, 0)$ in C^2
   (c) $A = \begin{bmatrix} i & 3i \\ 2i & 0 \end{bmatrix}$ and $B = \begin{bmatrix} 1 & 3 \\ 2 & 0 \end{bmatrix}$ in complex M_{22}

21. Which of the following sets of vectors in C^3 are linearly independent?

   (a) $\mathbf{u}_1 = (1-i, 1, 0)$, $\mathbf{u}_2 = (2, 1+i, 0)$, $\mathbf{u}_3 = (1+i, i, 0)$
   (b) $\mathbf{u}_1 = (1, 0, -i)$, $\mathbf{u}_2 = (1+i, 1, 1-2i)$, $\mathbf{u}_3 = (0, i, 2)$
   (c) $\mathbf{u}_1 = (i, 0, 2-i)$, $\mathbf{u}_2 = (0, 1, i)$, $\mathbf{u}_3 = (-i, -1-4i, 3)$

22. Let V be the vector space of all complex-valued functions of the real variable x. Show that the following vectors are linearly dependent.

$$\mathbf{f} = 3 + 3i \cos 2x, \quad \mathbf{g} = \sin^2 x + i \cos^2 x, \quad \mathbf{h} = \cos^2 x - i \sin^2 x$$

23. Explain why the following sets of vectors are not bases for the indicated vector spaces. (Solve this problem by inspection.)

   (a) $\mathbf{u}_1 = (i, 2i)$, $\mathbf{u}_2 = (0, 3i)$, $\mathbf{u}_3 = (1, 7i)$ for C^2
   (b) $\mathbf{u}_1 = (-1+i, 0, 2-i)$, $\mathbf{u}_2 = (1, -i, 1+i)$ for C^3

24. Which of the following sets of vectors are bases for C^2?

   (a) $(2i, -i)$, $(4i, 0)$ (b) $(1+i, 1)$, $(1+i, i)$
   (c) $(0, 0)$, $(1+i, 1-i)$ (d) $(2-3i, i)$, $(3+2i, -1)$

25. Which of the following sets of vectors are bases for C^3?

   (a) $(i, 0, 0)$, $(i, i, 0)$, (i, i, i) (b) $(1, 0, -i)$, $(1+i, 1, 1-2i)$, $(0, i, 2)$
   (c) $(i, 0, 2-i)$, $(0, 1, i)$, $(-i, -1-4i, 3)$ (d) $(1, 0, i)$, $(2-i, 1, 2+i)$, $(0, 3i, 3i)$

In Exercises 26–29, determine the dimension of and a basis for the solution space of the system.

26. $\quad x_1 + (1+i)x_2 = 0$
   $\quad (1-i)x_1 + \qquad 2x_2 = 0$

27. $\quad 2x_1 - (1+i)x_2 = 0$
   $\quad (-1+i)x_1 + \qquad x_2 = 0$

28. $x_1 + (2-i)x_2 \qquad\quad = 0$
   $\qquad\quad x_2 + 3ix_3 = 0$
   $ix_1 + (2+2i)x_2 + 3ix_3 = 0$

29. $x_1 + ix_2 - 2ix_3 + \quad x_4 = 0$
   $ix_1 + 3x_2 + \ 4x_3 - 2ix_4 = 0$

30. Prove: If $\mathbf{u}$ and $\mathbf{v}$ are vectors in complex Euclidean n-space, then

$$\mathbf{u} \cdot (k\mathbf{v}) = \bar{k}(\mathbf{u} \cdot \mathbf{v})$$

31. (a) Prove part (b) of Theorem 10.4.1.   (b) Prove part (c) of Theorem 10.4.1.

32. Establish the identity

$$\mathbf{u} \cdot \mathbf{v} = \frac{1}{4}\|\mathbf{u} + \mathbf{v}\|^2 - \frac{1}{4}\|\mathbf{u} - \mathbf{v}\|^2 + \frac{i}{4}\|\mathbf{u} + i\mathbf{v}\|^2 - \frac{i}{4}\|\mathbf{u} - i\mathbf{v}\|^2$$

for vectors in complex Euclidean *n*-space.

33. (*For readers who have studied calculus.*) Prove that complex $C(-\infty, \infty)$ is a subspace of the vector space of complex-valued functions of a real variable.

10.5 COMPLEX INNER PRODUCT SPACES

In Section 6.1 we defined the notion of an inner product on a real vector space by using the basic properties of the Euclidean inner product on R^n as axioms. In this section we shall define inner products on complex vector spaces by using the properties of the Euclidean inner product on C^n as axioms.

Unitary Spaces   Motivated by Theorem 10.4.1 we make the following definition.

> ### Definition
>
> An ***inner product on a complex vector space*** V is a function that associates a complex number $\langle \mathbf{u}, \mathbf{v} \rangle$ with each pair of vectors $\mathbf{u}$ and $\mathbf{v}$ in V in such a way that the following axioms are satisfied for all vectors $\mathbf{u}$, $\mathbf{v}$, and $\mathbf{w}$ in V and all scalars k.
>
> (1) $\langle \mathbf{u}, \mathbf{v} \rangle = \overline{\langle \mathbf{v}, \mathbf{u} \rangle}$
>
> (2) $\langle \mathbf{u} + \mathbf{v}, \mathbf{w} \rangle = \langle \mathbf{u}, \mathbf{w} \rangle + \langle \mathbf{v}, \mathbf{w} \rangle$
>
> (3) $\langle k\mathbf{u}, \mathbf{v} \rangle = k\langle \mathbf{u}, \mathbf{v} \rangle$
>
> (4) $\langle \mathbf{v}, \mathbf{v} \rangle \geq 0$ and $\langle \mathbf{v}, \mathbf{v} \rangle = 0$ if and only if $\mathbf{v} = \mathbf{0}$

A complex vector space with an inner product is called a ***complex inner product space*** or a ***unitary space***.

The following additional properties follow immediately from the four inner product axioms:

  (i) $\langle \mathbf{0}, \mathbf{v} \rangle = \langle \mathbf{v}, \mathbf{0} \rangle = 0$

 (ii) $\langle \mathbf{u}, \mathbf{v} + \mathbf{w} \rangle = \langle \mathbf{u}, \mathbf{v} \rangle + \langle \mathbf{u}, \mathbf{w} \rangle$

(iii) $\langle \mathbf{u}, k\mathbf{v} \rangle = \bar{k}\langle \mathbf{u}, \mathbf{v} \rangle$

Since only (iii) differs from the corresponding results for real inner products, we will prove it and leave the other proofs as exercises.

$$
\begin{aligned}
\langle \mathbf{u}, k\mathbf{v} \rangle &= \overline{\langle k\mathbf{v}, \mathbf{u} \rangle} && \text{[Axiom 5]} \\
&= \overline{k\langle \mathbf{v}, \mathbf{u} \rangle} && \text{[Axiom 5]} \\
&= \bar{k}\overline{\langle \mathbf{v}, \mathbf{u} \rangle} && \text{[Property of conjugates]} \\
&= \bar{k}\langle \mathbf{u}, \mathbf{v} \rangle && \text{[Axiom 5]}
\end{aligned}
$$

EXAMPLE 1 Inner Product on C^n

Let $\mathbf{u} = (u_1, u_2, \ldots, u_n)$ and $\mathbf{v} = (v_1, v_2, \ldots, v_n)$ be vectors in C^n. The Euclidean inner product $\langle \mathbf{u}, \mathbf{v} \rangle = \mathbf{u} \cdot \mathbf{v} = u_1 \bar{v}_1 + u_2 \bar{v}_2 + \cdots + u_n \bar{v}_n$ satisfies all the inner product axioms by Theorem 10.4.1. ◆

EXAMPLE 2 Inner Product on Complex M_{22}

If

$$U = \begin{bmatrix} u_1 & u_2 \\ u_3 & u_4 \end{bmatrix} \quad \text{and} \quad V = \begin{bmatrix} v_1 & v_2 \\ v_3 & v_4 \end{bmatrix}$$

are any 2×2 matrices with complex entries, then the following formula defines a complex inner product on complex M_{22} (verify):

$$\langle U, V \rangle = u_1 \bar{v}_1 + u_2 \bar{v}_2 + u_3 \bar{v}_3 + u_4 \bar{v}_4$$

For example, if

$$U = \begin{bmatrix} 0 & i \\ 1 & 1+i \end{bmatrix} \quad \text{and} \quad V = \begin{bmatrix} 1 & -i \\ 0 & 2i \end{bmatrix}$$

then

$$\begin{aligned}
\langle U, V \rangle &= (0)(\bar{1}) + i(\overline{-i}) + (1)(\bar{0}) + (1+i)(\overline{2i}) \\
&= (0)(1) + i(i) + (1)(0) + (1+i)(-2i) \\
&= 0 + i^2 + 0 - 2i - 2i^2 \\
&= 1 - 2i
\end{aligned}$$

◆

Calculus Required

EXAMPLE 3 Inner Product on Complex $C[a, b]$

If $f(x) = f_1(x) + if_2(x)$ is a complex-valued function of the real variable x, and if $f_1(x)$ and $f_2(x)$ are continuous on $[a, b]$, then we define

$$\int_a^b f(x)\, dx = \int_a^b [f_1(x) + if_2(x)]\, dx = \int_a^b f_1(x)\, dx + i \int_a^b f_2(x)\, dx$$

In words, *the integral of $f(x)$ is the integral of the real part of f plus i times the integral of the imaginary part of f.*

We leave it as an exercise to show that if the functions $\mathbf{f} = f_1(x) + if_2(x)$ and $\mathbf{g} = g_1(x) + ig_2(x)$ are vectors in complex $C[a, b]$, then the following formula defines an inner product on complex $C[a, b]$:

$$\begin{aligned}
\langle \mathbf{f}, \mathbf{g} \rangle &= \int_a^b [f_1(x) + if_2(x)]\overline{[g_1(x) + ig_2(x)]}\, dx \\
&= \int_a^b [f_1(x) + if_2(x)][g_1(x) - ig_2(x)]\, dx \\
&= \int_a^b [f_1(x)g_1(x) + f_2(x)g_2(x)]\, dx + i \int_a^b [f_2(x)g_1(x) - f_1(x)g_2(x)]\, dx
\end{aligned}$$

◆

In complex inner product spaces, as in real inner product spaces, the **_norm_** (or **_length_**) of a vector $\mathbf{u}$ is defined by

$$\|\mathbf{u}\| = \langle \mathbf{u}, \mathbf{u} \rangle^{1/2}$$

and the ***distance*** between two vectors **u** and **v** is defined by

$$d(\mathbf{u}, \mathbf{v}) = \|\mathbf{u} - \mathbf{v}\|$$

It can be shown that with these definitions Theorems 6.2.2 and 6.2.3 remain true in complex inner product spaces (Exercise 35).

EXAMPLE 4 Norm and Distance in C^n

If $\mathbf{u} = (u_1, u_2, \ldots, u_n)$ and $\mathbf{v} = (v_1, v_2, \ldots, v_n)$ are vectors in C^n with the Euclidean inner product, then

$$\|\mathbf{u}\| = \langle \mathbf{u}, \mathbf{u} \rangle^{1/2} = \sqrt{|u_1|^2 + |u_2|^2 + \cdots + |u_n|^2}$$

and

$$d(\mathbf{u}, \mathbf{v}) = \|\mathbf{u} - \mathbf{v}\| = \langle \mathbf{u} - \mathbf{v}, \mathbf{u} - \mathbf{v} \rangle^{1/2}$$
$$= \sqrt{|u_1 - v_1|^2 + |u_2 - v_2|^2 + \cdots + |u_n - v_n|^2}$$

Observe that these are just the formulas for the Euclidean norm and distance discussed in Section 10.4. ◆

Calculus Required

EXAMPLE 5 Norm of a Function in Complex $C[0, 2\pi]$

If complex $C[0, 2\pi]$ has the inner product of Example 3, and if $\mathbf{f} = e^{imx}$, where m is any integer, then with the help of Formula (15) of Section 10.3 we obtain

$$\|\mathbf{f}\| = \langle \mathbf{f}, \mathbf{f} \rangle^{1/2} = \left[\int_0^{2\pi} e^{imx}\overline{e^{imx}}\, dx \right]^{1/2}$$
$$= \left[\int_0^{2\pi} e^{imx} e^{-imx}\, dx \right]^{1/2} = \left[\int_0^{2\pi} dx \right]^{1/2} = \sqrt{2\pi} \qquad ◆$$

Orthogonal Sets

The definitions of such terms as ***orthogonal vectors, orthogonal set, orthonormal set***, and ***orthonormal basis*** carry over to unitary spaces without change. Moreover, Theorems 6.2.4, 6.3.1, 6.3.3, 6.3.4, 6.3.5, 6.3.6, and 6.5.4 remain valid in complex inner product spaces, and the Gram–Schmidt process can be used to convert an arbitrary basis for a complex inner product space into an orthonormal basis.

EXAMPLE 6 Orthogonal Vectors in C^2

The vectors

$$\mathbf{u} = (i, 1) \quad \text{and} \quad \mathbf{v} = (1, i)$$

in C^2 are orthogonal with respect to the Euclidean inner product, since

$$\mathbf{u} \cdot \mathbf{v} = (i)(\overline{1}) + (1)(\overline{i}) = (i)(1) + (1)(-i) = 0 \qquad ◆$$

EXAMPLE 7 Constructing an Orthonormal Basis for C^3

Consider the vector space C^3 with the Euclidean inner product. Apply the Gram–Schmidt process to transform the basis vectors $\mathbf{u}_1 = (i, i, i)$, $\mathbf{u}_2 = (0, i, i)$, $\mathbf{u}_3 = (0, 0, i)$ into an orthonormal basis.

Solution.

Step 1. $\mathbf{v}_1 = \mathbf{u}_1 = (i, i, i)$

Step 2. $\mathbf{v}_2 = \mathbf{u}_2 - \mathrm{proj}_{W_1}\mathbf{u}_2 = \mathbf{u}_2 - \dfrac{\langle \mathbf{u}_2, \mathbf{v}_1 \rangle}{\|\mathbf{v}_1\|^2}\mathbf{v}_1$

$$= (0, i, i) - \frac{2}{3}(i, i, i) = \left(-\frac{2}{3}i, \frac{1}{3}i, \frac{1}{3}i\right)$$

Step 3. $\mathbf{v}_3 = \mathbf{u}_3 - \mathrm{proj}_{W_2}\mathbf{u}_3 = \mathbf{u}_3 - \dfrac{\langle \mathbf{u}_3, \mathbf{v}_1 \rangle}{\|\mathbf{v}_1\|^2}\mathbf{v}_1 - \dfrac{\langle \mathbf{u}_3, \mathbf{v}_2 \rangle}{\|\mathbf{v}_2\|^2}\mathbf{v}_2$

$$= (0, 0, i) - \frac{1}{3}(i, i, i) - \frac{1/3}{2/3}\left(-\frac{2}{3}i, \frac{1}{3}i, \frac{1}{3}i\right)$$

$$= \left(0, -\frac{1}{2}i, \frac{1}{2}i\right)$$

Thus,

$$\mathbf{v}_1 = (i, i, i), \qquad \mathbf{v}_2 = \left(-\frac{2}{3}i, \frac{1}{3}i, \frac{1}{3}i\right), \qquad \mathbf{v}_3 = \left(0, -\frac{1}{2}i, \frac{1}{2}i\right)$$

form an orthogonal basis for C^3. The norms of these vectors are

$$\|\mathbf{v}_1\| = \sqrt{3}, \qquad \|\mathbf{v}_2\| = \frac{\sqrt{6}}{3}, \qquad \|\mathbf{v}_3\| = \frac{1}{\sqrt{2}}$$

so an orthonormal basis for C^3 is

$$\frac{\mathbf{v}_1}{\|\mathbf{v}_1\|} = \left(\frac{i}{\sqrt{3}}, \frac{i}{\sqrt{3}}, \frac{i}{\sqrt{3}}\right), \qquad \frac{\mathbf{v}_2}{\|\mathbf{v}_2\|} = \left(-\frac{2i}{\sqrt{6}}, \frac{i}{\sqrt{6}}, \frac{i}{\sqrt{6}}\right),$$

$$\frac{\mathbf{v}_3}{\|\mathbf{v}_3\|} = \left(0, -\frac{i}{\sqrt{2}}, \frac{i}{\sqrt{2}}\right) \qquad\qquad \blacklozenge$$

Calculus Required

EXAMPLE 8 Orthonormal Set in Complex $C[0, 2\pi]$

Let complex $C[0, 2\pi]$ have the inner product of Example 3, and let W be the set of vectors in $C[0, 2\pi]$ of the form

$$e^{imx} = \cos mx + i \sin mx$$

where m is an integer. The set W is orthogonal because if

$$\mathbf{f} = e^{ikx} \quad \text{and} \quad \mathbf{g} = e^{ilx}$$

are distinct vectors in W, then

$$\langle \mathbf{f}, \mathbf{g} \rangle = \int_0^{2\pi} e^{ikx}\overline{e^{ilx}}\, dx = \int_0^{2\pi} e^{ikx}e^{-ilx}\, dx = \int_0^{2\pi} e^{i(k-l)x}\, dx$$

$$= \int_0^{2\pi} \cos(k - l)x\, dx + i \int_0^{2\pi} \sin(k - l)x\, dx$$

$$= \left[\frac{1}{k - l}\sin(k - l)x\right]_0^{2\pi} - i\left[\frac{1}{k - l}\cos(k - l)x\right]_0^{2\pi}$$

$$= (0) - i(0) = 0$$

If we normalize each vector in the orthogonal set W, we obtain an orthonormal set. But in Example 5 we showed that each vector in W has norm $\sqrt{2\pi}$, so the vectors

$$\frac{1}{\sqrt{2\pi}}e^{imx}, \qquad m = 0, \pm 1, \pm 2, \ldots$$

form an orthonormal set in complex $C[0, 2\pi]$. ♦

Exercise Set 10.5

1. Let $\mathbf{u} = (u_1, u_2)$ and $\mathbf{v} = (v_1, v_2)$. Show that $\langle \mathbf{u}, \mathbf{v} \rangle = 3u_1\bar{v}_1 + 2u_2\bar{v}_2$ defines an inner product on C^2.

2. Compute $\langle \mathbf{u}, \mathbf{v} \rangle$ using the inner product in Exercise 1.

    (a) $\mathbf{u} = (2i, -i)$, $\mathbf{v} = (-i, 3i)$ (b) $\mathbf{u} = (0, 0)$, $\mathbf{v} = (1 - i, 7 - 5i)$
    (c) $\mathbf{u} = (1 + i, 1 - i)$, $\mathbf{v} = (1 - i, 1 + i)$ (d) $\mathbf{u} = (3i, -1 + 2i)$, $\mathbf{v} = (3i, -1 + 2i)$

3. Let $\mathbf{u} = (u_1, u_2)$ and $\mathbf{v} = (v_1, v_2)$. Show that

$$\langle \mathbf{u}, \mathbf{v} \rangle = u_1\bar{v}_1 + (1 + i)u_1\bar{v}_2 + (1 - i)u_2\bar{v}_1 + 3u_2\bar{v}_2$$

    defines an inner product on C^2.

4. Compute $\langle \mathbf{u}, \mathbf{v} \rangle$ using the inner product in Exercise 3.

    (a) $\mathbf{u} = (2i, -i)$, $\mathbf{v} = (-i, 3i)$ (b) $\mathbf{u} = (0, 0)$, $\mathbf{v} = (1 - i, 7 - 5i)$
    (c) $\mathbf{u} = (1 + i, 1 - i)$, $\mathbf{v} = (1 - i, 1 + i)$ (d) $\mathbf{u} = (3i, -1 + 2i)$, $\mathbf{v} = (3i, -1 + 2i)$

5. Let $\mathbf{u} = (u_1, u_2)$ and $\mathbf{v} = (v_1, v_2)$. Determine which of the following are inner products on C^2. For those that are not, list the axioms that do not hold.

    (a) $\langle \mathbf{u}, \mathbf{v} \rangle = u_1\bar{v}_1$ (b) $\langle \mathbf{u}, \mathbf{v} \rangle = u_1\bar{v}_1 - u_2\bar{v}_2$ (c) $\langle \mathbf{u}, \mathbf{v} \rangle = |u_1|^2|v_1|^2 + |u_2|^2|v_2|^2$
    (d) $\langle \mathbf{u}, \mathbf{v} \rangle = 2u_1\bar{v}_1 + iu_1\bar{v}_2 + iu_2\bar{v}_1 + 2u_2\bar{v}_2$ (e) $\langle \mathbf{u}, \mathbf{v} \rangle = 2u_1\bar{v}_1 + iu_1\bar{v}_2 - iu_2\bar{v}_1 + 2u_2\bar{v}_2$

6. Use the inner product of Example 2 to find $\langle U, V \rangle$ if

$$U = \begin{bmatrix} -i & 1 + i \\ 1 - i & i \end{bmatrix} \quad \text{and} \quad V = \begin{bmatrix} 3 & -2 - 3i \\ 4i & 1 \end{bmatrix}$$

7. Let $\mathbf{u} = (u_1, u_2, u_3)$ and $\mathbf{v} = (v_1, v_2, v_3)$. Does $\langle \mathbf{u}, \mathbf{v} \rangle = u_1\bar{v}_1 + u_2\bar{v}_2 + u_3\bar{v}_3 - iu_3\bar{v}_1$ define an inner product on C^3? If not, list all axioms that fail to hold.

8. Let V be the vector space of complex-valued functions of the real variable x, and let $\mathbf{f} = f_1(x) + if_2(x)$ and $\mathbf{g} = g_1(x) + ig_2(x)$ be vectors in V. Does

$$\langle \mathbf{f}, \mathbf{g} \rangle = (f_1(0) + if_2(0))\overline{(g_1(0) + ig_2(0))}$$

    define an inner product on V? If not, list all axioms that fail to hold.

9. Let C^2 have the inner product of Exercise 1. Find $\|\mathbf{w}\|$ if

    (a) $\mathbf{w} = (-i, 3i)$ (b) $\mathbf{w} = (1 - i, 1 + i)$ (c) $\mathbf{w} = (0, 2 - i)$ (d) $\mathbf{w} = (0, 0)$

10. For each vector in Exercise 9, use the Euclidean inner product to find $\|\mathbf{w}\|$.

11. Use the inner product of Exercise 3 to find $\|\mathbf{w}\|$ if

    (a) $\mathbf{w} = (1, -i)$ (b) $\mathbf{w} = (1 - i, 1 + i)$ (c) $\mathbf{w} = (3 - 4i, 0)$ (d) $\mathbf{w} = (0, 0)$

12. Use the inner product of Example 2 to find $\|A\|$ if

    (a) $A = \begin{bmatrix} -i & 7i \\ 6i & 2i \end{bmatrix}$ (b) $A = \begin{bmatrix} -1 & 1 + i \\ 1 - i & 3 \end{bmatrix}$

13. Let C^2 have the inner product of Exercise 1. Find $d(\mathbf{x}, \mathbf{y})$ if

    (a) $\mathbf{x} = (1, 1)$, $\mathbf{y} = (i, -i)$ (b) $\mathbf{x} = (1 - i, 3 + 2i)$, $\mathbf{y} = (1 + i, 3)$

14. Repeat the directions of Exercise 13 using the Euclidean inner product on C^2.

15. Repeat the directions of Exercise 13 using the inner product of Exercise 3.

16. Let complex M_{22} have the inner product of Example 2. Find $d(A, B)$ if

(a) $A = \begin{bmatrix} i & 5i \\ 8i & 3i \end{bmatrix}$ and $B = \begin{bmatrix} -5i & 0 \\ 7i & -3i \end{bmatrix}$ (b) $A = \begin{bmatrix} -1 & 1-i \\ 1+i & 2 \end{bmatrix}$ and $B = \begin{bmatrix} 2i & 2-3i \\ i & 1 \end{bmatrix}$

17. Let C^3 have the Euclidean inner product. For which complex values of k are $\mathbf{u}$ and $\mathbf{v}$ orthogonal?

(a) $\mathbf{u} = (2i, i, 3i)$, $\mathbf{v} = (i, 6i, k)$ (b) $\mathbf{u} = (k, k, 1+i)$, $\mathbf{v} = (1, -1, 1-i)$

18. Let M_{22} have the inner product of Example 2. Determine which of the following are orthogonal to

$$A = \begin{bmatrix} 2i & i \\ -i & 3i \end{bmatrix}$$

(a) $\begin{bmatrix} -3 & 1-i \\ 1-i & 2 \end{bmatrix}$ (b) $\begin{bmatrix} 1 & 1 \\ 0 & -1 \end{bmatrix}$ (c) $\begin{bmatrix} 0 & 0 \\ 0 & 0 \end{bmatrix}$ (d) $\begin{bmatrix} 0 & 1 \\ 3-i & 0 \end{bmatrix}$

19. Let C^3 have the Euclidean inner product. Show that for all values of the variable θ the vector
$\mathbf{x} = e^{i\theta} \left(\dfrac{i}{\sqrt{3}}, \dfrac{1}{\sqrt{3}}, \dfrac{1}{\sqrt{3}} \right)$ has norm 1 and is orthogonal to both $(1, i, 0)$ and $(0, i, -i)$.

20. Let C^2 have the Euclidean inner product. Which of the following form orthonormal sets?

(a) $(i, 0), (0, 1-i)$ (b) $\left(\dfrac{i}{\sqrt{2}}, -\dfrac{i}{\sqrt{2}} \right), \left(\dfrac{i}{\sqrt{2}}, \dfrac{i}{\sqrt{2}} \right)$ (c) $\left(\dfrac{i}{\sqrt{2}}, \dfrac{i}{\sqrt{2}} \right), \left(-\dfrac{i}{\sqrt{2}}, -\dfrac{i}{\sqrt{2}} \right)$ (d) $(i, 0), (0, 0)$

21. Let C^3 have the Euclidean inner product. Which of the following form orthonormal sets?

(a) $\left(\dfrac{i}{\sqrt{2}}, 0, \dfrac{i}{\sqrt{2}} \right), \left(\dfrac{i}{\sqrt{3}}, \dfrac{i}{\sqrt{3}}, -\dfrac{i}{\sqrt{3}} \right), \left(-\dfrac{i}{\sqrt{2}}, 0, \dfrac{i}{\sqrt{2}} \right)$ (b) $\left(\dfrac{2}{3}i, -\dfrac{2}{3}i, \dfrac{1}{3}i \right), \left(\dfrac{2}{3}i, \dfrac{1}{3}i, -\dfrac{2}{3}i \right), \left(\dfrac{1}{3}i, \dfrac{2}{3}i, \dfrac{2}{3}i \right)$

(c) $\left(\dfrac{i}{\sqrt{6}}, \dfrac{i}{\sqrt{6}}, -\dfrac{2i}{\sqrt{6}} \right), \left(\dfrac{i}{\sqrt{2}}, -\dfrac{i}{\sqrt{2}}, 0 \right)$

22. Let
$$\mathbf{x} = \left(\dfrac{i}{\sqrt{5}}, -\dfrac{i}{\sqrt{5}} \right) \quad \text{and} \quad \mathbf{y} = \left(\dfrac{2i}{\sqrt{30}}, \dfrac{3i}{\sqrt{30}} \right)$$

Show that $\{\mathbf{x}, \mathbf{y}\}$ is an orthonormal set if C^2 has the inner product
$$\langle \mathbf{u}, \mathbf{v} \rangle = 3u_1 \bar{v}_1 + 2u_2 \bar{v}_2$$
but is not orthonormal if C^2 has the Euclidean inner product.

23. Show that
$$\mathbf{u}_1 = (i, 0, 0, i), \quad \mathbf{u}_2 = (-i, 0, 2i, i), \quad \mathbf{u}_3 = (2i, 3i, 2i, -2i), \quad \mathbf{u}_4 = (-i, 2i, -i, i)$$
is an orthogonal set in C^4 with the Euclidean inner product. By normalizing each of these vectors, obtain an orthonormal set.

24. Let C^2 have the Euclidean inner product. Use the Gram–Schmidt process to transform the basis $\{\mathbf{u}_1, \mathbf{u}_2\}$ into an orthonormal basis.

(a) $\mathbf{u}_1 = (i, -3i)$, $\mathbf{u}_2 = (2i, 2i)$ (b) $\mathbf{u}_1 = (i, 0)$, $\mathbf{u}_2 = (3i, -5i)$

25. Let C^3 have the Euclidean inner product. Use the Gram–Schmidt process to transform the basis $\{\mathbf{u}_1, \mathbf{u}_2, \mathbf{u}_3\}$ into an orthonormal basis.

(a) $\mathbf{u}_1 = (i, i, i)$, $\mathbf{u}_2 = (-i, i, 0)$, $\mathbf{u}_3 = (i, 2i, i)$ (b) $\mathbf{u}_1 = (i, 0, 0)$, $\mathbf{u}_2 = (3i, 7i, -2i)$, $\mathbf{u}_3 = (0, 4i, i)$

26. Let C^4 have the Euclidean inner product. Use the Gram–Schmidt process to transform the basis $\{\mathbf{u}_1, \mathbf{u}_2, \mathbf{u}_3, \mathbf{u}_4\}$ into an orthonormal basis.

$$\mathbf{u}_1 = (0, 2i, i, 0), \quad \mathbf{u}_2 = (i, -i, 0, 0), \quad \mathbf{u}_3 = (i, 2i, 0, -i), \quad \mathbf{u}_4 = (i, 0, i, i)$$

27. Let C^3 have the Euclidean inner product. Find an orthonormal basis for the subspace spanned by $(0, i, 1-i)$ and $(-i, 0, 1+i)$.

28. Let C^4 have the Euclidean inner product. Express the vector $\mathbf{w} = (-i, 2i, 6i, 0)$ in the form $\mathbf{w} = \mathbf{w}_1 + \mathbf{w}_2$, where the vector $\mathbf{w}_1$ is in the space W spanned by $\mathbf{u}_1 = (-i, 0, i, 2i)$ and $\mathbf{u}_2 = (0, i, 0, i)$, and $\mathbf{w}_2$ is orthogonal to W.

29. (a) Prove: If k is a complex number and $\langle \mathbf{u}, \mathbf{v} \rangle$ is an inner product on a complex vector space, then $\langle \mathbf{u} - k\mathbf{v}, \mathbf{u} - k\mathbf{v} \rangle = \langle \mathbf{u}, \mathbf{u} \rangle - \overline{k}\langle \mathbf{u}, \mathbf{v} \rangle - k\overline{\langle \mathbf{u}, \mathbf{v} \rangle} + k\overline{k}\langle \mathbf{v}, \mathbf{v} \rangle$.
    (b) Use the result in (1) to prove that $0 \le \langle \mathbf{u}, \mathbf{u} \rangle - \overline{k}\langle \mathbf{u}, \mathbf{v} \rangle - k\overline{\langle \mathbf{u}, \mathbf{v} \rangle} + k\overline{k}\langle \mathbf{v}, \mathbf{v} \rangle$.

30. Prove that if $\mathbf{u}$ and $\mathbf{v}$ are vectors in a complex inner product space, then

$$|\langle \mathbf{u}, \mathbf{v} \rangle|^2 \le \langle \mathbf{u}, \mathbf{u} \rangle \langle \mathbf{v}, \mathbf{v} \rangle$$

This result, called the ***Cauchy–Schwarz inequality for complex inner product spaces***, differs from its real analog (Theorem 6.2.1) in that an absolute value sign must be included on the left side. [***Hint.*** Let $k = \langle \mathbf{u}, \mathbf{v} \rangle / \langle \mathbf{v}, \mathbf{v} \rangle$ in the inequality of Exercise 29(b).]

31. Prove: If $\mathbf{u} = (u_1, u_2, \ldots, u_n)$ and $\mathbf{v} = (v_1, v_2, \ldots, v_n)$ are vectors in C^n, then

$$|u_1\overline{v}_1 + u_2\overline{v}_2 + \cdots + u_n\overline{v}_n| \le (|u_1|^2 + |u_2|^2 + \cdots + |u_n|^2)^{1/2}(|v_1|^2 + |v_2|^2 + \cdots + |v_n|^2)^{1/2}$$

This is the complex version of Formula (4) in Theorem 4.1.3. [***Hint.*** Use Exercise 30.]

32. Prove that equality holds in the Cauchy–Schwarz inequality for complex vector spaces if and only if $\mathbf{u}$ and $\mathbf{v}$ are linearly dependent.

33. Prove that if $\langle \mathbf{u}, \mathbf{v} \rangle$ is an inner product on a complex vector space, then

$$\langle \mathbf{0}, \mathbf{v} \rangle = \langle \mathbf{v}, \mathbf{0} \rangle = 0$$

34. Prove that if $\langle \mathbf{u}, \mathbf{v} \rangle$ is an inner product on a complex vector space, then

$$\langle \mathbf{u}, \mathbf{v} + \mathbf{w} \rangle = \langle \mathbf{u}, \mathbf{v} \rangle + \langle \mathbf{u}, \mathbf{w} \rangle$$

35. Theorems 6.2.2 and 6.2.3 remain true in complex inner product spaces. In each part prove that this is so.
    (a) Theorem 6.2.2*a*   (b) Theorem 6.2.2*b*   (c) Theorem 6.2.2*c*   (d) Theorem 6.2.2*d*
    (e) Theorem 6.2.3*a*   (f) Theorem 6.2.3*b*   (g) Theorem 6.2.3*c*   (h) Theorem 6.2.3*d*

36. In Example 7 it was shown that the vectors

$$\mathbf{v}_1 = \left(\frac{i}{\sqrt{3}}, \frac{i}{\sqrt{3}}, \frac{i}{\sqrt{3}}\right), \quad \mathbf{v}_2 = \left(-\frac{2i}{\sqrt{6}}, \frac{i}{\sqrt{6}}, \frac{i}{\sqrt{6}}\right), \quad \mathbf{v}_3 = \left(0, -\frac{i}{\sqrt{2}}, \frac{i}{\sqrt{2}}\right)$$

form an orthonormal basis for C^3. Use Theorem 6.3.1 to express $\mathbf{u} = (1 - i, 1 + i, 1)$ as a linear combination of these vectors.

37. Prove that if $\mathbf{u}$ and $\mathbf{v}$ are vectors in a complex inner product space, then

$$\langle \mathbf{u}, \mathbf{v} \rangle = \frac{1}{4}\|\mathbf{u} + \mathbf{v}\|^2 - \frac{1}{4}\|\mathbf{u} - \mathbf{v}\|^2 + \frac{i}{4}\|\mathbf{u} + i\mathbf{v}\|^2 - \frac{i}{4}\|\mathbf{u} - i\mathbf{v}\|^2$$

38. Prove: If $\{\mathbf{v}_1, \mathbf{v}_2, \ldots, \mathbf{v}_n\}$ is an orthonormal basis for a complex inner product space V, and if $\mathbf{u}$ and $\mathbf{w}$ are any vectors in V, then

$$\langle \mathbf{u}, \mathbf{w} \rangle = \langle \mathbf{u}, \mathbf{v}_1 \rangle \overline{\langle \mathbf{w}, \mathbf{v}_1 \rangle} + \langle \mathbf{u}, \mathbf{v}_2 \rangle \overline{\langle \mathbf{w}, \mathbf{v}_2 \rangle} + \cdots + \langle \mathbf{u}, \mathbf{v}_n \rangle \overline{\langle \mathbf{w}, \mathbf{v}_n \rangle}$$

[***Hint.*** Use Theorem 6.3.1 to express $\mathbf{u}$ and $\mathbf{w}$ as linear combinations of the basis vectors.]

39. (***For readers who have studied calculus.***) Prove that if $\mathbf{f} = f_1(x) + if_2(x)$ and $\mathbf{g} = g_1(x) + ig_2(x)$ are vectors in complex $C[a, b]$, then the formula

$$\langle \mathbf{f}, \mathbf{g} \rangle = \int_a^b [f_1(x) + if_2(x)]\overline{[g_1(x) + ig_2(x)]}\, dx$$

defines a complex inner product on $C[a, b]$.

40. (***For readers who have studied calculus.***) Let $\mathbf{f} = x$ and $\mathbf{g} = 1 + ix$ be vectors in complex $C[0, 1]$ and let this space have the inner product defined in Exercise 39. Find
    (a) $\|\mathbf{g}\|$ (b) $\langle \mathbf{f}, \mathbf{g} \rangle$ (c) $\langle \mathbf{g}, \mathbf{f} \rangle$

41. (*For readers who have studied calculus.*) Let $\mathbf{f} = f_1(x) + if_2(x)$ and $\mathbf{g} = g_1(x) + ig_2(x)$ be vectors in complex $C[0, 1]$ and let this space have the inner product defined in Exercise 39. Show that the vectors $e^{2\pi imx}$, where $m = 0, \pm 1, \pm 2, \ldots$, form an orthonormal set.

10.6 UNITARY, NORMAL, AND HERMITIAN MATRICES

For matrices with real entries, the orthogonal matrices $(A^{-1} = A^T)$ and the symmetric matrices $(A = A^T)$ played an important role in the orthogonal diagonalization problem (Section 7.3). For matrices with complex entries, the orthogonal and symmetric matrices are of relatively little importance; they are superseded by two new classes of matrices, the unitary and Hermitian matrices, which we shall discuss in this section.

Unitary Matrices If A is a matrix with complex entries, then the ***conjugate transpose*** of A, denoted by A^*, is defined by

$$A^* = \overline{A}^T$$

where $\overline{A}$ is the matrix whose entries are the complex conjugates of the corresponding entries in A and $\overline{A}^T$ is the transpose of $\overline{A}$.

EXAMPLE 1 Conjugate Transpose

If

$$A = \begin{bmatrix} 1 + i & -i & 0 \\ 2 & 3 - 2i & i \end{bmatrix}, \quad \text{then} \quad \overline{A} = \begin{bmatrix} 1 - i & i & 0 \\ 2 & 3 + 2i & -i \end{bmatrix}$$

so

$$A^* = \overline{A}^T = \begin{bmatrix} 1 - i & 2 \\ i & 3 + 2i \\ 0 & -i \end{bmatrix} \qquad \blacklozenge$$

The following theorem shows that the basic properties of the conjugate transpose are similar to those of the transpose. The proofs are left as exercises.

Theorem 10.6.1 **Properties of the Conjugate Transpose**

If A and B are matrices with complex entries and k is any complex number, then:

(a) $(A^*)^* = A$ (b) $(A + B)^* = A^* + B^*$
(c) $(kA)^* = \overline{k}A^*$ (d) $(AB)^* = B^*A^*$

REMARK. Recall from Formula (7) of Section 4.1 that if $\mathbf{u}$ and $\mathbf{v}$ are column vectors in R^n, then the Euclidean inner product on R^n can be expressed as $\mathbf{u} \cdot \mathbf{v} = \mathbf{u}^T\mathbf{v}$. We leave it for you to confirm that if $\mathbf{u}$ and $\mathbf{v}$ are column vectors in C^n, then the Euclidean inner product on C^n can be expressed as $\mathbf{u} \cdot \mathbf{v} = \mathbf{u}^*\mathbf{v}$.

Recall that a matrix with real entries is called *orthogonal* if $A^{-1} = A^T$. The complex analogs of the orthogonal matrices are called *unitary* matrices. They are defined as follows:

Definition

A square matrix A with complex entries is called **unitary** if

$$A^{-1} = A^*$$

The following theorem parallels Theorem 6.5.1.

Theorem 10.6.2   **Equivalent Statements**

If A is an $n \times n$ matrix with complex entries, then the following are equivalent.

(*a*) *A is unitary.*

(*b*) *The row vectors of A form an orthonormal set in C^n with the Euclidean inner product.*

(*c*) *The column vectors of A form an orthonormal set in C^n with the Euclidean inner product.*

EXAMPLE 2 **A 2 × 2 Unitary Matrix**

The matrix

$$A = \begin{bmatrix} \dfrac{1+i}{2} & \dfrac{1+i}{2} \\[2mm] \dfrac{1-i}{2} & \dfrac{1+i}{2} \end{bmatrix} \tag{1}$$

has row vectors

$$\mathbf{r}_1 = \left(\frac{1+i}{2}, \frac{1+i}{2} \right), \qquad \mathbf{r}_2 = \left(\frac{1-i}{2}, \frac{-1+i}{2} \right)$$

Relative to the Euclidean inner product on C^n, we have

$$\|\mathbf{r}_1\| = \sqrt{\left| \frac{1+i}{2} \right|^2 + \left| \frac{1+i}{2} \right|^2} = \sqrt{\frac{1}{2} + \frac{1}{2}} = 1$$

$$\|\mathbf{r}_2\| = \sqrt{\left| \frac{1-i}{2} \right|^2 + \left| \frac{-1+i}{2} \right|^2} = \sqrt{\frac{1}{2} + \frac{1}{2}} = 1$$

and

$$\mathbf{r}_1 \cdot \mathbf{r}_2 = \left(\frac{1+i}{2} \right) \overline{\left(\frac{1-i}{2} \right)} + \left(\frac{1+i}{2} \right) \overline{\left(\frac{-1+i}{2} \right)}$$

$$= \left(\frac{1+i}{2} \right) \left(\frac{1+i}{2} \right) + \left(\frac{1+i}{2} \right) \left(\frac{-1-i}{2} \right)$$

$$= \frac{i}{2} - \frac{i}{2} = 0$$

so the row vectors form an orthonormal set in C^2. Thus, A is unitary and

$$A^{-1} = A^* = \begin{bmatrix} \dfrac{1-i}{2} & \dfrac{1+i}{2} \\ \dfrac{1-i}{2} & \dfrac{-1-i}{2} \end{bmatrix} \tag{2}$$

The reader should verify that matrix (2) is the inverse of matrix (1) by showing that $AA^* = A^*A = I$. ◆

Charles Hermite (1822–1901) was a French mathematician who made fundamental contributions to algebra, matrix theory, and various branches of analysis. He is noted for using integrals to solve a general fifth-degree polynomial equation. He also proved that the number e (the base for natural logarithms) is a transcendental number, that is, a number that is not the root of any polynomial equation with rational coefficients.

Recall that a square matrix A with real entries is called *orthogonally diagonalizable* if there is an orthogonal matrix P such that $P^{-1}AP \, (= P^TAP)$ is diagonal. For complex matrices we have an analogous concept.

> **Definition**
>
> A square matrix A with complex entries is called *unitarily diagonalizable* if there is a unitary P such that $P^{-1}AP \, (= P^*AP)$ is diagonal; the matrix P is said to *unitarily diagonalize* A.

We have two questions to consider:

- Which matrices are unitarily diagonalizable?
- How do we find a unitary matrix P to carry out the diagonalization?

Before pursuing these questions, we note that our earlier definitions of the terms *eigenvector, eigenvalue, eigenspace, characteristic equation,* and *characteristic polynomial* carry over without change to complex vector spaces.

Hermitian Matrices   In Section 7.3 we saw that the symmetric matrices played a fundamental role in the problem of orthogonally diagonalizing a matrix with real entries. The most natural complex analogs of the real symmetric matrices are the *Hermitian* matrices, which are defined as follows:

> **Definition**
>
> A square matrix A with complex entries is called *Hermitian* if
> $$A = A^*$$

EXAMPLE 3 A 3 × 3 Hermitian Matrix

If

$$A = \begin{bmatrix} 1 & i & 1+i \\ -i & -5 & 2-i \\ 1-i & 2+i & 3 \end{bmatrix}$$

then

$$\overline{A} = \begin{bmatrix} 1 & -i & 1-i \\ i & -5 & 2+i \\ 1+i & 2-i & 3 \end{bmatrix}, \quad \text{so} \quad A^* = \overline{A}^T = \begin{bmatrix} 1 & i & 1+i \\ -i & -5 & 2-i \\ 1-i & 2+i & 3 \end{bmatrix} = A$$

which means that A is Hermitian. ◆

It is easy to recognize Hermitian matrices by inspection: As seen in (3) the entries on the main diagonal are real numbers and the "mirror image" of each entry across the main diagonal is its complex conjugate.

$$\begin{bmatrix} 1 & i & 1+i \\ -i & -5 & 2-i \\ 1-i & 2+i & 3 \end{bmatrix} \tag{3}$$

Normal Matrices Hermitian matrices enjoy many but not all of the properties of real symmetric matrices. For example, just as the real symmetric matrices are orthogonally diagonalizable, so we shall see that the Hermitian matrices are unitarily diagonalizable. However, whereas the real symmetric matrices are the only matrices with real entries that are orthogonally diagonalizable (Theorem 7.3.1), the Hermitian matrices do not constitute the entire class of unitarily diagonalizable matrices; that is, there are unitarily diagonalizable matrices that are not Hermitian. To explain why this is so we shall need the following definition:

Definition

A square matrix A with complex entries is called **normal** if

$$AA^* = A^*A$$

EXAMPLE 4 Hermitian and Unitary Matrices

Every Hermitian matrix A is normal since $AA^* = AA = A^*A$, and every unitary matrix A is normal since $AA^* = I = A^*A$. ◆

The following two theorems are the complex analogs of Theorems 7.3.1 and 7.3.2. The proofs will be omitted.

Theorem 10.6.3 **Equivalent Statements**

If A is a square matrix with complex entries, then the following are equivalent:

(a) *A is unitarily diagonalizable.*
(b) *A has an orthonormal set of n eigenvectors.*
(c) *A is normal.*

Theorem 10.6.4

If A is a normal matrix, then eigenvectors from different eigenspaces of A are orthogonal.

Theorem 10.6.3 tells us that a square matrix A with complex entries is unitarily diagonalizable if and only if it is normal. Theorem 10.6.4 will be the key to constructing a matrix that unitarily diagonalizes a normal matrix.

Diagonalization Procedure We saw in Section 7.3 that a symmetric matrix A is orthogonally diagonalized by any orthogonal matrix whose column vectors are eigenvectors of A. Similarly, a normal matrix A is diagonalized by any unitary matrix whose column vectors are eigenvectors of A. The procedure for diagonalizing a normal matrix is as follows:

Step 1. Find a basis for each eigenspace of A.

Step 2. Apply the Gram–Schmidt process to each of these bases to obtain an orthonormal basis for each eigenspace.

Step 3. Form the matrix P whose columns are the basis vectors constructed in Step 2. This matrix unitarily diagonalizes A.

The justification of this procedure should be clear. Theorem 10.6.4 ensures that eigenvectors from *different* eigenspaces are orthogonal, and the application of the Gram–Schmidt process ensures that the eigenvectors within the *same* eigenspace are orthonormal. Thus, the *entire* set of eigenvectors obtained by this procedure is orthonormal. Theorem 10.6.3 ensures that this orthonormal set of eigenvectors is a basis.

EXAMPLE 5 Unitary Diagonalization

The matrix

$$A = \begin{bmatrix} 2 & 1+i \\ 1-i & 3 \end{bmatrix}$$

is unitarily diagonalizable because it is Hermitian and therefore normal. Find a matrix P that unitarily diagonalizes A.

Solution.

The characteristic polynomial of A is

$$\det(\lambda I - A) = \det \begin{bmatrix} \lambda - 2 & -1-i \\ -1+i & \lambda - 3 \end{bmatrix} = (\lambda - 2)(\lambda - 3) - 2 = \lambda^2 - 5\lambda + 4$$

so the characteristic equation is

$$\lambda^2 - 5\lambda + 4 = (\lambda - 1)(\lambda - 4) = 0$$

and the eigenvalues are $\lambda = 1$ and $\lambda = 4$.

By definition,

$$\mathbf{x} = \begin{bmatrix} x_1 \\ x_2 \end{bmatrix}$$

will be an eigenvector of A corresponding to λ if and only if $\mathbf{x}$ is a nontrivial solution of

$$\begin{bmatrix} \lambda - 2 & -1-i \\ -1+i & \lambda - 3 \end{bmatrix} \begin{bmatrix} x_1 \\ x_2 \end{bmatrix} = \begin{bmatrix} 0 \\ 0 \end{bmatrix} \tag{4}$$

To find the eigenvectors corresponding to $\lambda = 1$, we substitute this value in (4):

$$\begin{bmatrix} -1 & -1-i \\ -1+i & -2 \end{bmatrix} \begin{bmatrix} x_1 \\ x_2 \end{bmatrix} = \begin{bmatrix} 0 \\ 0 \end{bmatrix}$$

Solving this system by Gauss–Jordan elimination yields (verify)

$$x_1 = (-1-i)s, \qquad x_2 = s$$

Thus, the eigenvectors of A corresponding to $\lambda = 1$ are the nonzero vectors in C^2 of the form

$$\mathbf{x} = \begin{bmatrix} (-1-i)s \\ s \end{bmatrix} = s \begin{bmatrix} -1-i \\ 1 \end{bmatrix}$$

Thus, this eigenspace is one-dimensional with basis

$$\mathbf{u} = \begin{bmatrix} -1-i \\ 1 \end{bmatrix} \tag{5}$$

In this case the Gram–Schmidt process involves only one step: normalizing this vector. Since

$$\|\mathbf{u}\| = \sqrt{|-1-i|^2 + |1|^2} = \sqrt{2+1} = \sqrt{3}$$

the vector

$$\mathbf{p}_1 = \frac{\mathbf{u}}{\|\mathbf{u}\|} = \begin{bmatrix} \dfrac{-1-i}{\sqrt{3}} \\ \dfrac{1}{\sqrt{3}} \end{bmatrix}$$

is an orthonormal basis for the eigenspace corresponding to $\lambda = 1$.

To find the eigenvectors corresponding to $\lambda = 4$, we substitute this value in (4):

$$\begin{bmatrix} 2 & -1-i \\ -1+i & 1 \end{bmatrix} \begin{bmatrix} x_1 \\ x_2 \end{bmatrix} = \begin{bmatrix} 0 \\ 0 \end{bmatrix}$$

Solving this system by Gauss–Jordan elimination yields (verify)

$$x_1 = \left(\frac{1+i}{2}\right)s, \qquad x_2 = s$$

so the eigenvectors of A corresponding to $\lambda = 4$ are the nonzero vectors in C^2 of the form

$$\mathbf{x} = \begin{bmatrix} \left(\dfrac{1+i}{2}\right)s \\ s \end{bmatrix} = s \begin{bmatrix} \dfrac{1+i}{2} \\ 1 \end{bmatrix}$$

Thus, the eigenspace is one-dimensional with basis

$$\mathbf{u} = \begin{bmatrix} \dfrac{1+i}{2} \\ 1 \end{bmatrix}$$

Applying the Gram–Schmidt process (i.e., normalizing this vector) yields

$$\mathbf{p}_2 = \frac{\mathbf{u}}{\|\mathbf{u}\|} = \begin{bmatrix} \dfrac{1+i}{\sqrt{6}} \\ \dfrac{2}{\sqrt{6}} \end{bmatrix}$$

Thus,

$$P = [\mathbf{p}_1 \mid \mathbf{p}_2] = \begin{bmatrix} \dfrac{-1-i}{\sqrt{3}} & \dfrac{1+i}{\sqrt{6}} \\ \dfrac{1}{\sqrt{3}} & \dfrac{2}{\sqrt{6}} \end{bmatrix}$$

diagonalizes A and

$$P^{-1}AP = \begin{bmatrix} 1 & 0 \\ 0 & 4 \end{bmatrix} \qquad ♦$$

Eigenvalues of Hermitian and Symmetric Matrices In Theorem 7.3.2 it was stated that the eigenvalues of a symmetric matrix with real entries are real numbers. This important result is a corollary of the following more general theorem.

Theorem 10.6.5

The eigenvalues of a Hermitian matrix are real numbers.

Proof. If λ is an eigenvalue and $\mathbf{v}$ a corresponding eigenvector of an $n \times n$ Hermitian matrix A, then

$$A\mathbf{v} = \lambda \mathbf{v}$$

If we multiply each side of this equation on the left by $\mathbf{v}^*$ and then use the remark following Theorem 10.6.1 to write $\mathbf{v}^*\mathbf{v} = \|\mathbf{v}\|^2$ (with the Euclidean inner product on C^n), then we obtain

$$\mathbf{v}^*A\mathbf{v} = \mathbf{v}^*(\lambda\mathbf{v}) = \lambda\mathbf{v}^*\mathbf{v} = \lambda\|\mathbf{v}\|^2$$

But if we agree not to distinguish between the 1×1 matrix $\mathbf{v}^*A\mathbf{v}$ and its entry, and if we use the fact that eigenvectors are nonzero, then we can express λ as

$$\lambda = \frac{\mathbf{v}^*A\mathbf{v}}{\|\mathbf{v}\|^2} \tag{6}$$

Thus, to show that λ is a real number it suffices to show that the entry of $\mathbf{v}^*A\mathbf{v}$ is real. One way to do this is to show that the matrix $\mathbf{v}^*A\mathbf{v}$ is Hermitian, since we know that Hermitian matrices have real numbers on the main diagonal. But,

$$(\mathbf{v}^*A\mathbf{v})^* = \mathbf{v}^*A^*(\mathbf{v}^*)^* = \mathbf{v}^*A\mathbf{v}$$

which shows that $\mathbf{v}^*A\mathbf{v}$ is Hermitian and completes the proof. ∎

The proof of the following theorem is an immediate consequence of Theorem 10.6.5 and is left as an exercise.

Theorem 10.6.6

The eigenvalues of a symmetric matrix with real entries are real numbers.

Exercise Set 10.6

1. In each part find A^*.

(a) $A = \begin{bmatrix} 2i & 1-i \\ 4 & 3+i \\ 5+i & 0 \end{bmatrix}$
(b) $A = \begin{bmatrix} 2i & 1-i & -1+i \\ 4 & 5-7i & -i \\ i & 3 & 1 \end{bmatrix}$
(c) $A = \begin{bmatrix} 7i & 0 & -3i \end{bmatrix}$
(d) $A = \begin{bmatrix} a_{11} & a_{12} & a_{13} \\ a_{21} & a_{22} & a_{23} \end{bmatrix}$

2. Which of the following are Hermitian matrices?

(a) $\begin{bmatrix} 0 & i \\ i & 2 \end{bmatrix}$
(b) $\begin{bmatrix} 1 & 1+i \\ 1-i & -3 \end{bmatrix}$
(c) $\begin{bmatrix} i & i \\ -i & i \end{bmatrix}$
(d) $\begin{bmatrix} -2 & 1-i & -1+i \\ 1+i & 0 & 3 \\ -1-i & 3 & 5 \end{bmatrix}$
(e) $\begin{bmatrix} 1 & 0 & 0 \\ 0 & 1 & 0 \\ 0 & 0 & 1 \end{bmatrix}$

3. Find k, l, and m to make A a Hermitian matrix.

$$A = \begin{bmatrix} -1 & k & -i \\ 3-5i & 0 & m \\ l & 2+4i & 2 \end{bmatrix}$$

4. Use Theorem 10.6.2 to determine which of the following are unitary matrices.

(a) $\begin{bmatrix} i & 0 \\ 0 & i \end{bmatrix}$
(b) $\begin{bmatrix} \dfrac{i}{\sqrt{2}} & \dfrac{1}{\sqrt{2}} \\ -\dfrac{i}{\sqrt{2}} & \dfrac{1}{\sqrt{2}} \end{bmatrix}$
(c) $\begin{bmatrix} 1+i & 1+i \\ 1-i & -1+i \end{bmatrix}$
(d) $\begin{bmatrix} -\dfrac{i}{\sqrt{2}} & \dfrac{i}{\sqrt{6}} & \dfrac{i}{\sqrt{3}} \\ 0 & -\dfrac{i}{\sqrt{6}} & \dfrac{i}{\sqrt{3}} \\ \dfrac{i}{\sqrt{2}} & \dfrac{i}{\sqrt{6}} & \dfrac{i}{\sqrt{3}} \end{bmatrix}$

5. In each part verify that the matrix is unitary and find its inverse.

(a) $\begin{bmatrix} \dfrac{3}{5} & \dfrac{4}{5}i \\ -\dfrac{4}{5} & \dfrac{3}{5}i \end{bmatrix}$
(b) $\begin{bmatrix} \dfrac{1}{\sqrt{2}} & \dfrac{1}{\sqrt{2}} \\ -\dfrac{1+i}{2} & \dfrac{1+i}{2} \end{bmatrix}$

(c) $\begin{bmatrix} \dfrac{1}{2\sqrt{2}}(\sqrt{3}+i) & \dfrac{1}{2\sqrt{2}}(1-i\sqrt{3}) \\ \dfrac{1}{2\sqrt{2}}(1+i\sqrt{3}) & \dfrac{1}{2\sqrt{2}}(i-\sqrt{3}) \end{bmatrix}$
(d) $\begin{bmatrix} \dfrac{1+i}{2} & -\dfrac{1}{2} & \dfrac{1}{2} \\ \dfrac{i}{\sqrt{3}} & \dfrac{1}{\sqrt{3}} & -\dfrac{i}{\sqrt{3}} \\ \dfrac{3+i}{2\sqrt{15}} & \dfrac{4+3i}{2\sqrt{15}} & \dfrac{5i}{2\sqrt{15}} \end{bmatrix}$

6. Show that the matrix

$$\frac{1}{\sqrt{2}}\begin{bmatrix} e^{i\theta} & e^{-i\theta} \\ ie^{i\theta} & -ie^{-i\theta} \end{bmatrix}$$

is unitary for every real value of θ.

In Exercises 7–12 find a unitary matrix P that diagonalizes A, and determine $P^{-1}AP$.

7. $A = \begin{bmatrix} 4 & 1-i \\ 1+i & 5 \end{bmatrix}$
8. $A = \begin{bmatrix} 3 & -i \\ i & 3 \end{bmatrix}$
9. $A = \begin{bmatrix} 6 & 2+2i \\ 2-2i & 4 \end{bmatrix}$

10. $\begin{bmatrix} 0 & 3+i \\ 3-i & -3 \end{bmatrix}$
11. $A = \begin{bmatrix} 5 & 0 & 0 \\ 0 & -1 & -1+i \\ 0 & -1-i & 0 \end{bmatrix}$
12. $A = \begin{bmatrix} 2 & \dfrac{i}{\sqrt{2}} & -\dfrac{i}{\sqrt{2}} \\ -\dfrac{i}{\sqrt{2}} & 2 & 0 \\ \dfrac{i}{\sqrt{2}} & 0 & 2 \end{bmatrix}$

13. Show that the eigenvalues of the symmetric matrix

$$A = \begin{bmatrix} 1 & 4i \\ 4i & 3 \end{bmatrix}$$

are not real. Does this violate Theorem 10.6.6?

14. Find a 2×2 matrix that is both Hermitian and unitary and whose entries are not all real numbers.

15. Prove: If A is an $n \times n$ matrix with complex entries, then $\det(\overline{A}) = \overline{\det(A)}$. [**Hint.** First show that the signed elementary products from $\overline{A}$ are the conjugates of the signed elementary products from A.]

16. (a) Use the result of Exercise 15 to prove that if A is an $n \times n$ matrix with complex entries, then $\det(A^*) = \overline{\det(A)}$.

(b) Prove: If A is Hermitian, then $\det(A)$ is real.

(c) Prove: If A is unitary, then $|\det(A)| = 1$.

17. Prove that the entries on the main diagonal of a Hermitian matrix are real numbers.

18. Let

$$A = \begin{bmatrix} a_{11} & a_{12} & a_{13} \\ a_{21} & a_{22} & a_{23} \\ a_{31} & a_{32} & a_{33} \end{bmatrix} \quad \text{and} \quad B = \begin{bmatrix} b_{11} & b_{12} & b_{13} \\ b_{21} & b_{22} & b_{23} \\ b_{31} & b_{32} & b_{33} \end{bmatrix}$$

be matrices with complex entries. Show that

(a) $(A^*)^* = A$ (b) $(A+B)^* = A^* + B^*$ (c) $(kA)^* = \bar{k}A^*$ (d) $(AB)^* = B^*A^*$

19. Prove: If A is invertible, then so is A^*, in which case $(A^*)^{-1} = (A^{-1})^*$.

20. Show that if A is a unitary matrix, then A^* is also unitary.

21. Prove that an $n \times n$ matrix with complex entries is unitary if and only if its rows form an orthonormal set in C^n with the Euclidean inner product.

22. Use Exercises 20 and 21 to show that an $n \times n$ matrix is unitary if and only if its columns form an orthonormal set in C^n with the Euclidean inner product.

23. Let λ and μ be distinct eigenvalues of a Hermitian matrix A.

(a) Prove that if $\mathbf{x}$ is an eigenvector corresponding to λ and $\mathbf{y}$ an eigenvector corresponding to μ, then $\mathbf{x}^* A\mathbf{y} = \lambda \mathbf{x}^* \mathbf{y}$ and $\mathbf{x}^* A\mathbf{y} = \mu \mathbf{x}^* \mathbf{y}$.

(b) Prove Theorem 10.6.4.

Chapter 10 Supplementary Exercises

1. Let $\mathbf{u} = (u_1, u_2, \ldots, u_n)$ and $\mathbf{v} = (v_1, v_2, \ldots, v_n)$ be vectors in C^n, and let $\bar{\mathbf{u}} = (\bar{u}_1, \bar{u}_2, \ldots, \bar{u}_n)$ and $\bar{\mathbf{v}} = (\bar{v}_1, \bar{v}_2, \ldots, \bar{v}_n)$.

(a) Prove: $\overline{\mathbf{u} \cdot \mathbf{v}} = \bar{\mathbf{u}} \cdot \bar{\mathbf{v}}$.

(b) Prove: $\mathbf{u}$ and $\mathbf{v}$ are orthogonal if and only if $\bar{\mathbf{u}}$ and $\bar{\mathbf{v}}$ are orthogonal.

2. Show that if the matrix

$$\begin{bmatrix} a & b \\ -\bar{b} & \bar{a} \end{bmatrix}$$

is nonzero, then it is invertible.

3. Find a basis for the solution space of the system

$$\begin{bmatrix} -1 & -i & 1 \\ -i & 1 & i \\ 1 & i & -1 \end{bmatrix} \begin{bmatrix} x_1 \\ x_2 \\ x_3 \end{bmatrix} = \begin{bmatrix} 0 \\ 0 \\ 0 \end{bmatrix}$$

4. Prove: If a and b are complex numbers such that $|a|^2 + |b|^2 = 1$, and if θ is a real number, then

$$A = \begin{bmatrix} a & b \\ -e^{i\theta}\bar{b} & e^{i\theta}\bar{a} \end{bmatrix}$$

is a unitary matrix.

5. Find the eigenvalues of the matrix

$$\begin{bmatrix} 0 & 0 & 1 \\ 1 & 0 & \omega + 1 + \dfrac{1}{\omega} \\ 0 & 1 & -\omega - 1 - \dfrac{1}{\omega} \end{bmatrix}$$

where $\omega = e^{2\pi i/3}$.

6. (a) Prove that if z is a complex number other than 1, then

$$1 + z + z^2 + \cdots + z^n = \frac{1 - z^{n+1}}{1 - z}$$

[**Hint.** Let S be the sum on the left side of the equation and consider the quantity $S - zS$.]

(b) Use the result in (a) to prove that if $z^n = 1$ and $z \neq 1$, then $1 + z + z^2 + \cdots + z^{n-1} = 0$.

(c) Use the result in (a) to obtain Lagrange's trigonometric identity

$$1 + \cos\theta + \cos 2\theta + \cdots + \cos n\theta = \frac{1}{2} + \frac{\sin[(n + \frac{1}{2})\theta]}{2\sin(\theta/2)}$$

for $0 < \theta < 2\pi$. [**Hint.** Let $z = \cos\theta + i\sin\theta$.]

7. Let $\omega = e^{2\pi i/3}$. Show that the vectors $\mathbf{v}_1 = (1/\sqrt{3})(1, 1, 1)$, $\mathbf{v}_2 = (1/\sqrt{3})(1, \omega, \omega^2)$, and $\mathbf{v}_3 = (1/\sqrt{3})(1, \omega^2, \omega^4)$ form an orthonormal set in C^3. [**Hint.** Use part (b) of Exercise 6.]

8. Show that if U is an $n \times n$ unitary matrix and $|z_1| = |z_2| = \cdots = |z_n| = 1$, then the product

$$U \begin{bmatrix} z_1 & 0 & 0 & \cdots & 0 \\ 0 & z_2 & 0 & \cdots & 0 \\ \vdots & \vdots & \vdots & & \vdots \\ 0 & 0 & 0 & \cdots & z_n \end{bmatrix}$$

is also unitary.

9. Suppose that $A^* = -A$.

(a) Show that iA is Hermitian.

(b) Show that A is unitarily diagonalizable and has pure imaginary eigenvalues.

Chapter 10 Technology Exercises

The following exercises are designed to be solved using a technology utility. Typically, this will be MATLAB, *Mathematica*, Maple, Derive, or Mathcad, but it may also be some other type of linear algebra software or a scientific calculator with some linear algebra capabilities. For each exercise you will need to read the relevant documentation for the particular utility you are using. The goal of these exercises is to provide you with a basic proficiency with your technology utility. Once you have mastered the techniques in these exercises, you will be able to use your technology utility to solve many of the problems in the regular exercise sets.

Sections 10.1 and 10.2

T1. (*Complex numbers and numerical operations*) Read your documentation on entering and displaying complex numbers and for performing the basic arithmetic operations of addition, subtraction, multiplication, and division. Experiment with numbers of your own choosing until you feel you have mastered the operations.

T2. (*Matrices with complex entries*) For most technology utilities the procedures for adding, subtracting, multiplying, and inverting matrices with complex entries is the same as for matrices with real entries. Experiment with these operations on some matrices of your own choosing, and then try using your utility to solve some of the exercises in Sections 10.1 and 10.2.

T3. (*Complex conjugate*) Read your documentation on finding the conjugate of a complex number, and then use your utility to perform the computations in Example 1 of Section 10.2.

Section 10.3

T1. (*Modulus and argument*) Read your documentation on finding the modulus and argument of a complex number, and then use your utility to perform the computations in Example 1.

Section 10.6

T1. (*Conjugate transpose*) Read your documentation on finding the conjugate transpose of a matrix with complex entries, and then use your utility to perform the computations in Examples 1 and 3.

T2. (*Unitary diagonalization*) Use your technology utility to diagonalize the matrix A in Example 5 and to find a matrix P that unitarily diagonalizes A. (See Technology Exercise T1 of Section 7.2.)

Applications of Linear Algebra

INTRODUCTION: This chapter consists of 21 applications of linear algebra. With one clearly marked exception, each application is in its own independent section, so that sections can be deleted or permuted as desired. Each topic begins with a list of linear algebra prerequisites.

Because our primary objective in this chapter is to present applications of linear algebra, proofs are often omitted. Whenever results from other fields are needed, they are stated precisely, with motivation where possible, but usually without proof.

11.1 CONSTRUCTING CURVES AND SURFACES THROUGH SPECIFIED POINTS

In this section we describe a technique that uses determinants to construct lines, circles, and general conic sections through specified points in the plane. The procedure is also used to pass planes and spheres in 3-space through fixed points.

PREREQUISITES: Linear Systems
Determinants
Analytic Geometry

The following theorem follows from Theorem 2.3.6.

Theorem 11.1.1

A homogeneous linear system with as many equations as unknowns has a nontrivial solution if and only if the determinant of the coefficient matrix is zero.

We shall now show how this result can be used to determine equations of various curves and surfaces through specified points.

A Line Through Two Points   Suppose that (x_1, y_1) and (x_2, y_2) are two distinct points in the plane. There exists a unique line

$$c_1 x + c_2 y + c_3 = 0 \tag{1}$$

that passes through these two points (Figure 11.1.1). Notice that c_1, c_2, and c_3 are not all zero and that these coefficients are unique only up to a multiplicative constant. Because (x_1, y_1) and (x_2, y_2) lie on the line, substituting them in (1) gives the two equations

$$c_1 x_1 + c_2 y_1 + c_3 = 0 \tag{2}$$

$$c_1 x_2 + c_2 y_2 + c_3 = 0 \tag{3}$$

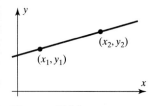

Figure 11.1.1

The three equations, (1), (2), and (3), can be grouped together and rewritten as

$$x c_1 + y c_2 + c_3 = 0$$
$$x_1 c_1 + y_1 c_2 + c_3 = 0$$
$$x_2 c_1 + y_2 c_2 + c_3 = 0$$

which is a homogeneous linear system of three equations for c_1, c_2, and c_3. Because c_1, c_2, and c_3 are not all zero, this system has a nontrivial solution, so that the determinant of the system must be zero. That is,

$$\begin{vmatrix} x & y & 1 \\ x_1 & y_1 & 1 \\ x_2 & y_2 & 1 \end{vmatrix} = 0 \tag{4}$$

Consequently, every point (x, y) on the line satisfies (4); conversely, it can be shown that every point (x, y) that satisfies (4) lies on the line.

EXAMPLE 1 Equation of a Line

Find the equation of the line that passes through the two points $(2, 1)$ and $(3, 7)$.

Solution.

Substituting the coordinates of the two points into Equation (4) gives

$$\begin{vmatrix} x & y & 1 \\ 2 & 1 & 1 \\ 3 & 7 & 1 \end{vmatrix} = 0$$

The cofactor expansion of this determinant along the first row then gives

$$-6x + y + 11 = 0 \qquad \blacklozenge$$

A Circle Through Three Points

Suppose that there are three distinct points in the plane, (x_1, y_1), (x_2, y_2), and (x_3, y_3), not all lying on a straight line. From analytic geometry we know that there is a unique circle, say,

$$c_1(x^2 + y^2) + c_2 x + c_3 y + c_4 = 0 \qquad (5)$$

which passes through them (Figure 11.1.2). Substituting the coordinates of the three points into this equation gives

$$c_1(x_1^2 + y_1^2) + c_2 x_1 + c_3 y_1 + c_4 = 0 \qquad (6)$$
$$c_1(x_2^2 + y_2^2) + c_2 x_2 + c_3 y_2 + c_4 = 0 \qquad (7)$$
$$c_1(x_3^2 + y_3^2) + c_2 x_3 + c_3 y_3 + c_4 = 0 \qquad (8)$$

As before, Equations (5) through (8) form a homogeneous linear system with a nontrivial solution for c_1, c_2, c_3, and c_4. Thus, the determinant of the coefficient matrix is zero:

$$\begin{vmatrix} x^2 + y^2 & x & y & 1 \\ x_1^2 + y_1^2 & x_1 & y_1 & 1 \\ x_2^2 + y_2^2 & x_2 & y_2 & 1 \\ x_3^2 + y_3^2 & x_3 & y_3 & 1 \end{vmatrix} = 0 \qquad (9)$$

This is a determinant form for the equation of the circle.

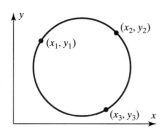

Figure 11.1.2

EXAMPLE 2 Equation of a Circle

Find the equation of the circle that passes through the three points $(1, 7)$, $(6, 2)$, and $(4, 6)$.

Solution.

Substituting the coordinates of the three points into Equation (9) gives

$$\begin{vmatrix} x^2 + y^2 & x & y & 1 \\ 50 & 1 & 7 & 1 \\ 40 & 6 & 2 & 1 \\ 52 & 4 & 6 & 1 \end{vmatrix} = 0$$

which reduces to

$$10(x^2 + y^2) - 20x - 40y - 200 = 0$$

In standard form this is

$$(x - 1)^2 + (y - 2)^2 = 5^2$$

Thus, the circle has center $(1, 2)$ and radius 5. ◆

A General Conic Section Through Five Points

The general equation of a conic section in the plane (a parabola, hyperbola, or ellipse, or degenerate forms of these curves) is given by

$$c_1x^2 + c_2xy + c_3y^2 + c_4x + c_5y + c_6 = 0$$

This equation contains six coefficients, but we can reduce the number to five if we divide through by any one of them that is not zero. Thus, only five coefficients must be determined, so that five distinct points in the plane are sufficient to determine the equation of the conic section (Figure 11.1.3). As before, the equation can be put in determinant form (see Exercise 6):

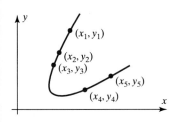

Figure 11.1.3

$$\begin{vmatrix} x^2 & xy & y^2 & x & y & 1 \\ x_1^2 & x_1y_1 & y_1^2 & x_1 & y_1 & 1 \\ x_2^2 & x_2y_2 & y_2^2 & x_2 & y_2 & 1 \\ x_3^2 & x_3y_3 & y_3^2 & x_3 & y_3 & 1 \\ x_4^2 & x_4y_4 & y_4^2 & x_4 & y_4 & 1 \\ x_5^2 & x_5y_5 & y_5^2 & x_5 & y_5 & 1 \end{vmatrix} = 0 \qquad (10)$$

EXAMPLE 3 Equation of an Orbit

An astronomer who wants to determine the orbit of an asteroid about the sun sets up a Cartesian coordinate system in the plane of the orbit with the sun at the origin. Astronomical units of measurement are used along the axes (1 astronomical unit = mean distance of earth to sun = 93 million miles). By Kepler's first law the orbit must be an ellipse, so the astronomer makes five observations of the asteroid at five different times and finds five points along the orbit to be

$(8.025, 8.310)$, $(10.170, 6.355)$, $(11.202, 3.212)$, $(10.736, 0.375)$, $(9.092, -2.267)$

Find the equation of the orbit.

Solution.

Substituting the coordinates of the five given points into (10) gives

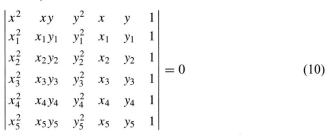

The cofactor expansion of this determinant along the first row is

$$386.799x^2 - 102.896xy + 446.026y^2 - 2476.409x - 1427.971y - 17109.378 = 0$$

Figure 11.1.4 is an accurate diagram of the orbit, together with the five given points. ◆

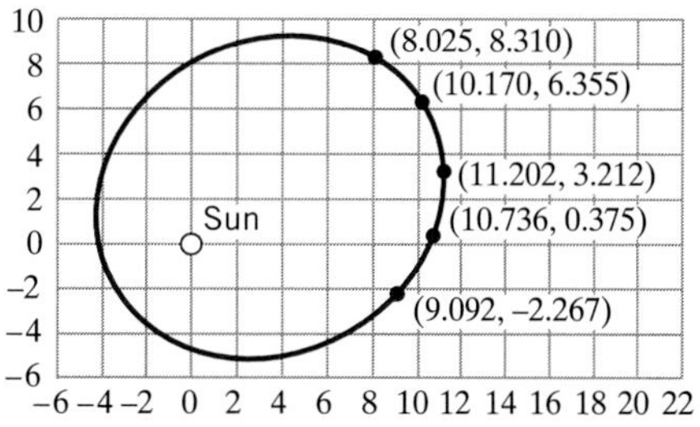

Figure 11.1.4

A Plane Through Three Points

In Exercise 7 we ask the reader to show the following: The plane in 3-space with equation

$$c_1 x + c_2 y + c_3 z + c_4 = 0$$

that passes through three noncollinear points (x_1, y_1, z_1), (x_2, y_2, z_2), and (x_3, y_3, z_3) is given by the determinant equation

$$\begin{vmatrix} x & y & z & 1 \\ x_1 & y_1 & z_1 & 1 \\ x_2 & y_2 & z_2 & 1 \\ x_3 & y_3 & z_3 & 1 \end{vmatrix} = 0 \tag{11}$$

EXAMPLE 4 Equation of a Plane

The equation of the plane that passes through the three noncollinear points $(1, 1, 0)$, $(2, 0, -1)$, and $(2, 9, 2)$ is

$$\begin{vmatrix} x & y & z & 1 \\ 1 & 1 & 0 & 1 \\ 2 & 0 & -1 & 1 \\ 2 & 9 & 2 & 1 \end{vmatrix} = 0$$

which reduces to

$$2x - y + 3z - 1 = 0 \qquad ♦$$

A Sphere Through Four Points

In Exercise 8 we ask the reader to show the following: The sphere in 3-space with equation

$$c_1(x^2 + y^2 + z^2) + c_2 x + c_3 y + c_4 z + c_5 = 0$$

that passes through four noncoplanar points (x_1, y_1, z_1), (x_2, y_2, z_2), (x_3, y_3, z_3), and (x_4, y_4, z_4) is given by the following determinant equation:

$$\begin{vmatrix} x^2 + y^2 + z^2 & x & y & z & 1 \\ x_1^2 + y_1^2 + z_1^2 & x_1 & y_1 & z_1 & 1 \\ x_2^2 + y_2^2 + z_2^2 & x_2 & y_2 & z_2 & 1 \\ x_3^2 + y_3^2 + z_3^2 & x_3 & y_3 & z_3 & 1 \\ x_4^2 + y_4^2 + z_4^2 & x_4 & y_4 & z_4 & 1 \end{vmatrix} = 0 \tag{12}$$

EXAMPLE 5 Equation of a Sphere

The equation of the sphere that passes through the four points $(0, 3, 2)$, $(1, -1, 1)$, $(2, 1, 0)$, and $(5, 1, 3)$ is

$$\begin{vmatrix} x^2 + y^2 + z^2 & x & y & z & 1 \\ 13 & 0 & 3 & 2 & 1 \\ 3 & 1 & -1 & 1 & 1 \\ 5 & 2 & 1 & 0 & 1 \\ 35 & 5 & 1 & 3 & 1 \end{vmatrix} = 0$$

This reduces to

$$x^2 + y^2 + z^2 - 4x - 2y - 6z + 5 = 0$$

which in standard form is

$$(x - 2)^2 + (y - 1)^2 + (z - 3)^2 = 9 \qquad \blacklozenge$$

Exercise Set 11.1

1. Find the equations of the lines that pass through the following points:

   (a) $(1, -1)$, $(2, 2)$ (b) $(0, 1)$, $(1, -1)$

2. Find the equations of the circles that pass through the following points:

   (a) $(2, 6)$, $(2, 0)$, $(5, 3)$ (b) $(2, -2)$, $(3, 5)$, $(-4, 6)$

3. Find the equation of the conic section that passes through the points $(0, 0)$, $(0, -1)$, $(2, 0)$, $(2, -5)$, and $(4, -1)$.

4. Find the equations of the planes in 3-space that pass through the following points:

   (a) $(1, 1, -3)$, $(1, -1, 1)$, $(0, -1, 2)$ (b) $(2, 3, 1)$, $(2, -1, -1)$, $(1, 2, 1)$

5. Find the equations of the spheres in 3-space that pass through the following points:

   (a) $(1, 2, 3)$, $(-1, 2, 1)$, $(1, 0, 1)$, $(1, 2, -1)$ (b) $(0, 1, -2)$, $(1, 3, 1)$, $(2, -1, 0)$, $(3, 1, -1)$

6. Show that Equation (10) is the equation of the conic section that passes through five given distinct points in the plane.

7. Show that Equation (11) is the equation of the plane in 3-space that passes through three given noncollinear points.

8. Show that Equation (12) is the equation of the sphere in 3-space that passes through four given noncoplanar points.

9. Find a determinant equation for the parabola of the form

   $$c_1 y + c_2 x^2 + c_3 x + c_4 = 0$$

   that passes through three given noncollinear points in the plane.

Technology Exercises 11.1

The following exercises are designed to be solved using a technology utility. Typically, this will be MATLAB, *Mathematica*, Maple, Derive, or Mathcad, but it may also be some other type of linear algebra software or a scientific calculator with some linear algebra capabilities. For each exercise you will need to read the relevant documentation for the particular utility you are using. The goal of these exercises is to provide you with a basic proficiency with your technology utility. Once you have mastered the techniques in these exercises, you will be able to use your technology utility to solve many of the problems in the regular exercise sets.

T1. The general equation of a quadric surface is given by

$$a_1x^2 + a_2y^2 + a_3z^2 + a_4xy + a_5xz + a_6yz + a_7x + a_8y + a_9z + a_{10} = 0$$

Given nine points on this surface, it may be possible to determine its equation.

(a) Show that if the nine points (x_i, y_i) for $i = 1, 2, 3, \ldots, 9$ lie on this surface, and if they determine uniquely the equation of this surface, then its equation can be written in determinant form as

$$
\begin{vmatrix}
x^2 & y^2 & z^2 & xy & xz & yz & x & y & z & 1 \\
x_1^2 & y_1^2 & z_1^2 & x_1y_1 & x_1z_1 & y_1z_1 & x_1 & y_1 & z_1 & 1 \\
x_2^2 & y_2^2 & z_2^2 & x_2y_2 & x_2z_2 & y_2z_2 & x_2 & y_2 & z_2 & 1 \\
x_3^2 & y_3^2 & z_3^2 & x_3y_3 & x_3z_3 & y_3z_3 & x_3 & y_3 & z_3 & 1 \\
x_4^2 & y_4^2 & z_4^2 & x_4y_4 & x_4z_4 & y_4z_4 & x_4 & y_4 & z_4 & 1 \\
x_5^2 & y_5^2 & z_5^2 & x_5y_5 & x_5z_5 & y_5z_5 & x_5 & y_5 & z_5 & 1 \\
x_6^2 & y_6^2 & z_6^2 & x_6y_6 & x_6z_6 & y_6z_6 & x_6 & y_6 & z_6 & 1 \\
x_7^2 & y_7^2 & z_7^2 & x_7y_7 & x_7z_7 & y_7z_7 & x_7 & y_7 & z_7 & 1 \\
x_8^2 & y_8^2 & z_8^2 & x_8y_8 & x_8z_8 & y_8z_8 & x_8 & y_8 & z_8 & 1 \\
x_9^2 & y_9^2 & z_9^2 & x_9y_9 & x_9z_9 & y_9z_9 & x_9 & y_9 & z_9 & 1
\end{vmatrix} = 0
$$

(b) Use the result in part (a) to determine the equation of the quadric surface that passes through the points $(1, 2, 3), (2, 1, 7), (0, 4, 6), (3, -1, 4), (3, 0, 11), (-1, 5, 8), (9, -8, 3),$ $(4, 5, 3),$ and $(-2, 6, 10)$.

(c) Use the methods of Section 9.7 to identify the resulting surface in part (b).

T2. (a) A hyperplane in the n-dimensional Euclidean space R^n has an equation of the form

$$a_1x_1 + a_2x_2 + a_3x_3 + \cdots + a_nx_n + a_{n+1} = 0$$

where $a_i, i = 1, 2, 3, \ldots, n + 1$, are constants, not all zero, and $x_i, i = 1, 2, 3, \ldots, n,$ are variables for which

$$(x_1, x_2, x_3, \ldots, x_n) \in R^n$$

A point

$$(x_{10}, x_{20}, x_{30}, \ldots, x_{n0}) \in R^n$$

lies on this hyperplane if

$$a_1x_{10} + a_2x_{20} + a_3x_{30} + \cdots + a_nx_{n0} + a_{n+1} = 0$$

Given that the n points $(x_{1i}, x_{2i}, x_{3i}, \ldots, x_{ni}), i = 1, 2, 3, \ldots, n,$ lie on this hyperplane and that they uniquely determine the equation of the hyperplane, show that the equation of the hyperplane can be written in determinant form as

$$
\begin{vmatrix}
x_1 & x_2 & x_3 & \cdots & x_n & 1 \\
x_{11} & x_{21} & x_{31} & \cdots & x_{n1} & 1 \\
x_{12} & x_{22} & x_{32} & \cdots & x_{n2} & 1 \\
x_{13} & x_{23} & x_{33} & \cdots & x_{n3} & 1 \\
\vdots & \vdots & \vdots & \ddots & \vdots & \vdots \\
x_{1n} & x_{2n} & x_{3n} & \cdots & x_{nn} & 1
\end{vmatrix} = 0
$$

(b) Determine the equation of the hyperplane in R^9 that goes through the following nine points:

$$(1, 2, 3, 4, 5, 6, 7, 8, 9) \quad (2, 3, 4, 5, 6, 7, 8, 9, 1) \quad (3, 4, 5, 6, 7, 8, 9, 1, 2)$$
$$(4, 5, 6, 7, 8, 9, 1, 2, 3) \quad (5, 6, 7, 8, 9, 1, 2, 3, 4) \quad (6, 7, 8, 9, 1, 2, 3, 4, 5)$$
$$(7, 8, 9, 1, 2, 3, 4, 5, 6) \quad (8, 9, 1, 2, 3, 4, 5, 6, 7) \quad (9, 1, 2, 3, 4, 5, 6, 7, 8)$$

11.2 ELECTRICAL NETWORKS

In this section basic laws of electrical circuits are discussed, and it is shown how these laws can be used to obtain systems of linear equations whose solutions yield the currents flowing in an electrical circuit.

PREREQUISITE: Linear Systems

The simplest electrical circuits consist of two basic components:

electrical sources   denoted by   ———+|⊢—————

resistors   denoted by   ————/\/\/\————

Electrical sources, such as batteries, create currents in an electrical circuit. Resistors, such as lightbulbs, limit the magnitudes of the currents.

There are three basic quantities associated with electrical circuits: ***electrical potential*** (E), ***resistance*** (R), and ***current*** (I). These are commonly measured in the following units:

$$E \quad \text{in volts} \quad (\text{V})$$
$$R \quad \text{in ohms} \quad (\Omega)$$
$$I \quad \text{in amperes} \quad (\text{A})$$

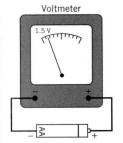

Voltmeter

1.5 V

Figure 11.2.1

Electrical potential is associated with two points in an electrical circuit and is measured in practice by connecting those points to a device called a *voltmeter*. For example, a common AA battery is rated at 1.5 volts, which means that this is the electrical potential across its positive and negative terminals (Figure 11.2.1).

In an electrical circuit the electrical potential between two points is called the ***voltage drop*** between these points. As we shall see, currents and voltage drops can be either positive or negative.

The flow of current in an electrical circuit is governed by three basic principles:

1. ***Ohm's Law*** The voltage drop across a resistor is the product of the current passing through it and its resistance; that is, $E = IR$.

2. ***Kirchhoff's Current Law*** The sum of the currents flowing into any point equals the sum of the currents flowing out from the point.

3. ***Kirchhoff's Voltage Law*** Around any closed loop the algebraic sum of the voltage drops is zero.

EXAMPLE 1 Finding Currents in a Circuit

Find the unknown currents I_1, I_2, and I_3 in the circuit shown in Figure 11.2.2.

Solution.

The flow directions for the currents I_1, I_2, and I_3 (marked by the arrowheads) were picked arbitrarily. Any of these currents that turn out to be negative actually flow opposite to the direction selected.

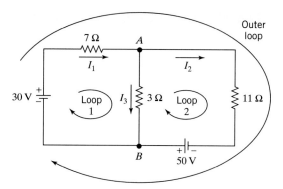

Figure 11.2.2

Applying Kirchhoff's current law to points A and B yields

$$I_1 = I_2 + I_3 \quad \text{(Point } A\text{)}$$
$$I_3 + I_2 = I_1 \qquad \text{(Point } B\text{)}$$

Since these equations both simplify to the same linear equation

$$I_1 - I_2 - I_3 = 0 \tag{1}$$

we still need two more equations to determine I_1, I_2, and I_3 uniquely. We will obtain them using Kirchhoff's voltage law.

To apply Kirchhoff's voltage law to a loop, select a positive direction around the loop (say clockwise) and make the following sign conventions:

• A current passing through a resistor produces a positive voltage drop if it flows in the positive direction of the loop and a negative voltage drop if it flows in the negative direction of the loop.

• A current passing through an electrical source produces a positive voltage drop if the positive direction of the loop is from $+$ to $-$ and a negative voltage drop if the positive direction of the loop is from $-$ to $+$.

Applying Kirchhoff's voltage law and Ohm's law to loop 1 in Figure 11.2.2 yields

$$7I_1 + 3I_3 - 30 = 0 \tag{2}$$

and to loop 2 yields

$$11I_2 - 3I_3 - 50 = 0 \tag{3}$$

Combining (1), (2), and (3) yields the linear system

$$
\begin{aligned}
I_1 - I_2 - I_3 &= 0 \\
7I_1 + 3I_3 &= 30 \\
11I_2 - 3I_3 &= 50
\end{aligned}
$$

Solving this linear system yields the following values for the currents:

$$I_1 = \tfrac{570}{131} \text{ (A)}, \qquad I_2 = \tfrac{590}{131} \text{ (A)}, \qquad I_3 = -\tfrac{20}{131} \text{ (A)}$$

Note that I_3 is negative, which means that this current flows opposite to the direction indicated in Figure 11.2.2. Also note that we could have applied Kirchhoff's voltage law to the outer loop of the circuit. However, this produces a redundant equation (try it). ♦

Exercise Set 11.2

In Exercises 1–4 find the currents in the circuits.

1.

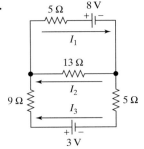

2.

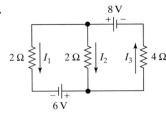

3.

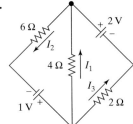

4.

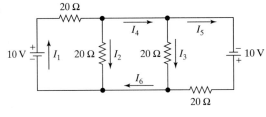

5. Show that if the current I_5 in the circuit of the accompanying figure is zero, then $R_4 = R_3 R_2 / R_1$.

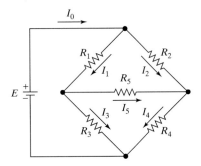

Figure Ex-5

[**Remark.** This circuit, called a Wheatstone bridge circuit, is used for the precise measurement of resistance. Here, R_4 is an unknown resistance and R_1, R_2, and R_3 are adjustable calibrated resistors. R_5 represents a galvanometer—a device for measuring current. After varying the resistances R_1, R_2, and R_3 until the galvanometer reading is zero, the formula $R_4 = R_3 R_2 / R_1$ determines the unknown resistance R_4.]

6. Show that if the two currents labeled I in the circuits of the accompanying figure are equal, then $R = \dfrac{1}{\dfrac{1}{R_1} + \dfrac{1}{R_2}}$.

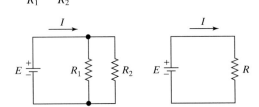

Figure Ex-6

Technology Exercises 11.2

The following exercises are designed to be solved using a technology utility. Typically, this will be MATLAB, *Mathematica*, Maple, Derive, or Mathcad, but it may also be some other type of linear algebra software or a scientific calculator with some linear algebra capabilities. For each exercise you will need to read the relevant documentation for the particular utility you are using. The goal of these exercises is to provide you with a basic proficiency with your technology utility. Once you have mastered the techniques in these exercises, you will be able to use your technology utility to solve many of the problems in the regular exercise sets.

T1. The accompanying figure shows a sequence of different circuits.

  (a) Solve for the current I_1, for the circuit in part (*a*) of the figure.
  (b) Solve for the currents I_1 through I_3, for the circuit in part (*b*) of the figure.
  (c) Solve for the currents I_1 through I_5, for the circuit in part (*c*) of the figure.
  (d) Continue this process until you discover a pattern in the values of $I_1, I_2, I_3, \ldots$.
  (e) Investigate the sequence of values for I_1 in each of the circuits in parts (a), (b), (c), and so on, and numerically show that the limit of this sequence approaches the value

$$\left(\frac{\sqrt{5}-1}{2}\right)\frac{E}{R} \approx (0.6180)\frac{E}{R}$$

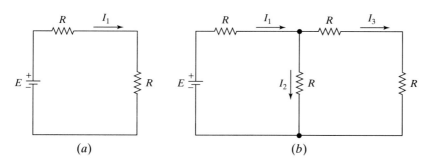

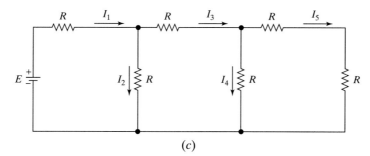

Figure Ex-T1

T2. The accompanying figure shows a sequence of different circuits.

  (a) Solve for the current I_1, for the circuit in part (*a*) of the figure.
  (b) Solve for the current I_1, for the circuit in part (*b*) of the figure.
  (c) Solve for the current I_1, for the circuit in part (*c*) of the figure.
  (d) Continue this process until you discover a pattern in the values of I_1.
  (e) Investigate the sequence of values for I_1 in each of the circuits in parts (a), (b), (c), and so on, and numerically show that the limit of this sequence approaches the value

$$\left(\frac{\sqrt{5}-1}{2}\right)\frac{E}{R} \approx (0.6180)\frac{E}{R}$$

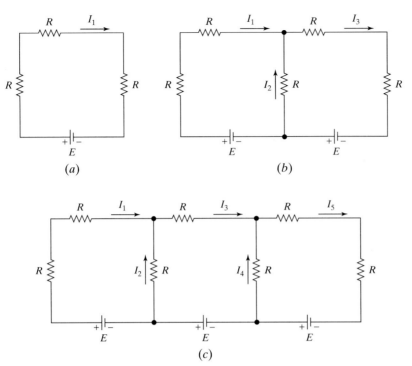

Figure Ex-T2

11.3 GEOMETRIC LINEAR PROGRAMMING

In this section a geometric technique is described for maximizing or minimizing a linear expression in two variables subject to a set of linear constraints.

PREREQUISITES: Linear Systems
Linear Inequalities

Linear Programming   The study of linear programming theory has expanded greatly since the pioneer work of George Dantzig in the late 1940s. Today, linear programming is applied to a wide variety of problems in industry and science. In this section we present a geometric approach to the solution of simple linear programming problems. Let us begin with some examples.

EXAMPLE 1 Maximizing Sales Revenue

A candy manufacturer has 130 pounds of chocolate-covered cherries and 170 pounds of chocolate-covered mints in stock. He decides to sell them in the form of two different mixtures. One mixture will contain half cherries and half mints by weight and will sell for $2.00 per pound. The other mixture will contain one-third cherries and two-thirds mints by weight and will sell for $1.25 per pound. How many pounds of each mixture should the candy manufacturer prepare in order to maximize his sales revenue?

Solution.

Let us first formulate this problem mathematically. Let the mixture of half cherries and half mints be called mix A, and let x_1 be the number of pounds of this mixture to be prepared. Let the mixture of one-third cherries and two-thirds mints be called mix B, and let x_2 be the number of pounds of this mixture to be prepared. Since mix A sells for \$2.00 per pound and mix B sells for \$1.25 per pound, the total sales z (in dollars) will be

$$z = 2.00x_1 + 1.25x_2$$

Since each pound of mix A contains $\frac{1}{2}$ pound of cherries and each pound of mix B contains $\frac{1}{3}$ pound of cherries, the total number of pounds of cherries used in both mixtures is

$$\tfrac{1}{2}x_1 + \tfrac{1}{3}x_2$$

Similarly, since each pound of mix A contains $\frac{1}{2}$ pound of mints and each pound of mix B contains $\frac{2}{3}$ pound of mints, the total number of pounds of mints used in both mixtures is

$$\tfrac{1}{2}x_1 + \tfrac{2}{3}x_2$$

Because the manufacturer can use at most 130 pounds of cherries and 170 pounds of mints, we must have

$$\tfrac{1}{2}x_1 + \tfrac{1}{3}x_2 \leq 130$$
$$\tfrac{1}{2}x_1 + \tfrac{2}{3}x_2 \leq 170$$

Furthermore, since x_1 and x_2 cannot be negative numbers, we must have

$$x_1 \geq 0 \quad \text{and} \quad x_2 \geq 0$$

The problem can therefore be formulated mathematically as follows: Find values of x_1 and x_2 that maximize

$$z = 2.00x_1 + 1.25x_2$$

subject to

$$\tfrac{1}{2}x_1 + \tfrac{1}{3}x_2 \leq 130$$
$$\tfrac{1}{2}x_1 + \tfrac{2}{3}x_2 \leq 170$$
$$x_1 \geq 0$$
$$x_2 \geq 0$$

Later in this section we shall show how to solve this type of mathematical problem geometrically. ◆

EXAMPLE 2   Maximizing Annual Yield

A woman has up to \$10,000 to invest. Her broker suggests investing in two bonds, A and B. Bond A is a rather risky bond with an annual yield of 10%, and bond B is a rather safe bond with an annual yield of 7%. After some consideration, she decides to invest at most \$6000 in bond A, at least \$2000 in bond B, and to invest at least as much in bond A as in bond B. How should she invest her \$10,000 in order to maximize her annual yield?

Solution.

To formulate this problem mathematically, let x_1 be the number of dollars to be invested in bond A and let x_2 be the number of dollars to be invested in bond B. Since each dollar invested in bond A earns \$.10 per year and each dollar invested in bond B earns \$.07 per year, the total dollar amount z earned each year by both bonds is

$$z = .10x_1 + .07x_2$$

The constraints imposed can be formulated mathematically as follows:

Invest no more than \$10,000:	$x_1 + x_2 \leq 10{,}000$
Invest at most \$6000 in bond A:	$x_1 \leq 6000$
Invest at least \$2000 in bond B:	$x_2 \geq 2000$
Invest at least as much in bond A as in bond B:	$x_1 \geq x_2$

We also have the implicit assumption that x_1 and x_2 are nonnegative:

$$x_1 \geq 0 \quad \text{and} \quad x_2 \geq 0$$

Thus, the complete mathematical formulation of the problem is as follows: Find values of x_1 and x_2 that maximize

$$z = .10x_1 + .07x_2$$

subject to

$$x_1 + x_2 \leq 10{,}000$$
$$x_1 \leq 6000$$
$$x_2 \geq 2000$$
$$x_1 - x_2 \geq 0$$
$$x_1 \geq 0$$
$$x_2 \geq 0$$

◆

EXAMPLE 3 Minimizing Cost

A student desires to design a breakfast of corn flakes and milk that is as economical as possible. On the basis of what he eats during his other meals, he decides that his breakfast should supply him with at least 9 grams of protein, at least $\frac{1}{3}$ the recommended daily allowance (RDA) of vitamin D, and at least $\frac{1}{4}$ the RDA of calcium. He finds the following nutrition information on the milk and corn flakes containers:

	Milk ($\frac{1}{2}$ cup)	Corn Flakes (1 ounce)
Cost	7.5 cents	5.0 cents
Protein	4 grams	2 grams
Vitamin D	$\frac{1}{8}$ of RDA	$\frac{1}{10}$ of RDA
Calcium	$\frac{1}{6}$ of RDA	None

In order not to have his mixture too soggy or too dry, the student decides to limit himself to mixtures that contain 1 to 3 ounces of corn flakes per cup of milk, inclusive. What quantities of milk and corn flakes should he use to minimize the cost of his breakfast?

Solution.

For the mathematical formulation of this problem, let x_1 be the quantity of milk used (measured in $\frac{1}{2}$-cup units), and let x_2 be the quantity of corn flakes used (measured in

1-ounce units). Then if z is the cost of the breakfast in cents, we may write the following.

Cost of breakfast:	$z = 7.5x_1 + 5.0x_2$
At least 9 grams protein:	$4x_1 + 2x_2 \geq 9$
At least $\frac{1}{3}$ RDA vitamin D:	$\frac{1}{8}x_1 + \frac{1}{10}x_2 \geq \frac{1}{3}$
At least $\frac{1}{4}$ RDA calcium:	$\frac{1}{6}x_1 \geq \frac{1}{4}$
At least 1 ounce corn flakes per cup (two $\frac{1}{2}$-cups) of milk:	$\dfrac{x_2}{x_1} \geq \dfrac{1}{2}$ (or $x_1 - 2x_2 \leq 0$)
At most 3 ounces corn flakes per cup (two $\frac{1}{2}$-cups) of milk:	$\dfrac{x_2}{x_1} \leq \dfrac{3}{2}$ (or $3x_1 - 2x_2 \geq 0$)

As before, we also have the implicit assumption that $x_1 \geq 0$ and $x_2 \geq 0$. Thus the complete mathematical formulation of the problem is as follows: Find values of x_1 and x_2 that minimize

$$z = 7.5x_1 + 5.0x_2$$

subject to

$$4x_1 + 2x_2 \geq 9$$
$$\tfrac{1}{8}x_1 + \tfrac{1}{10}x_2 \geq \tfrac{1}{3}$$
$$\tfrac{1}{6}x_1 \geq \tfrac{1}{4}$$
$$x_1 - 2x_2 \leq 0$$
$$3x_1 - 2x_2 \geq 0$$
$$x_1 \geq 0$$
$$x_2 \geq 0 \qquad\qquad \blacklozenge$$

Geometric Solution of Linear Programming Problems

Each of the preceding three examples is a special case of the following problem.

Problem.   Find values of x_1 and x_2 that either maximize or minimize

$$z = c_1x_1 + c_2x_2 \tag{1}$$

subject to

$$a_{11}x_1 + a_{12}x_2 \ (\leq)(\geq)(=) \ b_1$$
$$a_{21}x_1 + a_{22}x_2 \ (\leq)(\geq)(=) \ b_2$$
$$\vdots \qquad\quad \vdots \qquad\qquad \vdots \tag{2}$$
$$a_{m1}x_1 + a_{m2}x_2 \ (\leq)(\geq)(=) \ b_m$$

and

$$x_1 \geq 0, \qquad x_2 \geq 0 \tag{3}$$

In each of the m conditions of (2), any one of the symbols $\leq$, $\geq$, or $=$ may be used.

The problem above is called the ***general linear programming problem*** in two variables. The linear function z in (1) is called the ***objective function***. Equations (2) and (3) are called the ***constraints***; in particular, the equations in (3) are called the ***nonnegativity constraints*** on the variables x_1 and x_2.

We shall now show how to solve a linear programming problem in two variables graphically. A pair of values (x_1, x_2) that satisfy all of the constraints is called a ***feasible solution***. The set of all feasible solutions determines a subset of the x_1x_2-plane called the ***feasible region***. Our desire is to find a feasible solution that maximizes the objective function. Such a solution is called an ***optimal solution***.

To examine the feasible region of a linear programming problem, let us note that each constraint of the form

$$a_{i1}x_1 + a_{i2}x_2 = b_i$$

defines a line in the x_1x_2-plane, whereas each constraint of the form

$$a_{i1}x_1 + a_{i2}x_2 \le b_i \quad \text{or} \quad a_{i1}x_1 + a_{i2}x_2 \ge b_i$$

defines a half-plane that includes its boundary line

$$a_{i1}x_1 + a_{i2}x_2 = b_i$$

Thus, the feasible region is always an intersection of finitely many lines and half-planes. For example, the four constraints

$$\tfrac{1}{2}x_1 + \tfrac{1}{3}x_2 \le 130$$
$$\tfrac{1}{2}x_1 + \tfrac{2}{3}x_2 \le 170$$
$$x_1 \ge 0$$
$$x_2 \ge 0$$

of Example 1 define the half-planes illustrated in parts (a), (b), (c), and (d) of Figure 11.3.1. The feasible region of this problem is thus the intersection of these four half-planes, which is illustrated in Figure 11.3.1e.

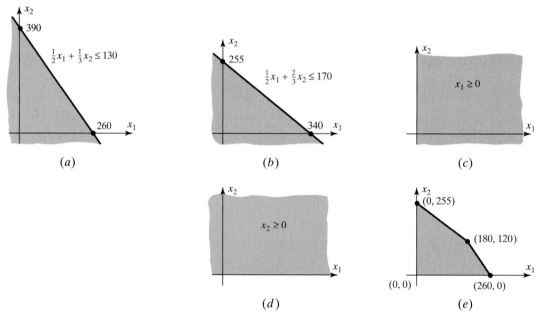

Figure 11.3.1

It can be shown that the feasible region of a linear programming problem has a boundary consisting of a finite number of straight line segments. If the feasible region can be enclosed in a sufficiently large circle, it is called **bounded** (Figure 11.3.1e); otherwise it is called **unbounded** (Figure 11.3.5). If the feasible region is *empty* (contains no points), then the constraints are inconsistent and the linear programming problem has no solution (Figure 11.3.6).

Those boundary points of a feasible region that are intersections of two of the straight line boundary segments are called **extreme points**. (They are also called *corner points* or *vertex points*.) For example, from Figure 11.3.1*e* the feasible region of Example 1 has four extreme points:

$$(0, 0), \quad (0, 255), \quad (180, 120), \quad (260, 0) \tag{4}$$

The importance of the extreme points of a feasible region is shown by the following theorem.

Theorem 11.3.1   **Maximum and Minimum Values**

If the feasible region of a linear programming problem is nonempty and bounded, then the objective function attains both a maximum and minimum value and these occur at extreme points of the feasible region. If the feasible region is unbounded, then the objective function may or may not attain a maximum or minimum value; however, if it attains a maximum or minimum value, it does so at an extreme point.

Figure 11.3.2 suggests the idea behind the proof of this theorem. Since the objective function

$$z = c_1 x_1 + c_2 x_2$$

of a linear programming problem is a linear function of x_1 and x_2, its level curves (the curves along which z has constant values) are straight lines. As we move in a direction perpendicular to these level curves, the objective function either increases or decreases monotonically. Within a bounded feasible region, the maximum and minimum values of z must therefore occur at extreme points, as Figure 11.3.2 indicates.

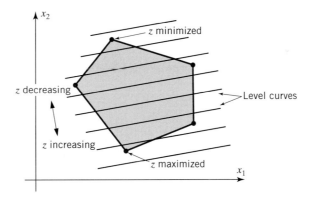

Figure 11.3.2

In the next few examples we use Theorem 11.3.1 to solve several linear programming problems and illustrate the variations in the nature of the solutions that may occur.

EXAMPLE 4 **Example 1 Revisited**

From Figure 11.3.1*e* we see that the feasible region of Example 1 is bounded. Consequently, from Theorem 11.3.1 the objective function

$$z = 2.00 x_1 + 1.25 x_2$$

attains both its minimum and maximum values at extreme points. The four extreme points and the corresponding values of z are given in the following table.

Extreme Point (x_1, x_2)	Value of $z = 2.00x_1 + 1.25x_2$
(0, 0)	0
(0, 255)	318.75
(180, 120)	510.00
(260, 0)	520.00

We see that the largest value of z is 520.00 and the corresponding optimal solution is (260, 0). Thus the candy manufacturer attains maximum sales of $520 when he produces 260 pounds of mixture A and none of mixture B. ◆

EXAMPLE 5 Using Theorem 11.3.1

Find values of x_1 and x_2 that maximize

$$z = x_1 + 3x_2$$

subject to

$$2x_1 + 3x_2 \leq 24$$
$$x_1 - x_2 \leq 7$$
$$x_2 \leq 6$$
$$x_1 \geq 0$$
$$x_2 \geq 0$$

Solution.

In Figure 11.3.3 we have drawn the feasible region of this problem. Since it is bounded, the maximum value of z is attained at one of the five extreme points. The values of the objective function at the five extreme points are given in the following table.

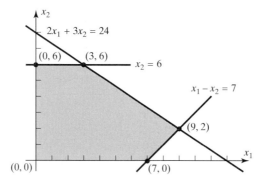

Extreme Point (x_1, x_2)	Value of $z = x_1 + 3x_2$
(0, 6)	18
(3, 6)	21
(9, 2)	15
(7, 0)	7
(0, 0)	0

Figure 11.3.3

From this table the maximum value of z is 21, which is attained at $x_1 = 3$ and $x_2 = 6$. ◆

EXAMPLE 6 Using Theorem 11.3.1

Find values of x_1 and x_2 that maximize

$$z = 4x_1 + 6x_2$$

subject to

$$2x_1 + 3x_2 \leq 24$$
$$x_1 - x_2 \leq 7$$
$$x_2 \leq 6$$
$$x_1 \geq 0$$
$$x_2 \geq 0$$

Solution.

The constraints in this problem are identical to the constraints in Example 5, so the feasible region of this problem is also given by Figure 11.3.3. The values of the objective function at the extreme points are given in the following table.

Extreme Point (x_1, x_2)	Value of $z = 4x_1 + 6x_2$
(0, 6)	36
(3, 6)	48
(9, 2)	48
(7, 0)	28
(0, 0)	0

We see that the objective function attains a maximum value of 48 at two adjacent extreme points, (3, 6) and (9, 2). This shows that an optimal solution to a linear programming problem need not be unique. As we ask the reader to show in Exercise 9, if the objective function has the same value at two adjacent extreme points, it has the same value at all points on the straight line boundary segment connecting the two extreme points. Thus, in this example the maximum value of z is attained at all points on the straight line segment connecting the extreme points (3, 6) and (9, 2). ♦

EXAMPLE 7   The Feasible Region Is a Line Segment

Find values of x_1 and x_2 that minimize

$$z = 2x_1 - x_2$$

subject to

$$2x_1 + 3x_2 = 12$$
$$2x_1 - 3x_2 \geq 0$$
$$x_1 \geq 0$$
$$x_2 \geq 0$$

Solution.

In Figure 11.3.4 we have drawn the feasible region of this problem. Because one of the constraints is an equality constraint, the feasible region is a straight line segment with two extreme points. The values of z at the two extreme points are given in the following table.

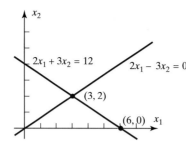

Figure 11.3.4

Extreme Point (x_1, x_2)	Value of $z = 2x_1 - x_2$
(3, 2)	4
(6, 0)	12

The minimum value of z is thus 4 and is attained at $x_1 = 3$ and $x_2 = 2$. ◆

EXAMPLE 8 Using Theorem 11.3.1

Find values of x_1 and x_2 that maximize

$$z = 2x_1 + 5x_2$$

subject to

$$2x_1 + x_2 \geq 8$$
$$-4x_1 + x_2 \leq 2$$
$$2x_1 - 3x_2 \leq 0$$
$$x_1 \geq 0$$
$$x_2 \geq 0$$

Solution.

The feasible region of this linear programming problem is illustrated in Figure 11.3.5. Since it is unbounded, we are not assured by Theorem 11.3.1 that the objective function attains a maximum value. In fact, it is easily seen that since the feasible region contains points for which both x_1 and x_2 are arbitrarily large and positive, the objective function

$$z = 2x_1 + 5x_2$$

can be made arbitrarily large and positive. This problem has no optimal solution. Instead, we say the problem has an ***unbounded solution***. ◆

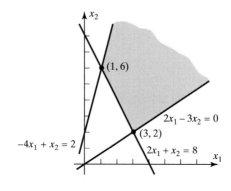

Figure 11.3.5

EXAMPLE 9 Using Theorem 11.3.1

Find values of x_1 and x_2 that maximize

$$z = -5x_1 + x_2$$

subject to

$$2x_1 + x_2 \geq 8$$
$$-4x_1 + x_2 \leq 2$$
$$2x_1 - 3x_2 \leq 0$$
$$x_1 \geq 0$$
$$x_2 \geq 0$$

Solution.

The above constraints are the same as those in Example 8, so the feasible region of this problem is also given by Figure 11.3.5. In Exercise 10 we ask the reader to show that the objective function of this problem attains a maximum within the feasible region. By Theorem 11.3.1, this maximum must be attained at an extreme point. The values of z at the two extreme points of the feasible region are given in the following table.

Extreme Point (x_1, x_2)	Value of $z = -5x_1 + x_2$
(1, 6)	1
(3, 2)	−13

The maximum value of z is thus 1 and is attained at the extreme point $x_1 = 1$, $x_2 = 6$.
◆

EXAMPLE 10 Inconsistent Constraints

Find values of x_1 and x_2 that minimize

$$z = 3x_1 - 8x_2$$

subject to

$$2x_1 - x_2 \leq 4$$
$$3x_1 + 11x_2 \leq 33$$
$$3x_1 + 4x_2 \geq 24$$
$$x_1 \geq 0$$
$$x_2 \geq 0$$

Solution.

As can be seen from Figure 11.3.6, the intersection of the five half-planes defined by the five constraints is empty. This linear programming problem has no feasible solutions since the constraints are inconsistent.
◆

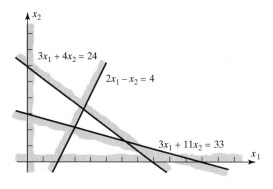

Figure 11.3.6 There are no points common to all five shaded half-planes.

Exercise Set 11.3

1. Find values of x_1 and x_2 that maximize

$$z = 3x_1 + 2x_2$$

subject to

$$2x_1 + 3x_2 \leq 6$$
$$2x_1 - x_2 \geq 0$$
$$x_1 \leq 2$$
$$x_2 \leq 1$$
$$x_1 \geq 0$$
$$x_2 \geq 0$$

2. Find values of x_1 and x_2 that minimize

$$z = 3x_1 - 5x_2$$

subject to

$$2x_1 - x_2 \leq -2$$
$$4x_1 - x_2 \geq 0$$
$$x_2 \leq 3$$
$$x_1 \geq 0$$
$$x_2 \geq 0$$

3. Find values of x_1 and x_2 that minimize

$$z = -3x_1 + 2x_2$$

subject to

$$3x_1 - x_2 \geq -5$$
$$-x_1 + x_2 \geq 1$$
$$2x_1 + 4x_2 \geq 12$$
$$x_1 \geq 0$$
$$x_2 \geq 0$$

4. Solve the linear programming problem posed in Example 2.

5. Solve the linear programming problem posed in Example 3.

6. A trucking firm ships the containers of two companies, A and B. Each container from company A weighs 40 pounds and is 2 cubic feet in volume. Each container from company B

weighs 50 pounds and is 3 cubic feet in volume. The trucking firm charges company A $2.20 for each container shipped and charges company B $3.00 for each container shipped. If one of the firm's trucks cannot carry more than 37,000 pounds and cannot hold more than 2000 cubic feet, how many containers from companies A and B should a truck carry to maximize the shipping charges?

7. Repeat Exercise 6 if the trucking firm raises its price for shipping a container from company A to $2.50.

8. A manufacturer produces sacks of chicken feed from two ingredients, A and B. Each sack is to contain at least 10 ounces of nutrient N_1, at least 8 ounces of nutrient N_2, and at least 12 ounces of nutrient N_3. Each pound of ingredient A contains 2 ounces of nutrient N_1, 2 ounces of nutrient N_2, and 6 ounces of nutrient N_3. Each pound of ingredient B contains 5 ounces of nutrient N_1, 3 ounces of nutrient N_2, and 4 ounces of nutrient N_3. If ingredient A costs 8 cents per pound and ingredient B costs 9 cents per pound, how much of each ingredient should the manufacturer use in each sack of feed to minimize his costs?

9. If the objective function of a linear programming problem has the same value at two adjacent extreme points, show that it has the same value at all points on the straight line segment connecting the two extreme points. [**Hint.** If (x_1', x_2') and (x_1'', x_2'') are any two points in the plane, a point (x_1, x_2) lies on the straight line segment connecting them if

$$x_1 = tx_1' + (1-t)x_1''$$

and

$$x_2 = tx_2' + (1-t)x_2''$$

where t is a number in the interval $[0, 1]$.]

10. Show that the objective function in Example 9 attains a maximum value in the feasible set. [**Hint.** Examine the level curves of the objective function.]

Technology Exercises 11.3

The following exercises are designed to be solved using a technology utility. Typically, this will be MATLAB, *Mathematica*, Maple, Derive, or Mathcad, but it may also be some other type of linear algebra software or a scientific calculator with some linear algebra capabilities. For each exercise you will need to read the relevant documentation for the particular utility you are using. The goal of these exercises is to provide you with a basic proficiency with your technology utility. Once you have mastered the techniques in these exercises, you will be able to use your technology utility to solve many of the problems in the regular exercise sets.

T1. Consider the feasible region consisting of $0 \le x, 0 \le y$ along with the set of inequalities

$$x \cos\left(\frac{(2k+1)\pi}{4n}\right) + y \sin\left(\frac{(2k+1)\pi}{4n}\right) \le \cos\left(\frac{\pi}{4n}\right)$$

for $k = 0, 1, 2, \ldots, n-1$. Maximize the objective function

$$z = 3x + 4y$$

assuming that (a) $n = 1$, (b) $n = 2$, (c) $n = 3$, (d) $n = 4$, (e) $n = 5$, (f) $n = 6$, (g) $n = 7$, (h) $n = 8$, (i) $n = 9$, (j) $n = 10$, and (k) $n = 11$. (l) Next, maximize this objective function using the nonlinear feasible region, $0 \le x, 0 \le y$, and

$$x^2 + y^2 \le 1$$

(m) Let the results of parts (a) through (k) begin a sequence of values for z_{max}. Do these values approach the value determined in part (l)? Explain.

T2. Repeat Exercise T1 using the objective function $z = x + y$.

11.4 THE ASSIGNMENT PROBLEM

A number of facilities are to be assigned different tasks. Each possible assignment results in a certain cost. In this section we describe an algorithm, called the Hungarian method, to find an assignment with minimum cost.

PREREQUISITE: Matrix Notation

A basic problem in operations research is to assign tasks to facilities on a one-to-one basis in some optimal way. For example, the problem may be to find the best assignment of workers to jobs, sports players to field positions, equipment to construction sites, and so forth. The assignment problem requires that there be as many facilities as tasks, say n of each. In this case, there are exactly $n!$ different ways to assign the tasks to the facilities on a one-to-one basis. This follows because there are n ways to assign the first task, $n - 1$ ways to assign the second, $n - 2$ ways to assign the third, and so on—a total of

$$n \cdot (n - 1) \cdot (n - 2) \cdot \; \cdots \; \cdot 3 \cdot 2 \cdot 1 = n!$$

possible assignments. Among these $n!$ possible assignments we are to find one that is optimal in some sense. To define the notion of an optimal assignment precisely, we introduce the following quantities. Let

$$c_{ij} = \text{cost of assigning the } i\text{th facility the } j\text{th task}$$

for $i, j = 1, 2, \ldots, n$. The units of c_{ij} might be dollars, miles, hours—whatever is appropriate to the problem. We define the *cost matrix* to be the $n \times n$ matrix

$$C = \begin{bmatrix} c_{11} & c_{12} & \cdots & c_{1n} \\ c_{21} & c_{22} & \cdots & c_{2n} \\ \vdots & \vdots & & \vdots \\ c_{n1} & c_{n2} & \cdots & c_{nn} \end{bmatrix}$$

The requirement that each facility be assigned a unique task on a one-to-one basis is equivalent to the condition that no two of the corresponding c_{ij}'s come from the same row or column. This leads to the following definition.

Definition

Given an $n \times n$ cost matrix C, an ***assignment*** is a set of n entry positions, no two of which lie in the same row or column.

An optimal assignment is then defined as follows.

Definition

The sum of the n entries of an assignment is called its ***cost***. An assignment with the smallest possible cost is called an ***optimal assignment***.

The ***assignment problem*** is to find an optimal assignment in a given cost matrix. For example, in assigning n pieces of equipment to n construction sites, c_{ij} could be the distance in miles between the ith piece of equipment and the jth construction site.

An optimal assignment is one for which the total distance traveled by the n pieces of equipment is a minimum.

EXAMPLE 1   Minimizing the Sum of Two Bids

A college intends to install air-conditioning in three of its buildings during a one-week spring break. It invites three contractors to submit separate bids for the work involved in each of the three buildings. The bids it receives (in 1000-dollar units) are listed in Table 1.

TABLE 1

	Bids		
	Bldg 1	**Bldg 2**	**Bldg 3**
Contractor 1	53	96	37
Contractor 2	47	87	41
Contractor 3	60	92	36

Each contractor can install the air-conditioning for only one building during the one-week period, so the college must assign a different contractor to each building. To which building should each contractor be assigned in order to minimize the sum of the corresponding bids?

Solution.

The cost matrix for this problem is the 3×3 matrix

$$\begin{bmatrix} 53 & 96 & 37 \\ 47 & 87 & 41 \\ 60 & 92 & 36 \end{bmatrix} \tag{1}$$

Because there are only six ($= 3!$) possible assignments, we may solve this problem by computing the cost of each of them. We have shaded the entries associated with each of the six assignments and computed their sums.

$$\begin{bmatrix} 53 & 96 & 37 \\ 47 & 87 & 41 \\ 60 & 92 & 36 \end{bmatrix}$$
$$53 + 87 + 36 = 176$$
(a)

$$\begin{bmatrix} 53 & 96 & 37 \\ 47 & 87 & 41 \\ 60 & 92 & 36 \end{bmatrix}$$
$$53 + 92 + 41 = 186$$
(b)

$$\begin{bmatrix} 53 & 96 & 37 \\ 47 & 87 & 41 \\ 60 & 92 & 36 \end{bmatrix}$$
$$47 + 96 + 36 = 179$$
(c)

$$\begin{bmatrix} 53 & 96 & 37 \\ 47 & 87 & 41 \\ 60 & 92 & 36 \end{bmatrix}$$
$$47 + 92 + 37 = 176$$
(d)

$$\begin{bmatrix} 53 & 96 & 37 \\ 47 & 87 & 41 \\ 60 & 92 & 36 \end{bmatrix}$$
$$60 + 96 + 41 = 197$$
(e)

$$\begin{bmatrix} 53 & 96 & 37 \\ 47 & 87 & 41 \\ 60 & 92 & 36 \end{bmatrix}$$
$$60 + 87 + 37 = 184$$
(f)

Note that the bid totals range from a minimum of \$176,000 to a maximum of \$197,000. Because the minimum bid total of \$176,000 is attained by both assignments (a) and (d), the college should assign the contractors to the buildings in one of the following two ways:

Contractor 1 to building 1	Contractor 1 to building 3
Contractor 2 to building 2 or	Contractor 2 to building 1
Contractor 3 to building 3	Contractor 3 to building 2

The method of brute force employed in this example quickly becomes impractical as the size of the cost matrix increases. For example, for a 10×10 cost matrix there are a total of 3,628,800 ($= 10!$) assignments possible. We shall now describe a practical method for solving any assignment problem. ◆

The Hungarian Method   Suppose that a particular assignment problem has the cost matrix

$$\begin{bmatrix} 0 & 5 & 7 & 0 & 3 \\ 0 & 4 & 0 & 2 & 5 \\ 6 & 0 & 3 & 7 & 7 \\ 7 & 9 & 4 & 0 & 0 \\ 3 & 2 & 0 & 0 & 1 \end{bmatrix} \tag{2}$$

Notice that all the entries of this cost matrix are nonnegative and that it contains many zero entries. Notice also that we can find an assignment consisting entirely of zero entries, namely,

$$\begin{bmatrix} \boxed{0} & 5 & 7 & 0 & 3 \\ 0 & 4 & \boxed{0} & 2 & 5 \\ 6 & \boxed{0} & 3 & 7 & 7 \\ 7 & 9 & 4 & 0 & \boxed{0} \\ 3 & 2 & 0 & \boxed{0} & 1 \end{bmatrix} \tag{3}$$

This assignment must be optimal, as its cost is zero and it is impossible to find an assignment with a cost less than zero if all the entries of the cost matrix are nonnegative.

Very few assignment problems are as easy to solve as the preceding one. However, the following theorem leads to a method of converting an arbitrary assignment problem to one that can be solved as easily.

Theorem 11.4.1   **Optimal Assignment**

If a number is added to or subtracted from all of the entries of any one row or column of a cost matrix, then an optimal assignment for the resulting cost matrix is also an optimal assignment for the original cost matrix.

To see why this theorem is true, suppose that five is added to each entry of the second row of a given cost matrix. Because each assignment contains exactly one entry from the second row, it follows that the cost of each assignment for the new matrix is exactly five more than the cost of the corresponding assignment for the original matrix. Thus, corresponding assignments preserve their ordering with respect to cost, so an optimal assignment of either matrix corresponds to an optimal assignment of the other. A similar argument holds if a number is added to any column of the cost matrix, or if subtraction rather than addition is used.

We now introduce the ***Hungarian method***, which is a five-step procedure for applying this theorem to a given cost matrix and obtaining one with nonnegative entries that contains an assignment consisting entirely of zero entries. Such an assignment (called an *optimal assignment of zeros*) will then be an optimal assignment for the original problem. The Hungarian method is outlined in Figure 11.4.1 for an $n \times n$ cost matrix. The first two steps use Theorem 11.4.1 to generate a cost matrix with nonnegative entries and with at least one zero entry in each row and column. The last three steps are applied iteratively as many times as necessary to generate a cost matrix that contains an optimal assignment of zeros. In Exercise 8 we ask the reader to show that each time Step 5 is applied, the sum of the entries of the new cost matrix generated is strictly less than the sum of the entries of the preceding cost matrix. This guarantees that the iterative process cannot continue indefinitely.

The Hungarian Method	
Steps	**Remarks**
1. Subtract the smallest entry in each row from all the entries of its row.	After this step, each row has at least one zero entry and all other entries are nonnegative.
2. Subtract the smallest entry in each column from all the entries of its column.	After this step, each row *and* column has at least one zero entry and all other entries are nonnegative.
3. Draw lines through appropriate rows and columns so that all the zero entries of the cost matrix are covered and the *minimum* number of such lines is used.	There may be several ways to do this. The important thing is that the *minimum* number of lines be used. Algorithms suitable for computer coding are available for this; however, for small values of n, trial and error will suffice.
4. *Test for Optimality.* (*i*) If the minimum number of covering lines is n, an optimal assignment of zeros is possible and we are finished. (*ii*) If the minimum number of covering lines is less than n, an optimal assignment of zeros is not yet possible. Proceed to Step 5.	See Exercise 7 for a justification of this test. If the test is affirmative, a judicious search will produce a set of n zero entries, no two of which lie in the same row or column. Algorithms, which we will not discuss, are available to systematically find such an optimal assignment of zeros.
5. Determine the smallest entry not covered by any line. Subtract this entry from all uncovered entries and then add it to all entries covered by both a horizontal and a vertical line. Return to Step 3.	This step is equivalent to applying Theorem 11.4.1 by subtracting the smallest uncovered entry from each uncovered row and then adding it to each covered column.

Figure 11.4.1

EXAMPLE 2 Minimizing Total Distance Traveled

A construction company has four large bulldozers located at four different garages. The bulldozers are to be moved to four different construction sites. The distances in miles between the bulldozers and construction sites are given in Table 2.

TABLE 2

		Construction Site			
		1	**2**	**3**	**4**
Bulldozer	**1**	90	75	75	80
	2	35	85	55	65
	3	125	95	90	105
	4	45	110	95	115

How should the bulldozers be moved to the construction sites in order to minimize the total distance traveled?

Solution.

We shall apply the Hungarian method to matrix (4), which is the cost matrix for the problem.

$$\begin{bmatrix} 90 & 75 & 75 & 80 \\ 35 & 85 & 55 & 65 \\ 125 & 95 & 90 & 105 \\ 45 & 110 & 95 & 115 \end{bmatrix} \tag{4}$$

Step 1. Subtract 75 from the first row of matrix (4), subtract 35 from its second row, subtract 90 from its third row, and subtract 45 from its fourth row to obtain matrix (5).

$$\begin{bmatrix} 15 & 0 & 0 & 5 \\ 0 & 50 & 20 & 30 \\ 35 & 5 & 0 & 15 \\ 0 & 65 & 50 & 70 \end{bmatrix} \tag{5}$$

Step 2. The first three columns of matrix (5) already contain zero entries; therefore, we need only subtract 5 from its fourth column. The result is matrix (6).

$$\begin{bmatrix} 15 & 0 & 0 & 0 \\ 0 & 50 & 20 & 25 \\ 35 & 5 & 0 & 10 \\ 0 & 65 & 50 & 65 \end{bmatrix} \tag{6}$$

Step 3. Cover the zero entries of matrix (6) with a minimum number of vertical and horizontal lines. This may be done by first trying to cover the zeros with one line, then with two, and finally with three. The indicated covering is not unique. (See Exercise 3 for an alternative covering.)

Step 4. Because the minimum number of lines used in Step 3 is three, an optimal assignment of zeros is not yet possible.

Step 5. Subtract 20, the smallest uncovered entry of matrix (6), from each of its uncovered entries and add it to the two entries covered twice with lines. The result is matrix (7).

$$\begin{bmatrix} 35 & 0 & 0 & 0 \\ 0 & 30 & 0 & 5 \\ 55 & 5 & 0 & 10 \\ 0 & 45 & 30 & 45 \end{bmatrix} \quad (7)$$

Step 6. (Step 3 of Figure 11.4.1.) Cover the zero entries of matrix (7) with a minimum number of vertical and horizontal lines.

Step 7. Because the minimum number of lines is still three, an optimal assignment of zeros is not yet possible.

Step 8. Subtract 5, the smallest uncovered entry of matrix (7), from each of its uncovered entries and add it to the two entries covered twice with lines. The result is matrix (8).

$$\begin{bmatrix} 40 & 0 & 5 & 0 \\ 0 & 25 & 0 & 0 \\ 55 & 0 & 0 & 5 \\ 0 & 40 & 30 & 40 \end{bmatrix} \quad (8)$$

Step 9. (Step 3 of Figure 11.4.1.) Cover the zero entries of matrix (8) with a minimum number of vertical and horizontal lines.

Step 10. Because the zero entries of matrix (8) cannot be covered with fewer than four lines, it must contain an optimal assignment of zeros.

By trial and error, we can find the following two optimal assignments of zeros in matrix (8):

$$\begin{bmatrix} 40 & 0 & 5 & \boxed{0} \\ 0 & 25 & \boxed{0} & 0 \\ 55 & \boxed{0} & 0 & 5 \\ \boxed{0} & 40 & 30 & 40 \end{bmatrix} \qquad \begin{bmatrix} 40 & \boxed{0} & 5 & 0 \\ 0 & 25 & 0 & \boxed{0} \\ 55 & 0 & \boxed{0} & 5 \\ \boxed{0} & 40 & 30 & 40 \end{bmatrix} \quad (9)$$

$$\text{(a)} \qquad\qquad\qquad\qquad \text{(b)}$$

Assignment (a) leads to the following movement of bulldozers to construction sites:

Bulldozer 1 to construction site 4

Bulldozer 2 to construction site 3

Bulldozer 3 to construction site 2

Bulldozer 4 to construction site 1

From Table 2 the corresponding minimum distance traveled is

$$80 + 55 + 95 + 45 = 275 \text{ miles}$$

Similarly, assignment (b) leads to the alternative solution

Bulldozer 1 to construction site 2

Bulldozer 2 to construction site 4

Bulldozer 3 to construction site 3

Bulldozer 4 to construction site 1

with the same minimum distance traveled

$$75 + 65 + 90 + 45 = 275 \text{ miles}$$ ◆

Before giving another example, we should mention that the assignment problems and associated cost matrices that can be solved by the Hungarian method must satisfy the following three conditions:

1. *The cost matrix must be square.* In the next example we shall discuss a procedure for handling assignment problems for which the cost matrix is not square.

2. *The entries of the cost matrix should be integers.* For hand calculations this is more a convenience than anything else. But for machine calculations this allows for use of exact integer arithmetic and avoids roundoff error. For practical problems, noninteger entries can always be converted to integer entries by multiplying the cost matrix by a suitable power of ten.

3. *The problem must be one of minimization.* The problem of maximizing the sum of entries of a cost matrix is easily converted to one of minimizing the sum by multiplying each entry of the cost matrix by -1.

EXAMPLE 3 The Bride–Groom Problem

A marriage broker has four female clients and five male clients who desire to be married. She ranks the possible matchings between her clients on a scale of zero to ten: zero for the poorest match and ten for the best match. Her rankings are given in Table 3.

TABLE 3

| | | \multicolumn{5}{c}{**Prospective Grooms**} |
		Bob	**Tom**	**Joe**	**Hal**	**Don**
Prospective Brides	**Sue**	7	4	7	3	10
	Ann	5	9	3	8	7
	Bea	3	5	6	2	9
	Fay	6	5	0	4	8

How should she match her clients in order to maximize the sum of the rankings of the matches?

Solution.

Because there is one more prospective groom than prospective bride, one of the grooms cannot be matched. Thus, the cost matrix is not square and the Hungarian method cannot be applied directly. To circumvent this problem we introduce a "dummy" prospective bride whose match with any of the five grooms has a rank of zero: The prospective groom matched with the dummy bride is then, in reality, the groom not matched. We therefore add a row of zeros to Table 3 corresponding to the dummy bride and in this way are led to the following square cost matrix:

$$\begin{bmatrix} 7 & 4 & 7 & 3 & 10 \\ 5 & 9 & 3 & 8 & 7 \\ 3 & 5 & 6 & 2 & 9 \\ 6 & 5 & 0 & 4 & 8 \\ 0 & 0 & 0 & 0 & 0 \end{bmatrix} \tag{10}$$

The problem as stated is one of maximizing a sum. We convert it to one of minimizing a sum by multiplying each entry of matrix (10) by -1 to obtain

$$\begin{bmatrix} -7 & -4 & -7 & -3 & -10 \\ -5 & -9 & -3 & -8 & -7 \\ -3 & -5 & -6 & -2 & -9 \\ -6 & -5 & 0 & -4 & -8 \\ 0 & 0 & 0 & 0 & 0 \end{bmatrix} \tag{11}$$

We now apply the Hungarian method to matrix (11) to find an optimal assignment.

Step 1. Subtract -10 from the first row of matrix (11), subtract -9 from its second row, subtract -9 from its third row, and subtract -8 from its fourth row to obtain matrix (12).

$$\begin{bmatrix} 3 & 6 & 3 & 7 & 0 \\ 4 & 0 & 6 & 1 & 2 \\ 6 & 4 & 3 & 7 & 0 \\ 2 & 3 & 8 & 4 & 0 \\ 0 & 0 & 0 & 0 & 0 \end{bmatrix} \tag{12}$$

Step 2. Because all the columns of matrix (12) contain zero entries, Step 2 in Figure 11.4.1 is not needed.

Step 3. Cover the zero entries of matrix (12) with a minimum number of vertical and horizontal lines.

Step 4. Because the minimum number of lines used in Step 3 is three, an optimal assignment of zeros is not possible.

Step 5. Subtract 2, the smallest uncovered entry of matrix (12), from each of its uncovered entries and add it to the two entries covered twice with lines. The result is matrix (13).

$$\begin{bmatrix} 1 & 4 & 1 & 5 & 0 \\ 4 & 0 & 6 & 1 & 4 \\ 4 & 2 & 1 & 5 & 0 \\ 0 & 1 & 6 & 2 & 0 \\ 0 & 0 & 0 & 0 & 2 \end{bmatrix} \tag{13}$$

Step 6. (Step 3 of Figure 11.4.1.) Cover the zero entries of matrix (13) with a minimum number of vertical and horizontal lines.

Step 7. Because the minimum number of lines is four, an optimal assignment of zeros is not yet possible.

Step 8. Subtract 1, the smallest uncovered entry of matrix (13), from each of its uncovered entries and add it to the three entries covered twice with lines. The result is matrix (14).

$$\begin{bmatrix} 0 & 3 & 0 & 4 & 0 \\ 4 & 0 & 6 & 1 & 5 \\ 3 & 1 & 0 & 4 & 0 \\ 0 & 1 & 6 & 2 & 1 \\ 0 & 0 & 0 & 0 & 3 \end{bmatrix} \tag{14}$$

Step 9. (Step 3 of Figure 11.4.1.) Cover the zero entries of matrix (14) with a minimum number of vertical and horizontal lines.

Step 10. Because the zero entries of matrix (14) cannot be covered with fewer than five lines, it must contain an optimal assignment of zeros. Such an assignment is given in matrix (15).

$$\begin{bmatrix} 0 & 3 & \boxed{0} & 4 & 0 \\ 4 & \boxed{0} & 6 & 1 & 5 \\ 3 & 1 & 0 & 4 & \boxed{0} \\ \boxed{0} & 1 & 6 & 2 & 1 \\ 0 & 0 & 0 & \boxed{0} & 3 \end{bmatrix} \tag{15}$$

The optimal assignment indicated in (15) leads to the following bride–groom matchings:

Sue–Joe	(rank $= 7$)
Ann–Tom	(rank $= 9$)
Bea–Don	(rank $= 9$)
Fay–Bob	(rank $= 6$)
Hal (unmatched)	

The resulting maximum ranking sum is $7 + 9 + 9 + 6 = 31$. In Exercise 4 we ask the reader to find an alternative optimal assignment of zeros in matrix (15) and to verify that it leads to the same maximum ranking sum. ◆

Exercise Set 11.4

1. Find an optimal assignment and corresponding cost for each of the following three cost matrices using the Hungarian method.

(a) $\begin{bmatrix} 17 & 4 & 10 \\ 15 & 5 & 8 \\ 18 & 7 & 11 \end{bmatrix}$ (b) $\begin{bmatrix} 3 & -2 & 0 & 1 \\ 5 & 3 & -3 & 4 \\ 2 & 7 & 5 & 3 \\ 5 & -2 & 0 & 1 \end{bmatrix}$ (c) $\begin{bmatrix} 12 & 9 & 7 & 7 & 10 \\ 15 & 11 & 8 & 13 & 14 \\ 9 & 6 & 5 & 12 & 12 \\ 6 & 9 & 13 & 7 & 10 \\ 8 & 13 & 12 & 9 & 13 \end{bmatrix}$

2. Solve Example 1 by the Hungarian method.

3. Find a set of three lines that covers the zero entries of matrix (6) in Example 2 different from the set given. Then use the Hungarian method and verify that it leads to the same optimal assignments of zeros given in (9).

4. Find an optimal assignment of zeros in matrix (15) of Example 3 different from the one given. Verify that the new assignment leads to the same cost as the one in the example.

5. A coin dealer is to sell four coins through a mail auction. Bids are received for each of the four coins from five bidders with instructions from each bidder that at most one of his bids is to be honored. The bids are given in Table 4.

TABLE 4

	Bids			
	Coin 1	Coin 2	Coin 3	Coin 4
Bidder 1	$150	$65	$210	$135
Bidder 2	175	75	230	155
Bidder 3	135	85	200	140
Bidder 4	140	70	190	130
Bidder 5	170	50	200	160

(a) How should the dealer assign the four coins in order to maximize the sum of the resulting bids? (Notice that a "dummy" coin must be added to produce a square cost matrix. Whichever bidder receives the dummy coin does not receive any real coin.)

(b) Suppose that bidder 2 instructed the dealer that at most two of his bids are to be honored. How can the problem be modified, and what is an optimal assignment for the new problem?

6. A Little League manager has nine players to assign to the nine positions of a baseball team. The manager ranks each of the players on a scale from 0 to 25 for each of the positions, taking into account both the ability of the player in the position and the importance of the position to the game. Referring to Table 5, how should the manager assign the nine players in order to maximize the sum of the rankings?

TABLE 5

	Player								
Position	Sam	Jill	John	Liz	Ann	Lois	Pete	Alex	Herb
P	20	15	10	10	17	23	25	5	15
C	10	10	12	15	9	7	8	7	8
1B	12	9	9	10	10	5	7	13	9
2B	13	14	10	15	15	5	8	20	10
3B	12	13	10	15	14	5	9	20	10
SS	15	14	15	16	15	5	10	20	10
LF	7	9	12	12	7	6	7	15	12
CF	5	6	8	8	5	4	5	10	7
RF	5	6	8	8	5	4	5	10	7

7. Prove part (*ii*) of the test for optimality in Step 4 of the Hungarian method (Figure 11.4.1) by showing that the n zero entries of an optimal assignment of zeros cannot be covered with fewer than n lines. [**Remark.** The proof of part (*i*), which was given in 1931 by two Hungarian mathematicians, D. König and E. Egerváry, is quite difficult. For completeness, the proof is given in the solutions manual for this text.]

8. (a) Let the zero entries of an $n \times n$ nonnegative cost matrix C be covered with m lines, with $m < n$. Let $a > 0$ be the smallest uncovered entry, and let C' be the cost matrix obtained by applying Step 5 of the Hungarian method to C. Show that

$$\text{(Sum of all entries of } C) - \text{(Sum of all entries of } C') = an(n - m)$$

[**Hint.** Notice that Step 5 is equivalent to adding a to every entry of C covered by one line, adding $2a$ to every entry covered by two lines, and subtracting a from every entry.]

   (b) Use the result of part (a) to show that the Hungarian method will produce a cost matrix with an optimal assignment of zeros in a finite number of steps.

Technology Exercises 11.4

The following exercises are designed to be solved using a technology utility. Typically, this will be MATLAB, *Mathematica*, Maple, Derive, or Mathcad, but it may also be some other type of linear algebra software or a scientific calculator with some linear algebra capabilities. For each exercise you will need to read the relevant documentation for the particular utility you are using. The goal of these exercises is to provide you with a basic proficiency with your technology utility. Once you have mastered the techniques in these exercises, you will be able to use your technology utility to solve many of the problems in the regular exercise sets.

T1. Use the Hungarian method to determine the optimal assignment for the 10×10 cost matrix

$$C = \begin{bmatrix} 2 & 3 & 4 & 5 & 6 & 7 & 8 & 9 & 10 & 1 \\ 3 & 4 & 5 & 6 & 7 & 8 & 9 & 10 & 11 & 2 \\ 4 & 5 & 6 & 7 & 8 & 9 & 10 & 11 & 12 & 3 \\ 5 & 6 & 7 & 8 & 9 & 10 & 11 & 12 & 13 & 4 \\ 6 & 7 & 8 & 9 & 10 & 11 & 12 & 13 & 14 & 5 \\ 7 & 8 & 9 & 10 & 11 & 12 & 13 & 14 & 15 & 6 \\ 8 & 9 & 10 & 11 & 12 & 13 & 14 & 15 & 16 & 7 \\ 9 & 10 & 11 & 12 & 13 & 14 & 15 & 16 & 17 & 8 \\ 10 & 11 & 12 & 13 & 14 & 15 & 16 & 17 & 18 & 9 \\ 1 & 2 & 3 & 4 & 5 & 6 & 7 & 8 & 9 & 0 \end{bmatrix}$$

What is interesting about your results?

T2. If you are working with a CAS, use it to determine the most general (a) 2×2, (b) 3×3, (c) 4×4, and (d) $n \times n$ matrices with the property that all assignments have the same cost, so that all assignments are optimal.

11.5   CUBIC SPLINE INTERPOLATION

In this section an artist's drafting aid is used as a physical model for the mathematical problem of finding a curve that passes through specified points in the plane. The parameters of the curve are determined by solving a linear system of equations.

PREREQUISITES: Linear Systems
Matrix Algebra
Differential Calculus

Curve Fitting   Fitting a curve through specified points in the plane is a common problem encountered in analyzing experimental data, in ascertaining the relations among variables, and in design work. In Figure 11.5.1 seven points in the xy-plane are displayed, and in Figure 11.5.2 a smooth curve has been drawn that passes through them. A curve that passes through a set of points in the plane is said to **interpolate** those points, and the curve is called an **interpolating curve** for those points. The interpolating curve in Figure 11.5.2 was drawn with the aid of a *drafting spline* (Figure 11.5.3). This drafting aid consists of a thin, flexible strip of wood or other material that is bent to pass through the points to be interpolated. Attached sliding weights hold the spline in position while the artist draws the interpolating curve. The drafting spline will serve as the physical model for a mathematical theory of interpolation that we will discuss in this section.

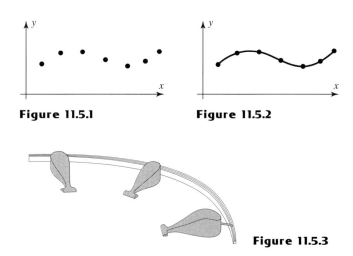

Figure 11.5.1   **Figure 11.5.2**

Figure 11.5.3

Statement of the Problem   Suppose that we are given n points in the xy-plane,

$$(x_1, y_1), (x_2, y_2), \ldots, (x_n, y_n)$$

which we wish to interpolate with a "well-behaved" curve (Figure 11.5.4). For convenience, we take the points to be equally spaced in the x-direction, although our results can easily be extended to the case of unequally spaced points. If we let the common distance between the x-coordinates of the points be h, then we have

$$x_2 - x_1 = x_3 - x_2 = \cdots = x_n - x_{n-1} = h$$

Let $y = S(x)$, $x_1 \leq x \leq x_n$ denote the interpolating curve that we seek. We assume that this curve describes the displacement of a drafting spline that interpolates the n points

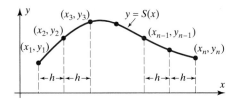

Figure 11.5.4

when the weights holding down the spline are situated precisely at the n points. It is known from linear beam theory that for small displacements the fourth derivative of the displacement of a beam is zero along any interval of the x-axis that contains no external forces acting on the beam. If we treat our drafting spline as a thin beam and realize that the only external forces acting on it arise from the weights at the n specified points, then it follows that

$$S^{(\text{iv})}(x) \equiv 0 \tag{1}$$

for values of x lying in the $n - 1$ open intervals

$$(x_1, x_2), (x_2, x_3), \ldots, (x_{n-1}, x_n)$$

between the n points.

We also need the result from linear beam theory which states that for a beam acted upon only by external forces the displacement must have two continuous derivatives. In the case of the interpolating curve $y = S(x)$ constructed by the drafting spline, this means that $S(x)$, $S'(x)$, and $S''(x)$ must be continuous for $x_1 \leq x \leq x_n$.

The condition that $S''(x)$ be continuous is what causes a drafting spline to produce a pleasing curve, as it results in continuous *curvature*. The eye can perceive sudden changes in curvature—that is, discontinuities in $S''(x)$—but sudden changes in higher derivatives are not discernible. Thus, the condition that $S''(x)$ be continuous is the minimal prerequisite for the interpolating curve to be perceptible as a single smooth curve, rather than as a series of separate curves pieced together.

To determine the mathematical form of the function $S(x)$, we observe that because $S^{(\text{iv})}(x) \equiv 0$ in the intervals between the n specified points, it follows by integrating this equation four times that $S(x)$ must be a *cubic polynomial* in x in each such interval. In general, however, $S(x)$ will be a different cubic polynomial in each interval, so $S(x)$ must have the form

$$S(x) = \begin{cases} S_1(x), & x_1 \leq x \leq x_2 \\ S_2(x), & x_2 \leq x \leq x_3 \\ \vdots \\ S_{n-1}(x), & x_{n-1} \leq x \leq x_n \end{cases} \tag{2}$$

where $S_1(x), S_2(x), \ldots, S_{n-1}(x)$ are cubic polynomials. For convenience, we will write these in the form

$$S_1(x) = a_1(x - x_1)^3 + b_1(x - x_1)^2 + c_1(x - x_1) + d_1, \qquad x_1 \leq x \leq x_2$$
$$S_2(x) = a_2(x - x_2)^3 + b_2(x - x_2)^2 + c_2(x - x_2) + d_2, \qquad x_2 \leq x \leq x_3$$
$$\vdots \tag{3}$$
$$S_{n-1}(x) = a_{n-1}(x - x_{n-1})^3 + b_{n-1}(x - x_{n-1})^2$$
$$+ c_{n-1}(x - x_{n-1}) + d_{n-1}, \qquad x_{n-1} \leq x \leq x_n$$

The a_i's, b_i's, c_i's, and d_i's constitute a total of $4n - 4$ coefficients that we must determine to completely specify $S(x)$. If we choose these coefficients so that $S(x)$ interpolates the n specified points in the plane and $S(x)$, $S'(x)$, and $S''(x)$ are continuous, then the resulting interpolating curve is called a ***cubic spline***.

Derivation of the Formula of a Cubic Spline   From Equations (2) and (3) we have

$$S(x) = S_1(x) = a_1(x - x_1)^3 + b_1(x - x_1)^2 + c_1(x - x_1) + d_1, \qquad x_1 \le x \le x_2$$
$$S(x) = S_2(x) = a_2(x - x_2)^3 + b_2(x - x_2)^2 + c_2(x - x_2) + d_2, \qquad x_2 \le x \le x_3$$
$$\vdots \qquad \vdots$$
$$S(x) = S_{n-1}(x) = a_{n-1}(x - x_{n-1})^3 + b_{n-1}(x - x_{n-1})^2$$
$$+ c_{n-1}(x - x_{n-1}) + d_{n-1}, \qquad x_{n-1} \le x \le x_n$$

$$(4)$$

so that

$$S'(x) = S_1'(x) = 3a_1(x - x_1)^2 + 2b_1(x - x_1) + c_1, \qquad x_1 \le x \le x_2$$
$$S'(x) = S_2'(x) = 3a_2(x - x_2)^3 + 2b_2(x - x_2) + c_2, \qquad x_2 \le x \le x_3$$
$$\vdots \qquad \vdots$$
$$S'(x) = S_{n-1}'(x) = 3a_{n-1}(x - x_{n-1})^2 + 2b_{n-1}(x - x_{n-1}) + c_{n-1}, \qquad x_{n-1} \le x \le x_n$$

$$(5)$$

and

$$S''(x) = S_1''(x) = 6a_1(x - x_1) + 2b_1, \qquad x_1 \le x \le x_2$$
$$S''(x) = S_2''(x) = 6a_2(x - x_2) + 2b_2, \qquad x_2 \le x \le x_3$$
$$\vdots \qquad \vdots$$
$$S''(x) = S_{n-1}''(x) = 6a_{n-1}(x - x_{n-1}) + 2b_{n-1}, \qquad x_{n-1} \le x \le x_n$$

$$(6)$$

We will now use these equations and the four properties of cubic splines stated below to express the unknown coefficients a_i, b_i, c_i, d_i, $i = 1, 2, \ldots, n - 1$, in terms of the known coordinates $y_1, y_2, \ldots, y_n$.

1. $S(x)$ *interpolates the points* (x_i, y_i), $i = 1, 2, \ldots, n$.

Because $S(x)$ interpolates the points (x_i, y_i), $i = 1, 2, \ldots, n$, we have

$$S(x_1) = y_1, \quad S(x_2) = y_2, \ldots, \quad S(x_n) = y_n \qquad (7)$$

From the first $n - 1$ of these equations and (4) we obtain

$$d_1 = y_1$$
$$d_2 = y_2$$
$$\vdots$$
$$d_{n-1} = y_{n-1}$$

$$(8)$$

From the last equation in (7), the last equation in (4), and the fact that $x_n - x_{n-1} = h$, we obtain

$$a_{n-1}h^3 + b_{n-1}h^2 + c_{n-1}h + d_{n-1} = y_n \qquad (9)$$

2. $S(x)$ *is continuous on* $[x_1, x_n]$.

Because $S(x)$ is continuous for $x_1 \le x \le x_n$, it follows that at each point x_i in the set $x_2, x_3, \ldots, x_{n-1}$ we must have

$$S_{i-1}(x_i) = S_i(x_i), \qquad i = 2, 3, \ldots, n - 1 \qquad (10)$$

Otherwise, the graphs of $S_{i-1}(x)$ and $S_i(x)$ would not join together to form a continuous curve at x_i. Using the interpolating property $S_i(x_i) = y_i$, it follows from

(10) that $S_{i-1}(x_i) = y_i$, $i = 2, 3, \ldots, n - 1$, or from (4) that

$$
\begin{aligned}
a_1 h^3 + b_1 h^2 + c_1 h + d_1 &= y_2 \\
a_2 h^3 + b_2 h^2 + c_2 h + d_2 &= y_3 \\
&\;\;\vdots \\
a_{n-2} h^3 + b_{n-2} h^2 + c_{n-2} h + d_{n-2} &= y_{n-1}
\end{aligned}
\tag{11}
$$

3. $S'(x)$ *is continuous on* $[x_1, x_n]$.

Because $S'(x)$ is continuous for $x_1 \leq x \leq x_n$, it follows that

$$
S'_{i-1}(x_i) = S'_i(x_i), \qquad i = 2, 3, \ldots, n - 1
$$

or from (5)

$$
\begin{aligned}
3a_1 h^2 + 2b_1 h + c_1 &= c_2 \\
3a_2 h^2 + 2b_2 h + c_2 &= c_3 \\
&\;\;\vdots \\
3a_{n-2} h^2 + 2b_{n-2} h + c_{n-2} &= c_{n-1}
\end{aligned}
\tag{12}
$$

4. $S''(x)$ *is continuous on* $[x_1, x_2]$.

Because $S''(x)$ is continuous for $x_1 \leq x \leq x_n$, it follows that

$$
S''_{i-1}(x_i) = S''_i(x_i), \qquad i = 2, 3, \ldots, n - 1
$$

or from (6)

$$
\begin{aligned}
6a_1 h + 2b_1 &= 2b_2 \\
6a_2 h + 2b_2 &= 2b_3 \\
&\;\;\vdots \\
6a_{n-2} h + 2b_{n-2} &= 2b_{n-1}
\end{aligned}
\tag{13}
$$

Equations (8), (9), (11), (12), and (13) constitute a system of $4n - 6$ linear equations in the $4n - 4$ unknown coefficients a_i, b_i, c_i, d_i, $i = 1, 2, \ldots, n - 1$. Consequently, we need two more equations to determine these coefficients uniquely. Before obtaining these additional equations, however, we can simplify our existing system by expressing the unknowns a_i, b_i, c_i, and d_i in terms of new unknown quantities

$$
M_1 = S''(x_1), \quad M_2 = S''(x_2), \ldots, \quad M_n = S''(x_n)
$$

and the known quantities

$$
y_1, y_2, \ldots, y_n
$$

For example, from (6) it follows that

$$
\begin{aligned}
M_1 &= 2b_1 \\
M_2 &= 2b_2 \\
&\;\;\vdots \\
M_{n-1} &= 2b_{n-1}
\end{aligned}
$$

so that

$$
b_1 = \tfrac{1}{2} M_1, \quad b_2 = \tfrac{1}{2} M_2, \ldots, \quad b_{n-1} = \tfrac{1}{2} M_{n-1}
$$

Moreover, we already know from (8) that

$$
d_1 = y_1, \quad d_2 = y_2, \ldots, \quad d_{n-1} = y_{n-1}
$$

We leave it as an exercise for the reader to derive the expressions for the a_i's and c_i's in terms of the M_i's and y_i's. The final result is as follows:

Theorem 11.5.1 **Cubic Spline Interpolation**

Given n points (x_1, y_1), (x_2, y_2), ..., (x_n, y_n) with $x_{i+1} - x_i = h$, $i = 1, 2, ..., n-1$, the cubic spline

$$S(x) = \begin{cases} a_1(x - x_1)^3 + b_1(x - x_1)^2 + c_1(x - x_1) + d_1, & x_1 \leq x \leq x_2 \\ a_2(x - x_2)^3 + b_2(x - x_2)^2 + c_2(x - x_2) + d_2, & x_2 \leq x \leq x_3 \\ \qquad \vdots \\ a_{n-1}(x - x_{n-1})^3 + b_{n-1}(x - x_{n-1})^2 \\ \qquad\qquad + c_{n-1}(x - x_{n-1}) + d_{n-1}, & x_{n-1} \leq x \leq x_n \end{cases}$$

that interpolates these points has coefficients given by

$$\begin{aligned} a_i &= (M_{i+1} - M_i)/6h \\ b_i &= M_i/2 \\ c_i &= (y_{i+1} - y_i)/h - [(M_{i+1} + 2M_i)h/6] \\ d_i &= y_i \end{aligned} \qquad (14)$$

for $i = 1, 2, ..., n - 1$, where $M_i = S''(x_i)$, $i = 1, 2, ..., n$.

From this result, we see that the quantities $M_1, M_2, ..., M_n$ uniquely determine the cubic spline. To find these quantities we substitute the expressions for a_i, b_i, and c_i given in (14) into (12). After some algebraic simplification, we obtain

$$\begin{aligned} M_1 + 4M_2 + M_3 &= 6(y_1 - 2y_2 + y_3)/h^2 \\ M_2 + 4M_3 + M_4 &= 6(y_2 - 2y_3 + y_4)/h^2 \\ &\vdots \\ M_{n-2} + 4M_{n-1} + M_n &= 6(y_{n-2} - 2y_{n-1} + y_n)/h^2 \end{aligned} \qquad (15)$$

or, in matrix form,

$$\begin{bmatrix} 1 & 4 & 1 & 0 & \cdots & 0 & 0 & 0 & 0 \\ 0 & 1 & 4 & 1 & \cdots & 0 & 0 & 0 & 0 \\ 0 & 0 & 1 & 4 & \cdots & 0 & 0 & 0 & 0 \\ \vdots & \vdots & \vdots & \vdots & & \vdots & \vdots & \vdots & \vdots \\ 0 & 0 & 0 & 0 & \cdots & 4 & 1 & 0 & 0 \\ 0 & 0 & 0 & 0 & \cdots & 1 & 4 & 1 & 0 \\ 0 & 0 & 0 & 0 & \cdots & 0 & 1 & 4 & 1 \end{bmatrix} \begin{bmatrix} M_1 \\ M_2 \\ M_3 \\ M_4 \\ \vdots \\ M_{n-3} \\ M_{n-2} \\ M_{n-1} \\ M_n \end{bmatrix} = \frac{6}{h^2} \begin{bmatrix} y_1 - 2y_2 + y_3 \\ y_2 - 2y_3 + y_4 \\ y_3 - 2y_4 + y_5 \\ \vdots \\ y_{n-4} - 2y_{n-3} + y_{n-2} \\ y_{n-3} - 2y_{n-2} + y_{n-1} \\ y_{n-2} - 2y_{n-1} + y_n \end{bmatrix}$$

This is a linear system of $n - 2$ equations for the n unknowns $M_1, M_2, ..., M_n$. Thus, we still need two additional equations to determine $M_1, M_2, ..., M_n$ uniquely. The reason for this is that there are infinitely many cubic splines that interpolate the given points, so that we simply do not have enough conditions to determine a unique cubic spline passing through the points. We discuss below three possible ways of specifying the two additional conditions required to obtain a unique cubic spline through the points. (The exercises present two more.) They are summarized in Table 1 shown on the following page.

TABLE 1

Natural Spline	The second derivative of the spline is zero at the endpoints.	$M_1 = 0$ $M_n = 0$	$$\begin{bmatrix} 4 & 1 & 0 & \cdots & 0 & 0 & 0 \\ 1 & 4 & 1 & \cdots & 0 & 0 & 0 \\ \vdots & \vdots & \vdots & & \vdots & \vdots & \vdots \\ 0 & 0 & 0 & \cdots & 1 & 4 & 1 \\ 0 & 0 & 0 & \cdots & 0 & 1 & 4 \end{bmatrix} \begin{bmatrix} M_2 \\ M_3 \\ \vdots \\ M_{n-2} \\ M_{n-1} \end{bmatrix} = \frac{6}{h^2} \begin{bmatrix} y_1 - 2y_2 + y_3 \\ y_2 - 2y_3 + y_4 \\ \vdots \\ y_{n-2} - 2y_{n-1} + y_n \end{bmatrix}$$
Parabolic Runout Spline	The spline reduces to a parabolic curve on the first and last intervals.	$M_1 = M_2$ $M_n = M_{n-1}$	$$\begin{bmatrix} 5 & 1 & 0 & \cdots & 0 & 0 & 0 \\ 1 & 4 & 1 & \cdots & 0 & 0 & 0 \\ \vdots & \vdots & \vdots & & \vdots & \vdots & \vdots \\ 0 & 0 & 0 & \cdots & 1 & 4 & 1 \\ 0 & 0 & 0 & \cdots & 0 & 1 & 5 \end{bmatrix} \begin{bmatrix} M_2 \\ M_3 \\ \vdots \\ M_{n-2} \\ M_{n-1} \end{bmatrix} = \frac{6}{h^2} \begin{bmatrix} y_1 - 2y_2 + y_3 \\ y_2 - 2y_3 + y_4 \\ \vdots \\ y_{n-2} - 2y_{n-1} + y_n \end{bmatrix}$$
Cubic Runout Spline	The spline is a single cubic curve on the first two and last two intervals.	$M_1 = 2M_2 - M_3$ $M_n = 2M_{n-1} - M_{n-2}$	$$\begin{bmatrix} 6 & 0 & 0 & \cdots & 0 & 0 & 0 \\ 1 & 4 & 1 & \cdots & 0 & 0 & 0 \\ \vdots & \vdots & \vdots & & \vdots & \vdots & \vdots \\ 0 & 0 & 0 & \cdots & 1 & 4 & 1 \\ 0 & 0 & 0 & \cdots & 0 & 0 & 6 \end{bmatrix} \begin{bmatrix} M_2 \\ M_3 \\ \vdots \\ M_{n-2} \\ M_{n-1} \end{bmatrix} = \frac{6}{h^2} \begin{bmatrix} y_1 - 2y_2 + y_3 \\ y_2 - 2y_3 + y_4 \\ \vdots \\ y_{n-2} - 2y_{n-1} + y_n \end{bmatrix}$$

The Natural Spline

The two simplest mathematical conditions we can impose are

$$M_1 = M_n = 0$$

These conditions together with (15) result in an $n \times n$ linear system for $M_1, M_2, \ldots, M_n$, which can be written in matrix form as

$$\begin{bmatrix} 1 & 0 & 0 & 0 & \cdots & 0 & 0 & 0 \\ 1 & 4 & 1 & 0 & \cdots & 0 & 0 & 0 \\ 0 & 1 & 4 & 1 & \cdots & 0 & 0 & 0 \\ \vdots & \vdots & \vdots & \vdots & & \vdots & \vdots & \vdots \\ 0 & 0 & 0 & 0 & \cdots & 1 & 4 & 1 \\ 0 & 0 & 0 & 0 & \cdots & 0 & 0 & 1 \end{bmatrix} \begin{bmatrix} M_1 \\ M_2 \\ M_3 \\ \vdots \\ M_{n-1} \\ M_n \end{bmatrix} = \frac{6}{h^2} \begin{bmatrix} 0 \\ y_1 - 2y_2 + y_3 \\ y_2 - 2y_3 + y_4 \\ \vdots \\ y_{n-2} - 2y_{n-1} + y_n \\ 0 \end{bmatrix}$$

For numerical calculations it is more convenient to eliminate M_1 and M_n from this system and write

$$\begin{bmatrix} 4 & 1 & 0 & 0 & \cdots & 0 & 0 & 0 \\ 1 & 4 & 1 & 0 & \cdots & 0 & 0 & 0 \\ 0 & 1 & 4 & 1 & \cdots & 0 & 0 & 0 \\ \vdots & \vdots & \vdots & \vdots & & \vdots & \vdots & \vdots \\ 0 & 0 & 0 & 0 & \cdots & 1 & 4 & 1 \\ 0 & 0 & 0 & 0 & \cdots & 0 & 1 & 4 \end{bmatrix} \begin{bmatrix} M_2 \\ M_3 \\ M_4 \\ \vdots \\ M_{n-2} \\ M_{n-1} \end{bmatrix} = \frac{6}{h^2} \begin{bmatrix} y_1 - 2y_2 + y_3 \\ y_2 - 2y_3 + y_4 \\ y_3 - 2y_4 + y_5 \\ \vdots \\ y_{n-3} - 2y_{n-2} + y_{n-1} \\ y_{n-2} - 2y_{n-1} + y_n \end{bmatrix} \qquad (16)$$

together with

$$M_1 = 0 \qquad (17)$$

$$M_n = 0 \qquad (18)$$

Thus, the $(n-2) \times (n-2)$ linear system can be solved for the $n-2$ coefficients $M_2, M_3, \ldots, M_{n-1}$, and M_1 and M_n are determined by (17) and (18).

Physically, the natural spline results when the ends of a drafting spline extend freely beyond the interpolating points without constraint. The end portions of the spline outside the interpolating points will fall on straight line paths, causing $S''(x)$ to vanish at the endpoints x_1 and x_n and resulting in the mathematical conditions $M_1 = M_n = 0$.

The natural spline tends to flatten the interpolating curve at the endpoints, which may be undesirable. Of course, if it is required that $S''(x)$ vanish at the endpoints, then the natural spline must be used.

The Parabolic Runout Spline
The two additional constraints imposed for this type of spline are

$$M_1 = M_2 \tag{19}$$

$$M_n = M_{n-1} \tag{20}$$

If we use the preceding two equations to eliminate M_1 and M_n from (15), we obtain the $(n-2) \times (n-2)$ linear system

$$
\begin{bmatrix}
5 & 1 & 0 & 0 & \cdots & 0 & 0 & 0 \\
1 & 4 & 1 & 0 & \cdots & 0 & 0 & 0 \\
0 & 1 & 4 & 1 & \cdots & 0 & 0 & 0 \\
\vdots & \vdots & \vdots & \vdots & & \vdots & \vdots & \vdots \\
0 & 0 & 0 & 0 & \cdots & 1 & 4 & 1 \\
0 & 0 & 0 & 0 & \cdots & 0 & 1 & 5
\end{bmatrix}
\begin{bmatrix}
M_2 \\ M_3 \\ M_4 \\ \vdots \\ M_{n-2} \\ M_{n-1}
\end{bmatrix}
= \frac{6}{h^2}
\begin{bmatrix}
y_1 - 2y_2 + y_3 \\
y_2 - 2y_3 + y_4 \\
y_3 - 2y_4 + y_5 \\
\vdots \\
y_{n-3} - 2y_{n-2} + y_{n-1} \\
y_{n-2} - 2y_{n-1} + y_n
\end{bmatrix}
\tag{21}
$$

for $M_2, M_3, \ldots, M_{n-1}$. Once these $n - 2$ values have been determined, then M_1 and M_n are determined from (19) and (20).

From (14) we see that $M_1 = M_2$ implies that $a_1 = 0$, and $M_n = M_{n-1}$ implies that $a_{n-1} = 0$. Thus, from (3) there are no cubic terms in the formula for the spline over the end intervals $[x_1, x_2]$ and $[x_{n-1}, x_n]$. Hence, as the name suggests, the parabolic runout spline reduces to a parabolic curve over these end intervals.

The Cubic Runout Spline
For this type of spline we impose the two additional conditions

$$M_1 = 2M_2 - M_3 \tag{22}$$

$$M_n = 2M_{n-1} - M_{n-2} \tag{23}$$

Using these two equations to eliminate M_1 and M_n from (15) results in the following $(n-2) \times (n-2)$ linear system for $M_2, M_3, \ldots, M_{n-1}$:

$$
\begin{bmatrix}
6 & 0 & 0 & 0 & \cdots & 0 & 0 & 0 \\
1 & 4 & 1 & 0 & \cdots & 0 & 0 & 0 \\
0 & 1 & 4 & 1 & \cdots & 0 & 0 & 0 \\
\vdots & \vdots & \vdots & \vdots & & \vdots & \vdots & \vdots \\
0 & 0 & 0 & 0 & \cdots & 1 & 4 & 1 \\
0 & 0 & 0 & 0 & \cdots & 0 & 0 & 6
\end{bmatrix}
\begin{bmatrix}
M_2 \\ M_3 \\ M_4 \\ \vdots \\ M_{n-2} \\ M_{n-1}
\end{bmatrix}
= \frac{6}{h^2}
\begin{bmatrix}
y_1 - 2y_2 + y_3 \\
y_2 - 2y_3 + y_4 \\
y_3 - 2y_4 + y_5 \\
\vdots \\
y_{n-3} - 2y_{n-2} + y_{n-1} \\
y_{n-2} - 2y_{n-1} + y_n
\end{bmatrix}
\tag{24}
$$

After we solve this linear system for $M_2, M_3, \ldots, M_{n-1}$, we can use (22) and (23) to determine M_1 and M_n.

If we rewrite (22) as

$$M_2 - M_1 = M_3 - M_2$$

it follows from (14) that $a_1 = a_2$. Because $S'''(x) = 6a_1$ on $[x_1, x_2]$ and $S'''(x) = 6a_2$ on $[x_2, x_3]$, we see that $S'''(x)$ is constant over the entire interval $[x_1, x_3]$. Consequently, $S(x)$ consists of a *single* cubic curve over the interval $[x_1, x_3]$ rather than two different cubic curves pieced together at x_2. [To see this, integrate $S'''(x)$ three times.] A similar analysis shows that $S(x)$ consists of a single cubic curve over the last two intervals.

Whereas the natural spline tends to produce an interpolating curve that is flat at the endpoints, the cubic runout spline has the opposite tendency: it produces a curve with pronounced curvature at the endpoints. If neither behavior is desired, the parabolic runout spline is a reasonable compromise.

EXAMPLE 1 Using a Parabolic Runout Spline

The density of water is well known to reach a maximum at a temperature slightly above freezing. Table 2, from the *Handbook of Chemistry and Physics* (Cleveland, Ohio: Chemical Rubber Publishing Company), gives the density of water in grams per cubic centimeter for five equally spaced temperatures from $-10°C$ to $30°C$. We will interpolate these five temperature–density measurements with a parabolic runout spline and attempt to find the maximum density of water in this range by finding the maximum value on this cubic spline. In the exercises we ask the reader to perform similar calculations using a natural spline and a cubic runout spline to interpolate the data points.

TABLE 2

Temperature (°C)	Density (g/cm³)
−10	.99815
0	.99987
10	.99973
20	.99823
30	.99567

Set

$$x_1 = -10, \quad y_1 = .99815$$
$$x_2 = 0, \quad y_2 = .99987$$
$$x_3 = 10, \quad y_3 = .99973$$
$$x_4 = 20, \quad y_4 = .99823$$
$$x_5 = 30, \quad y_5 = .99567$$

Then

$$6[y_1 - 2y_2 + y_3]/h^2 = -.0001116$$
$$6[y_2 - 2y_3 + y_4]/h^2 = -.0000816$$
$$6[y_3 - 2y_4 + y_5]/h^2 = -.0000636$$

and the linear system (21) for the parabolic runout spline becomes

$$\begin{bmatrix} 5 & 1 & 0 \\ 1 & 4 & 1 \\ 0 & 1 & 5 \end{bmatrix} \begin{bmatrix} M_2 \\ M_3 \\ M_4 \end{bmatrix} = \begin{bmatrix} -.0001116 \\ -.0000816 \\ -.0000636 \end{bmatrix}$$

Solving this system yields

$$M_2 = -.00001973$$
$$M_3 = -.00001293$$
$$M_4 = -.00001013$$

From (19) and (20) we have

$$M_1 = M_2 = -.00001973$$
$$M_5 = M_4 = -.00001013$$

Solving for the a_i's, b_i's, c_i's, and d_i's in (14), we obtain the following expression for the interpolating parabolic runout spline:

$$S(x) = \begin{cases} -.00000987(x+10)^2 + .0002707(x+10) + .99815, & -10 \le x \le 0 \\ .000000113(x-0)^3 - .00000987(x-0)^2 + .0000733(x-0) + .99987, & 0 \le x \le 10 \\ .000000047(x-10)^3 - .00000647(x-10)^2 - .0000900(x-10) + .99973, & 10 \le x \le 20 \\ -.00000507(x-20)^2 - .0002053(x-20) + .99823, & 20 \le x \le 30 \end{cases}$$

This spline is plotted in Figure 11.5.5. From that figure we see that the maximum is attained in the interval [0, 10]. To find this maximum, we set $S'(x)$ equal to zero in the interval [0, 10]:

$$S'(x) = .000000339x^2 - .0000197x + .0000733 = 0$$

To three significant digits the root of this quadratic in the interval [0, 10] is $x = 3.99$, and for this value of x, $S(3.99) = 1.00001$. Thus, according to our interpolated estimate, the maximum density of water is 1.00001 g/cm^3 attained at 3.99°C. This agrees well with the experimental maximum density of 1.00000 g/cm^3 attained at 3.98°C. (In the original metric system, the gram was *defined* as the mass of one cubic centimeter of water at its maximum density.)   ♦

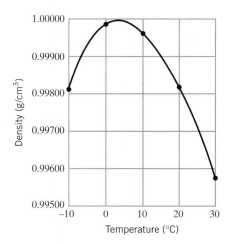

Figure 11.5.5

Closing Remarks   In addition to producing excellent interpolating curves, cubic splines and their generalizations are useful for numerical integration and differentiation, for the numerical solution of differential and integral equations, and in optimization theory.

Exercise Set 11.5

1. Derive the expressions for a_i and c_i in Equations (14) of Theorem 11.5.1.

2. The six points

$$(0, .00000), \quad (.2, .19867), \quad (.4, .38942), \quad (.6, .56464), \quad (.8, .71736), \quad (1.0, .84147)$$

lie on the graph of $y = \sin x$, where x is in radians.

(a) Find the parabolic runout spline that interpolates these six points for $.4 \le x \le .6$. Maintain an accuracy of five decimal places in your calculations.

(b) Calculate $S(.5)$ for the spline you found in part (a). What is the percentage error of $S(.5)$ with respect to the "exact" value of $\sin(.5) = .47943$?

3. The following five points

$$(0, 1), \quad (1, 7), \quad (2, 27), \quad (3, 79), \quad (4, 181)$$

lie on a single cubic curve.

(a) Which of the three types of cubic splines (natural, parabolic runout, or cubic runout) would agree exactly with the single cubic curve on which the five points lie?

(b) Determine the cubic spline you chose in part (a), and verify that it is a single cubic curve that interpolates the five points.

4. Repeat the calculations in Example 1 using a natural spline to interpolate the five data points.

5. Repeat the calculations in Example 1 using a cubic runout spline to interpolate the five data points.

6. (*The periodic spline*) If it is known or if it is desired that the n points (x_1, y_1), (x_2, y_2), ..., (x_n, y_n) to be interpolated lie on a single cycle of a periodic curve with period $x_n - x_1$, then an interpolating cubic spline $S(x)$ must satisfy

$$S(x_1) = S(x_n)$$
$$S'(x_1) = S'(x_n)$$
$$S''(x_1) = S''(x_n)$$

(a) Show that these three periodicity conditions require that

$$y_1 = y_n$$
$$M_1 = M_n$$
$$4M_1 + M_2 + M_{n-1} = 6(y_{n-1} - 2y_1 + y_2)/h^2$$

(b) Using the three equations in part (a) and Equations (15), construct an $(n-1) \times (n-1)$ linear system for $M_1, M_2, \ldots, M_{n-1}$ in matrix form.

7. (*The clamped spline*) Suppose that, in addition to the n points to be interpolated, we are given specific values y_1' and y_n' for the slopes $S'(x_1)$ and $S'(x_n)$ of the interpolating cubic spline at the endpoints x_1 and x_n.

(a) Show that

$$2M_1 + M_2 = 6(y_2 - y_1 - hy_1')/h^2$$
$$2M_n + M_{n-1} = 6(y_{n-1} - y_n + hy_n')/h^2$$

(b) Using the equations in part (a) and Equations (15), construct an $n \times n$ linear system for $M_1, M_2, \ldots, M_n$ in matrix form. [**Remark.** The clamped spline described in this exercise is the most accurate type of spline for interpolation work if the slopes at the endpoints are known or can be estimated.]

The following exercises are designed to be solved using a technology utility. Typically, this will be MATLAB, *Mathematica*, Maple, Derive, or Mathcad, but it may also be some other type of linear algebra software or a scientific calculator with some linear algebra capabilities. For each exercise you will need to read the relevant documentation for the particular utility you are using. The goal of these exercises is to provide you with a basic proficiency with your technology utility. Once you have mastered the techniques in these exercises, you will be able to use your technology utility to solve many of the problems in the regular exercise sets.

T1. In the solution of the natural cubic spline problem, it is necessary to solve a system of equations having coefficient matrix

$$A_n = \begin{bmatrix} 4 & 1 & 0 & \cdots & 0 & 0 & 0 \\ 1 & 4 & 1 & \cdots & 0 & 0 & 0 \\ \vdots & \vdots & \vdots & \ddots & \vdots & \vdots & \vdots \\ 0 & 0 & 0 & \cdots & 1 & 4 & 1 \\ 0 & 0 & 0 & \cdots & 0 & 1 & 4 \end{bmatrix}$$

If we could present a formula for the inverse of this matrix, then the solution for the natural cubic spline problem can be easily obtained. In this exercise and the next, we use a computer to discover this formula. Toward this end, we first determine an expression for the determinant of A_n, denoted by the symbol D_n. Given that

$$A_1 = [4] \quad \text{and} \quad A_2 = \begin{bmatrix} 4 & 1 \\ 1 & 4 \end{bmatrix}$$

we see that

$$D_1 = \det(A_1) = \det[4] = 4$$

and

$$D_2 = \det(A_2) = \det \begin{bmatrix} 4 & 1 \\ 1 & 4 \end{bmatrix} = 15$$

(a) Use the cofactor expansion of determinants to show that

$$D_n = 4D_{n-1} - D_{n-2}$$

for $n = 3, 4, 5, \ldots$. This says, for example, that

$$D_3 = 4D_2 - D_1 = 4(15) - 4 = 56$$
$$D_4 = 4D_3 - D_2 = 4(56) - 15 = 209$$

and so on. Using a computer, check this result for $5 \leq n \leq 10$.

(b) By writing

$$D_n = 4D_{n-1} - D_{n-2}$$

and the identity, $D_{n-1} = D_{n-1}$, in matrix form,

$$\begin{bmatrix} D_n \\ D_{n-1} \end{bmatrix} = \begin{bmatrix} 4 & -1 \\ 1 & 0 \end{bmatrix} \begin{bmatrix} D_{n-1} \\ D_{n-2} \end{bmatrix}$$

show that

$$\begin{bmatrix} D_n \\ D_{n-1} \end{bmatrix} = \begin{bmatrix} 4 & -1 \\ 1 & 0 \end{bmatrix}^{n-2} \begin{bmatrix} D_2 \\ D_1 \end{bmatrix} = \begin{bmatrix} 4 & -1 \\ 1 & 0 \end{bmatrix}^{n-2} \begin{bmatrix} 15 \\ 4 \end{bmatrix}$$

(c) Use the methods in Section 7.2 and a computer to show that

$$\begin{bmatrix} 4 & -1 \\ 1 & 0 \end{bmatrix}^{n-2} = \frac{\begin{bmatrix} (2+\sqrt{3})^{n-1} - (2-\sqrt{3})^{n-1} & (2-\sqrt{3})^{n-2} - (2+\sqrt{3})^{n-2} \\ (2+\sqrt{3})^{n-2} - (2-\sqrt{3})^{n-2} & (2-\sqrt{3})^{n-3} - (2+\sqrt{3})^{n-3} \end{bmatrix}}{2\sqrt{3}}$$

and hence

$$D_n = \frac{(2+\sqrt{3})^{n+1} - (2-\sqrt{3})^{n+1}}{2\sqrt{3}}$$

for $n = 1, 2, 3, \ldots$.

(d) Using a computer, check this result for $1 \leq n \leq 10$.

T2. In this exercise, we determine a formula for calculating A_n^{-1} from D_k for $k = 0, 1, 2, 3, \ldots, n$ assuming that D_0 is defined to be 1.

(a) Use a computer to compute A_k^{-1} for $k = 1, 2, 3, 4$, and 5.

(b) From your results in part (a), discover the conjecture that

$$A_n^{-1} = [\alpha_{ij}]$$

where $\alpha_{ij} = \alpha_{ji}$ and

$$\alpha_{ij} = (-1)^{i+j} \left(\frac{D_{n-j} D_{i-1}}{D_n} \right)$$

for $i \leq j$.

(c) Use the result in part (b) to compute A_7^{-1} and compare it to the result obtained using the computer.

11.6 MARKOV CHAINS

In this section we describe a general model of a system that changes from state to state. We then apply the model to several concrete problems.

PREREQUISITES: Linear Systems
Matrices
Intuitive Understanding of Limits

A Markov Process   Suppose a physical or mathematical system undergoes a process of change such that at any moment it can occupy one of a finite number of states. For example, the weather in a certain city could be in one of three possible states: sunny, cloudy, or rainy. Or an individual could be in one of four possible emotional states: happy, sad, angry, or apprehensive. Suppose that such a system changes with time from one state to another and at scheduled times the state of the system is observed. If the state of the system at any observation cannot be predicted with certainty, but the probability that a given state occurs can be predicted by just knowing the state of the system at the preceding observation, then the process of change is called a **Markov chain** or **Markov process**.

> ### Definition
>
> If a Markov chain has k possible states, which we label as $1, 2, \ldots, k$, then the probability that the system is in state i at any observation after it was in state j at the preceding observation is denoted by p_{ij} and is called the ***transition probability*** from state j to state i. The matrix $P = [p_{ij}]$ is called the ***transition matrix of the Markov chain***.

For example, in a three-state Markov chain, the transition matrix has the form

Preceding State

$$
\begin{array}{ccc}
1 & 2 & 3
\end{array}
$$

$$
\begin{bmatrix}
p_{11} & p_{12} & p_{13} \\
p_{21} & p_{22} & p_{23} \\
p_{31} & p_{32} & p_{33}
\end{bmatrix}
\begin{array}{l}
1 \\
2 \quad \textbf{New State} \\
3
\end{array}
$$

In this matrix p_{32} is the probability that the system will change from state 2 to state 3, p_{11} is the probability that the system will still be in state 1 if it was previously in state 1, and so forth.

EXAMPLE 1 Transition Matrix of the Markov Chain

A car rental agency has three rental locations, denoted by 1, 2, and 3. A customer may rent a car from any of the three locations and return the car to any of the three locations. The manager finds that customers return the cars to the various locations according to the following probabilities:

Rented from Location

$$
\begin{array}{ccc}
1 & 2 & 3
\end{array}
$$

$$
\begin{bmatrix}
.8 & .3 & .2 \\
.1 & .2 & .6 \\
.1 & .5 & .2
\end{bmatrix}
\begin{array}{l}
1 \quad \textbf{Returned} \\
2 \qquad \textbf{to} \\
3 \quad \textbf{Location}
\end{array}
$$

This matrix is the transition matrix of the system considered as a Markov chain. From this matrix the probability is .6 that a car rented from location 3 will be returned to location 2, the probability is .8 that a car rented from location 1 will be returned to location 1, and so forth. ◆

EXAMPLE 2 Transition Matrix of the Markov Chain

By reviewing its donation records, the alumni office of a college finds that 80% of its alumni who contribute to the annual fund one year will also contribute the next year, and 30% of those who do not contribute one year will contribute the next. This can be viewed as a Markov chain with two states: state 1 corresponds to an alumnus giving a donation in any one year, and state 2 corresponds to the alumnus not giving a donation in that year. The transition matrix is

$$
P = \begin{bmatrix} .8 & .3 \\ .2 & .7 \end{bmatrix}
$$

◆

In the examples above, the transition matrices of the Markov chains have the property that the entries in any column sum to 1. This is not accidental. If $P = [p_{ij}]$ is the transition matrix of any Markov chain with k states, then for each j we must have

$$p_{1j} + p_{2j} + \cdots + p_{kj} = 1 \tag{1}$$

because if the system is in state j at one observation, it is certain to be in one of the k possible states at the next observation.

A matrix with property (1) is called a **stochastic matrix**, a **probability matrix**, or a **Markov matrix**. From the preceding discussion, it follows that the transition matrix for a Markov chain must be a stochastic matrix.

In a Markov chain the state of the system at any observation time cannot generally be determined with certainty. The best one can usually do is specify probabilities for each of the possible states. For example, in a Markov chain with three states we might describe the possible state of the system at some observation time by a column vector

$$\mathbf{x} = \begin{bmatrix} x_1 \\ x_2 \\ x_3 \end{bmatrix}$$

in which x_1 is the probability that the system is in state 1, x_2 the probability that it is in state 2, and x_3 the probability that it is in state 3. In general we make the following definition.

Definition

The **state vector** for an observation of a Markov chain with k states is a column vector $\mathbf{x}$ whose ith component x_i is the probability that the system is in the ith state at that time.

Observe that the entries in any state vector for a Markov chain are nonnegative and have a sum of 1. (Why?) A column vector that has this property is called a **probability vector**.

Let us suppose now that we know the state vector $\mathbf{x}^{(0)}$ for a Markov chain at some initial observation. The following theorem will enable us to determine the state vectors

$$\mathbf{x}^{(1)}, \mathbf{x}^{(2)}, \ldots, \mathbf{x}^{(n)}, \ldots$$

at the subsequent observation times.

Theorem 11.6.1

If P is the transition matrix of a Markov chain and $\mathbf{x}^{(n)}$ is the state vector at the nth observation, then $\mathbf{x}^{(n+1)} = P\mathbf{x}^{(n)}$.

The proof of this theorem involves ideas from probability theory and will not be given here. From this theorem it follows that

$$\mathbf{x}^{(1)} = P\mathbf{x}^{(0)}$$
$$\mathbf{x}^{(2)} = P\mathbf{x}^{(1)} = P^2\mathbf{x}^{(0)}$$
$$\mathbf{x}^{(3)} = P\mathbf{x}^{(2)} = P^3\mathbf{x}^{(0)}$$
$$\vdots$$
$$\mathbf{x}^{(n)} = P\mathbf{x}^{(n-1)} = P^n\mathbf{x}^{(0)}$$

In this way, the initial state vector $\mathbf{x}^{(0)}$ and the transition matrix P determine $\mathbf{x}^{(n)}$ for $n = 1, 2, \ldots$.

EXAMPLE 3 Example 2 Revisited

The transition matrix in Example 2 was

$$P = \begin{bmatrix} .8 & .3 \\ .2 & .7 \end{bmatrix}$$

We now construct the probable future donation record of a new graduate who did not give a donation in the initial year after graduation. For such a graduate the system is initially in state 2 with certainty, so the initial state vector is

$$\mathbf{x}^{(0)} = \begin{bmatrix} 0 \\ 1 \end{bmatrix}$$

From Theorem 11.6.1 we then have

$$\mathbf{x}^{(1)} = P\mathbf{x}^{(0)} = \begin{bmatrix} .8 & .3 \\ .2 & .7 \end{bmatrix} \begin{bmatrix} 0 \\ 1 \end{bmatrix} = \begin{bmatrix} .3 \\ .7 \end{bmatrix}$$

$$\mathbf{x}^{(2)} = P\mathbf{x}^{(1)} = \begin{bmatrix} .8 & .3 \\ .2 & .7 \end{bmatrix} \begin{bmatrix} .3 \\ .7 \end{bmatrix} = \begin{bmatrix} .45 \\ .55 \end{bmatrix}$$

$$\mathbf{x}^{(3)} = P\mathbf{x}^{(2)} = \begin{bmatrix} .8 & .3 \\ .2 & .7 \end{bmatrix} \begin{bmatrix} .45 \\ .55 \end{bmatrix} = \begin{bmatrix} .525 \\ .475 \end{bmatrix}$$

Thus, after three years the alumnus can be expected to make a donation with probability .525. Beyond three years, we find the following state vectors (to three decimal places):

$$\mathbf{x}^{(4)} = \begin{bmatrix} .563 \\ .438 \end{bmatrix}, \quad \mathbf{x}^{(5)} = \begin{bmatrix} .581 \\ .419 \end{bmatrix}, \quad \mathbf{x}^{(6)} = \begin{bmatrix} .591 \\ .409 \end{bmatrix}, \quad \mathbf{x}^{(7)} = \begin{bmatrix} .595 \\ .405 \end{bmatrix}$$

$$\mathbf{x}^{(8)} = \begin{bmatrix} .598 \\ .402 \end{bmatrix}, \quad \mathbf{x}^{(9)} = \begin{bmatrix} .599 \\ .401 \end{bmatrix}, \quad \mathbf{x}^{(10)} = \begin{bmatrix} .599 \\ .401 \end{bmatrix}, \quad \mathbf{x}^{(11)} = \begin{bmatrix} .600 \\ .400 \end{bmatrix}$$

For all n beyond eleven, we have

$$\mathbf{x}^{(n)} = \begin{bmatrix} .600 \\ .400 \end{bmatrix}$$

to three decimal places. In other words, the state vectors converge to a fixed vector as the number of observations increases. (We shall discuss this further below.) ◆

EXAMPLE 4 Example 1 Revisited

The transition matrix in Example 1 was

$$\begin{bmatrix} .8 & .3 & .2 \\ .1 & .2 & .6 \\ .1 & .5 & .2 \end{bmatrix}$$

If a car is rented initially from location 2, then the initial state vector is

$$\mathbf{x}^{(0)} = \begin{bmatrix} 0 \\ 1 \\ 0 \end{bmatrix}$$

Using this vector and Theorem 11.6.1 one obtains the later state vectors listed in Table 1.

TABLE 1

n / $x^{(n)}$		0	1	2	3	4	5	6	7	8	9	10	11
$x_1^{(n)}$	0	.300	.400	.477	.511	.533	.544	.550	.553	.555	.556	.557
$x_2^{(n)}$	1	.200	.370	.252	.261	.240	.238	.233	.232	.231	.230	.230
$x_3^{(n)}$	0	.500	.230	.271	.228	.227	.219	.217	.215	.214	.214	.213

For all values of n greater than 11, all state vectors are equal to $\mathbf{x}^{(11)}$ to three decimal places.

Two things should be observed in this example. First, it was not necessary to know how long a customer kept the car. That is, in a Markov process the time period between observations need not be regular. Second, the state vectors approach a fixed vector as n increases, just as in the first example. ◆

EXAMPLE 5 Using Theorem 11.6.1

Figure 11.6.1

A traffic officer is assigned to control the traffic at the eight intersections indicated in Figure 11.6.1. She is instructed to remain at each intersection for an hour and then to either remain at the same intersection or move to a neighboring intersection. To avoid establishing a pattern, she is told to choose her new intersection on a random basis, with each possible choice equally likely. For example, if she is at intersection 5, her next intersection can be 2, 4, 5, or 8, each with probability $\frac{1}{4}$. Every day she starts at the location where she stopped the day before. The transition matrix for this Markov chain is

Old Intersection

$$
\begin{array}{cccccccc}
1 & 2 & 3 & 4 & 5 & 6 & 7 & 8
\end{array}
$$

$$
\begin{bmatrix}
\frac{1}{3} & \frac{1}{3} & 0 & \frac{1}{5} & 0 & 0 & 0 & 0 \\
\frac{1}{3} & \frac{1}{3} & 0 & 0 & \frac{1}{4} & 0 & 0 & 0 \\
0 & 0 & \frac{1}{3} & \frac{1}{5} & 0 & \frac{1}{3} & 0 & 0 \\
\frac{1}{3} & 0 & \frac{1}{3} & \frac{1}{5} & \frac{1}{4} & 0 & \frac{1}{4} & 0 \\
0 & \frac{1}{3} & 0 & \frac{1}{5} & \frac{1}{4} & 0 & 0 & \frac{1}{3} \\
0 & 0 & \frac{1}{3} & 0 & 0 & \frac{1}{3} & \frac{1}{4} & 0 \\
0 & 0 & 0 & \frac{1}{5} & 0 & \frac{1}{3} & \frac{1}{4} & \frac{1}{3} \\
0 & 0 & 0 & 0 & \frac{1}{4} & 0 & \frac{1}{4} & \frac{1}{3}
\end{bmatrix}
\begin{array}{l}
1 \\
2 \\
3 \\
4 \\
5 \\
6 \\
7 \\
8
\end{array}
\quad
\begin{array}{l}
\textbf{New} \\
\textbf{Intersection}
\end{array}
$$

If the traffic officer initially begins at intersection 5, her probable locations, hour by hour, are given by the state vectors given in Table 2. For all values of n greater than 22, all state vectors are equal to $\mathbf{x}^{(22)}$ to three decimal places. So, as with the first two examples, the state vectors approach a fixed vector as n increases. ◆

TABLE 2

$x^{(n)}$ \ n	0	1	2	3	4	5	10	15	20	22
$x_1^{(n)}$	0	.000	.133	.116	.130	.123	.113	.109	.108	.107
$x_2^{(n)}$	0	.250	.146	.163	.140	.138	.115	.109	.108	.107
$x_3^{(n)}$	0	.000	.050	.039	.067	.073	.100	.106	.107	.107
$x_4^{(n)}$	0	.250	.113	.187	.162	.178	.178	.179	.179	.179
$x_5^{(n)}$	1	.250	.279	.190	.190	.168	.149	.144	.143	.143
$x_6^{(n)}$	0	.000	.000	.050	.056	.074	.099	.105	.107	.107
$x_7^{(n)}$	0	.000	.133	.104	.131	.125	.138	.142	.143	.143
$x_8^{(n)}$	0	.250	.146	.152	.124	.121	.108	.107	.107	.107

Limiting Behavior of the State Vectors   In our examples we saw that the state vectors approached some fixed vector as the number of observations increased. We now ask whether the state vectors always approach a fixed vector in a Markov chain. A simple example shows that this is not the case.

EXAMPLE 6 System Oscillates Between Two State Vectors

Let

$$P = \begin{bmatrix} 0 & 1 \\ 1 & 0 \end{bmatrix} \quad \text{and} \quad \mathbf{x}^{(0)} = \begin{bmatrix} 1 \\ 0 \end{bmatrix}$$

Then, because $P^2 = I$ and $P^3 = P$, we have that

$$\mathbf{x}^{(0)} = \mathbf{x}^{(2)} = \mathbf{x}^{(4)} = \cdots = \begin{bmatrix} 1 \\ 0 \end{bmatrix}$$

and

$$\mathbf{x}^{(1)} = \mathbf{x}^{(3)} = \mathbf{x}^{(5)} = \cdots = \begin{bmatrix} 0 \\ 1 \end{bmatrix}$$

This system oscillates indefinitely between the two state vectors $\begin{bmatrix} 1 \\ 0 \end{bmatrix}$ and $\begin{bmatrix} 0 \\ 1 \end{bmatrix}$, and so does not approach any fixed vector.   ◆

However, if we impose a mild condition on the transition matrix, we can show that a fixed limiting state vector is approached. This condition is described by the following definition.

Definition

A transition matrix is *regular* if some integer power of it has all positive entries.

Thus, for a regular transition matrix P, there is some positive integer m such that all entries of P^m are positive. This is the case with the transition matrices of Examples 1 and 2 for $m = 1$. In Example 5 it turns out that P^4 has all positive entries. Consequently, in all three examples the transition matrices are regular.

A Markov chain that is governed by a regular transition matrix is called a *regular Markov chain*. We shall see that every regular Markov chain has a fixed state vector $\mathbf{q}$ such that $P^n\mathbf{x}^{(0)}$ approaches $\mathbf{q}$ as n increases for any choice of $\mathbf{x}^{(0)}$. This result is of major importance in the theory of Markov chains. It is based on the following theorem.

Theorem 11.6.2 **Behavior of P^n as $n \to \infty$**

If P is a regular transition matrix, then as $n \to \infty$

$$P^n \to \begin{bmatrix} q_1 & q_1 & \cdots & q_1 \\ q_2 & q_2 & \cdots & q_2 \\ \vdots & \vdots & & \vdots \\ q_k & q_k & \cdots & q_k \end{bmatrix}$$

where the q_i are positive numbers such that $q_1 + q_2 + \cdots + q_k = 1$.

We will not prove this theorem here. (The interested reader is referred to a more specialized text, such as J. Kemeny and J. Snell, *Finite Markov Chains*, New York: Springer-Verlag, 1976.)

Let us set

$$Q = \begin{bmatrix} q_1 & q_1 & \cdots & q_1 \\ q_2 & q_2 & \cdots & q_2 \\ \vdots & \vdots & & \vdots \\ q_k & q_k & \cdots & q_k \end{bmatrix} \quad \text{and} \quad \mathbf{q} = \begin{bmatrix} q_1 \\ q_2 \\ \vdots \\ q_k \end{bmatrix}$$

Thus, Q is a transition matrix, all of whose columns are equal to the probability vector $\mathbf{q}$. Q has the property that if $\mathbf{x}$ is any probability vector, then

$$Q\mathbf{x} = \begin{bmatrix} q_1 & q_1 & \cdots & q_1 \\ q_2 & q_2 & \cdots & q_2 \\ \vdots & \vdots & & \vdots \\ q_k & q_k & \cdots & q_k \end{bmatrix} \begin{bmatrix} x_1 \\ x_2 \\ \vdots \\ x_k \end{bmatrix} = \begin{bmatrix} q_1x_1 + q_1x_2 + \cdots + q_1x_k \\ q_2x_1 + q_2x_2 + \cdots + q_2x_k \\ \vdots & \vdots & & \vdots \\ q_kx_1 + q_kx_2 + \cdots + q_kx_k \end{bmatrix}$$

$$= (x_1 + x_2 + \cdots + x_k) \begin{bmatrix} q_1 \\ q_2 \\ \vdots \\ q_k \end{bmatrix} = (1)\mathbf{q} = \mathbf{q}$$

That is, Q transforms any probability vector $\mathbf{x}$ into the fixed probability vector $\mathbf{q}$. This result leads to the following theorem.

Theorem 11.6.3 **Behavior of $P^n\mathbf{x}$ as $n \to \infty$**

If P is a regular transition matrix and $\mathbf{x}$ is any probability vector, then as $n \to \infty$

$$P^n\mathbf{x} \to \begin{bmatrix} q_1 \\ q_2 \\ \vdots \\ q_k \end{bmatrix} = \mathbf{q}$$

where $\mathbf{q}$ is a fixed probability vector, independent of n, all of whose entries are positive.

This result holds since Theorem 11.6.2 implies that $P^n \to Q$ as $n \to \infty$. This in turn implies that $P^n \mathbf{x} \to Q\mathbf{x} = \mathbf{q}$ as $n \to \infty$. Thus, for a regular Markov chain, the system eventually approaches a fixed state vector $\mathbf{q}$. The vector $\mathbf{q}$ is called the ***steady-state vector*** of the regular Markov chain.

For systems with many states, usually the most efficient technique of computing the steady-state vector $\mathbf{q}$ is simply to calculate $P^n \mathbf{x}$ for some large n. Our examples illustrate this procedure. Each is a regular Markov process, so that convergence to a steady-state vector is ensured. Another way of computing the steady-state vector is to make use of the following theorem.

Theorem 11.6.4   Steady-State Vector

The steady-state vector $\mathbf{q}$ *of a regular transition matrix* P *is the unique probability vector that satisfies the equation* $P\mathbf{q} = \mathbf{q}$.

To see this, consider the matrix identity $PP^n = P^{n+1}$. By Theorem 11.6.2, both P^n and P^{n+1} approach Q as $n \to \infty$. Thus, we have $PQ = Q$. Any one column of this matrix equation gives $P\mathbf{q} = \mathbf{q}$. To show that $\mathbf{q}$ is the only probability vector that satisfies this equation, suppose $\mathbf{r}$ is another probability vector such that $P\mathbf{r} = \mathbf{r}$. Then also $P^n \mathbf{r} = \mathbf{r}$ for $n = 1, 2, \ldots$. Letting $n \to \infty$, Theorem 11.6.3 leads to $\mathbf{q} = \mathbf{r}$.

Theorem 11.6.4 can also be expressed by the statement that the homogeneous linear system

$$(I - P)\mathbf{q} = \mathbf{0}$$

has a unique solution vector $\mathbf{q}$ with nonnegative entries that satisfy the condition $q_1 + q_2 + \cdots + q_k = 1$. We can apply this technique to the computation of the steady-state vectors for our examples.

EXAMPLE 7   Example 2 Revisited

In Example 2 the transition matrix was

$$P = \begin{bmatrix} .8 & .3 \\ .2 & .7 \end{bmatrix}$$

so that the linear system $(I - P)\mathbf{q} = \mathbf{0}$ is

$$\begin{bmatrix} .2 & -.3 \\ -.2 & .3 \end{bmatrix} \begin{bmatrix} q_1 \\ q_2 \end{bmatrix} = \begin{bmatrix} 0 \\ 0 \end{bmatrix} \tag{2}$$

This leads to the single independent equation

$$.2q_1 - .3q_2 = 0$$

or

$$q_1 = 1.5q_2$$

Thus, setting $q_2 = s$, any solution of (2) is of the form

$$\mathbf{q} = s \begin{bmatrix} 1.5 \\ 1 \end{bmatrix}$$

where s is an arbitrary constant. To make the vector $\mathbf{q}$ a probability vector, we set $s = 1/(1.5 + 1) = .4$. Consequently,

$$\mathbf{q} = \begin{bmatrix} .6 \\ .4 \end{bmatrix}$$

is the steady-state vector of this regular Markov chain. This means that over the long run, 60% of the alumni will give a donation in any one year, and 40% will not. Observe that this agrees with the result obtained numerically in Example 3. ◆

EXAMPLE 8 Example 1 Revisited

In Example 1 the transition matrix was

$$P = \begin{bmatrix} .8 & .3 & .2 \\ .1 & .2 & .6 \\ .1 & .5 & .2 \end{bmatrix}$$

so that the linear system $(I - P)\mathbf{q} = \mathbf{0}$ is

$$\begin{bmatrix} .2 & -.3 & -.2 \\ -.1 & .8 & -.6 \\ -.1 & -.5 & .8 \end{bmatrix} \begin{bmatrix} q_1 \\ q_2 \\ q_3 \end{bmatrix} = \begin{bmatrix} 0 \\ 0 \\ 0 \end{bmatrix}$$

The reduced row-echelon form of the coefficient matrix is (verify)

$$\begin{bmatrix} 1 & 0 & -\frac{34}{13} \\ 0 & 1 & -\frac{14}{13} \\ 0 & 0 & 0 \end{bmatrix}$$

so that the original linear system is equivalent to the system

$$q_1 = \left(\tfrac{34}{13}\right)q_3$$
$$q_2 = \left(\tfrac{14}{13}\right)q_3$$

Setting $q_3 = s$, any solution of the linear system is of the form

$$\mathbf{q} = s \begin{bmatrix} \frac{34}{13} \\ \frac{14}{13} \\ 1 \end{bmatrix}$$

To make this a probability vector, we set

$$s = \frac{1}{\frac{34}{13} + \frac{14}{13} + 1} = \frac{13}{61}$$

Thus, the steady-state vector of the system is

$$\mathbf{q} = \begin{bmatrix} \frac{34}{61} \\ \frac{14}{61} \\ \frac{13}{61} \end{bmatrix} = \begin{bmatrix} .5573\ldots \\ .2295\ldots \\ .2131\ldots \end{bmatrix}$$

This agrees with the result obtained numerically in Table 1. The entries of $\mathbf{q}$ give the long-run probabilities that any one car will be returned to location 1, 2, or 3, respectively. If the car rental agency has a fleet of 1000 cars, it should design its facilities so that there are at least 558 spaces at location 1, at least 230 spaces at location 2, and at least 214 spaces at location 3. ◆

EXAMPLE 9 Example 5 Revisited

We will not give the details of the calculations but simply state that the unique probability vector solution of the linear system $(I - P)\mathbf{q} = \mathbf{0}$ is

$$\mathbf{q} = \begin{bmatrix} \frac{3}{28} \\ \frac{3}{28} \\ \frac{3}{28} \\ \frac{5}{28} \\ \frac{4}{28} \\ \frac{3}{28} \\ \frac{4}{28} \\ \frac{3}{28} \end{bmatrix} = \begin{bmatrix} .1071\ldots \\ .1071\ldots \\ .1071\ldots \\ .1785\ldots \\ .1428\ldots \\ .1071\ldots \\ .1428\ldots \\ .1071\ldots \end{bmatrix}$$

The entries in this vector indicate the proportion of time the traffic officer spends at each intersection over the long term. Thus, if the objective is for her to spend the same proportion of time at each intersection, then the strategy of random movement with equal probabilities from one intersection to another is not a good one. (See Exercise 5.) ◆

Exercise Set 11.6

1. For the transition matrix $P = \begin{bmatrix} .4 & .5 \\ .6 & .5 \end{bmatrix}$

(a) calculate $\mathbf{x}^{(n)}$ for $n = 1, 2, 3, 4, 5$, if $\mathbf{x}^{(0)} = \begin{bmatrix} 1 \\ 0 \end{bmatrix}$;

(b) state why P is regular and find its steady-state vector.

2. Consider the transition matrix

$$P = \begin{bmatrix} .2 & .1 & .7 \\ .6 & .4 & .2 \\ .2 & .5 & .1 \end{bmatrix}$$

(a) Calculate $\mathbf{x}^{(1)}$, $\mathbf{x}^{(2)}$, and $\mathbf{x}^{(3)}$ to three decimal places if

$$\mathbf{x}^{(0)} = \begin{bmatrix} 0 \\ 0 \\ 1 \end{bmatrix}$$

(b) State why P is regular and find its steady-state vector.

3. Find the steady-state vectors of the following regular transition matrices:

(a) $\begin{bmatrix} \frac{1}{3} & \frac{3}{4} \\ \frac{2}{3} & \frac{1}{4} \end{bmatrix}$ (b) $\begin{bmatrix} .81 & .26 \\ .19 & .74 \end{bmatrix}$ (c) $\begin{bmatrix} \frac{1}{3} & \frac{1}{2} & 0 \\ \frac{1}{3} & 0 & \frac{1}{4} \\ \frac{1}{3} & \frac{1}{2} & \frac{3}{4} \end{bmatrix}$

4. Let P be the transition matrix

$$\begin{bmatrix} \frac{1}{2} & 0 \\ \frac{1}{2} & 1 \end{bmatrix}$$

(a) Show that P is not regular.

(b) Show that as n increases, $P^n \mathbf{x}^{(0)}$ approaches $\begin{bmatrix} 0 \\ 1 \end{bmatrix}$ for any initial state vector $\mathbf{x}^{(0)}$.

(c) What conclusion of Theorem 11.6.3 is not valid for the steady state of this transition matrix?

5. Verify that if P is a $k \times k$ regular transition matrix all of whose row sums are equal to one, then the entries of its steady-state vector are all equal to $1/k$.

6. Show that the transition matrix

$$P = \begin{bmatrix} 0 & \frac{1}{2} & \frac{1}{2} \\ \frac{1}{2} & \frac{1}{2} & 0 \\ \frac{1}{2} & 0 & \frac{1}{2} \end{bmatrix}$$

is regular, and use Exercise 5 to find its steady-state vector.

7. John is either happy or sad. If he is happy one day, he is happy the next day four times out of five. If he is sad one day, he is sad the next day one time out of three. Over the long term, what are the chances that John is happy on any given day?

8. A country is divided into three demographic regions. It is found that each year 5% of the residents of region 1 move to region 2, and 5% move to region 3. Of the residents of region 2, 15% move to region 1 and 10% move to region 3. And of the residents of region 3, 10% move to region 1 and 5% move to region 2. What percentage of the population resides in each of the three regions after a long period of time?

Technology Exercises 11.6

The following exercises are designed to be solved using a technology utility. Typically, this will be MATLAB, *Mathematica*, Maple, Derive, or Mathcad, but it may also be some other type of linear algebra software or a scientific calculator with some linear algebra capabilities. For each exercise you will need to read the relevant documentation for the particular utility you are using. The goal of these exercises is to provide you with a basic proficiency with your technology utility. Once you have mastered the techniques in these exercises, you will be able to use your technology utility to solve many of the problems in the regular exercise sets.

T1. Consider the sequence of transition matrices

$$\{P_2, P_3, P_4, \ldots\}$$

with

$$P_2 = \begin{bmatrix} 0 & \frac{1}{2} \\ 1 & \frac{1}{2} \end{bmatrix}, \qquad P_3 = \begin{bmatrix} 0 & 0 & \frac{1}{3} \\ 0 & \frac{1}{2} & \frac{1}{3} \\ 1 & \frac{1}{2} & \frac{1}{3} \end{bmatrix},$$

$$P_4 = \begin{bmatrix} 0 & 0 & 0 & \frac{1}{4} \\ 0 & 0 & \frac{1}{3} & \frac{1}{4} \\ 0 & \frac{1}{2} & \frac{1}{3} & \frac{1}{4} \\ 1 & \frac{1}{2} & \frac{1}{3} & \frac{1}{4} \end{bmatrix}, \qquad P_5 = \begin{bmatrix} 0 & 0 & 0 & 0 & \frac{1}{5} \\ 0 & 0 & 0 & \frac{1}{4} & \frac{1}{5} \\ 0 & 0 & \frac{1}{3} & \frac{1}{4} & \frac{1}{5} \\ 0 & \frac{1}{2} & \frac{1}{3} & \frac{1}{4} & \frac{1}{5} \\ 1 & \frac{1}{2} & \frac{1}{3} & \frac{1}{4} & \frac{1}{5} \end{bmatrix},$$

and so on. Use a computer to show that each of these transition matrices is regular, and determine steady-state vectors $\mathbf{x}_k$ so that

$$P_k \mathbf{x}_k = \mathbf{x}_k$$

for $k = 2, 3, 4, \ldots, n$.

T2. A mouse is placed in a box with nine rooms as shown in the accompanying figure. Assume that it is equally likely that the mouse goes through any door in the room, or stays put in the room.

(a) Construct the 9 × 9 transition matrix for this problem and show that it is regular.

(b) Determine the steady-state vector for the matrix.

(c) Use a symmetry argument to show that this problem may be solved using only a 3 × 3 matrix.

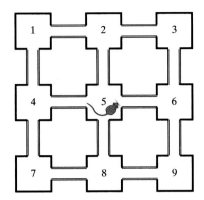

Figure Ex-T2

11.7   GRAPH THEORY

In this section we introduce matrix representations of relations among members of a set. We use matrix arithmetic to analyze these relationships.

PREREQUISITES: **Matrix Addition and Multiplication**

Relations Among Members of a Set   There are countless examples of sets with finitely many members in which some relation exists among members of the set. For example, the set could consist of a collection of people, animals, countries, companies, sports teams, or cities; and the relation between two members, *A* and *B*, of such a set could be that person *A* dominates person *B*, animal *A* feeds on animal *B*, country *A* militarily supports country *B*, company *A* sells its product to company *B*, sports team *A* consistently beats sports team *B*, or city *A* has a direct airline flight to city *B*.

We shall now show how the theory of *directed graphs* can be used to mathematically model relations such as those in the preceding examples.

Directed Graphs   A *directed graph* is a finite set of elements, $\{P_1, P_2, \ldots, P_n\}$, together with a finite collection of ordered pairs (P_i, P_j) of distinct elements of this set, with no ordered pair being repeated. The elements of the set are called *vertices*, and the ordered pairs are called *directed edges*, of the directed graph. We use the notation $P_i \rightarrow P_j$ (read "P_i is connected to P_j") to indicate that the directed edge (P_i, P_j) belongs to the directed graph. Geometrically, we can visualize a directed graph (Figure 11.7.1) by representing the vertices as points in the plane, and representing the directed edge $P_i \rightarrow P_j$ by drawing a line or arc from vertex P_i to vertex P_j, with an arrow pointing from P_i to P_j. If both $P_i \rightarrow P_j$ and $P_j \rightarrow P_i$ hold (denoted $P_i \leftrightarrow P_j$), we draw a single line between P_i and P_j with two oppositely pointing arrows (as with P_2 and P_3 in the figure).

As in Figure 11.7.1, for example, a directed graph may have separate "components" of vertices that are connected only among themselves; and some vertices, such as P_5,

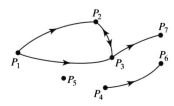

Figure 11.7.1

may not be connected with any other vertex. Also, because $P_i \rightarrow P_i$ is not permitted in a directed graph, a vertex cannot be connected with itself by a single arc that does not pass through any other vertex.

Figure 11.7.2 shows diagrams representing three more examples of directed graphs.

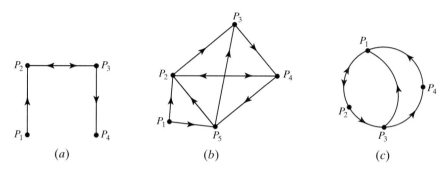

(a)　　　　　(b)　　　　　(c)

Figure 11.7.2

With a directed graph having n vertices, we may associate an $n \times n$ matrix $M = [m_{ij}]$, called the **vertex matrix** of the directed graph. Its elements are defined by

$$m_{ij} = \begin{cases} 1, & \text{if } P_i \rightarrow P_j \\ 0, & \text{otherwise} \end{cases}$$

for $i, j = 1, 2, \ldots, n$. For the three directed graphs in Figure 11.7.2, the corresponding vertex matrices are as follows:

Figure 11.7.2a:　$M = \begin{bmatrix} 0 & 1 & 0 & 0 \\ 0 & 0 & 1 & 0 \\ 0 & 1 & 0 & 1 \\ 0 & 0 & 0 & 0 \end{bmatrix}$

Figure 11.7.2b:　$M = \begin{bmatrix} 0 & 1 & 0 & 0 & 1 \\ 0 & 0 & 1 & 1 & 0 \\ 0 & 0 & 0 & 1 & 0 \\ 0 & 1 & 0 & 0 & 1 \\ 0 & 1 & 1 & 0 & 0 \end{bmatrix}$

Figure 11.7.2c:　$M = \begin{bmatrix} 0 & 1 & 0 & 0 \\ 1 & 0 & 1 & 0 \\ 1 & 0 & 0 & 1 \\ 1 & 0 & 0 & 0 \end{bmatrix}$

By their definition, vertex matrices have the following two properties:

(i) All entries are either 0 or 1.

(ii) All diagonal entries are 0.

Conversely, any matrix with these two properties determines a unique directed graph having the given matrix as its vertex matrix. For example, the matrix

$$M = \begin{bmatrix} 0 & 1 & 1 & 0 \\ 0 & 0 & 1 & 0 \\ 1 & 0 & 0 & 1 \\ 0 & 0 & 0 & 0 \end{bmatrix}$$

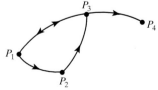

Figure 11.7.3

determines the directed graph in Figure 11.7.3.

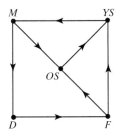

Figure 11.7.4

EXAMPLE 1 Influences Within a Family

A certain family consists of a mother, father, daughter, and two sons. The family members have influence, or power, over each other in the following ways: the mother can influence the daughter and the oldest son; the father can influence the two sons; the daughter can influence the father; the oldest son can influence the youngest son; and the youngest son can influence the mother. We may model this family influence pattern with a directed graph whose vertices are the five family members. If family member A influences family member B, we write $A \to B$. Figure 11.7.4 is the resulting directed graph, where we have used obvious letter designations for the five family members. The vertex matrix of this directed graph is

$$
\begin{array}{c}
\\
M \\
F \\
D \\
OS \\
YS
\end{array}
\begin{array}{ccccc}
M & F & D & OS & YS \\
\left[\begin{array}{ccccc}
0 & 0 & 1 & 1 & 0 \\
0 & 0 & 0 & 1 & 1 \\
0 & 1 & 0 & 0 & 0 \\
0 & 0 & 0 & 0 & 1 \\
1 & 0 & 0 & 0 & 0
\end{array}\right]
\end{array}
$$

♦

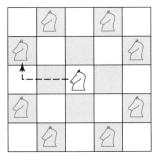

Figure 11.7.5

EXAMPLE 2 Vertex Matrix: Moves on a Chessboard

In chess the knight moves in an "L"-shaped pattern about the chessboard. For the board in Figure 11.7.5 it may move horizontally two squares and then vertically one square, or it may move vertically two squares and then horizontally one square. Thus, from the center square in the figure, the knight may move to any of the eight marked shaded squares. Suppose that the knight is restricted to the nine numbered squares in Figure 11.7.6. If by $i \to j$ we mean that the knight may move from square i to square j, the directed graph in Figure 11.7.7 illustrates all possible moves that the knight may make among these nine squares. In Figure 11.7.8 we have "unraveled" Figure 11.7.7 to make the pattern of possible moves clearer.

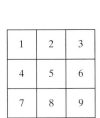

Figure 11.7.6

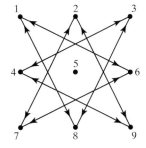

Figure 11.7.7

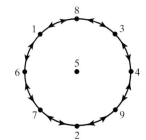

Figure 11.7.8

The vertex matrix of this directed graph is given by

$$
M = \begin{bmatrix}
0 & 0 & 0 & 0 & 0 & 1 & 0 & 1 & 0 \\
0 & 0 & 0 & 0 & 0 & 0 & 1 & 0 & 1 \\
0 & 0 & 0 & 1 & 0 & 0 & 0 & 1 & 0 \\
0 & 0 & 1 & 0 & 0 & 0 & 0 & 0 & 1 \\
0 & 0 & 0 & 0 & 0 & 0 & 0 & 0 & 0 \\
1 & 0 & 0 & 0 & 0 & 0 & 1 & 0 & 0 \\
0 & 1 & 0 & 0 & 0 & 1 & 0 & 0 & 0 \\
1 & 0 & 1 & 0 & 0 & 0 & 0 & 0 & 0 \\
0 & 1 & 0 & 1 & 0 & 0 & 0 & 0 & 0
\end{bmatrix}
$$

♦

In Example 1 the father cannot directly influence the mother; that is, $F \to M$ is not true. But he can influence the youngest son, who can then influence the mother. We write this as $F \to YS \to M$, and call it a **2-step connection** from F to M. Analogously, we call $M \to D$ a **1-step connection**, $F \to OS \to YS \to M$ a **3-step connection**, and so forth. Let us now consider a technique for finding the number of all possible r-step connections ($r = 1, 2, \ldots$) from one vertex P_i to another vertex P_j of an arbitrary directed graph. (This will include the case when P_i and P_j are the same vertex.) The number of 1-step connections from P_i to P_j is simply m_{ij}. That is, there is either zero or one 1-step connection from P_i to P_j, depending on whether m_{ij} is zero or one. For the number of 2-step connections, we consider the square of the vertex matrix. If we let $m_{ij}^{(2)}$ be the (i, j)-th element of M^2, we have

$$m_{ij}^{(2)} = m_{i1}m_{1j} + m_{i2}m_{2j} + \cdots + m_{in}m_{nj} \tag{1}$$

Now, if $m_{i1} = m_{1j} = 1$, there is a 2-step connection $P_i \to P_1 \to P_j$ from P_i to P_j. But if either m_{i1} or m_{1j} is zero, such a 2-step connection is not possible. Thus, $P_i \to P_1 \to P_j$ is a 2-step connection if and only if $m_{i1}m_{1j} = 1$. Similarly, for any $k = 1, 2, \ldots, n$, $P_i \to P_k \to P_j$ is a 2-step connection from P_i to P_j if and only if the term $m_{ik}m_{kj}$ on the right side of (1) is one; otherwise, the term is zero. Thus, the right side of (1) is the total number of two 2-step connections from P_i to P_j.

A similar argument will work for finding the number of 3-, 4-, $\ldots$, n-step connections from P_i to P_j. In general, we have the following result.

Theorem 11.7.1

Let M be the vertex matrix of a directed graph and let $m_{ij}^{(r)}$ *be the (i, j)-th element of M^r. Then* $m_{ij}^{(r)}$ *is equal to the number of r-step connections from P_i to P_j.*

EXAMPLE 3 Using Theorem 11.7.1

Figure 11.7.9 is the route map of a small airline that services the four cities P_1, P_2, P_3, P_4. As a directed graph, its vertex matrix is

$$M = \begin{bmatrix} 0 & 1 & 1 & 0 \\ 1 & 0 & 1 & 0 \\ 1 & 0 & 0 & 1 \\ 0 & 1 & 1 & 0 \end{bmatrix}$$

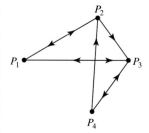

Figure 11.7.9

We have that

$$M^2 = \begin{bmatrix} 2 & 0 & 1 & 1 \\ 1 & 1 & 1 & 1 \\ 0 & 2 & 2 & 0 \\ 2 & 0 & 1 & 1 \end{bmatrix} \quad \text{and} \quad M^3 = \begin{bmatrix} 1 & 3 & 3 & 1 \\ 2 & 2 & 3 & 1 \\ 4 & 0 & 2 & 2 \\ 1 & 3 & 3 & 1 \end{bmatrix}$$

If we are interested in connections from city P_4 to city P_3, we may use Theorem 11.7.1 to find their number. Because $m_{43} = 1$, there is one 1-step connection; because $m_{43}^{(2)} = 1$, there is one 2-step connection; and because $m_{43}^{(3)} = 3$, there are three 3-step connections.

To verify this, from Figure 11.7.9 we find

$$1\text{-step connections from } P_4 \text{ to } P_3: \quad P_4 \rightarrow P_3$$
$$2\text{-step connections from } P_4 \text{ to } P_3: \quad P_4 \rightarrow P_2 \rightarrow P_3$$
$$3\text{-step connections from } P_4 \text{ to } P_3: \quad P_4 \rightarrow P_3 \rightarrow P_4 \rightarrow P_3$$
$$P_4 \rightarrow P_2 \rightarrow P_1 \rightarrow P_3$$
$$P_4 \rightarrow P_3 \rightarrow P_1 \rightarrow P_3 \qquad \blacklozenge$$

Cliques   In everyday language a "clique" is a closely knit group of people (usually three or more) that tends to communicate within itself and has no place for outsiders. In graph theory this concept is given a more precise meaning:

> ### Definition
>
> A subset of a directed graph is called a ***clique*** if it satisfies the following three conditions:
>
> (i) The subset contains at least three vertices.
> (ii) For each pair of vertices P_i and P_j in the subset, both $P_i \rightarrow P_j$ and $P_j \rightarrow P_i$ are true.
> (iii) The subset is as large as possible; that is, it is not possible to add another vertex to the subset and still satisfy condition (ii).

This definition suggests that cliques are maximal subsets that are in perfect "communication" with each other. For example, if the vertices represent cities, and $P_i \rightarrow P_j$ means that there is a direct airline flight from city P_i to city P_j, then there is a direct flight between any two cities within a clique in either direction.

EXAMPLE 4 A Directed Graph with Two Cliques

The directed graph illustrated in Figure 11.7.10 (which might represent the route map of an airline) has two cliques:

$$\{P_1, P_2, P_3, P_4\} \quad \text{and} \quad \{P_3, P_4, P_6\}$$

This example shows that a directed graph may contain several cliques and that a vertex may simultaneously belong to more than one clique. $\blacklozenge$

For simple directed graphs, cliques can be found by inspection. But for large directed graphs, it would be desirable to have a systematic procedure for detecting cliques. For this purpose it will be helpful to define a matrix $S = [s_{ij}]$ related to a given directed graph as follows:

$$s_{ij} = \begin{cases} 1, & \text{if } P_i \leftrightarrow P_j \\ 0, & \text{otherwise} \end{cases}$$

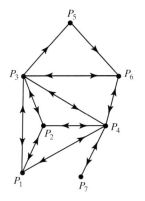

Figure 11.7.10

The matrix S determines a directed graph that is the same as the given directed graph, with the exception that the directed edges with only one arrow are deleted. For example, if the original directed graph is given by Figure 11.7.11a, the directed graph that has S as its vertex matrix is given in Figure 11.7.11b. The matrix S may be obtained from the

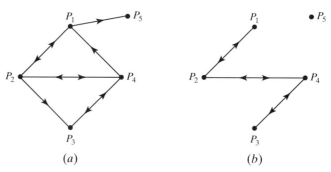

Figure 11.7.11

vertex matrix M of the original directed graph by setting $s_{ij} = 1$ if $m_{ij} = m_{ji} = 1$ and setting $s_{ij} = 0$ otherwise.

The following theorem, which uses the matrix S, is helpful for identifying cliques.

Theorem 11.7.2   **Identifying Cliques**

Let $s_{ij}^{(3)}$ be the (i, j)-th element of S^3. Then a vertex P_i belongs to some clique if and only if $s_{ii}^{(3)} \neq 0$.

Proof. If $s_{ii}^{(3)} \neq 0$, then there is at least one 3-step connection from P_i to itself in the modified directed graph determined by S. Suppose it is $P_i \rightarrow P_j \rightarrow P_k \rightarrow P_i$. In the modified directed graph, all directed relations are two-way, so that we also have the connections $P_i \leftrightarrow P_j \leftrightarrow P_k \leftrightarrow P_i$. But this means $\{P_i, P_j, P_k\}$ is either a clique or a subset of a clique. In either case, P_i must belong to some clique. The converse statement, if P_i belongs to a clique, then $s_{ij}^{(3)} \neq 0$, follows in a similar manner. ■

EXAMPLE 5 Using Theorem 11.7.2

Suppose that a directed graph has as its vertex matrix

$$M = \begin{bmatrix} 0 & 1 & 1 & 1 \\ 1 & 0 & 1 & 0 \\ 0 & 1 & 0 & 1 \\ 1 & 0 & 0 & 0 \end{bmatrix}$$

Then

$$S = \begin{bmatrix} 0 & 1 & 0 & 1 \\ 1 & 0 & 1 & 0 \\ 0 & 1 & 0 & 0 \\ 1 & 0 & 0 & 0 \end{bmatrix} \quad \text{and} \quad S^3 = \begin{bmatrix} 0 & 3 & 0 & 2 \\ 3 & 0 & 2 & 0 \\ 0 & 2 & 0 & 1 \\ 2 & 0 & 1 & 0 \end{bmatrix}$$

Because all diagonal entries of S^3 are zero, it follows from Theorem 11.7.2 that the directed graph has no cliques. ♦

EXAMPLE 6 Using Theorem 11.7.2

Suppose that a directed graph has as its vertex matrix

$$M = \begin{bmatrix} 0 & 1 & 0 & 1 & 1 \\ 1 & 0 & 0 & 1 & 0 \\ 1 & 1 & 0 & 1 & 0 \\ 1 & 1 & 0 & 0 & 0 \\ 1 & 0 & 0 & 1 & 0 \end{bmatrix}$$

Then

$$S = \begin{bmatrix} 0 & 1 & 0 & 1 & 1 \\ 1 & 0 & 0 & 1 & 0 \\ 0 & 0 & 0 & 0 & 0 \\ 1 & 1 & 0 & 0 & 0 \\ 1 & 0 & 0 & 0 & 0 \end{bmatrix} \quad \text{and} \quad S^3 = \begin{bmatrix} 2 & 4 & 0 & 4 & 3 \\ 4 & 2 & 0 & 3 & 1 \\ 0 & 0 & 0 & 0 & 0 \\ 4 & 3 & 0 & 2 & 1 \\ 3 & 1 & 0 & 1 & 0 \end{bmatrix}$$

The nonzero diagonal entries of S^3 are $s_{11}^{(3)}$, $s_{22}^{(3)}$, and $s_{44}^{(3)}$. Consequently, in the given directed graph, P_1, P_2, and P_4 belong to cliques. Because a clique must contain at least three vertices, the directed graph has only one clique, namely, $\{P_1, P_2, P_4\}$.   ◆

Dominance-Directed Graphs   In many groups of individuals or animals, there is a definite "pecking order" or dominance relation between any two members of the group. That is, given any two individuals A and B, either A dominates B or B dominates A, but not both. In terms of a directed graph in which $P_i \rightarrow P_j$ means P_i dominates P_j, this means that for all distinct pairs, either $P_i \rightarrow P_j$ or $P_j \rightarrow P_i$, but not both. In general, we have the following definition.

> ### Definition
> A **dominance-directed graph** is a directed graph such that for any distinct pair of vertices P_i and P_j, either $P_i \rightarrow P_j$ or $P_j \rightarrow P_i$, but not both.

An example of a directed graph satisfying this definition is a league of n sports teams that play each other exactly one time, as in one round of a round-robin tournament in which no ties are allowed. If $P_i \rightarrow P_j$ means that team P_i beat team P_j in their single match, it is easily seen that the definition of a dominance-directed group is satisfied. For this reason, dominance-directed graphs are sometimes called **tournaments**.

Figure 11.7.12 illustrates some dominance-directed graphs with three, four, and five vertices, respectively. In these three graphs, the circled vertices have the following interesting property: from each one there is either a 1-step or a 2-step connection to any other vertex in its graph. In a sports tournament these vertices would correspond to the most "powerful" teams in the sense that these teams either beat any given team or beat some other team that beat the given team. We can now state and prove a theorem that guarantees that any dominance-directed graph has at least one vertex with this property.

> ### Theorem 11.7.3   Connections in Dominance Directed Graphs
> *In any dominance-directed graph there is at least one vertex from which there is a 1-step or 2-step connection to any other vertex.*

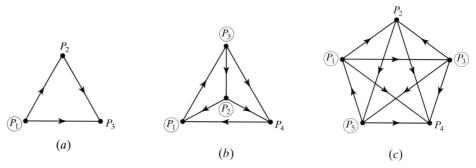

Figure 11.7.12

Proof. Consider a vertex (there may be several) with the largest total number of 1-step and 2-step connections to other vertices in the graph. By renumbering the vertices, we may assume that P_1 is such a vertex. Suppose there is some vertex P_i such that there is no 1-step or 2-step connection from P_1 to P_i. Then, in particular, $P_1 \rightarrow P_i$ is not true, so that by definition of a dominance-directed graph, it must be that $P_i \rightarrow P_1$. Next, let P_k be any vertex such that $P_1 \rightarrow P_k$ is true. Then we cannot have $P_k \rightarrow P_i$, as then $P_1 \rightarrow P_k \rightarrow P_i$ would be a 2-step connection from P_1 to P_i. Thus, it must be that $P_i \rightarrow P_k$. That is, P_i has 1-step connections to all the vertices to which P_1 has 1-step connections. The vertex P_i must then also have 2-step connections to all the vertices to which P_1 has 2-step connections. But because, in addition, we have that $P_i \rightarrow P_1$, this means that P_i has more 1-step and 2-step connections to other vertices than does P_1. However, this contradicts the way in which P_1 was chosen. Hence, there can be no vertex P_i to which P_1 has no 1-step or 2-step connection. ∎

This proof shows that a vertex with the largest total number of 1-step and 2-step connections to other vertices has the property stated in the theorem. There is a simple way of finding such vertices using the vertex matrix M and its square M^2. The sum of the entries in the ith row of M is the total number of 1-step connections from P_i to other vertices, and the sum of the entries of the ith row of M^2 is the total number of 2-step connections from P_i to other vertices. Consequently, the sum of the entries of the ith row of the matrix $A = M + M^2$ is the total number of 1-step and 2-step connections from P_i to other vertices. In other words, a row of $A = M + M^2$ with the largest row sum identifies a vertex having the property stated in Theorem 11.7.3.

EXAMPLE 7 Using Theorem 11.7.3

Suppose that five baseball teams play each other exactly once, and the results are as indicated in the dominance-directed graph of Figure 11.7.13. The vertex matrix of the

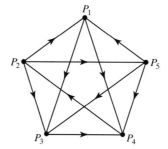

Figure 11.7.13

graph is

$$M = \begin{bmatrix} 0 & 0 & 1 & 1 & 0 \\ 1 & 0 & 1 & 0 & 1 \\ 0 & 0 & 0 & 1 & 0 \\ 0 & 1 & 0 & 0 & 0 \\ 1 & 0 & 1 & 1 & 0 \end{bmatrix}$$

And so

$$A = M + M^2 = \begin{bmatrix} 0 & 0 & 1 & 1 & 0 \\ 1 & 0 & 1 & 0 & 1 \\ 0 & 0 & 0 & 1 & 0 \\ 0 & 1 & 0 & 0 & 0 \\ 1 & 0 & 1 & 1 & 0 \end{bmatrix} + \begin{bmatrix} 0 & 1 & 0 & 1 & 0 \\ 1 & 0 & 2 & 3 & 0 \\ 0 & 1 & 0 & 0 & 0 \\ 1 & 0 & 1 & 0 & 1 \\ 0 & 1 & 1 & 2 & 0 \end{bmatrix} = \begin{bmatrix} 0 & 1 & 1 & 2 & 0 \\ 2 & 0 & 3 & 3 & 1 \\ 0 & 1 & 0 & 1 & 0 \\ 1 & 1 & 1 & 0 & 1 \\ 1 & 1 & 2 & 3 & 0 \end{bmatrix}$$

The row sums of A are

$$\text{1st row sum} = 4$$
$$\text{2nd row sum} = 9$$
$$\text{3rd row sum} = 2$$
$$\text{4th row sum} = 4$$
$$\text{5th row sum} = 7$$

Because the second row has the largest row sum, the vertex P_2 must have a 1-step or 2-step connection to any other vertex. This is easily verified from Figure 11.7.13.   ◆

We have informally suggested that a vertex with the largest number of 1-step and 2-step connections to other vertices is a "powerful" vertex. We can formalize this concept with the following definition.

Definition

The **power** of a vertex of a dominance-directed graph is the total number of 1-step and 2-step connections from it to other vertices. Alternatively, the power of a vertex P_i is the sum of the entries of the ith row of the matrix $A = M + M^2$, where M is the vertex matrix of the directed graph.

EXAMPLE 8 Example 7 Revisited

Let us rank the five baseball teams in Example 7 according to their powers. From the calculations for the row sums in that example we have

$$\text{Power of team } P_1 = 4$$
$$\text{Power of team } P_2 = 9$$
$$\text{Power of team } P_3 = 2$$
$$\text{Power of team } P_4 = 4$$
$$\text{Power of team } P_5 = 7$$

Hence, the ranking of the teams according to their powers would be

$$P_2 \text{ (first)}, \quad P_5 \text{ (second)}, \quad P_1 \text{ and } P_4 \text{ (tied for third)}, \quad P_3 \text{ (last)} \qquad ◆$$

Exercise Set 11.7

1. Construct the vertex matrix for each of the directed graphs illustrated in the accompanying figure.

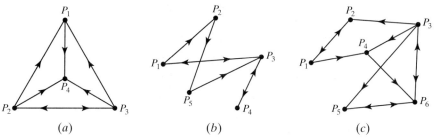

(a) (b) (c)

Figure Ex-1

2. Draw a diagram of the directed graph corresponding to each of the following vertex matrices.

(a) $\begin{bmatrix} 0 & 1 & 1 & 0 \\ 1 & 0 & 0 & 0 \\ 0 & 0 & 0 & 1 \\ 1 & 0 & 1 & 0 \end{bmatrix}$ (b) $\begin{bmatrix} 0 & 0 & 1 & 0 & 0 \\ 1 & 0 & 0 & 0 & 1 \\ 0 & 1 & 0 & 1 & 1 \\ 0 & 0 & 0 & 0 & 0 \\ 1 & 1 & 1 & 0 & 0 \end{bmatrix}$ (c) $\begin{bmatrix} 0 & 1 & 0 & 1 & 0 & 1 \\ 1 & 0 & 0 & 0 & 1 & 0 \\ 0 & 0 & 0 & 0 & 0 & 0 \\ 1 & 1 & 0 & 0 & 1 & 0 \\ 0 & 0 & 0 & 1 & 0 & 1 \\ 0 & 1 & 0 & 0 & 1 & 0 \end{bmatrix}$

3. Let M be the following vertex matrix of a directed graph:

$$\begin{bmatrix} 0 & 1 & 1 & 1 \\ 1 & 0 & 0 & 0 \\ 0 & 1 & 0 & 1 \\ 0 & 1 & 1 & 0 \end{bmatrix}$$

(a) Draw a diagram of the directed graph.
(b) Use Theorem 11.7.1 to find the number of 1-, 2-, and 3-step connections from the vertex P_1 to the vertex P_2. Verify your answer by listing the various connections as in Example 3.
(c) Repeat part (b) for the 1-, 2-, and 3-step connections from P_1 to P_4.

4. By inspection, locate all cliques in each of the directed graphs illustrated in the accompanying figure.

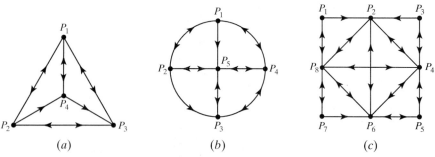

(a) (b) (c)

Figure Ex-4

5. For each of the following vertex matrices, use Theorem 11.7.2 to find all cliques in the corresponding directed graphs.

(a) $\begin{bmatrix} 0 & 1 & 0 & 1 & 0 \\ 1 & 0 & 1 & 0 & 1 \\ 0 & 1 & 0 & 1 & 1 \\ 1 & 0 & 0 & 0 & 1 \\ 1 & 0 & 1 & 1 & 0 \end{bmatrix}$ (b) $\begin{bmatrix} 0 & 1 & 0 & 1 & 1 & 0 \\ 1 & 0 & 1 & 0 & 1 & 1 \\ 0 & 1 & 0 & 1 & 0 & 1 \\ 1 & 0 & 1 & 0 & 1 & 1 \\ 0 & 1 & 0 & 1 & 0 & 0 \\ 0 & 0 & 1 & 1 & 1 & 0 \end{bmatrix}$

6. For the dominance-directed graph illustrated in the accompanying figure, construct the vertex matrix and find the power of each vertex.

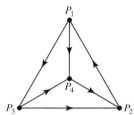

Figure Ex-6

7. Five baseball teams play each other one time with the following results:

> A beats B, C, D
> B beats C, E
> C beats D, E
> D beats B
> E beats A, D

Rank the five baseball teams in accordance with the powers of the vertices they correspond to in the dominance-directed graph representing the outcomes of the games.

Technology Exercises 11.7

The following exercises are designed to be solved using a technology utility. Typically, this will be MATLAB, *Mathematica*, Maple, Derive, or Mathcad, but it may also be some other type of linear algebra software or a scientific calculator with some linear algebra capabilities. For each exercise you will need to read the relevant documentation for the particular utility you are using. The goal of these exercises is to provide you with a basic proficiency with your technology utility. Once you have mastered the techniques in these exercises, you will be able to use your technology utility to solve many of the problems in the regular exercise sets.

T1. A graph having n vertices such that every vertex is connected to every other vertex has vertex matrix given by

$$M_n = \begin{bmatrix} 0 & 1 & 1 & 1 & 1 & \cdots & 1 \\ 1 & 0 & 1 & 1 & 1 & \cdots & 1 \\ 1 & 1 & 0 & 1 & 1 & \cdots & 1 \\ 1 & 1 & 1 & 0 & 1 & \cdots & 1 \\ 1 & 1 & 1 & 1 & 0 & \cdots & 1 \\ \vdots & \vdots & \vdots & \vdots & \vdots & \ddots & \vdots \\ 1 & 1 & 1 & 1 & 1 & \cdots & 0 \end{bmatrix}$$

In this problem we develop a formula for M_n^k whose (i, j)-th entry equals the number of k-step connections from P_i to P_j.

(a) Use a computer to compute the eight matrices M_n^k for $n = 2, 3$ and for $k = 2, 3, 4, 5$.

(b) Use the results in part (a) and symmetry arguments to show that M_n^k can be written as

$$M_n^k = \begin{bmatrix} 0 & 1 & 1 & 1 & 1 & \cdots & 1 \\ 1 & 0 & 1 & 1 & 1 & \cdots & 1 \\ 1 & 1 & 0 & 1 & 1 & \cdots & 1 \\ 1 & 1 & 1 & 0 & 1 & \cdots & 1 \\ 1 & 1 & 1 & 1 & 0 & \cdots & 1 \\ \vdots & \vdots & \vdots & \vdots & \vdots & \ddots & \vdots \\ 1 & 1 & 1 & 1 & 1 & \cdots & 0 \end{bmatrix}^k = \begin{bmatrix} \alpha_k & \beta_k & \beta_k & \beta_k & \beta_k & \cdots & \beta_k \\ \beta_k & \alpha_k & \beta_k & \beta_k & \beta_k & \cdots & \beta_k \\ \beta_k & \beta_k & \alpha_k & \beta_k & \beta_k & \cdots & \beta_k \\ \beta_k & \beta_k & \beta_k & \alpha_k & \beta_k & \cdots & \beta_k \\ \beta_k & \beta_k & \beta_k & \beta_k & \alpha_k & \cdots & \beta_k \\ \vdots & \vdots & \vdots & \vdots & \vdots & \ddots & \vdots \\ \beta_k & \beta_k & \beta_k & \beta_k & \beta_k & \cdots & \alpha_k \end{bmatrix}$$

(c) Using the fact that $M_n^k = M_n M_n^{k-1}$, show that

$$\begin{bmatrix} \alpha_k \\ \beta_k \end{bmatrix} = \begin{bmatrix} 0 & n-1 \\ 1 & n-2 \end{bmatrix} \begin{bmatrix} \alpha_{k-1} \\ \beta_{k-1} \end{bmatrix}$$

with

$$\begin{bmatrix} \alpha_1 \\ \beta_1 \end{bmatrix} = \begin{bmatrix} 0 \\ 1 \end{bmatrix}$$

(d) Using part (c), show that

$$\begin{bmatrix} \alpha_k \\ \beta_k \end{bmatrix} = \begin{bmatrix} 0 & n-1 \\ 1 & n-2 \end{bmatrix}^{k-1} \begin{bmatrix} 0 \\ 1 \end{bmatrix}$$

(e) Use the methods of Section 7.2 to compute

$$\begin{bmatrix} 0 & n-1 \\ 1 & n-2 \end{bmatrix}^{k-1}$$

and thereby obtain expressions for α_k and β_k, and eventually show that

$$M_n^k = \left(\frac{(n-1)^k - (-1)^k}{n} \right) U_n + (-1)^k I_n$$

where U_n is the $n \times n$ matrix all of whose entries are ones and I_n is the $n \times n$ identity matrix.

(f) Show that for $n > 2$, all vertices for these directed graphs belong to cliques.

T2. Consider a round-robin tournament between n players (labeled $a_1, a_2, a_3, \ldots, a_n$) where a_1 beats a_2, a_2 beats a_3, a_3 beats $a_4, \ldots, a_{n-1}$ beats a_n, and a_n beats a_1. Compute the "power" of each player, showing that they all have the same power; then determine that common power. [**Hint.** Use a computer to study the cases $n = 3, 4, 5, 6$; then make a conjecture and prove your conjecture to be true.]

11.8 GAMES OF STRATEGY

In this section we discuss a general game in which two competing players choose separate strategies to reach opposing objectives. The optimal strategy of each player is found in certain cases with the use of matrix techniques.

PREREQUISITES: Matrix Multiplication
Basic Probability Concepts

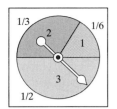

Row-wheel
of player *R*

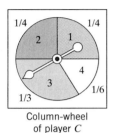

Column-wheel
of player *C*

Figure 11.8.1

Game Theory   To introduce the basic concepts in the theory of games, we will consider the following carnival-type game that two people agree to play. We will call the participants in the game *player R* and *player C*. Each player has a stationary wheel with a movable pointer on it as in Figure 11.8.1. For reasons that will become clear, we will call player *R*'s wheel the *row-wheel* and player *C*'s wheel the *column-wheel*. The row-wheel is divided into three sectors numbered 1, 2, and 3, and the column-wheel is divided into four sectors numbered 1, 2, 3, and 4. The fractions of the area occupied by the various sectors are indicated in the figure. To play the game each player spins the pointer of his or her wheel and lets it come to rest at random. The number of the sector in which each pointer comes to rest is called the *move* of that player. Thus, player *R* has three possible moves and player *C* has four possible moves. Depending on the move each player makes, player *C* then makes a payment of money to player *R* according to Table 1.

		Player *C*'s Move			
		1	**2**	**3**	**4**
Player *R*'s Move	**1**	$3	$5	−$2	−$1
	2	−$2	$4	−$3	−$4
	3	$6	−$5	$0	$3

For example, if the row-wheel pointer comes to rest in sector 1 (player *R* makes move 1), and the column-wheel pointer comes to rest in sector 2 (player *C* makes move 2), then player *C* must pay player *R* the sum of $5. Some of the entries in this table are negative, indicating that player *C* makes a negative payment to player *R*. By this we mean that player *R* makes a positive payment to player *C*. For example, if the row-wheel shows 2 and the column-wheel shows 4, then player *R* pays player *C* the sum of $4, because the corresponding entry in the table is −$4. In this way the positive entries of the table are the gains of player *R* and the losses of player *C*, and the negative entries are the gains of player *C* and the losses of player *R*.

In this game the players have no control over their moves; each move is determined by chance. However, if each player can decide whether he or she wants to play, then each would want to know how much he or she can expect to win or lose over the long term if he or she chooses to play. (Later in the section we will discuss this question and also consider a more complicated situation in which the players can exercise some control over their moves by varying the sectors of their wheels.)

Two-Person Zero-Sum Matrix Games   The game described above is an example of a ***two-person zero-sum matrix game***. The term "zero-sum" means that in each play of the game the positive gain of one player is equal to the negative gain (loss) of the other player. That is, the sum of the two gains is zero. The term "matrix game" is used to describe a two-person game in which each player has only a finite number of moves, so that all possible outcomes of each play, and the corresponding gains of the players, may be displayed in tabular or matrix form, as in Table 1.

In a general game of this type, let player *R* have *m* possible moves and let player *C* have *n* possible moves. In a play of the game, each player makes one of his or her possible moves, and then a *payoff* is made from player *C* to player *R*, depending on the moves. For $i = 1, 2, \ldots, m$, and $j = 1, 2, \ldots, n$, let us set

$$a_{ij} = \text{payoff that player } C \text{ makes to player } R \text{ if player } R$$
$$\text{makes move } i \text{ and player } C \text{ makes move } j$$

This payoff need not be money; it may be any type of commodity to which we can attach a numerical value. As before, if an entry a_{ij} is negative, we mean that player C receives a payoff of $|a_{ij}|$ from player R. We arrange these mn possible payoffs in the form of an $m \times n$ matrix

$$A = \begin{bmatrix} a_{11} & a_{12} & \cdots & a_{1n} \\ a_{21} & a_{22} & \cdots & a_{2n} \\ \vdots & \vdots & & \vdots \\ a_{m1} & a_{m2} & \cdots & a_{mn} \end{bmatrix}$$

which we will call the ***payoff matrix*** of the game.

Each player is to make his or her moves on a probabilistic basis. For example, for the game discussed in the introduction, the ratio of the area of a sector to the area of the wheel would be the probability that the player makes the move corresponding to that sector. Thus, from Figure 11.8.1 we see that player R would make move 2 with probability $\frac{1}{3}$, and player C would make move 2 with probability $\frac{1}{4}$. In the general case we make the following definitions:

$$p_i = \text{probability that player } R \text{ makes move } i \quad (i = 1, 2, \ldots, m)$$
$$q_j = \text{probability that player } C \text{ makes move } j \quad (j = 1, 2, \ldots, n)$$

It follows from these definitions that

$$p_1 + p_2 + \cdots + p_m = 1$$

and

$$q_1 + q_2 + \cdots + q_n = 1$$

With the probabilities p_i and q_j we form two vectors:

$$\mathbf{p} = [p_1 \quad p_2 \quad \cdots \quad p_m] \quad \text{and} \quad \mathbf{q} = \begin{bmatrix} q_1 \\ q_2 \\ \vdots \\ q_n \end{bmatrix}$$

We call the row vector $\mathbf{p}$ the ***strategy of player R*** and the column vector $\mathbf{q}$ the ***strategy of player C***. For example, from Figure 11.8.1 we have

$$\mathbf{p} = \begin{bmatrix} \frac{1}{6} & \frac{1}{3} & \frac{1}{2} \end{bmatrix} \quad \text{and} \quad \mathbf{q} = \begin{bmatrix} \frac{1}{4} \\ \frac{1}{4} \\ \frac{1}{3} \\ \frac{1}{6} \end{bmatrix}$$

for the carnival game described earlier.

From the theory of probability, if the probability that player R makes move i is p_i, and independently the probability that player C makes move j is q_j, then $p_i q_j$ is the probability that for any one play of the game, player R makes move i *and* player C makes move j. The payoff to player R for such a pair of moves is a_{ij}. If we multiply each possible payoff by its corresponding probability and sum over all possible payoffs, we obtain the expression

$$a_{11}p_1q_1 + a_{12}p_1q_2 + \cdots + a_{1n}p_1q_n + a_{21}p_2q_1 + \cdots + a_{mn}p_mq_n \quad (1)$$

Equation (1) is a weighted average of the payoffs to player R; each payoff is weighted according to the probability of its occurrence. In the theory of probability this weighted average is called the ***expected payoff*** to player R. It can be shown that if the game is played many times, the long-term average payoff per play to player R is given by this expression. We denote this expected payoff by $E(\mathbf{p}, \mathbf{q})$ to emphasize the fact that it depends on the strategies of the two players. From the definition of the payoff matrix A and the strategies $\mathbf{p}$ and $\mathbf{q}$, it can be verified that we may express the expected payoff in matrix notation as

$$E(\mathbf{p}, \mathbf{q}) = [p_1 \quad p_2 \quad \cdots \quad p_m] \begin{bmatrix} a_{11} & a_{12} & \cdots & a_{1n} \\ a_{21} & a_{22} & \cdots & a_{2n} \\ \vdots & \vdots & & \vdots \\ a_{m1} & a_{m2} & \cdots & a_{mn} \end{bmatrix} \begin{bmatrix} q_1 \\ q_2 \\ \vdots \\ q_n \end{bmatrix} = \mathbf{p}A\mathbf{q} \qquad (2)$$

Because $E(\mathbf{p}, \mathbf{q})$ is the expected payoff to player R, it follows that $-E(\mathbf{p}, \mathbf{q})$ is the expected payoff to player C.

EXAMPLE 1 Expected Payoff to Player *R*

For the carnival game described earlier, we have

$$E(\mathbf{p}, \mathbf{q}) = \mathbf{p}A\mathbf{q} = \begin{bmatrix} \frac{1}{6} & \frac{1}{3} & \frac{1}{2} \end{bmatrix} \begin{bmatrix} 3 & 5 & -2 & -1 \\ -2 & 4 & -3 & -4 \\ 6 & -5 & 0 & 3 \end{bmatrix} \begin{bmatrix} \frac{1}{4} \\ \frac{1}{4} \\ \frac{1}{4} \\ \frac{1}{3} \\ \frac{1}{6} \end{bmatrix} = \frac{13}{72} = .1805\ldots$$

Thus, in the long run player R can expect to receive an average of about 18 cents from player C in each play of the game. ◆

So far we have been discussing the situation in which each player has a predetermined strategy. We will now consider the more difficult situation in which both players may change their strategies independently. For example, in the game described in the introduction, we would allow both players to alter the areas of the sectors of their wheels and thereby control the probabilities of their respective moves. This qualitatively changes the nature of the problem and puts us firmly in the field of true game theory. It is understood that neither player knows the strategy the other will choose. It is also assumed that each player will make the best possible choice of strategy and that the other player knows this. Thus, player R attempts to choose a strategy $\mathbf{p}$ such that $E(\mathbf{p}, \mathbf{q})$ is as large as possible for the best strategy $\mathbf{q}$ that player C can choose; and similarly player C attempts to choose a strategy $\mathbf{q}$ such that $E(\mathbf{p}, \mathbf{q})$ is as small as possible for the best strategy $\mathbf{p}$ that player R can choose. To see that such choices are actually possible, we shall need the following theorem, called the ***Fundamental Theorem of Two-Person Zero-Sum Games***. (The general proof, which involves ideas from the theory of linear programming, will be omitted. However, later we shall prove two special cases of the theorem.)

Theorem 11.8.1   Fundamental Theorem of Zero-Sum Games

There exist strategies $\mathbf{p}^*$ and $\mathbf{q}^*$ such that

$$E(\mathbf{p}^*, \mathbf{q}) \geq E(\mathbf{p}^*, \mathbf{q}^*) \geq E(\mathbf{p}, \mathbf{q}^*) \qquad (3)$$

for all strategies $\mathbf{p}$ and $\mathbf{q}$.

The strategies $\mathbf{p}^*$ and $\mathbf{q}^*$ in this theorem are the best possible strategies for players R and C, respectively. To see why this is so let $v = E(\mathbf{p}^*, \mathbf{q}^*)$. The left-hand inequality of Equation (3) then reads

$$E(\mathbf{p}^*, \mathbf{q}) \geq v \quad \text{for all strategies } \mathbf{q}$$

This means that if player R chooses the strategy $\mathbf{p}^*$, then no matter what strategy $\mathbf{q}$ player C chooses, the expected payoff to player R will never be below v. Moreover, it is not possible for player R to achieve an expected payoff greater than v. To see why, suppose there is some strategy $\mathbf{p}^{**}$ that player R can choose such that

$$E(\mathbf{p}^{**}, \mathbf{q}) > v \quad \text{for all strategies } \mathbf{q}$$

Then, in particular,

$$E(\mathbf{p}^{**}, \mathbf{q}^*) > v$$

But this contradicts the right-hand inequality of Equation (3), which requires that $v \geq E(\mathbf{p}^{**}, \mathbf{q}^*)$. Consequently, the best player R can do is prevent his or her expected payoff from falling below the value v. Similarly, the best player C can do is ensure that player R's expected payoff does not exceed v, and this can be achieved by using strategy $\mathbf{q}^*$.

On the basis of this discussion, we arrive at the following definitions.

Definition

If $\mathbf{p}^*$ and $\mathbf{q}^*$ are strategies such that

$$E(\mathbf{p}^*, \mathbf{q}) \geq E(\mathbf{p}^*, \mathbf{q}^*) \geq E(\mathbf{p}, \mathbf{q}^*) \tag{4}$$

for all strategies $\mathbf{p}$ and $\mathbf{q}$, then

(i) $\mathbf{p}^*$ is called an *optimal strategy for player R*;
(ii) $\mathbf{q}^*$ is called an *optimal strategy for player C*;
(iii) $v = E(\mathbf{p}^*, \mathbf{q}^*)$ is called the *value* of the game.

The wording in this definition suggests that optimal strategies are not necessarily unique. This is indeed the case, and in Exercise 2 we ask the reader to show this. However, it can be proved that any two sets of optimal strategies always result in the same value v of the game. That is, if $\mathbf{p}^*, \mathbf{q}^*$ and $\mathbf{p}^{**}, \mathbf{q}^{**}$ are optimal strategies, then

$$E(\mathbf{p}^*, \mathbf{q}^*) = E(\mathbf{p}^{**}, \mathbf{q}^{**}) \tag{5}$$

The value of a game is thus the expected payoff to player R when both players choose any possible optimal strategies.

To find optimal strategies we must find vectors $\mathbf{p}^*$ and $\mathbf{q}^*$ that satisfy Equation (4). This is generally done by using linear programming techniques. Next, we discuss special cases for which optimal strategies may be found by more elementary techniques.

We now introduce the following definition.

Definition

An entry a_{rs} in a payoff matrix A is called a *saddle point* if

(i) a_{rs} is the smallest entry in its row, and
(ii) a_{rs} is the largest entry in its column.

A game whose payoff matrix has a saddle point is called *strictly determined*.

For example, the shaded element in each of the following payoff matrices is a saddle point:

$$\begin{bmatrix} 3 & 1 \\ -4 & 0 \end{bmatrix}, \qquad \begin{bmatrix} 30 & -50 & -5 \\ 60 & 90 & 75 \\ -10 & 60 & -30 \end{bmatrix}, \qquad \begin{bmatrix} 0 & -3 & 5 & -9 \\ 15 & -8 & -2 & 10 \\ 7 & 10 & 6 & 9 \\ 6 & 11 & -3 & 2 \end{bmatrix}$$

If a matrix has a saddle point a_{rs}, it turns out that the following strategies are optimal strategies for the two players:

$$\mathbf{p}^* = [0 \quad 0 \quad \cdots \quad \underset{\underset{r\text{th entry}}{\nearrow}}{1} \quad \cdots \quad 0], \qquad \mathbf{q}^* = \begin{bmatrix} 0 \\ 0 \\ \vdots \\ 1 \\ \vdots \\ 0 \end{bmatrix} \leftarrow s\text{th entry}$$

That is, an optimal strategy for player R is to always make the rth move, and an optimal strategy for player C is to always make the sth move. Such strategies for which only one move is possible are called ***pure strategies***. Strategies for which more than one move is possible are called ***mixed strategies***. To show that the above pure strategies are optimal, the reader may verify the following three equations (see Exercise 6):

$$E(\mathbf{p}^*, \mathbf{q}^*) = \mathbf{p}^* A \mathbf{q}^* = a_{rs} \tag{6}$$

$$E(\mathbf{p}^*, \mathbf{q}) = \mathbf{p}^* A \mathbf{q} \geq a_{rs} \quad \text{for any strategy } \mathbf{q} \tag{7}$$

$$E(\mathbf{p}, \mathbf{q}^*) = \mathbf{p} A \mathbf{q}^* \leq a_{rs} \quad \text{for any strategy } \mathbf{p} \tag{8}$$

Together, these three equations imply that

$$E(\mathbf{p}^*, \mathbf{q}) \geq E(\mathbf{p}^*, \mathbf{q}^*) \geq E(\mathbf{p}, \mathbf{q}^*)$$

for all strategies $\mathbf{p}$ and $\mathbf{q}$. Because this is exactly Equation (4), it follows that $\mathbf{p}^*$ and $\mathbf{q}^*$ are optimal strategies.

From Equation (6) the value of a strictly determined game is simply the numerical value of a saddle point a_{rs}. It is possible for a payoff matrix to have several saddle points, but then the uniqueness of the value of a game guarantees that the numerical values of all saddle points are the same.

EXAMPLE 2   Optimal Strategies to Maximize a Viewing Audience

Two competing television networks, R and C, are scheduling one-hour programs in the same time period. Network R can schedule one of three possible programs and network C can schedule one of four possible programs. Neither network knows which program the other will schedule. Both networks ask the same outside polling agency to give them an estimate of how all possible pairings of the programs will divide the viewing audience. The agency gives them each Table 2, whose (i, j)-th entry is the percentage of the viewing audience that will watch network R if network R's program i is paired against network C's program j. What program should each network schedule in order to maximize its viewing audience?

TABLE 2 *Audience Percentage for Network R*

		Network *C*'s Program			
		1	2	3	4
Network *R*'s Program	1	60	20	30	55
	2	50	75	45	60
	3	70	45	35	30

Solution.

Subtract 50 from each entry in the above table to construct the following matrix:

$$\begin{bmatrix} 10 & -30 & -20 & 5 \\ 0 & 25 & -5 & 10 \\ 20 & -5 & -15 & -20 \end{bmatrix}$$

This is the payoff matrix of the two-person zero-sum game in which each network is considered to start with 50% of the audience, and the (i, j)-th entry of the matrix is the percentage of the viewing audience that network C loses to network R if programs i and j are paired against each other. It is easily seen that the entry

$$a_{23} = -5$$

is a saddle point of the payoff matrix. Hence, the optimal strategy of network R is to schedule program 2 and the optimal strategy of network C is to schedule program 3. This will result in network R's receiving 45% of the audience and network C's receiving 55% of the audience. ◆

2 × 2 Matrix Games Another case in which the optimal strategies can be found by elementary means occurs when each player has only two possible moves. In this case, the payoff matrix is a 2×2 matrix

$$A = \begin{bmatrix} a_{11} & a_{12} \\ a_{21} & a_{22} \end{bmatrix}$$

If the game is strictly determined, at least one of the four entries of A is a saddle point, and the techniques discussed above can then be applied to determine optimal strategies for the two players. If the game is not strictly determined, we first compute the expected payoff for arbitrary strategies **p** and **q**:

$$E(\mathbf{p}, \mathbf{q}) = \mathbf{p}A\mathbf{q} = [p_1 \quad p_2] \begin{bmatrix} a_{11} & a_{12} \\ a_{21} & a_{22} \end{bmatrix} \begin{bmatrix} q_1 \\ q_2 \end{bmatrix}$$

$$= a_{11}p_1q_1 + a_{12}p_1q_2 + a_{21}p_2q_1 + a_{22}p_2q_2 \tag{9}$$

Because

$$p_1 + p_2 = 1 \quad \text{and} \quad q_1 + q_2 = 1 \tag{10}$$

we may substitute $p_2 = 1 - p_1$ and $q_2 = 1 - q_1$ into (9) to obtain

$$E(\mathbf{p}, \mathbf{q}) = a_{11}p_1q_1 + a_{12}p_1(1 - q_1) + a_{21}(1 - p_1)q_1 + a_{22}(1 - p_1)(1 - q_1) \tag{11}$$

If we rearrange the terms in Equation (11), we may write

$$E(\mathbf{p}, \mathbf{q}) = [(a_{11} + a_{22} - a_{12} - a_{21})p_1 - (a_{22} - a_{21})]q_1 + (a_{12} - a_{22})p_1 + a_{22} \tag{12}$$

By examining the coefficient of the q_1 term in (12), we see that if we set

$$p_1 = p_1^* = \frac{a_{22} - a_{21}}{a_{11} + a_{22} - a_{12} - a_{21}} \tag{13}$$

then that coefficient is zero, and (12) reduces to

$$E(\mathbf{p}^*, \mathbf{q}) = \frac{a_{11}a_{22} - a_{12}a_{21}}{a_{11} + a_{22} - a_{12} - a_{21}} \tag{14}$$

Equation (14) is independent of $\mathbf{q}$; that is, if player R chooses the strategy determined by (13), player C cannot change the expected payoff by varying his or her strategy.

In a similar manner, it may be verified that if player C chooses the strategy determined by

$$q_1 = q_1^* = \frac{a_{22} - a_{12}}{a_{11} + a_{22} - a_{12} - a_{21}} \tag{15}$$

then substituting in (12) gives

$$E(\mathbf{p}, \mathbf{q}^*) = \frac{a_{11}a_{22} - a_{12}a_{21}}{a_{11} + a_{22} - a_{12} - a_{21}} \tag{16}$$

Equations (14) and (16) show that

$$E(\mathbf{p}^*, \mathbf{q}) = E(\mathbf{p}^*, \mathbf{q}^*) = E(\mathbf{p}, \mathbf{q}^*) \tag{17}$$

for all strategies $\mathbf{p}$ and $\mathbf{q}$. Thus, the strategies determined by (13), (15), and (10) are optimal strategies for players R and C, respectively, and so we have the following result.

Theorem 11.8.2 **Optimal Strategies for a 2 × 2 Matrix Game**

For a 2 × 2 game that is not strictly determined, optimal strategies for players R and C are

$$\mathbf{p}^* = \left[\frac{a_{22} - a_{21}}{a_{11} + a_{22} - a_{12} - a_{21}} \quad \frac{a_{11} - a_{12}}{a_{11} + a_{22} - a_{12} - a_{21}} \right]$$

and

$$\mathbf{q}^* = \left[\begin{array}{c} \dfrac{a_{22} - a_{12}}{a_{11} + a_{22} - a_{12} - a_{21}} \\ \dfrac{a_{11} - a_{21}}{a_{11} + a_{22} - a_{12} - a_{21}} \end{array} \right]$$

The value of the game is

$$v = \frac{a_{11}a_{22} - a_{12}a_{21}}{a_{11} + a_{22} - a_{12} - a_{21}}$$

In order to be complete, we must show that the entries in the vectors $\mathbf{p}^*$ and $\mathbf{q}^*$ are numbers strictly between 0 and 1. In Exercise 7 we ask the reader to show that this is the case as long as the game is not strictly determined.

Equation (17) is interesting in that it implies that either player may force the expected payoff to be the value of the game by choosing his or her optimal strategy, regardless of which strategy the other player chooses. This is not true, in general, for games in which either player has more than two moves.

EXAMPLE 3 Using Theorem 11.8.2

The federal government desires to inoculate its citizens against a certain flu virus. The virus has two strains, and the proportions in which the two strains occur in the virus population is not known. Two vaccines have been developed. Vaccine 1 is 85% effective against strain 1 and 70% effective against strain 2. Vaccine 2 is 60% effective against strain 1 and 90% effective against strain 2. What inoculation policy should the government adopt?

Solution.

We may consider this a two-person game in which player R (the government) desires to make the payoff (the fraction of citizens resistant to the virus) as large as possible, and player C (the virus) desires to make the payoff as small as possible. The payoff matrix is

$$
\begin{array}{cc}
 & \textbf{Strain} \\
 & \begin{array}{cc} \textbf{1} & \textbf{2} \end{array} \\
\textbf{Vaccine}\ \begin{array}{c} \textbf{1} \\ \textbf{2} \end{array} & \begin{bmatrix} .85 & .70 \\ .60 & .90 \end{bmatrix}
\end{array}
$$

This matrix has no saddle points, so that Theorem 11.8.2 is applicable. Consequently,

$$
p_1^* = \frac{a_{22} - a_{21}}{a_{11} + a_{22} - a_{12} - a_{21}} = \frac{.90 - .60}{.85 + .90 - .70 - .60} = \frac{.30}{.45} = \frac{2}{3}
$$

$$
p_2^* = 1 - p_1^* = 1 - \frac{2}{3} = \frac{1}{3}
$$

$$
q_1^* = \frac{a_{22} - a_{12}}{a_{11} + a_{22} - a_{12} - a_{21}} = \frac{.90 - .70}{.85 + .90 - .70 - .60} = \frac{.20}{.45} = \frac{4}{9}
$$

$$
q_2^* = 1 - q_1^* = 1 - \frac{4}{9} = \frac{5}{9}
$$

$$
v = \frac{a_{11}a_{22} - a_{12}a_{21}}{a_{11} + a_{22} - a_{12} - a_{21}} = \frac{(.85)(.90) - (.70)(.60)}{.85 + .90 - .70 - .60} = \frac{.345}{.45} = .7666\ldots
$$

Thus, the optimal strategy for the government is to inoculate $\frac{2}{3}$ of the citizens with vaccine 1 and $\frac{1}{3}$ of the citizens with vaccine 2. This will guarantee that about 76.7% of the citizens will be resistant to a virus attack regardless of the distribution of the two strains.

In contrast, a virus distribution of $\frac{4}{9}$ of strain 1 and $\frac{5}{9}$ of strain 2 will result in the same 76.7% of resistant citizens, regardless of the inoculation strategy adopted by the government. ◆

Exercise Set 11.8

1. Suppose that a game has a payoff matrix

$$
A = \begin{bmatrix} -4 & 6 & -4 & 1 \\ 5 & -7 & 3 & 8 \\ -8 & 0 & 6 & -2 \end{bmatrix}
$$

(a) If players R and C use strategies

$$
\mathbf{p} = \begin{bmatrix} \frac{1}{2} & 0 & \frac{1}{2} \end{bmatrix} \quad \text{and} \quad \mathbf{q} = \begin{bmatrix} \frac{1}{4} \\ \frac{1}{4} \\ \frac{1}{4} \\ \frac{1}{4} \end{bmatrix}
$$

respectively, what is the expected payoff of the game?

(b) If player C keeps his strategy fixed as in part (a), what strategy should player R choose to maximize his expected payoff?

(c) If player R keeps her strategy fixed as in part (a), what strategy should player C choose to minimize the expected payoff to player R?

2. Construct a simple example to show that optimal strategies are not necessarily unique. For example, find a payoff matrix with several equal saddle points.

3. For the strictly determined games with the following payoff matrices, find optimal strategies for the two players, and find the values of the games.

(a) $\begin{bmatrix} 5 & 2 \\ 7 & 3 \end{bmatrix}$ (b) $\begin{bmatrix} -3 & -2 \\ 2 & 4 \\ -4 & 1 \end{bmatrix}$ (c) $\begin{bmatrix} 2 & -2 & 0 \\ -6 & 0 & -5 \\ 5 & 2 & 3 \end{bmatrix}$ (d) $\begin{bmatrix} -3 & 2 & -1 \\ -2 & -1 & 5 \\ -4 & 1 & 0 \\ -3 & 4 & 6 \end{bmatrix}$

4. For the 2×2 games with the following payoff matrices, find optimal strategies for the two players, and find the values of the games.

(a) $\begin{bmatrix} 6 & 3 \\ -1 & 4 \end{bmatrix}$ (b) $\begin{bmatrix} 40 & 20 \\ -10 & 30 \end{bmatrix}$ (c) $\begin{bmatrix} 3 & 7 \\ -5 & 4 \end{bmatrix}$ (d) $\begin{bmatrix} 3 & 5 \\ 5 & 2 \end{bmatrix}$ (e) $\begin{bmatrix} 7 & -3 \\ -5 & -2 \end{bmatrix}$

5. Player R has two playing cards: a black ace and a red four. Player C also has two cards: a black two and a red three. Each player secretly selects one of his or her cards. If both selected cards are the same color, player C pays player R the sum of the face values in dollars. If the cards are different colors, player R pays player C the sum of the face values. What are optimal strategies for both players, and what is the value of the game?

6. Verify Equations (6), (7), and (8).

7. Show that the entries of the optimal strategies $\mathbf{p}^*$ and $\mathbf{q}^*$ given in Theorem 11.8.2 are numbers strictly between zero and one.

Technology Exercises 11.8

The following exercises are designed to be solved using a technology utility. Typically, this will be MATLAB, *Mathematica*, Maple, Derive, or Mathcad, but it may also be some other type of linear algebra software or a scientific calculator with some linear algebra capabilities. For each exercise you will need to read the relevant documentation for the particular utility you are using. The goal of these exercises is to provide you with a basic proficiency with your technology utility. Once you have mastered the techniques in these exercises, you will be able to use your technology utility to solve many of the problems in the regular exercise sets.

T1. Consider a game between two players where each player can make up to n different moves ($n > 1$). If the ith move of player R and the jth move of player C are such that $i + j$ is even, then C pays R \$1. If $i + j$ is odd, then R pays C \$1. Assume that both players have the same strategy, that is, $\mathbf{p}_n = [\rho_i]_{1 \times n}$ and $\mathbf{q}_n = [\rho_i]_{n \times 1}$, where $\rho_1 + \rho_2 + \rho_3 + \cdots + \rho_n = 1$. Use a computer to show that

$$E(\mathbf{p}_2, \mathbf{q}_2) = (\rho_1 - \rho_2)^2$$
$$E(\mathbf{p}_3, \mathbf{q}_3) = (\rho_1 - \rho_2 + \rho_3)^2$$
$$E(\mathbf{p}_4, \mathbf{q}_4) = (\rho_1 - \rho_2 + \rho_3 - \rho_4)^2$$
$$E(\mathbf{p}_5, \mathbf{q}_5) = (\rho_1 - \rho_2 + \rho_3 - \rho_4 + \rho_5)^2$$

Using these results as a guide, prove in general that the expected payoff to player R is

$$E(\mathbf{p}_n, \mathbf{q}_n) = \left(\sum_{j=1}^{n} (-1)^{j+1} \rho_j \right)^2 \geq 0$$

which shows that in the long run, player R *will not lose* in this game.

T2. Consider a game between two players where each player can make up to n different moves ($n > 1$). If both players make the same move, then player C pays player R $\$(n - 1)$. However, if both players make different moves, then player R pays player C $\$1$. Assume that both players have the same strategy, that is, $\mathbf{p}_n = [\rho_i]_{1 \times n}$ and $\mathbf{q}_n = [\rho_i]_{n \times 1}$, where $\rho_1 + \rho_2 + \rho_3 + \cdots + \rho_n = 1$. Use a computer to show that

$$E(\mathbf{p}_2, \mathbf{q}_2) = \frac{1}{2}(\rho_1 - \rho_1)^2 + \frac{1}{2}(\rho_1 - \rho_2)^2 + \frac{1}{2}(\rho_2 - \rho_1)^2 + \frac{1}{2}(\rho_2 - \rho_2)^2$$

$$E(\mathbf{p}_3, \mathbf{q}_3) = \frac{1}{2}(\rho_1 - \rho_1)^2 + \frac{1}{2}(\rho_1 - \rho_2)^2 + \frac{1}{2}(\rho_1 - \rho_3)^2$$

$$+ \frac{1}{2}(\rho_2 - \rho_1)^2 + \frac{1}{2}(\rho_2 - \rho_2)^2 + \frac{1}{2}(\rho_2 - \rho_3)^2$$

$$+ \frac{1}{2}(\rho_3 - \rho_1)^2 + \frac{1}{2}(\rho_3 - \rho_2)^2 + \frac{1}{2}(\rho_3 - \rho_3)^2$$

$$E(\mathbf{p}_4, \mathbf{q}_4) = \frac{1}{2}(\rho_1 - \rho_1)^2 + \frac{1}{2}(\rho_1 - \rho_2)^2 + \frac{1}{2}(\rho_1 - \rho_3)^2 + \frac{1}{2}(\rho_1 - \rho_4)^2$$

$$+ \frac{1}{2}(\rho_2 - \rho_1)^2 + \frac{1}{2}(\rho_2 - \rho_2)^2 + \frac{1}{2}(\rho_2 - \rho_3)^2 + \frac{1}{2}(\rho_2 - \rho_4)^2$$

$$+ \frac{1}{2}(\rho_3 - \rho_1)^2 + \frac{1}{2}(\rho_3 - \rho_2)^2 + \frac{1}{2}(\rho_3 - \rho_3)^2 + \frac{1}{2}(\rho_3 - \rho_4)^2$$

$$+ \frac{1}{2}(\rho_4 - \rho_1)^2 + \frac{1}{2}(\rho_4 - \rho_2)^2 + \frac{1}{2}(\rho_4 - \rho_3)^2 + \frac{1}{2}(\rho_4 - \rho_4)^2$$

Using these results as a guide, prove in general that the expected payoff to player R is

$$E(\mathbf{p}_n, \mathbf{q}_n) = \frac{1}{2} \sum_{i=1}^{n} \sum_{j=1}^{n} (\rho_i - \rho_j)^2 \geq 0$$

which shows that in the long run, player R *will not lose* in this game.

11.9 LEONTIEF ECONOMIC MODELS

In this section we discuss two linear models for economic systems. Some results about nonnegative matrices are applied to determine equilibrium price structures and outputs necessary to satisfy demand.

PREREQUISITES: Linear Systems
Matrices

Economic Systems Matrix theory has been very successful in describing the interrelations among prices, outputs, and demands in economic systems. In this section we discuss some simple models based on the ideas of Nobel-laureate Wassily Leontief. We examine two different but related models: the closed or input–output model, and the open or production model. In each, we are given certain economic parameters that describe the interrelations between the "industries" in the economy under consideration. Using matrix theory, we then evaluate certain other parameters, such as prices or output levels, in order to satisfy a desired economic objective. We begin with the closed model.

Leontief Closed (Input–Output) Model

First we present a simple example, then we proceed to the general theory of the model.

EXAMPLE 1 An Input–Output Model

Three homeowners—a carpenter, an electrician, and a plumber—agree to make repairs in their three homes. They agree to work a total of 10 days each according to the following schedule:

	Work Performed by		
	Carpenter	**Electrician**	**Plumber**
Days of Work in Home of Carpenter	2	1	6
Days of Work in Home of Electrician	4	5	1
Days of Work in Home of Plumber	4	4	3

For tax purposes they must report and pay each other a reasonable daily wage, even for the work each does on his or her own home. Their normal daily wages are about \$100, but they agree to adjust their respective daily wages so that each homeowner will come out even, that is, so that the total amount paid out by each is the same as the total amount each receives. We can set

$$p_1 = \text{daily wage of carpenter}$$
$$p_2 = \text{daily wage of electrician}$$
$$p_3 = \text{daily wage of plumber}$$

To satisfy the "equilibrium" condition that each homeowner comes out even, we require that

$$\text{total expenditures} = \text{total income}$$

for each of the homeowners for the 10-day period. For example, the carpenter pays a total of $2p_1 + p_2 + 6p_3$ for the repairs in his own home, and receives a total income of $10p_1$ for the repairs that he performs on all three homes. Equating these two expressions then gives the first of the following three equations:

$$2p_1 + p_2 + 6p_3 = 10p_1$$
$$4p_1 + 5p_2 + p_3 = 10p_2$$
$$4p_1 + 4p_2 + 3p_3 = 10p_3$$

The remaining two equations are the equilibrium equations for the electrician and the plumber. Dividing these equations by 10 and rewriting them in matrix form yields

$$\begin{bmatrix} .2 & .1 & .6 \\ .4 & .5 & .1 \\ .4 & .4 & .3 \end{bmatrix} \begin{bmatrix} p_1 \\ p_2 \\ p_3 \end{bmatrix} = \begin{bmatrix} p_1 \\ p_2 \\ p_3 \end{bmatrix} \tag{1}$$

Equation (1) can be rewritten as a homogeneous system by subtracting the left side from the right side to obtain

$$
\begin{bmatrix}
.8 & -.1 & -.6 \\
-.4 & .5 & -.1 \\
-.4 & -.4 & .7
\end{bmatrix}
\begin{bmatrix}
p_1 \\
p_2 \\
p_3
\end{bmatrix}
=
\begin{bmatrix}
0 \\
0 \\
0
\end{bmatrix}
$$

The solution of this homogeneous system is found to be (verify)

$$
\begin{bmatrix}
p_1 \\
p_2 \\
p_3
\end{bmatrix}
= s
\begin{bmatrix}
31 \\
32 \\
36
\end{bmatrix}
$$

where s is an arbitrary constant. This constant is a scale factor, which the homeowners may choose for their convenience. For example, they may set $s = 3$ so that the corresponding daily wages—\$93, \$96, and \$108—are about \$100. ◆

This example illustrates the salient features of the Leontief input–output model of a closed economy. In the basic Equation (1) each column sum of the coefficient matrix is 1, corresponding to the fact that each of the homeowners' "output" of labor is completely distributed among these same homeowners in the proportions given by the entries in the column. Our problem is to determine suitable "prices" for these outputs so as to put the system in equilibrium, that is, so that each homeowner's total expenditures equal his or her total income.

In the general model we have an economic system consisting of a finite number of "industries," which we number as industries $1, 2, \ldots, k$. Over some fixed period of time, each industry produces an "output" of some good or service that is completely utilized in a predetermined manner by the k industries. An important problem is to find suitable "prices" to be charged for these k outputs so that for each industry, total expenditures equal total income. Such a price structure represents an equilibrium position for the economy.

For the fixed time period in question, let us set

p_i = price charged by the ith industry for its total output

e_{ij} = fraction of the total output of the jth industry purchased by the ith industry

for $i, j = 1, 2, \ldots, k$. By definition, we have

(i) $p_i \geq 0, \qquad i = 1, 2, \ldots, k$

(ii) $e_{ij} \geq 0, \qquad i, j = 1, 2, \ldots, k$

(iii) $e_{1j} + e_{2j} + \cdots + e_{kj} = 1, \qquad j = 1, 2, \ldots, k$

With these quantities, we form the ***price vector***

$$
\mathbf{p} =
\begin{bmatrix}
p_1 \\
p_2 \\
\vdots \\
p_k
\end{bmatrix}
$$

and the ***exchange*** or ***input–output matrix***

$$
E =
\begin{bmatrix}
e_{11} & e_{12} & \cdots & e_{1k} \\
e_{21} & e_{22} & \cdots & e_{2k} \\
\vdots & \vdots & & \vdots \\
e_{k1} & e_{k2} & \cdots & e_{kk}
\end{bmatrix}
$$

Condition (iii) expresses the fact that all the column sums of the exchange matrix are 1.

As in the example, in order that the expenditures of each industry be equal to its income, the following matrix equation must be satisfied [see (1)]:

$$E\mathbf{p} = \mathbf{p} \tag{2}$$

or

$$(I - E)\mathbf{p} = \mathbf{0} \tag{3}$$

Equation (3) is a homogeneous linear system for the price vector **p**. It will have a nontrivial solution if and only if the determinant of its coefficient matrix $I - E$ is zero. In Exercise 6 we ask the reader to show that this is the case for any exchange matrix E. Thus, (3) always has nontrivial solutions for the price vector **p**.

Actually, for our economic model to make sense we need more than just the fact that (3) has nontrivial solutions for **p**. We also need the prices p_i of the k outputs to be nonnegative numbers. We express this condition as $\mathbf{p} \geq 0$. (In general, if A is any vector or matrix, the notation $A \geq 0$ means that every entry of A is nonnegative, and the notation $A > 0$ means that every entry of A is positive. Similarly, $A \geq B$ means $A - B \geq 0$, and $A > B$ means $A - B > 0$.) To show that (3) has a nontrivial solution for which $\mathbf{p} \geq 0$ is a bit more difficult than showing merely that some nontrivial solution exists. But it is true, and we state this fact without proof in the following theorem.

Theorem 11.9.1

*If E is an exchange matrix, then $E\mathbf{p} = \mathbf{p}$ always has a nontrivial solution **p** whose entries are nonnegative.*

Let us consider a few simple examples of this theorem.

EXAMPLE 2 Using Theorem 11.9.1

Let

$$E = \begin{bmatrix} \frac{1}{2} & 0 \\ \frac{1}{2} & 1 \end{bmatrix}$$

Then $(I - E)\mathbf{p} = \mathbf{0}$ is

$$\begin{bmatrix} \frac{1}{2} & 0 \\ -\frac{1}{2} & 0 \end{bmatrix} \begin{bmatrix} p_1 \\ p_2 \end{bmatrix} = \begin{bmatrix} 0 \\ 0 \end{bmatrix}$$

which has the general solution

$$\mathbf{p} = s \begin{bmatrix} 0 \\ 1 \end{bmatrix}$$

where s is an arbitrary constant. We then have nontrivial solutions $\mathbf{p} \geq 0$ for any $s > 0$. ◆

EXAMPLE 3 Using Theorem 11.9.1

Let

$$E = \begin{bmatrix} 1 & 0 \\ 0 & 1 \end{bmatrix}$$

Then $(I - E)\mathbf{p} = \mathbf{0}$ has the general solution

$$\mathbf{p} = s \begin{bmatrix} 1 \\ 0 \end{bmatrix} + t \begin{bmatrix} 0 \\ 1 \end{bmatrix}$$

where s and t are independent arbitrary constants. Nontrivial solutions $\mathbf{p} \geq 0$ then result from any $s \geq 0$ and $t \geq 0$, not both zero. ◆

Example 2 indicates that in some situations one of the prices must be zero in order to satisfy the equilibrium condition. Example 3 indicates that there may be several linearly independent price structures available. Neither of these situations describes a truly interdependent economic structure. The following theorem gives sufficient conditions for both cases to be excluded.

Theorem 11.9.2

Let E be an exchange matrix such that for some positive integer m all the entries of E^m are positive. Then there is exactly one linearly independent solution of $(I - E)\mathbf{p} = \mathbf{0}$, and it may be chosen so that all its entries are positive.

We will not give a proof of this theorem. The reader who has read Section 11.6 on Markov chains may observe that this theorem is essentially the same as Theorem 11.6.4. What we are calling exchange matrices in this section were called stochastic or Markov matrices in Section 11.6.

EXAMPLE 4 Using Theorem 11.9.2

The exchange matrix in Example 1 was

$$E = \begin{bmatrix} .2 & .1 & .6 \\ .4 & .5 & .1 \\ .4 & .4 & .3 \end{bmatrix}$$

Because $E > 0$, the condition $E^m > 0$ in Theorem 11.9.2 is satisfied for $m = 1$. Consequently, we are guaranteed that there is exactly one linearly independent solution of $(I - E)\mathbf{p} = \mathbf{0}$, and it can be chosen so that $\mathbf{p} > 0$. In that example, we found that

$$\mathbf{p} = \begin{bmatrix} 31 \\ 32 \\ 36 \end{bmatrix}$$

is such a solution. ◆

Leontief Open (Production) Model
In contrast with the closed model, in which the outputs of k industries are distributed only among themselves, the open model attempts to satisfy an outside demand for the outputs. Portions of these outputs may still be distributed among the industries themselves, to keep them operating, but there is to be some excess, some net production, with which to satisfy the outside demand. In the closed model the outputs of the industries are fixed, and our objective is to determine prices for these outputs so that the equilibrium condition, that expenditures equal incomes, is satisfied. In the open model it is the prices that are fixed, and our objective is to determine levels of the outputs of the industries needed to satisfy the

outside demand. We will measure the levels of the outputs in terms of their economic values using the fixed prices. To be precise, over some fixed period of time, let

$x_i =$ monetary value of the total output of the ith industry

$d_i =$ monetary value of the output of the ith industry needed to satisfy the outside demand

$c_{ij} =$ monetary value of the output of the ith industry needed by the jth industry to produce one unit of monetary value of its own output

With these quantities, we define the ***production vector***

$$\mathbf{x} = \begin{bmatrix} x_1 \\ x_2 \\ \vdots \\ x_k \end{bmatrix}$$

the ***demand vector***

$$\mathbf{d} = \begin{bmatrix} d_1 \\ d_2 \\ \vdots \\ d_k \end{bmatrix}$$

and the ***consumption matrix***

$$C = \begin{bmatrix} c_{11} & c_{12} & \cdots & c_{1k} \\ c_{21} & c_{22} & \cdots & c_{2k} \\ \vdots & \vdots & & \vdots \\ c_{k1} & c_{k2} & \cdots & c_{kk} \end{bmatrix}$$

By their nature, we have that

$$\mathbf{x} \geq 0, \quad \mathbf{d} \geq 0, \quad \text{and} \quad C \geq 0$$

From the definition of c_{ij} and x_j, it can be seen that the quantity

$$c_{i1}x_1 + c_{i2}x_2 + \cdots + c_{ik}x_k$$

is the value of the output of the ith industry needed by all k industries to produce a total output specified by the production vector $\mathbf{x}$. Because this quantity is simply the ith entry of the column vector $C\mathbf{x}$, we can say further that the ith entry of the column vector

$$\mathbf{x} - C\mathbf{x}$$

is the value of the excess output of the ith industry available to satisfy the outside demand. The value of the outside demand for the output of the ith industry is the ith entry of the demand vector $\mathbf{d}$. Consequently, we are led to the following equation

$$\mathbf{x} - C\mathbf{x} = \mathbf{d}$$

or

$$(I - C)\mathbf{x} = \mathbf{d} \tag{4}$$

for the demand to be exactly met, without any surpluses or shortages. Thus, given C and $\mathbf{d}$, our objective is to find a production vector $\mathbf{x} \geq 0$ that satisfies Equation (4).

EXAMPLE 5 Production Vector for a Town

A town has three main industries: a coal-mining operation, an electric power-generating plant, and a local railroad. To mine $1 of coal, the mining operation must purchase $.25

of electricity to run its equipment and $.25 of transportation for its shipping needs. To produce $1 of electricity, the generating plant requires $.65 of coal for fuel, $.05 of its own electricity to run auxiliary equipment, and $.05 of transportation. To provide $1 of transportation, the railroad requires $.55 of coal for fuel and $.10 of electricity for its auxiliary equipment. In a certain week the coal-mining operation receives orders for $50,000 of coal from outside the town and the generating plant receives orders for $25,000 of electricity from outside. There is no outside demand for the local railroad. How much must each of the three industries produce in that week to exactly satisfy their own demand and the outside demand?

Solution.

For the one-week period let

$$x_1 = \text{value of total output of coal-mining operation}$$
$$x_2 = \text{value of total output of power-generating plant}$$
$$x_3 = \text{value of total output of local railroad}$$

From the information supplied the consumption matrix of the system is

$$C = \begin{bmatrix} 0 & .65 & .55 \\ .25 & .05 & .10 \\ .25 & .05 & 0 \end{bmatrix}$$

The linear system $(I - C)\mathbf{x} = \mathbf{d}$ is then

$$\begin{bmatrix} 1.00 & -.65 & -.55 \\ -.25 & .95 & -.10 \\ -.25 & -.05 & 1.00 \end{bmatrix} \begin{bmatrix} x_1 \\ x_2 \\ x_3 \end{bmatrix} = \begin{bmatrix} 50,000 \\ 25,000 \\ 0 \end{bmatrix}$$

The coefficient matrix on the left is invertible, and the solution is given by

$$\mathbf{x} = (I - C)^{-1}\mathbf{d} = \frac{1}{503} \begin{bmatrix} 756 & 542 & 470 \\ 220 & 690 & 190 \\ 200 & 170 & 630 \end{bmatrix} \begin{bmatrix} 50,000 \\ 25,000 \\ 0 \end{bmatrix} = \begin{bmatrix} 102,087 \\ 56,163 \\ 28,330 \end{bmatrix}$$

Thus, the total output of the coal-mining operation should be $102,087, the total output of the power-generating plant should be $56,163, and the total output of the railroad should be $28,330. ◆

Let us reconsider Equation (4):

$$(I - C)\mathbf{x} = \mathbf{d}$$

If the square matrix $I - C$ is invertible, we can write

$$\mathbf{x} = (I - C)^{-1}\mathbf{d} \tag{5}$$

In addition, if the matrix $(I - C)^{-1}$ has only nonnegative entries, then we are guaranteed that for any $\mathbf{d} \geq 0$, Equation (5) has a unique nonnegative solution for $\mathbf{x}$. This is a particularly desirable situation, as it means that any outside demand can be met. The terminology used to describe this case is given in the following definition.

Definition

A consumption matrix C is said to be ***productive*** if $(I - C)^{-1}$ exists and

$$(I - C)^{-1} \geq 0$$

We will now consider some simple criteria that guarantee that a consumption matrix is productive. The first is given in the following theorem.

Theorem 11.9.3   **Productive Consumption Matrix**

A consumption matrix C is productive if and only if there is some production vector **x** ≥ 0 *such that* **x** > C**x**.

(The proof is outlined in Exercise 8.) The condition **x** > C**x** means that there is some production schedule possible such that each industry produces more than it consumes.

Theorem 11.9.3 has two interesting corollaries. Suppose that all the row sums of *C* are less than one. If

$$\mathbf{x} = \begin{bmatrix} 1 \\ 1 \\ \vdots \\ 1 \end{bmatrix}$$

then C**x** is a column vector whose entries are these row sums. Therefore, **x** > C**x**, and the condition of Theorem 11.9.3 is satisfied. Thus, we arrive at the following corollary:

Corollary 11.9.4

A consumption matrix is productive if each of its row sums is less than 1.

As we ask the reader to show in Exercise 7, this corollary leads to the following:

Corollary 11.9.5

A consumption matrix is productive if each of its column sums is less than 1.

Recalling the definition of the entries of the consumption matrix *C*, we see that the *j*th column sum of *C* is the total value of the outputs of all *k* industries needed to produce one unit of value of output of the *j*th industry. The *j*th industry is thus said to be **profitable** if that *j*th column sum is less than 1. In other words, Corollary 11.9.5 says that a consumption matrix is productive if all *k* industries in the economic system are profitable.

EXAMPLE 6   Using Corollary 11.9.5

The consumption matrix in Example 5 was

$$C = \begin{bmatrix} 0 & .65 & .55 \\ .25 & .05 & .10 \\ .25 & .05 & 0 \end{bmatrix}$$

All three column sums in this matrix are less than 1 and so all three industries are profitable. Consequently, by Corollary 11.9.5 the consumption matrix *C* is productive. This can also be seen in the calculations in Example 5, as $(I - C)^{-1}$ is nonnegative. ◆

Exercise Set 11.9

1. For the following exchange matrices, find nonnegative price vectors that satisfy the equilibrium condition (3).

   (a) $\begin{bmatrix} \frac{1}{2} & \frac{1}{3} \\ \frac{1}{2} & \frac{2}{3} \end{bmatrix}$
   (b) $\begin{bmatrix} \frac{1}{2} & 0 & \frac{1}{2} \\ \frac{1}{3} & 0 & \frac{1}{2} \\ \frac{1}{6} & 1 & 0 \end{bmatrix}$
   (c) $\begin{bmatrix} .35 & .50 & .30 \\ .25 & .20 & .30 \\ .40 & .30 & .40 \end{bmatrix}$

2. Using Theorem 11.9.3 and its corollaries, show that each of the following consumption matrices is productive.

   (a) $\begin{bmatrix} .8 & .1 \\ .3 & .6 \end{bmatrix}$
   (b) $\begin{bmatrix} .70 & .30 & .25 \\ .20 & .40 & .25 \\ .05 & .15 & .25 \end{bmatrix}$
   (c) $\begin{bmatrix} .7 & .3 & .2 \\ .1 & .4 & .3 \\ .2 & .4 & .1 \end{bmatrix}$

3. Using Theorem 11.9.2, show that there is only one linearly independent price vector for the closed economic system with exchange matrix

   $$E = \begin{bmatrix} 0 & .2 & .5 \\ 1 & .2 & .5 \\ 0 & .6 & 0 \end{bmatrix}$$

4. Three neighbors have backyard vegetable gardens. Neighbor A grows tomatoes, neighbor B grows corn, and neighbor C grows lettuce. They agree to divide their crops among themselves as follows: A gets $\frac{1}{2}$ of the tomatoes, $\frac{1}{3}$ of the corn, and $\frac{1}{4}$ of the lettuce. B gets $\frac{1}{3}$ of the tomatoes, $\frac{1}{3}$ of the corn, and $\frac{1}{4}$ of the lettuce. C gets $\frac{1}{6}$ of the tomatoes, $\frac{1}{3}$ of the corn, and $\frac{1}{2}$ of the lettuce. What prices should the neighbors assign to their respective crops if the equilibrium condition of a closed economy is to be satisfied, and if the lowest-priced crop is to have a price of $100?

5. Three engineers—a civil engineer (CE), an electrical engineer (EE), and a mechanical engineer (ME)—each have a consulting firm. The consulting they do is of a multidisciplinary nature, so they buy a portion of each others' services. For each $1 of consulting the CE does, she buys $.10 of the EE's services and $.30 of the ME's services. For each $1 of consulting the EE does, she buys $.20 of the CE's services and $.40 of the ME's services. And for each $1 of consulting the ME does, she buys $.30 of the CE's services and $.40 of the EE's services. In a certain week the CE receives outside consulting orders of $500, the EE receives outside consulting orders of $700, and the ME receives outside consulting orders of $600. What dollar amount of consulting does each engineer perform in that week?

6. Using the fact that the column sums of an exchange matrix E are all 1, show that the column sums of $I - E$ are zero. From this, show that $I - E$ has zero determinant, and so $(I - E)\mathbf{p} = \mathbf{0}$ has nontrivial solutions for $\mathbf{p}$.

7. Show that Corollary 11.9.5 follows from Corollary 11.9.4. [**Hint.** Use the fact that $(A^T)^{-1} = (A^{-1})^T$ for any invertible matrix A.]

8. (**For readers who have studied calculus.**) Prove Theorem 11.9.3 as follows:

   (a) Prove the "only if" part of the theorem; that is, show that if C is a productive consumption matrix, then there is a vector $\mathbf{x} \geq 0$ such that $\mathbf{x} > C\mathbf{x}$.

   (b) Prove the "if" part of the theorem as follows:

   Step 1. Show that if there is a vector $\mathbf{x}^* \geq 0$ such that $C\mathbf{x}^* < \mathbf{x}^*$, then $\mathbf{x}^* > 0$.

   Step 2. Show that there is a number λ such that $0 < \lambda < 1$ and $C\mathbf{x}^* < \lambda \mathbf{x}^*$.

   Step 3. Show that $C^n \mathbf{x}^* < \lambda^n \mathbf{x}^*$ for $n = 1, 2, \ldots$.

   Step 4. Show that $C^n \to 0$ as $n \to \infty$.

Step 5. By multiplying out, show that

$$(I - C)(I + C + C^2 + \cdots + C^{n-1}) = I - C^n$$

for $n = 1, 2, \ldots$.

Step 6. By letting $n \to \infty$ in Step 5, show that the matrix infinite sum

$$S = I + C + C^2 + \cdots$$

exists and that $(I - C)S = I$.

Step 7. Show that $S \geq 0$ and that $S = (I - C)^{-1}$.

Step 8. Show that C is a productive consumption matrix.

Technology Exercises 11.9

The following exercises are designed to be solved using a technology utility. Typically, this will be MATLAB, *Mathematica*, Maple, Derive, or Mathcad, but it may also be some other type of linear algebra software or a scientific calculator with some linear algebra capabilities. For each exercise you will need to read the relevant documentation for the particular utility you are using. The goal of these exercises is to provide you with a basic proficiency with your technology utility. Once you have mastered the techniques in these exercises, you will be able to use your technology utility to solve many of the problems in the regular exercise sets.

T1. Consider a sequence of exchange matrices $\{E_2, E_3, E_4, E_5, \ldots, E_n\}$, where

$$E_2 = \begin{bmatrix} 0 & \frac{1}{2} \\ 1 & \frac{1}{2} \end{bmatrix}, \qquad E_3 = \begin{bmatrix} 0 & \frac{1}{2} & \frac{1}{3} \\ 1 & 0 & \frac{1}{3} \\ 0 & \frac{1}{2} & \frac{1}{3} \end{bmatrix},$$

$$E_4 = \begin{bmatrix} 0 & \frac{1}{2} & \frac{1}{3} & \frac{1}{4} \\ 1 & 0 & \frac{1}{3} & \frac{1}{4} \\ 0 & \frac{1}{2} & 0 & \frac{1}{4} \\ 0 & 0 & \frac{1}{3} & \frac{1}{4} \end{bmatrix}, \qquad E_5 = \begin{bmatrix} 0 & \frac{1}{2} & \frac{1}{3} & \frac{1}{4} & \frac{1}{5} \\ 1 & 0 & \frac{1}{3} & \frac{1}{4} & \frac{1}{5} \\ 0 & \frac{1}{2} & 0 & \frac{1}{4} & \frac{1}{5} \\ 0 & 0 & \frac{1}{3} & 0 & \frac{1}{5} \\ 0 & 0 & 0 & \frac{1}{4} & \frac{1}{5} \end{bmatrix},$$

and so on. Use a computer to show that $E_2^2 > 0_2$, $E_3^3 > 0_3$, $E_4^4 > 0_4$, $E_5^5 > 0_5$, and make the conjecture that although $E_n^n > 0_n$ is true, $E_n^k > 0_n$ is not true for $k = 1, 2, 3, \ldots, n - 1$. Next, use a computer to determine the vectors $\mathbf{p}_n$ such that $E_n \mathbf{p}_n = \mathbf{p}_n$ (for $n = 2, 3, 4, 5, 6$), and then see if you can discover a pattern that would allow you to easily compute $\mathbf{p}_{n+1}$ from $\mathbf{p}_n$. Test your discovery by first constructing $\mathbf{p}_8$ from

$$\mathbf{p}_7 = \begin{bmatrix} 2520 \\ 3360 \\ 1890 \\ 672 \\ 175 \\ 36 \\ 7 \end{bmatrix}$$

and then checking to see if $E_8 \mathbf{p}_8 = \mathbf{p}_8$.

T2. Consider an open production model having n industries with $n > 1$. In order to produce \$1 of its own output, the jth industry must spend \$$(1/n)$ for the output of the ith industry (for all $i \neq j$), but the jth industry (for all $j = 1, 2, 3, \ldots, n$) spends nothing for its own output. Construct the consumption matrix C_n, show that it is productive, and determine an expression for $(I_n - C_n)^{-1}$. In determining an expression for $(I_n - C_n)^{-1}$, use a computer to study the cases when $n = 2, 3, 4,$ and 5; then make a conjecture and prove your conjecture to be true. [**Hint.** If $F_n = [1]_{n \times n}$ (i.e., the $n \times n$ matrix with every entry equal to one), first show that

$$F_n^2 = nF_n$$

and then express your value of $(I_n - C_n)^{-1}$ in terms of n, I_n, and F_n.]

11.10 FOREST MANAGEMENT

In this section we discuss a matrix model for the management of a forest where trees are grouped into classes according to height. The optimal sustainable yield of a periodic harvest is calculated when the trees of different height classes can have different economic values.

PREREQUISITE: Matrix Operations

Optimal Sustainable Yield   Our objective is to introduce a simplified model for the sustainable harvesting of a forest whose trees are classified by height. The height of a tree is assumed to determine its economic value when it is cut down and sold. Initially, there is a distribution of trees of various heights. The forest is then allowed to grow for a certain period of time, after which some of the trees of various heights are harvested. The trees left unharvested are to be of the same height configuration as the original forest, so that the harvest is sustainable. As we will see, there are many such sustainable harvesting procedures. We want to find one for which the total economic value of all the trees removed is as large as possible. This determines the *optimal sustainable yield* of the forest and is the largest yield that can be attained continually without depleting the forest.

The Model   Suppose that a harvester has a forest of Douglas fir trees that are to be sold as Christmas trees year after year. Every December the harvester cuts down some of the trees to be sold. For each tree cut down a seedling is planted in its place. In this way the total number of trees in the forest is always the same. (In this simplified model, we will not take into account trees that die between harvests. We assume that every seedling planted survives and grows until it is harvested.)

In the marketplace, trees of different heights have different economic values. Suppose that there are n different price classes corresponding to certain height intervals, as shown in Table 1 and Figure 11.10.1. The first class consists of seedlings with heights in the interval $[0, h_1)$, and these seedlings are of no economic value. The nth class consists of trees with heights greater than or equal to h_{n-1}.

Let x_i $(i = 1, 2, \ldots, n)$ be the number of trees within the ith class that remain after each harvest. We form a column vector with the numbers and call it the *nonharvest vector*:

$$\mathbf{x} = \begin{bmatrix} x_1 \\ x_2 \\ \vdots \\ x_n \end{bmatrix}$$

TABLE 1

Class	Value (dollars)	Height Interval
1 (seedlings)	None	$[0, h_1)$
2	p_2	$[h_1, h_2)$
3	p_3	$[h_2, h_3)$
$\vdots$	$\vdots$	$\vdots$
$n - 1$	p_{n-1}	$[h_{n-2}, h_{n-1})$
n	p_n	$[h_{n-1}, \infty)$

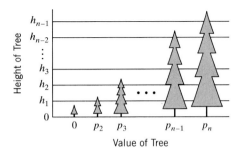

Figure 11.10.1

For a sustainable harvesting policy, the forest is to be returned after each harvest to the fixed configuration given by the nonharvest vector **x**. Part of our problem is to find those nonharvest vectors **x** for which sustainable harvesting is possible.

Because the total number of trees in the forest is fixed, we can set

$$x_1 + x_2 + \cdots + x_n = s \tag{1}$$

where s is predetermined by the amount of land available and the amount of space each tree requires. Referring to Figure 11.10.2, we have the following situation. The forest configuration is given by the vector **x** after each harvest. Between harvests the trees grow and produce a new forest configuration before each harvest. A certain number of trees are removed from each class at the harvest. Finally, a seedling is planted in place of each tree removed to return the forest again to the configuration **x**.

Consider first the growth of the forest between harvests. During this period a tree in the ith class may grow and move up to a higher height class. Or its growth may be retarded for some reason, and it will remain in the same class. We consequently define the following growth parameters g_i for $i = 1, 2, \ldots, n - 1$:

$g_i =$ the fraction of trees in the ith class that grow into
the $(i + 1)$-st class during a growth period

For simplicity we assume that a tree can move at most one height class upward in one growth period. With this assumption we have

$1 - g_i =$ the fraction of trees in the ith class that remain in
the ith class during a growth period

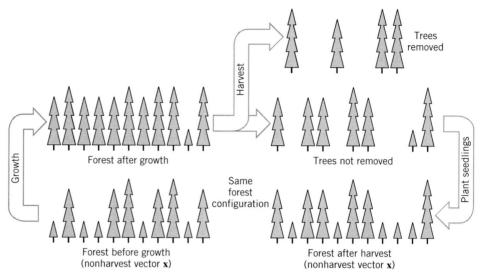

Figure 11.10.2

With these $n - 1$ growth parameters we form the following $n \times n$ **growth matrix**:

$$
G = \begin{bmatrix}
1 - g_1 & 0 & 0 & \cdots & 0 \\
g_1 & 1 - g_2 & 0 & \cdots & 0 \\
0 & g_2 & 1 - g_3 & \cdots & 0 \\
\vdots & \vdots & \vdots & \vdots & \vdots \\
0 & 0 & 0 & \cdots & 1 - g_{n-1} & 0 \\
0 & 0 & 0 & \cdots & g_{n-1} & 1
\end{bmatrix}
\tag{2}
$$

Because the entries of the vector $\mathbf{x}$ are the numbers of trees in the n classes before the growth period, the reader can verify that the entries of the vector

$$
G\mathbf{x} = \begin{bmatrix}
(1 - g_1)x_1 \\
g_1 x_1 + (1 - g_2)x_2 \\
g_2 x_2 + (1 - g_3)x_3 \\
\vdots \\
g_{n-2}x_{n-2} + (1 - g_{n-1})x_{n-1} \\
g_{n-1}x_{n-1} + x_n
\end{bmatrix}
\tag{3}
$$

are the numbers of trees in the n classes after the growth period.

Suppose that during the harvest we remove y_i $(i = 1, 2, \ldots, n)$ trees from the ith class. We will call the column vector

$$
\mathbf{y} = \begin{bmatrix} y_1 \\ y_2 \\ \vdots \\ y_n \end{bmatrix}
$$

the **harvest vector**. Thus, a total of

$$
y_1 + y_2 + \cdots + y_n
$$

trees are removed at each harvest. This is also the total number of trees added to the first class (the new seedlings) after each harvest. If we define the following $n \times n$ **replacement**

matrix

$$R = \begin{bmatrix} 1 & 1 & \cdots & 1 \\ 0 & 0 & \cdots & 0 \\ \vdots & \vdots & & \vdots \\ 0 & 0 & \cdots & 0 \end{bmatrix} \tag{4}$$

then the column vector

$$R\mathbf{y} = \begin{bmatrix} y_1 + y_2 + \cdots + y_n \\ 0 \\ 0 \\ \vdots \\ 0 \end{bmatrix} \tag{5}$$

specifies the configuration of trees planted after each harvest.

At this point we are ready to write the following equation, which characterizes a sustainable harvesting policy:

$$\begin{bmatrix} \text{configuration} \\ \text{at end of} \\ \text{growth period} \end{bmatrix} - [\text{harvest}] + \begin{bmatrix} \text{new seedling} \\ \text{replacement} \end{bmatrix} = \begin{bmatrix} \text{configuration} \\ \text{at beginning of} \\ \text{growth period} \end{bmatrix}$$

or, mathematically,

$$G\mathbf{x} - \mathbf{y} + R\mathbf{y} = \mathbf{x}$$

This equation can be rewritten as

$$(I - R)\mathbf{y} = (G - I)\mathbf{x} \tag{6}$$

or, more fully,

$$\begin{bmatrix} 0 & -1 & -1 & \cdots & -1 & -1 \\ 0 & 1 & 0 & \cdots & 0 & 0 \\ 0 & 0 & 1 & \cdots & 0 & 0 \\ \vdots & \vdots & \vdots & & \vdots & \vdots \\ 0 & 0 & 0 & \cdots & 1 & 0 \\ 0 & 0 & 0 & \cdots & 0 & 1 \end{bmatrix} \begin{bmatrix} y_1 \\ y_2 \\ y_3 \\ \vdots \\ y_{n-1} \\ y_n \end{bmatrix}$$

$$= \begin{bmatrix} -g_1 & 0 & 0 & \cdots & 0 & 0 \\ g_1 & -g_2 & 0 & \cdots & 0 & 0 \\ 0 & g_2 & -g_3 & \cdots & 0 & 0 \\ \vdots & \vdots & \vdots & & \vdots & \vdots \\ 0 & 0 & 0 & \cdots & -g_{n-1} & 0 \\ 0 & 0 & 0 & \cdots & g_{n-1} & 0 \end{bmatrix} \begin{bmatrix} x_1 \\ x_2 \\ x_3 \\ \vdots \\ x_{n-1} \\ x_n \end{bmatrix}$$

We will refer to Equation (6) as the ***sustainable harvesting condition***. Any vectors $\mathbf{x}$ and $\mathbf{y}$ with nonnegative entries, and such that $x_1 + x_2 + \cdots + x_n = s$, which satisfy this matrix equation, determine a sustainable harvesting policy for the forest. Note that if $y_1 > 0$, then the harvester is removing seedlings of no economic value and replacing them with new seedlings. As there is no point in doing this, we assume that

$$y_1 = 0 \tag{7}$$

With this assumption, it can be verified that (6) is the matrix form of the following set of equations:

$$y_2 + y_3 + \cdots + y_n = g_1 x_1$$
$$y_2 = g_1 x_1 - g_2 x_2$$
$$y_3 = g_2 x_2 - g_3 x_3$$
$$\vdots$$
$$y_{n-1} = g_{n-2} x_{n-2} - g_{n-1} x_{n-1}$$
$$y_n = g_{n-1} x_{n-1}$$

(8)

Notice that the first equation in (8) is the sum of the remaining $n - 1$ equations.

Because we must have $y_i \geq 0$ for $i = 2, 3, \ldots, n$, Equations (8) require that

$$g_1 x_1 \geq g_2 x_2 \geq \cdots \geq g_{n-1} x_{n-1} \geq 0 \tag{9}$$

Conversely, if $\mathbf{x}$ is a column vector with nonnegative entries that satisfy Equation (9), then (7) and (8) define a column vector $\mathbf{y}$ with nonnegative entries. Furthermore, $\mathbf{x}$ and $\mathbf{y}$ then satisfy the sustainable harvesting condition (6). In other words, a necessary and sufficient condition that a nonnegative column vector $\mathbf{x}$ determine a forest configuration that is capable of sustainable harvesting is that its entries satisfy (9).

Optimal Sustainable Yield   Because we remove y_i trees from the ith class ($i = 2, 3, \ldots, n$) and each tree in the ith class has an economic value of p_i, the total yield of the harvest, Yld, is given by

$$Yld = p_2 y_2 + p_3 y_3 + \cdots + p_n y_n \tag{10}$$

Using (8) we may substitute for the y_i's in (10) to obtain

$$Yld = p_2 g_1 x_1 + (p_3 - p_2) g_2 x_2 + \cdots + (p_n - p_{n-1}) g_{n-1} x_{n-1} \tag{11}$$

Combining (11), (1), and (9), we can now state the problem of maximizing the yield of the forest over all possible sustainable harvesting policies as follows:

Problem.   Find nonnegative numbers $x_1, x_2, \ldots, x_n$ that maximize

$$Yld = p_2 g_1 x_1 + (p_3 - p_2) g_2 x_2 + \cdots + (p_n - p_{n-1}) g_{n-1} x_{n-1}$$

subject to

$$x_1 + x_2 + \cdots + x_n = s$$

and

$$g_1 x_1 \geq g_2 x_2 \geq \cdots \geq g_{n-1} x_{n-1} \geq 0$$

As formulated above, this problem belongs to the field of linear programming. However, we shall illustrate the following result below, without linear programming theory, by actually exhibiting a sustainable harvesting policy.

Theorem 11.10.1   **Optimal Sustainable Yield**

The optimal sustainable yield is achieved by harvesting all the trees from one particular height class and none of the trees from any other height class.

Let us first set

$$Yld_k = \text{yield obtained by harvesting all of the } k\text{th}$$
$$\text{class and none of the other classes}$$

The largest value of Yld_k for $k = 2, 3, \ldots, n$ will then be the optimal sustainable yield, and the corresponding value of k will be the class that should be completely harvested to attain the optimal sustainable yield. Because no class but the kth is harvested, we have

$$y_2 = y_3 = \cdots = y_{k-1} = y_{k+1} = \cdots = y_n = 0 \tag{12}$$

In addition, because all of the kth class is harvested, no trees are left unharvested in the kth class, and no trees are ever present in the height classes above the kth class. Thus,

$$x_k = x_{k+1} = \cdots = x_n = 0 \tag{13}$$

Substituting (12) and (13) into the sustainable harvesting condition (8) gives

$$
\begin{aligned}
y_k &= g_1 x_1 \\
0 &= g_1 x_1 - g_2 x_2 \\
0 &= g_2 x_2 - g_3 x_3 \\
&\ \vdots \\
0 &= g_{k-2} x_{k-2} - g_{k-1} x_{k-1} \\
y_k &= g_{k-1} x_{k-1}
\end{aligned} \tag{14}
$$

Equations (14) can also be written as

$$y_k = g_1 x_1 = g_2 x_2 = \cdots = g_{k-1} x_{k-1} \tag{15}$$

from which it follows that

$$
\begin{aligned}
x_2 &= g_1 x_1 / g_2 \\
x_3 &= g_1 x_1 / g_3 \\
&\ \vdots \\
x_{k-1} &= g_1 x_1 / g_{k-1}
\end{aligned} \tag{16}
$$

If we substitute Equations (13) and (16) into

$$x_1 + x_2 + \cdots + x_n = s$$

[which is Equation (1)], we can solve for x_1 and obtain

$$x_1 = \frac{s}{1 + \dfrac{g_1}{g_2} + \dfrac{g_1}{g_3} + \cdots + \dfrac{g_1}{g_{k-1}}} \tag{17}$$

For the yield Yld_k, we combine (10), (12), (15), and (17) to obtain

$$
\begin{aligned}
Yld_k &= p_2 y_2 + p_3 y_3 + \cdots + p_n y_n \\
&= p_k y_k \\
&= p_k g_1 x_1 \\
&= \frac{p_k s}{\dfrac{1}{g_1} + \dfrac{1}{g_2} + \cdots + \dfrac{1}{g_{k-1}}}
\end{aligned} \tag{18}
$$

Equation (18) determines Yld_k in terms of the known growth and economic parameters for any $k = 2, 3, \ldots, n$. Thus, the optimal sustainable yield is found as follows.

Theorem 11.10.2   **Finding the Optimal Sustainable Yield**

The optimal sustainable yield is the largest value of

$$\frac{p_k s}{\dfrac{1}{g_1} + \dfrac{1}{g_2} + \cdots + \dfrac{1}{g_{k-1}}}$$

for $k = 2, 3, \ldots, n$. The corresponding value of k is the number of the class that is completely harvested.

In Exercise 4 we ask the reader to show that the nonharvest vector $\mathbf{x}$ for the optimal sustainable yield is

$$\mathbf{x} = \frac{s}{\dfrac{1}{g_1} + \dfrac{1}{g_2} + \cdots + \dfrac{1}{g_{k-1}}} \begin{bmatrix} 1/g_1 \\ 1/g_2 \\ \vdots \\ 1/g_{k-1} \\ 0 \\ 0 \\ \vdots \\ 0 \end{bmatrix} \qquad (19)$$

Theorem 11.10.2 implies that it is not necessarily the highest-priced class of trees that should be totally cropped. The growth parameters g_i must also be taken into account to determine the optimal sustainable yield.

EXAMPLE 1 Using Theorem 11.10.2

For a Scots pine forest in Scotland with a growth period of six years, the following growth matrix was found (see M. B. Usher, "A Matrix Approach to the Management of Renewable Resources, with Special Reference to Selection Forests," *Journal of Applied Ecology*, vol. 3, 1966, pp. 355–367):

$$G = \begin{bmatrix} .72 & 0 & 0 & 0 & 0 & 0 \\ .28 & .69 & 0 & 0 & 0 & 0 \\ 0 & .31 & .75 & 0 & 0 & 0 \\ 0 & 0 & .25 & .77 & 0 & 0 \\ 0 & 0 & 0 & .23 & .63 & 0 \\ 0 & 0 & 0 & 0 & .37 & 1.00 \end{bmatrix}$$

Suppose that the prices of trees in the five tallest height classes are

$$p_2 = \$50, \qquad p_3 = \$100, \qquad p_4 = \$150, \qquad p_5 = \$200, \qquad p_6 = \$250$$

Which class should be completely harvested to obtain the optimal sustainable yield, and what is the yield?

Solution.

From the matrix G we have that

$$g_1 = .28, \qquad g_2 = .31, \qquad g_3 = .25, \qquad g_4 = .23, \qquad g_5 = .37$$

Equation (18) then gives

$$Yld_2 = 50s/(.28^{-1}) = 14.0s$$
$$Yld_3 = 100s/(.28^{-1} + .31^{-1}) = 14.7s$$
$$Yld_4 = 150s/(.28^{-1} + .31^{-1} + .25^{-1}) = 13.9s$$
$$Yld_5 = 200s/(.28^{-1} + .31^{-1} + .25^{-1} + .23^{-1}) = 13.2s$$
$$Yld_6 = 250s/(.28^{-1} + .31^{-1} + .25^{-1} + .23^{-1} + .37^{-1}) = 14.0s$$

We see that Yld_3 is the largest of these five quantities, so from Theorem 11.10.2 the third class should be completely harvested every six years to maximize the sustainable yield. The corresponding optimal sustainable yield is $\$14.7s$, where s is the total number of trees in the forest. ◆

Exercise Set 11.10

1. A certain forest is divided into three height classes and has a growth matrix between harvests given by

$$G = \begin{bmatrix} \frac{1}{2} & 0 & 0 \\ \frac{1}{2} & \frac{1}{3} & 0 \\ 0 & \frac{2}{3} & 1 \end{bmatrix}$$

   If the price of trees in the second class is $30 and the price of trees in the third class is $50, which class should be completely harvested to attain the optimal sustainable yield? What is the optimal yield if there are 1000 trees in the forest?

2. In Example 1, to what level must the price of trees in the fifth class rise so that the fifth class is the one to completely harvest in order to attain the optimal sustainable yield?

3. In Example 1, what must the ratio of the prices $p_2: p_3: p_4: p_5: p_6$ be in order that the yields $Yld_k, k = 2, 3, 4, 5, 6$, all be the same? (In this case, any sustainable harvesting policy will produce the same optimal sustainable yield.)

4. Derive Equation (19) for the nonharvest vector **x** corresponding to the optimal sustainable harvesting policy described in Theorem 11.10.2.

5. For the optimal sustainable harvesting policy described in Theorem 11.10.2, how many trees are removed from the forest during each harvest?

6. If all the growth parameters $g_1, g_2, \ldots, g_{n-1}$ in the growth matrix G are equal, what should the ratio of the prices $p_2: p_3: \ldots : p_n$ be in order that any sustainable harvesting policy be an optimal sustainable harvesting policy? (See Exercise 3.)

Technology Exercises 11.10

The following exercises are designed to be solved using a technology utility. Typically, this will be MATLAB, *Mathematica*, Maple, Derive, or Mathcad, but it may also be some other type of linear algebra software or a scientific calculator with some linear algebra capabilities. For each exercise you will need to read the relevant documentation for the particular utility you are using. The goal of these exercises is to provide you with a basic proficiency with your technology utility. Once you have mastered the techniques in these exercises, you will be able to use your technology utility to solve many of the problems in the regular exercise sets.

T1. A particular forest has growth parameters given by

$$g_i = \frac{1}{i}$$

for $i = 1, 2, 3, \ldots, n - 1$, where n (the total number of height classes) can be chosen as large as needed. Suppose that the value of a tree in the kth height interval is given by

$$p_k = a(k - 1)^\rho$$

where a is a constant (in dollars) and ρ is a parameter satisfying $1 \le \rho \le 2$.

(a) Show that the yield Yld_k is given by

$$Yld_k = \frac{2a(k - 1)^{\rho-1}s}{k}$$

(b) For

$$\rho = 1.0, 1.1, 1.2, 1.3, 1.4, 1.5, 1.6, 1.7, 1.8, 1.9$$

   use a computer to determine the class number that should be completely harvested, and determine the optimal sustainable yield in each case. Make sure that you only allow k to take on integer values in your calculations.

(c) Repeat the calculations in part (b) using

$$\rho = 1.91, 1.92, 1.93, 1.94, 1.95, 1.96, 1.97, 1.98, 1.99$$

(d) Show that if $\rho = 2$, then the optimal sustainable yield can never be larger than $2as$.

(e) Compare the values of k determined in parts (b) and (c) to $1/(2 - \rho)$, and use some calculus to explain why

$$k \simeq \frac{1}{2 - \rho}$$

T2. A particular forest has growth parameters given by

$$g_i = \frac{1}{2^i}$$

for $i = 1, 2, 3, \dots, n - 1$, where n (the total number of height classes) can be chosen as large as needed. Suppose that the value of a tree in the kth height interval is given by

$$p_k = a(k - 1)^\rho$$

where a is a constant (in dollars) and ρ is a parameter satisfying $1 \le \rho$.

(a) Show that the yield Yld_k is given by

$$Yld_k = \frac{a(k - 1)^\rho s}{2^k - 2}$$

(b) For

$$\rho = 1, 2, 3, 4, 5, 6, 7, 8, 9, 10$$

use a computer to determine the class number that should be completely harvested in order to obtain an optimal yield, and determine the optimal sustainable yield in each case. Make sure that you only allow k to take on integer values in your calculations.

(c) Compare the values of k determined in part (b) to $1 + \rho / \ln(2)$ and use some calculus to explain why

$$k \simeq 1 + \frac{\rho}{\ln(2)}$$

11.11 COMPUTER GRAPHICS

In this section we assume that a view of a three-dimensional object is displayed on a video screen and show how matrix algebra can be used to obtain new views of the object by rotation, translation, and scaling.

PREREQUISITES: Matrix Algebra
Analytic Geometry

Visualization of a Three-Dimensional Object Suppose that we want to visualize a three-dimensional object by displaying various views of it on a video screen. The object we have in mind to display is to be determined by a finite number of straight line segments. As an example, consider the truncated right pyramid with hexagonal base illustrated in Figure 11.11.1. We first introduce an xyz-coordinate system in which to embed the object. As in Figure 11.11.1, we orient the coordinate system so that its origin is at the center of the video screen and the xy-plane coincides with the plane of the screen. Consequently, an observer will see only the projection of the view of the three-dimensional object onto the two-dimensional xy-plane.

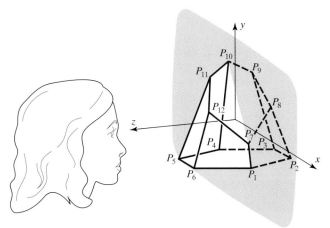

Figure 11.11.1

In the xyz-coordinate system, the endpoints $P_1, P_2, \ldots, P_n$ of the straight line segments that determine the view of the object will have certain coordinates, say,

$$(x_1, y_1, z_1), \quad (x_2, y_2, z_2), \quad \ldots, \quad (x_n, y_n, z_n)$$

These coordinates, together with a specification of which pairs are to be connected by straight line segments, are to be stored in the memory of the video display system. For example, assume that the 12 vertices of the truncated pyramid in Figure 11.11.1 have the following coordinates (the screen is 4 units wide by 3 units high):

$$P_1: (1.000, -.800, .000), \qquad P_2: (.500, -.800, -.866),$$
$$P_3: (-.500, -.800, -.866), \qquad P_4: (-1.000, -.800, .000),$$
$$P_5: (-.500, -.800, .866), \qquad P_6: (.500, -.800, .866),$$
$$P_7: (.840, -.400, .000), \qquad P_8: (.315, .125, -.546),$$
$$P_9: (-.210, .650, -.364), \qquad P_{10}: (-.360, .800, .000),$$
$$P_{11}: (-.210, .650, .364), \qquad P_{12}: (.315, .125, .546)$$

These 12 vertices are connected pairwise by 18 straight line segments as follows, where $P_i \leftrightarrow P_j$ denotes that point P_i is connected to point P_j:

$$P_1 \leftrightarrow P_2, \quad P_2 \leftrightarrow P_3, \quad P_3 \leftrightarrow P_4, \quad P_4 \leftrightarrow P_5, \quad P_5 \leftrightarrow P_6, \quad P_6 \leftrightarrow P_1,$$
$$P_7 \leftrightarrow P_8, \quad P_8 \leftrightarrow P_9, \quad P_9 \leftrightarrow P_{10}, \quad P_{10} \leftrightarrow P_{11}, \quad P_{11} \leftrightarrow P_{12}, \quad P_{12} \leftrightarrow P_7,$$
$$P_1 \leftrightarrow P_7, \quad P_2 \leftrightarrow P_8, \quad P_3 \leftrightarrow P_9, \quad P_4 \leftrightarrow P_{10}, \quad P_5 \leftrightarrow P_{11}, \quad P_6 \leftrightarrow P_{12}$$

In View 1 these 18 straight line segments are shown as they would appear on the video screen. It should be noticed that only the x- and y-coordinates of the vertices are needed by the video display system to draw the view, as only the projection of the object onto the xy-plane is displayed. However, we must keep track of the z-coordinates to carry out certain transformations discussed later.

We now show how to form new views of the object by scaling, translating, or rotating the initial view. We first construct a $3 \times n$ matrix P, referred to as the *coordinate matrix of the view*, whose columns are the coordinates of the n points of a view:

$$P = \begin{bmatrix} x_1 & x_2 & \cdots & x_n \\ y_1 & y_2 & \cdots & y_n \\ z_1 & z_2 & \cdots & z_n \end{bmatrix}$$

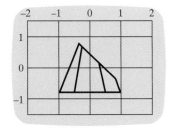

View 1

For example, the coordinate matrix P corresponding to View 1 is the 3×12 matrix

$$\begin{bmatrix} 1.000 & .500 & -.500 & -1.000 & -.500 & .500 & .840 & .315 & -.210 & -.360 & -.210 & .315 \\ -.800 & -.800 & -.800 & -.800 & -.800 & -.800 & -.400 & .125 & .650 & .800 & .650 & .125 \\ .000 & -.866 & -.866 & .000 & .866 & .866 & .000 & -.546 & -.364 & .000 & .364 & .546 \end{bmatrix}$$

We will show below how to transform the coordinate matrix P of a view to a new coordinate matrix P' corresponding to a new view of the object. The straight line segments connecting the various points move with the points as they are transformed. In this way each view is uniquely determined by its coordinate matrix once we have specified which pairs of points in the original view are to be connected by straight lines.

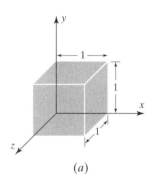

(a)

Scaling   The first type of transformation we consider consists of scaling a view along the x, y, and z directions by factors of α, β, and γ, respectively. By this we mean that if a point P_i has coordinates (x_i, y_i, z_i) in the original view, it is to move to a new point P_i' with coordinates $(\alpha x_i, \beta y_i, \gamma z_i)$ in the new view. This has the effect of transforming a unit cube in the original view to a rectangular parallelepiped of dimensions $\alpha \times \beta \times \gamma$ (Figure 11.11.2). Mathematically, this may be accomplished with matrix multiplication as follows. Define a 3×3 diagonal matrix

$$S = \begin{bmatrix} \alpha & 0 & 0 \\ 0 & \beta & 0 \\ 0 & 0 & \gamma \end{bmatrix}$$

Then, if a point P_i in the original view is represented by the column vector

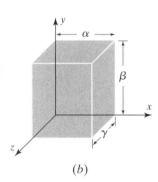

(b)

Figure 11.11.2

$$\begin{bmatrix} x_i \\ y_i \\ z_i \end{bmatrix}$$

the transformed point P_i' is represented by the column vector

$$\begin{bmatrix} x_i' \\ y_i' \\ z_i' \end{bmatrix} = \begin{bmatrix} \alpha & 0 & 0 \\ 0 & \beta & 0 \\ 0 & 0 & \gamma \end{bmatrix} \begin{bmatrix} x_i \\ y_i \\ z_i \end{bmatrix}$$

Using the coordinate matrix P, which contains the coordinates of all n points of the original view as its columns, these n points can be transformed simultaneously to produce the coordinate matrix P' of the scaled view, as follows:

$$SP = \begin{bmatrix} \alpha & 0 & 0 \\ 0 & \beta & 0 \\ 0 & 0 & \gamma \end{bmatrix} \begin{bmatrix} x_1 & x_2 & \cdots & x_n \\ y_1 & y_2 & \cdots & y_n \\ z_1 & z_2 & \cdots & z_n \end{bmatrix}$$

$$= \begin{bmatrix} \alpha x_1 & \alpha x_2 & \cdots & \alpha x_n \\ \beta y_1 & \beta y_2 & \cdots & \beta y_n \\ \gamma z_1 & \gamma z_2 & \cdots & \gamma z_n \end{bmatrix} = P'$$

The new coordinate matrix can then be entered into the video display system to produce the new view of the object. As an example, View 2 is View 1 scaled by setting $\alpha = 1.8$, $\beta = 0.5$, and $\gamma = 3.0$. Notice that the scaling $\gamma = 3.0$ along the z-axis is not visible in View 2, since we see only the projection of the object onto the xy-plane.

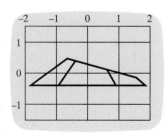

View 2

View 1 scaled by
$\alpha = 1.8$, $\beta = 0.5$, $\gamma = 3.0$.

Translation   We next consider the transformation of translating or displacing an object to a new position on the screen. Referring to Figure 11.11.3, suppose we desire

$P'_i(x_i + x_0, y_i + y_0, z_i + z_0)$

x

z

$P_i(x_i, y_i, z_i)$

Figure 11.11.3

to change an existing view so that each point P_i with coordinates (x_i, y_i, z_i) moves to a new point P'_i with coordinates $(x_i + x_0, y_i + y_0, z_i + z_0)$. The vector

$$\begin{bmatrix} x_0 \\ y_0 \\ z_0 \end{bmatrix}$$

is called the ***translation vector*** of the transformation. By defining a $3 \times n$ matrix T as

$$T = \begin{bmatrix} x_0 & x_0 & \cdots & x_0 \\ y_0 & y_0 & \cdots & y_0 \\ z_0 & z_0 & \cdots & z_0 \end{bmatrix}$$

all n points of the view determined by the coordinate matrix P can be translated by matrix addition by means of the equation

$$P' = P + T$$

The coordinate matrix P' then specifies the new coordinates of the n points. For example, if we wish to translate View 1 according to the translation vector

$$\begin{bmatrix} 1.2 \\ 0.4 \\ 1.7 \end{bmatrix}$$

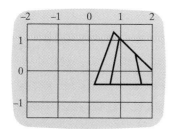

View 3
View 1 translated by
$x_0 = 1.2$, $y_0 = 0.4$, $z_0 = 1.7$.

the result is View 3. Notice, again, that the translation $z_0 = 1.7$ along the z-axis does not show up explicitly in View 3.

In Exercise 7, a technique of performing translations by matrix multiplication rather than by matrix addition is explained.

Rotation   A more complicated type of transformation is a rotation of a view about one of the three coordinate axes. We begin with a rotation about the z-axis (the axis perpendicular to the screen) through an angle θ. Given a point P_i in the original view with coordinates (x_i, y_i, z_i), we wish to compute the new coordinates (x'_i, y'_i, z'_i) of the rotated point P'_i. Referring to Figure 11.11.4 and using a little trigonometry, the reader should be able to derive the following:

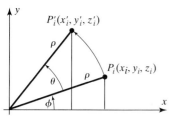

$P'_i(x'_i, y'_i, z'_i)$

ρ

θ ρ $P_i(x_i, y_i, z_i)$

ϕ x

Figure 11.11.4

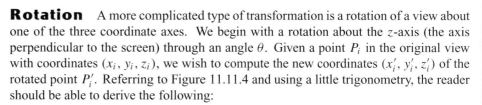

$$x'_i = \rho \cos(\phi + \theta) = \rho \cos \phi \cos \theta - \rho \sin \phi \sin \theta = x_i \cos \theta - y_i \sin \theta$$
$$y'_i = \rho \sin(\phi + \theta) = \rho \cos \phi \sin \theta + \rho \sin \phi \cos \theta = x_i \sin \theta + y_i \cos \theta$$
$$z'_i = z_i$$

These equations can be written in matrix form as

$$\begin{bmatrix} x_i' \\ y_i' \\ z_i' \end{bmatrix} = \begin{bmatrix} \cos\theta & -\sin\theta & 0 \\ \sin\theta & \cos\theta & 0 \\ 0 & 0 & 1 \end{bmatrix} \begin{bmatrix} x_i \\ y_i \\ z_i \end{bmatrix}$$

If we let R denote the 3×3 matrix in this equation, all n points can be rotated by the matrix product

$$P' = RP$$

to yield the coordinate matrix P' of the rotated view.

Rotations about the x and y axes can be accomplished analogously, and the resulting rotation matrices are given with Views 4, 5, and 6. These three new views of the truncated pyramid correspond to rotations of View 1 about the x, y, and z axes, respectively, each through an angle of $90°$.

Rotation about the x-axis

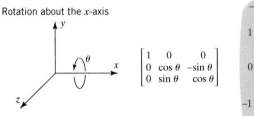

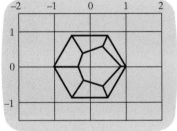

View 4 View 1 rotated $90°$ about the x-axis.

Rotation about the y-axis

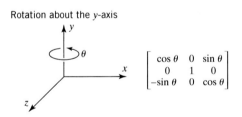

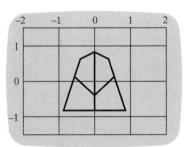

View 5 View 1 rotated $90°$ about the y-axis.

Rotation about the z-axis

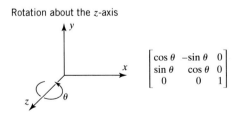

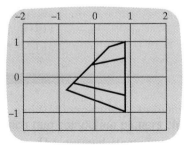

View 6 View 1 rotated $90°$ about the z-axis.

Rotations about three coordinate axes may be combined to give oblique views of an object. For example, View 7 is View 1 rotated first about the x-axis through $30°$, then about the y-axis through $-70°$, and finally about the z-axis through $-27°$. Mathe-

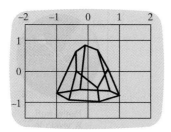

View 7
Oblique view of
truncated pyramid.

matically, these three successive rotations can be embodied in the single transformation equation $P' = RP$, where R is the product of three individual rotation matrices:

$$R_1 = \begin{bmatrix} 1 & 0 & 0 \\ 0 & \cos(30°) & -\sin(30°) \\ 0 & \sin(30°) & \cos(30°) \end{bmatrix}$$

$$R_2 = \begin{bmatrix} \cos(-70°) & 0 & \sin(-70°) \\ 0 & 1 & 0 \\ -\sin(-70°) & 0 & \cos(-70°) \end{bmatrix}$$

$$R_3 = \begin{bmatrix} \cos(-27°) & -\sin(-27°) & 0 \\ \sin(-27°) & \cos(-27°) & 0 \\ 0 & 0 & 1 \end{bmatrix}$$

in the order

$$R = R_3 R_2 R_1 = \begin{bmatrix} .305 & -.025 & -.952 \\ -.155 & .985 & -.076 \\ .940 & .171 & .296 \end{bmatrix}$$

As a final illustration, in View 8 we have two separate views of the truncated pyramid, which constitute a stereoscopic pair. They were produced by first rotating View 7 about the y-axis through an angle of $-3°$ and translating it to the right, then rotating the same View 7 about the y-axis through an angle of $+3°$ and translating it to the left. The translation distances were chosen so that the stereoscopic views are about $2\frac{1}{2}$ inches apart—the approximate distance between a pair of eyes.

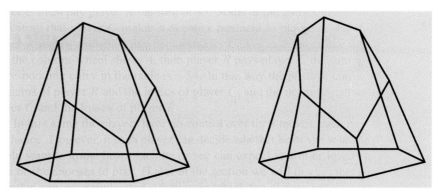

View 8 Stereoscopic figure of truncated pyramid. The three-dimensionality of the diagram can be seen by holding the book about one foot away and focusing on a distant object. Then by shifting gaze to View 8 without refocusing, the two views of the stereoscopic pair can be made to merge together and produce the desired effect.

Exercise Set 11.11

1. View 9 is a view of a square with vertices $(0, 0, 0)$, $(1, 0, 0)$, $(1, 1, 0)$, and $(0, 1, 0)$.

(a) What is the coordinate matrix of View 9?

(b) What is the coordinate matrix of View 9 after it is scaled by a factor $1\frac{1}{2}$ in the x-direction and $\frac{1}{2}$ in the y-direction? Draw a sketch of the scaled view.

(c) What is the coordinate matrix of View 9 after it is translated by the vector

$$\begin{bmatrix} -2 \\ -1 \\ 3 \end{bmatrix}?$$

Draw a sketch of the translated view.

(d) What is the coordinate matrix of View 9 after it is rotated through an angle of $-30°$ about the z-axis? Draw a sketch of the rotated view.

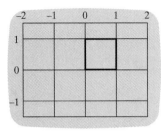

View 9 Square with vertices $(0, 0, 0)$, $(1, 0, 0)$, $(1, 1, 0)$, and $(0, 1, 0)$ (Exercises 1 and 2).

2. (a) If the coordinate matrix of View 9 is multiplied by the matrix

$$\begin{bmatrix} 1 & \frac{1}{2} & 0 \\ 0 & 1 & 0 \\ 0 & 0 & 1 \end{bmatrix}$$

the result is the coordinate matrix of View 10. Such a transformation is called a *shear in the x-direction with factor* $\frac{1}{2}$ *with respect to the y-coordinate*. Show that under such a transformation, a point with coordinates (x_i, y_i, z_i) has new coordinates $(x_i + \frac{1}{2}y_i, y_i, z_i)$.

(b) What are the coordinates of the four vertices of the shear square in View 10?

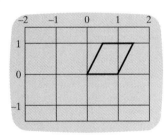

View 10 View 9 sheared along x-axis by $\frac{1}{2}$ with respect to the y-coordinate (Exercise 2).

(c) The matrix

$$\begin{bmatrix} 1 & 0 & 0 \\ .6 & 1 & 0 \\ 0 & 0 & 1 \end{bmatrix}$$

determines a *shear in the y-direction with factor .6 with respect to the x-coordinate* (e.g., View 11). Sketch a view of the square in View 9 after such a shearing transformation, and find the new coordinates of its four vertices.

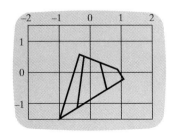

View 11 View 1 sheared along y-axis by .6 with respect to the x-coordinate (Exercise 2).

3. (a) The *reflection about the xz-plane* is defined as the transformation that takes a point (x_i, y_i, z_i) to the point $(x_i, -y_i, z_i)$ (e.g., View 12). If P and P' are the coordinate matrices of a view and its reflection about the xz-plane, respectively, find a matrix M such that $P' = MP$.

   (b) Analogous to part (a), define the *reflection about the yz-plane* and construct the corresponding transformation matrix. Draw a sketch of View 1 reflected about the yz-plane.

   (c) Analogous to part (a), define the *reflection about the xy-plane* and construct the corresponding transformation matrix. Draw a sketch of View 1 reflected about the xy-plane.

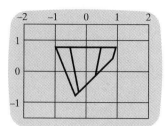

View 12 View 1 reflected about the xz-plane (Exercise 3).

4. (a) View 13 is View 1 subject to the following five transformations:

   1. Scale by a factor of $\frac{1}{2}$ in the x-direction, 2 in the y-direction, and $\frac{1}{3}$ in the z-direction.

   2. Translate $\frac{1}{2}$ unit in the x-direction.

   3. Rotate $20°$ about the x-axis.

   4. Rotate $-45°$ about the y-axis.

   5. Rotate $90°$ about the z-axis.

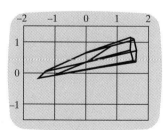

View 13 View 1 scaled, translated, and rotated (Exercise 4).

Construct the five matrices M_1, M_2, M_3, M_4, and M_5 associated with these five transformations.

   (b) If P is the coordinate matrix of View 1 and P' is the coordinate matrix of View 13, express P' in terms of M_1, M_2, M_3, M_4, M_5, and P.

5. (a) View 14 is View 1 subject to the following seven transformations:

   1. Scale by a factor of .3 in the x-direction and .5 in the y-direction.

   2. Rotate $45°$ about the x-axis.

   3. Translate 1 unit in the x-direction.

   4. Rotate $35°$ about the y-axis.

   5. Rotate $-45°$ about the z-axis.

   6. Translate 1 unit in the z-direction.

   7. Scale by a factor of 2 in the x-direction.

Construct the matrices M_1, M_2, ..., M_7 associated with these seven transformations.

(b) If P is the coordinate matrix of View 1 and P' is the coordinate matrix of View 14, express P' in terms of $M_1, M_2, \ldots, M_7$, and P.

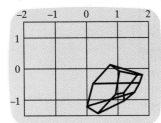

View 14 View 1 scaled, translated, and rotated (Exercise 5).

6. Suppose that a view with coordinate matrix P is to be rotated through an angle θ about an axis through the origin and specified by two angles α and β (see the accompanying figure). If P' is the coordinate matrix of the rotated view, find rotation matrices R_1, R_2, R_3, R_4, and R_5 such that

$$P' = R_5 R_4 R_3 R_2 R_1 P$$

[**Hint.** The desired rotation can be accomplished in the following five steps:

1. Rotate through an angle of β about the y-axis.
2. Rotate through an angle of α about the z-axis.
3. Rotate through an angle of θ about the y-axis.
4. Rotate through an angle of $-\alpha$ about the z-axis.
5. Rotate through an angle of $-\beta$ about the y-axis.]

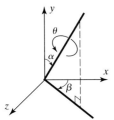

Figure Ex-6

7. This exercise illustrates a technique for translating a point with coordinates (x_i, y_i, z_i) to a point with coordinates $(x_i + x_0, y_i + y_0, z_i + z_0)$ by matrix multiplication rather than matrix addition.

(a) Let the point (x_i, y_i, z_i) be associated with the column vector

$$\mathbf{v}_i = \begin{bmatrix} x_i \\ y_i \\ z_i \\ 1 \end{bmatrix}$$

and let the point $(x_i + x_0, y_i + y_0, z_i + z_0)$ be associated with the column vector

$$\mathbf{v}'_i = \begin{bmatrix} x_i + x_0 \\ y_i + y_0 \\ z_i + z_0 \\ 1 \end{bmatrix}$$

Find a 4×4 matrix M such that $\mathbf{v}'_i = M\mathbf{v}_i$.

(b) Find the specific 4×4 matrix of the above form that will effect the translation of the point $(4, -2, 3)$ to the point $(-1, 7, 0)$.

8. For the three rotation matrices given with Views 4, 5, and 6, show that

$$R^{-1} = R^T$$

(A matrix with this property is called an **orthogonal matrix**. See Section 6.5.)

Technology Exercises 11.11

The following exercises are designed to be solved using a technology utility. Typically, this will be MATLAB, *Mathematica*, Maple, Derive, or Mathcad, but it may also be some other type of linear algebra software or a scientific calculator with some linear algebra capabilities. For each exercise you will need to read the relevant documentation for the particular utility you are using. The goal of these exercises is to provide you with a basic proficiency with your technology utility. Once you have mastered the techniques in these exercises, you will be able to use your technology utility to solve many of the problems in the regular exercise sets.

T1. Let (a, b, c) be a unit vector normal to the plane $ax + by + cz = 0$, and let $\mathbf{r} = (x, y, z)$ be a vector. It can be shown that the mirror image of the vector $\mathbf{r}$ through the above plane has coordinates $\mathbf{r}_m = (x_m, y_m, z_m)$, where

$$\begin{bmatrix} x_m \\ y_m \\ z_m \end{bmatrix} = M \begin{bmatrix} x \\ y \\ z \end{bmatrix}$$

with

$$M = I - 2\mathbf{n}\mathbf{n}^T = \begin{bmatrix} 1 & 0 & 0 \\ 0 & 1 & 0 \\ 0 & 0 & 1 \end{bmatrix} - 2 \begin{bmatrix} a \\ b \\ c \end{bmatrix} [a \quad b \quad c]$$

(a) Show that $M^2 = I$ and give a physical reason why this must be so. [***Hint.*** Use the fact that (a, b, c) is a unit vector to show that $\mathbf{n}^T\mathbf{n} = 1$.]

(b) Use a computer to show that $\det(M) = -1$.

(c) The eigenvectors of M satisfy the equation

$$\begin{bmatrix} x_m \\ y_m \\ z_m \end{bmatrix} = M \begin{bmatrix} x \\ y \\ z \end{bmatrix} = \lambda \begin{bmatrix} x \\ y \\ z \end{bmatrix}$$

and therefore correspond to those vectors whose direction is not affected by a reflection through the plane. Use a computer to determine the eigenvectors and eigenvalues of M and then give a physical argument to support your answer.

T2. A vector $\mathbf{v} = (x, y, z)$ is rotated by an angle θ about an axis having unit vector (a, b, c), thereby forming the rotated vector $\mathbf{v}_R = (x_R, y_R, z_R)$. It can be shown that

$$\begin{bmatrix} x_R \\ y_R \\ z_R \end{bmatrix} = R(\theta) \begin{bmatrix} x \\ y \\ z \end{bmatrix}$$

with

$$R(\theta) = \cos(\theta) \begin{bmatrix} 1 & 0 & 0 \\ 0 & 1 & 0 \\ 0 & 0 & 1 \end{bmatrix} + (1 - \cos(\theta)) \begin{bmatrix} a \\ b \\ c \end{bmatrix} [a \quad b \quad c]$$

$$+ \sin(\theta) \begin{bmatrix} 0 & -c & b \\ c & 0 & -a \\ -b & a & 0 \end{bmatrix}$$

(a) Use a computer to show that $R(\theta)R(\varphi) = R(\theta + \varphi)$, and then give a physical reason why this must be so. Depending on the sophistication of the computer you are using, you may have to experiment using different values of a, b, and

$$c = \sqrt{1 - a^2 - b^2}$$

(b) Show also that $R^{-1}(\theta) = R(-\theta)$ and explain physically why this must be so.

(c) Use a computer to show that $\det(R(\theta)) = +1$.

11.12 EQUILIBRIUM TEMPERATURE DISTRIBUTIONS

In this section we shall see that the equilibrium temperature distribution within a trapezoidal plate can be found when the temperatures around the edges of the plate are specified. The problem is reduced to solving a system of linear equations. Also, an interactive technique for solving the problem and a "random walk" approach to the problem are described.

> PREREQUISITES: Linear Systems
> Matrices
> Intuitive Understanding of Limits

Boundary Data   Suppose that the two faces of the thin trapezoidal plate shown in Figure 11.12.1a are insulated from heat. Suppose that we are also given the temperature along the four edges of the plate. For example, let the temperature be constant on each edge with values of $0°$, $0°$, $1°$, and $2°$, as in the figure. After a period of time the temperature inside the plate will stabilize. Our objective in this section is to determine this equilibrium temperature distribution at the points inside the plate. As we will see, the interior equilibrium temperature is completely determined by the **boundary data**, that is, the temperature along the edges of the plate.

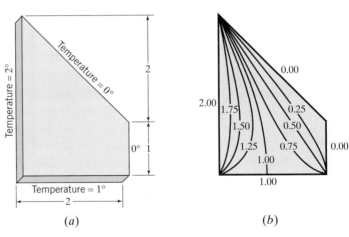

(a)

(b)

Figure 11.12.1

The equilibrium temperature distribution can be visualized by the use of curves that connect points of equal temperature. Such curves are called **isotherms** of the temperature distribution. In Figure 11.12.1b we have sketched a few isotherms using information we derive later in the chapter.

Although all our calculations will be for the trapezoidal plate illustrated, our techniques generalize easily to a plate of any practical shape. They also generalize to the problem of finding the temperature within a three-dimensional body. In fact, our "plate" could be the cross section of some solid object if the flow of heat perpendicular to the cross section is negligible. For example, Figure 11.12.1 could represent the cross section of a long dam. The dam is exposed to three different temperatures: the temperature of the ground at its base, the temperature of the water on one side, and the temperature of the air on the other side. A knowledge of the temperature distribution inside the dam is necessary to determine the thermal stresses to which it is subjected.

Next, we shall consider a certain thermodynamic principle that characterizes the temperature distribution we are seeking.

The Mean-Value Property
There are many different ways to obtain a mathematical model for our problem. The approach we use is based on the following property of equilibrium temperature distributions.

Figure 11.12.2

Theorem 11.12.1	**The Mean-Value Property**

Let a plate be in thermal equilibrium and let P be a point inside the plate. Then if C is any circle with center at P that is completely contained in the plate, the temperature at P is the average value of the temperature on the circle (Figure 11.12.2).

This property is a consequence of certain basic laws of molecular motion, and we will not attempt to derive it. Basically, this property states that in equilibrium, thermal energy tends to distribute itself as evenly as possible consistent with the boundary conditions. It can be shown that the mean-value property uniquely determines the equilibrium temperature distribution of a plate.

Unfortunately, determining the equilibrium temperature distribution from the mean-value property is not an easy matter. However, if we restrict ourselves to finding the temperature only at a finite set of points within the plate, the problem can be reduced to solving a linear system. We pursue this idea next.

Discrete Formulation of the Problem
We can overlay our trapezoidal plate with a succession of finer and finer square nets or meshes (Figure 11.12.3). In (*a*) we have a rather coarse net; in (*b*) we have a net with half the spacing as in (*a*); and

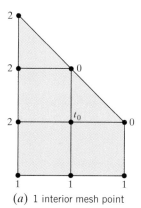

(*a*) 1 interior mesh point

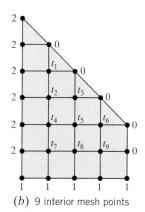

(*b*) 9 interior mesh points

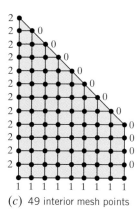

(*c*) 49 interior mesh points

Figure 11.12.3

in (*c*) we have a net with the spacing again reduced by half. The points of intersection of the net lines are called *mesh points*. We classify them as **boundary mesh points** if they fall on the boundary of the plate, or **interior mesh points** if they lie in the interior of the plate. For the three net spacings we have chosen, there are 1, 9, and 49 interior mesh points, respectively.

In the discrete formulation of our problem, we try to find the temperature only at the interior mesh points of some particular net. For a rather fine net, as in (*c*), this will provide an excellent picture of the temperature distribution throughout the entire plate.

At the boundary mesh points the temperature is given by the boundary data. (In Figure 11.12.3 we have labeled all the boundary mesh points with their corresponding temperatures.) At the interior mesh points, we shall apply the following discrete version of the mean-value property.

Theorem 11.12.2   **Discrete Mean-Value Property**

At each interior mesh point, the temperature is approximately the average of the temperatures at the four neighboring mesh points.

This discrete version is a reasonable approximation to the true mean-value property. But because it is only an approximation, it will provide only an approximation to the true temperatures at the interior mesh points. However, the approximations will get better as the mesh spacing decreases. In fact, as the mesh spacing approaches zero, the approximations approach the exact temperature distribution, a fact proved in advanced courses in numerical analysis. We will illustrate this convergence by computing the approximate temperatures at the mesh points for the three mesh spacings given in Figure 11.12.3.

Case (*a*) of Figure 11.12.3 is simple, as there is only one interior mesh point. If we let t_0 be the temperature at this mesh point, the discrete mean-value property immediately gives

$$t_0 = \tfrac{1}{4}(2 + 1 + 0 + 0) = .75$$

In case (*b*) we can label the temperatures at the nine interior mesh points $t_1, t_2, \ldots, t_9$, as in Figure 11.12.3*b*. (The particular ordering is not important.) By applying the discrete mean-value property successively to each of these nine mesh points, we obtain the following nine equations:

$$
\begin{aligned}
t_1 &= \tfrac{1}{4}(t_2 + 2 + 0 + 0) \\
t_2 &= \tfrac{1}{4}(t_1 + t_3 + t_4 + 2) \\
t_3 &= \tfrac{1}{4}(t_2 + t_5 + 0 + 0) \\
t_4 &= \tfrac{1}{4}(t_2 + t_5 + t_7 + 2) \\
t_5 &= \tfrac{1}{4}(t_3 + t_4 + t_6 + t_8) \\
t_6 &= \tfrac{1}{4}(t_5 + t_9 + 0 + 0) \\
t_7 &= \tfrac{1}{4}(t_4 + t_8 + 1 + 2) \\
t_8 &= \tfrac{1}{4}(t_5 + t_7 + t_9 + 1) \\
t_9 &= \tfrac{1}{4}(t_6 + t_8 + 1 + 0)
\end{aligned}
\tag{1}
$$

This is a system of nine linear equations in nine unknowns. We can rewrite it in matrix form as

$$\mathbf{t} = M\mathbf{t} + \mathbf{b} \tag{2}$$

where

$$
\mathbf{t} = \begin{bmatrix} t_1 \\ t_2 \\ t_3 \\ t_4 \\ t_5 \\ t_6 \\ t_7 \\ t_8 \\ t_9 \end{bmatrix}, \quad
M = \begin{bmatrix}
0 & \frac{1}{4} & 0 & 0 & 0 & 0 & 0 & 0 & 0 \\
\frac{1}{4} & 0 & \frac{1}{4} & \frac{1}{4} & 0 & 0 & 0 & 0 & 0 \\
0 & \frac{1}{4} & 0 & 0 & \frac{1}{4} & 0 & 0 & 0 & 0 \\
0 & \frac{1}{4} & 0 & 0 & \frac{1}{4} & 0 & \frac{1}{4} & 0 & 0 \\
0 & 0 & \frac{1}{4} & \frac{1}{4} & 0 & \frac{1}{4} & 0 & \frac{1}{4} & 0 \\
0 & 0 & 0 & 0 & \frac{1}{4} & 0 & 0 & 0 & \frac{1}{4} \\
0 & 0 & 0 & \frac{1}{4} & 0 & 0 & 0 & \frac{1}{4} & 0 \\
0 & 0 & 0 & 0 & \frac{1}{4} & 0 & \frac{1}{4} & 0 & \frac{1}{4} \\
0 & 0 & 0 & 0 & 0 & \frac{1}{4} & 0 & \frac{1}{4} & 0
\end{bmatrix}, \quad
\mathbf{b} = \begin{bmatrix} \frac{1}{2} \\ \frac{1}{2} \\ 0 \\ \frac{1}{2} \\ 0 \\ 0 \\ \frac{3}{4} \\ \frac{1}{4} \\ \frac{1}{4} \end{bmatrix}
$$

To solve Equation (2) we write it as

$$(I - M)\mathbf{t} = \mathbf{b}$$

The solution for **t** is thus

$$\mathbf{t} = (I - M)^{-1}\mathbf{b} \tag{3}$$

as long as the matrix $(I - M)$ is invertible. This is indeed the case, and the solution for **t** as calculated by (3) is

$$
\mathbf{t} = \begin{bmatrix}
0.7846 \\
1.1383 \\
0.4719 \\
1.2967 \\
0.7491 \\
0.3265 \\
1.2995 \\
0.9014 \\
0.5570
\end{bmatrix} \tag{4}
$$

Figure 11.12.4 is a diagram of the plate with the nine interior mesh points labeled with their temperatures as given by this solution.

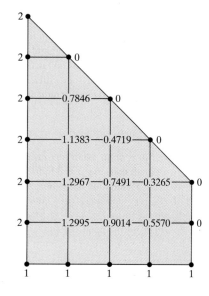

Figure 11.12.4

For case (*c*) of Figure 11.12.3, we repeat this same procedure. We label the temperatures at the 49 interior mesh points as $t_1, t_2, \ldots, t_{49}$ in some manner. For example, we may begin at the top of the plate and proceed from left to right along each row of mesh points. Applying the discrete mean-value property to each mesh point gives a system of 49 linear equations in 49 unknowns:

$$t_1 = \tfrac{1}{4}(t_2 + 2 + 0 + 0)$$
$$t_2 = \tfrac{1}{4}(t_1 + t_3 + t_4 + 2)$$
$$\vdots \tag{5}$$
$$t_{48} = \tfrac{1}{4}(t_{41} + t_{47} + t_{49} + 1)$$
$$t_{49} = \tfrac{1}{4}(t_{42} + t_{48} + 0 + 1)$$

In matrix form, Equations (5) are

$$\mathbf{t} = M\mathbf{t} + \mathbf{b}$$

where $\mathbf{t}$ and $\mathbf{b}$ are column vectors with 49 entries, and M is a 49×49 matrix. As in (3) the solution for $\mathbf{t}$ is

$$\mathbf{t} = (I - M)^{-1}\mathbf{b} \tag{6}$$

In Figure 11.12.5 we display the temperatures at the 49 mesh points found by Equation (6). The nine unshaded temperatures in this figure fall on the mesh points of Figure 11.12.4. In Table 1 we compare the temperatures at these nine common mesh points for the three different mesh spacings used.

TABLE 1

	Temperatures at Common Mesh Points		
	Case (*a*)	**Case (*b*)**	**Case (*c*)**
t_1	—	0.7846	0.8048
t_2	—	1.1383	1.1533
t_3	—	0.4719	0.4778
t_4	—	1.2967	1.3078
t_5	0.7500	0.7491	0.7513
t_6	—	0.3265	0.3157
t_7	—	1.2995	1.3042
t_8	—	0.9014	0.9032
t_9	—	0.5570	0.5554

Knowing that the temperatures of the discrete problem approach the exact temperatures as the mesh spacing decreases, we may surmise that the nine temperatures obtained in case (*c*) are closer to the exact values than those in case (*b*).

A Numerical Technique   To obtain the 49 temperatures in case (*c*) of Figure 11.12.3 it was necessary to solve a linear system with 49 unknowns. A finer net might involve a linear system with hundreds or even thousands of unknowns. Exact algorithms for the solutions of such large systems are impractical, and for this reason we now discuss a numerical technique for the practical solution of these systems.

To describe this technique, we look again at Equation (2):

$$\mathbf{t} = M\mathbf{t} + \mathbf{b} \tag{7}$$

The vector $\mathbf{t}$ we are seeking appears on both sides of this equation. We consider a way

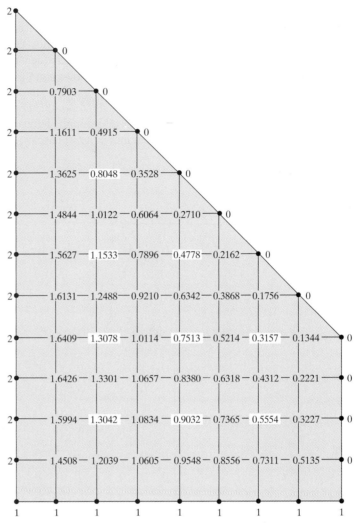

Figure 11.12.5

of generating better and better approximations to the vector solution $\mathbf{t}$. For the initial approximation $\mathbf{t}^{(0)}$ we may take $\mathbf{t}^{(0)} = \mathbf{0}$ if no better choice is available. If we substitute $\mathbf{t}^{(0)}$ into the right side of (7) and label the resulting left side as $\mathbf{t}^{(1)}$, we have

$$\mathbf{t}^{(1)} = M\mathbf{t}^{(0)} + \mathbf{b} \tag{8}$$

Usually $\mathbf{t}^{(1)}$ is a better approximation to the solution than is $\mathbf{t}^{(0)}$. If we substitute $\mathbf{t}^{(1)}$ into the right side of (7), we generate another approximation, which we label $\mathbf{t}^{(2)}$:

$$\mathbf{t}^{(2)} = M\mathbf{t}^{(1)} + \mathbf{b} \tag{9}$$

Continuing in this way, we generate a sequence of approximations as follows:

$$
\begin{aligned}
\mathbf{t}^{(1)} &= M\mathbf{t}^{(0)} + \mathbf{b} \\
\mathbf{t}^{(2)} &= M\mathbf{t}^{(1)} + \mathbf{b} \\
\mathbf{t}^{(3)} &= M\mathbf{t}^{(2)} + \mathbf{b} \\
&\ \ \vdots \\
\mathbf{t}^{(n)} &= M\mathbf{t}^{(n-1)} + \mathbf{b} \\
&\ \ \vdots
\end{aligned}
\tag{10}
$$

One would hope that this sequence of approximations $\mathbf{t}^{(0)}, \mathbf{t}^{(1)}, \mathbf{t}^{(2)}, \ldots$ converges to the exact solution of (7). We do not have the space here to go into the theoretical considerations necessary to show this. However, suffice it to say that for the particular problem we are considering, the sequence converges to the exact solution for any mesh size and for any initial approximation $\mathbf{t}^{(0)}$.

This technique of generating successive approximations to the solution of (7) is a variation of a technique called *Jacobi iteration*; the approximations themselves are called *iterates*. As a numerical example, let us apply Jacobi iteration to the calculation of the nine mesh point temperatures of case (b). Setting $\mathbf{t}^{(0)} = \mathbf{0}$, we have from Equation (2)

$$\mathbf{t}^{(1)} = M\mathbf{t}^{(0)} + \mathbf{b} = M\mathbf{0} + \mathbf{b} = \mathbf{b} = \begin{bmatrix} .5000 \\ .5000 \\ .0000 \\ .5000 \\ .0000 \\ .0000 \\ .7500 \\ .2500 \\ .2500 \end{bmatrix}$$

$$\mathbf{t}^{(2)} = M\mathbf{t}^{(1)} + \mathbf{b}$$

$$= \begin{bmatrix} 0 & \frac{1}{4} & 0 & 0 & 0 & 0 & 0 & 0 & 0 \\ \frac{1}{4} & 0 & \frac{1}{4} & \frac{1}{4} & 0 & 0 & 0 & 0 & 0 \\ 0 & \frac{1}{4} & 0 & 0 & \frac{1}{4} & 0 & 0 & 0 & 0 \\ 0 & \frac{1}{4} & 0 & 0 & \frac{1}{4} & 0 & \frac{1}{4} & 0 & 0 \\ 0 & 0 & \frac{1}{4} & \frac{1}{4} & 0 & \frac{1}{4} & 0 & \frac{1}{4} & 0 \\ 0 & 0 & 0 & 0 & \frac{1}{4} & 0 & 0 & 0 & \frac{1}{4} \\ 0 & 0 & 0 & \frac{1}{4} & 0 & 0 & 0 & \frac{1}{4} & 0 \\ 0 & 0 & 0 & 0 & \frac{1}{4} & 0 & \frac{1}{4} & 0 & \frac{1}{4} \\ 0 & 0 & 0 & 0 & 0 & \frac{1}{4} & 0 & \frac{1}{4} & 0 \end{bmatrix} \begin{bmatrix} .5000 \\ .5000 \\ .0000 \\ .5000 \\ .0000 \\ .0000 \\ .7500 \\ .2500 \\ .2500 \end{bmatrix} + \begin{bmatrix} .5000 \\ .5000 \\ .0000 \\ .5000 \\ .0000 \\ .0000 \\ .7500 \\ .2500 \\ .2500 \end{bmatrix} = \begin{bmatrix} .6250 \\ .7500 \\ .1250 \\ .8125 \\ .1875 \\ .0625 \\ .9375 \\ .5000 \\ .3125 \end{bmatrix}$$

Some additional iterates are

$$\mathbf{t}^{(3)} = \begin{bmatrix} 0.6875 \\ 0.8906 \\ 0.2344 \\ 0.9688 \\ 0.3750 \\ 0.1250 \\ 1.0781 \\ 0.6094 \\ 0.3906 \end{bmatrix}, \quad \mathbf{t}^{(10)} = \begin{bmatrix} 0.7791 \\ 1.1230 \\ 0.4573 \\ 1.2770 \\ 0.7236 \\ 0.3131 \\ 1.2848 \\ 0.8827 \\ 0.5446 \end{bmatrix}, \quad \mathbf{t}^{(20)} = \begin{bmatrix} 0.7845 \\ 1.1380 \\ 0.4716 \\ 1.2963 \\ 0.7486 \\ 0.3263 \\ 1.2992 \\ 0.9010 \\ 0.5567 \end{bmatrix}, \quad \mathbf{t}^{(30)} = \begin{bmatrix} 0.7846 \\ 1.1383 \\ 0.4719 \\ 1.2967 \\ 0.7491 \\ 0.3265 \\ 1.2995 \\ 0.9014 \\ 0.5570 \end{bmatrix}$$

All iterates beginning with the thirtieth are equal to $\mathbf{t}^{(30)}$ to four decimal places. Consequently, $\mathbf{t}^{(30)}$ is the exact solution to four decimal places. This agrees with our previous result given in Equation (4).

The Jacobi iteration scheme applied to the linear system (5) with 49 unknowns produces iterates that begin repeating to four decimal places after 119 iterations. Thus, $\mathbf{t}^{(119)}$ would provide the 49 temperatures of case (c) correct to four decimal places.

A Monte Carlo Technique

In this section we describe a so-called **Monte Carlo technique** for computing the temperature at a single interior mesh point of the discrete problem without having to compute the temperatures at the remaining interior mesh points. First we define a **discrete random walk** along the net. By this we mean a directed path along the net lines (Figure 11.12.6) that joins a succession of mesh points such that the direction of departure from each mesh point is chosen at random. Each of the four possible directions of departure from each mesh point along the path is to be equally probable.

By the use of random walks, we can compute the temperature at a specified interior mesh point on the basis of the following property.

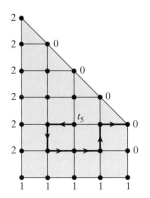

Figure 11.12.6

Theorem 11.12.3   Random Walk Property

Let $W_1, W_2, \ldots, W_n$ be a succession of random walks, all of which begin at a specified interior mesh point. Let $t_1^*, t_2^*, \ldots, t_n^*$ be the temperatures at the boundary mesh points first encountered along each of these random walks. Then the average value $(t_1^* + t_2^* + \cdots + t_n^*)/n$ of these boundary temperatures approaches the temperature at the specified interior mesh point as the number of random walks n increases without bound.

This property is a consequence of the discrete mean-value property that the mesh point temperatures satisfy. The proof of the random walk property involves elementary concepts from probability theory and we will not give it here.

In Table 2 we display the results of a large number of computer-generated random walks for the evaluation of the temperature t_5 of the nine-point mesh of case (b) in Figure 11.12.6. The first column lists the number n of the random walk. The second column lists the temperature t_n^* of the boundary point first encountered along the corresponding random walk. The last column contains the cumulative average of the boundary temperatures encountered along the n random walks. Thus, after 1000 random walks we have the approximation $t_5 \simeq .7550$. This compares with the exact value $t_5 = .7491$ that we had previously evaluated. As can be seen, the convergence to the exact value is not too rapid.

TABLE 2

n	t_n^*	$(t_1^* + \cdots + t_n^*)/n$	n	t_n^*	$(t_1^* + \cdots + t_n^*)/n$
1	1	1.0000	20	1	0.9500
2	2	1.5000	30	0	0.8000
3	1	1.3333	40	0	0.8250
4	0	1.0000	50	2	0.8400
5	2	1.2000	100	0	0.8300
6	0	1.0000	150	1	0.8000
7	2	1.1429	200	0	0.8050
8	0	1.0000	250	1	0.8240
9	2	1.1111	500	1	0.7860
10	0	1.0000	1000	0	0.7550

Exercise Set 11.12

1. A plate in the form of a circular disk has boundary temperatures of $0°$ on the left of its circumference and $1°$ on the right half of its circumference. A net with four interior mesh points is overlaid on the disk (see the accompanying figure).

   (a) Using the discrete mean-value property, write the 4×4 linear system $\mathbf{t} = M\mathbf{t} + \mathbf{b}$ that determines the approximate temperatures at the four interior mesh points.

   (b) Solve the linear system in part (a).

   (c) Use the Jacobi iteration scheme with $\mathbf{t}^{(0)} = \mathbf{0}$ to generate the iterates $\mathbf{t}^{(1)}, \mathbf{t}^{(2)}, \mathbf{t}^{(3)}, \mathbf{t}^{(4)}$, and $\mathbf{t}^{(5)}$ for the linear system in part (a). What is the "error vector" $\mathbf{t}^{(5)} - \mathbf{t}$, where $\mathbf{t}$ is the solution found in part (b)?

   (d) By certain advanced methods, it can be determined that the exact temperatures to four decimal places at the four mesh points are $t_1 = t_3 = .2871$ and $t_2 = t_4 = .7129$. What are the percentage errors in the values found in part (b)?

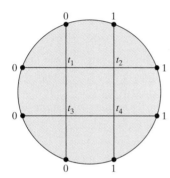

Figure Ex-1

2. Use Theorem 11.12.1 to find the exact equilibrium temperature at the center of the disk in Exercise 1.

3. Calculate the first two iterates $\mathbf{t}^{(1)}$ and $\mathbf{t}^{(2)}$ for case (b) of Figure 11.12.3 with nine interior mesh points [Equation (2)] when the initial iterate is chosen as

   $$\mathbf{t}^{(0)} = [1 \ \ 1 \ \ 1 \ \ 1 \ \ 1 \ \ 1 \ \ 1 \ \ 1 \ \ 1]^T$$

4. The random walk illustrated in Figure Ex-4a can be described by six arrows

   $$\leftarrow \downarrow \rightarrow \rightarrow \uparrow \rightarrow$$

   that specify the directions of departure from the successive mesh points along the path. Figure Ex-4b is an array of 100 computer-generated, randomly oriented arrows arranged in a 10×10 array. Use these arrows to determine random walks to approximate the temperature t_5, as in Table 2. Proceed as follows:

   (1) Take the last two digits of your telephone number. Use the last digit to specify a row and the other to specify a column.

   (2) Go to the arrow in the array with that row and column number.

   (3) Using this arrow as a starting point, move through the array of arrows as you would read a book (left to right and top to bottom). Beginning at the point labeled t_5 in Figure Ex-4a and using this sequence of arrows to specify a sequence of directions, move from mesh point to mesh point until you reach a boundary mesh point. This completes your first random walk. Record the temperature at the boundary mesh point. (If you reach the end of the arrow array, continue with the arrow in the upper left corner.)

   (4) Return to the interior mesh point labeled t_5 and begin where you left off in the arrow array; generate your next random walk. Repeat this process until you have completed 10 random walks and have recorded 10 boundary temperatures.

(5) Calculate the average of the 10 boundary temperatures recorded. (The exact value is $t_5 = .7491$.)

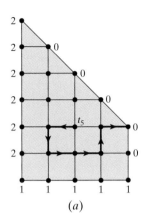

(a)

	0	1	2	3	4	5	6	7	8	9
0	↓	←	←	↓	↓	←	←	↓	↓	↑
1	←	←	→	→	←	↑	↑	→	→	↓
2	←	↑	←	←	←	→	←	↓	←	↓
3	↑	→	→	→	→	↑	↑	←	↓	↑
4	↑	↑	←	←	↓	↓	↓	→	←	↓
5	↓	←	→	↓	→	↓	↑	←	↑	↓
6	↑	↑	↑	↓	↓	↓	↑	←	↑	←
7	←	↓	↓	↑	→	→	↑	↓	↓	←
8	←	↑	→	↓	←	↓	↓	←	↓	←
9	↑	↑	→	→	↑	↓	↓	→	↓	→

(b)

Figure Ex-4

Technology Exercises 11.12

The following exercises are designed to be solved using a technology utility. Typically, this will be MATLAB, *Mathematica*, Maple, Derive, or Mathcad, but it may also be some other type of linear algebra software or a scientific calculator with some linear algebra capabilities. For each exercise you will need to read the relevant documentation for the particular utility you are using. The goal of these exercises is to provide you with a basic proficiency with your technology utility. Once you have mastered the techniques in these exercises, you will be able to use your technology utility to solve many of the problems in the regular exercise sets.

T1. Suppose that we have the square region described by

$$\mathcal{R} = \{(x, y) \mid 0 \le x \le 1, 0 \le y \le 1\}$$

and suppose that the equilibrium temperature distribution $u(x, y)$ along the boundary is given by $u(x, 0) = T_B$, $u(x, 1) = T_T$, $u(0, y) = T_L$, and $u(1, y) = T_R$. Suppose next that this region is partitioned into an $(n + 1) \times (n + 1)$ mesh using

$$x_i = \frac{i}{n} \quad \text{and} \quad y_j = \frac{j}{n}$$

for $i = 0, 1, 2, \ldots, n$ and $j = 0, 1, 2, \ldots, n$. If the temperatures of the interior mesh points are labeled by

$$u_{i,j} = u(x_i, y_i) = u(i/n, j/n)$$

then show that

$$u_{i,j} = \frac{1}{4}(u_{i-1,j} + u_{i+1,j} + u_{i,j-1} + u_{i,j+1})$$

for $i = 1, 2, 3, \ldots, n - 1$ and $j = 1, 2, 3, \ldots, n - 1$. To handle the boundary points, define

$$u_{0,j} = T_L, \quad u_{n,j} = T_R, \quad u_{i,0} = T_B, \quad \text{and} \quad u_{i,n} = T_T$$

for $i = 1, 2, 3, \ldots, n - 1$ and $j = 1, 2, 3, \ldots, n - 1$. Next let

$$F_{n+1} = \begin{bmatrix} 0 & I_n \\ 1 & 0 \end{bmatrix}$$

be the $(n + 1) \times (n + 1)$ matrix with the $n \times n$ identity matrix in the upper right-hand corner, a one in the lower left-hand corner, and zeros everywhere else. For example,

$$F_2 = \begin{bmatrix} 0 & 1 \\ 1 & 0 \end{bmatrix}, \qquad F_3 = \begin{bmatrix} 0 & 1 & 0 \\ 0 & 0 & 1 \\ 1 & 0 & 0 \end{bmatrix},$$

$$F_4 = \begin{bmatrix} 0 & 1 & 0 & 0 \\ 0 & 0 & 1 & 0 \\ 0 & 0 & 0 & 1 \\ 1 & 0 & 0 & 0 \end{bmatrix}, \qquad F_5 = \begin{bmatrix} 0 & 1 & 0 & 0 & 0 \\ 0 & 0 & 1 & 0 & 0 \\ 0 & 0 & 0 & 1 & 0 \\ 0 & 0 & 0 & 0 & 1 \\ 1 & 0 & 0 & 0 & 0 \end{bmatrix},$$

and so on. By defining the $(n+1) \times (n+1)$ matrix

$$M_{n+1} = F_{n+1} + F_{n+1}^T = \begin{bmatrix} 0 & I_n \\ 1 & 0 \end{bmatrix} + \begin{bmatrix} 0 & I_n \\ 1 & 0 \end{bmatrix}^T$$

show that if U_{n+1} is the $(n+1) \times (n+1)$ matrix with entries u_{ij}, then the set of equations

$$u_{i,j} = \frac{1}{4}(u_{i-1,j} + u_{i+1,j} + u_{i,j-1} + u_{i,j+1})$$

for $i = 1, 2, 3, \ldots, n-1$ and $j = 1, 2, 3, \ldots, n-1$ can be written as the matrix equation

$$U_{n+1} = \frac{1}{4}(M_{n+1}U_{n+1} + U_{n+1}M_{n+1})$$

where we consider only those elements of U_{n+1} with $i = 1, 2, 3, \ldots, n-1$ and $j = 1, 2, 3, \ldots, n-1$.

T2. The results of the preceding exercise and the discussion in the text suggest the following algorithm for solving for the equilibrium temperature in the square region

$$\mathcal{R} = \{(x, y) \mid 0 \le x \le 1, 0 \le y \le 1\}$$

given the boundary conditions

$$u(x, 0) = T_B, \qquad u(x, 1) = T_T, \qquad u(0, y) = T_L, \qquad u(1, y) = T_R$$

(1) Choose a value for n, and then choose an initial guess, say

$$\mathbf{U}_{n+1}^{(0)} = \begin{bmatrix} 0 & T_L & \cdots & T_L & 0 \\ T_B & 0 & \cdots & 0 & T_T \\ \vdots & \vdots & & \vdots & \vdots \\ T_B & 0 & \cdots & 0 & T_T \\ 0 & T_R & \cdots & T_R & 0 \end{bmatrix}$$

(2) For each value of $k = 0, 1, 2, 3, \ldots$, compute $U_{n+1}^{(k+1)}$ using

$$U_{n+1}^{(k+1)} = \frac{1}{4}(M_{n+1}U_{n+1}^{(k)} + U_{n+1}^{(k)}M_{n+1})$$

where M_{n+1} is as defined in Exercise T1. Then adjust $U_{n+1}^{(k+1)}$ by replacing all edge entries by the initial edge entries in $U_{n+1}^{(0)}$. [**Note.** The edge entries of a matrix are the entries in the first and last columns and first and last rows.]

(3) Continue this process until $U_{n+1}^{(k+1)} - U_{n+1}^{(k)}$ is approximately the zero matrix. This suggests that

$$U_{n+1} = \lim_{k \to \infty} U_{n+1}^{(k)}$$

Use a computer and this algorithm to solve for $u(x, y)$ given that

$$u(x, 0) = 0, \quad u(x, 1) = 0, \quad u(0, y) = 0, \quad u(1, y) = 2$$

Choose $n = 6$ and compute up to $U_{n+1}^{(30)}$. The exact solution can be expressed as

$$u(x, y) = \frac{8}{\pi} \sum_{m=1}^{\infty} \frac{\sinh((2m - 1)\pi x) \sin((2m - 1)\pi y)}{(2m - 1) \sinh((2m - 1)\pi)}$$

Use a computer to compute $u(i/6, j/6)$ for $i, j = 0, 1, 2, 3, 4, 5, 6$, and then compare your results to the values of $u(i/6, j/6)$ in $U_{n+1}^{(30)}$.

11.13   COMPUTED TOMOGRAPHY

In this section we shall see how constructing a cross-sectional view of a human body by analyzing X-ray scans leads to an inconsistent linear system. We present an iteration technique that provides an "approximate solution" of the linear system.

> PREREQUISITES: Linear Systems
> Natural Logarithms
> Euclidean Space R^n

The basic problem of computed tomography is to construct an image of a cross section of the human body using data collected from many individual beams of X rays that are passed through the cross section. These data are processed by a computer, and the computed cross section is displayed on a video monitor. Figure 11.13.1 is a diagram of General Electric's CT system showing a patient prepared to have a cross section of his head scanned by X-ray beams.

Such a system is also known as a **CAT scanner**, for *C*omputer-*A*ided *T*omography scanner. Figure 11.13.2 displays a typical cross section of a human head produced by the system.

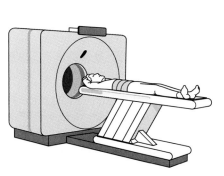

Figure 11.13.1

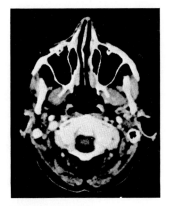

Figure 11.13.2

The first commercial system of computed tomography for medical use was developed in 1971 by G. N. Hounsfield of EMI, Ltd., in England. In 1979, Houndsfield and A. M. Cormack were awarded the Nobel prize for their pioneering work in the field. As we will see in this section, the construction of a cross section or tomograph requires the solution of a large linear system of equations. Certain algorithms, called algebraic reconstruction techniques (ARTs), can be used to solve these linear systems, whose solutions yield the cross sections in digital form.

Scanning Modes   Unlike conventional X-ray pictures that are formed by X rays that are projected *perpendicular* to the plane of the picture, tomographs are constructed from thousands of individual, hairline-thin X-ray beams that *lie in the plane* of the cross section. After they pass through the cross section, the intensities of the X-ray beams are measured by an X-ray detector, and these measurements are relayed to a computer when they are processed. Figures 11.13.3 and 11.13.4 illustrate two possible modes of scanning the cross section: the *parallel mode* and the *fan-beam mode*. In the parallel mode a single X-ray source and X-ray detector pair are translated across the field of view containing the cross section, and many measurements of the parallel beams are recorded. Then the source and detector pair are rotated through a small angle, and another set of measurements is taken. This is repeated until the desired number of beam measurements is completed. For example, in the original 1971 machine, 160 parallel measurements were taken through 180 angles spaced $1°$ apart: a total of $160 \times 180 = 28,800$ beam measurements. Each such scan took approximately $5\frac{1}{2}$ minutes.

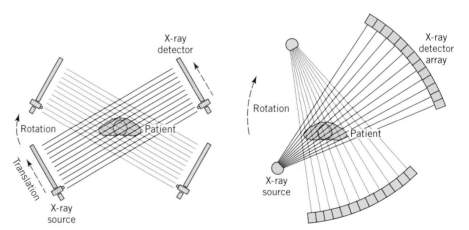

Figure 11.13.3 Parallel mode.   **Figure 11.13.4** Fan-beam mode.

In the fan-beam mode of scanning, a single X-ray tube generates a fan of collimated beams whose intensities are measured simultaneously by an array of detectors on the other side of the field of view. The X-ray tube and detector array are rotated through many angles, and a set of measurements is taken at each angle until the scan is completed. In the General Electric CT system, which uses the fan-beam mode, each scan takes 1 second.

Derivation of Equations   To see how the cross section is reconstructed from the many individual beam measurements, refer to Figure 11.13.5. Here the field of view in which the cross section is situated has been divided into many square *pixels*

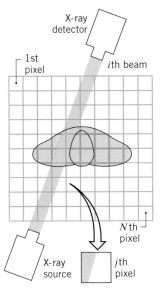

Figure 11.13.5

(picture elements) numbered 1 through N as indicated. It is our desire to determine the X-ray density of each pixel. In the EMI system, 6400 pixels were used, arranged in a square 80×80 array. The G.E. CT system uses 262,144 pixels in a 512×512 array, each pixel being about 1 mm on a side. After the densities of the pixels are determined by the method we will describe, they are reproduced on a video monitor, with each pixel shaded a level of gray proportional to its X-ray density. Because different tissues within the human body have different X-ray densities, the video display clearly distinguishes the various tissues and organs within the cross section.

Figure 11.13.6 shows a single pixel with an X-ray beam of roughly the same width as the pixel passing squarely through it. The photons constituting the X-ray beam are absorbed by the tissue within the pixel at a rate proportional to the X-ray density of the tissue. Quantitatively, the X-ray density of the jth pixel is denoted by x_j and is defined by

$$x_j = \ln\left(\frac{\text{number of photons entering the } j\text{th pixel}}{\text{number of photons leaving the } j\text{th pixel}}\right)$$

where "ln" denotes the natural logarithmic function. Using the logarithm property $\ln(a/b) = -\ln(b/a)$, we also have

$$x_j = -\ln\left(\begin{array}{l}\text{fraction of photons that pass through}\\\text{the } j\text{th pixel without being absorbed}\end{array}\right)$$

If the X-ray beam passes through an entire row of pixels (Figure 11.13.7), then the number of photons leaving one pixel is equal to the number of photons entering the next pixel in the row. If the pixels are numbered $1, 2, \ldots, n$, then the additive property of the logarithmic function gives

$$x_1 + x_2 + \cdots + x_n = \ln\left(\frac{\text{number of photons entering the first pixel}}{\text{number of photons leaving the } n\text{th pixel}}\right)$$

$$= -\ln\left(\begin{array}{l}\text{fraction of photons that pass}\\\text{through the row of } n \text{ pixels}\\\text{without being absorbed}\end{array}\right) \quad (1)$$

Thus, to determine the total X-ray density of a row of pixels, we simply sum the individual pixel densities.

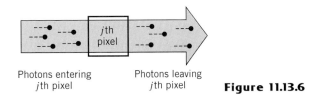

Photons entering jth pixel

Photons leaving jth pixel

Figure 11.13.6

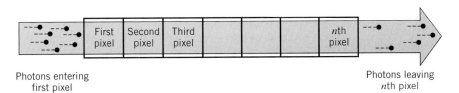

Photons entering first pixel

Photons leaving nth pixel

Figure 11.13.7

Next, consider the X-ray beam in Figure 11.13.5. By the **beam density** of the ith beam of a scan, denoted by b_i, we mean

$$b_i = \ln \left(\frac{\begin{array}{c} \textit{number of photons of the ith beam entering the detector} \\ \textit{without the cross section in the field of view} \end{array}}{\begin{array}{c} \textit{number of photons of the ith beam entering the detector} \\ \textit{with the cross section in the field of view} \end{array}} \right)$$

$$= -\ln \left(\begin{array}{c} \textit{fraction of photons of the ith beam that} \\ \textit{pass through the cross section without} \\ \textit{being absorbed} \end{array} \right) \tag{2}$$

The numerator in the first expression for b_i is obtained by performing a calibration scan without the cross section in the field of view. The resulting detector measurements are stored within the computer's memory. Then a clinical scan is performed with the cross section in the field of view, the b_i's of all the beams constituting the scan are computed, and the values are stored for further processing.

For each beam that passes squarely through a row of pixels, we must have

$$\left(\begin{array}{c} \textit{fraction of photons of the} \\ \textit{beam that pass through the} \\ \textit{row of pixels without being} \\ \textit{absorbed} \end{array} \right) = \left(\begin{array}{c} \textit{fraction of photons of the} \\ \textit{beam that pass through the} \\ \textit{cross section without being} \\ \textit{absorbed} \end{array} \right)$$

Thus, if the ith beam passes squarely through a row of n pixels, then it follows from Equations (1) and (2) that

$$x_1 + x_2 + \cdots + x_n = b_i$$

In this equation, b_i is known from the clinical and calibration measurements, and $x_1, x_2, \ldots, x_n$ are unknown pixel densities that must be determined.

More generally, if the ith beam passes squarely through a row (or column) of pixels with numbers $j_1, j_2, \ldots, j_i$, then we have

$$x_{j_1} + x_{j_2} + \cdots + x_{j_i} = b_i$$

If we set

$$a_{ij} = \begin{cases} 1, & \text{if } j = j_1, j_2, \ldots, j_i \\ 0, & \text{otherwise} \end{cases}$$

then we may write this equation as

$$a_{i1}x_1 + a_{i2}x_2 + \cdots + a_{iN}x_N = b_i \tag{3}$$

We shall refer to Equation (3) as the ith *beam equation*.

Referring to Figure 11.13.5, however, we see that the beams of a scan do not necessarily pass through a row or column of pixels squarely. Instead, a typical beam passes diagonally through each pixel in its path. There are many ways to take this into account. In Figure 11.13.8 we outline three methods of defining the quantities a_{ij} that appear in Equation (3), each of which reduces to our previous definition when the beam passes squarely through a row or column of pixels. Reading down the figure, each method is more exact than its predecessor, but with successively more computational difficulty.

Using any one of the three methods to define the a_{ij}'s in the ith beam equation, we can write the set of M beam equations in a complete scan as

$$\begin{array}{ccccccc} a_{11}x_1 & + & a_{12}x_2 & + \cdots + & a_{1N}x_N & = & b_1 \\ a_{21}x_1 & + & a_{22}x_2 & + \cdots + & a_{2N}x_N & = & b_2 \\ \vdots & & \vdots & & \vdots & & \vdots \\ a_{M1}x_1 & + & a_{M2}x_2 & + \cdots + & a_{MN}x_N & = & b_M \end{array} \tag{4}$$

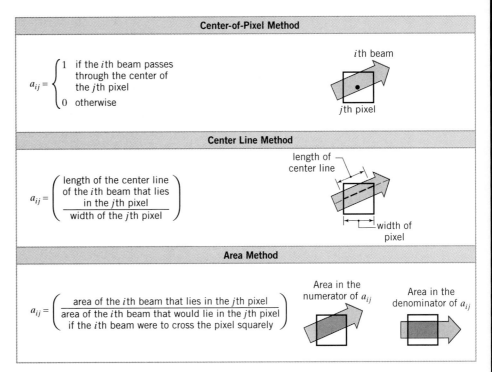

Figure 11.13.8

In this way we have a linear system of M equations (the M beam equations) in N unknowns (the N pixel densities).

Depending on the number of beams and pixels used, we may have $M > N$, $M = N$, or $M < N$. We will consider only the case $M > N$, the so-called *overdetermined case*, in which there are more beams in the scan than pixels in the field of view. Because of inherent modeling and experimental errors in the problem, we should not expect our linear system to have an exact mathematical solution for the pixel densities. In the next section we attempt to find an "approximate" solution to this linear system.

Algebraic Reconstruction Techniques   There have been many mathematical algorithms devised to treat the overdetermined linear system (4). The one we will describe belongs to the class of so-called **Algebraic Reconstruction Techniques** (ARTs). This method, which can be traced to an iterative technique originally introduced by S. Kaczmarz in 1937, was the one used in the first commercial machine. To introduce this technique, consider the following system of three equations in two unknowns:

$$
\begin{aligned}
L_1:& \quad x_1 + x_2 = 2 \\
L_2:& \quad x_1 - 2x_2 = -2 \\
L_3:& \quad 3x_1 - x_2 = 3
\end{aligned}
\tag{5}
$$

The lines L_1, L_2, L_3 determined by these three equations are plotted in the x_1x_2-plane. As shown in Figure 11.13.9a, the three lines do not have a common intersection, and so the three equations do not have an exact solution. However, the points (x_1, x_2) on the shaded triangle formed by the three lines are all situated "near" these three lines and can be thought of as constituting "approximate" solutions to our system. The following iterative procedure describes a geometric construction for generating points on the boundary of that triangular region (Figure 11.13.9b):

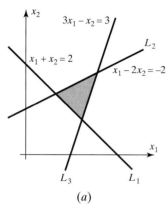
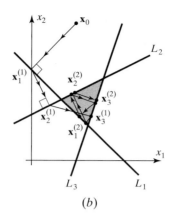
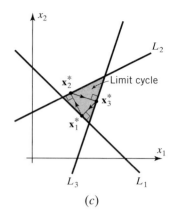

(a) (b) (c)

Figure 11.13.9

Algorithm 1

Step 0. Choose an arbitrary starting point $\mathbf{x}_0$ in the x_1x_2-plane.

Step 1. Project $\mathbf{x}_0$ orthogonally onto the first line L_1 and call the projection $\mathbf{x}_1^{(1)}$. The superscript (1) indicates that this is the first of several cycles through the steps.

Step 2. Project $\mathbf{x}_1^{(1)}$ orthogonally onto the second line L_2 and call the projection $\mathbf{x}_2^{(1)}$.

Step 3. Project $\mathbf{x}_2^{(1)}$ orthogonally onto the third line L_3 and call the projection $\mathbf{x}_3^{(1)}$.

Step 4. Take $\mathbf{x}_3^{(1)}$ as the new value of $\mathbf{x}_0$ and cycle through Steps 1 through 3 again. In the second cycle, label the projected points $\mathbf{x}_1^{(2)}$, $\mathbf{x}_2^{(2)}$, $\mathbf{x}_3^{(2)}$; in the third cycle label the projected points $\mathbf{x}_1^{(3)}$, $\mathbf{x}_2^{(3)}$, $\mathbf{x}_3^{(3)}$; and so forth.

This algorithm generates three sequences of points

$$L_1: \quad \mathbf{x}_1^{(1)}, \mathbf{x}_1^{(2)}, \mathbf{x}_1^{(3)}, \ldots$$

$$L_2: \quad \mathbf{x}_2^{(1)}, \mathbf{x}_2^{(2)}, \mathbf{x}_2^{(3)}, \ldots$$

$$L_3: \quad \mathbf{x}_3^{(1)}, \mathbf{x}_3^{(2)}, \mathbf{x}_3^{(3)}, \ldots$$

that lie on the three lines L_1, L_2, and L_3, respectively. It can be shown that as long as the three lines are not all parallel, then the first sequence converges to a point $\mathbf{x}_1^*$ on L_1, the second sequence converges to a point $\mathbf{x}_2^*$ on L_2, and the third sequence converges to a point $\mathbf{x}_3^*$ on L_3 (Figure 11.13.9c). These three limit points form what is called the **limit cycle** of the iterative process. It can be shown that the limit cycle is independent of the starting point $\mathbf{x}_0$.

We next discuss the specific formulas needed to effect the orthogonal projections in Algorithm 1. First, because the equation of a line in x_1x_2-space is

$$a_1 x_1 + a_2 x_2 = b$$

we can express it in vector form as

$$\mathbf{a}^T \mathbf{x} = b$$

where

$$\mathbf{a} = \begin{bmatrix} a_1 \\ a_2 \end{bmatrix} \quad \text{and} \quad \mathbf{x} = \begin{bmatrix} x_1 \\ x_2 \end{bmatrix}$$

The following theorem gives the necessary projection formula (Exercise 5).

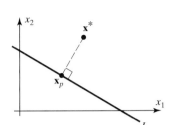

Figure 11.13.10

Theorem 11.13.1   **Orthogonal Projection Formula**

Let L be a line in R^2 with equation $\mathbf{a}^T\mathbf{x} = b$, and let $\mathbf{x}^*$ be any point in R^2 (Figure 11.13.10). Then the orthogonal projection, $\mathbf{x}_p$, of $\mathbf{x}^*$ onto L is given by

$$\mathbf{x}_p = \mathbf{x}^* + \frac{(b - \mathbf{a}^T\mathbf{x}^*)}{\mathbf{a}^T\mathbf{a}}\mathbf{a}$$

EXAMPLE 1 Using Algorithm 1

We can use Algorithm 1 to find an approximate solution of the linear system given in (5) and illustrated in Figure 11.13.9. If we write the equations of the three lines as

$$L_1: \quad \mathbf{a}_1^T\mathbf{x} = b_1$$
$$L_2: \quad \mathbf{a}_2^T\mathbf{x} = b_2$$
$$L_3: \quad \mathbf{a}_3^T\mathbf{x} = b_3$$

where

$$\mathbf{x} = \begin{bmatrix} x_1 \\ x_2 \end{bmatrix}, \quad \mathbf{a}_1 = \begin{bmatrix} 1 \\ 1 \end{bmatrix}, \quad \mathbf{a}_2 = \begin{bmatrix} 1 \\ -2 \end{bmatrix}, \quad \mathbf{a}_3 = \begin{bmatrix} 3 \\ -1 \end{bmatrix},$$
$$b_1 = 2, \quad b_2 = -2, \quad b_3 = 3$$

then, using Theorem 11.13.1, the iteration scheme in Algorithm 1 can be expressed as

$$\mathbf{x}_k^{(p)} = \mathbf{x}_{k-1}^{(p)} + \frac{(b_k - \mathbf{a}_k^T\mathbf{x}_{k-1}^{(p)})}{\mathbf{a}_k^T\mathbf{a}_k}\mathbf{a}_k, \quad k = 1, 2, 3$$

where $p = 1$ for the first cycle of iterates, $p = 2$ for the second cycle of iterates, and so forth. After each cycle of iterates [i.e., after $\mathbf{x}_3^{(p)}$ is computed], the next cycle of iterates is begun with $\mathbf{x}_0$ set equal to $\mathbf{x}_3^{(p)}$.

Table 1 gives the numerical results of six cycles of iterations starting with the initial point $\mathbf{x}_0 = (1, 3)$.

Using certain techniques that are impractical for large linear systems, the exact values of the points of the limit cycle in this example can be shown to be

$$\mathbf{x}_1^* = \left(\tfrac{12}{11}, \tfrac{10}{11}\right) = (1.09090\ldots, .90909\ldots)$$
$$\mathbf{x}_2^* = \left(\tfrac{46}{55}, \tfrac{78}{55}\right) = (.83636\ldots, 1.41818\ldots)$$
$$\mathbf{x}_3^* = \left(\tfrac{31}{22}, \tfrac{27}{22}\right) = (1.40909\ldots, 1.22727\ldots)$$

It can be seen that the sixth cycle of iterates provides an excellent approximation to the limit cycle. Any one of the three iterates $\mathbf{x}_1^{(6)}$, $\mathbf{x}_2^{(6)}$, $\mathbf{x}_3^{(6)}$ can be used as an approximate solution of the linear system. (The large discrepancies in the values of $\mathbf{x}_1^{(6)}$, $\mathbf{x}_2^{(6)}$, and $\mathbf{x}_3^{(6)}$ are due to the artificial nature of this illustrative example. In practical problems these discrepancies would be much smaller.) ◆

TABLE 1

	x_1	x_2
$\mathbf{x}_0$	1.00000	3.00000
$\mathbf{x}_1^{(1)}$	.00000	2.00000
$\mathbf{x}_2^{(1)}$	.40000	1.20000
$\mathbf{x}_3^{(1)}$	1.30000	.90000
$\mathbf{x}_1^{(2)}$	1.20000	.80000
$\mathbf{x}_2^{(2)}$	.88000	1.44000
$\mathbf{x}_3^{(2)}$	1.42000	1.26000
$\mathbf{x}_1^{(3)}$	1.08000	.92000
$\mathbf{x}_2^{(3)}$	.83200	1.41600
$\mathbf{x}_3^{(3)}$	1.40800	1.22400
$\mathbf{x}_1^{(4)}$	1.09200	.90800
$\mathbf{x}_2^{(4)}$	.83680	1.41840
$\mathbf{x}_3^{(4)}$	1.40920	1.22760
$\mathbf{x}_1^{(5)}$	1.09080	.90920
$\mathbf{x}_2^{(5)}$	.83632	1.41816
$\mathbf{x}_3^{(5)}$	1.40908	1.22724
$\mathbf{x}_1^{(6)}$	1.09092	.90908
$\mathbf{x}_2^{(6)}$	.83637	1.41818
$\mathbf{x}_3^{(6)}$	1.40909	1.22728

To generalize Algorithm 1 so that it applies to an overdetermined system of M equations in N unknowns,

$$\begin{aligned}
a_{11}x_1 + a_{12}x_2 + \cdots + a_{1N}x_N &= b_1 \\
a_{21}x_1 + a_{22}x_2 + \cdots + a_{2N}x_N &= b_2 \\
\vdots \qquad \vdots \qquad\qquad \vdots \qquad \vdots \\
a_{M1}x_1 + a_{M2}x_2 + \cdots + a_{MN}x_N &= b_M
\end{aligned} \qquad (6)$$

we introduce column vectors $\mathbf{x}$ and $\mathbf{a}_i$ as follows:

$$\mathbf{x} = \begin{bmatrix} x_1 \\ x_2 \\ \vdots \\ x_N \end{bmatrix}, \qquad \mathbf{a}_i = \begin{bmatrix} a_{i1} \\ a_{i2} \\ \vdots \\ a_{iN} \end{bmatrix}, \qquad i = 1, 2, \ldots, M$$

With these vectors, the M equations constituting our linear system (6) can be written in vector form as

$$\mathbf{a}_i^T \mathbf{x} = b_i, \qquad i = 1, 2, \ldots, M$$

Each of these M equations defines what is called a ***hyperplane*** in the N-dimensional Euclidean space R^N. In general these M hyperplanes have no common intersection, and so we seek instead some point in R^N that is reasonably "close" to all of them. Such a point will constitute an approximate solution of the linear system, and its N entries will determine approximate pixel densities with which to form the desired cross section.

As in the two-dimensional case, we will introduce an iterative process that generates cycles of successive orthogonal projections onto the M hyperplanes beginning with some arbitrary initial point in R^N. Our notation for these successive iterates is

$$\mathbf{x}_k^{(p)} = \begin{pmatrix} \textit{the iterate lying on the kth hyperplane} \\ \textit{generated during the pth cycle of iterations} \end{pmatrix}$$

The algorithm is as follows:

Algorithm 2

Step 0. Choose any point in R^N and label it $\mathbf{x}_0$.

Step 1. For the first cycle of iterates, set $p = 1$.

Step 2. For $k = 1, 2, \ldots, M$, compute

$$\mathbf{x}_k^{(p)} = \mathbf{x}_{k-1}^{(p)} + \frac{(b_k - \mathbf{a}_k^T \mathbf{x}_{k-1}^{(p)})}{\mathbf{a}_k^T \mathbf{a}_k} \mathbf{a}_k$$

Step 3. Set $\mathbf{x}_0^{(p+1)} = \mathbf{x}_M^{(p)}$.

Step 4. Increase the cycle number p by 1 and return to Step 2.

In Step 2 the iterate $\mathbf{x}_k^{(p)}$ is called the ***orthogonal projection*** of $\mathbf{x}_{k-1}^{(p)}$ onto the hyperplane $\mathbf{a}_k^T \mathbf{x} = b_k$. Consequently, as in the two-dimensional case, this algorithm determines a sequence of orthogonal projections from one hyperplane onto the next in which we cycle back to the first hyperplane after each projection onto the last hyperplane.

It can be shown that if the vectors $\mathbf{a}_1, \mathbf{a}_2, \ldots, \mathbf{a}_M$ span R^N, then the iterates $\mathbf{x}_M^{(1)}, \mathbf{x}_M^{(2)}, \mathbf{x}_M^{(3)}, \ldots$ lying on the Mth hyperplane will converge to a point $\mathbf{x}_M^*$ on that hyperplane which does not depend on the choice of the initial point $\mathbf{x}_0$. In computed tomography one of the iterates $\mathbf{x}_M^{(p)}$ for p sufficiently large is taken as an approximate solution of the linear system for the pixel densities.

Notice that for the center-of-pixel method, the scalar quantity $\mathbf{a}_k^T \mathbf{a}_k$ appearing in the equation in Step 2 of the algorithm is simply the number of pixels in which the kth beam passes through the center. Similarly, notice that the scalar quantity

$$b_k - \mathbf{a}_k^T \mathbf{x}_{k-1}^{(p)}$$

in that same equation can be interpreted as the *excess kth beam density* that results if the pixel densities are set equal to the entries of $\mathbf{x}_{k-1}^{(p)}$. This provides the following interpretation of our ART iteration scheme for the center-of-pixel method: *Generate the pixel densities of each iterate by distributing the excess beam density of successive beams in the scan evenly among those pixels in which the beam passes through the center. When the last beam in the scan has been reached, return to the first beam and continue.*

EXAMPLE 2 Using Algorithm 2

We can use Algorithm 2 to find the unknown pixel densities of the 9 pixels arranged in the 3×3 array illustrated in Figure 11.13.11. These 9 pixels are scanned using the parallel mode with 12 beams whose measured beam densities are indicated in the figure. We choose the center-of-pixel method to set up the 12 beam equations. (In Exercises 7 and 8 the reader is asked to set up the beam equations using the center line and area methods.) As the reader can verify, the beam equations are

$$x_7 + x_8 + x_9 = 13.00 \qquad x_3 + x_6 + x_9 = 18.00$$
$$x_4 + x_5 + x_6 = 15.00 \qquad x_2 + x_5 + x_8 = 12.00$$
$$x_1 + x_2 + x_3 = \;\;8.00 \qquad x_1 + x_4 + x_7 = \;\;6.00$$
$$x_6 + x_8 + x_9 = 14.79 \qquad x_2 + x_3 + x_6 = 10.51$$
$$x_3 + x_5 + x_7 = 14.31 \qquad x_1 + x_5 + x_9 = 16.13$$
$$x_1 + x_2 + x_4 = \;\;3.81 \qquad x_4 + x_7 + x_8 = \;\;7.04$$

Table 2 illustrates the results of the iteration scheme starting with an initial iterate $\mathbf{x}_0 = \mathbf{0}$. The table gives the values of each of the first cycle of iterates, $\mathbf{x}_1^{(1)}$ through $\mathbf{x}_{12}^{(1)}$, but thereafter gives the iterates $\mathbf{x}_{12}^{(p)}$ only for various values of p. The iterates $\mathbf{x}_{12}^{(p)}$ start repeating to two decimal places for $p \geq 45$, and so we take the entries of $\mathbf{x}_{12}^{(45)}$ as approximate values of the 9 pixel densities.   ◆

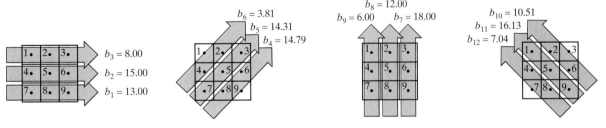

Figure 11.13.11

We close this section by noting that the field of computed tomography is presently a very active research area. In fact, the ART scheme discussed here has been replaced in commercial systems by more sophisticated techniques that are faster and provide a more accurate view of the cross section. However, all the new techniques address the same basic mathematical problem: finding a good approximate solution of a large overdetermined inconsistent linear system of equations.

TABLE 2

		Pixel Densities								
		x_1	x_2	x_3	x_4	x_5	x_6	x_7	x_8	x_9
	$\mathbf{x}_0$	.00	.00	.00	.00	.00	.00	.00	.00	.00
	$\mathbf{x}_1^{(1)}$	.00	.00	.00	.00	.00	.00	4.33	4.33	4.33
	$\mathbf{x}_2^{(1)}$	.00	.00	.00	5.00	5.00	5.00	4.33	4.33	4.33
	$\mathbf{x}_3^{(1)}$	2.67	2.67	2.67	5.00	5.00	5.00	4.33	4.33	4.33
	$\mathbf{x}_4^{(1)}$	2.67	2.67	2.67	5.00	5.00	5.37	4.33	4.71	4.71
First Cycle of Iterates	$\mathbf{x}_5^{(1)}$	2.67	2.67	3.44	5.00	5.77	5.37	5.10	4.71	4.71
	$\mathbf{x}_6^{(1)}$	.49	.49	3.44	2.83	5.77	5.37	5.10	4.71	4.71
	$\mathbf{x}_7^{(1)}$	.49	.49	4.93	2.83	5.77	6.87	5.10	4.71	6.20
	$\mathbf{x}_8^{(1)}$	.49	.84	4.93	2.83	6.11	6.87	5.10	5.05	6.20
	$\mathbf{x}_9^{(1)}$	−.31	.84	4.93	2.02	6.11	6.87	4.30	5.05	6.20
	$\mathbf{x}_{10}^{(1)}$	−.31	.13	4.22	2.02	6.11	6.16	4.30	5.05	6.20
	$\mathbf{x}_{11}^{(1)}$	1.06	.13	4.22	2.02	7.49	6.16	4.30	5.05	7.58
	$\mathbf{x}_{12}^{(1)}$	1.06	.13	4.22	.58	7.49	6.16	2.85	3.61	7.58
	$\mathbf{x}_{12}^{(2)}$	2.03	.69	4.42	1.34	7.49	5.39	2.65	3.04	6.61
	$\mathbf{x}_{12}^{(3)}$	1.78	.51	4.52	1.26	7.49	5.48	2.56	3.22	6.86
	$\mathbf{x}_{12}^{(4)}$	1.82	.52	4.62	1.37	7.49	5.37	2.45	3.22	6.82
	$\mathbf{x}_{12}^{(5)}$	1.79	.49	4.71	1.43	7.49	5.31	2.37	3.25	6.85
	$\mathbf{x}_{12}^{(10)}$	1.68	.44	5.03	1.70	7.49	5.03	2.04	3.29	6.96
	$\mathbf{x}_{12}^{(20)}$	1.49	.48	5.29	2.00	7.49	4.73	1.79	3.25	7.15
	$\mathbf{x}_{12}^{(30)}$	1.38	.55	5.34	2.11	7.49	4.62	1.74	3.19	7.26
	$\mathbf{x}_{12}^{(40)}$	1.33	.59	5.33	2.14	7.49	4.59	1.75	3.15	7.31
	$\mathbf{x}_{12}^{(45)}$	1.32	.60	5.32	2.15	7.49	4.59	1.76	3.14	7.32

Exercise Set 11.13

1. (a) Setting $\mathbf{x}_k^{(p)} = (x_{k1}^{(p)}, x_{k2}^{(p)})$, show that the three projection equations

$$\mathbf{x}_k^{(p)} = \mathbf{x}_{k-1}^{(p)} + \frac{(b_k - \mathbf{a}_k^T \mathbf{x}_{k-1}^{(p)})}{\mathbf{a}_k^T \mathbf{a}_k} \mathbf{a}_k, \qquad k = 1, 2, 3$$

for the three lines in Equation (5) can be written as

$$k = 1: \quad \begin{aligned} x_{11}^{(p)} &= \tfrac{1}{2}[2 + x_{01}^{(p)} - x_{02}^{(p)}] \\ x_{12}^{(p)} &= \tfrac{1}{2}[2 - x_{01}^{(p)} + x_{02}^{(p)}] \end{aligned}$$

$$k = 2: \quad \begin{aligned} x_{21}^{(p)} &= \tfrac{1}{5}[-2 + 4x_{11}^{(p)} + 2x_{12}^{(p)}] \\ x_{22}^{(p)} &= \tfrac{1}{5}[4 + 2x_{11}^{(p)} + x_{12}^{(p)}] \end{aligned}$$

$$k = 3: \quad \begin{aligned} x_{31}^{(p)} &= \tfrac{1}{10}[9 + x_{21}^{(p)} + 3x_{22}^{(p)}] \\ x_{32}^{(p)} &= \tfrac{1}{10}[-3 + 3x_{21}^{(p)} + 9x_{22}^{(p)}] \end{aligned}$$

where $(x_{01}^{(p+1)}, x_{02}^{(p+1)}) = (x_{31}^{(p)}, x_{32}^{(p)})$ for $p = 1, 2, \ldots$.

(b) Show that the three pairs of equations in part (a) can be combined to produce

$$\begin{aligned} x_{31}^{(p)} &= \tfrac{1}{20}[28 + x_{31}^{(p-1)} - x_{32}^{(p-1)}] \\ x_{32}^{(p)} &= \tfrac{1}{20}[24 + 3x_{31}^{(p-1)} - 3x_{32}^{(p-1)}] \end{aligned} \qquad p = 1, 2, \ldots$$

where $(x_{31}^{(0)}, x_{32}^{(0)}) = (x_{01}^{(1)}, x_{02}^{(1)}) = \mathbf{x}_0^{(1)}$. [**Note.** Using this pair of equations, one complete cycle of three orthogonal projections can be performed in a single step.]

(c) Because $\mathbf{x}_3^{(p)}$ tends to the limit point $\mathbf{x}_3^*$ as $p \to \infty$, the equations in part (b) become

$$\begin{aligned} x_{31}^* &= \tfrac{1}{20}[28 + x_{31}^* - x_{32}^*] \\ x_{32}^* &= \tfrac{1}{20}[24 + 3x_{31}^* - 3x_{32}^*] \end{aligned}$$

as $p \to \infty$. Solve this linear system for $\mathbf{x}_3^* = (x_{31}^*, x_{32}^*)$. [**Note.** The simplifications of the ART formulas described in this exercise are impractical for the large linear systems that arise in realistic computed tomography problems.]

2. Use the result of Exercise 1(b) to find $\mathbf{x}_3^{(1)}, \mathbf{x}_3^{(2)}, \ldots, \mathbf{x}_3^{(6)}$ to five decimal places in Example 1 using the following initial points:

(a) $\mathbf{x}_0 = (0, 0)$ (b) $\mathbf{x}_0 = (1, 1)$ (c) $\mathbf{x}_0 = (148, -15)$

3. (a) Show directly that the points of the limit cycle in Example 1,

$$\mathbf{x}_1^* = \left(\tfrac{12}{11}, \tfrac{10}{11}\right), \qquad \mathbf{x}_2^* = \left(\tfrac{46}{55}, \tfrac{78}{55}\right), \qquad \mathbf{x}_3^* = \left(\tfrac{31}{22}, \tfrac{27}{22}\right)$$

form a triangle whose vertices lie on the lines L_1, L_2, and L_3 and whose sides are perpendicular to these lines (Figure 11.13.9c).

(b) Using the equations derived in Exercise 1(a), show that if $\mathbf{x}_0^{(1)} = \mathbf{x}_3^* = \left(\tfrac{31}{22}, \tfrac{27}{22}\right)$, then

$$\begin{aligned} \mathbf{x}_1^{(1)} &= \mathbf{x}_1^* = \left(\tfrac{12}{11}, \tfrac{10}{11}\right) \\ \mathbf{x}_2^{(1)} &= \mathbf{x}_2^* = \left(\tfrac{46}{55}, \tfrac{78}{55}\right) \\ \mathbf{x}_3^{(1)} &= \mathbf{x}_3^* = \left(\tfrac{31}{22}, \tfrac{27}{22}\right) \end{aligned}$$

[**Note.** Either part of this exercise shows that successive orthogonal projections of any point on the limit cycle will move around the limit cycle indefinitely.]

4. The following three lines in the $x_1 x_2$-plane,

$$\begin{aligned} L_1: & \quad x_2 = 1 \\ L_2: & \quad x_1 - x_2 = 2 \\ L_3: & \quad x_1 - x_2 = 0 \end{aligned}$$

do not have a common intersection. Draw an accurate sketch of the three lines and graphically perform several cycles of the orthogonal projections described in Algorithm 1, beginning with the initial point $\mathbf{x}_0 = (0, 0)$. On the basis of your sketch, determine the three points of the limit cycle.

5. Prove Theorem 11.13.1 by verifying that

(a) the point $\mathbf{x}_p$ as defined in the theorem lies on the line $\mathbf{a}^T\mathbf{x} = b$ (i.e., $\mathbf{a}^T\mathbf{x}_p = b$);

(b) the vector $\mathbf{x}_p - \mathbf{x}^*$ is orthogonal to the line $\mathbf{a}^T\mathbf{x} = b$ (i.e., $\mathbf{x}_p - \mathbf{x}^*$ is parallel to $\mathbf{a}$).

6. As stated in the text, the iterates $\mathbf{x}_M^{(1)}, \mathbf{x}_M^{(2)}, \mathbf{x}_M^{(3)}, \ldots$ defined in Algorithm 2 will converge to a unique limit point $\mathbf{x}_M^*$ if the vectors $\mathbf{a}_1, \mathbf{a}_2, \ldots, \mathbf{a}_M$ span R^N. Show that if this is the case and if the center-of-pixel method is used, then the center of each of the N pixels in the field of view is crossed by at least one of the M beams in the scan.

7. Construct the 12 beam equations in Example 2 using the center line method. Assume that the distance between the center lines of adjacent beams is equal to the width of a single pixel.

8. Construct the 12 beam equations in Example 2 using the area method. Assume that the width of each beam is equal to the width of a single pixel and that the distance between the center lines of adjacent beams is also equal to the width of a single pixel.

Technology Exercises 11.13

The following exercises are designed to be solved using a technology utility. Typically, this will be MATLAB, *Mathematica*, Maple, Derive, or Mathcad, but it may also be some other type of linear algebra software or a scientific calculator with some linear algebra capabilities. For each exercise you will need to read the relevant documentation for the particular utility you are using. The goal of these exercises is to provide you with a basic proficiency with your technology utility. Once you have mastered the techniques in these exercises, you will be able to use your technology utility to solve many of the problems in the regular exercise sets.

T1. Given the set of equations

$$a_k x + b_k y = c_k$$

for $k = 1, 2, 3, \ldots, n$ (with $n > 2$), let us consider the following algorithm for obtaining an approximate solution to the system.

(1) Solve all possible pairs of equations

$$a_i x + b_i y = c_i \quad \text{and} \quad a_j x + b_j y = c_j$$

for $i, j = 1, 2, 3, \ldots, n$ and $i < j$ for their unique solutions. This leads to

$$\frac{1}{2} n(n-1)$$

solutions, which we label as

$$(x_{ij}, y_{ij})$$

for $i, j = 1, 2, 3, \ldots, n$ and $i < j$.

(2) Construct the geometric center of these points defined by

$$(x_C, y_C) = \left(\frac{2}{n(n-1)} \sum_{i=1}^{n-1} \sum_{j=i+1}^{n} x_{ij}, \ \frac{2}{n(n-1)} \sum_{i=1}^{n-1} \sum_{j=i+1}^{n} y_{ij} \right)$$

and use this as the approximate solution to the original system.

Use this algorithm to approximate the solution to the system

$$\begin{aligned} x + y &= 2 \\ x - 2y &= -2 \\ 3x - y &= 3 \end{aligned}$$

and compare your results to those in this section.

T2. (*For readers who have studied calculus.*) Given the set of equations

$$a_k x + b_k y = c_k$$

for $k = 1, 2, 3, \ldots, n$ (with $n > 2$), let us consider the following least squares algorithm for obtaining an approximate solution (x^*, y^*) to the system. Given a point (α, β) and the line $a_i x + b_i y = c_i$, the distance from this point to the line is given by

$$\frac{|a_i\alpha + b_i\beta - c_i|}{\sqrt{a_i^2 + b_i^2}}$$

If we define a function $f(x, y)$ by

$$f(x, y) = \sum_{i=1}^{n} \frac{(a_i x + b_i y - c_i)^2}{a_i^2 + b_i^2}$$

and then determine the point (x^*, y^*) that minimizes this function, we will determine the point that is *closest* to each of these lines in a summed least squares sense. Show that x^* and y^* are solutions to the system

$$\left(\sum_{i=1}^{n} \frac{a_i^2}{a_i^2 + b_i^2}\right) x^* + \left(\sum_{i=1}^{n} \frac{a_i b_i}{a_i^2 + b_i^2}\right) y^* = \sum_{i=1}^{n} \frac{a_i c_i}{a_i^2 + b_i^2}$$

and

$$\left(\sum_{i=1}^{n} \frac{a_i b_i}{a_i^2 + b_i^2}\right) x^* + \left(\sum_{i=1}^{n} \frac{b_i^2}{a_i^2 + b_i^2}\right) y^* = \sum_{i=1}^{n} \frac{b_i c_i}{a_i^2 + b_i^2}$$

Apply this algorithm to the system

$$\begin{aligned} x + \ y &= \ \ 2 \\ x - 2y &= -2 \\ 3x - \ y &= \ \ 3 \end{aligned}$$

and compare your results to those in this section.

11.14   FRACTALS

In this section we shall use certain classes of linear transformations to describe and generate intricate sets in the Euclidean plane. These sets, called fractals, are currently the focus of much mathematical and scientific research.

PREREQUISITES: Geometry of Linear Operators on R^2 (Section 9.2)
Euclidean Space R^n
Natural Logarithms
Intuitive Understanding of Limits

Fractals in the Euclidean Plane

At the end of the nineteenth century and beginning of the twentieth century, various bizarre and wild sets of points in the Euclidean plane began appearing in mathematics. Although they were initially mathematical curiosities, these sets, called *fractals*, are rapidly growing in importance. It is now recognized that they reveal a regularity in physical and biological phenomena previously dismissed as "random," "noisy," or "chaotic." For example, fractals are around us in the shapes of clouds, mountains, coastlines, trees, and ferns.

In this section we give a brief description of certain types of fractals in the Euclidean plane R^2. Much of this description is due to two mathematicians, Benoit B. Mandelbrot and Michael Barnsley, who are both active researchers in the field.

Self-Similar Sets

To start our study of fractals, we need to introduce some terminology about sets in R^2. We shall call a set in R^2 **bounded** if it can be enclosed by a suitably large circle (Figure 11.14.1) and **closed** if it contains all of its boundary points

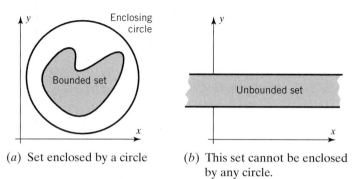

(a) Set enclosed by a circle   (b) This set cannot be enclosed by any circle.

Figure 11.14.1

(Figure 11.14.2). Two sets in R^2 will be called *congruent* if they can be made to coincide exactly by translating and rotating them appropriately within R^2 (Figure 11.14.3). We will also rely on the reader's intuitive concept of *overlapping* and *nonoverlapping sets*, as illustrated in Figure 11.14.4.

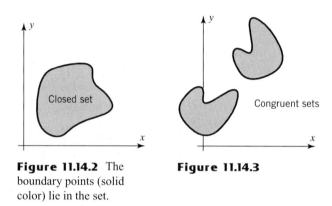

Figure 11.14.2 The boundary points (solid color) lie in the set.

Figure 11.14.3

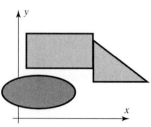

(a) Overlapping sets

(b) Nonoverlapping sets

Figure 11.14.4

If $T: R^2 \rightarrow R^2$ is the linear operator that scales by a factor of s (see Table 8 of Section 4.2), and if Q is a set in R^2, then the set $T(Q)$ (the set of images of points in Q under T) is called a *dilation* of the set Q if $s > 1$ and a *contraction* of Q if $0 < s < 1$ (Figure 11.14.5). In either case we say that $T(Q)$ is the set Q *scaled by the factor s*.

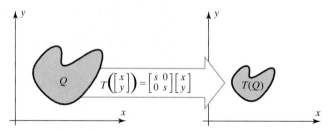

$$T\left(\begin{bmatrix} x \\ y \end{bmatrix}\right) = \begin{bmatrix} s & 0 \\ 0 & s \end{bmatrix}\begin{bmatrix} x \\ y \end{bmatrix}$$

Figure 11.14.5 A contraction of Q.

The types of fractals we shall consider first are called *self-similar*. In general, we define a self-similar set in R^2 as follows:

Definition

A closed and bounded subset of the Euclidean plane R^2 is said to be *self-similar* if it can be expressed in the form

$$S = S_1 \cup S_2 \cup S_3 \cup \cdots \cup S_k \tag{1}$$

where $S_1, S_2, S_3, \ldots, S_k$ are nonoverlapping sets, each of which is congruent to S scaled by the same factor s $(0 < s < 1)$.

If S is a self-similar set, then (1) is sometimes called a *decomposition* of S into nonoverlapping congruent sets.

EXAMPLE 1 Line Segment

A line segment in R^2 (Figure 11.14.6*a*) can be expressed as the union of two nonoverlapping congruent line segments (Figure 11.14.6*b*). In Figure 11.14.6*b* we have separated the two line segments slightly so that they can be seen more easily. Each of these two smaller line segments is congruent to the original line segment scaled by a factor of $\frac{1}{2}$. Hence, a line segment is a self-similar set with $k = 2$ and $s = \frac{1}{2}$. ◆

(a) (b)

Figure 11.14.6

EXAMPLE 2 Square

A square (Figure 11.14.7*a*) can be expressed as the union of four nonoverlapping congruent squares (Figure 11.14.7*b*), where we have again separated the smaller squares slightly. Each of the four smaller squares is congruent to the original square scaled by a factor of $\frac{1}{2}$. Hence, a square is a self-similar set with $k = 4$ and $s = \frac{1}{2}$. ◆

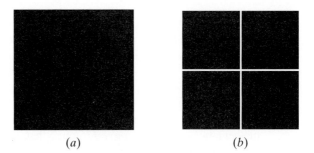

(a) (b) **Figure 11.14.7**

EXAMPLE 3 Sierpinski Carpet

The set suggested by Figure 11.14.8*a* was first described by the Polish mathematician Waclaw Sierpinski (1882–1969). It can be expressed as the union of eight nonoverlapping congruent subsets (Figure 11.14.8*b*), each of which is congruent to the original set scaled by a factor of $\frac{1}{3}$. Hence, it is a self-similar set with $k = 8$ and $s = \frac{1}{3}$. Notice that the intricate square-within-a-square pattern continues forever on a smaller and smaller scale (although this can only be suggested in a figure such as the one shown). ◆

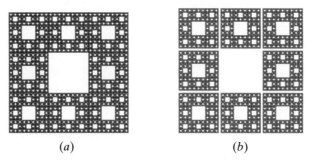

(a)   (b)   **Figure 11.14.8**

EXAMPLE 4 Sierpinski Triangle

Figure 11.14.9a illustrates another set due to Sierpinski. It is a self-similar set with $k = 3$ and $s = \frac{1}{2}$ (Figure 11.14.9b). As with the Sierpinski carpet, the intricate triangle-within-a-triangle pattern continues forever on a smaller and smaller scale. ◆

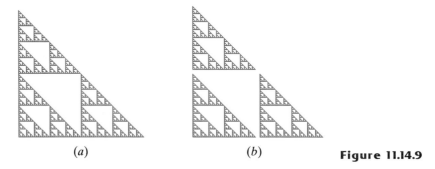

(a)   (b)   **Figure 11.14.9**

The Sierpinski carpet and triangle have a more intricate structure than the line segment and the square in that they exhibit a pattern that is repeated indefinitely. This difference will be explored more fully later in this section.

Topological Dimension of a Set   In Section 5.4 we defined the dimension of a *subspace* of a vector space to be the number of vectors in a basis, and we found that definition to coincide with our intuitive sense of dimension. For example, the origin of R^2 is zero-dimensional, lines through the origin are one-dimensional, and R^2 itself is two-dimensional. This definition of dimension is a special case of a more general concept, called ***topological dimension***, which is applicable to sets in R^n that are not necessarily subspaces. A precise definition of this concept is studied in a branch of mathematics called *topology*. Although that definition is beyond the scope of this text, we can state informally that

- a point in R^2 has topological dimension zero;
- a curve in R^2 has topological dimension one;
- a region in R^2 has topological dimension two.

It can be proved that the topological dimension of a set in R^n must be an integer between 0 and n, inclusive. In this text we shall denote the topological dimension of a set S by $d_T(S)$.

EXAMPLE 5 Topological Dimensions of Sets

Table 1 gives the topological dimensions of the sets studied in our earlier examples. The first two results in this table are intuitively obvious; however, the last two are not. Informally stated, both the Sierpinski carpet and triangle contain so many "holes" that those sets resemble web-like networks of lines rather than regions. Hence they have topological dimension one. The proofs are quite difficult. ◆

TABLE 1

Set S	$d_T(S)$
Line segment	1
Square	2
Sierpinski carpet	1
Sierpinski triangle	1

Hausdorff Dimension of a Self-Similar Set

In 1919 the German mathematician Felix Hausdorff (1868–1942) gave an alternative definition for the dimension of an arbitrary set in R^n. His definition is quite complicated, but for a self-similar set it reduces to something rather simple:

Definition

The **Hausdorff dimension** of a self-similar set S of form (1) is denoted by $d_H(S)$ and is defined by

$$d_H(S) = \frac{\ln k}{\ln(1/s)} \tag{2}$$

In this definition "ln" denotes the natural logarithm function. Equation (2) can also be expressed as

$$s^{d_H(S)} = \frac{1}{k} \tag{3}$$

in which the Hausdorff dimension $d_H(S)$ appears as an exponent. Formula (3) is more helpful for interpreting the concept of Hausdorff dimension; it states, for example, that if you scale a self-similar set by a factor of $s = \frac{1}{2}$, then its area (or, more properly, its *measure*) decreases by a factor of $\left(\frac{1}{2}\right)^{d_H(S)}$. Thus, scaling a line segment by a factor of $\frac{1}{2}$ reduces its measure (length) by a factor of $\left(\frac{1}{2}\right)^1 = \frac{1}{2}$, and scaling a square region by a factor of $\frac{1}{2}$ reduces its measure (area) by a factor of $\left(\frac{1}{2}\right)^2 = \frac{1}{4}$.

Before proceeding to some examples, a few facts about the Hausdorff dimension of a set are in order:

- The topological dimension and Hausdorff dimension of a set need not be the same.
- The Hausdorff dimension of a set need not be an integer.
- The topological dimension of a set is less than or equal to its Hausdorff dimension, that is, $d_T(S) \leq d_H(S)$.

EXAMPLE 6 Hausdorff Dimensions of Sets

Table 2 lists the Hausdorff dimensions of the sets studied in our earlier examples.

TABLE 2

Set S	s	k	$d_H(S) = \dfrac{\ln k}{\ln(1/s)}$
Line segment	$\frac{1}{2}$	2	$\ln 2 / \ln 2 = 1$
Square	$\frac{1}{2}$	4	$\ln 4 / \ln 2 = 2$
Sierpinski carpet	$\frac{1}{3}$	8	$\ln 8 / \ln 3 = 1.892\ldots$
Sierpinski triangle	$\frac{1}{2}$	3	$\ln 3 / \ln 2 = 1.584\ldots$

♦

Fractals   Comparing Tables 1 and 2, we see that the Hausdorff and topological dimensions are equal for both the line segment and square, but unequal for the Sierpinski carpet and triangle. In 1977 Benoit B. Mandelbrot suggested that sets for which the topological and Hausdorff dimensions differ must be quite complicated (as Hausdorff had earlier suggested in 1919). Mandelbrot proposed calling such sets *fractals* and he offered the following definition.

Definition

> A *fractal* is a subset of a Euclidean space whose Hausdorff dimension and topological dimension are not equal.

Mandelbrot has suggested that this definition is rather limited and probably will be replaced in the future. But in the meantime, it remains the formal definition of a fractal. According to this definition the Sierpinski carpet and Sierpinski triangle are fractals, whereas the line segment and square are not.

It follows from the preceding definition that a set whose Hausdorff dimension is not an integer must be a fractal (why?). However, we will see later that the converse is not true; that is, it is possible for a fractal to have an integer Hausdorff dimension.

Similitudes   We shall now show how some techniques from linear algebra can be used to generate fractals. This linear algebra approach also leads to algorithms that can be exploited to draw fractals on a computer. We begin with a definition.

Definition

> A *similitude* with scale factor s is a mapping of R^2 into R^2 of the form
> $$T\left(\begin{bmatrix} x \\ y \end{bmatrix}\right) = s \begin{bmatrix} \cos\theta & -\sin\theta \\ \sin\theta & \cos\theta \end{bmatrix} \begin{bmatrix} x \\ y \end{bmatrix} + \begin{bmatrix} e \\ f \end{bmatrix}$$
> where s, θ, e, and f are scalars.

Geometrically, a similitude is a composition of three simpler mappings: a scaling by a factor of s, a rotation about the origin through an angle θ, and a translation (e units

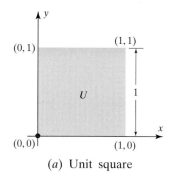

(a) Unit square

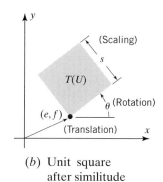

(b) Unit square
after similitude

Figure 11.14.10

in the x-direction and f units in the y-direction). Figure 11.14.10 illustrates the effect of a similitude on the unit square U.

For our application to fractals we shall only need similitudes that are **contractions**, by which we mean that the scale factor s is restricted to the range $0 < s < 1$. Consequently, when we refer to similitudes we shall always mean similitudes subject to this restriction.

Similitudes are important in the study of fractals because of the following fact:

> If $T: R^2 \to R^2$ is a similitude with scale factor s and S is a closed and bounded set in R^2, then the image $T(S)$ of the set S under T is congruent to S scaled by s.

Recall from the definition of a self-similar set in R^2 that a closed and bounded set S in R^2 is self-similar if it can be expressed in the form

$$S = S_1 \cup S_2 \cup S_3 \cup \cdots \cup S_k$$

where $S_1, S_2, S_3, \ldots, S_k$ are nonoverlapping sets each of which is congruent to S scaled by the same factor s ($0 < s < 1$) [see (1)]. In the following examples we will find similitudes that produce the sets $S_1, S_2, S_3, \ldots, S_k$ from S for the line segment, square, Sierpinski carpet, and Sierpinski triangle.

EXAMPLE 7 Line Segment

We shall take as our line segment the line segment S connecting the points $(0, 0)$ and $(1, 0)$ in the xy-plane (Figure 11.14.11a). Consider the two similitudes

$$T_1\left(\begin{bmatrix} x \\ y \end{bmatrix}\right) = \frac{1}{2}\begin{bmatrix} 1 & 0 \\ 0 & 1 \end{bmatrix}\begin{bmatrix} x \\ y \end{bmatrix}$$
$$T_2\left(\begin{bmatrix} x \\ y \end{bmatrix}\right) = \frac{1}{2}\begin{bmatrix} 1 & 0 \\ 0 & 1 \end{bmatrix}\begin{bmatrix} x \\ y \end{bmatrix} + \begin{bmatrix} \frac{1}{2} \\ 0 \end{bmatrix} \tag{4}$$

both of which have $s = \frac{1}{2}$ and $\theta = 0$. In Figure 11.14.11b we show how these two similitudes map the unit square U. The similitude T_1 maps U onto the smaller square $T_1(U)$, and the similitude T_2 maps U onto the smaller square $T_2(U)$. At the same time, T_1 maps the line segment S onto the smaller line segment $T_1(S)$, and T_2 maps S onto the smaller nonoverlapping line segment $T_2(S)$. The union of these two smaller nonoverlapping line segments is precisely the original line segment S; that is,

$$S = T_1(S) \cup T_2(S) \tag{5}$$

◆

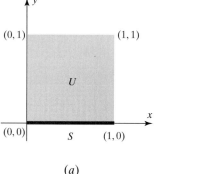

(a)

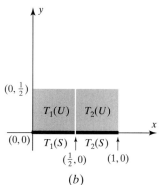

(b)

Figure 11.14.11

EXAMPLE 8 Square

Let us consider the unit square U in the xy-plane (Figure 11.14.12a) and the following four similitudes, all having $s = \frac{1}{2}$ and $\theta = 0$:

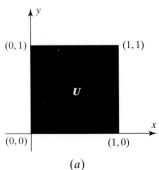

(a)

$$T_1 \left(\begin{bmatrix} x \\ y \end{bmatrix} \right) = \frac{1}{2} \begin{bmatrix} 1 & 0 \\ 0 & 1 \end{bmatrix} \begin{bmatrix} x \\ y \end{bmatrix}$$

$$T_2 \left(\begin{bmatrix} x \\ y \end{bmatrix} \right) = \frac{1}{2} \begin{bmatrix} 1 & 0 \\ 0 & 1 \end{bmatrix} \begin{bmatrix} x \\ y \end{bmatrix} + \begin{bmatrix} \frac{1}{2} \\ 0 \end{bmatrix}$$

$$T_3 \left(\begin{bmatrix} x \\ y \end{bmatrix} \right) = \frac{1}{2} \begin{bmatrix} 1 & 0 \\ 0 & 1 \end{bmatrix} \begin{bmatrix} x \\ y \end{bmatrix} + \begin{bmatrix} 0 \\ \frac{1}{2} \end{bmatrix}$$ (6)

$$T_4 \left(\begin{bmatrix} x \\ y \end{bmatrix} \right) = \frac{1}{2} \begin{bmatrix} 1 & 0 \\ 0 & 1 \end{bmatrix} \begin{bmatrix} x \\ y \end{bmatrix} + \begin{bmatrix} \frac{1}{2} \\ \frac{1}{2} \end{bmatrix}$$

The images of the unit square U under these four similitudes are the four squares shown in Figure 11.14.12b. Thus,

$$U = T_1(U) \cup T_2(U) \cup T_3(U) \cup T_4(U) \tag{7}$$

is a decomposition of U into four nonoverlapping squares that are congruent to U scaled by the same scale factor $\left(s = \frac{1}{2} \right)$. ♦

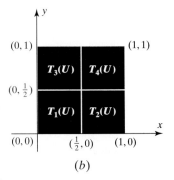

(b)

Figure 11.14.12

EXAMPLE 9 Sierpinski Carpet

Let us consider a Sierpinski carpet S over the unit square U of the xy-plane (Figure 11.14.13a) and the following eight similitudes, all having $s = \frac{1}{3}$ and $\theta = 0$:

$$T_i \left(\begin{bmatrix} x \\ y \end{bmatrix} \right) = \frac{1}{3} \begin{bmatrix} 1 & 0 \\ 0 & 1 \end{bmatrix} \begin{bmatrix} x \\ y \end{bmatrix} + \begin{bmatrix} e_i \\ f_i \end{bmatrix}, \qquad i = 1, 2, 3, \ldots, 8 \tag{8}$$

where the eight values of $\begin{bmatrix} e_i \\ f_i \end{bmatrix}$ are

$$\begin{bmatrix} 0 \\ 0 \end{bmatrix}, \quad \begin{bmatrix} \frac{1}{3} \\ 0 \end{bmatrix}, \quad \begin{bmatrix} \frac{2}{3} \\ 0 \end{bmatrix}, \quad \begin{bmatrix} 0 \\ \frac{1}{3} \end{bmatrix}, \quad \begin{bmatrix} \frac{2}{3} \\ \frac{1}{3} \end{bmatrix}, \quad \begin{bmatrix} 0 \\ \frac{2}{3} \end{bmatrix}, \quad \begin{bmatrix} \frac{1}{3} \\ \frac{2}{3} \end{bmatrix}, \quad \begin{bmatrix} \frac{2}{3} \\ \frac{2}{3} \end{bmatrix}$$

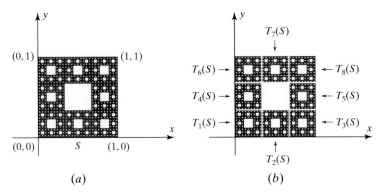

(a)   (b)

Figure 11.14.13

The images of S under these eight similitudes are the eight sets shown in Figure 11.14.13b. Thus,

$$S = T_1(S) \cup T_2(S) \cup T_3(S) \cup \cdots \cup T_8(S) \qquad (9)$$

is a decomposition of S into eight nonoverlapping sets that are congruent to S scaled by the same scale factor $\left(s = \frac{1}{3}\right)$. ◆

EXAMPLE 10 Sierpinski Triangle

Let us consider a Sierpinski triangle S fitted inside the unit square U of the xy-plane, as shown in Figure 11.14.14a, and the following three similitudes, all having $s = \frac{1}{2}$ and $\theta = 0$:

$$T_1\left(\begin{bmatrix} x \\ y \end{bmatrix}\right) = \frac{1}{2}\begin{bmatrix} 1 & 0 \\ 0 & 1 \end{bmatrix}\begin{bmatrix} x \\ y \end{bmatrix}$$

$$T_2\left(\begin{bmatrix} x \\ y \end{bmatrix}\right) = \frac{1}{2}\begin{bmatrix} 1 & 0 \\ 0 & 1 \end{bmatrix}\begin{bmatrix} x \\ y \end{bmatrix} + \begin{bmatrix} \frac{1}{2} \\ 0 \end{bmatrix} \qquad (10)$$

$$T_3\left(\begin{bmatrix} x \\ y \end{bmatrix}\right) = \frac{1}{2}\begin{bmatrix} 1 & 0 \\ 0 & 1 \end{bmatrix}\begin{bmatrix} x \\ y \end{bmatrix} + \begin{bmatrix} 0 \\ \frac{1}{2} \end{bmatrix}$$

The images of S under these three similitudes are the three sets in Figure 11.14.14b. Thus,

$$S = T_1(S) \cup T_2(S) \cup T_3(S) \qquad (11)$$

is a decomposition of S into three nonoverlapping sets that are congruent to S scaled by the same scale factor $\left(s = \frac{1}{2}\right)$. ◆

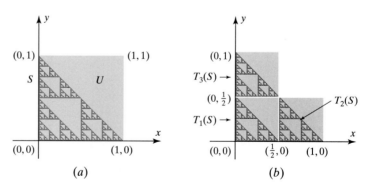

Figure 11.14.14

In the preceding examples we started with a specific set S and showed that it was self-similar by finding similitudes $T_1, T_2, T_3, \ldots, T_k$ with the same scale factor such that $T_1(S), T_2(S), T_3(S), \ldots, T_k(S)$ were nonoverlapping sets and such that

$$S = T_1(S) \cup T_2(S) \cup T_3(S) \cup \cdots \cup T_k(S) \qquad (12)$$

The following theorem addresses the converse problem of determining a self-similar set from a collection of similitudes.

> ### Theorem 11.14.1
>
> *If $T_1, T_2, T_3, \ldots, T_k$ are contracting similitudes with the same scale factor, then there is a unique nonempty closed and bounded set S in the Euclidean plane such that*
>
> $$S = T_1(S) \cup T_2(S) \cup T_3(S) \cup \cdots \cup T_k(S)$$
>
> *Furthermore, if the sets $T_1(S), T_2(S), T_3(S), \ldots, T_k(S)$ are nonoverlapping, then S is self-similar.*

Algorithms for Generating Fractals In general, there is no simple way to obtain the set S in the preceding theorem directly. We now describe an iterative procedure that will determine S from the similitudes that define it. We first give an example of the procedure and then give an algorithm for the general case.

EXAMPLE 11 Sierpinski Carpet

Figure 11.14.15 shows the unit square region S_0 in the xy-plane, which will serve as an "initial" set for an iterative procedure for the construction of the Sierpinski carpet. The set S_1 in the figure is the result of mapping S_0 with each of the eight similitudes T_i ($i = 1, 2, \ldots, 8$) in (8) that determine the Sierpinski carpet. It consists of eight square regions, each of side length $\frac{1}{3}$, surrounding an empty middle square. We next apply the eight similitudes to S_1 and arrive at the set S_2. Similarly, applying the eight similitudes to S_2 results in the set S_3. It we continue this process indefinitely, the sequence of sets $S_1, S_2, S_3, \ldots$ will "converge" to a set S, which is the Sierpinski carpet. ♦

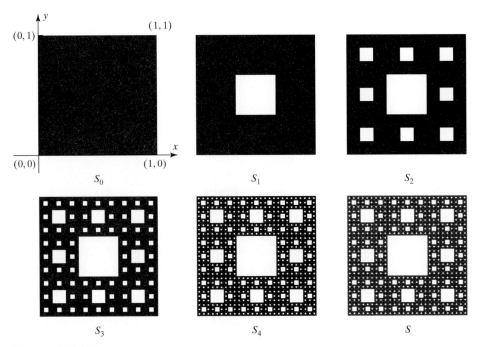

Figure 11.14.15

REMARK. Although we should properly give a definition of what it means for a sequence of sets to "converge" to a given set, an intuitive interpretation will suffice in this introductory treatment.

Although we started in Figure 11.14.15 with the unit square region to arrive at the Sierpinski carpet, we could have started with any nonempty set S_0. The only restriction is that the set S_0 be closed and bounded. For example, if we start with the particular set S_0 shown in Figure 11.14.16, then S_1 is the set obtained by applying each of the eight similitudes in (8). Applying the eight similitudes to S_1 results in the set S_2. As before, applying the eight similitudes indefinitely yields the Sierpinski carpet S as the limiting set.

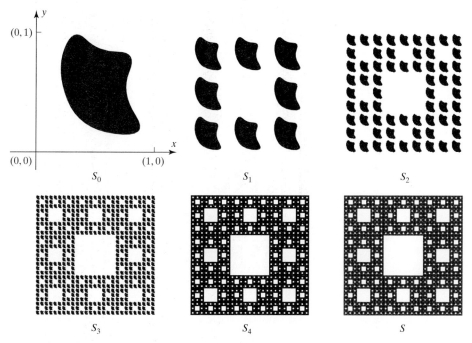

Figure 11.14.16

The general algorithm illustrated in the preceding example is as follows: Let $T_1, T_2, T_3, \ldots, T_k$ be contracting similitudes with the same scale factor, and for an arbitrary set Q in R^2 define the set $\mathfrak{J}(Q)$ by

$$\mathfrak{J}(Q) = T_1(Q) \cup T_2(Q) \cup T_3(Q) \cup \cdots \cup T_k(Q)$$

The following algorithm generates a sequence of sets $S_0, S_1, \ldots, S_n, \ldots$ that converges to the set S in Theorem 11.14.1.

Algorithm 1

Step 0. Choose an arbitrary nonempty closed and bounded set S_0 in R^2.

Step 1. Compute $S_1 = \mathfrak{J}(S_0)$.

Step 2. Compute $S_2 = \mathfrak{J}(S_1)$.

Step 3. Compute $S_3 = \mathfrak{J}(S_2)$.

$\vdots$

Step *n*. Compute $S_n = \mathfrak{J}(S_{n-1})$.

$\vdots$

EXAMPLE 12 Sierpinski Triangle

Let us construct the Sierpinski triangle determined by the three similitudes given in (10). The corresponding set mapping is $\Im(Q) = T_1(Q) \cup T_2(Q) \cup T_3(Q)$. Figure 11.14.17 shows an arbitrary closed and bounded set S_0, the first four iterates S_1, S_2, S_3, S_4, and the limiting set S (the Sierpinski triangle). ◆

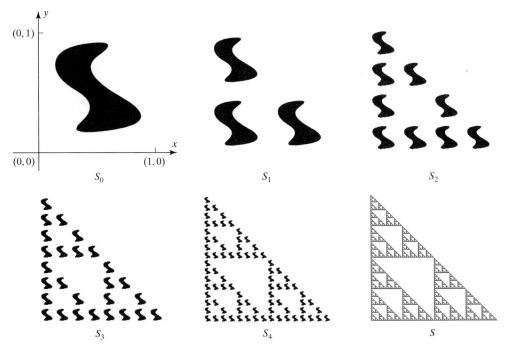

Figure 11.14.17

EXAMPLE 13 Using Algorithm 1

Consider the following two similitudes:

$$T_1\left(\begin{bmatrix} x \\ y \end{bmatrix}\right) = \frac{1}{2} \begin{bmatrix} 1 & 0 \\ 0 & 1 \end{bmatrix}$$

$$T_2\left(\begin{bmatrix} x \\ y \end{bmatrix}\right) = \frac{1}{2} \begin{bmatrix} \cos\theta & -\sin\theta \\ \sin\theta & \cos\theta \end{bmatrix} \begin{bmatrix} x \\ y \end{bmatrix} + \begin{bmatrix} .3 \\ .3 \end{bmatrix}$$

The actions of these two similitudes on the unit square U are illustrated in Figure 11.14.18.

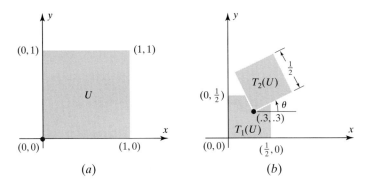

Figure 11.14.18

Here, the rotation angle θ is a parameter that we shall vary to generate different self-similar sets. The self-similar sets determined by these two similitudes are shown in Figure 11.14.19 for various values of θ. For simplicity, we have not drawn the xy axes, but in each case the origin is the lower left point of the set. These sets were generated on a computer using Algorithm 1 for the various values of θ. Because $k = 2$ and $s = \frac{1}{2}$, it follows from (2) that the Hausdorff dimension of these sets for any value of θ is 1. It can be shown that the topological dimension of these sets is 1 for $\theta = 0$ and 0 for all other values of θ. It follows that the self-similar set for $\theta = 0$ is not a fractal [it is the straight line segment from $(0, 0)$ to $(.6, .6)$], while the self-similar sets for all other values of θ are fractals. In particular, they are examples of fractals with integer Hausdorff dimension. ◆

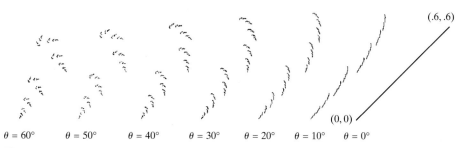

$\theta = 60°$ $\theta = 50°$ $\theta = 40°$ $\theta = 30°$ $\theta = 20°$ $\theta = 10°$ $\theta = 0°$

Figure 11.14.19

A Monte Carlo Approach
The set-mapping approach of constructing self-similar sets described in Algorithm 1 is rather time consuming on a computer because the similitudes involved must be applied to each of the many computer screen pixels in the successive iterated sets. In 1985 Michael Barnsley described an alternative, more practical, method of generating a self-similar set defined through its similitudes. It is a so-called **Monte Carlo method** that takes advantage of probability theory. Barnsley refers to it as the **Random Iteration Algorithm**.

Let $T_1, T_2, T_3, \ldots, T_k$ be contracting similitudes with the same scale factor. The following algorithm generates a sequence of points

$$\begin{bmatrix} x_0 \\ y_0 \end{bmatrix}, \begin{bmatrix} x_1 \\ y_1 \end{bmatrix}, \ldots, \begin{bmatrix} x_n \\ y_n \end{bmatrix}, \ldots$$

that collectively converge to the set S in Theorem 11.14.1.

Algorithm 2

Step 0. Choose an arbitrary point $\begin{bmatrix} x_0 \\ y_0 \end{bmatrix}$ in S.

Step 1. Choose one of the k similitudes at random, say T_{k_1}, and compute

$$\begin{bmatrix} x_1 \\ y_1 \end{bmatrix} = T_{k_1} \left(\begin{bmatrix} x_0 \\ y_0 \end{bmatrix} \right)$$

Step 2. Choose one of the k similitudes at random, say T_{k_2}, and compute

$$\begin{bmatrix} x_2 \\ y_2 \end{bmatrix} = T_{k_2} \left(\begin{bmatrix} x_1 \\ y_1 \end{bmatrix} \right)$$

⋮

Step *n*. Choose one of the k similitudes at random, say T_{k_n}, and compute

$$\begin{bmatrix} x_n \\ y_n \end{bmatrix} = T_{k_n}\left(\begin{bmatrix} x_{n-1} \\ y_{n-1} \end{bmatrix}\right)$$

$\vdots$

On a computer screen the pixels corresponding to the points generated by this algorithm will fill out the pixel representation of the limiting set S.

Figure 11.14.20 shows four stages of the Random Iteration Algorithm that generate the Sierpinski carpet, starting with the initial point $\begin{bmatrix} 0 \\ 0 \end{bmatrix}$.

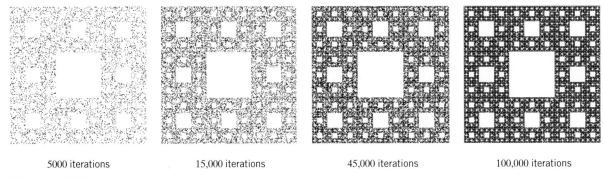

| 5000 iterations | 15,000 iterations | 45,000 iterations | 100,000 iterations |

Figure 11.14.20

REMARK. Although Step 0 in the preceding algorithm requires the selection of an initial point in the set S, which may not be known in advance, this is not a serious problem. In practice, one can usually start with any point in R^2 and after a few iterations (say ten or so) the point generated will be sufficiently close to S that the algorithm will work correctly from that point on.

More General Fractals   So far, we have discussed fractals that are self-similar sets according to the definition of a self-similar set in R^2. However, Theorem 11.14.1 remains true if the similitudes $T_1, T_2, \ldots, T_k$ are replaced by more general transformations, called *contracting affine transformations*. An affine transformation is defined as follows:

> ### Definition
>
> An *affine transformation* is a mapping of R^2 into R^2 of the form
> $$T\left(\begin{bmatrix} x \\ y \end{bmatrix}\right) = \begin{bmatrix} a & b \\ c & d \end{bmatrix}\begin{bmatrix} x \\ y \end{bmatrix} + \begin{bmatrix} e \\ f \end{bmatrix}$$
> where $a, b, c, d, e,$ and f are scalars.

Figure 11.14.21 shows how an affine transformation maps the unit square U onto a parallelogram $T(U)$. An affine transformation is said to be **contracting** if the Euclidean distance between any two points in the plane is strictly decreased after the two points are mapped by the transformation. It can be shown that any k contracting affine transformations $T_1, T_2, \ldots, T_k$ determine a unique closed and bounded set S satisfying the equation

$$S = T_1(S) \cup T_2(S) \cup T_3(S) \cup \cdots \cup T_k(S) \tag{13}$$

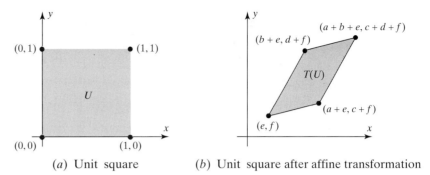

(a) Unit square   (b) Unit square after affine transformation

Figure 11.14.21

Equation (13) has the same form as Equation (12), which we used to find self-similar sets. Although Equation (13), which uses contracting affine transformations, does not determine a self-similar set S, the set it does determine has many of the features of self-similar sets. For example, Figure 11.14.22 shows how a set in the plane resembling a fern (an example made famous by Barnsley) can be generated through four contracting

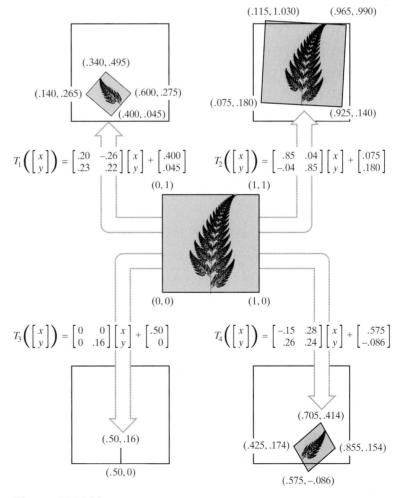

$$T_1\left(\begin{bmatrix} x \\ y \end{bmatrix}\right) = \begin{bmatrix} .20 & -.26 \\ .23 & .22 \end{bmatrix}\begin{bmatrix} x \\ y \end{bmatrix} + \begin{bmatrix} .400 \\ .045 \end{bmatrix}$$

$$T_2\left(\begin{bmatrix} x \\ y \end{bmatrix}\right) = \begin{bmatrix} .85 & .04 \\ -.04 & .85 \end{bmatrix}\begin{bmatrix} x \\ y \end{bmatrix} + \begin{bmatrix} .075 \\ .180 \end{bmatrix}$$

$$T_3\left(\begin{bmatrix} x \\ y \end{bmatrix}\right) = \begin{bmatrix} 0 & 0 \\ 0 & .16 \end{bmatrix}\begin{bmatrix} x \\ y \end{bmatrix} + \begin{bmatrix} .50 \\ 0 \end{bmatrix}$$

$$T_4\left(\begin{bmatrix} x \\ y \end{bmatrix}\right) = \begin{bmatrix} -.15 & .28 \\ .26 & .24 \end{bmatrix}\begin{bmatrix} x \\ y \end{bmatrix} + \begin{bmatrix} .575 \\ -.086 \end{bmatrix}$$

Figure 11.14.22

affine transformations. Notice how the middle fern is the slightly overlapping union of the four smaller affine-image ferns surrounding it. Notice also how T_3, because the determinant of its matrix part is zero, maps the entire fern onto the small straight line segment between the points $(.50, 0)$ and $(.50, .16)$. Figure 11.14.22 contains a wealth of information and should be studied carefully.

Michael Barnsley is actively pursuing an application of the above theory to the field of data compression and transmission. The fern, for example, is completely determined by the four affine transformations T_1, T_2, T_3, T_4. These four transformations in turn, are determined by the 24 numbers given in Figure 11.14.22 defining their corresponding values of a, b, c, d, e, and f. In other words, these 24 numbers completely *encode* the picture of the fern. Storing these 24 numbers in a computer requires considerably less memory space than storing a pixel-by-pixel description of the fern. In principle, any picture represented by a pixel map on a computer screen can be described through a finite number of affine transformations, although it is not easy to determine which transformations to use. Nevertheless, once encoded, the affine transformations generally require several orders of magnitude less computer memory than a pixel-by-pixel description of the pixel map.

Further Readings

Readers interested in learning more about fractals are referred to the following books, the first of which elaborates on the linear transformation approach of this section.

1. MICHAEL BARNSLEY, *Fractals Everywhere* (New York: Academic Press, 1993).

2. BENOIT B. MANDELBROT, *The Fractal Geometry of Nature* (New York: W. H. Freeman, 1982).

3. HEINZ-OTTO PEITGEN AND P. H. RICHTER, *The Beauty of Fractals* (New York: Springer-Verlag, 1986).

4. HEINZ-OTTO PEITGEN AND DIETMAR SAUPE, *The Science of Fractal Images* (New York: Springer-Verlag, 1988).

Exercise Set 11.14

1. The self-similar set in the accompanying figure has the sizes indicated. Given that its lower left corner is situated at the origin of the xy-plane, find the similitudes that determine the set. What is its Hausdorff dimension? Is it a fractal?

2. Find the Hausdorff dimension of the self-similar set shown in the accompanying figure. Use a ruler to measure the figure and determine an approximate value of the scale factor s. What are the rotation angles of the similitudes determining this set?

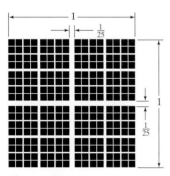

Figure Ex-1

Figure Ex-2

3. For each of the self-similar sets in the accompanying figure find: (i) the scale factor s of the similitudes describing the set; (ii) the rotation angles θ of all similitudes describing the set (all rotation angles are multiples of $90°$); and (iii) the Hausdorff dimension of the set. Which of the sets are fractals and why?

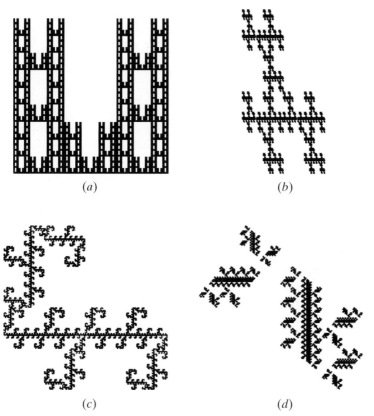

(a)

(b)

(c)

(d)

Figure Ex-3

4. Of the four affine transformations shown in Figure 11.14.22, only the transformation T_2 is a similitude. Determine its scale factor s and rotation angle θ.

5. Find the coordinates of the tip of the fern in Figure 11.14.22. [***Hint.*** The transformation T_2 maps the tip of the fern to itself.]

6. The square in Figure 11.14.7a was expressed as the union of four nonoverlapping squares as in Figure 11.14.7b. Suppose instead that it is expressed as the union of 16 nonoverlapping squares. Verify that its Hausdorff dimension is still two, as determined by Equation (2).

7. Show that the four similitudes

$$T_1\left(\begin{bmatrix} x \\ y \end{bmatrix}\right) = \frac{3}{4}\begin{bmatrix} 1 & 0 \\ 0 & 1 \end{bmatrix}\begin{bmatrix} x \\ y \end{bmatrix}$$

$$T_2\left(\begin{bmatrix} x \\ y \end{bmatrix}\right) = \frac{3}{4}\begin{bmatrix} 1 & 0 \\ 0 & 1 \end{bmatrix}\begin{bmatrix} x \\ y \end{bmatrix} + \begin{bmatrix} \frac{1}{4} \\ 0 \end{bmatrix}$$

$$T_3\left(\begin{bmatrix} x \\ y \end{bmatrix}\right) = \frac{3}{4}\begin{bmatrix} 1 & 0 \\ 0 & 1 \end{bmatrix}\begin{bmatrix} x \\ y \end{bmatrix} + \begin{bmatrix} 0 \\ \frac{1}{4} \end{bmatrix}$$

$$T_4\left(\begin{bmatrix} x \\ y \end{bmatrix}\right) = \frac{3}{4}\begin{bmatrix} 1 & 0 \\ 0 & 1 \end{bmatrix}\begin{bmatrix} x \\ y \end{bmatrix} + \begin{bmatrix} \frac{1}{4} \\ \frac{1}{4} \end{bmatrix}$$

express the unit square as the union of four *overlapping* squares. Evaluate the right-hand side of Equation (2) for the values of k and s determined by these similitudes, and show that the result is not the correct value of the Hausdorff dimension of the unit square. [**Note.** This exercise shows the necessity of the nonoverlapping condition in the definition of a self-similar set and its Hausdorff dimension.]

8. All of the results in this section can be extended to R^n. Compute the Hausdorff dimension of the unit cube in R^3 (see the accompanying figure). Given that the topological dimension of the unit cube is 3, determine whether it is a fractal. [**Hint.** Express the unit cube as the union of 8 smaller congruent nonoverlapping cubes.]

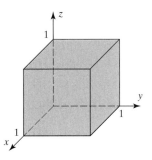

Figure Ex-8

9. The set in R^3 in the accompanying figure is called the **Menger sponge**. It is a self-similar set obtained by drilling out certain square holes from the unit cube. Notice that each face of the Menger sponge is a Sierpinski carpet and that the holes in the Sierpinski carpet now run all the way through the Menger sponge. Determine the values of k and s for the Menger sponge and find its Hausdorff dimension. Is the Menger sponge a fractal?

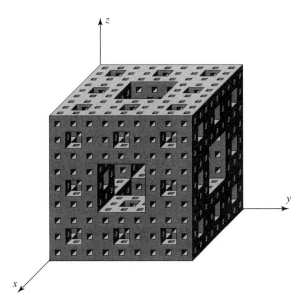

Figure Ex-9

10. The two similitudes

$$T_1\left(\begin{bmatrix} x \\ y \end{bmatrix}\right) = \frac{1}{3}\begin{bmatrix} 1 & 0 \\ 0 & 1 \end{bmatrix}\begin{bmatrix} x \\ y \end{bmatrix} \quad \text{and} \quad T_2\left(\begin{bmatrix} x \\ y \end{bmatrix}\right) = \frac{1}{3}\begin{bmatrix} 1 & 0 \\ 0 & 1 \end{bmatrix}\begin{bmatrix} x \\ y \end{bmatrix} + \begin{bmatrix} \frac{2}{3} \\ 0 \end{bmatrix}$$

determine a fractal known as the **Cantor set**. Starting with the unit square region U as an initial set, sketch the first four sets that Algorithm 1 determines. Also, find the Hausdorff dimension of the Cantor set. (This famous set was the first example that Hausdorff gave in his 1919 paper of a set whose Hausdorff dimension is not equal to its topological dimension.)

11. Compute the areas of the sets S_0, S_1, S_2, S_3, and S_4 in Figure 11.14.15.

Technology Exercises 11.14

The following exercises are designed to be solved using a technology utility. Typically, this will be MATLAB, *Mathematica*, Maple, Derive, or Mathcad, but it may also be some other type of linear algebra software or a scientific calculator with some linear algebra capabilities. For each exercise you will need to read the relevant documentation for the particular utility you are using. The goal of these exercises is to provide you with a basic proficiency with your technology utility. Once you have mastered the techniques in these exercises, you will be able to use your technology utility to solve many of the problems in the regular exercise sets.

T1. Use similitudes of the form

$$T_i\left(\begin{bmatrix} x \\ y \\ z \end{bmatrix}\right) = \frac{1}{3}\begin{bmatrix} 1 & 0 & 0 \\ 0 & 1 & 0 \\ 0 & 0 & 1 \end{bmatrix}\begin{bmatrix} x \\ y \\ z \end{bmatrix} + \begin{bmatrix} a_i \\ b_i \\ c_i \end{bmatrix}$$

to show that the Menger sponge (see Exercise 9) is the set S satisfying

$$S = \bigcup_{i=1}^{20} T_i(S)$$

for appropriately chosen similitudes T_i (for $i = 1, 2, 3, \ldots, 20$). Determine these similitudes by determining the collection of 3×1 matrices

$$\left\{ \begin{bmatrix} a_i \\ b_i \\ c_i \end{bmatrix} \middle\| \text{ for } i = 1, 2, 3, \ldots, 20 \right\}$$

T2. Generalize the ideas involved in the Cantor set (in R^1), the Sierpinski carpet (in R^2), and the Menger sponge (in R^3) to R^n by considering the set S satisfying

$$S = \bigcup_{i=1}^{m_n} T_i(S)$$

with

$$T_i\left(\begin{bmatrix} x_1 \\ x_2 \\ x_3 \\ \vdots \\ x_n \end{bmatrix}\right) = \frac{1}{3}\begin{bmatrix} 1 & 0 & 0 & \cdots & 0 \\ 0 & 1 & 0 & \cdots & 0 \\ 0 & 0 & 1 & \cdots & 0 \\ \vdots & \vdots & \vdots & \ddots & \vdots \\ 0 & 0 & 0 & \cdots & 1 \end{bmatrix}\begin{bmatrix} x_1 \\ x_2 \\ x_3 \\ \vdots \\ x_n \end{bmatrix} + \begin{bmatrix} a_{1i} \\ a_{2i} \\ a_{3i} \\ \vdots \\ a_{ni} \end{bmatrix}$$

where each a_{ki} equals either 0, $\frac{1}{3}$, or $\frac{2}{3}$, and no two of them ever equal $\frac{1}{3}$ at the same time. Use a computer to construct the set

$$\left\{ \begin{bmatrix} a_{1i} \\ a_{2i} \\ a_{3i} \\ \vdots \\ a_{ni} \end{bmatrix} \middle\| \text{ for } i = 1, 2, 3, \ldots, m_n \right\}$$

thereby determining the value of m_n for $n = 2, 3, 4$. Then develop an expression for m_n.

11.15 CHAOS

In this section we use a map of the unit square in the xy-plane onto itself to describe the concept of a chaotic mapping.

PREREQUISITES: Geometry of Linear Operators on R^2 (Section 9.2)
Eigenvalues and Eigenvectors
Intuitive Understanding of Limits and Continuity

Chaos   The word *chaos* was first used in a mathematical sense in 1975 by Tien-Yien Li and James Yorke in a paper entitled "Period Three Implies Chaos." The term is now used to describe the behavior of certain mathematical mappings and physical phenomena that at first glance seem to behave in a random or disorderly fashion but actually have an underlying element of order (e.g., random-number generation, shuffling cards, cardiac arrhythmia, fluttering airplane wings, changes in the red spot of Jupiter, and deviations in the orbit of Pluto). In this section we discuss a particular chaotic mapping called *Arnold's cat map*, after the Russian mathematician Vladimir I. Arnold who first described it using a diagram of a cat.

Arnold's Cat Map   To describe Arnold's cat map we need a few ideas about *modular arithmetic*. If x is a real number, then the notation x mod 1 denotes the unique number in the interval $[0, 1)$ that differs from x by an integer. For example,

$$2.3 \bmod 1 = 0.3, \quad 0.9 \bmod 1 = 0.9, \quad -3.7 \bmod 1 = 0.3, \quad 2.0 \bmod 1 = 0$$

Notice that if x is a nonnegative number, then x mod 1 is simply the fractional part of x. If (x, y) is an ordered pair of real numbers, then the notation (x, y) mod 1 denotes $(x \bmod 1, y \bmod 1)$. For example,

$$(2.3, -7.9) \bmod 1 = (0.3, 0.1)$$

Observe that for every real number x the point x mod 1 lies in the unit interval $[0, 1)$ and for every ordered pair (x, y) the point (x, y) mod 1 lies in the unit square

$$S = \{(x, y) \mid 0 \le x < 1, 0 \le y < 1\}$$

Also observe that the upper boundary and the right-hand boundary of the square are not included in S.

Arnold's cat map is the transformation $\Gamma: R^2 \to R^2$ defined by the formula

$$\Gamma: (x, y) \to (x + y, x + 2y) \bmod 1$$

or in matrix notation as

$$\Gamma\left(\begin{bmatrix} x \\ y \end{bmatrix}\right) = \begin{bmatrix} 1 & 1 \\ 1 & 2 \end{bmatrix} \begin{bmatrix} x \\ y \end{bmatrix} \bmod 1 \tag{1}$$

To understand the geometry of Arnold's cat map, it is helpful to write (1) in the factored form

$$\Gamma\left(\begin{bmatrix} x \\ y \end{bmatrix}\right) = \begin{bmatrix} 1 & 0 \\ 1 & 1 \end{bmatrix} \begin{bmatrix} 1 & 1 \\ 0 & 1 \end{bmatrix} \begin{bmatrix} x \\ y \end{bmatrix} \bmod 1$$

which expresses Arnold's cat map as the composition of a shear in the x-direction with factor 1, followed by a shear in the y-direction with factor 1. Because the computations are performed mod 1, Γ maps all points of R^2 into the unit square S.

We will illustrate the effect of Arnold's cat map on the unit square S, which is shaded in Figure 11.15.1a and contains a picture of a cat. It can be shown that it does not matter

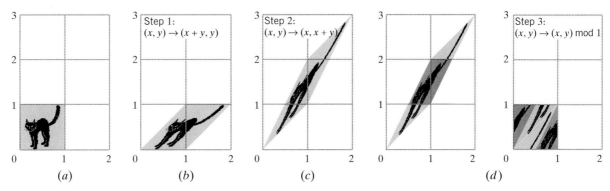

Figure 11.15.1

whether the mod 1 computations are carried out after each shear or at the very end. We will discuss both methods, first performing them at the end. The steps are as follows:

Step 1. Shear in the *x*-direction with factor 1 (Figure 11.15.1*b*):

$$(x, y) \rightarrow (x + y, y)$$

or in matrix notation

$$\begin{bmatrix} 1 & 1 \\ 0 & 1 \end{bmatrix} \begin{bmatrix} x \\ y \end{bmatrix} = \begin{bmatrix} x + y \\ y \end{bmatrix}$$

Step 2. Shear in the *y*-direction with factor 1 (Figure 11.15.1*c*):

$$(x, y) \rightarrow (x, x + y)$$

or in matrix notation

$$\begin{bmatrix} 1 & 0 \\ 1 & 1 \end{bmatrix} \begin{bmatrix} x \\ y \end{bmatrix} = \begin{bmatrix} x \\ x + y \end{bmatrix}$$

Step 3. Reassembly into *S* (Figure 11.15.1*d*):

$$(x, y) \rightarrow (x, y) \bmod 1$$

The geometric effect of the mod 1 arithmetic is to break up the parallelogram in Figure 11.15.1*c* and reassemble the pieces of *S* as shown in Figure 11.15.1*d*.

For computer implementation, it is more convenient to perform the mod 1 arithmetic at each step, rather than at the end. With this approach there is a reassembly at each step, but the net effect is the same. The steps are as follows:

Step 1. Shear in the *x*-direction with factor 1, followed by a reassembly into *S* (Figure 11.15.2*b*):

$$(x, y) \rightarrow (x + y, y) \bmod 1$$

Step 2. Shear in the *y*-direction with factor 1, followed by a reassembly into *S* (Figure 11.15.2*c*):

$$(x, y) \rightarrow (x, x + y) \bmod 1$$

Repeated Mappings   Chaotic mappings such as Arnold's cat map usually arise in physical models in which an operation is performed repeatedly. For example, cards are mixed by repeated shuffles, paint is mixed by repeated stirs, water in a tidal basin is mixed by repeated tidal changes, and so forth. Thus, we are interested in examining the

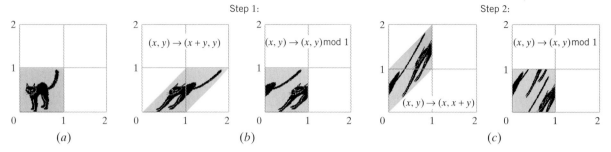

Figure 11.15.2

effect on S of repeated applications (or **iterations**) of Arnold's cat map. Figure 11.15.3, which was generated on a computer, shows the effect of 25 iterations of Arnold's cat map on the cat in the unit square S. Two interesting phenomena occur:

- The cat returns to its original form at the 25th iteration.
- At some of the intermediate iterations, the cat is decomposed into streaks that seem to have a specific direction.

Much of the remainder of this section is devoted to explaining these phenomena.

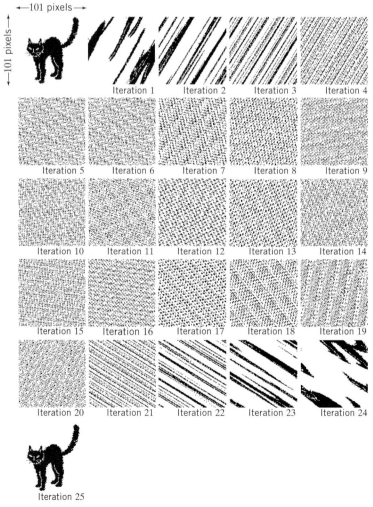

Figure 11.15.3

Periodic Points   Our first goal is to explain why the cat in Figure 11.15.3 returns to its original configuration at the 25th iteration. For this purpose it will be helpful to think of a ***picture*** in the xy-plane as an assignment of colors to the points in the plane. For pictures generated on a computer screen or other digital device, hardware limitations require that a picture be broken up into discrete squares, called ***pixels***. For example, in the computer-generated pictures in Figure 11.15.3 the unit square S is divided into a grid with 101 pixels on a side for a total of 10,201 pixels, each of which is black or white (Figure 11.15.4). An assignment of colors to pixels to create a picture is called a ***pixel map***.

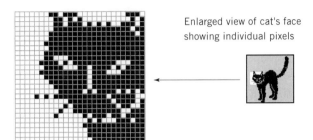

Enlarged view of cat's face showing individual pixels

Figure 11.15.4

As shown in Figure 11.15.5, each pixel in S can be assigned a unique pair of coordinates of the form $(m/101, n/101)$ that identifies its lower left-hand corner, where m and n are integers in the range $0, 1, 2, \ldots, 100$. We call these points ***pixel points*** because each such point identifies a unique pixel. Instead of restricting the discussion to the case where S is subdivided into an array with 101 pixels on a side, let us consider the more general case where there are p pixels per side. Thus, each pixel map in S consists of p^2 pixels uniformly spaced $1/p$ units apart in both the x and y directions. The pixel points in S have coordinates of the form $(m/p, n/p)$, where m and n are integers ranging from 0 to $p - 1$.

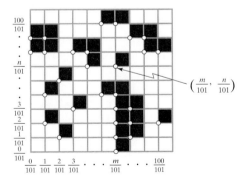

$$\left(\frac{m}{101}, \frac{n}{101}\right)$$

Figure 11.15.5

Under Arnold's cat map each pixel point of S is transformed into another pixel point of S. To see why this is so, observe that the image of the pixel point $(m/p, n/p)$ under Γ is given in matrix form by

$$\Gamma\left(\begin{bmatrix} \dfrac{m}{p} \\ \dfrac{n}{p} \end{bmatrix}\right) = \begin{bmatrix} 1 & 1 \\ 1 & 2 \end{bmatrix} \begin{bmatrix} \dfrac{m}{p} \\ \dfrac{n}{p} \end{bmatrix} \bmod 1 = \begin{bmatrix} \dfrac{m+n}{p} \\ \dfrac{m+2n}{p} \end{bmatrix} \bmod 1 \tag{2}$$

The ordered pair $((m + n)/p, (m + 2n)/p) \bmod 1$ is of the form $(m'/p, n'/p)$, where m' and n' lie in the range $0, 1, 2, \ldots, p - 1$. Specifically, m' and n' are the remainders when $m + n$ and $m + 2n$ are divided by p, respectively. Consequently, each point in S of the form $(m/p, n/p)$ is mapped onto another point of the same form.

Because Arnold's cat map transforms every pixel point of S into another pixel point of S, and because there are only p^2 different pixel points in S, it follows that any given pixel point must return to its original position after at most p^2 iterations of Arnold's cat map.

EXAMPLE 1 Using Formula [2]

If $p = 76$, then (2) becomes

$$\Gamma\left(\begin{bmatrix} \dfrac{m}{76} \\ \dfrac{n}{76} \end{bmatrix}\right) = \begin{bmatrix} \dfrac{m+n}{76} \\ \dfrac{m+2n}{76} \end{bmatrix} \bmod 1$$

In this case the successive iterates of the point $\left(\frac{27}{76}, \frac{58}{76}\right)$ are

0		1		2		3		4		5		6		7		8
$\begin{bmatrix} \frac{27}{76} \\ \frac{58}{76} \end{bmatrix}$	→	$\begin{bmatrix} \frac{9}{76} \\ \frac{67}{76} \end{bmatrix}$	→	$\begin{bmatrix} \frac{0}{76} \\ \frac{67}{76} \end{bmatrix}$	→	$\begin{bmatrix} \frac{67}{76} \\ \frac{58}{76} \end{bmatrix}$	→	$\begin{bmatrix} \frac{49}{76} \\ \frac{31}{76} \end{bmatrix}$	→	$\begin{bmatrix} \frac{4}{76} \\ \frac{35}{76} \end{bmatrix}$	→	$\begin{bmatrix} \frac{39}{76} \\ \frac{74}{76} \end{bmatrix}$	→	$\begin{bmatrix} \frac{37}{76} \\ \frac{35}{76} \end{bmatrix}$	→	$\begin{bmatrix} \frac{72}{76} \\ \frac{31}{76} \end{bmatrix}$

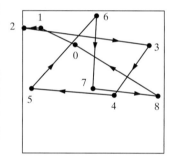

Figure 11.15.6

(verify). Because the point returns to its initial position on the ninth application of Arnold's cat map (but no sooner), the point is said to have period 9, and the set of nine distinct iterates of the point is called a 9-cycle. Figure 11.15.6 shows this 9-cycle with the initial point labeled 0 and its successive iterates labeled accordingly. ◆

In general, a point that returns to its initial position after n applications of Arnold's cat map, but does not return with fewer than n applications, is said to have *period n*, and its set of n distinct iterates is called an *n-cycle*. Arnold's cat map maps $(0, 0)$ into $(0, 0)$, so this point has period 1. Points with period 1 are also called *fixed points*. We leave it as an exercise (Exercise 11) to show that $(0, 0)$ is the only fixed point of Arnold's cat map.

Period Versus Pixel Width   If P_1 and P_2 are points with periods q_1 and q_2, respectively, then P_1 returns to its initial position in q_1 iterations (but no sooner), and P_2 returns to its initial position in q_2 iterations (but no sooner); thus, both points return to their initial positions in any number of iterations that is a multiple of both q_1 and q_2. In general, for a pixel map with p^2 pixel points of the form $(m/p, n/p)$, we let $\Pi(p)$ denote the least common multiple of the periods of all the pixel points in the map [i.e., $\Pi(p)$ is the smallest integer that is divisible by all of the periods]. It follows that the pixel map will return to its initial configuration in $\Pi(p)$ iterations of Arnold's cat map (but no sooner). For this reason, we call $\Pi(p)$ the *period of the pixel map*. In Exercise 4 we ask the reader to show that if $p = 101$, then all pixel points have period 1, 5, or 25, so that $\Pi(101) = 25$. This explains why the cat in Figure 11.15.3 returned to its initial configuration in 25 iterations.

Figure 11.15.7 shows how the period of a pixel map varies with p. While the general tendency is for the period to increase as p increases, there is a surprising amount of irregularity in the graph. Indeed, there is no simple function that specifies this relationship (see Exercise 1).

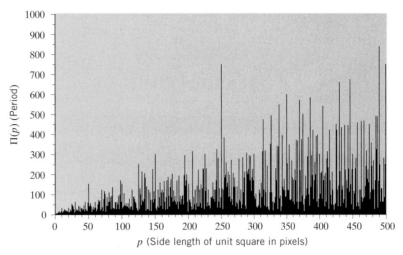

Figure 11.15.7

   Although a pixel map with p pixels on a side does not return to its initial configuration until $\Pi(p)$ iterations have occurred, various unexpected things can occur at intermediate iterations. For example, Figure 11.15.8 shows a pixel map with $p = 250$ of the famous Hungarian-American mathematician John von Neumann. It can be shown that $\Pi(250) = 750$, so that the pixel map will return to its initial configuration after 750 iterations of Arnold's cat map (but no sooner). However, after 375 iterations, the pixel map is turned upside down, and after another 375 iterations (for a total of 750) the pixel map is returned to its initial configuration. Moreover, there are so many pixel points with periods that

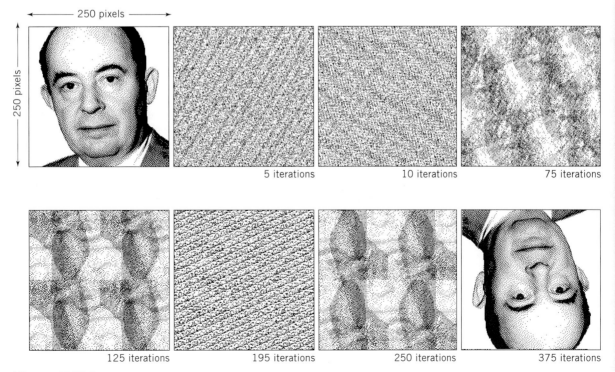

Figure 11.15.8

divide 750 that multiple ghostlike images of the original likeness occur at intermediate iterations; at 195 iterations numerous miniatures of the original likeness occur in diagonal rows.

The Tiled Plane   Our next objective is to explain the cause of the linear streaks that occur in Figure 11.15.3. For this purpose it will be helpful to view Arnold's cat map another way. As defined, Arnold's cat map is not a linear transformation because of the mod 1 arithmetic. However, there is an alternative way of defining Arnold's cat map that avoids the mod 1 arithmetic and results in a linear transformation. For this purpose imagine that the unit square S with its picture of the cat is a "tile" and suppose that the entire plane is covered with such tiles, as in Figure 11.15.9. We say that the xy-plane has been **tiled** with the unit square. If we apply the matrix transformation in (1) to the entire tiled plane without performing the mod 1 arithmetic, then it can be shown that the portion of the image within S will be identical to the image that we obtained using the mod 1 arithmetic (Figure 11.15.9). In short, the tiling results in the same pixel map in S as the mod 1 arithmetic, but in the tiled case Arnold's cat map is a linear transformation.

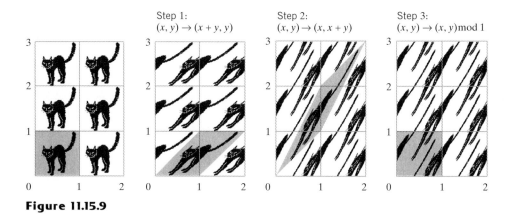

Figure 11.15.9

It is important to understand, however, that tiling and mod 1 arithmetic produce periodicity in different ways. If a pixel map in S has period n, then in the case of mod 1 arithmetic each point returns to its original position at the end of n iterations. In the case of tiling, points need not return to their original positions; rather, each point is replaced by a point of the same color at the end of n iterations.

Properties of Arnold's Cat Map   To understand the cause of the streaks in Figure 11.15.3, think of Arnold's cat map as a linear transformation on the tiled plane. Observe that the matrix

$$C = \begin{bmatrix} 1 & 1 \\ 1 & 2 \end{bmatrix}$$

that defines Arnold's cat map is symmetric and has a determinant of 1. The fact that the determinant is 1 means that multiplication by this matrix preserves areas; that is, the area of any figure in the plane and the area of its image are the same. This is also true for figures in S in the case of mod 1 arithmetic, since the effect of the mod 1 arithmetic is to cut up the figure and reassemble the pieces without any overlap, as shown in Figure 11.15.1d. Thus, in Figure 11.15.3 the area of the cat (whatever it is) is the same as the total area of the blotches in each iteration.

The fact that the matrix is symmetric means that its eigenvalues are real and the corresponding eigenvectors are perpendicular. We leave it for the reader to show that the eigenvalues and corresponding eigenvectors of C are

$$\lambda_1 = \frac{3 + \sqrt{5}}{2} = 2.6180\ldots, \qquad \lambda_2 = \frac{3 - \sqrt{5}}{2} = 0.3819\ldots,$$

$$\mathbf{v}_1 = \begin{bmatrix} 1 \\ \dfrac{1 + \sqrt{5}}{2} \end{bmatrix} = \begin{bmatrix} 1 \\ 1.6180\ldots \end{bmatrix}, \qquad \mathbf{v}_2 = \begin{bmatrix} \dfrac{-1 - \sqrt{5}}{2} \\ 1 \end{bmatrix} = \begin{bmatrix} -1.6180\ldots \\ 1 \end{bmatrix}$$

For each application of Arnold's cat map the eigenvalue λ_1 causes a stretching in the direction of the eigenvector $\mathbf{v}_1$ by a factor of $2.6180\ldots$, and the eigenvalue λ_2 causes a compression in the direction of the eigenvector $\mathbf{v}_2$ by a factor of $0.3819\ldots$. Figure 11.15.10 shows a square centered at the origin whose sides are parallel to the two eigenvector directions. Under the above mapping this square is deformed into the rectangle whose sides are also parallel to the two eigenvector directions. The area of the square and rectangle are the same.

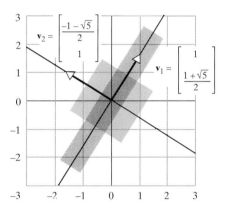

Figure 11.15.10

To explain the cause of the streaks in Figure 11.15.3, consider S to be part of the tiled plane, and let $\mathbf{p}$ be a point of S with period n. Because we are considering tiling, there is a point $\mathbf{q}$ in the plane with the same color as $\mathbf{p}$ that on successive iterations moves toward the position initially occupied by $\mathbf{p}$, reaching that position on the nth iteration. This point is $\mathbf{q} = (A^{-1})^n \mathbf{p} = A^{-n} \mathbf{p}$, since

$$A^n \mathbf{q} = A^n (A^{-n} \mathbf{p}) = \mathbf{p}$$

Thus, with successive iterations, points of S flow away from their initial positions, while at the same time other points in the plane (with corresponding colors) flow toward those initial positions, completing their trip on the final iteration of the cycle. Figure 11.15.11 illustrates this in the case where $n = 4$, $\mathbf{q} = \left(-\frac{8}{3}, \frac{5}{3}\right)$, and $\mathbf{p} = A^4 \mathbf{q} = \left(\frac{1}{3}, \frac{2}{3}\right)$. Notice that $\mathbf{p} \bmod 1 = \mathbf{q} \bmod 1 = \left(\frac{1}{3}, \frac{2}{3}\right)$, so both points occupy the same positions on their respective tiles. The outgoing point moves in the general direction of the eigenvector $\mathbf{v}_1$, as indicated by the arrows in Figure 11.15.11, and the incoming point moves in the general direction of eigenvector $\mathbf{v}_2$. It is the "flow lines" in the general directions of the eigenvectors that form the streaks in Figure 11.15.3.

Nonperiodic Points

Thus far we have considered the effect of Arnold's cat map on pixel points of the form $(m/p, n/p)$ for an arbitrary positive integer p. We know that all such points are periodic. We now consider the effect of Arnold's cat map on an

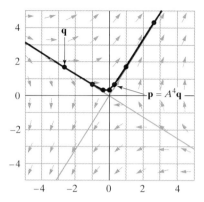

Figure 11.15.11

arbitrary point (a, b) in S. We classify such points as rational if the coordinates a and b are both rational numbers, and irrational if at least one of the coordinates is irrational. Every rational point is periodic, since it is a pixel point for a suitable choice of p. For example, the rational point $(r_1/s_1, r_2/s_2)$ can be written as $(r_1 s_2/s_1 s_2, r_2 s_1/s_1 s_2)$, so it is a pixel point with $p = s_1 s_2$. It can be shown (Exercise 13) that the converse is also true: Every periodic point must be a rational point.

It follows from the preceding discussion that the irrational points in S are nonperiodic, so that successive iterates of an irrational point (x_0, y_0) in S must all be distinct points in S. Figure 11.15.12, which was computer generated, shows an irrational point and selected iterates up to 100,000. For the particular irrational point that we selected, the iterates do not seem to cluster in any particular region of S; rather, they appear to be spread throughout S, becoming denser with successive iterations.

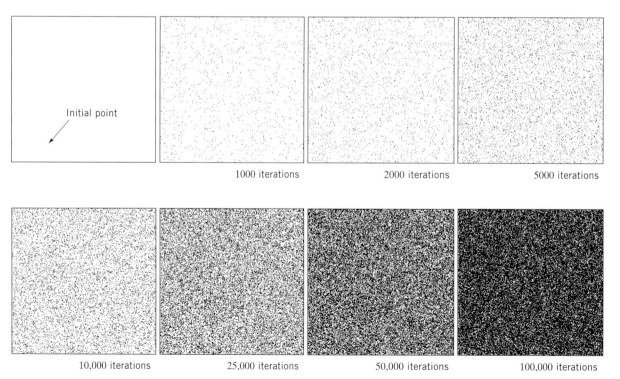

Figure 11.15.12

The behavior of the iterates in Figure 11.15.12 is sufficiently important that there is some terminology associated with it. We say that a set D of points in S is **dense in S** if every circle centered at any point of S encloses points of D, no matter how small the radius of the circle is taken (Figure 11.15.13). It can be shown that the rational points are dense in S and the iterates of most (but not all) of the irrational points are dense in S.

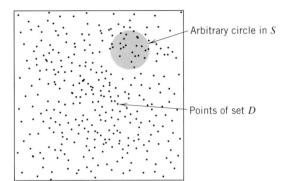

Arbitrary circle in S

Points of set D

Figure 11.15.13

Definition of Chaos   We know that under Arnold's cat map the rational points of S are periodic and dense in S and that some but not all of the irrational points have iterates that are dense in S. These are the basic ingredients of chaos. There are several definitions of chaos in current use, but the following one, which is an outgrowth of a definition introduced by Robert L. Devaney in 1986 in his book *An Introduction to Chaotic Dynamical Systems* (Benjamin/Cummings Publishing Co., Inc.), is most closely related to our work.

Definition

A mapping T of S onto itself is said to be **chaotic** if:

(i) S contains a dense set of periodic points of the mapping T.
(ii) There is a point in S whose iterates under T are dense in S.

Thus Arnold's cat map satisfies the definition of a chaotic mapping. What is noteworthy about this definition is that a chaotic mapping exhibits an element of order and an element of disorder—the periodic points move regularly in cycles, but the points with dense iterates move irregularly, often obscuring the regularity of the periodic points. This fusion of order and disorder characterizes chaotic mappings.

Dynamical Systems   Chaotic mappings arise in the study of **dynamical systems**. Informally stated, a dynamical system can be viewed as a system that has a specific state or configuration at each point of time, but that changes its state with time. Chemical systems, ecological systems, electrical systems, biological systems, economic systems, and so forth, can be looked at in this way. In a **discrete-time dynamical system**, the state changes at discrete points of time rather than at each instant. In a **discrete-time chaotic dynamical system** each state results from a chaotic mapping of the preceding state. For example, if one imagines that Arnold's cat map is applied at discrete points of time, then the pixel maps in Figure 11.15.3 can be viewed as the evolution of a discrete-time chaotic dynamical system from some initial set of states (each point of the cat is a single initial state) to successive sets of states.

One of the fundamental problems in the study of dynamical systems is to predict future states of the system from a known initial state. In practice, however, the exact initial state is rarely known because of errors in the devices used to measure the initial state. It was believed at one time that if the measuring devices were sufficiently accurate and the computers used to perform the iteration were sufficiently powerful, then one could predict the future states of the system to any degree of accuracy. But the discovery of chaotic systems shattered this belief because it was found that for such systems the slightest error in measuring the initial state or in the computation of the iterates becomes magnified exponentially, thereby preventing an accurate prediction of future states. Let us demonstrate this *sensitivity to initial conditions* with Arnold's cat map.

Suppose that P_0 is a point in the xy-plane whose exact coordinates are $(0.77837, 0.70904)$. A measurement error of 0.00001 is made in the y-coordinate, so that the point is thought to be located at $(0.77837, 0.70905)$, which we denote by Q_0. Both P_0 and Q_0 are pixel points with $p = 100,000$ (why?), and thus, since $\Pi(100,000) = 75,000$, both return to their initial positions after 75,000 iterations. In Figure 11.15.14 we show the first 50 iterates of P_0 under Arnold's cat map as crosses and the first 50 iterates of Q_0 as circles. Although P_0 and Q_0 are close enough that their symbols overlap initially, only their first eight iterates have overlapping symbols; from the ninth iteration on their iterates follow divergent paths.

It is possible to quantify the growth of the error from the eigenvalues and eigenvectors of Arnold's cat map. For this purpose we shall think of Arnold's cat map as a linear transformation on the tiled plane. Recall from Figure 11.15.10 and the related discussion that the projected distance between two points in S in the direction of the eigenvector $\mathbf{v}_1$ increases by a factor of $2.6180\ldots(= \lambda_1)$ with each iteration (Figure 11.15.15). After nine iterations this projected distance increases by a factor of $(2.6180\ldots)^9 = 5777.99\ldots$, and with an initial error of roughly $1/100,000$ in the direction of $\mathbf{v}_1$, this distance is $0.05777\ldots$ or about $\frac{1}{17}$ the width of the unit square S. After 12 iterations this small initial error grows to $(2.6180\ldots)^{12}/100,000 = 1.0368\ldots$, which is greater than the width of S. Thus, we lose complete track of the true iterates within S after 12 iterations because of the exponential growth of the initial error.

Although sensitivity to initial conditions limits the ability to predict the future evolution of dynamical systems, new techniques are presently being investigated to describe this future evolution in alternative ways.

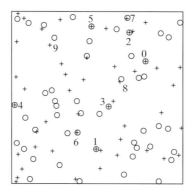

Figure 11.15.14

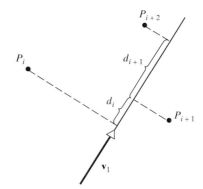

Figure 11.15.15

Exercise Set 11.15

1. In a journal article [F. J. Dyson and H. Falk, "Period of a Discrete Cat Mapping," *The American Mathematical Monthly*, vol. 99, August–September 1992, pp. 603–614] the following results concerning the nature of the function $\Pi(p)$ were established:

   (i) $\Pi(p) = 3p$ if and only if $p = 2 \cdot 5^k$ for $k = 1, 2, \ldots$.
   (ii) $\Pi(p) = 2p$ if and only if $p = 5^k$ for $k = 1, 2, \ldots$ or $p = 6 \cdot 5^k$ for $k = 0, 1, 2, \ldots$.
   (iii) $\Pi(p) \leq 12p/7$ for all other choices of p.

   Find $\Pi(250)$, $\Pi(25)$, $\Pi(125)$, $\Pi(30)$, $\Pi(10)$, $\Pi(50)$, $\Pi(3750)$, $\Pi(6)$, and $\Pi(5)$.

2. Find all the n-cycles that are subsets of the 36 points in S of the form $(m/6, n/6)$ with m and n in the range $0, 1, 2, 3, 4, 5$. Then find $\Pi(6)$.

3. (***Fibonacci Shift-Register Random-Number Generator***) A well-known method of generating a sequence of "pseudorandom" integers $x_0, x_1, x_2, x_3, \ldots$ in the interval from 0 to $p - 1$ is based on the following algorithm:

   (i) Pick any two integers x_0 and x_1 from the range $0, 1, 2, \ldots, p - 1$.
   (ii) Set $x_{n+1} = (x_n + x_{n-1}) \bmod p$ for $n = 1, 2, \ldots$.

   Here $x \bmod p$ denotes the number in the interval from 0 to $p - 1$ that differs from x by a multiple of p. For example, $35 \bmod 9 = 8$ (because $8 = 35 - 3 \cdot 9$); $36 \bmod 9 = 0$ (because $0 = 36 - 4 \cdot 9$); and $-3 \bmod 9 = 6$ (because $6 = -3 + 1 \cdot 9$).

   (a) Generate the sequence of pseudorandom numbers that results from the choices $p = 15$, $x_0 = 3$, and $x_1 = 7$ until the sequence starts repeating.
   (b) Show that the following formula is equivalent to step (ii) of the algorithm:

   $$\begin{bmatrix} x_{n+1} \\ x_{n+2} \end{bmatrix} = \begin{bmatrix} 1 & 1 \\ 1 & 2 \end{bmatrix} \begin{bmatrix} x_{n-1} \\ x_n \end{bmatrix} \bmod p \quad \text{for } n = 1, 2, 3, \ldots$$

   (c) Use the formula in part (b) to generate the sequence of vectors for the choices $p = 21$, $x_0 = 5$, and $x_1 = 5$ until the sequence starts repeating.

   REMARK. If we take $p = 1$ and pick x_0 and x_1 from the interval $[0, 1)$, then the above random-number generator produces pseudorandom numbers in the interval $[0, 1)$. The resulting scheme is precisely Arnold's cat map. Furthermore, if we eliminate the modular arithmetic in the algorithm and take $x_0 = x_1 = 1$, then the resulting sequence of integers is the famous Fibonacci sequence, $1, 1, 2, 3, 5, 8, 13, 21, 34, 55, 89, \ldots$, in which each number after the first two is the sum of the preceding two numbers.

4. For $C = \begin{bmatrix} 1 & 1 \\ 1 & 2 \end{bmatrix}$, it can be verified that

   $$C^{25} = \begin{bmatrix} 7{,}778{,}742{,}049 & 12{,}586{,}269{,}025 \\ 12{,}586{,}269{,}025 & 20{,}365{,}011{,}074 \end{bmatrix}$$

   It can also be verified that $12{,}586{,}269{,}025$ is divisible by 101 and that when $7{,}778{,}742{,}049$ and $20{,}365{,}011{,}074$ are divided by 101 the remainder is 1.

   (a) Show that every point in S of the form $(m/101, n/101)$ returns to its starting position after 25 iterations under Arnold's cat map.
   (b) Show that every point in S of the form $(m/101, n/101)$ has period 1, 5, or 25.
   (c) Show that the point $\left(\frac{1}{101}, 0\right)$ has period greater than 5 by iterating it five times.
   (d) Show that $\Pi(101) = 25$.

5. Show that for the mapping $T: S \to S$ defined by $T(x, y) = \left(x + \frac{5}{12}, y\right) \bmod 1$, every point in S is a periodic point. Why does this show that the mapping is not chaotic?

6. An *Anosov automorphism* on R^2 is a mapping from the unit square S onto S of the form

$$\begin{bmatrix} x \\ y \end{bmatrix} \rightarrow \begin{bmatrix} a & b \\ c & d \end{bmatrix} \begin{bmatrix} x \\ y \end{bmatrix} \bmod 1$$

in which (i) a, b, c, and d are integers, (ii) the determinant of the matrix is ± 1, and (iii) the eigenvalues of the matrix do not have magnitude 1. It can be shown that all Anosov automorphisms are chaotic mappings.

(a) Show that Arnold's cat map is an Anosov automorphism.

(b) Which of the following are the matrices of an Anosov automorphism?

$$\begin{bmatrix} 0 & 1 \\ 1 & 0 \end{bmatrix}, \quad \begin{bmatrix} 3 & 2 \\ 1 & 1 \end{bmatrix}, \quad \begin{bmatrix} 1 & 0 \\ 0 & 1 \end{bmatrix}, \quad \begin{bmatrix} 5 & 7 \\ 2 & 3 \end{bmatrix}, \quad \begin{bmatrix} 6 & 2 \\ 5 & 2 \end{bmatrix}$$

(c) Show that the following mapping of S onto S is not an Anosov automorphism.

$$\begin{bmatrix} x \\ y \end{bmatrix} \rightarrow \begin{bmatrix} 0 & 1 \\ -1 & 0 \end{bmatrix} \begin{bmatrix} x \\ y \end{bmatrix} \bmod 1$$

What is the geometric effect of this transformation on S? Use your observation to show that the mapping is not a chaotic mapping by showing that all points in S are periodic points.

7. Show that Arnold's cat map is one-to-one over the unit square S and that its range is S.

8. Show that the inverse of Arnold's cat map is given by

$$\Gamma^{-1}(x, y) = (2x - y, -x + y) \bmod 1$$

9. Show that the unit square S can be partitioned into four triangular regions on each of which Arnold's cat map is a transformation of the form

$$\begin{bmatrix} x \\ y \end{bmatrix} \rightarrow \begin{bmatrix} 1 & 1 \\ 1 & 2 \end{bmatrix} \begin{bmatrix} x \\ y \end{bmatrix} + \begin{bmatrix} a \\ b \end{bmatrix}$$

where a and b need not be the same for each region. [***Hint.*** Find the regions in S that map onto the four shaded regions of the parallelogram in Figure 11.15.1d.]

10. If (x_0, y_0) is a point in S and (x_n, y_n) is its nth iterate under Arnold's cat map, show that

$$\begin{bmatrix} x_n \\ y_n \end{bmatrix} = \begin{bmatrix} 1 & 1 \\ 1 & 2 \end{bmatrix}^n \begin{bmatrix} x_0 \\ y_0 \end{bmatrix} \bmod 1$$

This result implies that the modular arithmetic need only be performed once rather than after each iteration.

11. Show that $(0, 0)$ is the only fixed point of Arnold's cat map by showing that the only solution of the equation

$$\begin{bmatrix} x_0 \\ y_0 \end{bmatrix} = \begin{bmatrix} 1 & 1 \\ 1 & 2 \end{bmatrix} \begin{bmatrix} x_0 \\ y_0 \end{bmatrix} \bmod 1$$

with $0 \le x_0 < 1$ and $0 \le y_0 < 1$ is $x_0 = y_0 = 0$. ***Hint.*** For appropriate nonnegative integers, r and s, the preceding equation can be written as

$$\begin{bmatrix} x_0 \\ y_0 \end{bmatrix} = \begin{bmatrix} 1 & 1 \\ 1 & 2 \end{bmatrix} \begin{bmatrix} x_0 \\ y_0 \end{bmatrix} - \begin{bmatrix} r \\ s \end{bmatrix}$$

12. Find all 2-cycles of Arnold's cat map by finding all solutions of the equation

$$\begin{bmatrix} x_0 \\ y_0 \end{bmatrix} = \begin{bmatrix} 1 & 1 \\ 1 & 2 \end{bmatrix}^2 \begin{bmatrix} x_0 \\ y_0 \end{bmatrix} \bmod 1$$

with $0 \le x_0 < 1$ and $0 \le y_0 < 1$. ***Hint.*** For appropriate nonnegative integers, r and s, the preceding equation can be written as

$$\begin{bmatrix} x_0 \\ y_0 \end{bmatrix} = \begin{bmatrix} 2 & 3 \\ 3 & 5 \end{bmatrix} \begin{bmatrix} x_0 \\ y_0 \end{bmatrix} - \begin{bmatrix} r \\ s \end{bmatrix}$$

13. Show that every periodic point of Arnold's cat map must be a rational point by showing that for all solutions of the equation

$$\begin{bmatrix} x_0 \\ y_0 \end{bmatrix} = \begin{bmatrix} 1 & 1 \\ 1 & 2 \end{bmatrix}^n \begin{bmatrix} x_0 \\ y_0 \end{bmatrix} \bmod 1$$

the numbers x_0 and y_0 are quotients of integers.

Technology Exercises 11.15

The following exercises are designed to be solved using a technology utility. Typically, this will be MATLAB, *Mathematica*, Maple, Derive, or Mathcad, but it may also be some other type of linear algebra software or a scientific calculator with some linear algebra capabilities. For each exercise you will need to read the relevant documentation for the particular utility you are using. The goal of these exercises is to provide you with a basic proficiency with your technology utility. Once you have mastered the techniques in these exercises, you will be able to use your technology utility to solve many of the problems in the regular exercise sets.

T1. The methods of Exercise 4 show that for the cat map $\Pi(p)$ is the smallest integer satisfying the equation

$$\begin{bmatrix} 1 & 1 \\ 1 & 2 \end{bmatrix}^{\Pi(p)} \bmod p = \begin{bmatrix} 1 & 0 \\ 0 & 1 \end{bmatrix}$$

This suggests that one way to determine $\Pi(p)$ is to compute

$$\begin{bmatrix} 1 & 1 \\ 1 & 2 \end{bmatrix}^n \bmod p$$

starting with $n = 1$ and stopping when this produces the identity matrix. Use this idea to compute $\Pi(p)$ for $p = 2, 3, \ldots, 10$. Compare your results to the formulas given in Exercise 1, if they apply. What can you conjecture about

$$\begin{bmatrix} 1 & 1 \\ 1 & 2 \end{bmatrix}^{\frac{1}{2}\Pi(p)} \bmod p$$

when $\Pi(p)$ is even?

T2. The eigenvalues and eigenvectors for the cat map matrix

$$C = \begin{bmatrix} 1 & 1 \\ 1 & 2 \end{bmatrix}$$

are

$$\lambda_1 = \frac{3 + \sqrt{5}}{2}, \qquad \lambda_2 = \frac{3 - \sqrt{5}}{2},$$

$$\mathbf{v}_1 = \begin{bmatrix} 1 \\ \dfrac{1 + \sqrt{5}}{2} \end{bmatrix}, \qquad \mathbf{v}_2 = \begin{bmatrix} 1 \\ \dfrac{1 - \sqrt{5}}{2} \end{bmatrix}$$

Using these eigenvalues and eigenvectors we can define

$$D = \begin{bmatrix} \dfrac{3 + \sqrt{5}}{2} & 0 \\ 0 & \dfrac{3 - \sqrt{5}}{2} \end{bmatrix} \quad \text{and} \quad P = \begin{bmatrix} 1 & 1 \\ \dfrac{1 + \sqrt{5}}{2} & \dfrac{1 - \sqrt{5}}{2} \end{bmatrix}$$

and write $C = PDP^{-1}$; hence, $C^n = PD^nP^{-1}$. Use a computer to show that

$$C^n = \begin{bmatrix} c_{11}^{(n)} & c_{12}^{(n)} \\ c_{21}^{(n)} & c_{22}^{(n)} \end{bmatrix}$$

where

$$c_{11}^{(n)} = \left(\frac{1 + \sqrt{5}}{2\sqrt{5}} \right) \left(\frac{3 - \sqrt{5}}{2} \right)^n - \left(\frac{1 - \sqrt{5}}{2\sqrt{5}} \right) \left(\frac{3 + \sqrt{5}}{2} \right)^n$$

$$c_{22}^{(n)} = \left(\frac{1 + \sqrt{5}}{2\sqrt{5}} \right) \left(\frac{3 + \sqrt{5}}{2} \right)^n - \left(\frac{1 - \sqrt{5}}{2\sqrt{5}} \right) \left(\frac{3 - \sqrt{5}}{2} \right)^n$$

and

$$c_{12}^{(n)} = c_{21}^{(n)} = \frac{1}{\sqrt{5}} \left\{ \left(\frac{3 + \sqrt{5}}{2} \right)^n - \left(\frac{3 - \sqrt{5}}{2} \right)^n \right\}$$

How can you use these results and your conclusions in Exercise T1 to simplify the method for computing $\Pi(p)$?

11.16 CRYPTOGRAPHY

In this section we present a method of encoding and decoding messages. We also examine modular arithmetic and show how Gaussian elimination can sometimes be used to break an opponent's code.

> PREREQUISITES: Matrices
> Gaussian Elimination
> Matrix Operations
> Linear Independence
> Linear Transformations (Sections 8.1 and 8.2)

Ciphers The study of encoding and decoding secret messages is called ***cryptography***. Although secret codes date to the earliest days of written communication, there has been a recent surge of interest in the subject because of the need to maintain the privacy of information transmitted over public lines of communication. In the language of cryptography, codes are called ***ciphers***, uncoded messages are called ***plaintext***, and coded messages are called ***ciphertext***. The process of converting from plaintext to ciphertext is called ***enciphering***, and the reverse process of converting from ciphertext to plaintext is called ***deciphering***.

The simplest ciphers, called ***substitution ciphers***, are those which replace each letter of the alphabet by a different letter. For example, in the substitution cipher

Plain   A B C D E F G H I J K L M N O P Q R S T U V W X Y Z
Cipher D E F G H I J K L M N O P Q R S T U V W X Y Z A B C

the plaintext letter *A* is replaced by *D*, the plaintext letter *B* by *E*, and so forth. With this cipher the plaintext message

ROME WAS NOT BUILT IN A DAY

becomes

URPH ZDV QRW EXLOW LQ D GDB

Hill Ciphers   A disadvantage of substitution ciphers is that they preserve the frequencies of individual letters, making it relatively easy to break the code by statistical methods. One way to overcome this problem is to divide the plaintext into groups of letters and encipher the plaintext group by group, rather than one letter at a time. A system of cryptography in which the plaintext is divided into sets of *n* letters, each of which is replaced by a set of *n* cipher letters, is called a ***polygraphic system***. In this section we will study a class of polygraphic systems based on matrix transformations. (The ciphers that we will discuss are called ***Hill ciphers*** after Lester S. Hill, who introduced them in two papers: "Cryptography in an Algebraic Alphabet," *American Mathematical Monthly, 36*, June–July 1929, pp. 306–312; and "Concerning Certain Linear Transformation Apparatus of Cryptography," *American Mathematical Monthly, 38*, March 1931, pp. 135–154.)

In the discussion to follow we assume that each plaintext and ciphertext letter except *Z* is assigned the numerical value that specifies its position in the standard alphabet (Table 1). For reasons that will become clear later, *Z* is assigned a value of zero.

TABLE 1

A	*B*	*C*	*D*	*E*	*F*	*G*	*H*	*I*	*J*	*K*	*L*	*M*	*N*	*O*	*P*	*Q*	*R*	*S*	*T*	*U*	*V*	*W*	*X*	*Y*	*Z*
1	2	3	4	5	6	7	8	9	10	11	12	13	14	15	16	17	18	19	20	21	22	23	24	25	0

In the simplest Hill ciphers, successive *pairs* of plaintext are transformed into ciphertext by the following procedure:

Step 1. Choose a 2×2 matrix with integer entries

$$A = \begin{bmatrix} a_{11} & a_{12} \\ a_{21} & a_{22} \end{bmatrix}$$

to perform the encoding. Certain additional conditions on *A* will be imposed later.

Step 2. Group successive plaintext letters into pairs, adding an arbitrary "dummy" letter to fill out the last pair if the plaintext has an odd number of letters, and replace each plaintext letter by its numerical value.

Step 3. Successively convert each plaintext pair $p_1 p_2$ into a column vector

$$\mathbf{p} = \begin{bmatrix} p_1 \\ p_2 \end{bmatrix}$$

and form the product $A\mathbf{p}$. We will call $\mathbf{p}$ a ***plaintext vector*** and $A\mathbf{p}$ the corresponding ***ciphertext vector***.

Step 4. Convert each ciphertext vector into its alphabetic equivalent.

EXAMPLE 1 Hill Cipher of a Message

Use the matrix

$$\begin{bmatrix} 1 & 2 \\ 0 & 3 \end{bmatrix}$$

to obtain the Hill cipher for the plaintext message

I AM HIDING

Solution.

If we group the plaintext into pairs and add the dummy letter *G* to fill out the last pair, we obtain

IA   MH   ID   IN   GG

or equivalently from Table 1,

9 1   13 8   9 4   9 14   7 7

To encipher the pair *IA*, we form the matrix product

$$\begin{bmatrix} 1 & 2 \\ 0 & 3 \end{bmatrix} \begin{bmatrix} 9 \\ 1 \end{bmatrix} = \begin{bmatrix} 11 \\ 3 \end{bmatrix}$$

which, from Table 1, yields the ciphertext *KC*.

To encipher the pair *MH* we form the product

$$\begin{bmatrix} 1 & 2 \\ 0 & 3 \end{bmatrix} \begin{bmatrix} 13 \\ 8 \end{bmatrix} = \begin{bmatrix} 29 \\ 24 \end{bmatrix} \tag{1}$$

However, there is a problem here, because the number 29 has no alphabet equivalent (Table 1). To resolve this problem we make the following agreement:

> *Whenever an integer greater than* 25 *occurs, it will be replaced by the remainder that results when this integer is divided by* 26.

Because the remainder after division by 26 is one of the integers 0, 1, 2, . . . , 25, this procedure will always yield an integer with an alphabet equivalent.

Thus, in (1) we replace 29 by 3, which is the remainder after dividing 29 by 26. It now follows from Table 1 that the ciphertext for the pair *MH* is *CX*.

The computations for the remaining ciphertext vectors are

$$\begin{bmatrix} 1 & 2 \\ 0 & 3 \end{bmatrix} \begin{bmatrix} 9 \\ 4 \end{bmatrix} = \begin{bmatrix} 17 \\ 12 \end{bmatrix}$$

$$\begin{bmatrix} 1 & 2 \\ 0 & 3 \end{bmatrix} \begin{bmatrix} 9 \\ 14 \end{bmatrix} = \begin{bmatrix} 37 \\ 42 \end{bmatrix} \quad \text{or} \quad \begin{bmatrix} 11 \\ 16 \end{bmatrix}$$

$$\begin{bmatrix} 1 & 2 \\ 0 & 3 \end{bmatrix} \begin{bmatrix} 7 \\ 7 \end{bmatrix} = \begin{bmatrix} 21 \\ 21 \end{bmatrix}$$

These correspond to the ciphertext pairs *QL*, *KP*, and *UU*, respectively. In summary, the entire ciphertext message is

KC   CX   QL   KP   UU

which would usually be transmitted as a single string without spaces:

KCCXQLKPUU ◆

Because the plaintext was grouped in pairs and enciphered by a 2×2 matrix, the Hill cipher in Example 1 is referred to as a ***Hill 2-cipher***. It is obviously also possible to group the plaintext in triples and encipher by a 3×3 matrix with integer entries; this is called a ***Hill 3-cipher***. In general, for a ***Hill n-cipher***, plaintext is grouped into sets of n letters and enciphered by an $n \times n$ matrix with integer entries.

Modular Arithmetic   In Example 1 integers greater than 25 were replaced by their remainders after division by 26. This technique of working with remainders is at the core of a body of mathematics called *modular arithmetic*. Because of its importance in cryptography, we will digress for a moment to touch on some of the main ideas in this area.

In modular arithmetic we are given a positive integer m, called the ***modulus***, and any two integers whose difference is an integer multiple of the modulus are regarded as "equal" or "equivalent" with respect to the modulus. More precisely, we make the following definition.

Definition

If m is a positive integer and a and b are any integers, then we say that a is ***equivalent*** to b modulo m, written

$$a = b \quad (\mathrm{mod}\ m)$$

if $a - b$ is an integer multiple of m.

EXAMPLE 2 Various Equivalences

$$7 = 2 \quad (\mathrm{mod}\ 5)$$
$$19 = 3 \quad (\mathrm{mod}\ 2)$$
$$-1 = 25 \quad (\mathrm{mod}\ 26)$$
$$12 = 0 \quad (\mathrm{mod}\ 4)$$

◆

For any modulus m it can be proved that every integer a is equivalent, modulo m, to exactly one of the integers

$$0, 1, 2, \ldots, m - 1$$

We call this integer the ***residue*** of a modulo m, and we write

$$Z_m = \{0, 1, 2, \ldots, m - 1\}$$

to denote the set of residues modulo m.

If a is a *nonnegative* integer, then its residue modulo m is simply the remainder that results when a is divided by m. For an arbitrary integer a, the residue can be found using the following theorem.

Theorem 11.16.1

For any integer a and modulus m, let

$$R = remainder\ of\ \frac{|a|}{m}$$

Then the residue r of a modulo m is given by

$$r = \begin{cases} R & \text{if } a \geq 0 \\ m - R & \text{if } a < 0 \ \text{ and } \ R \neq 0 \\ 0 & \text{if } a < 0 \ \text{ and } \ R = 0 \end{cases}$$

EXAMPLE 3 Residues mod 26

Find the residue modulo 26 of (a) 87, (b) -38, and (c) -26.

Solution (a). Dividing $|87| = 87$ by 26 yields a remainder of $R = 9$, so $r = 9$. Thus,

$$87 = 9 \quad (\text{mod } 26)$$

Solution (b). Dividing $|-38| = 38$ by 26 yields a remainder of $R = 12$, so $r = 26 - 12 = 14$. Thus,

$$-38 = 14 \quad (\text{mod } 26)$$

Solution (c). Dividing $|-26| = 26$ by 26 yields a remainder of $R = 0$. Thus,

$$-26 = 0 \quad (\text{mod } 26) \qquad \blacklozenge$$

In ordinary arithmetic every nonzero number a has a *reciprocal* or *multiplicative inverse*, denoted by a^{-1}, such that

$$aa^{-1} = a^{-1}a = 1$$

In modular arithmetic we have the following corresponding concept:

Definition

If a is a number in Z_m, then a number a^{-1} in Z_m is called a ***reciprocal*** or ***multiplicative inverse*** of a modulo m if $aa^{-1} = a^{-1}a = 1$ (mod m).

It can be proved that if a and m have no common prime factors, then a has a unique reciprocal modulo m; conversely, if a and m have a common prime factor, then a has no reciprocal modulo m.

EXAMPLE 4 Reciprocal of 3 mod 26

The number 3 has a reciprocal modulo 26 because 3 and 26 have no common prime factors. This reciprocal can be obtained by finding the number x in Z_{26} that satisfies the modular equation

$$3x = 1 \quad (\text{mod } 26)$$

Although there are general methods for solving such modular equations, it would take us too far afield to study them. However, because 26 is relatively small, this equation can be solved by trying the possible solutions, 0 to 25, one at a time. With this approach we find that $x = 9$ is the solution, as

$$3 \cdot 9 = 27 = 1 \quad (\text{mod } 26)$$

Thus,

$$3^{-1} = 9 \quad (\text{mod } 26) \qquad \blacklozenge$$

EXAMPLE 5 A Number with No Reciprocal mod 26

The number 4 has no reciprocal modulo 26 because 4 and 26 have 2 as a common prime factor (see Exercise 8). $\blacklozenge$

For future reference, we provide the following table of reciprocals modulo 26:

TABLE 2 *Reciprocals Modulo 26*

a	1	3	5	7	9	11	15	17	19	21	23	25
a^{-1}	1	9	21	15	3	19	7	23	11	5	17	25

Deciphering   Every useful cipher must have a procedure for decipherment. In the case of a Hill cipher, decipherment uses the inverse (mod 26) of the enciphering matrix. To be precise, if m is a positive integer, then a square matrix A with entries in Z_m is said to be *invertible modulo m* if there is a matrix B with entries in Z_m such that

$$AB = BA = I \quad (\text{mod } m)$$

Suppose now that

$$A = \begin{bmatrix} a_{11} & a_{12} \\ a_{21} & a_{22} \end{bmatrix}$$

is invertible modulo 26 and this matrix is used in a Hill 2-cipher. If

$$\mathbf{p} = \begin{bmatrix} p_1 \\ p_2 \end{bmatrix}$$

is a plaintext vector, then

$$\mathbf{c} = A\mathbf{p} \quad (\text{mod } 26)$$

is the corresponding ciphertext vector and

$$\mathbf{p} = A^{-1}\mathbf{c} \quad (\text{mod } 26)$$

Thus, each plaintext vector can be recovered from the corresponding ciphertext vector by multiplying it on the left by A^{-1} (mod 26).

In cryptography it is important to know which matrices are invertible modulo 26 and how to obtain their inverses. We now investigate these questions.

In ordinary arithmetic, a square matrix A is invertible if and only if $\det(A) \neq 0$; or equivalently, if and only if $\det(A)$ has a reciprocal. The following theorem is the analog of this result in modular arithmetic.

Theorem 11.16.2

A square matrix A with entries in Z_m is invertible modulo m if and only if the residue of $\det(A)$ modulo m has a reciprocal modulo m.

Because the residue of $\det(A)$ modulo m will have a reciprocal modulo m if and only if this residue and m have no common prime factors, we have the following corollary.

Corollary 11.16.3

A square matrix A with entries in Z_m is invertible modulo m if and only if m and the residue of $\det(A)$ modulo m have no common prime factors.

Because the only prime factors of $m = 26$ are 2 and 13, we have the following corollary, which is useful in cryptography.

Corollary 11.16.4

A square matrix A with entries in Z_{26} is invertible modulo 26 if and only if the residue of $\det(A)$ modulo 26 is not divisible by 2 or 13.

We leave it for the reader to verify that if

$$A = \begin{bmatrix} a & b \\ c & d \end{bmatrix}$$

has entries in Z_{26} and the residue of $\det(A) = ad - bc$ modulo 26 is not divisible by 2 or 13, then the inverse of A (mod 26) is given by

$$A^{-1} = (ad - bc)^{-1} \begin{bmatrix} d & -b \\ -c & a \end{bmatrix} \quad (\bmod\ 26) \tag{2}$$

where $(ad - bc)^{-1}$ is the reciprocal of the residue of $ad - bc$ (mod 26).

EXAMPLE 6 Inverse of a Matrix mod 26

Find the inverse of

$$A = \begin{bmatrix} 5 & 6 \\ 2 & 3 \end{bmatrix}$$

modulo 26.

Solution.

$$\det(A) = ad - bc = 5 \cdot 3 - 6 \cdot 2 = 3$$

so from Table 2

$$(ad - bc)^{-1} = 3^{-1} = 9 \quad (\bmod\ 26)$$

Thus, from (2)

$$A^{-1} = 9 \begin{bmatrix} 3 & -6 \\ -2 & 5 \end{bmatrix} = \begin{bmatrix} 27 & -54 \\ -18 & 45 \end{bmatrix} = \begin{bmatrix} 1 & 24 \\ 8 & 19 \end{bmatrix} \quad (\bmod\ 26)$$

As a check,

$$AA^{-1} = \begin{bmatrix} 5 & 6 \\ 2 & 3 \end{bmatrix} \begin{bmatrix} 1 & 24 \\ 8 & 19 \end{bmatrix} = \begin{bmatrix} 53 & 234 \\ 26 & 105 \end{bmatrix} = \begin{bmatrix} 1 & 0 \\ 0 & 1 \end{bmatrix} \quad (\text{mod } 26)$$

Similarly, $A^{-1}A = I$.   ◆

EXAMPLE 7   Decoding a Hill 2-Cipher

Decode the following Hill 2-cipher, which was enciphered by the matrix in Example 6:

GTNKGKDUSK

Solution.

From Table 1 the numerical equivalent of this ciphertext is

7   20   14   11   7   11   4   21   19   11

To obtain the plaintext pairs, we multiply each ciphertext vector by the inverse of A (obtained in Example 6):

$$\begin{bmatrix} 1 & 24 \\ 8 & 19 \end{bmatrix} \begin{bmatrix} 7 \\ 20 \end{bmatrix} = \begin{bmatrix} 487 \\ 436 \end{bmatrix} = \begin{bmatrix} 19 \\ 20 \end{bmatrix} \quad (\text{mod } 26)$$

$$\begin{bmatrix} 1 & 24 \\ 8 & 19 \end{bmatrix} \begin{bmatrix} 14 \\ 11 \end{bmatrix} = \begin{bmatrix} 278 \\ 321 \end{bmatrix} = \begin{bmatrix} 18 \\ 9 \end{bmatrix} \quad (\text{mod } 26)$$

$$\begin{bmatrix} 1 & 24 \\ 8 & 19 \end{bmatrix} \begin{bmatrix} 7 \\ 11 \end{bmatrix} = \begin{bmatrix} 271 \\ 265 \end{bmatrix} = \begin{bmatrix} 11 \\ 5 \end{bmatrix} \quad (\text{mod } 26)$$

$$\begin{bmatrix} 1 & 24 \\ 8 & 19 \end{bmatrix} \begin{bmatrix} 4 \\ 21 \end{bmatrix} = \begin{bmatrix} 508 \\ 431 \end{bmatrix} = \begin{bmatrix} 14 \\ 15 \end{bmatrix} \quad (\text{mod } 26)$$

$$\begin{bmatrix} 1 & 24 \\ 8 & 19 \end{bmatrix} \begin{bmatrix} 19 \\ 11 \end{bmatrix} = \begin{bmatrix} 283 \\ 361 \end{bmatrix} = \begin{bmatrix} 23 \\ 23 \end{bmatrix} \quad (\text{mod } 26)$$

From Table 1 the alphabet equivalents of these vectors are

ST   RI   KE   NO   WW

which yields the message

STRIKE   NOW   ◆

Breaking a Hill Cipher

Because the purpose of enciphering messages and information is to prevent "opponents" from learning their contents, cryptographers are concerned with the *security* of their ciphers—that is, how readily they can be broken (deciphered by their opponents). We will conclude this section by discussing one technique for breaking Hill ciphers.

Suppose that you are able to obtain some corresponding plaintext and ciphertext from an opponent's message. For example, on examining some intercepted ciphertext, you may be able to deduce that the message is a letter that begins *DEAR SIR*. We will show that with a small amount of such data it may be possible to determine the deciphering matrix of a Hill code and consequently obtain access to the rest of the message.

It is a basic result in linear algebra that a linear transformation is completely determined by its values at a basis. This principle suggests that if we have a Hill *n*-cipher, and if

$$\mathbf{p}_1, \mathbf{p}_2, \ldots, \mathbf{p}_n$$

are linearly independent plaintext vectors whose corresponding ciphertext vectors

$$A\mathbf{p}_1, A\mathbf{p}_2, \ldots, A\mathbf{p}_n$$

are known, then there is enough information available to determine the matrix A, hence $A^{-1} \pmod{m}$.

The following theorem, whose proof is discussed in the exercises, provides a way to do this.

Theorem 11.16.5 **Determining the Deciphering Matrix**

Let $\mathbf{p}_1, \mathbf{p}_2, \ldots, \mathbf{p}_n$ be linearly independent plaintext vectors, and let $\mathbf{c}_1, \mathbf{c}_2, \ldots, \mathbf{c}_n$ be the corresponding ciphertext vectors in a Hill n-cipher. If

$$P = \begin{bmatrix} \mathbf{p}_1^T \\ \mathbf{p}_2^T \\ \vdots \\ \mathbf{p}_n^T \end{bmatrix}$$

is the $n \times n$ matrix with row vectors $\mathbf{p}_1^T, \mathbf{p}_2^T, \ldots, \mathbf{p}_n^T$ and if

$$C = \begin{bmatrix} \mathbf{c}_1^T \\ \mathbf{c}_2^T \\ \vdots \\ \mathbf{c}_n^T \end{bmatrix}$$

is the $n \times n$ matrix with row vectors $\mathbf{c}_1^T, \mathbf{c}_2^T, \ldots, \mathbf{c}_n^T$, then the sequence of elementary row operations that reduces C to I transforms P to $(A^{-1})^T$.

This theorem tells us that to find the transpose of the deciphering matrix A^{-1}, we must find a sequence of row operations that reduces C to I and then perform this same sequence of operations on P. The following example illustrates a simple algorithm for doing this.

EXAMPLE 8 Using Theorem 11.16.5

The following Hill 2-cipher is intercepted:

$$IOSBTGXESPXHOPDE$$

Decipher the message, given that it starts with the word *DEAR*.

Solution.

From Table 1, the numerical equivalent of the known plaintext is

	DE	*AR*
	4 5	1 18

and the numerical equivalent of the corresponding ciphertext is

	IO	*SB*
	9 15	19 2

so the corresponding plaintext and ciphertext vectors are

$$\mathbf{p}_1 = \begin{bmatrix} 4 \\ 5 \end{bmatrix} \leftrightarrow \mathbf{c}_1 = \begin{bmatrix} 9 \\ 15 \end{bmatrix}$$

$$\mathbf{p}_2 = \begin{bmatrix} 1 \\ 18 \end{bmatrix} \leftrightarrow \mathbf{c}_2 = \begin{bmatrix} 19 \\ 2 \end{bmatrix}$$

We want to reduce

$$C = \begin{bmatrix} \mathbf{c}_1^T \\ \mathbf{c}_2^T \end{bmatrix} = \begin{bmatrix} 9 & 15 \\ 19 & 2 \end{bmatrix}$$

to I by elementary row operations and simultaneously apply these operations to

$$P = \begin{bmatrix} \mathbf{p}_1^T \\ \mathbf{p}_2^T \end{bmatrix} = \begin{bmatrix} 4 & 5 \\ 1 & 18 \end{bmatrix}$$

to obtain $(A^{-1})^T$ (the transpose of the deciphering matrix). This can be accomplished by adjoining P to the right of C and applying row operations to the resulting matrix $[C \mid P]$ until the left side is reduced to I. The final matrix will then have the form $[I \mid (A^{-1})^T]$. The computations can be carried out as follows:

$$\begin{bmatrix} 9 & 15 & | & 4 & 5 \\ 19 & 2 & | & 1 & 18 \end{bmatrix} \qquad \longleftarrow \text{We formed the matrix } [C \mid P].$$

$$\begin{bmatrix} 1 & 45 & | & 12 & 15 \\ 19 & 2 & | & 1 & 18 \end{bmatrix} \qquad \longleftarrow \text{We multiplied the first row by } 9^{-1} = 3.$$

$$\begin{bmatrix} 1 & 19 & | & 12 & 15 \\ 19 & 2 & | & 1 & 18 \end{bmatrix} \qquad \longleftarrow \text{We replaced 45 by its residue modulo 26.}$$

$$\begin{bmatrix} 1 & 19 & | & 12 & 15 \\ 0 & -359 & | & -227 & -267 \end{bmatrix} \qquad \longleftarrow \text{We added } -19 \text{ times the first row to the second.}$$

$$\begin{bmatrix} 1 & 19 & | & 12 & 15 \\ 0 & 5 & | & 7 & 19 \end{bmatrix} \qquad \longleftarrow \text{We replaced the entries in the second row by their residues modulo 26.}$$

$$\begin{bmatrix} 1 & 19 & | & 12 & 15 \\ 0 & 1 & | & 147 & 399 \end{bmatrix} \qquad \longleftarrow \text{We multiplied the second row by } 5^{-1} = 21.$$

$$\begin{bmatrix} 1 & 19 & | & 12 & 15 \\ 0 & 1 & | & 17 & 9 \end{bmatrix} \qquad \longleftarrow \text{We replaced the entries in the second row by their residues modulo 26.}$$

$$\begin{bmatrix} 1 & 0 & | & -311 & -156 \\ 0 & 1 & | & 17 & 9 \end{bmatrix} \qquad \longleftarrow \text{We added } -19 \text{ times the second row to the first.}$$

$$\begin{bmatrix} 1 & 0 & | & 1 & 0 \\ 0 & 1 & | & 17 & 9 \end{bmatrix} \qquad \longleftarrow \text{We replaced the entries in the first row by their residues modulo 26.}$$

Thus,

$$(A^{-1})^T = \begin{bmatrix} 1 & 0 \\ 17 & 9 \end{bmatrix}$$

so the deciphering matrix is

$$A^{-1} = \begin{bmatrix} 1 & 17 \\ 0 & 9 \end{bmatrix}$$

To decipher the message, we first group the ciphertext into pairs and find the numerical equivalent of each letter:

IO	SB	TG	XE	SP	XH	OP	DE
9 15	19 2	20 7	24 5	19 16	24 8	15 16	4 5

Next, we multiply successive ciphertext vectors on the left by A^{-1} and find the alphabet equivalents of the resulting plaintext pairs:

$$\begin{bmatrix} 1 & 17 \\ 0 & 9 \end{bmatrix} \begin{bmatrix} 9 \\ 15 \end{bmatrix} = \begin{bmatrix} 4 \\ 5 \end{bmatrix} \quad \begin{matrix} D \\ E \end{matrix}$$

$$\begin{bmatrix} 1 & 17 \\ 0 & 9 \end{bmatrix} \begin{bmatrix} 19 \\ 2 \end{bmatrix} = \begin{bmatrix} 1 \\ 18 \end{bmatrix} \quad \begin{matrix} A \\ R \end{matrix}$$

$$\begin{bmatrix} 1 & 17 \\ 0 & 9 \end{bmatrix} \begin{bmatrix} 20 \\ 7 \end{bmatrix} = \begin{bmatrix} 9 \\ 11 \end{bmatrix} \quad \begin{matrix} I \\ K \end{matrix}$$

$$\begin{bmatrix} 1 & 17 \\ 0 & 9 \end{bmatrix} \begin{bmatrix} 24 \\ 5 \end{bmatrix} = \begin{bmatrix} 5 \\ 19 \end{bmatrix} \quad \begin{matrix} E \\ S \end{matrix} \quad \text{(mod 26)}$$

$$\begin{bmatrix} 1 & 17 \\ 0 & 9 \end{bmatrix} \begin{bmatrix} 19 \\ 16 \end{bmatrix} = \begin{bmatrix} 5 \\ 14 \end{bmatrix} \quad \begin{matrix} E \\ N \end{matrix}$$

$$\begin{bmatrix} 1 & 17 \\ 0 & 9 \end{bmatrix} \begin{bmatrix} 24 \\ 8 \end{bmatrix} = \begin{bmatrix} 4 \\ 20 \end{bmatrix} \quad \begin{matrix} D \\ T \end{matrix}$$

$$\begin{bmatrix} 1 & 17 \\ 0 & 9 \end{bmatrix} \begin{bmatrix} 15 \\ 16 \end{bmatrix} = \begin{bmatrix} 1 \\ 14 \end{bmatrix} \quad \begin{matrix} A \\ N \end{matrix}$$

$$\begin{bmatrix} 1 & 17 \\ 0 & 9 \end{bmatrix} \begin{bmatrix} 4 \\ 5 \end{bmatrix} = \begin{bmatrix} 11 \\ 19 \end{bmatrix} \quad \begin{matrix} K \\ S \end{matrix}$$

Finally, we construct the message from the plaintext pairs:

| DE | AR | IK | ES | EN | DT | AN | KS |

DEAR IKE SEND TANKS ◆

Further Readings

Readers interested in learning more about mathematical cryptography are referred to the following books, the first of which is elementary and the second more advanced.

1. ABRAHAM SINKOV, *Elementary Cryptanalysis, a Mathematical Approach* (Mathematical Association of America, Mathematical Library, 1966).

2. ALAN G. KONHEIM, *Cryptography, a Primer* (New York: Wiley-Interscience, 1981).

Exercise Set 11.16

1. Obtain the Hill cipher of the message

   DARK NIGHT

for each of the following enciphering matrices:

(a) $\begin{bmatrix} 1 & 3 \\ 2 & 1 \end{bmatrix}$ (b) $\begin{bmatrix} 4 & 3 \\ 1 & 2 \end{bmatrix}$

2. In each part determine whether the matrix is invertible modulo 26. If so, find its inverse modulo 26 and check your work by verifying that $AA^{-1} = A^{-1}A = I$ (mod 26).

(a) $A = \begin{bmatrix} 9 & 1 \\ 7 & 2 \end{bmatrix}$ (b) $A = \begin{bmatrix} 3 & 1 \\ 5 & 3 \end{bmatrix}$ (c) $A = \begin{bmatrix} 8 & 11 \\ 1 & 9 \end{bmatrix}$

(d) $A = \begin{bmatrix} 2 & 1 \\ 1 & 7 \end{bmatrix}$ (e) $A = \begin{bmatrix} 3 & 1 \\ 6 & 2 \end{bmatrix}$ (f) $A = \begin{bmatrix} 1 & 8 \\ 1 & 3 \end{bmatrix}$

3. Decode the message

   SAKNOXAOJX

given that it is a Hill cipher with enciphering matrix

$$\begin{bmatrix} 4 & 1 \\ 3 & 2 \end{bmatrix}$$

4. A Hill 2-cipher is intercepted that starts with the pairs

   SL   HK

Find the deciphering and enciphering matrices, given that the plaintext is known to start with the word *ARMY*.

5. Decode the following Hill 2-cipher if the last four plaintext letters are known to be *ATOM*.

   LNGIHGYBVRENJYQO

6. Decode the following Hill 3-cipher if the first nine plaintext letters are *IHAVECOME*:

   HPAFQGGDUGDDHPGODYNOR

7. If, in addition to the standard alphabet, a period, comma, and question mark are allowed, then 29 plaintext and ciphertext symbols are available and all matrix arithmetic would be done modulo 29. Under what conditions would a matrix with entries in Z_{29} be invertible modulo 29?

8. Show than the modular equation $4x = 1$ (mod 26) has no solution in Z_{26} by successively substituting the values $x = 0, 1, 2, \ldots, 25$.

9. (a) Let P and C be the matrices in Theorem 11.16.5. Show that $P = C(A^{-1})^T$.
   (b) To prove Theorem 11.16.5, let $E_1, E_2, \ldots, E_n$ be the elementary matrices that correspond to the row operations that reduce C to I, so

   $$E_n \cdots E_2 E_1 C = I$$

   Show that

   $$E_n \cdots E_2 E_1 P = (A^{-1})^T$$

   from which it follows that the same sequence of row operations that reduces C to I converts P to $(A^{-1})^T$.

10. (a) If A is the enciphering matrix of a Hill n-cipher, show that

    $$A^{-1} = (C^{-1}P)^T \quad (\text{mod } 26)$$

    where C and P are the matrices defined in Theorem 11.16.5.
    (b) Instead of using Theorem 11.16.5 as in the text, find the deciphering matrix A^{-1} of Example 8 by using the result in part (a) and Equation (2) to compute C^{-1}. [**Note.** Although this method is practical for Hill 2-ciphers, Theorem 11.16.5 is more efficient for Hill n-ciphers with $n > 2$.]

Technology Exercises 11.16

The following exercises are designed to be solved using a technology utility. Typically, this will be MATLAB, *Mathematica*, Maple, Derive, or Mathcad, but it may also be some other type of linear algebra software or a scientific calculator with some linear algebra capabilities. For each exercise you will need to read the relevant documentation for the particular utility you are using. The goal of these exercises is to provide you with a basic proficiency with your technology utility. Once you have mastered the techniques in these exercises, you will be able to use your technology utility to solve many of the problems in the regular exercise sets.

T1. Two integers that have no common factors (except 1) are said to be relatively prime. Given a positive integer n, let $S_n = \{a_1, a_2, a_3, \ldots, a_m\}$, where $a_1 < a_2 < a_3 < \cdots < a_m$, be the set of all positive integers less than n and relatively prime to n. For example, if $n = 9$, then

$$S_9 = \{a_1, a_2, a_3, \ldots, a_6\} = \{1, 2, 4, 5, 7, 8\}$$

(a) Construct a table consisting of n and S_n for $n = 2, 3, \ldots, 15$, and then compute

$$\sum_{k=1}^{m} a_k \quad \text{and} \quad \left(\sum_{k=1}^{m} a_k \right) \pmod{n}$$

in each case. Draw a conjecture for $n > 15$ and prove your conjecture to be true. [**Hint.** Use the fact that if a is relatively prime to n, then $n - a$ is also relatively prime to n.]

(b) Given a positive integer n and the set S_n, let P_n be the $m \times m$ matrix

$$P_n = \begin{bmatrix} a_1 & a_2 & a_3 & \cdots & a_{m-1} & a_m \\ a_2 & a_3 & a_4 & \cdots & a_m & a_1 \\ a_3 & a_4 & a_5 & \cdots & a_1 & a_2 \\ \vdots & \vdots & \vdots & \ddots & \vdots & \vdots \\ a_{m-1} & a_m & a_1 & \cdots & a_{m-3} & a_{m-2} \\ a_m & a_1 & a_2 & \cdots & a_{m-2} & a_{m-1} \end{bmatrix}$$

so that, for example,

$$P_9 = \begin{bmatrix} 1 & 2 & 4 & 5 & 7 & 8 \\ 2 & 4 & 5 & 7 & 8 & 1 \\ 4 & 5 & 7 & 8 & 1 & 2 \\ 5 & 7 & 8 & 1 & 2 & 4 \\ 7 & 8 & 1 & 2 & 4 & 5 \\ 8 & 1 & 2 & 4 & 5 & 7 \end{bmatrix}$$

Use a computer to compute $\det(P_n)$ and $\det(P_n) \pmod{n}$ for $n = 2, 3, \ldots, 15$, and then use these results to construct a conjecture.

(c) Use the results of part (a) to prove your conjecture to be true. [**Hint.** Add the first $m - 1$ rows of P_n to its last row and then use Theorem 2.2.3.] What do these results imply about the inverse of $P_n \pmod{n}$?

T2. Given a positive integer n greater than one, the number of positive integers less than n and relatively prime to n is called the **Euler phi function** of n and is denoted by $\varphi(n)$. For example, $\varphi(6) = 2$ since only two positive integers (i.e., 1 and 5) are less than 6 and have no common factor with 6.

(a) Using a computer, for each value of $n = 2, 3, \ldots, 25$ compute and print out all positive integers that are less than n and relatively prime to n. Then use these integers to determine the values of $\varphi(n)$ for $n = 2, 3, \ldots, 25$. Can you discover a pattern in the results?

(b) It can be shown that if $\{p_1, p_2, p_3, \ldots, p_m\}$ are all the distinct prime factors of n, then

$$\varphi(n) = n \left(1 - \frac{1}{p_1}\right) \left(1 - \frac{1}{p_2}\right) \left(1 - \frac{1}{p_3}\right) \cdots \left(1 - \frac{1}{p_m}\right)$$

For example, since $\{2, 3\}$ are the distinct prime factors of 12, we have

$$\varphi(12) = 12 \left(1 - \frac{1}{2}\right) \left(1 - \frac{1}{3}\right) = 4$$

which agrees with the fact that $\{1, 5, 7, 11\}$ are the only positive integers less than 12 and relatively prime to 12. Using a computer, print out all the prime factors of n for $n = 2, 3, \ldots, 25$. Then compute $\varphi(n)$ using the formula above and compare it to your results in part (a).

11.17 GENETICS

In this section we investigate the propagation of an inherited trait in successive generations by computing powers of a matrix.

> PREREQUISITES: Eigenvalues and Eigenvectors
> Diagonalization of a Matrix
> Intuitive Understanding of Limits

Inheritance Traits   In this section we examine the inheritance of traits in animals or plants. The inherited trait under consideration is assumed to be governed by a set of two genes, which we designate by A and a. Under *autosomal inheritance* each individual in the population of either gender possesses two of these genes, the possible pairings being designated AA, Aa, and aa. This pair of genes is called the individual's *genotype*, and it determines how the trait controlled by the genes is manifested in the individual. For example, in snapdragons a set of two genes determines the color of the flower. Genotype AA produces red flowers, genotype Aa produces pink flowers, and genotype aa produces white flowers. In humans, eye coloration is controlled through autosomal inheritance. Genotypes AA and Aa have brown eyes, and genotype aa has blue eyes. In this case we say that gene A *dominates* gene a, or that gene a is *recessive* to gene A, because genotype Aa has the same outward trait as genotype AA.

In addition to autosomal inheritance we will also discuss *X-linked inheritance*. In this type of inheritance the male of the species possesses only one of the two possible genes (A or a), and the female possesses a pair of the two genes (AA, Aa, or aa). In humans, color blindness, hereditary baldness, hemophilia, and muscular dystrophy, to name a few, are traits controlled by X-linked inheritance.

Below we explain the manner in which the genes of the parents are passed on to their offspring for the two types of inheritance. We construct matrix models that give the probable genotypes of the offspring in terms of the genotypes of the parents, and we use these matrix models to follow the genotype distribution of a population through successive generations.

Autosomal Inheritance   In autosomal inheritance an individual inherits one gene from each of its parents' pairs of genes to form its own particular pair. As far as we know, it is a matter of chance which of the two genes a parent passes on to the offspring. Thus, if one parent is of genotype Aa, it is equally likely that the offspring will inherit the A gene or the a gene from that parent. If one parent is of genotype aa and the other parent is of genotype Aa, the offspring will always receive an a gene from the aa parent, and will receive either an A gene or an a gene, with equal probability, from the Aa parent. Consequently, each of the offspring has equal probability of being genotype aa or Aa. In Table 1 we list the probabilities of the possible genotypes of the offspring for all possible combinations of the genotypes of the parents.

TABLE 1

Genotype of Offspring	Genotypes of Parents					
	AA–AA	AA–Aa	AA–aa	Aa–Aa	Aa–aa	aa–aa
AA	1	$\frac{1}{2}$	0	$\frac{1}{4}$	0	0
Aa	0	$\frac{1}{2}$	1	$\frac{1}{2}$	$\frac{1}{2}$	0
aa	0	0	0	$\frac{1}{4}$	$\frac{1}{2}$	1

EXAMPLE 1 Distribution of Genotypes in a Population

Suppose that a farmer has a large population of plants consisting of some distribution of all three possible genotypes AA, Aa, and aa. The farmer desires to undertake a breeding program in which each plant in the population is always fertilized with a plant of genotype AA and is then replaced by one of its offspring. We want to derive an expression for the distribution of the three possible genotypes in the population after any number of generations.

For $n = 0, 1, 2, \ldots$, let us set

$$a_n = \text{fraction of plants of genotype } AA \text{ in } n\text{th generation}$$
$$b_n = \text{fraction of plants of genotype } Aa \text{ in } n\text{th generation}$$
$$c_n = \text{fraction of plants of genotype } aa \text{ in } n\text{th generation}$$

Thus, a_0, b_0, and c_0 specify the initial distribution of the genotypes. We also have that

$$a_n + b_n + c_n = 1 \quad \text{for } n = 0, 1, 2, \ldots$$

From Table 1 we can determine the genotype distribution of each generation from the genotype distribution of the preceding generation by the following equations:

$$
\begin{aligned}
a_n &= a_{n-1} + \tfrac{1}{2} b_{n-1} \\
b_n &= c_{n-1} + \tfrac{1}{2} b_{n-1} \qquad n = 1, 2, \ldots \\
c_n &= 0
\end{aligned}
\tag{1}
$$

For example, the first of these three equations states that all the offspring of a plant of genotype AA will be of genotype AA under this breeding program, and half of the offspring of a plant of genotype Aa will be of genotype AA.

Equations (1) can be written in matrix notation as

$$\mathbf{x}^{(n)} = M\mathbf{x}^{(n-1)}, \qquad n = 1, 2, \ldots \tag{2}$$

where

$$\mathbf{x}^{(n)} = \begin{bmatrix} a_n \\ b_n \\ c_n \end{bmatrix}, \quad \mathbf{x}^{(n-1)} = \begin{bmatrix} a_{n-1} \\ b_{n-1} \\ c_{n-1} \end{bmatrix}, \quad \text{and} \quad M = \begin{bmatrix} 1 & \tfrac{1}{2} & 0 \\ 0 & \tfrac{1}{2} & 1 \\ 0 & 0 & 0 \end{bmatrix}$$

Notice that the three columns of the matrix M are the same as the first three columns of Table 1.

From Equation (2) it follows that

$$\mathbf{x}^{(n)} = M\mathbf{x}^{(n-1)} = M^2\mathbf{x}^{(n-2)} = \cdots = M^n\mathbf{x}^{(0)} \tag{3}$$

Consequently, if we can find an explicit expression for M^n, we can use (3) to obtain an explicit expression for $\mathbf{x}^{(n)}$. To find an explicit expression for M^n, we first diagonalize M. That is, we find an invertible matrix P and a diagonal matrix D such that

$$M = PDP^{-1} \tag{4}$$

With such a diagonalization, we then have (see Exercise 1)

$$M^n = PD^n P^{-1} \quad \text{for } n = 1, 2, \ldots$$

where

$$D^n = \begin{bmatrix} \lambda_1 & 0 & 0 & \cdots & 0 \\ 0 & \lambda_2 & 0 & \cdots & 0 \\ \vdots & \vdots & \vdots & & \vdots \\ 0 & 0 & 0 & \cdots & \lambda_k \end{bmatrix}^n = \begin{bmatrix} \lambda_1^n & 0 & 0 & \cdots & 0 \\ 0 & \lambda_2^n & 0 & \cdots & 0 \\ \vdots & \vdots & \vdots & & \vdots \\ 0 & 0 & 0 & \cdots & \lambda_k^n \end{bmatrix}$$

The diagonalization of M is accomplished by finding its eigenvalues and corresponding eigenvectors. These are as follows (verify):

Eigenvalues: $\qquad \lambda_1 = 1, \qquad \lambda_2 = \tfrac{1}{2}, \qquad \lambda_3 = 0$

Corresponding eigenvectors: $\quad \mathbf{v}_1 = \begin{bmatrix} 1 \\ 0 \\ 0 \end{bmatrix}, \quad \mathbf{v}_2 = \begin{bmatrix} 1 \\ -1 \\ 0 \end{bmatrix}, \quad \mathbf{v}_3 = \begin{bmatrix} 1 \\ -2 \\ 1 \end{bmatrix}$

Thus, in Equation (4) we have

$$D = \begin{bmatrix} \lambda_1 & 0 & 0 \\ 0 & \lambda_2 & 0 \\ 0 & 0 & \lambda_3 \end{bmatrix} = \begin{bmatrix} 1 & 0 & 0 \\ 0 & \tfrac{1}{2} & 0 \\ 0 & 0 & 0 \end{bmatrix}$$

and

$$P = [\mathbf{v}_1 \mid \mathbf{v}_2 \mid \mathbf{v}_3] = \begin{bmatrix} 1 & 1 & 1 \\ 0 & -1 & -2 \\ 0 & 0 & 1 \end{bmatrix}$$

Therefore,

$$\mathbf{x}^{(n)} = PD^n P^{-1}\mathbf{x}^{(0)} = \begin{bmatrix} 1 & 1 & 1 \\ 0 & -1 & -2 \\ 0 & 0 & 1 \end{bmatrix} \begin{bmatrix} 1 & 0 & 0 \\ 0 & \left(\tfrac{1}{2}\right)^n & 0 \\ 0 & 0 & 0 \end{bmatrix} \begin{bmatrix} 1 & 1 & 1 \\ 0 & -1 & -2 \\ 0 & 0 & 1 \end{bmatrix} \begin{bmatrix} a_0 \\ b_0 \\ c_0 \end{bmatrix}$$

or

$$\mathbf{x}^{(n)} = \begin{bmatrix} a_n \\ b_n \\ c_n \end{bmatrix} = \begin{bmatrix} 1 & 1 - \left(\tfrac{1}{2}\right)^n & 1 - \left(\tfrac{1}{2}\right)^{n-1} \\ 0 & \left(\tfrac{1}{2}\right)^n & \left(\tfrac{1}{2}\right)^{n-1} \\ 0 & 0 & 0 \end{bmatrix} \begin{bmatrix} a_0 \\ b_0 \\ c_0 \end{bmatrix}$$

$$= \begin{bmatrix} a_0 + b_0 + c_0 - \left(\tfrac{1}{2}\right)^n b_0 - \left(\tfrac{1}{2}\right)^{n-1} c_0 \\ \left(\tfrac{1}{2}\right)^n b_0 + \left(\tfrac{1}{2}\right)^{n-1} c_0 \\ 0 \end{bmatrix}$$

Using the fact that $a_0 + b_0 + c_0 = 1$, we thus have

$$a_n = 1 - \left(\tfrac{1}{2}\right)^n b_0 - \left(\tfrac{1}{2}\right)^{n-1} c_0$$
$$b_n = \left(\tfrac{1}{2}\right)^n b_0 + \left(\tfrac{1}{2}\right)^{n-1} c_0 \qquad n = 1, 2, \ldots \qquad (5)$$
$$c_n = 0$$

These are explicit formulas for the fractions of the three genotypes in the nth generation of plants in terms of the initial genotype fractions.

Because $\left(\tfrac{1}{2}\right)^n$ tends to zero as n approaches infinity, it follows from these equations that

$$a_n \to 1$$
$$b_n \to 0$$
$$c_n = 0$$

as n approaches infinity. That is, in the limit all plants in the population will be genotype AA. ◆

EXAMPLE 2 Modifying Example 1

We can modify Example 1 so that instead of fertilizing each plant with one of genotype AA, each plant is fertilized with a plant of its own genotype. Using the same notation as in Example 1, we then find

$$\mathbf{x}^{(n)} = M^n \mathbf{x}^{(0)}$$

where

$$M = \begin{bmatrix} 1 & \tfrac{1}{4} & 0 \\ 0 & \tfrac{1}{2} & 0 \\ 0 & \tfrac{1}{4} & 1 \end{bmatrix}$$

The columns of this new matrix M are the same as the columns of Table 1 corresponding to parents with genotypes AA–AA, Aa–Aa, and aa–aa.

The eigenvalues of M are (verify)

$$\lambda_1 = 1, \qquad \lambda_2 = 1, \qquad \lambda_3 = \tfrac{1}{2}$$

The eigenvalue $\lambda_1 = 1$ has multiplicity two and its corresponding eigenspace is two-dimensional. Picking two linearly independent eigenvectors $\mathbf{v}_1$ and $\mathbf{v}_2$ in that eigenspace, and a single eigenvector $\mathbf{v}_3$ for the simple eigenvalue $\lambda_3 = \tfrac{1}{2}$, we have (verify)

$$\mathbf{v}_1 = \begin{bmatrix} 1 \\ 0 \\ 0 \end{bmatrix}, \qquad \mathbf{v}_2 = \begin{bmatrix} 0 \\ 0 \\ 1 \end{bmatrix}, \qquad \mathbf{v}_3 = \begin{bmatrix} 1 \\ -2 \\ 1 \end{bmatrix}$$

The calculations for $\mathbf{x}^{(n)}$ are then

$$\mathbf{x}^{(n)} = M^n \mathbf{x}^{(0)} = PD^n P^{-1} \mathbf{x}^{(0)}$$

$$= \begin{bmatrix} 1 & 0 & 1 \\ 0 & 0 & -2 \\ 0 & 1 & 1 \end{bmatrix} \begin{bmatrix} 1 & 0 & 0 \\ 0 & 1 & 0 \\ 0 & 0 & \left(\tfrac{1}{2}\right)^n \end{bmatrix} \begin{bmatrix} 1 & \tfrac{1}{2} & 0 \\ 0 & \tfrac{1}{2} & 1 \\ 0 & -\tfrac{1}{2} & 0 \end{bmatrix} \begin{bmatrix} a_0 \\ b_0 \\ c_0 \end{bmatrix}$$

$$= \begin{bmatrix} 1 & \tfrac{1}{2} - \left(\tfrac{1}{2}\right)^{n+1} & 0 \\ 0 & \left(\tfrac{1}{2}\right)^n & 0 \\ 0 & \tfrac{1}{2} - \left(\tfrac{1}{2}\right)^{n+1} & 1 \end{bmatrix} \begin{bmatrix} a_0 \\ b_0 \\ c_0 \end{bmatrix}$$

Thus,

$$a_n = a_0 + \left[\tfrac{1}{2} - \left(\tfrac{1}{2} \right)^{n+1} \right] b_0$$
$$b_n = \left(\tfrac{1}{2} \right)^n b_0 \qquad\qquad n = 1, 2, \ldots \qquad (6)$$
$$c_n = c_0 + \left[\tfrac{1}{2} - \left(\tfrac{1}{2} \right)^{n+1} \right] b_0$$

In the limit, as n tends to infinity, $\left(\tfrac{1}{2} \right)^n \to 0$ and $\left(\tfrac{1}{2} \right)^{n+1} \to 0$ so that

$$a_n \to a_0 + \tfrac{1}{2} b_0$$
$$b_n \to 0$$
$$c_n \to c_0 + \tfrac{1}{2} b_0$$

Thus, fertilization of each plant with one of its own genotype produces a population that in the limit contains only genotypes AA and aa. ◆

Autosomal Recessive Diseases   There are many genetic diseases governed by autosomal inheritance in which a normal gene A dominates an abnormal gene a. Genotype AA is a normal individual; genotype Aa is a carrier of the disease but is not afflicted with the disease; and genotype aa is afflicted with the disease. In humans such genetic diseases are often associated with a particular racial group; for instance, cystic fibrosis (predominant among Caucasians), sickle-cell anemia (predominant among blacks), Cooley's anemia (predominant among people of Mediterranean origin), and Tay-Sachs disease (predominant among Eastern European Jews).

Suppose that an animal breeder has a population of animals that carries an autosomal recessive disease. Suppose, further, that those animals afflicted with the disease do not survive to maturity. One possible way to control such a disease is for the breeder to always mate a female, regardless of her genotype, with a normal male. In this way, all future offspring will either have a normal father and a normal mother (AA–AA matings) or a normal father and a carrier mother (AA–Aa matings). There can be no AA–aa matings since animals of genotype aa do not survive to maturity. Under this type of mating program no future offspring will be afflicted with the disease, although there will still be carriers in future generations. Let us now determine the fraction of carriers in future generations. We set

$$\mathbf{x}^{(n)} = \begin{bmatrix} a_n \\ b_n \end{bmatrix}, \qquad n = 1, 2, \ldots$$

where

a_n = fraction of population of genotype AA in nth generation

b_n = fraction of population of genotype Aa (carriers) in nth generation

Because each offspring has at least one normal parent, we may consider the controlled mating program as one of continual mating with genotype AA, as in Example 1. Thus, the transition of genotype distributions from one generation to the next is governed by the equation

$$\mathbf{x}^{(n)} = M\mathbf{x}^{(n-1)}, \qquad n = 1, 2, \ldots$$

where

$$M = \begin{bmatrix} 1 & \tfrac{1}{2} \\ 0 & \tfrac{1}{2} \end{bmatrix}$$

Knowing the initial distribution $\mathbf{x}^{(0)}$, the distribution of genotypes in the nth generation is thus given by

$$\mathbf{x}^{(n)} = M^n \mathbf{x}^{(0)}, \qquad n = 1, 2, \ldots$$

The diagonalization of M is easily carried out (see Exercise 4) and leads to

$$\mathbf{x}^{(n)} = PD^n P^{-1}\mathbf{x}^{(0)} = \begin{bmatrix} 1 & 1 \\ 0 & -1 \end{bmatrix} \begin{bmatrix} 1 & 0 \\ 0 & (\frac{1}{2})^n \end{bmatrix} \begin{bmatrix} 1 & 1 \\ 0 & -1 \end{bmatrix} \begin{bmatrix} a_0 \\ b_0 \end{bmatrix}$$

$$= \begin{bmatrix} 1 & 1 - (\frac{1}{2})^n \\ 0 & (\frac{1}{2})^n \end{bmatrix} \begin{bmatrix} a_0 \\ b_0 \end{bmatrix} = \begin{bmatrix} a_0 + b_0 - (\frac{1}{2})^n b_0 \\ (\frac{1}{2})^n b_0 \end{bmatrix}$$

Because $a_0 + b_0 = 1$, we have

$$\begin{aligned} a_n &= 1 - (\tfrac{1}{2})^n b_0 \\ b_n &= (\tfrac{1}{2})^n b_0 \end{aligned} \qquad n = 1, 2, \ldots \tag{7}$$

Thus, as n tends to infinity we have

$$a_n \to 1$$
$$b_n \to 0$$

so that in the limit there will be no carriers in the population.

From (7) we see that

$$b_n = \tfrac{1}{2}b_{n-1}, \qquad n = 1, 2, \ldots \tag{8}$$

That is, the fraction of carriers in each generation is one-half the fraction of carriers in the preceding generation. It would be of interest to also investigate the propagation of carriers under random mating, when two animals mate without regard to their genotypes. Unfortunately, such random mating leads to nonlinear equations and the techniques of this section are not applicable. However, by other techniques it can be shown that under random mating Equation (8) is replaced by

$$b_n = \frac{b_{n-1}}{1 + \tfrac{1}{2}b_{n-1}}, \qquad n = 1, 2, \ldots \tag{9}$$

As a numerical example, suppose that the breeder starts with a population in which 10% of the animals are carriers. Under the controlled mating program governed by Equation (8), the percentage of carriers can be reduced to 5% in one generation. But under random mating, Equation (9) predicts that 9.5% of the population will be carriers after one generation (i.e., $b_n = .095$ if $b_{n-1} = .10$). In addition, under controlled mating no offspring will ever be afflicted with the disease, but with random mating it can be shown that about 1 in 400 offspring will be born with the disease when 10% of the population are carriers.

X-Linked Inheritance

As mentioned in the introduction, in X-linked inheritance the male possesses one gene (A or a) and the female possesses two genes (AA, Aa, or aa). The term "X-linked" is used because such genes are found on the X-chromosome, of which the male has one and the female has two. The inheritance of such genes is as follows: A male offspring receives one of his mother's two genes with equal probability, and a female offspring receives the one gene of her father and one of her mother's two genes with equal probability. Readers familiar with basic probability can verify that this type of inheritance leads to the genotype probabilities in Table 2.

We will discuss a program of inbreeding in connection with X-linked inheritance. We begin initially with a male and female; select two of their offspring at random, one of each gender, and mate them; select two of the resulting offspring and mate them; and so forth. Such inbreeding is commonly performed with animals. (Among humans, such brother-sister marriages were used by the rulers of ancient Egypt to keep the royal line pure.)

TABLE 2

			Genotypes of Parents (Father, Mother)					
			(A, AA)	(A, Aa)	(A, aa)	(a, AA)	(a, Aa)	(a, aa)
Offspring	Male	A	1	$\frac{1}{2}$	0	1	$\frac{1}{2}$	0
		a	0	$\frac{1}{2}$	1	0	$\frac{1}{2}$	1
	Female	AA	1	$\frac{1}{2}$	0	0	0	0
		Aa	0	$\frac{1}{2}$	1	1	$\frac{1}{2}$	0
		aa	0	0	0	0	$\frac{1}{2}$	1

The original male-female pair can be one of the six types, corresponding to the six columns of Table 2:

$$(A, AA), \quad (A, Aa), \quad (A, aa), \quad (a, AA), \quad (a, Aa), \quad (a, aa)$$

The sibling-pairs mated in each successive generation have certain probabilities of being one of these six types. To compute these probabilities, for $n = 0, 1, 2, \ldots$, let us set

$a_n = $ probability sibling-pair mated in nth generation is type (A, AA)

$b_n = $ probability sibling-pair mated in nth generation is type (A, Aa)

$c_n = $ probability sibling-pair mated in nth generation is type (A, aa)

$d_n = $ probability sibling-pair mated in nth generation is type (a, AA)

$e_n = $ probability sibling-pair mated in nth generation is type (a, Aa)

$f_n = $ probability sibling-pair mated in nth generation is type (a, aa)

With these probabilities we form a column vector

$$\mathbf{x}^{(n)} = \begin{bmatrix} a_n \\ b_n \\ c_n \\ d_n \\ e_n \\ f_n \end{bmatrix}, \qquad n = 0, 1, 2, \ldots$$

From Table 2 it follows that

$$\mathbf{x}^{(n)} = M\mathbf{x}^{(n-1)}, \qquad n = 1, 2, \ldots \tag{10}$$

where

$$M = \begin{bmatrix} 1 & \frac{1}{4} & 0 & 0 & 0 & 0 \\ 0 & \frac{1}{4} & 0 & 1 & \frac{1}{4} & 0 \\ 0 & 0 & 0 & 0 & \frac{1}{4} & 0 \\ 0 & \frac{1}{4} & 0 & 0 & 0 & 0 \\ 0 & \frac{1}{4} & 1 & 0 & \frac{1}{4} & 0 \\ 0 & 0 & 0 & 0 & \frac{1}{4} & 1 \end{bmatrix} \begin{matrix} (A, AA) \\ (A, Aa) \\ (A, aa) \\ (a, AA) \\ (a, Aa) \\ (a, aa) \end{matrix}$$

$$\begin{matrix} (A, AA) & (A, Aa) & (A, aa) & (a, AA) & (a, Aa) & (a, aa) \end{matrix}$$

For example, suppose that in the $(n - 1)$-st generation the sibling-pair mated is type

(A, Aa). Then their male offspring will be either genotype A or a with equal probability, and their female offspring will be either genotype AA or Aa with equal probability. Because one of the male offspring and one of the female offspring are chosen at random for mating, the next sibling-pair will be one of type (A, AA), (A, Aa), (a, AA), (a, Aa) with equal probability. Thus, the second column of M contains "$\frac{1}{4}$" in each of the four rows corresponding to these four sibling-pairs. (See Exercise 9 for the remaining columns.)

As in our previous examples, it follows from (10) that

$$\mathbf{x}^{(n)} = M^n \mathbf{x}^{(0)}, \qquad n = 1, 2, \ldots \tag{11}$$

After lengthy calculations, the eigenvalues and eigenvectors of M turn out to be

$$\lambda_1 = 1, \quad \lambda_2 = 1, \quad \lambda_3 = \tfrac{1}{2}, \quad \lambda_4 = -\tfrac{1}{2}, \quad \lambda_5 = \tfrac{1}{4}(1 + \sqrt{5}), \quad \lambda_6 = \tfrac{1}{4}(1 - \sqrt{5})$$

$$\mathbf{v}_1 = \begin{bmatrix} 1 \\ 0 \\ 0 \\ 0 \\ 0 \\ 0 \end{bmatrix}, \qquad \mathbf{v}_2 = \begin{bmatrix} 0 \\ 0 \\ 0 \\ 0 \\ 0 \\ 1 \end{bmatrix}, \qquad \mathbf{v}_3 = \begin{bmatrix} -1 \\ 2 \\ -1 \\ 1 \\ -2 \\ 1 \end{bmatrix}, \qquad \mathbf{v}_4 = \begin{bmatrix} 1 \\ -6 \\ -3 \\ 3 \\ 6 \\ -1 \end{bmatrix},$$

$$\mathbf{v}_5 = \begin{bmatrix} \tfrac{1}{4}(-3 - \sqrt{5}) \\ 1 \\ \tfrac{1}{4}(-1 + \sqrt{5}) \\ \tfrac{1}{4}(-1 + \sqrt{5}) \\ 1 \\ \tfrac{1}{4}(-3 - \sqrt{5}) \end{bmatrix}, \qquad \mathbf{v}_6 = \begin{bmatrix} \tfrac{1}{4}(-3 + \sqrt{5}) \\ 1 \\ \tfrac{1}{4}(-1 - \sqrt{5}) \\ \tfrac{1}{4}(-1 - \sqrt{5}) \\ 1 \\ \tfrac{1}{4}(-3 + \sqrt{5}) \end{bmatrix}$$

The diagonalization of M then leads to

$$\mathbf{x}^{(n)} = PD^n P^{-1} \mathbf{x}^{(0)}, \qquad n = 1, 2, \ldots \tag{12}$$

where

$$P = \begin{bmatrix} 1 & 0 & -1 & 1 & \tfrac{1}{4}(-3 - \sqrt{5}) & \tfrac{1}{4}(-3 + \sqrt{5}) \\ 0 & 0 & 2 & -6 & 1 & 1 \\ 0 & 0 & -1 & -3 & \tfrac{1}{4}(-1 + \sqrt{5}) & \tfrac{1}{4}(-1 - \sqrt{5}) \\ 0 & 0 & 1 & 3 & \tfrac{1}{4}(-1 + \sqrt{5}) & \tfrac{1}{4}(-1 - \sqrt{5}) \\ 0 & 0 & -2 & 6 & 1 & 1 \\ 0 & 1 & 1 & -1 & \tfrac{1}{4}(-3 - \sqrt{5}) & \tfrac{1}{4}(-3 + \sqrt{5}) \end{bmatrix}$$

$$D^n = \begin{bmatrix} 1 & 0 & 0 & 0 & 0 & 0 \\ 0 & 1 & 0 & 0 & 0 & 0 \\ 0 & 0 & \left(\tfrac{1}{2}\right)^n & 0 & 0 & 0 \\ 0 & 0 & 0 & \left(-\tfrac{1}{2}\right)^n & 0 & 0 \\ 0 & 0 & 0 & 0 & \left[\tfrac{1}{4}(1 + \sqrt{5})\right]^n & 0 \\ 0 & 0 & 0 & 0 & 0 & \left[\tfrac{1}{4}(1 - \sqrt{5})\right]^n \end{bmatrix}$$

$$P^{-1} = \begin{bmatrix} 1 & \frac{2}{3} & \frac{1}{3} & \frac{2}{3} & \frac{1}{3} & 0 \\ 0 & \frac{1}{3} & \frac{2}{3} & \frac{1}{3} & \frac{2}{3} & 1 \\ 0 & \frac{1}{8} & -\frac{1}{4} & \frac{1}{4} & -\frac{1}{8} & 0 \\ 0 & -\frac{1}{24} & -\frac{1}{12} & \frac{1}{12} & \frac{1}{24} & 0 \\ 0 & \frac{1}{20}(5+\sqrt{5}) & \frac{1}{5}\sqrt{5} & \frac{1}{5}\sqrt{5} & \frac{1}{20}(5+\sqrt{5}) & 0 \\ 0 & \frac{1}{20}(5-\sqrt{5}) & -\frac{1}{5}\sqrt{5} & -\frac{1}{5}\sqrt{5} & \frac{1}{20}(5-\sqrt{5}) & 0 \end{bmatrix}$$

We will not write out the matrix product in (12), as it is rather unwieldy. However, if a specific vector $\mathbf{x}^{(0)}$ is given, the calculation for $\mathbf{x}^{(n)}$ is not too cumbersome (see Exercise 6).

Because the absolute values of the last four diagonal entries of D are less than 1, we see that as n tends to infinity

$$D^n \rightarrow \begin{bmatrix} 1 & 0 & 0 & 0 & 0 & 0 \\ 0 & 1 & 0 & 0 & 0 & 0 \\ 0 & 0 & 0 & 0 & 0 & 0 \\ 0 & 0 & 0 & 0 & 0 & 0 \\ 0 & 0 & 0 & 0 & 0 & 0 \\ 0 & 0 & 0 & 0 & 0 & 0 \end{bmatrix}$$

And so from Equation (12)

$$\mathbf{x}^{(n)} \rightarrow P \begin{bmatrix} 1 & 0 & 0 & 0 & 0 & 0 \\ 0 & 1 & 0 & 0 & 0 & 0 \\ 0 & 0 & 0 & 0 & 0 & 0 \\ 0 & 0 & 0 & 0 & 0 & 0 \\ 0 & 0 & 0 & 0 & 0 & 0 \\ 0 & 0 & 0 & 0 & 0 & 0 \end{bmatrix} P^{-1} \mathbf{x}^{(0)}$$

Performing the matrix multiplication on the right we obtain (verify)

$$\mathbf{x}^{(n)} \rightarrow \begin{bmatrix} a_0 + \frac{2}{3}b_0 + \frac{1}{3}c_0 + \frac{2}{3}d_0 + \frac{1}{3}e_0 \\ 0 \\ 0 \\ 0 \\ 0 \\ f_0 + \frac{1}{3}b_0 + \frac{2}{3}c_0 + \frac{1}{3}d_0 + \frac{2}{3}e_0 \end{bmatrix} \tag{13}$$

That is, in the limit all sibling-pairs will be either type (A, AA) or type (a, aa). For example, if the initial parents are type (A, Aa) (i.e., $b_0 = 1$ and $a_0 = c_0 = d_0 = e_0 = f_0 = 0$), then as n tends to infinity

$$\mathbf{x}^{(n)} \rightarrow \begin{bmatrix} \frac{2}{3} \\ 0 \\ 0 \\ 0 \\ 0 \\ \frac{1}{3} \end{bmatrix}$$

Thus, in the limit there is probability $\frac{2}{3}$ that the sibling-pairs will be (A, AA), and probability $\frac{1}{3}$ that they will be (a, aa).

Exercise Set 11.17

1. Show that if $M = PDP^{-1}$, then $M^n = PD^n P^{-1}$ for $n = 1, 2, \ldots$.

2. In Example 1 suppose that the plants are always fertilized with a plant of genotype Aa rather than one of genotype AA. Derive formulas for the fractions of the plants of genotypes AA, Aa, and aa in the nth generation. Also, find the limiting genotype distribution as n tends to infinity.

3. In Example 1 suppose that the initial plants are fertilized with genotype AA, the first generation is fertilized with genotype Aa, the second generation is fertilized with genotype AA, and this alternating pattern of fertilization is kept up. Find formulas for the fractions of the plants of genotypes AA, Aa, and aa in the nth generation.

4. In the section on autosomal recessive diseases, find the eigenvalues and eigenvectors of the matrix M and verify Equation (7).

5. Suppose that a breeder has an animal population in which 25% of the population are carriers of an autosomal recessive disease. If the breeder allows the animals to mate irrespective of their genotype, use Equation (9) to calculate the number of generations required for the percentage of carriers to fall from 25% to 10%. If the breeder instead implements the controlled mating program determined by Equation (8), what will the percentage of carriers be after the same number of generations?

6. In the section on X-linked inheritance, suppose that the initial parents are equally likely to be of any of the six possible genotype parents; that is,

$$\mathbf{x}^{(0)} = \begin{bmatrix} \frac{1}{6} \\ \frac{1}{6} \\ \frac{1}{6} \\ \frac{1}{6} \\ \frac{1}{6} \\ \frac{1}{6} \end{bmatrix}$$

Using Equation (12) calculate $\mathbf{x}^{(n)}$ and also calculate the limit of $\mathbf{x}^{(n)}$ as n tends to infinity.

7. From (13) show that under X-linked inheritance with inbreeding, the probability that the limiting sibling-pairs will be of type (A, AA) is the same as the proportion of A genes in the initial population.

8. In X-linked inheritance suppose that none of the females of genotype Aa survive to maturity. Under inbreeding the possible sibling-pairs are then

$$(A, AA), \quad (A, aa), \quad (a, AA), \quad \text{and} \quad (a, aa)$$

Find the transition matrix that describes how the genotype distribution changes in one generation.

9. Derive the matrix M in Equation (10) from Table 2.

Technology Exercises 11.17

The following exercises are designed to be solved using a technology utility. Typically, this will be MATLAB, *Mathematica*, Maple, Derive, or Mathcad, but it may also be some other type of linear algebra software or a scientific calculator with some linear algebra capabilities. For each exercise you will need to read the relevant documentation for the particular utility you are using. The goal of these exercises is to provide you with a basic proficiency with your technology utility. Once you have mastered the techniques in these exercises, you will be able to use your technology utility to solve many of the problems in the regular exercise sets.

T1. (a) Use a computer to verify that the eigenvalues and eigenvectors of

$$M = \begin{bmatrix} 1 & \frac{1}{4} & 0 & 0 & 0 & 0 \\ 0 & \frac{1}{4} & 0 & 1 & \frac{1}{4} & 0 \\ 0 & 0 & 0 & 0 & \frac{1}{4} & 0 \\ 0 & \frac{1}{4} & 0 & 0 & 0 & 0 \\ 0 & \frac{1}{4} & 1 & 0 & \frac{1}{4} & 0 \\ 0 & 0 & 0 & 0 & \frac{1}{4} & 1 \end{bmatrix}$$

as given in the text are correct.

(b) Starting with $\mathbf{x}^{(n)} = M\mathbf{x}^{(n-1)}$ and the assumption that

$$\lim_{n \to \infty} \mathbf{x}^{(n)} = \mathbf{x}$$

exists, we must have

$$\lim_{n \to \infty} \mathbf{x}^{(n)} = M \lim_{n \to \infty} \mathbf{x}^{(n-1)} \quad \text{or} \quad \mathbf{x} = M\mathbf{x}$$

This suggests that $\mathbf{x}$ can be solved directly using the equation $(M - I)\mathbf{x} = \mathbf{0}$. Use a computer to solve the equation $\mathbf{x} = M\mathbf{x}$, where

$$\mathbf{x} = \begin{bmatrix} a \\ b \\ c \\ d \\ e \\ f \end{bmatrix}$$

and $a + b + c + d + e + f = 1$; compare your results to Equation (13). Explain why the solution to $(M - I)\mathbf{x} = \mathbf{0}$ along with $a + b + c + d + e + f = 1$ is not specific enough to determine $\lim_{n \to \infty} \mathbf{x}^{(n)}$.

T2. (a) Given

$$P = \begin{bmatrix} 1 & 0 & -1 & 1 & \frac{1}{4}(-3-\sqrt{5}) & \frac{1}{4}(-3+\sqrt{5}) \\ 0 & 0 & 2 & -6 & 1 & 1 \\ 0 & 0 & -1 & -3 & \frac{1}{4}(-1+\sqrt{5}) & \frac{1}{4}(-1-\sqrt{5}) \\ 0 & 0 & 1 & 3 & \frac{1}{4}(-1+\sqrt{5}) & \frac{1}{4}(-1-\sqrt{5}) \\ 0 & 0 & -2 & 6 & 1 & 1 \\ 0 & 1 & 1 & -1 & \frac{1}{4}(-3-\sqrt{5}) & \frac{1}{4}(-3+\sqrt{5}) \end{bmatrix}$$

from Equation (12) and

$$\lim_{n \to \infty} D^n = \begin{bmatrix} 1 & 0 & 0 & 0 & 0 & 0 \\ 0 & 1 & 0 & 0 & 0 & 0 \\ 0 & 0 & 0 & 0 & 0 & 0 \\ 0 & 0 & 0 & 0 & 0 & 0 \\ 0 & 0 & 0 & 0 & 0 & 0 \\ 0 & 0 & 0 & 0 & 0 & 0 \end{bmatrix}$$

use a computer to show that

$$\lim_{n \to \infty} M^n = \begin{bmatrix} 1 & \frac{2}{3} & \frac{1}{3} & \frac{2}{3} & \frac{1}{3} & 0 \\ 0 & 0 & 0 & 0 & 0 & 0 \\ 0 & 0 & 0 & 0 & 0 & 0 \\ 0 & 0 & 0 & 0 & 0 & 0 \\ 0 & 0 & 0 & 0 & 0 & 0 \\ 0 & \frac{1}{3} & \frac{2}{3} & \frac{1}{3} & \frac{2}{3} & 1 \end{bmatrix}$$

(b) Use a computer to calculate M^n for $n = 10, 20, 30, 40, 50, 60, 70$, and then compare your results to the limit in part (a).

11.18 AGE-SPECIFIC POPULATION GROWTH

In this section we investigate, using the Leslie matrix model, the growth over time of a female population that is divided into age classes. We then determine the limiting age distribution and growth rate of the population.

PREREQUISITES: Eigenvalues and Eigenvectors
 Diagonalization of a Matrix
 Intuitive Understanding of Limits

One of the most common models of population growth used by demographers is the so-called Leslie model, developed in the 1940s. This model describes the growth of the female portion of a human or animal population. In this model the females are divided into age classes of equal duration. To be specific, suppose that the maximum age attained by any female in the population is L years (or some other time unit) and we divide the population into n age classes. Then each class is L/n years in duration. We label the age classes according to Table 1.

TABLE 1

Age Class	Age Interval
1	$[0, L/n)$
2	$[L/n, 2L/n)$
3	$[2L/n, 3L/n)$
⋮	⋮
$n - 1$	$[(n-2)L/n, (n-1)L/n)$
n	$[(n-1)L/n, L]$

Suppose that we know the number of females in each of the n classes at time $t = 0$. In particular, let there be $x_1^{(0)}$ females in the first class, $x_2^{(0)}$ females in the second class, and so forth. With these n numbers we form a column vector:

$$\mathbf{x}^{(0)} = \begin{bmatrix} x_1^{(0)} \\ x_2^{(0)} \\ \vdots \\ x_n^{(0)} \end{bmatrix}$$

We call this vector the ***initial age distribution vector.***

As time progresses, the number of females within each of the n classes changes because of three biological processes: birth, death, and aging. By describing these three processes quantitatively, we will see how to project the initial age distribution vector into the future.

The easiest way to study the aging process is to observe the population at discrete times, say, $t_0, t_1, t_2, \ldots, t_k, \ldots$. The Leslie model requires that the duration between any two successive observation times be the same as the duration of the age intervals. Therefore, we set

$$t_0 = 0$$
$$t_1 = L/n$$
$$t_2 = 2L/n$$
$$\vdots$$
$$t_k = kL/n$$
$$\vdots$$

With this assumption, all females in the $(i + 1)$-st class at time t_{k+1} were in the ith class at time t_k.

The birth and death processes between two successive observation times can be described by means of the following demographic parameters:

a_i $(i = 1, 2, \ldots, n)$	The average number of daughters born to each female during the time she is in the ith age class
b_i $(i = 1, 2, \ldots, n-1)$	The fraction of females in the ith age class that can be expected to survive and pass into the $(i + 1)$-st age class

By their definitions, we have that

(i) $a_i \geq 0$ for $i = 1, 2, \ldots, n$

(ii) $0 < b_i \leq 1$ for $i = 1, 2, \ldots, n-1$

Notice that we do not allow any b_i to equal zero, as then no females will survive beyond the ith age class. We also assume that at least one a_i is positive so that some births occur. Any age class for which the corresponding value of a_i is positive is called a ***fertile age class***.

We next define the age distribution vector $\mathbf{x}^{(k)}$ at time t_k by

$$\mathbf{x}^{(k)} = \begin{bmatrix} x_1^{(k)} \\ x_2^{(k)} \\ \vdots \\ x_n^{(k)} \end{bmatrix}$$

where $x_i^{(k)}$ is the number of females in the ith age class at time t_k. Now, at time t_k, the females in the first age class are just those daughters born between times t_{k-1} and t_k. Thus, we can write

$$\left\{\begin{array}{l}\text{number of} \\ \text{females} \\ \text{in class 1} \\ \text{at time } t_k\end{array}\right\} = \left\{\begin{array}{l}\text{number of} \\ \text{daughters} \\ \text{born to} \\ \text{females in} \\ \text{class 1} \\ \text{between times} \\ t_{k-1} \text{ and } t_k\end{array}\right\} + \left\{\begin{array}{l}\text{number of} \\ \text{daughters} \\ \text{born to} \\ \text{females in} \\ \text{class 2} \\ \text{between times} \\ t_{k-1} \text{ and } t_k\end{array}\right\} + \cdots + \left\{\begin{array}{l}\text{number of} \\ \text{daughters} \\ \text{born to} \\ \text{females in} \\ \text{class } n \\ \text{between times} \\ t_{k-1} \text{ and } t_k\end{array}\right\}$$

or, mathematically,

$$x_1^{(k)} = a_1 x_1^{(k-1)} + a_2 x_2^{(k-1)} + \cdots + a_n x_n^{(k-1)} \tag{1}$$

The females in the $(i+1)$-st age class $(i = 1, 2, \ldots, n-1)$ at time t_k are those females in the ith class at time t_{k-1} who are still alive at time t_k. Thus,

$$\left\{\begin{array}{l}\text{number of} \\ \text{females in} \\ \text{class } i+1 \\ \text{at time } t_k\end{array}\right\} = \left\{\begin{array}{l}\text{fraction of} \\ \text{females in} \\ \text{class } i \\ \text{who survive} \\ \text{and pass into} \\ \text{class } i+1\end{array}\right\} \left\{\begin{array}{l}\text{number of} \\ \text{females in} \\ \text{class } i \\ \text{at time } t_{k-1}\end{array}\right\}$$

or, mathematically,

$$x_{i+1}^{(k)} = b_i x_i^{(k-1)}, \qquad i = 1, 2, \ldots, n-1 \tag{2}$$

Using matrix notation, Equations (1) and (2) can be written as

$$\begin{bmatrix} x_1^{(k)} \\ x_2^{(k)} \\ x_3^{(k)} \\ \vdots \\ x_n^{(k)} \end{bmatrix} = \begin{bmatrix} a_1 & a_2 & a_3 & \cdots & a_{n-1} & a_n \\ b_1 & 0 & 0 & \cdots & 0 & 0 \\ 0 & b_2 & 0 & \cdots & 0 & 0 \\ \vdots & \vdots & \vdots & & \vdots & \vdots \\ 0 & 0 & 0 & \cdots & b_{n-1} & 0 \end{bmatrix} \begin{bmatrix} x_1^{(k-1)} \\ x_2^{(k-1)} \\ x_3^{(k-1)} \\ \vdots \\ x_n^{(k-1)} \end{bmatrix}$$

or, more compactly,

$$\mathbf{x}^{(k)} = L\mathbf{x}^{(k-1)}, \quad k = 1, 2, \ldots \tag{3}$$

where L is the **Leslie matrix**

$$L = \begin{bmatrix} a_1 & a_2 & a_3 & \cdots & a_{n-1} & a_n \\ b_1 & 0 & 0 & \cdots & 0 & 0 \\ 0 & b_2 & 0 & \cdots & 0 & 0 \\ \vdots & \vdots & \vdots & & \vdots & \vdots \\ 0 & 0 & 0 & \cdots & b_{n-1} & 0 \end{bmatrix} \tag{4}$$

From Equation (3) it follows that

$$\begin{aligned} \mathbf{x}^{(1)} &= L\mathbf{x}^{(0)} \\ \mathbf{x}^{(2)} &= L\mathbf{x}^{(1)} = L^2\mathbf{x}^{(0)} \\ \mathbf{x}^{(3)} &= L\mathbf{x}^{(2)} = L^3\mathbf{x}^{(0)} \\ &\vdots \\ \mathbf{x}^{(k)} &= L\mathbf{x}^{(k-1)} = L^k\mathbf{x}^{(0)} \end{aligned} \tag{5}$$

Thus, if we know the initial age distribution $\mathbf{x}^{(0)}$ and the Leslie matrix L, we can determine the female age distribution at any later time.

EXAMPLE 1 Female Age Distribution for Animals

Suppose that the oldest age attained by the females in a certain animal population is 15 years and we divide the population into three age classes with equal durations of five years. Let the Leslie matrix for this population be

$$L = \begin{bmatrix} 0 & 4 & 3 \\ \frac{1}{2} & 0 & 0 \\ 0 & \frac{1}{4} & 0 \end{bmatrix}$$

If there are initially 1000 females in each of the three age classes, then from Equation (3) we have

$$\mathbf{x}^{(0)} = \begin{bmatrix} 1,000 \\ 1,000 \\ 1,000 \end{bmatrix}$$

$$\mathbf{x}^{(1)} = L\mathbf{x}^{(0)} = \begin{bmatrix} 0 & 4 & 3 \\ \frac{1}{2} & 0 & 0 \\ 0 & \frac{1}{4} & 0 \end{bmatrix} \begin{bmatrix} 1,000 \\ 1,000 \\ 1,000 \end{bmatrix} = \begin{bmatrix} 7,000 \\ 500 \\ 250 \end{bmatrix}$$

$$\mathbf{x}^{(2)} = L\mathbf{x}^{(1)} = \begin{bmatrix} 0 & 4 & 3 \\ \frac{1}{2} & 0 & 0 \\ 0 & \frac{1}{4} & 0 \end{bmatrix} \begin{bmatrix} 7,000 \\ 500 \\ 250 \end{bmatrix} = \begin{bmatrix} 2,750 \\ 3,500 \\ 125 \end{bmatrix}$$

$$\mathbf{x}^{(3)} = L\mathbf{x}^{(2)} = \begin{bmatrix} 0 & 4 & 3 \\ \frac{1}{2} & 0 & 0 \\ 0 & \frac{1}{4} & 0 \end{bmatrix} \begin{bmatrix} 2,750 \\ 3,500 \\ 125 \end{bmatrix} = \begin{bmatrix} 14,375 \\ 1,375 \\ 875 \end{bmatrix}$$

Thus, after 15 years there are 14,375 females between 0 and 5 years of age, 1375 females between 5 and 10 years of age, and 875 females between 10 and 15 years of age. ◆

Limiting Behavior Although Equation (5) gives the age distribution of the population at any time, it does not immediately give a general picture of the dynamics of the growth process. For this we need to investigate the eigenvalues and eigenvectors of the Leslie matrix. The eigenvalues of L are the roots of its characteristic polynomial. As we ask the reader to verify in Exercise 2, this characteristic polynomial is

$$p(\lambda) = |\lambda I - L|$$
$$= \lambda^n - a_1\lambda^{n-1} - a_2b_1\lambda^{n-2} - a_3b_1b_2\lambda^{n-3} - \cdots - a_nb_1b_2\cdots b_{n-1}$$

To analyze the roots of this polynomial, it will be convenient to introduce the function

$$q(\lambda) = \frac{a_1}{\lambda} + \frac{a_2b_1}{\lambda^2} + \frac{a_3b_1b_2}{\lambda^3} + \cdots + \frac{a_nb_1b_2\cdots b_{n-1}}{\lambda^n} \qquad (6)$$

Using this function, the characteristic equation $p(\lambda) = 0$ can be written (verify)

$$q(\lambda) = 1 \quad \text{for } \lambda \neq 0 \qquad (7)$$

Because all the a_i and b_i are nonnegative, we see that $q(\lambda)$ is monotonically decreasing for λ greater than zero. Furthermore, $q(\lambda)$ has a vertical asymptote at $\lambda = 0$ and

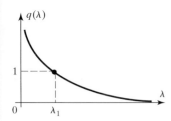

Figure 11.18.1

approaches zero as $\lambda \to \infty$. Consequently, as Figure 11.18.1 indicates, there is a unique λ, say $\lambda = \lambda_1$, such that $q(\lambda_1) = 1$. That is, the matrix L has a unique positive eigenvalue. It can also be shown (see Exercise 3) that λ_1 has multiplicity 1; that is, λ_1 is not a repeated root of the characteristic equation. Although we omit the computational details, the reader can verify that an eigenvector corresponding to λ_1 is

$$\mathbf{x}_1 = \begin{bmatrix} 1 \\ b_1/\lambda_1 \\ b_1 b_2/\lambda_1^2 \\ b_1 b_2 b_3/\lambda_1^3 \\ \vdots \\ b_1 b_2 \cdots b_{n-1}/\lambda_1^{n-1} \end{bmatrix} \tag{8}$$

Because λ_1 has multiplicity 1, its corresponding eigenspace has dimension 1 (Exercise 3), and so any eigenvector corresponding to it is some multiple of $\mathbf{x}_1$. We can summarize these results in the following theorem.

Theorem 11.18.1 **Existence of a Positive Eigenvalue**

A Leslie matrix L has a unique positive eigenvalue λ_1. This eigenvalue has multiplicity 1 and an eigenvector $\mathbf{x}_1$ all of whose entries are positive.

We will now show that the long-term behavior of the age distribution of the population is determined by the positive eigenvalue λ_1 and its eigenvector $\mathbf{x}_1$.

In Exercise 9 we ask the reader to prove the following result.

Theorem 11.18.2 **Eigenvalues of a Leslie Matrix**

If λ_1 is the unique positive eigenvalue of a Leslie matrix L and λ_k is any other real or complex eigenvalue of L, then $|\lambda_k| \leq \lambda_1$.

For our purposes the conclusion in Theorem 11.18.2 is not strong enough; we need λ_1 to satisfy $|\lambda_k| < \lambda_1$. In this case λ_1 would be called the **dominant eigenvalue** of L. However, as the following example shows, not all Leslie matrices satisfy this condition.

EXAMPLE 2 Leslie Matrix with No Dominant Eigenvalue

Let

$$L = \begin{bmatrix} 0 & 0 & 6 \\ \frac{1}{2} & 0 & 0 \\ 0 & \frac{1}{3} & 0 \end{bmatrix}$$

Then the characteristic polynomial of L is

$$p(\lambda) = |\lambda I - L| = \lambda^3 - 1$$

The eigenvalues of L are thus the solutions of $\lambda^3 = 1$; namely,

$$\lambda = 1, \qquad -\frac{1}{2} + \frac{\sqrt{3}}{2}i, \qquad -\frac{1}{2} - \frac{\sqrt{3}}{2}i$$

All three eigenvalues have absolute value 1, and so the unique positive eigenvalue $\lambda_1 = 1$ is not dominant. Note that this matrix has the property that $L^3 = I$. This means that for any choice of the initial age distribution $\mathbf{x}^{(0)}$, we have

$$\mathbf{x}^{(0)} = \mathbf{x}^{(3)} = \mathbf{x}^{(6)} = \cdots = \mathbf{x}^{(3k)} = \cdots$$

The age distribution vector thus oscillates with a period of three time units. Such oscillations (or **population waves**, as they are called) could not occur if λ_1 were dominant, as we will see below. ◆

It is beyond the scope of this book to discuss necessary and sufficient conditions for λ_1 to be a dominant eigenvalue. However, we will state the following sufficient condition without proof.

Theorem 11.18.3 **Dominant Eigenvalue**

If two successive entries a_i and a_{i+1} in the first row of a Leslie matrix L are nonzero, then the positive eigenvalue of L is dominant.

Thus, if the female population has two successive fertile age classes, then its Leslie matrix has a dominant eigenvalue. This is always the case for realistic populations if the duration of the age classes is sufficiently small. Notice that in Example 2 there is only one fertile age class (the third), so the condition of Theorem 11.18.3 is not satisfied. In what follows, we always assume that the condition of Theorem 11.18.3 is satisfied.

Let us assume that L is diagonalizable. This is not really necessary for the conclusions we will draw, but it does simplify the arguments. In this case, L has n eigenvalues, $\lambda_1, \lambda_2, \ldots, \lambda_n$, not necessarily distinct, and n linearly independent eigenvectors, $\mathbf{x}_1, \mathbf{x}_2, \ldots, \mathbf{x}_n$, corresponding to them. In this listing we place the dominant eigenvalue λ_1 first. We construct a matrix P whose columns are the eigenvectors of L:

$$P = [\mathbf{x}_1 \mid \mathbf{x}_2 \mid \mathbf{x}_3 \mid \cdots \mid \mathbf{x}_n]$$

The diagonalization of L is then given by the equation

$$L = P \begin{bmatrix} \lambda_1 & 0 & 0 & \cdots & 0 \\ 0 & \lambda_2 & 0 & \cdots & 0 \\ \vdots & \vdots & \vdots & & \vdots \\ 0 & 0 & 0 & \cdots & \lambda_n \end{bmatrix} P^{-1}$$

From this it follows that

$$L^k = P \begin{bmatrix} \lambda_1^k & 0 & 0 & \cdots & 0 \\ 0 & \lambda_2^k & 0 & \cdots & 0 \\ \vdots & \vdots & \vdots & & \vdots \\ 0 & 0 & 0 & \cdots & \lambda_n^k \end{bmatrix} P^{-1}$$

for $k = 1, 2, \ldots$. For any initial age distribution vector $\mathbf{x}^{(0)}$ we then have

$$L^k \mathbf{x}^{(0)} = P \begin{bmatrix} \lambda_1^k & 0 & 0 & \cdots & 0 \\ 0 & \lambda_2^k & 0 & \cdots & 0 \\ \vdots & \vdots & \vdots & & \vdots \\ 0 & 0 & 0 & \cdots & \lambda_n^k \end{bmatrix} P^{-1} \mathbf{x}^{(0)}$$

for $k = 1, 2, \dots$. Dividing both sides of this equation by λ_1^k and using the fact that $\mathbf{x}^{(k)} = L^k \mathbf{x}^{(0)}$, we have

$$\frac{1}{\lambda_1^k} \mathbf{x}^{(k)} = P \begin{bmatrix} 1 & 0 & 0 & \cdots & 0 \\ 0 & \left(\dfrac{\lambda_2}{\lambda_1}\right)^k & 0 & \cdots & 0 \\ \vdots & \vdots & \vdots & & \vdots \\ 0 & 0 & 0 & \cdots & \left(\dfrac{\lambda_n}{\lambda_1}\right)^k \end{bmatrix} P^{-1} \mathbf{x}^{(0)} \tag{9}$$

Because λ_1 is the dominant eigenvalue, we have $|\lambda_i / \lambda_1| < 1$ for $i = 2, 3, \dots, n$. It follows that

$$(\lambda_i / \lambda_1)^k \to 0 \text{ as } k \to \infty \quad \text{for } i = 2, 3, \dots, n$$

Using this fact, we may take the limit of both sides of (9) to obtain

$$\lim_{k \to \infty} \left\{ \frac{1}{\lambda_1^k} \mathbf{x}^{(k)} \right\} = P \begin{bmatrix} 1 & 0 & 0 & \cdots & 0 \\ 0 & 0 & 0 & \cdots & 0 \\ \vdots & \vdots & \vdots & & \vdots \\ 0 & 0 & 0 & \cdots & 0 \end{bmatrix} P^{-1} \mathbf{x}^{(0)} \tag{10}$$

Let us denote the first entry of the column vector $P^{-1} \mathbf{x}^{(0)}$ by the constant c. As we ask the reader to show in Exercise 4, the right side of (10) can be written as $c\mathbf{x}_1$, where c is a positive constant that depends only on the initial age distribution vector $\mathbf{x}^{(0)}$. Thus, (10) becomes

$$\lim_{k \to \infty} \left\{ \frac{1}{\lambda_1^k} \mathbf{x}^{(k)} \right\} = c\mathbf{x}_1 \tag{11}$$

Equation (11) gives us the approximation

$$\mathbf{x}^{(k)} \simeq c\lambda_1^k \mathbf{x}_1 \tag{12}$$

for large values of k. From (12) we also have

$$\mathbf{x}^{(k-1)} \simeq c\lambda_1^{k-1} \mathbf{x}_1 \tag{13}$$

Comparing Equations (12) and (13) we see that

$$\mathbf{x}^{(k)} \simeq \lambda_1 \mathbf{x}^{(k-1)} \tag{14}$$

for large values of k. This means that for large values of time, each age distribution vector is a scalar multiple of the preceding age distribution vector, the scalar being the positive eigenvalue of the Leslie matrix. Consequently, the *proportion* of females in each of the age classes becomes constant. As we will see in the following example, these limiting proportions can be determined from the eigenvector $\mathbf{x}_1$.

EXAMPLE 3 Example 1 Revisited

The Leslie matrix in Example 1 was

$$L = \begin{bmatrix} 0 & 4 & 3 \\ \frac{1}{2} & 0 & 0 \\ 0 & \frac{1}{4} & 0 \end{bmatrix}$$

Its characteristic polynomial is $p(\lambda) = \lambda^3 - 2\lambda - \frac{3}{8}$, and the reader can verify that the positive eigenvalue is $\lambda_1 = \frac{3}{2}$. From (8) the corresponding eigenvector $\mathbf{x}_1$ is

$$\mathbf{x}_1 = \begin{bmatrix} 1 \\ b_1/\lambda_1 \\ b_1 b_2/\lambda_1^2 \end{bmatrix} = \begin{bmatrix} 1 \\ \frac{1}{2} \\ \frac{3}{2} \\ \frac{\left(\frac{1}{2}\right)\left(\frac{1}{4}\right)}{\left(\frac{3}{2}\right)^2} \end{bmatrix} = \begin{bmatrix} 1 \\ \frac{1}{3} \\ \frac{1}{18} \end{bmatrix}$$

From (14) we have

$$\mathbf{x}^{(k)} \simeq \frac{3}{2}\mathbf{x}^{(k-1)}$$

for large values of k. Hence, every five years the number of females in each of the three classes will increase by about 50%, as will the total number of females in the population.

From (12) we have

$$\mathbf{x}^{(k)} \simeq c\left(\frac{3}{2}\right)^k \begin{bmatrix} 1 \\ \frac{1}{3} \\ \frac{1}{18} \end{bmatrix}$$

Consequently, eventually the females will be distributed among the three age classes in the ratios $1 : \frac{1}{3} : \frac{1}{18}$. This corresponds to a distribution of 72% of the females in the first age class, 24% of the females in the second age class, and 4% of the females in the third age class. ♦

EXAMPLE 4 Female Age Distribution for Humans

In this example we use birth and death parameters from the year 1965 for Canadian females. Because few women over 50 years of age bear children, we restrict ourselves to the portion of the female population between 0 and 50 years of age. The data are for 5-year age classes, so there are a total of 10 age classes. Rather than write out the 10×10 Leslie matrix in full, we list the birth and death parameters as follows:

Age Interval	a_i	b_i
[0, 5)	0.00000	0.99651
[5, 10)	0.00024	0.99820
[10, 15)	0.05861	0.99802
[15, 20)	0.28608	0.99729
[20, 25)	0.44791	0.99694
[25, 30)	0.36399	0.99621
[30, 35)	0.22259	0.99460
[35, 40)	0.10457	0.99184
[40, 45)	0.02826	0.98700
[45, 50)	0.00240	—

Using numerical techniques, the positive eigenvalue and corresponding eigenvector can be approximated by

$$\lambda_1 = 1.07622 \quad \text{and} \quad \mathbf{x}_1 = \begin{bmatrix} 1.00000 \\ 0.92594 \\ 0.85881 \\ 0.79641 \\ 0.73800 \\ 0.68364 \\ 0.63281 \\ 0.58482 \\ 0.53897 \\ 0.49429 \end{bmatrix}$$

Thus, if Canadian women continued to reproduce and die as they did in 1965, eventually every 5 years their numbers would increase by 7.622%. From the eigenvector $\mathbf{x}_1$, we see that, in the limit, for every 100,000 females between 0 and 5 years of age, there will be 92,594 females between 5 and 10 years of age, 85,881 females between 10 and 15 years of age, and so forth. ◆

Let us look again at Equation (12), which gives the age distribution vector of the population for large times:

$$\mathbf{x}^{(k)} \simeq c\lambda_1^k \mathbf{x}_1 \tag{15}$$

Three cases arise according to the value of the positive eigenvalue λ_1:

> (i) The population is eventually increasing if $\lambda_1 > 1$
>
> (ii) The population is eventually decreasing if $\lambda_1 < 1$
>
> (iii) The population eventually stabilizes if $\lambda_1 = 1$

The case $\lambda_1 = 1$ is particularly interesting as it determines a population that has *zero population growth*. For any initial age distribution, the population approaches a limiting age distribution that is some multiple of the eigenvector $\mathbf{x}_1$. From Equations (6) and (7) we see that $\lambda_1 = 1$ is an eigenvalue if and only if

$$a_1 + a_2 b_1 + a_3 b_1 b_2 + \cdots + a_n b_1 b_2 \cdots b_{n-1} = 1 \tag{16}$$

The expression

$$R = a_1 + a_2 b_1 + a_3 b_1 b_2 + \cdots + a_n b_1 b_2 \cdots b_{n-1} \tag{17}$$

is called the *net reproduction rate* of the population. (See Exercise 5 for a demographic interpretation of R.) Thus, we can say that a population has zero population growth if and only if its net reproduction rate is 1.

Exercise Set 11.18

1. Suppose that a certain animal population is divided into two age classes and has a Leslie matrix

$$L = \begin{bmatrix} 1 & \frac{2}{3} \\ \frac{1}{2} & 0 \end{bmatrix}$$

(a) Calculate the positive eigenvalue λ_1 of L and the corresponding eigenvector $\mathbf{x}_1$.

(b) Beginning with the initial age distribution vector

$$\mathbf{x}^{(0)} = \begin{bmatrix} 100 \\ 0 \end{bmatrix}$$

calculate $\mathbf{x}^{(1)}$, $\mathbf{x}^{(2)}$, $\mathbf{x}^{(3)}$, $\mathbf{x}^{(4)}$, and $\mathbf{x}^{(5)}$, rounding off to the nearest integer when necessary.

(c) Calculate $\mathbf{x}^{(6)}$ using the exact formula $\mathbf{x}^{(6)} = L\mathbf{x}^{(5)}$ and using the approximation formula $\mathbf{x}^{(6)} \simeq \lambda_1 \mathbf{x}^{(5)}$.

2. Find the characteristic polynomial of a general Leslie matrix given by Equation (4).

3. (a) Show that the positive eigenvalue λ_1 of a Leslie matrix is always simple. Recall that a root λ_0 of a polynomial $q(\lambda)$ is simple if and only if $q'(\lambda_0) \neq 0$.

(b) Show that the eigenspace corresponding to λ_1 has dimension 1.

4. Show that the right side of Equation (10) is $c\mathbf{x}_1$, where c is the first entry of the column vector $P^{-1}\mathbf{x}^{(0)}$.

5. Show that the net reproduction rate R, defined by (17), can be interpreted as the average number of daughters born to a single female during her expected lifetime.

6. Show that a population is eventually decreasing if and only if its net reproduction rate is less than 1. Similarly, show that a population is eventually increasing if and only if its net reproduction rate is greater than 1.

7. Calculate the net reproduction rate of the animal population in Example 1.

8. (*For readers with a hand calculator.*) Calculate the net reproduction rate of the Canadian female population in Example 4.

9. (*For readers who have read Sections 10.1–10.3.*) Prove Theorem 11.18.2. [**Hint.** Write $\lambda_k = re^{i\theta}$, substitute into (7), take the real parts of both sides, and show that $r \leq \lambda_1$.]

Technology Exercises 11.18

The following exercises are designed to be solved using a technology utility. Typically, this will be MATLAB, *Mathematica*, Maple, Derive, or Mathcad, but it may also be some other type of linear algebra software or a scientific calculator with some linear algebra capabilities. For each exercise you will need to read the relevant documentation for the particular utility you are using. The goal of these exercises is to provide you with a basic proficiency with your technology utility. Once you have mastered the techniques in these exercises, you will be able to use your technology utility to solve many of the problems in the regular exercise sets.

T1. Consider the sequence of Leslie matrices

$$L_2 = \begin{bmatrix} 0 & a \\ b_1 & 0 \end{bmatrix}, \qquad L_3 = \begin{bmatrix} 0 & 0 & a \\ b_1 & 0 & 0 \\ 0 & b_2 & 0 \end{bmatrix},$$

$$L_4 = \begin{bmatrix} 0 & 0 & 0 & a \\ b_1 & 0 & 0 & 0 \\ 0 & b_2 & 0 & 0 \\ 0 & 0 & b_3 & 0 \end{bmatrix}, \qquad L_5 = \begin{bmatrix} 0 & 0 & 0 & 0 & a \\ b_1 & 0 & 0 & 0 & 0 \\ 0 & b_2 & 0 & 0 & 0 \\ 0 & 0 & b_3 & 0 & 0 \\ 0 & 0 & 0 & b_4 & 0 \end{bmatrix}, \ldots$$

(a) Use a computer to show that

$$L_2^2 = I_2, \qquad L_3^3 = I_3, \qquad L_4^4 = I_4, \qquad L_5^5 = I_5, \ldots$$

for a suitable choice of a in terms of $b_1, b_2, \ldots, b_{n-1}$.

(b) From your results in part (a) conjecture a relationship between a and $b_1, b_2, \ldots, b_{n-1}$ that will make $L_n^n = I_n$, where

$$
L_n = \begin{bmatrix}
0 & 0 & 0 & \cdots & 0 & a \\
b_1 & 0 & 0 & \cdots & 0 & 0 \\
0 & b_2 & 0 & \cdots & 0 & 0 \\
0 & 0 & b_3 & \cdots & 0 & 0 \\
\vdots & \vdots & \vdots & \ddots & \vdots & \vdots \\
0 & 0 & 0 & \cdots & b_{n-1} & 0
\end{bmatrix}
$$

(c) Determine an expression for $p_n(\lambda) = |\lambda I_n - L_n|$ and use it to show that all eigenvalues of L_n satisfy $|\lambda| = 1$ when a and $b_1, b_2, \ldots, b_{n-1}$ are related by the equation determined in part (b).

T2. Consider the sequence of Leslie matrices

$$
L_2 = \begin{bmatrix} a & ap \\ b & 0 \end{bmatrix}, \qquad
L_3 = \begin{bmatrix} a & ap & ap^2 \\ b & 0 & 0 \\ 0 & b & 0 \end{bmatrix},
$$

$$
L_4 = \begin{bmatrix} a & ap & ap^2 & ap^3 \\ b & 0 & 0 & 0 \\ 0 & b & 0 & 0 \\ 0 & 0 & b & 0 \end{bmatrix}, \qquad
L_5 = \begin{bmatrix} a & ap & ap^2 & ap^3 & ap^4 \\ b & 0 & 0 & 0 & 0 \\ 0 & b & 0 & 0 & 0 \\ 0 & 0 & b & 0 & 0 \\ 0 & 0 & 0 & b & 0 \end{bmatrix}, \ldots
$$

$$
L_n = \begin{bmatrix}
a & ap & ap^2 & \cdots & ap^{n-2} & ap^{n-1} \\
b & 0 & 0 & \cdots & 0 & 0 \\
0 & b & 0 & \cdots & 0 & 0 \\
0 & 0 & b & \cdots & 0 & 0 \\
\vdots & \vdots & \vdots & \ddots & \vdots & \vdots \\
0 & 0 & 0 & \cdots & b & 0
\end{bmatrix}
$$

where $0 < p < 1$, $0 < b < 1$, and $1 < a$.

(a) Choose a value for n (say, $n = 8$). For various values of a, b, and p, use a computer to determine the dominant eigenvalue of L_n, and then compare your results to the value of $a + bp$.

(b) Show that

$$
p_n(\lambda) = |\lambda I_n - L_n| = \lambda^n - a \left(\frac{\lambda^n - (bp)^n}{\lambda - bp} \right)
$$

which means that the eigenvalues of L_n must satisfy

$$
\lambda^{n+1} - (a + bp)\lambda^n + a(bp)^n = 0
$$

(c) Can you now provide a rough proof to explain the fact that $\lambda_1 \simeq a + bp$?

11.19   HARVESTING OF ANIMAL POPULATIONS

In this section we employ the Leslie matrix model of population growth to model the sustainable harvesting of an animal population. We also examine the effect of harvesting different fractions of different age groups.

PREREQUISITE: Age-specific Population Growth (Section 11.18)

Harvesting   In Section 11.18 we used the Leslie matrix model to examine the growth of a female population that was divided into discrete age classes. In this section, we investigate the effects of harvesting an animal population growing according to such a model. By *harvesting* we mean the removal of animals from the population. (The word "harvesting" is not necessarily a euphemism for "slaughtering"; the animals may be removed from the population for other purposes.)

In this section we restrict ourselves to *sustainable harvesting policies*. By this we mean the following:

Definition

A harvesting policy in which an animal population is periodically harvested is said to be **sustainable** if the yield of each harvest is the same and the age distribution of the population remaining after each harvest is the same.

Thus, the animal population is not depleted by a sustainable harvesting policy; only the excess growth is removed.

As in Section 11.18 we will discuss only the females of the population. If the number of males in each age class is equal to the number of females—a reasonable assumption for many populations—then our harvesting policies will also apply to the male portion of the population.

The Harvesting Model   Figure 11.19.1 illustrates the basic idea of the model. We begin with a population having a particular age distribution. It undergoes

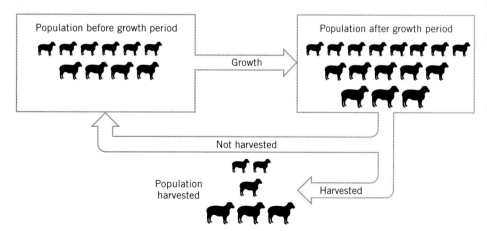

Figure 11.19.1

a growth period that will be described by the Leslie matrix. At the end of the growth period, a certain fraction of each age class is harvested in such a way that the unharvested population has the same age distribution as the original population. This cycle repeats after each harvest so that the yield is sustainable. The duration of the harvest is assumed to be short in comparison with the growth period so that any growth or change in the population during the harvest period can be neglected.

To describe this harvesting model mathematically, let

$$\mathbf{x} = \begin{bmatrix} x_1 \\ x_2 \\ \vdots \\ x_n \end{bmatrix}$$

be the age distribution vector of the population at the beginning of the growth period. Thus, x_i is the number of females in the ith class left unharvested. As in Section 11.18 we require that the duration of each age class be identical with the duration of the growth period. For example, if the population is harvested once a year, then the population is divided into 1-year age classes.

If L is the Leslie matrix describing the growth of the population, then the vector $L\mathbf{x}$ is the age distribution vector of the population at the end of the growth period, immediately before the periodic harvest. Let h_i, for $i = 1, 2, \ldots, n$, be the fraction of females from the ith class which is harvested. We use these n numbers to form an $n \times n$ diagonal matrix

$$H = \begin{bmatrix} h_1 & 0 & 0 & \cdots & 0 \\ 0 & h_2 & 0 & \cdots & 0 \\ 0 & 0 & h_3 & \cdots & 0 \\ \vdots & \vdots & \vdots & & \vdots \\ 0 & 0 & 0 & \cdots & h_n \end{bmatrix}$$

which we shall call the **harvesting matrix**. By definition, we have

$$0 \le h_i \le 1 \quad (i = 1, 2, \ldots, n)$$

That is, we may harvest none ($h_i = 0$), all ($h_i = 1$), or some fraction ($0 < h_i < 1$) of each of the n classes. Because the number of females in the ith class immediately before each harvest is the ith entry $(L\mathbf{x})_i$ of the vector $L\mathbf{x}$, it can be seen that the ith entry of the column vector

$$HL\mathbf{x} = \begin{bmatrix} h_1(L\mathbf{x})_1 \\ h_2(L\mathbf{x})_2 \\ \vdots \\ h_n(L\mathbf{x})_n \end{bmatrix}$$

is the number of females harvested from the ith class.

From the definition of a sustainable harvesting policy, we have

$$\begin{bmatrix} \text{age distribution} \\ \text{at end of} \\ \text{growth period} \end{bmatrix} - [\text{harvest}] = \begin{bmatrix} \text{age distribution} \\ \text{at beginning of} \\ \text{growth period} \end{bmatrix}$$

or, mathematically,

$$L\mathbf{x} - HL\mathbf{x} = \mathbf{x} \tag{1}$$

If we write Equation (1) in the form

$$(I - H)L\mathbf{x} = \mathbf{x} \tag{2}$$

we see that **x** must be an eigenvector of the matrix $(I - H)L$ corresponding to the eigenvalue 1. As we will now show, this places certain restrictions on the values of h_i and **x**.

Suppose that the Leslie matrix of the population is

$$
L = \begin{bmatrix}
a_1 & a_2 & a_3 & \cdots & a_{n-1} & a_n \\
b_1 & 0 & 0 & \cdots & 0 & 0 \\
0 & b_2 & 0 & \cdots & 0 & 0 \\
\vdots & \vdots & \vdots & & \vdots & \vdots \\
0 & 0 & 0 & \cdots & b_{n-1} & 0
\end{bmatrix}
\tag{3}
$$

Then the matrix $(I - H)L$ is (verify)

$$
(I-H)L = \begin{bmatrix}
(1-h_1)a_1 & (1-h_1)a_2 & (1-h_1)a_3 & \cdots & (1-h_1)a_{n-1} & (1-h_1)a_n \\
(1-h_2)b_1 & 0 & 0 & \cdots & 0 & 0 \\
0 & (1-h_3)b_2 & 0 & \cdots & 0 & 0 \\
\vdots & \vdots & \vdots & & \vdots & \vdots \\
0 & 0 & 0 & \cdots & (1-h_n)b_{n-1} & 0
\end{bmatrix}
$$

Thus, we see that $(I - H)L$ is a matrix with the same mathematical form as a Leslie matrix. In Section 11.18 we showed that a necessary and sufficient condition for a Leslie matrix to have 1 as an eigenvalue is that its net reproduction rate also be 1 [see Eq. (16) of Section 11.18]. Calculating the net reproduction rate of $(I - H)L$ and setting it equal to 1, we obtain (verify)

$$
\begin{aligned}
(1 - h_1)[a_1 + a_2 b_1 (1 - h_2) &+ a_3 b_1 b_2 (1 - h_2)(1 - h_3) + \cdots \\
&+ a_n b_1 b_2 \cdots b_{n-1} (1 - h_2)(1 - h_3) \cdots (1 - h_n)] = 1
\end{aligned}
\tag{4}
$$

This equation places a restriction on the allowable harvesting fractions. Only those values of $h_1, h_2, \ldots, h_n$ which satisfy (4) and which lie in the interval $[0, 1]$ can produce a sustainable yield.

If $h_1, h_2, \ldots, h_n$ do satisfy (4), then the matrix $(I - H)L$ has the desired eigenvalue $\lambda_1 = 1$; and, furthermore, this eigenvalue has multiplicity 1, as the positive eigenvalue of a Leslie matrix always has multiplicity 1 (Theorem 11.18.1). This means that there is only one linearly independent eigenvector **x** satisfying Equation (2). [See Exercise 3(b) of Section 11.18.] One possible choice for **x** is the following normalized eigenvector:

$$
\mathbf{x}_1 = \begin{bmatrix}
1 \\
b_1(1 - h_2) \\
b_1 b_2 (1 - h_2)(1 - h_3) \\
b_1 b_2 b_3 (1 - h_2)(1 - h_3)(1 - h_4) \\
\vdots \\
b_1 b_2 b_3 \cdots b_{n-1} (1 - h_2)(1 - h_3) \cdots (1 - h_n)
\end{bmatrix}
\tag{5}
$$

Any other solution **x** of (2) is a multiple of $\mathbf{x}_1$. Thus, the vector $\mathbf{x}_1$ determines the proportion of females within each of the n classes after a harvest under a sustainable harvesting policy. But there is an ambiguity in the total number of females in the population after each harvest. This can be determined by some auxiliary condition, such as an ecological or economic constraint. For example, for a population economically supported by the harvester, the largest population the harvester can afford to raise between harvests would determine the particular constant that $\mathbf{x}_1$ is multiplied by to produce the appropriate vector **x** in Equation (2). For a wild population the natural habitat of the population would determine how large the total population could be between harvests.

Summarizing our results so far, we see that there is a wide choice in the values of $h_1, h_2, \ldots, h_n$ that will produce a sustainable yield. But once these values are selected, the proportional age distribution of the population after each harvest is uniquely determined by the normalized eigenvector $\mathbf{x}_1$ defined by Equation (5). We now consider a few particular harvesting strategies of this type.

Uniform Harvesting

With many populations it is difficult to distinguish or catch animals of specific ages. If animals are caught at random, we can reasonably assume that the same fraction of each age class is harvested. We therefore set

$$h = h_1 = h_2 = \cdots = h_n$$

Equation (2) then reduces to (verify)

$$L\mathbf{x} = \left(\frac{1}{1-h}\right)\mathbf{x}$$

Hence, $1/(1-h)$ must be the unique positive eigenvalue λ_1 of the Leslie growth matrix L. That is,

$$\lambda_1 = \frac{1}{1-h}$$

Solving for the harvesting fraction h, we obtain

$$h = 1 - (1/\lambda_1) \tag{6}$$

The vector $\mathbf{x}_1$, in this case, is the same as the eigenvector of L corresponding to the eigenvalue λ_1. From Equation (8) of Section 11.18, this is

$$\mathbf{x}_1 = \begin{bmatrix} 1 \\ b_1/\lambda_1 \\ b_1 b_2/\lambda_1^2 \\ b_1 b_2 b_3/\lambda_1^3 \\ \vdots \\ b_1 b_2 \cdots b_{n-1}/\lambda_1^{n-1} \end{bmatrix} \tag{7}$$

From (6) we can see that the larger λ_1 is, the larger is the fraction of animals we can harvest without depleting the population. Notice that we need $\lambda_1 > 1$ in order that the harvesting fraction h lie in the interval $(0, 1)$. This is to be expected, as $\lambda_1 > 1$ is the condition that the population be increasing.

EXAMPLE 1 Harvesting Sheep

For a certain species of domestic sheep in New Zealand with a growth period of one year, the following Leslie matrix was found (see G. Caughley, "Parameters for Seasonally

Breeding Populations," *Ecology*, vol. 48, 1967, pp. 834–839):

$$
L = \begin{bmatrix}
.000 & .045 & .391 & .472 & .484 & .546 & .543 & .502 & .468 & .459 & .433 & .421 \\
.845 & 0 & 0 & 0 & 0 & 0 & 0 & 0 & 0 & 0 & 0 & 0 \\
0 & .975 & 0 & 0 & 0 & 0 & 0 & 0 & 0 & 0 & 0 & 0 \\
0 & 0 & .965 & 0 & 0 & 0 & 0 & 0 & 0 & 0 & 0 & 0 \\
0 & 0 & 0 & .950 & 0 & 0 & 0 & 0 & 0 & 0 & 0 & 0 \\
0 & 0 & 0 & 0 & .926 & 0 & 0 & 0 & 0 & 0 & 0 & 0 \\
0 & 0 & 0 & 0 & 0 & .895 & 0 & 0 & 0 & 0 & 0 & 0 \\
0 & 0 & 0 & 0 & 0 & 0 & .850 & 0 & 0 & 0 & 0 & 0 \\
0 & 0 & 0 & 0 & 0 & 0 & 0 & .786 & 0 & 0 & 0 & 0 \\
0 & 0 & 0 & 0 & 0 & 0 & 0 & 0 & .691 & 0 & 0 & 0 \\
0 & 0 & 0 & 0 & 0 & 0 & 0 & 0 & 0 & .561 & 0 & 0 \\
0 & 0 & 0 & 0 & 0 & 0 & 0 & 0 & 0 & 0 & .370 & 0
\end{bmatrix}
$$

The sheep have a lifespan of 12 years, so they are divided into 12 age classes of duration 1 year each. By the use of numerical techniques, the unique positive eigenvalue of L can be found to be

$$\lambda_1 = 1.176$$

From Equation (6) the harvesting fraction h is

$$h = 1 - (1/\lambda_1) = 1 - (1/1.176) = 0.150$$

Thus, the uniform harvesting policy is one in which 15.0% of the sheep from each of the 12 age classes is harvested every year. From (7) the age distribution vector of the sheep after each harvest is proportional to

$$
\mathbf{x}_1 = \begin{bmatrix}
1.000 \\
0.719 \\
0.596 \\
0.489 \\
0.395 \\
0.311 \\
0.237 \\
0.171 \\
0.114 \\
0.067 \\
0.032 \\
0.010
\end{bmatrix}
\tag{8}
$$

From (8) we see that for every 1000 sheep between 0 and 1 year of age that are not harvested, there are 719 sheep between 1 and 2 years of age, 596 sheep between 2 and 3 years of age, and so forth. ◆

Harvesting Only the Youngest Age Class

In some populations only the youngest females are of any economic value, so the harvester seeks to harvest only the females from the youngest age class. Accordingly, let us set

$$h_1 = h$$
$$h_2 = h_3 = \cdots = h_n = 0$$

Equation (4) then reduces to

$$(1 - h)(a_1 + a_2 b_1 + a_3 b_1 b_2 + \cdots + a_n b_1 b_2 \cdots b_{n-1}) = 1$$

or

$$(1 - h)R = 1$$

where R is the net reproduction rate of the population. [See Equation (17) of Section 11.18.] Solving for h, we obtain

$$h = 1 - (1/R) \tag{9}$$

Notice from this equation that a sustainable harvesting policy is possible only if $R > 1$. This is reasonable because only if $R > 1$ is the population increasing. From Equation (5) the age distribution vector after each harvest is proportional to the vector

$$\mathbf{x}_1 = \begin{bmatrix} 1 \\ b_1 \\ b_1 b_2 \\ b_1 b_2 b_3 \\ \vdots \\ b_1 b_2 b_3 \cdots b_{n-1} \end{bmatrix} \tag{10}$$

EXAMPLE 2 Sustainable Harvesting Policy

Let us apply this type of sustainable harvesting policy to the sheep population in Example 1. For the net reproduction rate of the population we find

$$\begin{aligned} R &= a_1 + a_2 b_1 + a_3 b_1 b_2 + \cdots + a_n b_1 b_2 \cdots b_{n-1} \\ &= (.000) + (.045)(.845) + \cdots + (.421)(.845)(.975) \cdots (.370) \\ &= 2.514 \end{aligned}$$

From Equation (9) the fraction of the first age class harvested is

$$h = 1 - (1/R) = 1 - (1/2.514) = .602$$

From Equation (10) the age distribution of the sheep population after the harvest is proportional to the vector

$$\mathbf{x}_1 = \begin{bmatrix} 1.000 \\ 0.845 \\ (.845)(.975) \\ (.845)(.975)(.965) \\ \\ \vdots \\ \\ (.845)(.975) \cdots (.370) \end{bmatrix} = \begin{bmatrix} 1.000 \\ 0.845 \\ 0.824 \\ 0.795 \\ 0.755 \\ 0.699 \\ 0.626 \\ 0.532 \\ 0.418 \\ 0.289 \\ 0.162 \\ 0.060 \end{bmatrix} \tag{11}$$

A direct calculation gives us the following (see also Exercise 3):

$$L\mathbf{x}_1 = \begin{bmatrix} 2.514 \\ 0.845 \\ 0.824 \\ 0.795 \\ 0.755 \\ 0.699 \\ 0.626 \\ 0.532 \\ 0.418 \\ 0.289 \\ 0.162 \\ 0.060 \end{bmatrix} \tag{12}$$

The vector $L\mathbf{x}_1$ is the age distribution vector immediately before the harvest. The total of all entries in $L\mathbf{x}_1$ is 8.520, so that the first entry 2.514 is 29.5% of the total. This means that immediately before each harvest, 29.5% of the population is in the youngest age class. Since 60.2% of this class is harvested, it follows that 17.8% (= 60.2% of 29.5%) of the entire sheep population is harvested each year. This can be compared with the uniform harvesting policy of Example 1, in which 15.0% of the sheep population is harvested each year.   ◆

Optimal Sustainable Yield   We saw in Example 1 that a sustainable harvesting policy in which the same fraction of each age class is harvested produces a yield of 15.0% of the sheep population. In Example 2 we saw that if only the youngest age class is harvested, the resulting yield is 17.8% of the population. There are many other possible sustainable harvesting policies and each will generally provide a different yield. It would be of interest to find a sustainable harvesting policy that produces the largest possible yield. Such a policy is called an ***optimal sustainable harvesting policy***, and the resulting yield is called the ***optimal sustainable yield***. However, determining the optimal sustainable yield requires linear programming theory, which we will not discuss here. We refer the reader to the following result, which appears in J. R. Beddington and D. B. Taylor, "Optimum Age Specific Harvesting of a Population," *Biometrics*, vol. 29, 1973, pp. 801–809.

Theorem 11.19.1   **Optimal Sustainable Yield**

An optimal sustainable harvesting policy is one in which either one or two age classes are harvested. If two age classes are harvested, then the older age class is completely harvested.

As an illustration, it can be shown that the optimal sustainable yield of the sheep population is attained when

$$\begin{aligned} h_1 &= 0.522 \\ h_9 &= 1.000 \end{aligned} \tag{13}$$

and all other values of h_i are zero. Thus, 52.2% of the sheep between 0 and 1 year of age and all the sheep between 8 and 9 years of age are harvested. As we ask the reader to show in Exercise 2, the resulting optimal sustainable yield is 19.9% of the population.

Exercise Set 11.19

1. Let a certain animal population be divided into three 1-year age classes and have as its Leslie matrix

$$L = \begin{bmatrix} 0 & 4 & 3 \\ \frac{1}{2} & 0 & 0 \\ 0 & \frac{1}{4} & 0 \end{bmatrix}$$

(a) Find the yield and the age distribution vector after each harvest if the same fraction of each of the three age classes is harvested every year.

(b) Find the yield and the age distribution vector after each harvest if only the youngest age class is harvested every year. Also, find the fraction of the youngest age class that is harvested.

2. For the optimal sustainable harvesting policy described by Equations (13), find the vector $\mathbf{x}_1$ that specifies the age distribution of the population after each harvest. Also calculate the vector $L\mathbf{x}_1$ and verify that the optimal sustainable yield is 19.9% of the population.

3. If only the first age class of an animal population is harvested, use Equation (10) to show that

$$L\mathbf{x}_1 - \mathbf{x}_1 = \begin{bmatrix} R - 1 \\ 0 \\ 0 \\ \vdots \\ 0 \end{bmatrix}$$

where R is the net reproduction rate of the population.

4. If only the Ith class of an animal population is to be periodically harvested ($I = 1, 2, \ldots, n$), find the corresponding harvesting fraction h_I.

5. Suppose that all of the Jth class and a certain fraction h_I of the Ith class of an animal population is to be periodically harvested ($1 \leq I < J \leq n$). Calculate h_I.

Technology Exercises 11.19

The following exercises are designed to be solved using a technology utility. Typically, this will be MATLAB, *Mathematica*, Maple, Derive, or Mathcad, but it may also be some other type of linear algebra software or a scientific calculator with some linear algebra capabilities. For each exercise you will need to read the relevant documentation for the particular utility you are using. The goal of these exercises is to provide you with a basic proficiency with your technology utility. Once you have mastered the techniques in these exercises, you will be able to use your technology utility to solve many of the problems in the regular exercise sets.

T1. The results of Theorem 11.19.1 suggest the following algorithm for determining the optimal sustainable yield.

(i) For each value of $i = 1, 2, \ldots, n$, set $h_i = h$ and $h_k = 0$ for $k \neq i$ and calculate the respective yields. These n calculations give the one-age-class results. Of course, any calculation leading to a value of h not between 0 and 1 is rejected.

(ii) For each value of $i = 1, 2, \ldots, n - 1$ and $j = i + 1, i + 2, \ldots, n$, set $h_i = h$, $h_j = 1$, and $h_k = 0$ for $k \neq i, j$ and calculate the respective yields. These $\frac{1}{2}n(n - 1)$ calculations give the two-age-class results. Of course, any calculation leading to a value of h not between 0 and 1 is again rejected.

(iii) Of the yields calculated in (i) and (ii), the largest is the optimal sustainable yield. Note that there will be at most

$$n + \frac{1}{2}n(n-1) = \frac{1}{2}n(n+1)$$

calculations in all. Once again, some of these may lead to a value of h not between 0 and 1 and must therefore be rejected.

If we use this algorithm for the sheep example in the text, there will be at most $\frac{1}{2}(12)(12+1) = 78$ calculations to consider. Use a computer to do the two-age-class calculations for $h_1 = h$, $h_j = 1$, and $h_k = 0$ for $k \neq 1$ or j for $j = 2, 3, \ldots, 12$. Construct a summary table consisting of the values of h_1 and the percentage yields using $j = 2, 3, \ldots, 12$, which will show that the largest of these yields occurs when $j = 9$.

T2. Using the algorithm in Exercise T1, do the one-age-class calculations for $h_i = h$ and $h_k = 0$ for $k \neq i$ for $i = 1, 2, \ldots, 12$. Construct a summary table consisting of the values of h_i and the percentage yields using $i = 1, 2, \ldots, 12$, which will show that the largest of these yields occurs when $i = 9$.

11.20   A LEAST SQUARES MODEL FOR HUMAN HEARING

In this section we apply the method of least squares approximation to a model for human hearing. The use of this method is motivated by energy considerations.

PREREQUISITES: Inner Product Spaces
                 Orthogonal Projection
                 Fourier Series (Section 9.4)

Anatomy of the Ear   We begin with a brief discussion of the nature of sound and human hearing. Figure 11.20.1 is a schematic diagram of the ear showing its three main components: the outer ear, middle ear, and inner ear. Sound waves enter the outer ear where they are channeled to the eardrum, causing it to vibrate. Three tiny bones in the middle ear mechanically link the eardrum with the snail-shaped cochlea within the inner ear. These bones pass on the vibrations of the eardrum to a fluid within the cochlea. The cochlea contains thousands of minute hairs that oscillate with the fluid. Those near the entrance of the cochlea are stimulated by high frequencies and those near the tip are stimulated by low frequencies. The movements of these hairs activate nerve

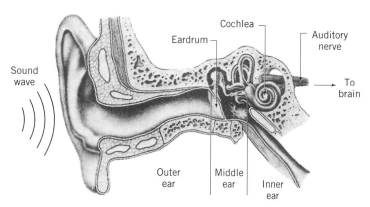

Figure 11.20.1

cells that send signals along various neural pathways to the brain, where the signals are interpreted as sound.

The sound waves themselves are variations in time of the air pressure. For the auditory system, the most elementary type of sound wave is a sinusoidal variation in the air pressure. This type of sound wave stimulates the hairs within the cochlea in such a way that a nerve impulse along a single neural pathway is produced (Figure 11.20.2). A sinusoidal sound wave can be described by a function of time

$$q(t) = A_0 + A \sin(\omega t - \delta) \tag{1}$$

where $q(t)$ is the atmospheric pressure at the eardrum, A_0 is the normal atmospheric pressure, A is the maximum deviation of the pressure from the normal atmospheric pressure, $\omega/2\pi$ is the frequency of the wave in cycles per second, and δ is the phase angle of the wave. To be perceived as sound, such sinusoidal waves must have frequencies within a certain range. For humans this range is roughly 20 cycles per second (cps) to 20,000 cps. Frequencies outside this range will not stimulate the hairs within the cochlea enough to produce nerve signals.

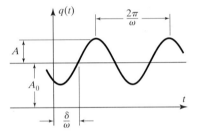
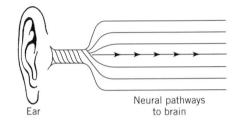

Figure 11.20.2

To a reasonable degree of accuracy, the ear is a linear system. This means that if a complex sound wave is a finite sum of sinusoidal components of different amplitudes, frequencies, and phase angles, say,

$$q(t) = A_0 + A_1 \sin(\omega_1 t - \delta_1) + A_2 \sin(\omega_2 t - \delta_2) + \cdots + A_n \sin(\omega_n t - \delta_n) \tag{2}$$

then the response of the ear consists of nerve impulses along the same neural pathways that would be stimulated by the individual components (Figure 11.20.3).

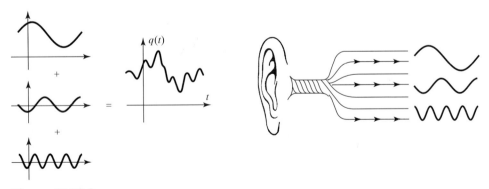

Figure 11.20.3

Let us now consider some periodic sound wave $p(t)$ with period T [i.e., $p(t) \equiv p(t + T)$] that is *not* a finite sum of sinusoidal waves. If we examine the response of the ear to such a periodic wave, we find that it is the same as the response to some wave that is the sum of sinusoidal waves. That is, there is some sound wave $q(t)$ as given by Equation (2) that produces the same response as $p(t)$, even though $p(t)$ and $q(t)$ are different functions of time.

We now want to determine the frequencies, amplitudes, and phase angles of the sinusoidal components of $q(t)$. Because $q(t)$ produces the same response as the periodic wave $p(t)$, it is reasonable to expect that $q(t)$ has the same period T as $p(t)$. This requires that each sinusoidal term in $q(t)$ have period T. Consequently, the frequencies of the sinusoidal components must be integer multiples of the basic frequency $1/T$ of the function $p(t)$. Thus, the ω_k in Equation (2) must be of the form

$$\omega_k = 2k\pi/T, \qquad k = 1, 2, \dots$$

But because the ear cannot perceive sinusoidal waves with frequencies greater than 20,000 cps, we may omit those values of k for which $\omega_k/2\pi = k/T$ is greater than 20,000. Thus, $q(t)$ is of the form

$$q(t) = A_0 + A_1 \sin\left(\frac{2\pi t}{T} - \delta_1\right) + \cdots + A_n \sin\left(\frac{2n\pi t}{T} - \delta_n\right) \tag{3}$$

where n is the largest integer such that n/T is not greater than 20,000.

We now turn our attention to the values of the amplitudes $A_0, A_1, \dots, A_n$ and the phase angles $\delta_1, \delta_2, \dots, \delta_n$ that appear in Equation (3). There is some criterion by which the auditory system "picks" these values so that $q(t)$ produces the same response as $p(t)$. To examine this criterion, let us set

$$e(t) = p(t) - q(t)$$

If we consider $q(t)$ as an approximation to $p(t)$, then $e(t)$ is the error in this approximation, an error that the ear cannot perceive. In terms of $e(t)$, the criterion for the determination of the amplitudes and the phase angles is that the quantity

$$\int_0^T [e(t)]^2 \, dt = \int_0^T [p(t) - q(t)]^2 \, dt \tag{4}$$

be as small as possible. We cannot go into the physiological reasons for this, but we note that this expression is proportional to the *acoustic energy* of the error wave $e(t)$ over one period. In other words, it is the energy of the difference between the two sound waves $p(t)$ and $q(t)$ that determines whether the ear perceives any difference between them. If this energy is as small as possible, then the two waves produce the same sensation of sound. Mathematically, the function $q(t)$ in (4) is the least squares approximation to $p(t)$ from the vector space $C[0, T]$ of continuous functions on the interval $[0, T]$. (See Section 9.4.)

Least squares approximations by continuous functions arise in a wide variety of engineering and scientific approximation problems. Apart from the acoustics problem just discussed, some other examples are as follows.

1. Let $S(x)$ be the axial strain distribution in a uniform rod lying along the x-axis from $x = 0$ to $x = l$ (Figure 11.20.4). The strain energy in the rod is proportional to the integral

$$\int_0^l [S(x)]^2 \, dx$$

The closeness of an approximation $q(x)$ to $S(x)$ can be judged according to the strain energy of the difference of the two strain distributions. That energy is proportional to

$$\int_0^l [S(x) - q(x)]^2 \, dx$$

which is a least squares criterion.

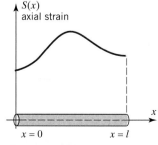

Figure 11.20.4

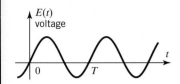

Figure 11.20.5

2. Let $E(t)$ be a periodic voltage across a resistor in an electrical circuit (Figure 11.20.5). The electrical energy transferred to the resistor during one period T is proportional to

$$\int_0^T [E(t)]^2\, dt$$

If $q(t)$ has the same period as $E(t)$ and is to be an approximation to $E(t)$, then the criterion of closeness might be taken as the energy of the difference voltage. This is proportional to

$$\int_0^T [E(t) - q(t)]^2\, dt$$

which is again a least squares criterion.

3. Let $y(x)$ be the vertical displacement of a uniform flexible string whose equilibrium position is along the x-axis from $x = 0$ to $x = l$ (Figure 11.20.6). The elastic potential energy of the string is proportional to

$$\int_0^l [y(x)]^2\, dx$$

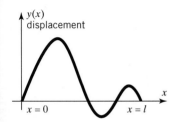

Figure 11.20.6

If $q(x)$ is to be an approximation to the displacement, then as before, the energy integral

$$\int_0^l [y(x) - q(x)]^2\, dx$$

determines a least squares criterion for the closeness of the approximation.

Least squares approximation is also used in situations where there is no a priori justification for its use, such as for approximating business cycles, population growth curves, sales curves, and so forth. It is used in these cases because of its mathematical simplicity. In general, if no other error criterion is immediately apparent for an approximation problem, the least squares criterion is the one most often chosen.

The following result was obtained in Section 9.4.

Theorem 11.20.1   **Minimizing Mean Square Error on $[0, 2\pi]$**

If $f(t)$ is continuous on $[0, 2\pi]$, the trigonometric function $g(t)$ of the form

$$g(t) = \tfrac{1}{2}a_0 + a_1 \cos t + \cdots + a_n \cos nt + b_1 \sin t + \cdots + b_n \sin nt$$

which minimizes the mean square error

$$\int_0^{2\pi} [f(t) - g(t)]^2\, dt$$

has coefficients

$$a_k = \frac{1}{\pi} \int_0^{2\pi} f(t) \cos kt\, dt, \qquad k = 0, 1, 2, \ldots, n$$

$$b_k = \frac{1}{\pi} \int_0^{2\pi} f(t) \sin kt\, dt, \qquad k = 1, 2, \ldots, n$$

If the original function $f(t)$ is defined over the interval $[0, T]$ instead of $[0, 2\pi]$, a change of scale will yield the following result (see Exercise 8):

> ## Theorem 11.20.2 | Minimizing Mean Square Error on $[0, T]$
>
> *If $f(t)$ is continuous on $[0, T]$, the trigonometric function $g(t)$ of the form*
>
> $$g(t) = \frac{1}{2}a_0 + a_1 \cos \frac{2\pi}{T}t + \cdots + a_n \cos \frac{2n\pi}{T}t + b_1 \sin \frac{2\pi}{T}t + \cdots + b_n \sin \frac{2n\pi}{T}t$$
>
> *which minimizes the mean square error*
>
> $$\int_0^T [f(t) - g(t)]^2 \, dt$$
>
> *has coefficients*
>
> $$a_k = \frac{2}{T} \int_0^T f(t) \cos \frac{2k\pi t}{T} \, dt, \quad k = 0, 1, 2, \dots, n$$
>
> $$b_k = \frac{2}{T} \int_0^T f(t) \sin \frac{2k\pi t}{T} \, dt, \quad k = 1, 2, \dots, n$$

EXAMPLE 1 Least Squares Approximation to a Sound Wave

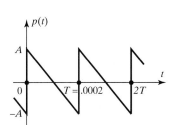

Figure 11.20.7

Let a sound wave $p(t)$ have a saw-tooth pattern with a basic frequency of 5000 cps (Figure 11.20.7). Assume units are chosen so that the normal atmospheric pressure is at the zero level and the maximum amplitude of the wave is A. The basic period of the wave is $T = 1/5000 = .0002$ second. From $t = 0$ to $t = T$, the function $p(t)$ has the equation

$$p(t) = \frac{2A}{T}\left(\frac{T}{2} - t\right)$$

Theorem 11.20.2 then yields the following (verify):

$$a_0 = \frac{2}{T}\int_0^T p(t)\, dt = \frac{2}{T}\int_0^T \frac{2A}{T}\left(\frac{T}{2} - t\right) dt = 0$$

$$a_k = \frac{2}{T}\int_0^T p(t)\cos \frac{2k\pi t}{T}\, dt = \frac{2}{T}\int_0^T \frac{2A}{T}\left(\frac{T}{2} - t\right)\cos \frac{2k\pi t}{T}\, dt = 0, \quad k = 1, 2, \dots$$

$$b_k = \frac{2}{T}\int_0^T p(t)\sin \frac{2k\pi t}{T}\, dt = \frac{2}{T}\int_0^T \frac{2A}{T}\left(\frac{T}{2} - t\right)\sin \frac{2k\pi t}{T}\, dt = \frac{2A}{k\pi}, \quad k = 1, 2, \dots$$

We can now investigate how the sound wave $p(t)$ is perceived by the human ear. We notice that $4/T = 20{,}000$ cps, so we need only go up to $k = 4$ in the formulas above. The least squares approximation to $p(t)$ is then

$$q(t) = \frac{2A}{\pi}\left[\sin \frac{2\pi}{T}t + \frac{1}{2}\sin \frac{4\pi}{T}t + \frac{1}{3}\sin \frac{6\pi}{T}t + \frac{1}{4}\sin \frac{8\pi}{T}t\right]$$

The four sinusoidal terms have frequencies of 5000, 10,000, 15,000, and 20,000 cps, respectively. In Figure 11.20.8 we have plotted $p(t)$ and $q(t)$ over one period. Although $q(t)$ is not a very good point-by-point approximation to $p(t)$, to the ear, both $p(t)$ and $q(t)$ produce the same sensation of sound. ◆

As discussed in Section 9.4, the least squares approximation becomes better as the number of terms in the approximating trigonometric polynomial becomes larger. More precisely,

$$\int_0^{2\pi}\left[f(t) - \frac{1}{2}a_0 - \sum_{k=1}^n (a_k \cos kt + b_k \sin kt)\right]^2 dt$$

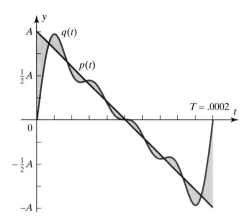

Figure 11.20.8

tends to zero as n approaches infinity. We denote this by writing

$$f(t) = \frac{1}{2}a_0 + \sum_{k=1}^{\infty}(a_k \cos kt + b_k \sin kt)$$

where the right side of this equation is the Fourier series of $f(t)$. Whether the Fourier series of $f(t)$ converges to $f(t)$ for each t is another question, and a more difficult one. For most continuous functions encountered in applications, the Fourier series does indeed converge to its corresponding function for each value of t.

Exercise Set 11.20

1. Find the trigonometric polynomial of order 3 that is the least squares approximation to the function $f(t) = (t - \pi)^2$ over the interval $[0, 2\pi]$.

2. Find the trigonometric polynomial of order 4 that is the least squares approximation to the function $f(t) = t^2$ over the interval $[0, T]$.

3. Find the trigonometric polynomial of order 4 that is the least squares approximation to the function $f(t)$ over the interval $[0, 2\pi]$, where

$$f(t) = \begin{cases} \sin t, & 0 \le t \le \pi \\ 0, & \pi < t \le 2\pi \end{cases}$$

4. Find the trigonometric polynomial of arbitrary order n that is the least squares approximation to the function $f(t) = \sin \frac{1}{2}t$ over the interval $[0, 2\pi]$.

5. Find the trigonometric polynomial of arbitrary order n that is the least squares approximation to the function $f(t)$ over the interval $[0, T]$, where

$$f(t) = \begin{cases} t, & 0 \le t \le \frac{1}{2}T \\ T - t, & \frac{1}{2}T < t \le T \end{cases}$$

6. For the inner product

$$\langle \mathbf{u}, \mathbf{v} \rangle = \int_0^{2\pi} u(t)v(t)\, dt$$

show that

(a) $\|1\| = \sqrt{2\pi}$ (b) $\| \cos kt \| = \sqrt{\pi}$ for $k = 1, 2, \ldots$ (c) $\| \sin kt \| = \sqrt{\pi}$ for $k = 1, 2, \ldots$

7. Show that the $2n + 1$ functions

$$1, \cos t, \cos 2t, \ldots, \cos nt, \sin t, \sin 2t, \ldots, \sin nt$$

are orthogonal over the interval $[0, 2\pi]$ relative to the inner product $\langle \mathbf{u}, \mathbf{v} \rangle$ defined in Exercise 6.

8. If $f(t)$ is defined and continuous on the interval $[0, T]$, show that $f(T\tau/2\pi)$ is defined and continuous for τ in the interval $[0, 2\pi]$. Use this fact to show how Theorem 11.20.2 follows from Theorem 11.20.1.

Technology Exercises 11.20

The following exercises are designed to be solved using a technology utility. Typically, this will be MATLAB, *Mathematica*, Maple, Derive, or Mathcad, but it may also be some other type of linear algebra software or a scientific calculator with some linear algebra capabilities. For each exercise you will need to read the relevant documentation for the particular utility you are using. The goal of these exercises is to provide you with a basic proficiency with your technology utility. Once you have mastered the techniques in these exercises, you will be able to use your technology utility to solve many of the problems in the regular exercise sets.

T1. Let g be the function

$$g(t) = \frac{3 + 4 \sin t}{5 - 4 \cos t}$$

for $0 \leq t \leq 2\pi$. Use a computer to determine the Fourier coefficients

$$\begin{Bmatrix} a_k \\ b_k \end{Bmatrix} = \frac{1}{\pi} \int_0^{2\pi} \left(\frac{3 + 4 \sin t}{5 - 4 \cos t} \right) \begin{Bmatrix} \cos kt \\ \sin kt \end{Bmatrix} dt$$

for $k = 0, 1, 2, 3, 4, 5$. From your results make a conjecture as to the general expressions for a_k and b_k. Test your conjecture by calculating

$$\frac{1}{2} a_0 + \sum_{k=1}^{\infty} (a_k \cos kt + b_k \sin kt)$$

on the computer and see if it converges to $g(t)$.

T2. Let g be the function

$$g(t) = e^{\cos t} [\cos(\sin t) + \sin(\sin t)]$$

for $0 \leq t \leq 2\pi$. Use a computer to determine the Fourier coefficients

$$\begin{Bmatrix} a_k \\ b_k \end{Bmatrix} = \frac{1}{\pi} \int_0^{2\pi} g(t) \begin{Bmatrix} \cos kt \\ \sin kt \end{Bmatrix} dt$$

for $k = 0, 1, 2, 3, 4, 5$. From your results make a conjecture as to the general expressions for a_k and b_k. Test your conjecture by calculating

$$\frac{1}{2} a_0 + \sum_{k=1}^{\infty} (a_k \cos kt + b_k \sin kt)$$

on the computer and see if it converges to $g(t)$.

11.21 WARPS AND MORPHS

Among the more interesting image-manipulation techniques available for computer graphics are warps and morphs. In this section we show how linear transformations can be used to distort a single picture to produce a warp, or to distort and blend two pictures to produce a morph.

PREREQUISITES: Geometry of Linear Operators on R^2 (Section 9.2)
Linear Independence
Bases in R^2

Most computer graphics software allows you to manipulate an image in various ways, such as by scaling, rotating, or slanting the image. Distorting an image by moving the corners of a rectangle containing the image is another basic image-manipulation technique. Distorting various pieces of an image in different ways is a more complicated procedure that results in a *warp* of the picture. In addition, warping two different images in complementary ways and blending the warps results in a *morph* of the two pictures (from the Greek root meaning *shape* or *form*). The main application of warping and morphing images has been the production of special effects in motion pictures and television or print advertisements. However, many scientific and technological applications for such techniques have also arisen—for example, studying the evolution of the shapes of living organisms, analyzing the growth and development of living organisms, assisting in reconstructive and cosmetic surgery, investigating variations in the design of a product, and "aging" photographs of missing people or police suspects.

Warps We begin by describing a simple warp of a triangular region in the plane. Let the three vertices of a triangle be given by the three noncollinear points $\mathbf{v}_1$, $\mathbf{v}_2$, and $\mathbf{v}_3$ (Figure 11.21.1a). We shall call this triangle the **begin-triangle**. If $\mathbf{v}$ is any point in the begin-triangle, then there are unique constants c_1 and c_2 such that

$$\mathbf{v} - \mathbf{v}_3 = c_1(\mathbf{v}_1 - \mathbf{v}_3) + c_2(\mathbf{v}_2 - \mathbf{v}_3) \tag{1}$$

Equation (1) expresses the vector $\mathbf{v} - \mathbf{v}_3$ as a (unique) linear combination of the two linearly independent vectors $\mathbf{v}_1 - \mathbf{v}_3$ and $\mathbf{v}_2 - \mathbf{v}_3$ with respect to an origin at $\mathbf{v}_3$. If we set $c_3 = 1 - c_1 - c_2$, then we can rewrite (1) as

$$\mathbf{v} = c_1\mathbf{v}_1 + c_2\mathbf{v}_2 + c_3\mathbf{v}_3 \tag{2}$$

where

$$c_1 + c_2 + c_3 = 1 \tag{3}$$

from the definition of c_3. We say that $\mathbf{v}$ is a **convex combination** of the vectors $\mathbf{v}_1$, $\mathbf{v}_2$, and $\mathbf{v}_3$ if (2) and (3) are satisfied and, in addition, the coefficients c_1, c_2, and c_3 are nonnegative. It can be shown (Exercise 6) that $\mathbf{v}$ lies in the triangle determined by $\mathbf{v}_1$, $\mathbf{v}_2$, and $\mathbf{v}_3$ if and only if it is a convex combination of those three vectors.

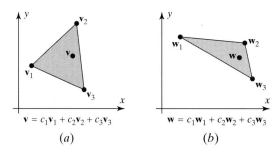

$$\mathbf{v} = c_1\mathbf{v}_1 + c_2\mathbf{v}_2 + c_3\mathbf{v}_3 \qquad\qquad \mathbf{w} = c_1\mathbf{w}_1 + c_2\mathbf{w}_2 + c_3\mathbf{w}_3$$

(a)   (b)   **Figure 11.21.1**

Next, given three noncollinear points $\mathbf{w}_1$, $\mathbf{w}_2$, and $\mathbf{w}_3$ of an ***end-triangle*** (Figure 11.21.1*b*), there is a unique ***affine transformation*** that maps $\mathbf{v}_1$ to $\mathbf{w}_1$, $\mathbf{v}_2$ to $\mathbf{w}_2$, and $\mathbf{v}_3$ to $\mathbf{w}_3$. That is, there is a unique 2×2 invertible matrix M and a unique vector $\mathbf{b}$ such that

$$\mathbf{w}_i = M\mathbf{v}_i + \mathbf{b} \quad \text{for } i = 1, 2, 3 \tag{4}$$

(See Exercise 5 for the evaluation of M and $\mathbf{b}$.) Moreover, it can be shown (Exercise 3) that the image $\mathbf{w}$ of the vector $\mathbf{v}$ in (2) under this affine transformation is

$$\mathbf{w} = c_1\mathbf{w}_1 + c_2\mathbf{w}_2 + c_3\mathbf{w}_3 \tag{5}$$

This is a basic property of affine transformations: They map a convex combination of vectors to the same convex combination of the images of the vectors.

Now suppose that the begin-triangle contains a picture within it (Figure 11.21.2*a*). That is, to each point in the begin-triangle we assign a gray level, say 0 for white and 100 for black, with any other gray level lying between 0 and 100. In particular, let a scalar-valued function ρ_0, called the ***picture-density*** of the begin-triangle, be defined so that $\rho_0(\mathbf{v})$ is the gray level at the point $\mathbf{v}$ in the begin-triangle. We can now define a picture in the end-triangle, called a ***warp*** of the original picture, with a picture-density ρ_1 by defining the gray level at the point $\mathbf{w}$ within the end-triangle to be the gray level of the point $\mathbf{v}$ in the begin-triangle that maps onto $\mathbf{w}$. In equation form, the picture-density ρ_1 is determined by

$$\rho_1(\mathbf{w}) = \rho_0(c_1\mathbf{v}_1 + c_2\mathbf{v}_2 + c_3\mathbf{v}_3) \tag{6}$$

In this way, as c_1, c_2, and c_3 vary over all nonnegative values that add to one, (5) generates all points $\mathbf{w}$ in the end-triangle and (6) generates the gray levels $\rho_1(\mathbf{w})$ of the warped picture at those points (Figure 11.21.2*b*).

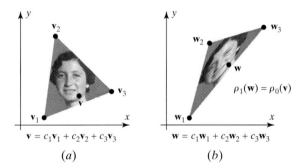

$$\mathbf{v} = c_1\mathbf{v}_1 + c_2\mathbf{v}_2 + c_3\mathbf{v}_3$$

$$\mathbf{w} = c_1\mathbf{w}_1 + c_2\mathbf{w}_2 + c_3\mathbf{w}_3$$

$$\rho_1(\mathbf{w}) = \rho_0(\mathbf{v})$$

(*a*)   (*b*)   **Figure 11.21.2**

Equation (6) determines a very simple warp of a picture within a single triangle. More generally, we can break up a picture into many triangular regions and warp each triangular region differently. This gives us much freedom in designing a warp through our choice of triangular regions and how we change them. To this end, suppose we are given a picture contained within some rectangular region of the plane. We choose n points $\mathbf{v}_1, \mathbf{v}_2, \ldots, \mathbf{v}_n$ within the rectangle, which we call ***vertex points***, so that they fall on key elements or features of the picture we wish to warp (Figure 11.21.3*a*). Once the

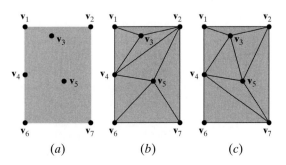

(*a*)   (*b*)   (*c*)   **Figure 11.21.3** vertex points are chosen, we complete a **_triangulation_** of the rectangular region; that is, we draw lines between the vertex points in such a way that we have the following conditions (Figure 11.21.3b):

1. The lines form the sides of a set of triangles.

2. The lines do not intersect.

3. Each vertex point is the vertex of at least one triangle.

4. The union of the triangles is the rectangle.

5. The set of triangles is maximal (i.e., no more vertices can be connected).

Notice that condition 4 requires that each corner of the rectangle containing the picture be a vertex point.

One can always form a triangulation from any n vertex points, but the triangulation is not necessarily unique. For example, Figures 11.21.3b and 11.21.3c are two different triangulations of the set of vertex points in Figure 11.21.3a. Since there are various computer algorithms that perform triangulations very quickly, it is not necessary to perform the tiresome triangulation task by hand; one need only specify the desired vertex points and let a computer generate a triangulation from them. If n is the number of vertex points chosen, it can be shown that the number of triangles m of any triangulation of those points is given by

$$m = 2n - 2 - k \tag{7}$$

where k is the number of vertex points lying on the boundary of the rectangle, including the four situated at the corner points.

The warp is specified by moving the n vertex points $\mathbf{v}_1, \mathbf{v}_2, \dots, \mathbf{v}_n$ to new locations $\mathbf{w}_1, \mathbf{w}_2, \dots, \mathbf{w}_n$ according to the changes in the picture we desire (Figures 11.21.4a and 11.21.4b). However, we impose two restrictions on the movements of the vertex points:

1. The four vertex points at the corners of the rectangle are to remain fixed, and any vertex point on a side of the rectangle is to remain fixed or move to another point on the same side of the rectangle. All other vertex points are to remain in the interior of the rectangle.

2. The triangles determined by the triangulation are not to overlap after their vertices have been moved.

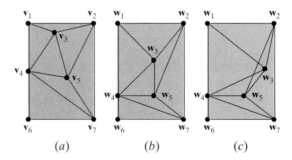

(a)   (b)   (c)   **Figure 11.21.4**

The first restriction guarantees that the rectangular shape of the begin-picture is preserved. The second restriction guarantees that the displaced vertex points still form a triangulation of the rectangle and that the new triangulation is similar to the original one. For example, Figure 11.21.4c is not an allowable movement of the vertex points shown in Figure 11.21.4a. Although a violation of this condition can be handled mathematically without too much additional effort, the resulting warps usually produce unnatural results and we shall not consider them here.

Figure 11.21.5 is a warp of a photograph of a woman using a triangulation with 94 vertex points and 179 triangles. Notice how the vertex points in the begin-triangulation

Begin-picture

Warped picture

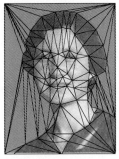

Begin-triangulation

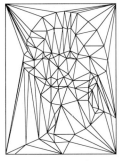

Begin-triangulation

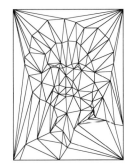

Warped triangulation

Warped triangulation **Figure 11.21.5**

are chosen to lie along key features of the picture (hairline, eyes, lips, etc.). These vertex points were moved to final positions corresponding to those same features in a picture of the women taken 20 years after the begin-picture. Thus, the warped picture represents the woman forced into her older shape but using her younger gray levels.

Time-Varying Warps   A *time-varying warp* is the set of warps generated when the vertex points of the begin-picture are moved continually in time from their original positions to specified final positions. This gives us a motion picture in which the begin-picture is continually warped to a final warp. Let us choose time units so that $t = 0$ corresponds to our begin-picture and $t = 1$ corresponds to our final warp. The simplest way of moving the vertex points from time 0 to time 1 is with constant velocity along straight-line paths from their initial positions to their final positions.

To describe such a motion, let $\mathbf{u}_i(t)$ denote the position of the ith vertex point at any time t between 0 and 1. Thus $\mathbf{u}_i(0) = \mathbf{v}_i$ (its given position in the begin-picture) and $\mathbf{u}_i(1) = \mathbf{w}_i$ (its given position in the final warp). In between, we determine its position by

$$\mathbf{u}_i(t) = (1 - t)\mathbf{v}_i + t\mathbf{w}_i \tag{8}$$

Notice that (8) expresses $\mathbf{u}_i(t)$ as a convex combination of $\mathbf{v}_i$ and $\mathbf{w}_i$ for each t in $[0, 1]$. Figure 11.21.6 illustrates a time-varying triangulation of a plain rectangular region with six vertex points. The lines connecting the vertex points at the different times are the space-time paths of these vertex points in this space-time diagram.

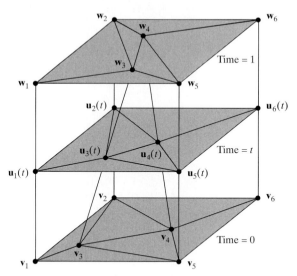

Figure 11.21.6

Once the positions of the vertex points are computed at time t, a warp is performed between the begin-picture and the triangulation at time t determined by the displaced vertex points at that time. Figure 11.21.7 shows a time-varying warp at five values of t generated from the warp between $t = 0$ and $t = 1$ shown in Figure 11.21.5.

$t = 0.00 \qquad t = 0.25 \qquad t = 0.50 \qquad t = 0.75 \qquad t = 1.00$

Figure 11.21.7

Morphs   A *time-varying morph* can be described as a blending of two time-varying warps of two different pictures using two triangulations that match corresponding features in the two pictures. One of the two pictures is designated as the begin-picture and the other as the end-picture. First, a time-varying warp from $t = 0$ to $t = 1$ is generated in which the begin-picture is warped into the shape of the end-picture. Then a time-varying warp from $t = 1$ to $t = 0$ is generated in which the end-picture is warped into the shape of the begin-picture. Finally, a weighted average of the gray levels of the two warps at each time t is produced to generate the morph of the two images at time t.

Figure 11.21.8 shows two photographs of a woman taken 20 years apart. Below the pictures are two corresponding triangulations in which corresponding features of the two photographs are matched. The time-varying morph between these two pictures for five values of t between 0 and 1 is shown in Figure 11.21.9.

The procedure for producing such a morph is outlined in the following nine steps (Figure 11.21.10):

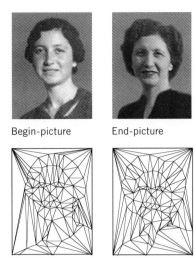

Begin-picture   End-picture

Begin-triangulation   End-triangulation   **Figure 11.21.8**

$t = 0.00$ $t = 0.25$ $t = 0.50$ $t = 0.75$ $t = 1.00$

Figure 11.21.9

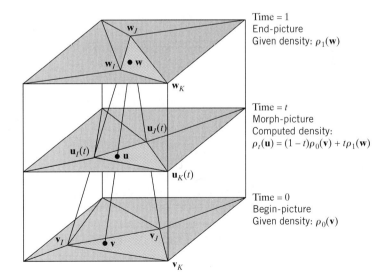

Time = 1
End-picture
Given density: $\rho_1(\mathbf{w})$

Time = t
Morph-picture
Computed density:
$\rho_t(\mathbf{u}) = (1 - t)\rho_0(\mathbf{v}) + t\rho_1(\mathbf{w})$

Time = 0
Begin-picture
Given density: $\rho_0(\mathbf{v})$

Figure 11.21.10

Step 1. Given a begin-picture with picture-density ρ_0 and an end-picture with picture-density ρ_1, position n vertex points $\mathbf{v}_1, \mathbf{v}_2, \ldots, \mathbf{v}_n$ in the begin-picture at key features of that picture.

Step 2. Position n corresponding vertex points $\mathbf{w}_1, \mathbf{w}_2, \ldots, \mathbf{w}_n$ in the end-picture at the corresponding key features of that picture.

Step 3. Triangulate the begin- and end-pictures in similar ways by drawing lines between corresponding vertex points in both pictures.

Step 4. For any time t between 0 and 1, find the vertex points $\mathbf{u}_1(t)$, $\mathbf{u}_2(t)$, ..., $\mathbf{u}_n(t)$ in the morph picture at that time using the formula

$$\mathbf{u}_i(t) = (1 - t)\mathbf{v}_i + t\mathbf{w}_i, \qquad i = 1, 2, \ldots, n \tag{9}$$

Step 5. Triangulate the morph picture at time t similar to the begin- and end-picture triangulations.

Step 6. For any point $\mathbf{u}$ in the morph picture at time t, find the triangle in the triangulation of the morph picture in which it lies and the vertices $\mathbf{u}_I(t)$, $\mathbf{u}_J(t)$, and $\mathbf{u}_K(t)$ of that triangle. (See Exercise 1 to determine whether a given point lies in a given triangle.)

Step 7. Express $\mathbf{u}$ as a convex combination of $\mathbf{u}_I(t)$, $\mathbf{u}_J(t)$, and $\mathbf{u}_K(t)$ by finding the constants c_I, c_J, and c_K such that

$$\mathbf{u} = c_I\mathbf{u}_I(t) + c_J\mathbf{u}_J(t) + c_K\mathbf{u}_K(t) \tag{10}$$

and

$$c_I + c_J + c_K = 1 \tag{11}$$

Step 8. Determine the locations of the point $\mathbf{u}$ in the begin- and end-pictures using

$$\mathbf{v} = c_I\mathbf{v}_I + c_J\mathbf{v}_J + c_K\mathbf{v}_K \quad \text{(in the begin-picture)} \tag{12}$$

and

$$\mathbf{w} = c_I\mathbf{w}_I + c_J\mathbf{w}_J + c_K\mathbf{w}_K \quad \text{(in the end-picture)} \tag{13}$$

Step 9. Finally, determine the picture-density $\rho_t(\mathbf{u})$ of the morph-picture at the point $\mathbf{u}$ using

$$\rho_t(\mathbf{u}) = (1 - t)\rho_0(\mathbf{v}) + t\rho_1(\mathbf{w}) \tag{14}$$

Step 9 is the key step in distinguishing a warp from a morph. Equation (14) takes weighted averages of the gray levels of the begin- and end-pictures to produce the gray levels of the morph-picture. The weights depend on the fraction of the distances that the vertex points have moved from their beginning positions to their ending positions. For example, if the vertex points have moved one-fourth of the way to their destinations (i.e., if $t = 0.25$), then we use one-fourth of the gray levels of the end-picture and three-fourths of the gray levels of the begin-picture. Thus, as time progresses, not only does the shape of the begin-picture gradually change into the shape of the end-picture (as in a warp) but the gray levels of the begin-picture also gradually change into the gray levels of the end-picture.

The procedure described above to generate a morph is cumbersome to perform by hand, but it is the kind of dull, repetitive procedure at which computers excel. A successful morph demands good preparation and requires more artistic ability than mathematical ability. (The software designer is required to have the mathematical ability.) The two photographs to be morphed should be carefully chosen so that they have matching features, and the vertex points in the two photographs also should be carefully chosen so that the triangles in the two resulting triangulations contain similar features of the two pictures. When done correctly, each frame of the morph should look just as "real" as the begin- and end-pictures.

The techniques we have discussed in this section can be generalized in numerous ways to produce much more elaborate warps and morphs. For example:

1. If the pictures are in color, the three components of the picture colors (red, green, and blue) can be morphed separately to produce a color morph.

2. Rather than follow straight-line paths to their destinations, the vertices of a triangulation can be directed separately along more complicated paths to produce a variety of results.

3. Rather than travel with constant speeds along their paths, the vertices of a triangulation can be directed to have different speeds at different times. For example, in a morph between two faces, the hairline can be made to change first, then the nose, and so forth.

4. Similarly, the gray-level mixing of the begin-picture and end-picture at different times and different vertices can be varied in a more complicated way than that in Equation (14).

5. One can morph two surfaces in three-dimensional space (representing two complete heads, for example) by triangulating the surfaces and using the techniques in this section.

6. One can morph two solids in three-dimensional space (for example, two three-dimensional tomographs of a beating human heart at two different times) by dividing the two solids into corresponding tetrahedral regions.

7. Two film strips can be morphed frame by frame by different amounts between each pair of frames to produce a morphed film strip in which, say, an actor walking along a set is gradually morphed into an ape walking along the set.

8. Instead of using straight lines to triangulate two pictures to be morphed, more complicated curves, such as spline curves, can be matched between the two pictures.

9. Three or more pictures can be morphed together by generalizing the formulas given in this section.

These and other generalizations have made warping and morphing two of the most active areas in computer graphics.

Exercise Set 11.21

1. Determine whether the vector $\mathbf{v}$ is a convex combination of the vectors $\mathbf{v}_1$, $\mathbf{v}_2$, and $\mathbf{v}_3$. Do this by solving Equations (1) and (3) for c_1, c_2, and c_3 and ascertaining whether these coefficients are nonnegative.

(a) $\mathbf{v} = \begin{bmatrix} 3 \\ 3 \end{bmatrix}$, $\mathbf{v}_1 = \begin{bmatrix} 1 \\ 1 \end{bmatrix}$, $\mathbf{v}_2 = \begin{bmatrix} 3 \\ 5 \end{bmatrix}$, $\mathbf{v}_3 = \begin{bmatrix} 4 \\ 2 \end{bmatrix}$

(b) $\mathbf{v} = \begin{bmatrix} 2 \\ 4 \end{bmatrix}$, $\mathbf{v}_1 = \begin{bmatrix} 1 \\ 1 \end{bmatrix}$, $\mathbf{v}_2 = \begin{bmatrix} 3 \\ 5 \end{bmatrix}$, $\mathbf{v}_3 = \begin{bmatrix} 4 \\ 2 \end{bmatrix}$

(c) $\mathbf{v} = \begin{bmatrix} 0 \\ 0 \end{bmatrix}$, $\mathbf{v}_1 = \begin{bmatrix} 3 \\ 3 \end{bmatrix}$, $\mathbf{v}_2 = \begin{bmatrix} -2 \\ -2 \end{bmatrix}$, $\mathbf{v}_3 = \begin{bmatrix} 3 \\ 0 \end{bmatrix}$

(d) $\mathbf{v} = \begin{bmatrix} 1 \\ 0 \end{bmatrix}$, $\mathbf{v}_1 = \begin{bmatrix} 3 \\ 3 \end{bmatrix}$, $\mathbf{v}_2 = \begin{bmatrix} -2 \\ -2 \end{bmatrix}$, $\mathbf{v}_3 = \begin{bmatrix} 3 \\ 0 \end{bmatrix}$

2. Verify Equation (7) for the two triangulations given in Figure 11.21.3.

3. Let an affine transformation be given by a 2×2 matrix M and a two-dimensional vector $\mathbf{b}$. Let $\mathbf{v} = c_1\mathbf{v}_1 + c_2\mathbf{v}_2 + c_3\mathbf{v}_3$, where $c_1 + c_2 + c_3 = 1$; let $\mathbf{w} = M\mathbf{v} + \mathbf{b}$; and let $\mathbf{w}_i = M\mathbf{v}_i + \mathbf{b}$ for $i = 1, 2, 3$. Show that $\mathbf{w} = c_1\mathbf{w}_1 + c_2\mathbf{w}_2 + c_3\mathbf{w}_3$. (This shows that an affine transformation maps a convex combination of vectors to the same convex combination of the images of the vectors.)

4. (a) Exhibit a triangulation of the points in Figure 11.21.3 in which the points $\mathbf{v}_3$, $\mathbf{v}_5$, and $\mathbf{v}_6$ form the vertices of a single triangle.

   (b) Exhibit a triangulation of the points in Figure 11.21.3 in which the points $\mathbf{v}_2$, $\mathbf{v}_5$, and $\mathbf{v}_7$ do *not* form the vertices of a single triangle.

5. Find the 2×2 matrix M and two-dimensional vector $\mathbf{b}$ that define the affine transformation that maps the three vectors $\mathbf{v}_1$, $\mathbf{v}_2$, and $\mathbf{v}_3$ to the three vectors $\mathbf{w}_1$, $\mathbf{w}_2$, and $\mathbf{w}_3$. Do this by setting up a system of six linear equations for the four entries of the matrix M and the two entries of the vector $\mathbf{b}$.

(a) $\mathbf{v}_1 = \begin{bmatrix} 1 \\ 1 \end{bmatrix}$, $\mathbf{v}_2 = \begin{bmatrix} 2 \\ 3 \end{bmatrix}$, $\mathbf{v}_3 = \begin{bmatrix} 2 \\ 1 \end{bmatrix}$, $\mathbf{w}_1 = \begin{bmatrix} 4 \\ 3 \end{bmatrix}$, $\mathbf{w}_2 = \begin{bmatrix} 9 \\ 5 \end{bmatrix}$, $\mathbf{w}_3 = \begin{bmatrix} 5 \\ 3 \end{bmatrix}$

(b) $\mathbf{v}_1 = \begin{bmatrix} -2 \\ 2 \end{bmatrix}$, $\mathbf{v}_2 = \begin{bmatrix} 0 \\ 0 \end{bmatrix}$, $\mathbf{v}_3 = \begin{bmatrix} 2 \\ 1 \end{bmatrix}$, $\mathbf{w}_1 = \begin{bmatrix} -8 \\ 1 \end{bmatrix}$, $\mathbf{w}_2 = \begin{bmatrix} 0 \\ 1 \end{bmatrix}$, $\mathbf{w}_3 = \begin{bmatrix} 5 \\ 4 \end{bmatrix}$

(c) $\mathbf{v}_1 = \begin{bmatrix} -2 \\ 1 \end{bmatrix}$, $\mathbf{v}_2 = \begin{bmatrix} 3 \\ 5 \end{bmatrix}$, $\mathbf{v}_3 = \begin{bmatrix} 1 \\ 0 \end{bmatrix}$, $\mathbf{w}_1 = \begin{bmatrix} 0 \\ -2 \end{bmatrix}$, $\mathbf{w}_2 = \begin{bmatrix} 5 \\ 2 \end{bmatrix}$, $\mathbf{w}_3 = \begin{bmatrix} 3 \\ -3 \end{bmatrix}$

(d) $\mathbf{v}_1 = \begin{bmatrix} 0 \\ 2 \end{bmatrix}$, $\mathbf{v}_2 = \begin{bmatrix} 2 \\ 2 \end{bmatrix}$, $\mathbf{v}_3 = \begin{bmatrix} -4 \\ -2 \end{bmatrix}$, $\mathbf{w}_1 = \begin{bmatrix} \frac{5}{2} \\ -1 \end{bmatrix}$, $\mathbf{w}_2 = \begin{bmatrix} \frac{7}{2} \\ 3 \end{bmatrix}$, $\mathbf{w}_3 = \begin{bmatrix} -\frac{7}{2} \\ -9 \end{bmatrix}$

6. (a) Let $\mathbf{a}$ and $\mathbf{b}$ be linearly independent vectors in the plane. Show that if c_1 and c_2 are nonnegative numbers such that $c_1 + c_2 = 1$, then the vector $c_1\mathbf{a} + c_2\mathbf{b}$ lies on the line segment connecting the tips of the vectors $\mathbf{a}$ and $\mathbf{b}$.

(b) Let $\mathbf{a}$ and $\mathbf{b}$ be linearly independent vectors in the plane. Show that if c_1 and c_2 are nonnegative numbers such that $c_1 + c_2 \leq 1$, then the vector $c_1\mathbf{a} + c_2\mathbf{b}$ lies in the triangle connecting the origin and the tips of the vectors $\mathbf{a}$ and $\mathbf{b}$. [*Hint.* First examine the vector $c_1\mathbf{a} + c_2\mathbf{b}$ multiplied by the scale factor $1/(c_1 + c_2)$.]

(c) Let $\mathbf{v}_1$, $\mathbf{v}_2$, and $\mathbf{v}_3$ be noncollinear points in the plane. Show that if c_1, c_2, and c_3 are nonnegative numbers such that $c_1 + c_2 + c_3 = 1$, then the vector $c_1\mathbf{v}_1 + c_2\mathbf{v}_2 + c_3\mathbf{v}_3$ lies in the triangle connecting the tips of the three vectors. [*Hint.* Let $\mathbf{a} = \mathbf{v}_1 - \mathbf{v}_3$ and $\mathbf{b} = \mathbf{v}_2 - \mathbf{v}_3$, and then use Equation (1) and part (b) of this exercise.]

7. (a) What can you say about the coefficients c_1, c_2, and c_3 that determine a convex combination $\mathbf{v} = c_1\mathbf{v}_1 + c_2\mathbf{v}_2 + c_3\mathbf{v}_3$ if $\mathbf{v}$ lies on one of the three vertices of the triangle determined by the three vectors $\mathbf{v}_1$, $\mathbf{v}_2$, and $\mathbf{v}_3$?

(b) What can you say about the coefficients c_1, c_2, and c_3 that determine a convex combination $\mathbf{v} = c_1\mathbf{v}_1 + c_2\mathbf{v}_2 + c_3\mathbf{v}_3$ if $\mathbf{v}$ lies on one of the three sides of the triangle determined by the three vectors $\mathbf{v}_1$, $\mathbf{v}_2$, and $\mathbf{v}_3$?

(c) What can you say about the coefficients c_1, c_2, and c_3 that determine a convex combination $\mathbf{v} = c_1\mathbf{v}_1 + c_2\mathbf{v}_2 + c_3\mathbf{v}_3$ if $\mathbf{v}$ lies in the interior of the triangle determined by the three vectors $\mathbf{v}_1$, $\mathbf{v}_2$, and $\mathbf{v}_3$?

8. The center of gravity of a triangle lies on the line segment connecting any one of the three vertices of the triangle with the midpoint of the opposite side. Its location on this line segment is two-thirds of the distance from the vertex. If the three vertices are given by the vectors $\mathbf{v}_1$, $\mathbf{v}_2$, and $\mathbf{v}_3$, write the center of gravity as a convex combination of these three vectors.

Technology Exercises 11.21

The following exercises are designed to be solved using a technology utility. Typically, this will be MATLAB, *Mathematica*, Maple, Derive, or Mathcad, but it may also be some other type of linear algebra software or a scientific calculator with some linear algebra capabilities. For each exercise you will need to read the relevant documentation for the particular utility you are using. The goal of these exercises is to provide you with a basic proficiency with your technology utility. Once you have mastered the techniques in these exercises, you will be able to use your technology utility to solve many of the problems in the regular exercise sets.

T1. To warp or morph a surface in R^3 we must be able to triangulate the surface. Let $\mathbf{v}_1 = \begin{bmatrix} v_{11} \\ v_{12} \\ v_{13} \end{bmatrix}$,

$\mathbf{v}_2 = \begin{bmatrix} v_{21} \\ v_{22} \\ v_{23} \end{bmatrix}$, and $\mathbf{v}_3 = \begin{bmatrix} v_{31} \\ v_{32} \\ v_{33} \end{bmatrix}$ be three noncollinear vectors on the surface. Then a vector

$\mathbf{v} = \begin{bmatrix} v_1 \\ v_2 \\ v_3 \end{bmatrix}$ lies in the triangle formed by these three vectors if and only if $\mathbf{v}$ is a convex

combination of the three vectors; that is, $\mathbf{v} = c_1\mathbf{v}_1 + c_2\mathbf{v}_2 + c_3\mathbf{v}_3$ for some nonnegative coefficients c_1, c_2, and c_3 whose sum is one.

(a) Show that in this case c_1, c_2, and c_3 are solutions of the following linear system:

$$\begin{bmatrix} v_{11} & v_{21} & v_{31} \\ v_{12} & v_{22} & v_{32} \\ v_{13} & v_{23} & v_{33} \\ 1 & 1 & 1 \end{bmatrix} \begin{bmatrix} c_1 \\ c_2 \\ c_3 \end{bmatrix} = \begin{bmatrix} v_1 \\ v_2 \\ v_3 \\ 1 \end{bmatrix}$$

In parts (b)–(d) determine whether the vector $\mathbf{v}$ is a convex combination of the vectors

$$\mathbf{v}_1 = \begin{bmatrix} 2 \\ 7 \\ -5 \end{bmatrix}, \mathbf{v}_2 = \begin{bmatrix} 3 \\ 0 \\ 9 \end{bmatrix}, \text{ and } \mathbf{v}_3 = \begin{bmatrix} 2 \\ 2 \\ -4 \end{bmatrix}.$$

(b) $\mathbf{v} = \dfrac{1}{4} \begin{bmatrix} 9 \\ 9 \\ 9 \end{bmatrix}$ (c) $\mathbf{v} = \dfrac{1}{4} \begin{bmatrix} 10 \\ 9 \\ 9 \end{bmatrix}$ (d) $\mathbf{v} = \dfrac{1}{4} \begin{bmatrix} 13 \\ -7 \\ 50 \end{bmatrix}$

T2. To warp or morph a solid object in R^3 we first partition the object into disjoint tetrahedrons.

Let $\mathbf{v}_1 = \begin{bmatrix} v_{11} \\ v_{12} \\ v_{13} \end{bmatrix}$, $\mathbf{v}_2 = \begin{bmatrix} v_{21} \\ v_{22} \\ v_{23} \end{bmatrix}$, $\mathbf{v}_3 = \begin{bmatrix} v_{31} \\ v_{32} \\ v_{33} \end{bmatrix}$, and $\mathbf{v}_4 = \begin{bmatrix} v_{41} \\ v_{42} \\ v_{43} \end{bmatrix}$ be four noncoplanar vectors.

Then a vector $\mathbf{v} = \begin{bmatrix} v_1 \\ v_2 \\ v_3 \end{bmatrix}$ lies in the solid tetrahedron formed by these four vectors if and only

if $\mathbf{v}$ is a convex combination of the three vectors; that is, $\mathbf{v} = c_1\mathbf{v}_1 + c_2\mathbf{v}_2 + c_3\mathbf{v}_3 + c_4\mathbf{v}_4$ for some nonnegative coefficients c_1, c_2, c_3, and c_4 whose sum is one.

(a) Show that in this case c_1, c_2, c_3, and c_4 are solutions of the following linear system:

$$\begin{bmatrix} v_{11} & v_{21} & v_{31} & v_{41} \\ v_{12} & v_{22} & v_{32} & v_{42} \\ v_{13} & v_{23} & v_{33} & v_{43} \\ 1 & 1 & 1 & 1 \end{bmatrix} \begin{bmatrix} c_1 \\ c_2 \\ c_3 \\ c_4 \end{bmatrix} = \begin{bmatrix} v_1 \\ v_2 \\ v_3 \\ 1 \end{bmatrix}$$

In parts (b)–(d) determine whether the vector $\mathbf{v}$ is a convex combination of the vectors

$$\mathbf{v}_1 = \begin{bmatrix} 2 \\ -6 \\ 1 \end{bmatrix}, \mathbf{v}_2 = \begin{bmatrix} -3 \\ 4 \\ 2 \end{bmatrix}, \mathbf{v}_3 = \begin{bmatrix} 7 \\ 2 \\ 3 \end{bmatrix}, \text{ and } \mathbf{v}_4 = \begin{bmatrix} -1 \\ 3 \\ 2 \end{bmatrix}.$$

(b) $\mathbf{v} = \begin{bmatrix} 5 \\ 0 \\ 7 \end{bmatrix}$ (c) $\mathbf{v} = \begin{bmatrix} 1 \\ 1 \\ 2 \end{bmatrix}$ (d) $\mathbf{v} = \begin{bmatrix} 1 \\ 2 \\ 2 \end{bmatrix}$

Answers to Exercises

EXERCISE SET 1.1 (page 6)

1. (a), (c), (f)   **2.** (a), (b), (c)

3. (a) $x = \frac{3}{7} + \frac{5}{7}t$ **(b)** $x_1 = \frac{5}{3}s - \frac{4}{3}t + \frac{7}{3}$ $x_1 = \frac{1}{4}r - \frac{5}{8}s + \frac{3}{4}t - \frac{1}{8}$ $v = \frac{8}{3}q - \frac{2}{3}r + \frac{1}{3}s - \frac{4}{3}t$

$\quad\quad\quad y = t$ $\quad\quad\quad x_2 = s$ $\quad\quad\quad x_2 = r$ $\quad\quad\quad\quad w = q$

$\quad\quad\quad\quad\quad\quad\quad\quad x_3 = t$ $\quad\quad\quad x_3 = s$ $\quad\quad\quad\quad x = r$

$\quad\quad\quad\quad\quad\quad\quad\quad\quad\quad\quad\quad\quad x_4 = t$ $\quad\quad\quad\quad y = s$

$\quad\quad\quad\quad\quad\quad\quad\quad\quad\quad\quad\quad\quad\quad\quad\quad\quad z = t$

4. (a) $\begin{bmatrix} 3 & -2 & -1 \\ 4 & 5 & 3 \\ 7 & 3 & 2 \end{bmatrix}$ **(b)** $\begin{bmatrix} 2 & 0 & 2 & 1 \\ 3 & -1 & 4 & 7 \\ 6 & 1 & -1 & 0 \end{bmatrix}$ **(c)** $\begin{bmatrix} 1 & 2 & 0 & -1 & 1 & 1 \\ 0 & 3 & 1 & 0 & -1 & 2 \\ 0 & 0 & 1 & 7 & 0 & 1 \end{bmatrix}$ **(d)** $\begin{bmatrix} 1 & 0 & 0 & 1 \\ 0 & 1 & 0 & 2 \\ 0 & 0 & 1 & 3 \end{bmatrix}$

5. (a) $2x_1 \qquad\qquad = 0$ **(b)** $3x_1 \qquad\quad - 2x_3 = 5$ **(c)** $7x_1 + 2x_2 + x_3 - 3x_4 = 5$ **(d)** $x_1 \qquad\qquad\qquad = 7$

$\quad\quad 3x_1 - 4x_2 = 0$ $\qquad 7x_1 + x_2 + 4x_3 = -3$ $\qquad\quad x_1 + 2x_2 + 4x_3 \qquad = 1$ $\qquad\qquad x_2 \qquad\qquad = -2$

$\quad\quad\quad\quad\quad\; x_2 = 1$ $\qquad\qquad - 2x_2 + x_3 = 7$ $\qquad\qquad\qquad\qquad\qquad\qquad\qquad\qquad\qquad x_3 \quad = 3$

$\qquad x_4 = 4$

6. (a) $x - 2y = 5$ **(b)** Let $x = t$; then $t - 2y = 5$. Solving for y yields $y = \frac{1}{2}t - \frac{5}{2}$.

10. $k = 6$: infinitely many solutions
$k \neq 6$: no solutions
No value of k yields one solution.

11. (a) The lines have no common point of intersection.
   (b) The lines intersect in exactly one point.
   (c) The three lines coincide.

EXERCISE SET 1.2 (page 19)

1. (a), (b), (c), (d), (h), (i), (j)   **2.** (a), (b), (d), (e)

3. (a) Both   **(b)** Neither   **(c)** Both
   (d) Row-echelon   **(e)** Neither   **(f)** Both

4. (a) $x_1 = -3, x_2 = 0, x_3 = 7$
   (b) $x_1 = 7t + 8, x_2 = -3t + 2, x_3 = -t - 5, x_4 = t$
   (c) $x_1 = 6s - 3t - 2, x_2 = s, x_3 = -4t + 7, x_4 = -5t + 8, x_5 = t$
   (d) Inconsistent

5. (a) $x_1 = -37, x_2 = -8, x_3 = 5$
  (b) $x_1 = 13t - 10, x_2 = 13t - 5, x_3 = -t + 2, x_4 = t$
  (c) $x_1 = -7s + 2t - 11, x_2 = s, x_3 = -3t - 4, x_4 = -3t + 9, x_5 = t$
  (d) Inconsistent

6. (a) $x_1 = 3, x_2 = 1, x_3 = 2$
  (b) $x_1 = -\frac{1}{7} - \frac{3}{7}t, x_2 = \frac{1}{7} - \frac{4}{7}, x_3 = t$
  (c) $x = t - 1, y = 2s, z = s, w = t$
  (d) Inconsistent

8. (a) Inconsistent
  (b) $x_1 = -4, x_2 = 2, x_3 = 7$
  (c) $x_1 = 3 + 2t, x_2 = t$
  (d) $x = \frac{8}{5} - \frac{3}{5}t - \frac{3}{5}s, y = \frac{1}{10} + \frac{2}{5}t - \frac{1}{10}s, z = t, w = s$

10. (a) $x_1 = 2 - 12t, x_2 = 5 - 27t, x_3 = t$
  (b) Inconsistent
  (c) $u = -2s - 3t - 6, v = s, w = -t - 2, x = t + 3, y = t$

12. (a), (c), (d)

13. (a) $x_1 = 0, x_2 = 0, x_3 = 0$
  (b) $x_1 = -s, x_2 = -t - s, x_3 = 4s, x_4 = t$
  (c) $w = t, x = -t, y = t, z = 0$

14. (a) Only the trivial solution
  (b) $u = 7s - 5t, v = -6s + 4t, w = 2s, x = 2t$
  (c) Only the trivial solution

15. (a) $I_1 = -1, I_2 = 0, I_3 = 1, I_4 = 2$ **(b)** $Z_1 = -s - t, Z_2 = s, Z_3 = -t, Z_4 = 0, Z_5 = t$

16. (a) $x = \frac{2}{3}a - \frac{1}{9}b, y = -\frac{1}{3}a + \frac{2}{9}b$ **(b)** $x_1 = a - \frac{1}{3}c, x_2 = a - \frac{1}{2}b, x_3 = -a + \frac{1}{2}b + \frac{1}{3}c$

17. $a = -4$, none; $a \ne \pm 4$, exactly one; $a = 4$, infinitely many

19. $\begin{bmatrix} 1 & 3 \\ 0 & 1 \end{bmatrix}$ and $\begin{bmatrix} 1 & 0 \\ 0 & 1 \end{bmatrix}$ are possible answers. **20.** $\alpha = \pi/2, \beta = \pi, \gamma = 0$ **22.** $\lambda = 4, \lambda = 2$

23. If $\lambda = 1$, then $x_1 = x_2 = -\frac{1}{2}s, x_3 = s$ **24.** $x = -13/7, y = 91/54, z = -91/8$
  If $\lambda = 2$, then $x_1 = -\frac{1}{2}s, x_2 = 0, x_3 = s$

25. $a = 1, b = -6, c = 2, d = 10$

26. $a = 1, b = -2, c = -4, d = -29$ **30. (a)** Three lines, at least two of which are distinct **(b)** Three identical lines

31. (a) False **(b)** True **(c)** False **(d)** False

32. (a) False **(b)** False **(c)** False **(d)** False

EXERCISE SET 1.3 [page 33]

1. (a) Undefined **(b)** 4×2 **(c)** Undefined **(d)** Undefined
  (e) 5×5 **(f)** 5×2 **(g)** Undefined **(h)** 5×2

2. $a = 5, b = -3, c = 4, d = 1$

3. (a) $\begin{bmatrix} 7 & 6 & 5 \\ -2 & 1 & 3 \\ 7 & 3 & 7 \end{bmatrix}$ **(b)** $\begin{bmatrix} -5 & 4 & -1 \\ 0 & -1 & -1 \\ -1 & 1 & 1 \end{bmatrix}$ **(c)** $\begin{bmatrix} 15 & 0 \\ -5 & 10 \\ 5 & 5 \end{bmatrix}$ **(d)** $\begin{bmatrix} -7 & -28 & -14 \\ -21 & -7 & -35 \end{bmatrix}$ **(e)** Undefined

(f) $\begin{bmatrix} 22 & -6 & 8 \\ -2 & 4 & 6 \\ 10 & 0 & 4 \end{bmatrix}$ **(g)** $\begin{bmatrix} -39 & -21 & -24 \\ 9 & -6 & -15 \\ -33 & -12 & -30 \end{bmatrix}$ **(h)** $\begin{bmatrix} 0 & 0 \\ 0 & 0 \\ 0 & 0 \end{bmatrix}$ **(i)** 5 **(j)** -25 **(k)** 168 **(l)** Undefined

4. (a) $\begin{bmatrix} 7 & 2 & 4 \\ 3 & 5 & 7 \end{bmatrix}$ **(b)** $\begin{bmatrix} -5 & 0 & -1 \\ 4 & -1 & 1 \\ -1 & -1 & 1 \end{bmatrix}$ **(c)** $\begin{bmatrix} -5 & 0 & -1 \\ 4 & -1 & 1 \\ -1 & -1 & 1 \end{bmatrix}$ **(d)** Undefined

(e) $\begin{bmatrix} -\frac{1}{4} & \frac{3}{2} \\ \frac{9}{4} & 0 \\ \frac{3}{4} & \frac{9}{4} \end{bmatrix}$ **(f)** $\begin{bmatrix} 0 & -1 \\ 1 & 0 \end{bmatrix}$ **(g)** $\begin{bmatrix} 9 & 1 & -1 \\ -13 & 2 & -4 \\ 0 & 1 & -6 \end{bmatrix}$ **(h)** $\begin{bmatrix} 9 & -13 & 0 \\ 1 & 2 & 1 \\ -1 & -4 & -6 \end{bmatrix}$

5. (a) $\begin{bmatrix} 12 & -3 \\ -4 & 5 \\ 4 & 1 \end{bmatrix}$ **(b)** Undefined **(c)** $\begin{bmatrix} 42 & 108 & 75 \\ 12 & -3 & 21 \\ 36 & 78 & 63 \end{bmatrix}$ **(d)** $\begin{bmatrix} 3 & 45 & 9 \\ 11 & -11 & 17 \\ 7 & 17 & 13 \end{bmatrix}$ **(e)** $\begin{bmatrix} 3 & 45 & 9 \\ 11 & -11 & 17 \\ 7 & 17 & 13 \end{bmatrix}$

(f) $\begin{bmatrix} 21 & 17 \\ 17 & 35 \end{bmatrix}$ **(g)** $\begin{bmatrix} 0 & -2 & 11 \\ 12 & 1 & 8 \end{bmatrix}$ **(h)** $\begin{bmatrix} 12 & 6 & 9 \\ 48 & -20 & 14 \\ 24 & 8 & 16 \end{bmatrix}$ **(i)** 61 **(j)** 35 **(k)** (28)

6. (a) $\begin{bmatrix} -6 & -3 \\ 36 & 0 \\ 4 & 7 \end{bmatrix}$ **(b)** Undefined **(c)** $\begin{bmatrix} 2 & -10 & 11 \\ 13 & 2 & 5 \\ 4 & -3 & 13 \end{bmatrix}$ **(d)** $\begin{bmatrix} 10 & -6 \\ -14 & 2 \\ -1 & -8 \end{bmatrix}$ **(e)** $\begin{bmatrix} 40 & 72 \\ 26 & 42 \end{bmatrix}$ **(f)** $\begin{bmatrix} 0 & 0 & 0 \\ 0 & 0 & 0 \\ 0 & 0 & 0 \end{bmatrix}$

7. (a) [67 41 41] **(b)** [63 67 57] **(c)** $\begin{bmatrix} 41 \\ 21 \\ 67 \end{bmatrix}$ **(d)** $\begin{bmatrix} 6 \\ 6 \\ 63 \end{bmatrix}$ **(e)** [24 56 97] **(f)** $\begin{bmatrix} 76 \\ 98 \\ 97 \end{bmatrix}$

8. (a) $\begin{bmatrix} 67 \\ 64 \\ 63 \end{bmatrix} = 6\begin{bmatrix} 3 \\ 6 \\ 0 \end{bmatrix} + 0\begin{bmatrix} -2 \\ 5 \\ 4 \end{bmatrix} + 7\begin{bmatrix} 7 \\ 4 \\ 9 \end{bmatrix}$ **(b)** $\begin{bmatrix} 6 \\ 6 \\ 63 \end{bmatrix} = 3\begin{bmatrix} 6 \\ 0 \\ 7 \end{bmatrix} + 6\begin{bmatrix} -2 \\ 1 \\ 7 \end{bmatrix} + 0\begin{bmatrix} 4 \\ 3 \\ 5 \end{bmatrix}$

$\begin{bmatrix} 41 \\ 21 \\ 67 \end{bmatrix} = -2\begin{bmatrix} 3 \\ 6 \\ 0 \end{bmatrix} + 1\begin{bmatrix} -2 \\ 5 \\ 4 \end{bmatrix} + 7\begin{bmatrix} 7 \\ 4 \\ 9 \end{bmatrix}$ $\begin{bmatrix} -6 \\ 17 \\ 41 \end{bmatrix} = -2\begin{bmatrix} 6 \\ 0 \\ 7 \end{bmatrix} + 5\begin{bmatrix} -2 \\ 1 \\ 7 \end{bmatrix} + 4\begin{bmatrix} 4 \\ 3 \\ 5 \end{bmatrix}$

$\begin{bmatrix} 41 \\ 59 \\ 57 \end{bmatrix} = 4\begin{bmatrix} 3 \\ 6 \\ 0 \end{bmatrix} + 3\begin{bmatrix} -2 \\ 5 \\ 4 \end{bmatrix} + 5\begin{bmatrix} 7 \\ 4 \\ 9 \end{bmatrix}$ $\begin{bmatrix} 70 \\ 31 \\ 122 \end{bmatrix} = 7\begin{bmatrix} 6 \\ 0 \\ 7 \end{bmatrix} + 4\begin{bmatrix} -2 \\ 1 \\ 7 \end{bmatrix} + 9\begin{bmatrix} 4 \\ 3 \\ 5 \end{bmatrix}$

10. (a) [67 41 41] = 3[6 −2 4] − 2[0 1 3] + 7[7 7 5]   **11.** 182
[64 21 59] = 6[6 −2 4] + 5[0 1 3] + 4[7 7 5]
[63 67 57] = 0[6 −2 4] + 4[0 1 3] + 9[7 7 5]

(b) [6 −6 70] = 6[3 −2 7] − 2[6 5 4] + 4[0 4 9]
[6 17 31] = 0[3 −2 7] + 1[6 5 4] + 3[0 4 9]
[63 41 122] = 7[3 −2 7] + 7[6 5 4] + 5[0 4 9]

13. (a) $A = \begin{bmatrix} 2 & -3 & 5 \\ 9 & -1 & 1 \\ 1 & 5 & 4 \end{bmatrix}$, $\mathbf{x} = \begin{bmatrix} x_1 \\ x_2 \\ x_3 \end{bmatrix}$, $\mathbf{b} = \begin{bmatrix} 7 \\ -1 \\ 0 \end{bmatrix}$ **(b)** $A = \begin{bmatrix} 4 & 0 & -3 & 1 \\ 5 & 1 & 0 & -8 \\ 2 & -5 & 9 & -1 \\ 0 & 3 & -1 & 7 \end{bmatrix}$, $\mathbf{x} = \begin{bmatrix} x_1 \\ x_2 \\ x_3 \\ x_4 \end{bmatrix}$, $\mathbf{b} = \begin{bmatrix} 1 \\ 3 \\ 0 \\ 2 \end{bmatrix}$

14. (a) $\begin{aligned} 3x_1 - x_2 + 2x_3 &= 2 \\ 4x_1 + 3x_2 + 7x_3 &= -1 \\ -2x_1 + x_2 + 5x_3 &= 4 \end{aligned}$ **(b)** $\begin{aligned} 3w - 2x \quad\ + z &= 0 \\ 5w \quad\ + 2y - 2z &= 0 \\ 3w + x + 4y + 7z &= 0 \\ -2w + 5x + y + 6z &= 0 \end{aligned}$

15. $\begin{bmatrix} -1 & 23 & -10 \\ 37 & -13 & 8 \\ 29 & 23 & 41 \end{bmatrix}$ (both parts) **16. (a)** $\begin{bmatrix} -3 & -15 & -11 \\ 21 & -15 & 44 \end{bmatrix}$ **(b)** $\begin{bmatrix} 4 & -7 & -19 & -43 \\ 2 & 2 & 18 & 17 \\ 0 & 5 & 25 & 35 \\ 2 & 3 & 23 & 24 \end{bmatrix}$ **(c)** $\begin{bmatrix} 3 & 3 \\ -1 & 4 \\ 1 & 5 \\ 4 & -4 \\ 0 & 14 \end{bmatrix}$

17. (a) A_{11} is a 2×3 matrix and B_{11} is a 2×2 matrix. $A_{11}B_{11}$ does not exist. **(b)** $\begin{bmatrix} -1 & 23 & -10 \\ 37 & -13 & 8 \\ 29 & 23 & 41 \end{bmatrix}$

21. (a) $\begin{bmatrix} a_{11} & 0 & 0 & 0 & 0 & 0 \\ 0 & a_{22} & 0 & 0 & 0 & 0 \\ 0 & 0 & a_{33} & 0 & 0 & 0 \\ 0 & 0 & 0 & a_{44} & 0 & 0 \\ 0 & 0 & 0 & 0 & a_{55} & 0 \\ 0 & 0 & 0 & 0 & 0 & a_{66} \end{bmatrix}$ **(b)** $\begin{bmatrix} a_{11} & a_{12} & a_{13} & a_{14} & a_{15} & a_{16} \\ 0 & a_{22} & a_{23} & a_{24} & a_{25} & a_{26} \\ 0 & 0 & a_{33} & a_{34} & a_{35} & a_{36} \\ 0 & 0 & 0 & a_{44} & a_{45} & a_{46} \\ 0 & 0 & 0 & 0 & a_{55} & a_{56} \\ 0 & 0 & 0 & 0 & 0 & a_{66} \end{bmatrix}$

(c) $\begin{bmatrix} a_{11} & 0 & 0 & 0 & 0 & 0 \\ a_{21} & a_{22} & 0 & 0 & 0 & 0 \\ a_{31} & a_{32} & a_{33} & 0 & 0 & 0 \\ a_{41} & a_{42} & a_{43} & a_{44} & 0 & 0 \\ a_{51} & a_{52} & a_{53} & a_{54} & a_{55} & 0 \\ a_{61} & a_{62} & a_{63} & a_{64} & a_{65} & a_{66} \end{bmatrix}$ **(d)** $\begin{bmatrix} a_{11} & a_{12} & 0 & 0 & 0 & 0 \\ a_{21} & a_{22} & a_{23} & 0 & 0 & 0 \\ 0 & a_{32} & a_{33} & a_{34} & 0 & 0 \\ 0 & 0 & a_{43} & a_{44} & a_{45} & 0 \\ 0 & 0 & 0 & a_{54} & a_{55} & a_{56} \\ 0 & 0 & 0 & 0 & a_{65} & a_{66} \end{bmatrix}$

22. (a) $\begin{bmatrix} 2 & 3 & 4 & 5 \\ 3 & 4 & 5 & 6 \\ 4 & 5 & 6 & 7 \\ 5 & 6 & 7 & 8 \end{bmatrix}$ **(b)** $\begin{bmatrix} 1 & 1 & 1 & 1 \\ 1 & 2 & 4 & 8 \\ 1 & 3 & 9 & 27 \\ 1 & 4 & 16 & 64 \end{bmatrix}$ **(c)** $\begin{bmatrix} -1 & -1 & 1 & 1 \\ -1 & -1 & -1 & 1 \\ 1 & -1 & -1 & -1 \\ 1 & 1 & -1 & -1 \end{bmatrix}$

25. One; namely, $A = \begin{bmatrix} 1 & 1 & 0 \\ 1 & -1 & 0 \\ 0 & 0 & 0 \end{bmatrix}$ **26.** None

27. (a) $\pm \begin{bmatrix} 1 & 1 \\ 1 & 1 \end{bmatrix}$ **(b)** Four; namely, $\begin{bmatrix} \pm\sqrt{5} & 0 \\ 0 & \pm 3 \end{bmatrix}$ **(c)** No; for example, $\begin{bmatrix} -1 & 0 \\ 0 & 1 \end{bmatrix}$

28. (a) Yes; for example, $\begin{bmatrix} 0 & 1 \\ 0 & 0 \end{bmatrix}$ **(b)** Yes; for example, $\begin{bmatrix} 1 & 0 \\ 0 & 0 \end{bmatrix}$

29. (a) True **(b)** True **(c)** False **(d)** True

30. (a) True **(b)** False; for example, $A = \begin{bmatrix} 1 & -1 \\ 1 & -1 \end{bmatrix}$ **(c)** True **(d)** True

EXERCISE SET 1.4 [page 47]

4. $A^{-1} = \begin{bmatrix} 2 & -1 \\ -5 & 3 \end{bmatrix}$, $B^{-1} = \begin{bmatrix} \frac{1}{5} & \frac{3}{20} \\ -\frac{1}{5} & \frac{1}{10} \end{bmatrix}$, $C^{-1} = \begin{bmatrix} -\frac{1}{2} & -2 \\ 1 & 3 \end{bmatrix}$, $D^{-1} = \begin{bmatrix} \frac{1}{2} & 0 \\ 0 & \frac{1}{3} \end{bmatrix}$

7. (a) $A = \begin{bmatrix} \frac{5}{13} & \frac{1}{13} \\ -\frac{3}{13} & \frac{2}{13} \end{bmatrix}$ **(b)** $A = \begin{bmatrix} \frac{2}{7} & 1 \\ \frac{1}{7} & \frac{3}{7} \end{bmatrix}$ **(c)** $A = \begin{bmatrix} -\frac{2}{5} & 1 \\ -\frac{1}{5} & \frac{3}{5} \end{bmatrix}$ **(d)** $A = \begin{bmatrix} -\frac{9}{13} & \frac{1}{13} \\ \frac{2}{13} & -\frac{6}{13} \end{bmatrix}$

8. $A^3 = \begin{bmatrix} 8 & 0 \\ 28 & 1 \end{bmatrix}$, $A^{-3} = \begin{bmatrix} \frac{1}{8} & 0 \\ -\frac{7}{2} & 1 \end{bmatrix}$, $A^2 - 2A + I = \begin{bmatrix} 1 & 0 \\ 4 & 0 \end{bmatrix}$

9. (a) $p(A) = \begin{bmatrix} 1 & 1 \\ 2 & -1 \end{bmatrix}$ **(b)** $p(A) = \begin{bmatrix} 20 & 7 \\ 14 & 6 \end{bmatrix}$ **(c)** $p(A) = \begin{bmatrix} 39 & 13 \\ 26 & 13 \end{bmatrix}$

11. $\begin{bmatrix} \cos\theta & -\sin\theta \\ \sin\theta & \cos\theta \end{bmatrix}$ **12.** $\begin{bmatrix} \frac{1}{2}(e^x + e^{-x}) & \frac{1}{2}(e^{-x} - e^x) \\ \frac{1}{2}(e^{-x} - e^x) & \frac{1}{2}(e^x + e^{-x}) \end{bmatrix}$

13. $A^{-1} = \begin{bmatrix} \frac{1}{a_{11}} & 0 & \cdots & 0 \\ 0 & \frac{1}{a_{22}} & \cdots & 0 \\ \vdots & \vdots & & \vdots \\ 0 & 0 & \cdots & \frac{1}{a_{nn}} \end{bmatrix}$ **16.** No **18.** $C = -A^{-1}BA^{-1}$

19. (a) $\begin{bmatrix} \frac{1}{2} & -\frac{1}{2} & 0 & 0 \\ \frac{1}{2} & \frac{1}{2} & 0 & 0 \\ 0 & 0 & \frac{1}{2} & -\frac{1}{2} \\ -1 & 0 & \frac{1}{2} & \frac{1}{2} \end{bmatrix}$ **(b)** $\begin{bmatrix} 1 & -1 & 0 & 0 \\ 0 & 1 & 0 & 0 \\ 0 & 0 & 1 & -1 \\ 0 & 0 & 0 & 1 \end{bmatrix}$

20. (a) One example is $\begin{bmatrix} 1 & 2 & 3 \\ 2 & 1 & 4 \\ 3 & 4 & 5 \end{bmatrix}$. **(b)** One example is $\begin{bmatrix} 0 & -1 & -1 \\ 1 & 0 & -1 \\ 1 & 1 & 0 \end{bmatrix}$. **22.** Yes **23.** $A^{-1} = \begin{bmatrix} \frac{1}{2} & \frac{1}{2} & -\frac{1}{2} \\ -\frac{1}{2} & \frac{1}{2} & \frac{1}{2} \\ \frac{1}{2} & -\frac{1}{2} & \frac{1}{2} \end{bmatrix}$

31. (a) For example, $A = \begin{bmatrix} 1 & 0 \\ 0 & 0 \end{bmatrix}, B = \begin{bmatrix} 0 & 1 \\ 0 & 0 \end{bmatrix}$ **(b)** $AB + BA$

32. (a) Same as $31(a)$ **(b)** $A^2 - AB + BA - B^2$ **33.** $\begin{bmatrix} \pm 1 & 0 & 0 \\ 0 & \pm 1 & 0 \\ 0 & 0 & \pm 1 \end{bmatrix}$

34. (a) If A is invertible, then A^T is invertible. **(b)** True

35. (a) False **(b)** True **(c)** True **(d)** False (they are sometimes equal)

EXERCISE SET 1.5 [page 56]

1. (a), (c), (d), (f)

2. (a) Add three times the first row to the second row.
(b) Multiply the third row by $\frac{1}{3}$.
(c) Interchange the first row and the fourth row.
(d) Add $\frac{1}{7}$ times third row to the first row.

3. (a) $\begin{bmatrix} 0 & 0 & 1 \\ 0 & 1 & 0 \\ 1 & 0 & 0 \end{bmatrix}$ **(b)** $\begin{bmatrix} 0 & 0 & 1 \\ 0 & 1 & 0 \\ 1 & 0 & 0 \end{bmatrix}$ **(c)** $\begin{bmatrix} 1 & 0 & 0 \\ 0 & 1 & 0 \\ -2 & 0 & 1 \end{bmatrix}$ **(d)** $\begin{bmatrix} 1 & 0 & 0 \\ 0 & 1 & 0 \\ 2 & 0 & 1 \end{bmatrix}$

4. No, since C cannot be obtained by performing a single row operation on B.

5. (a) $\begin{bmatrix} -7 & 4 \\ 2 & -1 \end{bmatrix}$ **(b)** $\begin{bmatrix} -\frac{5}{39} & \frac{2}{13} \\ \frac{4}{39} & \frac{1}{13} \end{bmatrix}$ **(c)** Not invertible

6. (a) $\begin{bmatrix} \frac{3}{2} & -\frac{11}{10} & -\frac{6}{5} \\ -1 & 1 & 1 \\ -\frac{1}{2} & \frac{7}{10} & \frac{2}{5} \end{bmatrix}$ **(b)** Not invertible **(c)** $\begin{bmatrix} \frac{1}{2} & -\frac{1}{2} & \frac{1}{2} \\ -\frac{1}{2} & \frac{1}{2} & \frac{1}{2} \\ \frac{1}{2} & \frac{1}{2} & -\frac{1}{2} \end{bmatrix}$ **(d)** $\begin{bmatrix} \frac{7}{2} & 0 & -3 \\ -1 & 1 & 0 \\ 0 & -1 & 1 \end{bmatrix}$ **(e)** $\begin{bmatrix} \frac{1}{2} & -\frac{1}{2} & \frac{1}{2} \\ 0 & 0 & 1 \\ \frac{1}{2} & \frac{1}{2} & -\frac{1}{2} \end{bmatrix}$

7. (a) $\begin{bmatrix} 1 & 3 & 1 \\ 0 & 1 & -1 \\ -2 & 2 & 0 \end{bmatrix}$ **(b)** $\begin{bmatrix} \frac{\sqrt{2}}{26} & \frac{-3\sqrt{2}}{26} & 0 \\ \frac{4\sqrt{2}}{26} & \frac{\sqrt{2}}{26} & 0 \\ 0 & 0 & 1 \end{bmatrix}$ **(c)** $\begin{bmatrix} 1 & 0 & 0 & 0 \\ -\frac{1}{3} & \frac{1}{3} & 0 & 0 \\ 0 & -\frac{1}{5} & \frac{1}{5} & 0 \\ 0 & 0 & -\frac{1}{7} & \frac{1}{7} \end{bmatrix}$ **(d)** Not invertible **(e)** $\begin{bmatrix} -\frac{4}{5} & \frac{3}{5} & \frac{1}{5} & \frac{1}{5} \\ \frac{3}{2} & 0 & -1 & 0 \\ \frac{1}{2} & 0 & 0 & 0 \\ \frac{4}{5} & \frac{2}{5} & -\frac{1}{5} & -\frac{1}{5} \end{bmatrix}$

8. (a) $\begin{bmatrix} \frac{1}{k_1} & 0 & 0 & 0 \\ 0 & \frac{1}{k_2} & 0 & 0 \\ 0 & 0 & \frac{1}{k_3} & 0 \\ 0 & 0 & 0 & \frac{1}{k_4} \end{bmatrix}$ **(b)** $\begin{bmatrix} 0 & 0 & 0 & \frac{1}{k_4} \\ 0 & 0 & \frac{1}{k_3} & 0 \\ 0 & \frac{1}{k_2} & 0 & 0 \\ \frac{1}{k_1} & 0 & 0 & 0 \end{bmatrix}$ **(c)** $\begin{bmatrix} \frac{1}{k} & 0 & 0 & 0 \\ -\frac{1}{k^2} & \frac{1}{k} & 0 & 0 \\ \frac{1}{k^3} & -\frac{1}{k^2} & \frac{1}{k} & 0 \\ -\frac{1}{k^4} & \frac{1}{k^3} & -\frac{1}{k^2} & \frac{1}{k} \end{bmatrix}$

9. (a) $E_1 = \begin{bmatrix} 1 & 0 \\ 5 & 1 \end{bmatrix}$, $E_2 = \begin{bmatrix} 1 & 0 \\ 0 & \frac{1}{2} \end{bmatrix}$ **(b)** $A^{-1} = E_2 E_1$ **(c)** $A = E_1^{-1} E_2^{-1}$

10. (a) $\begin{bmatrix} 1 & -4 & 7 \\ 4 & 5 & -3 \\ 2 & -1 & 0 \end{bmatrix}$ **(b)** $\begin{bmatrix} 2 & -1 & 0 \\ \frac{4}{3} & \frac{5}{3} & -1 \\ 1 & -4 & 7 \end{bmatrix}$ **(c)** $\begin{bmatrix} 10 & 9 & -6 \\ 4 & 5 & -3 \\ 1 & -4 & 7 \end{bmatrix}$

11. $\begin{bmatrix} 0 & 1 & 0 \\ 1 & 0 & 0 \\ 0 & 0 & 1 \end{bmatrix} \begin{bmatrix} 1 & 0 & 0 \\ 0 & 1 & 0 \\ -2 & 0 & 1 \end{bmatrix} \begin{bmatrix} 1 & 0 & 0 \\ 0 & 1 & 0 \\ 0 & 1 & 1 \end{bmatrix} \begin{bmatrix} 1 & 3 & 3 & 8 \\ 0 & 1 & 7 & 8 \\ 0 & 0 & 0 & 0 \end{bmatrix}$

16. (b) Add -1 times the first row to the second row.
Add -1 times the first row to the third row.
Add -1 times the second row to the first row.
Add the second row to the third row.

19. (a) False **(b)** False **(c)** True **(d)** True

20. (a) True **(b)** True **(c)** True **(d)** False

21. In general, no. Try $b = 1, a = c = d = 0$.

EXERCISE SET 1.6 [page 64]

1. $x_1 = 3, x_2 = -1$ **2.** $x_1 = -3, x_2 = -3$ **3.** $x_1 = -1, x_2 = 4, x_3 = -7$

4. $x_1 = 1, x_2 = -11, x_3 = 16$ **5.** $x_1 = 1, x_2 = 5, x_3 = -1$

6. $w = -6, x = 1, y = 10, z = -7$ **7.** $x_1 = 2b_1 - 5b_2, x_2 = -b_1 + 3b_2$

8. $x_1 = -\frac{15}{2}b_1 + \frac{1}{2}b_2 + \frac{5}{2}b_3, x_2 = \frac{1}{2}b_1 + \frac{1}{2}b_2 - \frac{1}{2}b_3, x_3 = \frac{5}{2}b_1 - \frac{1}{2}b_2 - \frac{1}{2}b_3$

9. (a) $x_1 = \frac{16}{3}, x_2 = -\frac{4}{3}, x_3 = -\frac{11}{3}$ **(b)** $x_1 = -\frac{5}{3}, x_2 = \frac{5}{3}, x_3 = \frac{10}{3}$
(c) $x_1 = 3, x_2 = 0, x_3 = -4$

11. (a) $x_1 = \frac{22}{17}, x_2 = \frac{1}{17}$ **(b)** $x_1 = \frac{21}{17}, x_2 = \frac{11}{17}$

12. (a) $x_1 = -18, x_2 = -1, x_3 = -14$ **(b)** $x_1 = -\frac{421}{2}, x_2 = -\frac{25}{2}, x_3 = -\frac{327}{2}$

13. (a) $x_1 = \frac{7}{15}, x_2 = \frac{4}{15}$ **(b)** $x_1 = \frac{34}{15}, x_2 = \frac{28}{15}$ **(c)** $x_1 = \frac{19}{15}, x_2 = \frac{13}{15}$ **(d)** $x_1 = -\frac{1}{5}, x_2 = \frac{3}{5}$

14. (a) $x_1 = 18, x_2 = -9, x_3 = 2$ **(b)** $x_1 = -23, x_2 = 11, x_3 = -2$
(c) $x_1 = 5, x_2 = -2, x_3 = 0$

15. (a) $x_1 = -12 - 3t, x_2 = -5 - t, x_3 = t$ **(b)** $x_1 = 7 - 3t, x_2 = 3 - t, x_3 = t$

16. $b_1 = 2b_2$ **17.** $b_1 = b_2 + b_3$ **18.** No restrictions **19.** $b_1 = b_3 + b_4, b_2 = 2b_3 + b_4$

21. $X = \begin{bmatrix} 11 & 12 & -3 & 27 & 26 \\ -6 & -8 & 1 & -18 & -17 \\ -15 & -21 & 9 & -38 & -35 \end{bmatrix}$

22. (a) Only the trivial solution $x_1 = x_2 = x_3 = x_4 = 0$; invertible
(b) Infinitely many solutions; not invertible

27. (a) $I - A$ is invertible.   **(b)** $\mathbf{x} = (I - A)^{-1}\mathbf{b}$

28. No. Try $A = I$.   **29.** Yes, for nonsquare matrices

EXERCISE SET 1.7 [page 71]

1. (a) $\begin{bmatrix} \frac{1}{2} & 0 \\ 0 & -\frac{1}{5} \end{bmatrix}$ **(b)** Not invertible   **(c)** $\begin{bmatrix} -1 & 0 & 0 \\ 0 & \frac{1}{2} & 0 \\ 0 & 0 & 3 \end{bmatrix}$ **2. (a)** $\begin{bmatrix} 6 & 3 \\ 4 & -1 \\ 4 & 10 \end{bmatrix}$ **(b)** $\begin{bmatrix} -24 & -10 & 12 \\ 3 & -10 & 0 \\ 60 & 20 & -16 \end{bmatrix}$

3. (a) $A^2 = \begin{bmatrix} 1 & 0 \\ 0 & 4 \end{bmatrix}$, $A^{-2} = \begin{bmatrix} 1 & 0 \\ 0 & \frac{1}{4} \end{bmatrix}$, $A^{-k} = \begin{bmatrix} 1 & 0 \\ 0 & 1/(-2)^k \end{bmatrix}$

(b) $A^2 = \begin{bmatrix} \frac{1}{4} & 0 & 0 \\ 0 & \frac{1}{9} & 0 \\ 0 & 0 & \frac{1}{16} \end{bmatrix}$, $A^{-2} = \begin{bmatrix} 4 & 0 & 0 \\ 0 & 9 & 0 \\ 0 & 0 & 16 \end{bmatrix}$, $A^{-k} = \begin{bmatrix} 2^k & 0 & 0 \\ 0 & 3^k & 0 \\ 0 & 0 & 4^k \end{bmatrix}$

4. (b), (c)   **5.** (a)   **6.** $a = 11, b = -9, c = -13$ **7.** $a = 2, b = -1$

8. (a) Do not commute   **(b)** Commute   **10. (a)** $\begin{bmatrix} 1 & 0 & 0 \\ 0 & -1 & 0 \\ 0 & 0 & -1 \end{bmatrix}$ **(b)** $\begin{bmatrix} \pm\frac{1}{3} & 0 & 0 \\ 0 & \pm\frac{1}{2} & 0 \\ 0 & 0 & \pm 1 \end{bmatrix}$

11. (a) $\begin{bmatrix} a_{11} & a_{12} & a_{13} \\ a_{21} & a_{22} & a_{23} \\ a_{31} & a_{32} & a_{33} \end{bmatrix} \begin{bmatrix} 3 & 0 & 0 \\ 0 & 5 & 0 \\ 0 & 0 & 7 \end{bmatrix}$ **(b)** No   **16. (b)** Yes   **17.** Yes

19. $\begin{bmatrix} 4 & 0 & 0 \\ 0 & 4 & 0 \\ 0 & 0 & 4 \end{bmatrix}, \begin{bmatrix} 4 & 0 & 0 \\ 0 & 4 & 0 \\ 0 & 0 & -1 \end{bmatrix}, \begin{bmatrix} 4 & 0 & 0 \\ 0 & -1 & 0 \\ 0 & 0 & 4 \end{bmatrix}, \begin{bmatrix} -1 & 0 & 0 \\ 0 & 4 & 0 \\ 0 & 0 & 4 \end{bmatrix},$

$\begin{bmatrix} -1 & 0 & 0 \\ 0 & -1 & 0 \\ 0 & 0 & 4 \end{bmatrix}, \begin{bmatrix} -1 & 0 & 0 \\ 0 & 4 & 0 \\ 0 & 0 & -1 \end{bmatrix}, \begin{bmatrix} 4 & 0 & 0 \\ 0 & -1 & 0 \\ 0 & 0 & -1 \end{bmatrix}, \begin{bmatrix} -1 & 0 & 0 \\ 0 & -1 & 0 \\ 0 & 0 & -1 \end{bmatrix}$

20. (a) Yes   **(b)** No (unless $n = 1$)   **(c)** Yes   **(d)** No (unless $n = 1$)

23. No   **24. (a)** $x_1 = \frac{7}{4}, x_2 = 1, x_3 = -\frac{1}{2}$ **(b)** $x_1 = -8, x_2 = -4, x_3 = 3$

25. $A = \begin{bmatrix} 1 & 10 \\ 0 & -2 \end{bmatrix}$ **26.** $\frac{n}{2}(1 + n)$ **27.** Multiply corresponding diagonal entries.

28. A is diagonal.   **30. (a)** True   **(b)** False   **(c)** True   **(d)** False

SUPPLEMENTARY EXERCISES [page 74]

1. $x' = \frac{3}{5}x + \frac{4}{5}y, y' = -\frac{4}{5}x + \frac{3}{5}y$ **2.** $x' = x\cos\theta + y\sin\theta, y' = -x\sin\theta + y\cos\theta$

3. One possible answer is   **4.** 3 pennies, 4 nickels, 6 dimes
$x_1 - 2x_2 - \ x_3 - x_4 = 0$
$x_1 + 5x_2 + 2x_4 \qquad = 0$

5. $x = 4, y = 2, z = 3$ **6.** Infinitely many if $a = 2$ or $a = -\frac{3}{2}$; none otherwise

7. (a) $a \neq 0, b \neq 2$ **(b)** $a \neq 0, b = 2$ **(c)** $a = 0, b = 2$ **(d)** $a = 0, b \neq 2$

8. $x = \frac{5}{9}, y = 9, z = \frac{1}{3}$ **9.** $K = \begin{bmatrix} 0 & 2 \\ 1 & 1 \end{bmatrix}$ **10.** $a = 2, b = -1, c = 1$

11. (a) $X = \begin{bmatrix} -1 & 3 & -1 \\ 6 & 0 & 1 \end{bmatrix}$ **(b)** $X = \begin{bmatrix} 1 & -2 \\ 3 & 1 \end{bmatrix}$ **(c)** $X = \begin{bmatrix} -\frac{113}{37} & -\frac{160}{37} \\ -\frac{20}{37} & -\frac{46}{37} \end{bmatrix}$

12. (a) $Z = \begin{bmatrix} -1 & -7 & 11 \\ 14 & 10 & -26 \end{bmatrix} X$ **(b)** $z_1 = -x_1 - 7x_2 + 11x_3$
$z_2 = 14x_1 + 10x_2 - 26x_3$

13. mpn multiplications and $mp(n-1)$ additions **15.** $a = 1, b = -2, c = 3$ **16.** $a = 1, b = -4, c = -5$

26. $A = -\frac{7}{5}, B = \frac{4}{5}, C = \frac{3}{5}$ **29. (b)** $\begin{bmatrix} a^n & 0 & 0 \\ 0 & b^n & 0 \\ d & 0 & c^n \end{bmatrix}$, where $d = \begin{cases} \dfrac{a^n - c^n}{a - c} & \text{if } a \neq c \\ na^{n-1} & \text{if } a = c \end{cases}$

EXERCISE SET 2.1 [page 87]

1. (a) 5 **(b)** 9 **(c)** 6 **(d)** 10 **(e)** 0 **(f)** 2 **2. (a)** Odd **(b)** Odd **(c)** Even **(d)** Even **(e)** Even **(f)** Even

3. 22 **4.** 0 **5.** 52 **6.** $-3\sqrt{6}$ **7.** $a^2 - 5a + 21$ **8.** 0 **9.** -65 **10.** -4 **11.** -123

12. $-c^4 + c^3 - 16c^2 + 8c - 2$ **13. (a)** $\lambda = 1, \lambda = -3$ **(b)** $\lambda = -2, \lambda = 3, \lambda = 4$ **16.** 275

17. (a) $= -120$ **(b)** $= -120$ **18.** $x = \dfrac{3 \pm \sqrt{33}}{4}$ **22.** Equals 0 if $n > 1$

24. The determinant is equal to the product of the diagonal entries.

25. The determinant is equal to the product of the diagonal entries.

EXERCISE SET 2.2 [page 94]

2. (a) -30 **(b)** -2 **(c)** 0 **(d)** 0 **3. (a)** -5 **(b)** -1 **(c)** 1

4. 30 **5.** 5 **6.** -17 **7.** 33 **8.** 39 **9.** 6 **10.** $-\frac{1}{6}$

11. -2 **12. (a)** -6 **(b)** 72 **(c)** -6 **(d)** 18

16. (a) $\det(A) = -1$ **(b)** $\det(A) = 1$ **17.** $x = 0, -1, \frac{1}{2}$ **18.** $x = 1, -3$

EXERCISE SET 2.3 [page 102]

1. (a) $\det(2A) = -40 = 2^2 \det(A)$ **(b)** $\det(-2A) = -448 = (-2)^3 \det(A)$

2. $\det AB = -170 = (\det A)(\det B)$

4. (a) Invertible **(b)** Not invertible **(c)** Not invertible **(d)** Not invertible

5. (a) -189 **(b)** $-\frac{1}{7}$ **(c)** $-\frac{8}{7}$ **(d)** $-\frac{1}{56}$ **(e)** 7

6. If $x = 0$, the first and third rows are proportional. **12. (a)** $k = \dfrac{5 \pm \sqrt{17}}{2}$ **(b)** $k = -1$
If $x = 2$, the first and second rows are proportional.

14. (a) $\begin{bmatrix} \lambda - 1 & -2 \\ -2 & \lambda - 1 \end{bmatrix} \begin{bmatrix} x_1 \\ x_2 \end{bmatrix} = \begin{bmatrix} 0 \\ 0 \end{bmatrix}$ **(b)** $\begin{bmatrix} \lambda - 2 & -3 \\ -4 & \lambda - 3 \end{bmatrix} \begin{bmatrix} x_1 \\ x_2 \end{bmatrix} = \begin{bmatrix} 0 \\ 0 \end{bmatrix}$ **(c)** $\begin{bmatrix} \lambda - 3 & -1 \\ 5 & \lambda + 3 \end{bmatrix} \begin{bmatrix} x_1 \\ x_2 \end{bmatrix} = \begin{bmatrix} 0 \\ 0 \end{bmatrix}$

15. (i) $\lambda^2 - 2\lambda - 3 = 0$ (ii) $\lambda = -1, \lambda = 3$ (iii) $\begin{bmatrix} -t \\ t \end{bmatrix}, \begin{bmatrix} t \\ t \end{bmatrix}$

(i) $\lambda^2 - 5\lambda - 6 = 0$ (ii) $\lambda = -1, \lambda = 6$ (iii) $\begin{bmatrix} -t \\ t \end{bmatrix}, \begin{bmatrix} \frac{3}{4}t \\ t \end{bmatrix}$

(i) $\lambda^2 - 4 = 0$ (ii) $\lambda = -2, \lambda = 2$ (iii) $\begin{bmatrix} -\frac{t}{5} \\ t \end{bmatrix}, \begin{bmatrix} -t \\ t \end{bmatrix}$

20. No   **21.** AB is singular.

22. **(a)** False   **(b)** True   **(c)** False   **(d)** True

23. **(a)** True   **(b)** True   **(c)** False   **(d)** True

EXERCISE SET 2.4 [page 112]

1. **(a)** $M_{11} = 29, M_{12} = 21, M_{13} = 27, M_{21} = -11, M_{22} = 13, M_{23} = -5, M_{31} = -19, M_{32} = -19, M_{33} = 19$
    (b) $C_{11} = 29, C_{12} = -21, C_{13} = 27, C_{21} = 11, C_{22} = 13, C_{23} = 5, C_{31} = -19, C_{32} = 19, C_{33} = 19$

2. **(a)** $M_{13} = 0, \ C_{13} = 0$ **(b)** $M_{23} = -96, \ C_{23} = 96$ **3.** 152
    (c) $M_{22} = -48, \ C_{22} = -48$ **(d)** $M_{21} = 72, \ C_{21} = -72$

4. **(a)** $\text{adj}(A) = \begin{bmatrix} 29 & 11 & -19 \\ -21 & 13 & 19 \\ 27 & 5 & 19 \end{bmatrix}$ **(b)** $A^{-1} = \begin{bmatrix} \frac{29}{152} & \frac{11}{152} & -\frac{19}{152} \\ -\frac{21}{152} & \frac{13}{152} & \frac{19}{152} \\ \frac{27}{152} & \frac{5}{152} & \frac{19}{152} \end{bmatrix}$

5. -40 **6.** -66 **7.** 0 **8.** $k^3 - 8k^2 - 10k + 95$ **9.** -240 **10.** 0

11. $A^{-1} = \begin{bmatrix} 3 & -5 & -5 \\ -3 & 4 & 5 \\ 2 & -2 & -3 \end{bmatrix}$ **12.** $A^{-1} = \begin{bmatrix} 2 & 0 & \frac{3}{2} \\ \frac{2}{3} & \frac{1}{3} & \frac{2}{3} \\ -1 & 0 & -1 \end{bmatrix}$ **13.** $A^{-1} = \begin{bmatrix} \frac{1}{2} & \frac{3}{2} & 1 \\ 0 & 1 & \frac{3}{2} \\ 0 & 0 & \frac{1}{2} \end{bmatrix}$

14. $A^{-1} = \begin{bmatrix} \frac{1}{2} & 0 & 0 \\ -4 & 1 & 0 \\ \frac{29}{12} & -\frac{1}{2} & \frac{1}{6} \end{bmatrix}$ **15.** $A^{-1} = \begin{bmatrix} -4 & 3 & 0 & -1 \\ 2 & -1 & 0 & 0 \\ -7 & 0 & -1 & 8 \\ 6 & 0 & 1 & -7 \end{bmatrix}$

16. $x_1 = 1, x_2 = 2$ **17.** $x = \frac{3}{11}, y = \frac{2}{11}, z = -\frac{1}{11}$

18. $x = -\frac{144}{55}, y = -\frac{61}{55}, z = \frac{46}{11}$ **19.** $x_1 = -\frac{30}{11}, x_2 = -\frac{38}{11}, x_3 = -\frac{40}{11}$

20. $x_1 = 5, x_2 = 8, x_3 = 3, x_4 = -1$ **21.** Cramer's rule does not apply.

22. $A^{-1} = \begin{bmatrix} \cos\theta & -\sin\theta & 0 \\ \sin\theta & \cos\theta & 0 \\ 0 & 0 & 1 \end{bmatrix}$ **23.** $y = 0$ **24.** $x = 1, y = 0, z = 2, w = 0$

31. $\det(A) = 10 \times (-108) = -1080$ **33.** 12   **34.** One

35. **(a)** True   **(b)** False   **(c)** True   **(d)** False

SUPPLEMENTARY EXERCISES [page 115]

1. $x' = \frac{3}{5}x + \frac{4}{5}y, y' = -\frac{4}{5}x + \frac{3}{5}y$ **2.** $x' = x\cos\theta + y\sin\theta, y' = -x\sin\theta + y\cos\theta$

4. 2   **5.** $\cos\beta = \dfrac{c^2 + a^2 - b^2}{2ac}, \cos\gamma = \dfrac{a^2 + b^2 - c^2}{2ab}$ **10.** **(b)** $\frac{19}{2}$

12. $\det(B) = (-1)^{n(n-1)/2} \det(A)$

13. **(a)** The ith and jth columns will be interchanged.
    (b) The ith column will be divided by c.
    (c) $-c$ times the jth column will be added to the ith column.

15. **(a)** $\lambda^3 + (-a_{11} - a_{22} - a_{33})\lambda^2 + (a_{11}a_{22} + a_{11}a_{33} + a_{22}a_{33} - a_{12}a_{21} - a_{13}a_{31} - a_{23}a_{32})\lambda +$
    $(a_{11}a_{23}a_{32} + a_{12}a_{21}a_{33} + a_{13}a_{22}a_{31} - a_{11}a_{22}a_{33} - a_{12}a_{23}a_{31} - a_{13}a_{21}a_{32})$

18. **(a)** $\lambda = -5, \lambda = 2, \lambda = 4; \begin{bmatrix} -2t \\ t \\ t \end{bmatrix}, \begin{bmatrix} 5t \\ t \\ t \end{bmatrix}, \begin{bmatrix} 7t \\ 19t \\ t \end{bmatrix}$ **(b)** $\lambda = 1; \begin{bmatrix} \frac{1}{2}t \\ -\frac{1}{2}t \\ t \end{bmatrix}$

EXERCISE SET 3.1 [page 125]

3. (a) $\overrightarrow{P_1P_2} = (-1,-1)$ (b) $\overrightarrow{P_1P_2} = (-7,-2)$ (c) $\overrightarrow{P_1P_2} = (2,1)$ (d) $\overrightarrow{P_1P_2} = (a,b)$
  (e) $\overrightarrow{P_1P_2} = (-5,12,-6)$ (f) $\overrightarrow{P_1P_2} = (1,-1,-2)$ (g) $\overrightarrow{P_1P_2} = (-a,-b,-c)$ (h) $\overrightarrow{P_1P_2} = (a,b,c)$

4. (a) $Q(5,10,-8)$ is one possible answer.   (b) $Q(-7,-4,-2)$ is one possible answer.

5. (a) $P(-1,2,-4)$ is one possible answer.   (b) $P(7,-2,-6)$ is one possible answer.

6. (a) $(-2,1,-4)$ (b) $(-10,6,4)$ (c) $(-7,1,10)$
  (d) $(80,-20,-80)$ (e) $(132,-24,-72)$ (f) $(-77,8,94)$

7. $x = (-\frac{8}{3},\frac{1}{2},\frac{8}{3})$ 8. $c_1 = 2, c_2 = -1, c_3 = 2$ 10. $c_1 = c_2 = c_3 = 0$

11. (a) $(\frac{9}{2},-\frac{1}{2},-\frac{1}{2})$ (b) $(\frac{23}{4},-\frac{9}{4},\frac{1}{4})$ 12. (a) $x' = 5,\ y' = 8$ (b) $x = -1,\ y = 3$

14. $u = \left(\frac{\sqrt{3}}{2},\frac{1}{2}\right), v = \left(-\frac{1}{2},-\frac{\sqrt{3}}{2}\right), u+v = \left(\frac{\sqrt{3}-1}{2},\frac{1-\sqrt{3}}{2}\right), u-v = \left(\frac{\sqrt{3}+1}{2},\frac{\sqrt{3}+1}{2}\right)$

EXERCISE SET 3.2 [page 128]

1. (a) 5   (b) $\sqrt{13}$ (c) 5   (d) $2\sqrt{3}$ (e) $3\sqrt{6}$ (f) 6   2. (a) $\sqrt{13}$ (b) $2\sqrt{26}$ (c) $\sqrt{209}$ (d) $3\sqrt{2}$

3. (a) $\sqrt{83}$ (b) $\sqrt{17}+\sqrt{26}$ (c) $4\sqrt{17}$ (d) $\sqrt{466}$ (e) $\left(\frac{3}{\sqrt{61}},\frac{6}{\sqrt{61}},-\frac{4}{\sqrt{61}}\right)$ (f) 1   4. $k = \pm\frac{4}{\sqrt{30}}$

6. (b) $(\frac{3}{5},\frac{4}{5})$ (c) $(\frac{2}{7},-\frac{3}{7},\frac{6}{7})$ 7. (b) $(6\sqrt{3}+5\sqrt{2},6-5\sqrt{2})$ 8. A sphere of radius 1 centered at (x_0,y_0,z_0)

12. Yes   13. (a) $a = c = 0$ (b) At least one of a or c is not zero, that is, $a^2 + c^2 > 0$

14. (a) The distance from x to the origin is less than 1.   (b) $\|x - x_0\| > 1$

EXERCISE SET 3.3 [page 136]

1. (a) -11 (b) -24 (c) 0   (d) 0   2. (a) $-\frac{11}{\sqrt{13}\sqrt{74}}$ (b) $-\frac{3}{\sqrt{10}}$ (c) 0   (d) 0

3. (a) Orthogonal   (b) Obtuse   (c) Acute   (d) Obtuse

4. (a) $(0,0)$ (b) $(\frac{8}{13},-\frac{12}{13})$ (c) $(-\frac{16}{13},0,-\frac{80}{13})$ (d) $(\frac{16}{89},\frac{12}{89},\frac{32}{89})$

5. (a) $(6,2)$ (b) $(-\frac{21}{13},-\frac{14}{13})$ (c) $(\frac{55}{13},1,-\frac{11}{13})$ (d) $(\frac{73}{89},-\frac{12}{89},-\frac{32}{89})$

6. (a) $\frac{2}{5}$ (b) $\frac{4\sqrt{5}}{5}$ (c) $\frac{18}{\sqrt{22}}$ (d) $\frac{43}{\sqrt{54}}$ 8. (b) $(3k,2k)$ for any scalar k (c) $(\frac{4}{5},\frac{3}{5}),(-\frac{4}{5},-\frac{3}{5})$

9. (a) 102   (b) $125\sqrt{2}$ (c) 170   (d) 170   10. For example, $(2,-5,0),(-3,0,5),(0,3,2),(1,-5,-5),(-3,3,7)$

11. $\cos\theta_1 = \frac{\sqrt{10}}{10},\ \cos\theta_2 = \frac{3\sqrt{10}}{10},\ \cos\theta_3 = 0$ 12. The right angle is at B.

13. $\pm(1/\sqrt{3},1/\sqrt{3},-1/\sqrt{3})$ 14. (a) $\frac{10}{3}$ (b) $-\frac{6}{5}$ (c) $\frac{-60+34\sqrt{3}}{33}$ (d) $\frac{1}{2}$

15. (a) 1   (b) $\frac{1}{\sqrt{17}}$ (c) $\frac{6}{\sqrt{10}}$ 18. $\cos^{-1}\left(\frac{2}{\sqrt{6}}\right)$

19. (b) $\cos\beta = \frac{b}{\|v\|},\ \cos\gamma = \frac{c}{\|v\|}$ 20. $\theta_1 \approx 71°, \theta_2 \approx 61°, \theta_3 \approx 36°$

24. (a) The vector u is dotted with a scalar.   (b) A scalar is added to the vector w.
  (c) Scalars do not have norms.   (d) The scalar k is dotted with a vector.

25. Yes; for example, if a and u are orthogonal   26. No; it merely says that u is orthogonal to $v - w$.

27. $r = (u \cdot r)\frac{u}{\|u\|^2} + (v \cdot r)\frac{v}{\|v\|^2} + (w \cdot r)\frac{w}{\|w\|^2}$ 28. Theorem of Pythagoras

EXERCISE SET 3.4 [page 147]

1. **(a)** $(32, -6, -4)$ **(b)** $(-14, -20, -82)$ **(c)** $(27, 40, -42)$
 (d) $(0, 176, -264)$ **(e)** $(-44, 55, -22)$ **(f)** $(-8, -3, -8)$

2. **(a)** $(18, 36, -18)$ **(b)** $(-3, 9, -3)$ 3. **(a)** $\sqrt{59}$ **(b)** $\sqrt{101}$ **(c)** 0

4. **(a)** $\dfrac{\sqrt{374}}{2}$ **(b)** $\sqrt{285}$ 7. For example, $(1, 1, 1) \times (2, -3, 5) = (8, -3, -5)$

8. **(a)** -10 **(b)** -110 9. **(a)** -3 **(b)** 3   **(c)** 3   **(d)** -3 **(e)** -3 **(f)** 0

10. **(a)** 16   **(b)** 45   11. **(a)** No   **(b)** Yes   **(c)** No   12. $\pm\left(0, \dfrac{2}{\sqrt{5}}, \dfrac{1}{\sqrt{5}}\right)$

13. $\left(\dfrac{6}{\sqrt{61}}, -\dfrac{3}{\sqrt{61}}, \dfrac{4}{\sqrt{61}}\right), \left(-\dfrac{6}{\sqrt{61}}, \dfrac{3}{\sqrt{61}}, -\dfrac{4}{\sqrt{61}}\right)$ 15. $2(\mathbf{v} \times \mathbf{u})$ 16. $\dfrac{12\sqrt{13}}{49}$

17. **(a)** $\dfrac{\sqrt{26}}{2}$ **(b)** $\dfrac{\sqrt{26}}{3}$ 19. **(a)** $\dfrac{2\sqrt{141}}{\sqrt{29}}$ **(b)** $\dfrac{\sqrt{137}}{3}$ 21. **(a)** $\sqrt{122}$ **(b)** $\theta \approx 40°19''$

22. Any scalar multiple of $(2, 2, 1)$ 23. **(a)** $\mathbf{m} = (0, 1, 0)$ and $\mathbf{n} = (1, 0, 0)$ **(b)** $(-1, 0, 0)$ **(c)** $(0, 0, -1)$

28. $(-8, 0, -8)$ 31. **(a)** $\frac{2}{3}$ **(b)** $\frac{1}{2}$ 35. **(b)** $\mathbf{u} \cdot \mathbf{w} \ne 0, \mathbf{v} \cdot \mathbf{w} = 0$

36. No, the equation is equivalent to $\mathbf{u} \times (\mathbf{v} - \mathbf{w}) = 0$ and hence to $\mathbf{v} - \mathbf{w} = k\mathbf{u}$ for some scalar k.

37. $\mathbf{u} \times (\mathbf{v} \times \mathbf{w}) \ne (\mathbf{u} \times \mathbf{v}) \times \mathbf{w}$, in general   38. They are collinear.

39. For example, $ab = ba$, $(ab)c = a(bc)$, and $ab = 0$ implies $a = 0$ or $b = 0$.

EXERCISE SET 3.5 [page 155]

1. **(a)** $-2(x + 1) + (y - 3) - (z + 2) = 0$ **(b)** $(x - 1) + 9(y - 1) + 8(z - 4) = 0$
 (c) $2z = 0$ **(d)** $x + 2y + 3z = 0$

2. **(a)** $-2x + y - z - 7 = 0$ **(b)** $x + 9y + 8z - 42 = 0$
 (c) $2z = 0$ **(d)** $x + 2y + 3y = 0$

3. **(a)** $(0, 0, 5)$ is a point in the plane and $\mathbf{n} = (-3, 7, 2)$ is a normal vector so that
 $-3(x - 0) + 7(y - 0) + 2(z - 5) = 0$ is a point-normal form; other points and normals
 yield other correct answers.   **(b)** $(x - 0) + 0(y - 0) - 4(z - 0) = 0$ is a possibility

4. **(a)** $2y - z + 1 = 0$ **(b)** $x + 9y - 5z - 26 = 0$

5. **(a)** Not parallel   **(b)** Parallel   **(c)** Parallel   6. **(a)** Parallel   **(b)** Not parallel

7. **(a)** Not perpendicular   **(b)** Perpendicular   8. **(a)** Perpendicular   **(b)** Not perpendicular

9. **(a)** $x = 3 + 2t, y = -1 + t, z = 2 + 3t$ **(b)** $x = -2 + 6t, y = 3 - 6t, z = -3 - 2t$
 (c) $x = 2, y = 2 + t, z = 6$ **(d)** $x = t, y = -2t, z = 3t$

10. **(a)** $x = 5 + t, y = -2 + 2t, z = 4 - 4t$ **(b)** $x = 2t, y = -t, z = -3t$

11. **(a)** $x = -12 - 7t, y = -41 - 23t, z = t$ **(b)** $x = \frac{5}{2}t, y = 0, z = t$

12. **(a)** $(-2, 4, 1) \cdot (x + 1, y - 2, z - 4) = 0$ **(b)** $(-1, 4, 3) \cdot (x - 2, y, z + 5) = 0$
 (c) $(-1, 0, 0) \cdot (x - 5, y + 2, z - 1) = 0$ **(d)** $(a, b, c) \cdot (x, y, z) = 0$

13. **(a)** Parallel   **(b)** Not parallel   14. **(a)** Perpendicular   **(b)** Not perpendicular

15. **(a)** $(x, y, z) = (-1, 2, 3) + t(7, -1, 5)$ $(-\infty < t < +\infty)$
 (b) $(x, y, z) = (2, 0, -1) + t(1, 1, 1)$ $(-\infty < t < +\infty)$
 (c) $(x, y, z) = (2, -4, 1) + t(0, 0, -2)$ $(-\infty < t < +\infty)$
 (d) $(x, y, z) = (0, 0, 0) + t(a, b, c)$ $(-\infty < t < +\infty)$

17. $2x + 3y - 5z + 36 = 0$ **18. (a)** $z = 0$ **(b)** $y = 0$ **(c)** $x = 0$

19. (a) $z - z_0 = 0$ **(b)** $x - x_0 = 0$ **(c)** $y - y_0 = 0$ **20.** $7x + 4y - 2z = 0$

21. $5x - 2y + z - 34 = 0$ **22.** $\left(-\frac{173}{3}, -\frac{43}{3}, \frac{49}{3}\right)$ **23.** $y + 2z - 9 = 0$

24. $x - y - 4z - 2 = 0$ **26.** $x = \frac{11}{5}t - 2, y = -\frac{2}{5}t + 5, z = t$

27. $x + 5y + 3z - 18 = 0$ **28.** $(x - 2) + (y + 1) - 3(z - 4) = 0$

29. $4x + 13y - z - 17 = 0$ **30.** $3x + 10y + 4z - 53 = 0$ **31.** $3x - y - z - 2 = 0$

32. $5x - 3y + 2z - 5 = 0$ **33.** $2x + 4y + 8z + 13 = 0$ **36.** $x - 4y + 4z + 9 = 0$

37. (a) $x = \frac{11}{23} + \frac{7}{23}t, y = -\frac{41}{23} - \frac{1}{23}t, z = t$ **(b)** $x = -\frac{2}{5}t, y = 0, z = t$

39. (a) $\frac{5}{3}$ **(b)** $\dfrac{1}{\sqrt{29}}$ **(c)** $\dfrac{4}{\sqrt{3}}$ **40. (a)** $\dfrac{1}{2\sqrt{26}}$ **(b)** 0 **(c)** $\dfrac{2}{\sqrt{6}}$

42. (a) $\dfrac{x - 3}{2} = y + 1 = \dfrac{z - 2}{3}$ **(b)** $\dfrac{x + 2}{6} = -\dfrac{y - 3}{6} = -\dfrac{z + 3}{2}$

43. (a) $x - 2y - 17 = 0$ and $x + 4z - 27 = 0$ is one possible answer.
   (b) $x - 2y = 0$ and $-7y + 2z = 0$ is one possible answer.

44. (a) $\theta \approx 35°$ **(b)** $\theta \approx 79°$ **45.** $\theta \approx 75°$ **46.** They are identical.

47. They are perpendicular.   **48.** It is the line segment joining P_1 to P_2.

49. For example, $x = x_0 + 2t, y = y_0 + 3t, z = z_0 + 5t$ and $x = x_0 + t$,
   $y = y_0 - 4t, z = z_0 + 2t$

EXERCISE SET 4.1 [page 170]

1. (a) $(-1, 9, -11, 1)$ **(b)** $(22, 53, -19, 14)$ **(c)** $(-13, 13, -36, -2)$
   (d) $(-90, -114, 60, -36)$ **(e)** $(-9, -5, -5, -3)$ **(f)** $(27, 29, -27, 9)$

2. $\left(\frac{6}{5}, \frac{2}{3}, \frac{2}{3}, \frac{2}{5}\right)$ **3.** $c_1 = 1, c_2 = 1, c_3 = -1, c_4 = 1$

5. (a) $\sqrt{29}$ **(b)** 3 **(c)** 13 **(d)** $\sqrt{31}$

6. (a) $\sqrt{133}$ **(b)** $\sqrt{30} + \sqrt{77}$ **(c)** $4\sqrt{30}$ **(d)** $\sqrt{1811}$ **(e)** $\left(\dfrac{1}{\sqrt{2}}, \dfrac{1}{3\sqrt{2}}, \dfrac{2}{3\sqrt{2}}, \dfrac{2}{3\sqrt{2}}\right)$ **(f)** 1

8. $k = \pm\frac{5}{7}$ **9. (a)** 7 **(b)** 14 **(c)** 7 **(d)** 11 **10. (a)** $\left(\dfrac{1}{\sqrt{10}}, \dfrac{3}{\sqrt{10}}\right), \left(-\dfrac{1}{\sqrt{10}}, -\dfrac{3}{\sqrt{10}}\right)$

11. (a) $\sqrt{10}$ **(b)** $\sqrt{56}$ **(c)** $\sqrt{59}$ **(d)** 10

14. (a) Yes **(b)** No **(c)** Yes **(d)** No **(e)** No **(f)** Yes

15. (a) $k = -3$ **(b)** $k = -2, k = -3$

16. $\pm\frac{1}{57}(-34, 44, -6, 11)$ **19.** $x_1 = 1, x_2 = -1, x_3 = 2$ **20.** -6

22. The component in the **a** direction is $\text{proj}_\mathbf{a}\mathbf{u} = \frac{4}{15}(-1, 1, 2, 3)$; the orthogonal component is
   $\frac{1}{15}(34, 11, 52, -27)$.

23. They do not intersect.

33. (a) Euclidean measure of "box" in R^n: $a_1 a_2 \cdots a_n$ **(b)** Length of diagonal: $\sqrt{a_1^2 + a_2^2 + \cdots + a_n^2}$

34. (b) The parallelogram law: The sum of the squares of the lengths of the four sides of a parallelogram is equal to the sum of the squares of the lengths of the two diagonals.

35. (a) $d(\mathbf{u}, \mathbf{v}) = \sqrt{2}$ **36.** Yes, since any two vectors lie on a plane

37. (a) True **(b)** True **(c)** False **(d)** True **(e)** True, unless $\mathbf{u} = \mathbf{0}$

EXERCISE SET 4.2 [page 185]

1. (a) Linear; $R^3 \to R^2$ **(b)** Nonlinear; $R^2 \to R^3$ **(c)** Linear; $R^3 \to R^3$ **(d)** Nonlinear; $R^4 \to R^2$

2. (a) $\begin{bmatrix} 2 & -3 & 0 & 1 \\ 3 & 5 & 0 & -1 \end{bmatrix}$ **(b)** $\begin{bmatrix} 7 & 2 & -8 \\ 0 & -1 & 5 \\ 4 & 7 & -1 \end{bmatrix}$ **(c)** $\begin{bmatrix} -1 & 1 \\ 3 & -2 \\ 5 & -7 \end{bmatrix}$ **(d)** $\begin{bmatrix} 1 & 0 & 0 & 0 \\ 1 & 1 & 0 & 0 \\ 1 & 1 & 1 & 0 \\ 1 & 1 & 1 & 1 \end{bmatrix}$

3. $\begin{bmatrix} 3 & 5 & -1 \\ 4 & -1 & 1 \\ 3 & 2 & -1 \end{bmatrix}$; $T(-1, 2, 4) = (3, -2, -3)$

4. (a) $\begin{bmatrix} 2 & -1 \\ 1 & 1 \end{bmatrix}$ **(b)** $\begin{bmatrix} 1 & 0 \\ 0 & 1 \end{bmatrix}$ **(c)** $\begin{bmatrix} 1 & 2 & 1 \\ 1 & 5 & 0 \\ 0 & 0 & 1 \end{bmatrix}$ **(d)** $\begin{bmatrix} 4 & 0 & 0 \\ 0 & 7 & 0 \\ 0 & 0 & -8 \end{bmatrix}$

5. (a) $\begin{bmatrix} 0 & 1 \\ -1 & 0 \\ 1 & 3 \\ 1 & -1 \end{bmatrix}$ **(b)** $\begin{bmatrix} 7 & 2 & -1 & 1 \\ 0 & 1 & 1 & 0 \\ -1 & 0 & 0 & 0 \end{bmatrix}$ **(c)** $\begin{bmatrix} 0 & 0 & 0 \\ 0 & 0 & 0 \\ 0 & 0 & 0 \\ 0 & 0 & 0 \\ 0 & 0 & 0 \end{bmatrix}$ **(d)** $\begin{bmatrix} 0 & 0 & 0 & 1 \\ 1 & 0 & 0 & 0 \\ 0 & 0 & 1 & 0 \\ 0 & 1 & 0 & 0 \\ 1 & 0 & -1 & 0 \end{bmatrix}$

6. (a) $\begin{bmatrix} -1 \\ 1 \end{bmatrix}$ **(b)** $\begin{bmatrix} 3 \\ 13 \end{bmatrix}$ **(c)** $\begin{bmatrix} -2x_1 + x_2 + 4x_3 \\ 3x_1 + 5x_2 + 7x_3 \\ 6x_1 \quad\ - x_3 \end{bmatrix}$ **(d)** $\begin{bmatrix} -x_1 + x_2 \\ 2x_1 + 4x_2 \\ 7x_1 + 8x_2 \end{bmatrix}$

7. (a) $T(-1, 4) = (5, 4)$ **(b)** $T(2, 1, -3) = (0, -2, 0)$

8. (a) $(-1, -2)$ **(b)** $(1, 2)$ **(c)** $(2, -1)$

9. (a) $(2, -5, -3)$ **(b)** $(2, 5, 3)$ **(c)** $(-2, -5, 3)$

10. (a) $(2, 0)$ **(b)** $(0, -5)$ **11. (a)** $(-2, 1, 0)$ **(b)** $(-2, 0, 3)$ **(c)** $(0, 1, 3)$

12. (a) $\left(\dfrac{3\sqrt{3}+4}{2}, \dfrac{3-4\sqrt{3}}{2} \right)$ **(b)** $\left(\dfrac{3-4\sqrt{3}}{2}, \dfrac{-3\sqrt{3}-4}{2} \right)$ **(c)** $\left(\dfrac{7\sqrt{2}}{2}, \dfrac{-\sqrt{2}}{2} \right)$ **(d)** $(4, 3)$

13. (a) $\left(-2, \dfrac{\sqrt{3}-2}{2}, \dfrac{1+2\sqrt{3}}{2} \right)$ **(b)** $(0, 1, 2\sqrt{2})$ **(c)** $(-1, -2, 2)$

14. (a) $\begin{bmatrix} 1 & 0 & 0 \\ 0 & 1/2 & \sqrt{3}/2 \\ 0 & -\sqrt{3}/2 & 1/2 \end{bmatrix}$ **(b)** $\begin{bmatrix} 1/2 & 0 & -\sqrt{3}/2 \\ 0 & 1 & 0 \\ \sqrt{3}/2 & 0 & 1/2 \end{bmatrix}$ **(c)** $\begin{bmatrix} 1/2 & \sqrt{3}/2 & 0 \\ -\sqrt{3}/2 & 1/2 & 0 \\ 0 & 0 & 1 \end{bmatrix}$

15. (a) $\left(-2, \dfrac{\sqrt{3}+2}{2}, \dfrac{-1+2\sqrt{3}}{2} \right)$ **(b)** $(-2\sqrt{2}, 1, 0)$ **(c)** $(1, 2, 2)$

16. (a) $\begin{bmatrix} 1 & 0 \\ 0 & -1 \end{bmatrix}$ **(b)** $\begin{bmatrix} 0 & 0 \\ 0 & \frac{1}{2} \end{bmatrix}$ **(c)** $\begin{bmatrix} 3 & 0 \\ 0 & -3 \end{bmatrix}$ **17. (a)** $\begin{bmatrix} 0 & 0 \\ 1/2 & -\sqrt{3}/2 \end{bmatrix}$ **(b)** $\begin{bmatrix} -\sqrt{2} & \sqrt{2} \\ \sqrt{2} & \sqrt{2} \end{bmatrix}$ **(c)** $\begin{bmatrix} -1 & 0 \\ 0 & -1 \end{bmatrix}$

18. (a) $\begin{bmatrix} -1 & 0 & 0 \\ 0 & 0 & 0 \\ 0 & 0 & 1 \end{bmatrix}$ **(b)** $\begin{bmatrix} 1 & 0 & 1 \\ 0 & \sqrt{2} & 0 \\ -1 & 0 & 1 \end{bmatrix}$ **(c)** $\begin{bmatrix} -1 & 0 & 0 \\ 0 & 1 & 0 \\ 0 & 0 & 0 \end{bmatrix}$

19. (a) $\begin{bmatrix} \sqrt{3}/8 & -\sqrt{3}/16 & 1/16 \\ 1/8 & 3/16 & -\sqrt{3}/16 \\ 0 & 1/8 & \sqrt{3}/8 \end{bmatrix}$ **(b)** $\begin{bmatrix} 0 & 0 & 0 \\ 0 & -1 & 0 \\ 0 & 0 & -1 \end{bmatrix}$ **(c)** $\begin{bmatrix} 0 & 1 & 0 \\ 0 & 0 & -1 \\ -1 & 0 & 0 \end{bmatrix}$

20. (a) Yes **(b)** Yes **(c)** No **21. (a)** Yes **(b)** No **22. (a)** $\begin{bmatrix} 1 & 0 & 0 \\ 0 & 0 & 0 \\ 0 & 0 & 0 \end{bmatrix} \begin{bmatrix} 0 & 0 & 0 \\ 0 & 1 & 0 \\ 0 & 0 & 0 \end{bmatrix} \begin{bmatrix} 0 & 0 & 0 \\ 0 & 0 & 0 \\ 0 & 0 & 1 \end{bmatrix}$

24. $\begin{bmatrix} \frac{1}{3}(1-\cos\theta)+\cos\theta & \frac{1}{3}(1-\cos\theta)-\dfrac{1}{\sqrt{3}}\sin\theta & \frac{1}{3}(1-\cos\theta)-\dfrac{1}{\sqrt{3}}\sin\theta \\[2mm] \frac{1}{3}(1-\cos\theta)-\dfrac{1}{\sqrt{3}}\sin\theta & \frac{1}{3}(1-\cos\theta)+\cos\theta & \frac{1}{3}(1-\cos\theta)-\dfrac{1}{\sqrt{3}}\sin\theta \\[2mm] \frac{1}{3}(1-\cos\theta)-\dfrac{1}{\sqrt{3}}\sin\theta & \frac{1}{3}(1-\cos\theta)-\dfrac{1}{\sqrt{3}}\sin\theta & \frac{1}{3}(1-\cos\theta)+\cos\theta \end{bmatrix}$

26. $135°$ **28. (c)** $90°$ **29. (a)** Twice the orthogonal projection on the x-axis
(b) Twice the reflection about the x-axis

30. (a) The x-coordinate is stretched by a factor of 2 and the y-coordinate is stretched by a factor of 3.
(b) Rotation through $30°$
31. Rotation through the angle 2θ **32.** Rotation through the angle $-\theta$

EXERCISE SET 4.3 [page 198]

1. (a) Not one-to-one **(b)** One-to-one **(c)** One-to-one **(d)** One-to-one
(e) One-to-one **(f)** One-to-one **(g)** One-to-one

2. (a) $\begin{bmatrix} 8 & 4 \\ 2 & 1 \end{bmatrix}$; not one-to-one **(b)** $\begin{bmatrix} 2 & -3 \\ 5 & 1 \end{bmatrix}$; one-to-one **(c)** $\begin{bmatrix} -1 & 3 & 2 \\ 2 & 0 & 4 \\ 1 & 3 & 6 \end{bmatrix}$; not one-to-one **(d)** $\begin{bmatrix} 1 & 2 & 3 \\ 2 & 5 & 3 \\ 1 & 0 & 8 \end{bmatrix}$; one-to-one

3. For example, the vector $(1, 3)$ is not in the range.

4. For example, the vector $(1, 6, 2)$ is not in the range.

5. (a) One-to-one; $\begin{bmatrix} \frac{1}{3} & -\frac{2}{3} \\ \frac{1}{3} & \frac{1}{3} \end{bmatrix}$; $T^{-1}(w_1, w_2) = \left(\frac{1}{3}w_1 - \frac{2}{3}w_2, \frac{1}{3}w_1 + \frac{1}{3}w_2\right)$ **(b)** Not one-to-one

(c) One-to-one; $\begin{bmatrix} 0 & -1 \\ -1 & 0 \end{bmatrix}$; $T^{-1}(w_1, w_2) = (-w_2, -w_1)$ **(d)** Not one-to-one

6. (a) One-to-one; $\begin{bmatrix} 1 & -2 & 4 \\ -1 & 2 & -3 \\ -1 & 3 & -5 \end{bmatrix}$; $T^{-1}(w_1, w_2, w_3) = (w_1 - 2w_2 + 4w_3, -w_1 + 2w_2 - 3w_3, -w_1 + 3w_2 - 5w_3)$

(b) Not one-to-one; $\begin{bmatrix} \frac{1}{2} & \frac{1}{2} & -\frac{1}{2} \\ -\frac{5}{14} & \frac{5}{14} & \frac{3}{14} \\ -\frac{1}{7} & \frac{1}{7} & \frac{1}{7} \end{bmatrix}$; $T^{-1}(w_1, w_2, w_3) = \left(\dfrac{w_1 + w_2 - w_3}{2}, \dfrac{-5w_1 + 5w_2 + 3w_3}{14}, \dfrac{-w_1 + w_2 + w_3}{7}\right)$

(c) One-to-one; $\begin{bmatrix} -\frac{3}{2} & -\frac{3}{2} & \frac{11}{2} \\ \frac{1}{2} & \frac{1}{2} & -\frac{3}{2} \\ -\frac{1}{2} & \frac{1}{2} & -\frac{1}{2} \end{bmatrix}$; $T^{-1}(w_1, w_2, w_3) = \left(\dfrac{-3w_1 - 3w_2 + 11w_3}{2}, \dfrac{w_1 + w_2 - 3w_3}{2}, \dfrac{-w_1 + w_2 - w_3}{2}\right)$

(d) Not one-to-one

7. (a) Reflection about the x-axis **(b)** Rotation through the angle $-\pi/4$ **(c)** Contraction by a factor of $\frac{1}{3}$
(d) Reflection about the yz-plane **(e)** Dilation by a factor of 5

8. (a) Linear **(b)** Nonlinear **(c)** Linear **(d)** Linear

9. (a) Linear **(b)** Nonlinear **(c)** Linear **(d)** Nonlinear

10. (a) Linear **(b)** Nonlinear **11. (a)** Linear **(b)** Linear

12. (a) For a reflection about the y-axis, $T(\mathbf{e}_1) = \begin{bmatrix} -1 \\ 0 \end{bmatrix}$ and $T(\mathbf{e}_2) = \begin{bmatrix} 0 \\ 1 \end{bmatrix}$.

Thus, $T = \begin{bmatrix} -1 & 0 \\ 0 & 1 \end{bmatrix}$.

(b) For a reflection about the xz-plane, $T(\mathbf{e}_1) = \begin{bmatrix} 1 \\ 0 \\ 0 \end{bmatrix}$, $T(\mathbf{e}_2) = \begin{bmatrix} 0 \\ -1 \\ 0 \end{bmatrix}$, and $T(\mathbf{e}_3) = \begin{bmatrix} 0 \\ 0 \\ 1 \end{bmatrix}$.

Thus, $T = \begin{bmatrix} 1 & 0 & 0 \\ 0 & -1 & 0 \\ 0 & 0 & 1 \end{bmatrix}$.

(c) For an orthogonal projection on the x-axis, $T(\mathbf{e}_1) = \begin{bmatrix} 1 \\ 0 \end{bmatrix}$ and $T(\mathbf{e}_2) = \begin{bmatrix} 0 \\ 0 \end{bmatrix}$.

Thus, $T = \begin{bmatrix} 1 & 0 \\ 0 & 0 \end{bmatrix}$.

(d) For an orthogonal projection on the yz-plane, $T(\mathbf{e}_1) = \begin{bmatrix} 0 \\ 0 \\ 0 \end{bmatrix}$, $T(\mathbf{e}_2) = \begin{bmatrix} 0 \\ 1 \\ 0 \end{bmatrix}$, and $T(\mathbf{e}_3) = \begin{bmatrix} 0 \\ 0 \\ 1 \end{bmatrix}$.

Thus, $T = \begin{bmatrix} 0 & 0 & 0 \\ 0 & 1 & 0 \\ 0 & 0 & 1 \end{bmatrix}$.

(e) For a rotation through a positive angle θ, $T(\mathbf{e}_1) = \begin{bmatrix} \cos\theta \\ \sin\theta \end{bmatrix}$ and $T(\mathbf{e}_2) = \begin{bmatrix} -\sin\theta \\ \cos\theta \end{bmatrix}$.

Thus, $T = \begin{bmatrix} \cos\theta & -\sin\theta \\ \sin\theta & \cos\theta \end{bmatrix}$.

(f) For a dilation by a factor $k \geq 1$, $T(\mathbf{e}_1) = \begin{bmatrix} k \\ 0 \\ 0 \end{bmatrix}$, $T(\mathbf{e}_2) = \begin{bmatrix} 0 \\ k \\ 0 \end{bmatrix}$, $T(\mathbf{e}_3) = \begin{bmatrix} 0 \\ 0 \\ k \end{bmatrix}$.

Thus, $T = \begin{bmatrix} k & 0 & 0 \\ 0 & k & 0 \\ 0 & 0 & k \end{bmatrix}$.

13. (a) $T(\mathbf{e}_1) = \begin{bmatrix} -1 \\ 0 \end{bmatrix}$ and $T(\mathbf{e}_2) = \begin{bmatrix} 0 \\ 0 \end{bmatrix}$. Thus, $T = \begin{bmatrix} -1 & 0 \\ 0 & 0 \end{bmatrix}$.

(b) $T(\mathbf{e}_1) = \begin{bmatrix} 0 \\ -1 \end{bmatrix}$ and $T(\mathbf{e}_2) = \begin{bmatrix} 1 \\ 0 \end{bmatrix}$. Thus, $T = \begin{bmatrix} 0 & 1 \\ -1 & 0 \end{bmatrix}$.

(c) $T(\mathbf{e}_1) = \begin{bmatrix} 0 \\ 3 \end{bmatrix}$ and $T(\mathbf{e}_2) = \begin{bmatrix} 0 \\ 0 \end{bmatrix}$. Thus, $T = \begin{bmatrix} 0 & 0 \\ 3 & 0 \end{bmatrix}$.

14. (a) $T(\mathbf{e}_1) = \begin{bmatrix} \frac{1}{5} \\ 0 \\ 0 \end{bmatrix}$, $T(\mathbf{e}_2) = \begin{bmatrix} 0 \\ -\frac{1}{5} \\ 0 \end{bmatrix}$, and $T(\mathbf{e}_3) = \begin{bmatrix} 0 \\ 0 \\ \frac{1}{5} \end{bmatrix}$. Thus, $T = \begin{bmatrix} \frac{1}{5} & 0 & 0 \\ 0 & -\frac{1}{5} & 0 \\ 0 & 0 & \frac{1}{5} \end{bmatrix}$.

(b) $T(\mathbf{e}_1) = \begin{bmatrix} 1 \\ 0 \\ 0 \end{bmatrix}$, $T(\mathbf{e}_2) = \begin{bmatrix} 0 \\ 0 \\ 0 \end{bmatrix}$, and $T(\mathbf{e}_3) = \begin{bmatrix} 0 \\ 0 \\ 0 \end{bmatrix}$. Thus, $T = \begin{bmatrix} 1 & 0 & 0 \\ 0 & 0 & 0 \\ 0 & 0 & 0 \end{bmatrix}$.

(c) $T(\mathbf{e}_1) = \begin{bmatrix} -1 \\ 0 \\ 0 \end{bmatrix}$, $T(\mathbf{e}_2) = \begin{bmatrix} 0 \\ -1 \\ 0 \end{bmatrix}$, and $T(\mathbf{e}_3) = \begin{bmatrix} 0 \\ 0 \\ -1 \end{bmatrix}$. Thus, $T = \begin{bmatrix} -1 & 0 & 0 \\ 0 & -1 & 0 \\ 0 & 0 & -1 \end{bmatrix}$.

15. **(a)** $T_A(\mathbf{e}_1) = \begin{bmatrix} -1 \\ 2 \\ 4 \end{bmatrix}, T_A(\mathbf{e}_2) = \begin{bmatrix} 3 \\ 1 \\ 5 \end{bmatrix}$, and $T_A(\mathbf{e}_3) = \begin{bmatrix} 0 \\ 2 \\ -3 \end{bmatrix}$

(b) $T_A(\mathbf{e}_1 + \mathbf{e}_2 + \mathbf{e}_3) = T_A(\mathbf{e}_1) + T_A(\mathbf{e}_2) + T_A(\mathbf{e}_3) = \begin{bmatrix} 2 \\ 5 \\ 6 \end{bmatrix}$

(c) $T_A(7\mathbf{e}_3) = 7T_A(\mathbf{e}_3) = \begin{bmatrix} 0 \\ 14 \\ -21 \end{bmatrix}$

16. **(a)** Linear transformation from $R^2 \to R^3$; one-to-one
(b) Linear transformation from $R^3 \to R^2$; not one-to-one

17. **(a)** $\left(\frac{1}{2}, \frac{1}{2}\right)$ **(b)** $\left(\frac{3}{4}, \frac{\sqrt{3}}{4}\right)$ **(c)** $\left(\frac{1 - 5\sqrt{3}}{4}, \frac{15 - \sqrt{3}}{4}\right)$

18. **(a)** $\lambda = 1; \begin{bmatrix} t \\ 0 \end{bmatrix}$ **(b)** $\lambda = 1; \begin{bmatrix} t \\ t \end{bmatrix}$ **(c)** $\lambda = 1; \begin{bmatrix} t \\ 0 \end{bmatrix}$ **(d)** $\lambda = \frac{1}{2}$; all vectors in R^2 are eigenvectors

$\lambda = -1; \begin{bmatrix} 0 \\ t \end{bmatrix}$ $\lambda = -1; \begin{bmatrix} t \\ -t \end{bmatrix}$ $\lambda = 0; \begin{bmatrix} 0 \\ t \end{bmatrix}$

19. **(a)** $\lambda = 1; \begin{bmatrix} 0 \\ s \\ t \end{bmatrix}$ **(b)** $\lambda = 1; \begin{bmatrix} s \\ 0 \\ t \end{bmatrix}$

$\lambda = -1; \begin{bmatrix} t \\ 0 \\ 0 \end{bmatrix}$ $\lambda = 0; \begin{bmatrix} 0 \\ t \\ 0 \end{bmatrix}$

(c) $\lambda = 2$; all vectors in R^3 are eigenvectors **(d)** $\lambda = 1; \begin{bmatrix} 0 \\ 0 \\ t \end{bmatrix}$

20. **(a)** Yes **(b)** Yes **23.** **(a)** $\begin{bmatrix} \cos 2\theta & \sin 2\theta \\ \sin 2\theta & -\cos 2\theta \end{bmatrix}$ **(b)** $\left(\frac{1 + 5\sqrt{3}}{2}, \frac{\sqrt{3} - 5}{2}\right)$

25. **(a)** False **(b)** True **(c)** False, since $\mathbf{x}$ could be 0 **(d)** True

26. **(a)** The range of T is properly smaller than R^n. **(b)** T must map infinitely many vectors to 0.

EXERCISE SET 5.1 [page 208]

1. Not a vector space. Axiom 8 fails. **2.** Not a vector space. Axiom 10 fails.

3. Not a vector space. Axioms 9 and 10 fail.

4. The set is a vector space under the given operations.

5. The set is a vector space under the given operations.

6. Not a vector space. Axioms 5 and 6 fail.

7. The set is a vector space under the given operations.

8. Not a vector space. Axioms 7 and 8 fail.

9. Not a vector space. Axioms 1, 4, 5, and 6 fail.

10. The set is a vector space under the given operations.

11. The set is a vector space under the given operations.

12. The set is a vector space under the given operations.

13. The set is a vector space under the given operations.

14. The set is a vector space under the given operations.

15. The set is a vector space under the given operations.

16. Not a vector space. Axiom 7 fails. Also, assuming that $0 = (1, 1)$, Axiom 4 holds but Axiom 5 fails.

21. No. A vector space must have a zero element.

22. No. Axioms 1, 4, and 6 will fail.

23. Yes. Moon behaves like the zero vector.   **24.** No

25. (1) Axiom 7   (2) Axiom 4   (3) Axiom 5   (4) Follows from statement 2
      (5) Axiom 3   (6) Axiom 5   (7) Axiom 4

27. (1) Axiom 1   (2) Hypothesis   (3) Axiom 3, then Axiom 5 and Axiom 4
      (4) Axiom 3, then Axiom 5 and Axiom 4

28. No; $\mathbf{0}_1 = \mathbf{0}_1 + \mathbf{0}_2 = \mathbf{0}_2$

29. No; $(-\mathbf{u})_1 = (-\mathbf{u})_1 + [\mathbf{u} + (-\mathbf{u})_2] = [(-\mathbf{u})_1 + \mathbf{u}] + (-\mathbf{u})_2 = (-\mathbf{u})_2$

30. $(\mathbf{u} + \mathbf{v}) - (\mathbf{v} + \mathbf{u}) = (\mathbf{u} + \mathbf{v}) + (-1)(\mathbf{v} + \mathbf{u}) = \mathbf{u} + \mathbf{v} + (-1)\mathbf{v} + (-1)\mathbf{u} = \mathbf{u} + \mathbf{0} + (-1)\mathbf{u} = \mathbf{0}$

EXERCISE SET 5.2 [page 219]

1. (a), (c)   **2.** (b), (d)   **3.** (a), (b), (d)   **4.** (b), (d), (e)   **5.** (a), (b), (d)

6. **(a)** Line; $x = -\frac{1}{2}t, y = -\frac{3}{2}t, z = t$ **(b)** Line; $x = 2t, y = t, z = 0$ **(c)** Origin
   (d) Origin   **(e)** Line; $x = -3t, y = -2t, z = t$ **(f)** Plane; $x - 3y + z = 0$

7. (a), (b), (d)

8. **(a)** $(-9, -7, -15) = -2\mathbf{u} + \mathbf{v} - 2\mathbf{w}$ **(b)** $(6, 11, 6) = 4\mathbf{u} - 5\mathbf{v} + \mathbf{w}$
   (c) $(0, 0, 0) = 0\mathbf{u} + 0\mathbf{v} + 0\mathbf{w}$ **(d)** $(7, 8, 9) = 0\mathbf{u} - 2\mathbf{v} + 3\mathbf{w}$

9. **(a)** $-9 - 7x - 15x^2 = -2\mathbf{p}_1 + \mathbf{p}_2 - 2\mathbf{p}_3$ **(b)** $6 + 11x + 6x^2 = 4\mathbf{p}_1 - 5\mathbf{p}_2 + \mathbf{p}_3$
   (c) $0 = 0\mathbf{p}_1 + 0\mathbf{p}_2 + 0\mathbf{p}_3$ **(d)** $7 + 8x + 9x^2 = 0\mathbf{p}_1 - 2\mathbf{p}_2 + 3\mathbf{p}_3$

10. (a), (b), (c)   **11.** **(a)** The vectors span.   **(b)** The vectors do not span.
            (c) The vectors do not span.   **(d)** The vectors span.

12. (a), (c), (e)   **13.** No   **14.** (a), (b), (d)   **15.** $y = z$

16. $x = 3t,\ \ y = -2t,\ \ z = 5t$, where $-\infty < t < +\infty$

23. **(a)** False   **(b)** True   **(c)** True   **(d)** True   **(e)** False

24. **(a)** They span a line if they are collinear and not both 0. They span a plane if they are not collinear.
   (b) If $\mathbf{u} = a\mathbf{v}$ and $\mathbf{v} = b\mathbf{u}$ for some real numbers a, b
   (c) We must have $\mathbf{b} = \mathbf{0}$ since a subspace must contain $\mathbf{x} = \mathbf{0}$ and then $\mathbf{b} = A\mathbf{0} = \mathbf{0}$.

25. No; $W_1 \cup W_2$ is not a subspace.

26. **(a)** For example, $\begin{bmatrix} 1 & 0 \\ 0 & 0 \end{bmatrix}, \begin{bmatrix} 0 & 1 \\ 0 & 0 \end{bmatrix}, \begin{bmatrix} 0 & 0 \\ 1 & 0 \end{bmatrix}, \begin{bmatrix} 0 & 0 \\ 0 & 1 \end{bmatrix}$

   (b) The set of matrices having one entry equal to 1 and all other entries equal to 0

27. They are not coplanar.

EXERCISE SET 5.3 [page 229]

1. **(a)** u_2 is a scalar multiple of u_1. **(b)** The vectors are linearly dependent by Theorem 5.3.3.
   (c) p_2 is a scalar multiple of p_1. **(d)** B is a scalar multiple of A.

2. **(d)**   3. None   4. **(d)**   5. **(a)** They do not lie in a plane. **(b)** They do lie in a plane.

6. **(a)** They do not lie on the same line. **(b)** They do not lie on the same line.
   (c) They do lie on the same line.

7. **(b)** $v_1 = \frac{2}{7}v_2 - \frac{3}{7}v_3,\quad v_2 = \frac{7}{2}v_1 + \frac{3}{2}v_3,\quad v_3 = -\frac{7}{3}v_1 + \frac{2}{3}v_2$

8. $\lambda = -\frac{1}{2}, \lambda = 1$ 17. If and only if the vector is not zero

18. **(a)** They are linearly independent since v_1, v_2, and v_3 do not lie in the same plane when
   they are placed with their initial points at the origin.
   (b) They are not linearly independent since v_1, v_2, and v_3 lie in the same plane when they
   are placed with their initial points at the origin.

19. (a), (d), (e), (f)

23. **(a)** False **(b)** False **(c)** True **(d)** False   26. **(a)** Yes

EXERCISE SET 5.4 [page 243]

1. **(a)** A basis for R^2 has two linearly independent vectors.
   (b) A basis for R^3 has three linearly independent vectors.
   (c) A basis for P_2 has three linearly independent vectors.
   (d) A basis for M_{22} has four linearly independent vectors.

2. (a), (b)   3. (a), (b)   4. (c), (d)   6. **(b)** Any two of the vectors v_1, v_2, v_3

7. **(a)** $(w)_S = (3, -7)$ **(b)** $(w)_S = \left(\frac{5}{28}, \frac{3}{14}\right)$ **(c)** $(w)_S = \left(a, \dfrac{b-a}{2}\right)$

8. **(a)** $(v)_S = (3, -2, 1)$ **(b)** $(v)_S = (-2, 0, 1)$

9. **(a)** $(p)_S = (4, -3, 1)$ **(b)** $(p)_S = (0, 2, -1)$

10. $(A)_S = (-1, 1, -1, 3)$ 11. Basis: $(1, 0, 1)$; dimension $= 1$

12. Basis: $\left(-\frac{1}{4}, -\frac{1}{4}, 1, 0\right), (0, -1, 0, 1)$; dimension $= 2$

13. Basis: $(4, 1, 0, 0), (-3, 0, 1, 0), (1, 0, 0, 1)$; dimension $= 3$

14. Basis: $(3, 1, 0), (-1, 0, 1)$; dimension $= 2$ 15. No basis; dimension $= 0$

16. Basis: $(4, -5, 1)$; dimension $= 1$

17. **(a)** $\left(\frac{2}{3}, 1, 0\right), \left(-\frac{5}{3}, 0, 1\right)$ **(b)** $(1, 1, 0), (0, 0, 1)$ **(c)** $(2, -1, 4)$ **(d)** $(1, 1, 0), (0, 1, 1)$

18. **(a)** 3-dimensional **(b)** 2-dimensional **(c)** 1-dimensional   19. 3-dimensional

20. **(a)** $\{v_1, v_2, e_1\}$ or $\{v_1, v_2, e_2\}$ **(b)** $\{v_1, v_2, e_1\}$ or $\{v_1, v_2, e_2\}$ or $\{v_1, v_2, e_3\}$

21. $\{v_1, v_2, e_2, e_3\}$ or $\{v_1, v_2, e_2, e_4\}$ or $\{v_1, v_2, e_3, e_4\}$

26. **(a)** One possible answer is $\{-1 + x - 2x^2, 3 + 3x + 6x^2, 9\}$.
   (b) One possible answer is $\{1 + x, x^2, -2 + 2x^2\}$.
   (c) One possible answer is $\{1 + x - 3x^2\}$.

27. **(a)** $(0, \sqrt{2})$ **(b)** $(1, 0)$ **(c)** $(-1, \sqrt{2})$ **(d)** $(a - b, \sqrt{2}b)$

28. **(a)** $(2, 0)$ **(b)** $\left(\dfrac{2}{\sqrt{3}}, -\dfrac{1}{\sqrt{3}}\right)$ **(c)** $(0, 1)$ **(d)** $\left(\dfrac{2}{\sqrt{3}}a, b - \dfrac{a}{\sqrt{3}}\right)$

30. Yes; for example, $\begin{bmatrix} 1 & 0 \\ 0 & \pm 1 \end{bmatrix}$, $\begin{bmatrix} 0 & 1 \\ \pm 1 & 0 \end{bmatrix}$ **31. (a)** n **(b)** $n(n+1)/2$ **(c)** $n(n+1)/2$

32. (a) $10 > 9 = \dim(M_{33})$ **(b)** The set $I_n, A, A^2, \ldots, A^{n^2}$ is linearly dependent.

33. (a) A linearly dependent set has at most n vectors.   **(b)** A spanning set has at least n vectors.

34. (a) The dimension is $n - 1$.
   (b) $(1, 0, 0, \ldots, 0, -1), (0, 1, 0, \ldots, 0, -1), (0, 0, 1, \ldots, 0, -1), \ldots, (0, 0, 0, \ldots, 1, -1)$
   is a basis of size $n - 1$.

35. (b) $\dim W = 2$ **(c)** $x - 1, x^2 - 1$ is a basis for W.

EXERCISE SET 5.5 [page 257]

1. $\mathbf{r}_1 = (2, -1, 0, 1), \mathbf{r}_2 = (3, 5, 7, -1), \mathbf{r}_3 = (1, 4, 2, 7)$;

$$\mathbf{c}_1 = \begin{bmatrix} 2 \\ 3 \\ 1 \end{bmatrix}, \mathbf{c}_2 = \begin{bmatrix} -1 \\ 5 \\ 4 \end{bmatrix}, \mathbf{c}_3 = \begin{bmatrix} 0 \\ 7 \\ 2 \end{bmatrix}, \mathbf{c}_4 = \begin{bmatrix} 1 \\ -1 \\ 7 \end{bmatrix}$$

2. (a) $1\begin{bmatrix} 2 \\ -1 \end{bmatrix} + 2\begin{bmatrix} 3 \\ 4 \end{bmatrix} = \begin{bmatrix} 8 \\ 7 \end{bmatrix}$ **(b)** $-2\begin{bmatrix} 4 \\ 3 \\ 0 \end{bmatrix} + 3\begin{bmatrix} 0 \\ 6 \\ -1 \end{bmatrix} + 5\begin{bmatrix} -1 \\ 2 \\ 4 \end{bmatrix} = \begin{bmatrix} -13 \\ 22 \\ 17 \end{bmatrix}$

(c) $-1\begin{bmatrix} -3 \\ 5 \\ 2 \\ 1 \end{bmatrix} + 2\begin{bmatrix} 6 \\ -4 \\ 3 \\ 8 \end{bmatrix} + 5\begin{bmatrix} 2 \\ 0 \\ -1 \\ 3 \end{bmatrix} = \begin{bmatrix} 25 \\ -13 \\ -1 \\ 30 \end{bmatrix}$ **(d)** $3\begin{bmatrix} 2 \\ 6 \end{bmatrix} + 0\begin{bmatrix} 1 \\ 3 \end{bmatrix} + (-5)\begin{bmatrix} 5 \\ -8 \end{bmatrix} = \begin{bmatrix} -19 \\ 58 \end{bmatrix}$

3. (a) $\begin{bmatrix} -2 \\ 10 \end{bmatrix} = \begin{bmatrix} 1 \\ 4 \end{bmatrix} - \begin{bmatrix} 3 \\ -6 \end{bmatrix}$ **(b)** $\mathbf{b}$ is not in the column space of A.   **(c)** $\begin{bmatrix} 1 \\ 9 \\ 1 \end{bmatrix} - 3\begin{bmatrix} -1 \\ 3 \\ 1 \end{bmatrix} + \begin{bmatrix} 1 \\ 1 \\ 1 \end{bmatrix} = \begin{bmatrix} 5 \\ 1 \\ -1 \end{bmatrix}$

(d) $\begin{bmatrix} 2 \\ 0 \\ 0 \end{bmatrix} = \begin{bmatrix} 1 \\ 1 \\ -1 \end{bmatrix} + (t-1)\begin{bmatrix} -1 \\ 1 \\ -1 \end{bmatrix} + t\begin{bmatrix} 1 \\ -1 \\ 1 \end{bmatrix}$ **(e)** $\begin{bmatrix} 4 \\ 3 \\ 5 \\ 7 \end{bmatrix} = -26\begin{bmatrix} 1 \\ 0 \\ 1 \\ 0 \end{bmatrix} + 13\begin{bmatrix} 2 \\ 1 \\ 2 \\ 1 \end{bmatrix} - 7\begin{bmatrix} 0 \\ 2 \\ 1 \\ 2 \end{bmatrix} + 4\begin{bmatrix} 1 \\ 1 \\ 3 \\ 2 \end{bmatrix}$

4. (a) $r\begin{bmatrix} -3 \\ 1 \\ 1 \\ 0 \end{bmatrix} + s\begin{bmatrix} 4 \\ -1 \\ 0 \\ 1 \end{bmatrix}$ **(b)** $\begin{bmatrix} -1 \\ 2 \\ 4 \\ -3 \end{bmatrix} + r\begin{bmatrix} -3 \\ 1 \\ 1 \\ 0 \end{bmatrix} + s\begin{bmatrix} 4 \\ -1 \\ 0 \\ 1 \end{bmatrix}$

5. (a) $\begin{bmatrix} 1 \\ 0 \end{bmatrix} + t\begin{bmatrix} 3 \\ 1 \end{bmatrix}; t\begin{bmatrix} 3 \\ 1 \end{bmatrix}$ **(b)** $\begin{bmatrix} -2 \\ 7 \\ 0 \end{bmatrix} + t\begin{bmatrix} -1 \\ -1 \\ 1 \end{bmatrix}; t\begin{bmatrix} -1 \\ -1 \\ 1 \end{bmatrix}$

(c) $\begin{bmatrix} -1 \\ 0 \\ 0 \\ 0 \end{bmatrix} + r\begin{bmatrix} 2 \\ 1 \\ 0 \\ 0 \end{bmatrix} + s\begin{bmatrix} -1 \\ 0 \\ 1 \\ 0 \end{bmatrix} + t\begin{bmatrix} -2 \\ 0 \\ 0 \\ 1 \end{bmatrix}; r\begin{bmatrix} 2 \\ 1 \\ 0 \\ 0 \end{bmatrix} + s\begin{bmatrix} -1 \\ 0 \\ 1 \\ 0 \end{bmatrix} + t\begin{bmatrix} -2 \\ 0 \\ 0 \\ 1 \end{bmatrix}$

(d) $\begin{bmatrix} \frac{6}{5} \\ \frac{7}{5} \\ 0 \\ 0 \end{bmatrix} + s\begin{bmatrix} \frac{7}{5} \\ \frac{4}{5} \\ 1 \\ 0 \end{bmatrix} + t\begin{bmatrix} \frac{1}{5} \\ -\frac{3}{5} \\ 0 \\ 1 \end{bmatrix}; s\begin{bmatrix} \frac{7}{5} \\ \frac{4}{5} \\ 1 \\ 0 \end{bmatrix} + t\begin{bmatrix} \frac{1}{5} \\ -\frac{3}{5} \\ 0 \\ 1 \end{bmatrix}$

6. (a) $\begin{bmatrix} 16 \\ 19 \\ 1 \end{bmatrix}$ **(b)** $\begin{bmatrix} 1 \\ 0 \\ 2 \end{bmatrix}, \begin{bmatrix} 0 \\ 1 \\ 0 \end{bmatrix}$ **(c)** $\begin{bmatrix} -1 \\ -1 \\ 1 \\ 0 \end{bmatrix}, \begin{bmatrix} 2 \\ -4 \\ 0 \\ 7 \end{bmatrix}$ **(d)** $\begin{bmatrix} -1 \\ -1 \\ 1 \\ 0 \\ 0 \end{bmatrix}, \begin{bmatrix} -2 \\ -1 \\ 0 \\ 1 \\ 0 \end{bmatrix}, \begin{bmatrix} -1 \\ -2 \\ 0 \\ 0 \\ 1 \end{bmatrix}$ **(e)** $\begin{bmatrix} -2 \\ 0 \\ 0 \\ 1 \\ 0 \end{bmatrix}, \begin{bmatrix} -16 \\ 2 \\ 5 \\ 0 \\ 12 \end{bmatrix}$

7. (a) $\mathbf{r}_1 = [1 \ \ 0 \ \ 2], \ \mathbf{r}_2 = [0 \ \ 0 \ \ 1], \ \mathbf{c} = \begin{bmatrix} 1 \\ 0 \\ 0 \end{bmatrix}, \ \mathbf{c}_2 = \begin{bmatrix} 2 \\ 1 \\ 0 \end{bmatrix}$

(b) $\mathbf{r}_1 = [1 \ \ -3 \ \ 0 \ \ 0], \ \mathbf{r}_2 = [0 \ \ 1 \ \ 0 \ \ 0], \ \mathbf{c}_1 = \begin{bmatrix} 1 \\ 0 \\ 0 \\ 0 \end{bmatrix}, \mathbf{c}_2 = \begin{bmatrix} -3 \\ 1 \\ 0 \\ 0 \end{bmatrix}$

(c) $\mathbf{r}_1 = [1 \ \ 2 \ \ 4 \ \ 5], \ \mathbf{r}_2 = [0 \ \ 1 \ \ -3 \ \ 0], \ \mathbf{r}_3 = [0 \ \ 0 \ \ 1 \ \ -3], \ \mathbf{r}_4 = [0 \ \ 0 \ \ 0 \ \ 1],$

$\mathbf{c}_1 = \begin{bmatrix} 1 \\ 0 \\ 0 \\ 0 \\ 0 \end{bmatrix}, \mathbf{c}_2 = \begin{bmatrix} 2 \\ 1 \\ 0 \\ 0 \\ 0 \end{bmatrix}, \mathbf{c}_3 = \begin{bmatrix} 4 \\ -3 \\ 1 \\ 0 \\ 0 \end{bmatrix}, \mathbf{c}_4 = \begin{bmatrix} 5 \\ 0 \\ -3 \\ 1 \\ 0 \end{bmatrix}$

(d) $\mathbf{r}_1 = [1 \ \ 2 \ \ -1 \ \ 5], \ \mathbf{r}_2 = [0 \ \ 1 \ \ 4 \ \ 3], \ \mathbf{r}_3 = [0 \ \ 0 \ \ 1 \ \ -7], \ \mathbf{r}_4 = [0 \ \ 0 \ \ 0 \ \ 1],$

$\mathbf{c}_1 = \begin{bmatrix} 1 \\ 0 \\ 0 \\ 0 \end{bmatrix}, \ \mathbf{c}_2 = \begin{bmatrix} 2 \\ 1 \\ 0 \\ 0 \end{bmatrix}, \ \mathbf{c}_3 = \begin{bmatrix} -1 \\ 4 \\ 1 \\ 0 \end{bmatrix}, \ \mathbf{c}_4 = \begin{bmatrix} 5 \\ 3 \\ -7 \\ 1 \end{bmatrix}$

8. (a) $(1, -1, 3), (0, 1, -19)$ **(b)** $\left(1, 0, -\frac{1}{2}\right)$ **(c)** $(1, 4, 5, 2), \left(0, 1, 1, \frac{4}{7}\right)$
(d) $(1, 4, 5, 6, 9), (0, 1, 1, 1, 2)$ **(e)** $(1, -3, 2, 2, 1), (0, 1, 2, 0, -1), \left(0, 0, 1, 0, -\frac{5}{12}\right)$

9. (a) $\begin{bmatrix} 1 \\ 5 \\ 7 \end{bmatrix}, \begin{bmatrix} -1 \\ -4 \\ -6 \end{bmatrix}$ **(b)** $\begin{bmatrix} 2 \\ 4 \\ 0 \end{bmatrix}$ **(c)** $\begin{bmatrix} 1 \\ 2 \\ -1 \end{bmatrix}, \begin{bmatrix} 4 \\ 1 \\ 3 \end{bmatrix}$ **(d)** $\begin{bmatrix} 1 \\ 3 \\ -1 \\ 2 \end{bmatrix}, \begin{bmatrix} 4 \\ -2 \\ 0 \\ 3 \end{bmatrix}$ **(e)** $\begin{bmatrix} 1 \\ 0 \\ 2 \\ 3 \\ -2 \end{bmatrix}, \begin{bmatrix} -3 \\ 3 \\ -3 \\ -6 \\ 9 \end{bmatrix}, \begin{bmatrix} 2 \\ 6 \\ -2 \\ 0 \\ 2 \end{bmatrix}$

10. (a) $(1, -1, 3), (5, -4, -4)$ **(b)** $(2, 0, -1)$ **(c)** $(1, 4, 5, 2), (2, 1, 3, 0)$
(d) $(1, 4, 5, 6, 9), (3, -2, 1, 4, -1)$ **(e)** $(1, -3, 2, 2, 1), (0, 3, 6, 0, -3), (2, -3, -2, 4, 4)$

11. (a) $(1, 1, -4, -3), (0, 1, -5, -2), \left(0, 0, 1, -\frac{1}{2}\right)$
(b) $\left(1, -1, 2, 0\right), (0, 1, 0, 0), \left(0, 0, 1, -\frac{1}{6}\right)$
(c) $(1, 1, 0, 0), (0, 1, 1, 1), (0, 0, 1, 1), (0, 0, 0, 1)$

12. (a) $\{\mathbf{v}_1, \mathbf{v}_2\}; \mathbf{v}_3 = 2\mathbf{v}_1 + \mathbf{v}_2, \mathbf{v}_4 = -2\mathbf{v}_1 + \mathbf{v}_2$
(b) $\{\mathbf{v}_1, \mathbf{v}_3\}; \mathbf{v}_2 = 2\mathbf{v}_1, \mathbf{v}_4 = \mathbf{v}_1 + \mathbf{v}_3$
(c) $\{\mathbf{v}_1, \mathbf{v}_2, \mathbf{v}_4\}; \mathbf{v}_3 = 2\mathbf{v}_1 - \mathbf{v}_2, \mathbf{v}_5 = -\mathbf{v}_1 + 3\mathbf{v}_2 + 2\mathbf{v}_4$

14. (b) $\begin{bmatrix} 0 & 0 & 0 \\ 0 & 1 & 0 \\ 0 & 0 & 1 \end{bmatrix}$ **15. (a)** True **(b)** True **(c)** False **(d)** True **(e)** False

16. $\begin{bmatrix} 3a & -5a \\ 3b & -5b \end{bmatrix}$ for all real numbers a, b not both 0. **17.** $\begin{bmatrix} 1 \\ 0 \\ 0 \end{bmatrix} + r \begin{bmatrix} -1 \\ 1 \\ 0 \end{bmatrix} + s \begin{bmatrix} -1 \\ 0 \\ 1 \end{bmatrix}$

18. The row spaces of AB and of B are the same since A is a product of elementary matrices.

EXERCISE SET 5.6 [page 269]

1. Rank $(A) = \text{rank}(A^T) = 2$

2. **(a)** Nullity $= 1$, rank $= 2$; $n = 3$ **(b)** Nullity $= 2$, rank $= 1$; $n = 3$
   (c) Nullity $= 2$, rank $= 2$; $n = 4$ **(d)** Nullity $= 3$, rank $= 2$; $n = 5$
   (e) Nullity $= 2$, rank $= 3$; $n = 5$

3. **(a)** $2; 1$ **(b)** $1; 2$ **(c)** $2; 2$ **(d)** $2; 3$ **(e)** $3; 2$

4. **(a)** $3; 3; 0; 0$ **(b)** $2; 2; 1; 1$ **(c)** $1; 1; 2; 2$ **(d)** $2; 2; 7; 3$
   (e) $2; 2; 3; 7$ **(f)** $0; 0; 4; 4$ **(g)** $2; 2; 0; 4$

5. **(a)** Rank $= 4$, nullity $= 0$ **(b)** Rank $= 3$, nullity $= 2$ **(c)** Rank $= 3$, nullity $= 0$

6. Rank $= \min(m, n)$, nullity $= n - \min(m, n)$

7. **(a)** Yes, 0 **(b)** No **(c)** Yes, 2 **(d)** Yes, 7 **(e)** No **(f)** Yes, 4 **(g)** Yes, 0

8. **(a)** Nullity $= 0$, number of parameters $= 0$ **(b)** Nullity $= 1$, number of parameters $= 1$
   (c) Nullity $= 2$, number of parameters $= 2$ **(d)** Nullity $= 7$, number of parameters $= 7$
   (e) Nullity $= 7$, number of parameters $= 7$ **(f)** Nullity $= 4$, number of parameters $= 4$
   (g) Nullity $= 0$, number of parameters $= 0$

9. $b_1 = r, b_2 = s, b_3 = 4s - 3r, b_4 = 2r - s, b_5 = 8s - 7r$ **11.** No

12. **(a)** Rank$(A) = 1$ if $t = 1$; rank$(A) = 2$ if $t = -2$; rank$(A) = 3$ if $t \neq 1, -2$
    (b) Rank$(A) = 2$ if $t = 1, \frac{3}{2}$; rank$(A) = 3$ if $t \neq 1, \frac{3}{2}$

13. Rank is 2 if $r = 2$ and $s = 1$; the rank is never 1.

16. **(a)** $\begin{bmatrix} 1 & 0 & 0 \\ 0 & 1 & 0 \\ 0 & 0 & 0 \end{bmatrix}$ **(b)** A line through the origin **(c)** A plane through the origin

    (d) The nullspace is a line through the origin and the row space is a plane through the origin.

17. **(a)** False **(b)** False **(c)** True **(d)** False

18. **(a)** 3 **(b)** 5 **(c)** 3 **(d)** 3 **19.** **(a)** 3 **(b)** 5 **(c)** 3 **(d)** 3

SUPPLEMENTARY EXERCISES [page 271]

1. **(a)** All of R^3 **2.** A line through the origin: $s = -2$
   (b) Plane: $2x - 3y + z = 0$ A plane through the origin: $s = 1$
   (c) Line: $x = 2t, y = t, z = 0$ The origin only: $s \neq 1, -2$
   (d) The origin: $(0, 0, 0)$ All of R^3: no value of s

3. **(a)** $a(4, 1, 1) + b(0, -1, 2)$ **(b)** $(a + c)(3, -1, 2) + b(1, 4, 1)$
   (c) $a(2, 3, 0) + b(-1, 0, 4) + c(4, -1, 1)$

5. **(a)** $\mathbf{v} = (-1 + r)\mathbf{v}_1 + \left(\frac{2}{3} - r\right)\mathbf{v}_2 + r\mathbf{v}_3$; r arbitrary **6.** A must be invertible. **7.** No

8. **(a)** Rank $= 2$, nullity $= 1$ **(b)** Rank $= 2$, nullity $= 2$
   (c) Rank $= 2$, nullity $= n - 2$ if $n \geq 2$; rank $= 1$, nullity $= 0$ if $n = 1$

9. **(a)** Rank $= 2$, nullity $= 1$ **(b)** Rank $= 3$, nullity $= 2$ **(c)** Rank $= n + 1$, nullity $= n$

11. $\{1, x^2, x^3, x^4, x^5, x^6, \ldots, x^n\}$ **12.** **(a)** 2 **(b)** 1 **(c)** 2 **(d)** 3 **13.** $0, 1,$ or 2

EXERCISE SET 6.1 [page 284]

1. (a) 2 **(b)** 11 **(c)** −13 **(d)** −8 **(e)** 0

2. (a) −2 **(b)** 62 **(c)** −74 **(d)** 8 **(e)** 0   **3. (a)** 3 **(b)** 56

4. (a) −29 **(b)** −15   **5. (b)** 29   **6. (b)** −42   **7. (a)** $\begin{bmatrix} \sqrt{3} & 0 \\ 0 & \sqrt{5} \end{bmatrix}$ **(b)** $\begin{bmatrix} 2 & 0 \\ 0 & \sqrt{6} \end{bmatrix}$

9. (a) No. Axiom 4 fails. **(b)** No. Axioms 2 and 3 fail. **(c)** Yes **(d)** No. Axiom 4 fails.

10. (a) $\sqrt{10}$ **(b)** $\sqrt{21}$ **(c)** $5\sqrt{5}$ **11. (a)** $3\sqrt{2}$ **(b)** $3\sqrt{5}$ **(c)** $3\sqrt{13}$

12. (a) $\sqrt{17}$ **(b)** 5   **13. (a)** $\sqrt{74}$ **(b)** 0   **14.** $3\sqrt{2}$ **15. (a)** $\sqrt{105}$ **(b)** $\sqrt{47}$

16. (a) 8 **(b)** −113 **(c)** −40 **(d)** 3 **(e)** $2\sqrt{53}$ **(f)** $\sqrt{881}$

17. (a) $\sqrt{2}, \frac{1}{3}\sqrt{6}, \frac{1}{5}\sqrt{10}$ **(b)** $\frac{2}{3}\sqrt{6}$

18. (a) $\dfrac{x^2}{4} + \dfrac{y^2}{16} = 1$ **(b)** $\dfrac{x^2}{\frac{1}{2}} + \dfrac{y^2}{1} = 1$

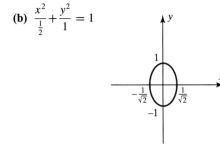

19. $\langle \mathbf{u}, \mathbf{v} \rangle = \frac{1}{9}u_1 v_1 + u_2 v_2$ **22.** Axiom 4 fails.

23. No for P_3, since $\mathbf{p} = x\left(x - \frac{1}{2}\right)(x - 1)$ satisfies $\langle \mathbf{p}, \mathbf{p} \rangle = 0$ **27. (a)** $-\frac{28}{15}$ **(b)** 0

28. (a) 0 **(b)** 1 **(c)** $\dfrac{2\ln 2}{\pi}$

32. (a) (1) Symmetry (2) Homogeneity (3) Symmetry   **34.** $a = 1/25, b = 1/16$

EXERCISE SET 6.2 [page 294]

1. (a) Yes **(b)** No **(c)** Yes **(d)** No **(e)** No **(f)** Yes   **2.** No

3. (a) $-\dfrac{1}{\sqrt{2}}$ **(b)** $-\dfrac{3}{\sqrt{73}}$ **(c)** 0 **(d)** $-\dfrac{20}{9\sqrt{10}}$ **(e)** $-\dfrac{1}{\sqrt{2}}$ **(f)** $\dfrac{2}{\sqrt{55}}$

4. (a) 0 **(b)** 0   **6. (a)** $\dfrac{19}{10\sqrt{7}}$ **(b)** 0

7. (a) Orthogonal **(b)** Orthogonal **(c)** Orthogonal **(d)** Not orthogonal

8. (a) $k = -3$ **(b)** $k = -2, k = -3$ **9.** $\pm\frac{1}{57}(-34, 44, -6, 11)$ **12.** $y = -\frac{1}{2}x$

13. (a) $x = t, y = -2t, z = -3t$ **(b)** $2x - 5y + 4z = 0$ **(c)** $x - z = 0$

14. (a) $(1, 2, -1, 2), (0, 1, -3, 2);$ $\begin{bmatrix} -5 \\ 3 \\ 1 \\ 0 \end{bmatrix}, \begin{bmatrix} 2 \\ -2 \\ 0 \\ 1 \end{bmatrix}$ **15. (a)** $\begin{bmatrix} 1 \\ 3 \\ 1 \end{bmatrix}, \begin{bmatrix} 0 \\ 1 \\ 1 \end{bmatrix}; \begin{bmatrix} 2 \\ -1 \\ 1 \end{bmatrix}$

16. **(a)** $(16, 19, 1)$ **(b)** $(0, 1, 0), \left(\frac{1}{2}, 0, 1\right)$ **(c)** $(-1, -1, 1, 0), \left(\frac{2}{7}, -\frac{4}{7}, 0, 1\right)$
    (d) $(-1, -1, 1, 0, 0), (-2, -1, 0, 1, 0), (-1, -2, 0, 0, 1)$

30. $\langle \mathbf{u}, \mathbf{v} \rangle = \frac{1}{2} u_1 v_1 + \frac{1}{6} u_2 v_2$

33. **(a)** The line $y = -x$ **(b)** The xz-plane **(c)** The x-axis

34. **(a)** A plane through the origin **(b)** A plane through the origin **(c)** Both equal R^3

35. **(a)** False **(b)** True **(c)** True **(d)** False

36. **(a)** All matrices with diagonal entries equal to 0 **(b)** All skew-symmetric matrices

EXERCISE SET 6.3 [page 308]

1. (a), (b), (d)   2. (b)   3. (b), (d)   4. (b), (d)   5. (a)   6. (a)

7. **(a)** $\left(-\frac{1}{\sqrt{5}}, \frac{2}{\sqrt{5}}\right), \left(\frac{2}{\sqrt{5}}, \frac{1}{\sqrt{5}}\right)$ **(b)** $\left(\frac{1}{\sqrt{2}}, 0, -\frac{1}{\sqrt{2}}\right), \left(\frac{1}{\sqrt{2}}, 0, \frac{1}{\sqrt{2}}\right), (0, 1, 0)$

    (c) $\left(\frac{1}{\sqrt{3}}, \frac{1}{\sqrt{3}}, \frac{1}{\sqrt{3}}\right), \left(-\frac{1}{\sqrt{2}}, \frac{1}{\sqrt{2}}, 0\right), \left(\frac{1}{\sqrt{6}}, \frac{1}{\sqrt{6}}, -\frac{2}{\sqrt{6}}\right)$

8. $\left(\frac{1}{2}, 0\right), (0, 1)$ 9. **(a)** $-\frac{7}{5} \mathbf{v}_1 + \frac{1}{5} \mathbf{v}_2 + 2 \mathbf{v}_3$ **(b)** $-\frac{37}{5} \mathbf{v}_1 - \frac{9}{5} \mathbf{v}_2 + 4 \mathbf{v}_3$ **(c)** $-\frac{3}{7} \mathbf{v}_1 - \frac{1}{7} \mathbf{v}_2 + \frac{5}{7} \mathbf{v}_3$

10. **(a)** $\frac{1}{7} \mathbf{v}_1 + \frac{5}{21} \mathbf{v}_2 + \frac{1}{3} \mathbf{v}_3 + \mathbf{v}_4$ **(b)** $\frac{15\sqrt{2}}{7} \mathbf{v}_1 + \frac{5\sqrt{2}}{21} \mathbf{v}_2 - \frac{2\sqrt{2}}{3} \mathbf{v}_3$ **(c)** $-\frac{3}{7} \mathbf{v}_1 + \frac{11}{63} \mathbf{v}_2 - \frac{1}{18} \mathbf{v}_3 + \frac{1}{2} \mathbf{v}_4$

11. **(a)** $(\mathbf{w})_S = (-2\sqrt{2}, 5\sqrt{2})$ **(b)** $(\mathbf{w})_S = (0, -2, 1)$

12. **(a)** $\mathbf{u} = \left(\frac{7}{5}, -\frac{1}{5}\right), \mathbf{v} = \left(\frac{13}{5}, \frac{16}{5}\right)$ **(b)** $\|\mathbf{u}\| = \sqrt{2}, d(\mathbf{u}, \mathbf{v}) = \sqrt{13}, \langle \mathbf{u}, \mathbf{v} \rangle = 3$

13. **(a)** $\mathbf{u} = \left(1, \frac{14}{5}, -\frac{2}{5}\right), \mathbf{v} = \left(0, -\frac{17}{5}, \frac{6}{5}\right), \mathbf{w} = \left(-4, -\frac{11}{5}, \frac{23}{5}\right)$ **(b)** $\|\mathbf{v}\| = \sqrt{13}, d(\mathbf{u}, \mathbf{v}) = 5\sqrt{3}, \langle \mathbf{w}, \mathbf{v} \rangle = 13$

14. **(a)** $\|\mathbf{u}\| = \sqrt{15}, \|\mathbf{v} - \mathbf{w}\| = 5, \|\mathbf{v} + \mathbf{w}\| = \sqrt{105}, \langle \mathbf{v}, \mathbf{w} \rangle = 20$
    (b) $\|\mathbf{u}\| = \sqrt{2}, \|\mathbf{v} - \mathbf{w}\| = \sqrt{34}, \|\mathbf{v} + \mathbf{w}\| = \sqrt{118}, \langle \mathbf{v}, \mathbf{w} \rangle = 21$

15. **(b)** $\mathbf{u} = -\frac{4}{5} \mathbf{v}_1 - \frac{11}{10} \mathbf{v}_2 + 0 \mathbf{v}_3 + \frac{1}{2} \mathbf{v}_4$

16. **(a)** $\left(\frac{1}{\sqrt{10}}, -\frac{3}{\sqrt{10}}\right), \left(\frac{3}{\sqrt{10}}, \frac{1}{\sqrt{10}}\right)$ **(b)** $(1, 0), (0, -1)$

17. **(a)** $\left(\frac{1}{\sqrt{3}}, \frac{1}{\sqrt{3}}, \frac{1}{\sqrt{3}}\right), \left(-\frac{1}{\sqrt{2}}, \frac{1}{\sqrt{2}}, 0\right), \left(\frac{1}{\sqrt{6}}, \frac{1}{\sqrt{6}}, -\frac{2}{\sqrt{6}}\right)$

    (b) $(1, 0, 0), \left(0, \frac{7}{\sqrt{53}}, -\frac{2}{\sqrt{53}}\right), \left(0, \frac{2}{\sqrt{53}}, \frac{7}{\sqrt{53}}\right)$

18. $\left(0, \frac{2}{\sqrt{5}}, \frac{1}{\sqrt{5}}, 0\right), \left(\frac{5}{\sqrt{30}}, -\frac{1}{\sqrt{30}}, \frac{2}{\sqrt{30}}, 0\right), \left(\frac{1}{\sqrt{10}}, \frac{1}{\sqrt{10}}, -\frac{2}{\sqrt{10}}, -\frac{2}{\sqrt{10}}\right), \left(\frac{1}{\sqrt{15}}, \frac{1}{\sqrt{15}}, -\frac{2}{\sqrt{15}}, \frac{3}{\sqrt{15}}\right)$

19. $\left(0, \frac{1}{\sqrt{5}}, \frac{2}{\sqrt{5}}\right), \left(-\frac{\sqrt{5}}{\sqrt{6}}, -\frac{2}{\sqrt{30}}, \frac{1}{\sqrt{30}}\right)$

20. $\left(\frac{1}{\sqrt{6}}, \frac{1}{\sqrt{6}}, \frac{1}{\sqrt{6}}\right), \left(\frac{1}{\sqrt{6}}, \frac{1}{\sqrt{6}}, -\frac{1}{\sqrt{6}}\right), \left(\frac{2}{\sqrt{6}}, -\frac{1}{\sqrt{6}}, 0\right)$

21. $\mathbf{w}_1 = \left(-\frac{4}{5}, 2, \frac{3}{5}\right), \mathbf{w}_2 = \left(\frac{9}{5}, 0, \frac{12}{5}\right)$

22. $\mathbf{w}_1 = \left(\frac{13}{14}, \frac{31}{14}, \frac{40}{14}\right), \mathbf{w}_2 = \left(\frac{1}{14}, -\frac{3}{14}, \frac{2}{14}\right)$

23. $\mathbf{w}_1 = \left(-\frac{5}{4}, -\frac{1}{4}, \frac{5}{4}, \frac{9}{4}\right), \mathbf{w}_2 = \left(\frac{1}{4}, \frac{9}{4}, \frac{19}{4}, -\frac{9}{4}\right)$

24. (a) $\begin{bmatrix} \dfrac{1}{\sqrt{5}} & -\dfrac{2}{\sqrt{5}} \\ \dfrac{2}{\sqrt{5}} & \dfrac{1}{\sqrt{5}} \end{bmatrix} \begin{bmatrix} \sqrt{5} & \sqrt{5} \\ 0 & \sqrt{5} \end{bmatrix}$ **(b)** $\begin{bmatrix} \dfrac{1}{\sqrt{2}} & -\dfrac{1}{\sqrt{3}} \\ 0 & \dfrac{1}{\sqrt{3}} \\ \dfrac{1}{\sqrt{2}} & \dfrac{1}{\sqrt{3}} \end{bmatrix} \begin{bmatrix} \sqrt{2} & 3\sqrt{2} \\ 0 & \sqrt{3} \end{bmatrix}$ **(c)** $\begin{bmatrix} \dfrac{1}{3} & \dfrac{8}{\sqrt{234}} \\ -\dfrac{2}{3} & \dfrac{11}{\sqrt{234}} \\ \dfrac{2}{3} & \dfrac{7}{\sqrt{234}} \end{bmatrix} \begin{bmatrix} 3 & \dfrac{1}{3} \\ 0 & \dfrac{\sqrt{26}}{3} \end{bmatrix}$

(d) $\begin{bmatrix} \dfrac{1}{\sqrt{2}} & -\dfrac{1}{\sqrt{3}} & \dfrac{1}{\sqrt{6}} \\ 0 & \dfrac{1}{\sqrt{3}} & \dfrac{2}{\sqrt{6}} \\ \dfrac{1}{\sqrt{2}} & \dfrac{1}{\sqrt{3}} & -\dfrac{1}{\sqrt{6}} \end{bmatrix} \begin{bmatrix} \sqrt{2} & \sqrt{2} & \sqrt{2} \\ 0 & \sqrt{3} & -\dfrac{1}{\sqrt{3}} \\ 0 & 0 & \dfrac{4}{\sqrt{6}} \end{bmatrix}$ **(e)** $\begin{bmatrix} \dfrac{1}{\sqrt{2}} & \dfrac{\sqrt{2}}{2\sqrt{19}} & -\dfrac{3}{\sqrt{19}} \\ \dfrac{1}{\sqrt{2}} & -\dfrac{\sqrt{2}}{2\sqrt{19}} & \dfrac{3}{\sqrt{19}} \\ 0 & \dfrac{3\sqrt{2}}{\sqrt{19}} & \dfrac{1}{\sqrt{19}} \end{bmatrix} \begin{bmatrix} \sqrt{2} & \dfrac{3}{\sqrt{2}} & \sqrt{2} \\ 0 & \dfrac{\sqrt{19}}{\sqrt{2}} & \dfrac{3\sqrt{2}}{\sqrt{19}} \\ 0 & 0 & \dfrac{1}{\sqrt{19}} \end{bmatrix}$

(f) Columns not linearly independent

29. $\mathbf{v}_1 = \dfrac{1}{\sqrt{2}}, \mathbf{v}_2 = \sqrt{\dfrac{3}{2}}x, \mathbf{v}_3 = \dfrac{\sqrt{5}}{2\sqrt{2}}(3x^2 - 1)$

30. (a) $1 + x + 4x^2 = \frac{7}{3}\sqrt{2}\mathbf{v}_1 + \frac{1}{3}\sqrt{6}\mathbf{v}_2 + \frac{8}{15}\sqrt{10}\,\mathbf{v}_3$

(b) $2 - 7x^2 = -\dfrac{\sqrt{2}}{3}\mathbf{v}_1 - \frac{28}{15}\sqrt{\dfrac{5}{2}}\,\mathbf{v}_3$ **(c)** $4 + 3x = 4\sqrt{2}\mathbf{v}_1 + \sqrt{6}\mathbf{v}_2$

31. $\mathbf{v}_1 = 1, \mathbf{v}_2 = \sqrt{3}(2x - 1), \mathbf{v}_3 = \sqrt{5}(6x^2 - 6x + 1)$

34. (a) Find two independent vectors in the plane and apply Gram–Schmidt.
(b) $(1/\sqrt{2}, 0, 1/\sqrt{2}), (-1/\sqrt{3}, 1/\sqrt{3}, 1/\sqrt{3})$

35. $(1/\sqrt{5}, 1/\sqrt{5}), (2/\sqrt{30}, -3/\sqrt{30})$

37. (a) True **(b)** False, unless the space has an inner product **(c)** True **(d)** True

EXERCISE SET 6.4 [page 318]

1. (a) $\begin{bmatrix} 21 & 25 \\ 25 & 35 \end{bmatrix}\begin{bmatrix} x_1 \\ x_2 \end{bmatrix} = \begin{bmatrix} 20 \\ 20 \end{bmatrix}$ **(b)** $\begin{bmatrix} 15 & -1 & 5 \\ -1 & 22 & 30 \\ 5 & 30 & 45 \end{bmatrix}\begin{bmatrix} x_1 \\ x_2 \\ x_3 \end{bmatrix} = \begin{bmatrix} -1 \\ 9 \\ 13 \end{bmatrix}$

2. (a) 0; column vectors are not linearly independent.
(b) 0; column vectors are not linearly independent.

3. (a) $x_1 = 5, x_2 = \frac{1}{2}$; $\begin{bmatrix} \frac{11}{2} \\ -\frac{9}{2} \\ -4 \end{bmatrix}$ **(b)** $x_1 = \frac{3}{7}, x_2 = -\frac{2}{3}$; $\begin{bmatrix} \frac{46}{21} \\ -\frac{5}{21} \\ \frac{13}{21} \end{bmatrix}$

(c) $x_1 = 12, x_2 = -3, x_3 = 9$; $\begin{bmatrix} 3 \\ 3 \\ 9 \\ 0 \end{bmatrix}$ **(d)** $x_1 = 14, x_2 = 30, x_3 = 26$; $\begin{bmatrix} 2 \\ 6 \\ -2 \\ 4 \end{bmatrix}$

4. (a) $\left(\frac{2}{3}, \frac{7}{3}, \frac{5}{3}\right)$ **(b)** $(3, -4, -1)$ **5. (a)** $(7, 2, 9, 5)$ **(b)** $\left(-\frac{12}{5}, -\frac{4}{5}, \frac{12}{5}, \frac{16}{5}\right)$

6. $(0, -1, 1, 1)$ **7. (a)** $\begin{bmatrix} 1 & 0 \\ 0 & 0 \end{bmatrix}$ **(b)** $\begin{bmatrix} 0 & 0 \\ 0 & 1 \end{bmatrix}$ **8. (a)** $\begin{bmatrix} 1 & 0 & 0 \\ 0 & 0 & 0 \\ 0 & 0 & 1 \end{bmatrix}$ **(b)** $\begin{bmatrix} 0 & 0 & 0 \\ 0 & 1 & 0 \\ 0 & 0 & 1 \end{bmatrix}$

9. (a) $\mathbf{v}_1 = (1, 0, -5), \mathbf{v}_2 = (0, 1, 3)$ **(b)** $\begin{bmatrix} \frac{10}{35} & \frac{15}{35} & -\frac{5}{35} \\ \frac{15}{35} & \frac{26}{35} & \frac{3}{35} \\ -\frac{5}{35} & \frac{3}{35} & \frac{34}{35} \end{bmatrix}$ **(c)** $\begin{bmatrix} \frac{10}{35}x_0 + \frac{15}{35}y_0 - \frac{5}{35}z_0 \\ \frac{15}{35}x_0 + \frac{26}{35}y_0 + \frac{3}{35}z_0 \\ -\frac{5}{35}x_0 + \frac{3}{35}y_0 + \frac{34}{35}z_0 \end{bmatrix}$ **(d)** $\dfrac{15}{\sqrt{35}}$

10. (a) $\mathbf{v}_1 = (2, -1, 4)$ **(b)** $\begin{bmatrix} \frac{4}{21} & -\frac{2}{21} & \frac{8}{21} \\ -\frac{2}{21} & \frac{1}{21} & -\frac{4}{21} \\ \frac{8}{21} & -\frac{4}{21} & \frac{16}{21} \end{bmatrix}$ **(c)** $\begin{bmatrix} \frac{4}{21}x_0 - \frac{2}{21}y_0 + \frac{8}{21}z_0 \\ -\frac{2}{21}x_0 + \frac{1}{21}y_0 - \frac{4}{21}z_0 \\ \frac{8}{21}x_0 - \frac{4}{21}y_0 + \frac{16}{21}z_0 \end{bmatrix}$ **(d)** $\dfrac{\sqrt{497}}{7}$

15. $[P] = A^T(AA^T)^{-1}A$

16. (1) Since $A^T\mathbf{0} = \mathbf{0}$ **(2)** Since A^TA is invertible   **(3)** Since the nullspace of A is nonzero if and only if the columns of A are dependent

17. (a) $A(A^TA)^{-1}A^T\mathbf{b}$ **(b)** $(A^TA)^{-1}A^T\mathbf{b}$ **(c)** $\|\mathbf{b} - A(A^TA)^{-1}A^T\mathbf{b}\|$ **(d)** $A(A^TA)^{-1}A^T$

EXERCISE SET 6.5 [page 330]

1. (b) $\begin{bmatrix} \frac{4}{5} & -\frac{9}{25} & \frac{12}{25} \\ 0 & \frac{4}{5} & \frac{3}{5} \\ -\frac{3}{5} & -\frac{12}{25} & \frac{16}{25} \end{bmatrix}$ **2. (b)** $T(-2, 3, 5) = \left(\frac{14}{3}, -\frac{5}{3}, \frac{11}{3}\right)$

3. (a) $\begin{bmatrix} 1 & 0 \\ 0 & 1 \end{bmatrix}$ **(b)** $\begin{bmatrix} \frac{1}{\sqrt{2}} & \frac{1}{\sqrt{2}} \\ -\frac{1}{\sqrt{2}} & \frac{1}{\sqrt{2}} \end{bmatrix}$ **(d)** $\begin{bmatrix} -\frac{1}{\sqrt{2}} & 0 & \frac{1}{\sqrt{2}} \\ \frac{1}{\sqrt{6}} & -\frac{2}{\sqrt{6}} & \frac{1}{\sqrt{6}} \\ \frac{1}{\sqrt{3}} & \frac{1}{\sqrt{3}} & \frac{1}{\sqrt{3}} \end{bmatrix}$ **(e)** $\begin{bmatrix} \frac{1}{2} & \frac{1}{2} & \frac{1}{2} & \frac{1}{2} \\ \frac{1}{2} & -\frac{5}{6} & \frac{1}{6} & \frac{1}{6} \\ \frac{1}{2} & \frac{1}{6} & \frac{1}{6} & -\frac{5}{6} \\ \frac{1}{2} & \frac{1}{6} & -\frac{5}{6} & \frac{1}{6} \end{bmatrix}$

5. (a) $[\mathbf{w}]_S = \begin{bmatrix} 3 \\ -7 \end{bmatrix}$ **(b)** $[\mathbf{w}]_S = \begin{bmatrix} \frac{5}{28} \\ \frac{3}{14} \end{bmatrix}$ **(c)** $[\mathbf{w}]_S = \begin{bmatrix} a \\ \frac{b-a}{2} \end{bmatrix}$

6. (a) $(\mathbf{v})_S = (3, -2, 1)$, $[\mathbf{v}]_S = \begin{bmatrix} 3 \\ -2 \\ 1 \end{bmatrix}$ **(b)** $(\mathbf{v})_S = (-2, 0, 1)$, $[\mathbf{v}]_S = \begin{bmatrix} -2 \\ 0 \\ 1 \end{bmatrix}$

7. (a) $(\mathbf{p})_S = (4, -3, 1)$, $[\mathbf{p}]_S = \begin{bmatrix} 4 \\ -3 \\ 1 \end{bmatrix}$ **(b)** $(\mathbf{p})_S = (0, 2, -1)$, $[\mathbf{p}]_S = \begin{bmatrix} 0 \\ 2 \\ -1 \end{bmatrix}$

8. $(A)_S = (-1, 1, -1, 3)$, $[A]_S = \begin{bmatrix} -1 \\ 1 \\ -1 \\ 3 \end{bmatrix}$

9. (a) $\mathbf{w} = (16, 10, 12)$ **(b)** $\mathbf{q} = 3 + 4x^2$ **(c)** $B = \begin{bmatrix} 15 & -1 \\ 6 & 3 \end{bmatrix}$

10. (a) $\begin{bmatrix} 2 & -3 \\ 1 & 4 \end{bmatrix}$ **(b)** $\begin{bmatrix} \frac{4}{11} & \frac{3}{11} \\ -\frac{1}{11} & \frac{2}{11} \end{bmatrix}$ **(c)** $[\mathbf{w}]_B = \begin{bmatrix} 3 \\ -5 \end{bmatrix}$, $[\mathbf{w}]_{B'} = \begin{bmatrix} -\frac{3}{11} \\ -\frac{13}{11} \end{bmatrix}$

11. (a) $\begin{bmatrix} \frac{13}{10} & -\frac{1}{2} \\ -\frac{2}{5} & 0 \end{bmatrix}$ **(b)** $\begin{bmatrix} 0 & -\frac{5}{2} \\ -2 & -\frac{13}{2} \end{bmatrix}$ **(c)** $[\mathbf{w}]_B = \begin{bmatrix} -\frac{17}{10} \\ \frac{8}{5} \end{bmatrix}$, $[\mathbf{w}]_{B'} = \begin{bmatrix} -4 \\ -7 \end{bmatrix}$

12. (a) $\begin{bmatrix} \frac{3}{4} & \frac{3}{4} & \frac{1}{12} \\ -\frac{3}{4} & -\frac{17}{12} & -\frac{17}{12} \\ 0 & \frac{2}{3} & \frac{2}{3} \end{bmatrix}$ **(b)** $\begin{bmatrix} \frac{19}{12} \\ -\frac{43}{12} \\ \frac{4}{3} \end{bmatrix}$ **13. (a)** $\begin{bmatrix} 3 & 2 & \frac{5}{2} \\ -2 & -3 & -\frac{1}{2} \\ 5 & 1 & 6 \end{bmatrix}$ **(b)** $\begin{bmatrix} -\frac{7}{2} \\ \frac{23}{2} \\ 6 \end{bmatrix}$

14. (a) $\begin{bmatrix} -\frac{2}{9} & \frac{7}{9} \\ \frac{1}{3} & -\frac{1}{6} \end{bmatrix}$ (b) $\begin{bmatrix} \frac{3}{4} & \frac{7}{2} \\ \frac{3}{2} & 1 \end{bmatrix}$ (c) $[\mathbf{p}]_B = \begin{bmatrix} 1 \\ -1 \end{bmatrix}$ (d) $[\mathbf{p}]_{B'} = \begin{bmatrix} -\frac{11}{4} \\ \frac{1}{2} \end{bmatrix}$

15. (b) $\begin{bmatrix} 2 & 0 \\ 1 & 3 \end{bmatrix}$ (c) $\begin{bmatrix} \frac{1}{2} & 0 \\ -\frac{1}{6} & \frac{1}{3} \end{bmatrix}$ (d) $[\mathbf{h}]_B = \begin{bmatrix} 2 \\ -5 \end{bmatrix}$, $[\mathbf{h}]_{B'} = \begin{bmatrix} 1 \\ -2 \end{bmatrix}$

16. (a) $(4\sqrt{2}, -2\sqrt{2})$ (b) $\left(-\frac{7}{2}\sqrt{2}, \frac{3}{2}\sqrt{2}\right)$

17. (a) $(-1 + 3\sqrt{3}, 3 + \sqrt{3})$ (b) $\left(\frac{5}{2} - \sqrt{3}, \frac{5}{2}\sqrt{3} + 1\right)$

18. (a) $\left(\frac{1}{2}\sqrt{2}, \frac{3}{2}\sqrt{2}, 5\right)$ (b) $\left(-\frac{5}{2}\sqrt{2}, \frac{7}{2}\sqrt{2}, -3\right)$

19. (a) $\left(-\frac{1}{2} - \frac{5}{2}\sqrt{3}, 2, \frac{5}{2} - \frac{1}{2}\sqrt{3}\right)$ (b) $\left(\frac{1}{2} - \frac{3}{2}\sqrt{3}, 6, -\frac{3}{2} - \frac{1}{2}\sqrt{3}\right)$

20. (a) $\left(-1, \frac{3}{2}\sqrt{2}, -\frac{7}{2}\sqrt{2}\right)$ (b) $\left(1, -\frac{3}{2}\sqrt{2}, \frac{9}{2}\sqrt{3}\right)$

21. (a) $A = \begin{bmatrix} \cos\theta & 0 & -\sin\theta \\ 0 & 1 & 0 \\ \sin\theta & 0 & \cos\theta \end{bmatrix}$ (b) $A = \begin{bmatrix} 1 & 0 & 0 \\ 0 & \cos\theta & \sin\theta \\ 0 & -\sin\theta & \cos\theta \end{bmatrix}$

22. $\begin{bmatrix} \frac{\sqrt{2}}{4} & \frac{\sqrt{6}}{4} & -\frac{\sqrt{2}}{2} \\ -\frac{\sqrt{3}}{2} & \frac{1}{2} & 0 \\ \frac{\sqrt{2}}{4} & \frac{\sqrt{6}}{4} & \frac{\sqrt{2}}{2} \end{bmatrix}$ 23. $a^2 + b^2 = \frac{1}{2}$ 26. (a) Rotation (b) Rotation followed by a reflection

27. (a) Rotation followed by a reflection (b) Rotation

30. (a) Rotation and reflection (b) Rotation and dilation
(c) Any rigid operator is angle preserving. Any dilation or contraction with $k \neq 0, 1$ is angle preserving but not rigid.

31. $\det(A) = \pm 1$ 32. $a = 0, b = \sqrt{2/3}, c = -\sqrt{1/3}$ or $a = 0, b = -\sqrt{2/3}, c = \sqrt{1/3}$

SUPPLEMENTARY EXERCISES [page 334]

1. (a) $(0, a, a, 0)$ with $a \neq 0$ (b) $\pm\left(0, \frac{2}{\sqrt{5}}, \frac{1}{\sqrt{5}}, 0\right)$ 6. $\pm\left(\frac{1}{\sqrt{2}}, 0, \frac{1}{\sqrt{2}}\right)$

7. $w_k = \frac{1}{k}, k = 1, 2, \ldots, n$ 8. No 11. (b) θ approaches $\frac{\pi}{2}$

12. (b) The diagonals of a parallelogram are perpendicular if and only if its sides have the same length.

EXERCISE SET 7.1 [page 344]

1. (a) $\lambda^2 - 2\lambda - 3 = 0$ (b) $\lambda^2 - 8\lambda + 16 = 0$ (c) $\lambda^2 - 12 = 0$
(d) $\lambda^2 + 3 = 0$ (e) $\lambda^2 = 0$ (f) $\lambda^2 - 2\lambda + 1 = 0$

2. (a) $\lambda = 3, \lambda = -1$ (b) $\lambda = 4$ (c) $\lambda = \sqrt{12}, \lambda = -\sqrt{12}$
(d) No real eigenvalues (e) $\lambda = 0$ (f) $\lambda = 1$

3. **(a)** Basis for eigenspace corresponding to $\lambda = 3$: $\begin{bmatrix} \frac{1}{2} \\ 1 \end{bmatrix}$; basis for eigenspace corresponding to $\lambda = -1$: $\begin{bmatrix} 0 \\ 1 \end{bmatrix}$

(b) Basis for eigenspace corresponding to $\lambda = 4$: $\begin{bmatrix} \frac{3}{2} \\ 1 \end{bmatrix}$

(c) Basis for eigenspace corresponding to $\lambda = \sqrt{12}$: $\begin{bmatrix} \frac{3}{\sqrt{12}} \\ 1 \end{bmatrix}$; basis for eigenspace corresponding to $\lambda = -\sqrt{12}$: $\begin{bmatrix} -\frac{3}{\sqrt{12}} \\ 1 \end{bmatrix}$

(d) There are no eigenspaces.

(e) Basis for eigenspace corresponding to $\lambda = 0$: $\begin{bmatrix} 1 \\ 0 \end{bmatrix}, \begin{bmatrix} 0 \\ 1 \end{bmatrix}$

(f) Basis for eigenspace corresponding to $\lambda = 1$: $\begin{bmatrix} 1 \\ 0 \end{bmatrix}, \begin{bmatrix} 0 \\ 1 \end{bmatrix}$

4. **(a)** $\lambda^3 - 6\lambda^2 + 11\lambda - 6 = 0$ **(b)** $\lambda^3 - 2\lambda = 0$ **(c)** $\lambda^3 + 8\lambda^2 + \lambda + 8 = 0$
 (d) $\lambda^3 - \lambda^2 - \lambda - 2 = 0$ **(e)** $\lambda^3 - 6\lambda^2 + 12\lambda - 8 = 0$ **(f)** $\lambda^3 - 2\lambda^2 - 15\lambda + 36 = 0$

5. **(a)** $\lambda = 1$, $\lambda = 2$, $\lambda = 3$ **(b)** $\lambda = 0$, $\lambda = \sqrt{2}$, $\lambda = -\sqrt{2}$ **(c)** $\lambda = -8$
 (d) $\lambda = 2$ **(e)** $\lambda = 2$ **(f)** $\lambda = -4$, $\lambda = 3$

6. **(a)** $\lambda = 1$: basis $\begin{bmatrix} 0 \\ 1 \\ 0 \end{bmatrix}$; $\lambda = 2$: basis $\begin{bmatrix} -\frac{1}{2} \\ 1 \\ 1 \end{bmatrix}$; $\lambda = 3$: basis $\begin{bmatrix} -1 \\ 1 \\ 1 \end{bmatrix}$

(b) $\lambda = 0$: basis $\begin{bmatrix} \frac{5}{3} \\ \frac{1}{3} \\ 1 \end{bmatrix}$; $\lambda = \sqrt{2}$: basis $\begin{bmatrix} \frac{1}{7}(15 + 5\sqrt{2}) \\ \frac{1}{7}(-1 + 2\sqrt{2}) \\ 1 \end{bmatrix}$; $\lambda = -\sqrt{2}$: basis $\begin{bmatrix} \frac{1}{7}(15 - 5\sqrt{2}) \\ \frac{1}{7}(-1 - 2\sqrt{2}) \\ 1 \end{bmatrix}$

(c) $\lambda = -8$: basis $\begin{bmatrix} -\frac{1}{6} \\ -\frac{1}{6} \\ 1 \end{bmatrix}$ **(d)** $\lambda = 2$: basis $\begin{bmatrix} \frac{1}{3} \\ \frac{1}{3} \\ 1 \end{bmatrix}$ **(e)** $\lambda = 2$: basis $\begin{bmatrix} -\frac{1}{3} \\ -\frac{1}{3} \\ 1 \end{bmatrix}$

(f) $\lambda = -4$: basis $\begin{bmatrix} -2 \\ \frac{8}{3} \\ 1 \end{bmatrix}$; $\lambda = 3$: basis $\begin{bmatrix} 5 \\ -2 \\ 1 \end{bmatrix}$

7. **(a)** $(\lambda - 1)^2(\lambda + 2)(\lambda + 1) = 0$ **(b)** $(\lambda - 4)^2(\lambda^2 + 3) = 0$

8. **(a)** $\lambda = 1$, $\lambda = -2$, $\lambda = -1$ **(b)** $\lambda = 4$

9. **(a)** $\lambda = 1$: basis $\begin{bmatrix} 0 \\ 0 \\ 0 \\ 1 \end{bmatrix}$ and $\begin{bmatrix} 2 \\ 3 \\ 1 \\ 0 \end{bmatrix}$; $\lambda = -2$; basis $\begin{bmatrix} -1 \\ 0 \\ 1 \\ 0 \end{bmatrix}$; $\lambda = -1$: basis $\begin{bmatrix} -2 \\ 1 \\ 1 \\ 0 \end{bmatrix}$ **(b)** $\lambda = 4$: basis $\begin{bmatrix} \frac{3}{2} \\ 1 \\ 0 \\ 0 \end{bmatrix}$

10. **(a)** $\lambda = -1$, $\lambda = 5$ **(b)** $\lambda = 3$, $\lambda = 7$, $\lambda = 1$ **(c)** $\lambda = -\frac{1}{3}$, $\lambda = 1$, $\lambda = \frac{1}{2}$

11. $\lambda = 1$, $\lambda = \frac{1}{512}$, $\lambda = 512$, $\lambda = 0$

12. For A^{25}, $\lambda = 1, -1$; basis for $\lambda = 1$: $\begin{bmatrix} -1 \\ 1 \\ 0 \end{bmatrix}, \begin{bmatrix} -1 \\ 0 \\ 1 \end{bmatrix}$; basis for $\lambda = -1$: $\begin{bmatrix} 2 \\ -1 \\ 1 \end{bmatrix}$

13. **(a)** $y = x$ and $y = 2x$ **(b)** No lines **(c)** $y = 0$ **14. (a)** -5 **(b)** 7

22. (a) $\lambda_1 = 1: \begin{bmatrix} 1 \\ 0 \\ 1 \end{bmatrix}; \lambda_2 = \frac{1}{2}: \begin{bmatrix} \frac{1}{2} \\ 1 \\ 0 \end{bmatrix}; \lambda_3 = \frac{1}{3}: \begin{bmatrix} 1 \\ 1 \\ 1 \end{bmatrix}$ **(b)** $\lambda_1 = -2: \begin{bmatrix} 1 \\ 0 \\ 1 \end{bmatrix}; \lambda_2 = -1: \begin{bmatrix} \frac{1}{2} \\ 1 \\ 0 \end{bmatrix}; \lambda_3 = 0: \begin{bmatrix} 1 \\ 1 \\ 1 \end{bmatrix}$

(c) $\lambda_1 = 3: \begin{bmatrix} 1 \\ 0 \\ 1 \end{bmatrix}; \lambda_2 = 4: \begin{bmatrix} \frac{1}{2} \\ 1 \\ 0 \end{bmatrix}; \lambda_3 = 5: \begin{bmatrix} 1 \\ 1 \\ 1 \end{bmatrix}$

24. (a) False **(b)** True **(c)** True **(d)** True

25. (a) A is 6×6. **(b)** A is invertible. **(c)** A has three eigenspaces.

EXERCISE SET 7.2 [page 354]

1. $\lambda = 0 : 1$ or 2; $\lambda = 1 : 1$; $\lambda = 2 : 1, 2$, or 3

2. (a) $\lambda = 3$, $\lambda = 5$
(b) For $\lambda = 3$, the rank of $3I - A$ is 1. For $\lambda = 5$, the rank of $5I - A$ is 2.
(c) A is diagonalizable since the eigenspaces produce a total of three basis vectors.

3. Not diagonalizable **4.** Not diagonalizable **5.** Not diagonalizable

6. Not diagonalizable **7.** Not diagonalizable

8. $P = \begin{bmatrix} \frac{4}{5} & \frac{3}{4} \\ 1 & 1 \end{bmatrix}; P^{-1}AP = \begin{bmatrix} 1 & 0 \\ 0 & 2 \end{bmatrix}$ **9.** $P = \begin{bmatrix} \frac{1}{3} & 0 \\ 1 & 1 \end{bmatrix}; P^{-1}AP = \begin{bmatrix} 1 & 0 \\ 0 & -1 \end{bmatrix}$

10. $P = \begin{bmatrix} 0 & 1 & 0 \\ 1 & 0 & 1 \\ -1 & 0 & 1 \end{bmatrix}; P^{-1}AP = \begin{bmatrix} 0 & 0 & 0 \\ 0 & 1 & 0 \\ 0 & 0 & 2 \end{bmatrix}$ **11.** $P = \begin{bmatrix} -2 & 0 & 1 \\ 0 & 1 & 0 \\ 1 & 0 & 0 \end{bmatrix}; P^{-1}AP = \begin{bmatrix} 3 & 0 & 0 \\ 0 & 3 & 0 \\ 0 & 0 & 2 \end{bmatrix}$

12. Not diagonalizable **13.** $P = \begin{bmatrix} 1 & 2 & 1 \\ 1 & 3 & 3 \\ 1 & 3 & 4 \end{bmatrix}; P^{-1}AP = \begin{bmatrix} 1 & 0 & 0 \\ 0 & 2 & 0 \\ 0 & 0 & 3 \end{bmatrix}$

14. Not diagonalizable **15.** $P = \begin{bmatrix} -\frac{1}{3} & 0 & 0 \\ 0 & 1 & 0 \\ 1 & 0 & 1 \end{bmatrix}; P^{-1}AP = \begin{bmatrix} 0 & 0 & 0 \\ 0 & 0 & 0 \\ 0 & 0 & 1 \end{bmatrix}$ **16.** Not diagonalizable

17. $P = \begin{bmatrix} 1 & 1 & 0 & 0 \\ 0 & 1 & 1 & 0 \\ 0 & 0 & 1 & 1 \\ 0 & 0 & 0 & 1 \end{bmatrix}; P^{-1}AP = \begin{bmatrix} -2 & 0 & 0 & 0 \\ 0 & -2 & 0 & 0 \\ 0 & 0 & 3 & 0 \\ 0 & 0 & 0 & 3 \end{bmatrix}$ **18.** $\begin{bmatrix} 1 & 0 \\ -1023 & 1024 \end{bmatrix}$ **19.** $\begin{bmatrix} -1 & 10237 & -2047 \\ 0 & 1 & 0 \\ 0 & 10245 & -2048 \end{bmatrix}$

20. (a) $\begin{bmatrix} 1 & 0 & 0 \\ 0 & 1 & 0 \\ 0 & 0 & 1 \end{bmatrix}$ **(b)** $\begin{bmatrix} 1 & 0 & 0 \\ 0 & 1 & 0 \\ 0 & 0 & 1 \end{bmatrix}$ **(c)** $\begin{bmatrix} 1 & -2 & 8 \\ 0 & -1 & 0 \\ 0 & 0 & -1 \end{bmatrix}$ **(d)** $\begin{bmatrix} 1 & -2 & 8 \\ 0 & -1 & 0 \\ 0 & 0 & -1 \end{bmatrix}$

21. $A^n = PD^nP^{-1} = \begin{bmatrix} 1 & 1 & 1 \\ 2 & 0 & -1 \\ 1 & -1 & 1 \end{bmatrix} \begin{bmatrix} 1^n & 0 & 0 \\ 0 & 3^n & 0 \\ 0 & 0 & 4^n \end{bmatrix} \begin{bmatrix} \frac{1}{6} & \frac{1}{3} & \frac{1}{6} \\ \frac{1}{2} & 0 & -\frac{1}{2} \\ \frac{1}{3} & -\frac{1}{3} & \frac{1}{3} \end{bmatrix}$

One possibility is $P = \begin{bmatrix} -b & -b \\ a - \lambda_1 & a - \lambda_2 \end{bmatrix}$ where λ_1 and λ_2 are as in Exercise 18 of Section 7.1.

25. (a) False **(b)** False **(c)** True **(d)** True

26. (a) Possible dimensions are 1 for $\lambda = 1$; 1 or 2 for $\lambda = 3$; 1, 2 or 3 for $\lambda = 4$.
  (b) Dimensions are 1 for $\lambda = 1$; 2 for $\lambda = 3$; 3 for $\lambda = 4$.
  (c) $\lambda = 4$

27. (a) Eigenvalues λ must satisfy $-1 < \lambda \le 1$.
  (b) If $A = PDP^{-1}$ with D diagonal, then $\lim_{k \to +\infty} A^k = PD'P^{-1}$, where D' is obtained from D by setting all diagonal entries that are not 1 to 0.

28. (a) Need $|x| < 1$, in which case the sum is $(1 - x)^{-1}$
  (b) Need all eigenvalues to have absolute value less than 1
  (c) $(I - A)^{-1}$

EXERCISE SET 7.3 [page 360]

1. (a) $\lambda^2 - 5\lambda = 0$; $\lambda = 0$: one-dimensional; $\lambda = 5$: one-dimensional
  (b) $\lambda^3 - 27\lambda - 54 = 0$; $\lambda = 6$: one-dimensional; $\lambda = -3$: two-dimensional
  (c) $\lambda^3 - 3\lambda^2 = 0$; $\lambda = 3$: one-dimensional; $\lambda = 0$: two-dimensional
  (d) $\lambda^3 - 12\lambda^2 + 36\lambda - 32 = 0$; $\lambda = 2$: two-dimensional; $\lambda = 8$: one-dimensional
  (e) $\lambda^4 - 8\lambda^3 = 0$; $\lambda = 0$: three-dimensional; $\lambda = 8$: one-dimensional
  (f) $\lambda^4 - 8\lambda^3 + 22\lambda^2 - 24\lambda + 9 = 0$; $\lambda = 1$: two-dimensional;
$\lambda = 3$: two-dimensional

2. $P = \begin{bmatrix} \dfrac{1}{\sqrt{2}} & -\dfrac{1}{\sqrt{2}} \\ \dfrac{1}{\sqrt{2}} & \dfrac{1}{\sqrt{2}} \end{bmatrix}$; $P^{-1}AP = \begin{bmatrix} 4 & 0 \\ 0 & 2 \end{bmatrix}$ **3.** $P = \begin{bmatrix} -\dfrac{2}{\sqrt{7}} & \dfrac{\sqrt{3}}{\sqrt{7}} \\ \dfrac{\sqrt{3}}{\sqrt{7}} & \dfrac{2}{\sqrt{7}} \end{bmatrix}$; $P^{-1}AP = \begin{bmatrix} 3 & 0 \\ 0 & 10 \end{bmatrix}$

4. $P = \begin{bmatrix} -\dfrac{2}{\sqrt{5}} & \dfrac{1}{\sqrt{5}} \\ \dfrac{1}{\sqrt{5}} & \dfrac{2}{\sqrt{5}} \end{bmatrix}$; $P^{-1}AP = \begin{bmatrix} 7 & 0 \\ 0 & 2 \end{bmatrix}$ **5.** $P = \begin{bmatrix} -\frac{4}{5} & 0 & \frac{3}{5} \\ 0 & 1 & 0 \\ \frac{3}{5} & 0 & \frac{4}{5} \end{bmatrix}$; $P^{-1}AP = \begin{bmatrix} 25 & 0 & 0 \\ 0 & -3 & 0 \\ 0 & 0 & -50 \end{bmatrix}$

6. $P = \begin{bmatrix} \dfrac{1}{\sqrt{2}} & \dfrac{1}{\sqrt{2}} & 0 \\ \dfrac{1}{\sqrt{2}} & -\dfrac{1}{\sqrt{2}} & 0 \\ 0 & 0 & 1 \end{bmatrix} \begin{bmatrix} 2 & 0 & 0 \\ 0 & 0 & 0 \\ 0 & 0 & 0 \end{bmatrix}$ **7.** $P = \begin{bmatrix} \dfrac{1}{\sqrt{3}} & \dfrac{1}{\sqrt{6}} & \dfrac{1}{\sqrt{2}} \\ \dfrac{1}{\sqrt{3}} & -\dfrac{2}{\sqrt{6}} & 0 \\ \dfrac{1}{\sqrt{3}} & \dfrac{1}{\sqrt{6}} & -\dfrac{1}{\sqrt{2}} \end{bmatrix} \begin{bmatrix} 0 & 0 & 0 \\ 0 & 3 & 0 \\ 0 & 0 & 3 \end{bmatrix}$

8. $P = \begin{bmatrix} 0 & 0 & \dfrac{1}{\sqrt{2}} & \dfrac{1}{\sqrt{2}} \\ 0 & 0 & \dfrac{1}{\sqrt{2}} & -\dfrac{1}{\sqrt{2}} \\ 1 & 0 & 0 & 0 \\ 0 & 1 & 0 & 0 \end{bmatrix} \begin{bmatrix} 0 & 0 & 0 & 0 \\ 0 & 0 & 0 & 0 \\ 0 & 0 & 4 & 0 \\ 0 & 0 & 0 & 2 \end{bmatrix}$ **9.** $P = \begin{bmatrix} -\frac{4}{5} & \frac{3}{5} & 0 & 0 \\ \frac{3}{5} & \frac{4}{5} & 0 & 0 \\ 0 & 0 & -\frac{4}{5} & \frac{3}{5} \\ 0 & 0 & \frac{3}{5} & \frac{4}{5} \end{bmatrix}$; $P^{-1}AP = \begin{bmatrix} -25 & 0 & 0 & 0 \\ 0 & 25 & 0 & 0 \\ 0 & 0 & -25 & 0 \\ 0 & 0 & 0 & 25 \end{bmatrix}$

10. $\begin{bmatrix} \dfrac{1}{\sqrt{2}} & -\dfrac{1}{\sqrt{2}} \\ \dfrac{1}{\sqrt{2}} & \dfrac{1}{\sqrt{2}} \end{bmatrix}$ **12. (b)** $\begin{bmatrix} \dfrac{1}{\sqrt{2}} & 0 & \dfrac{1}{\sqrt{2}} \\ 0 & 1 & 0 \\ -\dfrac{1}{\sqrt{2}} & 0 & \dfrac{1}{\sqrt{2}} \end{bmatrix}$ **14. (a)** True **(b)** True **(c)** False **(d)** True

15. Yes; take $A = \begin{bmatrix} 3 & 0 & 0 \\ 0 & 3 & 4 \\ 0 & 4 & 3 \end{bmatrix}$.

SUPPLEMENTARY EXERCISES [page 361]

1. (b) The transformation rotates vectors through the angle θ; therefore, if $0 < \theta < \pi$, then no nonzero vector is transformed into a vector in the same or opposite direction.

2. $\lambda = k$ with multiplicity 3.   **3. (c)** $\begin{bmatrix} 1 & 1 & 0 \\ 0 & 2 & 1 \\ 0 & 0 & 3 \end{bmatrix}$

9. $A^2 = \begin{bmatrix} 15 & 30 \\ 5 & 10 \end{bmatrix}$, $A^3 = \begin{bmatrix} 75 & 150 \\ 25 & 50 \end{bmatrix}$, $A^4 = \begin{bmatrix} 375 & 750 \\ 125 & 250 \end{bmatrix}$, $A^5 = \begin{bmatrix} 1875 & 3750 \\ 625 & 1250 \end{bmatrix}$

10. $A^3 = \begin{bmatrix} 1 & -3 & 3 \\ 3 & -8 & 6 \\ 6 & -15 & 10 \end{bmatrix}$, $A^4 = \begin{bmatrix} 3 & -8 & 6 \\ 6 & -15 & 10 \\ 10 & -24 & 15 \end{bmatrix}$ **11.** 0 and tr(A)

12. (b) $\begin{bmatrix} 0 & 0 & 0 & -1 \\ 1 & 0 & 0 & 2 \\ 0 & 1 & 0 & -1 \\ 0 & 0 & 1 & -3 \end{bmatrix}$ **15.** $\begin{bmatrix} 1 & 0 & 0 \\ -1 & -\frac{1}{2} & -\frac{1}{2} \\ 1 & -\frac{1}{2} & -\frac{1}{2} \end{bmatrix}$ **16. (a)** 18   **(b)** -1

17. They are all 0, 1, or -1.

EXERCISE SET 8.1 [page 373]

3. Nonlinear   **4.** Linear   **5.** Linear   **6.** Linear   **7.** Linear   **8. (a)** Linear   **(b)** Nonlinear

9. (a) Linear   **(b)** Nonlinear   **10. (a)** Nonlinear   **(b)** Linear

12. $T(x_1, x_2) = (-4x_1 + 5x_2, x_1 - 3x_2); T(5, -3) = (-35, 14)$

13. $T(x_1, x_2) = \frac{1}{7}(3x_1 - x_2, -9x_1 - 4x_2, 5x_1 + 10x_2); T(2, -3) = (\frac{9}{7}, -\frac{6}{7}, -\frac{20}{7})$

14. $T(x_1, x_2, x_3) = (-x_1 + 4x_2 - x_3, 5x_1 - 5x_2 - x_3, x_1 + 3x_3); T(2, 4, -1) = (15, -9, -1)$

15. $T(x_1, x_2, x_3) = (-41x_1 + 9x_2 + 24x_3, 14x_1 - 3x_2 - 8x_3); T(7, 13, 7) = (-2, 3)$

16. $T(2\mathbf{v}_1 - 3\mathbf{v}_2 + 4\mathbf{v}_3) = (-10, -7, 6)$

17. (a) Domain: R^2; codomain: R^2; $(T_2 \circ T_1)(x, y) = (2x - 3y, 2x + 3y)$
   (b) Domain: R^2; codomain: R^2; $(T_2 \circ T_1)(x, y) = (4x - 12y, 3x - 9y)$
   (c) Domain: R^2; codomain: R^2; $(T_2 \circ T_1)(x, y) = (2x + 3y, x - 2y)$
   (d) Domain: R^2; codomain: R^2; $(T_2 \circ T_1)(x, y) = (0, 2x)$

18. (a) Domain: R^2; codomain: R^2; $(T_3 \circ T_2 \circ T_1)(x, y) = (3x - 2y, x)$
   (b) Domain: R^2; codomain: R^2; $(T_3 \circ T_2 \circ T_1)(x, y) = (4y, 6y)$

19. (a) $a + d$ **(b)** $(T_2 \circ T_1)(A)$ does not exist since $T_1(A)$ is not a 2×2 matrix.

20. $(T_1 \circ T_2)(p(x)) = p(x); (T_2 \circ T_1)(p(x)) = p(x)$

21. $T_2(\mathbf{v}) = \frac{1}{4}\mathbf{v}$ **22.** $(T_2 \circ T_1)(a_0 + a_1x + a_2x^2) = (a_0 + a_1 + a_2)x + (a_1 + 2a_2)x^2 + a_2x^3$

23. (b) P_{mn} **26. (b)** $(3T)(x_1, x_2) = (6x_1 - 3x_2, 3x_2 + 3x_1)$

27. (b) $(T_1 + T_2)(x, y) = (3y, 4x); (T_2 - T_1)(x, y) = (y, 2x)$

28. (b) No   **31. (a)** $x^2 + 3x$ **(b)** $\sin x$ **(c)** $e^x - 1$

32. Yes   **33. (a)** True   **(b)** False   **(c)** True   **(d)** False   **34.** n^n

EXERCISE SET 8.2 [page 380]

1. (a), (c)   **2.** (a)   **3.** (a), (b), (c)   **4.** (a)   **5.** (b)   **6.** (a)

7. (a) $\left(\frac{1}{2}, 1\right)$ **(b)** $\left(\frac{3}{2}, -4, 1, 0\right)$ **(c)** No basis exists.

8. (a) $(1, -4)$ **(b)** $(4, 2, 6), (1, 1, 0), (-3, -4, 9)$ **(c)** (x, x^2, x^3)

10. (a) $\begin{bmatrix} 1 \\ 5 \\ 7 \end{bmatrix}, \begin{bmatrix} 0 \\ 1 \\ 1 \end{bmatrix}$ **(b)** $\begin{bmatrix} -\frac{14}{11} \\ \frac{19}{11} \\ 1 \end{bmatrix}$ **(c)** $\text{Rank}(T) = 2, \text{nullity}(T) = 1$

(d) $\text{Rank}(A) = 2, \text{nullity}(A) = 1$

11. (a) $\begin{bmatrix} 1 \\ 2 \\ 0 \end{bmatrix}$ **(b)** $\begin{bmatrix} \frac{1}{2} \\ 0 \\ 1 \end{bmatrix}, \begin{bmatrix} 0 \\ 1 \\ 0 \end{bmatrix}$ **(c)** $\text{Rank}(T) = 1, \text{nullity}(T) = 2$

(d) $\text{Rank}(A) = 1, \text{nullity}(A) = 2$

12. (a) $\begin{bmatrix} 1 \\ \frac{1}{4} \end{bmatrix}, \begin{bmatrix} 0 \\ 1 \end{bmatrix}$ **(b)** $\begin{bmatrix} -1 \\ -1 \\ 1 \\ 0 \end{bmatrix}, \begin{bmatrix} -\frac{4}{7} \\ \frac{2}{7} \\ 0 \\ 1 \end{bmatrix}$ **(c)** $\text{Rank}(T) = 2, \text{nullity}(T) = 2$

(d) $\text{Rank}(A) = 2, \text{nullity}(A) = 2$

13. (a) $\begin{bmatrix} 1 \\ 3 \\ -1 \\ 2 \end{bmatrix}, \begin{bmatrix} 0 \\ 1 \\ -\frac{2}{7} \\ \frac{5}{14} \end{bmatrix}, \begin{bmatrix} 0 \\ 0 \\ 0 \\ 1 \end{bmatrix}$ **(b)** $\begin{bmatrix} -1 \\ -1 \\ 1 \\ 0 \\ 0 \end{bmatrix}, \begin{bmatrix} -1 \\ -2 \\ 0 \\ 0 \\ 1 \end{bmatrix}$ **(c)** $\text{Rank}(T) = 3, \text{nullity}(T) = 2$

(d) $\text{Rank}(A) = 3, \text{nullity}(A) = 2$

14. (a) Range: xz-plane; nullspace: y-axis
(b) Range: yz-plane; nullspace: x-axis
(c) Range: plane $y = x$; nullspace: the line $x = -t, y = t, z = 0$

15. $\ker(T) = \{\mathbf{0}\}; R(T) = V$

16. (a) $\text{Nullity}(T) = 2$ **(b)** $\text{Nullity}(T) = 4$ **(c)** $\text{Nullity}(T) = 3$ **(d)** $\text{Nullity}(T) = 1$

17. $\text{Nullity}(T) = 0, \text{rank}(T) = 6$

18. (a) Dimension $= \text{nullity}(T) = 3$
(b) No. In order for $A\mathbf{x} = \mathbf{b}$ to be consistent for all $\mathbf{b}$ in R^5, we must have $R(T) = R^5$. But $R(T) \neq R^5$, since $\text{rank}(T) = \dim R(T) = 4$.

21. (a) $x = -t, y = -t, z = t, -\infty < t < +\infty$ **(b)** $14x - 8y - 5z = 0$

24. $\ker(D)$ consists of all constant polynomials.

25. $\ker(J)$ consists of all polynomials of the form kx.

26. $\ker(D \circ D)$ consists of all functions of the form $ax + b$; $\ker(D \circ D \circ D)$ consists of all functions of the form $ax^2 + bx + c$.

27. (a) Kernel, range **(b)** $(1, 1, 1)$ **(c)** The dimension of V **(d)** Three

28. (a) A plane through the origin **(b)** A line through the origin

29. (a) $D \circ D \circ D \circ D$, where D is differentiation **(b)** $D \circ D \circ \cdots \circ D$ ($n + 1$ times)

EXERCISE SET 8.3 [page 388]

1. **(a)** $\ker(T) = \{\mathbf{0}\}$; T is one-to-one.
   (b) $\ker(T) = \left\{k\left(-\frac{3}{2}, 1\right)\right\}$; T is not one-to-one.
   (c) $\ker(T) = \{\mathbf{0}\}$; T is one-to-one.
   (d) $\ker(T) = \{\mathbf{0}\}$; T is one-to-one.
   (e) $\ker(T) = \{k(1, 1)\}$; T is not one-to-one.
   (f) $\ker(T) = \{k(0, 1, -1)\}$; T is not one-to-one.

2. **(a)** $T^{-1}\begin{bmatrix} x_1 \\ x_2 \end{bmatrix} = \begin{bmatrix} x_1 - 2x_2 \\ -2x_1 + 5x_2 \end{bmatrix}$ **(b)** T has no inverse. **(c)** $T^{-1}\begin{bmatrix} x_1 \\ x_2 \end{bmatrix} = \begin{bmatrix} \frac{3}{19}x_1 - \frac{7}{19}x_2 \\ \frac{1}{19}x_1 + \frac{4}{19}x_2 \end{bmatrix}$

3. **(a)** T has no inverse. **(b)** $T^{-1}\begin{bmatrix} x_1 \\ x_2 \\ x_3 \end{bmatrix} = \begin{bmatrix} \frac{1}{8}x_1 + \frac{1}{8}x_2 - \frac{3}{4}x_3 \\ \frac{1}{8}x_1 + \frac{1}{8}x_2 + \frac{1}{4}x_3 \\ -\frac{3}{8}x_1 + \frac{5}{8}x_2 + \frac{1}{4}x_3 \end{bmatrix}$

   (c) $T^{-1}\begin{bmatrix} x_1 \\ x_2 \\ x_3 \end{bmatrix} = \begin{bmatrix} \frac{1}{2}x_1 - \frac{1}{2}x_2 + \frac{1}{2}x_3 \\ -\frac{1}{2}x_1 + \frac{1}{2}x_2 + \frac{1}{2}x_3 \\ \frac{1}{2}x_1 + \frac{1}{2}x_2 - \frac{1}{2}x_3 \end{bmatrix}$ **(d)** $T^{-1}\begin{bmatrix} x_1 \\ x_2 \\ x_3 \end{bmatrix} = \begin{bmatrix} 3x_1 + 3x_2 - x_3 \\ -2x_1 - 2x_2 + x_3 \\ -4x_1 - 5x_2 + 2x_3 \end{bmatrix}$

4. **(a)** Not one-to-one **(b)** Not one-to-one **(c)** One-to-one

5. **(a)** $\ker(T) = \{k(-1, 1)\}$ **(b)** T is not one-to-one since $\ker(T) \neq \{\mathbf{0}\}$.

6. **(a)** $\ker(T) = \{\mathbf{0}\}$ **(b)** T is one-to-one from Theorem 8.3.2.

7. **(a)** T is one-to-one. **(b)** T is not one-to-one. **(c)** T is not one-to-one. **(d)** T is one-to-one.

8. **(a)** T is one-to-one. **(b)** T is one-to-one. 9. No. A is not invertible.

10. **(a)** T is not one-to-one. **(b)** $T^{-1}(x_1, x_2, x_3, \ldots, x_n) = (x_n, x_{n-1}, x_{n-2}, \ldots, x_1)$
    (c) $T^{-1}(x_1, x_2, x_3, \ldots, x_n) = (x_n, x_1, x_2, \ldots, x_{n-1})$

11. **(a)** $a_i \neq 0$ for $i = 1, 2, 3, \ldots, n$ **(b)** $T^{-1}(x_1, x_2, x_3, \ldots, x_n) = \left(\dfrac{1}{a_1}x_1, \dfrac{1}{a_2}x_2, \dfrac{1}{a_3}x_3, \ldots, \dfrac{1}{a_n}x_n\right)$

12. **(b)** $T_1^{-1}(x, y) = \left(\frac{1}{2}x + \frac{1}{2}y, \frac{1}{2}x - \frac{1}{2}y\right)$; $T_2^{-1}(x, y) = \left(\frac{2}{5}x + \frac{1}{5}y, \frac{1}{5}x - \frac{2}{5}y\right)$; $(T_2 \circ T_1)^{-1}(x, y) = \left(\frac{3}{10}x - \frac{1}{10}y, \frac{1}{10}x + \frac{3}{10}y\right)$

13. **(a)** $T_1^{-1}(p(x)) = \dfrac{p(x)}{x}$; $T_2^{-1}(p(x)) = p(x - 1)$; $(T_2 \circ T_1)^{-1}(p(x)) = \dfrac{1}{x}p(x - 1)$

15. **(a)** $(1, -1)$ **(d)** $T^{-1}(2, 3) = 2 + x$

17. **(a)** T is not one-to-one. **(b)** T is one-to-one. $T^{-1}\begin{bmatrix} a & b \\ c & d \end{bmatrix} = \begin{bmatrix} a & c \\ b & d \end{bmatrix}$

    (c) T is one-to-one. $T^{-1}\begin{bmatrix} a & b \\ c & d \end{bmatrix} = \begin{bmatrix} d & -b \\ -c & a \end{bmatrix}$

20. J is not one-to-one since $J(x) = J(x^3)$.

21. T is not one-to-one since, for example, $f(x) = x^2(x - 1)^2$ is in its kernel.

22. **(a)** False, since T^{-1} does not exist. **(b)** True **(c)** True

23. Yes; it is one-to-one. 24. Yes; it is one-to-one.

25. No; it is not one-to-one.

EXERCISE SET 8.4 [page 399]

1. (a) $\begin{bmatrix} 0 & 0 & 0 \\ 1 & 0 & 0 \\ 0 & 1 & 0 \\ 0 & 0 & 1 \end{bmatrix}$
   2. (a) $\begin{bmatrix} 1 & 1 & 0 \\ 0 & -2 & -3 \end{bmatrix}$
   3. (a) $\begin{bmatrix} 1 & -1 & 1 \\ 0 & 1 & -2 \\ 0 & 0 & 1 \end{bmatrix}$

4. (a) $\begin{bmatrix} 2 & -1 \\ 2 & 0 \end{bmatrix}$
   5. (a) $\begin{bmatrix} 0 & 0 \\ -\frac{1}{2} & 1 \\ \frac{8}{3} & \frac{4}{3} \end{bmatrix}$
   6. (a) $\begin{bmatrix} 1 & -\frac{3}{2} & \frac{1}{2} \\ -1 & \frac{1}{2} & \frac{1}{2} \\ 0 & \frac{1}{2} & -\frac{1}{2} \end{bmatrix}$

7. (a) $\begin{bmatrix} 1 & 1 & 1 \\ 0 & 2 & 4 \\ 0 & 0 & 4 \end{bmatrix}$
   (b) $3 + 10x + 16x^2$
   8. (a) $\begin{bmatrix} 0 & 0 & 0 \\ 1 & -3 & 9 \\ 0 & 1 & -6 \\ 0 & 0 & 1 \end{bmatrix}$
   (b) $-11x + 7x^2 - x^3$

9. (a) $[T(\mathbf{v}_1)]_B = \begin{bmatrix} 1 \\ -2 \end{bmatrix}$, $[T(\mathbf{v}_2)]_B = \begin{bmatrix} 3 \\ 5 \end{bmatrix}$ **(b)** $T(\mathbf{v}_1) = \begin{bmatrix} 3 \\ -5 \end{bmatrix}$, $T(\mathbf{v}_2) = \begin{bmatrix} -2 \\ 29 \end{bmatrix}$

(c) $T\left(\begin{bmatrix} x_1 \\ x_2 \end{bmatrix}\right) = \begin{bmatrix} \frac{18}{7} & \frac{1}{7} \\ -\frac{107}{7} & \frac{24}{7} \end{bmatrix} \begin{bmatrix} x_1 \\ x_2 \end{bmatrix}$ **(d)** $\begin{bmatrix} \frac{19}{7} \\ -\frac{83}{7} \end{bmatrix}$

10. (a) $[T(\mathbf{v}_1)]_{B'} = \begin{bmatrix} 3 \\ 1 \\ -3 \end{bmatrix}$, $[T(\mathbf{v}_2)]_{B'} = \begin{bmatrix} -2 \\ 6 \\ 0 \end{bmatrix}$, $[T(\mathbf{v}_3)]_{B'} = \begin{bmatrix} 1 \\ 2 \\ 7 \end{bmatrix}$, $[T(\mathbf{v}_4)]_{B'} = \begin{bmatrix} 0 \\ 1 \\ 1 \end{bmatrix}$

(b) $T(\mathbf{v}_1) = \begin{bmatrix} 11 \\ 5 \\ 22 \end{bmatrix}$, $T(\mathbf{v}_2) = \begin{bmatrix} -42 \\ 32 \\ -10 \end{bmatrix}$, $T(\mathbf{v}_3) = \begin{bmatrix} -56 \\ 87 \\ 17 \end{bmatrix}$, $T(\mathbf{v}_4) = \begin{bmatrix} -13 \\ 17 \\ 2 \end{bmatrix}$

(c) $T\left(\begin{bmatrix} x_1 \\ x_2 \\ x_3 \\ x_4 \end{bmatrix}\right) = \begin{bmatrix} -\frac{253}{10} & \frac{49}{5} & \frac{241}{10} & -\frac{229}{10} \\ \frac{115}{2} & -39 & -\frac{65}{2} & \frac{153}{2} \\ 66 & -60 & -9 & 91 \end{bmatrix} \begin{bmatrix} x_1 \\ x_2 \\ x_3 \\ x_4 \end{bmatrix}$ **(d)** $\begin{bmatrix} -31 \\ 37 \\ 12 \end{bmatrix}$

11. (a) $[T(\mathbf{v}_1)]_B = \begin{bmatrix} 1 \\ 2 \\ 6 \end{bmatrix}$, $[T(\mathbf{v}_2)]_B = \begin{bmatrix} 3 \\ 0 \\ -2 \end{bmatrix}$, $[T(\mathbf{v}_3)]_B = \begin{bmatrix} -1 \\ 5 \\ 4 \end{bmatrix}$

(b) $T(\mathbf{v}_1) = 16 + 51x + 19x^2$, $T(\mathbf{v}_2) = -6 - 5x + 5x^2$, $T(\mathbf{v}_3) = 7 + 40x + 15x^2$

(c) $T(a_0 + a_1x + a_2x^2) = \dfrac{239a_0 - 161a_1 + 289a_2}{24} + \dfrac{201a_0 - 111a_1 + 247a_2}{8}x + \dfrac{61a_0 - 31a_1 + 107a_2}{12}x^2$

(d) $T(1 + x^2) = 22 + 56x + 14x^2$

12. (a) $[T_2 \circ T_1]_{B',B} = \begin{bmatrix} 1 & 1 \\ 2 & 4 \\ 0 & 4 \end{bmatrix}$, $[T_2]_{B'} = \begin{bmatrix} 1 & 1 & 1 \\ 0 & 2 & 4 \\ 0 & 0 & 4 \end{bmatrix}$, $[T_1]_{B',B} = \begin{bmatrix} 0 & 0 \\ 1 & 0 \\ 0 & 1 \end{bmatrix}$

(b) $[T_2 \circ T_1]_{B',B} = [T_2]_{B'}[T_1]_{B',B}$

13. (a) $[T_2 \circ T_1]_{B',B} = \begin{bmatrix} 0 & 0 \\ 6 & 0 \\ 0 & 0 \\ 0 & -9 \end{bmatrix}$, $[T_2]_{B',B''} = \begin{bmatrix} 0 & 0 & 0 \\ 3 & 0 & 0 \\ 0 & 3 & 0 \\ 0 & 0 & 3 \end{bmatrix}$, $[T_1]_{B'',B} = \begin{bmatrix} 2 & 0 \\ 0 & 0 \\ 0 & -3 \end{bmatrix}$

(b) $[T_2 \circ T_1]_{B',B} = [T_2]_{B',B''}[T_1]_{B'',B}$

16. $\begin{bmatrix} 0 & 0 & 0 & 1 \\ 1 & 0 & 0 & 0 \\ 0 & 1 & 0 & 0 \\ 0 & 0 & 1 & 0 \end{bmatrix}$ **18. (a)** $\begin{bmatrix} 0 & 1 & 0 \\ 0 & 0 & 2 \\ 0 & 0 & 0 \end{bmatrix}$ **(b)** $\begin{bmatrix} 0 & -\frac{3}{2} & \frac{23}{6} \\ 0 & 0 & -\frac{16}{3} \\ 0 & 0 & 0 \end{bmatrix}$ **(c)** $-6 + 48x$

19. (a) $\begin{bmatrix} 0 & 0 & 0 \\ 0 & 0 & -1 \\ 0 & 1 & 0 \end{bmatrix}$ **(b)** $\begin{bmatrix} 0 & 0 & 0 \\ 0 & 1 & 0 \\ 0 & 0 & 2 \end{bmatrix}$ **(c)** $\begin{bmatrix} 2 & 1 & 0 \\ 0 & 2 & 2 \\ 0 & 0 & 2 \end{bmatrix}$

(d) $14e^{2x} - 8xe^{2x} - 20x^2e^{2x}$ since $\begin{bmatrix} 2 & 1 & 0 \\ 0 & 2 & 2 \\ 0 & 0 & 2 \end{bmatrix} \begin{bmatrix} 4 \\ 6 \\ -10 \end{bmatrix} = \begin{bmatrix} 14 \\ -8 \\ -20 \end{bmatrix}$

20. Top row $V \to W$, bottom row $R^4 \to R^7$ **21. (a)** B', B'' **(b)** B', B'''

22. We can easily compute kernels, ranges, and compositions of linear transformations.

EXERCISE SET 8.5 [page 411]

1. $[T]_B = \begin{bmatrix} 1 & -2 \\ 0 & -1 \end{bmatrix}$, $[T]_{B'} = \begin{bmatrix} -\frac{3}{11} & -\frac{56}{11} \\ -\frac{2}{11} & \frac{3}{11} \end{bmatrix}$ **2.** $[T]_B = \begin{bmatrix} \frac{4}{5} & \frac{61}{10} \\ \frac{18}{5} & -\frac{19}{5} \end{bmatrix}$, $[T]_{B'} = \begin{bmatrix} -\frac{31}{2} & \frac{9}{2} \\ -\frac{75}{2} & \frac{25}{2} \end{bmatrix}$

3. $[T]_B = \begin{bmatrix} \dfrac{1}{\sqrt{2}} & -\dfrac{1}{\sqrt{2}} \\ \dfrac{1}{\sqrt{2}} & \dfrac{1}{\sqrt{2}} \end{bmatrix}$, $[T]_{B'} = \begin{bmatrix} \dfrac{13}{11\sqrt{2}} & -\dfrac{25}{11\sqrt{2}} \\ \dfrac{5}{11\sqrt{2}} & \dfrac{9}{11\sqrt{2}} \end{bmatrix}$

4. $[T]_B = \begin{bmatrix} 1 & 2 & -1 \\ 0 & -1 & 0 \\ 1 & 0 & 7 \end{bmatrix}$, $[T]_{B'} = \begin{bmatrix} 1 & 4 & 3 \\ -1 & -2 & -9 \\ 1 & 1 & 8 \end{bmatrix}$

5. $[T]_B = \begin{bmatrix} 1 & 0 & 0 \\ 0 & 1 & 0 \\ 0 & 0 & 0 \end{bmatrix}$, $[T]_{B'} = \begin{bmatrix} 1 & 0 & 0 \\ 0 & 1 & 1 \\ 0 & 0 & 0 \end{bmatrix}$ **6.** $[T]_B = \begin{bmatrix} 5 & 0 \\ 0 & 5 \end{bmatrix}$, $[T]_{B'} = \begin{bmatrix} 5 & 0 \\ 0 & 5 \end{bmatrix}$

7. $[T]_B = \begin{bmatrix} \frac{2}{3} & -\frac{2}{9} \\ \frac{1}{2} & \frac{4}{3} \end{bmatrix}$, $[T]_{B'} = \begin{bmatrix} 1 & 1 \\ 0 & 1 \end{bmatrix}$ **8. (a)** $\det(T) = 17$
(b) $\det(T) = 0$
(c) $\det(T) = 1$

10. (a) $[T]_B = \begin{bmatrix} 1 & 1 & 1 & 1 & 1 \\ 0 & 2 & 4 & 6 & 8 \\ 0 & 0 & 4 & 12 & 24 \\ 0 & 0 & 0 & 8 & 32 \\ 0 & 0 & 0 & 0 & 16 \end{bmatrix}$, where B is the standard basis for P_4; rank $(T) = 5$ and nullity $(T) = 0$.

(b) T is one-to-one.

11. (a) $\mathbf{u}_1' = \begin{bmatrix} -1 \\ 1 \end{bmatrix}$, $\mathbf{u}_2' = \begin{bmatrix} 1 \\ -2 \end{bmatrix}$ **(b)** $\mathbf{u}_1' = \begin{bmatrix} 1 \\ \dfrac{3 - \sqrt{21}}{2} \end{bmatrix}$, $\mathbf{u}_2' = \begin{bmatrix} 1 \\ \dfrac{3 + \sqrt{21}}{2} \end{bmatrix}$

12. (a) $\mathbf{u}_1' = \begin{bmatrix} -1 \\ 1 \\ 0 \end{bmatrix}$, $\mathbf{u}_2' = \begin{bmatrix} 1 \\ 0 \\ 1 \end{bmatrix}$, $\mathbf{u}_3' = \begin{bmatrix} -1 \\ -1 \\ 1 \end{bmatrix}$ **(b)** $\mathbf{u}_1' = \begin{bmatrix} -1 \\ 1 \\ 0 \end{bmatrix}$, $\mathbf{u}_2' = \begin{bmatrix} 1 \\ 0 \\ 1 \end{bmatrix}$, $\mathbf{u}_3' = \begin{bmatrix} -1 \\ -1 \\ 1 \end{bmatrix}$

(c) $\mathbf{u}_1' = \begin{bmatrix} 1 \\ 2 \\ 1 \end{bmatrix}$, $\mathbf{u}_2' = \begin{bmatrix} 0 \\ 1 \\ 0 \end{bmatrix}$, $\mathbf{u}_3' = \begin{bmatrix} -1 \\ 0 \\ 1 \end{bmatrix}$

13. (a) $\lambda = -4$, $\lambda = 3$

(b) Basis for eigenspace corresponding to $\lambda = -4$: $-2 + \frac{8}{3}x + x^2$;
basis for eigenspace corresponding to $\lambda = 3$: $5 - 2x + x^2$

14. (a) $\lambda = 1, \lambda = -2, \lambda = -1$

(b) Basis for eigenspace corresponding to $\lambda = 1$: $\begin{bmatrix} 0 & 0 \\ 0 & 1 \end{bmatrix}$ and $\begin{bmatrix} 2 & 3 \\ 1 & 0 \end{bmatrix}$;

basis for eigenspace corresponding to $\lambda = -2$: $\begin{bmatrix} -1 & 0 \\ 1 & 0 \end{bmatrix}$;

basis for eigenspace corresponding to $\lambda = -1$: $\begin{bmatrix} -2 & 1 \\ 1 & 0 \end{bmatrix}$

19. (a) False   **(b)** True   **(c)** True   **(d)** True

20. For example, $\begin{bmatrix} 1 & 0 \\ 0 & 2 \end{bmatrix}$, $\begin{bmatrix} 3 & 0 \\ 0 & 4 \end{bmatrix}$ since they have different eigenvalues

21. (1) $B = P^{-1}AP$ is similar to A.
(2) $I = P^{-1}P$
(3) The distributive law for matrices
(4) The determinant of a product is the product of the determinants.
(5) The commutative law for real multiplication   (6) $\det(P^{-1}) = 1/\det(P)$

22. $P^{-1}x$ is an eigenvector for B with eigenvalue λ.

23. The choice of an appropriate basis can yield a better understanding of the linear operator.

SUPPLEMENTARY EXERCISES (page 413)

1. No. $T(\mathbf{x}_1 + \mathbf{x}_2) = A(\mathbf{x}_1 + \mathbf{x}_2) + B \neq (A\mathbf{x}_1 + B) + (A\mathbf{x}_2 + B) = T(\mathbf{x}_1) + T(\mathbf{x}_2)$, and if $c \neq 1$, then $T(c\mathbf{x}) = cA\mathbf{x} + B \neq c(A\mathbf{x} + B) = cT(\mathbf{x})$.

2. (b) $A^n = \begin{bmatrix} \cos n\theta & -\sin n\theta \\ \sin n\theta & \cos n\theta \end{bmatrix}$

5. (a) $T(\mathbf{e}_3)$ and any two of $T(\mathbf{e}_1)$, $T(\mathbf{e}_2)$, and $T(\mathbf{e}_4)$ form bases for the range; $(-1, 1, 0, 1)$ is a basis for the kernel.
(b) Rank $= 3$, nullity $= 1$

6. (a) $(1, 1, -1)$ **(b)** $(-1, 2, 4)$ and $(3, 0, 1)$

7. (a) Rank $(T) = 2$ and nullity $(T) = 2$ **(b)** T is not one-to-one.

11. Rank $= 3$, nullity $= 1$ **13.** $\begin{bmatrix} 1 & 0 & 0 & 0 \\ 0 & 0 & 1 & 0 \\ 0 & 1 & 0 & 0 \\ 0 & 0 & 0 & 1 \end{bmatrix}$

14. (a) $\mathbf{v}_1 = 2\mathbf{u}_1 + \mathbf{u}_2$, $\mathbf{v}_2 = -\mathbf{u}_1 + \mathbf{u}_2 + \mathbf{u}_3$, $\mathbf{v}_3 = 3\mathbf{u}_1 + 4\mathbf{u}_2 + 2\mathbf{u}_3$
(b) $\mathbf{u}_1 = -2\mathbf{v}_1 - 2\mathbf{v}_2 + \mathbf{v}_3$, $\mathbf{u}_2 = 5\mathbf{v}_1 + 4\mathbf{v}_2 - 2\mathbf{v}_3$, $\mathbf{u}_3 = -7\mathbf{v}_1 - 5\mathbf{v}_2 + 3\mathbf{v}_3$

15. $[T]_{B'} = \begin{bmatrix} -4 & 0 & 9 \\ 1 & 0 & -2 \\ 0 & 1 & 1 \end{bmatrix}$ **17.** $[T]_B = \begin{bmatrix} 1 & -1 & 1 \\ 0 & 1 & 0 \\ 1 & 0 & -1 \end{bmatrix}$ **19. (b)** 1 and x **(c)** e^x and e^{-x}

20. (a) $\begin{bmatrix} 2 \\ 6 \\ 12 \end{bmatrix}$ **(e)**

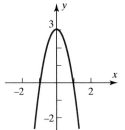

(d) $-3x^2 + 3$

21. The points are on the graph.

24. $\begin{bmatrix} 0 & 1 & 0 & 0 & \cdots & 0 \\ 0 & 0 & 1 & 0 & \cdots & 0 \\ 0 & 0 & 0 & 1 & \cdots & 0 \\ \vdots & \vdots & \vdots & \vdots & & \vdots \\ 0 & 0 & 0 & 0 & \cdots & 1 \\ 0 & 0 & 0 & 0 & \cdots & 0 \end{bmatrix}$

25. $\begin{bmatrix} 0 & 0 & 0 & \cdots & 0 \\ 1 & 0 & 0 & \cdots & 0 \\ 0 & \frac{1}{2} & 0 & \cdots & 0 \\ 0 & 0 & \frac{1}{3} & \cdots & 0 \\ \vdots & \vdots & \vdots & & \vdots \\ 0 & 0 & 0 & \cdots & \frac{1}{n+1} \end{bmatrix}$

EXERCISE SET 9.1 [page 424]

1. (a) $y_1 = c_1 e^{5x} - 2c_2 e^{-x}$ **(b)** $y_1 = 0$ **2. (a)** $y_1 = c_1 e^{7x} - 3c_2 e^{-x}$ **(b)** $y_1 = -\frac{1}{40} e^{7x} + \frac{81}{40} e^{-x}$
$\quad\quad y_2 = c_1 e^{5x} + c_2 e^{-x}$ $\quad\quad y_2 = 0$ $\quad\quad y_2 = 2c_1 e^{7x} + 2c_2 e^{-x}$ $\quad\quad y_2 = -\frac{1}{20} e^{7x} - \frac{27}{20} e^{-x}$

3. (a) $y_1 = -c_2 e^{2x} + c_3 e^{3x}$ **(b)** $y_1 = e^{2x} - 2e^{3x}$ **4.** $y_1 = (c_1 + c_2) e^{2x} + c_3 e^{8x}$
$\quad\quad y_2 = c_1 e^x + 2c_2 e^{2x} - c_3 e^{3x}$ $\quad\quad y_2 = e^x - 2e^{2x} + 2e^{3x}$ $\quad\quad y_2 = -c_2 e^{2x} + c_3 e^{8x}$
$\quad\quad y_3 = 2c_2 e^{2x} - c_3 e^{3x}$ $\quad\quad y_3 = -2e^{2x} + 2e^{3x}$ $\quad\quad y_3 = -c_1 e^{2x} + c_3 e^{8x}$

7. $y = c_1 e^{3x} + c_2 e^{-2x}$ **8.** $y = c_1 e^x + c_2 e^{2x} + c_3 e^{3x}$

EXERCISE SET 9.2 [page 434]

1. (a) $\begin{bmatrix} 0 & -1 \\ -1 & 0 \end{bmatrix}$ **(b)** $\begin{bmatrix} -1 & 0 \\ 0 & -1 \end{bmatrix}$ **(c)** $\begin{bmatrix} 1 & 0 \\ 0 & 0 \end{bmatrix}$ **(d)** $\begin{bmatrix} 0 & 0 \\ 0 & 1 \end{bmatrix}$

2. (a) $(-1, -2)$ **(b)** $(-2, -1)$ **(c)** $(2, 0)$ **(d)** $(0, 1)$

3. (a) $\begin{bmatrix} 1 & 0 & 0 \\ 0 & 1 & 0 \\ 0 & 0 & -1 \end{bmatrix}$ **(b)** $\begin{bmatrix} 1 & 0 & 0 \\ 0 & -1 & 0 \\ 0 & 0 & 1 \end{bmatrix}$ **(c)** $\begin{bmatrix} -1 & 0 & 0 \\ 0 & 1 & 0 \\ 0 & 0 & 1 \end{bmatrix}$

4. (a) $(1, 1, -1)$ **(b)** $(1, -1, 1)$ **(c)** $(-1, 1, 1)$

5. (a) $\begin{bmatrix} 0 & -1 & 0 \\ 1 & 0 & 0 \\ 0 & 0 & 1 \end{bmatrix}$ **(b)** $\begin{bmatrix} 1 & 0 & 0 \\ 0 & 0 & -1 \\ 0 & 1 & 0 \end{bmatrix}$ **(c)** $\begin{bmatrix} 0 & 0 & 1 \\ 0 & 1 & 0 \\ -1 & 0 & 0 \end{bmatrix}$

6. (a) Rectangle with vertices at $(0, 0), (1, 0), (1, -2), (0, -2)$ **(b)** Rectangle with vertices at $(0, 0), (-1, 0), (-1, 2), (0, 2)$
(c) Rectangle with vertices at $(0, 0), (1, 0), (1, \frac{1}{2}), (0, \frac{1}{2})$ **(d)** Square with vertices at $(0, 0), (2, 0), (2, 2), (0, 2)$
(e) Parallelogram with vertices at $(0, 0), (1, 0), (7, 2), (6, 2)$ **(f)** Parallelogram with vertices at $(0, 0), (1, -2), (1, 0), (0, 2)$

7. Rectangle with vertices at $(0, 0), (-3, 0), (0, 1), (-3, 1)$

8. (a) $\begin{bmatrix} \frac{1}{\sqrt{2}} & -\frac{1}{\sqrt{2}} \\ \frac{1}{\sqrt{2}} & \frac{1}{\sqrt{2}} \end{bmatrix}$ **(b)** $\begin{bmatrix} 0 & -1 \\ 1 & 0 \end{bmatrix}$ **(c)** $\begin{bmatrix} -1 & 0 \\ 0 & -1 \end{bmatrix}$ **(d)** $\begin{bmatrix} 0 & 1 \\ -1 & 0 \end{bmatrix}$ **(e)** $\begin{bmatrix} \frac{\sqrt{3}}{2} & \frac{1}{2} \\ -\frac{1}{2} & \frac{\sqrt{3}}{2} \end{bmatrix}$

9. (a) $\begin{bmatrix} 1 & 0 \\ 4 & 1 \end{bmatrix}$ **(b)** $\begin{bmatrix} 1 & -2 \\ 0 & 1 \end{bmatrix}$ **10. (a)** $\begin{bmatrix} 1 & 0 \\ 0 & \frac{1}{3} \end{bmatrix}$ **(b)** $\begin{bmatrix} 6 & 0 \\ 0 & 1 \end{bmatrix}$

11. (a) Expansion by a factor of 3 in the x-direction
(b) Expansion by a factor of -5 in the y-direction
(c) Shear by a factor of 4 in the x-direction

12. (a) $\begin{bmatrix} 2 & 0 \\ 0 & 1 \end{bmatrix}\begin{bmatrix} 3 & 0 \\ 0 & 1 \end{bmatrix}$; expansion in the y-direction by a factor of 3, then expansion in the x-direction by a factor of 2

(b) $\begin{bmatrix} 1 & 0 \\ 2 & 1 \end{bmatrix}\begin{bmatrix} 1 & 4 \\ 0 & 1 \end{bmatrix}$; shear in the x-direction by a factor of 4, then shear in the y- direction by a factor of 2

(c) $\begin{bmatrix} 0 & 1 \\ 1 & 0 \end{bmatrix}\begin{bmatrix} 4 & 0 \\ 0 & 1 \end{bmatrix}\begin{bmatrix} 1 & 0 \\ 0 & -2 \end{bmatrix}$; expansion in the y-direction by a factor of -2, then expansion in the x-direction by a factor of 4, then reflection about $y = x$

(d) $\begin{bmatrix} 1 & 0 \\ 4 & 1 \end{bmatrix}\begin{bmatrix} 1 & 0 \\ 1 & 18 \end{bmatrix}\begin{bmatrix} 1 & -3 \\ 0 & 1 \end{bmatrix}$; shear in the x-direction by a factor of -3, then expansion in the y-direction by a factor of 18, then shear in the y-direction by a factor of 4

13. (a) $\begin{bmatrix} \frac{1}{2} & 0 \\ 0 & 5 \end{bmatrix}$ **(b)** $\begin{bmatrix} 1 & 0 \\ 2 & 5 \end{bmatrix}$ **(c)** $\begin{bmatrix} 0 & -1 \\ -1 & 0 \end{bmatrix}$ **14. (a)** $\begin{bmatrix} 0 & 1 \\ -5 & 0 \end{bmatrix}$ **(b)** $\frac{1}{2}\begin{bmatrix} \sqrt{3} & -1 \\ -6\sqrt{3}+3 & 6+3\sqrt{3} \end{bmatrix}$

16. $16y - 11x - 3 = 0$ **17. (a)** $y = \frac{2}{7}x$ **(b)** $y = x$ **(c)** $y = \frac{1}{2}x$ **(d)** $y = -2x$

18. $\begin{bmatrix} 1 & -2 \\ 0 & 1 \end{bmatrix}$ **19. (b)** No. A is not invertible.

22. (a) $\begin{bmatrix} 1 & 0 & 0 \\ 0 & 0 & 1 \\ 0 & 1 & 0 \end{bmatrix}$ **(b)** $\begin{bmatrix} 0 & 0 & 1 \\ 0 & 1 & 0 \\ 1 & 0 & 0 \end{bmatrix}$ **(c)** $\begin{bmatrix} 0 & 1 & 0 \\ 1 & 0 & 0 \\ 0 & 0 & 1 \end{bmatrix}$

23. (a) $\begin{bmatrix} 1 & 0 & k \\ 0 & 1 & k \\ 0 & 0 & 1 \end{bmatrix}$ **(b)** xz-direction: $\begin{bmatrix} 1 & k & 0 \\ 0 & 1 & 0 \\ 0 & k & 1 \end{bmatrix}$; yz-direction: $\begin{bmatrix} 1 & 0 & 0 \\ k & 1 & 0 \\ k & 0 & 1 \end{bmatrix}$

24. (a) $\lambda_1 = 1$: $\begin{bmatrix} 1 \\ 0 \end{bmatrix}$; $\lambda_2 = -1$: $\begin{bmatrix} 0 \\ 1 \end{bmatrix}$ **(b)** $\lambda_1 = 1$: $\begin{bmatrix} 0 \\ 1 \end{bmatrix}$; $\lambda_2 = -1$: $\begin{bmatrix} 1 \\ 0 \end{bmatrix}$

(c) $\lambda_1 = 1$: $\begin{bmatrix} 1 \\ 1 \end{bmatrix}$; $\lambda_2 = -1$: $\begin{bmatrix} -1 \\ 1 \end{bmatrix}$ **(d)** $\lambda = 1$: $\begin{bmatrix} 1 \\ 0 \end{bmatrix}$

(e) $\lambda = 1$: $\begin{bmatrix} 0 \\ 1 \end{bmatrix}$ **(f)** (θ an odd integer multiple of π) $\lambda = -1$: $(1, 0), (0, 1)$
(θ an even integer multiple of π) $\lambda = 1$: $(1, 0), (0, 1)$
(θ not an integer multiple of π) no real eigenvalues

EXERCISE SET 9.3 [page 441]

1. $y = -\frac{1}{2} + \frac{7}{2}x$ **2.** $y = \frac{2}{3} + \frac{1}{6}x$ **3.** $y = 2 + 5x - 3x^2$ **4.** $y = -5 + 3x - 4x^2 + 2x^3$

8. $y = 4 - .2x + .2x^2$; if $x = 12$, then $y = 30.4$ ($\$30.4$ thousand)

EXERCISE SET 9.4 [page 447]

1. (a) $(1+\pi) - 2\sin x - \sin 2x$ **(b)** $(1+\pi) - 2\left[\sin x + \dfrac{\sin 2x}{2} + \dfrac{\sin 3x}{3} + \cdots + \dfrac{\sin nx}{n}\right]$

2. (a) $\frac{4}{3}\pi^2 + 4\cos x + \cos 2x + \frac{4}{9}\cos 3x - 4\pi\sin x - 2\pi\sin 2x - \dfrac{4\pi}{3}\sin 3x$

(b) $\frac{4}{3}\pi^2 + 4\displaystyle\sum_{k=1}^{n}\dfrac{\cos kx}{k^2} - 4\pi\sum_{k=1}^{n}\dfrac{\sin kx}{k}$

3. **(a)** $-\dfrac{1}{2} + \dfrac{1}{e-1}e^x$ **(b)** $\dfrac{1}{12} - \dfrac{3-e}{2e-2}$ 4. **(a)** $(4e-10) + (18-6e)x$ **(b)** $\dfrac{(3-e)(7e-19)}{2}$

5. **(a)** $\dfrac{3}{\pi}x$ **(b)** $1 - \dfrac{6}{\pi^2}$ 8. $\displaystyle\sum_{k=1}^{\infty} \dfrac{2}{k}\sin(kx)$

EXERCISE SET 9.5 [page 453]

1. (a), (c), (e), (g), (h)

2. **(a)** $A = \begin{bmatrix} 3 & 0 \\ 0 & 7 \end{bmatrix}$ **(b)** $A = \begin{bmatrix} 4 & -3 \\ -3 & -9 \end{bmatrix}$ **(c)** $A = \begin{bmatrix} 5 & \frac{5}{2} \\ \frac{5}{2} & 0 \end{bmatrix}$ **(d)** $A = \begin{bmatrix} 0 & -\frac{7}{2} \\ -\frac{7}{2} & 0 \end{bmatrix}$

3. **(a)** $A = \begin{bmatrix} 9 & 3 & -4 \\ 3 & -1 & \frac{1}{2} \\ -4 & \frac{1}{2} & 4 \end{bmatrix}$ **(b)** $\begin{bmatrix} 1 & -\frac{5}{2} & \frac{9}{2} \\ -\frac{5}{2} & 1 & 0 \\ \frac{9}{2} & 0 & -3 \end{bmatrix}$ **(c)** $A = \begin{bmatrix} 0 & \frac{1}{2} & \frac{1}{2} \\ \frac{1}{2} & 0 & \frac{1}{2} \\ \frac{1}{2} & \frac{1}{2} & 0 \end{bmatrix}$

(d) $A = \begin{bmatrix} \sqrt{2} & \sqrt{2} & -4\sqrt{3} \\ \sqrt{2} & 0 & 0 \\ -4\sqrt{3} & 0 & -\sqrt{3} \end{bmatrix}$ **(e)** $A = \begin{bmatrix} 1 & 1 & 0 & -5 \\ 1 & 1 & 0 & 0 \\ 0 & 0 & -1 & 2 \\ -5 & 0 & 2 & -1 \end{bmatrix}$

4. **(a)** $2x^2 + 5y^2 - 6xy$ **(b)** $7x_1^2 + 5x_1x_2$ **(c)** $x^2 - 3y^2 + 5z^2$
(d) $-2x_1^2 + 3x_3^2 + 7x_1x_2 + x_1x_3 + 12x_2x_3$ **(e)** $2x_1x_2 + 2x_1x_3 + 2x_1x_4 + 2x_2x_3 + 2x_2x_4 + 2x_3x_4$

5. **(a)** max value $= 5$ at $\pm(1, 0)$; min value $= -1$ at $\pm(0, 1)$

(b) max value $= \dfrac{11 + \sqrt{10}}{2}$ at $\pm\left(\dfrac{1}{\sqrt{20 - 6\sqrt{10}}}, \dfrac{1}{\sqrt{20 + 6\sqrt{10}}}\right)$;

min value $= \dfrac{11 - \sqrt{10}}{2}$ at $\pm\left(\dfrac{-1}{\sqrt{20 + 6\sqrt{10}}}, \dfrac{1}{\sqrt{20 - 6\sqrt{10}}}\right)$

(c) max value $= \dfrac{7 + \sqrt{10}}{2}$ at $\pm\left(\dfrac{1}{\sqrt{20 - 6\sqrt{10}}}, \dfrac{-1}{\sqrt{20 - 6\sqrt{10}}}\right)$;

min value $= \dfrac{7 - \sqrt{10}}{2}$ at $\pm\left(\dfrac{1}{\sqrt{20 + 6\sqrt{10}}}, \dfrac{1}{\sqrt{20 - 6\sqrt{10}}}\right)$

(d) max value $= \dfrac{3 + \sqrt{10}}{2}$ at $\pm\left(\dfrac{3}{\sqrt{20 - 2\sqrt{10}}}, \dfrac{3}{\sqrt{20 + 2\sqrt{10}}}\right)$; min value $= \dfrac{3 - \sqrt{10}}{2}$ at $\pm\left(\dfrac{3}{\sqrt{20 + 2\sqrt{10}}}, \dfrac{-3}{\sqrt{20 - 2\sqrt{10}}}\right)$

6. **(a)** max value $= 4$ at $\pm\left(\dfrac{1}{\sqrt{6}}, \dfrac{1}{\sqrt{6}}, \dfrac{2}{\sqrt{6}}\right)$; min value $= -2$ at $\pm\left(-\dfrac{1}{\sqrt{3}}, -\dfrac{1}{\sqrt{3}}, \dfrac{1}{\sqrt{3}}\right)$

(b) max value $= 3$ at $\left(\dfrac{2}{\sqrt{6}}, \dfrac{1}{\sqrt{6}}, \dfrac{1}{\sqrt{6}}\right)$; min value $= 0$ at $\left(\dfrac{1}{\sqrt{3}}, -\dfrac{1}{\sqrt{3}}, -\dfrac{1}{\sqrt{3}}\right)$

(c) max value $= 4$ at $\pm\left(\dfrac{1}{\sqrt{2}}, 0, \dfrac{1}{\sqrt{2}}\right)$; min value $= 2$ at all points on the unit sphere for which $x_3 = -x_1$

7. **(b)** 9. **(a)**

11. **(a)** Positive definite **(b)** Negative definite **(c)** Positive semidefinite
(d) Negative semidefinite **(e)** Indefinite **(f)** Indefinite

12. **(a)** Indefinite **(b)** Indefinite **(c)** Positive semidefinite **(d)** Indefinite
(e) Positive and negative semidefinite **(f)** Positive definite

13. **(c)** No. $T(k\mathbf{x}) \neq kT(\mathbf{x})$, unless $k = 0$ or 1.

14. **(a)** $k > 4$ **(b)** $k > 2$ **(c)** $-\frac{1}{3}\sqrt{15} < k < \frac{1}{3}\sqrt{15}$

15. $A = \begin{bmatrix} c_1^2 & c_1c_2 & c_1c_3 & \cdots & c_1c_n \\ c_1c_2 & c_2^2 & c_2c_3 & \cdots & c_2c_n \\ \vdots & \vdots & \vdots & & \vdots \\ c_1c_n & c_2c_n & c_3c_n & \cdots & c_n^2 \end{bmatrix}$

16. **(a)** $A = \begin{bmatrix} \dfrac{1}{n} & \dfrac{-1}{n(n-1)} & \dfrac{-1}{n(n-1)} & \cdots & \dfrac{-1}{n(n-1)} \\ \dfrac{-1}{n(n-1)} & \dfrac{1}{n} & \dfrac{-1}{n(n-1)} & \cdots & \dfrac{-1}{n(n-1)} \\ \vdots & \vdots & \vdots & & \vdots \\ \dfrac{-1}{n(n-1)} & \dfrac{-1}{n(n-1)} & \dfrac{-1}{n(n-1)} & \cdots & \dfrac{1}{n} \end{bmatrix}$ **(b)** Positive semidefinite

EXERCISE SET 9.6 [page 462]

1. **(a)** $\begin{bmatrix} x_1 \\ x_2 \end{bmatrix} = \begin{bmatrix} \dfrac{1}{\sqrt{2}} & \dfrac{1}{\sqrt{2}} \\ \dfrac{1}{\sqrt{2}} & -\dfrac{1}{\sqrt{2}} \end{bmatrix} \begin{bmatrix} y_1 \\ y_2 \end{bmatrix}; \ y_1^2 + 3y_2^2$ **(b)** $\begin{bmatrix} x_1 \\ x_2 \end{bmatrix} = \begin{bmatrix} \dfrac{1}{\sqrt{5}} & \dfrac{2}{\sqrt{5}} \\ -\dfrac{2}{\sqrt{5}} & \dfrac{1}{\sqrt{5}} \end{bmatrix} \begin{bmatrix} y_1 \\ y_2 \end{bmatrix}; \ y_1^2 + 6y_2^2$

   (c) $\begin{bmatrix} x_1 \\ x_2 \end{bmatrix} = \begin{bmatrix} \dfrac{1}{\sqrt{2}} & \dfrac{1}{\sqrt{2}} \\ \dfrac{1}{\sqrt{2}} & -\dfrac{1}{\sqrt{2}} \end{bmatrix} \begin{bmatrix} y_1 \\ y_2 \end{bmatrix}; \ y_1^2 - y_2^2$

   (d) $\begin{bmatrix} x_1 \\ x_2 \end{bmatrix} = \begin{bmatrix} \dfrac{\sqrt{17}-4}{\sqrt{34-8\sqrt{17}}} & \dfrac{\sqrt{17}+4}{\sqrt{34+8\sqrt{17}}} \\ \dfrac{1}{\sqrt{34-8\sqrt{17}}} & \dfrac{-1}{\sqrt{34+8\sqrt{17}}} \end{bmatrix} \begin{bmatrix} y_1 \\ y_2 \end{bmatrix}; \ (1+\sqrt{17})y_1^2 + (1-\sqrt{17})y_2^2$

2. **(a)** $\begin{bmatrix} x_1 \\ x_2 \\ x_3 \end{bmatrix} = \begin{bmatrix} \frac{2}{3} & \frac{1}{3} & \frac{2}{3} \\ -\frac{2}{3} & \frac{2}{3} & \frac{1}{3} \\ -\frac{1}{3} & -\frac{2}{3} & \frac{2}{3} \end{bmatrix} \begin{bmatrix} y_1 \\ y_2 \\ y_3 \end{bmatrix}; \ y_1^2 + 7y_2^2 + 4y_3^2$ **(b)** $\begin{bmatrix} x_1 \\ x_2 \\ x_3 \end{bmatrix} = \begin{bmatrix} \frac{1}{3} & \frac{2}{3} & \frac{2}{3} \\ \frac{2}{3} & \frac{1}{3} & -\frac{2}{3} \\ -\frac{2}{3} & \frac{2}{3} & -\frac{1}{3} \end{bmatrix} \begin{bmatrix} y_1 \\ y_2 \\ y_3 \end{bmatrix}; \ 7y_1^2 + 4y_2^2 + y_3^2$

   (c) $\begin{bmatrix} x_1 \\ x_2 \\ x_3 \end{bmatrix} = \begin{bmatrix} \dfrac{1}{\sqrt{14}} & \dfrac{1}{\sqrt{6}} & -\dfrac{4}{\sqrt{21}} \\ -\dfrac{2}{\sqrt{14}} & \dfrac{2}{\sqrt{6}} & \dfrac{1}{\sqrt{21}} \\ \dfrac{3}{\sqrt{14}} & \dfrac{1}{\sqrt{6}} & \dfrac{2}{\sqrt{21}} \end{bmatrix} \begin{bmatrix} y_1 \\ y_2 \\ y_3 \end{bmatrix}; \ 2y_2^2 - 7y_3^2$

   (d) $\begin{bmatrix} x_1 \\ x_2 \\ x_3 \end{bmatrix} = \begin{bmatrix} \dfrac{3}{\sqrt{10}} & \dfrac{1}{\sqrt{20}} & \dfrac{1}{\sqrt{20}} \\ -\dfrac{1}{\sqrt{10}} & \dfrac{3}{\sqrt{20}} & \dfrac{3}{\sqrt{20}} \\ 0 & \dfrac{1}{\sqrt{2}} & -\dfrac{1}{\sqrt{2}} \end{bmatrix} \begin{bmatrix} y_1 \\ y_2 \\ y_3 \end{bmatrix}; \ \sqrt{10}y_2^2 - \sqrt{10}y_3^2$

3. **(a)** $2x^2 - 3xy + 4y^2$ **(b)** $x^2 - xy$ **(c)** $5xy$ **(d)** $4x^2 - 2y^2$ **(e)** y^2

4. **(a)** $\begin{bmatrix} 2 & -\frac{3}{2} \\ -\frac{3}{2} & 4 \end{bmatrix}$ **(b)** $\begin{bmatrix} 1 & -\frac{1}{2} \\ -\frac{1}{2} & 0 \end{bmatrix}$ **(c)** $\begin{bmatrix} 0 & \frac{5}{2} \\ \frac{5}{2} & 0 \end{bmatrix}$ **(d)** $\begin{bmatrix} 4 & 0 \\ 0 & -2 \end{bmatrix}$ **(e)** $\begin{bmatrix} 0 & 0 \\ 0 & 1 \end{bmatrix}$

5. (a) $[x \quad y]\begin{bmatrix} 2 & -\frac{3}{2} \\ -\frac{3}{2} & 4 \end{bmatrix}\begin{bmatrix} x \\ y \end{bmatrix} + [-7 \quad 2]\begin{bmatrix} x \\ y \end{bmatrix} + 7 = 0$ **(b)** $[x \quad y]\begin{bmatrix} 1 & -\frac{1}{2} \\ -\frac{1}{2} & 0 \end{bmatrix}\begin{bmatrix} x \\ y \end{bmatrix} + [5 \quad 8]\begin{bmatrix} x \\ y \end{bmatrix} - 3 = 0$

(c) $[x \quad y]\begin{bmatrix} 0 & \frac{5}{2} \\ \frac{5}{2} & 0 \end{bmatrix}\begin{bmatrix} x \\ y \end{bmatrix} - 8 = 0$ **(d)** $[x \quad y]\begin{bmatrix} 4 & 0 \\ 0 & -2 \end{bmatrix}\begin{bmatrix} x \\ y \end{bmatrix} - 7 = 0$

(e) $[x \quad y]\begin{bmatrix} 0 & 0 \\ 0 & 1 \end{bmatrix}\begin{bmatrix} x \\ y \end{bmatrix} + [7 \quad -8]\begin{bmatrix} x \\ y \end{bmatrix} - 5 = 0$

6. (d) Hyperbola **(a)** Ellipse **(b)** Ellipse **(c)** Hyperbola **(e)** Circle
(f) Parabola **(g)** Parabola **(h)** Parabola **(i)** Parabola **(j)** Circle

7. (a) $9x'^2 + 4y'^2 = 36$, ellipse **(b)** $x'^2 - 16y'^2 = 16$, hyperbola
(c) $y'^2 = 8x'$, parabola **(d)** $x'^2 + y'^2 = 16$, circle
(e) $18y'^2 - 12x'^2 = 419$, hyperbola **(f)** $y' = -\frac{1}{7}x'^2$, parabola

8. (a) Hyperbola; possible equations are **(b)** Ellipse; possible equations are
$\quad 3x'^2 - 2y'^2 + 8 = 0, -2x'^2 + 3y'^2 + 8 = 0$ $\quad 7x'^2 + 3y'^2 = 9, 3x'^2 + 7y'^2 = 9$
(c) Hyperbola; possible equations are $4x'^2 - y'^2 = 3, 4y'^2 - x'^2 = 3$

9. $2x''^2 + y''^2 = 6$, ellipse **10.** $13y''^2 - 4x''^2 = 81$, hyperbola **11.** $2x''^2 - 3y''^2 = 24$, hyperbola

12. $6x''^2 + 11y''^2 = 66$, ellipse **13.** $4y''^2 - x''^2 = 0$, hyperbola **14.** $\sqrt{29}x'^2 - 3y' = 0$, parabola

15. (a) Two intersecting lines, $y = x$ and $y = -x$ **(b)** No graph
(c) The graph is the single point $(0, 0)$. **(d)** The graph is the line $y = x$.
(e) The graph consists of two parallel lines $\dfrac{3}{\sqrt{13}}x + \dfrac{2}{\sqrt{13}}y = \pm 2$. **(f)** The graph is the single point $(1, 2)$.

EXERCISE SET 9.7 [page 467]

1. (a) $x^2 + 2y^2 - z^2 + 4xy - 5yz$ **(b)** $3x^2 + 7z^2 + 2xy - 3xz + 4yz$ **(c)** $xy + xz + yz$
(d) $x^2 + y^2 - z^2$ **(e)** $3z^2 + 3xz$ **(f)** $2z^2 + 2xz + y^2$

2. (a) $\begin{bmatrix} 1 & 2 & 0 \\ 2 & 2 & -\frac{5}{2} \\ 0 & -\frac{5}{2} & -1 \end{bmatrix}$ **(b)** $\begin{bmatrix} 3 & 1 & -\frac{3}{2} \\ 1 & 0 & 2 \\ -\frac{3}{2} & 2 & 7 \end{bmatrix}$ **(c)** $\begin{bmatrix} 0 & \frac{1}{2} & \frac{1}{2} \\ \frac{1}{2} & 0 & \frac{1}{2} \\ \frac{1}{2} & \frac{1}{2} & 0 \end{bmatrix}$ **(d)** $\begin{bmatrix} 1 & 0 & 0 \\ 0 & 1 & 0 \\ 0 & 0 & -1 \end{bmatrix}$ **(e)** $\begin{bmatrix} 0 & 0 & \frac{3}{2} \\ 0 & 0 & 0 \\ \frac{3}{2} & 0 & 3 \end{bmatrix}$ **(f)** $\begin{bmatrix} 0 & 0 & 1 \\ 0 & 1 & 0 \\ 1 & 0 & 2 \end{bmatrix}$

3. (a) $[x \quad y \quad z]\begin{bmatrix} 1 & 2 & 0 \\ 2 & 2 & -\frac{5}{2} \\ 0 & -\frac{5}{2} & -1 \end{bmatrix}\begin{bmatrix} x \\ y \\ z \end{bmatrix} + [7 \quad 0 \quad 2]\begin{bmatrix} x \\ y \\ z \end{bmatrix} - 3 = 0$

(b) $[x \quad y \quad z]\begin{bmatrix} 3 & 1 & -\frac{3}{2} \\ 1 & 0 & 2 \\ -\frac{3}{2} & 2 & 7 \end{bmatrix}\begin{bmatrix} x \\ y \\ z \end{bmatrix} + [-3 \quad 0 \quad 0]\begin{bmatrix} x \\ y \\ z \end{bmatrix} - 4 = 0$

(c) $[x \quad y \quad z]\begin{bmatrix} 0 & \frac{1}{2} & \frac{1}{2} \\ \frac{1}{2} & 0 & \frac{1}{2} \\ \frac{1}{2} & \frac{1}{2} & 0 \end{bmatrix}\begin{bmatrix} x \\ y \\ z \end{bmatrix} - 1 = 0$ **(d)** $[x \quad y \quad z]\begin{bmatrix} 1 & 0 & 0 \\ 0 & 1 & 0 \\ 0 & 0 & -1 \end{bmatrix}\begin{bmatrix} x \\ y \\ z \end{bmatrix} - 7 = 0$

(e) $[x \quad y \quad z]\begin{bmatrix} 0 & 0 & \frac{3}{2} \\ 0 & 0 & 0 \\ \frac{3}{2} & 0 & 3 \end{bmatrix}\begin{bmatrix} x \\ y \\ z \end{bmatrix} + [0 \quad -14 \quad 0]\begin{bmatrix} x \\ y \\ z \end{bmatrix} + 9 = 0$

(f) $[x \quad y \quad z]\begin{bmatrix} 0 & 0 & 1 \\ 0 & 1 & 0 \\ 1 & 0 & 2 \end{bmatrix}\begin{bmatrix} x \\ y \\ z \end{bmatrix} + [2 \quad -1 \quad 3]\begin{bmatrix} x \\ y \\ z \end{bmatrix} = 0$

4. **(a)** Ellipsoid   **(b)** Hyperboloid of one sheet   **(c)** Hyperboloid of two sheets   **(d)** Elliptic cone
   (e) Elliptic paraboloid   **(f)** Hyperbolic paraboloid   **(g)** Sphere

5. **(a)** $9x'^2 + 36y'^2 + 4z'^2 = 36$, ellipsoid   **(b)** $6x'^2 + 3y'^2 - 2z'^2 = 18$, hyperboloid of one sheet
   (c) $3x'^2 - 3y'^2 - z'^2 = 3$, hyperboloid of two sheets   **(d)** $4x'^2 + 9y'^2 - z'^2 = 0$, elliptic cone
   (e) $x'^2 + 16y'^2 - 16z' = 0$, elliptic paraboloid   **(f)** $7x'^2 - 3y'^2 + z' = 0$, hyperbolic paraboloid
   (g) $x'^2 + y'^2 + z'^2 = 25$, sphere

6. **(a)** $25x'^2 - 3y'^2 - 50z'^2 - 150 = 0$, hyperboloid of two sheets
   (b) $2x'^2 + 2y'^2 + 8z'^2 - 5 = 0$, ellipsoid
   (c) $9x'^2 + 4y'^2 - 36z' = 0$, elliptic paraboloid
   (d) $x'^2 - y'^2 + z' = 0$, hyperbolic paraboloid

7. $x''^2 + y''^2 - 2z''^2 = -1$, hyperboloid of two sheets   **8.** $x''^2 + y''^2 + 2z''^2 = 4$, ellipsoid

9. $x''^2 - y''^2 + z'' = 0$, hyperbolic paraboloid

10. $6x''^2 + 3y''^2 - 8\sqrt{2}z'' = 0$, elliptic paraboloid

EXERCISE SET 9.8 [page 475]

1. Multiplications: mpn; additions: $mp(n - 1)$

2. Multiplications: $(k - 1)n^3$; additions: $(k - 1)(n^3 - n^2)$

3.

	$n = 5$	$n = 10$	$n = 100$	$n = 1000$
Solve $A\mathbf{x} = \mathbf{b}$ by Gauss–Jordan elimination	+: 50 ×: 65	+: 375 ×: 430	+: 383,250 ×: 343,300	+: 333,283,500 ×: 334,333,000
Solve $A\mathbf{x} = \mathbf{b}$ by Gaussian elimination	+: 50 ×: 65	+: 375 ×: 430	+: 383,250 ×: 343,300	+: 333,283,500 ×: 334,333,000
Find A^{-1} by reducing $[A \mid I]$ to $[I \mid A^{-1}]$	+: 80 ×: 125	+: 810 ×: 1000	+: 980,100 ×: 1,000,000	+: 998,001,000 ×: 1,000,000,000
Solve $A\mathbf{x} = \mathbf{b}$ as $\mathbf{x} = A^{-1}\mathbf{b}$	+: 100 ×: 150	+: 900 ×: 1100	+: 990,000 ×: 1,010,000	+: 999,000,000 ×: 1,001,000,000
Find $\det(A)$ by row reduction	+: 30 ×: 44	+: 285 ×: 339	+: 328,350 ×: 333,399	+: 332,833,500 ×: 333,333,999
Solve $A\mathbf{x} = \mathbf{b}$ by Cramer's Rule	+: 180 ×: 264	+: 3135 ×: 3729	+: 33,163,350 ×: 33,673,399	+: $33,316,633 \times 10^4$ ×: $33,366,733 \times 10^4$

4.

	$n = 5$ Execution Time (sec)	$n = 10$ Execution Time (sec)	$n = 100$ Execution Time (sec)	$n = 1000$ Execution Time (sec)
Solve $A\mathbf{x} = \mathbf{b}$ by Gauss–Jordan elimination	1.55×10^{-4}	1.05×10^{-3}	.878	836
Solve $A\mathbf{x} = \mathbf{b}$ by Gaussian elimination	1.55×10^{-4}	1.05×10^{-3}	.878	836
Find A^{-1} by reducing $[A \mid I]$ to $[I \mid A^{-1}]$	2.84×10^{-4}	2.41×10^{-3}	2.49	2499
Solve $A\mathbf{x} = \mathbf{b}$ as $\mathbf{x} = A^{-1}\mathbf{b}$	3.50×10^{-4}	2.65×10^{-3}	2.52	2502
Find $\det(A)$ by row reduction	1.03×10^{-4}	8.21×10^{-4}	.831	833
Solve $A\mathbf{x} = \mathbf{b}$ by Cramer's Rule	6.18×10^{-4}	90.3×10^{-4}	83.9	834×10^3

EXERCISE SET 9.9 [page 484]

1. $x_1 = 2, x_2 = 1$ 2. $x_1 = -2, x_2 = 1, x_3 = -3$ 3. $x_1 = 3, x_2 = -1$

4. $x_1 = 4, x_2 = -1$ 5. $x_1 = -1, x_2 = 1, x_3 = 0$ 6. $x_1 = 1, x_2 = -2, x_3 = 1$

7. $x_1 = -1, x_2 = 1, x_3 = 0$ 8. $x_1 = -1, x_2 = 1, x_3 = 1$

9. $x_1 = -3, x_2 = 1, x_3 = 2, x_4 = 1$ 10. $x_1 = 2, x_2 = -1, x_3 = 0, x_4 = 0$

11. (a) $A = LU = \begin{bmatrix} 2 & 0 & 0 \\ -2 & 1 & 0 \\ 2 & 1 & 1 \end{bmatrix} \begin{bmatrix} 1 & \frac{1}{2} & -\frac{1}{2} \\ 0 & 0 & 1 \\ 0 & 0 & 0 \end{bmatrix}$

(b) $A = L_1 DU = \begin{bmatrix} 1 & 0 & 0 \\ -1 & 1 & 0 \\ 1 & 1 & 1 \end{bmatrix} \begin{bmatrix} 2 & 0 & 0 \\ 0 & 1 & 0 \\ 0 & 0 & 1 \end{bmatrix} \begin{bmatrix} 1 & \frac{1}{2} & -\frac{1}{2} \\ 0 & 0 & 1 \\ 0 & 0 & 0 \end{bmatrix}$

(c) $A = L_2 U_2 = \begin{bmatrix} 1 & 0 & 0 \\ -1 & 1 & 0 \\ 1 & 1 & 1 \end{bmatrix} \begin{bmatrix} 2 & 1 & -1 \\ 0 & 0 & 1 \\ 0 & 0 & 0 \end{bmatrix}$

13. (b) $\begin{bmatrix} a & b \\ c & d \end{bmatrix} = \begin{bmatrix} 1 & 0 \\ \frac{c}{a} & 1 \end{bmatrix} \begin{bmatrix} a & b \\ 0 & \frac{ad - bc}{a} \end{bmatrix}$

14. Additions: $\frac{n^3}{3} + \frac{n^2}{2} - \frac{5n}{6}$; multiplications: $\frac{n^3}{3} + n^2 - \frac{n}{3}$

18. $A = PLU = \begin{bmatrix} 1 & 0 & 0 \\ 0 & 0 & 1 \\ 0 & 1 & 0 \end{bmatrix} \begin{bmatrix} 3 & 0 & 0 \\ 0 & 2 & 0 \\ 3 & 0 & 1 \end{bmatrix} \begin{bmatrix} 1 & -\frac{1}{3} & 0 \\ 0 & 1 & \frac{1}{2} \\ 0 & 0 & 1 \end{bmatrix}$

EXERCISE SET 10.1 [page 492]

1. (a–d)   2. (a) $(2, 3)$ (b) $(-4, 0)$ (c) $(-3, -2)$ (d) $(0, -5)$

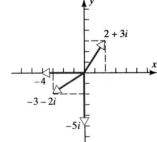

3. (a) $x = -2, y = -3$ (b) $x = 2, y = 1$

4. (a) $5 + 3i$ (b) $-3 - 7i$ (c) $4 - 8i$ (d) $-4 - 5i$ (e) $19 + 14i$ (f) $-\frac{11}{2} - \frac{17}{2}i$

5. (a) $2 + 3i$ (b) $-1 - 2i$ (c) $-2 + 9i$

6.

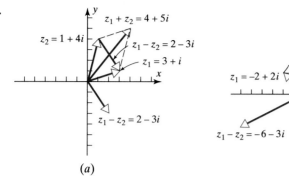

(a)

(b)

7.

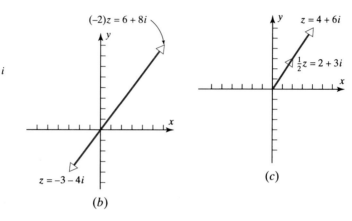

(a)

(b)

(c)

8. **(a)** $k_1 = -5, k_2 = 3$ **(b)** $k_1 = 3, k_2 = 1$

9. **(a)** $z_1 z_2 = 3 + 3i, z_1^2 = -9, z_2^2 = -2i$
   (b) $z_1 z_2 = 26, z_1^2 = -20 + 48i, z_2^2 = -5 - 12i$
   (c) $z_1 z_2 = \frac{11}{3} - i, z_1^2 = \frac{4}{9}(-3 + 4i), z_2^2 = -6 - \frac{5}{2}i$

10. **(a)** $9 - 8i$ **(b)** $-63 + 16i$ **(c)** $-32 - 24i$ **(d)** $22 + 19i$ 11. $76 - 88i$ 12. $26 - 18i$

13. $-26 + 18i$ 14. $-1 - 11i$ 15. $-\frac{63}{16} + i$ 16. $(2 + \sqrt{2}) + i(1 - \sqrt{2})$ 17. 0 18. $-24i$

19. **(a)** $\begin{bmatrix} 1 + 6i & -3 + 7i \\ 3 + 8i & 3 + 12i \end{bmatrix}$ **(b)** $\begin{bmatrix} 3 - 2i & 6 + 5i \\ 3 - 5i & 13 + 3i \end{bmatrix}$ **(c)** $\begin{bmatrix} 3 + 3i & 2 + 5i \\ 9 - 5i & 13 - 2i \end{bmatrix}$ **(d)** $\begin{bmatrix} 9 + i & 12 + 2i \\ 18 - 2i & 13 + i \end{bmatrix}$

20. **(a)** $\begin{bmatrix} 13 + 13i & -8 + 12i & -33 - 22i \\ 1 + i & 0 & i \\ 7 + 9i & -6 + 6i & -16 - 16i \end{bmatrix}$ **(b)** $\begin{bmatrix} 6 + 2i & -11 + 19i \\ -1 + 6i & -9 - 5i \end{bmatrix}$ **(c)** $\begin{bmatrix} 6i & 1 + i \\ -6 - i & 5 - 9i \end{bmatrix}$ **(d)** $\begin{bmatrix} 22 - 7i & 2 + 10i \\ -5 - 4i & 6 - 8i \\ 9 - i & -1 - i \end{bmatrix}$

22. **(a)** $z = -1 \pm i$ **(b)** $z = \frac{1}{2} \pm \frac{\sqrt{3}}{2}i$ 23. **(b)** i

EXERCISE SET 10.2 [page 498]

1. **(a)** $2 - 7i$ **(b)** $-3 + 5i$ **(c)** $-5i$ **(d)** i **(e)** -9 **(f)** 0

2. **(a)** 1 **(b)** 7 **(c)** 5 **(d)** $\sqrt{2}$ **(e)** 8 **(f)** 0

4. **(a)** $-\frac{17}{25} - \frac{19}{25}i$ **(b)** $\frac{23}{25} + \frac{11}{25}i$ **(c)** $\frac{23}{25} - \frac{11}{25}i$ **(d)** $-\frac{17}{25} + \frac{19}{25}i$ **(e)** $\frac{1}{5} - i$ **(f)** $\frac{\sqrt{26}}{5}$

5. **(a)** $-i$ **(b)** $\frac{1}{26} + \frac{5}{26}i$ **(c)** $7i$ 6. **(a)** $\frac{6}{5} + \frac{2}{5}i$ **(b)** $-\frac{2}{5} + \frac{1}{5}i$ **(c)** $\frac{3}{5} + \frac{11}{5}i$ **(d)** $\frac{3}{5} + \frac{1}{5}i$

7. $\frac{1}{2} + \frac{1}{2}i$ 8. $\frac{2}{5} + \frac{1}{5}i$ 9. $-\frac{7}{625} - \frac{24}{625}i$ 10. $-\frac{11}{25} + \frac{2}{25}i$ 11. $\frac{1 - \sqrt{3}}{4} + \frac{1 + \sqrt{3}}{4}i$

12. $-\frac{1}{26} - \frac{5}{26}i$ **13.** $-\frac{1}{10} + \frac{1}{10}i$ **14.** $-\frac{2}{5}$ **15. (a)** $-1 - 2i$ **(b)** $-\frac{3}{25} - \frac{4}{25}i$

17. (a)   **(b)**   **(c)**   **(d)**

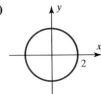 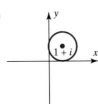 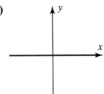 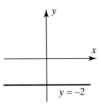

18. (a)   **(b)**   **(c)**   **(d)**

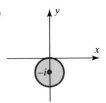 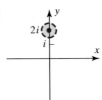 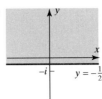

19. (a) $-y$ **(b)** $-x$ **(c)** y **(d)** x **20. (b)** $-i$ **23. (a)** $\dfrac{x_1 x_2 + y_1 y_2}{x_2^2 + y_2^2}$ **(b)** $\dfrac{x_2 y_1 - x_1 y_2}{x_2^2 + y_2^2}$

27. (c) Yes, if $z \ne 0$.   **28.** $x_1 = i, x_2 = -i$ **29.** $x_1 = 1 + i, x_2 = 1 - i$

30. $x_1 = \frac{1}{2} + i, x_2 = 2, x_3 = \frac{1}{2} - i$ **31.** $x_1 = i, x_2 = 0, x_3 = -i$

32. $x_1 = -(1 + i)t, x_2 = t$ **33.** $x_1 = (1 + i)t, x_2 = 2t$

34. $x_1 = -(1 - i)t, x_2 = -it, x_3 = t$ **35. (a)** $\begin{bmatrix} i & 2 \\ -1 & i \end{bmatrix}$ **(b)** $\begin{bmatrix} 0 & 1 \\ -i & 2i \end{bmatrix}$

39. (a) $\begin{bmatrix} -i & -2 - 2i & -1 + i \\ 1 & 2 & -i \\ i & i & 1 \end{bmatrix}$ **(b)** $\begin{bmatrix} 1 + i & -i & 1 \\ -7 + 6i & 5 - i & 1 + 4i \\ 1 + 2i & -i & 1 \end{bmatrix}$

41. (a) $|z_1 - z_2| = \sqrt{(a_1 - a_2)^2 + (b_1 - b_2)^2}$
  (b) $|(8 + 8i) - 12|^2 = 80 = 40 + 40 = |12 - (6 + 2i)|^2 + |(6 + 2i) - (8 + 8i)|^2$

EXERCISE SET 10.3 [page 505]

1. (a) 0 **(b)** $\pi/2$ **(c)** $-\pi/2$ **(d)** $\pi/4$ **(e)** $2\pi/3$ **(f)** $-\pi/4$

2. (a) $5\pi/3$ **(b)** $-\pi/3$ **(c)** $5\pi/3$

3. (a) $2\left[\cos\left(\dfrac{\pi}{2}\right) + i \sin\left(\dfrac{\pi}{2}\right)\right]$ **(b)** $4[\cos \pi + i \sin \pi]$ **(c)** $5\sqrt{2}\left[\cos\left(\dfrac{\pi}{4}\right) + i \sin\left(\dfrac{\pi}{4}\right)\right]$

  (d) $12\left[\cos\left(\dfrac{2\pi}{3}\right) + i \sin\left(\dfrac{2\pi}{3}\right)\right]$ **(e)** $3\sqrt{2}\left[\cos\left(-\dfrac{3\pi}{4}\right) + i \sin\left(-\dfrac{3\pi}{4}\right)\right]$ **(f)** $4\left[\cos\left(-\dfrac{\pi}{6}\right) + i \sin\left(-\dfrac{\pi}{6}\right)\right]$

4. (a) $6\left[\cos\left(\dfrac{5\pi}{12}\right) + i \sin\left(\dfrac{5\pi}{12}\right)\right]$ **(b)** $\dfrac{2}{3}\left[\cos\left(\dfrac{\pi}{12}\right) + i \sin\left(\dfrac{\pi}{12}\right)\right]$

  (c) $\dfrac{3}{2}\left[\cos\left(-\dfrac{\pi}{12}\right) + i \sin\left(-\dfrac{\pi}{12}\right)\right]$ **(d)** $\dfrac{32}{9}\left[\cos\left(\dfrac{11\pi}{12}\right) + i \sin\left(\dfrac{11\pi}{12}\right)\right]$

5. 1 **6. (a)** -64 **(b)** $-i$ **(c)** $-64\sqrt{3} - 64i$ **(d)** $-\dfrac{1 + \sqrt{3}i}{2048}$

7. (a)

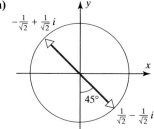

(b)

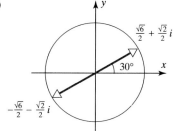

(c)

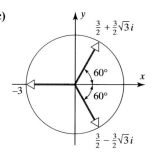

(d)

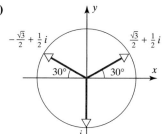

(e)

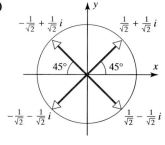

(f)

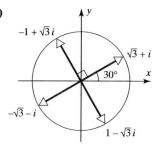

8.

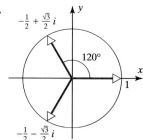

9.

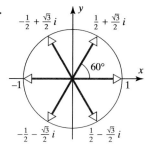

10. $\sqrt[4]{2}\left[\cos\left(\dfrac{\pi}{8}\right) + i\sin\left(\dfrac{\pi}{8}\right)\right]$, $\sqrt[4]{2}\left[\cos\left(\dfrac{9\pi}{8}\right) + i\sin\left(\dfrac{9\pi}{8}\right)\right]$ **11.** $\pm 2, \pm 2i$

12. The roots are $\pm(2^{1/4} + 2^{1/4}i)$, $\pm(2^{1/4} - 2^{1/4}i)$ and the factorization is
$z^4 + 8 = (z^2 - 2^{5/4}z + 2^{3/2}) \cdot (z^2 + 2^{5/4}z + 2^{3/2})$.

13. Rotates z clockwise by $90°$ **14. (a)** 16 **(b)** $\dfrac{i}{4^9}$

15. (a) $\mathrm{Re}(z) = -3$, $\mathrm{Im}(z) = 0$ **(b)** $\mathrm{Re}(z) = -3$, $\mathrm{Im}(z) = 0$
(c) $\mathrm{Re}(z) = 0$, $\mathrm{Im}(z) = -\sqrt{2}$ **(d)** $\mathrm{Re}(z) = -3$, $\mathrm{Im}(z) = 0$

20. $\cos 2\theta = \cos^2\theta - \sin^2\theta$, $\sin 2\theta = 2\sin\theta\cos\theta$
$\cos 3\theta = \cos^3\theta - 3\sin^2\theta\cos\theta$, $\sin 3\theta = 3\sin\theta\cos^2\theta - \sin^3\theta$

EXERCISE SET 10.4 [page 510]

1. (a) $(3i, -i, -2 - i, 4)$ **(b)** $(3 + 2i, -1 - 2i, -3 + 5i, -i)$
(c) $(-1 - 2i, 2i, 2 - i, -1)$ **(d)** $(-3 + 9i, 3 - 3i, -3 - 6i, 12 + 3i)$
(e) $(-3 + 2i, 3, -3 - 3i, i)$ **(f)** $(-1 - 5i, 3i, 4, -5)$

2. $(2 + i, 0, -3 + i, -4i)$ **3.** $c_1 = -2 - i, c_2 = 0, c_3 = 2 - i$

5. (a) $\sqrt{2}$ **(b)** $2\sqrt{3}$ **(c)** $\sqrt{10}$ **(d)** $\sqrt{37}$

6. (a) $\sqrt{43}$ **(b)** $\sqrt{10} + \sqrt{29}$ **(c)** $\sqrt{10} + \sqrt{10}i$ **(d)** $\sqrt{699}$ **(e)** $\left(\dfrac{1 + i}{\sqrt{6}}, \dfrac{2i}{\sqrt{6}}, 0\right)$ **(f)** 1

8. All k such that $|k| = \frac{1}{5}$ **9. (a)** 3 **(b)** $2 - 27i$ **(c)** $-5 - 10i$

10. The set is a vector space under the given operations.

11. Not a vector space. Axiom 6 fails; that is, the set is not closed under scalar multiplication. (Multiply by i, for example.)

12. (a) **13.** ker T is all multiples of $\begin{bmatrix} 1+3i \\ 1+i \\ -2 \end{bmatrix}$; nullity of $T = 1$

14. (b) **15.** (a), (d) **16.** (a), (b), (d)

17. (a) $(-3-2i)\mathbf{u} + (3-i)\mathbf{v} + (1+2i)\mathbf{w}$ (b) $(2+i)\mathbf{u} + (-1+i)\mathbf{v} + (-1-i)\mathbf{w}$
(c) $0\mathbf{u} + 0\mathbf{v} + 0\mathbf{w}$ (d) $(-5-4i)\mathbf{u} + (5+2i)\mathbf{v} + (2+4i)\mathbf{w}$

18. (a) (b) No (c) Yes (d) No **19.** (a), (b), (c)

20. (a) $\mathbf{u}_2 = i\mathbf{u}_1$ (b) Three vectors in a two-dimensional space
(c) A is a scalar multiple of B.

21. (b), (c) **22.** $\mathbf{f} - 3\mathbf{g} - 3\mathbf{h} = \mathbf{0}$

23. (a) Three vectors in a two-dimensional space (b) Two vectors in a three-dimensional space

24. (a), (b) **25.** (a), (b), (c), (d) **26.** $(-1-i, 1)$; dimension $= 1$

27. $(1, 1-i)$; dimension $= 1$ **28.** $(3+6i, -3i, 1)$; dimension $= 1$

29. $(\frac{5}{2}i, -\frac{1}{2}, 1, 0), (-\frac{1}{4}, \frac{3}{4}i, 0, 1)$; dimension $= 2$

EXERCISE SET 10.5 [page 517]

2. (a) -12 (b) 0 (c) $2i$ (d) 37 **4.** (a) $-4+5i$ (b) 0 (c) $4-4i$ (d) 42

5. (a) Axiom 4 fails. (b) Axiom 4 fails. (c) Axioms 2 and 3 fail.
(d) Axioms 1 and 4 fail. (e) This is an inner product.

6. $-9-5i$ **7.** No. Axioms 1 and 4 fail. **8.** No. Axiom 4 fails. **9.** (a) $\sqrt{21}$ (b) $\sqrt{10}$ (c) $\sqrt{10}$ (d) 0

10. (a) $\sqrt{10}$ (b) 2 (c) $\sqrt{5}$ (d) 0 **11.** (a) $\sqrt{2}$ (b) $2\sqrt{3}$ (c) 5 (d) 0

12. (a) $3\sqrt{10}$ (b) $\sqrt{14}$ **13.** (a) $\sqrt{10}$ (b) $2\sqrt{5}$ **14.** (a) 2 (b) $2\sqrt{2}$

15. (a) $2\sqrt{3}$ (b) $2\sqrt{2}$ **16.** (a) $7\sqrt{2}$ (b) $2\sqrt{3}$ **17.** (a) $-\frac{8}{3}i$ (b) None

18. (a), (b), (c) **20.** (b) **21.** (b), (c)

23. $\left(\dfrac{i}{\sqrt{2}}, 0, 0, \dfrac{i}{\sqrt{2}} \right), \left(-\dfrac{i}{\sqrt{6}}, 0, \dfrac{2i}{\sqrt{6}}, \dfrac{i}{\sqrt{6}} \right), \left(\dfrac{2i}{\sqrt{21}}, \dfrac{3i}{\sqrt{21}}, \dfrac{2i}{\sqrt{21}}, \dfrac{-2i}{\sqrt{21}} \right), \left(-\dfrac{i}{\sqrt{7}}, \dfrac{2i}{\sqrt{7}}, -\dfrac{i}{\sqrt{7}}, \dfrac{i}{\sqrt{7}} \right)$

24. (a) $\mathbf{v}_1 = \left(\dfrac{i}{\sqrt{10}}, -\dfrac{3i}{\sqrt{10}} \right), \mathbf{v}_2 = \left(\dfrac{3i}{\sqrt{10}}, \dfrac{i}{\sqrt{10}} \right)$ (b) $\mathbf{v}_1 = (i, 0), \mathbf{v}_2 = (0, -i)$

25. (a) $\mathbf{v}_1 = \left(\dfrac{i}{\sqrt{3}}, \dfrac{i}{\sqrt{3}}, \dfrac{i}{\sqrt{3}} \right), \mathbf{v}_2 = \left(-\dfrac{i}{\sqrt{2}}, \dfrac{i}{\sqrt{2}}, 0 \right), \mathbf{v}_3 = \left(\dfrac{i}{\sqrt{6}}, \dfrac{i}{\sqrt{6}}, -\dfrac{2i}{\sqrt{6}} \right)$

(b) $\mathbf{v}_1 = (i, 0, 0), \mathbf{v}_2 = \left(0, \dfrac{7i}{\sqrt{53}}, \dfrac{-2i}{\sqrt{53}} \right), \mathbf{v}_3 = \left(0, \dfrac{2i}{\sqrt{53}}, \dfrac{7i}{\sqrt{53}} \right)$

26. $\left(0, \dfrac{2i}{\sqrt{5}}, \dfrac{i}{\sqrt{5}}, 0 \right), \left(\dfrac{5i}{\sqrt{30}}, -\dfrac{i}{\sqrt{30}}, \dfrac{2i}{\sqrt{30}}, 0 \right), \left(\dfrac{i}{\sqrt{10}}, \dfrac{i}{\sqrt{10}}, -\dfrac{2i}{\sqrt{10}}, -\dfrac{2i}{\sqrt{10}} \right), \left(\dfrac{i}{\sqrt{15}}, \dfrac{i}{\sqrt{15}}, -\dfrac{2i}{\sqrt{15}}, \dfrac{3i}{\sqrt{15}} \right)$

27. $\mathbf{v}_1 = \left(0, \dfrac{i}{\sqrt{3}}, \dfrac{1-i}{\sqrt{3}} \right), \mathbf{v}_2 = \left(-\dfrac{3i}{\sqrt{15}}, \dfrac{2}{\sqrt{15}}, \dfrac{1+i}{\sqrt{15}} \right)$

28. $\mathbf{w}_1 = \left(-\dfrac{5i}{4}, -\dfrac{i}{4}, \dfrac{5i}{4}, \dfrac{9i}{4} \right), \mathbf{w}_2 = \left(\dfrac{i}{4}, \dfrac{9i}{4}, \dfrac{19i}{4}, -\dfrac{9i}{4} \right)$

36. $\mathbf{u} = -\sqrt{3}i\mathbf{v}_1 + \dfrac{3}{\sqrt{6}}\mathbf{v}_2 - \dfrac{1}{\sqrt{2}}\mathbf{v}_3$ **40. (a)** $\dfrac{2\sqrt{3}}{3}$ **(b)** $\dfrac{1}{2} - \dfrac{i}{3}$ **(c)** $\dfrac{1}{2} + \dfrac{i}{3}$

EXERCISE SET 10.6 [page 526]

1. (a) $\begin{bmatrix} -2i & 4 & 5-i \\ 1+i & 3-i & 0 \end{bmatrix}$ **(b)** $\begin{bmatrix} -2i & 4 & -i \\ 1+i & 5+7i & 3 \\ -1-i & i & 1 \end{bmatrix}$ **(c)** $\begin{bmatrix} -7i \\ 0 \\ 3i \end{bmatrix}$ **(d)** $\begin{bmatrix} \bar{a}_{11} & \bar{a}_{21} \\ \bar{a}_{12} & \bar{a}_{22} \\ \bar{a}_{13} & \bar{a}_{23} \end{bmatrix}$

2. (b), (d), (e)   **3.** $k = 3 + 5i,\ l = i,\ m = 2 - 4i$ **4.** (a), (b)

5. (a) $A^{-1} = \begin{bmatrix} \dfrac{3}{5} & -\dfrac{4}{5} \\ -\dfrac{4}{5}i & -\dfrac{3}{5}i \end{bmatrix}$ **(b)** $A^{-1} = \begin{bmatrix} \dfrac{1}{\sqrt{2}} & \dfrac{-1+i}{2} \\ \dfrac{1}{\sqrt{2}} & \dfrac{1-i}{2} \end{bmatrix}$

(c) $A^{-1} = \begin{bmatrix} \dfrac{1}{2\sqrt{2}}(\sqrt{3} - i) & \dfrac{1}{2\sqrt{2}}(1 - \sqrt{3}i) \\ \dfrac{1}{2\sqrt{2}}(1 + \sqrt{3}i) & \dfrac{1}{2\sqrt{2}}(-\sqrt{3} - i) \end{bmatrix}$ **(d)** $A^{-1} = \begin{bmatrix} \dfrac{1-i}{2} & -\dfrac{i}{\sqrt{3}} & \dfrac{3-i}{2\sqrt{15}} \\ -\dfrac{1}{2} & \dfrac{1}{\sqrt{3}} & \dfrac{4-3i}{2\sqrt{15}} \\ \dfrac{1}{2} & \dfrac{i}{\sqrt{3}} & -\dfrac{5i}{2\sqrt{15}} \end{bmatrix}$

7. $P = \begin{bmatrix} \dfrac{-1+i}{\sqrt{3}} & \dfrac{1-i}{\sqrt{6}} \\ \dfrac{1}{\sqrt{3}} & \dfrac{2}{\sqrt{6}} \end{bmatrix}$; $P^{-1}AP = \begin{bmatrix} 3 & 0 \\ 0 & 6 \end{bmatrix}$ **8.** $P = \begin{bmatrix} -\dfrac{i}{\sqrt{2}} & \dfrac{i}{\sqrt{2}} \\ \dfrac{1}{\sqrt{2}} & \dfrac{1}{\sqrt{2}} \end{bmatrix}$; $P^{-1}AP = \begin{bmatrix} 4 & 0 \\ 0 & 2 \end{bmatrix}$

9. $P = \begin{bmatrix} -\dfrac{1+i}{\sqrt{6}} & \dfrac{1+i}{\sqrt{3}} \\ \dfrac{2}{\sqrt{6}} & \dfrac{1}{\sqrt{3}} \end{bmatrix}$; $P^{-1}AP = \begin{bmatrix} 2 & 0 \\ 0 & 8 \end{bmatrix}$ **10.** $P = \begin{bmatrix} -\dfrac{2}{\sqrt{14}} & \dfrac{5}{\sqrt{35}} \\ \dfrac{3-i}{\sqrt{14}} & \dfrac{3-i}{\sqrt{35}} \end{bmatrix}$; $P^{-1}AP = \begin{bmatrix} -5 & 0 \\ 0 & 2 \end{bmatrix}$

11. $P = \begin{bmatrix} 0 & 1 & 0 \\ -\dfrac{1-i}{\sqrt{6}} & 0 & \dfrac{1-i}{\sqrt{3}} \\ \dfrac{2}{\sqrt{6}} & 0 & \dfrac{1}{\sqrt{3}} \end{bmatrix}$; $P^{-1}AP = \begin{bmatrix} 1 & 0 & 0 \\ 0 & 5 & 0 \\ 0 & 0 & -2 \end{bmatrix}$

12. $P = \begin{bmatrix} \dfrac{i}{\sqrt{2}} & 0 & -\dfrac{i}{\sqrt{2}} \\ -\dfrac{1}{2} & \dfrac{1}{\sqrt{2}} & -\dfrac{1}{2} \\ \dfrac{1}{2} & \dfrac{1}{\sqrt{2}} & \dfrac{1}{2} \end{bmatrix}$; $P^{-1}AP = \begin{bmatrix} 1 & 0 & 0 \\ 0 & 2 & 0 \\ 0 & 0 & 3 \end{bmatrix}$

13. $\lambda = 2 \pm i\sqrt{15}$; no, since A has complex entries.   **14.** $\begin{bmatrix} 0 & i \\ -i & 0 \end{bmatrix}$ is one possibility.

SUPPLEMENTARY EXERCISES [page 528]

3. $\begin{bmatrix} -i \\ 1 \\ 0 \end{bmatrix}, \begin{bmatrix} 1 \\ 0 \\ 1 \end{bmatrix}$ is one possibility.   **5.** $\lambda = 1,\ \omega,\ \omega^2\ (= \bar{\omega})$

EXERCISE SET 11.1 (page 536)

1. **(a)** $y = 3x - 4$ **(b)** $y = -2x + 1$

2. **(a)** $x^2 + y^2 - 4x - 6y + 4 = 0$ or $(x - 2)^2 + (y - 3)^2 = 9$
   (b) $x^2 + y^2 + 2x - 4y - 20 = 0$ or $(x + 1)^2 + (y - 2)^2 = 25$

3. $x^2 + 2xy + y^2 - 2x + y = 0$ (a parabola) 4. **(a)** $x + 2y + z = 0$ **(b)** $-x + y - 2z + 1 = 0$

5. **(a)** $x^2 + y^2 + z^2 - 2x - 4y - 2z = -2$ or $(x - 1)^2 + (y - 2)^2 + (z - 1)^2 = 4$
   (b) $x^2 + y^2 + z^2 - 2x - 2y = 3$ or $(x - 1)^2 + (y - 1)^2 + z^2 = 5$

9. $\begin{vmatrix} y & x^2 & x & 1 \\ y_1 & x_1^2 & x_1 & 1 \\ y_2 & x_2^2 & x_2 & 1 \\ y_3 & x_3^2 & x_3 & 1 \end{vmatrix} = 0$

EXERCISE SET 11.2 (page 540)

1. $I_1 = \frac{255}{317}$, $I_2 = \frac{97}{317}$, $I_3 = \frac{158}{317}$ 2. $I_1 = \frac{13}{5}$, $I_2 = -\frac{2}{5}$, $I_3 = \frac{11}{5}$

3. $I_1 = -\frac{5}{22}$, $I_2 = \frac{7}{22}$, $I_3 = \frac{6}{11}$ 4. $I_1 = \frac{1}{2}$, $I_2 = 0$, $I_3 = 0$, $I_4 = \frac{1}{2}$, $I_5 = \frac{1}{2}$, $I_6 = \frac{1}{2}$

EXERCISE SET 11.3 (page 552)

1. $x_1 = 2, x_2 = \frac{2}{3}$; maximum value of $z = \frac{22}{3}$ 2. No feasible solutions 3. Unbounded solution

4. Invest $6000 in bond A and $4000 in bond B; the annual yield is $880.

5. $\frac{7}{9}$ cup of milk, $\frac{25}{18}$ ounces of corn flakes; minimum cost $= \frac{335}{18} \approx 18.6¢$

6. 550 containers from company A and 300 containers from company B;
   maximum shipping charges $= \$2110$

7. 925 containers from company A and no containers from company B;
   maximum shipping charges $= \$2312.50$

8. 0.4 pound of ingredient A and 2.4 pounds of ingredient B; minimum cost $= 24.8¢$

EXERCISE SET 11.4 (page 562)

1. **(a)** $\begin{bmatrix} 17 & 4 & 10 \\ 15 & 5 & 8 \\ 18 & 7 & 11 \end{bmatrix}$ or $\begin{bmatrix} 17 & 4 & 10 \\ 15 & 5 & 8 \\ 18 & 7 & 11 \end{bmatrix}$ **(b)** $\begin{bmatrix} 3 & -2 & 0 & 1 \\ 5 & 3 & -3 & 4 \\ 2 & 7 & 5 & 3 \\ 5 & -2 & 0 & 1 \end{bmatrix}$ or $\begin{bmatrix} 3 & -2 & 0 & 1 \\ 5 & 3 & -3 & 4 \\ 2 & 7 & 5 & 3 \\ 5 & -2 & 0 & 1 \end{bmatrix}$
   cost $= 30$ cost $= 30$ cost $= -2$ cost $= -2$

   (c) $\begin{bmatrix} 12 & 9 & 7 & 7 & 10 \\ 15 & 11 & 8 & 13 & 14 \\ 9 & 6 & 5 & 12 & 12 \\ 6 & 9 & 13 & 7 & 10 \\ 8 & 13 & 12 & 9 & 13 \end{bmatrix}$ or $\begin{bmatrix} 12 & 9 & 7 & 7 & 10 \\ 15 & 11 & 8 & 13 & 14 \\ 9 & 6 & 5 & 12 & 12 \\ 6 & 9 & 13 & 7 & 10 \\ 8 & 13 & 12 & 9 & 13 \end{bmatrix}$ or $\begin{bmatrix} 12 & 9 & 7 & 7 & 10 \\ 15 & 11 & 8 & 13 & 14 \\ 9 & 6 & 5 & 12 & 12 \\ 6 & 9 & 13 & 7 & 10 \\ 8 & 13 & 12 & 9 & 13 \end{bmatrix}$
   cost $= 39$ cost $= 39$ cost $= 39$

3. $\begin{bmatrix} 15 & 0 & 0 & 0 \\ 0 & 50 & 20 & 25 \\ 35 & 5 & 0 & 10 \\ 0 & 65 & 50 & 65 \end{bmatrix}$ 4. $\begin{bmatrix} 0 & 3 & 0 & 4 & 0 \\ 4 & 0 & 6 & 1 & 5 \\ 3 & 1 & 0 & 4 & 0 \\ 0 & 1 & 6 & 2 & 1 \\ 0 & 0 & 0 & 0 & 3 \end{bmatrix}$

5. (a) Bidder 1 receives coin 3 ($210).
   Bidder 2 receives coin 1 ($175).
   Bidder 3 receives coin 2 ($85).
   Bidder 4 receives nothing.
   Bidder 5 receives coin 4 ($160).

   Sum of bids = $630

(b) Repeat bidder 2's row twice in the cost matrix; add two dummy coins to end up with a 6 × 6 cost matrix. Optimal assignment is
   Bidder 1 receives nothing.
   Bidder 2 receives coins 1 and 3 ($175 + $230).
   Bidder 3 receives coin 2 ($85).
   Bidder 4 receives nothing.
   Bidder 5 receives coin 4 ($160).

   Sum of bids = $650

6. P—Pete   2B—Jill or Ann   LF—Herb
   C—Liz   3B—Alex   CF—John or Lois
   1B—Sam   SS—Ann or Jill   RF—Lois or John

EXERCISE SET 11.5 (page 574)

2. (a) $S(x) = -.12643(x - .4)^3 - .20211(x - .4)^2 + .92158(x - .4) + .38942$
(b) $S(.5) = .47943$; error = 0%

3. (a) The cubic runout spline   **(b)** $S(x) = 3x^3 - 2x^2 + 5x + 1$

4. $S(x) = \begin{cases} -.00000042(x + 10)^3 & + .000214(x + 10) + .99815, & -10 \le x \le 0 \\ .00000024(x)^3 - .0000126(x)^2 & + .000088(x) + .99987, & 0 \le x \le 10 \\ -.00000004(x - 10)^3 - .0000054(x - 10)^2 & - .000092(x - 10) + .99973, & 10 \le x \le 20 \\ .00000022(x - 20)^3 - .0000066(x - 20)^2 & - .000212(x - 20) + .99823, & 20 \le x \le 30 \end{cases}$

Maximum at $(x, S(x)) = (3.93, 1.00004)$

5. $S(x) = \begin{cases} .00000009(x + 10)^3 - .0000121(x + 10)^2 + .000282(x + 10) + .99815, & -10 \le x \le 0 \\ .00000009(x)^3 - .0000093(x)^2 + .000070(x) + .99987, & 0 \le x \le 10 \\ .00000004(x - 10)^3 - .0000066(x - 10)^2 - .000087(x - 10) + .99973, & 10 \le x \le 20 \\ .00000004(x - 20)^3 - .0000053(x - 20)^2 - .000207(x - 20) + .99823, & 20 \le x \le 30 \end{cases}$

Maximum at $(x, S(x)) = (4.00, 1.00001)$

6. (b)
$$\begin{bmatrix} 4 & 1 & 0 & 0 & \cdots & 0 & 0 & 0 & 1 \\ 1 & 4 & 1 & 0 & \cdots & 0 & 0 & 0 & 0 \\ 0 & 1 & 4 & 1 & \cdots & 0 & 0 & 0 & 0 \\ \vdots & \vdots & \vdots & \vdots & & \vdots & \vdots & \vdots & \vdots \\ 0 & 0 & 0 & 0 & \cdots & 0 & 1 & 4 & 1 \\ 1 & 0 & 0 & 0 & \cdots & 0 & 0 & 1 & 4 \end{bmatrix} \begin{bmatrix} M_1 \\ M_2 \\ M_3 \\ \vdots \\ M_{n-2} \\ M_{n-1} \end{bmatrix} = \frac{6}{h^2} \begin{bmatrix} y_{n-1} - 2y_1 + y_2 \\ y_1 - 2y_2 + y_3 \\ y_2 - 2y_3 + y_4 \\ \vdots \\ y_{n-3} - 2y_{n-2} + y_{n-1} \\ y_{n-2} - 2y_{n-1} + y_1 \end{bmatrix}$$

7. (b)
$$\begin{bmatrix} 2 & 1 & 0 & 0 & \cdots & 0 & 0 & 0 & 1 \\ 1 & 4 & 1 & 0 & \cdots & 0 & 0 & 0 & 0 \\ 0 & 1 & 4 & 1 & \cdots & 0 & 0 & 0 & 0 \\ \vdots & \vdots & \vdots & \vdots & & \vdots & \vdots & \vdots & \vdots \\ 0 & 0 & 0 & 0 & \cdots & 0 & 1 & 4 & 1 \\ 0 & 0 & 0 & 0 & \cdots & 0 & 0 & 1 & 2 \end{bmatrix} \begin{bmatrix} M_1 \\ M_2 \\ M_3 \\ \vdots \\ M_{n-1} \\ M_n \end{bmatrix} = \frac{6}{h^2} \begin{bmatrix} -hy_1' - y_1 + y_2 \\ y_1 - 2y_2 + y_3 \\ y_2 - 2y_3 + y_4 \\ \vdots \\ y_{n-2} - 2y_{n-1} + y_n \\ y_{n-1} - y_n + hy_n' \end{bmatrix}$$

EXERCISE SET 11.6 (page 585)

1. (a) $\mathbf{x}^{(1)} = \begin{bmatrix} .4 \\ .6 \end{bmatrix}$, $\mathbf{x}^{(2)} = \begin{bmatrix} .46 \\ .54 \end{bmatrix}$, $\mathbf{x}^{(3)} = \begin{bmatrix} .454 \\ .546 \end{bmatrix}$, $\mathbf{x}^{(4)} = \begin{bmatrix} .4546 \\ .5454 \end{bmatrix}$, $\mathbf{x}^{(5)} = \begin{bmatrix} .45454 \\ .54546 \end{bmatrix}$

(b) P is regular since all entries of P are positive; $\mathbf{q} = \begin{bmatrix} \frac{5}{11} \\ \frac{6}{11} \end{bmatrix}$

2. (a) $\mathbf{x}^{(1)} = \begin{bmatrix} .7 \\ .2 \\ .1 \end{bmatrix}$, $\mathbf{x}^{(2)} = \begin{bmatrix} .23 \\ .52 \\ .25 \end{bmatrix}$, $\mathbf{x}^{(3)} = \begin{bmatrix} .273 \\ .396 \\ .331 \end{bmatrix}$

3. (a) $\begin{bmatrix} \frac{9}{17} \\ \frac{8}{17} \end{bmatrix}$ **(b)** $\begin{bmatrix} \frac{26}{45} \\ \frac{19}{45} \end{bmatrix}$ **(c)** $\begin{bmatrix} \frac{3}{19} \\ \frac{4}{19} \\ \frac{12}{19} \end{bmatrix}$

(b) P is regular, since all entries of P are positive: $\mathbf{q} = \begin{bmatrix} \frac{22}{72} \\ \frac{29}{72} \\ \frac{21}{72} \end{bmatrix}$

4. (a) $P^n = \begin{bmatrix} \left(\frac{1}{2}\right)^n & 0 \\ 1 - \left(\frac{1}{2}\right)^n & 1 \end{bmatrix}$, $n = 1, 2, \ldots$. Thus, no integer power of P has all positive entries.

(b) $P^n \to \begin{bmatrix} 0 & 0 \\ 1 & 1 \end{bmatrix}$ as n increases, so $P^n \mathbf{x}^{(0)} \to \begin{bmatrix} 0 \\ 1 \end{bmatrix}$ for any $\mathbf{x}^{(0)}$ as n increases.

(c) The entries of the limiting vector $\begin{bmatrix} 0 \\ 1 \end{bmatrix}$ are not all positive.

6. $P^2 = \begin{bmatrix} \frac{1}{2} & \frac{1}{4} & \frac{1}{4} \\ \frac{1}{4} & \frac{1}{2} & \frac{1}{4} \\ \frac{1}{4} & \frac{1}{4} & \frac{1}{2} \end{bmatrix}$ has all positive entries; $\mathbf{q} = \begin{bmatrix} \frac{1}{3} \\ \frac{1}{3} \\ \frac{1}{3} \end{bmatrix}$ **7.** $\frac{10}{13}$

8. $54\frac{1}{6}\%$ in region 1, $16\frac{2}{3}\%$ in region 2, and $29\frac{1}{6}\%$ in region 3

EXERCISE SET 11.7 [page 596]

1. (a) $\begin{bmatrix} 0 & 0 & 0 & 1 \\ 1 & 0 & 1 & 1 \\ 1 & 1 & 0 & 1 \\ 0 & 0 & 0 & 0 \end{bmatrix}$ **(b)** $\begin{bmatrix} 0 & 1 & 1 & 0 & 0 \\ 0 & 0 & 0 & 0 & 1 \\ 1 & 0 & 0 & 1 & 0 \\ 0 & 0 & 1 & 0 & 0 \\ 0 & 0 & 1 & 0 & 0 \end{bmatrix}$ **(c)** $\begin{bmatrix} 0 & 1 & 0 & 1 & 0 & 0 \\ 1 & 0 & 0 & 0 & 0 & 0 \\ 0 & 1 & 0 & 1 & 1 & 1 \\ 0 & 0 & 0 & 0 & 0 & 1 \\ 0 & 0 & 0 & 0 & 0 & 1 \\ 0 & 0 & 1 & 0 & 1 & 0 \end{bmatrix}$

2. (a) 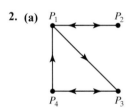 **(b)** 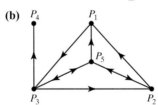 **(c)**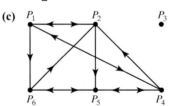

3. (a)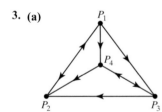

(b) 1-step: $P_1 \to P_2$

2-step: $P_1 \to P_4 \to P_2$
$P_1 \to P_3 \to P_2$

3-step: $P_1 \to P_2 \to P_1 \to P_2$
$P_1 \to P_3 \to P_4 \to P_2$
$P_1 \to P_4 \to P_3 \to P_2$

(c) 1-step: $P_1 \to P_4$

2-step: $P_1 \to P_3 \to P_4$

3-step: $P_1 \to P_2 \to P_1 \to P_4$
$P_1 \to P_4 \to P_3 \to P_4$

4. (a) $\{P_1, P_2, P_3\}$ **(b)** $\{P_3, P_4, P_5\}$ **(c)** $\{P_2, P_4, P_6, P_8\}$ and $\{P_4, P_5, P_6\}$

5. (a) None **(b)** $\{P_3, P_4, P_6\}$

6. $\begin{bmatrix} 0 & 0 & 1 & 1 \\ 1 & 0 & 0 & 0 \\ 0 & 1 & 0 & 1 \\ 0 & 1 & 0 & 0 \end{bmatrix}$ Power of $P_1 = 5$
Power of $P_2 = 3$
Power of $P_3 = 4$
Power of $P_4 = 2$

7. First, A; second, B and E (tie); fourth, C; fifth, D

EXERCISE SET 11.8 (page 606)

1. (a) $-5/8$ **(b)** $[0 \ 1 \ 0]$ **(c)** $[1 \ 0 \ 0 \ 0]^T$ **2.** Let $A = \begin{bmatrix} 1 & 1 \\ 1 & 1 \end{bmatrix}$, for example.

3. (a) $\mathbf{p}^* = [0 \ 1]$, $\mathbf{q}^* = \begin{bmatrix} 0 \\ 1 \end{bmatrix}$, $v = 3$ **(b)** $\mathbf{p}^* = [0 \ 1 \ 0]$, $\mathbf{q}^* = \begin{bmatrix} 1 \\ 0 \end{bmatrix}$, $v = 2$

(c) $\mathbf{p}^* = [0 \ 0 \ 1]$, $\mathbf{q}^* = \begin{bmatrix} 0 \\ 1 \\ 0 \end{bmatrix}$, $v = 2$ **(d)** $\mathbf{p}^* = [0 \ 1 \ 0 \ 0]$, $\mathbf{q}^* = \begin{bmatrix} 1 \\ 0 \\ 0 \end{bmatrix}$, $v = -2$

4. (a) $\mathbf{p}^* = \begin{bmatrix} \frac{5}{8} & \frac{3}{8} \end{bmatrix}$, $\mathbf{q}^* = \begin{bmatrix} \frac{1}{8} \\ \frac{7}{8} \end{bmatrix}$, $v = \frac{27}{8}$ **(b)** $\mathbf{p}^* = \begin{bmatrix} \frac{2}{3} & \frac{1}{3} \end{bmatrix}$, $\mathbf{q}^* = \begin{bmatrix} \frac{1}{6} \\ \frac{5}{6} \end{bmatrix}$, $v = \frac{70}{3}$

(c) $\mathbf{p}^* = [1 \ 0]$, $\mathbf{q}^* = \begin{bmatrix} 1 \\ 0 \end{bmatrix}$, $v = 3$ **(d)** $\mathbf{p}^* = \begin{bmatrix} \frac{3}{5} & \frac{2}{5} \end{bmatrix}$, $\mathbf{q}^* = \begin{bmatrix} \frac{3}{5} \\ \frac{2}{5} \end{bmatrix}$, $v = \frac{19}{5}$

(e) $\mathbf{p}^* = \begin{bmatrix} \frac{3}{13} & \frac{10}{13} \end{bmatrix}$, $\mathbf{q}^* = \begin{bmatrix} \frac{1}{13} \\ \frac{12}{13} \end{bmatrix}$, $v = -\frac{29}{13}$

5. $\mathbf{p}^* = \begin{bmatrix} \frac{13}{20} & \frac{7}{20} \end{bmatrix}$, $\mathbf{q}^* = \begin{bmatrix} \frac{11}{20} \\ \frac{9}{20} \end{bmatrix}$, $v = -\frac{3}{20}$

EXERCISE SET 11.9 (page 616)

1. (a) $\begin{bmatrix} 2 \\ 3 \end{bmatrix}$ **(b)** $\begin{bmatrix} 6 \\ 5 \\ 6 \end{bmatrix}$ **(c)** $\begin{bmatrix} 78 \\ 54 \\ 79 \end{bmatrix}$

2. (a) Use Corollary 11.9.4; all row sums are less than one. **3.** E^2 has all positive entries.
(b) Use Corollary 11.9.5; all column sums are less than one.

(c) Use Theorem 11.9.3, with $\mathbf{x} = \begin{bmatrix} 2 \\ 1 \\ 1 \end{bmatrix} > C\mathbf{x} = \begin{bmatrix} 1.9 \\ .9 \\ .9 \end{bmatrix}$.

4. Price of tomatoes, \$120.00; price of corn, \$100.00; price of lettuce, \$106.67

5. \$1256 for the CE, \$1448 for the EE, \$1556 for the ME

EXERCISE SET 11.10 (page 625)

1. The second class; \$15,000 **2.** \$223 **3.** $1 : 1.90 : 3.02 : 4.24 : 5.00$

5. $s/(g_1^{-1} + g_2^{-1} + \cdots + g_{k-1}^{-1})$ **6.** $1 : 2 : 3 : \cdots : n - 1$

EXERCISE SET 11.11 [page 631]

1. **(a)** $\begin{bmatrix} 0 & 1 & 1 & 0 \\ 0 & 0 & 1 & 1 \\ 0 & 0 & 0 & 0 \end{bmatrix}$ **(b)** $\begin{bmatrix} 0 & \frac{3}{2} & \frac{3}{2} & 0 \\ 0 & 0 & \frac{1}{2} & \frac{1}{2} \\ 0 & 0 & 0 & 0 \end{bmatrix}$

(c) $\begin{bmatrix} -2 & -1 & -1 & -2 \\ -1 & -1 & 0 & 0 \\ 3 & 3 & 3 & 3 \end{bmatrix}$ **(d)** $\begin{bmatrix} 0 & .866 & 1.366 & .500 \\ 0 & -.500 & .366 & .866 \\ 0 & 0 & 0 & 0 \end{bmatrix}$

2. **(b)** $(0,0,0)$, $(1,0,0)$, $\left(1\frac{1}{2},1,0\right)$, and $\left(\frac{1}{2},1,0\right)$
(c) $(0,0,0)$, $(1,.6,0)$, $(1,1.6,0)$, $(0,1,0)$

3. **(a)** $\begin{bmatrix} 1 & 0 & 0 \\ 0 & -1 & 0 \\ 0 & 0 & 1 \end{bmatrix}$ **(b)** $\begin{bmatrix} -1 & 0 & 0 \\ 0 & 1 & 0 \\ 0 & 0 & 1 \end{bmatrix}$ **(c)** $\begin{bmatrix} 1 & 0 & 0 \\ 0 & 1 & 0 \\ 0 & 0 & -1 \end{bmatrix}$

4. **(a)** $M_1 = \begin{bmatrix} \frac{1}{2} & 0 & 0 \\ 0 & 2 & 0 \\ 0 & 0 & \frac{1}{3} \end{bmatrix}$, $M_2 = \begin{bmatrix} \frac{1}{2} & \frac{1}{2} & \cdots & \frac{1}{2} \\ 0 & 0 & \cdots & 0 \\ 0 & 0 & \cdots & 0 \end{bmatrix}$, $M_3 = \begin{bmatrix} 1 & 0 & 0 \\ 0 & \cos 20° & -\sin 20° \\ 0 & \sin 20° & \cos 20° \end{bmatrix}$,

$M_4 = \begin{bmatrix} \cos(-45°) & 0 & \sin(-45°) \\ 0 & 1 & 0 \\ -\sin(-45°) & 0 & \cos(-45°) \end{bmatrix}$, $M_5 = \begin{bmatrix} 0 & -1 & 0 \\ 1 & 0 & 0 \\ 0 & 0 & 1 \end{bmatrix}$

(b) $P' = M_5 M_4 M_3 (M_1 P + M_2)$

5. **(a)** $M_1 = \begin{bmatrix} .3 & 0 & 0 \\ 0 & .5 & 0 \\ 0 & 0 & 1 \end{bmatrix}$, $M_2 = \begin{bmatrix} 1 & 0 & 0 \\ 0 & \cos 45° & -\sin 45° \\ 0 & \sin 45° & \cos 45° \end{bmatrix}$, $M_3 = \begin{bmatrix} 1 & 1 & \cdots & 1 \\ 0 & 0 & \cdots & 0 \\ 0 & 0 & \cdots & 0 \end{bmatrix}$,

$M_4 = \begin{bmatrix} \cos 35° & 0 & \sin 35° \\ 0 & 1 & 0 \\ -\sin 35° & 0 & \cos 35° \end{bmatrix}$, $M_5 = \begin{bmatrix} \cos(-45°) & -\sin(-45°) & 0 \\ \sin(-45°) & \cos(-45°) & 0 \\ 0 & 0 & 1 \end{bmatrix}$,

$M_6 = \begin{bmatrix} 0 & 0 & \cdots & 0 \\ 0 & 0 & \cdots & 0 \\ 1 & 1 & \cdots & 1 \end{bmatrix}$, $M_7 = \begin{bmatrix} 2 & 0 & 0 \\ 0 & 1 & 0 \\ 0 & 0 & 1 \end{bmatrix}$

(b) $P' = M_7(M_5 M_4 (M_2 M_1 P + M_3) + M_6)$

6. $R_1 = \begin{bmatrix} \cos \beta & 0 & \sin \beta \\ 0 & 1 & 0 \\ -\sin \beta & 0 & \cos \beta \end{bmatrix}$, $R_2 = \begin{bmatrix} \cos \alpha & -\sin \alpha & 0 \\ \sin \alpha & \cos \alpha & 0 \\ 0 & 0 & 1 \end{bmatrix}$, $R_3 = \begin{bmatrix} \cos \theta & 0 & \sin \theta \\ 0 & 1 & 0 \\ -\sin \theta & 0 & \cos \theta \end{bmatrix}$,

$R_4 = \begin{bmatrix} \cos \alpha & \sin \alpha & 0 \\ -\sin \alpha & \cos \alpha & 0 \\ 0 & 0 & 1 \end{bmatrix}$, $R_5 = \begin{bmatrix} \cos \beta & 0 & -\sin \beta \\ 0 & 1 & 0 \\ \sin \beta & 0 & \cos \beta \end{bmatrix}$

7. (a) $M = \begin{bmatrix} 1 & 0 & 0 & x_0 \\ 0 & 1 & 0 & y_0 \\ 0 & 0 & 1 & z_0 \\ 0 & 0 & 0 & 1 \end{bmatrix}$ **(b)** $\begin{bmatrix} 1 & 0 & 0 & -5 \\ 0 & 1 & 0 & 9 \\ 0 & 0 & 1 & -3 \\ 0 & 0 & 0 & 1 \end{bmatrix}$

EXERCISE SET 11.12 [page 644]

1. (a) $\begin{bmatrix} t_1 \\ t_2 \\ t_3 \\ t_4 \end{bmatrix} = \begin{bmatrix} 0 & \frac{1}{4} & \frac{1}{4} & 0 \\ \frac{1}{4} & 0 & 0 & \frac{1}{4} \\ \frac{1}{4} & 0 & 0 & \frac{1}{4} \\ 0 & \frac{1}{4} & \frac{1}{4} & 0 \end{bmatrix} \begin{bmatrix} t_1 \\ t_2 \\ t_3 \\ t_4 \end{bmatrix} + \begin{bmatrix} 0 \\ \frac{1}{2} \\ 0 \\ \frac{1}{2} \end{bmatrix}$ **(b)** $\mathbf{t} = \begin{bmatrix} \frac{1}{4} \\ \frac{3}{4} \\ \frac{1}{4} \\ \frac{3}{4} \end{bmatrix}$

(c) $\mathbf{t}^{(1)} = \begin{bmatrix} 0 \\ \frac{1}{2} \\ 0 \\ \frac{1}{2} \end{bmatrix}$, $\mathbf{t}^{(2)} = \begin{bmatrix} \frac{1}{8} \\ \frac{5}{8} \\ \frac{1}{8} \\ \frac{5}{8} \end{bmatrix}$, $\mathbf{t}^{(3)} = \begin{bmatrix} \frac{3}{16} \\ \frac{11}{16} \\ \frac{3}{16} \\ \frac{11}{16} \end{bmatrix}$, $\mathbf{t}^{(4)} = \begin{bmatrix} \frac{7}{32} \\ \frac{23}{32} \\ \frac{7}{32} \\ \frac{23}{32} \end{bmatrix}$, $\mathbf{t}^{(5)} = \begin{bmatrix} \frac{15}{64} \\ \frac{47}{64} \\ \frac{15}{64} \\ \frac{47}{64} \end{bmatrix}$, $\mathbf{t}^{(5)} - \mathbf{t} = \begin{bmatrix} -\frac{1}{64} \\ -\frac{1}{64} \\ -\frac{1}{64} \\ -\frac{1}{64} \end{bmatrix}$

(d) for t_1, 4.5%; for t_2, -1.8%

2. $\frac{1}{2}$ **3.** $\mathbf{t}^{(1)} = [\frac{3}{4} \quad \frac{5}{4} \quad \frac{2}{4} \quad \frac{5}{4} \quad \frac{4}{4} \quad \frac{2}{4} \quad \frac{5}{4} \quad \frac{4}{4} \quad \frac{3}{4}]^T$

$\mathbf{t}^{(2)} = [\frac{13}{16} \quad \frac{18}{16} \quad \frac{9}{16} \quad \frac{22}{16} \quad \frac{13}{16} \quad \frac{7}{16} \quad \frac{21}{16} \quad \frac{16}{16} \quad \frac{10}{16}]^T$

EXERCISE SET 11.13 [page 656]

1. (c) $x_3^* = (\frac{31}{22}, \frac{27}{22})$

2. (a) $\mathbf{x}_3^{(1)} = (1.40000, 1.20000)$ **(b)** Same as part (a) **(c)** $\mathbf{x}_3^{(1)} = (9.55000, 25.65000)$
$\mathbf{x}_3^{(2)} = (1.41000, 1.23000)$ $\mathbf{x}_3^{(2)} = (.59500, -1.21500)$
$\mathbf{x}_3^{(3)} = (1.40900, 1.22700)$ $\mathbf{x}_3^{(3)} = (1.49050, 1.47150)$
$\mathbf{x}_3^{(4)} = (1.40910, 1.22730)$ $\mathbf{x}_3^{(4)} = (1.40095, 1.20285)$
$\mathbf{x}_3^{(5)} = (1.40909, 1.22727)$ $\mathbf{x}_3^{(5)} = (1.40991, 1.22972)$
$\mathbf{x}_3^{(6)} = (1.40909, 1.22727)$ $\mathbf{x}_3^{(6)} = (1.40901, 1.22703)$

4. $\mathbf{x}_1^* = (1, 1), \mathbf{x}_2^* = (2, 0), \mathbf{x}_3^* = (1, 1)$

7.
$$x_7 + x_8 + x_9 = 13.00$$
$$x_4 + x_5 + x_6 = 15.00$$
$$x_1 + x_2 + x_3 = 8.00$$
$$.82843(x_6 + x_8) + .58579x_9 = 14.79$$
$$1.41421(x_3 + x_5 + x_7) = 14.31$$
$$.82843(x_2 + x_4) + .58579x_1 = 3.81$$
$$x_3 + x_6 + x_9 = 18.00$$
$$x_2 + x_5 + x_8 = 12.00$$
$$x_1 + x_4 + x_7 = 6.00$$
$$.82843(x_2 + x_6) + .58579x_3 = 10.51$$
$$1.41421(x_1 + x_5 + x_9) = 16.13$$
$$.82843(x_4 + x_8) + .58579x_7 = 7.04$$

8.
$$x_7 + x_8 + x_9 = 13.00$$
$$x_4 + x_5 + x_6 = 15.00$$
$$x_1 + x_2 + x_3 = 8.00$$
$$.04289(x_3 + x_5 + x_7) + .75000(x_6 + x_8) + .61396x_9 = 14.79$$
$$.91421(x_3 + x_5 + x_7) + .25000(x_2 + x_4 + x_6 + x_8) = 14.31$$
$$.04289(x_3 + x_5 + x_7) + .75000(x_2 + x_4) + .61396x_1 = 3.81$$
$$x_3 + x_6 + x_9 = 18.00$$
$$x_2 + x_5 + x_8 = 12.00$$
$$x_1 + x_4 + x_7 = 6.00$$
$$.04289(x_1 + x_5 + x_9) + .75000(x_2 + x_6) + .61396x_3 = 10.51$$
$$.91421(x_1 + x_5 + x_9) + .25000(x_2 + x_4 + x_6 + x_8) = 16.13$$
$$.04289(x_1 + x_5 + x_9) + .75000(x_4 + x_8) + .61396x_7 = 7.04$$

EXERCISE SET 11.14 [page 674]

1. $T_i\left(\begin{bmatrix} x \\ y \end{bmatrix}\right) = \dfrac{12}{25}\begin{bmatrix} 1 & 0 \\ 0 & 1 \end{bmatrix}\begin{bmatrix} x \\ y \end{bmatrix} + \begin{bmatrix} e_i \\ f_i \end{bmatrix}$, $i = 1, 2, 3, 4$, where the four values of $\begin{bmatrix} e_i \\ f_i \end{bmatrix}$ are $\begin{bmatrix} 0 \\ 0 \end{bmatrix}$, $\begin{bmatrix} \frac{13}{25} \\ 0 \end{bmatrix}$, $\begin{bmatrix} 0 \\ \frac{13}{25} \end{bmatrix}$, and $\begin{bmatrix} \frac{13}{25} \\ \frac{13}{25} \end{bmatrix}$; $d_H(S) = \ln(4)/\ln\left(\frac{25}{12}\right) = 1.888\ldots$

2. $s \approx .47$; $d_H(S) \approx \ln(4)/\ln(1/.47) = 1.8\ldots$. Rotation angles: $0°$ (upper left); $90°$ (upper right); $180°$ (lower left); $180°$ (lower right)

3. **(a)** (i) $s = \frac{1}{3}$; (ii) all rotation angles are $0°$; (iii) $d_H(S) = \ln(7)/\ln(3) = 1.771\ldots$. This set is a fractal.
   (b) (i) $s = \frac{1}{2}$; (ii) all rotation angles are $180°$; (iii) $d_H(S) = \ln(3)/\ln(2) = 1.584\ldots$. This set is a fractal.
   (c) (i) $s = \frac{1}{2}$; (ii) rotation angles: $90°$ (top); $180°$ (lower left); $180°$ (lower right);
   (iii) $d_H(S) = \ln(3)/\ln(2) = 1.584\ldots$. This set is a fractal.
   (d) (i) $s = \frac{1}{2}$; (ii) rotation angles: $90°$ (upper left); $180°$ (upper right); $180°$ (lower right);
   (iii) $d_H(S) = \ln(3)/\ln(2) = 1.584\ldots$. This set is a fractal.

4. $s = .8509\ldots, \theta = -2.69°\ldots$ 5. $(0.766, 0.996)$ rounded to three decimal places

6. $d_H(S) = \ln(16)/\ln(4) = 2$ 7. $\ln(4)/\ln\left(\frac{4}{3}\right) = 4.818\ldots$

8. $d_H(S) = \ln(8)/\ln(2) = 3$; the cube is not a fractal.

9. $k = 20$; $s = \frac{1}{3}$; $d_H(S) = \ln(20)/\ln(3) = 2.726\ldots$; the set is a fractal.

10.

Initial set

First iterate

Second iterate

 Third iterate

Fourth iterate

$d_H(S) = \ln(2)/\ln(3) = 0.6309\ldots$

11. Area of $S_0 = 1$; area of $S_1 = \frac{8}{9} = 0.888\ldots$; area of $S_2 = \left(\frac{8}{9}\right)^2 = 0.790\ldots$; area of $S_3 = \left(\frac{8}{9}\right)^3 = 0.702\ldots$; area of $S_4 = \left(\frac{8}{9}\right)^4 = 0.624\ldots$

EXERCISE SET 11.15 [page 689]

1. $\Pi(250) = 750$, $\Pi(25) = 50$, $\Pi(125) = 250$, $\Pi(30) = 60$, $\Pi(10) = 30$, $\Pi(50) = 150$, $\Pi(3750) = 7500$, $\Pi(6) = 12$, $\Pi(5) = 10$

2. One 1-cycle: $\{(0,0)\}$; one 3-cycle: $\left\{\left(\frac{3}{6}, 0\right), \left(\frac{3}{6}, \frac{3}{6}\right), \left(0, \frac{3}{6}\right)\right\}$; two 4-cycles:
   $\left\{\left(\frac{4}{6}, 0\right), \left(\frac{4}{6}, \frac{4}{6}\right), \left(\frac{2}{6}, 0\right), \left(\frac{2}{6}, \frac{2}{6}\right)\right\}$ and $\left\{\left(0, \frac{2}{6}\right), \left(\frac{2}{6}, \frac{4}{6}\right), \left(0, \frac{4}{6}\right), \left(\frac{4}{6}, \frac{2}{6}\right)\right\}$; two 12-cycles:
   $\left\{\left(0, \frac{1}{6}\right), \left(\frac{1}{6}, \frac{2}{6}\right), \left(\frac{3}{6}, \frac{5}{6}\right), \left(\frac{2}{6}, \frac{1}{6}\right), \left(\frac{3}{6}, \frac{4}{6}\right), \left(\frac{1}{6}, \frac{5}{6}\right), \left(0, \frac{5}{6}\right), \left(\frac{5}{6}, \frac{4}{6}\right), \left(\frac{3}{6}, \frac{1}{6}\right), \left(\frac{4}{6}, \frac{5}{6}\right), \left(\frac{3}{6}, \frac{2}{6}\right), \left(\frac{5}{6}, \frac{1}{6}\right)\right\}$
   and $\left\{\left(\frac{1}{6}, 0\right), \left(\frac{1}{6}, \frac{1}{6}\right), \left(\frac{2}{6}, \frac{3}{6}\right), \left(\frac{5}{6}, \frac{2}{6}\right), \left(\frac{1}{6}, \frac{3}{6}\right), \left(\frac{4}{6}, \frac{1}{6}\right), \left(\frac{5}{6}, 0\right), \left(\frac{5}{6}, \frac{5}{6}\right), \left(\frac{4}{6}, \frac{3}{6}\right), \left(\frac{1}{6}, \frac{4}{6}\right), \left(\frac{5}{6}, \frac{3}{6}\right), \left(\frac{2}{6}, \frac{5}{6}\right)\right\}$.
   $\Pi(6) = 12$

3. **(a)** 3, 7, 10, 2, 12, 14, 11, 10, 6, 1, 7, 8, 0, 8, 8, 1, 9, 10, 4, 14, 3, 2, 5, 7, 12, 4, 1, 5, 6, 11, 2, 13, 0, 13, 13, 11, 9, 5, 14, 4, 3, 7, $\ldots$
   (c) $(5, 5), (10, 15), (4, 19), (2, 0), (2, 2), (4, 6), (10, 16), (5, 0), (5, 5), \ldots$

4. (c) The first five iterates of $\left(\frac{1}{101}, 0\right)$ are $\left(\frac{1}{101}, \frac{1}{101}\right), \left(\frac{2}{101}, \frac{3}{101}\right), \left(\frac{5}{101}, \frac{8}{101}\right), \left(\frac{13}{101}, \frac{21}{101}\right),$ and $\left(\frac{34}{101}, \frac{55}{101}\right)$.

6. (b) The matrices of Anosov automorphisms are $\begin{bmatrix} 3 & 2 \\ 1 & 1 \end{bmatrix}$ and $\begin{bmatrix} 5 & 7 \\ 2 & 3 \end{bmatrix}$.

(c) The transformation affects a rotation of S through $90°$ in the clockwise direction.

9.

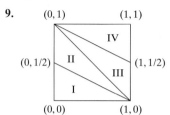
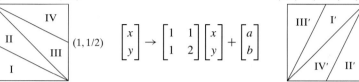
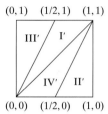

In region I: $\begin{bmatrix} a \\ b \end{bmatrix} = \begin{bmatrix} 0 \\ 0 \end{bmatrix}$; in region II: $\begin{bmatrix} a \\ b \end{bmatrix} = \begin{bmatrix} 0 \\ -1 \end{bmatrix}$; in region III:

$\begin{bmatrix} a \\ b \end{bmatrix} = \begin{bmatrix} -1 \\ -1 \end{bmatrix}$; in region IV: $\begin{bmatrix} a \\ b \end{bmatrix} = \begin{bmatrix} -1 \\ -2 \end{bmatrix}$

12. $\left(\frac{1}{5}, \frac{3}{5}\right)$ and $\left(\frac{4}{5}, \frac{2}{5}\right)$ form one 2-cycle, and $\left(\frac{2}{5}, \frac{1}{5}\right)$ and $\left(\frac{3}{5}, \frac{4}{5}\right)$ form another 2-cycle.

EXERCISE SET 11.16 [page 702]

1. (a) *GIYUOKEVBH*   **(b)** *SFANEFZWJH*

2. (a) $A^{-1} = \begin{bmatrix} 12 & 7 \\ 23 & 15 \end{bmatrix}$ **(b)** Not invertible   **(c)** $A^1 = \begin{bmatrix} 1 & 19 \\ 23 & 24 \end{bmatrix}$

(d) Not invertible   **(e)** Not invertible   **(f)** $A^{-1} = \begin{bmatrix} 15 & 12 \\ 21 & 5 \end{bmatrix}$

3. *WE LOVE MATH*

4. Deciphering matrix $= \begin{bmatrix} 7 & 15 \\ 6 & 5 \end{bmatrix}$; enciphering matrix $= \begin{bmatrix} 7 & 5 \\ 2 & 15 \end{bmatrix}$

5. *THEY SPLIT THE ATOM*   **6.** *I HAVE COME TO BURY CAESAR*

7. A is invertible modulo 29 if and only if $\det(A) \neq 0 \pmod{29}$.

EXERCISE SET 11.17 [page 714]

2. $\left.\begin{array}{l} a_n = \frac{1}{4} + \left(\frac{1}{2}\right)^{n+1}(a_0 - c_0) \\ b_n = \frac{1}{2} \\ c_n = \frac{1}{4} - \left(\frac{1}{2}\right)^{n+1}(a_0 - c_0) \end{array}\right\}$ $n = 1, 2, \ldots$ $\left.\begin{array}{l} a_n \to \frac{1}{4} \\ b_n = \frac{1}{2} \\ c_n \to \frac{1}{4} \end{array}\right\}$ as $n \to \infty$

3. $\left.\begin{array}{l} a_{2n+1} = \dfrac{2}{3} + \dfrac{1}{6(4)^n}(2a_0 - b_0 - 4c_0) \\[2mm] b_{2n+1} = \dfrac{1}{3} - \dfrac{1}{6(4)^n}(2a_0 - b_0 - 4c_0) \\[2mm] c_{2n+1} = 0 \end{array}\right\}$ $n = 0, 1, 2, \ldots$ $\left.\begin{array}{l} a_{2n} = \dfrac{5}{12} + \dfrac{1}{6(4)^n}(2a_0 - b_0 - 4c_0) \\[2mm] b_{2n} = \dfrac{1}{2} \\[2mm] c_{2n} = \dfrac{1}{12} - \dfrac{1}{6(4)^n}(2a_0 - b_0 - 4c_0) \end{array}\right\}$ $n = 1, 2, \ldots$

4. Eigenvalues: $\lambda_1 = 1, \lambda_2 = \frac{1}{2}$; eigenvectors: $\mathbf{e}_1 = \begin{bmatrix} 1 \\ 0 \end{bmatrix}, \mathbf{e}_2 = \begin{bmatrix} 1 \\ -1 \end{bmatrix}$

5. 12 generations; .006%

6. $\mathbf{x}^{(n)} = \begin{bmatrix} \dfrac{1}{2} + \dfrac{1}{2^{2n+3}}[(-3 - \sqrt{5})(1 + \sqrt{5})^{n+1} + (-3 + \sqrt{5})(1 - \sqrt{5})^{n+1}] \\[2mm] \dfrac{1}{2^{2n+1}}[(1 + \sqrt{5})^{n+1} + (1 - \sqrt{5})^{n+1}] \\[2mm] \dfrac{1}{2^{2n+1}}[(1 + \sqrt{5})^{n} + (1 - \sqrt{5})^{n}] \\[2mm] \dfrac{1}{2^{2n+1}}[(1 + \sqrt{5})^{n} + (1 - \sqrt{5})^{n}] \\[2mm] \dfrac{1}{2^{2n+1}}[(1 + \sqrt{5})^{n+1} + (1 - \sqrt{5})^{n+1}] \\[2mm] \dfrac{1}{2} + \dfrac{1}{2^{2n+3}}[(-3 - \sqrt{5})(1 + \sqrt{5})^{n+1} + (-3 + \sqrt{5})(1 - \sqrt{5})^{n+1}] \end{bmatrix}$; $\mathbf{x}^{(n)} \to \begin{bmatrix} \frac{1}{2} \\ 0 \\ 0 \\ 0 \\ 0 \\ \frac{1}{2} \end{bmatrix}$ as $n \to \infty$

8. $\begin{bmatrix} 1 & 0 & 0 & 0 \\ 0 & 0 & 0 & 0 \\ 0 & 0 & 0 & 0 \\ 0 & 0 & 0 & 1 \end{bmatrix}$

EXERCISE SET 11.18 (page 724)

1. (a) $\lambda_1 = \frac{3}{2}$, $\mathbf{x}_1 = \begin{bmatrix} 1 \\ \frac{1}{3} \end{bmatrix}$

(b) $\mathbf{x}^{(1)} = \begin{bmatrix} 100 \\ 50 \end{bmatrix}$, $\mathbf{x}^{(2)} = \begin{bmatrix} 175 \\ 50 \end{bmatrix}$, $\mathbf{x}^{(3)} = \begin{bmatrix} 250 \\ 88 \end{bmatrix}$, $\mathbf{x}^{(4)} = \begin{bmatrix} 382 \\ 125 \end{bmatrix}$, $\mathbf{x}^{(5)} = \begin{bmatrix} 570 \\ 191 \end{bmatrix}$

(c) $\mathbf{x}^{(6)} = L\mathbf{x}^{(5)} = \begin{bmatrix} 857 \\ 285 \end{bmatrix}$, $\mathbf{x}^{(6)} \simeq \lambda_1 \mathbf{x}^{(5)} = \begin{bmatrix} 855 \\ 287 \end{bmatrix}$

7. 2.375   **8.** 1.49611

EXERCISE SET 11.19 (page 734)

1. (a) Yield $= 33\frac{1}{3}\%$ of population; $\mathbf{x}_1 = \begin{bmatrix} 1 \\ \frac{1}{3} \\ \frac{1}{18} \end{bmatrix}$ **(b)** Yield $= 45.8\%$ of population; $\mathbf{x}_1 = \begin{bmatrix} 1 \\ \frac{1}{2} \\ \frac{1}{8} \end{bmatrix}$; harvest 57.9% of youngest age class

2. $\mathbf{x}_1 = \begin{bmatrix} 1.000 \\ .845 \\ .824 \\ .795 \\ .755 \\ .699 \\ .626 \\ .532 \\ 0 \\ 0 \\ 0 \\ 0 \end{bmatrix}$, $L\mathbf{x}_1 = \begin{bmatrix} 2.090 \\ .845 \\ .824 \\ .795 \\ .755 \\ .699 \\ .626 \\ .532 \\ .418 \\ 0 \\ 0 \\ 0 \end{bmatrix}$, $\dfrac{1.089 + .418}{7.584} = .199$

4. $h_I = (R - 1)/(a_1 b_1 b_2 \cdots b_{I-1} + \cdots + a_n b_1 b_2 \cdots b_{n-1})$

5. $h_I = \dfrac{a_1 + a_2 b_1 + \cdots + (a_{J-1} b_1 b_2 \cdots b_{J-2}) - 1}{a_1 b_1 b_2 \cdots b_{I-1} + \cdots + a_{J-1} b_1 b_2 \cdots b_{J-2}}$

EXERCISE SET 11.20 [page 740]

1. $\dfrac{\pi^2}{3} + 4\cos t + \cos 2t + \dfrac{4}{9}\cos 3t$

2. $\dfrac{T^2}{3} + \dfrac{T^2}{\pi^2}\left(\cos\dfrac{2\pi}{T}t + \dfrac{1}{2^2}\cos\dfrac{4\pi}{T}t + \dfrac{1}{3^2}\cos\dfrac{6\pi}{T}t + \dfrac{1}{4^2}\cos\dfrac{8\pi}{T}t\right) - \dfrac{T^2}{\pi}\left(\sin\dfrac{2\pi}{T}t + \dfrac{1}{2}\sin\dfrac{4\pi}{T}t + \dfrac{1}{3}\sin\dfrac{6\pi}{T}t + \dfrac{1}{4}\sin\dfrac{8\pi}{T}t\right)$

3. $\dfrac{1}{\pi} + \dfrac{1}{2}\sin t - \dfrac{2}{3\pi}\cos 2t - \dfrac{2}{15\pi}\cos 4t$

4. $\dfrac{4}{\pi}\left(\dfrac{1}{2} - \dfrac{1}{1\cdot 3}\cos t - \dfrac{1}{3\cdot 5}\cos 2t - \dfrac{1}{5\cdot 7}\cos 3t - \cdots - \dfrac{1}{(2n-1)(2n+1)}\cos nt\right)$

5. $\dfrac{T}{4} - \dfrac{8T}{\pi^2}\left(\dfrac{1}{2^2}\cos\dfrac{2\pi t}{T} + \dfrac{1}{6^2}\cos\dfrac{6\pi t}{T} + \dfrac{1}{10^2}\cos\dfrac{10\pi t}{T} + \cdots + \dfrac{1}{(2n)^2}\cos\dfrac{2n\pi t}{T}\right)$

EXERCISE SET 11.21 [page 749]

1. (a) Yes; $\mathbf{v} = \tfrac{1}{5}\mathbf{v}_1 + \tfrac{2}{5}\mathbf{v}_2 + \tfrac{2}{5}\mathbf{v}_3$ (b) No; $\mathbf{v} = \tfrac{2}{5}\mathbf{v}_1 + \tfrac{4}{5}\mathbf{v}_2 - \tfrac{1}{5}\mathbf{v}_3$

   (c) Yes; $\mathbf{v} = \tfrac{2}{5}\mathbf{v}_1 + \tfrac{3}{5}\mathbf{v}_2 + 0\mathbf{v}_3$ (d) Yes; $\mathbf{v} = \tfrac{4}{15}\mathbf{v}_1 + \tfrac{6}{15}\mathbf{v}_2 + \tfrac{5}{15}\mathbf{v}_3$

2. $m = $ number of triangles $= 7, n = $ number of vertex points $= 7$,
   $k = $ number of boundary vertex points $= 5$; Equation (7) is $7 = 2(7) - 2 - 5$.

3. $\mathbf{w} = M\mathbf{v} + \mathbf{b} = M(c_1\mathbf{v}_1 + c_2\mathbf{v}_2 + c_3\mathbf{v}_3) + (c_1 + c_2 + c_3)\mathbf{b} = c_1(M\mathbf{v}_1 + \mathbf{b}) + c_2(M\mathbf{v}_2 + \mathbf{b}) + c_3(M\mathbf{v}_3 + \mathbf{b}) = c_1\mathbf{w}_1 + c_2\mathbf{w}_2 + c_3\mathbf{w}_3$

4. (a) 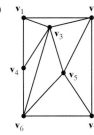   (b)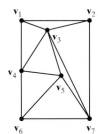

5. (a) $M = \begin{bmatrix} 1 & 2 \\ 0 & 1 \end{bmatrix}, \mathbf{b} = \begin{bmatrix} 1 \\ 2 \end{bmatrix}$ (b) $M = \begin{bmatrix} 3 & -1 \\ 1 & 1 \end{bmatrix}, \mathbf{b} = \begin{bmatrix} 0 \\ 1 \end{bmatrix}$

   (c) $M = \begin{bmatrix} 1 & 0 \\ 0 & 1 \end{bmatrix}, \mathbf{b} = \begin{bmatrix} 2 \\ -3 \end{bmatrix}$ (d) $M = \begin{bmatrix} \frac{1}{2} & 1 \\ 2 & 0 \end{bmatrix}, \mathbf{b} = \begin{bmatrix} \frac{1}{2} \\ -1 \end{bmatrix}$

7. (a) Two of the coefficients are zero.   (b) At least one of the coefficients is zero.
   (c) None of the coefficients are zero.

8. $\tfrac{1}{3}\mathbf{v}_1 + \tfrac{1}{3}\mathbf{v}_2 + \tfrac{1}{3}\mathbf{v}_3$

Photo Credits

Chapter 1. Page 13 (top): Granger Collection. Page 13 (bottom): Courtesy Deutsches Museum, München.

Chapter 2. Page 111: Granger Collection.

Chapter 3. Page 139: Granger Collection.

Chapter 4. Page 165 (top): Courtesy Columbia University. Page 165 (bottom): Courtesy of the University of Berlin.

Chapter 6. Page 304: Detail from a painting by P. S. Kroyer, 1897. Courtesy the Royal Danish Academy of Sciences and Letters. Page 305: Photo by Gerda Schimpf, Courtesy of the University of Berlin.

Chapter 9. Page 446: Science Photo Library/Photo Researchers.

Chaper 10. Pages 495 and 503: Granger Collection. Page 522: ©Corbis-Bettmann.

color insert photos:

Figure 11.1: Isaac Newton; Florian Cajori, ed., *Mathematical Principles of Natural Philosophy and His System of the World*, trans./ed. by Andrew Motte, Regents of the University of California Press, ©1934, 1962.

Figure 11.2: Gregory Heisler/The Image Bank.

Figure 11.3: Roger Tully/Tony Stone Images, Inc.

Figure 11.4 (left): Ulf E. Wallin/The Image Bank.

Figure 11.4 (right): Simon Wilkinson/The Image Bank.

Figure 11.5: Chris Rorres.

Figure 11.6: Bruce Coleman, Inc.

Figure 11.7: Roy P. Fontaine/Photo Researchers.

Figure 11.8: Hank DeLespinasse/The Image Bank.

Figure 11.9 (left): Tony Stone Images, Inc.

Figure 11.9 (center): Gary Withey/Bruce Coleman, Inc.

Figure 11.9 (right): Lou Jones/The Image Bank.

Figure 11.10 (left): Wolf Von Dem Bussche/The Image Bank.

Figure 11.10 (right): Michael Melford/The Image Bank.

Figure 11.11: James King-Holmes/Science Photo Library/Photo Researchers.

Figure 11.12: Charles Thatcher/Tony Stone Images, Inc.

Figure 11.13: David York/Medichrome.

Figure 11.14: Chris Rorres.

Figure 11.15: Chris Rorres.

Figure 11.16: UPI/Bettmann.

Figure 11.17: Erich Lessing/Art Resource.

Figure 11.18: Karen Kasmauski/Woodfin Camp & Associates.

Figure 11.19: Clyde H. Smith/Peter Arnold, Inc.

Figure 11.20: Network Graphics/Blaize Zito Associates, Inc.

Figure 11.21: Chris Rorres/Pictures of Panagiotitsa Rorres.

Index